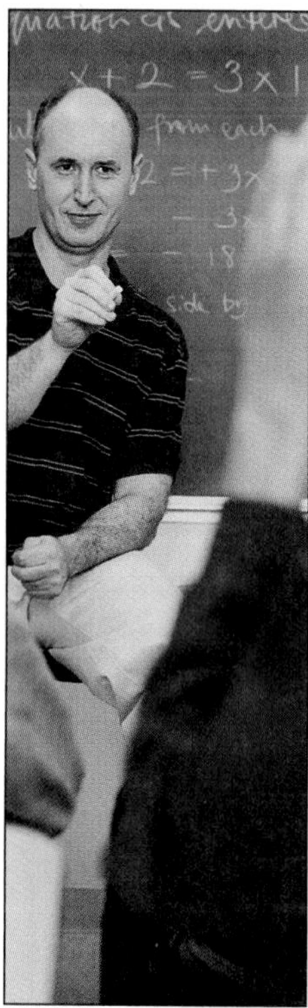

Every one of your students has the potential to make a difference. And realizing that potential starts right here, in your course.

When students succeed in your course—when they stay on-task and make the breakthrough that turns confusion into confidence—they are empowered to realize the possibilities for greatness that lie within each of them. We know your goal is to create an environment where students reach their full potential and experience the exhilaration of academic success that will last them a lifetime. *WileyPLUS* can help you reach that goal.

Wiley**PLUS** is a suite of resources—including the complete online text—that will help your students:

- come to class better prepared for your lectures
- get immediate feedback and context-sensitive help on assignments and quizzes
- track their progress throughout the course

www.wileyplus.com

88% of students surveyed said it improved their understanding of the material.*

WileyPLUS is built around the activities you perform

Prepare & Present

Create outstanding class presentations using a wealth of resources, such as PowerPoint™ slides, image galleries, interactive learningware, and more. Plus you can easily upload any materials you have created into your course, and combine them with the resources Wiley provides you with.

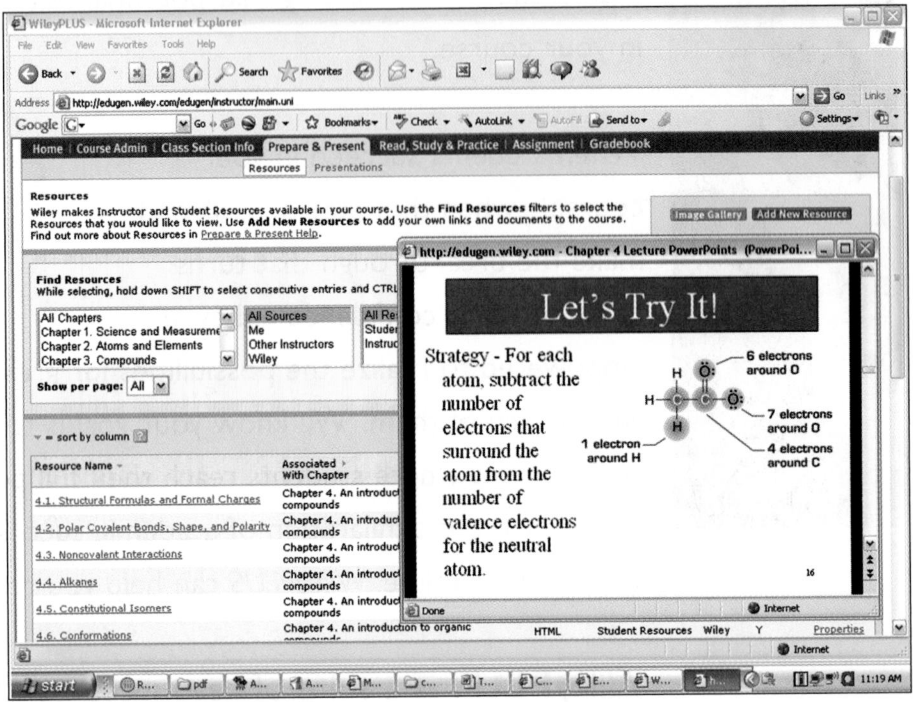

catalyst

CATALYST

With the 2nd edition of *Raymond GOB*, we are introducing an innovative online learning program called CATALYST. The CATALYST assignments ask students to consider the key concepts and topics at hand from different perspectives; with different givens/desired responses required each time a new question is presented.

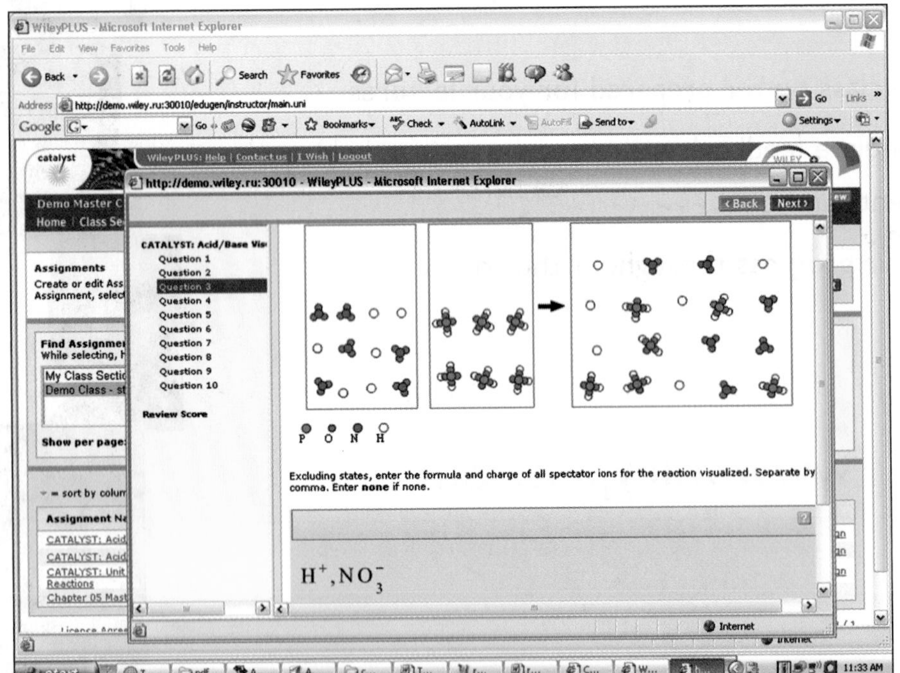

Create Assignments

Automate the assigning and grading of homework or quizzes by using the provided question banks. Student results will be automatically graded and recorded in your gradebook. *WileyPLUS* also links homework problems to relevant sections of the online text, hints, or solutions—context-sensitive help where students need it most!

* Based on 7000 survey responses from student users of *WileyPLUS* in academic year 2006-2007.

in your class each day. With Wiley**PLUS** you can:

Track Student Progress

Keep track of your students' progress via an instructor's gradebook, which allows you to analyze individual and overall class results. This gives you an accurate and realistic assessment of your students' progress and level of understanding.

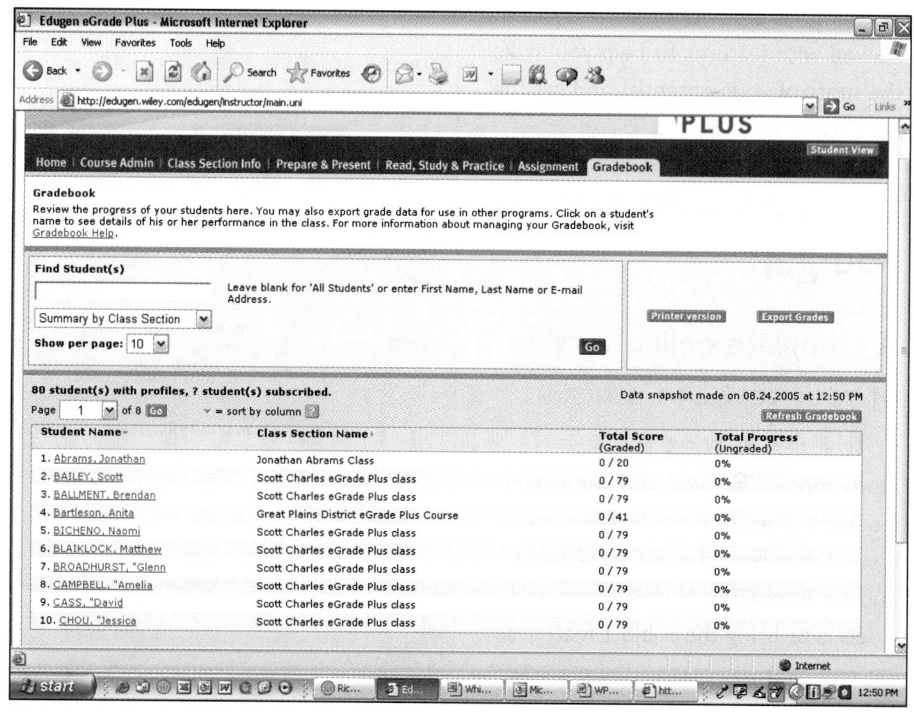

Now Available with WebCT and eCollege!

Now you can seamlessly integrate all of the rich content and resources available with *WileyPLUS* with the power and convenience of your WebCT or eCollege course. You and your students get the best of both worlds with single sign-on, an integrated gradebook, list of assignments and roster, and more. If your campus is using another course management system, contact your local Wiley Representative.

eCollege®

"I studied more for this class than I would have without *WileyPLUS*."

Melissa Lawler, *Western Washington Univ.*

For more information on what *WileyPLUS* can do to help your students reach their potential, please visit

www.wileyplus.com/experience

84% of students would recommend *WileyPLUS* to their next instructors.*

TO THE STUDENT

You have the potential to make a difference!

WileyPLUS is a powerful online system packed with features to help you make the most of your potential, and get the best grade you can!

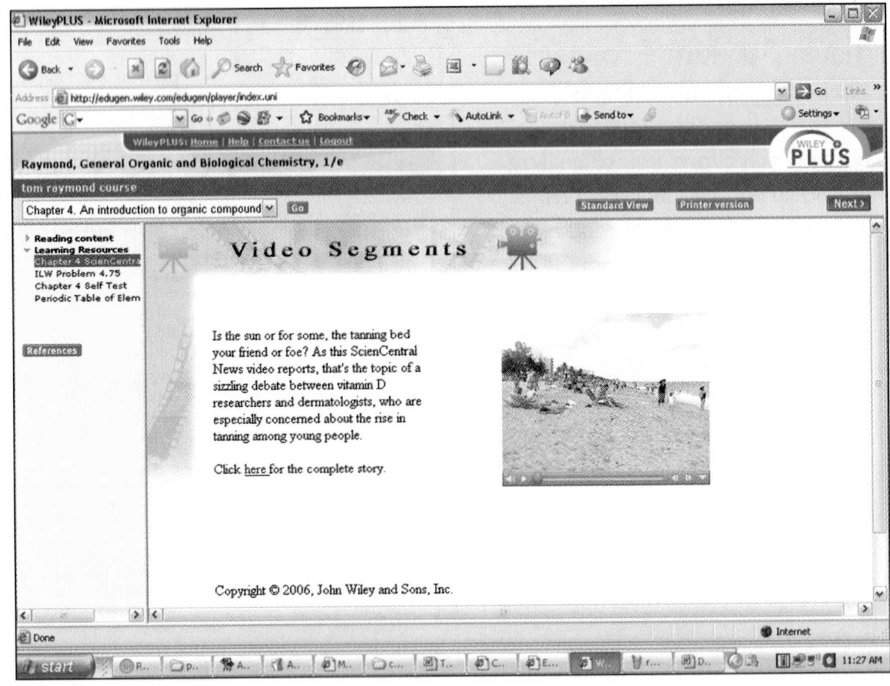

With Wiley**PLUS** you get:

A complete online version of your text and other study resources

Study more effectively and get instant feedback when you practice on your own. Resources like self-assessment quizzes, interactive learningware, and video clips bring the subject matter to life, and help you master the material.

Problem-solving help, instant grading, and feedback on your homework and quizzes

You can keep all of your assigned work in one location, making it easy for you to stay on task. Plus, many homework problems contain direct links to the relevant portion of your text to help you deal with problem-solving obstacles at the moment they come up.

The ability to track your progress and grades throughout the term.

A personal gradebook allows you to monitor your results from past assignments at any time. You'll always know exactly where you stand.

If your instructor uses *WileyPLUS*, you will receive a URL for your class. If not, your instructor can get more information about *WileyPLUS* by visiting www.wileyplus.com

"It has been a great help, and I believe it has helped me to achieve a better grade."
Michael Morris, *Columbia Basin College*

74% of students surveyed said it helped them get a better grade.*

GENERAL, ORGANIC, AND BIOLOGICAL CHEMISTRY

An Integrated Approach

SECOND EDITION

THE WILEY BICENTENNIAL–KNOWLEDGE FOR GENERATIONS

Each generation has its unique needs and aspirations. When Charles Wiley first opened his small printing shop in lower Manhattan in 1807, it was a generation of boundless potential searching for an identity. And we were there, helping to define a new American literary tradition. Over half a century later, in the midst of the Second Industrial Revolution, it was a generation focused on building the future. Once again, we were there, supplying the critical scientific, technical, and engineering knowledge that helped frame the world. Throughout the 20th Century, and into the new millennium, nations began to reach out beyond their own borders and a new international community was born. Wiley was there, expanding its operations around the world to enable a global exchange of ideas, opinions, and know-how.

For 200 years, Wiley has been an integral part of each generation's journey, enabling the flow of information and understanding necessary to meet their needs and fulfill their aspirations. Today, bold new technologies are changing the way we live and learn. Wiley will be there, providing you the must-have knowledge you need to imagine new worlds, new possibilities, and new opportunities.

Generations come and go, but you can always count on Wiley to provide you the knowledge you need, when and where you need it!

William J. Pesce
President and Chief Executive Officer

Peter Booth Wiley
Chairman of the Board

GENERAL, ORGANIC, AND BIOLOGICAL CHEMISTRY

An Integrated Approach

SECOND EDITION

Kenneth W. Raymond

Eastern Washington University

John Wiley & Sons, Inc.

VICE PRESIDENT AND EXECUTIVE PUBLISHER	Kaye Pace
PROJECT EDITOR	Jennifer Yee
PRODUCTION SERVICES MANAGER	Dorothy Sinclair
PRODUCTION EDITOR	Janet Foxman
EXECUTIVE MARKETING MANAGER	Amanda Wygal
CREATIVE DIRECTOR	Harry Nolan
ART DIRECTOR	Hope Miller
DESIGNER	Brian Salisbury
SENIOR PHOTO EDITOR	Lisa Gee
EDITORIAL ASSISTANT	Catherine Donovan
SENIOR MEDIA EDITOR	Thomas Kulesa
PRODUCTION MANAGEMENT	Suzanne Ingrao/Ingrao Associates
COVER ILLUSTRATION/DESIGN	Norm Christiansen
BICENTENNIAL LOGO DESIGN	Richard J. Pacifico
COVER PHOTOS	Image of orange based on photo © Stock Food/Punchstock (top) and © Photodisc/Superstock (bottom)

This volume contains selected illustrations from the following texts, reprinted with permission by John Wiley and Sons, Inc.

- Boyer, Rodney, Concepts in Biochemistry, Second Edition, © 2002.
- Hein, Morris; Best, Leo R.; Pattison, Scott; Arena, Susan, Introduction to General, Organic, and Biochemistry, Eighth Edition, © 2005.
- Holum, John R., Fundamentals of General, Organic, and Biological Chemistry, Sixth Edition, © 1998.
- Pratt, Charlotte W.; Cornely, Kathleen, Essential Biochemistry, © 2004.
- Voet, Donald; Voet, Judith G.; Pratt, Charlotte W., Fundamentals of Biochemistry: Life at the Molecular Level, 2nd Edition, © 2006.

This book was set in 10.5/12 Adobe Garamond by Preparé and printed and bound by Courier/Kendallville. The cover was printed by Courier/Kendallville.

This book is printed on acid-free paper. ∞

To order books or for customer service, please call 1-800-CALL WILEY (225-5945).

ISBN: 978-0-470-12927-2

Printed in the United States of America

10 9 8 7 6 5 4 3 2

PREFACE

This second edition of General, Organic, and Biochemistry: an Integrated Approach has been written for students preparing for careers in health-related fields such as nursing, dental hygiene, nutrition, medical technology, and occupational therapy. The text is also suitable for students majoring in other fields where it is important to have an understanding of chemistry and its relationship to living things. Students need have no previous background in chemistry, but should possess basic math skills. For those whose math is a bit rusty, the text provides reviews of the important material. While designed for use in one-semester or two-quarter General, Organic, and Biochemistry (GOB) courses, instructors have found that it also works well for one-year courses, especially when combined with the supplement *Chemistry Case Studies For Allied Health Students* by Colleen Kelley and Wendy Weeks.

In a GOB course it is essential to show how the subject matter relates to the students' future careers. For this reason, this text makes extensive use of real-life examples from the health sciences.

ORGANIZATION

In most GOB texts, a group of chapters on general chemistry is followed by a series of organic chemistry chapters, which is followed by chapters devoted to biochemistry. Years of experience in teaching health science courses have shown the author that there is a drawback to this approach: in many cases there is a long time interval between when a topic is first presented and when it is used again—enough of a break that students' familiarity with the subject matter has often lapsed.

In introducing GOB material, this text uses an integrated approach in which related general chemistry, organic chemistry, and biochemistry topics are presented in adjacent chapters. This approach helps students see the strong connections that exist between these three branches of chemistry and allows instructors to discuss these interrelationships while the material is still fresh in students' minds. This integration involves the following sets of chapters:

- **Chapter 3 (Compounds) and Chapter 4 (An Introduction to Organic Compounds).** An introduction to bonding and compounds is followed by a look at the members of a few key organic families.
- **Chapters 3 and 4 and Chapter 6 (Reactions).** A study of inorganic and organic compounds is followed (after a look at gases, liquids, and solids in Chapter 5) by an introduction to their reactions.
- **Chapter 7 (Solutions) and Chapter 8 (Lipids and Membranes).** A discussion of solubility is followed by a look at the importance of solubility in biochemistry. Some reactions from Chapter 6 are reintroduced.
- **Chapter 9 (Acids and Bases) and Chapter 10 (Carboxylic Acids, Phenols, and Amines).** Principles of acid/base chemistry from an inorganic perspective are followed by a chapter on the organic and biochemical aspects of this topic.

- **Chapter 11 (Alcohols, Aldehydes, and Ketones) and Chapter 12 (Carbohydrates).** An introduction to the chemistry of alcohols, aldehydes and ketones is followed by a presentation of related biochemical applications.

KEY FEATURES OF THE SECOND EDITION

The many reviewers of this text have made helpful suggestions. The major changes to the second edition are:

- Newly introduced material: Molar mass (Chapter 2), step-by-step instructions for drawing molecules and polyatomic ions (Chapter 4), ionic equations and net ionic equations (Chapter 7), and naming esters and amides (Chapter 10).
- Revised material: Changed group naming from IA, IIA format to 1A, 2A format (Chapter 2), added details on naming binary molecules and introduced a table of prefixes used for naming them (Chapter 2), modified treatment of the combined gas law and the ideal gas law (Chapter 5), and updated discussion of waxes, fatty acids, phospholipids, and eicosanoids (Chapter 8).
- New Appendicies: Important families of organic compounds (Appendix B) and Naming ions, ionic compounds, binary molecules, and organic compounds (Appendix C).
- Addition of new chapter sections: Reaction types (Section 6.2), Maintaining the pH of blood serum (Section 9.10), and Oxidation of phenols (Section 10.5).

PROBLEM SOLVING

Learning to do anything requires practice, and in chemistry this practice involves solving problems. This text offers students ample opportunities to do so.

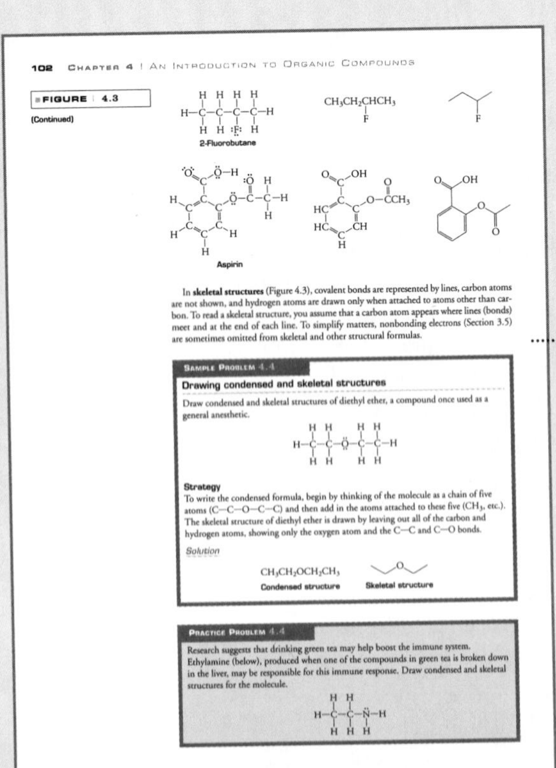
102 CHAPTER 4 | AN INTRODUCTION TO ORGANIC COMPOUNDS

FIGURE 4.3 (Continued)

2-Fluorobutane

$CH_3CH_2CHCH_3$ (F)

Aspirin

In **skeletal structures** (Figure 4.3), covalent bonds are represented by lines, carbon atoms are not shown, and hydrogen atoms are drawn only when attached to atoms other than carbon. To read a skeletal structure, you assume that a carbon atom appears where lines (bonds) meet and at the end of each line. To simplify matters, nonbonding electrons (Section 3.5) are sometimes omitted from skeletal and other structural formulas.

SAMPLE PROBLEM 4.4

Drawing condensed and skeletal structures

Draw condensed and skeletal structures of diethyl ether, a compound once used as a general anesthetic.

Strategy

To write the condensed formula, begin by thinking of the molecule as a chain of five atoms (C—C—O—C—C) and then add in the atoms attached to these five (CH_3, etc.). The skeletal structure of diethyl ether is drawn by leaving out all of the carbon and hydrogen atoms, showing only the oxygen atom and the C—C and C—O bonds.

Solution

$CH_3CH_2OCH_2CH_3$

Condensed structure

Skeletal structure

PRACTICE PROBLEM 4.4

Research suggests that drinking green tea may help boost the immune system. Ethylamine (below), produced when one of the compounds in green tea is broken down in the liver, may be responsible for this immune response. Draw condensed and skeletal structures for the molecule.

- **Sample Problems and Practice Problems.**
Each major topic is followed by a sample problem and a related practice problem. The solution to each sample problem is accompanied by a strategy to use when solving the problem. The answers to practice problems are given at the end of the chapter.

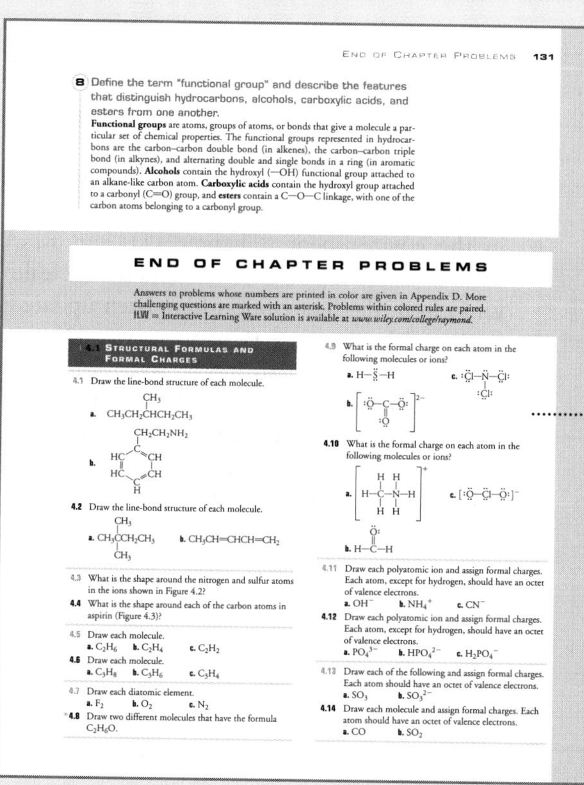

• **End of Chapter Problems.** A large number of problems can be found and the end of each chapter, and Appendix D provides answers for the odd-numbered ones. Many of the end of chapter problems are paired and some, marked with an asterisk, are more challenging than others. Each chapter includes a set of *Thinking It Through* problems that ask students to go a bit further with one or more of the concepts presented in the chapter. Over two hundred and fifty new problems have been added to the second edition. Of these, more than half are multi-part questions and many are challenging.

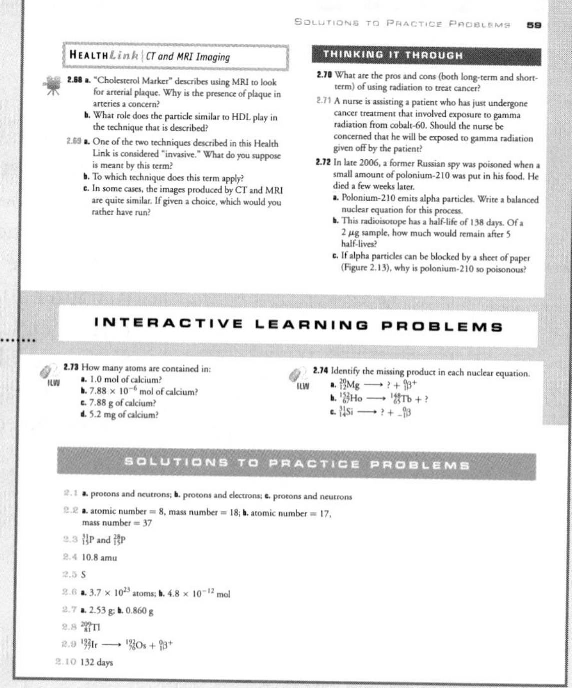

• **Interactive Learning Ware Problems.** The text's website (http://www.wiley.com/college/raymond) features Interactive Learning Ware (ILW), a step-by-step problem solving tutorial program that guides students through selected problems from the book. The ILW problems are representative of those that students frequently find most difficult and they reinforce students' critical thinking and problem solving skills. An ILW icon identifies each ILW problem in the end-of-chapter question section.

ILW

HEALTH LINKS AND BIOCHEMISTRY LINKS

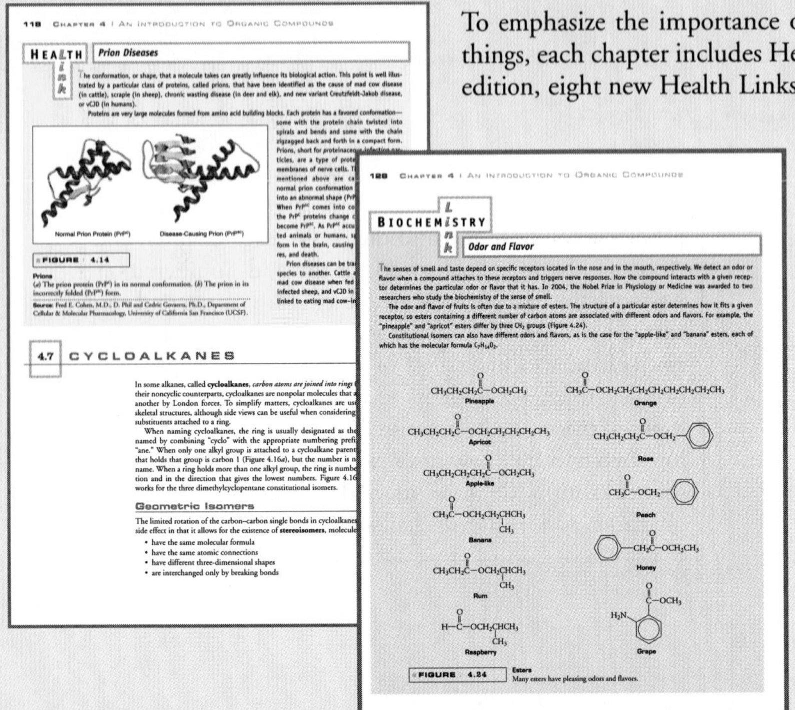

118 CHAPTER 4 | AN INTRODUCTION TO ORGANIC COMPOUNDS

HEALTH Link — Prion Diseases

The conformation, or shape, that a molecule takes can greatly influence its biological action. This point is well illustrated by a particular class of proteins, called prions, that have been identified as the cause of mad cow disease (in cattle), scrapie (in sheep), chronic wasting disease (in deer and elk), and new variant Creutzfeldt-Jakob disease, or vCJD (in humans).

Proteins are very large molecules formed from amino acid building blocks. Each protein has a favored conformation—some with the protein chain twisted into spirals and bends and some with the chain zigzagged back and forth in a compact form. Prions, short for *proteinaceous infection par-* ticles, are a type of prote membranes of nerve cells. T mentioned above are ca normal prion conformation into an abnormal shape (PrP When PrP^{SC} comes into co the PrP^{C} proteins change c become PrP^{SC}. As PrP^{SC} accu ted animals or humans, s form in the brain, causing res, and death.

Prion diseases can be tra species to another. Cattle a mad cow disease when fed infected sheep, and vCJD in linked to eating mad cow-in

FIGURE 4.14

Prions
(*a*) The prion protein (PrP^{C}) in its normal conformation. (*b*) The prion in its incorrectly folded (PrP^{SC}) form.

Source: Fred E. Cohen, M.D., D. Phil and Cedric Govaerts, Ph.D., Department of Cellular & Molecular Pharmacology, University of California San Francisco (UCSF).

4.7 CYCLOALKANES

In some alkanes, called **cycloalkanes**, *carbon atoms are joined into rings* their noncyclic counterparts, cycloalkanes are nonpolar molecules that another by London forces. To simplify matters, cycloalkanes are us skeletal structures, although side views can be useful when considering substituents attached to a ring.

When naming cycloalkanes, the ring is usually designated as the named by combining "cyclo" with the appropriate numbering prefi "ane." When only one alkyl group is attached to a cycloalkane parent that holds that group is carbon 1 (Figure 4.16*a*), but the number is n name. When a ring holds more than one alkyl group, the ring is numbe tion and in the direction that gives the lowest numbers. Figure 4.16 works for the three dimethylcyclopentane constitutional isomers.

Geometric Isomers

The limited rotation of the carbon–carbon single bonds in cycloalkanes side effect in that it allows for the existence of **stereoisomers**, molecule

- have the same molecular formula
- have the same atomic connections
- have different three-dimensional shapes
- are interchanged only by breaking bonds

128 CHAPTER 4 | AN INTRODUCTION TO ORGANIC COMPOUNDS

BIOCHEMISTRY Link — Odor and Flavor

The senses of smell and taste depend on specific receptors located in the nose and in the mouth, respectively. We detect an odor or flavor when a compound attaches to these receptors and triggers nerve responses. How the compound interacts with a given receptor determines the particular odor or flavor that it has. In 2004, the Nobel Prize in Physiology or Medicine was awarded to two researchers who study the biochemistry of the sense of smell.

The odor and flavor of fruits is often due to a mixture of esters. The structure of a particular ester determines how it fits a given receptor, so esters containing a different number of carbon atoms are associated with different odors and flavors. For example, the "pineapple" and "apricot" esters differ by three CH_2 groups (Figure 4.24).

Constitutional isomers can also have different odors and flavors, as is the case for the "apple-like" and "banana" esters, each of which has the molecular formula $C_7H_{14}O_2$.

FIGURE 4.24 **Esters**
Many esters have pleasing odors and flavors.

To emphasize the importance of chemistry to the health sciences and to living things, each chapter includes Health Links and Biochemistry Links. In the second edition, eight new Health Links have been added: Body Mass Index (Chapter 1), Dental Fillings (Chapter 3), Making Weight (Chapter 5), Saliva (Chapter 7), Omega-3 fatty acids (Chapter 8), Biofilms (Chapter 10), Drugs in the environment (Chapter 11), and RNA Interference (Chapter 14). In addition, six Health Links from the first edition have been updated: Body Temperature (Chapter 1), Dietary Reference Intakes (formerly Recommended Daily Allowances) (Chapter 2), CT and MRI (formerly CT and PET) (Chapter 2), Sunscreens (Chapter 4), *Trans* fats (Chapter 8), and Proteins in Medicine (formerly Enzymes in Medicine) (Chapter 13).

ON-LINE VIDEOS

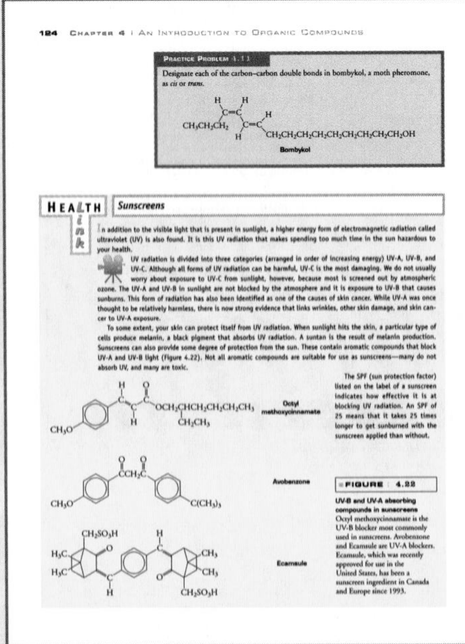

124 CHAPTER 4 | AN INTRODUCTION TO ORGANIC COMPOUNDS

PRACTICE PROBLEM 4.11

Designate each of the carbon–carbon double bonds in bombykol, a moth pheromone, as *cis* or *trans.*

HEALTH Link — Sunscreens

In addition to the visible light that is present in sunlight, a higher energy form of electromagnetic radiation called ultraviolet (UV) is also found. It is this UV radiation that makes spending too much time in the sun hazardous to your health.

UV radiation is divided into three categories (arranged in order of increasing energy) UV-A, UV-B, and UV-C. Although all forms of UV radiation can be harmful, UV-C is the most damaging. We do not usually worry about exposure to UV-C from sunlight, however, because most is screened out by atmospheric ozone. The UV-A and UV-B in sunlight are not blocked by the atmosphere and it is exposure to UV-B that causes sunburns. This form of radiation has also been identified as one of the causes of skin cancer. While UV-A was once thought to be relatively harmless, there is now strong evidence that links wrinkles, other skin damage, and skin cancer to UV-A exposure.

To some extent, your skin can protect itself from UV radiation. When sunlight hits the skin, a particular type of cells produce melanin, a black pigment that absorbs UV radiation. A suntan is the result of melanin production. Sunscreens can also provide some degree of protection from the sun. These contain aromatic compounds that block UV-A and UV-B light (Figure 4.22). Not all aromatic compounds are suitable for use as sunscreens—many do not absorb UV, and many are toxic.

The SPF (sun protection factor) listed on the label of a sunscreen indicates how effective it is at blocking UV radiation. An SPF of 25 means that it takes 25 times longer to get sunburned with the sunscreen applied than without.

FIGURE 4.22

UV-B and UV-A absorbing compounds in sunscreens
Octyl methoxycinnamate is the UV-B blocker most commonly used in sunscreens. Avobenzone and Ecamsule are UV-A blockers. Ecamsule, which was recently approved for use in the United States, has been a sunscreen ingredient in Canada and Europe since 1993.

For one topic in each chapter, a special icon is used to indicate that an on-line ScienCentral article and video clip are available for viewing.

These articles and video clips are of interest because they show how the chemistry being presented pertains to current events. Each set of end of chapter problems includes some related to the video content. The video titles include: Pork on the Run (Chapter 1), Cholesterol Marker (Chapter 2), Glowing Fish (Chapter 3), Tanning & Health (Chapter 4), Breath Of Life (Chapter 5), Hydrogen Cars (Chapter 6), Kidney Bones (Chapter 7), Teen Steroids (Chapter 8), Cystic Fibrosis Mucus (Chapter 9), Meth and the Brain (Chapter 10), Toxin Eaters (Chapter 11), Sweet Spot (Chapter 12), Young Hearts (Chapter 13), Cancer Screening (Chapter 14), Exercise Gene (Chapter 15).

OTHER TEXT FEATURES

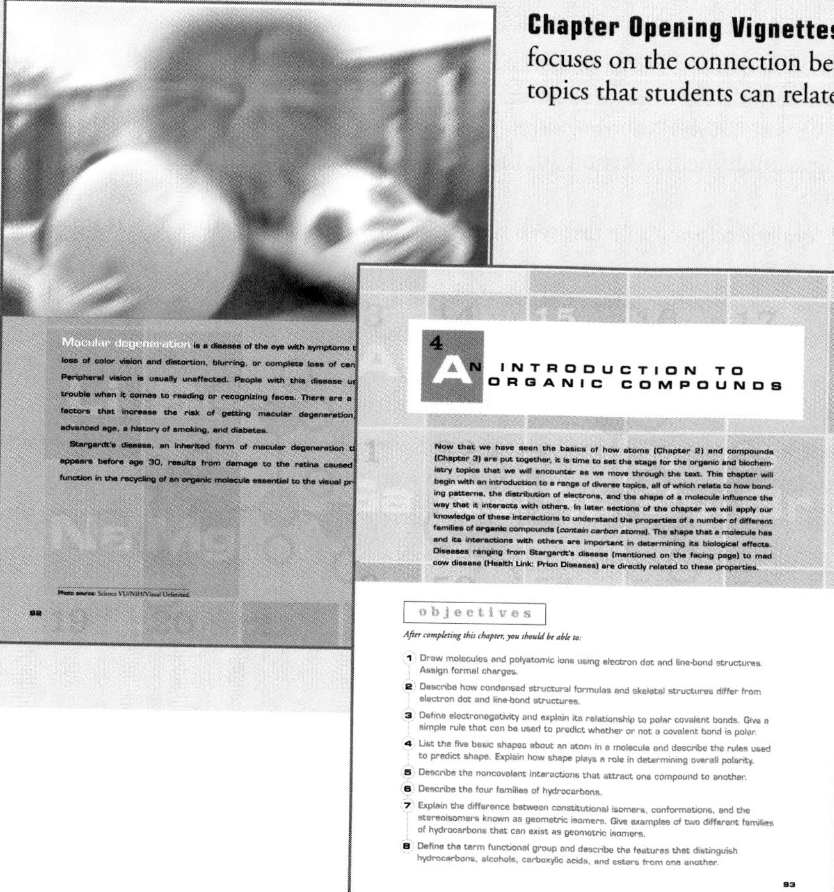
Macular degeneration is a disease of the eye with symptoms t
loss of color vision and distortion, blurring, or complete loss of cen
Peripheral vision is usually unaffected. People with this disease us
trouble when it comes to reading or recognizing faces. There are a
factors that increase the risk of getting macular degeneration,
advanced age, a history of smoking, and diabetes.
Stargardt's disease, an inherited form of macular degeneration t
appears before age 30, results from damage to the retina caused
function in the recycling of an organic molecule essential to the visual pr

Photo source: Science VU/NIH/Visual Unlimited.

92

4

AN INTRODUCTION TO ORGANIC COMPOUNDS

Now that we have seen the basics of how atoms (Chapter 2) and compounds (Chapter 3) are put together, it is time to set the stage for the organic and biochemistry topics that we will encounter as we move through the text. This chapter will begin with an introduction to a range of diverse topics, all of which relate to how bonding patterns, the distribution of electrons, and the shape of a molecule influence the way that it interacts with others. In later sections of the chapter we will apply our knowledge of these interactions to understand the properties of a number of different families of **organic** compounds (*contain carbon atoms*). The shape that a molecule has and its interactions with others are important in determining its biological effects. Diseases ranging from Stargardt's disease (mentioned on the facing page) to mad cow disease (Health Link: Prion Diseases) are directly related to these properties.

objectives

After completing this chapter, you should be able to:

1. Draw molecules and polyatomic ions using electron dot and line-bond structures. Assign formal charges.
2. Describe how condensed structural formulas and skeletal structures differ from electron dot and line-bond structures.
3. Define electronegativity and explain its relationship to polar covalent bonds. Give a simple rule that can be used to predict whether or not a covalent bond is polar.
4. List the five basic shapes about an atom in a molecule and describe the rules used to predict shape. Explain how shape plays a role in determining overall polarity.
5. Describe the noncovalent interactions that attract one compound to another.
6. Describe the four families of hydrocarbons.
7. Explain the difference between constitutional isomers, conformations, and the stereoisomers known as geometric isomers. Give examples of two different families of hydrocarbons that can exist as geometric isomers.
8. Define the term functional group and describe the features that distinguish hydrocarbons, alcohols, carboxylic acids, and esters from one another.

93

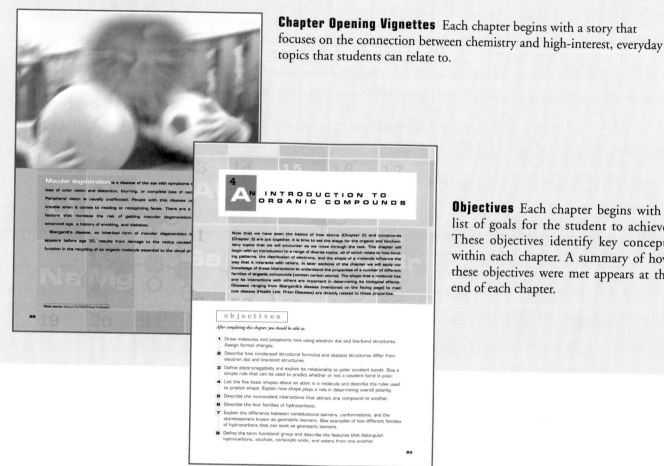

Chapter Opening Vignettes Each chapter begins with a story that focuses on the connection between chemistry and high-interest, everyday topics that students can relate to.

Objectives Each chapter begins with a list of goals for the student to achieve. These objectives identify key concepts within each chapter. A summary of how these objectives were met appears at the end of each chapter.

SUPPORT PACKAGE FOR STUDENTS

Student Web Site Within each chapter you will find an icon a indicating that there is a ScienCentral ScienCentral, Inc. news video at the website with a news story related to chapter content. The web site also features practice quizzes and Interactive Learningware problems.

Student Solutions Manual and Study Guide Written by Byron Howell, Tyler Junior College and Adeliza Flores, of Las Positas College, this supplement contains worked out solutions to the odd-numbered text problems; chapter summaries; sample problems, and practice problems.

Laboratory Manual Written by David Macaulay, Joseph Bauer, and Molly Bloomfield. This lab manual is written for the one or two-term chemistry lab course for students in the allied health sciences and related fields. These experiments are presented in an integrated table of contents and contain chapter references from General, Organic, and Biological Chemistry: An Integrated Approach, Second Edition.

Chemistry Case Studies For Allied Health Students Written by Colleen Kelley and Wendy Weeks. This manual was designed to bring relevance and critical thinking skills to the allied health chemistry course. Students are encouraged to become "diagnosticians" and apply their newly-acquired chemistry knowledge to solving real life health and environmental cases. The case manual also encourages a holistic approach by asking students to synthesize information across topics.

SUPPORT PACKAGE FOR INSTRUCTORS

PowerPoint Lecture Slides Created by Colleen Kelley, Pima Community College, these slides contain lecture outlines and key topics from each chapter of the text, along with supporting artwork and figures from the text. The slides also contain assessment questions and questions for in-class discussion.

WileyPlus Homework Management System WileyPlus is a powerful on-line tool that provides a completely integrated suite of teaching and learning resources in one easy-to-use web site. WileyPlus integrates Wiley's world-renowned content with media, including a multimedia version of the text, PowerPoint slides, on-line assessment, and more.

Digital Image Archive The text web site includes downloadable files of text images in JPEG format.

Test Bank Written by John Singer, Jackson Community College, the test bank includes multiple choice, true/false, and short answer questions.

Computerized Test Bank The IBM and Macintosh compatible version of the entire Test Bank has full editing features to help the instructor customize tests.

Instructor's Manual Written by Colleen Kelley, Pima Community College, this supplement provides Chapter Summaries and lecture outlines.

Instructor's Solutions Manual Written by Adeliza Flores, of Las Positas College, this supplement contains worked-out solutions to all of the end-of-chapter problems.

ACKNOWLEDGMENTS

I wish to thank my wife Susan and my son William for their encouragement, support, and patience.

It is with great appreciation that I acknowledge the important contributions made by Jennifer Yee, Catherine Donovan, Suzanne Ingrao, Janet Foxman, Lisa Gee, Hope Miller, Thomas Kulesa, Gabriel Dillon, Amanda Wygal, and all of the others at Wiley who were involved in helping to prepare this second edition.

Finally, I wish to acknowledge the important contributions made by the following reviewers of this text:

Jeannie Collins
University of Southern Indiana

Myriam Cotton
Pacific Lutheran University

Stephen Dunham
Moravian College

Joseph Fassler
Santa Rosa Junior College

Louis Giacinti
Milwaukee Area Technical College

Christina Goode
California State University, Fullerton

Saifunnissa B. Hassam
Santa Rosa Junior College

Colleen Kelley
Pima Community College

Ira Krull
Northeastern University

Paul Moggach
Georgian College

Michael Myers
California State University, Long Beach

Lynette Rushton
South Puget Sound Community College

Shaun Schmidt
Washburn University

John Singer
Jackson Community College

Carnetta Skipworth
Bowling Green Community College

Steven Stefanides
Wenatchee Valley College

Maria Vogt
Bloomfield College

William Wagener
West Liberty State College

Wendy Weeks
Pima Community College

Lynda Peebles
Texas Woman's University

Jennifer Powers
Kennesaw State University

Parris Powers
Volunteer State Community College

Rita Rhodes
University of Tulsa

Lenore Rodicio
Miami-Dade College

Lynette Rushton
South Puget Sound Community College

Sean Schmidt
Washburn University

Sara Selfe
Edmonds Community College

Anna Sequeira
Forsyth Tech Community College

Sonja Siewert
West Shore Community College

Carnetta Skipworth
Western Kentucky University

Robert St. Amand
Union County College

Richard Tarkka
University of Central Arkansas

Thottumkara Vinod
Western Illinois University

Tracy Whitehead
Henderson State University

Todd Wimpfheimer
Salem State College

ABOUT THE AUTHOR

Ken Raymond received a B.S. in Chemistry from Central Washington University in 1975 and a Ph.D. in Organic Chemistry from the University of Washington in 1981. Since joining the faculty of Eastern Washington University in 1982, his primary teaching responsibilities have been in the general, organic, and biochemistry series for the health sciences and in the upper-division organic chemistry lecture and laboratory series. In 1990 he received EWU's annual award for excellence in teaching. He has been chair of the Department of Chemistry and Biochemistry since 2000. When not grading papers, he plays mandolin and button accordion in a local folk band.

BRIEF CONTENTS

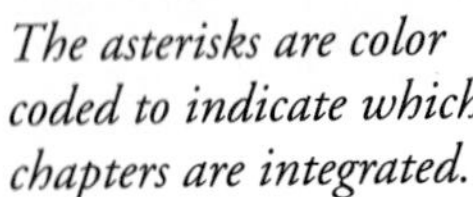

The asterisks are color coded to indicate which chapters are integrated.

CONTENTS

GENERAL, ORGANIC, AND BIOLOGICAL CHEMISTRY

An Integrated Approach

SECOND EDITION

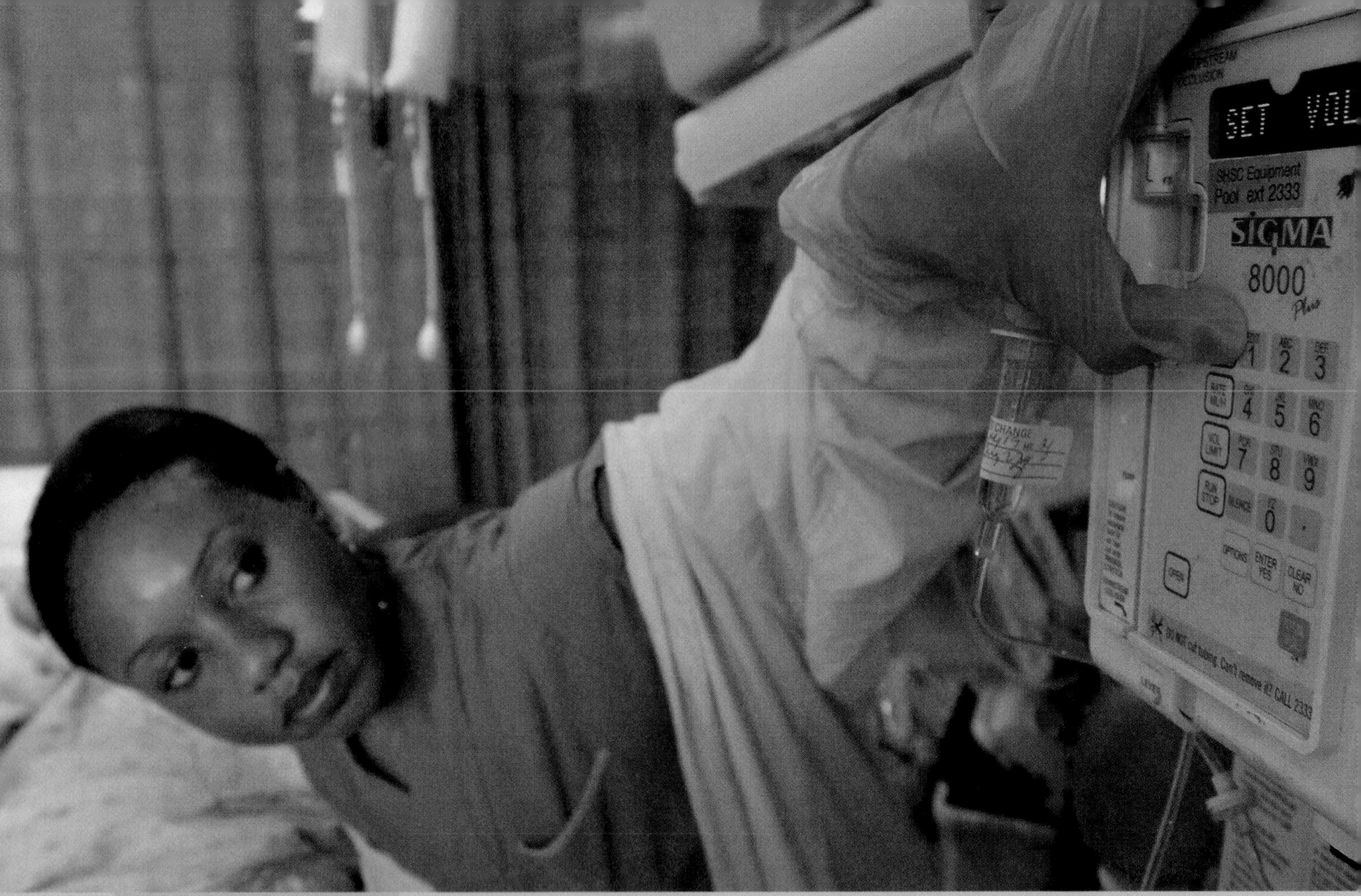

After their first chemistry lecture a group of students walks across campus together. One student says, “I want to be a nurse. I can see how we might need to study chemistry, but why does the textbook have to start with a chapter on math? What does math have to do with the health sciences?” The rest of the group voices their agreement. A nursing student walking past the group in the other direction overhears this comment and smiles. “They have no idea how much math they will use,” she thinks to herself.

Photo source: Rick Madonik/Toronto Star/Zuma Press.

1 SCIENCE AND MEASUREMENTS

In this first chapter of the text we will take a look at science, chemistry, and mathematics, and will see the important role that each plays in the health sciences.

objectives

After completing this chapter, you should be able to:

1. Explain the terms scientific method, law, theory, hypothesis, and experiment.
2. Define the terms matter and energy. Describe the three states of matter and the two forms of energy.
3. Describe and give examples of physical properties and physical change.
4. Convert from one unit of measurement into another.
5. Express values using scientific notation and metric prefixes.
6. Explain the difference between the terms accurate and precise.
7. Use the correct number of significant figures to report the results of calculations involving measured quantities.

1.1 THE SCIENTIFIC METHOD

■ Experiments test hypotheses.

Science is an approach that is used to try to make sense out of how the universe operates, ranging in scale from the very large (understanding how stars form) to the very small (understanding the behavior of the tiny particles from which everything is made). The knowledge gained from scientific studies has impacted our lives in many positive ways, including our ever-improving ability to treat diseases. For example, the medical scanners (including CT and MRI) and many of the therapeutic drugs (including antibiotics and anti-cancer drugs) used today are available as a result of the careful work of scientists (Figure 1.1).

In doing science, the **scientific method** is the process used to gather and interpret information. Observation is part of this process. One well-known story regarding the importance of observation involves the English scientist Isaac Newton (1642–1727). Reportedly, seeing an apple fall out of a tree led him to formulate the law of gravity, which states that *there is an attractive force between any two objects* (in this case, between the earth and an apple). This and other scientific **laws** are statements that *describe things that are consistently and reproducibly observed.* While a law does not explain why things happen, it can be used to predict what might happen in the future. For example, the law of gravity does not explain why things fall, but it does allow you to predict what will happen if you jump off of a ladder.

Explaining observations is a key component of the scientific method. The process begins with the construction of a **hypothesis**, *a tentative explanation (educated guess) that is based on presently known facts.* Clinicians, for example, make educated guesses when treating patients. If a patient complains of stomach pains, the clinician will ask a few questions and make a few observations before coming up with a hypothesis (diagnosis) as to the nature of the problem. This hypothesis is based on knowledge of symptoms and diseases.

The most important part of the scientific method is what happens once a hypothesis has been constructed—it must be tested by doing careful **experiments**. To test a hypothesis, a clinician might call for a series of medical tests (experiments) to be run. If the test results support the diagnosis, treatment can begin. If they invalidate the diagnosis, the clinician must revise the hypothesis and look for another cause of the illness.

Experiments must be designed so that the observations made are directly related to the question at hand. For example, if a patient has stomach pains, taking an x ray of his or her big toe will probably not help find the cause of the illness.

Once a hypothesis has survived repeated testing, it may become a **theory**—*an experimentally tested explanation of an observed behavior.* For a theory to survive, it must be consistent with existing experimental evidence, must accurately predict the results of future experiments, and must explain future observations.

Figure 1.2 shows the interconnections of the various parts of the scientific method—making an observation, forming a hypothesis, performing experiments, and creating a theory. Scientists do not necessarily follow these steps in order, nor do they always use all of the steps. It may be that an existing law suggests a new experiment or that a set of published experiments suggests a radically new hypothesis. Creativity is an important part of science; sometimes new theories arise when someone discovers an entirely new way of interpreting experimental results that hundreds of others had looked at before, but could not explain. In addition to creativity, a scientist must have sufficient knowledge of the field to be able to interpret experimental results and to evaluate hypotheses and experiments.

The fact that theories are based on experimental observations means that they sometimes change. In Section 2.1 two theories of the atom, the fundamental particle from which matter is created, are discussed. One of these theories dates back to the early 1800s, when technology was not very advanced and experiments provided much less information than is obtainable today (Figure 1.3). While the earliest theory of the atom accounted for the observations made up until the early 1800s, once better experimental results were obtained, errors were revealed.

Whether scientists study atoms or inherited diseases, theories must be continually reevaluated and, if necessary, revised as new experiments provide additional information. This change is an expected part of science.

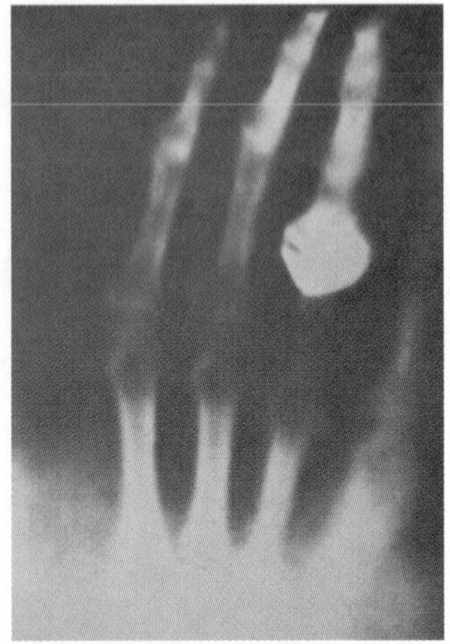

(*a*)

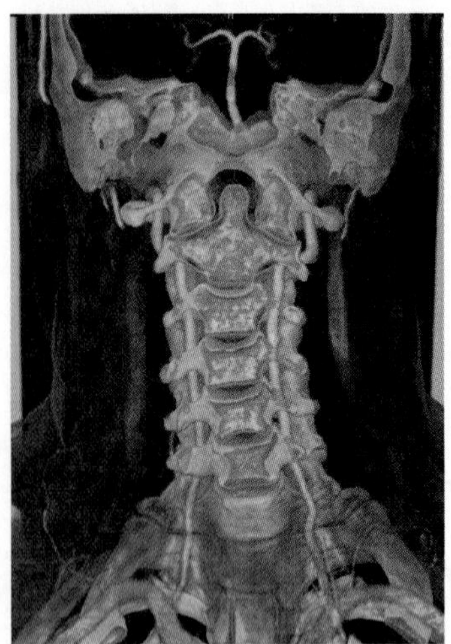

(*b*)

■ FIGURE | 1.1

Medical imaging
(*a*) The first x ray of the human body was taken in 1895 by Wilhelm Roentgen, the discoverer of x rays. In this x ray, you can see the bones of his wife's hand and her wedding ring. (*b*) With the improvements that have been made to x-ray equipment, clinicians can now obtain sharper and more detailed images, as in this scan of a patient's vertebrae.

Source: (*a*) SPL/Photoresearchers; (*b*) Gondelon/Photo Researchers, Inc.

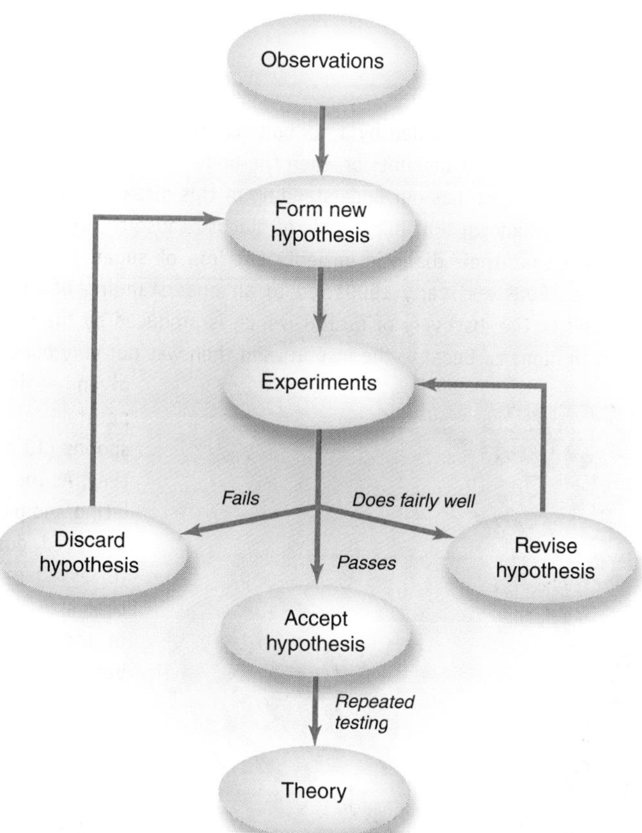

FIGURE 1.2

The scientific method
In the scientific method, experiments provide the information used to discard, revise, or accept hypotheses.

Source: From Biochemistry: A Foundation 1st edition by Ritter. © 1996. Reprinted with permission of Brooks/Cole, a division of Thomson Learning: *www.thomsonrights.com.* Fax 800 730 2215.

(*a*)

(*b*)

FIGURE 1.3

Modifying theories
Theories are sometimes revised when improved scientific equipment allows better experimental results to be obtained.

Source: (*a*) © Hulton-Deutsch Collection/Corbis; (*b*) James Holmes/Thomson Laboratories/Photo Researchers Inc.

HEALTH Link: Science and Medicine

The level of glucose (blood sugar) in the body is controlled by a hormone called insulin. Diabetes is the disease that occurs when insulin is not produced in sufficient amounts or when the body is not sensitive to its effects. As science has progressed over the years, so has our understanding of this disease and our ability to treat it. In the mid 1800s, before it was known that high levels of glucose cause the symptoms of diabetes, some physicians recommended that their diabetic patients eat lots of sugar. Others recommended starvation. Scientific studies in the late 1800s and early 1900s led to an understanding of the role that the pancreas plays in glucose metabolism and to the discovery of insulin, which is produced by the pancreas. Insulin was first used in 1922 to treat diabetes in humans. Because the insulin used then was not very pure, patients were given injections—often painful—of up to 2 teaspoons (10 milliliters) at a time. As the science of isolating and purifying insulin improved, dosages dropped to less than one-tenth of that size. Other advances in the treatment of diabetes included the use of oral drugs to control insulin levels (introduced in 1955), the use of genetically engineered human insulin (introduced in 1982) in place of that isolated from cattle and pigs, and the development of new methods to test blood glucose levels (Figure 1.4).

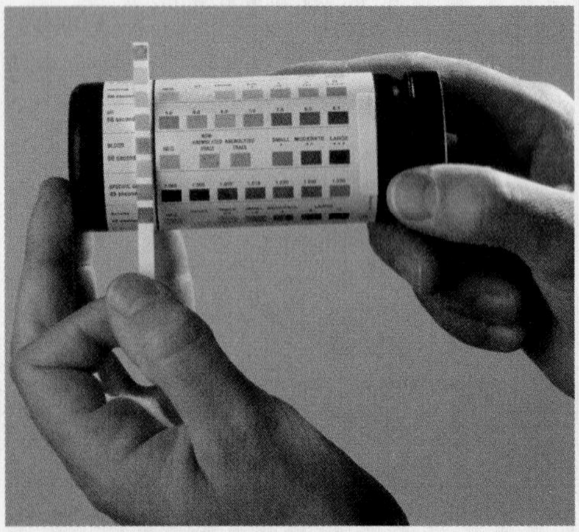

(*a*)

(*b*)

FIGURE 1.4

Glucose testing
It is important for diabetics to monitor their blood glucose levels. (*a*) When these test strips are dipped in a urine sample, the array of colors produced indicates the amount of glucose present. (*b*) Blood glucose monitors that require just a small drop of blood are alternatives to test strips.

Source: (*a*) Saturn Stills/Photo Researchers, Inc.; (*b*) Yoav Levy/Phototake.

1.2 MATTER AND ENERGY

Now that we have been introduced to some of the basic aspects of science, let us see how chemistry fits into the picture. **Chemistry** is *the study of matter and the changes it undergoes.* **Matter** is defined as *anything that has mass and occupies space.* In everyday terms, this definition includes your body, the air that you breathe, this book, and all of the other material around you.

We can describe matter in terms of **physical properties**, those characteristics that can be determined without changing the **chemical composition** of matter (what it is made of). For example, a cube of sugar is white, tastes sweet, and is odorless. The act of measuring these and other physical properties, including melting point (melting temperature), does not change the sugar into anything else—it is still sugar.

Matter is typically found in one of three different physical states or phases—as a **solid**, a **liquid**, or a **gas**. From our direct experience we know that

- *Solids have fixed shapes and volumes.*
- *Liquids have variable shapes and fixed volumes.*
- *Gases have variable shapes and volumes.*

FIGURE 1.5

Gas, liquid, and solid
A paint can (a solid), paint (a liquid), and paint fumes (a gas) illustrate the three physical states of matter.

Think about what happens if an opened can of paint gets spilled (Figure 1.5). Whether it is standing upright or lying on its side, the can (a solid) has the same shape and occupies the same volume of space. The paint (a liquid) keeps its original 1 gallon volume but changes its shape as it spreads out across the floor. The paint fumes (a gas) quickly change their shape and volume as they spread through the air in the room.

- Matter can exist in the solid, liquid, and gas states.

Converting matter between each of these states is an example of **physical change**, change in which the chemical composition of matter is not altered (Figure 1.6). Crushing a cube of sugar, boiling water to make steam, and melting an iron rod are examples of physical change.

Any time that matter is changed in any way, **work** has been done. This includes the physical changes just mentioned, as well as walking, running, or turning the pages of this book. All of these activities involve **energy**, which is defined as *the ability to do work and to transfer heat.*

Energy can be found in two forms, as **potential energy** *(stored energy)* or as **kinetic energy** *(the energy of motion)*. The water sitting behind a dam has potential energy. When the floodgates are opened and the water begins to pour through, potential energy is converted into kinetic energy.

- Potential energy is stored energy. Kinetic energy is the energy of motion.

All matter contains energy, so changes in matter (work) and changes in energy (potential or kinetic) are connected to one another. For example, if you drive a car, some of the potential energy of gasoline is converted into the kinetic energy used to move the pistons in the engine (doing work) and some is converted into heat, a form of kinetic energy related to the motion of the particles from which things are made.

FIGURE 1.6

Physical change
When snow melts in the spring, rivers fill with water. The conversion of snow into water is a physical change.

Source: Peter Van Rijn/Superstock.

SAMPLE PROBLEM 1.1

Potential versus kinetic energy

Which are mainly examples of potential energy and which are mainly examples of kinetic energy?

a. A mountain climber sits at the top of a peak.
b. A mountain climber rappels down a cliff.
c. A hamburger sits on a plate.
d. A nurse inflates a blood pressure cuff.

Strategy

Recall that potential energy is stored energy and that kinetic energy is the energy of motion.

Solution

a. potential energy
b. kinetic energy
c. potential energy
d. kinetic energy

PRACTICE PROBLEM 1.1

a. Describe changes in the kinetic energy and the potential energy of a barbell when a weightlifter picks it up.
b. Describe changes in the kinetic energy and the potential energy of a weightlifter when a barbell is picked up.

1.3 UNITS OF MEASUREMENT

FIGURE 1.7

Measuring hand strength
A dynamometer is used to measure a patient's hand strength.

Source: Keith Brofsky/Photodisc Green/Getty Images.

Making measurements is part of our everyday lives. Every time that you look at your watch to see how many minutes of class remain, tell a friend about your 5 mile run this morning, or save money by buying products with the lowest unit price, you are using measurements. Measurements are also a key part of the job of health professionals. A nurse might measure your pulse, blood pressure, and temperature, a dental hygienist might measure the depth of your gum pockets, or an occupational therapist might measure your hand strength to gauge the degree of recovery from an injury (Figure 1.7).

Measurements consist of two parts: a number and a **unit**. Saying that you swam for 3 is not very informative—was it 3 minutes, 3 hours, or 3 miles? The number must be accompanied by a unit, a quantity that is used as a standard of measurement (of time, of length, of volume, etc.). The **metric system** is the measurement system used most often worldwide. In this text we will use metric units and the **common units** used in the United States (Table 1.1). Occasionally, **SI units** (an international system of units related to the metric system) will be introduced. Table 1.2 lists some of the additional units that are commonly used in medical applications.

Mass

Mass is related to the amount of matter in a material—the more matter that it contains, the greater its mass. Units commonly used to measure mass are the gram (g), kilogram (kg), and pound (lb). A mass of 454 g is equivalent to 1 lb and 2.205 lb is equivalent to 1 kg (Figure 1.8).

(*a*)

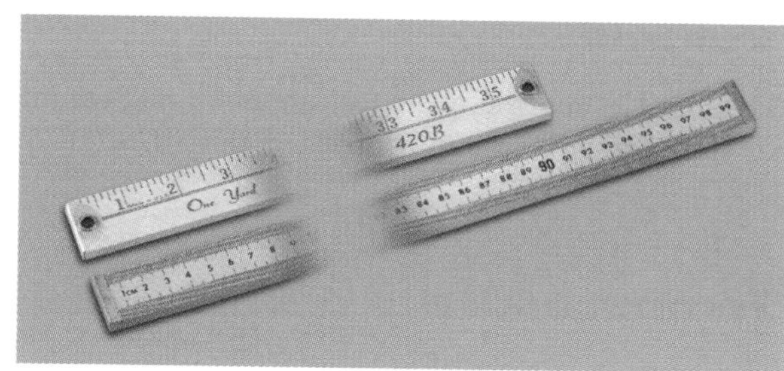

(*b*)

(*c*)

FIGURE 1.8 **Comparing units**
(*a*) One kilogram weighs a little more than two pounds. (*b*) One meter (bottom) is slightly longer than one yard (top). (*c*) One quart is slightly smaller than one liter.

Source: (*a*), (*b*) Andy Washnik/Wiley Archive; (*c*) Michael Dalton/Fundamental Photographs.

TABLE 1.1 MEASUREMENT UNITS

Quantity	Common Unit	Metric Unit	SI Unit	Relationships
Mass	Pound (lb)	Gram (g)	Kilogram (kg)	1 kg = 2.205 lb 1 kg = 1000 g
Length	Foot (ft)	Meter (m)	Meter (m)	1 m = 3.281 ft
Volume	Quart (qt)	Liter (L)	Cubic meter (m^3)	0.946 L = 1 qt 1 m^3 = 1000 L
Energy	calorie (cal)	calorie (cal)	Joule (J)	4.184 J = 1 cal
Temperature	Degree Fahrenheit (°F)	Degree Celsius (°C)	Kelvin (K)	$°F = (1.8 \times °C) + 32$ $°C = \frac{°F - 32}{1.8}$ $K = °C + 273.15$

TABLE 1.2 SOME MEASUREMENT UNITS USED IN MEDICINE

Quantity	Relationship
Mass	1 milligram (mg) = 1000 micrograms (mcg or μg)[a]
	1 grain (gr) = 65 milligrams (mg)
Volume	1 cubic centimeter (cc or cm^3) = 1 milliliter (mL)
	15 drops (gtt) = 1 milliliter (mL)
	1 teaspoon (tsp) = 5 milliliters (mL)
	1 tablespoon (T or tbsp) = 15 milliliters (mL)
	2 tablespoons (T or tbsp) = 1 ounce (oz)

[a] The prefixes micro and milli are explained in Section 1.4.

■ 454 g = 1 lb; 2.205 lb = 1 kg
1 m = 3.281 ft = 39.37 in
0.946 L = 1 qt
32°F = 0°C = 273.15 K
1 cal = 4.184 J

Length

The meter (m), which is slightly longer than a yard, is the metric unit of length. One meter is equal to 3.281 ft and 39.37 in.

Volume

The liter (L) and the quart (qt) are two units commonly used to measure volume (the amount of space occupied by a material). One quart is a bit smaller than a liter (1 qt = 0.946 L).

Temperature

The metric system uses the Celsius (°C) scale to measure temperature. On this scale, water freezes at 0°C and boils at 100°C. On the Fahrenheit (°F) scale, still used in the United States, water freezes at 32°F and boils at 212°F (Figure 1.9). Besides having different numerical values for the freezing and boiling points of water, these two temperature scales have degrees of different sizes. On the Fahrenheit scale there are 180 degrees between the temperatures where water boils and freezes (212°F − 32°F = 180°F). On the Celsius scale, however, there are only 100 degrees over this same range (100°C − 0°C = 100°C). This means that the boiling to freezing range for water has almost twice as many Fahrenheit degrees as Celsius degrees (180/100 = 1.8).

Scientists often measure temperature using the SI unit called the kelvin (K). A temperature of 0 K, known as absolute zero, is the temperature at which all heat energy has been removed from a sample. On the Kelvin temperature scale, the difference between the freezing point (273.15 K) and the boiling point (373.15 K) of water is 100 degrees, the same as that for the Celsius scale, so a kelvin is the same size as a Celsius degree.

Energy

The metric unit for energy, the calorie (cal), is defined as the amount of energy required to raise the temperature of 1 g of water from 14.5°C to 15.5°C. The SI energy unit, the joule (J), which is approximately equal to the energy expended by a human heart each time that it beats, is about one-fourth as large as a calorie (1 cal = 4.184 J).

When you hear the word "calorie," it might bring food to mind. One food Calorie (Cal) is equal to 1000 cal, which means that an 80 Cal cookie contains 80,000 cal of potential energy.

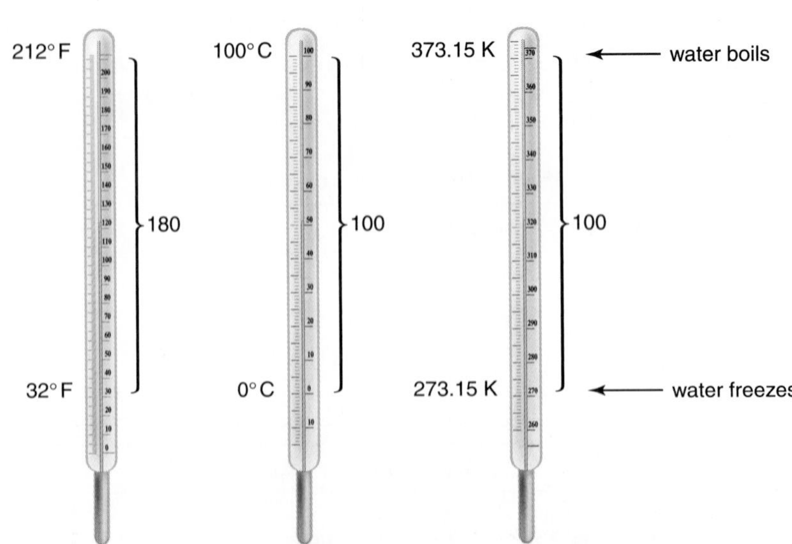

FIGURE 1.9

Temperature units
Water freezes and boils at different temperature values in the Fahrenheit, Celsius, and Kelvin scales.

SAMPLE PROBLEM 1.2

Comparing units

Which is larger?

a. 1 g or 1 lb **b.** 1 lb or 1 kg **c.** 1 L or 1 qt **d.** 1 cal or 1 J

Strategy

Refer to the "Relationships" column in Table 1.1 to get a feel for the relative sizes of these units.

Solution

a. 1 lb **b.** 1 kg **c.** 1 L **d.** 1 cal

PRACTICE PROBLEM 1.2

Which is a warmer temperature?

a. 273°C or 273 K? **b.** 32°F or 32°C? **c.** 0°F or 0 K?

1.4 SCIENTIFIC NOTATION AND METRIC PREFIXES

Scientific Notation

When making measurements, particularly in the sciences, there are many times when you must deal with very large or very small numbers. For example, a typical red blood cell has a diameter of about 0.0000075 m. In **scientific notation** (exponential notation) this diameter is written 7.5×10^{-6} m. Values expressed in scientific notation are written as a number between 1 and 10 multiplied by a power of ten. The superscripted number to the right of the ten is called an exponent.

$$7.5 \times 10^{-6}$$

A number between 1 and 10 — Exponent

An exponent with a positive value tells you how many times to multiply a number by 10,

$$3.5 \times 10^4 = 3.5 \times 10 \times 10 \times 10 \times 10 = 35000$$

$$6.22 \times 10^2 = 6.22 \times 10 \times 10 = 622$$

while an exponent with a negative value tells you how many times to divide a number by 10.

$$3.5 \times 10^{-4} = \frac{3.5}{10 \times 10 \times 10 \times 10} = 0.00035$$

$$6.22 \times 10^{-2} = \frac{6.22}{10 \times 10} = 0.0622$$

An easy way to convert a number into scientific notation is to shift the decimal point. For a number that is equal to or greater than 10, shift the decimal point to the left until you get a number between 1 and 10. The number of spaces that you moved the decimal place is the new exponent (see Table 1.3).

$$35000 = 3.5 \times 10^4$$

$$285.2 = 2.852 \times 10^2$$

$$8300000 = 8.3 \times 10^6$$

TABLE 1.3 SCIENTIFIC NOTATION

Number	Scientific Notation	Exponent
0.0001	1×10^{-4}	−4
0.001	1×10^{-3}	−3
0.01	1×10^{-2}	−2
0.1	1×10^{-1}	−1
1	1×10^{0}	0
10	1×10^{1}	1
100	1×10^{2}	2
1000	1×10^{3}	3
10000	1×10^{4}	4

For a number less than 1, shift the decimal point to the right until you get a number between 1 and 10. Put a negative sign in front of the number of spaces that you moved the decimal place and make this the new exponent.

$$0.00035 = 3.5 \times 10^{-4}$$

$$0.0445 = 4.45 \times 10^{-2}$$

$$0.00000003554 = 3.554 \times 10^{-8}$$

Metric Prefixes

Units larger and smaller than the metric units introduced in Section 1.3 can be created by attaching a prefix that indicates how the new unit relates to the original (see Table 1.4). For example, drugs are often administered in milliliter (mL) volumes. The prefix *milli* indicates that the original unit, in this case the liter, has been multiplied by 10^{-3}.

$$1 \text{ milliliter (mL)} = 1 \times 10^{-3} \text{ L}$$

Similarly, distance can be measured in kilometers. The prefix *kilo* indicates that the meter unit of length has been multiplied by 10^{3}.

$$1 \text{ kilometer (km)} = 1 \times 10^{3} \text{ m}$$

TABLE 1.4 SI AND METRIC PREFIXES

Prefix	Symbol	Multiplier	
giga	G	1,000,000,000	$= 10^{9}$
mega	M	1,000,000	$= 10^{6}$
kilo	k	1000	$= 10^{3}$
hecto	h	100	$= 10^{2}$
deka	da	10	$= 10^{1}$
		1	$= 10^{0}$
deci	d	0.1	$= 10^{-1}$
centi	c	0.01	$= 10^{-2}$
milli	m	0.001	$= 10^{-3}$
micro	μ	0.000001	$= 10^{-6}$
nano	n	0.000000001	$= 10^{-9}$

SAMPLE PROBLEM 1.3

Using scientific notation

Convert each number into scientific notation.

a. 0.0144 **b.** 144 **c.** 36.32 **d.** 0.0000098

Strategy

The decimal point is shifted to the left for numbers equal to or greater than 10 and shifted to the right for numbers less than 1.

Solution

a. 1.44×10^{-2} **b.** 1.44×10^{2} **c.** 3.632×10^{1} **d.** 9.8×10^{-6}

PRACTICE PROBLEM 1.3

One-millionth of a liter of blood contains about 5 million red blood cells. Express this volume of blood using a metric prefix and this number of cells using scientific notation.

1.5 MEASUREMENTS AND SIGNIFICANT FIGURES

We have just examined some of the units used to report the measured properties of a material. In this section we will address three of the important factors to consider when making measurements: accuracy, precision, and significant figures.

Accuracy is related to how close a measured value is to a true value. Suppose that a patient's temperature is taken twice and values of 98°F and 102°F are obtained. If the patient's actual temperature is 103°F, the second measurement is more accurate because it is closer to the true value.

■ Accurate measurements fall near the true value.

Precision is a measure of reproducibility. The closer that separate measurements come to one another, the more precise they are. Suppose that a patient's temperature is taken three times and values of 98°F, 99°F, and 97°F are obtained. Another set of temperature measurements gives 90°F, 100°F, and 96°F. The first three measurements are more in agreement with one another, so they are more precise than the second set.

■ Precise measurements are grouped together.

A set of precise measurements is not necessarily accurate and a set of accurate measurements is not necessarily precise. This is illustrated in Figure 1.10, using the game of darts as an example. Figure 1.10*a* shows the results of three shots that are precise, but not accurate—the shots fall close together, but they not are centered on the bull's-eye. In Figure 1.10*b*, the shots are accurate, but not precise, because the shots fall near the bull's-eye but not close together. Figure 1.10*c* shows three shots that are both accurate and precise.

(*a*)

(*b*)

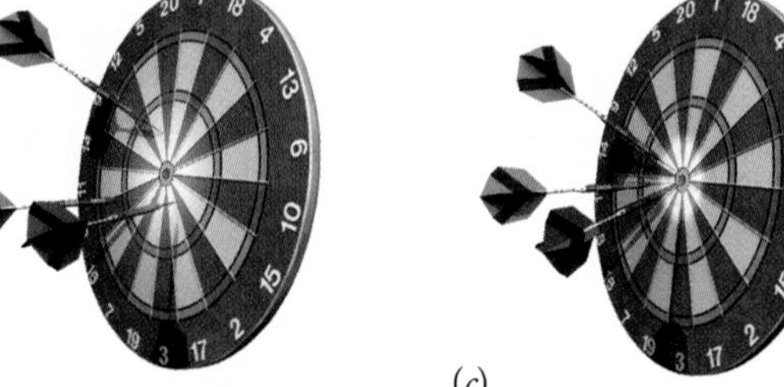

(*c*)

■ **FIGURE 1.10** **Accuracy and precision**
(*a*) The darts were thrown precisely (they are all close to one another) but not accurately. (*b*) The darts were thrown accurately (they fall near the bull's-eye) but not precisely. (*c*) The darts were thrown precisely and accurately.

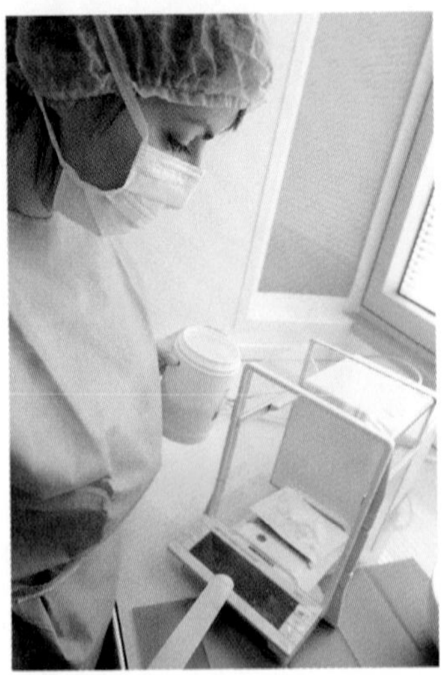

FIGURE | 1.11

Balances
Top-loading balances give a digital readout of the mass of an object.

Source: BSIP/Photo Researchers, Inc.

Significant Figures

The quality of the equipment used to make a measurement is one factor in obtaining accurate and precise results. For example, balances similar to the one shown in Figure 1.11 come in different models. A lower priced model might report masses to within ±0.1 g, and a higher priced one to within ±0.001 g.

Suppose that the precision of a balance is such that repeated measurements always agree to within ±0.1 g. On this balance, a U.S. quarter (25 cent coin) might have a reported mass of 5.6 g. This number, 5.6, has two **significant figures** (*those digits in a measurement that are reproducible when the measurement is repeated, plus the first doubtful digit*). Here the "6" in 5.6 is doubtful, because the balance reports mass with an error of ±0.1 g. Assuming that the balance is accurate, the actual mass of the quarter may be a little bit more or a little bit less than 5.6 grams.

On a different balance that reports masses with a precision of ±0.001 g, the reported mass of the same quarter might be 5.563 g. Using this measuring device, the mass of the quarter is reported with four significant figures.

For the numbers above (5.6 and 5.563), determining significant figures is straightforward: all of the digits written are significant. Things get a bit trickier when zeros are involved, because zeros that are part of the measurement are significant, while those that only specify the position of the decimal point are not. Table 1.5 summarizes the rules for determining when a digit is significant.

It is important to note that significant figures apply only to measurements, because measurements always contain some degree of error. Numbers have no error when they are obtained by an *exact count* (there are seven patients sitting in the waiting room) or are *defined* (12 eggs = 1 dozen, 1 km = 1000 m). These **exact numbers** have an unlimited number of significant figures.

TABLE | 1.5 SIGNIFICANT FIGURES

	Examples	Number of Significant Figures
Numbers are significant if they are:		
a. Nonzero digits	3.4	2
	25.85	4
b. Zeros between nonzero digits	308	3
	97.0002	6
c. Zeros at the end of a number when a decimal point is written	5.010	4
	200.	3
d. Digits in a number written in scientific notation (not including the power of ten)	7.0×10^{-5}	2
	6.02×10^{23}	3
Numbers are not significant if they are:		
a. Zeros at the beginning of a number	0.543	3
	0.0006	1
b. Zeros at the end of a number when no decimal point is written	200	1
	1,500,000	2

SAMPLE PROBLEM 1.4

Determining significant figures

Specify the number of significant figures in each measured value.

a. 30.1°C **b.** 0.00730 m **c.** 7.30×10^3 m **d.** 44.50 mL

Strategy

All nonzero digits are significant. Zeros, however, are significant only under certain conditions (see Table 1.5).

Solution

a. 3 **b.** 3 **c.** 3 **d.** 4

PRACTICE PROBLEM 1.4

Write each measured value in exponential notation, being sure to give the correct number of significant figures.

a. 7032 cal **b.** 88.0 L **c.** 0.00005 g **d.** 0.06430 lb

HEALTH Link: Body Mass Index

In early 2006, the Harris Poll® released a report titled, "Obesity Epidemic Continues to Worsen in the United States." According to this report, the percentage of Americans who are overweight or obese continues to rise. Based on Body Mass Index (discussed below), 66% of U.S. adults are overweight, and 27% are also obese. These numbers, which echo those given in other reports, are of concern to health professionals because being overweight or obese increases a person's risk of developing health problems. Among the identified overweight- and obesity-related diseases are type II diabetes, heart disease, high blood pressure, high cholesterol levels, stroke, cancer (colon, breast, and endometrial), and asthma.

For years, determining whether someone was overweight involved using height and weight charts. These charts were of limited usefulness because, when it comes to assessing the risk of overweight- and obesity-related disease, body weight is not the main the issue. The primary factor to consider is the percentage of body weight that is due to fat.

What body fat percentage is considered healthy? There is no single answer to this question because recommended body fat levels depend on a variety of factors. Table 1.6 lists the gender- and age-based recommendations for adults.

Health professionals can determine percentage body fat using a variety of techniques. One of these is the skin-fold measurement, in which calipers are used to test the thickness of folds of skin at various places on the body

TABLE 1.6 RECOMMENDED PERCENT BODY FAT

	Age	Underweight	Recommended weight	Overweight	Obese
Female	20–39	<23	23–33	34–38	>38
	40–59	<23	23–35	36–39	>39
	60–79	<25	25–35	36–41	>41
Male	20–39	<10	10–20	21–26	>26
	40–59	<11	11–22	23–27	>27
	60–79	<13	13–23	24–29	>29

Values also depend upon ethnicity. The values reported here are averages for the ethnic groups studied in Gallagher et al., *American Journal of Clinical Nutrition*, 2000; 72:694–701.

(Figure 1.12). A calculation using the measured values gives body fat percentage. Underwater weighing is another method used to determine body fat levels. Because fat has a lower density than muscle and bone, the more fat that a person has, the less the person will weigh underwater. Once measurements have been made, a set of equations is used to calculate percent body fat. In a different technique called bioelectrical impedance, electrodes are placed on different parts of the body and a low electrical current is applied. Fat is a poorer conductor of electricity than muscle and bone, so the higher the percent body fat, the greater the resistance or impedance to the current.

Although it is not based on direct measurements of percent body fat, Body Mass Index (BMI) is a good predictor of an individual's risk for overweight- or obesity-related disease. BMI is calculated from a person's weight and height, using the equation

$$\text{BMI} = 703 \times \frac{\text{weight (lb)}}{\text{height (in)}^2}$$

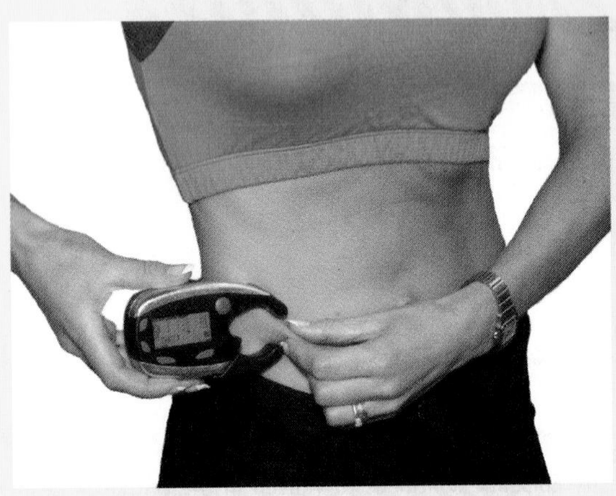

FIGURE 1.12

Skin fold calipers
Skinfold measurements can be used to determine percent body fat.

Source: Courtesy of Accu-Measure, LLC.

A 5 foot 1 inch (61.0 inch) tall person weighing 145 lbs is calculated to have a BMI of 27.4.

$$\text{BMI} = 703 \times \frac{\text{weight (lb)}}{\text{height (in)}^2} = 703 \times \frac{145}{61.0^2} = 27.4$$

According to adult BMI standards (Table 1.7), a person with a BMI of 27.4 is overweight. BMI is calculated in the same way for children and teens, but the interpretation of BMI differs.

TABLE 1.7 ADULT BODY MASS INDEX

BMI	Condition
Below 18.5	Underweight
18.5–24.9	Recommended weight
25.0–29.9	Overweight
30.0 or higher	Obese

Calculations Involving Significant Figures

Reporting answers with too many or too few significant figures is a problem commonly encountered with calculations involving measured values. The important thing to remember is that *calculations should not change the degree of uncertainty in a value.*

When doing multiplication or division with measured values, *the answer should have the same number of significant figures as the quantity with the fewest.* Suppose that you are asked to determine the area of a rectangle. According to your measurements, its width is 5.3 cm and its length is 6.1 cm. Since area = width × length, you use your calculator to multiply the two, and obtain

5.3 cm	Two significant figures
× 6.1 cm	Two significant figures
32.33 cm^2	Calculator answer (four significant figures)
(32 cm^2)	Correct answer (two significant figures)

The result given by your calculator has too many significant figures. Each of the original numbers (5.3 and 6.1) has just two significant figures, but the calculator has given an answer with four. Rewriting a number with the proper number of significant figures means that we have to drop the digits that are not significant (in this case, the two to the right of the decimal point) and round off the last digit of the number. We will use the following rules when rounding numbers:

- If the first digit to be removed is 0, 1, 2, 3, or 4, leave the last reported digit unchanged. (57.42 rounds off to 57.4 if three significant figures are needed and to 57 if two significant figures are needed.)
- If the first digit to be removed is 5, 6, 7, 8, or 9, increase the last reported digit by 1. (57.69 rounds off to 57.7 if three significant figures are needed and to 58 if two significant figures are needed.)

When doing addition or subtraction with measured values, *the answer should have the same number of decimal places as the quantity with the fewest decimal places.* Suppose that you are given three mass measurements and are asked to calculate the total mass:

13.5	g	One decimal place
2.335	g	Three decimal places
+653	g	Zero decimal places
668.835	g	Calculator answer (three decimal places)
(669	g)	Correct answer (Zero decimal places)

Sample Problem 1.5

Calculations involving significant figures

Each of the numbers below is measured. Solve the calculations and give the answer with the correct number of significant figures.

a. 0.12×1.77
b. $690.4 \div 12$
c. $5.444 - 0.44$
d. $16.5 + 0.114 + 3.55$

Strategy

You must apply a different rule to carry significant figures through multiplication or division than to carry them through addition or subtraction.

Solution

a. 0.21 **b.** 58 **c.** 5.00 **d.** 20.2

Practice Problem 1.5

Each of the numbers below is measured. Solve the calculations and give the answer with the correct number of significant figures.

a. 53.4×489.6
b. $6.333 \times 10^{-4} \times 5.77 \times 10^{3}$
c. $(5 \times 989.5) \div 16.3$
d. $(0.45 \times 6) + 3.14$

1.6 CONVERSION FACTORS AND THE FACTOR LABEL METHOD

What is your height in inches and in centimeters? What is the volume of a cup of coffee in milliliters? Answering these questions requires that you convert from one unit into another.

Some **unit conversions** are simple enough that you can probably do them in your head—six eggs are half a dozen and twenty-four inches are two feet. Solving other conversions may require a systematic approach called the **factor label method**, which uses **conversion factors** to transform one unit into another. Conversion factors are derived from the numerical relationship between two units.

Suppose that a 185 lb patient is prescribed a drug whose recommended dosage is listed in terms of kilograms of body weight. To administer the correct dose, you must convert

■ Although the terms "mass" and "weight" are often used interchangeably, they do not mean the same thing. Mass is related to the amount of matter in an object, while weight depends on gravity. An astronaut weighs much less in space than on the surface of the earth, but his or her mass does not change.

HEALTH *Link* Body Temperature

When you go in for a medical checkup, a health professional will almost always begin by taking your temperature. This is done because running a fever is a sign of illness. What *should* your temperature be? A temperature of 98.6°F (37.0°C), measured orally, is considered normal. This normal temperature is actually an average of the typical range of oral body temperatures (97.2 – 99.9°F) recorded for healthy people.

The human body is divided into two different temperature zones, the core and the shell, so temperature readings will vary depending on which part of your body is measured. The body's internal core, which holds the organs of the abdomen, chest, and head, is held at a constant temperature. The outer shell, that part of the body nearest the skin, is used to insulate the core. Shell temperatures fluctuate, depending on the whether the body is trying to keep or to lose heat, and typically shell temperatures run about 1°F lower than core temperatures.

Rectal temperature measurements are a good way to determine the core body temperature. While oral measurements can indicate core temperature, readings may be incorrect if the thermometer is not placed correctly in the mouth. Hot or cold drinks can also affect the results of oral temperature measurements.

Tympanic membrane (eardrum) measurements give an indication of the core temperature of the brain, while *axillary* (armpit) and *temporal artery* (an artery in the head that runs near the temple) give the shell temperature. Like oral measurements, these three methods are prone to error.

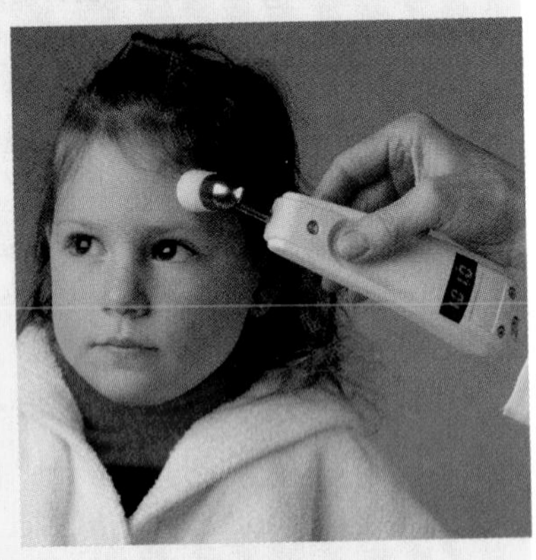

FIGURE 1.13

Temporal artery thermometer
A temporal artery thermometer measures temperature by detecting infrared (IR) energy released at the temporal artery.

Source: Courtesy Exergen Corporation.

How are temperatures measured?

A variety of methods can be used to take someone's temperature. The "low tech" method used by countless parents is touch—does your child's forehead feel hot? As you might expect, this is not the most reliable technique.

For centuries the *mercury thermometer* has been used to measure temperature. Its operation is based on the fact that mercury expands as it gets warmer—the higher the temperature, the longer the column of mercury in a thermometer. These thermometers, used for rectal, oral, and axillary temperature measurements, have fallen out of favor because they can be difficult to read, can transmit infection when not cleaned properly, and, if broken, can expose people to toxic mercury.

The *digital thermometer* is one alternative to the mercury thermometer. The operation of this thermometer is based on a thermistor, a device that conducts electricity better the higher the temperature. Digital thermometers, like mercury thermometers, are used to measure rectal, oral, and axillary temperatures. A digital thermometer pacifier has been developed for infant use.

A third type of thermometer measures temperature by detecting infrared (IR) energy (Section 2.6), a form of energy that is associated with heat. *Tympanic membrane* and *temporal artery thermometers* (Figure 1.13), which operate using IR energy, are quick to use but, in the case of the tympanic thermometers, can give false readings.

In an interesting application of IR-based temperature measurement, some airports have installed automated remote temperature sensors to scan people for fever (Figure 1.14). These devices, originally installed to help control the spread of SARS (severe acute respiratory syndrome), may be used to help fight avian flu.

FIGURE 1.14

Remote temperature sensing
In an attempt to slow the spread of SARS (severe acute respiratory syndrome), some airports have installed automated IR sensors to identify travelers who are running a fever.

Source: Courtesy of Infrared Solutions Inc. www.infraredsolutions.com

the patient's pound weight into kilograms. Converting from pounds to kilograms makes use of the equality 2.205 lb = 1 kg (Table 1.1). Two different conversion factors can be created from this relationship, the first of which is produced by dividing both sides of the equality by 1 kg. This and all other conversion factors are equal to 1.

$$2.205 \text{ lb} = 1 \text{ kg} \qquad \frac{2.205 \text{ lb}}{1 \text{ kg}} = \frac{1 \text{ kg}}{1 \text{ kg}} = 1 \qquad \text{conversion factor: } \frac{2.205 \text{ lb}}{1 \text{ kg}}$$

The second conversion factor is created by dividing both sides of the equality by 2.205 lb

$$1 \text{ kg} = 2.205 \text{ lb} \qquad \frac{1 \text{ kg}}{2.205 \text{ lb}} = \frac{2.205 \text{ lb}}{2.205 \text{ lb}} = 1 \qquad \text{conversion factor: } \frac{1 \text{ kg}}{2.205 \text{ lb}}$$

What is the kilogram weight of a 185 lb patient? To answer this question using the factor label method, we multiply 185 lb by the appropriate conversion factor (equal to 1). In this case the conversion factor to use is the one that has the desired new unit in the numerator. This allows the original units to cancel one another (Figure 1.15).

$$185 \cancel{\text{lb}} \times \frac{1 \text{ kg}}{2.205 \cancel{\text{lb}}} = 83.9 \text{ kg}$$

Looking at this answer, you might wonder why it is reported with three significant figures. In the equality 1 kg = 2.205 lb, the "1" is an exact number and has an unlimited number of significant figures. The value with the fewest significant figures is 185 lb.

Let us try another one. A vial contains 15 mL of blood serum. Convert this volume into liters. Converting from mL into L uses the equality 1 mL = 1×10^{-3} L (Table 1.4). The two conversion factors derived from this relationship are

$$\frac{1 \text{ mL}}{1 \times 10^{-3} \text{ L}} \quad \text{and} \quad \frac{1 \times 10^{-3} \text{ L}}{1 \text{ mL}}$$

and the conversion factor to use is the one with the new unit (L) in the numerator.

$$15 \cancel{\text{mL}} \times \frac{1 \times 10^{-3} \text{ L}}{1 \cancel{\text{mL}}} = 1.5 \times 10^{-2} \text{ L}$$

FIGURE 1.15

A problem with unit conversions

In September of 1999, the Mars Climate Orbiter (a $168 million weather satellite) fired its main engine to drop into orbit around Mars. Unfortunately, due to a mix-up in units—the computer on the orbiter used SI units, but NASA scientists sent it information in common units—the orbiter crashed into the planet.

Source: NASA.

If you do not have access to a direct relationship between two different units, making a unit conversion may require more than one step. Suppose that you are asked to convert the average volume of blood pumped by one beat of your heart (0.070 L) from liters into cups. What is this volume in cups? You may not know the direct relationship between liters and cups, but 0.946 L = 1 qt (Table 1.1). This gives the conversion factors

$$\frac{1\ \text{qt}}{0.946\ \text{L}} \quad \text{and} \quad \frac{0.946\ \text{L}}{1\ \text{qt}}$$

which means that

$$0.070\ \cancel{\text{L}} \times \frac{1\ \text{qt}}{0.946\ \cancel{\text{L}}} = 0.074\ \text{qt}$$

Knowing that there are 4 cups in one quart gives the equation

$$0.074\ \cancel{\text{qt}} \times \frac{4\ \text{cups}}{1\ \cancel{\text{qt}}} = 0.30\ \text{cup}$$

so 0.070 L is the same volume as 0.30 cup.

The two steps of this conversion can be taken care of at once by incorporating both conversion factors into one equation.

$$0.070\ \cancel{\text{L}} \times \frac{1\ \cancel{\text{qt}}}{0.946\ \cancel{\text{L}}} \times \frac{4\ \text{cups}}{1\ \cancel{\text{qt}}} = 0.30\ \text{cup}$$

SAMPLE PROBLEM 1.6

Unit conversions

a. A bottle of cough syrup lists the recommended adult dosage as 1.5 tablespoons. Convert this dosage into milliliters.

b. An aspirin tablet contains 5.0 grains of aspirin. Convert this mass into grams.

Strategy

To solve each part of this problem you must find the appropriate conversion factor. In part a, the conversion factor is built on the relationship between tablespoons and milliliters (see Table 1.2).

Solution

a. $1.5\ \cancel{\text{T}} \times \frac{15\ \text{mL}}{1\ \cancel{\text{T}}} = 23\ \text{mL}$

b. $5.0\ \cancel{\text{gr}} \times \frac{65\ \cancel{\text{mg}}}{1\ \cancel{\text{gr}}} \times \frac{1 \times 10^{-3}\ \text{g}}{1\ \cancel{\text{mg}}} = 0.33\ \text{g}$

PRACTICE PROBLEM 1.6

a. Naloxone is a drug used in emergency rooms to treat narcotic overdoses. For children, the recommended dosage of naloxone is 0.010 mg/kg of body weight. What dosage (in mg) should be prescribed for a 12 kg child?

b. Naloxone is sold in quantities of 0.40 mg/mL. How many milliliters of the drug should be administered to a 12 kg child?

Temperature Conversions

To convert between degrees Fahrenheit and degrees Celsius, one of the two equations below is used.

$$°F = (1.8 \times °C) + 32 \qquad °C = \frac{°F - 32}{1.8}$$

Between the freezing point (32°F, 0°C) and boiling point (212°F, 100°C) of water there are 180°F and 100°C. The ratio 180/100 equals 1.8, which is the source of this term in the equations—a degree Fahrenheit is 1.8 times smaller than a degree Celsius. The 32 comes from the different freezing point for water on the two temperature scales.

The relationship between degrees Celsius and kelvins is

$$K = °C + 273.15 \qquad °C = K - 273.15$$

These equations are simpler than the ones used to relate °F and °C because kelvins and Celsius degrees are the same size.

Let us see how a temperature conversion would be carried out. On a warm summer day, the temperature reaches 85°F. What is this temperature in °C?

$$°C = \frac{°F - 32}{1.8} = \frac{85 - 32}{1.8} = 29°C$$

SAMPLE PROBLEM 1.7

Temperature conversions

Liquid nitrogen (N_2), which has a freezing point of −210°C, is often used to remove warts and to treat precancerous skin lesions. Convert this temperature into kelvins.

Strategy

To solve this problem you must use an equation that relates °C and K.

Solution

$K = °C + 273.15 = -210 + 273.15 = 63\ K$

PRACTICE PROBLEM 1.7

The liquid helium used in magnetic resonance imagers (MRIs) has a temperature of 4.1 K. Convert this temperature into °C and °F.

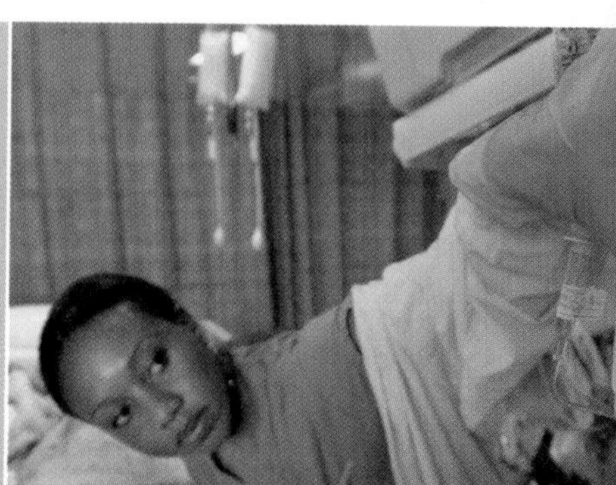

Why does the first chapter of a chemistry textbook for the health sciences include a discussion of math? Because math is an essential tool for those who plan a career in the health care field. Nurses, for example, might be called upon to calculate drug dosages (if 30 mg is the prescribed dose, how many milliliters should be drawn from a vial containing 50 mg/mL?) or IV drip rates (how many gtt/min to administer 50 cc over a 1 hr period?). Understanding how to do calculations of this sort is vital: the well-being of patients depends on getting the correct answer every time.

summary of objectives

1 Explain the terms scientific method, law, theory, hypothesis, and experiment.

Scientists use the **scientific method** to collect and interpret information. Scientific **laws** describe observations but do not attempt to explain them. A scientific **theory** is a **hypothesis** (tentative explanation) that has survived repeated testing by **experimentation**.

2 Define the terms matter and energy. Describe the three states of matter and the two forms of energy.

Matter has mass and occupies space, while **energy** is the capacity to do **work** and transfer heat. The three physical states of matter are **solid** (has a fixed shape and volume), **liquid** (has a variable shape and a fixed volume), and **gas** (has a variable shape and volume). Energy is found in two forms, as **potential energy** (stored energy) or **kinetic energy** (energy of motion).

3 Describe and give examples of physical properties and physical change.

The **physical properties** of matter, including odor and melting point, are those that can be determined without affecting the **chemical composition** of matter. **Physical changes**, including boiling, melting, and crushing, are those that do not affect chemical composition.

4 Convert from one unit of measurement into another.

Converting between units may require the use of a **conversion factor**, which is based on the relationship between two units. For example, the two conversion factors that may be formed from the equality 2.205 lb = 1 kg are

$$\frac{2.205\ \text{lb}}{1\ \text{kg}} \quad \text{and} \quad \frac{1\ \text{kg}}{2.205\ \text{lb}}$$

In the **factor label method**, converting 95 lb into kilograms involves multiplying by the conversion factor that cancels the original (lb) unit.

$$95\ \cancel{\text{lb}} \times \frac{1\ \text{kg}}{2.205\ \cancel{\text{lb}}} = 43\ \text{kg}$$

5 Express values using scientific notation and metric prefixes.

In **scientific notation**, values are expressed as a number between 1 and 10, multiplied by a power of ten. For example, in scientific notation 0.025 is written 2.5×10^{-2} and 6401 is written 6.401×10^{3}. **Metric prefixes** (kilo, centi, milli, etc.) can be used to create units of different sizes. For example, 2503 meters (m) can be expressed as 2.503 kilometers (km) and 4×10^{-2} m as 4 centimeters (cm).

6 Explain the difference between the terms accurate and precise.

Accurate measurements are close to the true value. **Precise** measurements are grouped closely together.

7 Use the correct number of significant figures to report the results of calculations involving measured quantities.
When reporting the results of measurements, values should be given with the correct number of **significant figures** (all certain digits plus the first uncertain one). **Exact numbers** (which come from an exact count or a definition) have no uncertain digits. When doing calculations involving measured values, the answer must be **rounded off** to the correct number of significant figures. For example, 5.5 g + 23.44 g + 0.225 g = 29.2 g and 88 cm × 62.33 cm = 5.5×10^3 cm^2.

END OF CHAPTER PROBLEMS

Answers to problems whose numbers are printed in color are given in Appendix D. More challenging questions are marked with an asterisk. Problems within color rules are paired. **ILW** = Interactive Learning Ware solution is available at *www.wiley.com/college/raymond.*

1.1 The Scientific Method

1.1 Is the statement "What goes up must come down" a scientific law or a scientific theory? Explain.

1.2 Centuries before any experiments had been carried out, philosophers had proposed the existence of the atom. Why are the proposals of these philosophers not considered theories?

1.3 How is a theory different from a hypothesis?

1.2 Matter and Energy

1.4 Define the terms "matter" and "energy."

1.5 What are the three states of matter?

1.6 A battery-powered remote control toy car sits at the bottom of a hill. The car begins to move and is steered up the hill.
- **a.** Describe the changes in kinetic energy that take place.
- **b.** Describe the changes in potential energy that take place.
- **c.** What work is done?

1.7 A gymnast runs across a mat and does a series of flips.
- **a.** Describe the changes in kinetic energy that take place.
- **b.** Describe the changes in potential energy that take place.
- **c.** What work is done?

1.8 On a hot day, a glass of iced tea is placed on a table.
- **a.** What are some of the physical properties of the ice?
- **b.** What change in physical state would you expect to take place if the iced tea sits in the sun for a while?

1.9
- **a.** List some of the physical properties of a piece of copper wire.
- **b.** Give examples of some of the physical changes that a piece of copper wire could undergo.

1.10 Suppose that you are camping in the winter. To obtain drinking water, you use a propane-fueled camp stove to melt snow.
- **a.** Is the melting of snow a physical process or a chemical process?
- **b.** Is the burning of propane a physical process or a chemical process?
- **c.** Describe the potential energy change that takes place for propane as it burns in the stove.
- **d.** Describe the kinetic energy change that takes place for water as the snow melts.

1.11 Rather than melting snow on a camp stove as described in the previous problem, you decide to eat handfuls of snow.
- **a.** Describe the change in the potential energy of your body as the snow that you have swallowed is melted.
- **b.** If you are stranded in the woods during the winter, why is it better to obtain water by melting snow than eating it?

1.3 Units of Measurement

1.12 Which is larger?
- **a.** 1 yd or 1 m
- **b.** 1 lb or 1 g
- **c.** 1 cup or 1 mL

1.13 Which is larger?
- **a.** 1 pint or 1 L
- **b.** 1°F or 1°C
- **c.** 1 gal or 1 L

1.14 In many countries, the energy content of food is listed in kilojoules (kJ), where 1 kJ = 1×10^3 J. Which is larger, one food Calorie or one kilojoule?

1.15 Which is larger?
a. 1 mg or 1 μg
b. 1 grain or 1 mg
c. 1 T or 1 t
d. 1 T or 1 oz

1.16 Which is larger?
a. 1 gtt or 1 mL
b. 1 T or 1 mL
c. 1 t or 1 oz
d. 1 mg or 1 gr

1.4 Scientific Notation and Metric Prefixes

1.17 Express each number using metric prefixes (example: one-tenth = deci or d). Give both the name and the abbreviation for each prefix.
a. one-thousandth
b. one million
c. one-hundredth

1.18 Express each number using metric prefixes (example: one-tenth = deci or d). Give both the name and the abbreviation for each prefix.
a. one thousandth
b. one-millionth
c. one-billionth

1.19 Express each distance in scientific notation and ordinary (decimal) notation, without using metric prefixes (example: 6.2 cm = 6.2×10^{-2} m = 0.062 m).
a. 1.5 km
b. 5.67 mm
c. 5.67 nm
d. 0.3 cm

1.20 Express each mass in scientific notation and ordinary (decimal) notation, without using metric prefixes (example: 5 μg = 5×10^{-6} g = 0.000005 g).
a. 24 μg
b. 0.716 mg
c. 15 dg
d. 412 kg

1.21 Which is the greater amount of energy?
a. 1 kcal or 1 kJ
b. 4.184 cal or 1 J

1.22 Which is the greater amount of energy?
a. 100 cal or 100 J
b. 100 Cal or 100 kJ

1.23 **a.** How many meters are in 1 km?
b. How many meters are in 5 km?
c. How many millimeters are in 1 m?

1.24 **a.** How many grams are in 2 kg?
b. How many grams are in 0.25 mg?
c. How many micrograms are in 0.25 g?

1.25 **a.** How many millimeters are in 2 km?
b. How many centiliters are in 3 dL?
c. How many kilograms are in 4 mg?

1.26 **a.** How many micrometers are in 5 mm?
b. How many milliliters are in 6 μL?
c. How many decigrams are in 9 ng?

1.5 Measurements and Significant Figures

1.27 How many significant figures does each number have? Assume that each is a measured value.
a. 1000000.5
b. 887.60
c. 0.668
d. 45
e. 0.00045

1.28 How many significant figures does each number have? Assume that each is a measured value.
a. 1.466
b. 3.5895
c. 600.2
d. 4.55×10^3
e. 0.001
f. 2×10^1
g. 2.0×10^1

1.29 Solve each calculation, reporting each answer with the correct number of significant figures. Assume that each value is a measured quantity.
a. 14×3.6
b. $0.0027 \div 6.7784$
c. $12.567 + 34$
d. $(1.2 \times 10^3 \times 0.66) + 1.0$

1.30 Solve each calculation, reporting each answer with the correct number of significant figures. Assume that each value is a measured quantity.
a. 0.114×5.2377
b. 3.11×14.5
c. $123.667 - 78.9$
d. $(6.21 + 0.04) \times 16.72$

1.31 A microbiologist wants to know the circumference of a cell being viewed through a microscope. Estimating the diameter of the cell to be 11 μm and knowing that circumference = π × diameter (we will assume that the cell is round, even though that is usually not the case), the microbiologist uses a calculator and gets the answer 34.55751919 μm. Taking significant figures into account, what answer should actually be reported? (π = 3.141592654 . . . .)

1.32 Given that area = π × radius^2 and radius = diameter/2, what is the area of the cell described in Problem 1.31? Report your answer with the correct number of significant figures.

*__1.33__ **a.** Calculate the area of the floor for the room that you are in by pacing off its length and width. Report your answer in units of paces^2 and be sure to report your answer with the correct number of significant figures.
b. Using a yardstick, measure the length of your stride. Convert your answer to part a into units of feet^2.

c. Use the yardstick to measure the length and width of the room. Calculate the area in units of feet^2, reporting your answer with the correct number of significant figures.

d. Compare your answers to parts b and c. Account for any difference.

1.34 You are at the state fair and pay a dollar for the chance to throw three baseballs in an attempt to knock over a pyramid of bowling pins. After your three tosses, the pins remain standing. Which of the following statements about your throws might be correct?

a. They were precise and accurate.
b. They were neither precise nor accurate.
c. They were precise but not accurate.

1.6 Conversion Factors and the Factor Label Method

1.35 Give the two conversion factors that are based on each equality.

a. 12 eggs = 1 dozen
b. 1×10^3 m = 1 km
c. 0.946 L = 1 qt

1.36 Give the two conversion factors that are based on each equality.

a. 2 T = 1 oz
b. 15 gtt = 1 mL
c. 1 mg = 1000 μg

1.37 a. 17 ft is how many yards?
b. 36.8 ft is how many inches?

1.38 a. 124 in. is how many feet?
b. 124 in. is how many yards?

1.39 Convert

a. 92 μg into grams
b. 27.2 ng into milligrams
c. 0.33 kg into milligrams
d. 7.27 mg into micrograms

1.40 Convert

a. 81.2 g into kilograms
b. 81.2 kg into grams
c. 29 μg into milligrams
d. 47.66 μg into decigrams

1.41 Convert your weight from pounds to kilograms.

1.42 Convert

a. 91°F into degrees Celsius
b. 53°C into degrees Fahrenheit
c. 0°C into kelvins
d. 309 K into degrees Celsius

1.43 Convert

a. 103°F into degrees Celsius
b. 25°C into degrees Fahrenheit
c. 35°C into kelvins
d. 405 K into degrees Fahrenheit

1.44 In 2006, 40,809 people completed the 12 km Bloomsday race in Spokane, WA. What is the distance of this race in miles?

1.45 It is estimated that an accordion player expends 9.2 kJ of energy per minute of playing time. Convert this value into calories (1 food Calorie = 1000 calories).

***1.46** Stavudine is an antiviral drug that has been tested as a treatment for AIDS. The daily recommended dosage of stavudine is 1.0 mg/kg of body weight. How many grams of this drug should be administered to a 150 lb patient?

***1.47** As an alternative to ear tags and lip tattoos, tetracycline (an antibiotic) is used to mark polar bears. The advantages of using tetracycline in this fashion are that it leaves a detectable deposit on the bears' teeth, it can be administered remotely, and using it doesn't require that the animal be sedated. If 25 mg/kg is an effective dose, how much tetracycline is needed (in grams) to mark a 1000 kg polar bear?

***1.48** Ivermectin is used to treat dogs that have intestinal parasites. The effective dosage of this drug is 10.5 μg/kg of body weight. How much ivermectin should be given to a 9.0 kg dog?

***1.49** Chloroquine is used to treat malaria. Studies have shown that an effective dose for children is 3.5 mg per kilogram (3.5 mg/kg) of body weight, every 6 hours. If a child weighs 12 kg, how many milligrams of this drug should be given in a 24 hour period?

1.50 Scientists have been able to cool matter to temperatures as low as 0.25 nK, which is only a fraction of a degree above absolute zero (0 K). Using scientific notation, express 0.25 nK in kelvins.

1.51 The tranquilizer Valium is sold in 2.0 mL syringes that contain 50.0 mg of drug per 1.0 mL of liquid (50.0 mg/1.0 mL). If a physician prescribes 25 mg of this drug, how many milliliters should be administered?

1.52 An antibiotic is sold in 3.0 mL ampoules that contain 60.0 mg of drug (60.0 mg/3.0 mL). How many milliliters of the antibiotic should be withdrawn from the ampoule if 45 mg are to be administered to a patient?

1.53 a. A vial contains 25 mg/mL of a particular drug. To administer 15 mg of the drug, how many milliliters should be drawn from the vial?
b. A patient is to receive 50 cc of a drug mixture intravenously over a 1 hr time period. What is the appropriate IV drip rate in gtt/min?

1.54 **a.** A prescription of antibiotics for a 30 lb child says to give 100 mg three times daily. Is this dosage safe if the proper pediatric dosage range for this drug is 10–30 mg per kilogram of body weight per day?
b. A patient's cough syrup prescription comes in a 250 mL bottle. For how long will the cough syrup last if he takes two teaspoons three times a day?

HEALTH*Link* | Science and Medicine

1.55 In the past 200 years, in what ways have scientific discoveries led to changes in the treatment of diabetes?

1.56 Until the late 1980s, what was the source of the insulin used to treat diabetes?

1.57 **a.** According to "Pork on the Run," how can coronary heart disease be prevented in diabetic pigs? Would you expect this to apply to diabetic people, as well?
b. Why were pigs selected for the study described in this video-clip?

HEALTH*Link* | Body Mass Index

1.58 **a.** A 6′ 2″ tall adult weighs 180 lbs. What is his BMI? Based on this value, what is his status: underweight, normal, overweight, or obese?
b. Answer part a, but using your height and weight.
c. A woman stands 1.65 m tall and weighs 72.7 kg. What is her BMI and what is her status?

1.59 September 2006 was the first time that models were banned from a top-level fashion show for being too thin. The organizers of the Madrid Fashion Week defined "too thin" as having a BMI of less than 18. How much would a 5′ 2″ model weigh if she had a BMI of 16?

HEALTH*Link* | Body Temperature

1.60 A patient has a temperature of 31°C. Should her clinician be concerned?

1.61 Suppose that you take your temperature orally and see that it is 99.1°F. Does this necessarily mean that you are running a fever? Explain.

1.62 Why do you suppose that the core body temperature is usually higher than the shell temperature?

1.63 If your temperature is taken rectally, one measurement is usually sufficient. If taken at the eardrum, however, more than one measurement is recommended. Why?

THINKING IT THROUGH

1.64 How do you feel about the possibility that your body temperature may be measured without your knowledge when you walk through some airports? Is this an invasion of your privacy? If so, is it justified to prevent the spread of disease? (Health Link: Body Temperature.)

INTERACTIVE LEARNING PROBLEMS

ILW

1.65 A bullet is fired at a speed of 2235 ft/s. What is the speed expressed in kilometers per hour?

SOLUTIONS TO PRACTICE PROBLEMS

1.1 **a.** Sitting on the floor, the barbell contains some potential energy. When picked up off of the floor, its potential energy increases. As the weightlifter moves the barbell its kinetic energy increases. **b.** From the food that he or she has eaten, the weightlifter contains some potential energy. As the barbell is picked up, some of this potential energy is converted into kinetic energy (movement of the barbell as well as the movement of muscles).

1.2 **a.** 273°C; **b.** 32°C; **c.** 0°F

1.3 One-millionth of a liter = 1 microliter = 1 μL; 5 million cells = 5×10^6 cells

1.4 **a.** 7.032×10^3 cal; **b.** 8.80×10^1 L; **c.** 5×10^{-5} g; **d.** 6.430×10^{-2} lb

1.5 **a.** 2.61×10^4; **b.** 3.65; **c.** 3×10^2; **d.** 6

1.6 **a.** 0.12 mg; **b.** 0.30 mL

1.7 **a.** −269.1°C; **b.** −452.4°F

A friend of yours has taken the day off from classes to take her grandmother to the hospital for a PET scan. The scan has been ordered to see whether she has Alzheimer's disease and, if so, how far it has progressed.

After completing the initial paperwork, the two sit in the waiting room for over an hour. Finally a technician comes over to explain that things are running behind schedule because of airport delays. He explains, "We have to fly in the radioactively tagged sugar molecules from Seattle. Once the airport courier arrives we can get started." Just then a heavy box labeled with radiation stickers is wheeled through the clinic doors.

Photo source: Stockbyte/Punchstock.

2 ATOMS AND ELEMENTS

In this chapter we turn our attention to the structure and properties of atoms—the fundamental particles from which matter is constructed. This discussion will include a look at elements, the periodic table, and the nature of nuclear radiation.

objectives

After completing this chapter, you should be able to:

1. Describe the subatomic structure of an atom.
2. Define the terms element and atomic symbol.
3. Explain how elements are arranged in the periodic table.
4. Explain how atomic number and mass number are used to indicate the makeup of an atom's nucleus. Describe how isotopes of an element differ from one another.
5. Define the terms mole and molar mass.
6. Define the term nuclear radiation and describe the four common types of radiation emitted by radioisotopes. Explain how exposure to radiation can be controlled.
7. Describe how fission differs from fusion. Define the terms chain reaction and critical mass.

2.1 ATOMS

In one of the views held by ancient Greek philosophers, matter was thought to be made from tiny, indestructible building blocks called **atoms** (from *atomos*, Greek for *indivisible*). The first scientific description of atoms, one based on careful measurements and experiments, was proposed in the early 1800s by the English scientist John Dalton. According to **Dalton's atomic theory**, *the atom is the basic unit from which matter is constructed.* He imagined that if you took a piece of iron and divided it again and again, you would eventually end up with an indivisible unit of matter called an iron atom (Figure 2.1). This iron atom would be different from other kinds of atoms, such as carbon, oxygen, and hydrogen.

As often happens in science, further studies into the nature of the atom gave new results that led to revisions of Dalton's atomic theory. According to current theory, normal matter is made from atoms, which are built from smaller **subatomic particles** called **protons, neutrons**, and **electrons**. Protons and neutrons form the compact core, or **nucleus**, of an atom, while electrons are dispersed in clouds around the nucleus (Figure 2.1). Atoms are mostly empty space. If an atom were the size of a classroom, for example, its nucleus would be about the size of the period at the end of this sentence and electrons would move throughout the remaining space.

■ Atoms consist of protons, neutrons, and electrons.

Mass and charge are the two characteristics of subatomic particles that are important for understanding atomic structure (Table 2.1). Atoms and the subatomic particles from which they are constructed are so small that it is usually inconvenient to describe their mass in grams. The mass of one proton, for example, is only 0.000000000000000000000001673 g or 1.673×10^{-24} g. Instead, the mass of atoms and subatomic particles is usually represented in terms of **atomic mass units** (amu). Protons and neutrons each have a mass of about 1 amu, while the much smaller electrons have a mass of about only 1/2000 amu. Because protons and neutrons are the only subatomic particles in an atom that have significant mass, the mass of an atom sits mostly in the nucleus, which is made up of protons and neutrons.

Studies have shown that protons are positively charged, neutrons are neutral (uncharged), and electrons are negatively charged (Table 2.1). This gives an atomic nucleus (contains protons and neutrons) a positive charge and gives the electron clouds that surround the nucleus a negative charge.

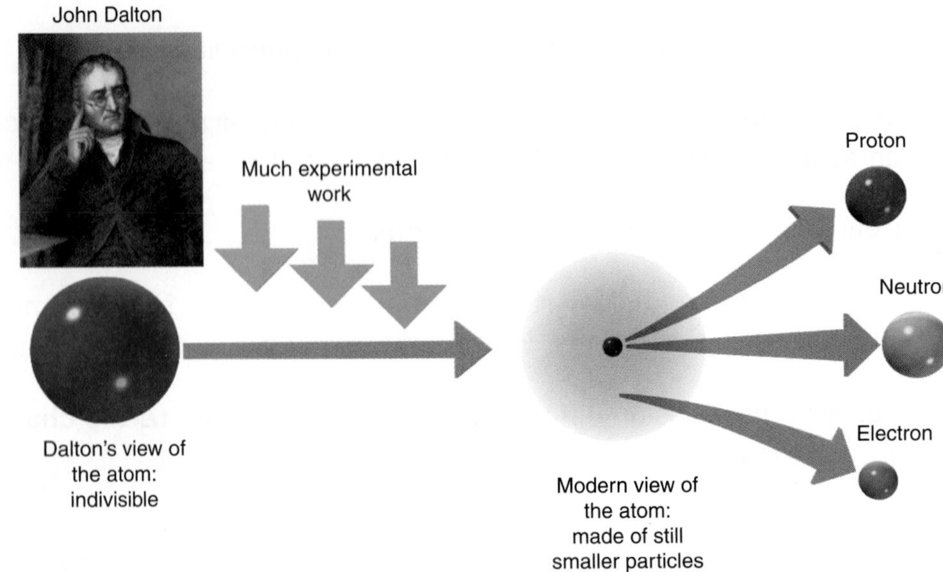

■ FIGURE | 2.1

The evolving theory of the atom John Dalton's proposal that atoms are the indivisible fundamental particle of matter fit the observations available at the time. Now we have a more complex view of atoms (based on experimental results obtained since Dalton's time), which proposes that atoms are composed of subatomic particles.

Source: Shelia Terry/Photo Researchers Inc.

TABLE 2.1 SUBATOMIC PARTICLES

Subatomic Particle	Mass (amu)	Charge	Location in Atom
Proton	1.00728	1+	Nucleus
Neutron	1.00867	0	Nucleus
Electron	0.00055	1−	Outside of nucleus

SAMPLE PROBLEM 2.1

Properties of subatomic particles

To which subatomic particle does each statement refer?

a. has a positive charge
b. is found outside of the nucleus
c. has the smallest mass
d. has a negative charge

Strategy

There are three subatomic particles—the proton, the neutron, and the electron. Review the paragraphs above to determine the mass, charge, and location of each.

Solution

a. proton **b.** electron **c.** electron **d.** electron

PRACTICE PROBLEM 2.1

a. Which two subatomic particles are found in the nucleus?
b. Which two subatomic particles have opposite charges?
c. Which two subatomic particles have nearly the same mass?

2.2 ELEMENTS

The term **element** is used to describe a substance that contains only one type of atom. An iron atom, a tiny sliver of iron, and a bar of iron are all examples of the element called iron. A carbon atom, a diamond, and a piece of graphite all belong to the element called carbon (Figure 2.2).

Over the years, the number of known elements has steadily grown. Before 1800, only 31 elements had been discovered, but by 2005 this number had grown to 113, of which 91 occur naturally. The remaining ones are artificial—they have been produced from the natural ones.

Elements are not named according to any particular set of rules. Some are named for their color (iodine from *iodes*, Greek for *violet*), their place of discovery (berkelium for Berkeley, California), or a person (einsteinium for Albert Einstein) (Figure 2.3). In addition to being identified by name, each element has a unique **atomic symbol**, a one or two letter abbreviation of the element name (Table 2.2). The symbols used for some elements are based on their older Latin names, as is the case for potassium (K from *kalium*) and sodium (Na from *natrium*).

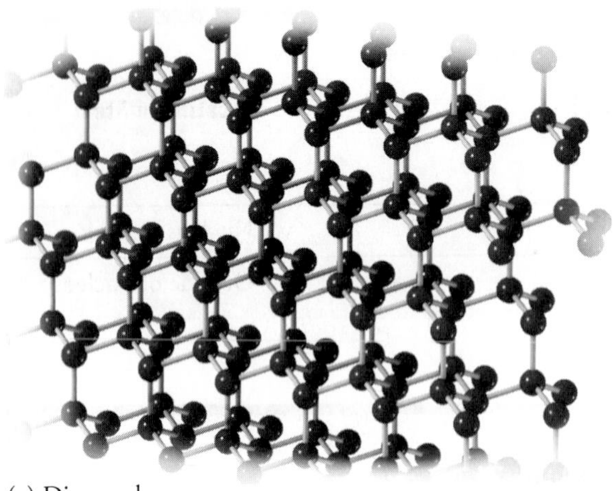

(*a*) Diamond

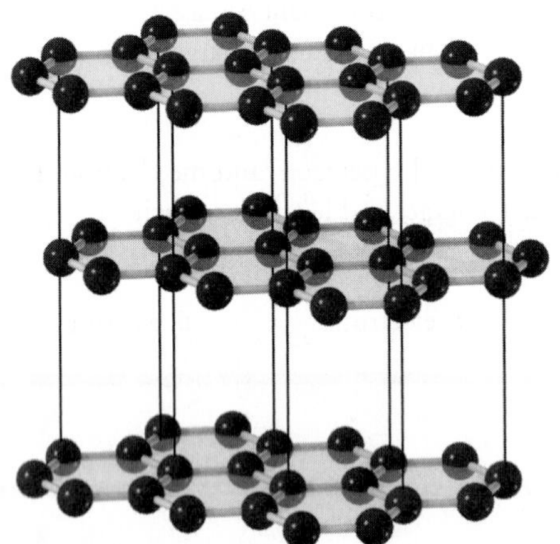

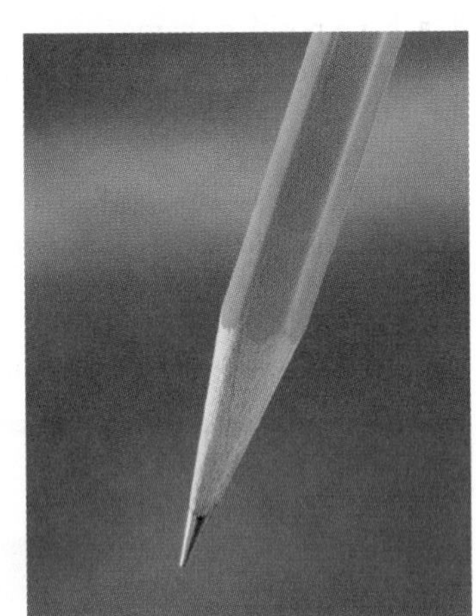

(*b*) Graphite

FIGURE | 2.2

Elements
A diamond and the graphite in a pencil consist of different arrangements of carbon atoms.

Source (photos): © Royalty-Free/Corbis. **(line art):** BROWN, THEODORE E.; LEMAY, H. EUGENE; BURSTEN, BRUCE E.; BURDGE, JULIA R., CHEMISTRY: THE CENTRAL SCIENCE, 10th edition, © 2006. Reprinted by permission of Pearson Education, Inc., Upper Saddle River, NJ.

FIGURE | 2.3

Some element names are based on odor or a location
Bromine (left) gets its name from *bromos*, Greek for *stench*, and copper gets its name from *cuprum*, Latin for *from the island of Cyprus*.

Source: Ken Karp/Wiley Archive.

TABLE | 2.2 Names and Symbols for Some Common Elements

Name	Symbol	Name	Symbol
Calcium	Ca	Mercury (hydragyrum)	Hg
Carbon	C	Neon	Ne
Chlorine	Cl	Nitrogen	N
Copper (cuprum)	Cu	Oxygen	O
Gold (aurum)	Au	Phosphorus	P
Helium	He	Potassium (kalium)	K
Hydrogen	H	Radon	Rn
Iodine	I	Silver (argentum)	Ag
Iron (ferrum)	Fe	Sodium (natrium)	Na
Lead (plumbum)	Pb	Sulfur	S
Lithium	Li	Tin (stannum)	Sn
Magnesium	Mg	Zinc	Zn

Although only four elements—oxygen (O), carbon (C), hydrogen (H), and nitrogen (N)—make up most of the mass of a human body, about two dozen elements play critical roles in the body's chemistry and are essential for life (Figure 2.4). The bulk elements are present in greater amounts than the trace elements.

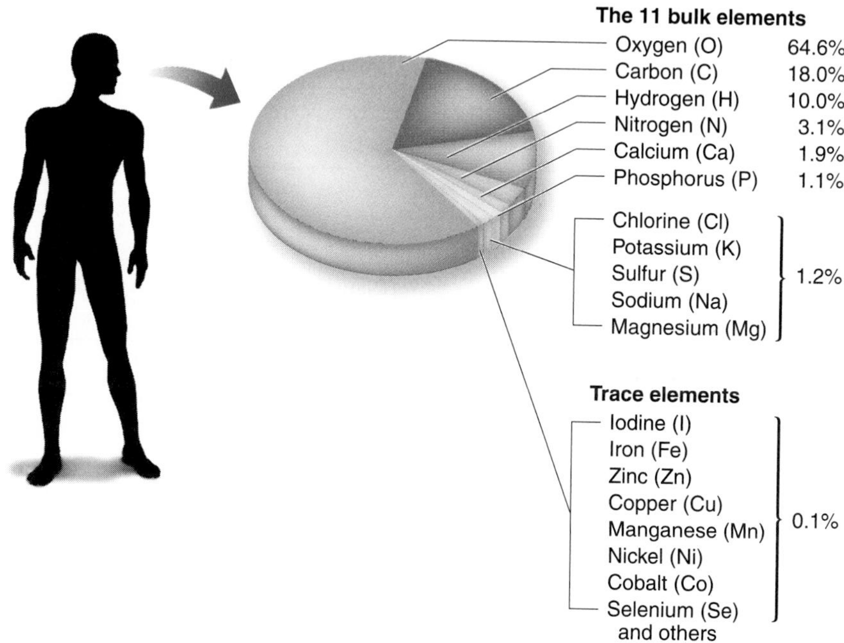

FIGURE 2.4 **The major elements in the human body, by mass** The bulk elements are present in substantial amounts; the trace elements are present in tiny amounts but are equally necessary for life.

HEALTH Link: *Dietary Reference Intakes (DRIs)*

In 1941 the National Academy of Sciences (NAS) created **Recommended Dietary Allowances** (RDAs). RDAs are the daily intake of nutrients that are sufficient to meet the needs of 97 to 98% of healthy people. Originally, RDAs were designed to prevent diseases caused by nutritional deficiencies.

As scientific understanding of health and nutrition progressed, a new set of guidelines became necessary. In the 1990s, the NAS created **Dietary Reference Intakes** (DRIs) to help health professionals make better nutritional recommendations. DRI reference values include the RDAs just mentioned, Adequate Intake (AI), and Tolerable Upper Intake Level (UL). **Adequate Intake**, which is used when an RDA is not available, is the daily intake of a nutrient believed to be sufficient to meet the needs of people. Due to unreliability of the experimental data, the percentage of people covered by a particular AI is uncertain. The **Tolerable Upper Intake Level** of a nutrient is the highest intake that can be safely consumed. Table 2.3 gives DRIs for some essential elements, as specified by the NAS Food and Nutrition Board.

TABLE 2.3 DIETARY REFERENCE INTAKES (DRIs) FOR SOME ESSENTIAL ELEMENTS[a]

Element	RDA[b] (mg/day)	AI[c] (mg/day)	UL[d] (mg/day)	Biological Function	Adverse Health Effect of Overconsumption
Ca		1300	2500	Formation of bones and teeth	Kidney stones
Fe	18		45	Component of hemoglobin	Heart, bladder, and pancreatic problems
I	0.15		1.1	Component of thyroid hormones	Thyroid problems
Mn		2.3	11	Helps some enzymes to function properly	Heart rhythm problems
P	700		4000	Tissue growth	Skeletal problems
Se	0.055		0.40	Helps to regulate the thyroid	Nerve damage
Zn	11		40	Helps some enzymes to function properly	Nausea, vomiting, and dizziness

[a]Some RDA and AI values differ with age and/or gender. Values reported here are the highest listed for adults.
[b]Recommended Dietary Allowance.
[c]Adequate Intake.
[d]Tolerable Upper Intake Level.

2.3 ATOMIC NUMBER AND MASS NUMBER

- Atomic number is the number of protons in an atom's nucleus.
- Mass number is the total number of protons and neutrons in an atom's nucleus.

All atoms of a particular element have the same number of protons in their nucleus. For example, all sulfur (S) atoms have 16 protons and all calcium (Ca) atoms have 20 protons. *The number of protons that an atom has in its nucleus* is known as the **atomic number**.

The total number of protons and neutrons in the nucleus of an atom is called the **mass number**. A sulfur atom containing 16 protons and 16 neutrons has a mass number of 32. An aluminum (Al) atom containing 13 protons and 14 neutrons has a mass number of 27.

To specify the number of protons and neutrons in the nucleus of a particular atom, **atomic notation** is commonly used. In atomic notation, an atom's atomic number is shown as a subscript and its mass number as a superscript, both written to the left of the atomic symbol. $^{14}_{7}N$ describes a nitrogen atom whose nucleus consists of 7 protons and 7 neutrons.

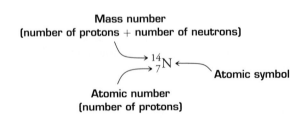

Mass number (number of protons + number of neutrons) → $^{14}_{7}N$ ← Atomic symbol

Atomic number (number of protons)

SAMPLE PROBLEM 2.2

Determining atomic and mass number

Give the atomic number and mass number of

a. a sulfur atom with 18 neutrons
b. an aluminum atom with 12 neutrons

Strategy

To solve this problem you must first know how many protons a sulfur atom and an aluminum atom have. Although this information has not yet been provided for all of the elements, in this case it can be found in the three preceding paragraphs. Once you know the number of protons for each, you can solve the problem—atomic number is equal to the number of protons and mass number is equal to the number of protons plus the number of neutrons.

Solution

a. atomic number = 16, mass number = 34
b. atomic number = 13, mass number = 25

PRACTICE PROBLEM 2.2

Give the atomic number and mass number of

a. an atom with 8 protons and 10 neutrons
b. an atom with 17 protons and 20 neutrons

Isotopes

While all atoms of a particular element always have the same number of protons, the number of neutrons that they carry can be different. *Atoms of an element that have different numbers of neutrons* are called **isotopes**. Hydrogen, the atoms of which always have one proton, exists as three isotopes. The atoms of one of these isotopes have no neutrons, the atoms of another have one neutron, and the atoms of the third have two neutrons (Figure 2.5). These isotopes can be represented using atomic notation ($^{1}_{1}H$, $^{2}_{1}H$, and $^{3}_{1}H$) or simply by giving the name of the element, followed by the mass number (hydrogen-1, hydrogen-2, and hydrogen-3). The isotopes of hydrogen are also known by other names: hydrogen-1 is protium, hydrogen-2 is deuterium, and hydrogen-3 is tritium.

■ Isotopes of an element have the same number of protons and a different number of neutrons.

All elements consist of more than one isotope and the isotopes of an element are usually not present in nature in equal amounts. Hydrogen, for example, is found mainly as the $^{1}_{1}H$ isotope. The most abundant of the three naturally occurring isotopes of oxygen ($^{16}_{8}O$, $^{17}_{8}O$, and $^{18}_{8}O$) is $^{16}_{8}O$ and the most abundant of the two chlorine isotopes ($^{35}_{17}Cl$ and $^{37}_{17}C$) is $^{35}_{17}Cl$.

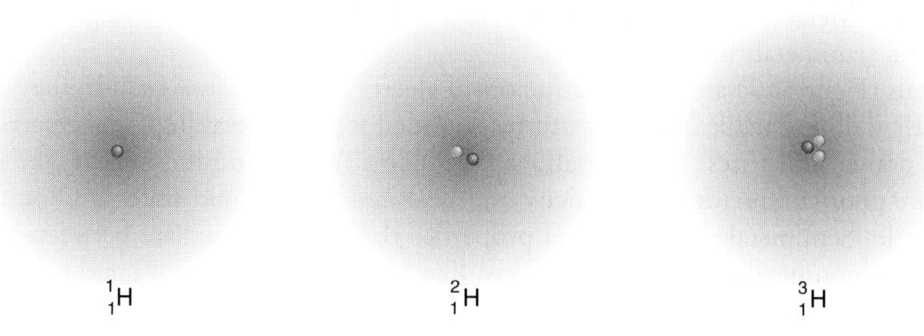

■ **FIGURE 2.5**

Hydrogen isotopes
All hydrogen atoms have one proton. The difference between the three hydrogen isotopes is in the number of neutrons present in the nucleus of each. $^{1}_{1}H$ has zero neutrons, $^{2}_{1}H$ has one, and $^{3}_{1}H$ has two.

SAMPLE PROBLEM 2.3

Identifying isotopes

Iodine-131 ($^{131}_{53}I$) is used for the detection and treatment of thyroid disorders. Which is an isotope of iodine-131, $^{131}_{52}Te$ or $^{132}_{53}I$?

Strategy

If you look back at the definition of "isotope," you will see that it mentions *atoms of an element.* Isotopes are atoms of a particular element (same atomic number) that have a different number of neutrons (different mass number).

Solution

$^{131}_{53}I$ and $^{132}_{53}I$ are isotopes—they have the same atomic number and different mass numbers.

PRACTICE PROBLEM 2.3

Which two atoms are isotopes? $^{31}_{15}P$, $^{31}_{14}Si$, $^{28}_{15}P$

Up to this point, we have dealt only with counting the number of protons and neutrons present in the nucleus of an atom. The number of electrons carried by an atom can be determined from the number of protons that it has. A neutral atom (net charge = 0) contains an equal number of protons (positive charges) and electrons (negative charges). For example, all neutral oxygen atoms (atomic number = 8) have 8 protons and 8 electrons and all neutral iron (Fe) atoms (atomic number = 26) have 26 protons and 26 electrons. Further discussion of electrons will be postponed until Chapter 3.

2.4 PERIODIC TABLE

■ The atomic weight of an element is the average mass of the atoms in a naturally occurring sample of the element.

The **periodic table of the elements** (Figure 2.6) is a complete list of the elements, arranged in order from smallest to largest atomic number. The atomic symbol, atomic number, and **atomic weight** of each element are included in the table. The term atomic weight refers to *the average mass of the atoms of an element, as it is found in nature.* The atomic weight of bromine is close to 80 amu, for example, because in nature elemental bromine consists of about 50% $^{79}_{35}Br$ and about 50% $^{81}_{35}Br$.

Notice that the periodic table shows the atomic weights of some elements surrounded by brackets. These elements (Po, Rn, Cf, etc.) do not have stable isotopes, so the reported atomic weight is the mass of the isotope that stays in existence for the longest time. The stability of isotopes is discussed in Sections 2.6–2.8.

Metals, Nonmetals, and Semimetals

The elements can be divided into three categories—**metals**, **nonmetals**, and **semimetals** (or metalloids). In Figure 2.6, the heavy line that starts to the left of boron (B) and then zigzags down to the right marks the area where the transition from metals to nonmetals takes place. Metals are to the left, nonmetals (with the exception of hydrogen) are to the right, and semimetals border this line.

Metals (Figure 2.7) are good conductors of electricity and heat. As solids, they are lustrous (they shine), malleable (they can be pounded without breaking), and ductile (they can be drawn into wires). Nonmetals, as their name implies, do not behave like metals. They are poor conductors of electricity and heat and, in the solid state, are nonlustrous and brittle. Semimetals have physical properties that are intermediate between those of metals and nonmetals. Silicon (Si) is shiny, like a metal, but is a poor conductor of heat and electricity, like a nonmetal.

Representative elements | Transition metals | Representative elements

Metals | Semimetals | Nonmetals

Period	1A	2A	3B	4B	5B	6B	7B	8B	8B	8B	1B	2B	3A	4A	5A	6A	7A	8A
1	1 **H** 1.00794																	2 **He** 4.00260
2	3 **Li** 6.941	4 **Be** 9.012182											5 **B** 10.811	6 **C** 12.0107	7 **N** 14.0067	8 **O** 15.9994	9 **F** 18.99840	10 **Ne** 20.1797
3	11 **Na** 22.98977	12 **Mg** 24.3050											13 **Al** 26.98154	14 **Si** 28.0855	15 **P** 30.97376	16 **S** 32.065	17 **Cl** 35.453	18 **Ar** 39.948
4	19 **K** 39.0983	20 **Ca** 40.078	21 **Sc** 44.95591	22 **Ti** 47.867	23 **V** 50.9415	24 **Cr** 51.9961	25 **Mn** 54.93805	26 **Fe** 55.845	27 **Co** 58.9332	28 **Ni** 58.6934	29 **Cu** 63.546	30 **Zn** 65.39	31 **Ga** 69.723	32 **Ge** 72.64	33 **As** 74.92160	34 **Se** 78.96	35 **Br** 79.904	36 **Kr** 83.80
5	37 **Rb** 85.4678	38 **Sr** 87.62	39 **Y** 88.9058	40 **Zr** 91.224	41 **Nb** 92.9063	42 **Mo** 95.94	43 **Tc** [98]	44 **Ru** 101.07	45 **Rh** 102.9055	46 **Pd** 106.42	47 **Ag** 107.868	48 **Cd** 112.411	49 **In** 114.818	50 **Sn** 118.710	51 **Sb** 121.760	52 **Te** 127.60	53 **I** 126.9045	54 **Xe** 131.293
6	55 **Cs** 132.9054	56 **Ba** 137.327	57 ***La** 138.9055	72 **Hf** 178.49	73 **Ta** 180.9479	74 **W** 183.84	75 **Re** 186.207	76 **Os** 190.23	77 **Ir** 192.217	78 **Pt** 195.078	79 **Au** 196.9665	80 **Hg** 200.59	81 **Tl** 204.383	82 **Pb** 207.2	83 **Bi** [208.9804]	84 **Po** [209]	85 **At** [210]	86 **Rn** [222]
7	87 **Fr** [223]	88 **Ra** [226]	89 **†Ac** [227]	104 **Rf** [261]	105 **Db** [262]	106 **Sg** [266]	107 **Bh** [264]	108 **Hs** [269]	109 **Mt** [268]	110 **Ds** [281]	111 **Rg** [272]	112 **Uub** [285]		114 **Uuq** [289]				

*Lanthanide elements	58 **Ce** 140.116	59 **Pr** 140.90765	60 **Nd** 144.24	61 **Pm** [145]	62 **Sm** 150.36	63 **Eu** 151.964	64 **Gd** 157.25	65 **Tb** 158.92534	66 **Dy** 162.50	67 **Ho** 164.93032	68 **Er** 167.259	69 **Tm** 168.93421	70 **Yb** 173.04	71 **Lu** 174.967
†Actinide series	90 **Th** 232.0381	91 **Pa** 231.0359	92 **U** 238.0289	93 **Np** [237]	94 **Pu** [244]	95 **Am** [243]	96 **Cm** [247]	97 **Bk** [247]	98 **Cf** [251]	99 **Es** [252]	100 **Fm** [257]	101 **Md** [258]	102 **No** [259]	103 **Lr** [262]

For elements that do not have stable isotopes, the mass of the most stable isotope is given in parentheses. Elements 112 and 114 have been discovered but have not been given official names. The production of elements 113, 115, 116, and 118 has been reported, but at the time of this writing the experiments have not been reproduced by other researchers.

FIGURE 2.6 **The periodic table of the elements**
The periodic table is an organized list of the known elements. The names and atomic symbols of the elements are listed inside the front cover of this text.

SAMPLE PROBLEM 2.4

Relating atomic weight and the percentages of isotopes found in nature

Silicon, which appears in nature as $^{28}_{14}Si$, $^{29}_{14}Si$, and $^{30}_{14}Si$, has an atomic weight of 28.1 amu. Which isotope predominates? (When used in problems in this book, atomic weight will always be rounded to one decimal place.)

Strategy

The key to solving this problem is to remember that the atomic weight of an element is the average mass of all of its isotopes as found in nature. If one isotope predominates, then its mass number will be near the average.

Solution

For the atomic weight (average mass) of silicon to be nearly 28 amu, the $^{28}_{14}Si$ isotope must predominate.

PRACTICE PROBLEM 2.4

In nature, boron is found as two isotopes, $^{10}_{5}B$ and $^{11}_{5}B$, but there is much more of the $^{11}_{5}B$ isotope. Based on this information, which of the three atomic weights is nearest the correct value for boron—10.1, 10.5, or 10.8 amu?

(a)

(b)

(c)

FIGURE | 2.7

Metals, nonmetals, and semimetals
(*a*) Silver and other metals are shiny and are good conductors of electricity and heat. (*b*) Sulfur and other solid nonmetals have a dull appearance and are poor conductors. (*c*) Silicon is a semimetal. Its shiny appearance is characteristic of a metal, but its poor conduction of heat and electricity is characteristic of a nonmetal.

Source: (*a*) © 1991 Paul Silverman-Fundamental Photographs; (*b*), (*c*) © 1994, 1990 Richard Megna-Fundamental Photographs.

Groups

■ Elements in the same column belong to the same group.

The *elements in the same vertical column* of the periodic table belong to the same **group**. Three different systems are commonly used to identify chemical groups. In this text we will use a system in which groups are identified by the numbers 1–8 and by the letter A or B. H, Li, Na, K, Rb, Cs, and Fr belong to group 1A and Sc, Y, La, and Ac belong to group 3B (Figure 2.6). Some chemical groups have been given names. Of these, the important ones to learn are the **alkali metals** (the group 1A metals), the **alkaline earth metals** (group 2A), the **halogens** (group 7A), and the **inert gases** (group 8A).

The periodic table is also divided into larger groupings of elements. Groups 1A, 2A, 3A, 4A, 5A, 6A, 7A, and 8A are called the **representative elements**, while groups 1B, 2B, 3B, 4B, 5B, 6B, 7B, and 8B are known as the transition elements or the **transition metals**. Elements with the atomic numbers 58 through 71 (found toward the bottom of the periodic table) are the **lanthanide elements** and those with atomic numbers 90 through 103 are the **actinide elements**.

Periods

Elements in the same horizontal row of the periodic table belong to the same **period**. The elements H and He make up the first period, while Li, Be, B, C, N, O, F, and Ne are the second period. Moving down the periodic table, the periods are numbered from 1 to 7 beginning with the first row. The term "period" relates to the periodic (regular) changes that take place in the properties of elements as you move through them in order of atomic number.

■ Elements in the same row belong to the same period.

Atomic size is one of the properties that shows periodic change. Moving across a period from left to right, the atoms of an element usually get smaller. Moving down a group, they usually get larger.

Metallic and nonmetallic behavior is another property that displays a periodic trend. Elements on the far left side of a period are the most metallic, while those on the far right side are the most nonmetallic. In the third period, Na, Mg, and Al are metals, Si is a semimetal, and P, S, Cl, and Ar are nonmetals. Moving down a group, atoms become more metallic—compare the nonmetal carbon (C) with the semimetals silicon (Si) and germanium (Ge) and the metals tin (Sn) and lead (Pb).

SAMPLE PROBLEM 2.5

Predicting the properties of elements

Which of the following is the best conductor of electricity?
Li, B, C, N, or F

Strategy

Metals are the elements that have the ability to conduct electricity. To solve this problem you must determine which is the most metallic element.

Solution

Lithium (Li) is the only metal listed, so it is the best conductor of electricity.

PRACTICE PROBLEM 2.5

Which of the following is likely to be the most brittle? Na, Al, Si, or S

2.5 THE MOLE

Atoms are so small that the only time they are dealt with one by one is in experiments that rely on very specialized scientific instruments. In most situations, such as filling a balloon with helium gas or using a mercury-filled thermometer to take someone's temperature, a very large number of atoms is involved. For this reason, atoms are best represented using a counting unit called the **mole.**

A **counting unit** is any term that *refers to a specific number of things.* If you say that you are going to the store to buy a dozen (12) eggs, you are using a counting unit. If a friend talks about a couple (2) of people that he met, a counting unit is being used. Other counting units include quartet (4) and gross (144 or 12 dozen).

None of the counting units just mentioned represent a large enough number to be useful for counting atoms. This is because even a very small sample of an element contains a huge number of atoms. For example, a cube of gold that is one inch on a side contains nearly 1×10^{24} (1,000,000,000,000,000,000,000,000) atoms. To deal with numbers of this size, the mole counting unit is used. *One mole corresponds to 6.02×10^{23} items* (Figure 2.8). This number is known as **Avogadro's number**, after the Italian physicist Amedeo Avogadro, whose work in the early 1800s contributed to our understanding of the atom.

■ One mole equals 6.02×10^{23} items.

Just how large a number is 6.02×10^{23}? Consider this: A machine capable of counting 1 billion pennies per second would require 19 million years to count one mole of pennies. Even though a mole is an incredibly large number of items, atoms are so small that one mole of them in a solid or liquid state does not occupy all that much space (see Figure 2.9).

Because the mole is the standard counting unit used to indicate the number of atoms present in a sample, it is useful to be able to convert back and forth from moles to number of atoms. Doing so depends on defining an appropriate conversion factor and using the factor label method (Section 1.6) to switch units. The number of carbon (C) atoms in 0.500 mol (moles) can be calculated as follows.

Given that 1 mol of atoms is equal to 6.02×10^{23} atoms, two conversion factors can be written.

$$\frac{6.02 \times 10^{23}\ \text{C atoms}}{1\ \text{mol C atoms}} \quad \text{and} \quad \frac{1\ \text{mol C atoms}}{6.02 \times 10^{23}\ \text{C atoms}}$$

To convert from number of moles to number of atoms, the conversion factor that has the number of atoms in the numerator is used. Setting up the equation and solving gives an answer of 3.01×10^{23} C atoms.

$$0.500\ \cancel{\text{mol C atoms}} \times \frac{6.02 \times 10^{23}\ \text{C atoms}}{1\ \cancel{\text{mol C atoms}}} = 3.01 \times 10^{23}\ \text{C atoms}$$

FIGURE | 2.8

Counting units
Dozen (12) and mole (6.02×10^{23}) are counting units that refer to a certain number of items.

This stack of pennies would stretch more than 20,000 times farther than the distance from the sun to the nearest star system, Alpha Centauri.
1 dozen pennies
1 mole of pennies

FIGURE | 2.9

The mole
One mole of mercury, carbon, cobalt, and zinc atoms (6.02×10^{23} atoms).

Source: Ken Karp/Wiley Archive.

SAMPLE PROBLEM 2.6

Working with conversion factors

How many items are present in the following?

a. 1/2 dozen eggs **b.** 2.00 gross of pencils **c.** 0.100 mol of C atoms

Strategy

Each part of the problem requires a different conversion factor. In part a, converting dozens of eggs into number of eggs uses a conversion factor with number of eggs in the numerator. One dozen eggs equals 12 eggs, so the conversion factor to use is

$$\frac{12 \text{ eggs}}{1 \text{ dozen eggs}}$$

Using a similar approach, conversion factors for parts b and c can be derived.

Solution

a. 6 eggs

$$0.5 \cancel{\text{dozen eggs}} \times \frac{12 \text{ eggs}}{1 \cancel{\text{dozen eggs}}} = 6 \text{ eggs}$$

b. 288 pencils

$$2.00 \cancel{\text{gross of pencils}} \times \frac{144 \text{ pencils}}{1 \cancel{\text{gross of pencils}}} = 288 \text{ pencils}$$

c. 6.02×10^{22} C atoms

$$0.100 \cancel{\text{mol C atoms}} \times \frac{6.02 \times 10^{23} \text{ C atoms}}{1 \cancel{\text{mol C atoms}}} = 6.02 \times 10^{22} \text{ C atoms}$$

PRACTICE PROBLEM 2.6

a. How many helium (He) atoms are present in 0.62 mol of He?
b. How many moles are 2.9×10^{12} fluorine (F) atoms?

The Mole and Mass

Counting units such as dozen and mole are good for indicating the number of a particular item, but are not useful for indicating mass, unless you know something about the mass of the individual items. For example, you know that one dozen people weigh more than one dozen doughnuts because one person weighs more than one doughnut. The same idea holds true for atoms. One mole of oxygen atoms has a greater mass than one mole of nitrogen atoms because one carbon atom (16.0 amu) has a greater mass than one nitrogen atom (14.0 amu).

The periodic table allows us to predict *the mass (in grams) of one mole of the atoms of an element*. This mass, called the **molar mass**, is numerically equal to the atomic weight of the element (in amu). For example, the atomic weight of lithium, rounded to one decimal place, is 6.9 amu and the molar mass of Li is 6.9 g/mol (1 mol Li = 6.9 g). Similarly, the atomic weight of argon is 39.9 amu and its molar mass is 39.9 g/mol.

■ The molar mass of an element (the mass in grams of one mole of the element) is equal to its atomic weight in amu.

Knowing this relationship allows the factor label method to be used to carry out a number of useful conversions.

- A sample containing 0.770 mol of carbon has a mass of 9.24 g.

$$0.770 \cancel{\text{mol C}} \times \frac{12.0 \text{ g C}}{1 \cancel{\text{mol C}}} = 9.24 \text{ g C}$$

- 1.25 g of carbon atoms is 0.104 mol.

$$1.25 \cancel{\text{g C}} \times \frac{1 \text{ mol C}}{12.0 \cancel{\text{g C}}} = 0.104 \text{ mol C}$$

- 10.9 g of carbon atoms is 5.47×10^{23} atoms.

$$10.9 \cancel{\text{g C}} \times \frac{1 \cancel{\text{mol C}}}{12.0 \cancel{\text{g C}}} \times \frac{6.02 \times 10^{23} \text{ atoms}}{1 \cancel{\text{mol C}}} = 5.47 \times 10^{23} \text{ atoms}$$

SAMPLE PROBLEM 2.7

Mass to mole conversions

50.0 g of lead (Pb) is how many moles?

Strategy

The problem asks you to convert from grams of Pb to moles of Pb, so a relationship between these two units is used to produce a conversion factor. The atomic weight of Pb is 207.2 amu, which means that its molar mass is 207.2 g/mol (1 mol Pb = 207.2 g). To convert from grams into moles, moles must be in the numerator of the conversion factor, so

$$\frac{1 \text{ mol Pb}}{207.2 \text{ g Pb}}$$

Solution

0.241 mol Pb

$$50.0 \cancel{\text{g Pb}} \times \frac{1 \text{ mol Pb}}{207.2 \cancel{\text{g Pb}}} = 0.241 \text{ mol Pb}$$

PRACTICE PROBLEM 2.7

a. What is the mass of 0.0122 mol of Pb?
b. What is the mass of 2.50×10^{21} Pb atoms?

2.6 RADIOACTIVE ISOTOPES

- Radioisotopes emit nuclear radiation (high energy particles and electromagnetic radiation).

For the 91 elements that occur naturally on the earth, more than a total of 300 isotopes have been identified. Over 1000 additional artificial isotopes have been produced. Studies of naturally occurring and artificial isotopes have shown that some have unstable nuclei that spontaneously disintegrate to become more stable, releasing high energy particles (individual or groups of subatomic particles) and/or high energy electromagnetic radiation (discussed in the paragraphs that follow). The particles and energy released in this nuclear change are called **nuclear radiation**, and atoms that emit nuclear radiation are called **radioactive isotopes** or **radioisotopes**. Of the three hydrogen isotopes, ${}^{1}_{1}H$ and ${}^{2}_{1}H$ are stable and emit no nuclear radiation, while ${}^{3}_{1}H$ is a radioisotope.

In Section 1.2 we were introduced to the idea of physical change, a change in which nothing new is created (melting ice, crushing sugar, etc.). As we will see below, in the **nuclear change** that takes place when nuclear radiation is released by a radioisotope, something new is usually created—when an atomic nucleus changes, so does the identity of the atom involved.

Common Forms of Radiation

Radioisotopes typically emit one or more of the four common types of radiation: alpha particles, beta particles, positrons, and gamma rays.

An **alpha particle** (${}^{4}_{2}\alpha$) is identical to the nucleus of a helium-4 atom (${}^{4}_{2}He$), in that each has two protons, two neutrons, and a 2+ charge. The alpha particle has a much greater energy, however, due to its high speed—a typical alpha particle is ejected from the nucleus of a radioisotope at 5–10% of the speed of light (Table 2.4).

When a radioisotope emits an alpha particle, the new atom that is formed has two fewer protons and two fewer neutrons than the original. For example, when thorium-230 emits an alpha particle (Figure 2.10*a*), radium-226 is formed, as is represented by the following **nuclear equation**.

$${}^{230}_{90}Th \longrightarrow {}^{226}_{88}Ra + {}^{4}_{2}\alpha$$

TABLE 2.4 COMMON FORMS OF RADIOACTIVITY

Name	Symbol	Makeup	Charge	Velocity[a]	Penetrating Ability
Alpha	α	2 protons + 2 neutrons	2+	5–10% of light speed	Low
Beta	β	Electron	1−	Up to 90% of light speed	Moderate
Positron	β^+	Positively charged electron	1+	Up to 90% of light speed	Moderate
Gamma	γ	Electromagnetic radiation	0	Light speed	High

[a] Light speed = 3×10^8 m/s (meters/second) = 186,000 mps (miles per second).

In a nuclear equation an arrow separates the starting radioisotope (shown on the left) from the products (shown on the right). To be useful, nuclear equations must be **balanced**, which means that *the sum of the mass numbers and the sum of the charges on atomic nuclei and subatomic particles must be the same on both sides of the equation.* In the balanced equation above, thorium's atomic number indicates a net nuclear charge of 90+ (from the 90 protons in its nucleus). The products also have a charge of 90+, due to radium's 88 protons and the alpha particle's 2 protons. The sum of the mass numbers is 230 on each side of the equation.

SAMPLE PROBLEM 2.8

Balancing nuclear equations

In targeted alpha therapy, a carrier substance is used to transport particular radioisotopes directly to cancer cells, which are then destroyed by the alpha radiation. Balance the nuclear equation for the loss of an alpha particle from actinium-225, one radioisotope used in this procedure.

$$^{225}_{89}\text{Ac} \longrightarrow ? + {}^{4}_{2}\alpha$$

Strategy

To balance this nuclear equation you must come up with an atomic number for the missing atom that will make the sum of the atomic numbers the same on both sides of the reaction arrow (hint: $89 = x + 2$). The same must be true of the mass numbers ($225 = y + 4$). Once you have decided on the atomic number for the missing element, you can refer to the periodic table to discover its identity.

Solution

$$^{225}_{89}\text{Ac} \longrightarrow {}^{221}_{87}\text{Fr} + {}^{4}_{2}\alpha$$

Balancing the charges on the nuclei and subatomic particles requires a sum of 89 on each side of the equation, so the missing product must have an atomic number of 87 ($89 = 87 + 2$), which corresponds to francium (Fr). Balancing the mass numbers requires a sum of 225 on each side of the equation, so the missing product must have a mass number of 221 ($225 = 221 + 4$).

PRACTICE PROBLEM 2.8

Bismuth-213 is another radioisotope used in targeted alpha therapy. Balance the nuclear equation for the loss of an α particle from this radioisotope.

$$^{213}_{83}\text{Bi} \longrightarrow ? + {}^{4}_{2}\alpha$$

A **beta particle** ($^{0}_{-1}\beta$) is an electron that is ejected from the nucleus of a radioisotope at up to 90% of the speed of light (Table 2.4). Since a beta particle is just a fast moving electron, it has the same charge and mass as an electron. On release of a beta particle, the nucleus of the atom formed has one more proton and one less neutron than the original radioisotope. Boron-12, for example, releases a beta particle to form carbon-12 (Figure 2.10*b*), as shown in the nuclear equation

$$^{12}_{5}\text{B} \longrightarrow {}^{12}_{6}\text{C} + {}^{0}_{-1}\beta$$

FIGURE | 2.10

Radioactive decay

(*a*) Loss of an alpha particle from thorium-230. (*b*) Loss of a beta particle from boron-12. (*c*) Loss of a positron from fluorine-18. (*d*) Loss of a gamma ray from an unstable nucleus.

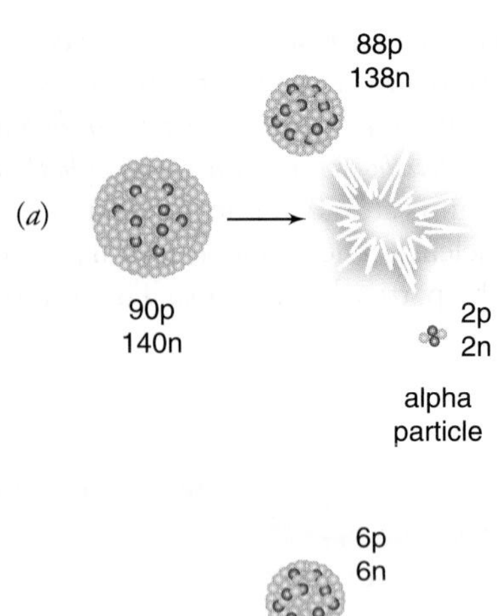

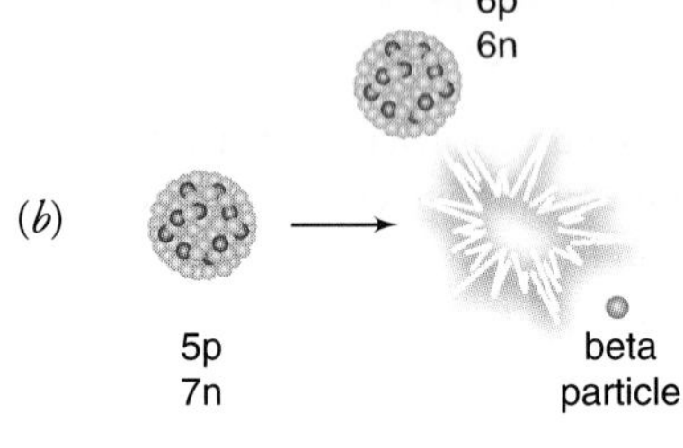

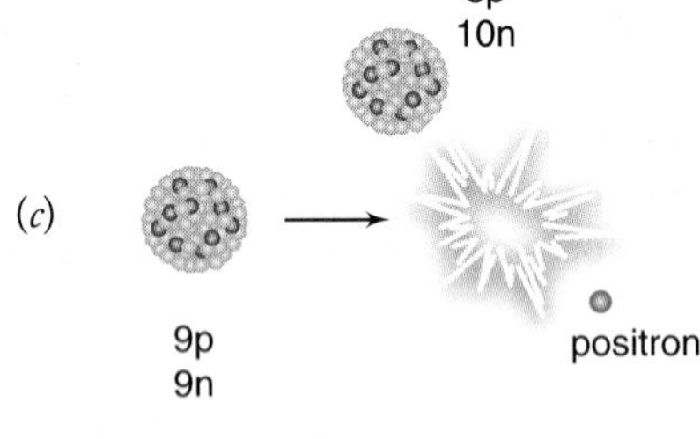

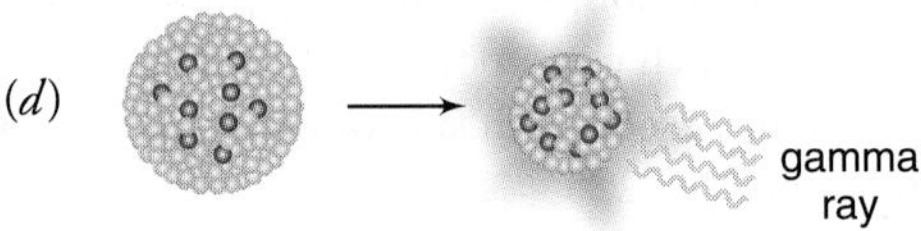

When using atomic notation to describe certain types of nuclear radiation, charge is written in the place normally reserved for atomic number. Examples include ${}^{0}_{-1}\beta$, ${}^{0}_{1}\beta^+$, and ${}^{0}_{0}\gamma$.

A beta particle carries a 1− charge and since it has no protons or neutrons, its mass number is 0. The nuclear equation shown above is balanced, because the sum of the charges on the nuclei and subatomic particles (5) and the sum of the mass numbers (12) is the same on both sides of the arrow.

It may be puzzling to think that a nucleus composed of protons and neutrons can eject an electron. What happens is that a neutron transforms into a proton and electron, and the electron is ejected from the nucleus at high speed as a beta particle.

A **positron** (${}^{0}_{1}\beta^+$) is a subatomic particle that has the same mass as a beta particle but carries a 1+ charge. Like beta particles, positrons are ejected from the nucleus of a radioisotope at speeds of up to 90% of the speed of light. Positron radiation has an important use in a medical procedure called positron emission tomography (see the paragraphs directly before the summary of chapter objectives).

The nucleus of an atom formed by release of a positron has one less proton and one more neutron than the original radioisotope. For example, when fluorine-18 emits a positron (Figure 2.10*c*), oxygen-18 is produced, as is shown by the nuclear equation,

$${}^{18}_{9}F \longrightarrow {}^{18}_{8}O + {}^{0}_{1}\beta^+$$

The nuclear equation above is balanced because the sum of the charges on the nuclei and subatomic particles (9) and the sum of the mass numbers (18) are the same on both sides of the equation.

The fourth common form of radiation, the **gamma ray** ($^{0}_{0}\gamma$), is **electromagnetic radiation**, a type of energy that travels as waves. The more well-known types of electromagnetic radiation (arranged in order of increasing energy) are radiowave, microwave, infrared, visible light, ultraviolet, x ray, and gamma ray.

■ Four common forms of radiation are the alpha particle ($^{4}_{2}\alpha$), the beta particle ($^{0}_{-1}\beta$), the positron ($^{0}_{1}\beta^{+}$), and the gamma ray ($^{0}_{0}\gamma$).

Gamma rays are a very high energy form of electromagnetic radiation. The release of alpha, beta, or positron radiation is often accompanied by the release of gamma rays because loss of any of these particles can leave the newly produced nucleus in an unstable state. When this unstable nucleus rearranges to a more stable form, energy is given off as gamma rays (Figure 2.10*d*). Iodine-131 is an example of a beta and gamma emitting radioisotope.

$$^{131}_{53}\text{I} \longrightarrow {}^{131}_{54}\text{Xe} + {}^{0}_{-1}\beta + {}^{0}_{0}\gamma$$

A gamma ray has no charge and, since it has no protons or neutrons, its mass number is 0.

SAMPLE PROBLEM 2.9

Balancing nuclear equations

To treat some forms of cancer, wire containing iridium-192 is surgically implanted in a tumor and removed at a later date. Beta particles emitted by this radioisotope kill the cancer cells. Balance the nuclear equation for the loss of a beta particle from this radioisotope.

$$^{192}_{77}\text{Ir} \longrightarrow ? + {}^{0}_{-1}\beta$$

Strategy

To balance this nuclear equation you must find an atomic number for the missing atom that will make the sum of the charges on nuclei and subatomic particles the same on both sides of the reaction arrow (hint: $77 = x + -1$). The same must be true of the mass numbers ($192 = y + 0$). Once you have decided on the atomic number for the missing element, you can refer to the periodic table to discover its identity.

Solution

$$^{192}_{77}\text{Ir} \longrightarrow {}^{192}_{78}\text{Pt} + {}^{0}_{-1}\beta$$

PRACTICE PROBLEM 2.9

Instead of releasing a beta particle, an iridium-192 nucleus may release a positron. Write a balanced nuclear equation for this process.

2.7 RADIOISOTOPES IN MEDICINE

When nuclear radiation comes into contact with matter, including living tissue, its kinetic energy is transferred to the surrounding atoms. The changes that follow can alter water and important biochemical substances (proteins, DNA, lipids, and others) involved in regulating processes that take place within cells, disrupting normal cellular functions (Figure 2.11). This accounts for the health risks associated with exposure to radiation.

■ FIGURE | 2.11

Radiation effects
Nuclear radiation can disrupt normal cellular functions by damaging proteins, DNA, lipids, and other biochemical substances important to life. (*a*) Normal cells. (*b*) Radiation-damaged cells.

Source: SPL/Photo Researchers, Inc.

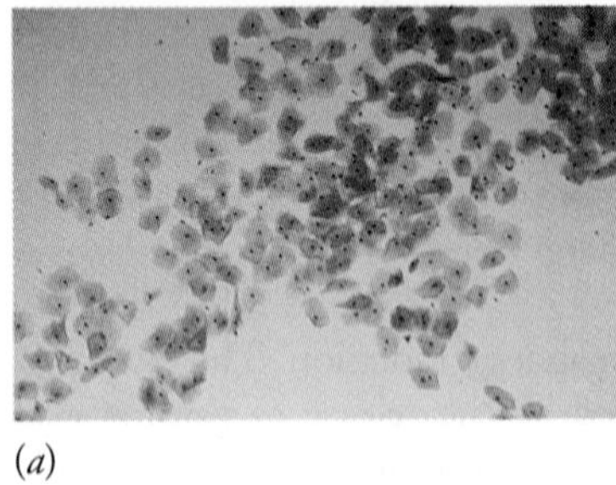

(*a*)

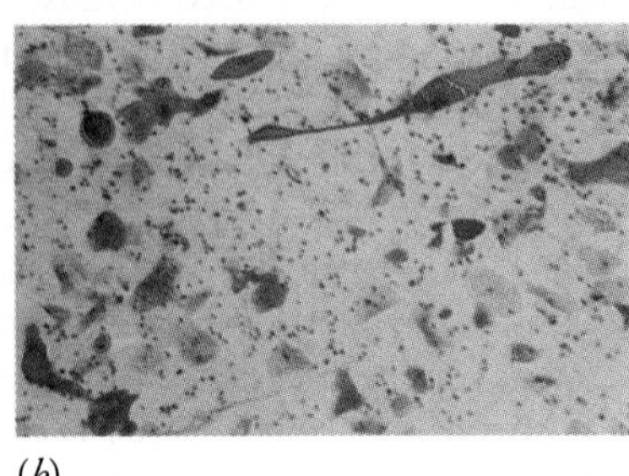

(*b*)

■ FIGURE | 2.12

Chernobyl
In attempting to repair the damaged reactor at the Chernobyl power plant, some workers received lethal doses of radiation.

Source: © Vladimir Repik/Reuters/Corbis.

The short-term effects of a single exposure to a high dose of nuclear radiation range from nausea to death within a few weeks, depending on the dose. Some of the workers who tried to control the 1986 Chernobyl power plant disaster in the Soviet Union were exposed to extremely high levels of radiation in a very short time (Figure 2.12), and more than 30 of them died soon after. The delayed effects of radiation exposure include an increased risk of cancer, cataracts, and mutations—changes in egg and sperm cells that can transmit genetic disorders to offspring.

The effects of long-term exposure to radiation at low levels are not well understood. One factor that makes estimating the effects of low doses of radiation difficult is that we are continually exposed to naturally occurring radioisotopes that are present in and around us. These levels of background radiation vary depending on where you live. For example, in regions of Brazil, India, and China, where monazite (a mineral containing radioactive thorium-232 and radium-226) is found, the background radiation level is hundreds of times higher than the average level found in the United States. The altitude that you live at determines how much exposure you will have to background radiation from the cosmic rays (atomic nuclei) that constantly bombard the earth from outer space. The higher the altitude, the less atmosphere there is above you to block the cosmic rays.

Controlling Exposure to Radiation

While we can't do much about background radiation, we can control our exposure to other sources of nuclear radiation. One way to do so is to use proper shielding (blocks radiation) when working with radioisotopes. Although fast moving, alpha particles travel only 4 to 5 cm in air before losing their energy, and they can only slightly penetrate surrounding matter (Figure 2.13 and Table 2.4). Gloves, clothing, or a sheet of paper are usually sufficient to block alpha particles. Being faster moving and smaller than alpha particles, beta particles and positrons can travel more than 1 meter in air and are able to

■ FIGURE | 2.13

Shielding
Alpha particles are blocked by paper, beta particles and positrons by a thin piece of plastic, and gamma rays by concrete.

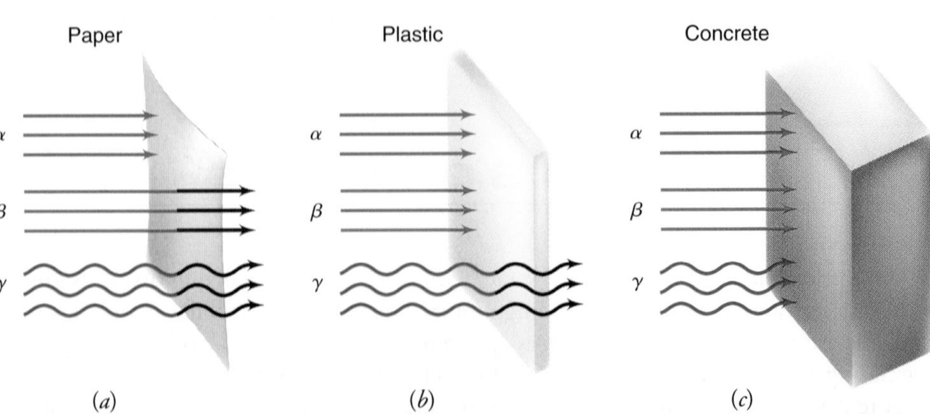

penetrate matter to a greater depth. Blocking these particles requires more substantial shielding than needed for alpha particles, typically a thin sheet of plastic or metal. Gamma rays are so energetic that they have a very high penetrating power—a thick slab of concrete or lead is required to stop them.

Half-life

Knowing how long a radioisotope will remain hazardous is also an important factor in controlling exposure to radiation. This is determined from the **half-life** of a radioisotope, *the time required for one-half of the atoms in a sample to decay (fall apart)* (Figure 2.14). Take, for example, the radioactive decay of radon-222 into polonium-218 by emission of an alpha particle.

$$^{222}_{86}\text{Rn} \longrightarrow {}^{218}_{84}\text{Po} + {}^{4}_{2}\alpha$$

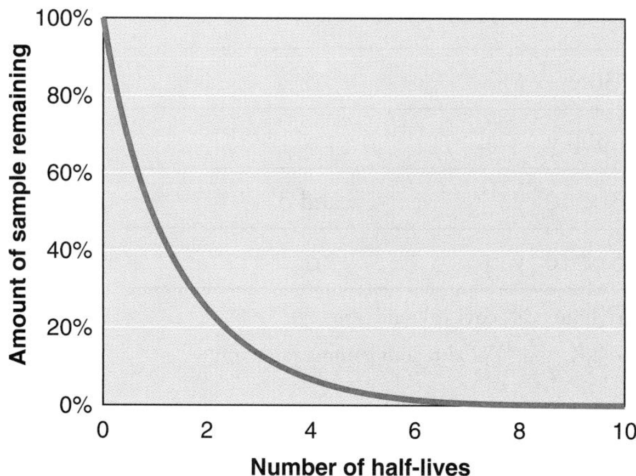

FIGURE 2.14

Half-life
After one half-life, a sample will have decayed to 50% of its original amount. After two half-lives 25% will remain and after three half-lives 12.5% will remain.

Radon-222 has a half-life of 3.8 days, which means that beginning with a 2 mg sample of this radioisotope, only 1 mg remains after 3.8 days. After another 3.8 days, 0.5 mg (one-half of 1 mg) of radon-222 remains. The half-life for a radioactive decay is unrelated to the starting amount of a sample, so beginning with 10 mg of $^{222}_{86}\text{Rn}$, 5 mg will remain after 3.8 days, and beginning with 0.01 mg, 0.005 mg will remain after 3.8 days.

■ After one half-life, half of the atoms in a radioisotope will have decayed.

Like radon-222, plutonium-239 is an alpha emitter, but its half-life of 24,000 years is considerably longer.

$$^{239}_{94}\text{Pu} \longrightarrow {}^{235}_{92}\text{U} + {}^{4}_{2}\alpha$$

Beginning with 1 g of plutonium-239 and waiting for one half-life (24,000 years), 0.5 g would remain. Over this same period of time, 1 g of radon-222 would disappear. After 71 half-lives (about 9 months) only one atom would remain. As Table 2.5 shows, half-lives range from small fractions of a second to over a billion years.

Half-life is taken into consideration when radioisotopes are selected for treating and diagnosing diseases. Radioisotopes with half-lives ranging from hours to weeks are generally preferred, because they do not decay before serving their purpose and they do not pose problems by being present long after their use.

Half-life also plays a role in determining whether a particular isotope is long-lived enough to be found in nature. Radioisotopes with extremely short half-lives, such as polonium-214 (0.0016 sec), decay so rapidly that they can be observed only immediately after their formation. Those with very long half-lives, such as uranium-238 (4.5 billion years), decay very slowly and are therefore present in nature.

TABLE 2.5 HALF-LIVES AND DECAY TYPES FOR SELECTED RADIOISOTOPES

Isotope	Half-life[a]	Decay Type[b]
$^{214}_{84}Po$	0.0016 s	α
$^{11}_{6}C$	20 m	β^+
$^{24}_{11}Na$	15 h	β
$^{222}_{86}Rn$	3.8 d	α
$^{131}_{53}I$	8.1 d	β
$^{32}_{15}P$	14.3 d	β
$^{3}_{1}H$	12.3 y	β
$^{90}_{38}Sr$	28.1 y	β
$^{14}_{6}C$	5730 y	β
$^{239}_{94}Pu$	24,000 y	α
$^{40}_{19}K$	1.3×10^9 y	β and β^+
$^{238}_{92}U$	4.5×10^9 y	α

[a] Units: seconds (s), minutes (m), hours (h), days (d), and years (y).

[b] $^{214}_{84}Po$, $^{24}_{11}Na$, $^{222}_{86}Rn$, $^{131}_{53}I$, $^{239}_{94}Pu$, $^{40}_{19}K$, and $^{238}_{92}U$ also emit gamma radiation.

SAMPLE PROBLEM 2.10

Calculations involving half-life

Rubidium-84, used to monitor cardiac output, has a half-life of 33 days. How many milligrams of a 10.0 mg sample of this radioisotope remain after 99 days?

Strategy

To solve this problem it helps to remember that a half-life of 33 days means that whatever the starting amount of rubidium-84, after 33 days only half remains. Of that remaining amount, in another 33 days, only half remains.

Solution

1.25 mg

After 1 half-life (33 days), 5.00 mg remain
After 2 half-lives (66 days), 2.50 mg remain
After 3 half-lives (99 days), 1.25 mg remain

PRACTICE PROBLEM 2.10

How many days are required for a 1.00 mg sample of $^{84}_{37}Rb$ to decay to 0.0625 mg?

Diagnosis and Therapy

Radioisotopes allow clinicians to diagnose a wide range of health problems (Table 2.6). Iodine-131 ($^{131}_{53}I$) can be used to monitor the state of the thyroid gland, because cancer, hyperthyroidism (an overactive thyroid), and other thyroid disorders cause characteristic changes in the rate at which iodine is taken up by this gland. After a patient is given a dose of an $^{131}_{53}I$-containing substance, external radiation detectors are used to measure the speed at which this radioisotope accumulates in the thyroid (Figure 2.15).

TABLE 2.6 Some Uses of Radioisotopes in Medicine

Radioisotope	Use
$^{24}_{11}Na$	Detecting blood vessel obstruction
$^{32}_{15}P$	Treating leukemia; detecting eye tumors
$^{51}_{24}Cr$	Imaging the spleen; detecting gastrointestinal disorders
$^{59}_{26}Fe$ and $^{52}_{26}Fe$	Detecting bone marrow disorders
$^{60}_{27}Co$	Treating cancer
$^{67}_{31}Ga$	Treating lymphomas; whole-body tumor scans
$^{75}_{34}Se$	Pancreas scans
$^{84}_{37}Rb$	Measuring cardiac output
$^{85}_{38}Sr$	Bone scans
$^{125}_{53}I$	Treating prostate and brain cancer
$^{131}_{53}I$	Treating and detecting thyroid disorders
$^{133}_{54}Xe$	Detecting lung malfunctions
$^{137}_{55}Cs$	Treating cancer
$^{167}_{69}Tm$	Bone and tumor scans
$^{186}_{75}Re$	Pain relief for bone, prostate, and breast cancer
$^{192}_{77}Ir$	Internal radiation therapy
$^{197}_{80}Hg$	Brain scans
$^{213}_{83}Bi$	Targeted alpha therapy
$^{225}_{89}Ac$	Targeted alpha therapy

Just as iodine-131 can be used for diagnosing thyroid problems, it can also be used therapeutically. To treat hyperthyroidism or thyroid cancer, patients are given a higher dose of $^{131}_{53}I$ than they would receive for diagnosis. As this radioisotope accumulates in normal and cancerous thyroid cells, it emits radiation that damages or kills the cells.

Radiologists can use a beam of gamma rays emitted by cobalt-60 ($^{60}_{27}Co$) or cesium-137 ($^{137}_{55}Cs$) to kill cancerous cells (Figure 2.16). This procedure is gradually being phased out and replaced by one that uses high-energy x rays, a form of treatment that does not require the presence of radioisotopes. Either way, even though attempts are made to limit the radiation damage to normal cells, side effects are common. Patients undergoing extensive radiation therapy routinely experience hair loss, nausea, and loss of white blood cells.

FIGURE | 2.15

A thyroid scan
After a patient is given a dose of iodine-131, a scanner measures radiation emitted by the thyroid gland.

Source: CNRI/Photo Researchers, Inc.

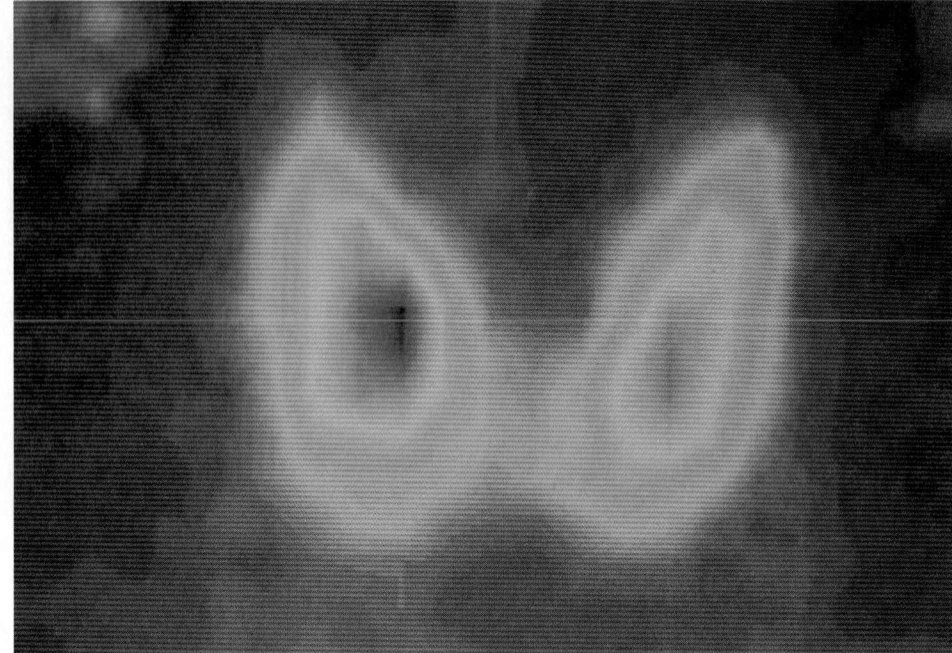

FIGURE | 2.16

Radiation cancer therapy
The radiation source, a gamma emitter, is moved through a circular path so only the tumor receives continuous radiation.

Source: From Biochemistry: A Foundation 1st edition by Ritter. © 1996. Reprinted with permission of Brooks/Cole, a division of Thomson Learning: *www.thomsonrights.com*. Fax 800 730-2215.

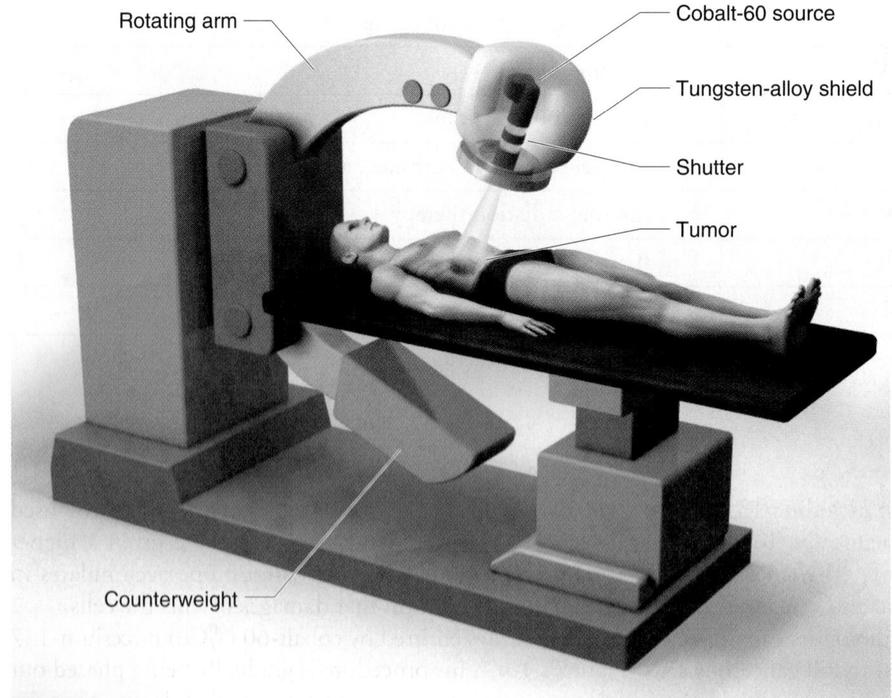

HEALTH Link: CT and MRI Imaging

To help diagnose injuries or diseases, clinicians sometimes find it useful to have images of various organs and tissues. This medical imaging commonly makes use of x rays or radio waves.

X rays are a form of electromagnetic radiation that has slightly less energy than gamma rays. The medical use of x rays involves placing a patient between an x-ray source and x-ray (photographic) film. X rays are absorbed to a different extent by various tissues, and only those x rays that pass through the body expose the x-ray film (Figure 2.17).

Contrast media, substances that completely block x rays, can be used to make specific structures stand out. For example, barium-containing substances are often administered orally or as an enema to allow a close look at the gastrointestinal tract.

Tomography, named after the Greek word *tomos*, meaning *a cut*, is a group of techniques that produce images of various two-dimensional slices of an object. Computed tomography (CT), also known as computed axial tomography (CAT), couples the use of computers with x-ray technology. To obtain a CT scan, a narrow beam of x rays is rotated around a patient, while detectors connected to a computer measure the location and strength of x rays that pass through the patient. This information can be processed to provide three-dimensional views of the body (Figure 2.18).

Electromagnetic radiation exists not only as high-energy x rays and gamma rays, but also as low-energy radio waves. Like gamma rays (see the paragraphs directly before the summary of chapter objectives) and x rays (see above), radio waves can play a role in medical imaging. **Magnetic Resonance Imaging** (MRI) is based on the effect that radio waves have on certain atomic nuclei, including hydrogen-1, which have a property known as nuclear spin. In the presence of a strong magnetic field, a radio wave with the proper energy can "tickle" a spinning hydrogen nucleus in a way that provides information about the atom's chemical environment. Software converts these atomic signals into images such as that in Figure 2.19.

In a typical MRI imager, the magnetic field is 25,000 times greater than the earth's magnetic field and about 150 times stronger than a refrigerator magnet. The strength of the MRI's magnetic field requires that care be taken with metal objects. This is why patients must remove rings and jewelry before having an MRI scan performed.

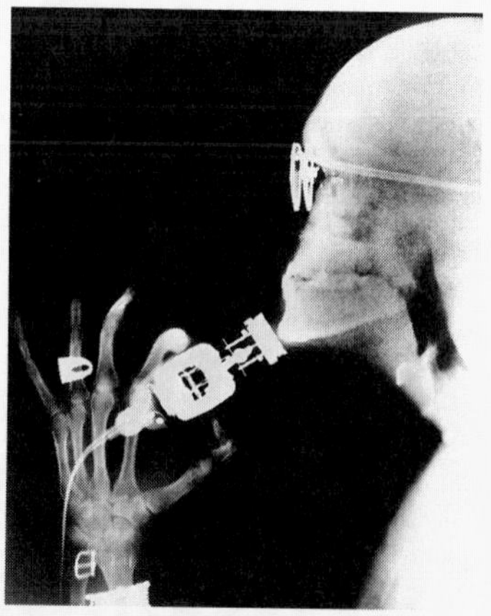

FIGURE 2.17

X rays
This x-ray image shows details of bone structure. X rays do not penetrate the glasses, ring, watch, or electric shaver.

Source: © Bettmann/Corbis.

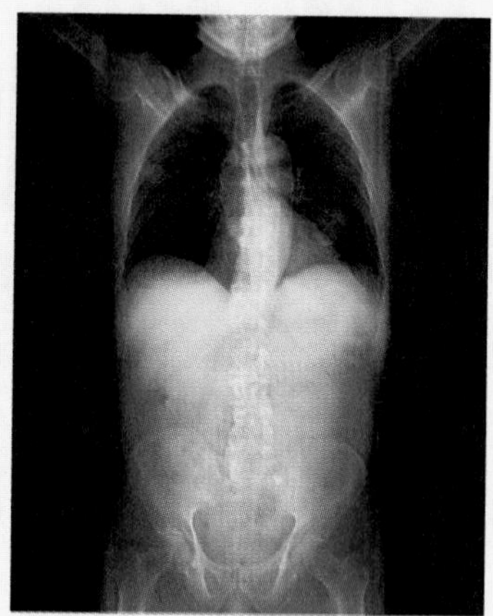

FIGURE 2.18

Computed tomography (CT)
This CT image shows a patient's crooked spine.

Source: Ben Edwards/Stone/Getty.

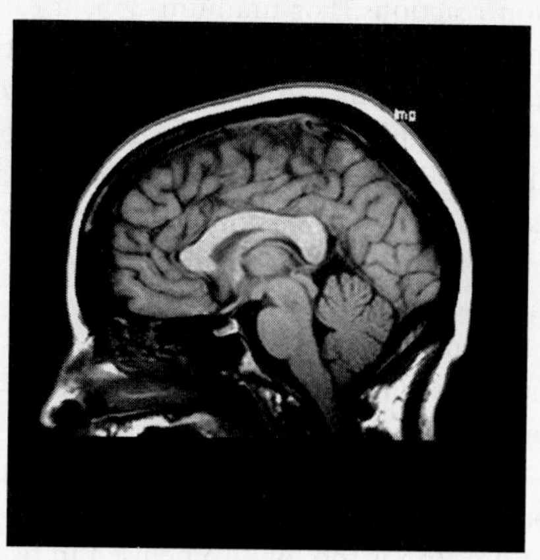

FIGURE 2.19

Magnetic Resonance Imaging
This MRI shows a patient's brain.

Living Art Enterprises, LLC/Photo Researchers, Inc.

2.8 FISSION AND FUSION

If you asked someone to tell you the first scientific equation that came to mind, the response that you would probably get is $E = mc^2$. This famous equation of Albert Einstein relates energy (E) to mass (m) and the speed of light (c). The value of the speed of light, 3.0×10^8 meters per second, is a very large number, so this equation predicts that a very small change in mass is associated with a very large change in energy. For the usual changes that go on around us—burning propane gas in a camp stove, boiling water, etc.—the changes in mass and energy are extremely small. When radioisotopes undergo decay, however, the change in mass is significant and considerable amounts of energy are released. For example, when one mole of uranium-238 undergoes α decay, 200,000 times more energy is released than when one mole of propane is burned. In the case of uranium-235, which has a half-life of 4.5 billion years, this energy is released very slowly. Other nuclear reactions (fission and fusion) release energy more quickly.

Fission

In **fission**, an atom's nucleus splits to produce two smaller nuclei, a number of neutrons, and energy. Fission reactions of the natural radioisotope uranium-235 and the artificial ones uranium-233 and plutonium-239 are useful as energy sources. For atoms of these radioisotopes, fission takes place when they are hit by neutrons. One of the fission reactions that uranium-235 undergoes releases energy and produces barium-142, krypton-91, and 3 neutrons.

$$^{235}_{92}\text{U} + {}^{1}_{0}\text{n} \longrightarrow {}^{142}_{36}\text{Ba} + {}^{91}_{36}\text{Kr} + 3\,{}^{1}_{0}\text{n}$$

This is an example of a **chain reaction**, one in which *one or more products* (neutrons in this case) *can initiate another cycle of the reaction*. Because neutrons must collide with atomic nuclei for fission to take place, the uranium-235 chain reaction will continue only if there are sufficient amounts of this radioisotope present. There must be enough uranium atoms that, at a minimum, one of the neutrons produced in each reaction is used to begin another fission reaction. *This minimum amount of radioisotope needed for a nuclear chain reaction to continue* is called the **critical mass.**

If a **supercritical mass** of a radioisotope is present, most of the neutrons produced will be used in initiating another cycle of fission reactions. For uranium-235, if each of the 3 released neutrons encounters other uranium-235 nuclei, 3 more fission reactions will take place. The 9 neutrons produced in these reactions can take part in 9 more fission reactions. As the number of fission reactions multiplies, so do the amount of energy released and the number of neutrons produced (Figure 2.20). This can lead to a nuclear explosion and is the basis for nuclear weapons. Of the two atomic bombs dropped on Japan during World War II, one contained a supercritical amount of uranium-235 and the other, a supercritical amount of plutonium-239.

In addition to rapidly releasing great amounts of energy, nuclear weapons produce radioisotopes, many of which have long half-lives. This radioactive fallout can have long-term adverse effects on living things and on the environment.

Uranium, as found in nature, consists of 0.72% uranium-235 and 99.28% uranium-238. No amount of natural uranium can be a critical mass, because uranium-238 does not undergo fission and the percentage of uranium-235 is too low. To be used in fission reactions, uranium must be enriched, a process that involves separating the two isotopes from one another. For nuclear reactors (see below), uranium must be enriched to about 3% uranium-235. Nuclear weapons require about 90% uranium-235.

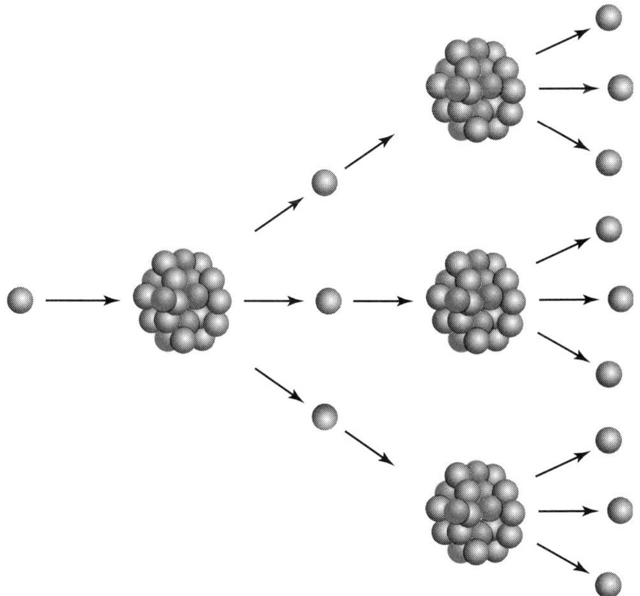

FIGURE 2.20

Fission

Collision with a neutron splits a uranium-235 nucleus. Neutrons produced in this chain reaction can initiate additional fission reactions.

http://www.mbe.doe.gov/me70/manhattan/images/FissionChainReaction.gif

Nuclear Reactors

There are a number of different types of nuclear reactors. What they all have in common are fuel rods, control rods, coolant, and a moderator. The core of a reactor contains **fuel rods**, which *contain the radioisotope that undergoes fission.* In many cases, uranium is the radioisotope of choice. The amount of energy released by the fuel rods is determined by **control rods**, which are constructed from boron or other *elements that absorb neutrons.* When the control rods are pushed into the core, neutrons are absorbed and fission decreases. If the control rods are pulled out of the core, more neutrons are exchanged between fuel rods and fission increases. To prevent the escape of radiation, the core of a nuclear reactor is shielded using steel and concrete. This absorbs neutrons and any alpha, beta, or gamma radiation that is produced.

The **coolant** in a nuclear reactor *removes the heat generated by the fission process.* Without coolant, the control rods could get so hot that they would melt (undergo a meltdown). The **moderator** is a substance that *slows neutrons.* Moderators are used because, in the case of uranium-235, slow-moving neutrons are absorbed more effectively than fast-moving ones. In many reactors, water is both a coolant and a moderator.

Nuclear power plants use the energy released through fission to generate electricity. At its most fundamental level, nuclear fission generates electricity in the same way that coal-fired plants do. The energy released (by fission, by burning coal) is used to produce steam that spins a turbine (Figure 2.21).

There are pros and cons to any form of electricity generation. The burning of coal, for example, gives products that contribute to air pollution, acid rain, and global warming. In this sense, nuclear reactors are clean—no air pollution is generated. The major concerns about nuclear power plants are related to radiation. One issue is that fuel rods eventually become less efficient as the fuel is consumed. The fuel rods can't simply be thrown away, because many fission products are themselves radioactive. One of the fission products, strontium-90, has a half-life of 28.8 years. It is generally agreed that it takes 20 half-lives for radiation released by radioisotopes to drop to safe levels, so strontium-90 generated today must be stored for 576 years (28.8 years $\times$ 20 = 576 years). Plutonium-239, also formed in fuel rods, has a half-life of 24,000 years and must be stored for considerably longer. Currently there is no final solution for storing these radioactive wastes. One promising approach is to encase the substances in glass or ceramic and bury them deep underground.

Sometimes spent fuel rods are **reprocessed**. This term refers to *chemical treatment of used nuclear fuel to separate fissile material—substances that can undergo fission.*

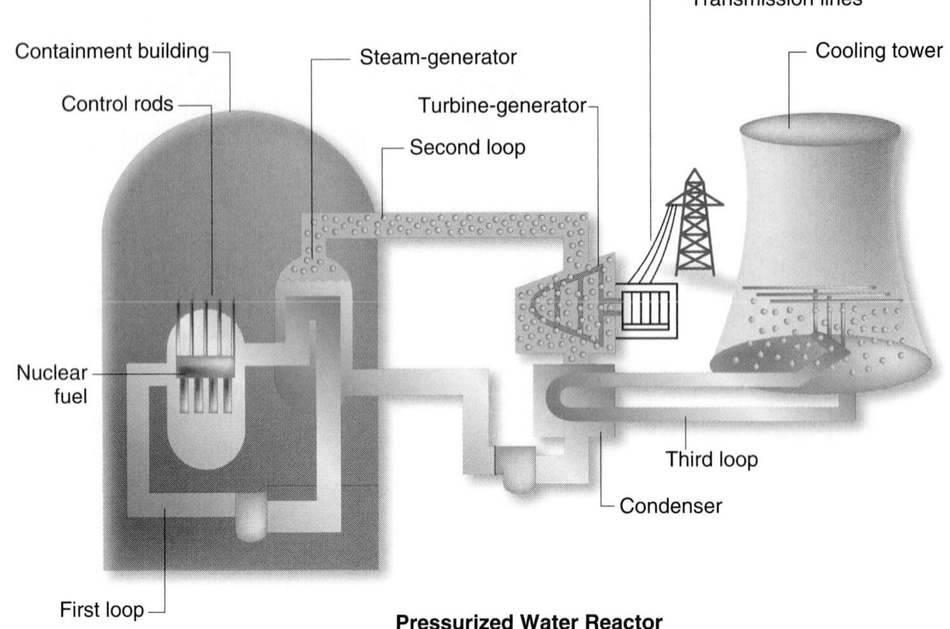

FIGURE 2.21

A nuclear reactor
In a nuclear reactor, heat generated by fission produces steam that turns an electricity-generating turbine. Cooling water from an external source converts the steam back to a liquid, which is then recycled through the reactor core.

http://www.nrc.gov/reading-rm/basic-ref/teachers/images/pwr-schematic.gif

Reprocessing can be carried out to obtain radioisotopes for use as fuel in nuclear reactors. For those working to limit the spread of nuclear weapons technology, reprocessing can be a concern. By reprocessing spent nuclear fuel rods, it is possible to obtain plutonium-239 with the purity and in quantities needed to produce nuclear weapons.

Fusion

We have seen that some large nuclei, such as uranium-235, undergo fission and release energy. On the other end of the spectrum, some small nuclei, such as hydrogen, can undergo **fusion**. In fusion, *nuclei combine (fuse)*, releasing large amounts of energy in the process. Our sun is a giant fusion reactor. One of the fusion reactions taking place in the sun is believed to be the combination of two hydrogen-1 nuclei to produce a deuterium nucleus and a positron.

$$2\,^{1}_{1}\text{H} \longrightarrow \,^{2}_{1}\text{H} + \,^{0}_{1}\beta^{+}$$

Currently, an internationally funded experiment is underway to study the possibility of using fusion as a power source, in much the same way that fission is currently used in nuclear reactors. This 30-year project, to be built in France, is known as ITER (International Thermonuclear Experimental Reactor). The fuel to be used in ITER is a combination of the hydrogen isotopes deuterium and tritium. When nuclei of these atoms undergo a fusion reaction, the products are helium-4, a neutron, and energy.

$$^{2}_{1}\text{H} + \,^{3}_{1}\text{H} \longrightarrow \,^{4}_{2}\text{He} + \,^{1}_{0}\text{n}$$

Because deuterium and tritium nuclei have an identical 1+ charge and repel one another, getting them to take part in a fusion reaction requires the input of a tremendous amount of energy—temperatures near 100,000,000 K. Anything heated to this temperature has a huge kinetic energy. To keep the nuclei near one another, ITER will use magnetic confinement (Figure 2.22).

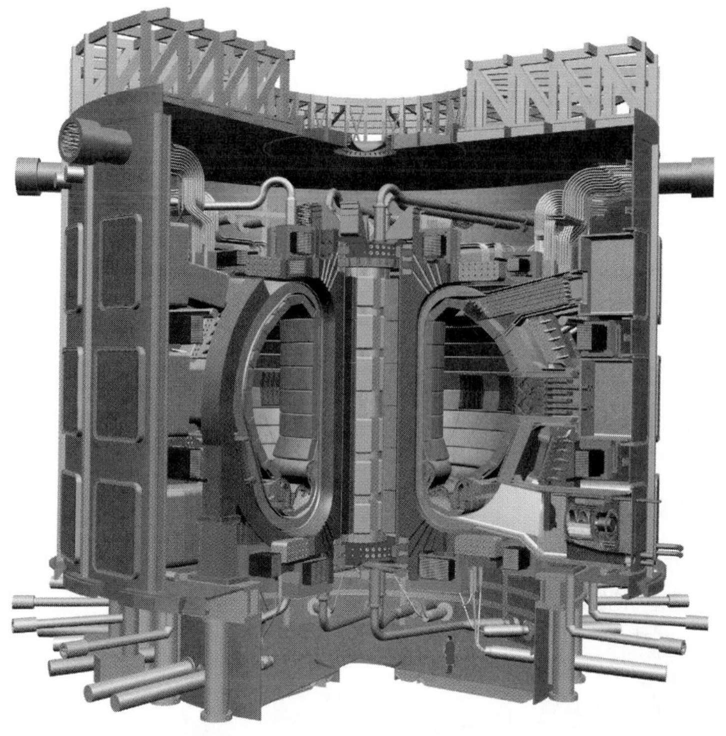

FIGURE 2.22

A cutaway view of ITER
ITER (International Thermonuclear Experimental Reactor) will explore the use of fusion as a power source.

Courtesy of ITER Organization.

In positron emission tomography (PET) a positron-emitting radioisotope is introduced into the body. Fluorine-18, one of the radioisotopes used in this technique, decays to form oxygen-18 and a positron.

$$^{18}_{9}\text{F} \longrightarrow {}^{18}_{8}\text{O} + {}^{0}_{1}\beta^{+}$$

When the released positron collides with an electron ($^{0}_{-1}\text{e}^{-}$) from a nearby atom, they destroy one another, producing two gamma rays in the process.

$$^{0}_{1}\beta^{+} + {}^{0}_{-1}\text{e}^{-} \longrightarrow 2\ {}^{0}_{0}\gamma$$

The gamma rays are detected by the PET scanner, and a computer processes the information to construct an array of two-dimensional images.

Fluorine-18 is an artificial radioisotope with a half-life of 110 minutes. To be used in PET, this rapidly decaying radioisotope must, within about three hours of being produced, be incorporated into glucose, delivered to a clinic, and used. Some large medical centers have their own cyclotrons (used to produce radioisotopes), but in many cases ^{18}F-labeled glucose is produced in one city and delivered to a clinic in a different one.

Being radioactive, the labeled glucose is transported in heavily shielded containers. In the clinic, a patient is moved to a shielded room, where the glucose is administered intravenously. Following this, the patient sits quietly for 30 to 90 minutes to allow it to move throughout his or her body. This radioactively tagged sugar will be at its highest levels in areas where energy demands are the greatest—in the brain for example. After a PET scan of the brain is run, a health professional can diagnose Alzheimer's and other diseases by seeing how glucose usage varies. The PET image shown here is of a normal brain (left) and the brain of a patient with Alzheimer's disease (right).

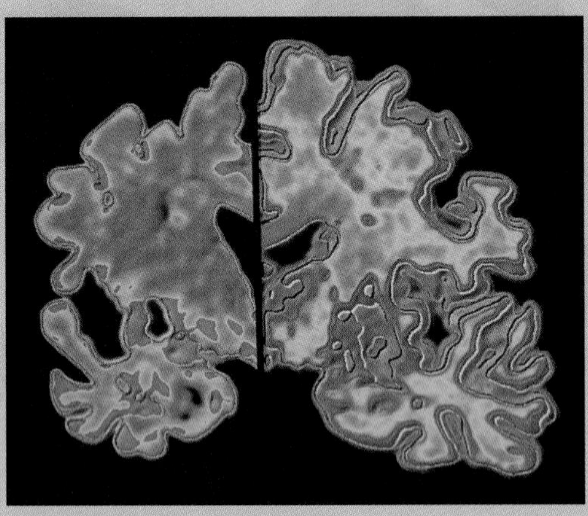

A. Pakieka/Photo Researchers, Inc.

summary of objectives

1 Describe the subatomic structure of an atom.

Matter is constructed from **atoms**, which are made from various combinations of the subatomic particles called **protons, neutrons**, and **electrons**. The nucleus of an atom consists of protons (mass $\approx$ 1 **amu**, charge = 1+) and neutrons (mass $\approx$ 1 amu, charge = 0). Electrons (mass $\approx$ 1/2000 amu, charge = 1−) are dispersed around the nucleus.

2 Define the terms element and atomic symbol.

The word **element** is used to describe matter that consists of just one type of atom. All atoms of a given element have the same number of protons in their nucleus. Each element has an **atomic symbol**, which is a one or two letter abbreviation of the element name.

3 Explain how elements are arranged in the periodic table.

The **periodic table** provides the atomic symbol, atomic number, and **atomic weight** (average mass in amu of the naturally occurring atoms of an element) of each of the elements. A line that starts to the left of boron (B) marks the location where the transition from **metals** (elements to the left of the line) to **nonmetals** (to the right of the line) takes place. **Semimetals**, which fall along the dividing line between metals and nonmetals, have properties that are intermediate between the metals and nonmetals. Elements that fall in the same vertical column of the periodic table belong to the same **group**, identified by a combination of numbers and letters (1A, 2A, etc.) and by names (alkali metals, halogens, etc.). Larger groupings of elements also exist, including the **representative elements**, the **transition metals**, and the **lanthanide** and **actinide elements**. Elements in the same horizontal row of the periodic table belong to the same **period**.

4 Explain how atomic number and mass number are used to indicate the makeup of an atom's nucleus. Describe how isotopes of an element differ from one another.

Atomic number is the number of protons in an atom's nucleus and **mass number** is the total number of protons and neutrons in its nucleus. Atoms of an element that have different numbers of neutrons are called isotopes. Isotopes of an element have the same atomic number and different mass numbers.

5 Define the terms mole and molar mass.

The **mole** is a counting unit that equals 6.02×10^{23} items. **Molar mass,** the mass (in grams) of one mole of an element, is equal to the atomic weight of the element (in amu).

6 Define the term nuclear radiation and describe the four common types of radiation emitted by radioisotopes. Explain how exposure to radiation can be controlled.

Radioisotopes contain unstable atomic nuclei that emit **nuclear radiation**, which commonly consists of particles (**alpha**, **beta**, or **positron**) or **electromagnetic radiation** (**gamma rays**). An alpha (α) particle consists of two protons and two neutrons, a beta (β) particle of an electron, a positron (β^+) of a positively charged electron, and a gamma (γ) ray of high-energy electromagnetic radiation.

Limiting exposure to radiation requires using appropriate shielding for the radiation type and knowing the **half-life** of the radioisotope, the time required for one-half of the atoms in a sample to decay.

7 Describe how fission differs from fusion. Define the terms chain reaction and critical mass.

In a **fission** reaction, a nucleus splits to produce two smaller nuclei and a number of neutrons. In a **fusion** reaction, nuclei combine to produce a larger nucleus. In each type of reaction or energy is released. A fission reaction is an example of a **chain reaction**, one in which one or more of the products can initiate a recurrence of the reaction. The minimum amount of a radioisotope needed for a nuclear chain reaction to be self-sustaining is called a **critical mass**.

END OF CHAPTER PROBLEMS

Answers to problems whose numbers are printed in color are given in Appendix B. More challenging questions are marked with an asterisk. Problems within color rules are paired. **ILW** = Interactive Learning Ware solution is available at *www.wiley.com/college/raymond.*

2.1 Atoms

2.1 Why does the nucleus of an atom have a positive charge?

2.2 Why is most of the mass of an atom located in the nucleus?

2.3 Describe the structure of an atom.

***2.4** One amu equals 1.66054×10^{-24} g.

a. 1.00728 g of protons is how many amu?

b. 1.00728 g of protons is how many protons?

*2.5 One amu equals 1.66054×10^{-24} g.
a. 1.00867 g of neutrons is how many amu?
b. 1.00867 g of neutrons is how many neutrons?

2.2 Elements

2.6 Give the atomic symbol for each element.
a. hydrogen **b.** helium **c.** chlorine **d.** oxygen

2.7 Give the atomic symbol for each element.
a. lithium **b.** bromine **c.** boron **d.** aluminum **e.** fluorine

2.8 Give the name of each element.
a. N **b.** C **c.** Sn **d.** Ag

2.9 Give the name of each element.
a. Be **b.** Ne **c.** Mg **d.** P

2.10 The following elements are among those that are required in the diet. Give the name of each.
a. I **b.** Na **c.** Fe

2.3 Atomic Number and Mass Number

2.11 How many protons and neutrons are present in the nucleus of each?
a. $^{19}_{9}F$ **b.** $^{23}_{11}Na$ **c.** $^{238}_{92}U$

2.12 How many protons and neutrons are present in the nucleus of each?
a. $^{133}_{54}Xe$ (used to detect lung malfunctions)
b. $^{75}_{34}Se$ (used in pancreas scans)
c. $^{84}_{37}Rb$ (used to measure cardiac output)

2.13 Which of the following statements do not accurately describe the isotopes of an element?
a. same number of protons
b. same mass number
c. same atomic number

2.14 Which of the following statements do not accurately describe the isotopes of an element?
a. same atomic symbol
b. same number of neutrons
c. same number of protons and neutrons

2.15 In nature, the element chlorine exists as two different isotopes, $^{35}_{17}Cl$ and $^{37}_{17}Cl$. The atomic weight of chlorine is 35.5 amu. Which chlorine isotope predominates?

2.16 In nature, the element neon exists as three different isotopes, $^{20}_{10}Ne$, $^{21}_{10}Ne$, and $^{22}_{10}Ne$. The atomic weight of neon is 20.2 amu. Which neon isotope predominates?

2.17 Name and give the atomic notation for an isotope that has 65 protons and 83 neutrons.

2.18 Calculate the number of protons, neutrons, and electrons in each neutral atom.
a. $^{4}_{2}He$ **b.** $^{37}_{17}Cl$ **c.** $^{222}_{86}Rn$ **d.** $^{40}_{20}Ca$

2.19 Calculate the number of protons, neutrons, and electrons in each neutral atom.
a. $^{24}_{12}Mg$ **b.** $^{55}_{25}Mn$ **c.** $^{64}_{30}Zn$ **d.** $^{74}_{34}Se$

2.20 Complete the table.

Name	Calcium		Copper
Atomic symbol		$^{13}_{6}C$	
Number of protons			29
Number of neutrons			63
Atomic number	20		
Mass number	40		

2.21 Complete the table.

Name	Helium		Iron
Atomic symbol		$^{34}_{17}Cl$	
Number of protons	2		
Number of neutrons			30
Atomic number			26
Mass number	3		

2.4 Periodic Table

2.22 Which of the two atoms is the most metallic?
a. Na and Cl **b.** O and Te

2.23 Which of the two atoms is the most metallic?
a. Al and Si **b.** Ca and Mg

2.24 Arrange each set of three atoms in order of size (largest to smallest).
a. Na, P, and Al **b.** O, Se, and S

2.25 Arrange each set of three atoms in order of size (largest to smallest).
a. I, F, and Br **b.** Ne, F, and O

2.26 List some of the physical properties of metals.

2.27 List some of the physical properties of nonmetals.

2.28 Name an element that belongs to each group.
a. a halogen **b.** an inert gas **c.** an alkaline earth metal **d.** an alkali metal

2.29 **a.** List the elements in group 5A by name and atomic symbol.
b. Specify whether each is a metal, semimetal, or nonmetal.

2.30 **a.** List the elements in the second period by name and atomic symbol.
b. Specify whether each is a metal, semimetal, or nonmetal.

*2.31 In nature, vanadium (V) is found as two isotopes.

Isotope	Mass (amu)	Percentage
vanadium-50	50.9440	99.75
vanadium-51	49.9472	0.25

Use this information to calculate the atomic weight of this element.

*2.32 In nature, nickel (Ni) is found as five isotopes.

Isotope	Mass (amu)	Percentage
nickel-58	57.9353	68.27
nickel-60	58.9302	26.10
nickel-61	60.9310	1.13
nickel-62	61.9283	3.59
nickel-64	63.9280	0.91

Use this information to calculate the atomic weight of this element.

2.5 The Mole

2.33 What is the
- a. atomic weight of helium (He)?
- b. molar mass of helium?
- c. mass (in grams) of 5.00 mol of helium?
- d. mass (in grams) of 0.100 mol of helium?
- e. mass (in grams) of 6.02×10^{23} helium atoms?

2.34 What is the
- a. atomic weight of magnesium (Mg)?
- b. molar mass of magnesium?
- c. mass (in amu) of 100 Mg atoms?
- d. mass (in grams) of 6.02×10^{23} Mg atoms?
- e. mass (in grams) of 3.00 mol of Mg?

2.35 How many atoms are present in 2.00 mol of aluminum?

2.36 How many atoms are present in 0.47 mol of uranium?

2.37 a. What is the atomic weight of sulfur (S)?
b. How many sulfur atoms are contained in 32.1 g of sulfur?

2.38 a. What is the atomic weight of carbon (C)?
b. How many carbon atoms are contained in 32.1 g of carbon?

2.39 How many atoms are contained in the following?
- a. 1.0 mol of carbon
- b. 1.22×10^{-9} mol of carbon
- c. 12.0 g of carbon
- d. 4.5 ng of carbon

2.40 How many atoms are contained in the following?
- a. 1.0 mol of fluorine
- b. 1.22×10^{-9} mol of fluorine
- c. 12.0 g of fluorine
- d. 4.5 ng of fluorine

2.41 In 3.45 mg of Fe there are how many
- a. grams of Fe?
- b. moles of Fe?
- c. Fe atoms?

2.42 In 7.71 μg of Zn there are how many
- a. grams of Zn?
- b. moles of Fe?
- c. Fe atoms?

2.6 Radioactive Isotopes

2.43 Write a balanced nuclear equation for each process.
- a. $^{187}_{80}Hg$ emits an alpha particle
- b. $^{226}_{88}Ra$ emits an alpha particle
- c. $^{238}_{92}U$ emits an alpha particle

2.44 Identify the missing product in each nuclear equation.
- a. $^{32}_{15}P \longrightarrow ? + ^{0}_{1}\beta^{+}$
- b. $^{40}_{19}K \longrightarrow ? + ^{0}_{-1}\beta$
- c. $^{40}_{19}K \longrightarrow ^{40}_{18}Ar + ?$

2.45 Identify the missing product in each nuclear equation.
- a. $^{14}_{8}O \longrightarrow ? + ^{0}_{1}\beta^{+}$
- b. $^{3}_{1}H \longrightarrow ^{3}_{2}He + ?$
- c. $^{14}_{6}C \longrightarrow ? + ^{0}_{-1}\beta$

2.46 a. Write the balanced nuclear equation for the loss of an alpha particle from $^{203}_{83}Bi$.
b. Write the balanced nuclear equation for the loss of a positron from $^{17}_{9}F$.

2.47 a. Write the balanced nuclear equation for the loss of an alpha particle from $^{35}_{16}S$.
b. Write the balanced nuclear equation for the loss of a beta particle from $^{27}_{12}Mg$.

2.48 $^{197}_{80}Hg$, a beta and gamma emitter, is used for brain scans. Write a balanced equation for the loss of 1 beta particle and 1 gamma ray from this radioisotope.

2.49 a. In the type of nuclear decay called electron capture, an electron from an atom's electron cloud falls into the nucleus. Iodine-128 undergoes electron capture to produce an x ray and another product. Write a balanced nuclear equation for this process.

$$^{128}_{53}I + ^{0}_{-1}e^{-} \longrightarrow ? + ^{0}_{0}x \text{ ray}$$

b. Write a balanced nuclear equation for loss of a positron from iodine-128.
c. Looking only at the new atom formed from the reactions in parts a and b of this question, is it possible to tell electron capture from beta emission?

2.50 In a radioactive decay series, one radioisotope decays into another, which decays into another, and so on. For example, in fourteen steps uranium-238 is converted to lead-206. Starting with uranium-238, the first decay in this series releases an alpha particle, the second decay releases a beta particle, and the third releases a beta particle. Write balanced nuclear equations for these three reactions.

2.7 Radioisotopes in Medicine

*2.51 Smoke detectors contain an alpha emitter. Considering the type of radiation released and the usual placement of a smoke detector, do these detectors pose a radiation risk? Explain.

2.52 A nurse is assisting a patient who has just undergone radiation therapy in which a radioisotope was administered intravenously. To ensure her own safety, what information might the nurse want to have regarding the radioisotope?

*__2.53__ Radioisotopes used for diagnosis are beta, gamma, or positron emitters. Why are alpha emitters not used for diagnostic purposes?

2.54 ${}^{59}_{26}Fe$, a beta emitter with a half-life of 45 days, is used to monitor iron metabolism.

a. Write a balanced nuclear equation for this radioactive decay.

b. How much time must elapse before a patient contains just 25% of an administered dose of ${}^{59}_{26}Fe$, assuming that this radioisotope is eliminated from the body only by radioactive decay?

2.55 ${}^{197}_{80}Hg$, a radioisotope used in brain scans, has a half-life of 66 hours.

a. Beginning with a 1.00 mg sample of this isotope, how many milligrams of the isotope will remain after 264 hours?

b. This isotope decays by emitting 1 neutron (${}^{1}_{0}n$) and 1 gamma ray. Write a balanced nuclear reaction for this decay process.

2.56 Given a 16 mg sample of ${}^{222}_{86}Rn$ and a 4 mg sample of ${}^{131}_{53}I$, which will decay to just 1 mg the fastest? Refer to Table 2.5 for the half-life of each isotope.

2.57 ${}^{52}_{26}Fe$, a positron emitter with a half-life of 8.2 hours, is used for PET bone marrow scans.

a. Write a balanced nuclear equation for this radioactive decay.

b. The detector used for this scan measures gamma rays. How is the release of positrons connected to the formation of gamma rays?

c. Assuming that this radioisotope is eliminated from the body only by radioactive decay, how much time must elapse before a patient contains just 25% of an administered dose of ${}^{52}_{26}Fe$?

d. The product of ${}^{52}_{26}Fe$ positron decay is a radioisotope that emits positrons. Write a balanced nuclear equation for the decay of this product.

2.8 Fission and Fusion

2.58 Uranium-235 can take part in other fission reactions than the one shown earlier in the chapter. Two of these nuclear reactions are shown below. Supply the missing product for each.

$${}^{235}_{92}U + {}^{1}_{0}n \longrightarrow {}^{103}_{42}Mo + ? + 2\,{}^{1}_{0}n$$

$${}^{235}_{92}U + {}^{1}_{0}n \longrightarrow {}^{91}_{36}Kr + ? + 3\,{}^{1}_{0}n$$

2.59 Explain the italicized terms in the following sentence. A *fission chain reaction* is self-sustaining when a *critical mass* of *fissile material* is present.

2.60 Plutonium-239 has a half-life of 24,000 years. If nuclear waste containing this radioisotope is stored today, what year will it be when the required 20 half-lives have passed and the waste is considered safe?

2.61 **a.** The United States stopped reprocessing fuel rods in the 1970s because of safety concerns. Since that time, spent fuel rods have been stored at nuclear reactor sites. What might those safety concerns have been?

b. In an energy plan released in 2006, it was proposed that reprocessing be restarted in the United States and the United States would accept spent fuel rods from other countries. What would be some benefits of this proposed plan?

2.62 In a fission reactor fueled by uranium, energy is released when a neutron encounters uranium-235. Although uranium-238, which makes up about 97% of the uranium in fuel rods, does not undergo fission, in three steps it can be transformed into plutonium-239, a fissile material. Complete each nuclear equation.

$${}^{238}_{92}U + ? \longrightarrow {}^{239}_{92}U$$

$${}^{239}_{92}U \longrightarrow {}^{239}_{93}Np + ?$$

(half-life = 23.5 minutes)

$${}^{239}_{93}Np \longrightarrow {}^{239}_{94}Pu + ?$$

(half-life = 2.36 days)

2.63 One plan for ITER is to use neutrons produced by fusion to "breed" tritium from lithium-6. The tritium, in turn, can be used as a fuel for fission. Write a balanced nuclear equation for this reaction of lithium-6 with a neutron to produce tritium.

2.64 Complete the fusion nuclear equations.

a. ${}^{2}_{1}H + {}^{2}_{1}H \longrightarrow {}^{1}_{1}p + ?$

b. ${}^{3}_{2}He + ? \longrightarrow {}^{4}_{2}He + 2\,{}^{1}_{1}p$

c. ${}^{13}_{6}C + {}^{4}_{2}He \longrightarrow {}^{1}_{0}n + ?$

d. ${}^{3}_{1}H + ? \longrightarrow {}^{4}_{2}He + 2\,{}^{1}_{0}n$

HEALTH*Link* | *Dietary Reference Intakes (DRIs)*

2.65 **a.** Table 2.3 lists the RDA for some elements and the AI for others. Why?

b. Which is a better measure of the amount of nutrient needed to meet a person's dietary needs, RDA or AI? Explain.

2.66 In NAS dietary tables the dietary reference intakes for selenium and iodine are reported in micrograms per day, rather than the milligrams per day given in Table 2.3. For these two elements, convert the values in Table 2.3 into micrograms per day.

2.67 The nutrition facts label on a box of macaroni and cheese says that one serving contains 20% of the daily recommended amount of calcium. Assuming that this value refers to adults, to how many milligrams of calcium does this correspond? (See Table 2.3.)

HEALTH*Link* | *CT and MRI Imaging*

2.68 a. "Cholesterol Marker" describes using MRI to look for arterial plaque. Why is the presence of plaque in arteries a concern?

b. What role does the particle similar to HDL play in the technique that is described?

2.69 a. One of the two techniques described in this Health Link is considered "invasive." What do you suppose is meant by this term?

b. To which technique does this term apply?

c. In some cases, the images produced by CT and MRI are quite similar. If given a choice, which would you rather have run?

THINKING IT THROUGH

2.70 What are the pros and cons (both long-term and short-term) of using radiation to treat cancer?

2.71 A nurse is assisting a patient who has just undergone cancer treatment that involved exposure to gamma radiation from cobalt-60. Should the nurse be concerned that he will be exposed to gamma radiation given off by the patient?

2.72 In late 2006, a former Russian spy was poisoned when a small amount of polonium-210 was put in his food. He died a few weeks later.

a. Polonium-210 emits alpha particles. Write a balanced nuclear equation for this process.

b. This radioisotope has a half-life of 138 days. Of a 2 μg sample, how much would remain after 5 half-lives?

c. If alpha particles can be blocked by a sheet of paper (Figure 2.13), why is polonium-210 so poisonous?

INTERACTIVE LEARNING PROBLEMS

ILW

2.73 How many atoms are contained in:

a. 1.0 mol of calcium?

b. 7.88×10^{-6} mol of calcium?

c. 7.88 g of calcium?

d. 5.2 mg of calcium?

ILW

2.74 Identify the missing product in each nuclear equation.

a. $^{20}_{12}\text{Mg} \longrightarrow ? + ^{0}_{1}\beta^{+}$

b. $^{152}_{67}\text{Ho} \longrightarrow ^{148}_{65}\text{Tb} + ?$

c. $^{31}_{14}\text{Si} \longrightarrow ? + ^{0}_{-1}\beta$

SOLUTIONS TO PRACTICE PROBLEMS

2.1 **a.** protons and neutrons; **b.** protons and electrons; **c.** protons and neutrons

2.2 **a.** atomic number = 8, mass number = 18; **b.** atomic number = 17, mass number = 37

2.3 $^{31}_{15}\text{P}$ and $^{28}_{15}\text{P}$

2.4 10.8 amu

2.5 S

2.6 **a.** 3.7×10^{23} atoms; **b.** 4.8×10^{-12} mol

2.7 **a.** 2.53 g; **b.** 0.860 g

2.8 $^{209}_{81}\text{Tl}$

2.9 $^{192}_{77}\text{Ir} \longrightarrow ^{192}_{76}\text{Os} + ^{0}_{1}\beta^{+}$

2.10 132 days

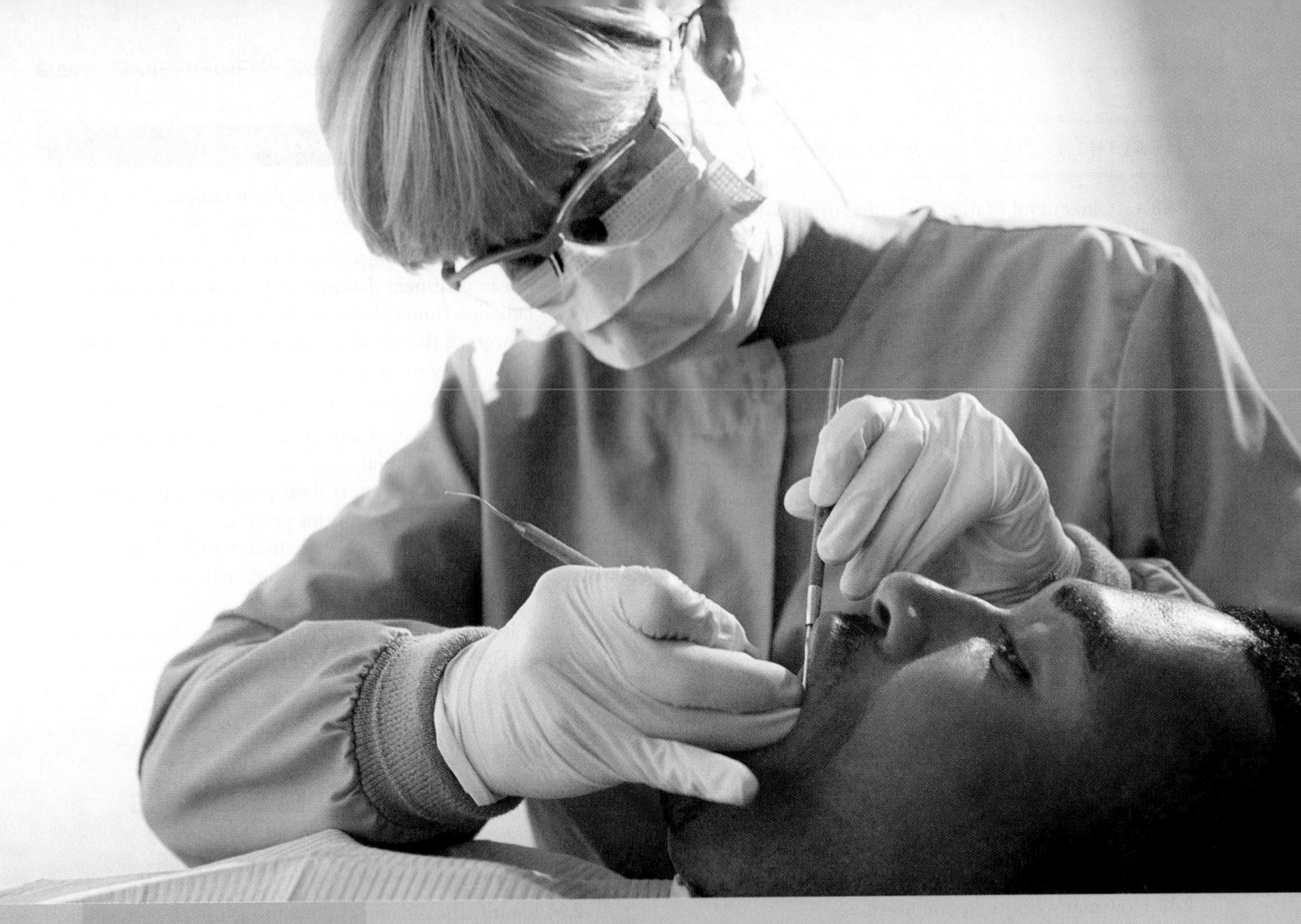

You have arrived at the dentist's office for a routine checkup. The dental hygienist shows you to a dental chair and proceeds to x ray your teeth and clean them. Then she asks, "Which flavor of fluoride do you want, mint, berry, or tropical?" As you swish the fluoride rinse around in your mouth you wonder what the fluoride is and how it helps your teeth.

Photo source: Thinkstock/Punchstock.

3 COMPOUNDS

The helium in a balloon, the nitrogen and oxygen gas in air, and the mercury in a thermometer are pure elements. These are among the relatively few examples of pure elements that you are likely to encounter. The chances are much greater that you will come into contact with compounds (matter constructed of two or more chemically combined elements). To understand the different ways that atoms interact with one another to form compounds—including those present in your teeth—we must first see how electrons arrange themselves about an atom's nucleus.

objectives

After completing this chapter, you should be able to:

1. Describe the differences in how the Bohr model and the quantum mechanical model of the atom view electrons as being arranged about an atom's nucleus.
2. Define the term valence electron and describe electron dot structures.
3. Define the term ion and explain how the electron dot structure of a representative element atom (groups 1A–8A) can be used to predict the charge on the monoatomic ion that it forms.
4. Describe the naming of monoatomic and polyatomic cations and anions.
5. Explain the difference between an ionic bond and a covalent bond.
6. Name and write the formulas of simple ionic compounds and binary molecules.
7. Define the terms formula weight and molecular weight. Use the molar mass of a compound to carry out conversions involving moles and mass.

3.1 IONS

Chapter 2 described atoms as consisting of a nucleus made up of protons and neutrons surrounded by clouds of electrons. The number of protons that an atom has is specified by its atomic number and the number of neutrons can be determined from its mass number, the total number of protons and neutrons. In a neutral (uncharged) atom, the number of electrons equals the number of protons, because the positive and negative charges cancel one another.

When an atom or a group of atoms gains or loses electrons, an **ion** results. Ions have *an unequal number of protons and electrons.* We will begin our look at ions with a few examples. The explanation of why these particular ions form will be delayed until the pertinent background material has been covered in Section 3.2.

If Li loses an electron, an ion with a 1+ charge, Li^+, is created (Figure 3.1*a*). This **monoatomic ion**—it is formed from a single atom—has a 1+ charge because it contains three positively charged protons and only two negatively charged electrons. (In atomic notation the *charge on a monoatomic ion is shown as a superscript to the right of the atomic symbol.*)

	Protons	Electrons	Charge	Name
Li	3	3	0	Lithium
Li^+	3	2	1+	Lithium ion

■ Ions have an unequal number of protons and electrons.

Ions that carry a *positive charge,* like Li^+, are called **cations.** Monoatomic cations formed from representative elements (groups 1A–8A) are given the same name as the original element, so Li^+ is a lithium ion.

Removing two electrons from a neutral calcium atom produces a calcium ion with a 2+ charge, because it has two more protons than electrons.

	Protons	Electrons	Charge	Name
Ca	20	20	0	Calcium
Ca^{2+}	20	18	2+	Calcium ion

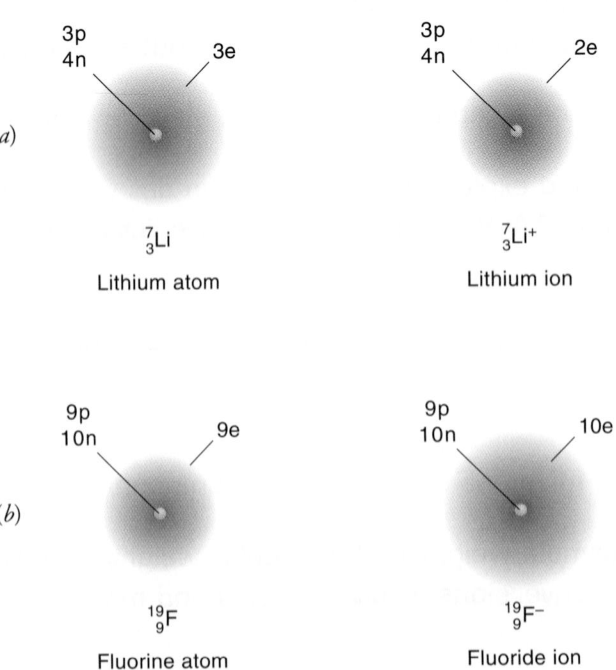

■ FIGURE 3.1

Ions
(*a*) $^{7}_{3}Li$ contains 3 protons, 4 neutrons, and 3 electrons and $^{7}_{3}Li^+$ has 3 protons, 4 neutrons, and 2 electrons. Monoatomic cations are smaller than the atoms from which they are formed.
(*b*) $^{19}_{9}F$ contains 9 protons, 10 neutrons, and 9 electrons and $^{19}_{9}F^-$ has 9 protons, 10 neutrons, and 10 electrons. Monoatomic anions are larger than the atoms from which they are formed.

Many transition metals (groups 1B–8B) are able to form several different cations. Iron ions, for example, can have a 2+ or a 3+ charge.

	Protons	Electrons	Charge	Name
Fe	26	26	0	Iron
Fe^{2+}	26	24	2+	Iron(II) ion
Fe^{3+}	26	23	3+	Iron(III) ion

When naming transition metal ions, a Roman numeral (in parentheses) is used to indicate charge—Fe^{2+} is iron(II) ion, Fe^{3+} is iron(III) ion, Cu^{+} is copper(I) ion, and Cu^{2+} is copper(II) ion.

SAMPLE PROBLEM 3.1

Naming monoatomic cations

Name each ion.

a. Mg^{2+} **b.** Co^{2+} **c.** Mn^{2+}

Strategy

There may be more to this problem than you expect. To name these monoatomic cations you must check to see if they are transition metal ions. If so, the charge on the ion should be part of the name.

Solution

a. magnesium ion **b.** cobalt(II) ion **c.** manganese(II) ion

PRACTICE PROBLEM 3.1

Name each ion.

a. Sc^{3+} **b.** Ti^{2+} **c.** Sr^{2+}

In an alternative naming system used for transition metal ions, the suffix *ous* is used for the ion with the smaller charge and the suffix *ic* for that with the greater charge (Table 3.1). Using this system, Fe^{2+} is ferrous ion, Fe^{3+} is ferric ion, Cu^{+} is cuprous ion, and Cu^{2+} is cupric ion. This naming system is not as easy to use as the Roman numeral–based system, because you cannot determine the charge on the ion directly from its name. Another challenge to using this system is that, for some ions, it makes use of the older Latin names for the elements, such as *ferrum* for iron and *cuprum* for copper.

TABLE 3.1 SOME TRANSITION METAL IONS

Atom	Ion	Name	Alternative Name
Chromium	Cr^{2+}	Chromium(II) ion	Chromous ion
	Cr^{3+}	Chromium(III) ion	Chromic ion
Copper	Cu^{+}	Copper(I) ion	Cuprous ion
	Cu^{2+}	Copper(II) ion	Cupric ion
Iron	Fe^{2+}	Iron(II) ion	Ferrous ion
	Fe^{3+}	Iron(III) ion	Ferric ion
Tin	Sn^{2+}	Tin(II) ion	Stannous ion
	Sn^{4+}	Tin(IV) ion	Stannic ion

■ Cations have a positive charge and anions have a negative charge.

Some atoms gain electrons to produce *negatively charged ions* called **anions.** When F gains an electron, an ion with a 1− charge, F^-, is formed. Monoatomic anions are named by changing the ending on the name of the element name to *ide*, so the anion formed from fluorine is called fluoride ion (Figure 3.1*b*).

	Protons	Electrons	Charge	Name
F	9	9	0	Fluorine
F^-	9	10	1−	Fluoride ion

When an oxygen atom gains two electrons, it is transformed into an oxide ion.

	Protons	Electrons	Charge	Name
O	8	8	0	Oxygen
O^{2-}	8	10	2−	Oxide ion

■ Polyatomic ions contain two or more atoms.

Many ions are **polyatomic**, which means that they are built from *two or more atoms.* Examples include ammonium ion (NH_4^+) and carbonate ion (CO_3^{2-}). As is the case for monoatomic ions, the charge on polyatomic ions is due to an imbalance in the total number of protons and electrons that are present. NH_4^+ contains 1 nitrogen atom, 4 hydrogen atoms, and a total of 10 electrons.

	Protons	Electrons	Charge
NH_4^+	7 from N + $\underbrace{4 \times 1 \text{ from each H}}$		
	7 + 4 = 11	10	1+

CO_3^{2-} is a combination of 1 carbon atom, 3 oxygen atoms, and a total of 32 electrons.

	Protons	Electrons	Charge
CO_3^{2-}	6 from C + $\underbrace{3 \times 8 \text{ from each O}}$		
	6 + 24 = 30	32	2−

In Chapter 4 we will take a look at how the atoms in polyatomic ions are connected to one another and where the charge resides.

Table 3.2 lists a number of the more common polyatomic ions. Some differ only in the number of hydrogen atoms that they contain, as is the case for PO_4^{3-}, HPO_4^{2-}, and $H_2PO_4^-$.

TABLE 3.2 COMMON POLYATOMIC IONS

	Formula	Name	Formula	Name
Cations				
	H_3O^+	Hydronium ion	NH_4^+	Ammonium ion
Anions				
	OH^-	Hydroxide ion	HSO_4^-	Hydrogensulfate (bisulfate) ion
	CO_3^{2-}	Carbonate ion	PO_4^{3-}	Phosphate ion
	HCO_3^-	Hydrogencarbonate (bicarbonate) ion	HPO_4^{2-}	Hydrogenphosphate ion
	NO_2^-	Nitrite ion	$H_2PO_4^-$	Dihydrogenphosphate ion
	NO_3^-	Nitrate ion	$Cr_2O_7^{2-}$	Dichromate ion
	SO_3^{2-}	Sulfite ion	$CH_3CO_2^-$	Acetate ion
	SO_4^{2-}	Sulfate ion	CN^-	Cyanide ion

The names of these ions reflect the number of hydrogen atoms that are present. Phosphate ion (PO_4^{3-}) has no hydrogen atoms, hydrogenphosphate ion (HPO_4^{2-}) has one hydrogen atom, and dihydrogenphosphate ion ($H_2PO_4^-$) has two. When an element can form polyatomic ions by combining with oxygen in two different ways, the suffixes *ate* and *ite* are used to indicate the relative number of oxygen atoms: Nitrate ion (NO_3^-) has one more oxygen atom than nitrite ion (NO_2^-), and sulfate ion (SO_4^{2-}) has one more oxygen atom than sulfite ion (SO_3^{2-}).

Most of the polyatomic ions in Table 3.2 are found in significant amounts in all living things and many are essential to survival. Hydrogencarbonate ion (HCO_3^-), for example, is involved in the transport of carbon dioxide from your tissues to your lungs.

3.2 THE ARRANGEMENT OF ELECTRONS

To be able to predict the type of ion that a particular atom or group of atoms will form requires an understanding of how electrons are placed around the nucleus of an atom. We will begin by looking at the **Bohr model** of the atom, an early attempt to describe electron arrangements.

When an electric current is passed through a tube containing hydrogen gas, the light emitted by the hydrogen atoms can be separated by a prism and viewed as an emission spectrum—a series of colored lines (Figure 3.2). At the beginning of the last century, the Danish physicist Niels Bohr proposed that the colors of light in the emission spectrum of hydrogen are directly related to the movement of a hydrogen atom's electron between different energy levels.

In Bohr's model of the atom, *electrons circle the nucleus in specific orbits, with each orbit corresponding to a different energy level* (Figure 3.3). An atom is in its **ground state** (*most stable electron arrangement*) when its electrons are in energy levels as near as possible to the nucleus. For a hydrogen atom, the ground state has the lone electron sitting in energy level 1. If a ground state hydrogen atom absorbs energy, its electron is pushed to *an orbit farther from the nucleus*, placing the atom into an **excited state**.

■ The ground state is an atom's most stable electron arrangement.

According to Bohr, the emission spectrum of hydrogen is produced when hydrogen atoms move from various excited states back to more stable states, releasing their energy in the process. This energy is given off in the form of electromagnetic radiation (Section 2.6). Transitions back to the ground state from energy level 2 ($2 \rightarrow 1$) or other energy levels (e.g., $3 \rightarrow 1$, $4 \rightarrow 1$) are of high enough energy that ultraviolet light is released, while movement of electrons to energy level 2 ($3 \rightarrow 2$, $4 \rightarrow 2$, etc.) causes visible frequencies of light to be emitted (Figure 3.3). The different colors of visible light have energies ranging from violet (highest energy) to red (lowest energy). Other transitions ($4 \rightarrow 3$, $6 \rightarrow 4$, etc.) release lower energy infrared light.

Bohr's model was very good at explaining the emission spectrum of hydrogen, but not those of other elements. As work in this field progressed, experiments provided new

■ FIGURE 3.2

Hydrogen emission spectrum Light from a hydrogen lamp, when passed through a prism, gives an emission spectrum that consists of just a few different colors of visible light.

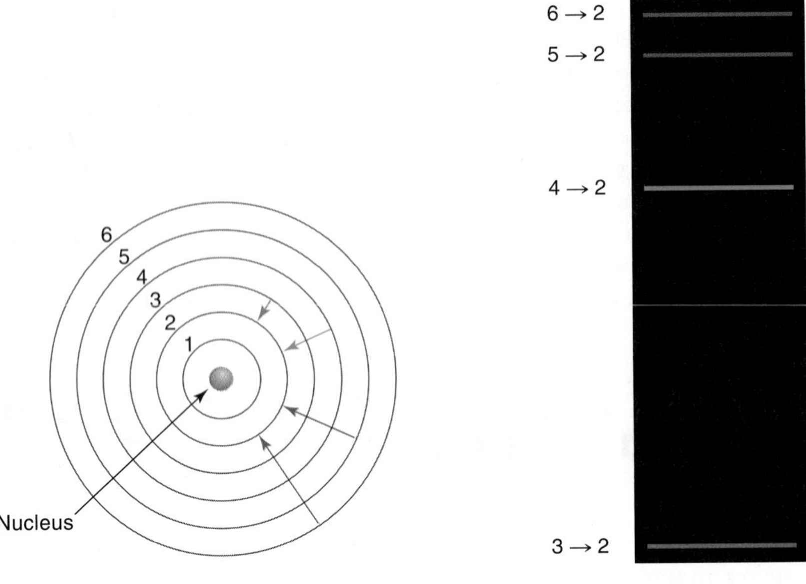

FIGURE | 3.3

The Bohr atomic model
In the Bohr model, electrons orbit the nucleus, with each orbit corresponding to a different energy level. When an electron jumps to an orbit nearer the nucleus, energy is released.

information regarding the behavior of electrons. One remarkable piece of evidence was that, under certain experimental conditions, electrons can behave as energy waves, rather than particles. In 1926, the Austrian physicist Erwin Schrödinger used these wave-like properties of electrons to devise a mathematical equation that described electron energy levels in a new way. Like the Bohr model, this new approach, called **quantum mechanics**, viewed an atom as having a series of energy levels. Instead of picturing electrons in fixed orbits about the nucleus, however, quantum mechanics assigns them to various **atomic orbitals**. These orbitals, *three-dimensional regions of space where there is a high probability of finding an electron,* are the electron clouds referred to when atomic structure was discussed in Section 2.1. Examples of these atomic orbitals include *s* and *p* orbitals (Figure 3.4).

FIGURE | 3.4

Orbitals
According to quantum mechanics, a variety of orbitals make up the energy levels of atoms. Among them are *s* orbitals (spherically shaped) and *p* orbitals (each with two teardrop-shaped lobes).

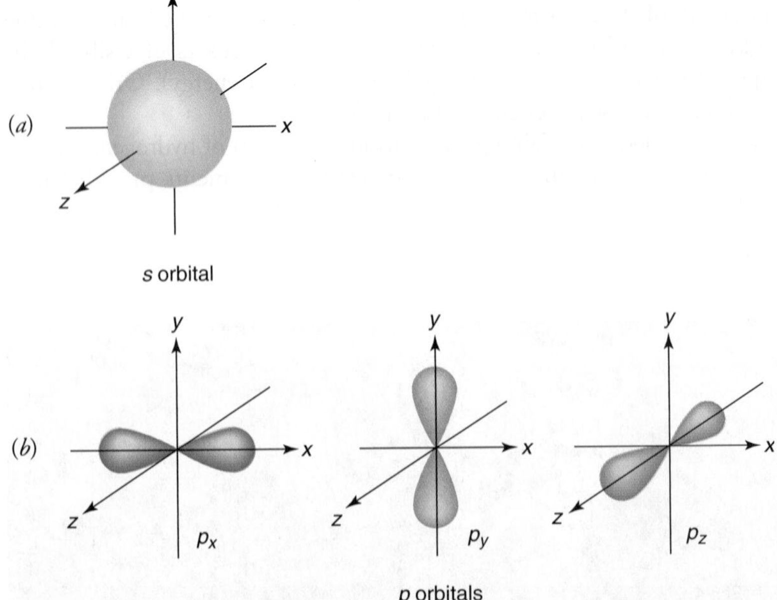

For our purposes, quantum mechanics is useful because it lets us calculate the maximum number of electrons that any particular energy level can hold. The equation used to carry out this calculation is

$$\textit{Maximum number of electrons per energy level} = 2n^2$$

where n is the number of the energy level. For example, the first energy level ($n = 1$) holds just two electrons ($2 \times 1^2 = 2$) and the second energy level ($n = 2$) holds up to eight electrons ($2 \times 2^2 = 8$). Table 3.3 shows the maximum number of electrons that can be carried by energy levels 1 through 4.

TABLE 3.3 The Maximum Number of Electrons Held in the First Four Electron Energy Levels

Energy Level (n)	Maximum Number of Electrons ($2n^2$)
1	2
2	8
3	18
4	32

Knowing the maximum number of electrons that an energy level can hold allows ground state electron arrangements to be predicted. The ground state of a hydrogen atom has one electron in the $n = 1$ energy level, while the ground state of a helium atom has two electrons in this energy level. Any other arrangement for a helium atom, say, one electron in the $n = 1$ energy level and one electron in the $n = 2$ energy level, is an excited state. The ground state of a lithium atom has two electrons in the $n = 1$ energy level and one electron in the $n = 2$ energy level (Figure 3.5).

Table 3.4 provides the ground state electron arrangements for the first 20 elements, and shows that an energy level is not always filled when the next energy level begins filling with electrons. In a potassium (K) atom the $n = 4$ energy level has 1 electron even though the $n = 3$ energy level holds only 8 of the allowed 18 electrons. As we will see, atoms are particularly stable when an energy level holds 8 electrons (Section 3.3).

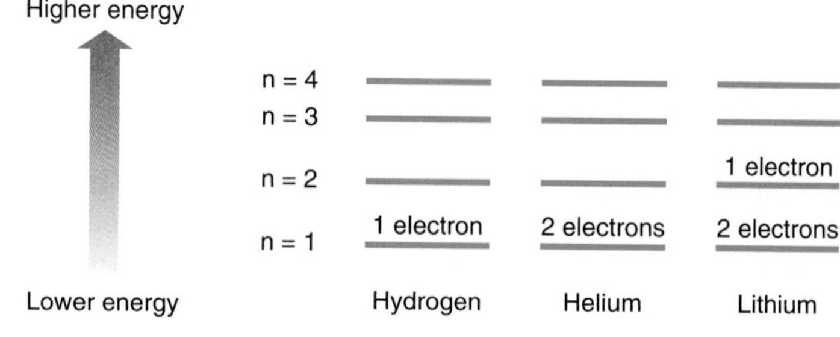

FIGURE 3.5 **Ground state electron arrangements for H, He, and Li**
The $n = 1$ energy level holds a maximum of two electrons and all of the electrons of hydrogen and helium are held in this first energy level. There is not room for all three of lithium's electrons in the $n = 1$ energy level, so one electron is placed in the $n = 2$ energy level.

TABLE 3.4 The Ground State Electron Distribution for the First 20 Elements[a]

		Number of Electrons in Energy Level			
Element	**Group**	**n=1**	**n=2**	**n=3**	**n=4**
H	1A	**1**			
He	8A	**2**			
Li	1A	2	**1**		
Be	2A	2	**2**		
B	3A	2	**3**		
C	4A	2	**4**		
N	5A	2	**5**		
O	6A	2	**6**		
F	7A	2	**7**		
Ne	8A	2	**8**		
Na	1A	2	8	**1**	
Mg	2A	2	8	**2**	
Al	3A	2	8	**3**	
Si	4A	2	8	**4**	
P	5A	2	8	**5**	
S	6A	2	8	**6**	
Cl	7A	2	8	**7**	
Ar	8A	2	8	**8**	
K	1A	2	8	8	**1**
Ca	2A	2	8	8	**2**

[a] Valence electrons are listed in bold.

Valence Electrons

Understanding the arrangement of electrons about an atom gives us some insight as to how members of a given group or period in the periodic table are related. Note in Table 3.4 that atoms of elements belonging to the same group have an identical number of electrons in their **valence shell** (the *highest numbered, occupied energy level*). For example, all group 1A atoms have one valence shell electron, or **valence electron**. For a hydrogen (H) atom the valence electron is in the $n = 1$ energy level, for a lithium (Li) atom it is in the $n = 2$ energy level, for a sodium (Na) atom it is in the $n = 3$ energy level, and for a potassium (K) atom, it is in the $n = 4$ energy level.

■ Valence electrons are held in an atom's outermost occupied energy level.

The number of valence electrons for the atoms of the representative elements follows a periodic trend (Figure 3.6)—group 1A with one valence electron, group 2A with two valence electrons, group 3A with three valence electrons, and so on through group 8A with eight valence electrons. The one exception in group 8A, the inert gases, is the helium atom, which has just two valence electrons.

1A	2A	3A	4A	5A	6A	7A	8A
H·							He:
Li·	Be:	·B·	·C·	·N·	·O·	:F:	:Ne:
Na·	Mg:	·Al·	·Si·	·P·	·S·	:Cl:	:Ar:
K·	Ca:						

FIGURE 3.6

Valence electrons
Representative elements in the same group have the same number of valence electrons. In the electron dot structures used here, valence electrons are shown as dots.

A correlation exists between the periods of the periodic table and energy levels. For atoms of the two elements in period 1 (H and He), valence electrons are held in energy level 1, which holds just two electrons. Atoms of period 2 elements (Li, Be, . . ., Ne—there are eight in all) hold their valence electrons in energy level 2, which holds up to eight electrons. This same principle holds true for periods 3, 4, and so on.

SAMPLE PROBLEM 3.2

Valence electrons

For each atom, give the total number of electrons and the number of valence electrons.

a. calcium (Ca)
b. silicon (Si)
c. argon (Ar)
d. radium (Ra)

Strategy
For these neutral atoms the total number of electrons is the same as the total number of protons (atomic number), which can be obtained from a periodic table. The number of valence electrons in these atoms (which all belong to the representative elements) can be determined from the group number.

Solution

a. 20 total electrons, 2 valence electrons
b. 14 total electrons, 4 valence electrons
c. 18 total electrons, 8 valence electrons
d. 88 total electrons, 2 valence electrons

PRACTICE PROBLEM 3.2

For each of the atoms in Sample Problem 3.2 above, which energy level holds the valence electrons?

Electron Dot Structures

In the early 1900s, the American chemist Gilbert N. Lewis developed **electron dot structures** to show the number of valence electrons that an atom carries. In these structures, *valence electrons are represented by dots.* Lithium has one valence electron, so its electron dot structure is the symbol Li with a dot next to it (Figure 3.6). The electron dot structure of beryllium (Be) has two electron dots and that of fluorine (F) has seven.

SAMPLE PROBLEM 3.3

Drawing electron dot structures of atoms

Draw the electron dot structure of

a. a bromine (Br) atom

b. a rubidium (Rb) atom

Strategy

To draw the electron dot structure of these atoms, you must know the number of valence electrons that they carry.

Solution

a. $:\ddot{\underset{\cdot}{\mathrm{Br}}}:$

b. Rb·

PRACTICE PROBLEM 3.3

Draw the electron dot structure of

a. a krypton (Kr)atom

b. a barium (Ba) atom

BIOCHEMISTRY Link

Bioluminescence

Experience tells us that a lightbulb gets hot when it is switched on. This happens because the production of light is usually accompanied by the release of heat. Sometimes, however, *light can be produced without heat,* in a process called luminescence. When luminescence takes place in a living thing, it is called bioluminescence.

A number of bacteria, sponges, jellyfish, clams, insects, and fish are bioluminescent. This production of light serves a number of different functions. For fireflies, the ability to emit light plays a key role in attracting mates. One species of squid ejects a glowing cloud of bioluminescent material that hides it from its attackers. The anglerfish, which lives deep in the ocean, attracts prey using a bioluminescent appendage (Figure 3.7).

The substance that all of these organisms use to produce light is called luciferin, after the Latin name *Lucifer* (bearer of light). Bioluminescence occurs when luciferin is acted on by a particular enzyme in the presence of oxygen gas and ATP (a biochemical energy source). Electrons in luciferin are pushed into an excited state and, when they return to the ground state, light is emitted.

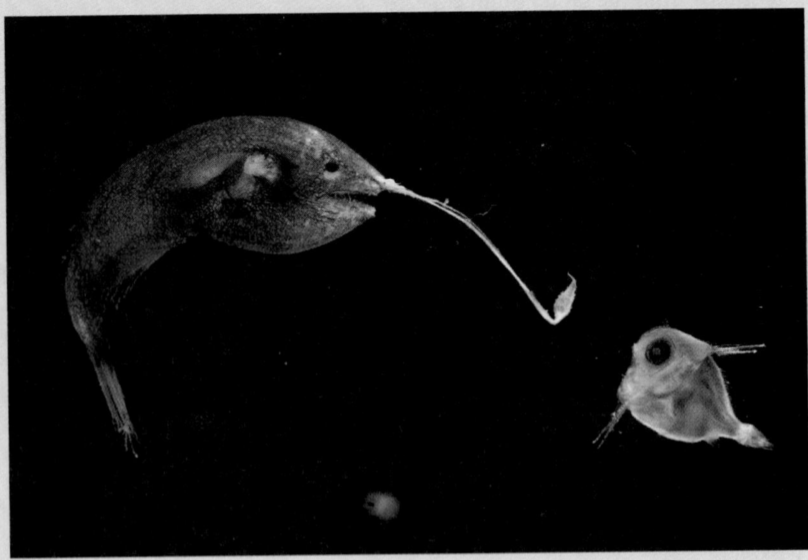

FIGURE 3.7 **The anglerfish**
In the dark ocean depths, a bioluminescent appendage attracts prey toward the anglerfish.

Source: Norbert Wu/Peter Arnold, Inc.

3.3 THE OCTET RULE

The elements helium (He), neon (Ne), argon (Ar), krypton (Kr), xenon (Xe), and radon (Rn) belong to the inert gas family, which gets its name from the fact that atoms of these elements are resistant to change and, with few exceptions, do not lose or gain electrons. This resistance to change (stability) is related to the number of valence electrons held by atoms of these elements. For helium and neon, the stability may have to do with the fact that these atoms have full valence shells—a helium atom's $n = 1$ energy level is filled with two electrons and a neon atom's $n = 2$ energy level is filled with eight electrons. Atoms of the other inert gases do not have full valence shells but, like the neon atom, carry eight valence electrons.

According to the **octet rule**, *atoms gain, lose, or share valence electrons in order to end up with eight valence electrons.* The effect that this has is of providing the same stable electron arrangement as found in inert gases.

■ An octet (8 valence electrons) is a stable electron arrangement.

An ion formed from fluorine always has a 1− charge, while one formed from sodium always has a 1+ charge. Fluoride ion (F^-) and sodium ion (Na^+) *have the same electron arrangement as* (are **isoelectronic** with) neon (Figure 3.8).

IA	IIA	IIIA	IVA	VA	VIA	VIIA	Inert gases VIIIA
							He:
Li^+				$:\ddot{N}:^{3-}$	$:\ddot{O}:^{2-}$	$:\ddot{F}:^-$	$:\ddot{Ne}:$
Na^+	Mg^{2+}	Al^{3+}		$:\ddot{P}:^{3-}$	$:\ddot{S}:^{2-}$	$:\ddot{Cl}:^-$	$:\ddot{Ar}:$
K^+	Ca^{2+}						

■ **FIGURE 3.8**

Some common ions of representative elements
Atoms lose or gain the number of electrons necessary to have an octet (to become isoelectronic with the nearest inert gas). Here, isoelectronic ions and atoms are shown with the same color.

Monoatomic Ions

Using electron dot structures makes it very easy to predict the ion that a particular atom will form. The electron dot structure of a neutral fluorine atom, for example, has seven valence electrons. Adding one more electron gives a fluoride ion with an octet and a 1− charge. The other halogen atoms behave the same way—chlorine, bromine, and iodine each have seven valence electrons and require just one more to reach an octet.

$:\ddot{F}:$ Fluorine atom	$:\ddot{F}:^-$ Fluoride ion
$\cdot\ddot{O}\cdot$ Oxygen atom	$:\ddot{O}:^{2-}$ Oxide ion
$\cdot\dot{N}\cdot$ Nitrogen atom	$:\ddot{N}:^{3-}$ Nitride ion

Oxygen and the other group 6A nonmetal atoms (S and Se) always form monoatomic anions with a charge of 2−. Each has six valence electrons when neutral and must gain exactly two more to achieve an octet. The group 5A nonmetals, nitrogen and phosphorus, have five valence electrons and require three more for an octet, giving ions with a charge of 3−. Carbon, with four valence electrons, might be expected to accept four more to make an octet. Only rarely, however, will atoms gain or lose more than three electrons, since this places too great a charge on the ion that would be produced. Carbon atoms reach an octet by other means, as we will see in Section 3.5.

■ Nonmetals gain electrons and metals lose electrons.

As the two previous paragraphs showed, nonmetal atoms gain electrons to reach an octet. In contrast, metal atoms lose electrons. The 11 electrons of a sodium atom are distributed among three energy levels: 2 electrons in the $n = 1$ energy level, 8 electrons in the $n = 2$ energy level, and 1 electron in the $n = 3$ energy level (Table 3.4). When a sodium atom loses the valence electron held in its $n = 3$ energy level, the resulting cation has 8 electrons (an octet) in its new valence shell ($n = 2$).

The valance electron shown in the electron dot structure of a sodium atom represents the lone electron in energy level 3 (Table 3.4). When this electron is lost from a sodium ion, energy level 3 is empty. The electron dot structure of Na^+ does not show any electron dots to reinforce the idea that the original valence shell has been emptied.

Na·	Na^+
Sodium atom	Sodium ion
Mg:	Mg^{2+}
Magnesium atom	Magnesium ion
·Ȧl·	Al^{3+}
Aluminum atom	Aluminum ion

Like sodium, other alkali metal atoms form cations with a 1+ charge. Alkaline earth metal atoms (two valence electrons) and group 3A metal atoms (three valence electrons) form cations with charges of, respectively, 2+ and 3+.

Using the octet rule and electron dot structures to predict the charge on an ion does not always work for transition metal atoms, because most transition metal atoms cannot lose the number of electrons required to reach an octet. For example, chromium (Cr), with 24 total electrons, would need to lose 6 electrons to end up with just 8 in its outermost filled shell. This does not happen because, as just mentioned, atoms rarely gain or lose more than 3 electrons. Table 3.1 shows the cations observed for some of the transition metals.

SAMPLE PROBLEM 3.4

Drawing electron dot structures of ions

Without referring to Figure 3.8, draw the electron dot structure of each atom and of the ion that it is expected to form.

a. Cs **b.** S

Strategy

To solve this problem you need to know how many valence electrons each atom has. You should be able to determine this by taking a look at a periodic table. Once you know the number of valence electrons and have drawn an electron dot structure, you need to remember that metals lose electrons and nonmetals gain electrons—enough in each case to reach an octet.

Solution

a. Cs· **b.** ·S̤̈·

Cs^+ $:\ddot{\underset{..}{S}}:^{2-}$

PRACTICE PROBLEM 3.4

Draw the electron dot structure of each atom and of the ion that it is expected to form.

a. Cl **b.** Se **c.** Ca

BIOCHEMISTRY Link

Ionophores and Biological Ion Transport

The cell membrane that surrounds each living cell is a barrier that prevents unwanted substances from entering and desired substances from leaving. The differing amount of ions within cells (in intracellular fluid) and outside of cells (in blood plasma) can be maintained because cell membranes control the passage of ions (Figure 3.9). Some bacteria produce substances called **ionophores**, whose job it is to transport ions across cell membranes. *Streptomyces* bacteria make valinomycin, a doughnut-shaped ionophore whose center is the correct size to hold a potassium ion (K^+) (Figure 3.10). This and other ionophores used as antibiotics (bacteria-killing drugs) destroy bacteria by transporting ions across their cell membranes, which upsets the balance of ions and disrupts key biological processes.

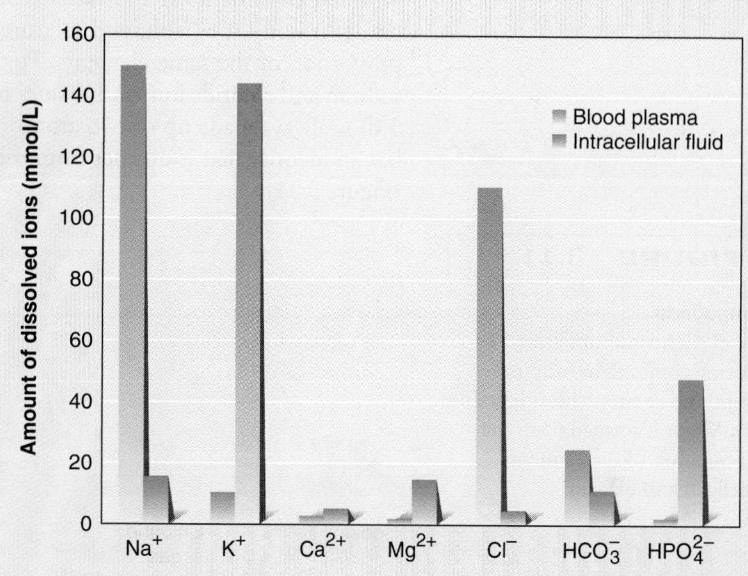

FIGURE 3.9 **The major dissolved ions in blood plasma and in intracellular fluid** Amounts of dissolved ions are expressed as mmol/L (millimoles of ion per liter of fluid).

Source: Figure 26.2, p. 1036 from HUMAN ANATOMY AND PHYSIOLOGY, 6th ed. By Elaine N. Marieb. Copyright © 2004 by Pearson Education, Inc. Reprinted by permission.

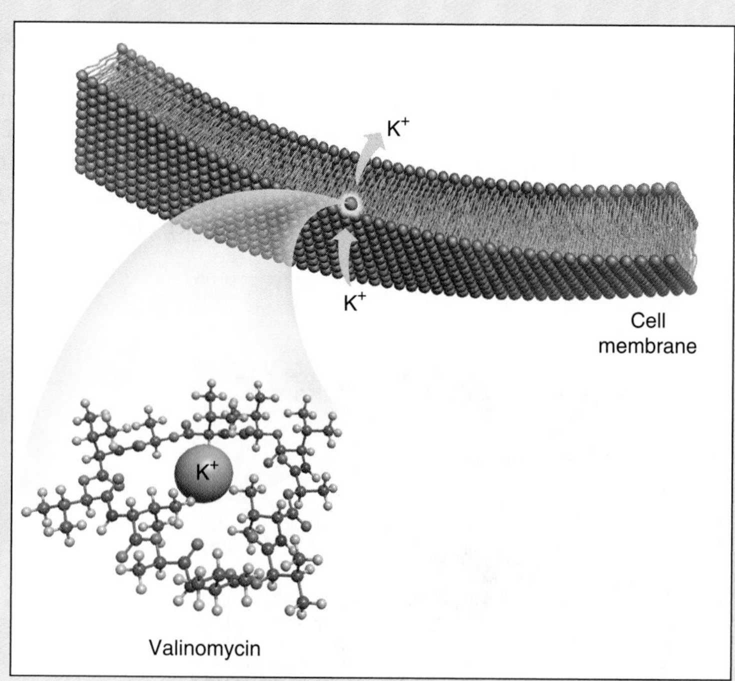

FIGURE 3.10 **Valinomycin** This ionophore transports K^+ across cell membranes.

3.4 IONIC COMPOUNDS

■ Compounds contain two or more elements.

As we saw in the introduction to this chapter, a **compound** is defined as *matter constructed of two or more chemically combined elements.* Examples include table salt (contains sodium ions and chloride ions) and water (contains hydrogen atoms and oxygen atoms). A compound is not just a haphazard mixture of elements—each compound always has the same proportion of the same elements. For example, table salt always has an equal number of sodium and chloride ions and a water molecule (a particular type of compound; see Section 3.6) is always made up of two atoms of hydrogen and one atom of oxygen. A compound has an identity that is distinct from the identities of the elements that went into making it (Figure 3.11).

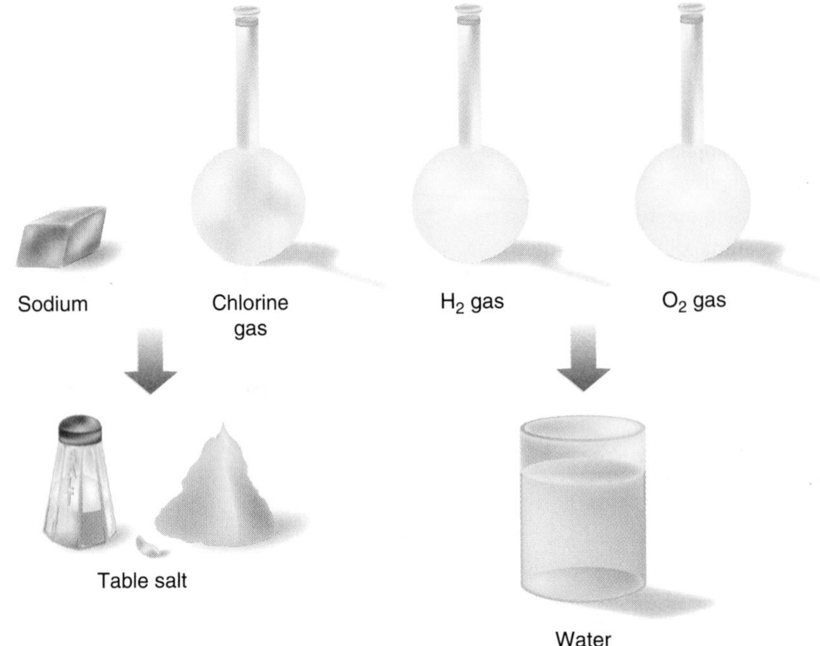

■ FIGURE | 3.11

Compounds
The elements sodium and chlorine combine to form the compound sodium chloride (table salt). Water is formed from the combination of the elements hydrogen and oxygen.

When the atoms in an element or compound combine to form new compounds, **chemical change** has taken place. For example, when a piece of iron rusts (Figure 3.12), a chemical change has occurred because iron (an element) combines with oxygen (a different element) to produce rust (a compound that is made from iron ions and oxygen ions). *The chemical changes that an element or a compound undergo* are called **chemical properties.**

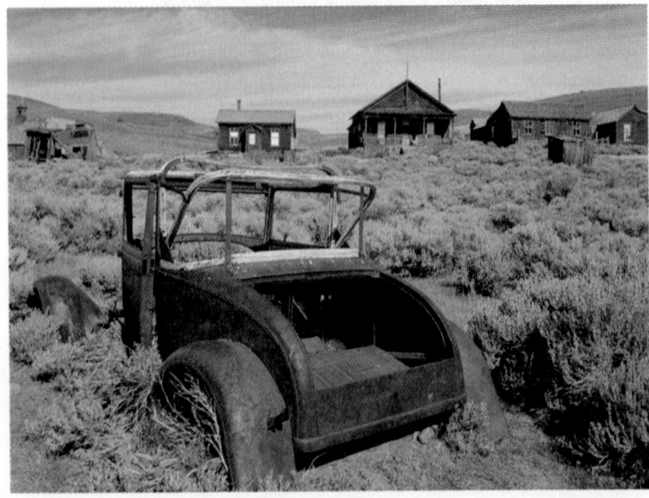

■ FIGURE | 3.12

Chemical change
Iron that is exposed to air and water soon rusts. This is a chemical change.

Source: Tony Gervis/Robert Harding World Imagery/Getty.

Some compounds are constructed from cations (positively charged ions) and anions (negatively charged ions). Sodium chloride (table salt), for example, contains sodium ions (Na^+) and chloride ions (Cl^-). The opposite charge on the Na^+ and Cl^- ions provides the **ionic bond** (*the attractive force*) that holds sodium chloride together. Figure 3.13 shows the structure of sodium chloride, which consists of a **crystal lattice** (array) of alternating cations and anions. In this lattice, each ion is surrounded by others of opposite charge.

Sodium chloride is an example of an **ionic compound**, a compound that forms when cations and anions interact. The simplest ionic compounds, such as sodium chloride, are **binary** (contain just two elements) and contain monoatomic ions.

The formula of any ionic compound is a listing of the relative number of each type of ion that is present, with the cation listed before the anion. Sodium chloride, whose formula is NaCl, contains an equal number of Na^+ and Cl^- ions. $MgCl_2$ is a binary ionic compound that has 2 Cl^- for each Mg^{2+}. Like all ionic compounds, $MgCl_2$ is neutral—the 2+ charge on each magnesium ion is countered by two 1− charges, one from each of the chloride ions. Note that the formulas NaCl and $MgCl_2$ do not specify the charges on the ions. It is assumed that these can be determined, based on knowledge of the periodic table and the octet rule.

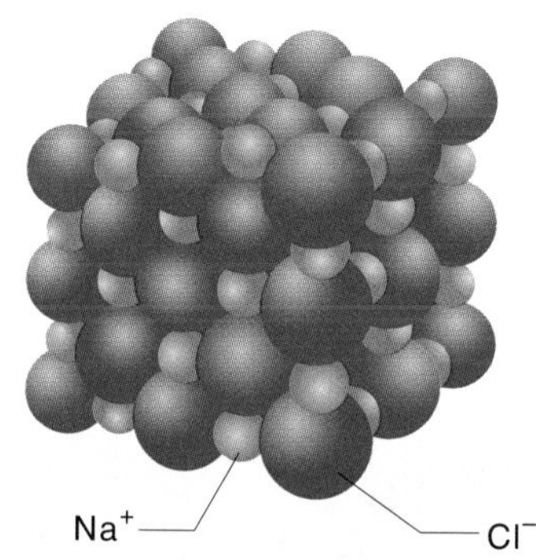

FIGURE 3.13

Sodium chloride
A sodium chloride crystal is a lattice of alternating Na^+ and Cl^- ions. Each ion is surrounded by ions of opposite charge.

- Cations and anions are held to one another by ionic bonds.

SAMPLE PROBLEM 3.5

Predicting formulas of ionic compounds

Write the formula of the ionic compound that forms between

a. sodium ions and fluoride ions
b. calcium ions and oxide ions
c. aluminum ions and oxide ions
d. sodium ions and sulfide ions

Strategy

You must first determine the charge expected on each of the ions and then decide how many of each will be required to produce a neutral compound. Once you have done this, the formula lists the cation followed by the anion, and includes the relative numbers of each.

Solution

a. NaF (used to prevent cavities)
b. CaO (lime—used in plaster, stucco, and mortar)
c. Al_2O_3 (used in the manufacture of dental cements)
d. Na_2S (used in the manufacture of rubber and in ore refining)

PRACTICE PROBLEM 3.5

Write the formula of the ionic compound that forms between

a. lithium ions and chloride ions
b. lithium ions and oxide ions
c. lithium ions and nitride ions

Ionic Compounds Containing Polyatomic Ions

Polyatomic ions also form ionic compounds. The food preservative sodium sulfite (Na_2SO_3) is the combination of sodium ions (Na^+) and sulfite ions (SO_3^{2-}). To produce a neutral ionic compound, each sulfite ion (2− charge) requires two sodium ions (1− charge from each).

When more than one copy of a particular polyatomic ion is present in an ionic compound, the formula of the ion is usually surrounded by parentheses, as is the case for the laxative $Mg(OH)_2$ and the expectorant $(NH_4)_2CO_3$.

When an ionic compound contains polyatomic ions, interpreting its formula depends on being familiar with the formulas of the polyatomic ions involved. For example, $Mg(OH)_2$ contains Mg^{2+} and OH^- ions and $(NH_4)_2CO_3$ contains NH_4^+ and CO_3^{2-} ions. It is important to note that in ionic compounds, polyatomic ions "act as one." The compound $NaNO_2$ consists of Na^+ and NO_2^- ions, not some combination of ions formed from Na, N, and O atoms.

FIGURE 3.14

Ionic compounds

Pictured here, clockwise from the upper left, are iron(II) sulfate, iron(III) sulfate, copper(II) sulfate, copper(II) carbonate, and sodium chloride. Ionic compounds containing transition metal ions, such as the first four mentioned here, are often brightly colored.

Source: Andrew Lambert Photography/Photo Researchers, Inc.

Naming Ionic Compounds

When naming ionic compounds, the cation name is placed before the anion name. Lithium ions (Li^+) combine with bromide ions (Br^-) to form lithium bromide (LiBr) and ammonium ions (NH_4^+) combine with nitrate ions (NO_3^-) to form ammonium nitrate (NH_4NO_3).

The number of times that an ion appears in the formula of an ionic compound is *not* specified in the name, so $BaCl_2$ is called barium chloride, not barium dichloride. Similarly, Na_2SO_4 is sodium sulfate and $Mg(HCO_3)_2$ is magnesium hydrogencarbonate. It is assumed that the formula can be determined from the name, because the charges on the various ions are known. For example, calcium bromide must have the formula $CaBr_2$ because calcium ions always have a charge of 2+ and bromide ions always have a charge of 1−. This means that two bromide ions will combine with one calcium ion to create a neutral ionic compound.

Assigning names works the same way when an ionic compound contains transition metal ions (Figure 3.14). CuCl, the combination of copper(I) ion (Cu^+) and chloride ion (Cl^-), is called copper(I) chloride, and $CuCl_2$ is named copper(II) chloride. Copper(I) chloride and copper(II) chloride are also known, respectively, as cuprous chloride and cupric chloride (Table 3.1). Iron(II) hydroxide has the formula $Fe(OH)_2$—one Fe^{2+} ion requires two OH^- ions to form a neutral compound.

Ionic compounds are widely used in medicine, by industry, and around the house. Table 3.5 lists some of their common uses.

TABLE 3.5 The Uses of Some Ionic Compounds

Name	Formula	Use
Ammonium carbonate	$(NH_4)_2CO_3$	Smelling salts
Barium sulfate	$BaSO_4$	Compound used to help view internal organs in x-ray studies
Calcium carbonate	$CaCO_3$	Antacid
Calcium sulfate	$CaSO_4$	Plaster casts
Lithium carbonate	Li_2CO_3	Treatment for manic depression
Magnesium hydroxide	$Mg(OH)_2$	Milk of magnesia
Magnesium sulfate	$MgSO_4$	Laxative
Silver nitrate	$AgNO_3$	Prevention of eye infections in newborns
Sodium bicarbonate	$NaHCO_3$	Baking soda and antacid
Sodium hydroxide	NaOH	Drain cleaner
Sodium iodide	NaI	Source of iodide ion for the thyroid
Sodium nitrate	$NaNO_3$	Food preservative
Sodium nitrite	$NaNO_2$	Meat preservative
Sodium acetate	CH_3CO_2Na[a]	Foot and hand warmers

[a] For ionic compounds involving polyatomic ions with an organic or biochemical source (such as $CH_3CO_2^-$), the formula sometimes lists the anion before the cation.

3.5 COVALENT BONDS

To reach an octet, metals lose electrons (Na becomes Na^+) and nonmetals gain electrons (Cl becomes Cl^-). For nonmetals, a second option is available for attaining an octet—valence electrons can be shared.

An example of this is what happens when two F atoms, each of which has seven valence electrons, interact with one another. When the atoms reach an appropriate distance, one pair of electrons is shared and each atom ends up with an octet. This *shared pair of valence electrons* is called a **covalent bond**.

■ In a covalent bond a pair of valence electrons is shared between two atoms.

This F atom has an octet

Shared electron pair

:F̤̈· ·F̤̈: ⇒ :F̤̈:F̤̈:

This F atom has an octet

Individual atoms

Atoms joined by a covalent bond

Generally, *the number of covalent bonds that a nonmetal atom forms is the same as the number of electrons that it needs to gain to have an octet.* Fluorine atoms, with seven valence electrons, form one covalent bond because the extra electron gained by sharing is enough to complete an octet. Atoms of the other second period nonmetals, oxygen (six valence electrons), nitrogen (five valence electrons), and carbon (four valence electrons) form, respectively, two, three, and four covalent bonds. Hydrogen atoms form just one covalent bond. Figure 3.15 shows the covalent bonding that can take place between H atoms and F, O, N, and C atoms.

The drawings in Figure 3.15*a* are electron dot structures—valence electrons are shown using dots. In an alternative approach, called the **line-bond method** (Figure 3.15*b*), *each pair of shared bonding electrons is represented by a line.* In these drawings, the *valence electrons not involved in bonds* are **nonbonding electrons**.

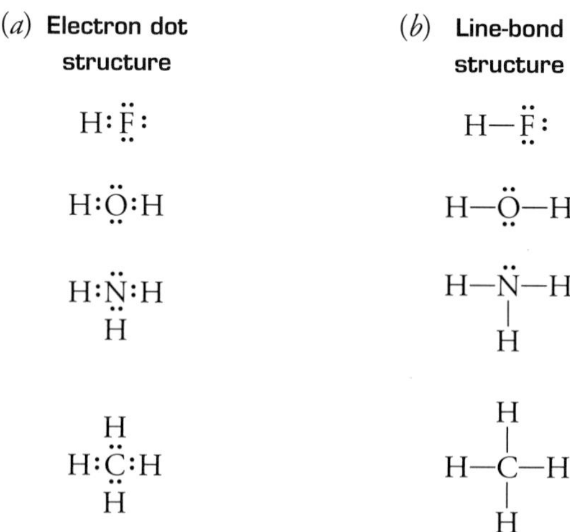

■ **FIGURE 3.15** **Covalent bonds**
Nonmetal atoms can satisfy the octet rule by forming single covalent bonds—generally one single bond for each valence electron required to complete an octet. To represent covalent bonds, (*a*) electron dot structures use pairs of electron dots and (*b*) line-bond structures use lines.

Sample Problem 3.6

Drawing molecules

Using the line-bond method, draw ethanethiol, the compound added to natural gas to give it a detectable odor.

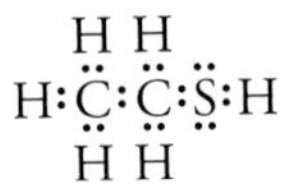

```
  H H
H:C:C:S:H
  H H
```

Strategy

Electron dot and line-bond structures differ in that each pair of valence electrons involved in a covalent bond is shown as a pair of dots in an electron dot structure and as a line in a line-bond structure.

Solution

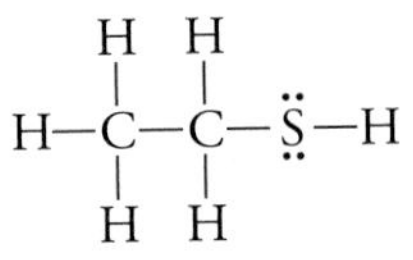

```
   H  H
   |  |   ..
H—C—C—S—H
   |  |   ..
   H  H
```

Practice Problem 3.6

Draw the electron dot structure of 2-propanol (rubbing alcohol).

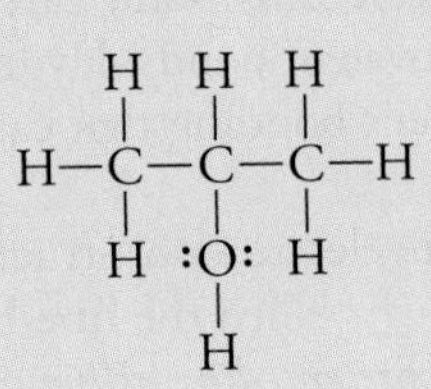

```
   H  H  H
   |  |  |
H—C—C—C—H
   |  |  |
   H :O: H
      |
      H
```

3.6 MOLECULES

■ The atoms in molecules are held together by covalent bonds.

The drawings in Figure 3.15 represent **molecules**—*uncharged groups of atoms connected to one another by covalent bonds.* An alternative to drawing the structure of a molecule is to give its **molecular formula**, which lists the number of each type of atom that is present. The molecular formulas of the molecules shown in Figure 3.15 are HF, H_2O, NH_3, and CH_4.

Most molecules are compounds, because they contain atoms of two or more different elements. Some molecules, however, are elements, because they contain just one type of atom. Seven elements (H_2, N_2, O_2, F_2, Cl_2, Br_2, and I_2) appear as **diatomic** *(two atom)* molecules. Oxygen is also found as the triatomic molecule called ozone(O_3). In Figure 3.15, atoms in the molecules are joined by **single bonds** (single covalent bonds), in which *one pair of electrons is shared* by two atoms. These are not the only covalent bonding patterns known, however. Under normal circumstances it is possible for two atoms to share up to three pairs of electrons. Atoms involved in a **double bond** share two pairs of electrons and atoms involved in a **triple bond** share three pairs of electrons.

An oxygen atom needs two electrons to gain an octet, so a given O atom is able to form either two single bonds or one double bond. A carbon atom, which requires four electrons to obtain an octet, has a number of covalent bonding options. It can form four single bonds, two double bonds, or various combinations of single, double, and triple bonds, as long as the total number of bonds is four (Figure 3.16). Chapter 4 will present an approach to use for going from a molecular formula to molecular structures similar to those shown below.

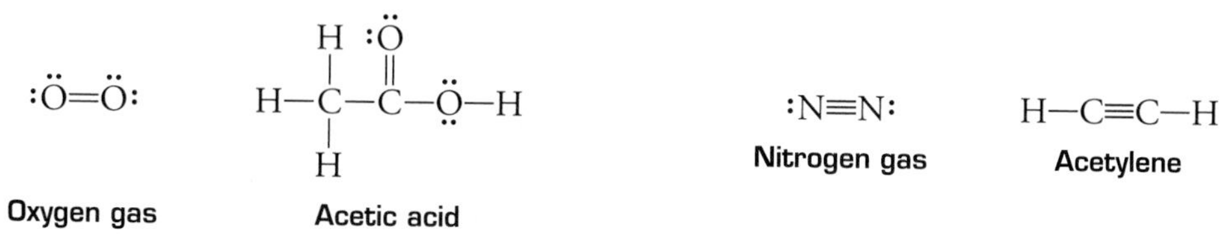

FIGURE 3.16 **Single, double, and triple bonds**
To reach an octet, oxygen atoms form two covalent bonds, nitrogen atoms form three covalent bonds, and carbon atoms form four covalent bonds.

Naming Binary Molecules

Molecules come in all shapes and sizes. How they are assigned names usually depends on the type of molecule that they are. Chapter 4, for example, will introduce a set of rules used to name organic molecules, a very large class of molecules that contain carbon atoms. Here, the relatively simple procedure used to name **binary molecules**—those that *contain just two different elements*—will be presented. Binary molecules are named by listing the elements in order of appearance in the molecular formula and changing the ending of the name of the second element to *ide*. For example, HF is hydrogen fluoride and HCl is hydrogen chloride.

■ Binary molecules contain two different elements.

When naming ionic compounds, the number of each type of ion is not specified ($CaCl_2$ is calcium chloride, not calcium dichloride), because ions always combine in fixed ratios to form a neutral compound. In binary molecules, however, atoms can sometimes combine in several different ways. Sulfur and oxygen atoms can bond to form two different molecules, SO_2 and SO_3. To distinguish such molecules by name, prefixes (Table 3.6) are added to specify the number of each type of atom that is present: SO_2 is named sulfur dioxide and SO_3 is named sulfur trioxide.

Other examples include:

- CO carbon monoxide (a poisonous, odorless gas)
- CO_2 carbon dioxide (a product of human metabolism)
- $SiCl_4$ silicon tetrachloride (used to prepare smoke screens in warfare)
- N_2O_5 dinitrogen pentoxide (used in the synthesis of certain organic compounds)

Sometimes binary molecules are better known by other names. Among these are H_2O (water, instead of dihydrogen oxide) and H_2S (hydrogen sulfide, instead of dihydrogen sulfide).

TABLE 3.6 PREFIXES USED FOR NAMING BINARY MOLECULES.[a]

Prefix	Number of atoms	Prefix	Number of atoms
mono	1	hexa	6
di	2	hepta	7
tri	3	octa	8
tetra	4	nona	9
penta	5	deca	10

[a]Names should not begin with "mono" (CO_2 is carbon dioxide, not monocarbon dioxide). When adding a prefix places two vowels together, the "a" or "o" ending on the prefix is dropped (CO is carbon monoxide, not carbon monooxide).

SAMPLE PROBLEM 3.7

Naming binary molecules

Name each binary molecule.

a. SiO_2 (used in glass manufacture)
b. SF_6 (used in electrical circuits)
c. P_2O_5 (a drying agent)

Strategy

When naming binary molecules, the element names are listed in the same order as given in the formula, the ending on the name of last element is changed to *ide*, and the number of times each appears is specified (see Table 3.6).

Solution

a. silicon dioxide
b. sulfur hexafluoride
c. diphosphorus pentoxide

PRACTICE PROBLEM 3.7

Name each binary molecule.

a. $SiBr_4$ **b.** P_2O_3 **c.** P_4Se_3

HEALTH Link: Dental Fillings

If you need to have a tooth filled, your experience will be very different from that of your ancestors. As recently as the mid-1800s it was standard practice to press pellets of lead, tin, or gold into dental cavities. Prior to that, anything that would plug the hole in the tooth (cork, resin, and others) was used. One problem with all of these filling materials was that they were not durable and tended to break or fall out. Today, the two most commonly used dental filling materials are amalgam and tooth-colored composites.

Any mixture of mercury and one or more other metals is called an amalgam. In dental amalgam, mercury is combined with silver and lesser amounts of tin, copper, and zinc. This mixture, which is soft to begin with and can be pressed into a tooth, sets quickly to form a hard filling. Composites are a special type of plastic made from organic (carbon-containing) compounds. They are soft and pliable until being hardened by exposure to an intense blue or ultraviolet light.

There are pros and cons to choosing either type of filling. Amalgam fillings are the stronger of the two and can last for 10 to 15 years. By comparison, composite fillings last an average of 5 years. While amalgam fillings are more durable, composite fillings can actually strengthen teeth because, unlike amalgam fillings, they bond directly to tooth material. Because of this, less of a tooth needs to be drilled away to prepare for a composite filling than for an amalgam one. Composite fillings are more expensive, but for those concerned with the appearance of their teeth, the extra cost may be worthwhile—composite fillings match tooth color, while amalgam does not (Figure 3.17).

Currently, one of the biggest issues related to dental fillings is whether amalgam is safe to have in your mouth. The concern is that some scientific studies have shown that amalgam fillings release trace amounts of mercury, a toxic element that at high enough levels can cause sometimes fatal neurological and brain damage. Opponents of amalgam use claim that no level of mercury is safe and that anyone with amalgam fillings should have them replaced with composite ones. Those in the pro-amalgam camp say that a person's daily exposure to mercury from amalgam fillings is not a concern because it is much lower than the average daily exposure to the mercury present in food and water because of pollution. Which is best for you should you need a filling? That is for you and your dental professional to decide.

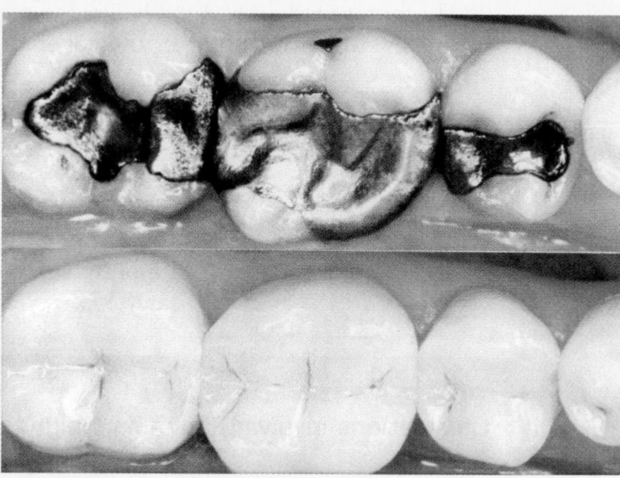

FIGURE 3.17

Dental fillings. Fillings are commonly made from amalgam (top) or composites (bottom)

Source: Courtesy of Dr. Joe Armel.

3.7 FORMULA WEIGHT, MOLECULAR WEIGHT, AND MOLAR MASS

Chapter 2 introduced atomic weight, the average mass of the naturally occurring atoms of an element. When dealing with an ionic compound, it can be helpful to know its **formula weight**, *the sum of the atomic weights of the elements in the formula*. Sodium chloride (NaCl) has a formula weight of 58.5 amu, which is determined by adding the atomic weights of sodium and chlorine.

$$\begin{array}{ccccc} Na^+ & & Cl^- & & NaCl \\ 23.0 \text{ amu} & + & 35.5 \text{ amu} & = & 58.5 \text{ amu} \end{array}$$

In this calculation, the atomic weights of Na and Cl were used, even though NaCl is composed of Na^+ and Cl^- ions. Compared to the mass of the protons and neutrons that make up the nucleus of Na and Cl atoms, electron mass is negligible (Section 2.1), so losing or gaining electrons to form ions has no effect on atomic weight—Na and Na^+ have the same atomic weight, as do Cl and Cl^-.

For more complex ionic compounds, calculation of formula weight works the same way. The formula weight of copper(II) nitrate, $Cu(NO_3)_2$, is 187.5 amu.

$$\underset{63.5\ \text{amu}}{Cu^{2+}} + \overbrace{\underset{2 \times 14.0\ \text{amu}}{2\ N} + \underset{6 \times 16.0\ \text{amu}}{6\ O}}^{2\ NO_3^-} = \underset{187.5\ \text{amu}}{Cu(NO_3)_2}$$

■ The molar mass of an ionic compound (the mass in grams of one mole) is equal to its formula weight in amu.

The mass in grams of one mole of an ionic compound (its molar mass) is numerically equivalent to its formula weight (in amu). For example, $Cu(NO_3)_2$ has a formula weight of 187.5 amu, so its molar mass is 187.5 g/mol.

This relationship allows conversions of the following type to be carried out.

- The formula weight of $AgNO_3$, used to prevent eye infection in newborns, is 169.9 amu. A sample containing 0.500 mol of $AgNO_3$ has a mass of 85.0 g.

$$0.500\ \cancel{\text{mol AgNO}_3} \times \frac{169.9\ \text{g AgNO}_3}{1\ \cancel{\text{mol AgNO}_3}} = 85.0\ \text{g AgNO}_3$$

- A 7.28 g sample of $AgNO_3$ is 4.28×10^{-2} mol.

$$7.28\ \cancel{\text{g AgNO}_3} \times \frac{1\ \text{mol AgNO}_3}{169.9\ \cancel{\text{g AgNO}_3}} = 4.28 \times 10^{-2}\ \text{mol AgNO}_3$$

SAMPLE PROBLEM 3.8

Calculations involving formula weight

a. What is the formula weight of the food preservative sodium sulfite (Na_2SO_3)?
b. What is the mass of 1.50 mol of Na_2SO_3?

Strategy

In part a, you must add up the individual atomic weights of each element in the formula. Solving part b involves a conversion factor that uses the molar mass of Na_2SO_3.

Solution

a. 126.1 amu

$$\underset{2 \times 23.0\ \text{amu}}{2\ Na^+} + \overbrace{\underset{32.1\ \text{amu}}{S} + \underset{3 \times 16.0\ \text{amu}}{3\ O}}^{SO_3^-} = \underset{126.1\ \text{amu}}{Na_2SO_3}$$

b. 189 g Na_2SO_3

$$1.50\ \cancel{\text{mol Na}_3\text{SO}_3} \times \frac{126.1\ \text{g Na}_2\text{SO}_3}{1\ \cancel{\text{mol Na}_2\text{SO}_3}} = 189\ \text{g Na}_2\text{SO}_3$$

PRACTICE PROBLEM 3.8

a. What is the formula weight of baking soda ($NaHCO_3$)?
b. What is the mass of 0.315 mol of baking soda?

Molecular Weight

Just as elements have an atomic weight and ionic compounds have a formula weight, molecules have a **molecular weight**—*the sum of the atomic weights of the elements in the formula.* The molecular weight of water (H_2O) is 18.0 amu, which is determined from the atomic weights of hydrogen and oxygen.

$$\begin{array}{ccccc} 2\,H & & O & & H_2O \\ 2 \times 1.0\text{ amu} & + & 16.0\text{ amu} & = & 18.0\text{ amu} \end{array}$$

The molecular weight of sulfur trioxide (SO_3) is 80.1 amu.

$$\begin{array}{ccccc} S & & 3O & & SO_3 \\ 32.1\text{ amu} & + & 3 \times 16.0\text{ amu} & = & 80.1\text{ amu} \end{array}$$

Since the molecular weight of sulfur trioxide is 80.1 amu, its molar mass is 80.1 g/mol and 0.0210 mol has a mass of 1.68 g.

■ The molar mass of a molecule (the mass in grams of one mole) is equal to its molecular weight in amu.

$$0.0210\ \cancel{\text{mol } SO_3} \times \frac{80.1\text{ g } SO_3}{1\ \cancel{\text{mol } SO_3}} = 1.68\text{ g } SO_3$$

SAMPLE PROBLEM 3.9

Calculations involving molecular weight

a. What is the molecular weight of chloroform ($CHCl_3$)?
b. What is the mass of 2.50 mol of chloroform?

Strategy
You can calculate the molecular weight of chloroform by adding up the atomic weights of carbon, hydrogen, and chlorine. (Remember to add in the atomic weight of chlorine three times, since it appears three times in the formula.) Part b of the problem can be solved by using a conversion factor related to the molar mass of $CHCl_3$.

Solution

a. 119.5 amu

$$\begin{array}{ccccccc} C & & H & & 3\,Cl & & CHCl_3 \\ 12.0\text{ amu} & + & 1.0\text{ amu} & + & 3 \times 35.5 & = & 119.5\text{ amu} \end{array}$$

b. 299 g

$$2.50\ \cancel{\text{mol } CHCl_3} \times \frac{119.5\text{ g } CHCl_3}{1\ \cancel{\text{mol } CHCl_3}} = 299\text{ g } CHCl_3$$

PRACTICE PROBLEM 3.9

a. What is the molecular weight of glycine ($C_2H_5NO_2$), one of the amino acids used to build proteins?
b. What is the mass of 4.00 mol of glycine?
c. How many glycine molecules are present in 0.00552 g of glycine?

BIOCHEMISTRY Link

Ethylene, a Plant Hormone

Many of the chemical changes that take place within cells are controlled by compounds called hormones, one example of which is ethylene, (C_2H_4) a plant hormone that stimulates the ripening of some fruits.

$$\begin{array}{c} H-C=C-H \\ \;\;\; | \quad\; | \\ \;\;\; H \quad H \end{array}$$

Ethylene

You can see the effects of this molecule by doing a simple experiment. Take two unripened tomatoes and place one of them in a plastic bag. If you watch them over the course of several days, you will find that the tomato in the plastic bag ripens more quickly than the other. Each tomato produces small amounts of ethylene gas, but, since the ethylene is unable to escape through plastic, the tomato in the bag is exposed to higher levels of this hormone and ripens more quickly.

Food distributors control ripening in the same way. Bananas, for example, are picked green and stored in a well-ventilated (ethylene-free) environment. This allows them to be shipped without spoiling. Once the bananas have reached their destination, they can be quickly ripened by exposure to ethylene gas (Figure 3.18).

FIGURE 3.18

Ethylene promotes ripening Bananas are shipped in well-ventilated containers. By not allowing ethylene levels to build up, the bananas can reach their destination before they ripen.

Source: Spencer Grant/Photo Edit.

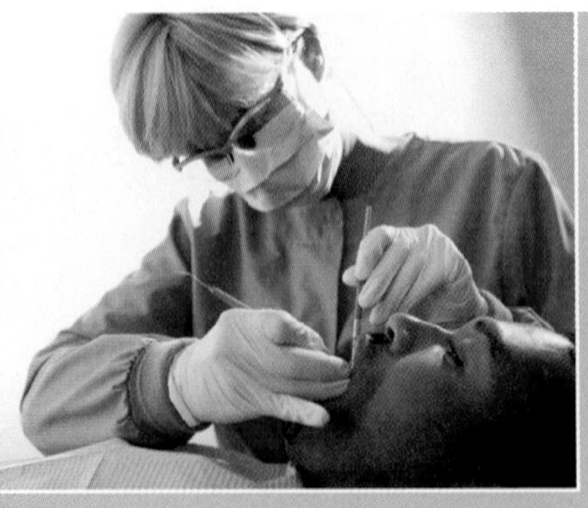

Tooth enamel is composed mostly of a mineral called hydroxyapatite, an ionic compound with the formula $Ca_5(OH)(PO_4)_3$. Tooth decay is what happens when enamel is damaged by the breakdown of hydroxyapatite. This demineralization takes place when tooth enamel is exposed to acids produced by bacteria present in dental plaque.

Fluoride ion (F^-) can be used to prevent tooth decay. When children are given fluoride, it is incorporated into their developing teeth through the formation of fluorapatite, $Ca_5(F)(PO_4)_3$. This mineral is stronger than hydroxyapatite and is not broken down as easily by acids. Adults also benefit from the use of fluoride, because existing hydroxyapatite can be converted into the more durable fluorapatite.

Fluoride has other benefits as well. It can reverse some demineralization damage through remineralization, the formation of new fluorapatite. Some studies have also shown that fluoride reduces the ability of bacterial plaque to produce the acids that cause cavities.

A number of options are available for administering fluoride. In some areas, fluoride is naturally present in the water. In others, water is fluoridated by addition of sodium fluoride or another fluoride-containing ionic compound. Many types of toothpaste contain fluoride, and fluoride tablets are available by prescription. The fluoride rinses or gels used at the dentist office contain higher levels of this anion.

summary of objectives

1 Describe the differences in how the Bohr model and the quantum mechanical model of the atom view electrons as being arranged about an atom's nucleus.

In the **Bohr model** of the atom, electrons are described as circling the nucleus in fixed orbits. In the **quantum mechanical model** of the atom, electrons reside in **orbitals** that surround the nucleus. $2n^2$ electrons (n is the energy level number) is the theoretical maximum allowed per energy level.

2 Define the term valence electron and describe electron dot structures.

Electrons in the highest numbered, occupied energy level of an atom are its **valence electrons**. In **electron dot** structures valence electrons are represented by dots.

3 Define the term ion and explain how the electron dot structure of a representative element atom (groups 1A–8A) can be used to predict the charge on the monoatomic ion that it forms.

Atoms or groups of atoms that have an unequal number of protons and electrons are called **ions**. Atoms of representative elements form **monoatomic** ions (formed from a single atom) in order to reach the stable electron arrangement consisting of eight valence electrons—an **octet**. Nonmetals gain electrons to reach an octet. (For example, N has five valence electrons and gains three more to become N^{3-}, Br has seven valence electrons and gains one more to become Br^-.) Metals lose valence electrons to reach an octet in the next lowest energy level. (For example, Na has one valence electron and loses it to become Na^+, Al has three valence electrons and loses them to become Al^{3+}.) Most transition metals are unable to lose enough electrons to attain an octet.

4 Describe the naming of monoatomic and polyatomic cations and anions.

Ions with a positive charge are **cations** and those with a negative charge are **anions**. Monoatomic cations are given the name of the original atom (Na^+ is sodium ion), while monoatomic anions are named by changing the ending of the name of the atom to *ide* (Cl^- is chloride ion). Names of ions formed from transition metal atoms specify the charge on the ion [Fe^{2+} is iron(II) ion and Fe^{3+} is iron(III) ion]. Alternatively, Fe^{2+} is ferrous ion and Fe^{3+} is ferric ion. While the names of polyatomic ions must be memorized, there is a pattern to how some names are assigned. A polyatomic anion whose name ends in "ite" has one less O atom than the related ion whose name ends in "ate" (SO_3^{2-} is sulfite ion and SO_4^{2-} is sulfate ion, NO_2^- is nitrite ion and NO_3^- is nitrate ion). Additionally, use of "hydrogen" specifies the relative number of H atoms that related ions have (PO_4^{3-} is phosphate ion, HPO_4^{2-} is hydrogenphosphate ion, and $H_2PO_4^-$ is dihydrogen phosphate ion).

5 Explain the difference between an ionic bond and a covalent bond.

An **ionic bond** is the attraction between ions of opposite charge (cations and anions are attracted to one another). Nonmetal atoms share electrons in **covalent bonds** as a way to obtain an octet of valence electrons.

6 Name and write the formulas of simple ionic compounds and binary molecules.

Ionic compounds are named by combining the name of the cation with the name of the anion, without specifying the relative number of each ion [NaCl is sodium chloride, $MgBr_2$ is magnesium bromide, and $Fe(OH)_2$ is iron(II) hydroxide]. **Molecules** (uncharged groups of atoms joined by covalent bonds) are **binary** when they contain atoms of just two different elements. Binary molecules are named by listing the names of the elements in the order that they appear in the formula, with the ending of the second element's name changed to *ide*. Prefixes, such as *mono*, *di*, and *tri*, are used to specify the number of each type of atom present (SO_2 is sulfur dioxide, SO_3 is sulfur trioxide).

7 Define the terms formula weight and molecular weight. Use the molar mass of a compound to carry out conversions involving moles and mass.

The **formula weight** of an ionic compound is the sum of the atomic weights of the elements in its formula. The **molecular weight** of a molecule is the sum of the atomic weights of the elements in its formula. The molar mass of an ionic compound or molecule (the mass in grams of one mole) is equal to its formula weight or molecular weight (in amu). The formula weight of LiF is 25.9 amu (6.9 amu for Li plus 19.0 amu for F), its molar mass is 25.9g/mol, and 0.50 mol of LiF has a mass of 13 g.

$$0.50\ \cancel{\text{mol LiF}} \times \frac{25.9\text{ g LiF}}{1\ \cancel{\text{mol LiF}}} = 13\text{ g LiF}$$

The molecular weight of Br_2 is 159.8 amu (79.9 amu for each Br), its molar mass is 159.8 g/mol, and 198 g of Br_2 is 1.24 mol.

$$198\ \cancel{\text{g Br}_2} \times \frac{1\text{ mol Br}_2}{159.8\ \cancel{\text{g Br}_2}} = 1.24\text{ mol Br}_2$$

END OF CHAPTER PROBLEMS

Answers to problems whose numbers are printed in color are given in Appendix D. More challenging questions are marked with an asterisk. Problems within color rules are paired. **ILW** = Interactive Learning Ware solution is available at *www.wiley.com/college/raymond.*

3.1 Ions

3.1 Give the total number of protons and electrons in each ion.
a. K^+ **b.** Mg^{2+} **c.** P^{3-}

3.2 Give the total number of protons and electrons in each ion.
a. Li^+ **b.** Al^{3+} **c.** S^{2-}

3.3 Give the total number of protons, neutrons, and electrons in each ion.
a. ${}^{63}_{29}Cu^+$ **b.** ${}^{19}_{9}F^-$ **c.** ${}^{37}_{17}Cl^-$

3.4 Give the total number of protons, neutrons, and electrons in each ion.
a. ${}^{23}_{11}Na^+$ **b.** ${}^{16}_{8}O^{2-}$ **c.** ${}^{35}_{17}Cl^-$

3.5 Give the name of each ion.
a. F^- **b.** O^{2-} **c.** Cl^- **d.** Br^-

3.6 Give the name of each ion.
a. Sr^{2+} **c.** S^{2-}
b. Cl^- **d.** Se^{2-}

3.7 Give the name of each ion.
a. $CO_3{}^{2-}$ **c.** $SO_3{}^{2-}$
b. $NO_3{}^-$ **d.** $CH_3CO_2{}^-$

3.8 Give the name of each ion.
a. H_3O^+ **c.** $HPO_4{}^{2-}$
b. OH^- **d.** $H_2PO_4{}^-$

3.9 Write the formula of each ion.
a. hydrogencarbonate ion
b. nitrite ion
c. sulfate ion

3.10 Write the formula of each ion.
a. hydrogensulfate ion
b. phosphate ion
c. dichromate ion

3.11 Draw the electron dot structure of a hydrogen ion.

3.12 Hydride ion has a 1− charge. Draw the electron dot structure of a hydride ion.

3.2 The Arrangement of Electrons

*__3.13__ For a helium atom, the energy separation between the ground state and excited state electron energy levels is different than that for a hydrogen atom. Does this cause helium to have a different emission spectrum than hydrogen? Explain.

3.14 Specify the number of electrons held in energy levels 1-4 of each atom.
a. Be **b.** N **c.** F

3.15 Specify the number of electrons held in energy levels 1-4 of each atom.
a. B **b.** C **c.** Mg

3.16 Specify the number of valence electrons for each atom.
a. H **d.** Br
b. Be **e.** Ne
c. C

3.17 Specify the number of valence electrons for each atom.
a. Li **d.** Kr
b. Si **e.** P
c. Al

3.18 For each, give the total number electrons, the number of valence electrons, and the number of the energy level that holds the valence electrons.
a. He **c.** Te
b. Xe **d.** Pb

3.19 For each, give the total number electrons, the number of valence electrons, and the number of the energy level that holds the valence electrons.
a. Br **c.** As
b. Kr **d.** I

3.20 For a neutral atom of element 114
a. how many total electrons are present?
b. how many valence electrons are present?
c. which energy level holds the valence electrons?
d. is the valence energy level full?

3.21 In October 2006, scientists reported making three atoms of element 118. For a neutral atom of this element
a. how many total electrons are present?
b. how many valence electrons are present?

c. which energy level holds the valence electrons?
d. is the valence energy level full?

3.3 The Octet Rule

3.22 When a nitrogen atom is converted into an ion, it becomes isoelectronic with neon.
a. What is the name of this ion?
b. How many electrons does nitrogen gain when it forms the ion?
c. What is the charge on the ion?

3.23 When a potassium atom is converted into an ion, it becomes isoelectronic with argon.
a. What is the name of this ion?
b. How many electrons does potassium lose when it forms the ion?
c. What is the charge on the ion?

3.24 Draw the electron dot structure of each atom and of the ion that it is expected to form.
a. Na **c.** Ar
b. Cl **d.** S

3.25 Draw the electron dot structure of each atom and of the ion that it is expected to form.
a. K **c.** Ca
b. Se **d.** O

3.4 Ionic Componds

3.26 To fight tooth decay, some toothpastes contain stannous fluoride, also known as tin(II) fluoride. Write the formula of this compound.

3.27 Name each ionic compound.
a. MgO **c.** CaF_2
b. Na_2SO_4 **d.** $FeCl_2$

3.28 Name each ionic compound.
a. Li_2O **c.** $Al(OH)_3$
b. $NaHSO_4$ **d.** $FeCl_3$

3.29 Write the formula of each ionic compound.
a. calcium hydrogenphosphate
b. copper(II) bromide
c. copper(II) sulfate
d. sodium hydrogensulfate

3.30 Write the formula of each ionic compound.
a. iron(III) hydroxide
b. ammonium bromide
c. copper(I) sulfate
d. tin(II) oxide

3.31 Write the formula of each ionic compound.
a. lithium sulfate (an antidepressant)
b. calcium dihydrogenphosphate (used in foods as a mineral supplement)
c. barium carbonate (used as a rat poison)

3.32 Write the formula of each ionic compound.
a. aluminum phosphate (used in some dental cements)
b. magnesium hydrogenphosphate (a laxative)
c. strontium bromide (an anticonvulsant)

3.33 Write the formula of the ionic compound that forms between
a. magnesium ions and fluoride ions
b. potassium ions and bromide ions
c. potassium ions and sulfide ions
d. aluminum ions and sulfide ions

3.34 In addition to sugars, citric acid, and other ingredients, the sports drink called Powerade contains potassium phosphate and potassium dihydrogenphosphate. Write the formula of each of these ionic compounds.

3.35 The following ionic compounds are improperly named. Give the correct name of each.
a. $SnCl_2$, tin chloride
b. FeO, iron oxide
c. $CrPO_4$, chromium phosphate

3.36 Give two different names for each ionic compound.
a. Cu_2SO_4 **d.** $Fe(NO_2)_3$
b. $CuSO_4$ **e.** $SnSO_4$
c. $Fe(NO_2)_2$ **f.** $Sn(SO_4)_2$

3.5 Covalent Bonds

3.37 Predict the number of covalent bonds formed by each nonmetal atom.
a. N **b.** Cl **c.** P

3.38 Predict the number of covalent bonds formed by each nonmetal atom.
a. C **b.** O **c.** Br

3.39 Draw the electron dot structure of each molecule.

```
       H     H
       |  ..  |
a. H—C—O—C—H
       |  ..  |
       H     H

       H       H  H :Ö
       |       |  |  ||
b. H—C———C—C—C—H
       |       |  |
       H  H—N: H
            |
            H
```

3.40 Draw the electron dot structure of each molecule.

a. 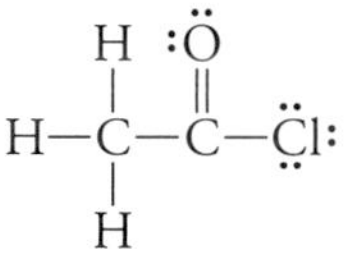
```
     H :Ö
     |  ‖   ..
  H—C—C—Cl:
     |      ..
     H
```

b. 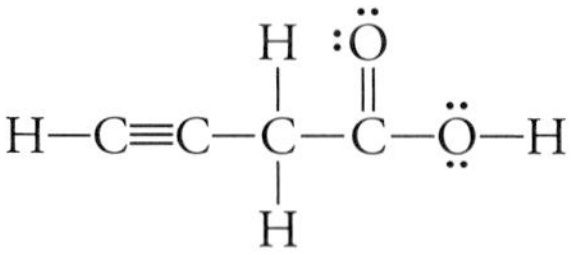
```
            H :Ö
            |  ‖
  H—C≡C—C—C—Ö—H
            |     ..
            H
```

3.41 Draw the line-bond structure of pyruvic acid, a compound formed during the breakdown of sugars by the body.

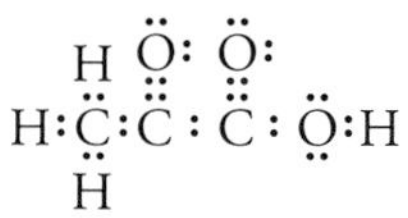

```
   H Ö: Ö:
  H:C:C : C : Ö:H
   H
```

Pyruvic acid

3.42 Draw the line-bond structure of lactic acid, a compound formed during anaerobic exercise.

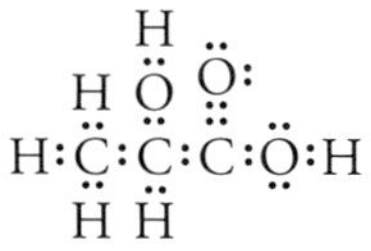

```
      H
    H Ö Ö:
  H:C:C:C:Ö:H
    H H
```

Lactic acid

3.43 Draw the electron dot structure of ethylene glycol (used as an antifreeze).

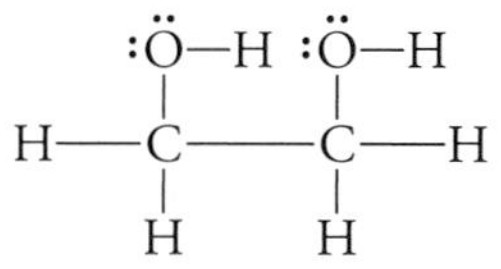

```
      :Ö—H :Ö—H
       |      |
  H——C———C——H
       |      |
       H      H
```

Ethylene glycol

3.44 Draw the electron dot structure of propylene glycol, an environmentally safe alternative to ordinary antifreeze.

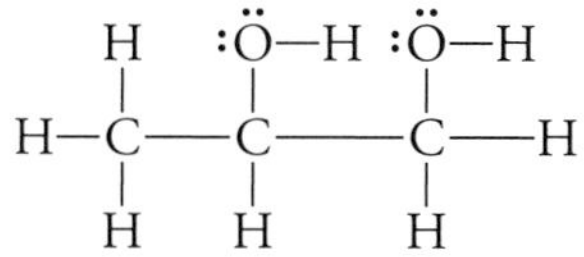

```
     H   :Ö—H :Ö—H
     |     |      |
  H—C——C———C——H
     |     |      |
     H     H      H
```

Propylene glycol

3.6 Molecules

3.45 Name each molecule.
a. NCl_3 **b.** PCl_3 **c.** PCl_5

3.46 Name each molecule.
a. CS_2 **b.** N_2O_3 **c.** NF_3

3.47 Phosphine (PH_3) is a poisonous gas that has the odor of decaying fish. Give another name for this binary molecule.

3.48 Methane (CH_4) is a major component of natural gas. Give another name for this binary molecule.

3.49 Dentists use nitrous oxide (N_2O) as an anesthetic. Give another name for this binary molecule.

*__3.50__ Draw the electron dot structure of the molecule formed when sufficient H atoms are added to give each atom an octet of valence electrons.
a. C **b.** N **c.** O

*__3.51__ Draw the electron dot structure of the molecule formed when sufficient H atoms are added to give each atom an octet of valence electrons.
a. F **b.** P **c.** Br

3.52 Indicate whether each is an ionic compound or a binary molecule.
a. $BaCl_2$ **d.** HgO
b. OCl_2 **e.** N_2O_3
c. CS_2 **f.** Cu_2O

3.53 Name each of the ionic compounds or binary molecules in the previous problem.

3.54 Indicate whether each is an ionic compound or a binary molecule.
a. SF_2 **d.** PF_5
b. MgF_2 **e.** NO_2
c. $SnCl_2$ **f.** SnO_2

3.55 Name each of the ionic compounds or binary molecules in the previous problem.

3.56 Are diatomic molecules considered compounds? Explain.

3.57 Do any elements exist as binary molecules? Explain.

3.7 Formula Weight and Molecular Weight

3.58 **a.** What is the formula weight of ammonium hydroxide?
b. What is the mass of 0.950 mol of ammonium hydroxide?
c. How many moles of ammonium hydroxide are present in 0.475 g?

3.59 **a.** What is the formula weight of Li_2CO_3?
b. What is the mass of 1.33×10^{-4} mol of Li_2CO_3?
c. How many moles of CO_3^{2-} ions are present in 73.5 g of Li_2CO_3?

3.60 **a.** What is the formula weight of sodium oxide?
b. How many oxide ions are present in 0.25 mol of sodium oxide?
c. How many sodium ions are present in 0.25 mol of sodium oxide?
d. How many oxide ions are present in 2.30 g of sodium oxide?
e. How many sodium ions are present in 2.30 g of sodium oxide?

3.61 **a.** What is the formula weight of magnesium iodide?
b. How many magnesium ions are present in 7.5×10^{-6} mol of magnesium iodide?
c. How many iodide ions are present in 7.5×10^{-6} mol of magnesium iodide?
d. How many magnesium ions are present in 4.5 mg of magnesium iodide?
e. How many iodide ions are present in 4.5 mg of magnesium iodide?

***3.62** The food additive potassium sorbate, $K(C_6H_7O_2)$, is a mold and yeast inhibitor.
a. What is the charge on the sorbate ion?
b. How many C atoms are present in 0.0150 g of potassium sorbate?

***3.63** To control manic-depressive behavior, some patients are administered up to 2000 mg of lithium carbonate per day. Convert this dosage into millimoles.

3.64 What is the molecular weight of each?
a. carbon monoxide
b. dinitrogen pentoxide
c. glucose ($C_6H_{12}O_6$)

3.65 **a.** What is the molecular weight of CCl_4?
b. What is the mass of 61.3 mol of CCl_4?
c. How many moles of CCl_4 are present in 0.465 g of CCl_4?
d. How many molecules of CCl_4 are present in 5.50×10^{-3} g of CCl_4?

3.66 **a.** What is the molecular weight of aspirin ($C_9H_8O_4$)?
b. What is the mass of 0.00225 mol of aspirin?
c. How many moles of aspirin are present in 500 mg of aspirin?
d. How many molecules of aspirin are present in 1.00 g of aspirin?

***3.67** One tablet of a particular analgesic contains 250 mg of acetaminophen ($C_8H_9NO_2$). How many acetaminophen molecules are contained in the tablet?

***3.68** A vitamin tablet contains 500 mg of vitamin C ($C_6H_8O_6$). How many vitamin C molecules are contained in the tablet?

3.69 Isovaleric acid ($C_5H_{10}O_2$) is the molecule responsible for foot odor. To smell this compound, it must be present in the air at a minimum of 250 parts per billion. This means that a 0.50 L volume of air would hold 0.00013 g of isovaleric acid.
a. To how many moles of isovaleric acid does this correspond?
b. To how many molecules of isovaleric acid does this correspond?

3.70 2-Isobutyl-3-methoxypyriazine, $C_9H_{14}N_2O$, has an "earthy" smell and is one of the compounds responsible for the aroma of coffee.
a. If 1.00 L of coffee contains 8.3×10^{-2} mg of 2-isobutyl-3-methoxypyriazine, how many molecules of this compound are present?
b. How many molecules of the compound would be present in 1 cup of this coffee?

BIOCHEMISTRY*Link* | *Bioluminescence*

3.71 How are electrons involved in the production of light by luciferin?

3.72 What are some of the biological functions of bioluminescence?

3.73 According to "Glowing Fish", what percentage of ocean life bioluminesces?

BIOCHEMISTRY*Link* | *Ionophores and Biological Ion Transport*

3.74 Explain how ionophores act as antibiotics.

3.75 K^+ attaches more strongly to valinomycin than Na^+ because of the size of the cavity (binding site) in the center of this compound. Is this cavity too large for Na^+ or is it too small?

BIOCHEMISTRY*Link* | *Dental Fillings*

3.76 Sometimes a filling can make a tooth sensitive to temperature changes. Given that amalgam is made from metals and that composites are made from nonmetals, which type of filling would you expect is most likely to cause thermal sensitivity?

3.77 Although amalgam is sometimes referred to as a compound, it does not fit the definition of this term. Explain.

BIOCHEMISTRY*Link* | *Ethylene, a Plant Hormone*

3.78 During ripening, bananas produce small amounts of ethylene. When bananas are shipped, why should they not be shipped in closed containers?

THINKING IT THROUGH

3.79 In many cities, there is great debate about the issue of fluoridating the water supply. What are the pros and cons of doing so?

INTERACTIVE LEARNING PROBLEMS

ILW

3.80 Name the following compounds

a. $Pd(NO_3)_4$
b. Si_3N_4
c. Ba_3N_2
d. SiS_2
e. How many grams of O are combined with 7.14×10^{21} atoms of N in the compound N_2O_5?

SOLUTIONS TO PRACTICE PROBLEMS

3.1 **a.** scandium(III) ion; **b.** titanium(II) ion; **c.** strontium ion

3.2 **a.** 4; **b.** 3; **c.** 3; **d.** 7

3.3 **a.** $:\ddot{\underset{..}{Kr}}:$; **b.** Ba:

3.4 **a.** $:\ddot{\underset{.}{Cl}}:$ $\quad :\ddot{\underset{..}{Cl}}:^{-}$
b. $\cdot\ddot{\underset{..}{Se}}\cdot$ $\quad :\ddot{\underset{..}{Se}}:^{2-}$
c. Ca: $\quad Ca^{2+}$

3.5 **a.** LiCl; **b.** Li_2O; **c.** Li_3N

3.6
```
   H  H  H
H:C : C : C:H
   H :O: H
      H
```

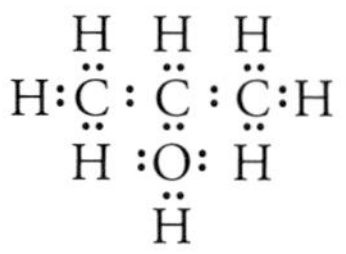

3.7 **a.** silicon tetrabromide; **b.** diphosphorus trioxide; **c.** tetraphosphorus triselenide

3.8 **a.** 84.0 amu; **b.** 26.5 g

3.9 **a.** 75.0 amu; **b.** 300 g; **c.** 4.43×10^{19} molecules

Macular degeneration is a disease of the eye with symptoms that include loss of color vision and distortion, blurring, or complete loss of central vision. Peripheral vision is usually unaffected. People with this disease usually have trouble when it comes to reading or recognizing faces. There are a number of factors that increase the risk of getting macular degeneration, including advanced age, a history of smoking, and diabetes.

Stargardt's disease, an inherited form of macular degeneration that usually appears before age 30, results from damage to the retina caused by a malfunction in the recycling of an organic molecule essential to the visual process.

Photo source: Science VU/NIH/Visuals Unlimited.

4 An Introduction to Organic Compounds

Now that we have seen the basics of how atoms (Chapter 2) and compounds (Chapter 3) are put together, it is time to set the stage for the organic and biochemistry topics that we will encounter as we move through the text. This chapter will begin with an introduction to a range of diverse topics, all of which relate to how bonding patterns, the distribution of electrons, and the shape of a molecule influence the way that it interacts with others. In later sections of the chapter we will apply our knowledge of these interactions to understand the properties of a number of different families of **organic** compounds (*contain carbon atoms*). The shape that a molecule has and its interactions with others are important in determining its biological effects. Diseases ranging from Stargardt's disease (mentioned on the facing page) to mad cow disease (Health Link: Prion Diseases) are directly related to these properties.

objectives

After completing this chapter, you should be able to:

1. Draw molecules and polyatomic ions using electron dot and line-bond structures. Assign formal charges.
2. Describe how condensed structural formulas and skeletal structures differ from electron dot and line-bond structures.
3. Define electronegativity and explain its relationship to polar covalent bonds. Give a simple rule that can be used to predict whether or not a covalent bond is polar.
4. List the five basic shapes about an atom in a molecule and describe the rules used to predict shape. Explain how shape plays a role in determining overall polarity.
5. Describe the noncovalent interactions that attract one compound to another.
6. Describe the four families of hydrocarbons.
7. Explain the difference between constitutional isomers, conformations, and the stereoisomers known as geometric isomers. Give examples of two different families of hydrocarbons that can exist as geometric isomers.
8. Define the term functional group and describe the features that distinguish hydrocarbons, alcohols, carboxylic acids, and esters from one another.

4.1 STRUCTURAL FORMULAS AND FORMAL CHARGES

Before we can consider organic molecules, we must expand on the discussion of covalent bonds and molecules that was initiated in Chapter 3. Section 3.6 introduced molecules, the uncharged groups of nonmetal atoms that are connected to one another by covalent bonds, and showed that the structure of a given molecule can be represented by an electron dot structure (all valence electrons are shown using dots) or a line-bond structure (each pair of shared bonding electrons is represented by a line). In the structures of 2-propanol (rubbing alcohol) shown here, the carbon atoms and oxygen atom have formed the number of covalent bonds required to reach an octet—four bonds for carbon atoms and two bonds for oxygen atoms.

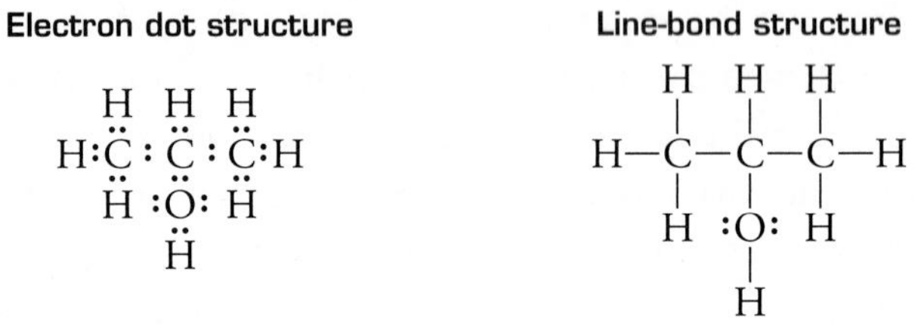

Electron dot structure

```
 H  H  H
H:C : C : C:H
 H :O: H
    H
```

Line-bond structure

```
   H   H   H
   |   |   |
H—C—C—C—H
   |   |   |
   H  :O:  H
       |
       H
```

2-Propanol

Either of these **structural formulas** provides more information about the 2-propanol than does the molecular formula (C_3H_8O), because molecular formulas tell nothing about how atoms are attached to one another. Suppose, for example, that a toxicologist is reporting on the health problems associated with exposure to high levels of the compound C_3H_8O. Because this molecular formula does not tell us how the molecule is put together, it is not clear whether she is talking about 2-propanol or one of the other two molecules that have the formula C_3H_8O. Representing the molecule using a structural formula (Figure 4.1) clarifies the matter.

Given a molecular formula, how do we go about drawing a structural formula? For small molecules, knowing the number of covalent bonds that an atom is expected to form can be a good place to start. As we saw in Section 3.5, the number of covalent bonds that a nonmetal atom forms is generally the same as the number of electrons that it needs to gain in order to have an octet. An oxygen atom has six electrons, needs two more to gain an octet, and can form two covalent bonds. A nitrogen atom has five valence electrons, needs three more to gain an octet, and can form three covalent bonds. The expected bonding patterns for hydrogen and the period two nonmetals are:

- A carbon atom can form 4 covalent bonds.
- A nitrogen atom can form 3 covalent bonds.
- An oxygen atom can form 2 covalent bonds.
- A halogen atom can form 1 covalent bond.
- A hydrogen atom can form 1 covalent bond.

```
   H   H   H              H   H   H                  H        H   H
   |   |   |              |   |   |                  |        |   |
H—C—C—C—H           H—C—C—C—Ö—H           H—C—Ö—C—C—H
   |   |   |              |   |   |                  |        |   |
   H  :O:  H              H   H   H                  H        H   H
       |
       H
```

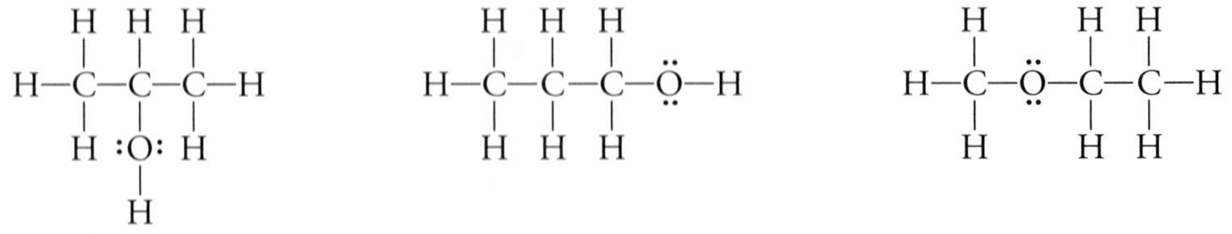

FIGURE 4.1 **Structural formulas**
Structural formulas, like the line-bond formulas shown here for the three molecules with the formula C_3H_8O, indicate the relative position of each atom within a molecule.

To draw the structural formula for HF, join H and F by a covalent bond to give each its required number of bonds. For NH_3, N becomes the central atom by forming three covalent bonds, one to each H atom. Note that in the structures of HF and NH_3, the F and N atoms each carry enough nonbonding electrons to have an octet. Figure 3.15 shows other structural formulas that can be drawn using this approach.

```
   ..        ..
H—F:      H—N—H
   ..        |
             H
```

Not all structural formulas are this easy to come up with. Fortunately, for more challenging situations there is a systematic approach that can be used. This approach will be introduced by showing how NCl_3 is drawn.

Drawing Line-Bond Structures

1. ***Count the total number of valence electrons.*** In a molecule, an atom's valence electrons end up either in covalent bonds or as nonbonding electrons. The structural formula of NCl_3 should show 26 electrons:

1 N atom	+	3 Cl atoms	
5 valence electrons	+	$\underbrace{3 \times 7 \text{ valence electrons}}$	
5	+	21	= 26 valence electrons

2. ***Use single covalent bonds to connect the atoms to one another.*** The one bit of guesswork here will be deciding which atom to attach to another. Although not always the case, central atoms often come first in the molecular formula. The structure of NCl_3 follows this pattern: N is the central atom. Note that although hydrogen is sometimes listed first in molecular formulas, it can never be a central atom. Hydrogen forms just one covalent bond.

   In NCl_3, nitrogen is the central atom, so we will connect it to the chlorine atoms using single bonds.

```
Cl—N—Cl
   |
   Cl
```

3. ***Beginning with the atoms attached to the central atom, add the remaining electrons in order to complete octets.*** Six valence electrons have been used in the NCl_3 structure above: two for each of the three covalent bonds. Of the original 26 electrons, 20 must still be added. Drawing enough electrons to give each chlorine atom an octet gives

```
 ..      ..
:Cl—N—Cl:
 ..  |   ..
    :Cl:
     ..
```

   This structure is not complete, because only 24 electrons are shown (3 bonds and 9 pairs of nonbonding electrons). The two remaining electrons are added to the nitrogen atom.

```
 ..  ..  ..
:Cl—N—Cl:
 ..  |   ..
    :Cl:
     ..
```

4. ***If the central atom does not have an octet, move pairs of nonbonding electrons from attached atoms to form multiple bonds with the central atom. Do so until the central atom has an octet.*** The NCl_3 drawn in step 3 is completed and need not be modified. The nitrogen atom has an octet (3 bonds and 1 pair of nonbonding electrons), as do the chlorine atoms (1 bond and 3 pairs of nonbonding electrons each).

Let us repeat this process for formaldehyde, which has the formula CH_2O. In this molecule, C is the central atom and is attached to each of the other three atoms.

1. ***Count the total number of valence electrons.*** CH_2O has 12 valence electrons available.

1 C atom	+	2 H atoms	+	1 O atom	
4 valence electrons	+	2 × 1 valence electron	+	6 valence electrons	
4	+	2	+ 6		= 12 valence electrons

2. ***Use single covalent bonds to connect the atoms to one another.*** As the hint above explained, C is the central atom.

H—C—O
 |
 H

3. ***Beginning with the atoms attached to the central atom, add the remaining electrons in order to complete octets.*** The drawing above shows three single bonds, so 6 electrons are represented. That leaves 6 more to add. Hydrogen atoms form just one covalent bond and can accept no more electrons, so the remaining electrons must go to the oxygen atom.

H—C—Ö:
 |
 H

Additional nonbonding electrons are not added to the carbon atom because all of the 12 valence electrons have been used.

4. ***If the central atom does not have an octet, move pairs of nonbonding electrons from attached atoms to form multiple bonds with the central atom. Do so until the central atom has an octet.*** The carbon atom does not have an octet, so a pair of nonbonding electrons is moved from the oxygen atom and used to form a double bond with the carbon atom. This gives the carbon atom an octet and the drawing is complete.

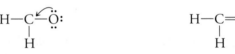

Carbon has 6 valence electrons **Carbon has an octet**

SAMPLE PROBLEM 4.1

Drawing line-bond structures

Draw the line-bond structure of hydrogen cyanide, HCN. In this molecule, carbon is the central atom.

Strategy

After completing step 3 of the procedure just described, you will find that the carbon atom does not have an octet. Move as many pairs of nonbonding electrons as needed to give carbon an octet.

Solution

1. The molecular formula of HCN indicates that 10 valence electrons are available, 1 from H, 4 from C, and 5 from N.

2. Carbon is the central atom, so the initial structure is

H—C—N

3. Having two covalent bonds, the structure above uses only 4 of the available 10 valence electrons. The hydrogen atom cannot accept any more electrons, so the remaining 6 are added to nitrogen.

$$\mathrm{H{-}C{-}\ddot{\underset{..}{N}}:}$$

4. While the nitrogen atom has an octet, carbon does not. Nonbonding electrons from N are moved to form multiple bonds with C, until C has an octet.

$\mathrm{H{-}C{-}\ddot{\underset{..}{N}}:}$	$\mathrm{H{-}C{=}\ddot{N}:}$	$\mathrm{H{-}C{\equiv}N:}$
Carbon has 4 valence electrons	**Carbon has 6 valence electrons**	**Carbon has an octet**

PRACTICE PROBLEM 4.1

Draw the line-bond structure of carbon disulfide, CS_2. In this molecule, C is the central atom.

In addition to describing the structure of molecules, structural formulas also provide details about the makeup of polyatomic ions. Nitrate ion (NO_3^-), for example, consists of a nitrogen atom covalently bonded to three oxygen atoms (Figure 4.2*a*). The charge on this ion is shown as a superscript to the right of the square brackets that surround the structural formula. As is always the case, the charge on the ion is due to an imbalance in the total number of protons and electrons in the structure.

To understand the behavior of molecules and polyatomic ions (including those present in cell membranes and those responsible for the cleansing action of soap), it is necessary to know *which atom or atoms carry a charge*. This charge, called **formal charge**, is determined by comparing the number of electrons that surround an atom in a compound to the number of valence electrons that it carries as a neutral atom. A formula to use when calculating formal charge is

$$\text{Formal charge} = \begin{matrix}\text{number of valence electrons}\\ \text{for a neutral atom}\end{matrix} - \begin{matrix}\text{number of electrons around the atom}\\ \text{in the compound}\end{matrix}$$

Using this formula, you assume that the atom "sees" its nonbonding electrons and the bonding electrons nearest to it (imagine that each covalent bond is split down the middle with each atom getting one electron).

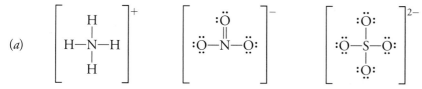

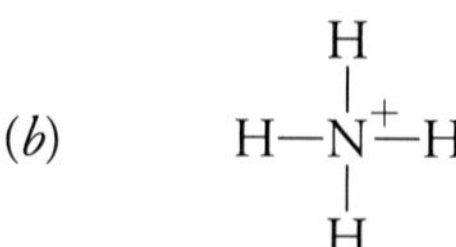

FIGURE 4.2 **Polyatomic ions**
Polyatomic ions are charged groups of nonmetal atoms that are connected by covalent bonds. (*a*) The net charge on the ion is shown as a superscript to the right. (*b*) The formal charge on each atom is shown.

To see examples of how formal charge is calculated, let us consider two of the ions in Figure 4.2*a*. In an ammonium ion, the nitrogen atom sees four electrons (the four bonding electrons nearest it). Because a neutral nitrogen atom has five valence electrons, the formal charge on this nitrogen atom is 1+.

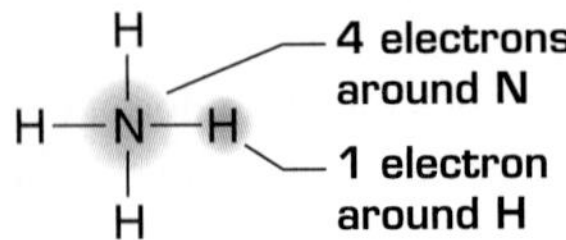

$$\text{Formal charge on N} = \frac{\text{number of valence electrons}}{\text{for a neutral atom}} - \frac{\text{number of electrons around}}{\text{the atom in the compound}}$$

$$= 5 - 4 = 1+$$

In an ammonium ion, each hydrogen atom sees one electron. Because a neutral hydrogen atom has one valence electron, the formal charge on each hydrogen atom is 0.

$$\text{Formal charge on H} = \frac{\text{number of valence electrons}}{\text{for a neutral atom}} - \frac{\text{number of electrons around}}{\text{the atom in the compound}}$$

$$= 1 - 1 = 0$$

The structure of the ammonium ion in Figure 4.2*b* shows the nitrogen atom carrying a formal charge of 1+.

In a nitrate ion, the nitrogen atom sees four electrons and has a formal charge of 1+.

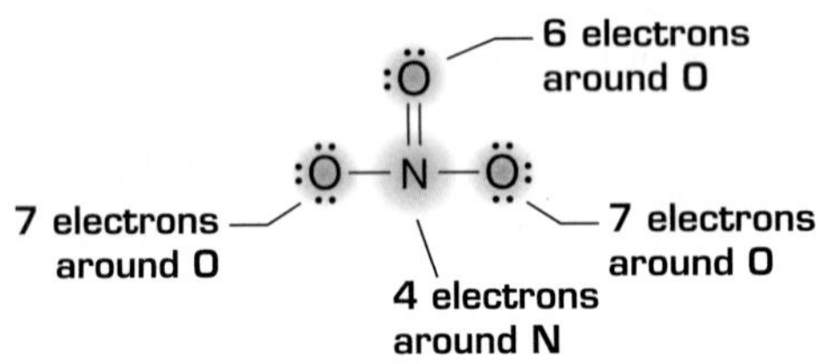

$$\text{Formal charge on N} = \frac{\text{number of valence electrons}}{\text{for a neutral atom}} - \frac{\text{number of electrons around}}{\text{the atom in the compound}}$$

$$= 5 - 4 = 1+$$

Two of the oxygen atoms see seven electrons and have a formal charge of 1−.

$$\text{Formal charge on O} = \frac{\text{number of valence electrons}}{\text{for a neutral atom}} - \frac{\text{number of electrons around}}{\text{the atom in the compound}}$$

$$= 6 - 7 = 1-$$

The third oxygen atom sees six electrons and has a formal charge of 0.

$$\text{Formal charge on O} = \frac{\text{number of valence electrons}}{\text{for a neutral atom}} - \frac{\text{number of electrons around}}{\text{the atom in the compound}}$$

$$= 6 - 6 = 0$$

■ Formal charges help keep track of the distribution of electrons in a compound.

An interesting thing to note for nitrate ion is that, while the structure shown in Figure 4.2*a* indicates a net charge of 1− on the ion, the formal charges in Figure 4.2*b* indicate that this overall charge is due to two negative charges and one positive charge. The sum of the individual formal charges in a compound will always equal the overall charge on the compound. In sulfate ion (Figure 4.2*b*), the overall charge of 2− is the sum of the formal charges on the S and O atoms (2+, 1−, 1−, 1−, and 1−).

SAMPLE PROBLEM 4.2

Assigning formal charges

The line-bond structure of acetate ion (Table 3.2) is shown below. Assign formal charges to the atoms in this ion.

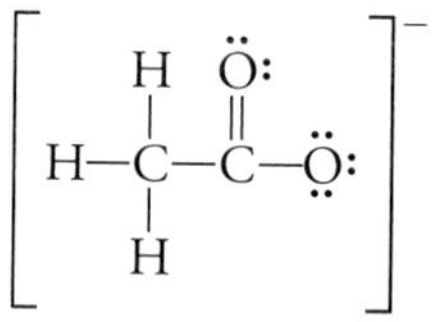

$$\left[H_3C{-}C({=}\ddot{O}{:}){-}\ddot{\underset{..}{O}}{:} \right]^-$$

Strategy

For each atom, subtract the number of electrons that surround the atom from the number of valence electrons for the neutral atom.

Solution

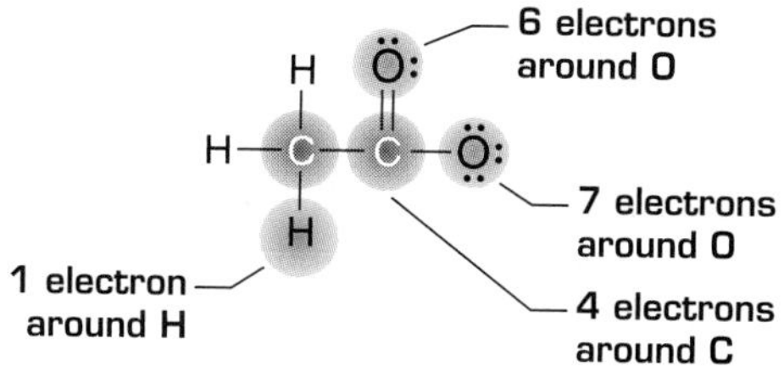

6 electrons around O

7 electrons around O

4 electrons around C

1 electron around H

Each carbon atom sees four electrons and has a formal charge of 0.

$$\text{Formal charge on C} = \frac{\text{number of valence electrons}}{\text{for a neutral atom}} - \frac{\text{number of electrons around}}{\text{the atom in the compound}}$$
$$= 4 - 4 = 0$$

Each hydrogen atom sees one electron and has a formal charge of 0.

$$\text{Formal charge on H} = \frac{\text{number of valence electrons}}{\text{for a neutral atom}} - \frac{\text{number of electrons around}}{\text{the atom in the compound}}$$
$$= 1 - 1 = 0$$

The double-bonded oxygen atom sees six electrons and has a formal charge of 0.

$$\text{Formal charge on O} = \frac{\text{number of valence electrons}}{\text{for a neutral atom}} - \frac{\text{number of electrons around}}{\text{the atom in the compound}}$$
$$= 6 - 6 = 0$$

The remaining oxygen atom sees seven electrons and has a formal charge of 1−.

$$\text{Formal charge on O} = \frac{\text{number of valence electrons}}{\text{for a neutral atom}} - \frac{\text{number of electrons around}}{\text{the atom in the compound}}$$
$$= 6 - 7 = 1-$$

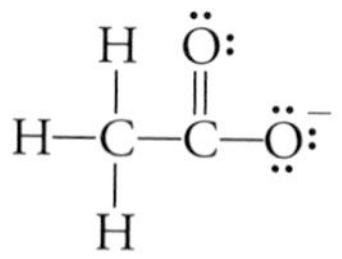

$$H_3C{-}C({=}\ddot{O}{:}){-}\ddot{\underset{..}{O}}{:}^-$$

PRACTICE PROBLEM 4.2

The line-bond structure of one form of glycine, an amino acid present in many proteins, is shown below. Assign formal charges to the atoms in this compound.

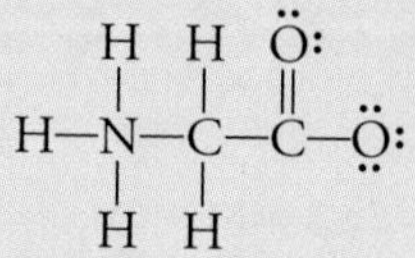

$$H_3N{-}CH_2{-}C({=}\ddot{O}{:}){-}\ddot{\underset{..}{O}}{:}$$

We have just seen how to assign formal charges when given a structural formula. The structural formulas for polyatomic ions can be drawn using the same set of four guidelines that we used to draw line-bond structures. Let us use those guidelines to draw nitrite ion (NO_2^-).

1. ***Count the total number of valence electrons.*** NO_2^- has 18 electrons available. In the case of polyatomic anions, one electron must be added for each negative charge. For polyatomic cations, one electron must be subtracted for each positive charge.

1 N atom	+	2 O atoms	+	1 negative charge
5 valence electrons	+	2 × 6 valence electrons	+	1 electron
5	+	12	+	1 = 18 electrons

2. ***Use single covalent bonds to connect the atoms to one another.*** Nitrogen, which is listed first in the formula, is the central atom.

O—N—O

3. ***Beginning with the atoms attached to the central atom, add the remaining electrons in order to complete octets.*** In the drawing above, 4 of the 18 available electrons have been used. The remaining 14 will be added, beginning with the O atoms and then moving to the N atom.

:Ö—N—Ö:

4. ***If the central atom does not have an octet, move pairs of nonbonding electrons from attached atoms to form multiple bonds with the central atom. Do so until the central atom has an octet.*** The nitrogen atom does not have an octet, so a pair of nonbonding electrons are moved from one of the oxygen atoms (it does not matter which one). This gives the nitrogen atom an octet.

:Ö—N—Ö: **Nitrogen has 6 valence electrons**

:O=N—Ö: **Nitrogen has an octet**

Formal charges must be added to complete the structure of NO_2^-. The formal charge on the nitrogen atom is 0,

$$\text{Formal charge on N} = \frac{\text{number of valence electrons}}{\text{for a neutral atom}} - \frac{\text{number of electrons around}}{\text{the atom in the compound}}$$

$$= 5 - 5 = 0$$

The formal charge on the oxygen atom on the left is 0.

$$\text{Formal charge on O} = \frac{\text{number of valence electrons}}{\text{for a neutral atom}} - \frac{\text{number of electrons around}}{\text{the atom in the compound}}$$

$$= 6 - 0 = 0$$

And the formal charge on oxygen atom on the right is 1-.

$$\text{Formal charge on O} = \frac{\text{number of valence electrons}}{\text{for a neutral atom}} - \frac{\text{number of electrons around}}{\text{the atom in the compound}}$$

$$= 6 - 7 = 1-$$

The final, correct structure is

:O=N—Ö:$^-$

SAMPLE PROBLEM 4.3

Drawing line-bond structures of polyatomic ions

Draw the line-bond structure of NO_2^+, assigning formal charges.

Strategy

As in NO_2^-, nitrogen is the central atom in NO_2^+. When counting available electrons in polyatomic cations, subtract one electron for each positive charge.

Solution

1. A nitrogen atom has 5 valence electrons and each oxygen atom has 6. This gives an initial total of 17 available electrons. Because of the positive charge on the ion, 1 electron is subtracted, giving a total of 16 available electrons for NO_2^+.
2. Nitrogen is the central atom, so the initial structure is

O—N—O

3. Having two covalent bonds, the structure above uses only 4 of the available 16 valence electrons. The remaining 12 are added to the oxygen atoms.

:Ö—N—Ö:

4. To give the nitrogen atom an octet, one pair of nonbonding electrons from each oxygen atom is used to make a multiple bond.

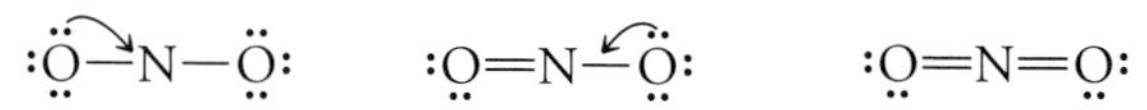

:Ö—N—Ö:	:O=N—Ö:	:O=N=O:
Nitrogen has 4 valence electrons	Nitrogen has 6 valence electrons	Nitrogen has an octet

In this drawing the nitrogen atom has a formal charge of 1+ and the oxygen atoms each have a formal charge of zero. The final structure is

:O=N⁺=O:

PRACTICE PROBLEM 4.3

Draw the line-bond structure of PO_3^{3-}.

Many compounds, especially those encountered in organic and biochemistry, are quite large, and drawing their electron dot or line-bond structural formulas can be a time-consuming task. For this reason, chemists have devised more abbreviated methods for representing structure. A **condensed structural formula** describes the attachment of atoms to one another, without showing all of the bonds. For example, a carbon atom with three attached hydrogen atoms can be written CH_3 and one with two attached hydrogen atoms can be written CH_2. Several examples of condensed structural formulas are shown in Figure 4.3.

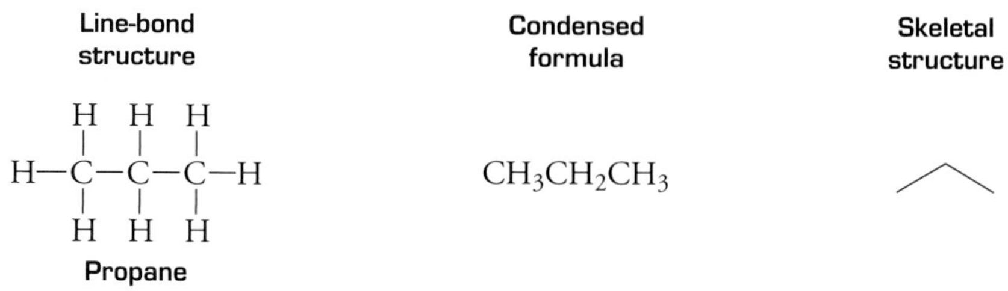

Line-bond structure	Condensed formula	Skeletal structure
H–C(H)(H)–C(H)(H)–C(H)(H)–H Propane	$CH_3CH_2CH_3$	^

FIGURE 4.3

Condensed and skeletal structural formulas
Condensed formulas show the attachment of atoms without always showing covalent bonds and nonbonding electrons. In skeletal structures carbon atoms are not shown and hydrogen atoms appear only when attached to atoms other than carbon.
(continues on the next page)

FIGURE 4.3

(Continued)

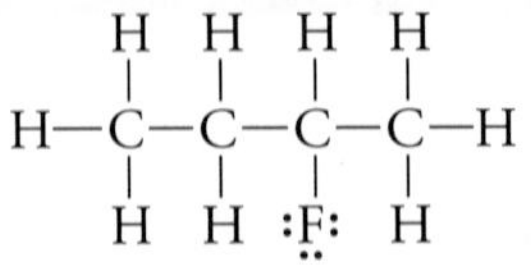

$CH_3CH_2CHCH_3$ (with F on C-2)

2-Fluorobutane

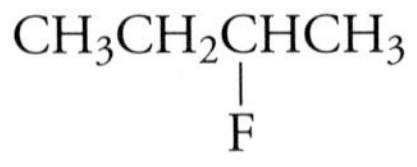

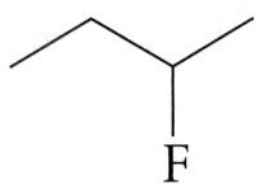

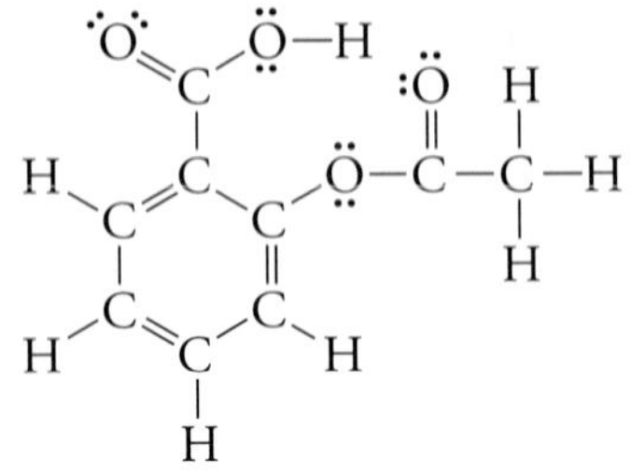

Aspirin

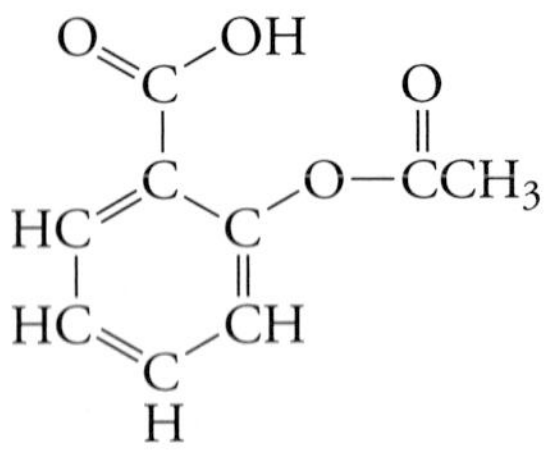

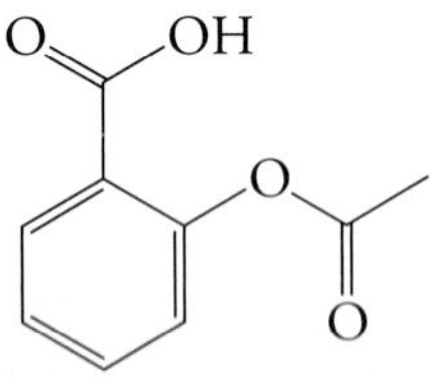

In **skeletal structures** (Figure 4.3), covalent bonds are represented by lines, carbon atoms are not shown, and hydrogen atoms are drawn only when attached to atoms other than carbon. To read a skeletal structure, you assume that a carbon atom appears where lines (bonds) meet and at the end of each line. To simplify matters, nonbonding electrons (Section 3.5) are sometimes omitted from skeletal and other structural formulas.

SAMPLE PROBLEM 4.4

Drawing condensed and skeletal structures

Draw condensed and skeletal structures of diethyl ether, a compound once used as a general anesthetic.

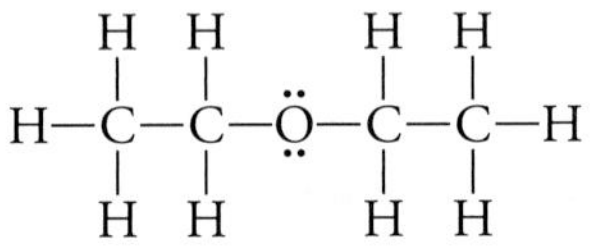

Strategy

To write the condensed formula, begin by thinking of the molecule as a chain of five atoms (C—C—O—C—C) and then add in the atoms attached to these five (CH_3, etc.). The skeletal structure of diethyl ether is drawn by leaving out all of the carbon and hydrogen atoms, showing only the oxygen atom and the C—C and C—O bonds.

Solution

$CH_3CH_2OCH_2CH_3$

Condensed structure **Skeletal structure**

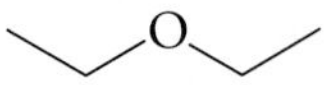

PRACTICE PROBLEM 4.4

Research suggests that drinking green tea may help boost the immune system. Ethylamine (below), produced when one of the compounds in green tea is broken down in the liver, may be responsible for this immune response. Draw condensed and skeletal structures for the molecule.

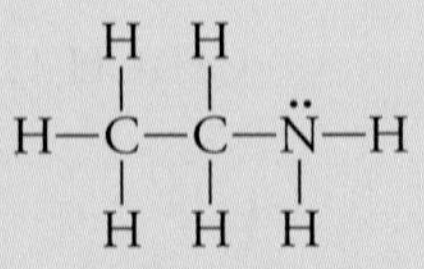

4.2 POLAR COVALENT BONDS, SHAPE, AND POLARITY

We have described a covalent bond as consisting of a shared pair of valence electrons. This is not the complete story, however, because the electrons in a covalent bond are not always shared equally between the two atoms. An unequal sharing of electrons is attributed to differences in **electronegativity**, *the ability of an atom to attract bonding electrons.* As shown in Figure 4.4, electronegativity displays a periodic trend—moving to the right across a period or up a group, electronegativity generally increases, with fluorine atoms being the most electronegative.

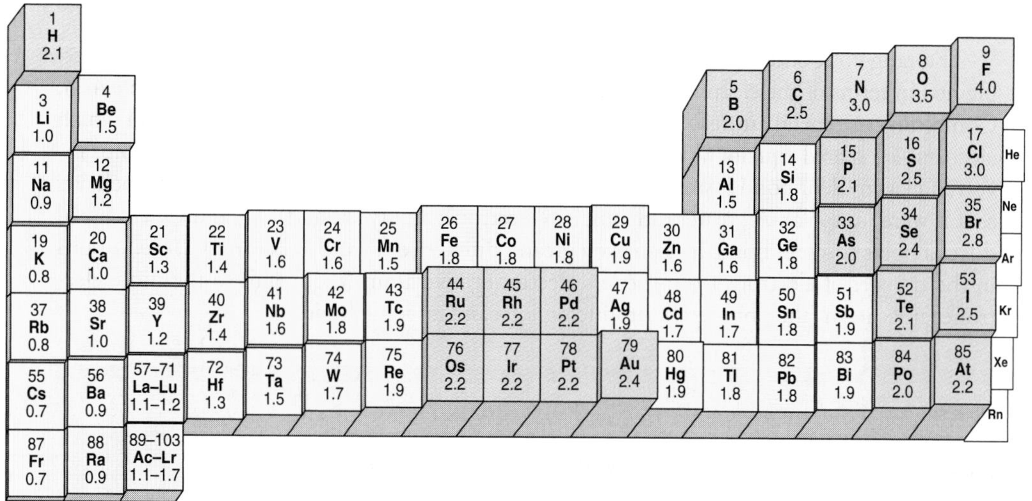

FIGURE 4.4

Electronegativity
Electronegativity reflects the ability of an atom to attract bonding electrons. Fluorine (F) is the most electronegative atom, and cesium (Cs) and francium (Fr) are the least electronegative. Inert gases do not attract bonding electrons and are not assigned an electronegativity value.

When the electrons in covalent bonds are shared by atoms with different electronegativities, a **polar covalent bond** results. In a polar covalent bond, *the unequal sharing of electrons gives the bond a partially positive end and a partially negative end (pole).* Fluorine is much more electronegative than hydrogen, so when a fluorine atom and a hydrogen atom combine to form a covalent bond, the fluorine atom pulls the shared pair of bonding electrons toward itself. As a result, the fluorine atom carries a partial negative charge, represented by δ− (delta minus), the hydrogen atom carries a partial positive charge (δ+), and the bond is polar (Figure 4.5).

Any covalent bond that forms between atoms of different electronegativity is polar covalent to some extent, although some levels of polarity are so small that the bonds behave as if they were nonpolar (not polar). For the compounds that we are most likely to encounter in organic and biochemistry, the important polar covalent bonds are those in which *either hydrogen or carbon atoms are covalently attached to nitrogen, oxygen, fluorine, or chlorine atoms.*

■ Bonds between carbon or hydrogen atoms and nitrogen, oxygen, fluorine, or chlorine atoms are polar covalent.

```
δ+  δ−
H—F:

δ+  δ−  δ+
H—O—H

   H
   |δ+ δ−
H—C—Cl:
   |
   H

   H
   |δ+ δ−  δ+
H—C—N—H
   |   |
   H   H δ+
```

FIGURE 4.5

Polar covalent bonds
Electrons are shared unequally in polar covalent bonds, which sets up partial positive (δ+) and partial negative (δ−) charges.

SAMPLE PROBLEM 4.5

Identifying polar covalent bonds

The compound below has been used to kill any insect larvae present in cereal and dried fruit. Label the polar covalent bonds, using the symbols δ+ and δ−.

```
   :O:     H
   ||      |
H—C—N—C—H
       |   |
       H   H
```

Strategy

We have defined polar covalent bonds as those between H or C atoms and N, O, F, or Cl atoms. In a polar covalent bond, the δ+ belongs with the less electronegative atom (H or C) and the δ− with the more electronegative one (N, O, F, or Cl).

Solution

```
          δ−
          :Ö        H
          ||δ+ δ−  δ+|
       H—C——N̈——C—H
           δ+|     |
             H     H
```

PRACTICE PROBLEM 4.5

Label the polar covalent bonds in aspirin (see Figure 4.3), using the symbols δ+ and δ−.

The range of bond types that we have seen in this and the previous chapter can be explained in terms of the electronegativity differences between the atoms involved (Table 4.1). Combining nonmetal atoms of identical electronegativity produces covalent bonds in which electrons are shared equally. When the electronegativity difference between the two atoms involved in a covalent bond is small (C and H, etc.), the bond is only slightly polar and behaves as if it were nonpolar. A greater difference in electronegativity (H and F, C and O, etc.) produces a polar covalent bond, and an even greater difference (Li and F, Na and Cl, etc.) results in the transfer of electrons from the less electronegative atom (the metal) to the more electronegative atom (the nonmetal), producing ions and an ionic bond.

TABLE 4.1 BOND TYPES

Bond	Ionic	Polar Covalent	Covalent
Characteristics	Attraction of opposite charges	Unequal sharing of an electron pair	Equal sharing of an electron pair
Example	Na^+ Cl^-	$^{\delta+}C—O^{\delta-}$	C—C
Electronegativity difference	Very large	Large	None

Electron dot, line-bond, condensed, and skeletal structures are designed to show how the atoms in a compound are attached to one another, but are not intended to necessarily represent its actual shape. Shape can be predicted, however, based on these structural formulas.

Consider the way that four groups of electrons arrange themselves around an atom. The carbon atom in methane (Table 4.2) has four single bonds, one to each of the four hydrogen atoms. The best way for these four groups of electrons to be as far apart from one another as

TABLE 4.2 COMMON MOLECULAR SHAPES

Number of Groups of Electrons[a]	Shape[b] (Bond Angles)	Example	Line-Bond Structure	Three-dimensional View[c]
4	Tetrahedral (110°)	Methane		

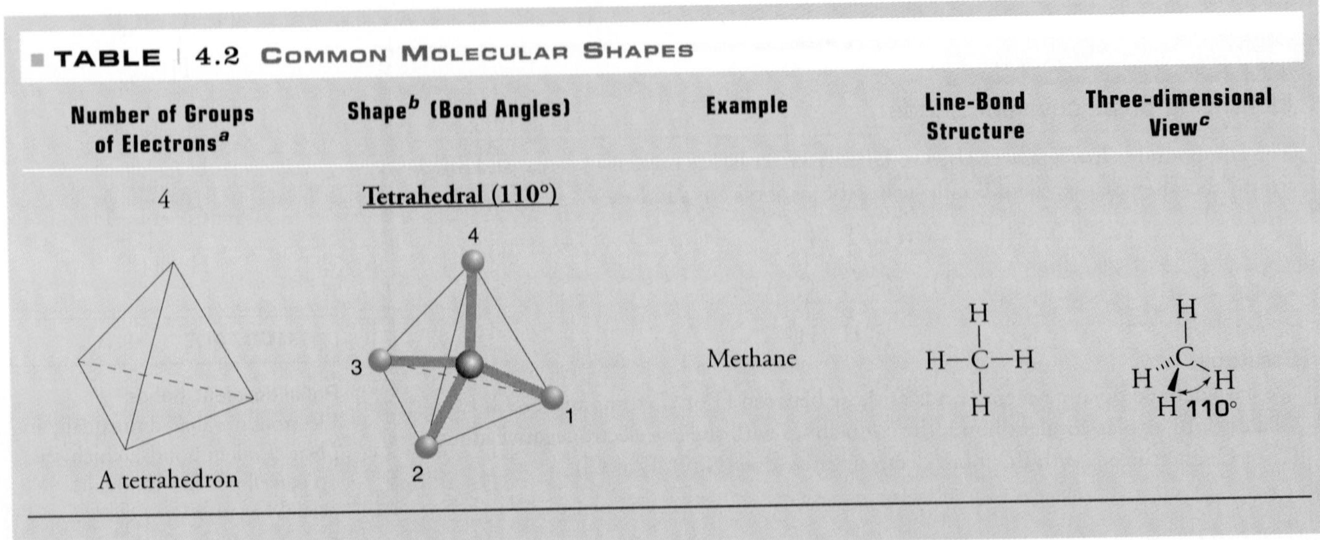

Number of Groups of Electrons[a]	Shape[b] (Bond Angles)	Example	Line-Bond Structure	Three-dimensional View[c]
	Pyramidal (110°) (nonbonding electron pair) 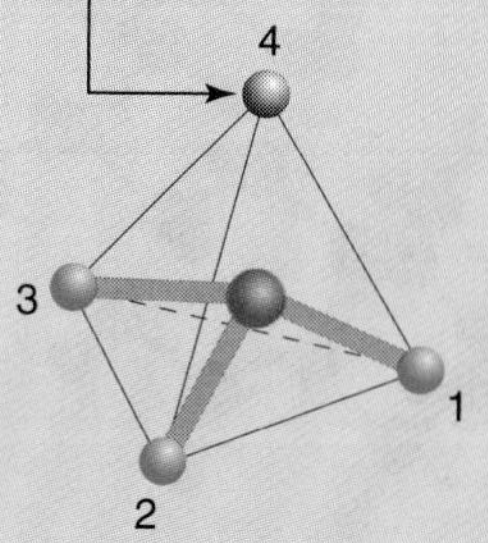	Ammonia	H—N̈—N | H	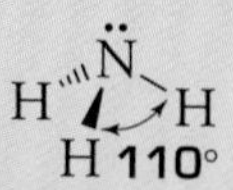 H 110°
	Bent (110°) (nonbonding electron pairs) 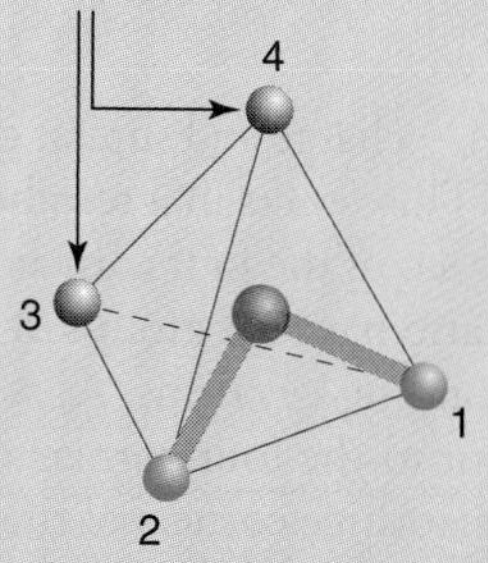	Water	H—Ö—H	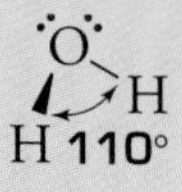 H 110°
3 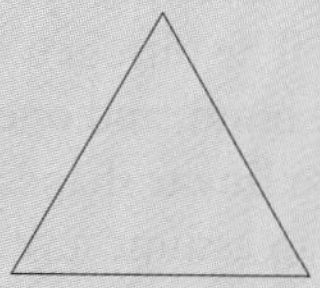 A triangle	**Trigonal planar (120°)** 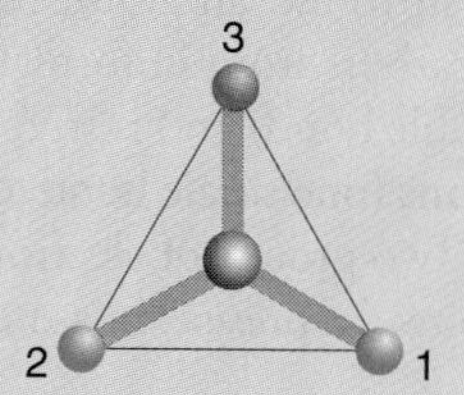	Formaldehyde	Ö: ‖ H—C—H	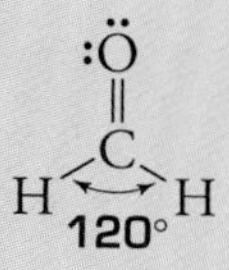 :Ö ‖ C H 120° H
	Bent (120°) (nonbonding electron pair) 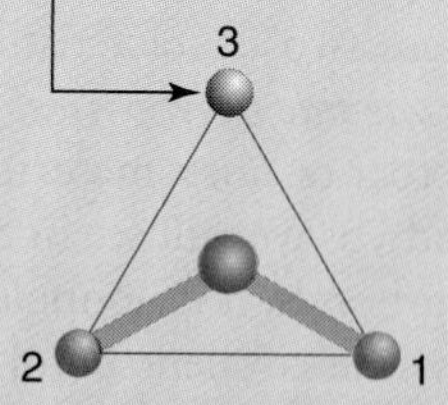	Nitrite ion	[Ö=N̈—Ö:]⁻	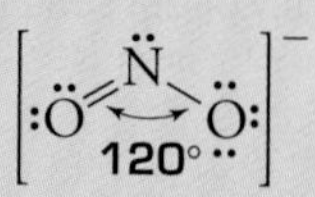 [:Ö=N̈—Ö:]⁻ 120°
2  A line	**Linear (180°)** 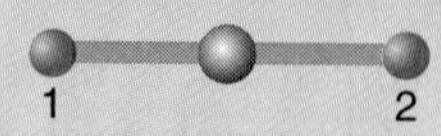	Carbon dioxide	:Ö=C=Ö:	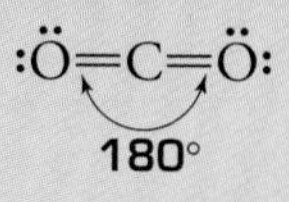 :Ö=C=Ö: 180°

[a]The number of groups of electrons around an atom is the sum of the attached atoms (the single, double, or triple bond to an atom counts as one group of electrons) and attached pairs of nonbonding electrons.

[b]Molecular shape is based on the relative placement of atoms. Nonbonding electron pairs are ignored.

[c]Wedge and dashed line notation can be used to show three-dimensional shapes. A solid wedge points out of the plane of the paper and a dashed wedge points in to the paper.

FIGURE | 4.6

Complex shapes
Large molecules, like cholesterol, can assume complex shapes.

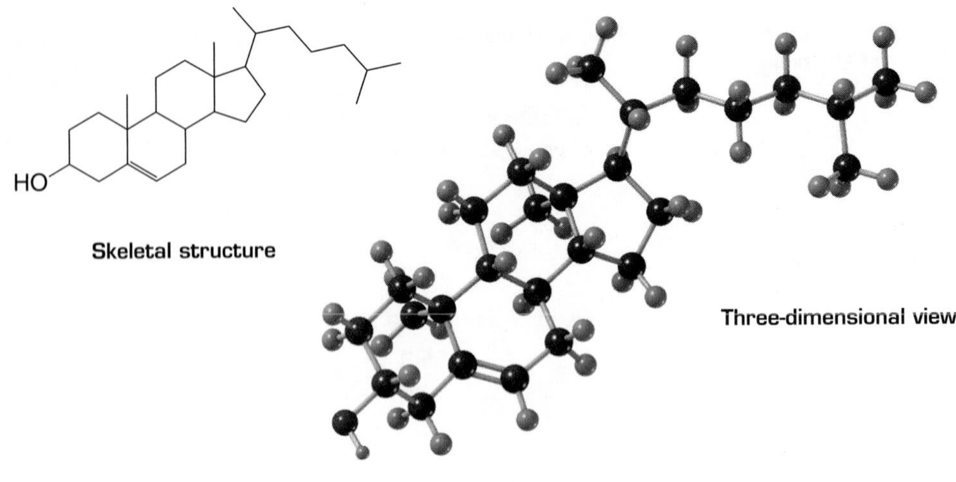

possible (electrons all have the same 1− charge and similar charges repel) is for them to point to the four corners of a tetrahedron. This results in a **tetrahedral** shape, in which the hydrogen atoms of methane are placed at each of the four corners of the same tetrahedron.

In ammonia, the nitrogen atom also has four groups of electrons surrounding it—three single bonds and one pair of nonbonding electrons. As with methane, the four groups of electrons arrange themselves to point to the four corners of a tetrahedron. With the three hydrogen atoms of ammonia sitting at three corners of the tetrahedron, the molecule has a **pyramidal** shape. Using a similar argument, the oxygen atom in water has four groups of electrons (two single bonds and two nonbonding pairs) and the molecule has a **bent** shape.

When three groups of electrons surround an atom, the two common shapes observed are **trigonal planar** and bent. The carbon atom in formaldehyde has three groups of electrons (two single bonds and one double bond) and is trigonal planar. The nitrogen atom in nitrite ion has three groups of electrons (one single bond, one double bond, and one pair of nonbonding electrons) and is bent. Two groups of electrons assume a **linear** shape, as seen for carbon dioxide. When many atoms are present in a molecule it can assume a very complex shape (Figure 4.6).

Polarity

Now that we have seen how to identify polar covalent bonds and to determine the shape about an atom, we can see how these two characteristics determine if a molecule, as a whole, is polar. In a **polar** molecule, *one side has a partial positive charge and the other has a partial negative charge.* Knowing whether or not a molecule is polar helps us to understand how it interacts with other compounds, as we will see in Section 4.3.

■ In a polar molecule, one side is partially positive and the other is partially negative.

If a molecule has no polar covalent bonds, it is nonpolar. The molecule F_2, for example, is nonpolar because it contains only one covalent bond—a nonpolar one (Figure 4.7*a*). Although the carbon and hydrogen atoms in ethane (CH_3CH_3) have a slightly different electronegativity, the C—H bonds behave as if they are nonpolar. For this reason ethane molecules are also nonpolar (Figure 4.7*b*).

For a molecule to be polar, it must contain one or more polar covalent bonds. In methylene chloride (CH_2Cl_2) each of the C—Cl bonds is polar covalent. The molecule has a tetrahedral shape, so the partial negative charges are focused on one side of the molecule and the partial positive charge on the other; thus the molecule is polar (Figure 4.7*c*). For polar molecules, an alternative to showing the partial charges ($\delta+$ and $\delta-$) for each individual polar covalent bond is to use an arrow that represents the overall contribution of each. This arrow points toward the partially negative end of the molecule. The other end of the arrow, which looks like a plus sign, is directed toward the partially positive end.

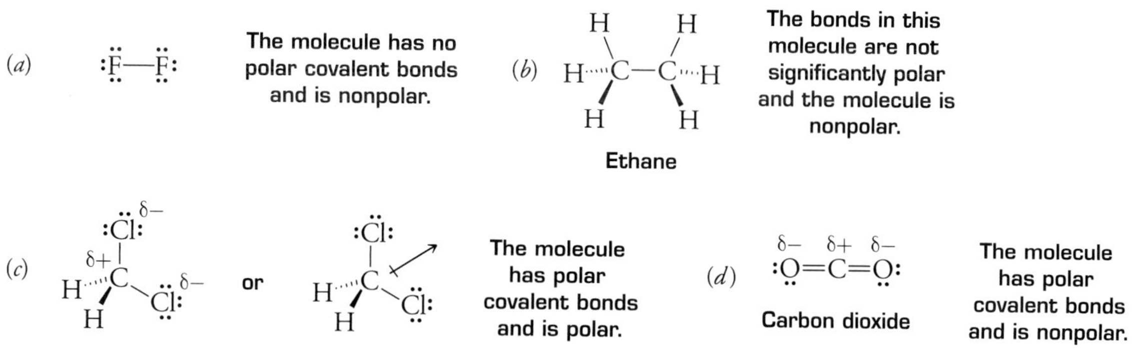

(*a*) :F—F: The molecule has no polar covalent bonds and is nonpolar.

(*b*) Ethane. The bonds in this molecule are not significantly polar and the molecule is nonpolar.

(*c*) or The molecule has polar covalent bonds and is polar.

(*d*) δ− δ+ δ− :O=C=O: Carbon dioxide. The molecule has polar covalent bonds and is nonpolar.

FIGURE 4.7

Polarity

Molecules that contain only nonpolar bonds are nonpolar. Depending on their shape, molecules that contain one or more polar covalent bonds may be polar or nonpolar.

Not all molecules with polar covalent bonds are polar, however. Carbon dioxide (CO_2) contains two polar covalent C=O bonds but is nonpolar because the linear shape of the CO_2 molecule causes the two polar covalent bonds to oppose and cancel one another (Figure 4.7*d*). Although each end of the molecule has a partial negative charge, one end is no more negative than the other. This shows why it is important to consider both the polarity of covalent bonds and the molecular shape when predicting the overall polarity of a molecule.

SAMPLE PROBLEM 4.6

Identifying polar molecules

Chloroform ($CHCl_3$) was once used as a general anesthetic. Its use for this purpose has been discontinued due to toxicity concerns. Is a chloroform molecule polar? (Hint: Carbon is the central atom.)

Strategy

To solve this problem you must first look for polar covalent bonds—for a molecule to be polar it must have at least one polar covalent bond. Then determine its shape—for a molecule to be polar, the shape must not cancel the contributions of the polar covalent bonds.

Solution

Yes. The molecule has a tetrahedral shape, and the three polar covalent bonds are arranged such that the partial positive and negative charges are on opposite sides of the molecule.

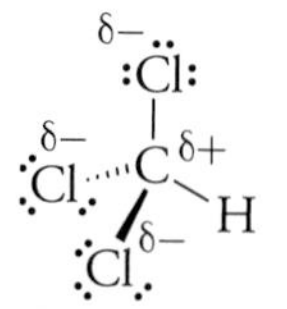

PRACTICE PROBLEM 4.6

Which of the following molecules are polar? Each is drawn showing its actual three-dimensional shape.

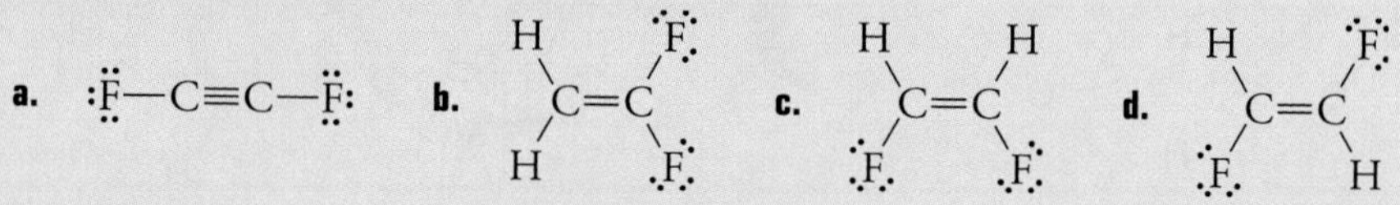

a. :F—C≡C—F: **b.** **c.** **d.**

4.3 NONCOVALENT INTERACTIONS

Up to this point in the chapter we have seen that atoms involved in polar covalent bonds carry a partial charge and that atoms can be assigned formal charges. We have also seen that, while the presence of polar covalent bonds is a requirement for a molecule to be polar, not all molecules that contain polar covalent bonds are polar; the shape of the molecule is a deciding factor. In this section we will see how all of these properties help determine the forces that attract molecules or ions to one another, which allows us to understand their properties (why some vitamins are water soluble and others are not, why some drugs remain in the body for months, etc.).

When neighboring molecules or ions (or remote parts of the same molecule or ion) interact with one another, they do so through **noncovalent interactions**—*interactions that do not involve the sharing of valence electrons.* Noncovalent interactions can be divided into two broad categories, those due to the attraction of permanent charges (hydrogen bonds, salt bridges, dipole–dipole interactions, ion–dipole interactions, and coordinate-covalent bonds) and those due to the attraction of temporary partial charges (London forces).

Noncovalent Interactions Due to Permanent Charges

■ In a hydrogen bond, an H atom attached to N, O, or F is attracted to a different N, O, or F.

A **hydrogen bond** is the interaction of a nitrogen, oxygen, or fluorine atom with a hydrogen atom that is covalently bonded to a different nitrogen, oxygen, or fluorine atom (Figure 4.8*a*).

When a hydrogen atom is covalently bonded to a highly electronegative atom (N, O, or F), a polar covalent bond exists and the hydrogen atom carries a partial positive charge (Section 4.2). A hydrogen bond is the interaction of this partially charged hydrogen atom with a pair of electrons on a different highly electronegative atom (N, O, or F). While

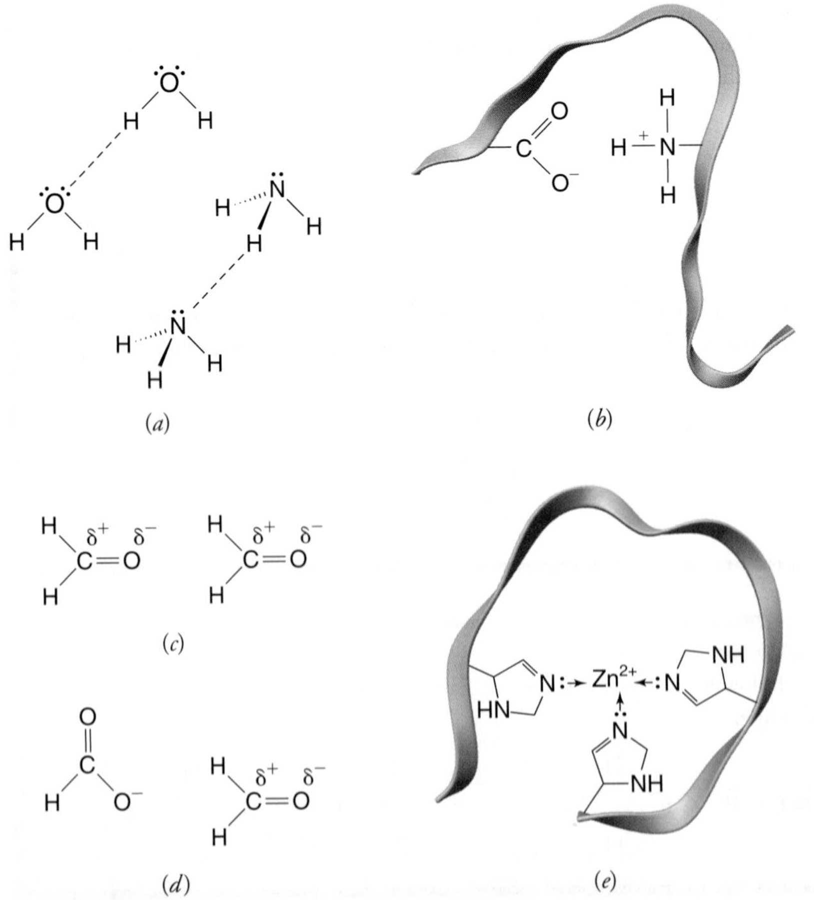

FIGURE 4.8

Noncovalent interactions and permanent charges
(*a*) Hydrogen bonds (dashed lines) form when a hydrogen atom, covalently bonded to an N, O, or F atom, is attracted to a different N, O, or F atom. (*b*) Salt bridges (ionic bonds) help proteins to maintain their correct three-dimensional shape. (*c*) Dipole–dipole forces, which are due to the attraction of opposite partial charges, hold neighboring polar groups to one another. (*d*) Ion–dipole interactions occur when an ion is attracted to the partially (and oppositely) charged end of a polar group. (*e*) The activity of many enzymes requires that a trace element be attached to the enzyme molecule. Here, Zn^{2+} is held in place by coordinate-covalent bonds (represented by arrows).

hydrogen bonds vary in strength, the average hydrogen bond is about ten times weaker than a typical covalent bond. As we will see in later chapters, hydrogen bonding makes an important contribution to the structure and function of proteins, nucleic acids, and other biologically important compounds.

A **salt bridge** is another name for an ionic bond. The term is used by biochemists to describe ionic bonds that form between charged groups in protein molecules (Figure 4.8*b*).

Dipole–dipole forces are the attraction of neighboring polar groups for one another. When polar compounds interact they do so by orienting themselves with the partially negative end of one pointing to the partially positive end of the other (Figure 4.8*c*).

Ion–dipole interactions occur between ions and atoms with a partial charge, as shown in Figure 4.8*d*. In a related interaction called a **coordinate-covalent bond**, the nonbonding electrons of a nonmetal atom (N, S, etc.) associate with metal cations (Fe^{3+}, Zn^{2+}, etc.). Many enzymes function properly only when a specific ion is held in place by a coordinate-covalent bond (Figure 4.8*e*).

Noncovalent Interactions Due to Temporary Partial Charges

Individual atoms, nonpolar molecules, or nonpolar parts of molecules or ions are attracted to one another by **London forces**. Named after the German physicist Fritz London, this force is the result of the continuous motion of electrons. Electron movement within a substance creates a temporary dipole, which tends to produce temporary dipoles on its neighbors as electrons in the neighbor become attracted to or repelled by the temporary charges. As the temporary dipole shifts, so do the dipoles on neighbors (Figure 4.9*a*), so London forces are a fleeting attraction between temporary dipoles.

The strength of London forces depends on the surface area of the substance involved. Greater surface area leads to stronger London force attraction (Figure 4.9*b*).

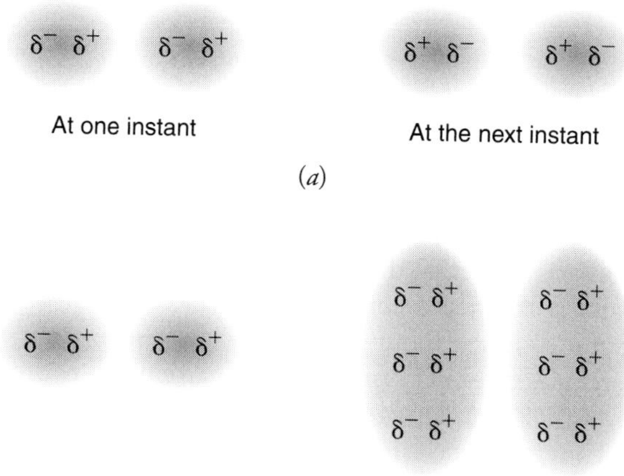

δ⁻ δ⁺ δ⁻ δ⁺
At one instant
δ⁺ δ⁻ δ⁺ δ⁻
At the next instant
(a)
A weak London force exists between small molecules.
A stronger London force exists between larger molecules.
(b)

FIGURE 4.9

Noncovalent interactions and temporary charges

London forces are attractive forces that result from temporary dipoles. (*a*) The movement of electrons within an atom, nonpolar molecule, or nonpolar parts of a compound creates a momentary dipole, which attracts or repels electrons in a neighbor to create another temporary dipole and a fleeting attraction between the oppositely charged ends of each. At the next instant, the electrons may have shifted to create an entirely new set of dipoles. (*b*) London forces depend on surface area. Small, spherical molecules have a minimal area of surface contact, so they are attracted to one another by relatively weak London forces. Large, nonspherical molecules have greater surface contact, and therefore, a stronger London force attraction for one another.

SAMPLE PROBLEM 4.7

Noncovalent interactions

Which noncovalent interactions can take place between the molecule shown below and another identical to it?

$$CH_3CH_2CH_2CH_2CH_2CH_2\overset{\overset{\Large \ddot{O}:}{\|}}{C}CH_3$$

Strategy

Possible answers include: hydrogen bonding, salt bridges, dipole–dipole forces, ion–dipole interactions, coordinate–covalent bonds, and London forces. Refer to Figures 4.8 and 4.9 for help deciding which noncovalent interactions can occur and which cannot.

Solution

The answers are dipole–dipole forces (the C=O bond is polar) and London forces. Hydrogen bonding is not possible because the molecule does not have a hydrogen atom covalently bonded to an N, O, or F atom. Salt bridges, ion-dipole interactions, and coordinate–covalent bonds do not occur because no ion is present.

PRACTICE PROBLEM 4.7

The molecule in Sample Problem 4.7 cannot form hydrogen bonds with another identical to it. Can the molecule hydrogen bond with a water molecule? Explain.

4.4 ALKANES

- Alkanes are hydrocarbons that contain only single bonds.
- Hydrocarbons contain only carbon and hydrogen atoms.

For the remainder of this chapter, we will turn our attention to organic compounds. As part of our discussion, we will revisit the topics presented above: structural formulas, polar covalent bonds, shape, polarity, and noncovalent interactions.

The first family of organic compounds that we will consider is the **alkanes**. The members of this family are molecules that consist only of carbon and hydrogen atoms and contain only single bonds. Examples include methane (CH_4), ethane (CH_3CH_3), and propane ($CH_3CH_2CH_3$) (Figure 4.10). The alkanes in this figure are ***normal*** alkanes, which means that *their carbon chains are unbranched* (are joined in one continuous line).

Alkanes belong to a larger grouping of organic compounds called **hydrocarbons**. While, as the name implies, all hydrocarbons contain only carbon and hydrogen atoms, they differ in the types of covalent bonds that they contain (Table 4.3). Later in this chapter the three other families of hydrocarbons—alkenes, alkynes, and aromatic compounds—will be introduced.

Each carbon atom in an alkane molecule has four covalent bonds that are arranged in a tetrahedral shape about the atom. The molecules in Figure 4.10 are represented using wedge and dashed line notation (see the footnote to Table 4.2 for an explanation).

Alkanes contain only nonpolar covalent bonds, are nonpolar molecules, and are attracted to one another by London forces. Table 4.4 shows that the more carbon atoms in a *normal* alkane, the higher its boiling point. The reason for this is that the longer the alkane, the greater its surface area and the stronger the London force that holds it to other molecules. Boiling points increase with strengthened intermolecular forces, because this

Name	Condensed structural formula	Three-dimensional structure
Methane	CH_4	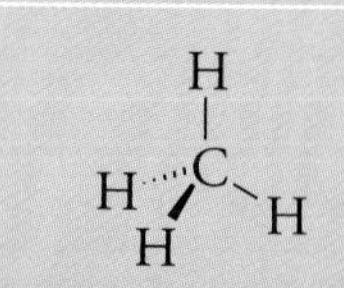
Ethane	CH_3CH_3	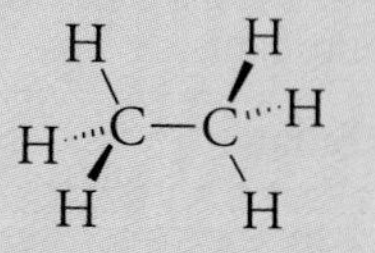
Propane	$CH_3CH_2CH_3$	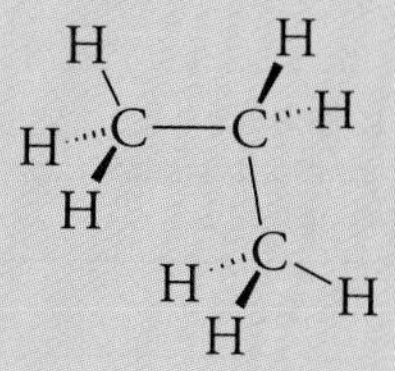

FIGURE 4.10 Alkanes
Methane, ethane, and propane are alkanes. A condensed structural formula and three-dimensional view of each are shown.

TABLE 4.3 The Hydrocarbons

Family	Key Feature	Class	Example	Name
Alkanes	Atoms joined by single bonds only	Saturated	$CH_3CH_2CH_3$	Propane
Alkenes	At least one carbon–carbon double bond	Unsaturated	$CH_2{=}CHCH_3$	Propene
Alkynes	At least one carbon–carbon triple bond	Unsaturated	$HC{\equiv}CCH_3$	Propyne
Aromatic compounds	Contains a ring of alternating single and double bonds	Unsaturated	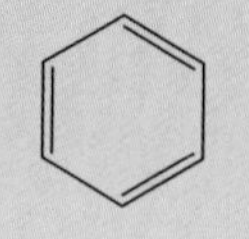 or 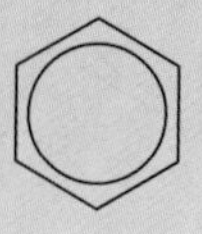	Benzene

physical change involves pulling molecules in the liquid phase apart from one another and moving them into the gas phase. An increase in London forces is also associated with a rise in melting point, because more energy (a higher temperature) is required to loosen the attraction between molecules in the solid phase. (More details about the transitions between solid, liquid, and gas phases will be presented in Chapter 5.)

The boiling points of the first four *normal* alkanes, methane through butane, are lower than room temperature (about 25°C), so at this temperature each exists as a gas. Larger *normal* alkanes, pentane through heptadecane (17 carbon atoms), are liquids at room temperature, while those larger than heptadecane are solids.

TABLE 4.4 Structure, Name, and Properties of Selected Hydrocarbons

Structural Formula	IUPAC Name	Common Name	Boiling Point (°C)
Alkanes			
CH_4	Methane		−164
CH_3CH_3	Ethane		−89
$CH_3CH_2CH_3$	Propane		−42
$CH_3CH_2CH_2CH_3$	Butane		0
$CH_3CH_2CH_2CH_2CH_3$	Pentane		36
Alkenes			
$CH_2{=}CH_2$	Ethene	Ethylene	−102
$CH_2{=}CHCH_3$	Propene	Propylene	−98
$CH_2{=}CHCH_2CH_3$	1-Butene		−6.5
$CH_2{=}CHCH_2CH_2CH_3$	1-Pentene		30
Alkynes			
$HC{\equiv}CH$	Ethyne	Acetylene	−75
$HC{\equiv}CCH_3$	Propyne		−23
$HC{\equiv}CCH_2CH_3$	1-Butyne		9
$HC{\equiv}CCH_2CH_2CH_3$	1-Pentyne		40
Aromatic compounds			
	Benzene		80
	Naphthalene		218
	Anthracene		340

In Chapter 3 we saw that binary molecules are named by listing the number of each element that is present in the formula (CO_2 is carbon dioxide, N_2O_5 is dinitrogen pentoxide, and so on). Alkanes cannot be named this way because, in most cases, a particular molecular formula refers to more than one molecular structure. (Section 4.5 will expand on this idea.) For this reason, the names of alkanes and other organic molecules are based on structure, rather than on molecular formula.

IUPAC rules, devised by the International Union of Pure and Applied Chemistry, are a widely used method of naming organic compounds. Using the IUPAC rules to name an alkane involves identifying the **parent chain** (*the longest continuous chain of carbon atoms in the molecule*) and **substituents** (*atoms or groups of atoms attached to the parent chain*). In alkanes the substituents, called **alkyl groups**, *are constructed solely of carbon and hydrogen atoms.* The paragraphs that follow will briefly introduce the use of the IUPAC rules for naming alkanes.

IUPAC Rules for Naming Alkanes

1. ***Name the parent chain.*** The parent chain is named by combining a numbering prefix (Table 4.5), which specifies the number of carbon atoms in the parent, with "ane," which identifies the molecule as an alkane. In Table 4.4, for example, propane gets its name from the fact that its parent chain has three carbon atoms ("prop") and is an alkane ("ane"). Similarly, the parent chain of butane has four carbon atoms and that of pentane has five.
2. ***Name any alkyl groups attached to the parent chain.*** Table 4.6 gives the names and structural formulas for the most commonly encountered alkyl groups, some branched and others unbranched. Note that the same numbering prefixes used to name parent chains are used to indicate the total number of carbon atoms in an alkyl group. The simplest alkyl groups are the methyl group, which has one carbon atom, and the ethyl group, which has two. A three-carbon substituent can be attached by an end carbon atom (propyl group) or by the middle carbon atom (isopropyl group). It may help to think of the "iso" prefix as referring to an alkyl group that branches into two methyl ($—CH_3$) groups.

   The alkyl groups with four carbon atoms are butyl, isobutyl, *sec*-butyl, and *tert*-butyl. A butyl group consists of an unbranched chain of four carbon atoms and an isobutyl group is a chain that branches into two methyl groups. A *sec*-butyl group has a chain of carbon atoms that branches into a methyl group and an ethyl group ($—CH_2CH_3$). In a *tert*-butyl group, the carbon atom attached directly to the parent chain branches into three methyl groups.
3. ***Determine the point of attachment of alkyl groups to the parent chain.*** The parent chain is numbered from the end nearer the first alkyl group. The molecule in Figure 4.11*a* is numbered from the right end of the parent, the end nearer the alkyl group, which places the methyl group on carbon 3. Numbering from the left would place the methyl group on carbon 4.

TABLE 4.5 The First Ten Numbering Prefixes for IUPAC Naming

Number of Carbon Atoms	Prefix
1	Meth
2	Eth
3	Prop
4	But
5	Pent
6	Hex
7	Hept
8	Oct
9	Non
10	Dec

TABLE 4.6 Formulas and Names of Alkyl Groups

Formula[a]	Name
$—CH_3$	Methyl
$—CH_2CH_3$	Ethyl
$—CH_2CH_2CH_3$	Propyl
$—CHCH_3$ with CH_3 branching from the CH	Isopropyl
$—CH_2CH_2CH_2CH_3$	Butyl
$—CH_2CHCH_3$ with CH_3 branching from the CH	Isobutyl
$—CHCH_2CH_3$ with CH_3 branching from the CH	*sec*-Butyl or *s*-butyl
$—C—CH_3$ with CH_3 above and CH_3 below the C	*tert*-Butyl or *t*-butyl

[a] The bond drawn to the left of each formula is the point of attachment to the parent chain.

FIGURE | 4.11

IUPAC names of alkanes

(*a*) The parent chain is numbered from the end nearer the first substituent. (*b* and *c*) Substituents are listed in alphabetical order (ignoring italicized letters), along with a number indicating their position of attachment to the parent chain. (*d*) If a substituent appears more than once, labels such as di, tri, and tetra are added to the substituent name. (*e*) The labels di, tri, and tetra are not used in alphabetizing.

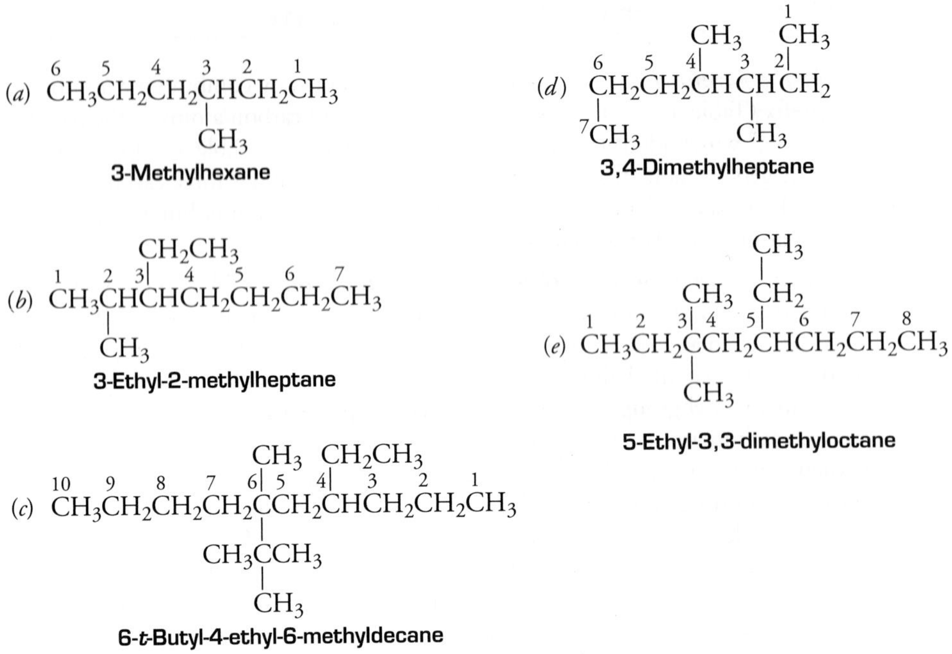

```
     6   5   4   3  2   1
(a) CH3CH2CH2CHCH2CH3
                |
               CH3
```
3-Methylhexane

```
                          1
             CH3     CH3
      6   5   4|   3  2|
(d)  CH2CH2CH CHCH2
      |           |
     7CH3        CH3
```
3,4-Dimethylheptane

```
         CH2CH3
      1  2 3|  4   5   6   7
(b) CH3CHCHCH2CH2CH2CH3
         |
        CH3
```
3-Ethyl-2-methylheptane

```
                  CH3
                   |
              CH3 CH2
      1   2   3| 4  5|  6   7   8
(e) CH3CH2CCH2CHCH2CH2CH3
              |
             CH3
```
5-Ethyl-3,3-dimethyloctane

```
                 CH3  CH2CH3
     10  9   8   7  6| 5   4|  3   2   1
(c) CH3CH2CH2CH2CCH2CHCH2CH2CH3
                   |
               CH3CCH3
                   |
                  CH3
```
6-*t*-Butyl-4-ethyl-6-methyldecane

4. ***Construct the name of the alkane by placing the alkyl groups in alphabetical order and specifying their position numbers, followed by the name of the parent chain. The labels di, tri, tetra, etc., are added if two or more identical substituents are present.*** The molecule in Figure 4.11*a* is named 3-methylhexane. Ethyl takes priority over methyl, so the molecule in Figure 4.11*b* is called 3-ethyl-2-methylheptane, not 2-methyl-3-ethylheptane. When alphabetizing substituents, italicized letters are not considered—*t*-butyl is listed before ethyl and methyl (Figure 4.11*c*). The molecule in Figure 4.11*d*, which has methyl groups on carbons 3 and 4, is called 3,4-dimethylheptane, not 3-methyl-4-methylheptane. A separate number must be used each time that a substituent appears, so the molecule in Figure 4.11*e* is called 5-ethyl-3,3-dimethyloctane. The prefixes di, tri, tetra, etc., are ignored when alphabetizing the names of substituents.

As seen in the names just given, the IUPAC rules call for separating numbers from numbers by using a comma, and numbers from letters by using a hyphen.

SAMPLE PROBLEM 4.8

Using the IUPAC rules

Give the correct IUPAC name for the molecule.

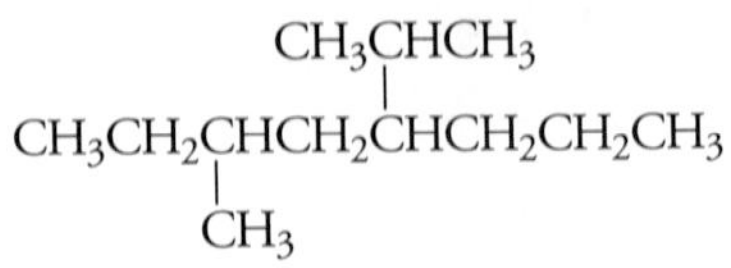

```
            CH3CHCH3
                |
CH3CH2CHCH2CHCH2CH2CH3
      |
     CH3
```

Strategy

The best way to approach the naming of this molecule is to apply the IUPAC rules step by step, as described above. Find and name the parent chain and then assign a name and position to each substituent.

Solution

5-isopropyl-3-methyloctane

The first step in determining the IUPAC name involves finding the parent chain. Here, the parent has eight carbon atoms, which makes it "octane." One methyl group and one isopropyl group are attached to the parent, which is numbered from the left side, nearer the first alkyl group. In the name 5-isopropyl-3-methyloctane, the alkyl groups are arranged alphabetically (isopropyl before methyl).

$$\overset{1}{CH_3}\overset{2}{CH_2}\overset{3}{\underset{\underset{CH_3}{|}}{C}H}\overset{4}{CH_2}\overset{\overset{CH_3CHCH_3}{5|}}{C}H\overset{6}{CH_2}\overset{7}{CH_2}\overset{8}{CH_3}$$

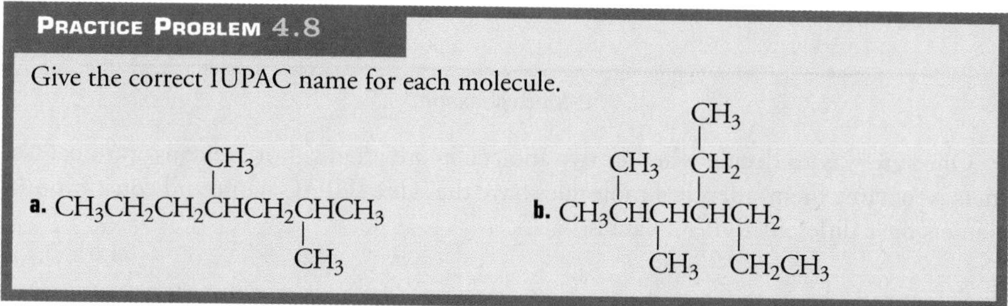

PRACTICE PROBLEM 4.8

Give the correct IUPAC name for each molecule.

4.5 CONSTITUTIONAL ISOMERS

As has already been mentioned in this chapter, different molecules can share the same molecular formula. This is the case for butane and 2-methylpropane, each of which has the molecular formula C_4H_{10}.

$$CH_3CH_2CH_2CH_3 \qquad CH_3\overset{\overset{CH_3}{|}}{C}HCH_3$$

Butane **2-Methylpropane**

Molecules, like the two shown above, that have the *same molecular formula but different atomic connections*, are called **constitutional isomers.**

■ Constitutional isomers are molecules that have the same molecular formula, but whose atoms are connected differently.

As the number of atoms in a molecular formula increases, so does the number of possible constitutional isomers. The molecular formula C_6H_{14}, for example, gives rise to five constitutional isomers: hexane, 2-methylpentane, 3-methylpentane, 2,2-dimethylbutane, and 2,3-dimethylbutane (Figure 4.12). The number of constitutional isomers grows

$$CH_3CH_2CH_2CH_2CH_2CH_3$$

Hexane

$$\overset{1}{CH_3}\overset{2}{\underset{\underset{CH_3}{|}}{C}H}\overset{3}{CH_2}\overset{4}{CH_2}\overset{5}{CH_3}$$

2-Methylpentane

$$\overset{1}{CH_3}\overset{2}{CH_2}\overset{3}{\underset{\underset{CH_3}{|}}{C}H}\overset{4}{CH_2}\overset{5}{CH_3}$$

3-Methylpentane

$$\overset{1}{CH_3}\overset{\overset{CH_3}{2|}}{\underset{\underset{CH_3}{|}}{C}}\overset{3}{CH_2}\overset{4}{CH_3}$$

2,2-Dimethylbutane

$$\overset{1}{CH_3}\overset{2}{\underset{\underset{CH_3}{|}}{C}H}\overset{\overset{CH_3}{3|}}{C}H\overset{4}{CH_3}$$

2,3-Dimethylbutane

■ **FIGURE 4.12**

Constitutional isomers of C_6H_{14} The five molecules shown are the constitutional isomers with the molecular formula C_6H_{14}. Constitutional isomers always have different names from one another.

rapidly with an increase in alkane size. C_8H_{18} has 18 constitutional isomers, $C_{15}H_{32}$ has 4,347 isomers, and $C_{30}H_{62}$ has 4,111,846,763.

In some cases, as is true for the formulas in Figure 4.12, identifying constitutional isomers is a simple process. Their structures look different enough that there is no question that they are constitutional isomers. At other times, making this identification can be more difficult. For example, although the molecules below might, at first, appear to be different, all structures represent the same molecule, 2-methylhexane. It is important to remember that line-bond structures do not show the true shape of a molecule, just the bonding arrangements.

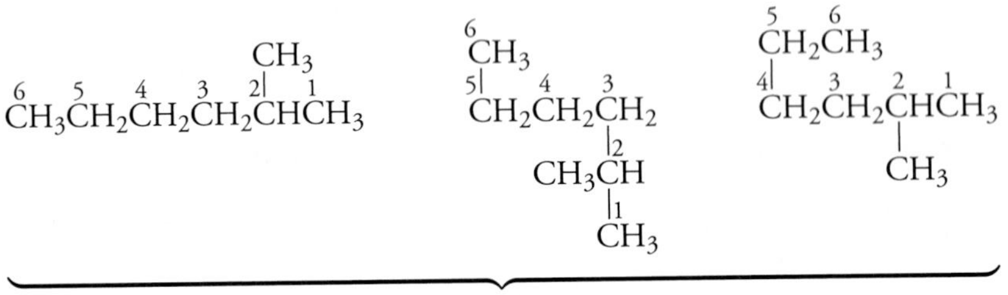

CH_3 (on C2)
$\overset{6}{C}H_3\overset{5}{C}H_2\overset{4}{C}H_2\overset{3}{C}H_2\overset{2}{C}H\overset{1}{C}H_3$

$\overset{6}{C}H_3$ (on C5)
$\overset{5}{C}H_2\overset{4}{C}H_2\overset{3}{C}H_2$
$CH_3\overset{2}{C}H$
$\overset{1}{C}H_3$

$\overset{5}{C}H_2\overset{6}{C}H_3$ (on C4)
$\overset{4}{C}H_2\overset{3}{C}H_2\overset{2}{C}H\overset{1}{C}H_3$
CH_3 (on C2)

2-Methylhexane

One sure way to decide whether two molecules are identical or are constitutional isomers is to name them. Identical molecules have the same IUPAC name and constitutional isomers have different IUPAC names.

SAMPLE PROBLEM 4.9

Identifying constitutional isomers

Are the two alkanes constitutional isomers or are they identical?

CH_3CH_2
$\quad CH_2CHCH_3$
$\qquad CH_3$

CH_3
CH_3CH
$\quad CH_2CH_2$
$\qquad CH_3$

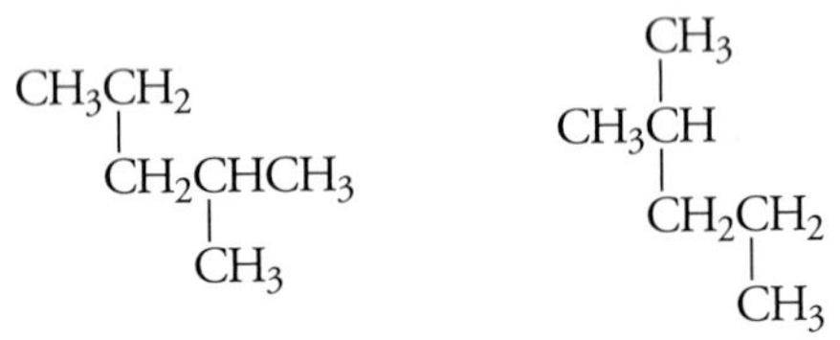

Strategy

Constitutional isomers have the same molecular formula but different structures (and names).

Solution

Identical molecules. They both have the same formula (C_6H_{14}) and the same name (2-methylpentane).

PRACTICE PROBLEM 4.9

Are the two alkanes constitutional isomers or are they identical?

CH_3CH_2
$CH_3CHCHCH_3$
$\quad CH_3$

CH_3CH_2
$CH_3CHCH_2CH_2CH_3$

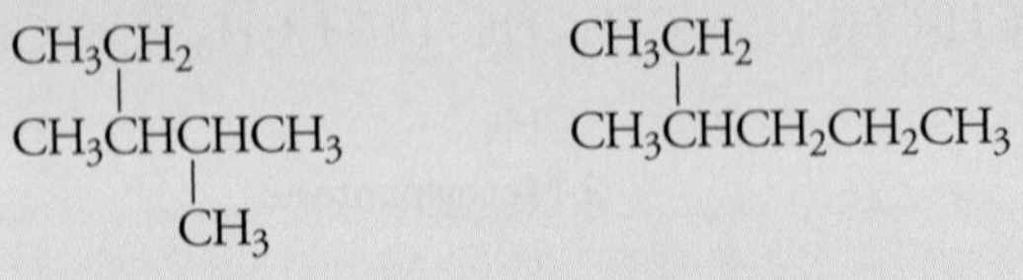

While on the subject of alkane constitutional isomers, it is worth discussing how we obtain and use some alkanes. The natural gas used to heat homes is mostly methane, mixed with smaller amounts of ethane, propane, and butane. Propane, by itself, is used as a fuel for camp stoves and barbecues and butane is used in disposable lighters.

Petroleum (crude oil), a mixture of hydrocarbons and lesser amounts of organic molecules that contain oxygen, nitrogen, and sulfur, is the source of most of the other alkanes that we use. Gasoline is a mixture of alkanes, having between 5 and 12 carbon atoms, other hydrocarbons, and additional compounds that are added to improve the performance of automobile engines. Gasoline contains a greater number of different alkanes than you might expect. There being three alkane constitutional isomers with the formula C_5H_{12}, five isomers with the formula C_6H_{14}, and many more with the formulas C_7H_{16} through $C_{12}H_{26}$, hundreds of different alkanes can be found in gasoline. Other important mixtures of alkanes include diesel fuel (contains alkanes with between 12 and 18 carbon atoms), motor oil (alkanes with more than 15 carbon atoms), and asphalt (alkanes with more than 35 carbon atoms).

4.6 CONFORMATIONS

When dealing with organic or biomolecules, it is often not enough to know only which constitutional isomer is present. Sometimes we must also know something about which of a number of possible three-dimensional shapes a molecule is found in. Among other things, the three-dimensional shape that a molecule has can affect the transport of certain compounds across cell membranes, the ability of the light-absorbing pigment present in the retina to absorb light, and the proper functioning of proteins (Health Link: Prion Diseases).

Rotation about single bonds allows most molecules to assume a number of different three-dimensional shapes. *The shapes that a molecule can take because of bond rotations* are called **conformations**. The different conformations of a molecule

- have the same molecular formula
- have the same atomic connections
- have different three-dimensional shapes
- are interchanged by the rotation of single bonds

■ Conformations are interchanged by rotation around single bonds.

Switching from one conformation to another always involves single bond rotation, never bond breaking. In the case of butane, pictured in Figure 4.13, three of the many possible conformations resulting from rotations about the bond between carbons 2 and 3 are shown. Of these three, the one shown in Figure 4.13*c* is the most stable (most favored) because it minimizes crowding by placing all of the atoms and bonds as far apart as possible.

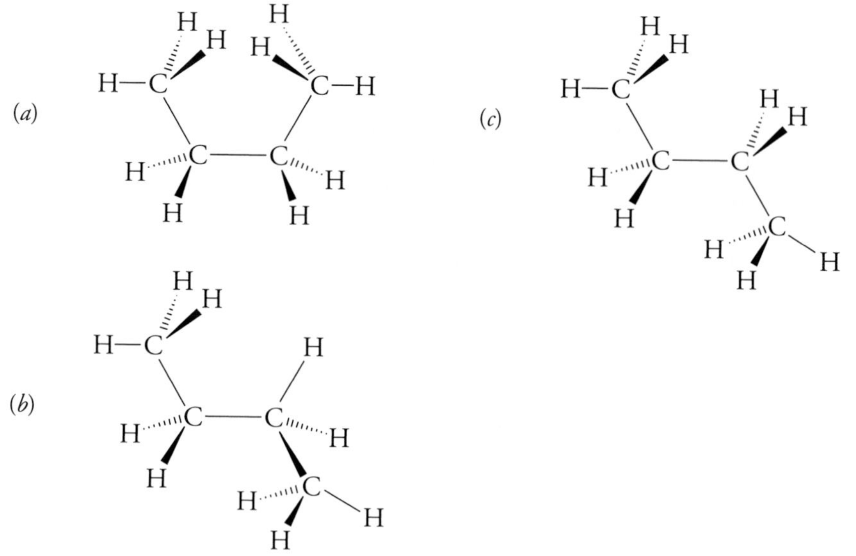

FIGURE 4.13

Conformations of butane
Rotation about the bond between carbons 2 and 3 in butane gives rise to different conformations (three-dimensional shapes) for the molecule.

HEALTH Link

Prion Diseases

The conformation, or shape, that a molecule takes can greatly influence its biological action. This point is well illustrated by a particular class of proteins, called prions, that have been identified as the cause of mad cow disease (in cattle), scrapie (in sheep), chronic wasting disease (in deer and elk), and new variant Creutzfeldt-Jakob disease, or vCJD (in humans).

Proteins are very large molecules formed from amino acid building blocks. Each protein has a favored conformation—some with the protein chain twisted into spirals and bends and some with the chain zigzagged back and forth in a compact form. Prions, short for *pro*teinaceous infect*ion* particles, are a type of protein found in the membranes of nerve cells. The prion diseases mentioned above are caused when the normal prion conformation (PrP^{C}) is twisted into an abnormal shape (PrP^{SC}) (Figure 4.14). When PrP^{SC} comes into contact with PrP^{C}, the PrP^{C} proteins change conformation and become PrP^{SC}. As PrP^{SC} accumulates in affected animals or humans, sponge-like holes form in the brain, causing dizziness, seizures, and death.

Prion diseases can be transferred from one species to another. Cattle are known to get mad cow disease when fed parts of scrapie-infected sheep, and vCJD in humans has been linked to eating mad cow–infected beef.

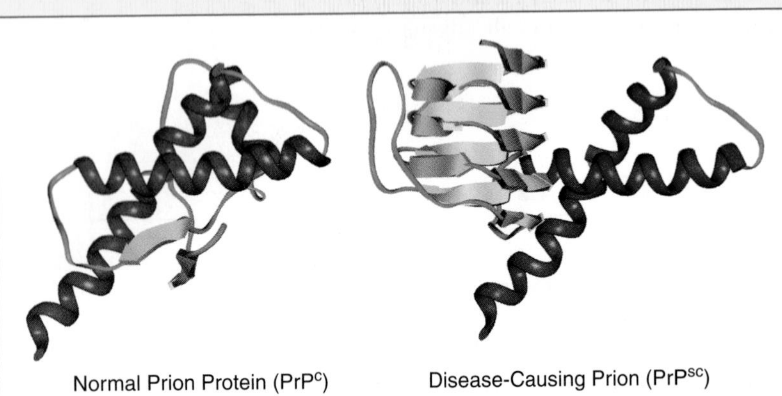

FIGURE 4.14

Prions
(*a*) The prion protein (PrP^{c}) in its normal conformation. (*b*) The prion in its incorrectly folded (PrP^{sc}) form.

Source: Fred E. Cohen, M.D., D. Phil and Cedric Govaerts, Ph.D., Department of Cellular & Molecular Pharmacology, University of California San Francisco (UCSF).

4.7 CYCLOALKANES

In some alkanes, called **cycloalkanes**, *carbon atoms are joined into rings* (Figure 4.15). Like their noncyclic counterparts, cycloalkanes are nonpolar molecules that are attracted to one another by London forces. To simplify matters, cycloalkanes are usually drawn using skeletal structures, although side views can be useful when considering the orientation of substituents attached to a ring.

When naming cycloalkanes, the ring is usually designated as the parent, which is named by combining "cyclo" with the appropriate numbering prefix (Table 4.5) and "ane." When only one alkyl group is attached to a cycloalkane parent, the carbon atom that holds that group is carbon 1 (Figure 4.16*a*), but the number is not included in the name. When a ring holds more than one alkyl group, the ring is numbered from the position and in the direction that gives the lowest numbers. Figure 4.16*b* shows how this works for the three dimethylcyclopentane constitutional isomers.

Geometric Isomers

The limited rotation of the carbon–carbon single bonds in cycloalkanes has an interesting side effect in that it allows for the existence of **stereoisomers**, molecules that

- have the same molecular formula
- have the same atomic connections
- have different three-dimensional shapes
- are interchanged only by breaking bonds

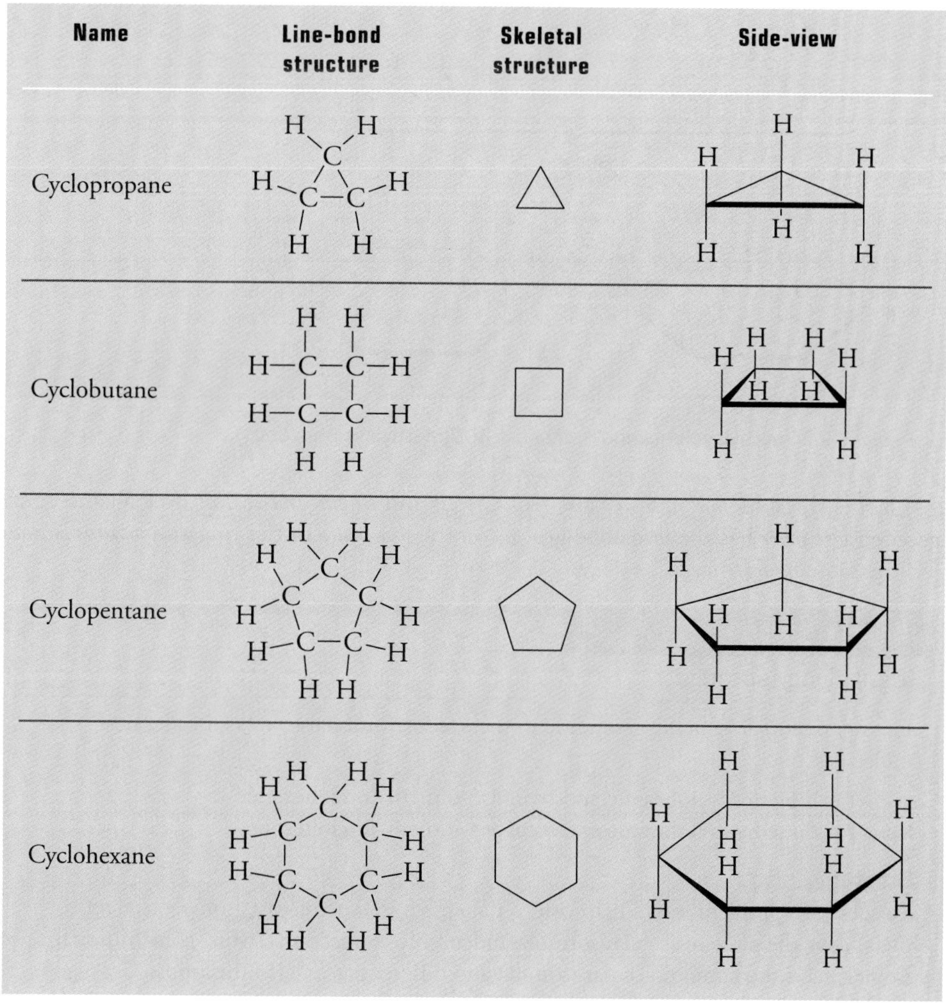

Name	Line-bond structure	Skeletal structure	Side-view
Cyclopropane			
Cyclobutane			
Cyclopentane			
Cyclohexane			

FIGURE 4.15

Cycloalkanes
Cycloalkanes are alkanes whose carbon atoms are joined in a ring.

Except for the last item in the list (*are interchanged only by breaking bonds*), the description above is identical to that given for conformations (Section 4.6). When stereoisomers exist because of restricted bond rotation, the stereoisomers are called **geometric isomers.**

■ Stereoisomers are interchanged only by breaking bonds.

To explore this concept, let us compare the three-dimensional structure of a noncyclic alkane with that of a cycloalkane. Figure 4.17*a* shows two conformations of butane, interchanged by rotation about the carbon–carbon single bond between carbons 2 and 3. The two 1,2-dimethylcyclohexane molecules in Figure 4.17*b* are stereoisomers. Rotation of the ring carbon–carbon bonds is limited, so the molecule that has methyl groups on the same face of the ring can never be rotated enough to look like the molecule with the methyl groups on opposite faces. Transforming one of these molecules into the other requires the breaking of covalent bonds, which makes them geometric isomers.

(*a*)

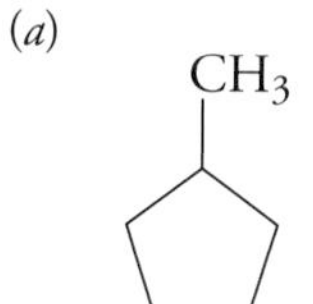

Methylcyclopentane

(*b*)

H_3C CH_3

1,1-Dimethylcyclopentane

CH_3 CH_3

1,2-Dimethylcyclopentane

CH_3 CH_3

1,3-Dimethylcyclopentane

FIGURE 4.16

IUPAC names of cycloalkanes
(*a*) With one substituent present, the parent ring is numbered starting at the substituent. The position number is omitted from the name, however. (*b*) When more than one substituent is present, the parent ring is numbered from the position and in the direction that gives the lowest numbers.

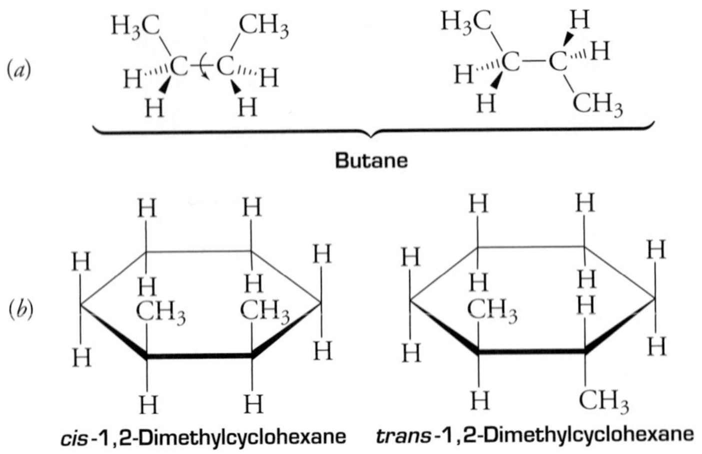

FIGURE | 4.17

Geometric isomers
(*a*) Two conformations of butane are interchanged by single bond rotation. (*b*) Two geometric isomers of 1,2-dimethylcyclohexane are interchanged only by bond breaking. In *cis*-1,2-dimethylcyclohexane the methyl groups are on the same face of the ring. In *trans*-1,2-dimethylcyclohexane they are on opposite sides.

Geometric isomers come in pairs—one is ***cis*** and one is ***trans***. For cycloalkanes, a *cis* geometric isomer has the two alkyl groups on the *same face* of the ring and a *trans* isomer has them on *opposite faces*.

SAMPLE PROBLEM 4.10

Identifying isomers

Identify each pair as being geometric isomers, constitutional isomers, or identical molecules.

a. *cis*-1,4-dimethylcyclohexane and *trans*-1,2-dimethylcyclohexane
b. *cis*-1,4-dimethylcyclohexane and *trans*-1,4-dimethylcyclohexane

Strategy
To solve this problem it will help you to remember that geometric isomers (stereoisomers) have the same molecular formula and atomic connections, while constitutional isomers have the same molecular formula and different atomic connections.

Solution

a. Constitutional isomers. While *cis* and *trans* molecules are geometric isomers, these two molecules are not geometric isomers of the same molecule. They are constitutional isomers because they have different atomic connections (1,4-dimethyl versus 1,2-dimethyl).
b. Geometric isomers. These two molecules have the same atomic connections (both are 1,4-dimethyl), so they are not constitutional isomers. They are geometric isomers of 1,4-dimethylcyclohexane.

PRACTICE PROBLEM 4.10

Draw and name the geometric isomers of 1-ethyl-2-methylcyclohexane.

4.8 ALKENES, ALKYNES, AND AROMATIC COMPOUNDS

■ Unsaturated molecules contain double or triple bonds.

Alkanes and cycloalkanes are **saturated** hydrocarbons, which is another way of saying that *they contain only single bonded carbon atoms.* The three other families of hydrocarbons (Table 4.3) are **unsaturated**, which means that *they contain double or triple bonds.* This use of the terms saturated and unsaturated is related to a reaction involving hydrogen gas (H_2) that will be introduced in Section 6.3.

- Alkenes contain carbon–carbon double bonds.
- Alkynes contain carbon–carbon triple bonds.

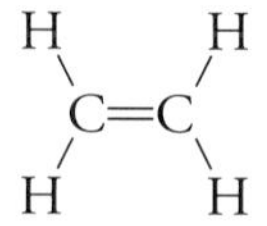

Alkenes *contain at least one carbon–carbon double bond.* In ethylene, the smallest alkene, the carbon atoms carry three groups of electrons—one double bond and two single bonds. This results in a trigonal planar shape around each carbon atom (Table 4.2). **Alkynes** *contain at least one carbon–carbon triple bond* and the smallest member of this family is acetylene. Each carbon atom in acetylene has two groups of electrons arranged in a linear shape.

$H_2C{=}CH_2$

Ethylene

$H{-}C{\equiv}C{-}H$

Acetylene

Some alkenes and alkynes can be obtained from petroleum, but many are produced by living things. The alkene called muscalure is a sex-attractant pheromone (chemical messenger) produced by houseflies (Figure 4.18).

The third family of unsaturated hydrocarbons is **aromatic hydrocarbons**. A well-known member of this family, benzene, is a ring of six carbon atoms that can be drawn with each carbon atom having one single and one double bond to neighboring carbon atoms, and one bond to a hydrogen atom.

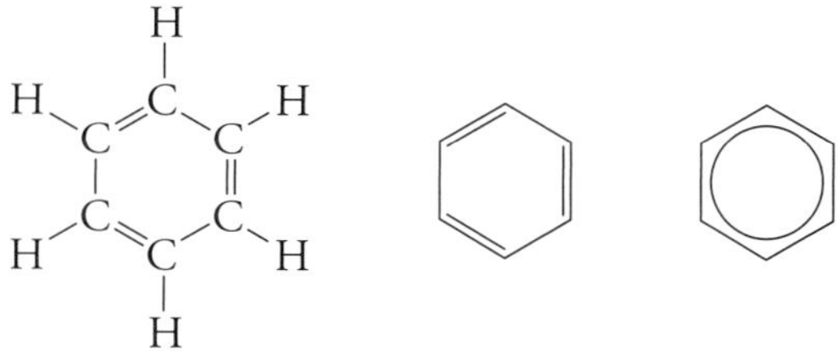

Benzene

Although the double bonds give benzene the appearance of being an alkene, it and other aromatic compounds do not behave as alkenes. To de-emphasize their alkene-like appearance, aromatic compounds are often drawn with a circle replacing the double bonds.

Like alkanes, the unsaturated hydrocarbons (alkenes, alkynes, and aromatic compounds) have only nonpolar covalent bonds and are nonpolar molecules. London forces hold members of these hydrocarbon families to one another, and increasing size leads to stronger London force attractions between molecules and higher melting and boiling points (Table 4.4).

Naming

When using the IUPAC rules to name alkenes or alkynes, the parent is the longest chain of carbon atoms that has the carbon–carbon double or triple bond. Numbering of the parent begins at the end nearer the double or triple bond, and the position of the bond is indicated using the lower number assigned to the multiply bonded carbon atoms. As for naming alkanes, the ending on the name indicates the organic family—IUPAC names of alkenes end with "ene" and those of alkynes end with "yne." Once the parent chain has been determined, alkyl groups are identified by name, position, and number of appearances (Figure 4.19*a*). Sometimes organic compounds are known by **common names**, names that are not assigned according to the IUPAC rules. Ethylene (for ethene), propylene (for propene), and acetylene (for ethyne) are examples.

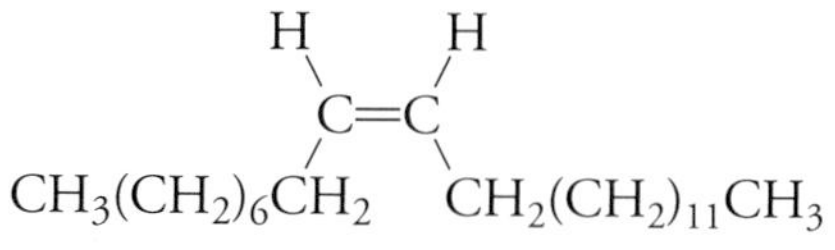

$CH_3(CH_2)_6CH_2{-}CH{=}CH{-}CH_2(CH_2)_{11}CH_3$

Muscalure

FIGURE 4.18

Muscalure
This alkene, produced by the female housefly, is a pheromone (a chemical messenger) that is used to attract a mate. In this condensed structure, $(CH_2)_6$ indicates six repeats of CH_2 and $(CH_2)_{11}$ indicates eleven repeats.

FIGURE 4.19

Naming unsaturated hydrocarbons
(*a*) Alkenes and alkynes. When assigning IUPAC names, parent chains, which must contain the carbon–carbon double or triple bond, are numbered from the end nearer these groups. For alkenes and alkynes containing two or three carbon atoms, the position of the double or triple bond is not specified. Alkene names end in "ene" and alkyne names end in "yne." (*b*) Aromatic compounds. Methylbenzene is also known as toluene. In common names, a 1,2 arrangement of substituents on a benzene ring is an *ortho* arrangement and dimethylbenzenes are called xylene, so 1,2-dimethylbenzene is *ortho*-xylene or *o*-xylene. Benzyl alcohol, benzoic acid, and ethyl benzoate are aromatic compounds that are not pure hydrocarbons.

(*a*)

$CH_2{=}CHCH_2CHCH_3$ (with CH_3 on C4)
4-Methyl-1-pentene

$CH_3CHCH_2CH{=}CHCH_2CH_2CH_3$ (with CH_3 on C2)
2-Methyl-4-octene

$CH_3CHCHC{\equiv}CH$ (with CH_3 on C3 and C4)
3,4-Dimethyl-1-pentyne

$CH_3C{\equiv}CCH_2CHCH_2CH_2CH_3$ (with CH_3CHCH_3 on C5)
5-Isopropyl-2-octyne

(*b*)

Benzene

CH_3
Methylbenzene (toluene)

CH_3, CH_3
1,2-Dimethylbenzene (*o*-xylene)

CH_3, CH_3
1,3-Dimethylbenzene (*m*-xylene)

CH_3, CH_3
1,4-Dimethylbenzene (*p*-xylene)

CH_2OH
Benzyl alcohol

$C(=O){-}OH$
Benzoic acid

$C(=O){-}OCH_2CH_3$
Ethyl benzoate

"Benzene" is an accepted IUPAC name, so benzene-related compounds are named using benzene as the parent ring. When only one substituent is attached, its position is not specified in the name, as is the case for methylbenzene (Figure 4.19*b*). In the presence of more than one substituent, the ring is numbered from the position and in the direction to give the lowest substituent numbers possible. Methylbenzene and 1,2-dimethylbenzene are also known, respectively, by the common names toluene and *ortho*-xylene. For common names of benzene-based compounds that have only two substituents attached to the ring, *ortho* or *o* refers to a 1,2 arrangement, *meta* or *m* to a 1,3 arrangement, and *para* or *p* to a 1,4 arrangement.

- When only two substituents are attached to benzene, they will be *ortho*, *meta*, or *para* to one another.

An aromatic ring may be part of a larger molecule, so not all aromatic compounds are pure hydrocarbons. This includes benzyl alcohol, benzoic acid, and ethyl benzoate, examples from three of the organic families that we will encounter in the section that follows.

Some aromatic compounds, called **polycyclic aromatic hydrocarbons** (PAHs), *contain benzene rings that are fused to one another (they share atoms and bonds).* Naphthalene and anthracene (Table 4.4), and phenanthrene (Figure 4.20), all of which can be obtained from coal, are simple examples. Benzo[*a*]pyrene, a large PAH formed during the burning of tobacco, coal, gasoline, and many other substances, is known to cause cancer.

FIGURE 4.20

Polycyclic aromatic hydrocarbons
Phenanthrene, benzo[*a*]pyrene, and other PAHs contain fused benzene rings.

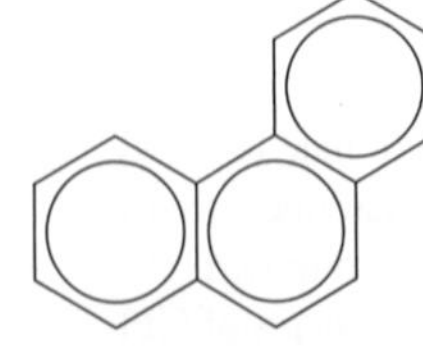
Phenanthrene

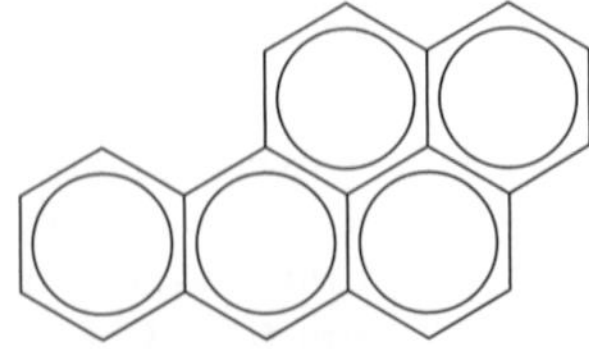
Benzo[a]pyrene

Geometric Isomers

Unlike single bonds, which rotate freely (except when they are in rings; see Section 4.7), double bonds cannot rotate. With this restricted bond rotation comes the possibility of some alkenes existing as *cis* and *trans* stereoisomers. In Figure 4.21*a*, the two geometric isomers of 2-butene are shown. In the *cis* isomer, the two methyl groups ($—CH_3$) are on the same side of a line connecting the two double-bonded carbon atoms, and in the *trans* isomer these groups are on opposite sides of that line.

Not all alkenes are found as *cis* and *trans* isomers. For geometric isomers to exist, neither carbon atom of the double bond may carry two identical attached atoms or groups of atoms. The alkene in Figure 4.21*b*, 2-methyl-2-butene, has two identical groups ($—CH_3$) attached to the carbon atom to the left side of the double bond. If the molecule drawn on the right is flipped top to bottom, it is identical to the one on the left and *cis* and *trans* isomers do not exist. In Figure 4.21*c* each of the carbons in the double bond has two different atoms or groups of atoms attached to it. If flipped, the 2-pentene on the right cannot be superimposed on the molecule to the left—this alkene has *cis* and *trans* isomers.

Alkynes, having a linear shape across the carbon–carbon triple bond, do not exist as *cis* and *trans* isomers.

(*a*)

H H C=C H_3C CH_3 — ***cis*-2-Butene**

H CH_3 C=C H_3C H — ***trans*-2-Butene**

(*c*)

H H C=C H_3C CH_2CH_3 — ***cis*-2-Pentene**

H_3C H C=C H CH_2CH_3 — ***trans*-2-Pentene**

(*b*)

H_3C H C=C H_3C CH_3 — H_3C CH_3 C=C H_3C H

2-Methyl-2-butene

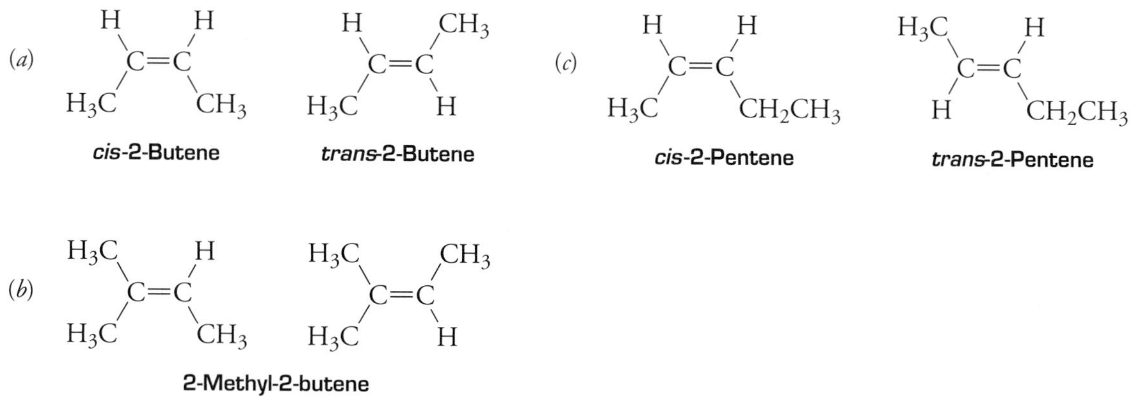

FIGURE 4.21 **Geometric isomers of alkenes**
(*a*) In the *cis* isomer the $—CH_3$ groups are on the same side of a line that connects the two double-bonded carbon atoms, and in the *trans* isomer they are on opposite sides. (*b*) 2-Methyl-2-butene has no *cis* or *trans* isomers, because the double-bonded carbon atom on the left is attached to identical groups of atoms ($—CH_3$ groups). When flipped top to bottom, the molecule on the right is the same as the one on the left. (*c*) In the *cis* isomer of 2-pentene, the alkyl groups are on the same side of the double bond.

SAMPLE PROBLEM 4.11

Identifying geometric isomers

Which geometric isomer of muscalure is shown in Figure 4.18?

Strategy

To distinguish the *cis* from the *trans* isomer, compare the relative positions of the carbon atoms attached to the carbon–carbon double bond.

Solution

The *cis* isomer. Both ends of the carbon chain are on the same side of a line that connects the double-bonded carbon atoms.

PRACTICE PROBLEM 4.11

Designate each of the carbon–carbon double bonds in bombykol, a moth pheromone, as *cis* or *trans*.

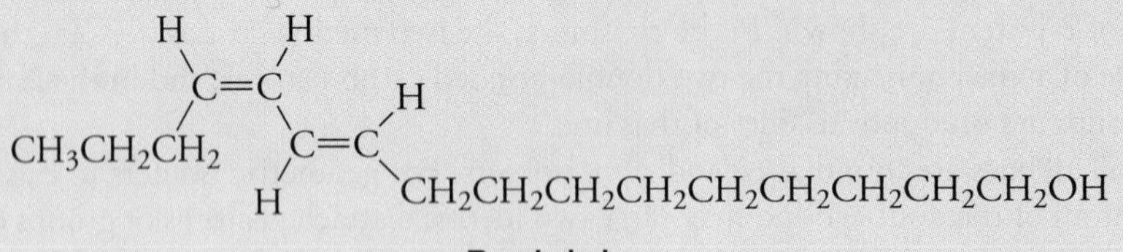

Bombykol

HEALTH Link — *Sunscreens*

In addition to the visible light that is present in sunlight, a higher energy form of electromagnetic radiation called ultraviolet (UV) is also found. It is this UV radiation that makes spending too much time in the sun hazardous to your health.

UV radiation is divided into three categories (arranged in order of increasing energy) UV-A, UV-B, and UV-C. Although all forms of UV radiation can be harmful, UV-C is the most damaging. We do not usually worry about exposure to UV-C from sunlight, however, because most is screened out by atmospheric ozone. The UV-A and UV-B in sunlight are not blocked by the atmosphere and it is exposure to UV-B that causes sunburns. This form of radiation has also been identified as one of the causes of skin cancer. While UV-A was once thought to be relatively harmless, there is now strong evidence that links wrinkles, other skin damage, and skin cancer to UV-A exposure.

To some extent, your skin can protect itself from UV radiation. When sunlight hits the skin, a particular type of cells produce melanin, a black pigment that absorbs UV radiation. A suntan is the result of melanin production. Sunscreens can also provide some degree of protection from the sun. These contain aromatic compounds that block UV-A and UV-B light (Figure 4.22). Not all aromatic compounds are suitable for use as sunscreens—many do not absorb UV, and many are toxic.

The SPF (sun protection factor) listed on the label of a sunscreen indicates how effective it is at blocking UV radiation. An SPF of 25 means that it takes 25 times longer to get sunburned with the sunscreen applied than without.

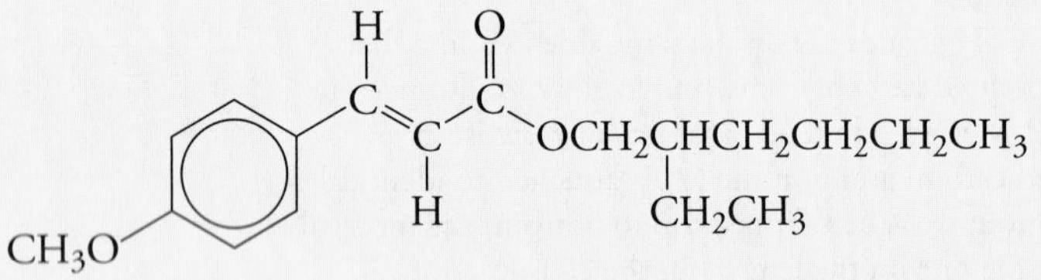

Octyl methoxycinnamate

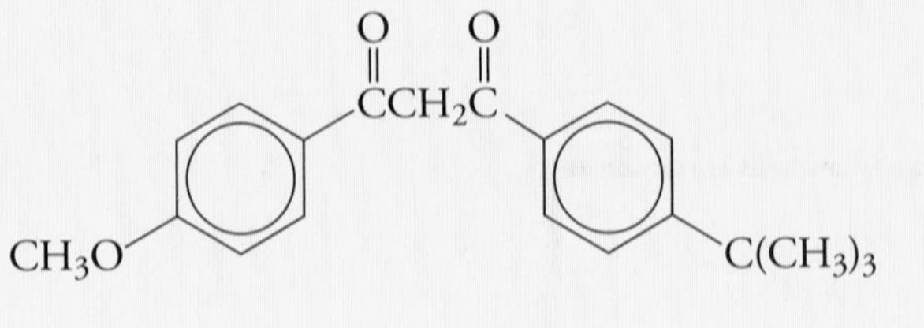

Avobenzone

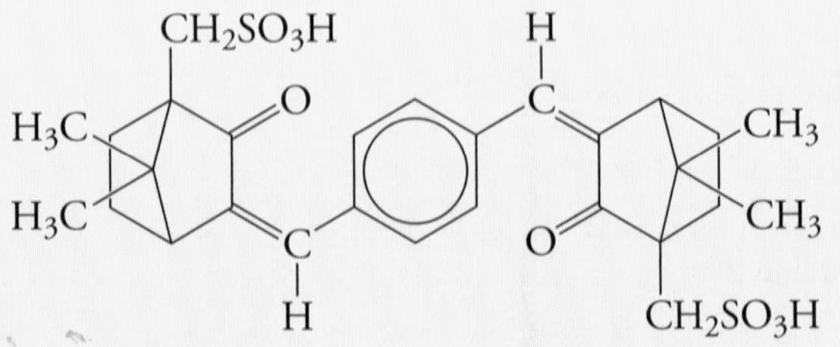

Ecamsule

FIGURE 4.22

UV-B and UV-A absorbing compounds in sunscreens
Octyl methoxycinnamate is the UV-B blocker most commonly used in sunscreens. Avobenzone and Ecamsule are UV-A blockers. Ecamsule, which was recently approved for use in the United States, has been a sunscreen ingredient in Canada and Europe since 1993.

4.9 ALCOHOLS, CARBOXYLIC ACIDS, AND ESTERS

In addition to the saturated and unsaturated hydrocarbons introduced in this chapter, there are a large number of other important families of organic molecules. In this section three of these families will be briefly considered: alcohols, carboxylic acids, and esters (Table 4.7).

TABLE 4.7 Structure, Name, and Boiling Point of Selected Alcohols, Carboxylic Acids, and Esters

Formula	IUPAC Name[a]	Boiling Point (°C)
Alcohols		
CH_3OH	Methanol	65.0
CH_3CH_2OH	Ethanol	78.5
$CH_3CH_2CH_2OH$	1-Propanol	97.4
$CH_3CH_2CH_2CH_2OH$	1-Butanol	117.3
$CH_3CH_2CH_2CH_2CH_2OH$	1-Pentanol	138
Carboxylic acids		
$HC(=O){-}OH$	Methanoic acid	100
$CH_3C(=O){-}OH$	Ethanoic acid	118
$CH_3CH_2C(=O){-}OH$	Propanoic acid	141
$CH_3CH_2CH_2C(=O){-}OH$	Butanoic acid	164
$CH_3CH_2CH_2CH_2C(=O){-}OH$	Pentanoic acid	187
Esters		
$HC(=O){-}OCH_2CH_3$	Ethyl methanoate	53
$CH_3C(=O){-}OCH_2CH_3$	Ethyl ethanoate	77
$CH_3CH_2C(=O){-}OCH_2CH_3$	Ethyl propanoate	99
$CH_3CH_2CH_2C(=O){-}OCH_2CH_3$	Ethyl butanoate	120

[a] Names are given here for reference purposes. The IUPAC rules for naming alcohols, carboxylic acids, and esters will be presented in later chapters.

These particular families have been selected for two reasons. First, they provide examples of compounds that interact by noncovalent interactions (Section 4.3) other than the London forces that come into play for hydrocarbons. Second, having some knowledge of the members of these classes of compounds will help us understand material presented in some of the chapters soon to follow.

■ Functional groups determine to which family an organic molecule belongs.

What distinguishes alcohols, carboxylic acids, and esters from the hydrocarbons is that their **functional groups** contain atoms other than carbon and hydrogen. A functional group is *an atom, group of atoms, or bond that gives a molecule a particular set of chemical properties,* and each organic family of organic compounds is defined by the functional group that its members contain. We have already encountered three functional groups in this chapter, the carbon–carbon double bond of alkenes, the carbon–carbon triple bond of alkynes, and the ring with alternating double and single bonds of aromatic compounds.

All **alcohols** contain a **hydroxyl** (—OH) functional group that is attached to an **alkane-type carbon atom**. An alkane-type carbon atom is singly bonded to carbon or hydrogen atoms.

```
    H   H
    |   |   ..
H—C—C—O—H
    |   |   ..
    H   H
```

An alkane-type carbon atom

Carboxylic acids contain a **carboxyl** functional group, which is the combination of a hydroxyl (—OH) group and a **carbonyl** (C═O) group. In carboxylic acids, the carbon atom of the carboxyl group is attached to a hydrogen atom, an alkane-type carbon atom, or an aromatic ring.

```
      :Ö
       ||   ..
H—C—O—H
            ..
```

```
     H   :Ö
     |     ||   ..
H—C—C—O—H
     |          ..
     H
```

```
               :Ö
                 ||  ..
(benzene ring)—C—O—H
                     ..
```

Although **carboxylic acids** contain an —OH group, they are not considered alcohols. The chemistry of carboxyl —OH groups and alcohol —OH groups is very different.

Esters contain a C—O—C linkage in which one of the carbon atoms belongs to a carbonyl group. Esters can be formed by reacting carboxylic acids with alcohols, and part of each original molecule can be seen in the ester structure.

$$CH_3CH_2\overset{\overset{\displaystyle O}{\|}}{C}—OCH_2CH_3$$

Carboxylic acid part | Alcohol part

Physical Properties

While the functional groups present in alcohols, carboxylic acids, and esters all contain one or more oxygen atoms, there are some differences in the intermolecular forces that each can experience. The hydroxyl group in alcohols, for example, contains an oxygen–hydrogen covalent bond, which allows alcohols to participate in hydrogen bonding (Figure 4.23*a*).

While the primary force that holds one alcohol molecule to another is a hydrogen bond, this is not the only intermolecular force that operates. Note in Table 4.7 that the longer the carbon chain for an alcohol, the higher its boiling point. Since all alcohols form hydrogen bonds with one another, the increase in boiling points is attributed to a greater contribution by London forces between the hydrocarbon parts of the alcohols, the larger they become.

Carboxylic acids contain a hydrogen–oxygen covalent bond and, like alcohols, they can form hydrogen bonds with one another. With the presence of an oxygen atom in the C=O group, each carboxylic acid molecule can be involved in two hydrogen bonds with another carboxylic acid, one involving the H of the —OH group and the other involving the O of the C=O (Figure 4.23*b*). As with alcohols, an increase in the number of carbon atoms in a carboxylic acid leads to stronger London force attractions and a higher boiling point (Table 4.7).

Esters are unable to form hydrogen bonds because they have no covalent bond between a hydrogen atom and an oxygen atom. Therefore, the forces that hold one ester to another are dipole–dipole interactions and London forces (Figure 4.23*c*). The more carbon atoms in an ester molecule, the greater the contribution of London forces and the higher its boiling point.

The effect that intermolecular forces have on boiling points is clearly demonstrated by comparing molecules of similar molecular weight. Molecules listed in Tables 4.4 and 4.7 with molecular weights ranging between 72 and 74 amu (pentane, ethyl methanoate, 1-butanol, and propanoic acid) have very different boiling points. Pentane has the lowest boiling point because its molecules are held to one another by the relatively weak London forces. The next highest boiling point is that of ethyl methanoate (the naming of alcohols, carboxylic acids, and esters will be covered in later chapters). These ester molecules are attracted to one another by the relatively stronger dipole–dipole forces. The molecules able to form hydrogen bonds—1-butanol and propanoic acid—have even higher boiling points.

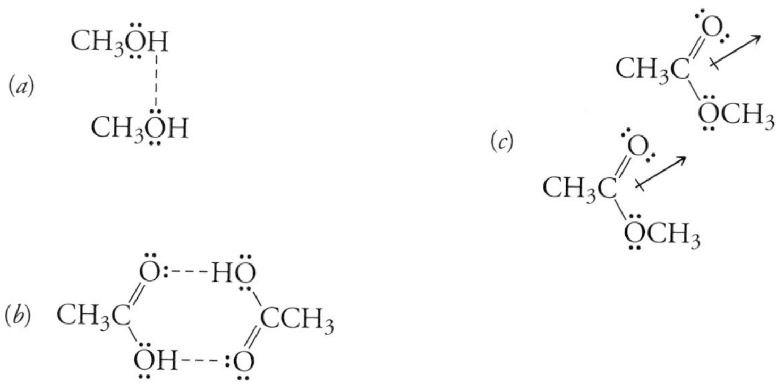

(*a*) CH_3OH ⋯ CH_3OH

(*b*) $CH_3C(=O)OH$ ⋯ $HO(O=)CCH_3$

(*c*) $CH_3C(=O)OCH_3$ / $CH_3C(=O)OCH_3$

FIGURE 4.23 **Noncovalent forces of alcohols, carboxylic acids, and esters**
(*a*) Alcohol molecules interact primarily through hydrogen bonds. (*b*) Carboxylic acid molecules can form two hydrogen bonds with other carboxylic acids. (*c*) Ester molecules can interact via dipole–dipole attractions.

BIOCHEMISTRY Link

Odor and Flavor

The senses of smell and taste depend on specific receptors located in the nose and in the mouth, respectively. We detect an odor or flavor when a compound attaches to these receptors and triggers nerve responses. How the compound interacts with a given receptor determines the particular odor or flavor that it has. In 2004, the Nobel Prize in Physiology or Medicine was awarded to two researchers who study the biochemistry of the sense of smell.

The odor and flavor of fruits is often due to a mixture of esters. The structure of a particular ester determines how it fits a given receptor, so esters containing a different number of carbon atoms are associated with different odors and flavors. For example, the "pineapple" and "apricot" esters differ by three CH_2 groups (Figure 4.24).

Constitutional isomers can also have different odors and flavors, as is the case for the "apple-like" and "banana" esters, each of which has the molecular formula $C_7H_{14}O_2$.

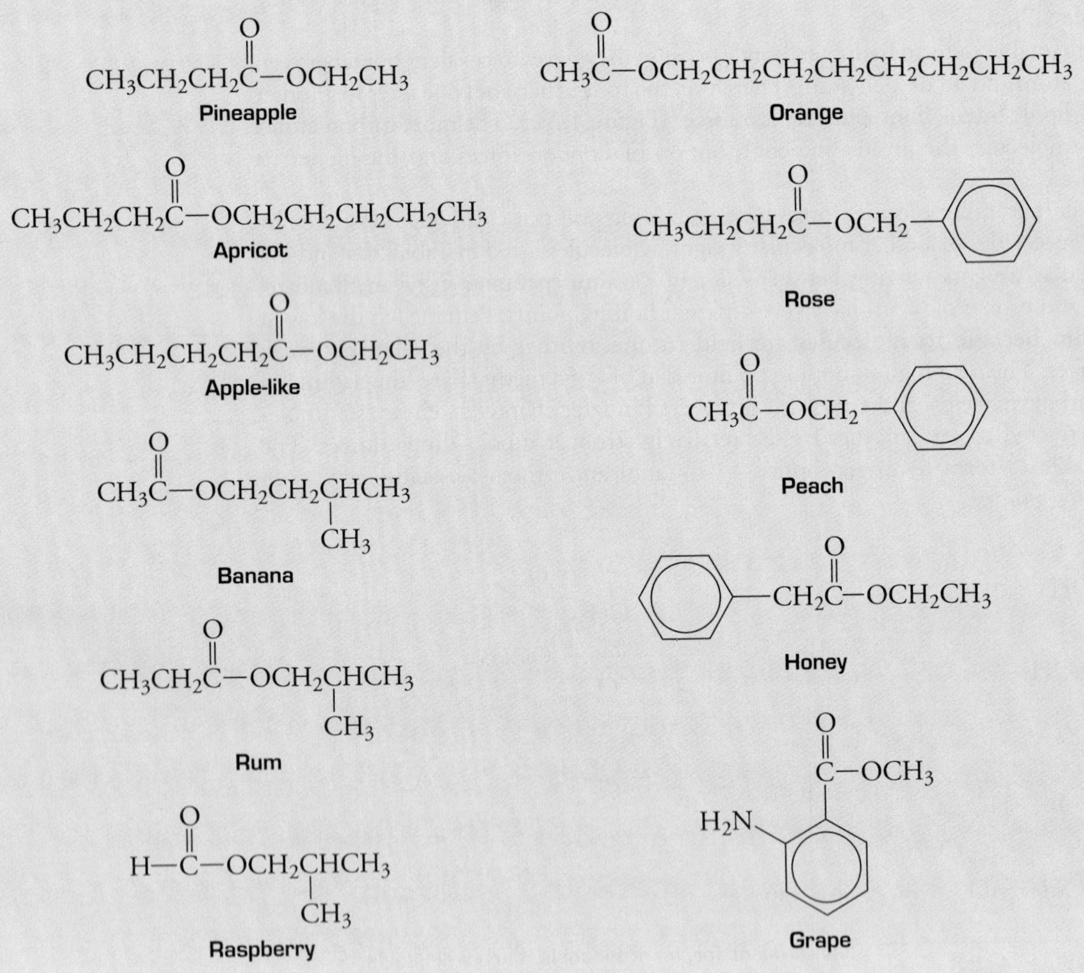

$CH_3CH_2CH_2C(=O)-OCH_2CH_3$

Pineapple

$CH_3C(=O)-OCH_2CH_2CH_2CH_2CH_2CH_2CH_2CH_3$

Orange

$CH_3CH_2CH_2C(=O)-OCH_2CH_2CH_2CH_2CH_3$

Apricot

$CH_3CH_2CH_2C(=O)-OCH_2-C_6H_5$

Rose

$CH_3CH_2CH_2CH_2C(=O)-OCH_2CH_3$

Apple-like

$CH_3C(=O)-OCH_2-C_6H_5$

Peach

$CH_3C(=O)-OCH_2CH_2CH(CH_3)CH_3$

Banana

$C_6H_5-CH_2C(=O)-OCH_2CH_3$

Honey

$CH_3CH_2C(=O)-OCH_2CH(CH_3)CH_3$

Rum

$H-C(=O)-OCH_2CH(CH_3)CH_3$

Raspberry

$H_2N-C_6H_4-C(=O)-OCH_3$ (ortho)

Grape

FIGURE 4.24 **Esters**
Many esters have pleasing odors and flavors.

The light-absorbing molecule present in the rods and cones of the retina is called rhodopsin. It is formed by the combination of a particular protein with an unsaturated molecule called 11-*cis*-retinal (derived from vitamin A; see Problem 4.74). When light enters the eye and hits rhodopsin, the 11-*cis*-retinal part of rhodopsin is converted to all-*trans*-retinal (another vitamin A derivative) and the change in rhodopsin's shape triggers a nerve response that tells the brain that light has reached the eye.

After this *cis* to *trans* change, rhodopsin is broken apart and all-*trans*-retinal is transported to a layer of cells behind the retina, where it is recycled back to 11-*cis*-retinal. In people with Stargardt's disease, the recycling reactions do not take place properly. When retinal remains in the *trans* isomer shape, it reacts with a cell membrane component called phosphatidyl ethanolamine (Sectioon 8.4) to produce a molecule linked to the degeneration of vision associated with the disease.

summary of objectives

1 Draw molecules and polyatomic ions using electron dot and line-bond structures. Assign formal charges.
Drawing the structure of a molecule or polyatomic ion involves counting the number of available valence electrons, joining the atoms by covalent bonds, and then adding the remaining electrons in a way that gives each atom an octet. Formal charge is determined by subtracting the number of electrons around a given atom in a molecule or ion from the number of valence electrons for the neutral, individual atom.

2 Describe how condensed structural formulas and skeletal structures differ from electron dot and line-bond structures.
In an electron dot structure, all valence electrons (nonbonding electrons, as well as those in bonds) are represented by dots. In the line-bond method, each pair of shared bonding electrons is represented by a line. A **condensed structural formula** describes the attachment of atoms to one another, without showing all of the bonds. A carbon atom with three attached hydrogen atoms is written CH_3 and one with two attached hydrogen atoms is written CH_2. In a **skeletal structure**, covalent bonds are represented by lines, carbon atoms are not shown, and hydrogen atoms are drawn only when attached to atoms other than carbon. To simplify matters, nonbonding electrons are sometimes omitted from structural formulas.

3 Define electronegativity and explain its relationship to polar covalent bonds. Give a simple rule that can be used to predict whether or not a covalent bond is polar.
Electronegativity is the ability of an atom to attract electrons in covalent bonds. When the two atoms in a covalent bond have different electronegativities, the bonding

electrons are pulled nearer the atom with the higher electronegativity (stronger attraction). This puts partial charges on the atoms involved (δ+ and δ−) and a polar covalent bond exists. A bond will be polar covalent if it consists of a hydrogen or carbon atom covalently attached to a nitrogen, oxygen, fluorine, or chlorine atom.

4 List the five basic shapes about an atom in a molecule and describe the rules used to predict shape. Explain how shape plays a role in determining overall polarity.

The basic shapes are **tetrahedral**, **pyramidal**, **trigonal planar**, **bent**, and **linear**. The shape about a particular atom depends on how many groups of electrons it holds (a group of electrons is either a pair of nonbonding electrons or a single, double, or triple covalent bond) and on how many of those groups are nonbonding pairs. An atom attached to four groups of electrons will be surrounded by a tetrahedral shape if it has no nonbonding pairs, a pyramidal shape if it has just one nonbonding pair, and a bent shape if it has two nonbonding pairs. An atom attached to three groups of electrons will be trigonal planar if it has no nonbonding pairs and bent if it has one nonbonding pair. An atom attached to just two groups of electrons is linear.

Molecules, as a whole, are **polar** when polar covalent bonds and the shapes around atoms combine to give one side of the molecule a partial negative charge and the other side a partial positive charge.

5 Describe the noncovalent interactions that attract one compound to another.

Compounds (or remote parts of the same compound) are attracted to one another through **noncovalent interactions**. A **hydrogen bond** is the attraction of a nitrogen, oxygen, or fluorine atom to a hydrogen atom covalently bonded to a different nitrogen, oxygen, or fluorine atom. **Salt bridge** is the term used by biochemists to describe an ionic bond. A **dipole–dipole interaction** is the attraction between opposite partial charges in polar groups and an **ion–dipole interaction** is the attraction of an ion for a polar group. A **coordinate-covalent interaction** involves the attraction between a cation and the nonbonding electrons of a nonmetal ion, and **London forces** arise from temporary dipoles.

6 Describe the four families of hydrocarbons.

Hydrocarbons contain only carbon and hydrogen atoms. Hydrocarbons that contain only single covalent bonds are called **alkanes**. **Alkenes** and **alkynes** are hydrocarbons that, respectively, have at least one carbon–carbon double bond and at least one carbon–carbon triple bond. **Aromatic compounds** consist of rings of carbon atoms joined by alternating double and single bonds.

7 Explain the difference between constitutional isomers, conformations, and the stereoisomers known as geometric isomers. Give examples of two different families of hydrocarbons that can exist as geometric isomers.

Constitutional isomers are molecules that have the same molecular formula, but different atomic connections. **Conformations** are the different three-dimensional shapes that a molecule can assume through rotation of single bonds. **Stereoisomers** are molecules that have the same molecular formula and the same atomic connections, but different three-dimensional shapes that can be interchanged only by bond breaking. Stereoisomers that result due to limited rotation about covalent bonds are called **geometric isomers**. The limited rotation of single bonds in the ring structure of **cycloalkanes** can produce ***cis*** (the substituents are on the same face of the ring) and ***trans*** (substituents are on opposite faces of the ring) geometric isomers. Because double bonds do not rotate, **alkenes** can also exist as *cis* and *trans* isomers—in *cis* alkenes the groups being compared are on the same side of a line connecting the two double-bonded carbon atoms, and in *trans* isomers they are on opposite sides.

8 **Define the term "functional group" and describe the features that distinguish hydrocarbons, alcohols, carboxylic acids, and esters from one another.**

Functional groups are atoms, groups of atoms, or bonds that give a molecule a particular set of chemical properties. The functional groups represented in hydrocarbons are the carbon–carbon double bond (in alkenes), the carbon–carbon triple bond (in alkynes), and alternating double and single bonds in a ring (in aromatic compounds). **Alcohols** contain the hydroxyl (—OH) functional group attached to an alkane-like carbon atom. **Carboxylic acids** contain the hydroxyl group attached to a carbonyl (C═O) group, and **esters** contain a C—O—C linkage, with one of the carbon atoms belonging to a carbonyl group.

END OF CHAPTER PROBLEMS

Answers to problems whose numbers are printed in color are given in Appendix D. More challenging questions are marked with an asterisk. Problems within colored rules are paired. **ILW** = Interactive Learning Ware solution is available at *www.wiley.com/college/raymond.*

4.1 Structural Formulas and Formal Charges

4.1 Draw the line-bond structure of each molecule.

a. 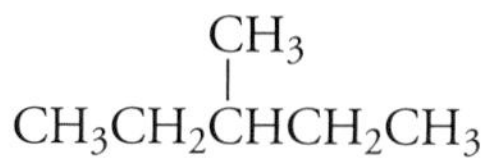 $CH_3CH_2CH(CH_3)CH_2CH_3$ (CH₃ drawn above the central CH)

b. 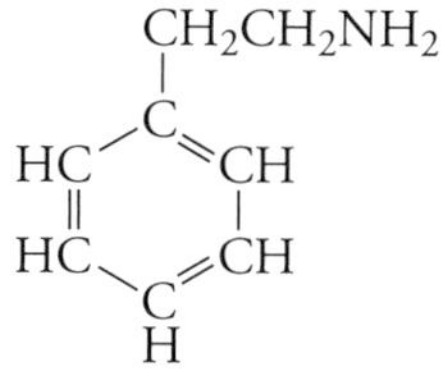 Ring structure: C bearing $CH_2CH_2NH_2$ at top, HC═ / CH, HC / CH, C–H at bottom

4.2 Draw the line-bond structure of each molecule.

a. 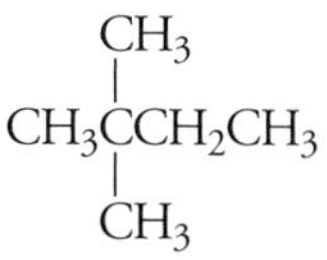 $CH_3C(CH_3)_2CH_2CH_3$ (CH₃ above and below the second C) **b.** $CH_3CH{=}CHCH{=}CH_2$

4.3 What is the shape around the nitrogen and sulfur atoms in the ions shown in Figure 4.2?

4.4 What is the shape around each of the carbon atoms in aspirin (Figure 4.3)?

4.5 Draw each molecule.
a. C_2H_6 **b.** C_2H_4 **c.** C_2H_2

4.6 Draw each molecule.
a. C_3H_8 **b.** C_3H_6 **c.** C_3H_4

4.7 Draw each diatomic element.
a. F_2 **b.** O_2 **c.** N_2

*__4.8__ Draw two different molecules that have the formula C_2H_6O.

4.9 What is the formal charge on each atom in the following molecules or ions?

a. 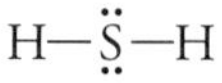 H—S̈—H (S with two lone pairs)

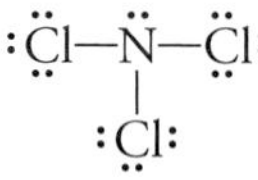

b. 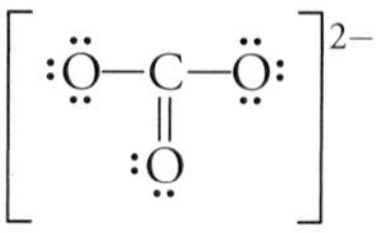 $[:\ddot{O}—C(=\ddot{O}:)—\ddot{O}:]^{2-}$

c. :C̈l—N̈(—C̈l:)—C̈l:

4.10 What is the formal charge on each atom in the following molecules or ions?

a. 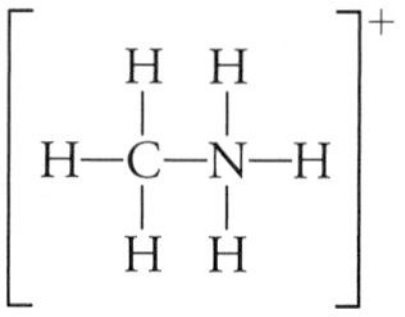 $[H—CH_2—NH_2—H]^+$ (H—C(H)(H)—N(H)(H)—H)

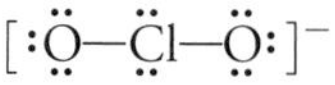

b. 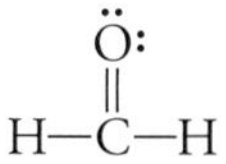 H—C(═Ö:)—H

c. $[:\ddot{O}—\ddot{Cl}—\ddot{O}:]^-$

4.11 Draw each polyatomic ion and assign formal charges. Each atom, except for hydrogen, should have an octet of valence electrons.
a. OH^- **b.** $NH_4{}^+$ **c.** CN^-

4.12 Draw each polyatomic ion and assign formal charges. Each atom, except for hydrogen, should have an octet of valence electrons.
a. $PO_4{}^{3-}$ **b.** $HPO_4{}^{2-}$ **c.** $H_2PO_4{}^-$

4.13 Draw each of the following and assign formal charges. Each atom should have an octet of valence electrons.
a. SO_3 **b.** $SO_3{}^{2-}$

4.14 Draw each molecule and assign formal charges. Each atom should have an octet of valence electrons.
a. CO **b.** SO_2

4.2 Polar Covalent Bonds, Shape, and Polarity

4.15 Label any polar covalent bond(s) in the molecules and ions that appear in Problem 4.9.

4.16 Label any polar covalent bond(s) in the molecules and ions that appear in Problem 4.10.

4.17 Specify the shape around each specified atom in Problem 4.9.

a. The S atom in H_2S.
b. The C atom in CO_3^{2-}.
c. The N atom in NCl_3.

4.18 Specify the shape around each specified atom ion in Problem 4.10.

a. The N atom in $CH_3NH_3^+$.
b. The C atom in CH_2O.
c. The Cl atom in ClO_2^-.

4.19 Which of the molecules are polar?

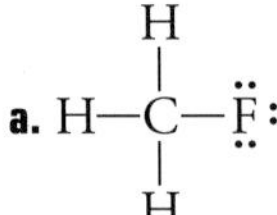

b. $:\ddot{F}-\overset{H}{\overset{|}{\underset{H}{\underset{|}{C}}}}-\ddot{F}:$ **c.** $:\ddot{F}-\overset{:\ddot{F}:}{\overset{|}{\underset{:\ddot{F}:}{\underset{|}{C}}}}-\ddot{F}:$

4.20 Which of the molecules are polar?

a. $H-\overset{H}{\overset{|}{\underset{H}{\underset{|}{C}}}}-\ddot{O}-H$ **c.** $H-\overset{H}{\overset{|}{\underset{H}{\underset{|}{C}}}}-\overset{\ddot{O}:}{\overset{\|}{C}}-\overset{H}{\overset{|}{\underset{H}{\underset{|}{C}}}}-H$

b. $:\ddot{Cl}-C\equiv C-\ddot{Cl}:$

4.21 True or false? A bent molecular shape always has a bond angle near 110°.

4.22 True or false? An atom surrounded by four groups of electrons always has a tetrahedral arrangement of atoms.

4.3 Noncovalent Interactions

4.23 Do hydrogen bonds form between formaldehyde molecules (Table 4.2)?

4.24 Do hydrogen bonds form between methanol molecules (Table 4.7)?

4.25 Which pairs of molecules can form a hydrogen bond with one another?

a. CH_3CH_3 and CH_3CH_3

b. $CH_3\overset{O}{\overset{\|}{C}}-H$ and $CH_3\overset{O}{\overset{\|}{C}}-H$

c. CH_3CH_2OH and CH_3CH_2OH

d. $CH_3\overset{O}{\overset{\|}{C}}-OH$ and $CH_3\overset{O}{\overset{\|}{C}}-OH$

4.26 Which pairs of molecules can form a hydrogen bond with one another?

a. $CH_2{=}CHCH_3$ and $CH_2{=}CHCH_3$
b. CH_3CH_2SH and CH_3CH_2SH
c. CH_3OH and CH_3OH
d. CH_3OCH_3 and CH_3OCH_3

4.27 Which of the molecules in Problem 4.25 can form a hydrogen bond with a water molecule?

4.28 Which of the molecules in Problem 4.26 can form a hydrogen bond with a water molecule?

4.29 A protein contains the following groups. Which can form salt bridges with one another?

a. $-CH_2OH$ and $HOCH_2-$

b. $-CH_2\overset{O}{\overset{\|}{C}}-O^-$ and $^+NH_3CH_2CH_2CH_2CH_2-$

c. $-CH_2\overset{O}{\overset{\|}{C}}-OH$ and $^+NH_3CH_2CH_2CH_2CH_2-$

d. $-\underset{CH_3}{\underset{|}{C}}HCH_3$ and CH_3-

4.30 **a.** Which of the pairs in Problem 4.29 interact primarily through hydrogen bonds?
b. Which interact primarily through London forces?

4.31 Which share the stronger London force interactions, two $CH_3CH_2CH_2CH_2CH_3$ molecules or two $CH_3CH(CH_3)CH_2CH_3$ molecules?

4.32 Which share the stronger London force interactions, two $CH_3CH_2CH_2CH_2CH_3$ molecules or two $CH_3CH_2CH_2CH_2CH_2CH_2CH_2CH_3$ molecules?

4.33 In ancient Greece, Socrates and others were executed by being forced to drink hemlock. The major poisonous ingredient in hemlock is an organic molecule called coniine.

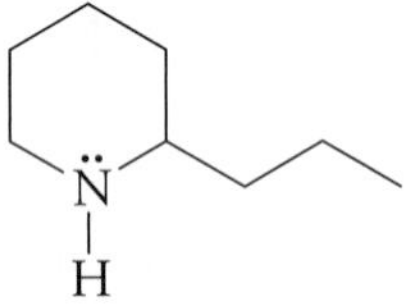

Coniine

Show two ways that a coniine molecule can form a hydrogen bond to a water molecule.

4.34 Can two coniine molecules (Problem 4.33) interact by London forces? Explain.

4.4 Alkanes

4.35 Which has the higher boiling point, $CH_3CH_2CH_2CH_2CH_3$ or $CH_3CH(CH_3)CH_2CH_3$? Explain.

4.36 Which has the higher boiling point, $CH_3CH_2CH_2CH_2CH_3$ or $CH_3CH_2CH_2CH_2CH_2CH_2CH_2CH_3$? Explain.

4.37 Arrange the molecules in order, from highest boiling point to lowest boiling point: decane, propane, butane.

4.38 Arrange the molecules in order, from highest boiling point to lowest boiling point: pentane, octane, methane.

4.39 Draw a line-bond structure of each alkane.

a. $CH_3CH_2C(CH_3)_3$

b. $CH_3CH_2CH(CH_3)CH(CH_3)CH_2CH_3$

4.40 Draw a line-bond structure of each alkane.

a. $CH_3CH(CH_2CH_2CH_3)CH_2CH_3$

b. $CH_3C(CH_2CH_3)_2CH(CH_3)CH_3$

4.41 Find and name the parent chain for each molecule, then give the complete IUPAC name for each.

a. $CH_3CH_2CHCH_2CH_3$
CH_3 (on the middle CH)

b. $CH_3CHCH_2CH_3$
$CH_2CH_2CH_3$ (on the CH)

c. $CH_2CH_2CH_3$ (above the C)
$CH_3CCH_2CH_3$
$CH_2CH_2CH_3$ (below the C)

4.42 Find and name the parent chain for each molecule, then give the complete IUPAC name for each.

a. $CH_3CH_2CH_2$
CH_3 (on the last CH_2)

b. CH_2CH_3 (above the second CH)
$CH_3CHCHCH_2CH_2CH_2CH_3$
$CH_3CH_2CH_2$ (below the first CH)

c. CH_3 (above the second CH)
$CH_3CH_2CH_2CH_2CHCH_2CHCH_3$
CH_2CH_3 (below the first CH)

4.5 Constitutional Isomers

***4.43** Draw and name six of the constitutional isomers with the formula C_7H_{16}.

***4.44** Draw and name six of the constitutional isomers with the formula C_8H_{18}.

4.45 Which pairs of molecules are constitutional isomers? Which are identical?

a. $CH_3CH_2CH_2CH_2CH_2CH_3$ and
CH_3CHCH_3
$CH_3CH_2CH_2$ (attached to the CH)

b. $CH_2CH_2CH_2CH_3$ with CH_2CH_3 (below the first CH_2) and
$CH_2CH_2CH_3$ with $CH_3CH_2CH_2$ (below the first CH_2)

c. $CH_3CHCH_2CH_2CH_3$ with CH_3 (on the CH) and
$CH_3CHCHCH_3$ with CH_3 (above the first CH) and CH_3 (below the second CH)

4.46 Which pairs of molecules are constitutional isomers? Which are identical?

a. $CH_3CH_2CH_2CH_3$ and
CH_3
$CH_3CH_2CH_2$ (the CH_3 attached to the last CH_2)

b. CH_3CHCH_3 with CH_3 (on the CH) and
$CH_3CH_2CH_2$
CH_3 (attached to the last CH_2)

c. $CH_3CHCHCH_3$ with H_3C and CH_3 (below the two CH groups) and
$CH_3CHCHCH_3$ with CH_3 (above the first CH) and CH_3 (below the second CH)

4.47 Which are constitutional isomers?

a. pentane and 2-methylpentane

b. 2-methylpentane and 3-methylpentane

c. 2,2-dimethylpropane and pentane

d. 2,2-dimethylpropane and cyclopentane

4.48 Which are constitutional isomers?

a. hexane and 3-methylpentane

b. hexane and 2,3-dimethylpentane

c. hexane and 2,3-dimethylbutane

d. hexane and cylcohexane

4.6 Conformations

***4.49** Which pairs of molecules are constitutional isomers? Which are different conformations of the same molecule?

a. H, H, H_3C on C–C with CH_3, H, H and H_3C, H, H on C–C with CH_3, H, H

b. H, H, H_3C on C–C with CH_3, H, H and H_3C, H, H_3C on C–C with H, H, H

***4.50** Which pairs of molecules are constitutional isomers? Which are different conformations of the same molecule?

a. H, H, H_3C on C–C with CH_3, H, CH_2CH_3 and H, H, H on C–C with CH_2CH_3, CH_3, CH_3

b. H, H, H_3C on C–C with CH_3, H, CH_2CH_3 and H_3C, H, H on C–C with CH_2CH_3, CH_3, H

4.51 **a.** How are constitutional isomers and conformations similar?

b. How are constitutional isomers and conformations different?

4.52 True or false?

a. Constitutional isomers have the same IUPAC name.

b. The different conformations of an alkane have the same IUPAC name.

4.7 Cycloalkanes

4.53 Draw and name the three ethylmethylcyclobutane constitutional isomers.

4.54 Draw and name three ethyldimethylcyclohexane constitutional isomers.

4.55 Give the correct IUPAC name for each molecule.

a. 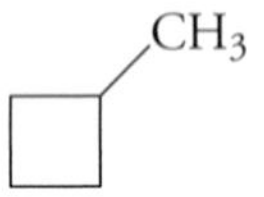 CH_3

b. 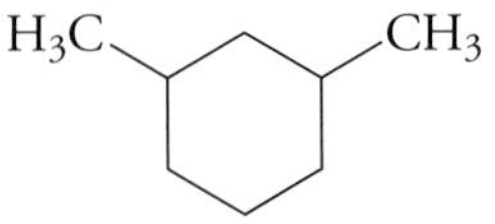 H_3C CH_3

c. 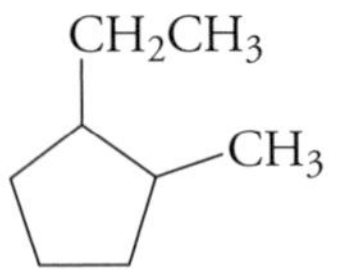 CH_2CH_3 CH_3

d. 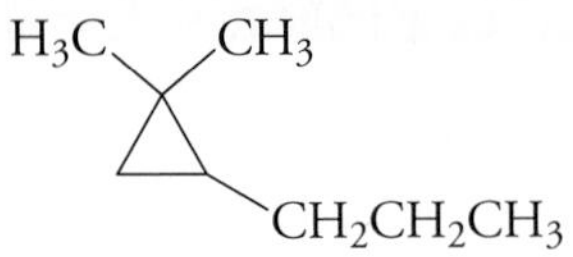 H_3C CH_3 $CH_2CH_2CH_3$

4.56 Give the correct IUPAC name for each molecule.

a. 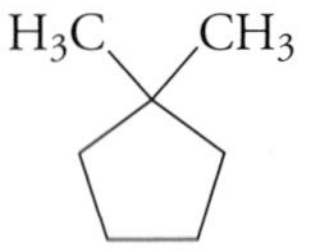 H_3C CH_3

b. 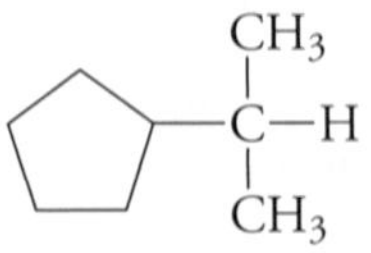 CH_3 C—H CH_3

c. $CH_2CH_2CH_3$ $CH_2CH_2CH_2CH_3$

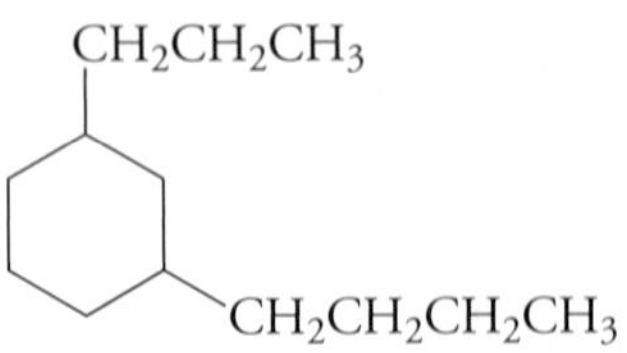

d. 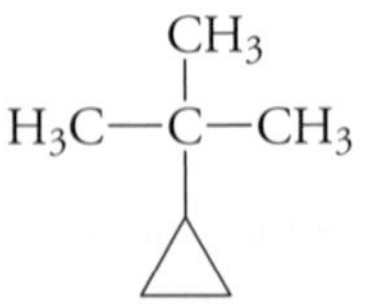 CH_3 H_3C—C—CH_3

4.57 Which molecule(s) in Problem 4.55 can exist as *cis* and *trans* isomers?

4.58 Which molecule(s) in Problem 4.56 can exist as *cis* and *trans* isomers?

4.59 Draw a side view of each cycloalkane. (See Figure 4.15 for examples.)

a. *trans*-1,2-dimethylcyclohexane

b. *trans*-1-ethyl-2-methylcyclohexane

c. *cis*-1,3-diethylcyclopentane

4.60 Draw a side view of each cycloalkane. (See Figure 4.15 for examples.)

a. *cis*-1,2-dimethylcyclopentane

b. *trans*-1,2-dimethylcyclopentane

c. *cis*-1-ethyl-2-isopropylcyclohexane

4.61 Give the complete IUPAC name (including the use of the term *cis* or *trans*) for each molecule.

a. 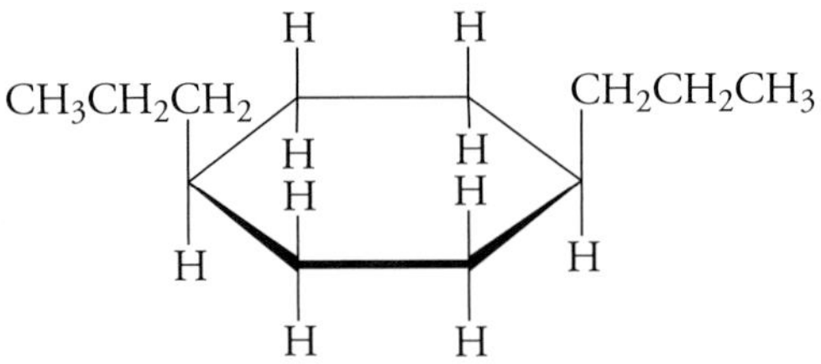 $CH_3CH_2CH_2$ $CH_2CH_2CH_3$

b.

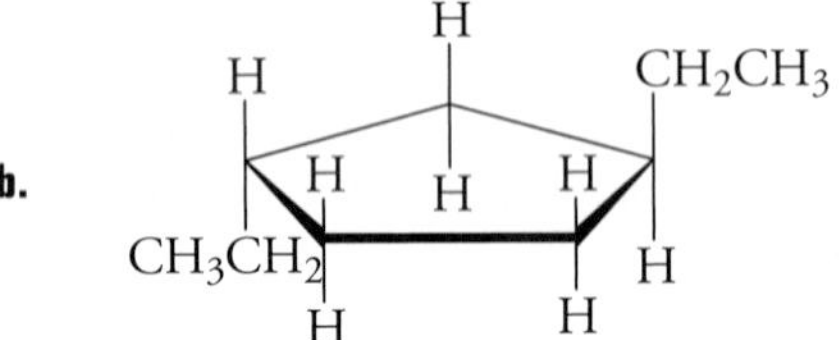

4.62 Give the complete IUPAC name (including the use of the term *cis* or *trans*) for each molecule.

a. 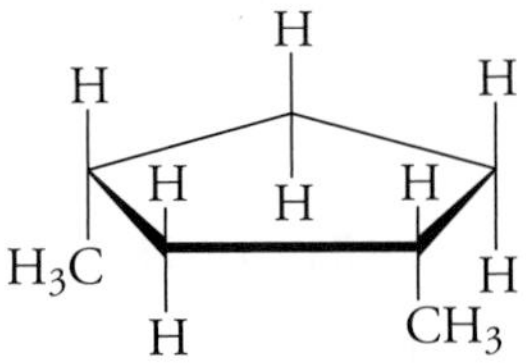 H_3C CH_3

b. 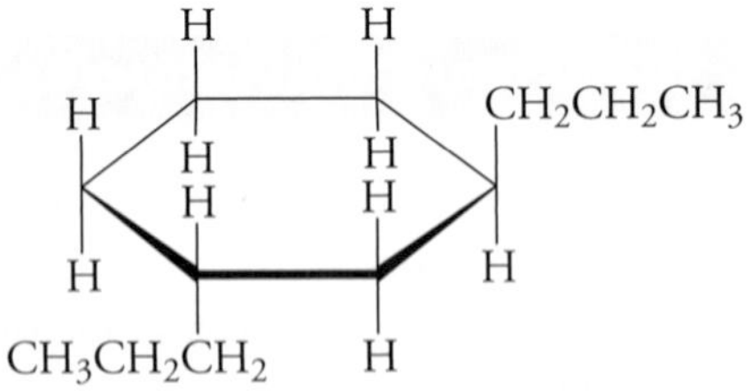 $CH_2CH_2CH_3$ $CH_3CH_2CH_2$

4.8 Alkenes, Alkynes, and Aromatic Compounds

4.63 Draw propene, showing the proper three-dimensional shape about each atom.

4.64 Draw propyne showing the proper three-dimensional shape about each atom.

4.65 Name each molecule.

a. $CH_2{=}CHCH_2CH_3$

b. $CH_3CHCH{=}CCH_3$ (with CH_3 on C2 and CH_3 on C4)

c. $CH_3C{\equiv}CCHCH_3$ (with CH_3 on C4)

4.66 Name each molecule.

a. $CH_3CH{=}CCH_3$ (with CH_3 on C3)

b. $HC{\equiv}CCCH_3$ (with two CH_3 on C3)

c. $CH_3CHCH{=}CHCH_3$ (with CH_3 on C2)

4.67 Draw each molecule.
a. 3-isopropyl-1-heptene
b. 2,3-dimethyl-2-butene
c. 5-*sec*-butyl-3-nonyne

4.68 Draw each molecule.
a. 2,3-dimethyl-3-hexene
b. 3,4-dimethyl-3-hexene
c. 4,5-dimethyl-1-hexyne

4.69 The molecule shown is a termite trail marking pheromone. Which double bonds are *cis* and which are *trans*?

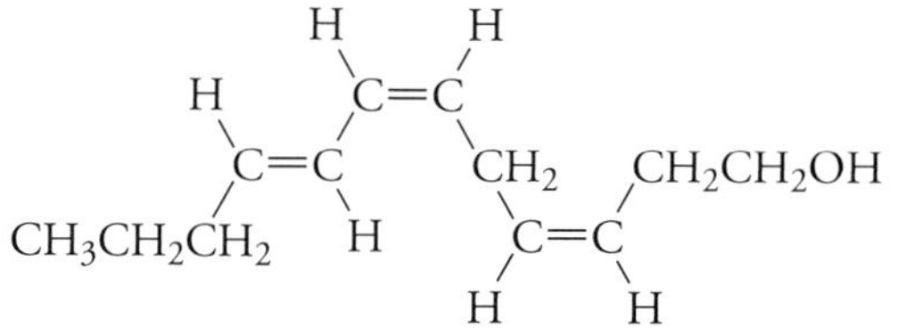

H H H C=C C=C CH₂ CH₂CH₂OH CH₃CH₂CH₂ H C=C H H

4.70 Name each molecule.

a. (benzene ring with H_3C and $CH_2CH_2CH_3$)

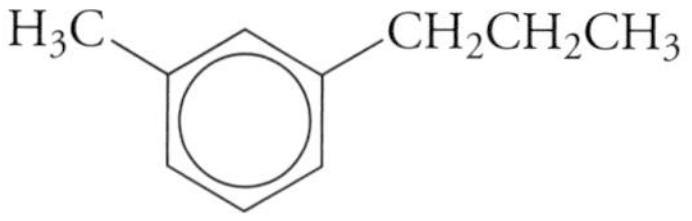

b. (benzene ring with CH_3CHCH_3)

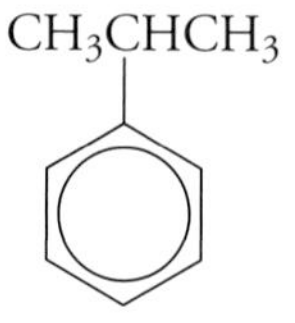

c. (benzene ring with H_3C, CH_3, CH_3)

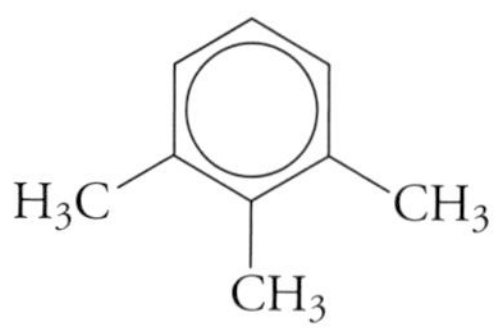

4.71 Name each molecule.

a. (benzene ring with H_3C and $CH_2CH_2CH_3$)

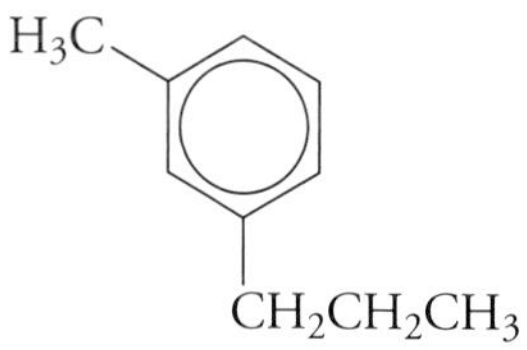

b. (benzene ring with CH_3, CH_3, CH_3)

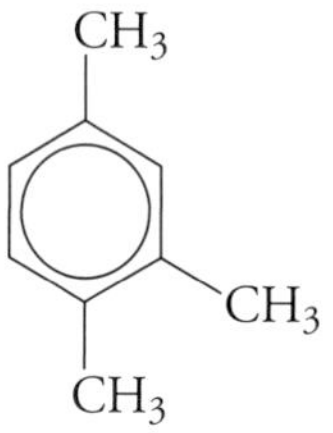

c. (benzene ring with CH_3CH_2, CH_3CH_2, and $C(CH_3)_3$: CH_3, CH_3, H_3C)

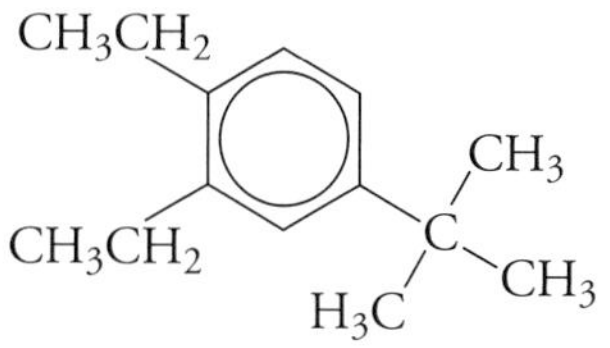

4.72 Draw each molecule.
a. 1,3-dipropylbenzene
b. *p*-diethylbenzene
c. 4-isobutyl-1,2-dimethylbenzene

4.73 Draw each molecule
a. 1,2-dipropylbenzene
b. *m*-diisopropylbenzene
c. 1,2,4-trimethylbenzene

*__4.74__ Label the indicated carbon–carbon double bonds as *cis* or *trans*.

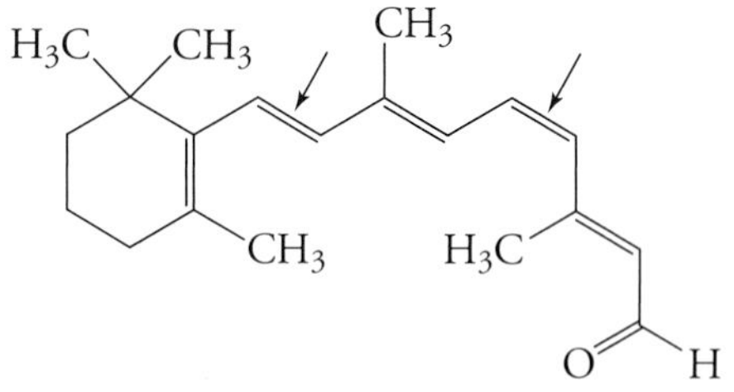

H_3C CH_3 CH_3 CH_3 H_3C O H

11-*cis*-Retinal
(a derivative of vitamin A)

4.75 Is the double bond in octyl methoxycinnamate (Figure 4.22) *cis*, *trans*, or neither?

4.76 Does β-myrcene, a compound present in lemon grass, bay, and hops, contain any *cis* or *trans* double bonds?

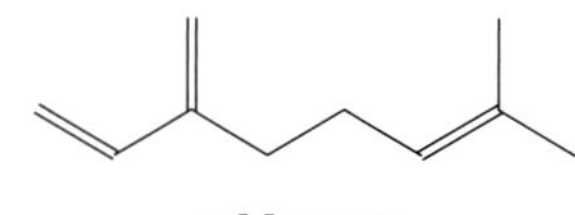

β-Myrcene

4.9 Alcohols, Carboxylic Acids, and Esters

4.77 List the primary noncovalent attraction between each pair of molecules.

a. $CH_3CH_2CH_2OH$ and $CH_3CH_2CH_2OH$

b. $CH_3\overset{O}{\overset{\|}{C}}OH$ and $CH_3\overset{O}{\overset{\|}{C}}OH$

c. $H-\overset{O}{\overset{\|}{C}}-OCH_3$ and $H-\overset{O}{\overset{\|}{C}}-OCH_3$

4.78 This molecule is an elephant sex pheromone.

$$\begin{array}{c} H \quad\; H \\ \diagdown\;\diagup \\ C{=}C \\ \diagup\quad\diagdown \end{array}$$
$$CH_3CH_2CH_2CH_2 \qquad CH_2CH_2CH_2CH_2CH_2CH_2O-\overset{O}{\overset{\|}{C}}CH_3$$

a. Which functional groups are present in this molecule?

b. Which geometric isomer is present?

c. Is the molecule saturated or unsaturated?

4.79 Dovonex is a prescription drug used to treat psoriasis.

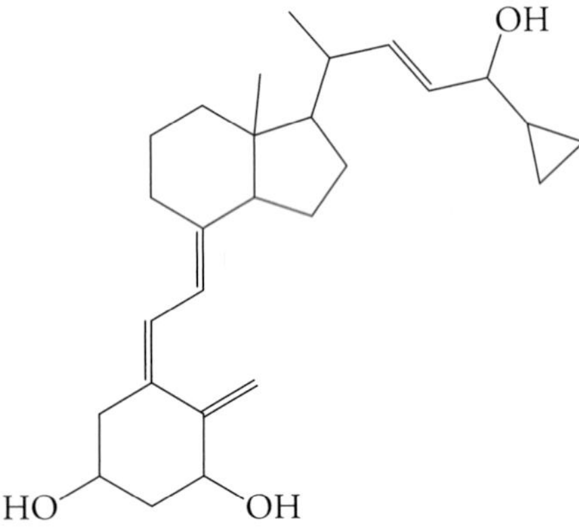

Dovonex

a. Which functional groups are present in this molecule?

b. Indicate which geometric isomer is present for the alkene group at the top of the structure.

c. Are any fused rings present in this molecule?

d. Three different types of cycloalkane rings are present. Identify them.

e. Is this molecule saturated or unsaturated?

HEALTH*Link* | *Prion Diseases*

4.80 Are covalent bonds broken when PrP^{C} is converted into PrP^{SC}? Explain.

4.81 Suggest a way to reduce the spread of mad cow disease between cattle.

HEALTH*Link* | *Sunscreens*

4.82 What properties are important for molecules used as sunscreens?

4.83 When applied to the skin of mice, Forskolin, a compound present in an Asian plant, was shown to increase the production of melanin. Which, do you suppose, were the results of this scientific study?

a. The mice tanned more quickly.

b. The mice did not sunburn as easily.

c. The mice were less susceptible to skin cancer.

4.84 **a.** "Tanning and Health" presented the results of a study in which mice were exposed to UV radiation. Based on this study, which leads to more skin cancer, a single dose of UV that causes sunburn or small doses of UV over several weeks?

b. While this video-clip describes hazards associated with exposure to UV radiation, it also presents an alternative view which contends that UV exposure is a good thing. How might UV exposure be beneficial to your health?

BIOCHEMISTRY*Link* | *Odor and Flavor*

*__4.85__ Although esters are generally known to have pleasant odors, many large esters have no odor. Explain.

4.86 Would you expect esters that are constitutional isomers of one another to have the same odor? Explain.

THINKING IT THROUGH

4.87 The term "organic" can have different meanings. What are two different ways to interpret a sign at the grocery store that reads "organic foods"?

4.88 If you were given the assignment to come up with a few likely candidates for new esters that might have the odor and flavor of grapes, where would you begin? (Refer to Figure 4.24.)

INTERACTIVE LEARNING PROBLEMS

ILW **4.89** What is the IUPAC name for:

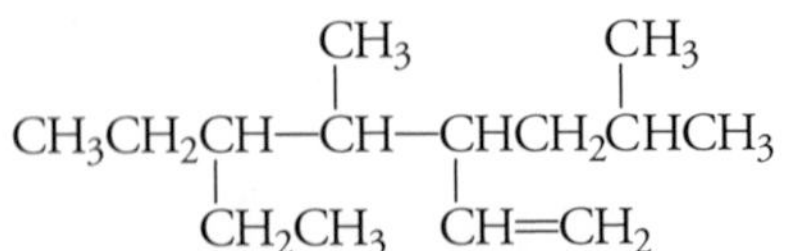

SOLUTIONS TO PRACTICE PROBLEMS

4.1 :S̈=C=S̈:

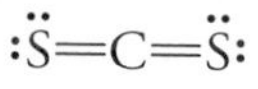

4.2 H_3N^+–CH₂–C(=O)–O⁻

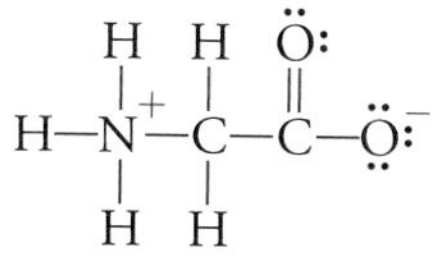

4.3 ⁻:Ö–P̈–Ö:⁻ with :Ö:⁻ on P

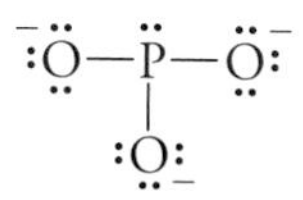

4.4 $CH_3CH_2NH_2$ NH₂

4.5 δ– O, δ– O, δ+ H, δ+, δ–, δ+ O δ+, O δ–

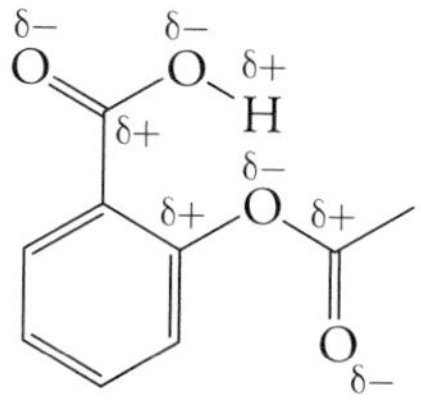

4.6 **b.** and **c.**

4.7 Yes. The O atom can form a hydrogen bond with an H atom in a water molecule.

4.8 **a.** 2,4-dimethylheptane; **b.** 4-ethyl-2,3-dimethylheptane

4.9 Constitutional isomers.

4.10

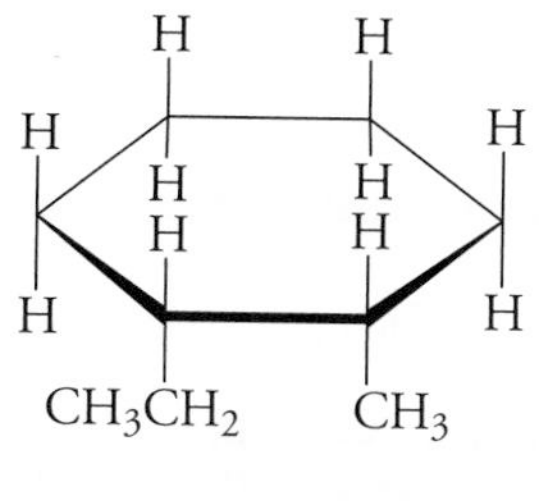

cis-1-ethyl-2-methylcyclohexane

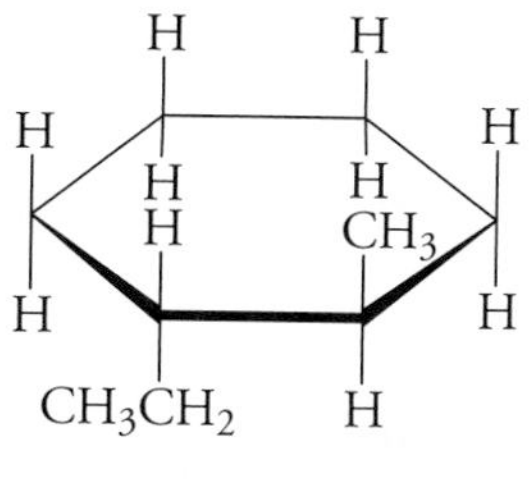

trans-1-ethyl-2-methylcyclohexane

4.11 $CH_3CH_2CH_2$–CH=CH–CH=CH–$CH_2CH_2CH_2CH_2CH_2CH_2CH_2CH_2CH_2OH$ (cis, trans)

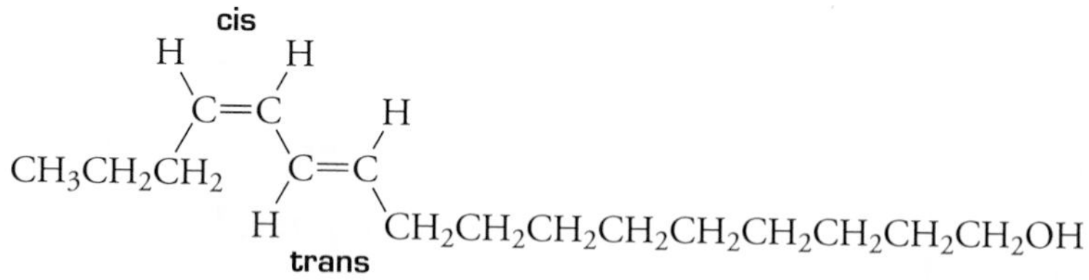

One day at the gym a student runs into a friend who is a nurse practitioner. When asked how she is doing, the friend excitedly tells about her invitation to join the medical support staff for a Himalayan climbing expedition. One of her jobs will be to keep records of the oxygen levels in the climbers' blood as they reach high altitudes. "Why are those records kept?" the student asks. His friend replies that at high altitudes, many climbers begin experiencing headaches, nausea, shortness of breath, and other symptoms of acute mountain sickness. In an attempt to find out whether this sickness and other climbing-related diseases are related to a drop in blood oxygen levels, many climbing expeditions monitor the climbers.

Photo source: © Atalante/Gamma.

5 GASES, LIQUIDS, AND SOLIDS

Air is a mixture of gases, and each breath that you take contains about 1×10^{22} atoms and molecules, about 99% of which are either N_2 or O_2. As we saw in Section 1.2, matter is typically found in three different physical states, as a solid, a liquid, or a gas. From direct experience we know that solids have fixed shapes and volumes, that liquids have variable shapes and fixed volumes, and that gases have variable shapes and volumes. In this chapter we will see how these differences can be explained by studying the group behavior of atoms and molecules. We will also take a look at various properties of solids, liquids, and gases, including the processes involved in getting oxygen into the blood and distributed throughout the body.

objectives

After completing this chapter, you should be able to:

1. Define the terms specific heat, heat of fusion, and heat of vaporization. Show how each can be used in calculations involving energy.
2. Describe the meaning of the terms enthalpy change, entropy change, and free energy change. Explain how the value of free energy change can be used to predict whether a process is spontaneous or nonspontaneous.
3. Convert between common pressure units.
4. List the variables that describe the condition of a gas and give the equations for the various gas laws.
5. Explain Dalton's law of partial pressure.
6. Define the terms density and specific gravity.
7. Describe the relationship between atmospheric pressure and the boiling point of a liquid.
8. Describe the difference between amorphous and crystalline solids and describe the makeup of the four classes of crystalline solids.

5.1 STATES OF MATTER AND ENERGY

In our earlier discussion of matter we saw that it is typically found in one of three different phases or states—as a solid, a liquid, or a gas. Having also learned about some of the properties of atoms and compounds, we can now take a closer look at the differences between these three phases. One major factor that is responsible for the varied behavior of solids, liquids, and gases is the nature of the interaction that attracts one particle (atom, ion, or molecule) to another.

Let us take a look at the three phases of water: $H_2O(s)$ (ice), $H_2O(l)$ (water), and $H_2O(g)$ (steam). Here, the notations (*s*), (*l*), and (*g*) refer, respectively, to solid, liquid, and gas. In ice, the water molecules are held fairly rigidly in place relative to one another (for water the intermolecular force doing so is primarily hydrogen bonding), giving this solid a fixed shape and volume. If ice is heated until it melts, liquid water is formed in which the molecules have a greater kinetic energy than in ice. (The higher the temperature of something, the greater the kinetic energy of the particles from which it is made.) Although the water molecules still interact with one another through hydrogen bonds, their increased motion allows them to slip and slide past one another (Figure 5.1). If water is heated until it boils, gaseous water is formed. The even greater kinetic energy allows the water molecules to separate completely from one another and move freely throughout the container that holds them.

Moving from any one of these phases to another (solid to liquid, for example) is a physical change. In terms of the language of Section 1.2, this is because the *chemical composition of the matter does not change.* When ice melts to form liquid water, individual water molecules are not broken apart—only the nature of the interactions between molecules is changed.

Let us fill in a few more details about phase changes. Beginning with ice at a temperature of −20°C (−4°F), for example, and gradually adding heat energy to warm it, we will see an increase in the temperature. When the temperature reaches 0°C (32°F), the melting point of ice or the freezing point of water, the temperature remains constant—even as more heat is added—until all of the ice has melted (Figure 5.2). The energy put in during this melting process is called the **heat of fusion**. With the continued addition of heat energy, water temperature rises until it reaches 100°C (212°F), the boiling point of water at standard atmospheric pressure (Section 5.2). As the water begins boiling, the temperature remains constant as heat is added, until all of the water has been converted to steam. The energy that goes into converting water from the liquid to the gas phase is called the **heat of vaporization**. Once the water has all boiled, the addition of more heat causes the temperature of the steam to rise.

- Heat of fusion is the heat required to melt a solid.
- Heat of vaporization is the heat required to evaporate a liquid.

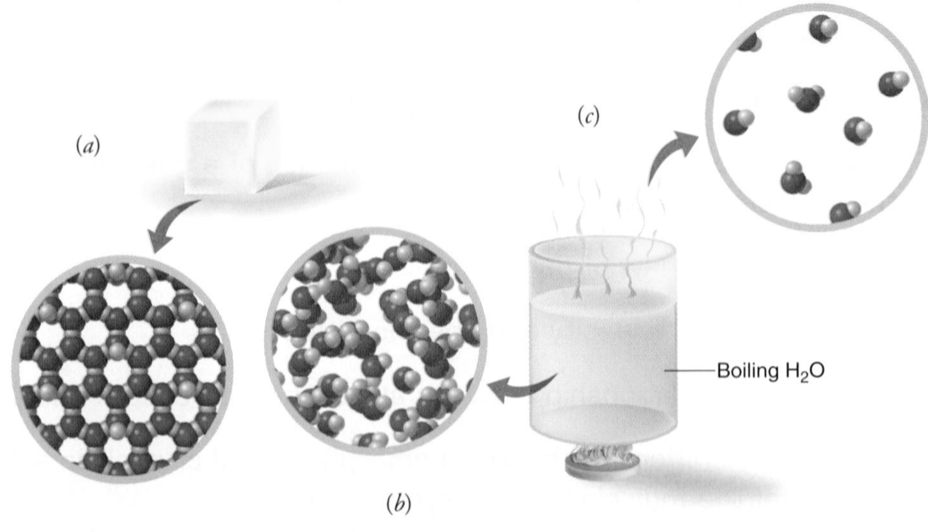

FIGURE 5.1

Phases of water
(*a*) In ice, water molecules are locked into place by hydrogen bonds to neighboring molecules. (*b*) In liquid water these noncovalent interactions are not as effective, due to increased motion associated with higher temperatures. Water molecules can slip past one another. (*c*) In steam, the water molecules do not interact with one another.

Source: BROWN, THEODORE E; LEMAY, H. EUGENE; BURSTEN, BRUCE E, CHEMISTRY; THE CENTRAL SCIENCE, 10th edition, © 2006, p.445. Adapted by permission of Pearson Education, Inc., Upper Saddle River, NJ.

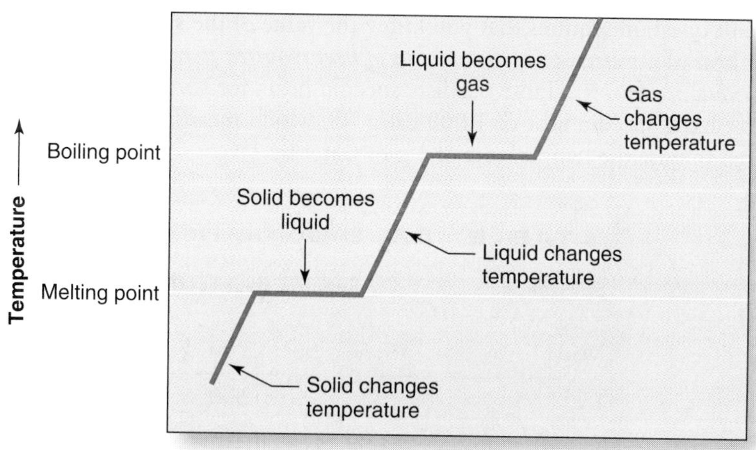

FIGURE 5.2

Phase changes of water
The energy required to convert ice into water is called the heat of fusion. The energy required to convert water into steam is the heat of vaporization.

This process can be reversed. As heat energy is removed from steam, its temperature drops. At a temperature of 100°C, where steam condenses to form liquid water, the temperature remains constant until only water is present. Further loss of heat energy lowers the temperature of water until, at 0°C, water begins to freeze. Again, the temperature remains at 0°C until all of the water has been converted into ice. Removal of more heat energy lowers the temperature of the ice.

Under certain conditions, some substances will skip the liquid phase and jump directly between the liquid and gas phases. The conversion of a solid directly into a gas is called **sublimation** and the reverse of this process is called **deposition**. Dry ice, $CO_2(s)$, is a common example of a compound that undergoes sublimation (Figure 5.3).

FIGURE 5.3

Sublimation
The sublimation of dry ice involves the direct conversion of $CO_2(s)$ into $CO_2(g)$.

Source: Charles D. Winters/ Photo Researchers, Inc.

SAMPLE PROBLEM 5.1

Energy and changes in physical state

To reduce a fever, rubbing alcohol (2-propanol) can be applied to the skin. As the alcohol evaporates (liquid becomes gas), the skin cools. Explain the changes in heat energy as this process takes place. Note: 2-Propanol vapors are flammable, so care must be taken when using this technique.

Strategy

To answer this question you must decide whether heat energy must be put into or removed from rubbing alcohol to convert it into a gas.

Solution

The heat energy required to convert rubbing alcohol from a liquid to a gas is provided by the heat in the skin. As heat moves from the skin into the rubbing alcohol, the skin cools.

PRACTICE PROBLEM 5.1

Refrigerators contain tubes that are filled with a liquid that is converted into a gas at a relatively low temperature. Which process do you suppose helps to maintain cool temperatures inside of a refrigerator, converting the liquid into a gas or converting the gas back into a liquid?

We can also look at the physical changes just described from a mathematical standpoint. For example, we might wonder exactly how many calories of heat energy are required to raise the temperature of 2.5 g of water from 20.0°C to 25.0°C. Answering this type of question requires that you know the value of the **specific heat** of water. The specific heat of a substance is *the amount of heat required to raise the temperature of 1 gram of the substance by 1°C.* Table 5.1 lists specific heats for several elements and compounds. Water has a specific heat of 1.000 cal/g °C, which means that 1.000 cal of energy is required to raise the temperature of 1 g of water by 1°C.

TABLE | 5.1 SPECIFIC HEAT

Substance	Specific Heat (cal/g °C)
Water	1.000
Ice[a]	0.500
Steam[b]	0.480
Ethanol$_{(\ell)}$	0.586
Copper$_{(s)}$	0.0924
Aluminum$_{(s)}$	0.0215
Gold$_{(s)}$	0.0310

[a] From −10°C to 0°C.
[b] At constant pressure.

To calculate the energy needed to increase the temperature of 2.5 g of water by 5.0°C, we can apply the factor label method, using specific heat as a conversion factor (Section 1.6). The conversion factors for specific heat calculations involving water are

$$\frac{1.000 \text{ cal}}{\text{g °C}} \quad \text{and} \quad \frac{\text{g °C}}{1.000 \text{ cal}}$$

Selecting the conversion factor which allows all units but calories to be canceled gives

$$2.5 \text{ g} \times 5.0 \text{ °C} \times \frac{1.000 \text{ cal}}{\text{g °C}} = 13 \text{ cal}$$

This calculation shows that an input of 13 cal of heat energy will raise the temperature of 2.5 g of water by 5.0°C.

Energy calculations may also be carried out using values in Table 5.2 for heat of fusion (the heat required to melt a solid) and heat of vaporization (the heat required to vaporize a liquid). For example, to determine the number of calories needed to melt 155 g of ice at 0°C, we use the heat of fusion of water (79.7 cal/g) to create two conversion factors.

$$\frac{79.7 \text{ cal}}{\text{g}} \quad \text{and} \quad \frac{\text{g}}{79.7 \text{ cal}}$$

The conversion factor to use is the one with the desired unit (cal) in the numerator.

$$155 \text{ g} \times \frac{79.7 \text{ cal}}{\text{g}} = 1.24 \times 10^4 \text{ cal} = 12.4 \text{ kcal}$$

TABLE | 5.2 HEAT OF FUSION AND HEAT OF VAPORIZATION

Substance	Heat of Fusion (cal/g)	Heat of Vaporization (cal/g)
Water	79.7	540
Ethanol	26.05	230
2-Propanol	21.37	159

SAMPLE PROBLEM 5.2

Calculations involving heat energy

A patient with a fever is sponged with 75 g of 2-propanol. How much heat energy is drawn from the patient when the alcohol vaporizes?

Strategy

This problem deals with the vaporization of 2-propanol, so you will need to get the appropriate heat of vaporization value from Table 5.2. Then, heat of vaporization can be used as a conversion factor to convert from grams into calories.

Solution

The heat of vaporization for 2-propanol is 159 cal/g.

$$75 \text{ g} \times \frac{159 \text{ cal}}{\text{g}} = 1.2 \times 10^4 \text{ cal} = 12 \text{ kcal}$$

PRACTICE PROBLEM 5.2

A 55 g block of aluminum and a 55 g block of copper are placed side by side on the burner of a stove. When the stove is turned on, the temperature of each metal increases from 25°C to 150°C. Which metal has absorbed more heat energy?

Enthalpy, Entropy, and Free Energy

Section 1.2 introduced the idea that energy (*the ability to do work and to transfer heat*) is involved whenever matter undergoes changes, including the phase changes just described. We will now explore this connection between energy and phase changes in more detail.

An important question to ask is *why* some changes are **spontaneous** (continue to occur once they are started) and others are **nonspontaneous** (will not run by themselves unless something keeps them going). Energy is a key factor in determining which term applies—a process is spontaneous *if it releases energy*. For example, if a rock sitting at the top of a hill is given a push, it spontaneously rolls down the hill and releases its stored potential energy as it does so. If a rock sitting at the bottom of a hill is given a push, it will not roll up the hill by itself (a nonspontaneous process), because this change requires the continual input of energy (Figure 5.4).

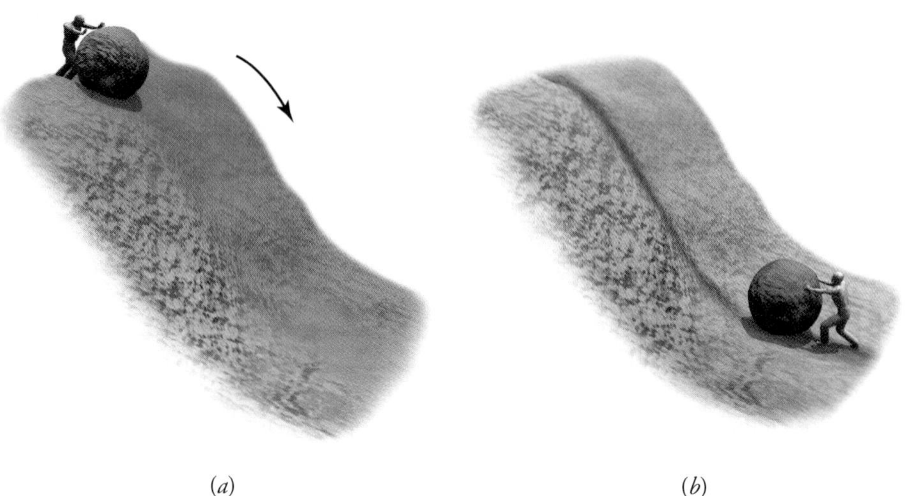

(*a*) (*b*)

FIGURE 5.4

Spontaneous and nonspontaneous processes
(*a*) *Spontaneous.* Once it begins moving, a rock will continue to roll down a hill.
(*b*) *Nonspontaneous.* A rock will not spontaneously roll up a hill.

FIGURE 5.5

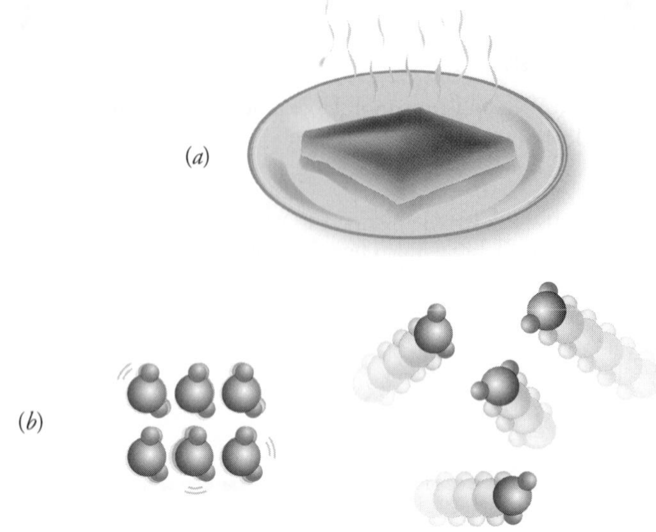

Enthalpy and entropy
(*a*) The loss of heat energy (enthalpy) from a hot piece of toast is spontaneous. (*b*) Gas molecules (left) will move away from one another. This increase in the distribution of energy (entropy) is spontaneous.

■ Enthalpy (H) is a measure of heat.

■ Entropy (S) is a measure of randomness.

From the standpoint of chemistry, there are three energy-related factors that determine whether or not a process is spontaneous. The first of these is heat, also known as **enthalpy**, which is represented by the symbol H. When discussing the change in heat (ΔH) for a process (the symbol Δ stands for *change*), we call a release of heat **exothermic** (exo means "out") and a gain of heat **endothermic** (endo means "in"). How is ΔH related to whether or not a change is spontaneous? Experience tells you that if you pull a hot piece of toast from the toaster and lay it on a plate, the toast will cool down (exothermic process) rather than get warmer (endothermic process). Heat always flows from matter at a higher temperature (the toast) to matter at a lower temperature (the surrounding air). Another way of saying this is that *the loss of heat is spontaneous and the gain of heat is nonspontaneous* (Figure 5.5*a*).

The second factor in determining whether or not a process is spontaneous is **entropy**. The entropy change (ΔS) for a process is related to how energy is distributed. To see what this means, consider two drawings in Figure 5.5*b*. If you were told that they represent the "before" and "after" of a collection of gas phase water molecules, which would you guess is the "after"? In fact, the drawing on the left is the initial state and that on the right is the final state. When the water molecules are packed together (left side), their overall energy is concentrated into one small area. For our purposes, assume that they vibrate to some extent, but are not free to move throughout the container. When the water molecules have spread out (right side), their energy is distributed in a greater number of ways—now they vibrate, rotate, and move through the container. In each case the total energy is the same. The drawing on the right, however, represents a state with greater entropy, because the energy is more dispersed. *An increase in entropy (more energy dispersal) is spontaneous and a decrease in entropy (less energy dispersal) is nonspontaneous.*

The third factor in determining spontaneous versus nonspontaneous processes is temperature (T). A change in temperature can convert a spontaneous process into a nonspontaneous one and vice versa. Before discussing the effect of temperature, however, let us reconsider the water–ice phase change. Under which conditions is ice spontaneously converted to water and under which conditions is the reverse (water to ice) spontaneous?

Heat must be added to melt ice, so looking only at the enthalpy change for the process, the ice to water conversion is endothermic and nonspontaneous. Heat must be removed to freeze water, so in terms of enthalpy, the water to ice phase change is exothermic and spontaneous.

The water molecules in liquid water have more freedom to move (their energy is more dispersed) than in ice, so looking just at entropy change, the ice to water conversion is spontaneous. The reverse process (water to ice) is nonspontaneous in terms of entropy change.

How does temperature enter into the picture? Suppose that you pour a glass of water on the sidewalk and place an ice cube in the puddle (Figure 5.6). On a warm day (25°C) you would expect the ice cube to melt (spontaneous). For the ice to melt spontaneously,

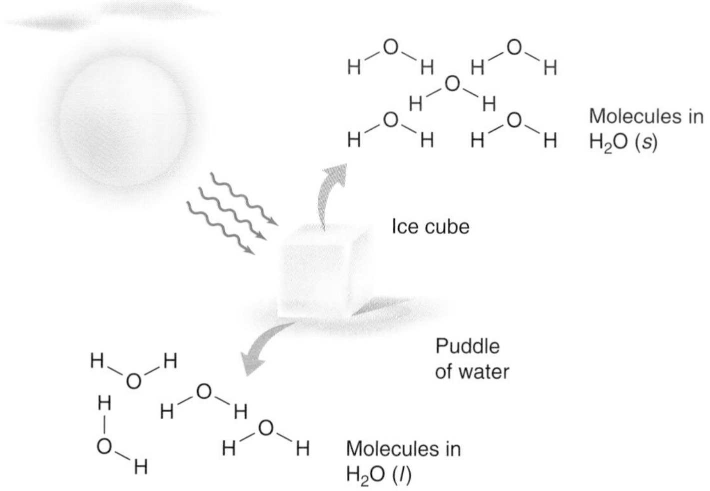

Molecules in H_2O (*s*)

Ice cube

Puddle of water

Molecules in H_2O (*l*)

FIGURE 5.6

Ice and water

The melting of an ice cube is endothermic (heat is required) so, in terms of enthalpy, the process is nonspontaneous. The energy of the molecules in ice is less dispersed than in liquid water so, in terms of entropy, the melting of ice is spontaneous. At 25°C ice spontaneously melts (ΔG is negative). At −25°C water spontaneously freezes (ΔG is negative).

at this temperature the contribution of entropy (favors melting) must be more significant than the contribution of enthalpy (favors freezing). On a cold day (−25°C), you would expect the water to freeze. At this temperature the contribution of enthalpy (favors freezing) must be more significant than that of entropy (favors melting).

Rather than trying to keep track of the separate effects of enthalpy, entropy, and temperature, all three are grouped into what is called **free energy** (G), also known as *Gibbs free energy*, after the chemist Willard Gibbs who made important contributions to our understanding of energy changes. Mathematically, these terms are related by the equation

$$\Delta G = \Delta H - T\Delta S$$

While we will not do any calculations involving this equation, it is useful for us to see the roles that enthalpy, entropy, and temperature play in determining ΔG. When the temperature (T) is high enough, the $T\Delta S$ term can become more important than ΔH in determining the overall ΔG. At low temperatures the ΔH term can be more important.

The last thing to mention about free energy at this point is how the value of ΔG is used to indicate whether a process is spontaneous or nonspontaneous. By convention, *a spontaneous process has a negative* ΔG and *a nonspontaneous process has a positive* ΔG. At 25°C, for example, ΔG for the conversion of ice into water has a value of −130 calories per mole of water (−130 cal/mol). The negative sign indicates that the process is spontaneous and allows us to predict that the ice should melt at this temperature. At a temperature of −25°C, ΔG for the conversion of ice to water equals +130 cal/mol. This process is nonspontaneous and we would not expect the ice to melt. The reverse process (water freezing) at this same temperature is spontaneous, however, and ΔG equals −130 cal/mol.

SAMPLE PROBLEM 5.3

Enthalpy, entropy, and free energy

a. In terms of enthalpy, is the conversion of water into steam spontaneous or nonspontaneous?

b. In terms of entropy, is the conversion of water into steam spontaneous or nonspontaneous?

c. At a temperature of 50°C and at normal atmospheric pressure is the conversion of water into steam spontaneous or nonspontaneous?

Strategy

To answer part a, you must decide whether heat must be added or removed to convert water into steam. To answer part b, think in terms of which phase (liquid or gas) allows molecules the greatest energy dispersion. To answer part c consider the boiling point of water.

Solution

a. Nonspontaneous. To boil water (convert water into steam), heat energy must be added.

b. Spontaneous. When water (a liquid) is converted into steam (a gas), the water molecules become separated from one another and their energy is more widely distributed. This process is similar to what happens when a pile of closely stacked marbles collapses and the marbles roll across the floor.

c. Nonspontaneous. At normal atmospheric pressure water boils at 100°C. At 50°C water will not boil spontaneously. At this temperature, the contribution of enthalpy ΔH outweighs the contribution of $T\Delta S$ to the overall ΔG for the process.

PRACTICE PROBLEM 5.3

Before going on a picnic on a hot summer day, you stop by the store and pick up a block of dry ice, $CO_2(s)$.

a. In terms of ΔH alone, is the sublimation of dry ice spontaneous or nonspontaneous?

b. In terms of ΔS alone, is the sublimation of dry ice spontaneous or nonspontaneous?

c. At 50°C does ΔG for the sublimation of dry ice have a negative or a positive value?

d. When placed in an ice chest, dry ice will keep food cold. Where does the heat go that is removed from the food?

5.2 GASES AND PRESSURE

In dealing with gases, one of the important properties to consider is **pressure**. We saw above that the particles that make up a gas do not interact with one another and are free to move about any container that holds them. Gas pressure is the force of collisions that take place between these particles and an object (the walls of a container that holds the gas, this book, your skin, etc.)

When a weather reporter says that a low pressure system is developing over the Pacific Ocean, he or she is referring to atmospheric pressure. The source of this pressure is the mass of all of the air above us in the atmosphere. For example, a column of air 1 meter in diameter that extends upward from the surface of the earth 32 km (20 miles), to the edge of the atmosphere, has a mass of about 10,000 kg (Figure 5.7). If you stand at sea level, this atmospheric mass pushes on all parts of your body with a standard (typical) pressure of

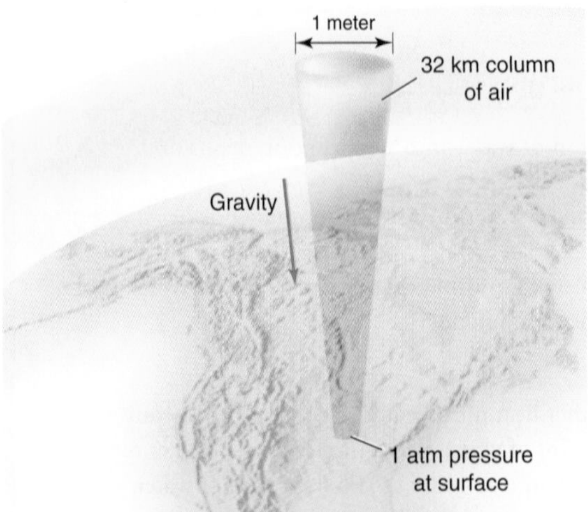

FIGURE 5.7

Atmospheric pressure
A-32-km tall column of air that is 1 meter in diameter has a mass of about 10,000 kg. The mass of the air above and around you is the cause of the air pressure that you experience.

Source: BROWN, THEODORE E; LEMAY, H. EUGENE; BURSTEN, BRUCE E, CHEMISTRY: THE CENTRAL SCIENCE, 10th edition, © 2006, p. 401. Adapted by permission of Pearson Education, Inc., Upper Saddle River, NJ.

(a)

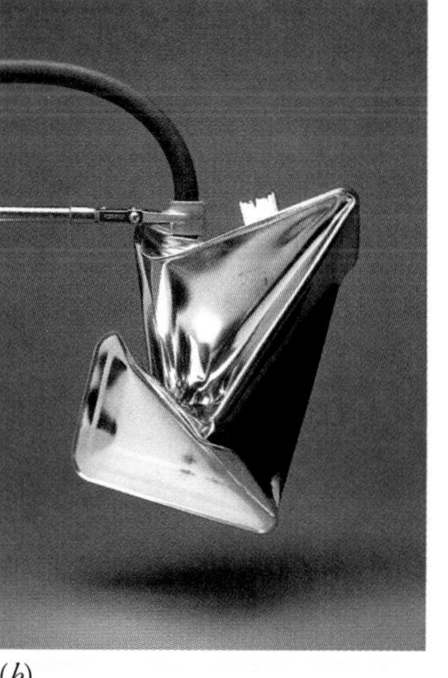
(b)

FIGURE 5.8

The effect of atmospheric pressure
(*a*) When the pressure inside a metal can is equal or nearly equal to the external atmospheric pressure, the can retains its shape. (*b*) When the pressure inside the can is reduced by evacuating the air, the external atmospheric pressure collapses the can.

Source: Ken Karp/Wiley Archive.

about 14.7 pounds per square inch (psi), defined also as 1 atmosphere (1 atm) (Figure 5.8). The less air (fewer atoms and molecules) above you in the atmosphere, the lower the atmospheric pressure (Figure 5.9). Denver, Colorado, at an elevation of 1.6 km (1.0 mi), has a typical atmospheric pressure of 0.74 atm (11 psi). The summit of Mt. Everest, at 9.3 km (5.6 mi), has an atmospheric pressure near 0.26 atm (3.8 psi).

- Standard atmospheric pressure is defined as 1 atm, 14.7 psi, or 760 torr.

Atmospheric pressure is measured using a device called a **barometer**, which is constructed by filling a long glass tube with mercury and carefully inverting the tube into a mercury-filled dish (Figure 5.10). The mercury does not drain out of the inverted tube because the pressure of the atmosphere pushes down on the mercury in the dish with enough force to hold a column of mercury inside the tube. At a pressure of 1 atm and a temperature of 0°C, the mercury column will have a height of 760 mm. The greater the atmospheric pressure, the higher the mercury level in the column, and the less the atmospheric pressure, the lower the mercury level.

The height of a barometer's mercury column is the source of a common pressure unit called the **torr**, named after Evangelista Torricelli (1608–1647), the Italian scientist who

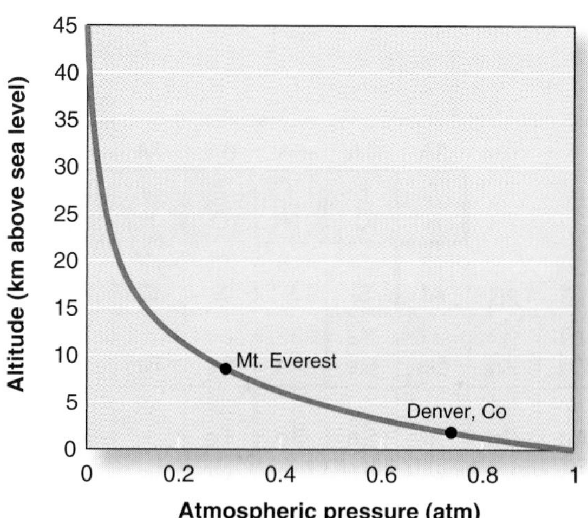

FIGURE 5.9

Altitude and atmospheric pressure
The higher the altitude, the lower the atmospheric pressure.

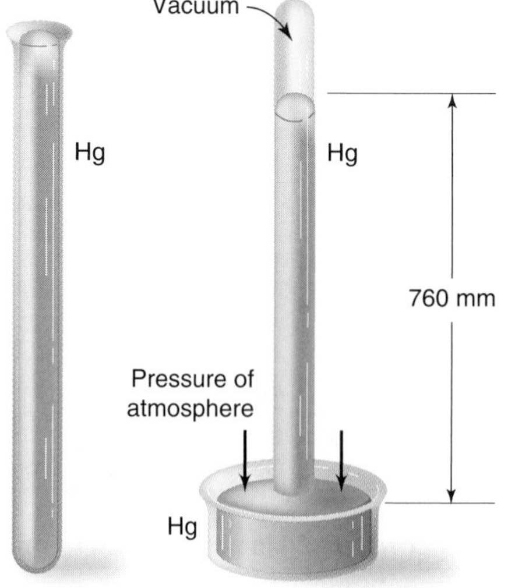

FIGURE 5.10

A barometer
At standard temperature (0°C) and pressure (1 atm) the pressure of the atmosphere supports a column of mercury that is 760 mm tall.

invented the barometer. Each millimeter of height of a mercury column corresponds to 1 torr, so a pressure of 1 atm (holds a mercury column at a height of 760 mm) equals 760 torr.

SAMPLE PROBLEM 5.4

Converting between pressure units

A pressure of 695 torr is how many atmospheres?

Strategy

This problem is solved by finding a conversion factor that relates torr and atmospheres.

Solution

0.914 atm

$$695\ \cancel{\text{torr}} \times \frac{1\ \text{atm}}{760\ \cancel{\text{torr}}} = 0.914\ \text{atm}$$

PRACTICE PROBLEM 5.4

a. A pressure of 35.0 psi is how many torr?
b. A pressure of 750 torr is how many atmospheres?

At a temperature of 90°C and a pressure of 1.0 atm, water is a liquid. At this same temperature, but at a pressure of 0.6 atm, water is a gas (Section 5.5). Because of the effect that temperature and pressure can have on the physical state of a substance, scientists often talk in terms of **standard temperature and pressure (STP)**, which corresponds to a temperature of 0°C and a pressure of 1 atm. Eleven elements exist as gases at STP (Figure 5.11). Of these, the group 8A elements (inert gases) consist of individual atoms, while the other gaseous elements, H_2, N_2, O_2, F_2, and Cl_2, are diatomic molecules. Some covalent compounds are gases at STP as well, including carbon dioxide (CO_2), ammonia (NH_3), and nitrous oxide (N_2O), all of which are small, low molecular weight compounds with low boiling points (Table 5.3). Ionic compounds and most metals are solids at STP. The only two elements that appear as liquids at STP are bromine (Br_2) and mercury (Hg).

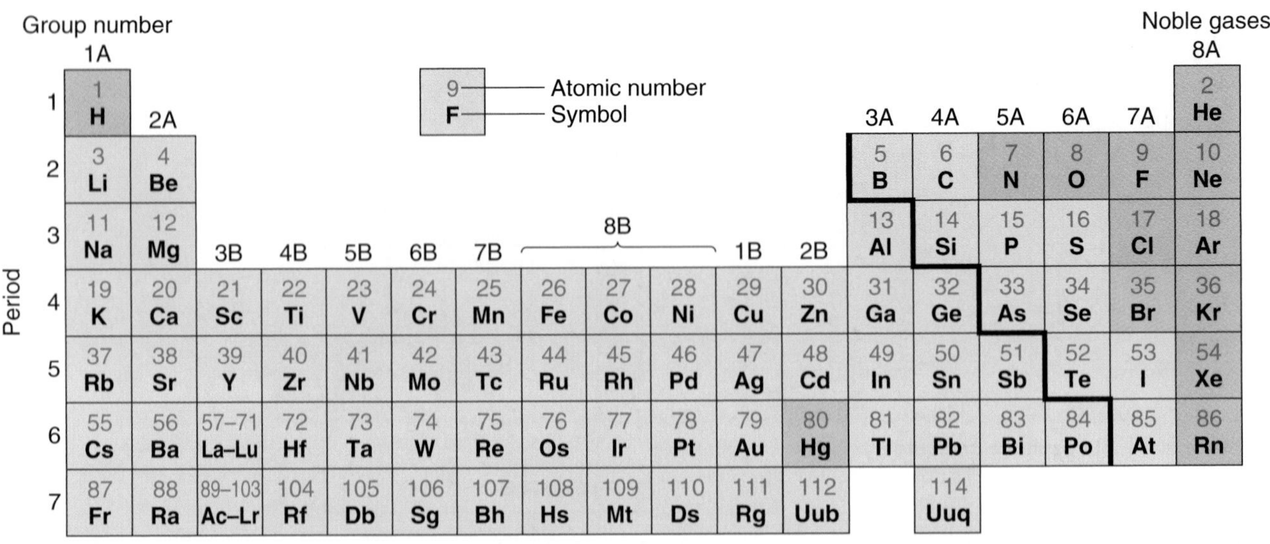

FIGURE 5.11 **Gaseous elements**
The elements shown in orange are gases at standard temperature and pressure (0°C and 1 atm). Elements shown in purple are solids at STP and those in green are liquids.

TABLE 5.3 Some Covalent Compounds That Are Gases at STP (0°C and 1 atm)

Compound	Name
CH_4	Methane
CO	Carbon monoxide
CO_2	Carbon dioxide
HCl	Hydrogen chloride
HF	Hydrogen fluoride
H_2S	Hydrogen sulfide
NH_3	Ammonia
NO	Nitric oxide
N_2O	Nitrous oxide
NO_2	Nitrogen dioxide
SO_2	Sulfur dioxide

When dealing with enclosed gases (in bottles as opposed to the open atmosphere, for example), a **manometer** can be used to measure pressure. An open-end manometer (Figure 5.12), one of several different styles of manometers, consists of a curved tube filled with mercury. By leaving one end of the tube open to the atmosphere and attaching the other end to a flask, the mercury in the manometer is acted on by two competing pressures, that of the atmosphere and that of the gas inside the flask. If the pressure inside the flask is equal to the atmospheric pressure, then an equal force is applied to the mercury column in each arm of the manometer and the mercury levels are the same. If the gas pressure in the flask is greater than atmospheric pressure, mercury is pushed away from the flask and toward the open end of the tube. Conversely, if the pressure in the flask is less than atmospheric pressure, mercury is pushed away from the open end of the tube. Each millimeter of height difference between the two mercury columns corresponds to a pressure difference of 1 torr between the atmosphere and the gas in the flask. The **sphygmomanometer** (from *sphygmos*, Greek for "pulse") used to measure blood pressure consists of an inflatable cuff that is attached to a manometer (Health Link: Blood Pressure).

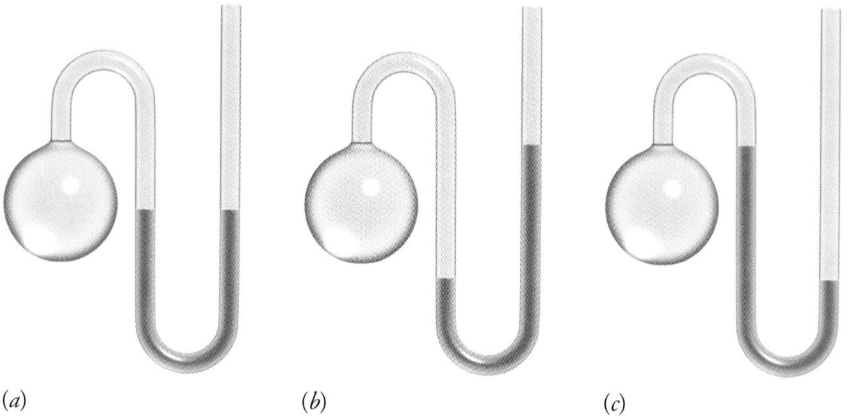

FIGURE 5.12

An open-end manometer
The relative height of the mercury in each arm of the column is related to the difference in pressure between the atmosphere and the gas inside the flask. (*a*) Internal pressure equals atmospheric pressure. (*b*) Internal pressure is greater than atmospheric pressure. (*c*) Internal pressure is lower than atmospheric pressure.

HEALTH Link

Blood Pressure

Blood pressure is the pressure that blood exerts on the walls of blood vessels. With each beat of the heart, blood is pushed into the arteries, causing a temporary increase in blood pressure called systolic pressure. As blood moves down the arteries and the heart prepares for another beat, the blood pressure drops temporarily and the lowest pressure reached is called diastolic pressure.

Blood pressure can be measured at the brachial artery in the arm, using a sphygmomanometer (Figure 5.13). The cuff of the sphygmomanometer is placed over the arm just above the elbow and inflated until the pressure of the cuff constricts blood flow through the brachial artery. As the pressure of the cuff is gradually reduced, a stethoscope is used to listen for the sound of blood as it first begins to flow through the constricted artery. The manometer reading at the first of these sounds is the systolic pressure. The cuff's pressure is reduced further until the blood flows freely through the artery. The pressure at this point, where the sounds of blood being pushed through a constricted artery are no longer heard, is the diastolic pressure.

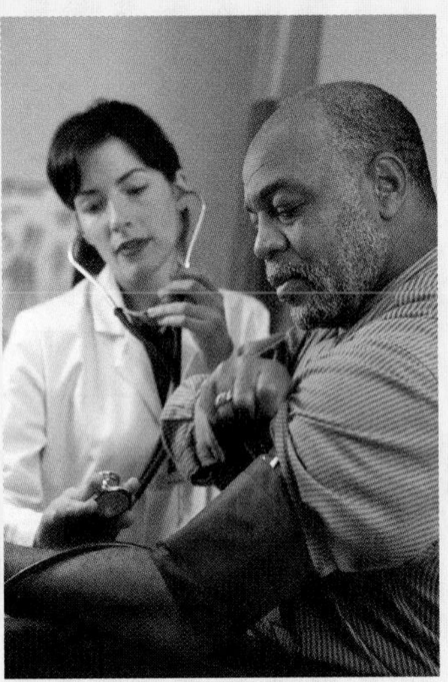

FIGURE | 5.13

Sphygmomanometer
A nurse uses a sphygmomanometer to measure blood pressure.

Source: Charles Thatcher/Stone/Getty.

Orthostatic hypotension (low blood pressure), which is especially common in the elderly, causes the dizziness that sometimes occurs when someone suddenly stands after sitting or lying down for a prolonged period. If the body does not respond quickly enough to this change in posture, blood pools in the lower limbs and a temporary blood pressure drop takes place. As a result, less blood is pumped to the brain and dizziness occurs.

Chronic hypertension (high blood pressure), caused by constricted movement of blood through the blood vessels, is a serious disease that increases the workload of the heart and can damage blood vessels, speeding the onset of atherosclerosis (hardening of the arteries).

In 2003 a new set of guidelines pertaining to blood pressure was issued by the National Heart, Lung and Blood Institute. These guidelines (Table 5.4) group blood pressure ranges into four categories: normal, prehypertension, stage 1 hypertension, and stage 2 hypertension. The higher the blood pressure, the greater the risk of heart disease. Studies have shown that each 20 point rise in systolic pressure or 10 point rise in diastolic pressure doubles the risk of heart disease.

TABLE | 5.4 Blood Pressure Guidelines

	Blood Pressure (mm Hg)			
Category	**Systolic**		**Diastolic**	**What You Should Do**
Normal	<120	and	<80	Maintain a healthy lifestyle[a]
Prehypertension	120–139	or	80–89	Adopt a healthy lifestyle
Stage 1 hypertension	140–159	or	90–99	Adopt a healthy lifestyle; take medication
Stage 2 hypertension	≥160	or	≥100	Adopt a healthy lifestyle; take more than one medication

[a] Maintain a healthy weight, exercise, drink alcohol in moderation, and reduce dietary sodium levels.

5.3 THE GAS LAWS

Whether we are dealing with inhaled anesthetic gases, the helium in a balloon, or the air in a tire, all gases behave in a similar manner. In this section we will take a look at the set of laws that describe their behavior.

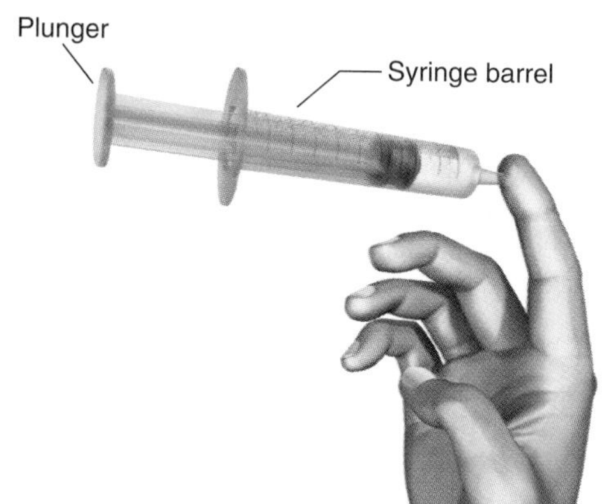

FIGURE 5.14

The relationship between pressure and volume
For a given sample of gas at a constant temperature, an increase in volume produces a drop in pressure.

Pressure and Volume

Suppose that you pick up a syringe barrel, put your finger over the opening, and pull back the plunger (Figure 5.14). As the plunger moves back, you will feel your skin being sucked into the syringe. This happens because as the volume inside the syringe barrel increases, the pressure of the air inside drops.

In the late 1600s, Robert Boyle studied this relationship between pressure and volume, and came to the conclusion that as volume increases, pressure decreases. Stated more carefully, **Boyle's law** says that, for a sample of a gas at a fixed temperature, *pressure and volume are inversely proportional*, as shown in Figure 5.15. Mathematically, Boyle's law can be expressed by the equation

■ As the volume of a gas increases, the pressure decreases.

$$P_1V_1 = P_2V_2$$

where P is pressure, V is volume, and the subscripts 1 and 2 refer to the initial and final states, respectively.

Let us see an example of how this equation can be used. If the syringe barrel in Figure 5.14 has an initial volume of 0.50 mL and the air inside is at a pressure of 1.0 atm, what is the pressure inside the barrel if your finger is placed over the opening and the plunger is pulled back to give a volume of 1.0 mL? In this problem we are given the initial volume (0.50 mL) and pressure (1.0 atm) and the final volume (1.0 mL), so rearranging the Boyle's law equation and solving for the final pressure gives

$$P_2 = \frac{P_1V_1}{V_2} = \frac{1.0 \text{ atm} \times 0.50 \text{ mL}}{1.0 \text{ mL}} = 0.50 \text{ atm}$$

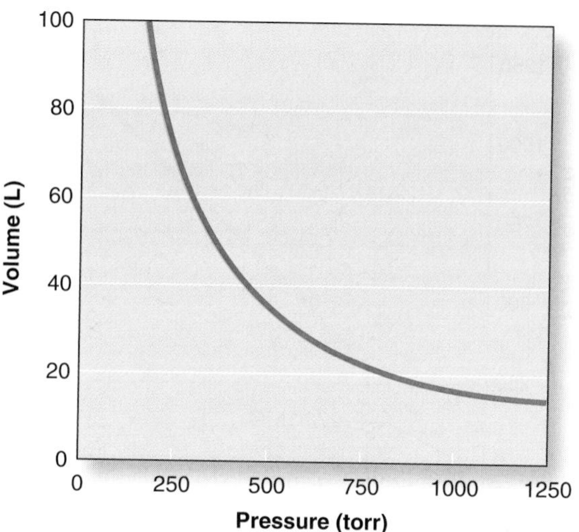

FIGURE 5.15

Boyle's law
For 1 mol of gas at constant temperature (0°C), an increase in volume leads to a decrease in pressure.

SAMPLE PROBLEM 5.5

Pressure–volume relationships

At an ocean depth of 33 ft, where the pressure is 2.0 atm, a scuba diver releases a bubble of air with a volume of 6.0 mL. What is the volume of the air bubble when it reaches the surface, where the pressure is 1.0 atm? (Assume that the temperature and the amount of gas in the bubble remains constant.)

Strategy

This problem gives you values for three of the four variables in the $P_1V_1 = P_2V_2$ equation. You must solve for the fourth.

Solution

$$V_2 = \frac{P_1V_1}{P_2} = \frac{2.0\ \cancel{\text{atm}} \times 6.0\ \text{mL}}{1.0\ \cancel{\text{atm}}} = 12\ \text{mL}$$

PRACTICE PROBLEM 5.5

If you take a piece of bubble wrap and push hard enough on one of the bubbles, it will pop. Explain how this relates to Boyle's law.

Pressure and Temperature

To maximize gas mileage and to help your auto tires last as long as possible, it is important to keep them inflated to the correct pressure. The inflation pressure listed on tires is the maximum recommended pressure and is based on measurements made when the tire is cold. As you drive your car, the tires heat up and for every 10°F increase in temperature, the tire pressure increases by about 1 psi. This means that if you drive to a gas station and check tire pressure, the reading you get will be higher than you would have obtained if you had checked the pressure before leaving home.

The French chemist Gay-Lussac (1778–1850) studied the pressure–temperature relationship for gases. **Gay-Lussac's law** states that, for a sample of a gas with a constant volume, *pressure and temperature are directly related.* As the temperature of a gas increases, so does its pressure, as shown in Figure 5.16. In this figure, note that when the pressure is

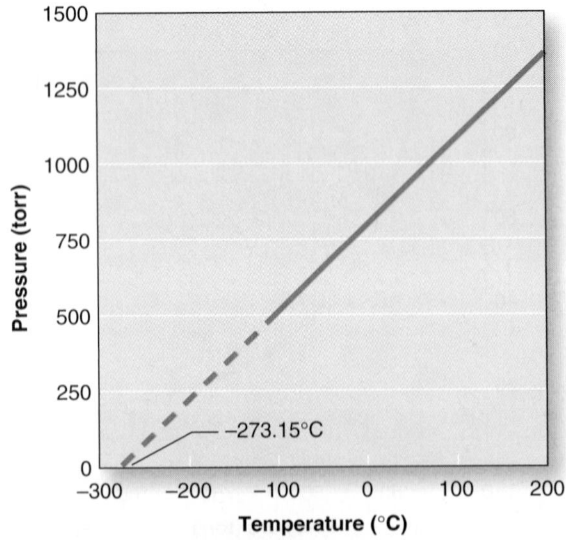

FIGURE 5.16

Gay-Lussac's law
A plot of pressure versus temperature for 1 mol of a gas with a constant volume of 22.4 L is a straight line. The temperature at which the pressure equals 0 torr (−273.15°C) is absolute zero (0 K).

extrapolated to 0 torr (gases never actually reach this pressure because they turn into liquids or solids long before reaching this point), the corresponding temperature is −273.15°C. This is the basis for the Kelvin temperature scale (0 K = −273.15°C).

The equation that mathematically describes Gay-Lussac's law is

■ As the temperature of a gas increases, so does the pressure.

$$\frac{P_1}{T_1} = \frac{P_2}{T_2}$$

where P is pressure and T is temperature (in kelvins). Let us use the example of a steel tank fitted with a pressure gauge to demonstrate the use of this equation (Figure 5.17). Suppose that at a temperature of 25°C the pressure in the tank is 1.5 atm. If the temperature is raised to 50°C, what is the new pressure? To solve this problem we begin by converting the temperatures from degrees Celsius to kelvins. Then, the equation above can be rearranged and the value for the new pressure (P_2) can be calculated.

$$\text{K} = °\text{C} + 273.15 = 25°\text{C} + 273.15 = 298\ \text{K}$$
$$\text{K} = °\text{C} + 273.15 = 50°\text{C} + 273.15 = 323\ \text{K}$$
$$P_2 = \frac{P_1 T_2}{T_1} = \frac{1.5\ \text{atm} \times 323\ \cancel{\text{K}}}{298\ \cancel{\text{K}}} = 1.6\ \text{atm}$$

To get the correct answer, temperatures must be expressed in kelvins. The Gay-Lussac equation predicts that doubling the temperature (in K) will double the pressure (see Sample Problem 5.6). In the example that we just worked, doubling the temperature (in °C) increased the pressure only from 1.5 to 1.6 atm.

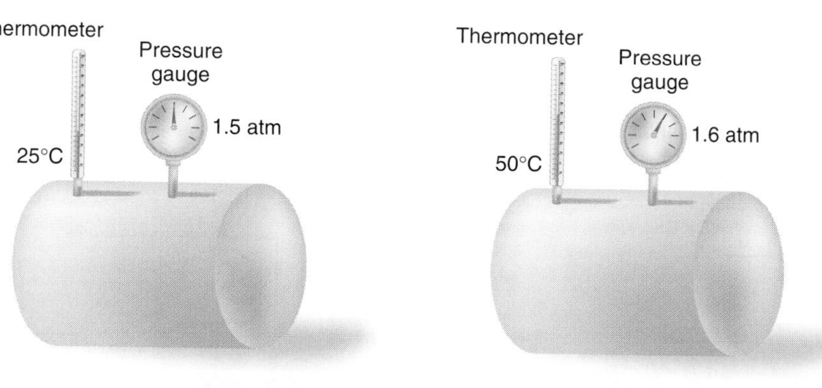

■ **FIGURE 5.17**

The relationship between temperature and pressure
For a given sample of gas with a fixed volume, an increase in temperature results in an increase in pressure.

SAMPLE PROBLEM 5.6

Pressure–temperature relationships

For the pressure tank shown in Figure 5.17, if the initial temperature is 25°C and the initial pressure is 1.5 atm, what is the new pressure if the temperature is increased to 323°C?

Strategy

As in the example worked directly above, the temperatures must be converted into kelvins before the Gay-Lussac equation is used to calculate the new pressure.

Solution

$$\text{K} = °\text{C} + 273.15 = 25°\text{C} + 273.15 = 298\ \text{K}$$
$$\text{K} = °\text{C} + 273.15 = 323°\text{C} + 273.15 = 596\ \text{K}$$
$$P_2 = \frac{P_1 T_2}{T_1} = \frac{1.5\ \text{atm} \times 596\ \cancel{\text{K}}}{298\ \cancel{\text{K}}} = 3.0\ \text{atm}$$

Increasing the temperature from 25°C to 323°C, which doubles the temperature in kelvins (298 K × 2 = 596 K), doubles the pressure.

PRACTICE PROBLEM 5.6

A visiting home care nurse checks the pressure gauge on a patient's portable oxygen tank and sees that it contains 815 psi of O_2. This reading may be inaccurate because the patient has been sitting in the sun and the tank has warmed to 35°C. What would the pressure reading be at a temperature of 25°C?

Volume and Temperature

Try this experiment: Blow up a balloon, put it in the freezer, and leave it there for a few minutes. When you return, the balloon will have shrunk. This happens because there is a direct relationship between volume and temperature.

■ As the temperature of a gas increases, so does the volume.

Jacques Charles (1746–1823) studied this relationship and formulated **Charles' law**, which states that, for a sample of a gas at a fixed pressure, *volume and temperature are directly related.* As the temperature of a gas increases, so does its volume, as can be shown using an apparatus in which a plug of mercury is suspended in a glass tube by the air trapped inside (Figure 5.18). The amount of gas remains constant because air cannot escape, and the pressure exerted on the air sample does not change because the amount of mercury and the atmospheric pressure remain constant. As the temperature of the trapped air increases, so does its volume, as indicated by the upward movement of the mercury plug. A direct relationship exists between temperature and volume (Figure 5.19). If the line in Figure 5.19 is extended to a volume of 0 L, the corresponding temperature is −273.15°C (0 K).

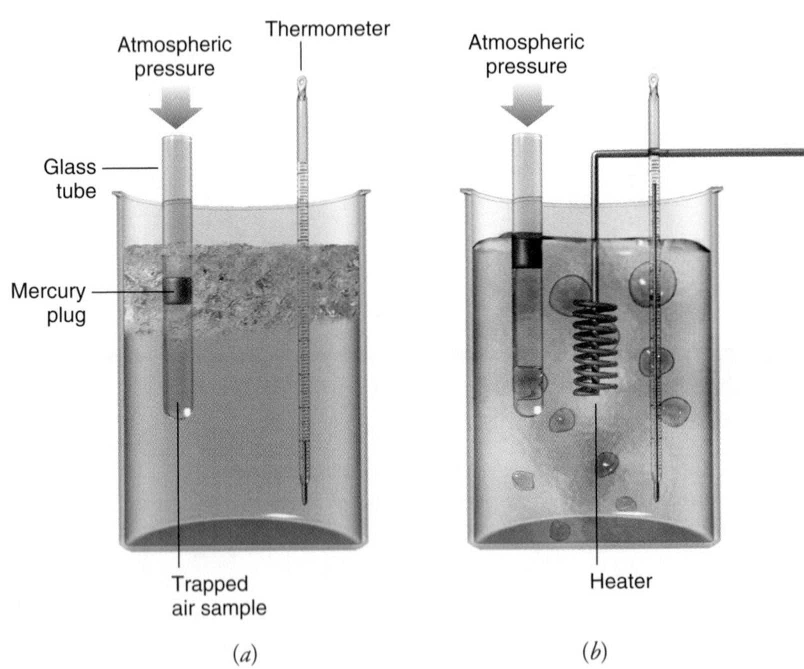

FIGURE 5.18

The relationship between temperature and volume
For a given sample of gas at fixed pressure, an increase in temperature results in an increase in volume. (*a*) An ice water bath at 0°C. (*b*) A boiling water bath at 100°C.

Source: From CHEMISTRY: THE MOLECULAR NATURE OF MATTER AND CHANGE, 4th edition, © 2006. Reproduced by permission of The McGraw-Hill Companies.

Mathematically, Charles' law can be written

$$\frac{V_1}{T_1} = \frac{V_2}{T_2}$$

where V is volume and T is temperature (in kelvins). Let us use this equation to predict the volume change for a balloon that is placed in a freezer. If the balloon has a volume of 2.0 L at room temperature (25°C), what is its volume after sitting in a −10°C freezer?

To solve this problem, the temperature must first be converted into kelvins.

$$K = °C + 273.15 = 25°C + 273.15 = 298\ K$$
$$K = °C + 273.15 = -10°C + 273.15 = 263\ K$$
$$V_2 = \frac{V_1 T_2}{T_1} = \frac{2.0\ L \times 263\ \cancel{K}}{298\ \cancel{K}} = 1.8\ L$$

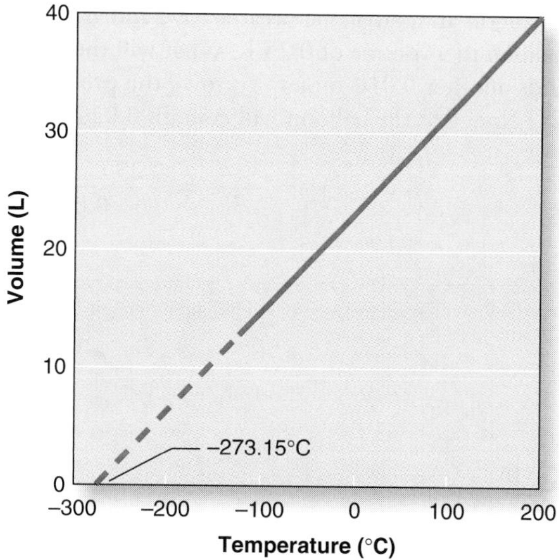

FIGURE 5.19

Charles' law
A plot of volume versus temperature for 1 mol of a gas at a pressure of 760 torr is a straight line. The temperature at which the volume equals 0 L (−273.15°C) is absolute zero (0 K).

SAMPLE PROBLEM 5.7

Volume–temperature relationships

Outside on a day when the temperature is 32°C a balloon has a volume of 2.5 L. What is the new volume of the balloon after it is brought indoors, where the temperature is 24°C?

Strategy

For calculations involving gases, all temperatures must be in kelvins. Once this conversion has been made, the Charles' law equation can be used to calculate the new volume.

Solution

$$K = °C + 273.15 = 32°C + 273.15 = 305\ K$$
$$K = °C + 273.15 = 24°C + 273.15 = 297\ K$$
$$V_2 = \frac{V_1 T_2}{T_1} = \frac{2.5\ L \times 297\ \cancel{K}}{305\ \cancel{K}} = 2.4\ L$$

PRACTICE PROBLEM 5.7

When handling containers of compressed gases, why is it important to keep them away from fires even if the gases are not flammable? Answer this question in terms of Charles' law.

Moles and Volume

To prepare for a birthday party, you are assigned the job of blowing up balloons. The more air that you blow into each balloon, the larger it gets. This relationship between the amount of gas in a sample and the volume that it occupies was studied in the early 1800s by Amedeo Avogadro. **Avogadro's law** states that, at a given temperature and pressure, *volume and the number of moles of gas are directly related* (Figure 5.20).

The equation that describes Avogadro's law is

$$\frac{V_1}{n_1} = \frac{V_2}{n_2}$$

where V is volume and n is the number of moles of gas. Suppose that you are blowing up a balloon and, after one breath, have added 0.010 mol of air molecules and inflated the balloon to a volume of 0.25 L. What will the volume of the balloon be if a second breath adds another 0.010 moles? To solve the problem the equation is rearranged to solve for V_2. Note that the balloon will contain 0.020 moles after the second breath.

$$V_2 = \frac{V_1 n_2}{n_1} = \frac{0.25\text{ L} \times 0.020\text{ mol}}{0.010\text{ mol}} = 0.50\text{ L}$$

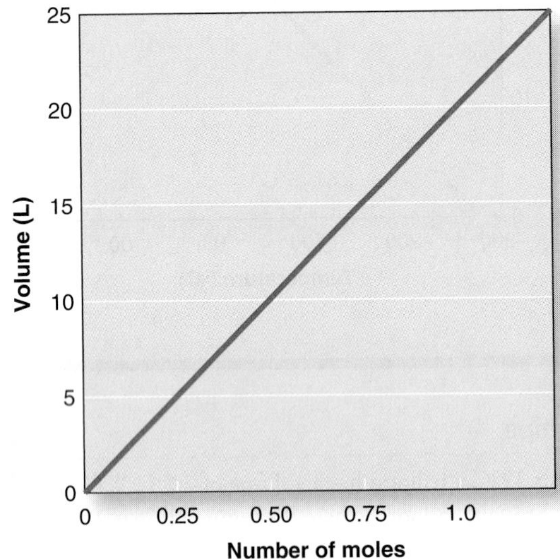

FIGURE 5.20

Avogadro's law
A plot of volume versus number of moles for a gas at standard temperature and pressure is a straight line.

SAMPLE PROBLEM 5.8

An inflated balloon that contains 0.045 mol of helium (He) gas has a volume of 1.1 L. If the pressure and temperature stay the same, what is the new volume if 0.080 g of He is added?

Strategy

To solve this problem you must solve the Avogadro's law equation for the new volume (V_2). Remember that the variables n_1 and n_2 refer to moles.

Solution

$$0.080\text{ g He} \times \frac{1\text{ mol He}}{2.0\text{ g He}} = 0.040\text{ mol He}$$

$$0.045\text{ mol He initially} + 0.040\text{ mol He added} = 0.085\text{ mol He}$$

$$V_2 = \frac{V_1 n_2}{n_1} = \frac{1.1\text{ L} \times 0.085\text{ mol}}{0.045\text{ mol}} = 2.1\text{ L}$$

PRACTICE PROBLEM 5.8

At STP, one mole of any gas occupies a volume of 22.4 L. Under these same conditions, how many moles of gas are present in a volume of 15.7 L?

The Combined Gas Law and the Ideal Gas Law

When dealing with a particular sample of gas, where the number of moles present is held constant, the three variables that describe its condition are pressure, temperature, and volume. Two at a time, these variables are described by Boyle's, Gay-Lussac's, and Charles' law.

Boyle's law $P_1V_1 = P_2V_2$

Gay-Lussac's law $\frac{P_1}{T_1} = \frac{P_2}{T_2}$

Charles' law $\frac{V_1}{T_1} = \frac{V_2}{T_2}$

A limitation of these laws is that they are useful only when one of the three variables is held constant. Temperature must be held constant to use Boyle's law, volume must be held constant to use Gay-Lussac's law, and pressure must be held constant to use Charles' law.

These three gas laws can be merged to create the **combined gas law**, which takes the form

Combined gas law $\frac{P_1V_1}{T_1} = \frac{P_2V_2}{T_2}$

where T is the temperature in kelvins. This law is useful because it allows the properties of a gas to be predicted over a wider range of conditions. For example, suppose that you are standing at sea level (atmospheric pressure = 1.0 atm) on a day when the temperature is 22°C (295 K) and are holding a balloon (volume = 2.9 L) filled with helium gas (He). If you release the balloon, what will its new volume be at a higher altitude (atmospheric pressure = 0.85 atm) where the temperature is 15°C (288 K)? There is quite of bit of information here, so let us begin solving this problem by listing all of the variables, then solving the combined gas law equation for the unknown.

$P_1 = 1.0$ atm $V_1 = 2.9$ L $T_1 = 295$ K

$P_2 = 0.85$ atm $V_2 =$ unknown $T_2 = 288$ K

$$\frac{P_1V_1}{T_1} = \frac{P_2V_2}{T_2}$$

$$V_2 = \frac{P_1V_1T_2}{P_2T_1} = \frac{1.0\ \cancel{\text{atm}} \times 2.9\ \text{L} \times 288\ \cancel{\text{K}}}{0.85\ \cancel{\text{atm}} \times 295\ \cancel{\text{K}}} = 3.3\ \text{L}$$

If, in addition to pressure, temperature, and volume, the number of moles of gas in a sample may change, then the combined gas law can be modified by bringing in Avogadro's law (relates volume and number of moles of a gas).

$$\frac{P_1V_1}{n_2T_1} = \frac{P_2V_2}{n_2T_2}$$

This new equation can be interpreted as saying that, for a given gas, pressure times volume divided by number of moles times temperature is a constant. This constant is called the **gas constant** (R). Mathematically,

$$\frac{PV}{nT} = R$$

where T is the temperature in kelvins. For **ideal gases**, *those that completely obey all of the gas laws that have been presented in this section*, the gas constant has a value of 0.0821 L atm/mol K. The value of R is different when other pressure units are used—here R is defined in terms of atmospheres. Except at very high pressures or very low temperatures, most real gases come fairly close to ideal behavior, so we will treat gases as being ideal.

The equation above, called the **ideal gas law**, is usually written in the form

$$PV = nRT$$

With this equation, we can solve for the value of any one variable (P, V, n, or T) if given values for the other three. For example, using the initial pressure, volume, and temperature values from the helium-filled balloon example above, we can calculate the number of moles of gas in the balloon.

$$P = 1.0 \text{ atm}$$
$$V = 2.9 \text{ L}$$
$$n = \text{unknown}$$
$$R = 0.0821 \text{ L atm/mol K}$$
$$T = 295 \text{ K}$$

$$PV = nRT$$

$$n = \frac{PV}{RT} = \frac{1.0 \cancel{\text{atm}} \times 2.9 \cancel{\text{L}}}{0.0821 \cancel{\text{L atm}}/\text{mol} \cancel{\text{K}} \times 295 \cancel{\text{K}}} = 0.12 \text{ mol}$$

SAMPLE PROBLEM 5.9

The ideal gas law

A flask filled with 1.00 mol of Ne(g) at a temperature of 25°C has a pressure of 1.00 atm. What is its volume?

Strategy

In calculations involving the ideal gas law, temperature should be expressed in K.

Solution

The problem gives values for n, T, and P. Before solving this problem, °C must be converted to K.

$$K = °C + 273.15 = 25°C + 273.15 = 298 \text{ K}$$

Rearranging the ideal gas equation and solving gives an answer of 24.5 L.

$$PV = nRT$$

$$V = \frac{nRT}{P} = \frac{1.00 \cancel{\text{mol}} \times 0.0821 \text{ L} \cancel{\text{atm}}/\cancel{\text{mol K}} \times 298 \cancel{\text{K}}}{1.00 \cancel{\text{atm}}} = 24.5 \text{ L}$$

PRACTICE PROBLEM 5.9

a. A flask is filled with 0.25 mol of gas at standard temperature and pressure. What is the volume of the flask in liters?

b. What is the pressure (in torr) in the flask if the temperature is reduced by 25°C?

5.4 PARTIAL PRESSURE

Besides predicting how a given sample of gas responds to changes in pressure, volume, temperature, and number of moles, the gas laws are useful for predicting the behavior of mixtures of gases. This is the case because when ideal gases are mixed, each exerts the same pressure that it would if it were alone in a container. Consider, for example, two interconnected 22.4 L cylinders at a temperature of 273 K and a pressure of 0.50 atm, one containing 0.50 mol of He(g) and the other containing 0.50 mol of Ne(g) (Figure 5.21a). If the valve between the two cylinders is opened, and the He is forced into the cylinder containing Ne, what is the new pressure?

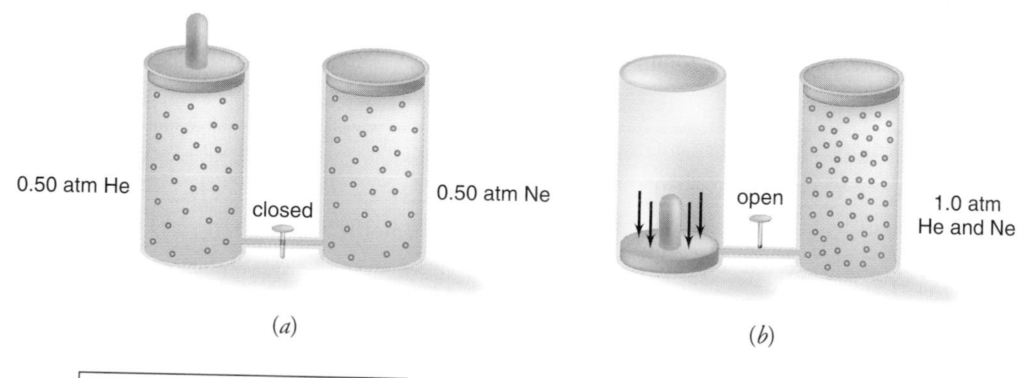

FIGURE 5.21 **Total pressure is the sum of partial pressures** (*a*) Cylinders of He and Ne gas with pressures of 0.50 atm. (*b*) When the gases are combined into one cylinder the total pressure is 1.0 atm, the sum of the partial pressure of each gas.

We can use the ideal gas law to solve this problem. Once all of the gas has been pushed into the right-hand cylinder, we know the final volume (22.4 L), the final number of moles (0.50 + 0.50 = 1.00 mol), and the temperature (298 K). This leaves pressure as the only unknown quantity in the $PV = nRT$ equation.

$$P = \text{unknown}$$
$$V = 22.4\ \text{L}$$
$$n = 1.00\ \text{mol}$$
$$R = 0.0821\ \text{L atm/mol K}$$
$$T = 298\ \text{K}$$

Rearranging the ideal gas law to solve for P gives

$$P = \frac{nRT}{V} = \frac{1.00\ \cancel{\text{mol}} \times 0.0821\ \cancel{\text{L}}\ \text{atm}/\cancel{\text{mol K}} \times 273\ \cancel{\text{K}}}{22.4\ \cancel{\text{L}}} = 1.00\ \text{atm}$$

The pressure that each gas in a mixture would exert, if alone, is called its **partial pressure. Dalton's law of partial pressure** states that *the total pressure of a mixture of gases is the sum of the partial pressures of its components.* For this example, the total pressure of 1.00 atm is the sum of the partial pressures of He (0.50 atm) and Ne (0.50 atm).

■ The partial pressure of a gas in a mixture is the pressure that the gas would exert if alone.

Dalton's law can be described by a mathematical equation. If the partial pressure of gases A, B, and C are written P_A, P_B, and P_C, then the total pressure (P_{total}) of a mixture containing these three gases is

$$P_{total} = P_A + P_B + P_C$$

SAMPLE PROBLEMS 5.10

Dalton's law of partial pressure

a. A 1.00 L flask contains 5.00×10^{-2} mol of neon. What is the pressure (in atmospheres) inside this flask at a temperature of 30°C?

b. An identical flask contains 5.00×10^{-2} mol of neon and 5.00×10^{-3} mol of argon. At 30°C, what is the partial pressure of each gas in atmospheres and what is the total pressure?

Strategy

The pressure due to any particular gas can be calculated using the ideal gas law. Since volume, number of moles, and temperature are given, the only unknown is the pressure. In part b, the total pressure is the sum of the pressures due to Ne and Ar (their partial pressures).

Solution

a. K = °C + 273.15 = 30°C + 273.15 = 303 K

$$P = \frac{nRT}{V} = \frac{5.00 \times 10^{-2}\ \cancel{\text{mol}} \times 0.0821\ \cancel{\text{L}}\ \text{atm}/\cancel{\text{mol K}} \times 303\ \cancel{\text{K}}}{1.00\ \cancel{\text{L}}} = 1.24\ \text{atm}$$

b. $$P_{Ne} = \frac{nRT}{V} = \frac{5.00 \times 10^{-2}\ \cancel{\text{mol}} \times 0.0821\ \cancel{\text{L}}\ \text{atm}/\cancel{\text{mol K}} \times 303\ \cancel{\text{K}}}{1.00\ \cancel{\text{L}}} = 1.24\ \text{atm}$$

$$P_{Ar} = \frac{nRT}{V} = \frac{5.00 \times 10^{-3}\ \cancel{\text{mol}} \times 0.0821\ \cancel{\text{L}}\ \text{atm}/\cancel{\text{mol K}} \times 303\ \cancel{\text{K}}}{1.00\ \cancel{\text{L}}} = 0.124\ \text{atm}$$

$$P_{total} = P_{Ne} + P_{Ar} = 1.24\ \text{atm} + 0.124\ \text{atm} = 1.36\ \text{atm}$$

PRACTICE PROBLEM 5.10

A 35 L gas cylinder contains a mixture of 7.5 g of N_2, 2.5 g of O_2, and 1.5 g of He. At 35°C, what is the partial pressure (in atm) of each gas and what is the total pressure?

Air is a mixture of N_2, O_2, and other gases, including Ar, CO_2, He, Kr, Xe, H_2, and CH_4. In dry air [no $H_2O(g)$ is present] at STP the partial pressure of N_2 is about 593 torr (760 torr = 1 atm), the partial pressure of O_2 is about 160 torr, and that of the remaining gases is about 7 torr.

$$P_{total} = P_{N_2} + P_{O_2} + P_{others} = 593\ \text{torr} + 160\ \text{torr} + 7\ \text{torr} = 760\ \text{torr}$$

From these partial pressures we can show that N_2 makes up 78.0% of air, O_2 makes up 21.1%, and the other gases 0.9%.

$$N_2 \qquad \frac{593\ \text{torr}}{760\ \text{torr}} \times 100 = 78.0\%$$

$$O_2 \qquad \frac{160\ \text{torr}}{760\ \text{torr}} \times 100 = 21.1\%$$

$$\text{Others} \qquad \frac{7\ \text{torr}}{760\ \text{torr}} \times 100 = 0.9\%$$

Because air is a uniform (evenly distributed) mixture, these percentages remain constant, regardless of the total air pressure. At the top of Mt. Everest, where a typical atmospheric pressure might be 200 torr, dry air still consists of 78.0% N_2, 21.1% O_2, and 0.9% other gases.

$$P_{N_2} = 0.780 \times 200\ \text{torr} = 156\ \text{torr}$$

$$P_{O_2} = 0.211 \times 200\ \text{torr} = 42.2\ \text{torr}$$

$$P_{others} = 0.009 \times 200\ \text{torr} = 2\ \text{torr}$$

$$P_{total} = P_{N_2} + P_{O_2} + P_{others} = 156\ \text{torr} + 42.2\ \text{torr} + 2\ \text{torr} = 200\ \text{torr}$$

Many climbers make up for the low partial pressure of oxygen at high altitudes by using supplemental oxygen (Figure 5.22).

FIGURE 5.22

High altitude climbing
The partial pressure of O_2 is low at high altitudes, so climbers often carry their own oxygen supplies.

Source: © Chamoux Initiative/Gamma.

SAMPLE PROBLEMS 5.11

Calculating the partial pressure of O_2 in dry air

On a sunny day in July, the air pressure in Spokane, Washington (altitude 0.6 km), is 696 torr. What is the partial pressure of O_2?

Strategy
The partial pressure of O_2 in dry air will always be 21.1% of the total pressure.

Solution

Air is 21.1% O_2, so at a pressure of 696 torr the partial pressure of O_2 is 21.1% of 696 torr.

$$P_{O_2} = 0.211 \times 696 \text{ torr} = 147 \text{ torr}$$

PRACTICE PROBLEM 5.11

In Spokane, Louisiana (altitude 0.21 km), the air pressure is 755 torr. What is the partial pressure of O_2?

5.5 LIQUIDS

Having looked at various properties of gases, let us now turn our attention to liquids. One of the important properties of liquids for us to consider is their thickness or **viscosity** (resistance to flow). The more viscous a liquid, the slower it pours (Figure 5.23). Viscosity is related to the strength of the noncovalent interactions between the molecules that make up a liquid—the stronger the attractions, the thicker the liquid. For example, molecules of 2-propanol are attracted to one another by hydrogen bonds that form between the —OH groups. Glycerol molecules, with three —OH groups each, can form more hydrogen bonds. Glycerol has a higher viscosity than 2-propanol.

Temperature has an effect on viscosity. As temperature rises, the increased kinetic energy of the molecules in a liquid helps them pull away from one another—higher temperature produces lower viscosity. You may have direct experience with this behavior. If you store maple syrup in the refrigerator, you will find that it pours very slowly when cold. If the syrup

FIGURE | 5.23

Viscosity
Glycerol molecules are able to form more hydrogen bonds with one another than are 2-propanol molecules, so glycerol is thicker (more viscous) than 2-propanol.

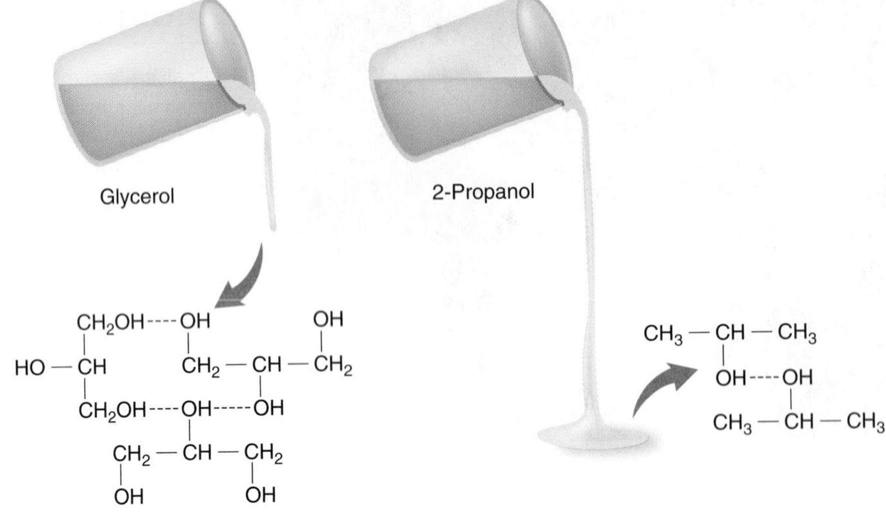

is warmed up, however, it pours as easily as water. This dependence of viscosity on temperature has applications in medicine. In some instances, whole body hypothermia (cooling) is used during surgery to slow bleeding because as blood is cooled, its viscosity increases.

■ Density is mass divided by volume.

Density and specific gravity are two other important properties of liquids. The **density** of a liquid (or any other substance) is *the amount of mass contained in a given volume.* At 20°C, ethanol has a density of 0.791 g/mL (0.791 g = 1 mL). Knowing the density of a liquid allows you to calculate the volume associated with a given mass. This is useful because it is usually easier to pour out a certain volume of liquid than to weigh out a particular mass. For example, suppose that a lab experiment calls for using 9.0 g of ethanol. At 20°C, how many milliliters of ethanol correspond to 9.0 g? To solve this problem, we use the density of ethanol as a conversion factor.

$$9.0 \text{ g ethanol} \times \frac{1 \text{ mL}}{0.791 \text{ g}} = 11 \text{ mL ethanol}$$

The density of a substance usually drops as the temperature rises because, while its mass is constant, its volume usually increases. Because density depends on temperature, temperature should be specified whenever density values are reported (see Table 5.5). The use of mercury (Hg) thermometers depends on the change in the density of mercury with temperature. At a temperature of 0°C, where the density of Hg is 13.60 g/mL, 100 g of Hg occupies a volume of 7.35 mL. At 100°C, where the density is 13.35 g/mL, the same 100 g of Hg has a volume of 7.49 mL. This increase in volume is what causes the mercury in a thermometer to rise.

TABLE | 5.5 DENSITY OF COMMON SUBSTANCES

Solids (at 20°C)	Density (g/mL)	Liquids (at 20°C)	Density (g/mL)	Gases[a] (at 0°C)	Density[b] (g/L)
Cork	0.25	Ethanol	0.791	Hydrogen (H_2)	0.0899
Fat (human)	0.94	Kerosene	0.82	Helium (He)	0.179
Bone	1.90	Water	1.00	Air	1.29
Salt	2.17	Whole blood	1.06	Oxygen (O_2)	1.43
Lead	11.3	Bromine (Br_2)	3.1	Chlorine (Cl_2)	3.21

[a] At standard temperature and pressure.
[b] Gas density is so low that values are usually reported in g/L, rather than g/mL.

Specific gravity relates the density of a substance to that of water.

$$\text{Specific gravity} = \frac{\text{density of substance}}{\text{density of water}}$$

Here is an example of how specific gravity can be calculated. At 25°C, the density of 2-propanol is 0.798 g/mL and that of water is 1.00 g/mL. Therefore, the specific gravity of 2-propanol is 0.798.

$$\text{Specific gravity} = \frac{\text{density of substance}}{\text{density of water}} = \frac{0.798 \text{ g/mL}}{1.00 \text{ g/mL}} = 0.798$$

With the density of water being 1.00 g/mL at 25°C, the specific gravity of 2-propanol or any other substance at this temperature is equal to its density. Specific gravity is likely to vary with temperature because a change in temperature affects the density of water and of the substance being tested, but not necessarily to the same extent.

Specific gravity measurements, which are made using refractometers, hydrometers, or test strips (Figure 5.24), can help to determine the acid levels in car batteries, the antifreeze levels in car radiators, and the alcohol content in beer and wine. The specific gravity of urine (a mixture of water and waste products excreted by the kidneys) can be used to diagnose kidney problems. Urine with a high specific gravity has too many waste products dissolved in it, which can indicate dehydration or overproduction of antidiuretic hormone (ADH), which regulates the amount of water in blood serum. High ADH levels can be indicative of stress or trauma and often follow major sugery. Urine with a low specific gravity may be an indication of kidney disease, excess fluid intake, or underproduction of ADH.

(*a*)

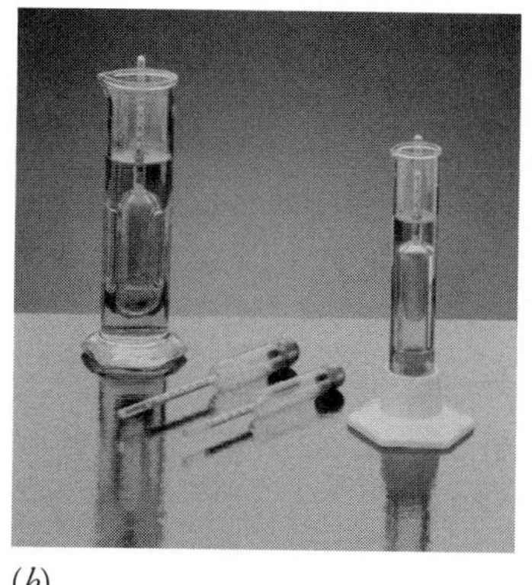

(*b*)

FIGURE 5.24

Measuring specific gravity

(a) Refractometers measure the extent to which a light beam is bent by a liquid. This refraction is related to specific gravity. (b) The level at which the bulb of a hydrometer floats in a liquid is determined by specific gravity.

Source: (a) Courtesy of MISCO; (b) Courtesy of Becton, Dickinson and Company.

SAMPLE PROBLEM 5.12

Density and specific gravity

a. At 20°C, 9.41 g of chloroform ($CHCl_3$) occupies a volume of 6.32 mL. What is the density of chloroform at this temperature?

b. At 20°C water has a density of 1.00 g/mL. Based on your answer to part a, what is the specific gravity of chloroform at this temperature?

Strategy

The density of a substance is obtained by dividing the mass (in grams) of a sample by the volume (in mL) that it occupies. The specific gravity of a substance at a given temperature is the density of the substance divided by the density of water at the same temperature.

Solution

a. $\text{Density} = \frac{9.41\text{ g}}{6.32\text{ mL}} = 1.49\text{ g/mL}$

b. $\text{Specific gravity} = \frac{\text{density of chloroform}}{\text{density of water}} = \frac{1.49\text{ g/}\cancel{\text{mL}}}{1.00\text{ g/}\cancel{\text{mL}}} = 1.49$

PRACTICE PROBLEM 5.12

At 0°C methanol (CH_3OH) has a density of 0.810 g/mL. At this temperature how many milliliters of methanol does 17.3 g occupy?

HEALTH Link: Making Weight

In wrestling, boxing, weight lifting, judo, and many other sports, athletes compete with others who have a similar weight. "Making weight" refers to the custom of rapidly losing weight to become eligible to compete in a particular weight classification (Figure 5.25). This rapid weight reduction typically involves loss of water (one cup of water weighs about 1/2 pound). Rapid water loss techniques include restricting fluid intake and increasing sweat production by sitting in a sauna or exercising in a hot room while wearing a rubber suit. After qualifying for a lower weight division, the idea is to rehydrate during the time that elapses between the weighing in and competition. An athlete doing this will be heavier and, presumably, stronger than his or her competitors.

It is not clear that making weight puts an athlete at a competitive advantage. Dehydration is known to adversely affect endurance, strength, energy, and motivation. Extreme dehydration can result in kidney and heart failure. In 1997, three college wrestlers died while trying to rapidly lose weight through water loss.

In response to these deaths, the NCAA created new rules to discourage rapid weight loss. These rules include banning the use of saunas and rubber suits for this purpose. A certification process that requires athletes to be hydrated before being eligible for weighing was also put in place. Athletes with a urine specific gravity greater than 1.025, which is at the high end of the normal range (1.002–1.0035), are considered to be dehydrated and ineligible for official weigh in.

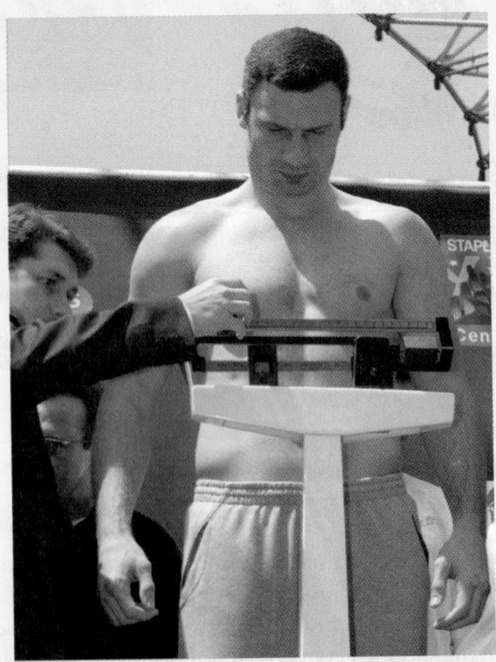

FIGURE 5.25

Making weight

Rapid weight reduction, achieved by loss of water from the body, can qualify an athlete for a lower weight division.

Vapor Pressure

Due to collisions that take place between the particles (atoms or molecules) that make up a liquid, particles at the surface are continually evaporating—being bounced off into the gas phase. At the same time, gas phase molecules are being trapped and converted to liquid (Figure 5.26).

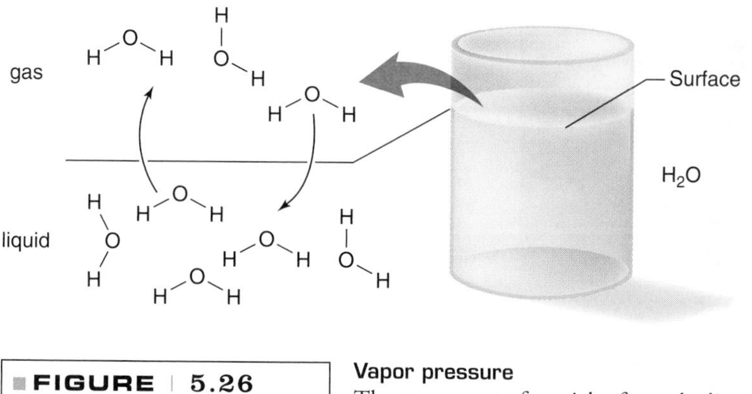

FIGURE 5.26 **Vapor pressure** The movement of particles from the liquid to the gas phase gives rise to vapor pressure.

If water, $H_2O(l)$, is placed in a closed container, water evaporates until the partial pressure of $H_2O(g)$ reaches a maximum level, called the **vapor pressure**. When water's vapor pressure is reached, any other gas that is present (air, for example) is saturated with water vapor because this gas holds the maximum amount of water vapor possible. For example, the air in a closed container is saturated with water vapor if, at 20°C, the partial pressure of water is 17.54 torr. This is the vapor pressure of water at this temperature. As Table 5.6 shows, the vapor pressure of water varies with temperature: the higher the temperature, the greater the vapor pressure.

The **boiling point** of a liquid is the temperature at which the vapor pressure of the liquid equals the atmospheric pressure. At this temperature, vapor forms in pockets (bubbles) throughout the liquid. Table 5.6 shows that the vapor pressure of water is 760 torr at 100°C, so at an atmospheric pressure of 760 torr (1 atm), water boils at 100°C. At other atmospheric pressures, the boiling point of water varies. In Denver, Colorado, at an altitude of 1.6 km and an air pressure of 560 torr, water boils at 92°C, and on Mt. Everest, at an altitude of 9.3 km and an air pressure of 200 torr, water boils at 67°C.

■ Liquids boil when their vapor pressure equals the pressure of the air above them.

TABLE 5.6 THE VAPOR PRESSURE OF WATER AT VARIOUS TEMPERATURES

Temperature (°C)	Vapor Pressure (torr)	Temperature (°C)	Vapor Pressure (torr)
0	4.58	80	355.10
10	9.21	90	525.76
20	17.54	100	760.00
30	31.82	110	1074.56
37[a]	47.07	125	1740.93
40	55.32	150	3570.48
50	92.51	175	6694.08
60	149.38	200	11659.16
70	233.70	300	64432.80

[a] Body temperature.

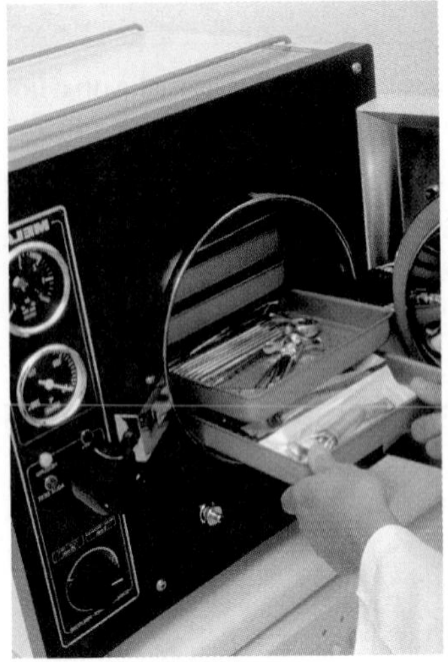

FIGURE 5.27

An autoclave
An autoclave produces pressures greater than 1 atm, so water boils at temperatures above 100°C. At these temperatures, most infectious agents are destroyed.

Source: Ulrich Sapountsis/ Photo Researchers Inc.

Increasing the pressure above 760 torr will push the boiling point of water to temperatures greater than 100°C. For example, when heat is applied to a pressure cooker, the pressure begins to climb, as does the temperature required for water to boil. At a pressure of 1520 torr, double the standard atmospheric pressure, water boils at 120°C. The high pressures reached in autoclaves, used for sterilization purposes, allow water to boil at temperatures high enough to kill most infectious agents (Figure 5.27).

HEALTH Link: Breathing

The process of supplying your cells with the O_2 that they need and removing the CO_2 that is produced depends on many of the principles discussed in this chapter (Figure 5.28).

Inhaling

When you inhale, your intercostal muscles lift your rib cage and your diaphragm contracts. The resulting increase in the volume of your thoracic cavity (holds the lungs and heart) expands your lung volume by about 0.5 L. This increase in lung volume at constant temperature results in a pressure decrease in the lungs (Boyle's law) to about 1 torr below atmospheric pressure and, since the air in your lungs is at a lower pressure than the atmosphere that surrounds you, air moves into your lungs until the internal and external pressures match.

Moving oxygen gas from the lungs to the tissues

At the alveoli, the chambers in your lungs where gas exchange with capillary blood takes place, the partial pressure of O_2 (P_{O_2}, Section 5.4) is about 104 torr, while blood entering the capillaries of the alveoli normally has a P_{O_2} of approximately 40 torr (the partial pressure of a gas above a liquid is directly related to the amount of gas dissolved in the liquid; Section 7.3).

The difference in O_2 partial pressures between air in the alveoli and blood entering the alveoli leads to a net movement of O_2 into the blood, so blood leaves the alveoli with a P_{O_2} of about 104 torr, enters the heart, and is pumped throughout the body. Much of the O_2 is carried by hemoglobin. The P_{O_2} in the remote tissues is less than 40 torr, so a net movement of O_2 from the blood (higher pressure) to the tissues (lower pressure) takes place. Blood leaves the tissues with a P_{O_2} of 40 torr and is cycled back to the lungs where it picks up more O_2.

Moving carbon dioxide from the tissues to the lungs

Blood in your tissues has a partial pressure of CO_2 (P_{CO_2}) above 45 torr, while blood entering the tissues has a P_{CO_2} of only 40 torr. A net movement of CO_2 from tissues (higher pressure) to the blood (lower pressure) takes place, and blood

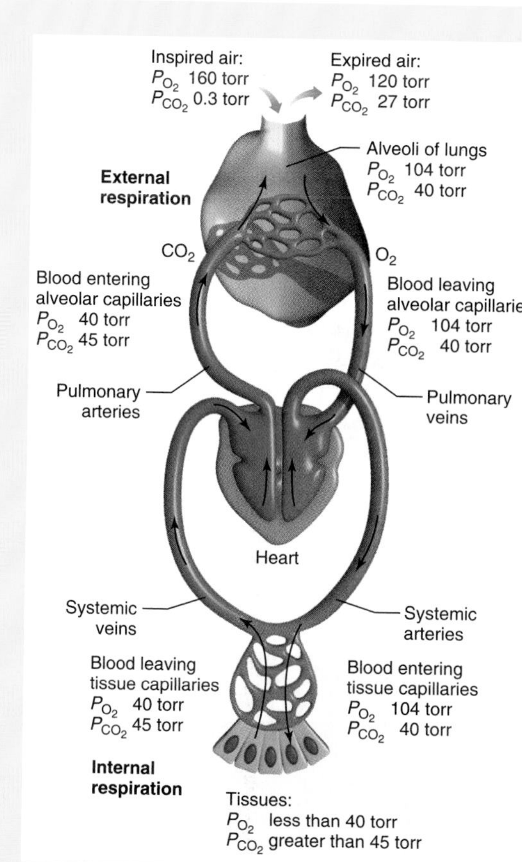

that moves from your tissues to your lungs has a P_{CO_2} of 45 torr. Much of the CO_2 is carried as HCO_3^-. On entering the capillaries of the alveoli, where the P_{CO_2} is 40 torr, CO_2 moves from the blood into the alveoli, leaving the blood with a P_{CO_2} of 40 torr.

Exhaling

When you exhale, your intercostal muscles and diaphragm relax. This decreases the volume of your thoracic cavity and reduces your lung volume (a decrease in volume at constant temperature results in a pressure increase) and, subsequently, the pressure in your lungs is raised to about 1 torr above atmospheric pressure. Since the air in your lungs is at a higher pressure than the atmosphere that surrounds you, air moves out of your lungs, carrying with it much of the CO_2 delivered to the lungs from the tissues.

FIGURE 5.28

Breathing
The movement of O_2 from the lungs to the tissues and of CO_2 from the tissues to the lungs is driven by pressure differences. The O_2 and CO_2 move from regions of high partial pressure to regions of low partial pressure.

Source: Figure 22.17, p. 854 from HUMAN ANATOMY AND PHYSIOLOGY, 6th ed. By Elaine N. Marieb. Copyright 2004 by Pearson Education, Inc. Reprinted by permission.

5.6 SOLIDS

As we have previously discussed, the particles (atoms, ions, or molecules) that make up a solid are held close to one another and have a limited ability to move around. Solids can be classified based on whether the arrangement of these particles is ordered (in **crystalline solids**) or not (in **amorphous solids**).

We will begin by considering crystalline solids, of which there are four types: **ionic**, **covalent**, **molecular**, and **metallic** (Table 5.7). Ionic solids consist of oppositely charged

TABLE 5.7 Solids

Type	Particles	Forces	Examples
Crystalline			
Ionic	Cations and anions	Ionic bonds	NaCl, KBr, $MgCl_2$
Covalent	Atoms	Covalent bonds	Diamond, graphite
Molecular	Molecules	Various noncovalent interactions	H_2O, CH_3CH_2OH
Metallic	Metal ion	Metallic bond	Na, Fe, Cu, Au
Amorphous	Molecules, atoms, or ions	Covalent or noncovalent	Rubber, glass

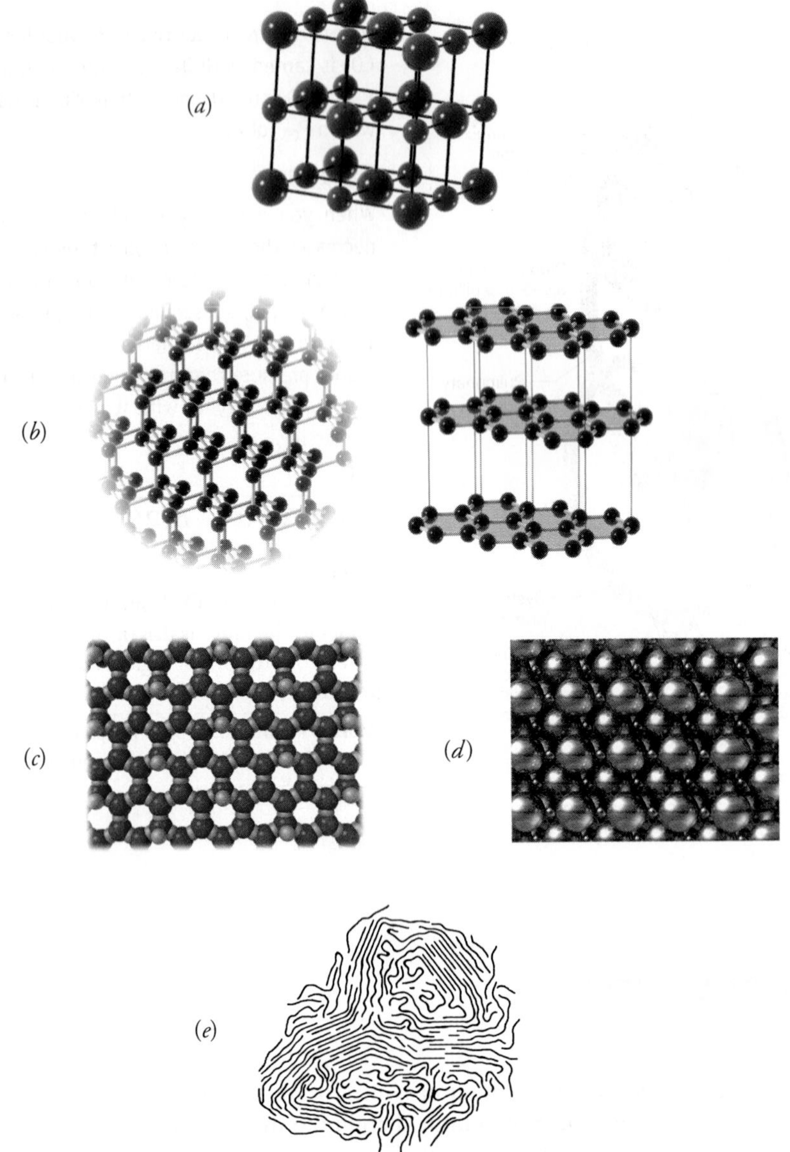

FIGURE | 5.29

Solids
In crystalline solids the particles have an ordered arrangement. The categories of crystalline solids are (*a*) ionic, (*b*) covalent, (*c*) molecular, and (*d*) metallic. In an amorphous solid (*e*) the arrangement of particles is not ordered.

Source: (*b*) and (*c*) BROWN, THEODORE E; LEMAY, H. EUGENE; BURSTEN, BRUCE E, CHEMISTRY: THE CENTRAL SCIENCE, 10th edition, © 2006, p. 472. Reprinted by permission of Pearson Education, Inc., Upper Saddle River, NJ.

ions held to one another by ionic bonds. The ordered arrangement in this type of solid, as we saw in Section 3.4, consists of the alternating cations and ions (Figure 5.29*a*). Most ionic solids are hard and have high melting points.

In covalent solids atoms are held to one another by an arrangement of covalent bonds that extends throughout the solid. Most melt or sublime at high temperatures. Examples include diamond and graphite, shown on the left and right, respectiviely, in Figure 5.29*b*. In diamond, one of the hardest substances known, each carbon atom is interlocked with four others in a tetrahedral arrangement. In graphite, the carbon atoms are arranged into many layered sheets of carbon atoms. Within a sheet each carbon atom is strongly connected to three other carbon atoms by covalent bonds, and individual sheets are able to move across one another, which gives graphite its slipperiness.

Molecular solids consist of an ordered arrangement of molecules attracted to one another by noncovalent interactions (hydrogen bonding, London forces, and so on). Because the forces that hold the molecules to one another are weaker than in ionic and covalent solids, melting points of molecular solids tend to be low by comparison. In ice, an example of a molecular solid (Figure 5.29*c*), the water molecules are attracted to one another by hydrogen bonds and the molecules are arranged in honeycomb-like shapes.

The structure of metallic solids (metals) is an array of metal cations immersed in a cloud of electrons that spans the entire crystalline structure (Figure 5.29*d*). The freely moving electrons—former valence electrons of the metal atoms—are shared in a metallic bond by all of the metal cations. The melting point and hardness exhibited by metallic solids vary, depending on the metal.

In contrast to all of the crystalline solids just mentioned, the particles in amorphous solids are not arranged in an ordered fashion. Rubber is an example of an amorphous solid—the molecules that make up its structure are arranged in a haphazard fashion (Figure 5.29*e*).

Each red blood cell contains about 300 million hemoglobin molecules, each of which transports up to four O_2 molecules from the lungs to the tissues. Hemoglobin bound to O_2 is called oxyhemoglobin and that without O_2 is called deoxyhemoglobin. To measure the percentage of oxyhemoglobin in a climber's blood, the probe of a pulse oximeter is clipped to a finger. The probe emits two different energies of electromagnetic radiation—infrared radiation absorbed by oxyhemoglobin and red light absorbed by deoxyhemoglobin. Analysis of the amount of each type of light that is absorbed allows the percentage of oxyhemoglobin to be calculated and displayed by the oximeter. After making a blood O_2 measurement, the climber radios the result to base camp.

The percentage of oxygen saturation is related to the partial pressure of O_2 (P_{O_2}) in blood, as shown to the right. A 97% O_2 saturation corresponds roughly with the normal arterial P_{O_2} of about 104 torr. A 90% O_2 saturation (P_{O_2} = 60 torr) or lower is dangerous.

summary of objectives

1 Define the terms specific heat, heat of fusion, and heat of vaporization. Show how each can be used in calculations involving energy.

The **specific heat** of a substance, the amount of heat required to raise the temperature of 1 gram of the substance by 1°C, can be used to calculate the changes in heat energy associated with a particular temperature change. Water, for example, has a specific heat of 1.000 cal/g °C, so to increase the temperature of 15.0 g of water by 25°C, 375 cal must be input.

$$15.0\text{ g} \times 25.0\ \cancel{^\circ\text{C}} \times \frac{1.000\text{ cal}}{\text{g}\ \cancel{^\circ\text{C}}} = 375\text{ cal}$$

Heat of fusion is the heat energy required to melt a solid and **heat of vaporization** is the heat energy required to vaporize a liquid. Water has a heat of fusion of 79.7 cal/g, so melting 15.0 g of ice requires 1.20×10^3 cal.

$$15.0\text{ g} \times \frac{79.7\text{ cal}}{\text{g}} = 1.20 \times 10^3\text{ cal}$$

Water has a heat of vaporization of 540 cal/g, so vaporizing 15.0 g of water requires 8.10×10^3 cal.

$$15.0\ \text{g} \times \frac{540\ \text{cal}}{\text{g}} = 8.10 \times 10^3\ \text{cal}$$

2 Describe the meaning of the terms enthalpy change, entropy change, and free energy change. Explain how the value of free energy change can be used to predict whether a process is spontaneous or nonspontaneous.

Enthalpy change (ΔH) is the heat released or absorbed during a process and **entropy change** (ΔS) is the change in how energy is distributed. **Free energy change** (ΔG) is a combination of enthalpy change, entropy change, and temperature ($\Delta G = \Delta H - T\Delta S$) that can be used to indicate whether a process is **spontaneous** (will continue on its own, once started) or **nonspontaneous** (will not run by itself). For a spontaneous process ΔG has a negative value and for a nonspontaneous process ΔG has a positive value.

3 Convert between common pressure units.

The **pressure** of a gas is due to collision of the gas particles with an object (the walls of a container, for example). Standard atmospheric pressure corresponds to 1 atmosphere (**atm**), 14.7 pounds per square inch (**psi**), and 760 **torr**. The relationships between these units can be used as conversion factors. For example, 225 torr equals 0.296 atm and 4.35 psi.

$$225\ \cancel{\text{torr}} \times \frac{1\ \text{atm}}{760\ \cancel{\text{torr}}} = 0.296\ \text{atm}$$

$$225\ \cancel{\text{torr}} \times \frac{14.7\ \text{psi}}{760\ \cancel{\text{torr}}} = 4.35\ \text{psi}$$

4 List the variables that describe the condition of a gas and give the equations for the various gas laws.

Pressure, temperature, volume, and number of moles are the variables that determine the condition of a gas sample. The mathematical relationships that relate these variables are **Boyle's law** ($P_1V_1 = P_2V_2$), **Gay-Lussac's law** ($P_1/T_1 = P_2/T_2$), **Charles' law** ($V_1/T_1 = V_2/T_2$), **Avogadro's law** ($V_1/n_1 = V_2/n_2$), the **combined gas law** ($P_1V_1/T_1 = P_2V_2/T_2$), and the **ideal gas law** ($PV = nRT$).

5 Explain Dalton's law of partial pressure.

Partial pressure, a term used when dealing with mixtures of gases, is the pressure that each of the gases in the mixture would exert if it were the only one present. **Dalton's law of partial pressure** states that the total pressure of a mixture is the sum of the partial pressures ($P_{total} = P_A + P_B + P_C + \ldots$).

6 Define the terms density and specific gravity.

The **density** of a substance is the mass of a given volume and **specific gravity** is the density of a substance at a particular temperature divided by the density of water at the same temperature.

7 Describe the relationship between atmospheric pressure and the boiling point of a liquid.

When a liquid evaporates, its maximum pressure above the liquid is called the **vapor pressure**. A liquid boils when its vapor pressure is equal to the atmospheric pressure.

8 Describe the difference between amorphous and crystalline solids and describe the makeup of the four classes of crystalline solids.

In a **crystalline solid** the arrangement of particles is ordered, while in an **amorphous** one it is not. The four classes of crystalline solids are **ionic** (consists of arrays of oppositely charged ions held to one another by ionic bonds), **covalent** (atoms are held to one another by an array of covalent bonds), **molecular** (consists of a ordered arrangement of molecules held to one another by noncovalent interactions), and **metallic** (consists of an array of metal cations immersed in a cloud of electrons that spans the entire crystalline structure).

END OF CHAPTER PROBLEMS

Answers to problems whose numbers are printed in color are given in Appendix D. More challenging questions are marked with an asterisk. Problems within color rules are paired. **ILW** = Interactive Learning Ware solution is available at *www.wiley.com/college/raymond.*

5.1 States of Matter and Energy

5.1 **a.** What is heat of fusion?
b. What is heat of vaporization?

5.2 **a.** What change in physical state takes place during sublimation?
b. Give an example of an element or compound that sublimes.

5.3 If you immerse your arm in a bucket of ice water, your arm gets cold. Where does the heat energy from your arm go and what process is the energy used for?

5.4 Some over-the-counter (nonprescription) wart removers contain ether. When a few drops are placed on a wart, it feels cold as the ether rapidly evaporates. Where does the heat energy from the wart go and what process is the energy used for?

5.5 Calculate the energy required to
a. warm 325 g of ice from −10°C to 0°C
b. warm 325 g of water from 10°C to 20°C

5.6 Calculate the energy released when
a. 750 g of ethanol are cooled from 25°C to 5.0°C
b. 750 g of water are cooled from 25°C to 5.0°C

5.7 Calculate the heat required to vaporize
a. 15 g of water **b.** 15 g of ethanol

5.8 Calculate the heat required to melt
a. 15 g of ice **b.** 15 g of ethanol (solid)

5.9 Steam can cause severe burns because of the heat released to the skin when the steam condenses to water. How much energy is released when 25 g of steam condense?

***5.10** An 11 g piece of gold has a temperature of 23°C. What will the new temperature of the gold be if 45 cal of heat are added?

***5.11** An 11 g piece of copper has a temperature of 23°C. What will the new temperature of the copper be if 45 cal of heat are added?

5.12 In chemical terms, what does the term *spontaneous* mean?

5.13 In chemical terms, what does the term *nonspontaneous* mean?

5.14 **a.** In terms of enthalpy, is the melting of a snowball spontaneous or nonspontaneous?
b. In terms of entropy, is the melting of a snowball spontaneous or nonspontaneous?
c. If a snowball sits outside at a temperature of −10°C, is the melting of the snowball spontaneous or nonspontaneous? Is the value of ΔG for this process negative or positive?
d. If a snowball sits inside at a temperature of 22°C, is the melting of the snowball spontaneous or nonspontaneous? Is the value of ΔG for this process negative or positive?

5.15 **a.** In terms of enthalpy, is the condensation of steam spontaneous or nonspontaneous?
b. In terms of entropy, is this process spontaneous or nonspontaneous?
c. At 90°C and 1 atm of atmospheric pressure is the condensation of steam spontaneous or nonspontaneous? Is the value of ΔG for this process negative or positive?
d. At 110°C and 1 atm of atmospheric pressure is this process spontaneous or nonspontaneous? Is the value of ΔG for this process negative or positive?

5.16 Some homes are cooled by swamp coolers (evaporative coolers), which consist of a box containing a fan and water soaked pads. When hot outside air is drawn into the box, heat in the air evaporates water from the pads. The cooled air is blown into the house.
a. Describe the enthalpy change for the air.
b. Describe the kinetic energy change for the air.
c. How much energy does it take to evaporate 1 mol of water?
d. Air has a heat capacity of 0.242 cal/g°C. How many calories of energy must be removed to cool 5.00 g of air by 20°C?
e. How many grams of water must evaporate to cool 7.50 g of air by 15.0 K?

***5.17** How much energy is required to convert
a. 35.5 g of ice at −10°C to water at 95.2°C?
b. 3.10 mol of ice from −8°C to steam at a 100°C?

***5.18** A coil of hot copper wire is placed in 50.0 grams of water that has an initial temperature of 18.0°C. The copper cools by 100.0°C, and the water warms up to 28.0°C. How many grams of copper are present?

***5.19** A hot piece of aluminum is placed in 50.0 grams of water that has an initial temperature of 18.0°C. The aluminum cools by 100.0°C, and the water warms up to 28.0°C. How many grams of aluminum are present?

5.2 GASES AND PRESSURE

5.20 Define STP.

5.21 **a.** Give an example of an element or compound that is a solid at STP.
b. Give an example of an element or compound that is a liquid at STP.
c. Give an example of an element or compound that is a gas at STP.

5.22 A pressure of 9.2 psi is how many
a. atmospheres? **b.** torr?

5.23 A pressure of 13.6 psi is how many
a. atmospheres? **b.** torr?

5.24 **a.** Estimate the atmospheric pressure, in atmospheres, at an altitude of 20 km (see Figure 5.9).
b. Convert your answer to part a into psi and torr.

5.25 **a.** Estimate the atmospheric pressure, in atmospheres, at an altitude of 5 km (see Figure 5.9).
b. Convert your answer to part a into psi and torr.

5.26 Is the average atmospheric pressure higher in Denver, Colorado (1.6 km above sea level), or in Denver, Oklahoma (0.35 km above sea level)? Explain.

5.27 At an atmospheric pressure of 760 torr and a temperature of 0°C, the mercury level in the right arm of the manometer pictured in Figure 5.12 is 5 mm higher than the mercury in the left arm. What is the gas pressure inside the flask?

5.28 Explain how a barometer works.

5.29 **a.** Listening to the weather report, you hear that the barometer is falling. What part of the barometer is falling?
b. What is happening to the air pressure?

5.30 In the SI measurement system, the unit of pressure is called the kilopascal (kPa). One atmosphere equals 101.3 kPa.
a. Convert 1.29 atm into kilopascals.
b. Convert 87.2 kPa into atmospheres.
c. Convert 612 torr into kilopascals.
d. Convert 277 kPa into torr.
e. Convert 9.22 psi into kilopascals.
f. Convert 583 kPa into pounds per square inch.

5.31 At altitudes above 14,000 feet, most people get at least a mild case of mountain sickness (see the first page of this chapter). Using Figure 5.9, estimate the air pressure in kilopascals (see Problem 5.30) at this altitude.

5.3 THE GAS LAWS

***5.32** Why is Boyle's law a law and not a theory? What is the difference?

5.33 At a pressure of 760 torr a balloon has a volume of 1.50 L. If the balloon is put into a container and the pressure is increased to 2500 torr (at constant temperature), what is the new volume of the balloon?

5.34 At a pressure of 1.5 atm a balloon has a volume of 3.25 L. If the pressure is decreased to 1.0 atm (at constant temperature), what is the new volume of the balloon?

5.35 At a temperature of 30°C, a balloon has a volume of 1.50 L. If the temperature is increased to 60°C (at constant pressure), what is the new volume of the balloon?

5.36 At a temperature of 22°C, a balloon has a volume of 8.0 L. If the temperature is increased to 55°C (at constant pressure), what is the new volume of the balloon?

5.37 At a temperature of 30°C, a gas inside a 1.50 L metal canister has a pressure of 760 torr. If the temperature is increased to 60°C (at constant volume), what is the new pressure of the gas?

5.38 At a temperature of 75°C, a gas inside a 5.00 L metal canister has a pressure of 2500 torr. If the temperature is decreased to 45°C (at constant volume), what is the new pressure of the gas?

5.39 A 2.0 L balloon contains 0.35 mol of $Cl_2(g)$. At constant pressure and temperature, what is the new volume of the balloon if 0.20 mol of gas is removed?

5.40 A 9.1 L balloon contains 1.25 mol of He(g). At constant pressure and temperature, what is the new volume of the balloon if 0.75 mol of He is added?

5.41 A balloon with a volume of 1.50 L is at a pressure of 760 torr and a temperature of 30°C. If the pressure is increased to 2500 torr and the temperature is raised to 60°C, what is the new volume of the balloon?

5.42 A balloon with a volume of 2.00 L is at a pressure of 1.5 atm and a temperature of 20°C. If the pressure is increased to 2.0 atm and the temperature is raised to 25°C, what is the new volume of the balloon?

5.43 A 575 mL metal can contains 2.50×10^{-2} mol of He at a temperature of 298 K. What is the pressure (in atm) inside the can? Is this pressure greater than or less than standard atmospheric pressure?

5.44 A 1.00 qt metal can contains 0.150 g of He at a temperature of 20°C. What is the pressure (in atm) inside the can? Is this pressure greater than or less than standard atmospheric pressure?

*5.45 The label on a can of spray paint warns you to keep it away from high temperatures. From the perspective of the ideal gas law, explain why.

5.46 A 250.0 mL flask contains 0.350 mol of O_2 at 40°C.
a. What is the pressure in atm?
b. What is the pressure in torr?
c. What is the pressure in psi?

5.47 A 100.0 mL flask contains 400 g of N_2 at 0°C.
a. What is the pressure in atm?
b. What is the pressure in torr?
c. What is the pressure in psi?

***5.48** A 250 mL flask contains He at a pressure of 760 torr and a temperature of 25°C. What mass of He is present?

*5.49 A 750 mL flask contains O_2 at a pressure of 0.75 atm and a temperature of 20°C. What mass of O_2 is present?

***5.50 a.** How many moles of Ar are present in a 2.0 L flask that has a pressure of 1.05 atm at a temperature of 25°C?
b. What is the mass of this Ar?

*5.51 **a.** How many moles of Ne are present in a 1.0 qt flask that has a pressure of 850 torr at a temperature of 35°C?
b. What is the mass of this Ne?

5.52 Use concepts discussed in this chapter to explain why an empty plastic milk bottle collapses if the air is pumped out of it.

5.53 A sample of a gas at a temperature of 25.0°C has a pressure of 815 torr and occupies a volume of 9.92 L.
a. Use Boyle's law to calculate the new pressure of the gas if the temperature is held constant and the volume is decreased to 5.92 L.
b. Use Gay-Lussac's law to calculate the new pressure of the gas if the volume is held constant and the temperature is increased to 125.0°C.
c. Use Charles' law to calculate the new volume of the gas if the pressure is held constant and the temperature is increased to 125.0°C.
d. Use the combined gas law to calculate the new pressure if the temperature is increased to 125.0°C and the volume is decreased to 5.92 L.
e. Use the ideal gas law to calculate the number of moles of gas that are present.

***5.54 a.** The cover of the rock band Led Zeppelin's first album pictured the Hindenburg, a hydrogen-filled dirigible, going down in flames. At a temperature of 25°C and a pressure of 760 torr, the Hindenburg held 7.062×10^6 cubic feet of hydrogen gas. How many grams of hydrogen did this represent?
b. If the Hindenburg had been filled with helium instead of hydrogen, assuming the same temperature, pressure, and volume, how many grams of helium would have been present?

5.55 True or false? Ten liters of helium gas at STP has the same mass as ten liters of neon gas at STP. Explain.

5.56 True or false? Two moles of helium gas at STP occupies the same volume as two moles of neon gas at STP. Explain.

5.4 Partial Pressure

5.57 A mixture of gases contains 0.75 mol of N_2, 0.25 mol of O_2, and 0.25 mol of He.
a. What is the partial pressure of each gas (in atm and in torr) in a 25 L cylinder at 350 K?
b. What is the total pressure?

5.58 A mixture of gases contains 0.75 mol of N_2, 0.25 mol of O_2, and 0.25 mol of He.
a. What is the partial pressure of each gas (in atm and in torr) in a 25 L cylinder at 250 K?
b. What is the total pressure?

5.59 For the mixture of gases in Problem 5.57, what are the partial pressures and the total pressure if 0.50 mol of $CO_2(g)$ is added?

5.60 For the mixture of gases in Problem 5.58, what are the partial pressures and the total pressure if 0.50 mol of $CO_2(g)$ is added?

5.61 In which town does dry air contain the highest partial pressure of O_2, Hot Coffee, Mississippi (altitude 279 ft) or Pie Town, New Mexico (altitude 7778 ft)? Explain.

***5.62** Some runners train at high altitudes. Why?

*5.63 Scuba divers sometimes breathe a gas mixture called trimix, which consists of helium, oxygen, and nitrogen gases. If a scuba tank with a volume of 2.5 L, a pressure of 2700 psi, and a temperature of 45°F contains 27% O_2, 12% He, and 8% N_2,
a. what is the partial pressure of each gas?
b. how many grams of each gas are present?
c. how many molecules of O_2 and N_2 and atoms of He are present?

***5.64** If the scuba tank in problem 5.63 is used until the total pressure drops to 1300 psi, assuming no change in temperature,
a. what is the partial pressure of each gas?
b. how many grams of each gas are present?
c. how many molecules of O_2 and N_2 and atoms of He are present?

5.5 Liquids

5.65 A patient has 25.0 mL of blood drawn and this volume of blood has a mass of 26.5 g. What is the density of the blood?

5.66 A patient has 0.050 L of blood drawn and this volume of blood has a mass of 55.0 g. What is the density of the blood?

5.67 At 20°C, how many milliliters of ethanol correspond to 35.2 g? (See Table 5.5.)

5.68 At 20°C, how many milliliters of ethanol correspond to 50.0 g? (See Table 5.5.)

5.69 At 20°C, what volume does 11.2 g of human fat occupy? (See Table 5.5.)

5.70 Why do gases have a lower density than solids and liquids?

5.71 What is the density, in g/mL, of $O_2(g)$ at 0°? (See Table 5.5.)

5.72 What is the specific gravity of whole blood at 20°C? (See Table 5.5.)

5.73 What is the specific gravity of kerosene at 20°C? (See Table 5.5.)

5.74 **a.** What is the specific gravity of water?
b. Does this value change with varying temperature?

5.75 If water is placed in a flask and a pump is used to reduce the atmospheric pressure above the water to 17.54 torr, at what temperature does the water boil? (See Table 5.6.)

5.76 Estimate the boiling point of water at a pressure of 920 torr.

5.77 Estimate the boiling point of water at a pressure of 2.2 atm.

*__5.78__ If you go camping in the mountains, why does it take longer to cook a pot of noodles than it does at sea level?

*__5.79__ Calculate the energy required for each. (Hint: Use information provided in Tables 5.1 and 5.5.)
a. warm 35.0 mL of water from 21°C to 29°C
b. warm 17.5 mL of water from 18°C to 54°C

*__5.80__ Calculate the energy required for each. (Hint: Use information provided in Tables 5.1 and 5.5.)
a. warm 2.60 mL of ethanol from 15°C to 35°C
b. warm 17.5 mL of water from 32°C to 87°C

5.81 Swamp coolers (see Problem 5.16) are not as effective in humid areas as they are in dry areas. Explain.

5.82 If you hang wet sheets outside on a clothesline on a sunny winter day when the temperature is 0°C, they will eventually dry. Explain.

5.6 Solids

5.83 Why do ionic solids typically have a higher melting point than molecular solids?

5.84 Describe the difference in the forces that hold the particles to one another in each of the four types of crystalline solid.

5.85 How is the structure of an amorphous solid different from the structure of a crystalline one?

*__5.86__ Crystalline solids tend to melt over a narrower temperature range than amorphous solids. Explain why.

5.87 At a microscopic level, molecular solids can have the same appearance as covalent solids. Why do molecular solids tend to have lower melting points?

5.88 Molecular solids and covalent solids each contain covalent bonds. Why do molecular solids tend to be harder?

HEALTH*Link* | Blood Pressure

*__5.89__ During severe bleeding, ADH (a hormone released by the hypothalamus) causes vasoconstriction (shrinking of the blood vessels) to take place. What effect does a decrease in blood vessel volume have on blood pressure?

5.90 Ethyl alcohol, the alcohol present in alcoholic beverages, inhibits the release of ADH (see the previous problem). What effect does drinking alcohol have on blood pressure?

HEALTH*Link* | Making Weight

5.91 One of the rule changes that the NCAA made to discourage rapid weight loss was to shorten the time between weigh in and competition from 24 hours to just 2 hours. Why would this discourage athletes from trying to make weight?

5.92 Why does a high urine specific gravity indicate dehydration?

5.93 **a.** Use the density of water (1.00 g/mL) to derive a conversion factor for water that has the units lb/cup.
b. If an athlete reduces her body's water volume by 5.5 cups through restricting fluid intake and sweating in a sauna, how much weight has she lost? Is this a good idea? Explain.

HEALTH*Link* | Breathing

5.94 The chest compressions given during cardiopulmonary resuscitation (CPR) cause the injured person to exhale. Explain why, in terms of Boyle's law.

5.95 If your diet is low in iron-containing foods, your body may not be able to produce normal amounts of hemoglobin and you may develop iron-deficiency anemia. The symptoms of anemia include shortness of breath and fatigue. Explain why insufficient hemoglobin can produce these symptoms.

5.96 According to CPR guidelines, when giving CPR you should administer 100 chest compressions per minute and 12 to 16 breaths per minute. According to "Breath of Life," what mistake is often made when CPR is given and to what is this mistake attributed?

THINKING IT THROUGH

5.97 Many sports organizations ban the use of and test for the presence of higher than normal levels of erythropoietin (EPO), a naturally occurring compound that stimulates the production of red blood cells. How might EPO enhance athletic performance?

5.98 If a patient is anemic (see Problem 5.95) will the % O_2 saturation reading given by a pulse oximeter be a good indication of the P_{O_2} in the patient's blood? Explain why or why not.

INTERACTIVE LEARNING PROBLEMS

5.99 A sample of helium at 740 torr and in a volume of 2.58 L was heated from 24.0 to 75.0°C. The volume of the container expanded to 2.81 L. What was the final pressure (in torr) of the helium?

5.100 A sample of carbon monoxide was prepared and collected over water at a temperature of 20°C and a total pressure of 754 torr. It occupied a volume of 268 mL. How many grams of CO were in the sample?

SOLUTIONS TO PRACTICE PROBLEMS

5.1 Converting the liquid into a gas. As the heat required to cause this physical change moves from the refrigerator into the liquid, the refrigerator cools.

5.2 Copper.

5.3 **a.** nonspontaneous; **b.** spontaneous; **c.** negative; **d.** The heat energy goes into subliming the dry ice.

5.4 **a.** 1.81×10^3 torr; **b.** 0.99 atm

5.5 Volume and pressure are inversely proportional. As you push on the bubble, its volume decreases. As the volume decreases, the pressure inside the bubble increases until it becomes so high that the bubble pops.

5.6 789 psi

5.7 If a container of compressed gas is exposed to a fire or any other intense heat source, its temperature will increase. As the temperature rises, so does the pressure of the gas. At a high enough pressure, the container might explode.

5.8 0.701 mol

5.9 **a.** 5.6 L; **b.** 690 torr (two significant figures)

5.10 $P_{N_2} = 0.19$ atm; $P_{O_2} = 0.056$ atm; $P_{He} = 0.27$ atm; $P_{total} = 0.52$ atm

5.11 159 torr

5.12 21.4 mL

RASPBERRY
diet
Snapple
ICED TEA
Protein 0g
Not a significant source of other nutrients.
*Percent Daily Values are based on a 2,000 calorie diet.
OUR REAL TEA STARTS WITH THE FINEST TEA LEAVES AND I
MADE FROM: WATER, CITRIC ACID, TEA, NATURAL FLAVORS,
ASPARTAME**, POTASSIUM CITRATE, MALIC ACID.
**PHENYLKETONURICS: CONTAINS PHENYLALANINE
DISTRIBUTED U
RYE BROOK, NY

You stop by the grocery store one afternoon to get something to drink. Diet Snapple looks good, so you pick up a bottle and read the label. One of the ingredients, the artificial sweetener called aspartame, comes with the warning, "Phenylketonurics: contains phenylalanine." Recalling from your chemistry class that phenylalanine is a naturally occurring amino acid, you wonder why it causes a problem for some people.

6 REACTIONS

The story behind this warning label found on foods and beverages that contain aspartame is related to the biochemistry of phenylalanine, one product formed when this sweetener is broken down in the body. In phenylketonurics, phenylalanine does not undergo a chemical change (Section 3.4) that occurs in the rest of the population, and the resulting buildup of phenylalanine produces harmful side effects.

In this chapter we will take our first detailed look at chemical changes, known also as **chemical reactions**. We will focus on the reactions of relatively small compounds, this being the necessary starting point if we are to understand the chemistry of the larger compounds (including aspartame) that are important to biochemistry.

objectives

After completing this chapter, you should be able to:

1. Interpret and balance chemical equations.
2. Classify reactions as involving synthesis, decomposition, single replacement, or double replacement.
3. Identify oxidation, reduction, combustion, and hydrogenation reactions.
4. Identify hydrolysis, hydration, and dehydration reactions of organic compounds.
5. Determine the limiting reactant, theoretical yield, and percent yield of a reaction.
6. Describe the difference in energy changes for spontaneous and nonspontaneous reactions, and list the factors that affect the rate of a chemical reaction.

6.1 CHEMICAL EQUATIONS

Chapter 2 described nuclear reactions, processes that involve the loss of particles and high energy electromagnetic radiation from atomic nuclei. In **chemical reactions**, the covalent and ionic bonds that hold elements and compounds together are broken and new bonds are formed. For example, the fuel cell technology being developed to generate electricity is based on a reaction between hydrogen gas (H_2) and oxygen gas (O_2). In this process, covalent H—H and O=O bonds are broken and the atoms recombine to form water (Figure 6.1). The change in free energy (Section 5.1) associated with this process is the source of the power made available by fuel cells.

■ In a chemical equation, an arrow separates reactants from products.

Chemical changes are represented using **chemical equations**, in which an arrow separates **reactants**, *the elements or compounds present at the start*, from **products**, *the new substances formed.* The chemical equation for the fuel cell reaction is

$$2H_2(g) + O_2(g) \longrightarrow 2H_2O(g)$$

Coefficients, the numbers placed in front of the formulas of reactants and products (for example, the 2's that appear in front of H_2 and H_2O in the reaction equation above), specify how many of each is used or produced during the reaction. When just one of a particular item appears in the equation, no coefficient is given. Recall from Chapter 5 that the physical state of elements and compounds can be indicated using *s* for solid, *l* for liquid, and *g* for gas. To this list we will add *aq* for aqueous (dissolved in water).

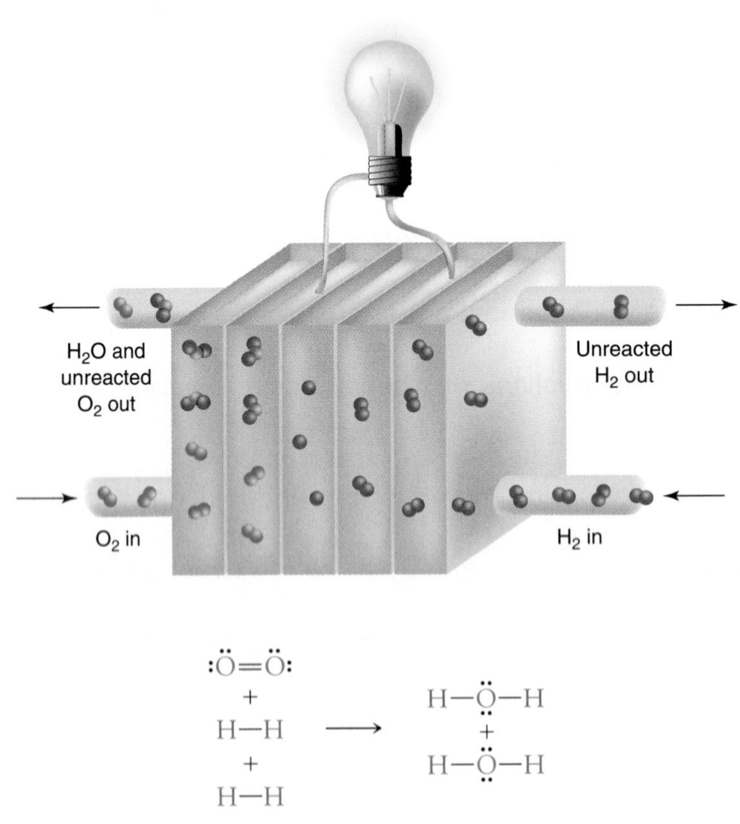

FIGURE 6.1 **Fuel cell reaction** In the conversion of hydrogen gas (H_2) and oxygen gas (O_2) into water, covalent bonds are broken and new ones are formed.

SAMPLE PROBLEM 6.1

Interpreting reaction equations

Potassium metal (K) must be handled very carefully because it reacts violently with water. What would you observe if potassium was placed in water?

$$2K(s) + 2H_2O(l) \longrightarrow H_2(g) + 2KOH(aq)$$

Strategy

The key word in this problem is "observe." What does the reaction equation tell you about the physical state of reactants and products?

Solution

Potassium (a solid) would react with water (a liquid) to produce bubbles (hydrogen gas).

PRACTICE PROBLEM 6.1

The presence of dissolved calcium ion (Ca^{2+}) is one of the causes of hard water. Hard water can be softened by adding sodium carbonate (Na_2CO_3). What would you observe if you carried out the following reaction?

$$Na_2CO_3(aq) + Ca^{2+}(aq) \longrightarrow CaCO_3(s) + 2Na^+(aq)$$

Balancing Chemical Equations

Chemical reactions involve the breaking and forming of bonds, not the destruction or creation of atoms. For this reason a **balanced** chemical equation has the same number of atoms of each element on both sides of the reaction arrow.

■ In a balanced equation, the same number of each atom appears on each side.

A simple example illustrates the importance of balanced equations. If explaining to a small child how to make a cheese sandwich, you might say, "Start with two pieces of bread and one slice of cheese." By explaining things in this way, you are making sure that all of the parts necessary for sandwich making are present. A balanced equation for making a cheese sandwich might be written

$$2 \text{ bread slices} + 1 \text{ cheese slice} \longrightarrow 1 \text{ sandwich}$$

In a similar fashion, writing balanced chemical equations helps us make sure that we have enough of each reactant for a reaction to take place. Let us take a look at a specific example. Nitrogen gas (N_2) can be reacted with oxygen gas (O_2) to produce the dental anesthetic nitrous oxide (N_2O). If asked for the balanced reaction equation we might begin by writing

$$N_2 + O_2 \longrightarrow N_2O$$

Not balanced

This equation is not balanced, because there are a different number of oxygen atoms on each side of the reaction arrow. This is clearly shown by creating a table that lists reactant atoms on one side and product atoms on the other.

Reactant atoms	Product atoms
2 N atoms	2 N atoms
2 O atoms	1 O atom

To balance the equation, coefficients that equalize the number of atoms on each side of the arrow must be found. How do we begin? For this reaction, we need to find a way to increase the number of oxygen atoms, since there are too few on the product side. Increasing the coefficient on N_2O adds oxygen atoms to the product side of the equation, but at the same time increases the number of nitrogen atoms.

$$N_2 + O_2 \longrightarrow 2N_2O$$

Not balanced

Reactant atoms	Product atoms
2 N atoms	4 N atoms
2 O atoms	2 O atoms

This problem is remedied by changing the coefficient for N_2. The resulting balanced equation indicates that two nitrogen molecules react with one oxygen molecule to form two nitrous oxide molecules. To run this reaction efficiently, you would need twice as many N_2 molecules as O_2 molecules.

$$2N_2 + O_2 \longrightarrow 2N_2O$$

Balanced equation

Reactant atoms	Product atoms
4 N atoms	4 N atoms
2 O atoms	2 O atoms

SAMPLE PROBLEM 6.2

Balancing reaction equations

When gasoline is burned in a car engine, some nitrogen gas (N_2) from the air also burns (reacts with O_2) to produce nitric oxide (NO), a colorless, toxic gas. Balance the chemical equation for this reaction.

$$N_2 + O_2 \longrightarrow NO$$

Strategy

To see if the reaction equation is already balanced, you should count the number of nitrogen and oxygen atoms on each side of the reaction arrow. If the equation is not balanced, change the coefficient on a reactant or the product and check the equation again.

Solution

As written, the equation is not balanced.

Reactant atoms	Product atoms
2 N atoms	1 N atom
2 O atoms	1 O atom

To balance the equation, more nitrogen and oxygen atoms must be added to the product side—the coefficient on NO is changed to 2.

$$N_2 + O_2 \longrightarrow 2NO$$

Balanced equation

Reactant atoms	Product atoms
2 N atoms	2 N atoms
2 O atoms	2 O atoms

Practice Problem 6.2

Nitric oxide (NO) reacts with oxygen gas to produce nitrogen dioxide, a brownish gas that is present in the smog that hovers over many large cities. Balance the corresponding reaction equation.

$$NO + O_2 \longrightarrow NO_2$$

Avoiding Common Errors

The examples above show how to balance a chemical equation by changing the coefficients on reactants and products. Now let us take a look at ways to avoid making the most common equation balancing errors.

- *Do not change the formula of a reactant or product.* For the chemical equation

  $$N_2 + O_2 \longrightarrow N_2O$$

  it might seem that changing the formula of N_2O to N_2O_2 would balance the equation, because there would be two nitrogen atoms and two oxygen atoms on each side of the reaction arrow.

  $$N_2 + O_2 \longrightarrow N_2O_2$$

  While this revised equation is balanced, it does not represent the original reaction, whose product is N_2O, not N_2O_2.

- *Do not add new reactants or products.* In the same reaction equation,

  $$N_2 + O_2 \longrightarrow N_2O$$

  it looks as if the number of N and O atoms can be equalized by adding an O atom to the product side.

  $$N_2 + O_2 \longrightarrow N_2O + O$$

  While this equation is also balanced, it no longer represents the original reaction.

- *Do not use multiples of the coefficients when writing the balanced equation.* The coefficients in a balanced equation give the relative numbers of each of the reactants and products, so, technically, each of the following equations is balanced. Of these, the first one is the best choice, because it uses the smallest whole number values for the coefficients.

  $$2N_2 + O_2 \longrightarrow 2N_2O$$
  $$4N_2 + 2O_2 \longrightarrow 4N_2O$$
  $$100N_2 + 50O_2 \longrightarrow 100N_2O$$

6.2 REACTION TYPES

It should be no surprise that, given the vast number of different compounds and elements that there are to work with, the number of possible chemical reactions is immense. In our introduction to reactions, it will be helpful to sort them in terms of similar characteristics. While there is no one correct way to classify reactions, a common approach identifies four broad reaction types: synthesis, decomposition, single replacement, and double replacement. As we continue through this chapter and into later chapters of the text, many reactions important to the health sciences will be introduced. Most, but not all, of these reactions will fit into one of these four categories.

- Many reactions involve synthesis, decomposition, single replacement, or double replacement.

In a **synthesis** reaction *two or more elements or compounds combine to form one more complex compound.* In general, a synthesis reaction can be written

$$A + B \longrightarrow AB$$

Section 6.1 presented two examples of this reaction type, each involving elements as reactants:

$$2N_2 + O_2 \longrightarrow 2N_2O$$

$$2H_2 + O_2 \longrightarrow 2H_2O$$

The reaction of carbon dioxide with water is an example of a synthesis reaction involving compounds.

$$CO_2 + H_2O \longrightarrow H_2CO_3$$

A **decomposition** reaction is the reverse of a synthesis reaction. It involves the *breakdown of one compound to form elements or simpler compounds.*

$$AB \longrightarrow A + B$$

The decompositions of water and calcium carbonate ($CaCO_3$) are examples.

$$2H_2O \longrightarrow 2H_2 + O_2$$

$$CaCO_3 \longrightarrow CaO + CO_2$$

In a **single replacement** reaction, *one element trades places with a different element in a compound.* In general terms this reaction appears as

$$A + BC \longrightarrow AC + B$$

In the reaction between elemental iron (Fe) and copper(II) sulfate ($CuSO_4$), for example, iron replaces copper. Similarly, when sodium (Na) reacts with hydrogen chloride (HCl), hydrogen is replaced.

$$Fe + CuSO_4 \longrightarrow FeSO_4 + Cu$$

$$2Na + 2HCl \longrightarrow 2NaCl + H_2$$

A **double replacement** reaction involves *parts of two compounds switching places.*

$$AB + CD \longrightarrow AD + CB$$

The reaction of sodium chloride (NaCl) with silver nitrate ($AgNO_3$) and of hydrogen chloride with sodium hydroxide (NaOH) are double replacement reactions.

$$NaCl + AgNO_3 \longrightarrow NaNO_3 + AgCl$$

$$HCl + NaOH \longrightarrow H_2O + NaCl$$

SAMPLE PROBLEM 6.3

Identifying reaction type

Classify each reaction as involving synthesis, decomposition, single replacement, or double replacement.

a. $MgO + H_2O \longrightarrow Mg(OH)_2$
b. $Mg + 2H_2O \longrightarrow Mg(OH)_2 + H_2$

Strategy
To determine reaction type, look to see whether there is just one product (synthesis), whether there is just one reactant (decomposition), whether one element replaces another (single replacement), or whether parts of two compounds are traded (double replacement).

Solution

a. This reaction forms just one product, so it involves synthesis.
b. In this reaction hydrogen atoms are displaced by magnesium, so the reaction involves single replacement.

PRACTICE PROBLEM 6.3

Classify the reaction as involving synthesis, decomposition, single replacement, or double replacement.

$$H_2 + 2AgNO_3 \longrightarrow 2Ag + 2HNO_3$$

- Many reactions involve synthesis, decomposition, single replacement, or double replacement.

6.3 OXIDATION AND REDUCTION

In Section 6.2 we saw that most chemical reactions fit one of four reaction types: synthesis ($A + B \longrightarrow AB$), decomposition ($AB \longrightarrow A + B$), single replacement ($A + BC \longrightarrow AC + B$), or double replacement ($AB + CD \longrightarrow AD + CB$). By focusing on a different set of characteristics than used to distinguish these four reaction types, certain reactions can be placed into a whole different set of categories. In this section we will look at one of these alternative reaction groupings: oxidation and reduction.

Section 3.4 showed us that NaCl consists of Na^+ and Cl^- ions. This compound can be created from elemental sodium (Na) and chlorine (Cl_2) by the reaction

$$2Na(s) + Cl_2(g) \longrightarrow 2NaCl(s)$$

In this synthesis reaction, neutral sodium atoms (charge = 0) lose an electron (are **oxidized**) to form Na^+ and each of the neutral chlorine atoms in Cl_2 gain an electron (are **reduced**) to form Cl^-. Since Cl_2 removes electrons from Na, Cl_2 is called an **oxidizing agent**. Similarly, because Na gives electrons to Cl_2, Na is the **reducing agent**.

- Oxidation involves the loss of electrons.
- Reduction involves the gain of electrons.

The single replacement reaction that takes places between copper metal (Cu) and aqueous silver nitrate ($AgNO_3$) also involves oxidation and reduction. Cu is oxidized to Cu^{2+} and Ag^+ is reduced to Ag (Figure 6.2).

$$Cu(s) + 2AgNO_3(aq) \longrightarrow Cu(NO_3)_2(aq) + 2Ag(s)$$

Oxidation and reduction always occur together because electrons move from the substance being oxidized to the one being reduced.

(*a*) (*b*)

FIGURE 6.2

An oxidation–reduction reaction
(*a*) Copper wire (Cu) is immersed in aqueous silver nitrate ($AgNO_3$). (*b*) After a few hours, a silver metal precipitate has formed and the liquid has turned blue, due to the presence of dissolved copper(II) ion. Copper has been oxidized ($Cu \longrightarrow Cu^{2+}$) and silver ion has been reduced ($Ag^+ \longrightarrow Ag$).

Source: (*a*) Andrew Lambert Photography/Photo Researchers, Inc.

SAMPLE PROBLEM 6.4

Identifying oxidation and reduction

If spilled on clothing, the iodine (I_2) present in Betadine (a disinfectant) can leave a stain. One way to remove an iodine stain is to treat it with sodium thiosulfate ($Na_2S_2O_3$).

$$\underset{\text{colored}}{I_2(s)} + 2Na_2S_2O_3(aq) \longrightarrow Na_2S_4O_6(aq) + \underset{\text{colorless}}{2NaI(aq)}$$

In this reaction is I_2 oxidized or is it reduced?

Strategy
To solve this problem you must decide whether the iodine atoms have lost electrons (oxidation) or gained electrons (reduction).

Solution
Iodine is reduced. The product NaI consists of Na^+ and I^- ions. The neutral iodine atoms in I_2 gain electrons when they are converted into I^-. Oxidation and reduction go together, so as I_2 is reduced, $Na_2S_2O_3$ is oxidized.

PRACTICE PROBLEM 6.4

Before bleach (NaOCl) is used on fabric treated with sodium thiosulfate to remove iodine stains, the fabric should be rinsed. Otherwise, any iodide ions (I^-) present will react with NaOCl and the stain will reappear.

$$2H^+ + NaOCl + 2I^- \longrightarrow NaCl + I_2 + H_2O$$

In this reaction is I^- oxidized or is it reduced?

FIGURE 6.3

Combustion
When wood is burned, the organic compounds that it is made from undergo oxidation to produce carbon dioxide (CO_2) and water.

Source: Photo Disc, Inc.

Combustion

When methane (CH_4), the major constituent of natural gas, reacts with oxygen gas (O_2), an oxidation reaction called **combustion** takes place. Combustion reactions, which always have O_2 as a reactant, are very rapid and result in the release of heat (Figure 6.3). When organic (carbon-containing) molecules such as methane are involved in a combustion reaction, the products are carbon dioxide and water. A chemical equation for the combustion of methane is

$$CH_4(g) + O_2(g) \longrightarrow CO_2(g) + H_2O(g)$$

Not balanced

By counting the number of C, H, and O atoms on each side of the equation we see that it is not balanced.

Reactant atoms	Product atoms
1 C atom	1 C atom
4 H atoms	2 H atoms
2 O atoms	3 O atoms

A good approach to use when balancing combustion reactions involving organic compounds is to balance the carbon atoms first, then the hydrogen atoms, and the oxygen atoms last. Looking at the table above, we see that the carbon atoms are balanced (1 on each side) but the hydrogen and oxygen atoms are not. To balance the hydrogen atoms, the coefficient on H_2O—the product molecule that hydrogen appears in—is increased.

$$CH_4(g) + O_2(g) \longrightarrow CO_2(g) + 2H_2O(g)$$

Not balanced

Reactant atoms	Product atoms
1 C atom	1 C atom
4 H atoms	4 H atoms
2 O atoms	4 O atoms

Now that the carbon and hydrogen atoms are balanced, we can deal with the oxygen atoms. The coefficient on O_2, the only reactant molecule that oxygen appears in, is increased to 2. This produces a balanced reaction equation.

$$CH_4(g) + 2O_2(g) \longrightarrow CO_2(g) + 2H_2O(g)$$

Balanced

Reactant atoms	Product atoms
1 C atom	1 C atom
4 H atoms	4 H atoms
4 O atoms	4 O atoms

Having balanced the reaction equation, let us turn our attention to its relationship to oxidation and reduction. Unlike the oxidation and reduction reactions considered earlier, ions are not involved here. With no charges to provide clues as to electron loss or gain, identifying oxidation or reduction in this and many other reactions involving organic compounds can be challenging. In this particular reaction a carbon atom is oxidized because it gains a partial positive charge (Figure 6.4). In CH_4 the bonds between the carbon and hydrogen atoms are nonpolar, so the carbon atom has no partial charge. In CO_2 the carbon atom is attached to the more electronegative oxygen atoms by polar covalent bonds, in which the carbon atom carries a partial positive charge and the oxygen atoms have partial negative charges. In going from no partial charge in CH_4 to a partial positive charge in CO_2 the carbon atom has been oxidized.

H
H—C—H
H

Bonds are nonpolar

δ− δ+ δ−
:Ö=C=Ö:

Bonds are polar covalent

FIGURE 6.4

Combustion of methane
When a methane (CH_4) molecule is converted into a carbon dioxide (CO_2) molecule, its carbon atom is oxidized. By gaining a partial positive charge, it has (partially) lost electrons.

Instead of analyzing each of the covalent bonds in organic reactants and products to look for evidence of oxidation or reduction, organic chemists have devised simple rules for identifying oxidation or reduction.

- An atom is oxidized if it
  - gains oxygen (is attached to more oxygen atoms in the product than in the reactant) and/or
  - loses hydrogen (is attached to fewer hydrogen atoms in the product than in the reactant).
- An atom is reduced if it
  - loses oxygen (is attached to fewer oxygen atoms) and/or
  - gains hydrogen (is attached to more hydrogen atoms).

■ An atom that gains oxygen and/or loses hydrogen has been oxidized.

■ An atom that loses oxygen and/or gains hydrogen has been reduced.

Interpreting the combustion reaction between CH_4 and O_2 just discussed, we see that carbon has been oxidized because, in the transformation of CH_4 into CO_2, the carbon atom loses hydrogen atoms and gains oxygen atoms. While our focus here is on the oxidation of carbon atoms in organic molecules, it should be noted that O_2 is reduced in combustion reactions.

SAMPLE PROBLEM 6.5

Balancing and identifying oxidation in combustion reactions

If not enough oxygen gas (O_2) is available when methane is burned, the incomplete combustion reaction that takes place produces carbon monoxide. A nonbalanced equation for this reaction is

$$2CH_4(g) + O_2(g) \longrightarrow CO(g) + 2H_2O(g)$$

a. Write the balanced equation.
b. Is carbon oxidized or is it reduced during this reaction?

Strategy

To balance the reaction equation, balance carbon first, then hydrogen, then oxygen. In determining if the carbon atom is oxidized or reduced, consider whether it loses or gains hydrogen or oxygen.

Solution

a.

$$2CH_4(g) + O_2(g) \longrightarrow CO(g) + 2H_2O(g)$$

Not balanced

Reactant atoms	Product atoms
2 C atoms	1 C atom
8 H atoms	4 H atoms
2 O atoms	3 O atoms

$$2CH_4(g) + O_2(g) \longrightarrow 2CO(g) + 2H_2O(g)$$

Not balanced

Reactant atoms	Product atoms
2 C atoms	2 C atoms
8 H atoms	4 H atoms
2 O atoms	4 O atoms

$$2CH_4(g) + O_2(g) \longrightarrow 2CO(g) + 4H_2O(g)$$

Not balanced

Reactant atoms	Product atoms
2 C atoms	2 C atoms
8 H atoms	8 H atoms
2 O atoms	6 O atoms

$$2CH_4(g) + 3O_2(g) \longrightarrow 2CO(g) + 4H_2O(g)$$

Balanced

Reactant atoms	Product atoms
2 C atom	2 C atom
8 H atoms	8 H atoms
6 O atoms	6 O atoms

b. When CH_4 is converted to CO, carbon is oxidized—it loses hydrogen and gains oxygen.

PRACTICE PROBLEM 6.5

The alkyne called acetylene (C_2H_2) is burned in the oxyacetylene torches used for welding. A nonbalanced equation is shown. Write the correct balanced equation.

$$2C_2H_2(g) + O_2(g) \longrightarrow CO_2(g) + H_2O(g)$$

Hydrogenation

Alkenes and other unsaturated hydrocarbons undergo a synthesis reaction called **catalytic hydrogenation**, in which hydrogen gas (H_2) acts as a reducing agent. The term "catalytic" comes from the required presence of platinum (Pt), a **catalyst** (*a substance that speeds up a reaction without itself being altered*) (Section 6.7). During catalytic hydrogenation, the carbon atoms in an alkene are reduced because they gain hydrogen atoms. This reaction is the origin of the terms saturated and unsaturated that were introduced in Section 4.8. A molecule is unsaturated (with respect to hydrogen atoms) when it can undergo hydrogenation and is saturated when no more H_2 can be added.

$$\underset{\text{Unsaturated}}{H_2C{=}CH_2} \xrightarrow[Pt]{H_2} \underset{\text{Saturated}}{H_3C{-}CH_3}$$

The reaction equation above shows an approach often used when writing organic reaction equations. The key organic reactant is shown to the left of the arrow and additional reactants, catalysts (if required), or conditions (heat, for example) are placed above or below the arrow. This approach often results in equations that are not balanced, a generally acceptable practice in organic chemistry.

A process similar to catalytic hydrogenation takes place in one of the series of reactions that cells use to manufacture the fatty acids (a type of carboxylic acid) that living things use for a variety of purposes. In this reaction, which is catalyzed by an **enzyme** (*a biochemical catalyst*), the carbon–carbon double bond of a particular compound (partial structure shown) is reduced by NADPH. When writing reaction equations, biochemists will sometimes give reactants separate reaction arrows, as is the case here (one for the carbon-containing reactant and one for NADPH and H^+).

$$-C{=}C{-}\overset{O}{\overset{\|}{C}}{-}S- \xrightarrow[\text{NADPH} + H^+ \;\rightarrow\; \text{NADP}^+]{\text{enzyme}} -\overset{H}{\overset{|}{C}}{-}\overset{H}{\overset{|}{C}}{-}\overset{O}{\overset{\|}{C}}{-}S-$$

In the chapters that follow, we will encounter other examples of oxidation and reduction in organic and biochemical compounds.

HEALTH Link: Antiseptics and Oxidation

Sepsis is a medical term related to infection, and antiseptics are compounds that prevent infections—usually by slowing the growth of harmful bacteria. In the late 1840s, an obstetrician working in a Viennese hospital noticed that when physicians washed their hands with soap and chlorine water before delivering babies, the death rate in the obstetrical ward decreased significantly. When he tried to promote this antiseptic practice, however, he was fired for upsetting the status quo. Later, in 1865, the surgeon Joseph Lister had better success at introducing the concepts of hand washing before doing surgery, of sterilizing bandages, and of using carbolic acid (phenol) as an antiseptic. Even though the amounts of phenol that Lister used were high enough to cause tissue damage, his ideas rapidly caught on (Figure 6.5).

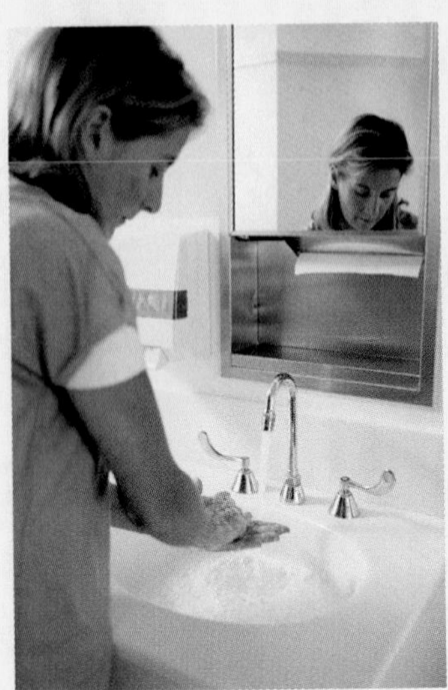

Phenol   Benzoyl peroxide

Many of the antiseptics used today are oxidizing agents. One example is tincture of iodine, a combination of I_2 and KI dissolved in a mixture of water and ethanol (CH_3CH_2OH). Another oxidizing agent, benzoyl peroxide, is the active ingredient in Oxy-10 and other acne medications. When a dilute mixture of benzoyl peroxide is applied to the skin, it kills bacteria that cause acne.

FIGURE 6.5 **Scrubbing before surgery**
Health professionals were not aware of the importance of hand washing and cleanliness until the late 1800s.

Source: Photo Disc, Inc.

SAMPLE PROBLEM 6.6

Oxidation and reduction of organic compounds

Write the chemical equation for the reaction that takes place when each alkene is mixed with H_2 and Pt.

a. 1-butene

b. *cis*-2-butene

c. *trans*-2-pentene

Strategy

In the catalytic hydrogenation of an alkene, a hydrogen atom is added to each of the double-bonded carbon atoms and the double bond becomes a single bond. For a review of alkene naming, see Section 4.8.

Solution

a. $CH_2{=}CHCH_2CH_3 \xrightarrow[Pt]{H_2} CH_3CH_2CH_2CH_3$

b. $(CH_3)HC{=}CH(CH_3)$ (*cis*: CH_3 groups on the same side) $\xrightarrow[Pt]{H_2} CH_3CH_2CH_2CH_3$

c. $(CH_3)HC{=}CH(CH_2CH_3)$ (*trans*: H and CH_2CH_3 on top, CH_3 and H on bottom) $\xrightarrow[Pt]{H_2} CH_3CH_2CH_2CH_2CH_3$

PRACTICE PROBLEM 6.6

Is the organic compound oxidized or reduced in each of the following reactions? (The reactants necessary for carrying out the reactions in parts b and c are not specified. These will be introduced in Chapter 11.)

a. 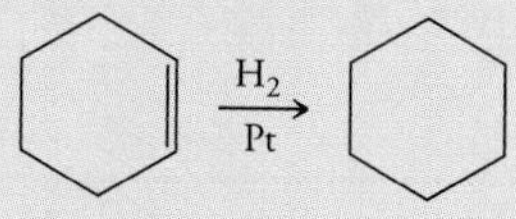 cyclohexene $\xrightarrow[\text{Pt}]{H_2}$ cyclohexane

b. $CH_3CH(OH)CH_3 \longrightarrow CH_3C(=O)CH_3$

c. $CH_3C(=O){-}H \longrightarrow CH_3C(=O){-}OH$

6.4 REACTIONS INVOLVING WATER

Water is a reactant or product in a number of reactions important to organic and biochemistry. In this section we will take a look at three of them: hydrolysis, hydration, and dehydration.

Hydrolysis

In a **hydrolysis** reaction, water (hydro) is used to split (lyse) a molecule. Esters are one class of molecules that undergo hydrolysis—when treated with water in the presence of hydroxide ion (OH^-) they split to form a **carboxylate ion** and an alcohol (Figure 6.6).

$$\underset{\text{An ester}}{CH_3CH_2CH_2C(=O){-}OCH_2CH_3} \xrightarrow[OH^-]{H_2O} \underset{\text{A carboxylate ion}}{CH_3CH_2CH_2C(=O){-}O^-} + \underset{\text{An alcohol}}{HOCH_2CH_3}$$

FIGURE 6.6 **Ester hydrolysis**
In the presence of OH^-, an ester reacts with water to form a carboxylate ion and an alcohol.

Carboxylate ions and carboxylic acids are closely related and differ only by the absence or presence of an H^+ ion. We will learn more about the relationship between carboxylic acids and carboxylate ions in chapters that follow.

Hydrolysis is one of the factors that determine the length of time that some drugs remain active. The local anesthetics procaine (also known as Novocain) and chloroprocaine are short acting, remaining effective for little more than an hour. A particular enzyme present in blood serum deactivates these two anesthetics by catalyzing hydrolysis of their ester groups to form carboxylate ions and alcohols (Figure 6.7).

■ Esters can be hydrolyzed, alkenes can be hydrated, and alcohols can be dehydrated.

Hydration

In a synthesis reaction known as **hydration**, water is added to a double bond. The formation of an alcohol by reacting water and an alkene, in the presence of H^+ (a catalyst), is a typical example (Figure 6.8). Hydration plays important roles in biochemistry. In Chapter 15, for example, we will discuss the citric acid cycle, a series of reactions important to metabolism. One step in this process, the conversion of citrate into isocitrate, involves the enzyme-catalyzed hydration of the intermediate compound *cis*-aconitate (Figure 6.9).

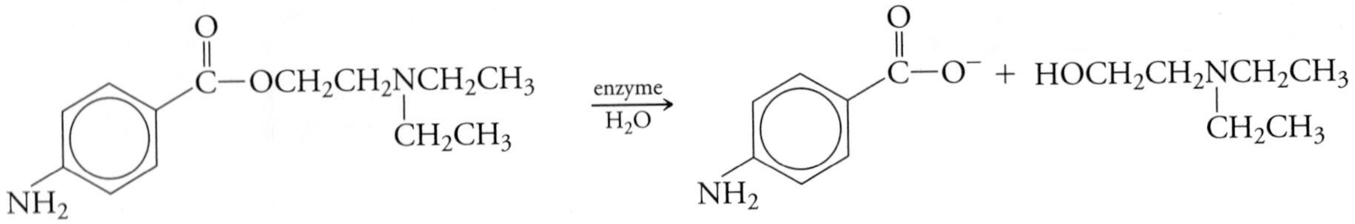

Procaine

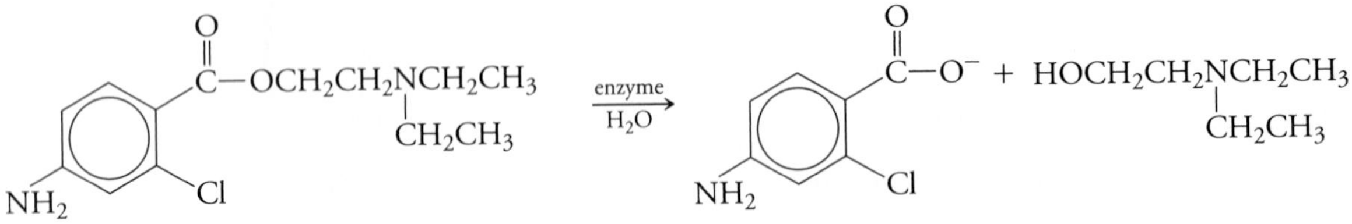

Chloroprocaine

FIGURE | 6.7

Local anesthetics
Procaine and chloroprocaine, two short-acting anesthetics, are hydrolyzed by the action of a particular blood serum enzyme.

FIGURE | 6.8

Adding water to an alkene
In the presence of an H^+ catalyst, H_2O adds across a carbon–carbon double bond to form an alcohol.

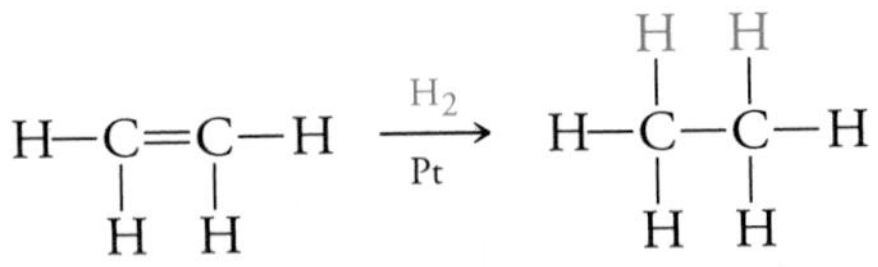

Unsaturated

Saturated

FIGURE | 6.9

Biochemical dehydration and hydration
The enzyme-catalyzed conversion of citrate into isocitrate involves the dehydration of citrate to form *cis*-aconitate and the hydration of *cis*-aconitate to form isocitrate.

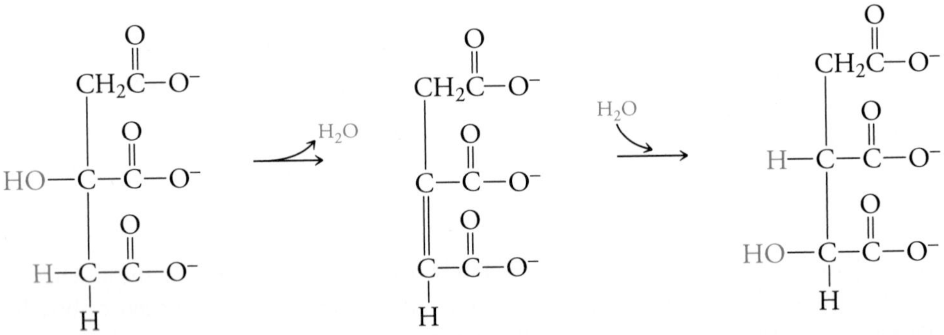

Citrate ⇒ Dehydration *cis*-Aconitate ⇒ Hydration Isocitrate

Dehydration

Dehydration is the reverse of hydration. When heated in the presence of H^+ (a catalyst), an alcohol splits to form an alkene plus water (Figure 6.10). Hydration and dehydration are the reverse of one another, because hydration converts an alkene into an alcohol, and dehydration converts an alcohol into an alkene. In Section 11.2 we will revisit hydration of alkenes and dehydration of alcohols.

FIGURE | 6.10

Alcohol dehydration
When heated in the presence of H^+, alcohols are converted to alkenes. In this reaction a water molecule is eliminated from the reactant.

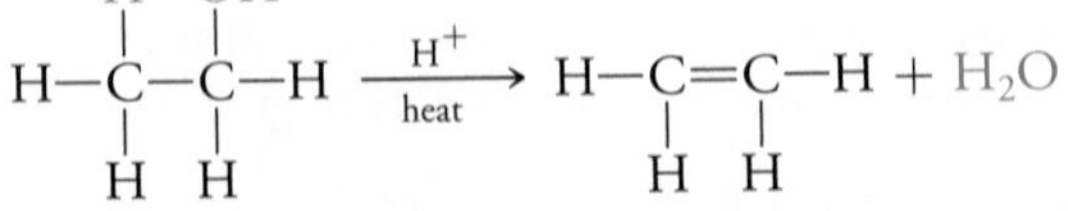

SAMPLE PROBLEM 6.7

Identifying the products of reactions involving water

Draw the product(s) of each reaction.

a. 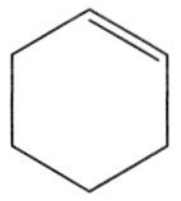 cyclohexene $\xrightarrow[H^+]{H_2O}$

c. $CH_3CH(OH)CH_3 \xrightarrow[heat]{H^+}$

b. $CH_3C(=O){-}OCH_3 \xrightarrow[OH^-]{H_2O}$

Strategy

To solve problems of this type, you must be able to match a particular reaction to the corresponding functional group—esters are hydrolyzed, alkenes are hydrated, and alcohols are dehydrated.

Solution

a. 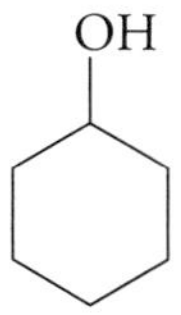 cyclohexanol (cyclohexane ring bearing OH)

c. $CH_2{=}CHCH_3 + H_2O$

b. $CH_3C(=O){-}O^- + HOCH_3$

PRACTICE PROBLEM 6.7

Draw the missing reactant for each reaction.

a. $\xrightarrow[OH^-]{H_2O}$ $C_6H_5C(=O){-}O^- + HOCH_2CH_3$

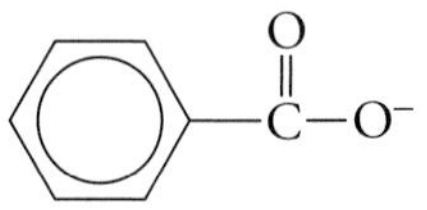

c. $\xrightarrow[H^+]{H_2O}$ cyclopentanol (cyclopentane ring bearing OH)

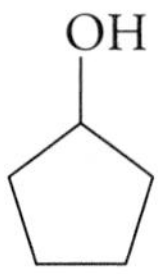

b. $\xrightarrow[heat]{H^+}$ $CH_3CH{=}CHCH_3$

6.5 MOLE AND MASS RELATIONSHIPS IN REACTIONS

We have seen that balanced equations accurately describe the number of reactants used and products formed in a reaction. Apart from letting us know about the changes that take place during a reaction, balanced equations have another important use—they allow us to predict which reactant (if any) will run out first and how much product can be expected to form.

To introduce this topic, let us go back to the sandwich example used earlier in the chapter. Suppose that you have invited friends over for lunch and want to make as many cheese sandwiches as possible. If you have 20 slices of bread, how many slices of cheese do you need? The answer is 10 and here is one approach to use to arrive at this answer. From the balanced equation for making a cheese sandwich

$$\text{2 bread slices} + \text{1 cheese slice} \longrightarrow \text{1 sandwich}$$

we can write a conversion factor (Section 1.6) based on the relationship between bread and cheese slices (2 bread slices require 1 cheese slice). The calculation shows that 20 bread slices require 10 cheese slices.

$$20\ \cancel{\text{bread slices}} \times \frac{\text{1 cheese slice}}{2\ \cancel{\text{bread slices}}} = \text{10 cheese slices}$$

Now let us use the same approach to deal with a chemical reaction. In the catalytic converter found on the exhaust system of cars, poisonous carbon monoxide produced when gasoline is burned gets oxidized to carbon dioxide. The balanced equation for this reaction

$$2CO(g) + O_2(g) \longrightarrow 2CO_2(g)$$

indicates that 2 molecules of CO react with 1 molecule of O_2 to form 2 molecules of CO_2. Since the equation describes relative numbers of reactants and products, it is also true that 2 dozen CO molecules react with 1 dozen O_2 molecules to form 2 dozen CO_2 molecules, and that 2 mol of CO react with 1 mol of O_2 to form 2 mol of CO_2. Recall that the mole is a counting unit that represents 6.02×10^{23} items.

Beginning with 6.0 mol of CO, how many moles of O_2 will be required to convert all of the CO to CO_2? The answer, 3.0 mol, can be obtained by using a conversion factor derived from the balanced equation (2 mol of CO completely reacts with 1 mol of O_2).

$$6.0\ \cancel{\text{mol CO}} \times \frac{1\ \text{mol}\ O_2}{2\ \cancel{\text{mol CO}}} = 3.0\ \text{mol}\ O_2$$

Beginning with the same 6.0 mol of CO, how many moles of CO_2 will be obtained if all of the CO is used up? The answer, 6.0 mol, can be calculated using the equation

$$6.0\ \cancel{\text{mol CO}} \times \frac{2\ \text{mol}\ CO_2}{2\ \cancel{\text{mol CO}}} = 6.0\ \text{mol}\ CO_2$$

SAMPLE PROBLEM 6.8

Interpreting reaction equations in terms of moles

In terms of the reaction equation above and starting with 5.8 mol of CO,

a. how many moles of O_2 are needed to convert all of the CO to CO_2?
b. how many moles of CO_2 are obtained when all of the CO is converted to CO_2?

Strategy

In part a, you are asked to convert from moles of CO into moles of O_2. The conversion factor to use is based on the relationship that 2 mol of CO require 1 mol of O_2 to completely react (see the balanced equation). A similar approach is used to solve part b.

Solution

a. Convert from moles of CO to moles of O_2

$$5.8\ \cancel{\text{mol CO}} \times \frac{1\ \text{mol}\ O_2}{2\ \cancel{\text{mol CO}}} = 2.9\ \text{mol}\ O_2$$

b. Convert from moles of CO to moles of CO_2

$$5.8\ \cancel{\text{mol CO}} \times \frac{2\ \text{mol}\ CO_2}{2\ \cancel{\text{mol CO}}} = 5.8\ \text{mol}\ CO_2$$

PRACTICE PROBLEM 6.8

Ethylene gas ($CH_2{=}CH_2$), which can be used to promote the ripening of fruits (Chapter 3, Biochemistry Link: Ethylene, a Plant Hormone), burns according to the equation

$$CH_2{=}CH_2 + 3O_2 \longrightarrow 2CO_2 + 2H_2O$$

Beginning with 5.8 mol of ethylene,

a. how many moles of O_2 are required?
b. how many moles of CO_2 are obtained when all of the ethylene is consumed?

Knowing how to use a balanced equation to determine the relative number of moles of reactants required and products obtained allows us to interpret reaction equations in terms of mass. Again, to introduce this idea, let us talk in terms of making cheese sandwiches. At 0.75 ounces per cheese slice, how many ounces of cheese are required to use up 24 pieces of bread in making sandwiches? Solving this problem requires two steps—calculating how many slices of cheese are needed and then determining how many ounces of cheese this corresponds to.

$$24\ \cancel{\text{bread slices}} \times \frac{1\ \text{cheese slice}}{2\ \cancel{\text{bread slices}}} = 12\ \text{cheese slices}$$

$$12\ \cancel{\text{cheese slices}} \times \frac{0.75\ \text{oz cheese}}{1\ \cancel{\text{cheese slice}}} = 9.0\ \text{oz cheese}$$

Let us try the same approach using a chemical reaction. The re-breather units used by some firefighters convert exhaled carbon dioxide (CO_2) into oxygen gas (O_2).

$$4KO_2(s) + 2CO_2(g) \longrightarrow 2K_2CO_3(s) + 3O_2(g)$$

How many grams of KO_2 are required to completely react with 0.400 mol of CO_2? Answering this question requires two steps. The first step is determining how many moles of KO_2 are needed and the second step is converting from moles of KO_2 to grams of KO_2.

1. *How many moles of KO_2 are required?* Using the approach outlined above,

$$0.400\ \cancel{\text{mol } CO_2} \times \frac{4\ \text{mol } KO_2}{2\ \cancel{\text{mol } CO_2}} = 0.800\ \text{mol } KO_2$$

2. *How many grams of KO_2 are needed?* The formula weight of KO_2 is 71.1 amu, so its molar mass is 71.1 g/mol.

$$0.800\ \cancel{\text{mol } KO_2} \times \frac{71.1\ \text{g } KO_2}{1\ \cancel{\text{mol } KO_2}} = 56.9\ \text{g } KO_2$$

Because you often measure out grams of a substance to run a reaction, calculations involving balanced equations sometimes involve gram to gram comparisons. For example, the balanced equation for the reaction that takes place when silver (Ag) becomes tarnished is

$$Ag(s) + S(s) \longrightarrow AgS(s)$$

How many grams of silver will completely react with 10.0 g of sulfur? Because reaction equations are balanced in terms of number (or moles) of reactants and products, not their mass, solving this problem requires that the equation be interpreted in terms of moles.

■ Reaction equations are balanced in terms of moles, not mass.

Solving this problem requires three steps.

1. *How many moles of S are present?* The atomic weight of sulfur is 32.1 amu, so its molar mass is 32.1 g/mol.

$$10.0 \text{ \sout{g S}} \times \frac{1 \text{ mol S}}{32.1 \text{ \sout{g S}}} = 0.312 \text{ mol S}$$

2. *How many moles of Ag are needed?* The conversion factor in this calculation is obtained from the balanced equation.

$$0.312 \text{ \sout{mol S}} \times \frac{1 \text{ mol Ag}}{1 \text{ \sout{mol S}}} = 0.312 \text{ mol Ag}$$

3. *How many grams of Ag are needed?* The final step makes use of the molar mass of Ag (107.9 g/mol).

$$0.312 \text{ \sout{mol Ag}} \times \frac{107.9 \text{ g Ag}}{1 \text{ \sout{mol Ag}}} = 33.7 \text{ g Ag}$$

It is important to note that the reaction equation above indicates that an equal number of silver and sulfur atoms react, not an equal mass of silver and sulfur atoms. Any calculation that involves a balanced reaction equation must use the number of moles of reactants and products, not their mass (Figure 6.11).

Grams of reactant

Use molar mass to convert

Moles of reactant

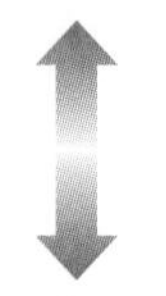

Use the balanced equation to convert

Moles of product

Use molar mass to convert

Grams of product

FIGURE 6.11

Interpreting reaction equations Calculations involving reaction equations must be done in terms of the number of moles of reactants and/or products. If the problem is stated in terms of grams, the appropriate conversions must be made.

Source: BROWN, THEODORE E; LEMAY, H. EUGENE; BURSTEN, BRUCE E, CHEMISTRY: THE CENTRAL SCIENCE, 10th edition, © 2006, p. 150. Adapted by permission of Pearson Education, Inc., Upper Saddle River, NJ.

SAMPLE PROBLEM 6.9

Interpreting reaction equations in terms of mass

In the following reaction, how many grams of water are needed to completely hydrolyze 61.5 g of the ester?

$$CH_3\overset{O}{\overset{\|}{C}}{-}OCH_3 + H_2O \xrightarrow{OH^-} CH_3\overset{O}{\overset{\|}{C}}{-}O^- + HOCH_3$$

Strategy

In any calculation that involves a balanced reaction equation, the interpretation must be done in terms of moles. Convert grams of ester into moles of ester, decide how many moles of water are required, and then convert from moles of water into grams of water.

Solution

The problem can be solved in three steps.

1. *How many moles of ester are present?* The molar mass of the ester ($C_3H_6O_2$) is 74.0 g/mol.

$$61.5 \text{ \sout{g ester}} \times \frac{1 \text{ mol ester}}{74.0 \text{ \sout{g ester}}} = 0.831 \text{ mol ester}$$

2. *How many moles of water are needed?* From the balanced equation, 1 mol of ester reacts with 1 mol of water.

$$0.831 \text{ \sout{mol ester}} \times \frac{1 \text{ mol } H_2O}{1 \text{ \sout{mol ester}}} = 0.831 \text{ mol } H_2O$$

3. *How many grams of water are needed?* The final step makes use of the molar mass of H_2O (18.0 g/mol).

$$0.831 \text{ \sout{mol H$_2$O}} \times \frac{18.0 \text{ g } H_2O}{1 \text{ \sout{mol H$_2$O}}} = 15.0 \text{ g } H_2O$$

PRACTICE PROBLEM 6.8

Ethylene gas ($CH_2{=}CH_2$), which can be used to promote the ripening of fruits (Chapter 3, Biochemistry Link: Ethylene, a Plant Hormone), burns according to the equation

$$CH_2{=}CH_2 + 3O_2 \longrightarrow 2CO_2 + 2H_2O$$

Beginning with 5.8 mol of ethylene,

a. how many moles of O_2 are required?
b. how many moles of CO_2 are obtained when all of the ethylene is consumed?

Knowing how to use a balanced equation to determine the relative number of moles of reactants required and products obtained allows us to interpret reaction equations in terms of mass. Again, to introduce this idea, let us talk in terms of making cheese sandwiches. At 0.75 ounces per cheese slice, how many ounces of cheese are required to use up 24 pieces of bread in making sandwiches? Solving this problem requires two steps—calculating how many slices of cheese are needed and then determining how many ounces of cheese this corresponds to.

$$24\ \text{\sout{bread slices}} \times \frac{1\ \text{cheese slice}}{2\ \text{\sout{bread slices}}} = 12\ \text{cheese slices}$$

$$12\ \text{\sout{cheese slices}} \times \frac{0.75\ \text{oz cheese}}{1\ \text{\sout{cheese slice}}} = 9.0\ \text{oz cheese}$$

Let us try the same approach using a chemical reaction. The re-breather units used by some firefighters convert exhaled carbon dioxide (CO_2) into oxygen gas (O_2).

$$4KO_2(s) + 2CO_2(g) \longrightarrow 2K_2CO_3(s) + 3O_2(g)$$

How many grams of KO_2 are required to completely react with 0.400 mol of CO_2? Answering this question requires two steps. The first step is determining how many moles of KO_2 are needed and the second step is converting from moles of KO_2 to grams of KO_2.

1. *How many moles of KO_2 are required?* Using the approach outlined above,

$$0.400\ \text{\sout{mol CO}}_2 \times \frac{4\ \text{mol KO}_2}{2\ \text{\sout{mol CO}}_2} = 0.800\ \text{mol KO}_2$$

2. *How many grams of KO_2 are needed?* The formula weight of KO_2 is 71.1 amu, so its molar mass is 71.1 g/mol.

$$0.800\ \text{\sout{mol KO}}_2 \times \frac{71.1\ \text{g KO}_2}{1\ \text{\sout{mol KO}}_2} = 56.9\ \text{g KO}_2$$

Because you often measure out grams of a substance to run a reaction, calculations involving balanced equations sometimes involve gram to gram comparisons. For example, the balanced equation for the reaction that takes place when silver (Ag) becomes tarnished is

$$Ag(s) + S(s) \longrightarrow AgS(s)$$

How many grams of silver will completely react with 10.0 g of sulfur? Because reaction equations are balanced in terms of number (or moles) of reactants and products, not their mass, solving this problem requires that the equation be interpreted in terms of moles.

■ Reaction equations are balanced in terms of moles, not mass.

Solving this problem requires three steps.

1. *How many moles of S are present?* The atomic weight of sulfur is 32.1 amu, so its molar mass is 32.1 g/mol.

$$10.0 \text{ g S} \times \frac{1 \text{ mol S}}{32.1 \text{ g S}} = 0.312 \text{ mol S}$$

2. *How many moles of Ag are needed?* The conversion factor in this calculation is obtained from the balanced equation.

$$0.312 \text{ mol S} \times \frac{1 \text{ mol Ag}}{1 \text{ mol S}} = 0.312 \text{ mol Ag}$$

3. *How many grams of Ag are needed?* The final step makes use of the molar mass of Ag (107.9 g/mol).

$$0.312 \text{ mol Ag} \times \frac{107.9 \text{ g Ag}}{1 \text{ mol Ag}} = 33.7 \text{ g Ag}$$

It is important to note that the reaction equation above indicates that an equal number of silver and sulfur atoms react, not an equal mass of silver and sulfur atoms. Any calculation that involves a balanced reaction equation must use the number of moles of reactants and products, not their mass (Figure 6.11).

Grams of reactant

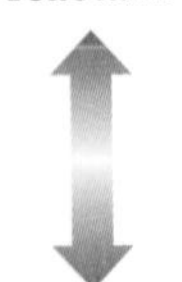

Use molar mass to convert

Moles of reactant

Use the balanced equation to convert

Moles of product

Use molar mass to convert

Grams of product

FIGURE 6.11

Interpreting reaction equations Calculations involving reaction equations must be done in terms of the number of moles of reactants and/or products. If the problem is stated in terms of grams, the appropriate conversions must be made.

Source: BROWN, THEODORE E; LEMAY, H. EUGENE; BURSTEN, BRUCE E, CHEMISTRY: THE CENTRAL SCIENCE, 10th edition, © 2006, p. 150. Adapted by permission of Pearson Education, Inc., Upper Saddle River, NJ.

SAMPLE PROBLEM 6.9

Interpreting reaction equations in terms of mass

In the following reaction, how many grams of water are needed to completely hydrolyze 61.5 g of the ester?

$$CH_3\overset{\overset{O}{\|}}{C}{-}OCH_3 + H_2O \xrightarrow{OH^-} CH_3\overset{\overset{O}{\|}}{C}{-}O^- + HOCH_3$$

Strategy

In any calculation that involves a balanced reaction equation, the interpretation must be done in terms of moles. Convert grams of ester into moles of ester, decide how many moles of water are required, and then convert from moles of water into grams of water.

Solution

The problem can be solved in three steps.

1. *How many moles of ester are present?* The molar mass of the ester ($C_3H_6O_2$) is 74.0 g/mol.

$$61.5 \text{ g ester} \times \frac{1 \text{ mol ester}}{74.0 \text{ g ester}} = 0.831 \text{ mol ester}$$

2. *How many moles of water are needed?* From the balanced equation, 1 mol of ester reacts with 1 mol of water.

$$0.831 \text{ mol ester} \times \frac{1 \text{ mol } H_2O}{1 \text{ mol ester}} = 0.831 \text{ mol } H_2O$$

3. *How many grams of water are needed?* The final step makes use of the molar mass of H_2O (18.0 g/mol).

$$0.831 \text{ mol } H_2O \times \frac{18.0 \text{ g } H_2O}{1 \text{ mol } H_2O} = 15.0 \text{ g } H_2O$$

PRACTICE PROBLEM 6.9

How many moles of H_2 are required to completely reduce 15.0 g of the unsaturated hydrocarbon?

$$CH_2{=}CHCH_2CH{=}CH_2 + 2\,H_2 \xrightarrow[\text{Pt}]{} CH_3CH_2CH_2CH_2CH_3$$

6.6 CALCULATING THE YIELD OF A REACTION

When reactions are actually carried out, the reactants are not always present in the exact ratios specified by the balanced chemical equation and products are not always obtained in the expected amounts. This makes it important to know how to determine the limiting reactant, the theoretical yield, and the percent yield of a reaction.

Limiting Reactant

The **limiting reactant** is the reactant that *determines (limits) the amount of product that can be formed* in a reaction (Figure 6.12). Identifying the limiting reactant is based on determining which of the reactants gets totally consumed.

■ The limiting reactant is the one that runs out first.

A reaction between nitrogen gas (N_2) and hydrogen gas (H_2) prepares ammonia (NH_3), which can be used as a fertilizer. In the reaction equation below, which reactant is limiting if 2.10 mol of N_2 are reacted with 5.70 mol of H_2?

$$N_2(g) + 3H_2(g) \longrightarrow 2NH_3(g)$$

To answer this question, the number of moles of H_2 required to react completely with 2.10 mol of N_2 must be calculated.

$$2.10\ \cancel{\text{mol N}_2} \times \frac{3\text{ mol H}_2}{1\ \cancel{\text{mol N}_2}} = 6.30\text{ mol H}_2$$

If 6.30 mol of H_2 are required to react with 2.10 mol of N_2, but only 5.70 mol of H_2 are available, then H_2 is the limiting reactant because it will be totally consumed.

What if 3.95 mol of N_2 are combined with 15.1 mol of H_2?

$$3.95\ \cancel{\text{mol N}_2} \times \frac{3\text{ mol H}_2}{1\ \cancel{\text{mol N}_2}} = 11.9\text{ mol H}_2$$

In this situation, 11.9 mol of H_2 are required to react completely with 3.95 mol of N_2, but 15.1 mol of H_2 are available. More than enough H_2 is present, so when the N_2 has been totally consumed, some H_2 will remain—N_2 is the limiting reactant.

■ FIGURE 6.12

Limiting reactant

The limiting reactant (the reactant that is completely consumed) determines the maximum amount of product that can be formed. To build bicycles, the mechanic pictured here has two "reactants," bicycle frames and wheels. Once the two wheels have been used, no more bicycles can be put together—wheels are the limiting reactant.

Sample Problem 6.10

Determining limiting reactant

Sulfur dioxide (SO_2), an air pollutant that can cause respiratory problems, is produced when sulfur or sulfur containing substances like coal or oil are burned. Some factories scrub SO_2 from exhaust gases by reacting it with calcium oxide (CaO).

$$CaO(s) + SO_2(g) \longrightarrow CaSO_3(s)$$

What is the limiting reactant if 40.0 g of CaO are reacted with 0.625 mol of SO_2?

Strategy

You must calculate the number of moles of SO_2 needed to completely react with 40.0 g of CaO. Convert grams of CaO into moles of CaO and then use the balanced equation to guide you.

Solution

Reaction equations are balanced in terms of moles, not grams, so solving this problem requires the conversion of grams of CaO into moles of CaO.

1. *How many moles of CaO are present?* Convert grams into moles by using the molar mass of CaO (56.1 g/mol).

$$40.0 \cancel{\text{g CaO}} \times \frac{1 \text{ mol CaO}}{56.1 \cancel{\text{g CaO}}} = 0.713 \text{ mol CaO}$$

2. *What is the limiting reactant?* To identify the limiting reactant, you must determine how many moles of SO_2 are needed to react completely with the 0.713 mol of CaO.

$$0.713 \cancel{\text{mol CaO}} \times \frac{1 \text{ mol } SO_2}{1 \cancel{\text{mol CaO}}} = 0.713 \text{ mol } SO_2$$

A total of 0.713 mol of SO_2 are required, but only 0.625 mol are actually present, so SO_2 is the limiting reactant.

Practice Problem 6.10

For the hydrolysis of the ester, an ingredient in some perfumes, what is the limiting reactant if 60.0 g of ester are reacted with 20.0 g of water?

$$C_6H_5-\overset{O}{\overset{\|}{C}}-OCH_3 + H_2O \xrightarrow{OH^-} C_6H_5-\overset{O}{\overset{\|}{C}}-O^- + HOCH_3$$

Theoretical Yield

The theoretical yield is determined by the amount of limiting reactant that is initially present.

When the limiting reactant has been used up a reaction will stop, even though other reactants (the nonlimiting ones) still remain. This means that the **theoretical yield**, *the maximum amount of product that can be obtained*, is determined by the amount of limiting reactant that is initially present. Let us see how theoretical yield is calculated, by considering the reaction that takes place when propane (C_3H_8) is used as a fuel.

$$C_3H_8(g) + 5O_2(g) \longrightarrow 3CO_2(g) + 4H_2O(g)$$

What is the theoretical yield of CO_2 (in moles and in grams) if 2.2 mol of propane are reacted with 14 mol of O_2? In this problem, calculating the theoretical yield requires two steps.

1. *What is the limiting reactant?* Identify the limiting reactant by calculating the number of moles of one reactant that are required to completely react with the other. As shown by the equation below, 2.2 mol of propane require 11 mol of O_2.

$$2.2 \, \cancel{\text{mol } C_3H_8} \times \frac{5 \text{ mol } O_2}{1 \, \cancel{\text{mol } C_3H_8}} = 11 \text{ mol } O_2$$

   If 11 mol of O_2 are required and 14 mol of O_2 are available, then C_3H_8 is the limiting reactant (when all of the C_3H_8 is consumed, O_2 still remains).

2. *What is the theoretical yield of CO_2?* The theoretical yield of CO_2 is the amount of CO_2 formed when all of the limiting reactant is converted to product.

$$2.2 \, \cancel{\text{mol } C_3H_8} \times \frac{3 \text{ mol } CO_2}{1 \, \cancel{\text{mol } C_3H_8}} = 6.6 \text{ mol } CO_2$$

$$6.6 \, \cancel{\text{mol } CO_2} \times \frac{44.0 \text{ g } CO_2}{1 \, \cancel{\text{mol } CO_2}} = 290 \text{ g } CO_2$$

Percent Yield

When a reaction is carried out, it is very unlikely that the theoretical yield of product will be obtained. The reactants might not completely react with one another or might combine to form unexpected or unwanted products. Sometimes, the product is difficult to separate from the reaction mixture, so it cannot all be recovered.

The *amount of product obtained* from a reaction is called the **actual yield**. To indicate how well the actual yield agrees with the theoretical yield, chemists report the **percent yield**, which is defined as

$$\text{Percent yield} = \frac{\text{actual yield}}{\text{theoretical yield}} \times 100$$

If the theoretical yield of product for a reaction is 35.7 g, but only 15.2 g are obtained, then a 42.6% yield was obtained.

$$\text{Percent yield} = \frac{15.2 \text{ g}}{35.7 \text{ g}} \times 100 = 42.6\%$$

SAMPLE PROBLEM 6.11

Calculating percent yield

A drug company runs a large-scale reaction to prepare the pain reliever acetaminophen. The theoretical yield of the reaction is 42 kg of acetaminophen. If 33 kg of acetaminophen are obtained, what is the percent yield of the reaction?

Strategy

You must identify the actual yield and the theoretical yield, and then use the equation above to solve for the percent yield.

Solution

$$\text{Percent yield} = \frac{\text{actual yield}}{\text{theoretical yield}} \times 100$$

$$= \frac{33 \text{ kg}}{42 \text{ kg}} \times 100$$

$$= 79\%$$

PRACTICE PROBLEM 6.11

One use of positron emission tomography, or PET (Chapter 2), is to study cerebral blood flow. In this particular application, fluorine-19 is used as the positron emitter. After being artificially produced, ^{19}F is incorporated into $CH_3{}^{19}F$, which is then administered to the patient undergoing the PET scan. The $CH_3{}^{19}F$ molecule is produced by the reaction

$$^{19}F^- + CH_3I \longrightarrow CH_3^{19}F + I^-$$

If the theoretical yield for a synthesis of $CH_3{}^{19}F$ is 3.66 μg, but 3.15 μg are obtained, what is the percent yield?

6.7 FREE ENERGY AND REACTION RATE

In Section 5.1 we discussed the spontaneous melting of ice at 25°C and the effects that enthalpy (ΔH), entropy (ΔS), and temperature (T) have on this process. We saw that the combined effects of ΔH, ΔS, and T are described by free energy change (ΔG). For a spontaneous process (energy is released) ΔG has a negative value and for a nonspontaneous process (energy must be added) ΔG has a positive value.

These same principles also apply to chemical reactions. Consider, for example, the combustion of propane gas ($CH_3CH_2CH_3$).

$$CH_3CH_2CH_3(g) + 5O_2(g) \longrightarrow 3CO_2(g) + 4H_2O(g) \qquad \Delta G = -496 \text{ kcal}$$

Here, the negative sign on the value of ΔG indicates that energy is released (496 kJ for each mole of propane burned) and that the reaction is spontaneous. As we know from experience, once the burner of a propane-fueled camp stove is lit, it will continue burning until the propane is gone.

- The ΔG values given here are for standard conditions: 1 atm and 25°C.
- ΔG is negative for a spontaneous reaction.

The reaction equation for the reverse of the combustion of propane is written

$$3CO_2(g) + 4H_2O(g) \longrightarrow CH_3CH_2CH_3(g) + 5O_2(g) \qquad \Delta G = 496 \text{ kcal}$$

With a ΔG of +496 kcal/mol (energy must be added), this reaction is nonspontaneous.

Energy Diagrams

In Figure 5.4, a rock rolling down a hill was used as an example of a spontaneous (energy releasing) process. Using a similar type of drawing, called a **reaction energy diagram**, we can represent the energy change that takes place during a chemical reaction (Figure 6.13). In such a diagram, energy is shown on the y-axis and progress of reaction is plotted on the x-axis, with reactants on the left and products on the right. The more potential energy stored in a molecule, the higher it appears on the y-axis, and the less energy that it contains, the lower it is on this axis. In a spontaneous reaction energy is released, so products contain less energy (are more stable than) reactants. In a nonspontaneous reaction, the opposite is true.

Just because a reaction releases energy does not guarantee that it will take place within a reasonable amount of time—a negative ΔG says only that a reaction has the potential of taking place. For example, a rock sitting at the top of a hill has the potential of rolling down the hill, but this will happen only if it is given a push.

Reaction rate, a measure of *how quickly products form*, is directly tied to **activation energy** (E_{act}), *the energy barrier that must be crossed to go from reactants to products*

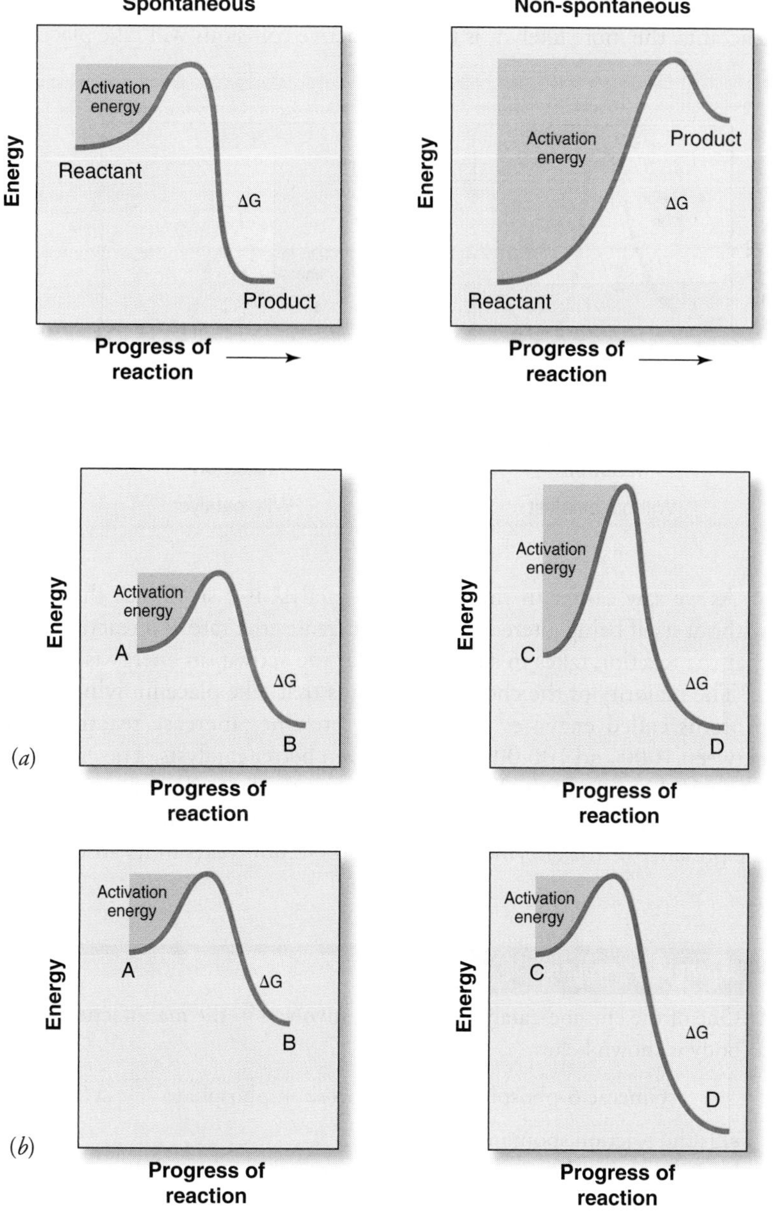

FIGURE 6.13

A reaction energy diagram
Energy diagrams show the progress of a reaction (reactant ⟶ product) and accompanying energy changes. In a spontaneous reaction products are lower in energy than reactants, while in a nonspontaneous one the opposite is true. The energy barrier between reactant and product is the activation energy for the reaction.

FIGURE 6.14

Activation energy
(*a*) The reactants and products for these two reactions have the same energy difference (ΔG), but the reactions have different activation energies—the second reaction will take place more slowly.
(*b*) These two reactions have the same activation energy, but do not have the same energy difference (ΔG) between reactants and products—the second reaction is more spontaneous.

(Figure 6.14). Activation energy is related to the energy that the reactant molecules must have in order to collide with sufficient force and in great enough numbers for bond breaking or bond making to occur. The lower the activation energy for a reaction, the faster it will take place. As shown in Figure 6.14, the activation energy and ΔG of a reaction are not related. Two spontaneous reactions with the same ΔG will take place at very different rates if their activation energies are different and reactions with the same activation energy will take place at the same rate, even if their ΔGs differ widely.

One way to alter the rate of a given chemical reaction is to change the temperature (Figure 6.15); the higher the temperature, the faster it will go. This is the case because the higher the temperature, the greater the kinetic energy of the reactants. With increased motion, collisions between reactants take place more frequently and with greater energy, so more reactants are able to find their way over the activation barrier to become products.

Another way that the rate of a reaction can sometimes be varied is to change the concentration of one or more reactants. (As we will see in Chapter 7, concentration refers to the amount of a substance that is present in a mixture.) A higher concentration of reactants

(*a*) (*b*)

FIGURE 6.15

The effect of temperature on reaction rate
Lowering the temperature leads to a drop in reaction rate. The reaction of an Alka Seltzer tablet is faster in (*a*) warm water than in (*b*) cold water.

Source: Andy Washnik.

can increase the rate of a reaction because the more crowded the reaction mixture is with reactants, the more likely it is that productive collisions will take place.

FIGURE | 6.16

Catalysts
Catalysts increase the rate of a reaction by lowering its activation energy.

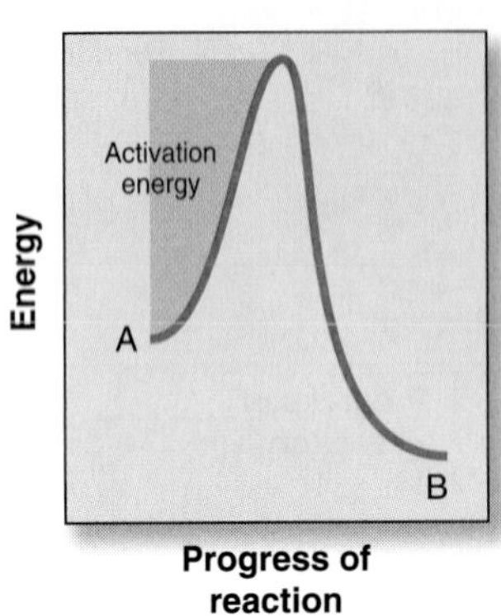

Without catalyst

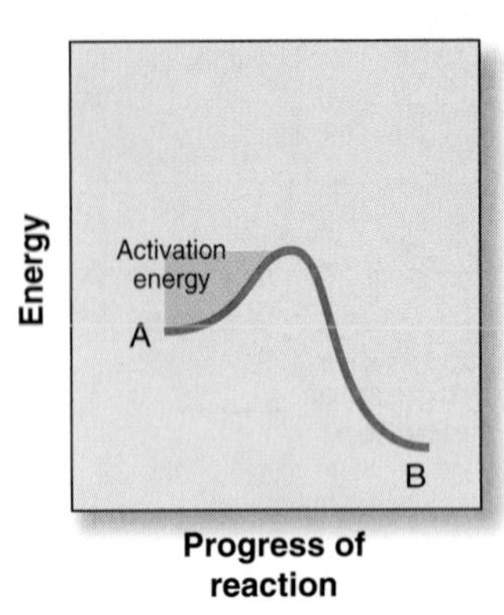

With catalyst

- Temperature, reactant concentration, and catalysts can alter the rate of a reaction.

As we saw earlier in this chapter, a catalyst is a substance that speeds up a reaction, without itself being altered. A catalyst increases the rate of a reaction by changing the path that the reaction takes in such a way that the activation energy is lowered (Figure 6.16).

The majority of the chemical reactions that take place in living things are catalyzed by proteins called enzymes. While most enzymes increase reaction rates by a factor of between 1000 and 100,000, some are even better catalysts. The enzyme OMP-decarboxylase, which catalyzes the breakdown of a particular carboxylic acid, increases the reaction rate by a factor of 10^{17}. This means that a process that takes place in 0.00001 second in the presence of the enzyme would require 32,000 years in its absence!

SAMPLE PROBLEM 6.12

One of the enzyme-catalyzed reactions involved in the manufacture of glucose in your body is shown below.

$$\text{Glucose 6-phosphate} \longrightarrow \text{glucose} + \text{phosphate} \qquad \Delta G = -3.3 \text{ kcal}$$

a. Is the reaction spontaneous?
b. How would you expect decreasing the concentration of glucose 6-phosphate to affect the reaction rate?
c. How would removing the enzyme affect the reaction rate?

Strategy
To decide whether or not the reaction is spontaneous, you must consider the value of ΔG.

Solution

a. Yes, because ΔG has a negative value.
b. Decreasing the reactant concentration should decrease the reaction rate.
c. The reaction would slow or stop completely—the enzyme acts as a catalyst.

PRACTICE PROBLEM 6.12

For this same reaction,

a. would removing the enzyme affect the value of ΔG?
b. would removing the enzyme affect whether or not the reaction is spontaneous?

HEALTH Link — Carbonic Anhydrase

Carbon dioxide, an end product of the metabolism that takes place in cells, is carried in the blood to the lungs, where it is exhaled. When they move into the blood about 95% of the CO_2 molecules end up in red blood cells, where some (about one-third) attach to hemoglobin, the oxygen-transporting molecules carried by red blood cells. The remaining two-thirds of the CO_2 molecules are converted into carbonic acid (H_2CO_3) in a hydration reaction catalyzed by the enzyme carbonic anhydrase (Figure 6.17). The 5% of CO_2 that does not go into red blood cells is dissolved directly in blood serum.

Carbonic anhydrase is an effective catalyst, with each enzyme molecule catalyzing the conversion of about one million CO_2 molecules each second. Once formed, H_2CO_3 can lose H^+ to produce bicarbonate ion (HCO_3^-), which moves from the red blood cells into the blood plasma (Figure 6.18).

At the lungs this process is reversed. Inside the red blood cells, carbonic anhydrase catalyzes the conversion of carbonic acid to CO_2. This CO_2, along with that released by hemoglobin and that dissolved in blood serum, moves into the lungs and is exhaled.

$$\mathrm{O{=}C{=}O} + \mathrm{H_2O} \underset{}{\overset{\text{carbonic anhydrase}}{\rightleftharpoons}} \mathrm{O{=}C(OH){-}OH}$$

Carbonic acid

$$\mathrm{O{=}C(OH){-}OH} \rightleftharpoons \mathrm{O{=}C(OH){-}O^-} + \mathrm{H^+}$$

Bicarbonate ion

FIGURE 6.17

Carbonic anhydrase
The enzyme carbonic anhydrase catalyzes the conversion of carbon dioxide into carbonic acid. This reaction involves hydration, the addition of water to a double bond. Carbonic acid rapidly loses H^+ to become bicarbonate ion. The double arrows indicate that the reactions run forward and backward.

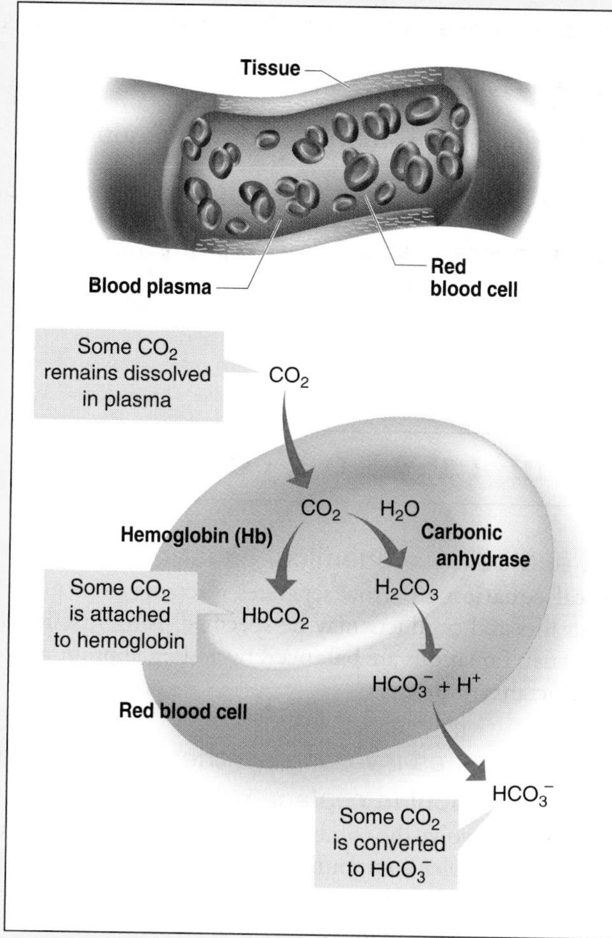

FIGURE 6.18

Transporting CO_2 in the blood
Carbon dioxide produced by the cells is carried by the blood in three different ways: directly dissolved in the blood, attached to the hemoglobin in red blood cells, and as HCO_3^- which is formed from the H_2CO_3 produced by the action of carbonic anhydrase. At the lungs, carbonic anhydrase catalyzes the conversion of H_2CO_3 to CO_2. This CO_2, the CO_2 released by hemoglobin, and that dissolved in serum move to the lungs and are exhaled.

Source: Figure 22.22, p. 860 from HUMAN ANATOMY AND PHYSIOLOGY, 6th ed. by Elaine N. Marieb. Copyright © 2004 by Pearson Education, Inc. Adapted by permission.

Protein 0g
Not a significant source of other nutrients.
*Percent Daily Values are based on a 2,000 calorie diet.
OUR REAL TEA STARTS WITH THE FINEST TEA LEAVES AND
MADE FROM: WATER, CITRIC ACID, TEA, NATURAL FLAVORS,
ASPARTAME**, POTASSIUM CITRATE, MALIC ACID.
**PHENYLKETONURICS: CONTAINS PHENYLALANINE
diet Snapple
ICED TEA

This chapter opened with a short discussion about the warning to phenylketonurics that is carried on foods that contain the artificial sweetener aspartame. Phenylketonuria (PKU) is an inherited disease that is caused by the inability to produce a properly working form of the enzyme called phenylalanine hydroxylase. The role of this enzyme is to serve as a catalyst in the transformation of one amino acid (phenylalanine) into another (tyrosine).

$$C_6H_5{-}CH_2CH(NH_2)C(=O){-}OH \xrightarrow{\text{phenylalanine hydroxylase}} HO{-}C_6H_4{-}CH_2CH(NH_2)C(=O){-}OH$$

Phenylalanine → Tyrosine

In the absence of phenylalanine hydroxylase, this oxidation reaction does not take place and phenylalanine is instead converted into other compounds that are toxic at high levels and, in developing brains, can cause irreversible damage. Within a few days of birth, newborns are usually tested for PKU. If this disease is detected and the infant is placed on a low-phenylalanine diet, normal brain development will take place.

Phenylalanine is one of the compounds used in the manufacture of aspartame. In the body, hydrolysis of this sweetener (one of the bonds hydrolyzed is an ester) releases phenylalanine. Those with PKU should limit their use of products that contain aspartame.

summary of objectives

1 Interpret and balance chemical equations.

In a **chemical equation** an arrow separates reactants from products. The physical states of reactants and products may be specified using *s* (solid), *l* (liquid), *g* (gas), and *aq* (aqueous). Equations are **balanced** by adding appropriate **coefficients** to the formulas of reactants or products.

2 Classify reactions as involving synthesis, decomposition, single replacement, or double replacement.

Synthesis reactions involve the combination of two or more elements or compounds to form one more complex compound. In **decomposition** reactions one compound is broken down to form elements or simpler compounds. In **single replacement** reactions, an element switches places with a different element in a compound. Parts of two compounds trade places in **double replacement** reactions.

3 Identify oxidation, reduction, combustion, and hydrogenation reactions.

Oxidation is the loss of electrons and **reduction** is the gain of electrons. An increase in the number of covalent bonds to oxygen and/or a decrease in the number of covalent bonds to hydrogen are an indication that an atom has been oxidized. Similarly, the loss of bonds to oxygen and/or the gain of bonds to hydrogen are an indication that reduction has taken place. In a **combustion reaction** the reactant is oxidized by O_2. **Catalytic hydrogenation** involves the reduction of a double bond by H_2 in the presence of Pt (a catalyst).

4 Identify hydrolysis, hydration, and dehydration reactions of organic compounds.

In a **hydrolysis reaction** water is used to split the other reactant. Hydrolysis of an ester in the presence of OH^- produces a carboxylate ion plus an alcohol. In a **hydration reaction**, water (with the assistance of the catalyst H^+) adds to the double bond of an alkene to form an alcohol. In a **dehydration reaction**, water (in the presence of H^+ and heat) is eliminated from an alcohol to produce an alkene.

5 Determine the limiting reactant, theoretical yield, and percent yield of a reaction.

Reaction equations are balanced in terms of numbers of reactants and products. Interpreting a balanced equation in terms of moles allows calculation of the relative amount of each reactant needed for the reaction to take place. Knowing the number of moles or the mass of reactants allows the **limiting reactant** (the reactant that runs out first) to be determined, which allows the **theoretical yield** (the amount of product expected from the limiting reactant) of a reaction to be calculated. Dividing the actual yield by the theoretical yield and then multiplying by 100 gives the **percent yield** of a reaction.

6 Describe the difference in energy changes for spontaneous and nonspontaneous reactions, and list the factors that affect the rate of a chemical reaction.

A reaction is spontaneous if ΔG is negative and nonspontaneous if ΔG is positive. **Activation energy**, **temperature**, **concentration**, and **catalysts** affect the rate of a reaction.

summary of reactions

Section 6.2

Synthesis

$$CO_2 + H_2O \longrightarrow H_2CO_3$$

Decomposition

$$CaCO_3 \longrightarrow CaO + CO_2$$

Single replacement

$$Fe + CuSO_4 \longrightarrow Cu + FeSO_4$$

Double replacement

$$NaCl + AgNO_3 \longrightarrow NaNO_3 + AgCl$$

Section 6.3

Oxidation and reduction

$$2Na + Cl_2 \longrightarrow 2NaCl$$

Combustion

$$CH_4 + 2O_2 \longrightarrow CO_2 + 2H_2O$$

Hydrogenation

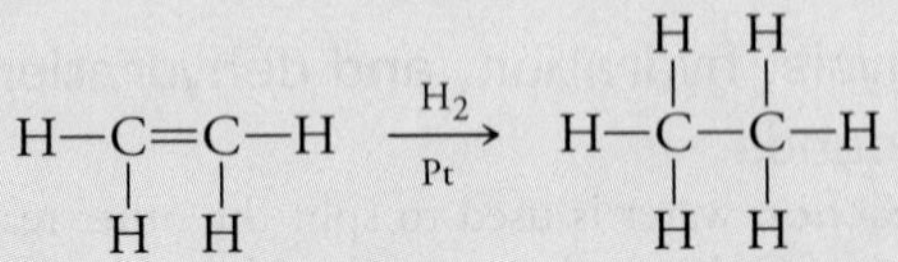

$$H{-}\underset{H}{C}{=}\underset{H}{C}{-}H \xrightarrow[Pt]{H_2} H{-}\underset{H}{\overset{H}{C}}{-}\underset{H}{\overset{H}{C}}{-}H$$

Section 6.4

Hydrolysis

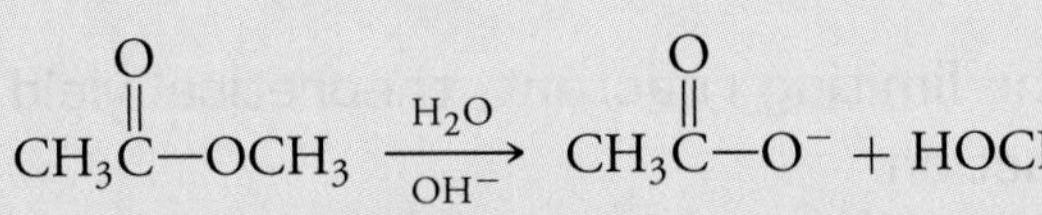

$$CH_3\overset{O}{\overset{\|}{C}}{-}OCH_3 \xrightarrow[OH^-]{H_2O} CH_3\overset{O}{\overset{\|}{C}}{-}O^- + HOCH_3$$

Hydration

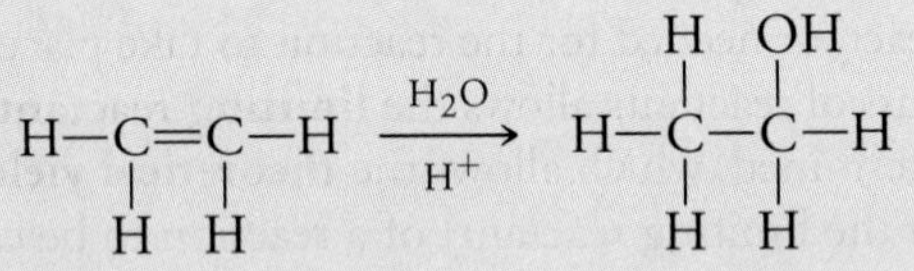

$$H{-}\underset{H}{C}{=}\underset{H}{C}{-}H \xrightarrow[H^+]{H_2O} H{-}\underset{H}{\overset{H}{C}}{-}\underset{H}{\overset{OH}{C}}{-}H$$

Dehydration

$$H{-}\underset{H}{\overset{H}{C}}{-}\underset{H}{\overset{OH}{C}}{-}H \xrightarrow[heat]{H^+} H{-}\underset{H}{C}{=}\underset{H}{C}{-}H + H_2O$$

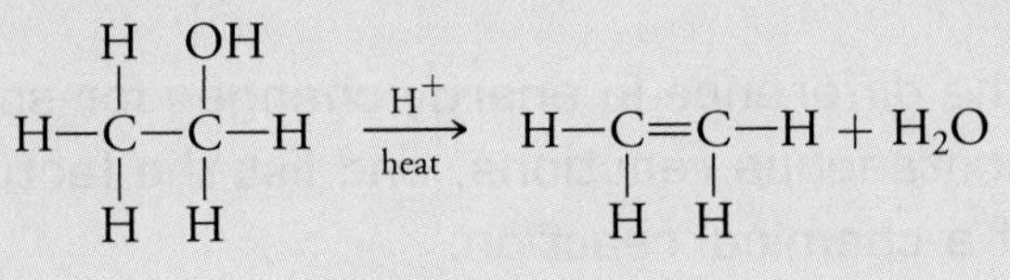

END OF CHAPTER PROBLEMS

Answers to problems whose numbers are printed in color are given in Appendix D. More challenging questions are marked with an asterisk. Problems within colored rules are paired. **ILW** = Interactive Learning Ware solution is available at *www.wiley.com/college/raymond*.

6.1 Chemical Equations

6.1 Write the following sentence as a balanced chemical equation. Phosphorus reacts with chlorine (Cl_2) to produce phosphorus trichloride.

6.2 Write the following sentence as a balanced chemical equation. Aqueous sodium chloride reacts with aqueous silver nitrate to form aqueous sodium nitrate and solid silver chloride.

6.3 What would you observe if you carried out the following reactions?

a. $NH_3(g) + HCl(g) \longrightarrow NH_4Cl(s)$

b. $HCO_3^-(aq) + H_3O^+(aq) \longrightarrow CO_2(g) + 2\ H_2O(l)$

6.4 What would you observe if you carried out the following reactions?

a. $AgNO_3(aq) + KCl(aq) \longrightarrow AgCl(s) + KNO_3(aq)$

b. $2\ Al(s) + Fe_2O_3(s) \longrightarrow 2\ Fe(l) + Al_2O_3(s)$

6.5 Balance the reaction equations.
a. $SO_2 + O_2 \longrightarrow SO_3$
b. $NO + O_2 \longrightarrow NO_2$

6.6 Balance the reaction equations.
a. $Mg + O_2 \longrightarrow MgO$
b. $K + O_2 \longrightarrow K_2O$

6.7 Balance the reaction equations.
a. $K + Cl_2 \longrightarrow KCl$
***b.** $CH_4 + Cl_2 \longrightarrow CH_2Cl_2 + HCl$

6.8 Balance the reaction equations.
a. $C_4H_6 + H_2 \longrightarrow C_4H_{10}$
***b.** $C_3H_8 + O_2 \longrightarrow CO + H_2O$

***6.9** Cave explorers can use lamps that contain calcium carbide (CaC_2) to light their way. When moistened, calcium carbide reacts to form acetylene gas (C_2H_2), which immediately burns to produce light. Balance the equation for the reaction of calcium carbide with water.

$$CaC_2(s) + H_2O(l) \longrightarrow Ca(OH)_2(aq) + C_2H_2(g)$$

6.2 Reaction Types

6.10 Classify each reaction as involving synthesis, decomposition, single replacement, or double replacement.
a. $2Na + Cl_2 \longrightarrow 2NaCl$
b. $2K + 2H_2O \longrightarrow H_2 + 2KOH$

6.11 Classify each reaction as involving synthesis, decomposition, single replacement, or double replacement.
a. $CaCO_3 + 2HCl \longrightarrow CaCl_2 + H_2CO_3$

b. $CH_3CH_2OH \xrightarrow[\text{heat}]{H^+} CH_2{=}CH_2 + H_2O$

6.12 Classify each reaction as involving synthesis, decomposition, single replacement, or double replacement.
a. $HBr + NaOH \longrightarrow NaBr + H_2O$
b. $Cu + 2AgNO_3 \longrightarrow Cu(NO_3)_2 + 2Ag$
c. $2S + 3O_2 \longrightarrow 2SO_3$
d. $H_2SO_4 \longrightarrow H_2O + SO_3$

6.13 Classify each reaction as involving synthesis, decomposition, single replacement, or double replacement.
a. $Cl_2 + 2NaI \longrightarrow 2NaCl + I_2$
b. $2K + 2HCl \longrightarrow 2KCl + H_2$
c. $N_2 + 3H_2 \longrightarrow 2NH_3$
d. $2KClO_3 \longrightarrow 2KCl + 3O_2$

6.14 The reaction shown here is one step in the fermentation process (Section 15.5). To which reaction type does the reaction belong?

$$CH_3\overset{\overset{O}{\|}}{C}-\overset{\overset{O}{\|}}{C}-OH \longrightarrow CH_3\overset{\overset{O}{\|}}{C}-OH + CO_2$$

Pyruvic acid — Acetal dehyde

6.15 The reaction shown here is one step in the citric acid cycle (Section 15.7) . To which reaction type does the reaction belong?

$$^{-}O-\overset{\overset{O}{\|}}{C}-CH{=}CH-\overset{\overset{O}{\|}}{C}-O^{-} + H_2O \longrightarrow$$

Fumarate

$$^{-}O-\overset{\overset{O}{\|}}{C}-CH_2-\overset{\overset{OH}{|}}{CH}-\overset{\overset{O}{\|}}{C}-O^{-}$$

Malate

6.16 As we will see in Section 10.4, an ester can be hydrolyzed in the following way. To which reaction type does the reaction belong?

$$CH_3\overset{\overset{O}{\|}}{C}-OCH_3 + HOH \xrightleftharpoons{H^+\text{ (catalyst)}} CH_3\overset{\overset{O}{\|}}{C}-OH + HOCH_3$$

6.17 As we will see in Section 11.2, an alcohol can be produced in the following way (Pt is a catalyst). To which reaction type does the reaction belong?

$$CH_3CH_2\overset{\overset{O}{\|}}{C}H + H_2 \xrightarrow{Pt} CH_2CH_2\overset{\overset{OH}{|}}{C}H_2$$

6.3 Oxidation and Reduction

6.18 Zinc reacts with copper(II) sulfate according to the equation

$$Zn(s) + CuSO_4(aq) \longrightarrow ZnSO_4(aq) + Cu(s)$$

a. Is zinc oxidized or is it reduced?
b. Is copper(II) ion oxidized or is it reduced?
c. What is the oxidizing agent?
d. What is the reducing agent?

6.19 Magnesium reacts with iron(II) chloride according to the equation

$$Mg(s) + FeCl_2(aq) \longrightarrow MgCl_2(aq) + Fe(s)$$

a. Is magnesium oxidized or is it reduced?
b. Is iron(II) ion oxidized or is it reduced?
c. What is the oxidizing agent?
d. What is the reducing agent?

6.20 Sodium metal reacts with oxygen gas (O_2) to form sodium oxide.

a. Write a balanced equation for this oxidation–reduction reaction. (In this problem you need not worry about the physical state of the reactants or product.)
b. Which reactant is oxidized?
c. Which reactant is reduced?
d. What is the oxidizing agent?
e. What is the reducing agent?

6.21 Aluminum metal reacts with oxygen gas (O_2) to form aluminum oxide.

a. Write a balanced equation for this oxidation–reduction reaction. (In this problem you need not worry about the physical state of the reactants or product.)
b. Which reactant is oxidized?
c. Which reactant is reduced?
d. What is the oxidizing agent?
e. What is the reducing agent?

6.22 Classify the reactions in Problems 6.18 and 6.20 as involving synthesis, decomposition, single replacement, or double replacement.

6.23 Classify the reactions in Problems 6.19 and 6.21 as involving synthesis, decomposition, single replacement, or double replacement.

6.24 When potassium metal reacts with sulfur, the following reaction takes place

$$K + S \longrightarrow K_2S$$

a. Balance the reaction equation.
b. Which reactant is oxidized?
c. Which reactant is reduced?

***6.25** Ethanol (CH_3CH_2OH) is mixed with gasoline to produce gasohol, a cleaner burning fuel than gasoline. Write the balanced equation for the complete combustion of ethanol by O_2 to produce CO_2 and H_2O.

***6.26** 1-Butene ($CH_2{=}CHCH_2CH_3$) burns in the presence of O_2 to produce CO_2 and H_2O. Write the balanced equation for this reaction.

6.27 a. "Hydrogen Cars" describes the possibility of using hydrogen as a fuel for cars, buses, and trucks. According to the video-clip, one of the positive aspects of this fuel is that when hydrogen is burned, water and energy are the only products. Write a balanced chemical equation for the reaction of H_2 and O_2 to form H_2O.
b. What products are formed when octane (C_8H_{18}), a typical component of gasoline, is burned?
c. Currently, what is the major disadvantage of using hydrogen as a fuel?

6.28 Many city water departments use chlorine gas (Cl_2) to disinfect their water supplies. When bubbled through water, Cl_2 is converted into hypochlorite ion (OCl^-), an oxidizing agent that kills harmful bacteria. In this reaction, chlorine atoms are both oxidized and reduced. Explain.

$$Cl_2(aq) + H_2O(l) \longrightarrow OCl^-(aq) + 2\,H^+(aq) + Cl^-(aq)$$

6.29 *p*-Phenylenediamine is a compound that is used in black hair dye. After being applied to and absorbed by hair, this compound is treated with H_2O_2 to produce a black color. Is *p*-phenylenediamine oxidized or is it reduced by H_2O_2?

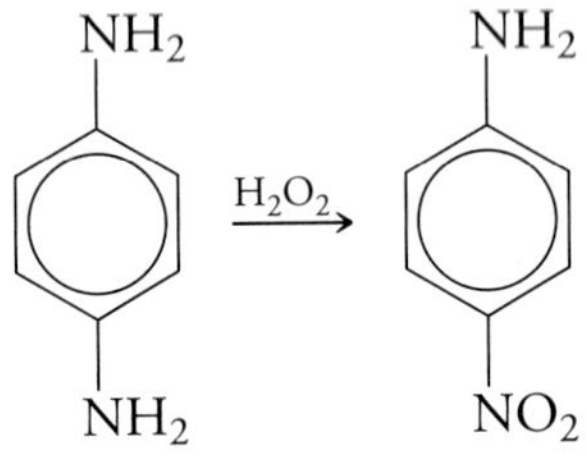

NH2 / NH2 —(H_2O_2)→ NH2 / NO2

p-Phenylenediamine

6.30 One of the reactions used to develop photographic film is shown below. Hydroquinone, a developer, reacts with silver ions to form silver metal, which gives the black color on a film negative.

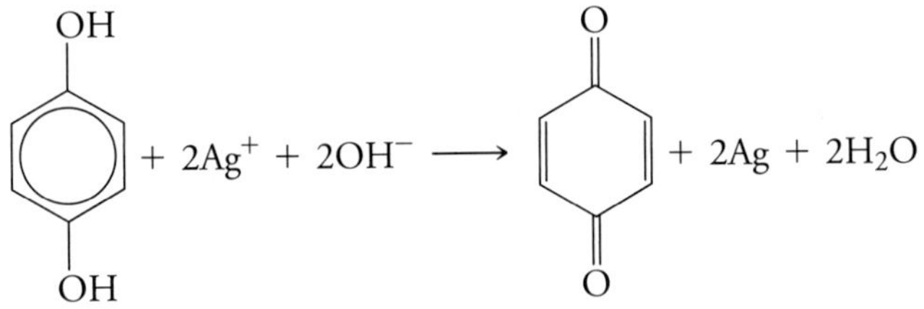

Hydroquinone (OH, OH) $+ 2Ag^+ + 2OH^- \longrightarrow$ Benzoquinone (O, O) $+ 2Ag + 2H_2O$

Hydroquinone **Benzoquinone**

a. Is hydroquinone oxidized or is it reduced?
b. Is Ag^+ oxidized or is it reduced?
c. What is the oxidizing agent?
d. What is the reducing agent?

6.31 Draw the saturated product expected when 3 mol of H_2, in the presence of Pt, are reacted with 1 mol of the termite trail marking pheromone shown below.

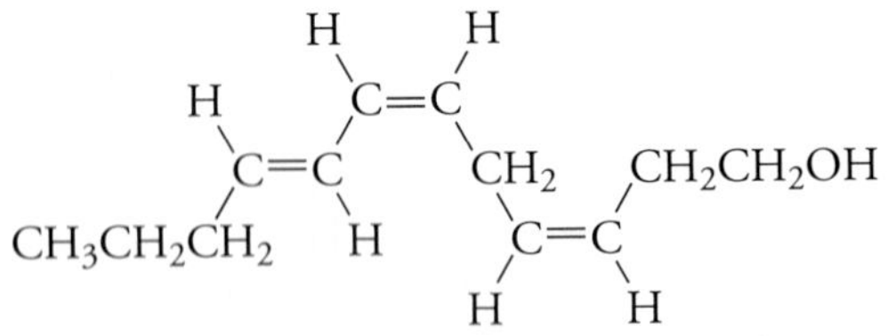

$CH_3CH_2CH_2(H)C{=}C(H)-CH{=}CH-CH_2-CH{=}CH-CH_2CH_2OH$

6.32 Draw the saturated product expected when 4 mol of H_2, in the presence of Pt, are reacted with 1 mol of α-farnesene, a compound present in the natural wax coating of apples.

α-Farnesene

6.33 Draw the skeletal structure of the product formed when each alkene is reacted with H_2, in the presence of Pt.

a. cyclopentene
b. 2-ethyl-3-methyl-1-pentene

6.34 Draw the skeletal structure of the product formed when each alkene is reacted with H_2, in the presence of Pt.

a. 3-methylcyclobutene
b. 2,3-dimethyl-2-butene

6.4 Reactions Involving Water

6.35 Draw the products of each hydrolysis reaction.

a. $H{-}C(=O){-}OCH_2CH_3 \xrightarrow[OH^-]{H_2O}$

b. $CH_3C(=O){-}OCH_2{-}C_6H_5 \xrightarrow[OH^-]{H_2O}$

c. $CH_3CH_2C(=O){-}OCH_2CH_2CH_3 \xrightarrow[OH^-]{H_2O}$

6.36 Draw the products of each hydrolysis reaction.

a. $CH_3CH(CH_3)C(=O){-}OCH_3 \xrightarrow[OH^-]{H_2O}$

b. $CH_3C(=O){-}OCH(CH_3)CH_3 \xrightarrow[OH^-]{H_2O}$

c. $C_6H_5{-}C(=O){-}OCH_2{-}C_6H_5 \xrightarrow[OH^-]{H_2O}$

6.37 Draw the hydration product formed when each alkene is reacted with H_2O in the presence of H^+.

a. *trans*-3-hexene
b. *cis*-3-hexene
c. 1,2-dimethylcyclopentene

6.38 Draw the hydration product formed when each alkene is reacted with H_2O in the presence of H^+.

a. *trans*-4-octene
b. *cis*-4-octene
c. 4,5-dimethylcyclohexene

6.39 Draw the organic dehydration product formed when each alcohol is heated in the presence of H^+.

a. $CH_3C(OH)(CH_3)CH_3$

b. (cyclohexanol)

c. $CH_3CH_2CH(OH)CH_2CH_3$

6.40 Draw the organic dehydration product formed when each alcohol is heated in the presence of H^+.

a. $CH_3CH_2C(OH)(CH_2CH_3)CH_2CH_3$

b. 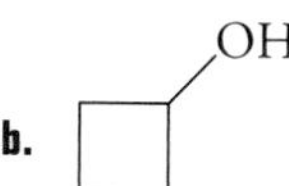

c. $CH_3CH(OH){-}C_6H_5$

6.41 Aspirin (acetylsalicylic acid) is an ester of acetic acid.

Aspirin

a. Circle the ester group in aspirin.
b. Aspirin is sold with cotton placed in the neck of the bottle to help keep moisture out. When exposed to water, the ester group of aspirin slowly hydrolyzes. Draw the hydrolysis products obtained if aspirin is reacted with H_2O in the presence of OH^-.

6.42 Triflusal is a drug used to inhibit blood clot formation. Draw the organic products obtained when triflusal is hydrolyzed in the presence of OH^-.

Triflusal

*6.43 The ester is reacted with H_2O and OH^- to form a carboxylate ion and molecule A. Molecule A is heated in the presence of H^+ to produce molecule B. Molecule B is treated with H_2O and H^+ to form molecule C. Draw A, B, and C.

$$CH_3CH_2\underset{CH_3}{\underset{|}{C}}H\overset{O}{\overset{\|}{C}}-O\underset{CH_3}{\underset{|}{C}}HCH_3 \longrightarrow CH_3CH_2\underset{CH_3}{\underset{|}{C}}H\overset{O}{\overset{\|}{C}}-O^- + A$$

$$A \xrightarrow[\text{heat}]{H^+} B$$

$$B \xrightarrow[H^+]{H_2O} C$$

6.44 The ester hydrolysis described in this chapter is actually a combination of two reactions. The reaction in Figure 6.6, for example, can be broken down into two steps. In the first, the ester is split to form a carboxylic acid and an alcohol. In step two, the carboxylic acid is transformed into a carboxylate ion through loss of H^+.

Step 1

$$CH_3CH_2CH_2\overset{O}{\overset{\|}{C}}-OCH_2CH_3 + H_2O \longrightarrow CH_3CH_2CH_2\overset{O}{\overset{\|}{C}}-OH + HOCH_2CH_3$$

Step 2

$$CH_3CH_2CH_2\overset{O}{\overset{\|}{C}}-OH + {}^-OH \longrightarrow CH_3CH_2CH_2\overset{O}{\overset{\|}{C}}-O^- + H_2O$$

To which reaction type (synthesis, decomposition, single replacement, or double replacement) does the first of the two steps belong?

6.5 Mole and Mass Relationships in Reactions

6.45 For the combustion of methane, beginning with 3.15 mol of CH_4,

a. how many moles of O_2 are required to completely consume the CH_4?

b. how many moles of CO_2 are obtained when the CH_4 is completely combusted?

c. how many moles of H_2O are obtained when the CH_4 is completely combusted?

$$CH_4(g) + 2O_2(g) \longrightarrow CO_2(g) + 2H_2O(g)$$

6.46 For the combustion of ethane, beginning with 0.27 mol of CH_3CH_3,

a. how many moles of O_2 are required to completely consume the CH_3CH_3?

b. how many moles of CO_2 are obtained when the CH_3CH_3 is completely combusted?

c. how many moles of H_2O are obtained when the CH_3CH_3 is completely combusted?

$$2CH_3CH_3(g) + 7O_2(g) \longrightarrow 4CO_2(g) + 6H_2O(g)$$

6.47 2-Propanol, also known as isopropyl alcohol or rubbing alcohol, can be produced by the reaction

$$\underset{\textbf{Propene}}{CH_3CH{=}CH_2} + H_2O \xrightarrow{H^+} \underset{\textbf{2-Propanol}}{CH_3\overset{OH}{\overset{|}{C}}HCH_3}$$

a. How many moles of H_2O are required to completely react with 55.7 mol of propene?

b. How many moles of H_2O are required to completely react with 1.66 mol of propene?

c. How many moles of 2-propanol are expected from the complete reaction of 47.2 g of propene?

d. How many grams of 2-propanol are expected from the complete reaction of 125 g of propene?

6.48 In the presence of H_2 and Pt, propene is reduced to propane.

$$\underset{\textbf{Propene}}{CH_3CH{=}CH_2} + H_2 \xrightarrow{Pt} \underset{\textbf{Propane}}{CH_3CH_2CH_3}$$

a. How many moles of H_2 are required to completely react with 14.3 mol of propene?

b. How many moles of H_2 are required to completely react with 0.88 mol of propene?

c. How many moles of propane are expected from the complete reaction of 27.7 g of propene?

d. How many grams of propane are expected from the complete reaction of 125 g of propene?

6.49 Consider the reaction

$$KOH(s) + CO_2(g) \longrightarrow KHCO_3(s)$$

a. How many grams of KOH are required to react completely with 5.00 mol of CO_2?

b. How many grams of $KHCO_3$ are produced from the complete reaction of 75.9 g of KOH?

6.50 Consider the reaction

$$2KClO_3(s) \longrightarrow 2KCl(s) + 3O_2(g)$$

a. How many grams of KCl are produced from the complete reaction of 6.60 mol of $KClO_3$?

b. How many grams of KCl are produced from the complete reaction of 47.5 g of $KClO_3$?

6.51 A reaction of iron metal with oxygen gas produces ferrous oxide (FeO).

$$Fe + O_2 \longrightarrow FeO$$

a. Balance the reaction equation.
b. When FeO forms, has Fe been oxidized or has it been reduced?
c. When FeO forms, has O_2 been oxidized or has it been reduced?

6.52 A reaction of iron metal with oxygen gas produces ferric oxide (Fe_2O_3).

$$Fe + O_2 \longrightarrow Fe_2O_3$$

a. Balance the reaction equation.
b. When Fe_2O_3 forms, has Fe been oxidized or has it been reduced?
c. When Fe_2O_3 forms, has O_2 been oxidized or has it been reduced?

6.6 Calculating the Yield of a Reaction

***6.53** One form of phosphorus, called white phosphorus, burns when exposed to air.

$$P_4(s) + O_2(g) \longrightarrow P_4O_{10}(s)$$

a. Balance the reaction equation.
b. What is the theoretical yield (in grams) of P_4O_{10} if 33.0 g of P_4 are reacted with 40.0 g of O_2?

***6.54** Mercury reacts with oxygen to produce mercury(II) oxide.

$$Hg(l) + O_2(g) \longrightarrow HgO(s)$$

a. Balance the reaction equation.
b. What is the theoretical yield (in grams) of HgO if 254 g of Hg are reacted with 254 g of O_2?

6.55 Many food colors and fabric dyes are manufactured from aniline. Aniline can be produced by the reaction below.

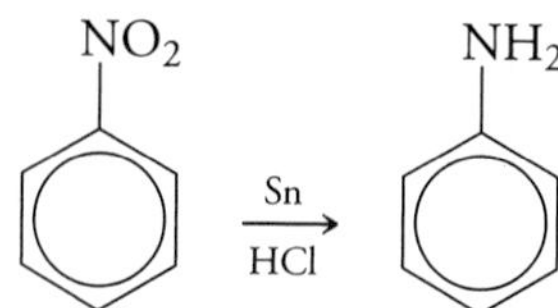

Nitrobenzene Aniline

a. What is the theoretical yield of aniline (in grams), beginning with 150 g of nitrobenzene? (Assume that nitrobenzene is the limiting reactant.)
b. Based on the theoretical yield from part a, what is the percent yield if 50.0 g of aniline are obtained?

6.56 In the presence of H_2 and Pt, ethylene is converted to ethane.

$$CH_2{=}CH_2 + H_2 \xrightarrow{Pt} CH_3CH_3$$

Ethylene Ethane

a. What is the theoretical yield of ethane (in moles) if 8.0 mol of ethylene are reacted with 4.0 mol of H_2?
b. What is the theoretical yield if 20.0 g of ethylene are reacted with 10.0 g of H_2?

***6.57** If a person drinks a glass of beer or wine, the first step in metabolizing the ethanol (CH_3CH_2OH) is its enzymatic conversion to acetaldehyde (CH_3CHO).

$$CH_3{-}CH(OH){-}H + NAD^+ \xrightarrow{\text{enzyme}} CH_3{-}C({=}O){-}H + NADH + H^+$$

Ethanol Acetaldehyde

a. In this reaction is ethanol oxidized or is it reduced?
b. Write this reaction equation by assigning the conversion of NAD^+ into NADH and H^+ its own arrow.
c. What theoretical yield of acetaldehyde (in grams) is expected from the reaction of 10.0 g of ethanol? (Assume that ethanol is the limiting reactant.)

***6.58** The toxic effects of methanol (CH_3OH) are due to its enzymatic conversion into formaldehyde (CH_2O) in the liver.

$$H{-}CH(OH){-}H + NAD^+ \xrightarrow{\text{enzyme}} H{-}C({=}O){-}H + NADH + H^+$$

Methanol Formaldehyde

a. In this reaction is methanol oxidized or is it reduced?
b. Write this reaction equation by assigning the conversion of NAD^+ into NADH and H^+ its own arrow.
c. What theoretical yield of methanol (in grams) is expected from the reaction of 10.0 g of methanol? (Assume that methanol is the limiting reactant.)

6.7 Free Energy and Reaction Rate

6.59 Which of the reactions are spontaneous?

a. $CO(g) + H_2O(g) \longrightarrow CO_2(g) + H_2(g)$
$\Delta G = -4.1$ kcal/mol

b. $2HI(g) \longrightarrow H_2(g) + I_2(g)$
$\Delta G = 0.6$ kcal/mol

6.60 Which of the reactions are spontaneous?

a. $3C_2H_2(g) \longrightarrow C_6H_6(g)$

$\Delta G = -119.0$ kcal/mol

b. fructose 1,6-bisphosphate + ADP $\longrightarrow$ fructose 6-phosphate + ATP

$\Delta G = 3.4$ kcal/mol

***6.61** Sample Problem 6.12 showed how glucose 6-phosphate is converted into glucose during the manufacture of glucose.

glucose 6-phosphate $\longrightarrow$ glucose + phosphate

$\Delta G = -3.3$ kcal/mol

In the breakdown of glucose, glucose is converted into glucose 6-phosphate.

glucose + ATP $\longrightarrow$ glucose 6-phosphate + ADP

$\Delta G = -4.0$ kcal/mol

a. Is the conversion of glucose 6-phosphate into glucose + phosphate spontaneous?
b. Is the conversion of glucose + ATP into glucose 6-phosphate + ADP spontaneous?
c. How can ΔG be negative for both the transformation of glucose 6-phosphate into glucose (part a) and the transformation of glucose into glucose 6-phosphate (part b)?

6.62 Draw two reaction energy diagrams that illustrate the type of results expected for the reaction below, one diagram for the reaction in the presence of Pt catalyst, and one diagram in the absence of Pt catalyst. In drawing the energy diagrams, assume that the reactions are spontaneous.

$$CH_3CH{=}CH_2 + H_2 \xrightarrow{Pt} CH_3CH_2CH_3$$

6.63 What is the likely effect on the rate of a reaction if
a. reactant concentration is decreased?
b. temperature is decreased?
c. a catalyst is added?

HEALTHLink | Antiseptics and Oxidation

6.64 When hydrogen peroxide is used as a disinfectant, it is the O_2 gas produced from the decomposition of H_2O_2 that kills bacteria. Balance the reaction equation.

$$H_2O_2(aq) \longrightarrow H_2O(l) + O_2(g)$$

6.65 When present in ointments and creams for treating acne, the oxidizing agent benzoyl peroxide ($C_{14}H_{10}O_4$) is safely handled. Pure benzoyl peroxide is quite hazardous to deal with, however. Any heat source can cause it to explode as the compound undergoes rapid oxidation. Complete the balancing of the equation for the combustion of benzoyl peroxide.

$$2\,C_{14}H_{10}O_4 + O_2 \longrightarrow CO_2 + H_2O$$

HEALTHLink | Carbonic Anhydrase

6.66 Explain the role that carbonic anhydrase has in the transport of CO_2 in the bloodstream.

6.67 Only 5% of the CO_2 that moves from cells into the blood is carried as dissolved CO_2. What happens to the other 95%?

THINKING IT THROUGH

6.68 As we will see in a later chapter, high temperatures can cause a protein to lose its normal biological function. This being the case, explain why an increase in temperature does not *always* lead to an increased reaction rate for biochemical reactions.

INTERACTIVE LEARNING PROBLEMS

ILW

6.69 Balance the following equations:

a. $Mg(OH)_2 + 2HBr \longrightarrow MgBr_2 + H_2O$
b. $HCl + Ca(OH)_2 \longrightarrow CaCl_2 + H_2O$
c. $Al_2O_3 + H_2SO_4 \longrightarrow Al_2(SO_4)_3 + H_2O$
d. $KHCO_3 + H_3PO_4 \longrightarrow K_2HPO_4 + H_2O + CO_2$
e. $C_9H_{20} + O_2 \longrightarrow CO_2 + H_2O$

ILW

6.70 In dilute nitric acid, HNO_3, copper metal dissolves according to the following equation:

$$3Cu(s) + 8HNO_3(aq) \longrightarrow 3Cu(NO_3)_2(aq) + 2NO(g) + 4H_2O$$

How many grams of HNO_3 are needed to dissolve 11.45 g of Cu according to this equation?

ILW

6.71 What is the product of 2-methyl-2-butene and

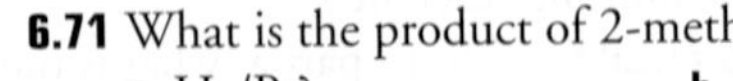

a. H_2/Pt? **b.** H_2O/H^+?

SOLUTIONS TO PRACTICE PROBLEMS

6.1 When the two liquids are mixed, a solid forms.

6.2 $2NO + O_2 \longrightarrow 2NO_2$

6.3 single displacement

6.4 oxidized

6.5 $2C_2H_2(g) + 5O_2(g) \longrightarrow 4CO_2(g) + 2H_2O(g)$

6.6 **a.** reduced; **b.** oxidized; **c.** oxidized

6.7 **a.** 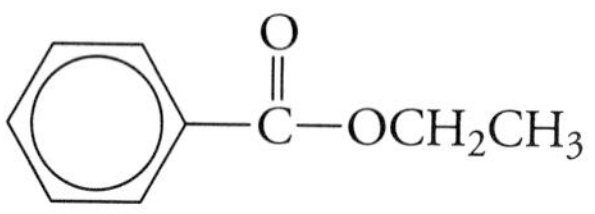 $C_6H_5-\overset{O}{\overset{\|}{C}}-OCH_2CH_3$ **b.** $CH_3CH_2\overset{OH}{\overset{|}{C}}HCH_3$ **c.** 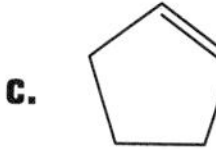

6.8 **a.** 17 mol O_2; **b.** 12 mol CO_2

6.9 0.442 mol H_2

6.10 ester

6.11 86.1%

6.12 **a.** no; **b.** no

A patient thumbs through an old magazine while waiting for her oncologist to come into the examination room. A knock sounds on the door and the doctor enters. After a quick exam, she looks at the results of the patient's blood test and then says, "As we discussed a few weeks ago, treatment just cannot be put off any longer. Ordinarily, I would start you on allopurinol a few days before beginning treatment to minimize a side effect of the chemotherapy drug. Since you are allergic to allopurinol, however, we will have to forgo its use and just watch carefully for signs of gout." The patient asks why gout is associated with use of this particular chemotherapy drug.

Photo source: Rick Gomez/Blend Images/Jupiter Images.

7 SOLUTIONS, COLLOIDS, AND SUSPENSIONS

In previous chapters we have looked at atoms and their grouping into elements and compounds. It turns out that most of the matter around us does not consist of pure elements or compounds. Instead, we are surrounded by mixtures of the two. In this chapter we will turn our attention to the properties of various mixtures, including those formed by some organic and biochemical compounds. Some disease states, including gout, are related to the behavior of mixtures.

objectives

After completing this chapter, you should be able to:

1. Define the term pure substance and describe what is meant by the terms homogeneous mixture and heterogeneous mixture. For a homogeneous mixture, explain what differentiates solutes from a solvent.
2. Describe the effect that temperature has on the solubility of gases, liquids, and solids in water.
3. Use solubility rules to predict whether or not a reaction between ionic compounds will produce a precipitate. For a precipitation reaction, be able to write the ionic and net ionic equation.
4. Explain Henry's law.
5. Explain the terms hydrophilic, hydrophobic, and amphipathic and give examples of compounds that belong to each category.
6. List some of the commonly encountered concentration units and perform calculations involving concentration.
7. Calculate the concentration of a solution after it has been diluted.
8. Distinguish between solutions, suspensions, and colloids.
9. Describe the principles of diffusion and osmosis.

7.1 SOLUTIONS

The kidneys are involved in maintaining a proper balance of **electrolytes** (*compounds that dissolve in water to form ions*) in your body. The consequences of having too much or too little of a particular electrolyte can have profound effects on your health. Included in the list of biochemically important ions are sodium ion (Na^+) and potassium ion (K^+), which are involved in the transmission of nerve impulses, and calcium ion (Ca^{2+}), which plays a role in muscle contraction and blood clotting. For us to be able to make sense out of this and the other chemistry that goes on in living things, we must understand how different compounds and elements interact with one another to form mixtures.

■ Pure substances can combine to form homogeneous or heterogeneous mixtures.

Before looking at the factors that determine how well two compounds mix with one another, let us define a few terms. A **pure substance** consists of just *one element or compound*. The elements mercury (Hg), silver (Ag), and oxygen gas (O_2) are pure substances, as are the compounds water (H_2O), sodium chloride (NaCl), and sucrose ($C_{12}H_{22}O_{11}$). A *combination of two or more pure substances* is called a **mixture**. Some mixtures are **heterogeneous** (*not evenly distributed*) as is illustrated by the reaction shown in Figure 7.1. When aqueous silver nitrate ($AgNO_3$) is mixed with aqueous sodium chloride (NaCl), a **precipitate** (*a solid reaction product*) of silver chloride (AgCl) forms. The two reactant mixtures combine to form a heterogeneous mixture.

(*a*)

(*b*)

$$AgNO_3(aq) + NaCl(aq) \longrightarrow AgCl(s) + NaNO_3(aq)$$

■ FIGURE | 7.1

A heterogeneous mixture
(*a*) Aqueous silver nitrate and aqueous sodium chloride are homogeneous mixtures. (*b*) When these two are mixed, silver chloride precipitates to produce a heterogeneous mixture.

Source: Ken Karp/Wiley Archive.

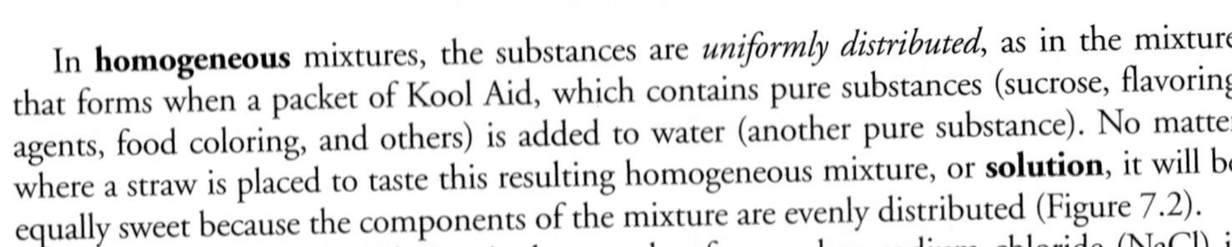

In **homogeneous** mixtures, the substances are *uniformly distributed*, as in the mixture that forms when a packet of Kool Aid, which contains pure substances (sucrose, flavoring agents, food coloring, and others) is added to water (another pure substance). No matter where a straw is placed to taste this resulting homogeneous mixture, or **solution**, it will be equally sweet because the components of the mixture are evenly distributed (Figure 7.2).

Another example of a solution is the one that forms when sodium chloride (NaCl) is dissolved in water. In this aqueous solution (Figure 7.1*a*), water is the **solvent** (*the solution component present in the greatest amount*) and Na^+ and Cl^- are the **solutes** (*components dissolved in the solvent*). The solutes and solvents that make up a solution can be solids, liquids, or gases. Carbonated water is a solution of the gas carbon dioxide (solute) dissolved in water (solvent). Air consists of oxygen gas (solute) and other gaseous solutes dissolved in nitrogen gas (solvent).

■ FIGURE | 7.2

Solutions
A Kool Aid solution is formed when the ingredients in the package (sugar, flavoring, and coloring) become uniformly distributed in water.

Source: Andy Washnik/Wiley Archive.

The formation of liquid solutions, such as Kool Aid or dissolved NaCl, requires that solute particles have two important characteristics:

- They must be about the same size as the solvent molecules.
- They must be able to interact with the solvent molecules through noncovalent interactions (see Section 4.3 for a review of these forces).

If solutes have each of these characteristics, then they will mix well with solvent and will not settle out of solution. In a solution, solutes and solvents stay uniformly mixed no matter how long they stand.

These characteristics (similar size and interacting through noncovalent interactions) are the source of a general guideline, ***like dissolves like***, that can be used to predict whether a substance is **soluble** (will dissolve) or **insoluble** (will not dissolve) in a particular solvent. Using this guideline we can understand why Na^+ and Cl^- ions are soluble in water. These ions are "like" water molecules—they are about the same size and can interact, in this case, by ion–dipole interactions. In Section 7.2 we will look in more detail at the ability of ionic compounds to dissolve in water.

Temperature plays a role in the formation of liquid solutions. For all gaseous solutes (O_2, CO_2, etc.), an increase in temperature leads to a decrease in **solubility** (*the amount of solute that will dissolve in a solvent at a given temperature*). At a pressure of 1 atm and at 20°C, 0.169 g of carbon dioxide (CO_2) will dissolve in 100 g of H_2O. Increasing the temperature to 60°C drops the solubility of CO_2 to 0.058 g/100 g H_2O. One way to explain this change in solubility is to think in terms of kinetic energy. As the temperature of a solvent (water, for example) is increased, its molecules acquire a greater kinetic energy. With the increased motion of the solvent molecules, the gaseous solute atoms or molecules get bumped out of the solution and into the air that sits above the liquid.

Most liquid and solid solutes become more soluble in water as temperature increases. At 20°C, for example, 35.9 g of sodium chloride will dissolve in 100 g of H_2O, and at 60°C, 37.1 g will dissolve. In some cases, the dependence of solubility on temperature is quite large. Ammonium chloride has a solubility in water of 37.2 g/100 g H_2O at 20°C and of 55.3 g/100 g H_2O at 60°C (Figure 7.3). As temperature is raised, the solvent molecules move more rapidly and are more effective at bouncing the liquid or solid solute particles apart from one another, allowing them to dissolve.

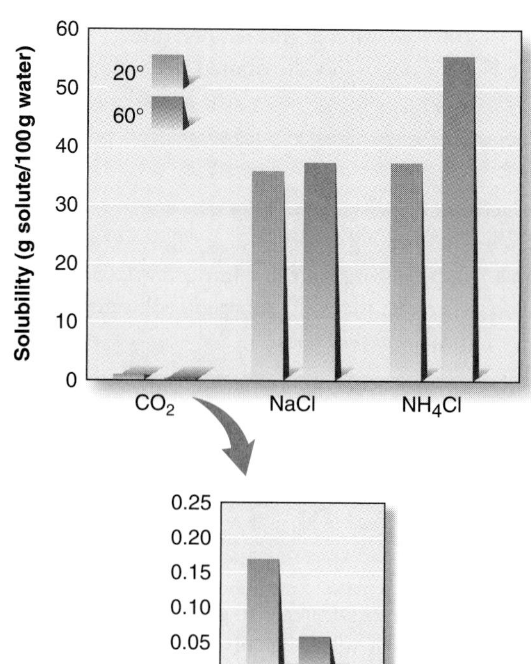

FIGURE 7.3

Temperature and solubility Gases become less soluble in water as the temperature increases, while liquids and solids typically become more soluble.

FIGURE | 7.4

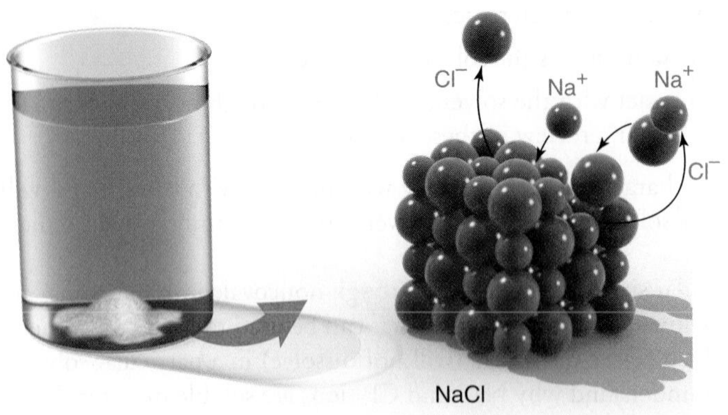

Solubility
Table salt (NaCl) is soluble in water, but when it reaches its solubility limit no more will dissolve. At this point, solute particles Na^+ and Cl^- enter and leave the solution at the same rate.

When a solute has reached its solubility limit at a given temperature, no additional solute will dissolve (Figure 7.4). While it may appear that nothing is happening in the resulting heterogeneous mixture, there is, in fact, some activity—solute particles continually dissolve in the solvent while others leave the solution. These two processes take place at the same speed, so no visible change takes place.

SAMPLE PROBLEM 7.1

Determining solubility

a. If 45.0 g of NaCl are added to 100 g of water at 20°C, will all of the NaCl dissolve?
b. Is the resulting mixture homogeneous or heterogeneous?

Strategy

NaCl has a solubility of 35.9 g per 100 g of water at 20°C. Based on this value, you can decide whether or not the NaCl is completely soluble and whether the mixture is homogeneous or heterogeneous.

Solution

a. No. Only 35.9 g of NaCl dissolve per 100 g of water at this temperature.
b. The mixture is heterogeneous. The NaCl is not evenly distributed since some sits on the bottom of the beaker.

PRACTICE PROBLEM 7.1

On a hot day, it is often easier to catch fish by casting your line into a deep, cool part of a lake than into a shallow, warm spot. One reason that fish gather in cool water may be related to the levels of oxygen dissolved in the water. Explain.

7.2 REACTIONS OF IONS IN AQUEOUS SOLUTIONS

In the previous section we used the guideline *like dissolves like* to account for the solubility of NaCl in water. To summarize, NaCl dissolves in water because Na^+ and Cl^- are similar in size to water molecules and can interact with them through ion–dipole interactions. There is one additional factor—the strength of the ionic crystal lattice—to consider when looking at the water solubility of ionic compounds. In some cases the ionic bonds that hold a crystal lattice together are so strong that the crystal will not dissolve in water, even if the individual ions would otherwise be soluble. As we just saw, NaCl dissolves in water to release the solutes Na^+ and Cl^-. Silver nitrate ($AgNO_3$), also soluble in water, dissolves to produce Ag^+ and NO_3^-. Silver chloride (AgCl), however, is not water soluble. This inability to dissolve in water has nothing directly to do with the ions involved. When paired with Na^+ in NaCl, Cl^- becomes a solute in water and, when paired with NO_3^- in $AgNO_3$, Ag^+ becomes a solute. In AgCl, the ionic bonds between Ag^+ and Cl^- ions are strong enough that they are not broken by water molecules and the ionic compound is insoluble. The solubility behavior of a number of ionic compounds in water is summarized in Table 7.1.

TABLE 7.1 The Solubility of Ionic Compounds in Water

Compound	Example	Exceptions	Example
Water soluble			
Nitrates	$AgNO_3$	None	None
Chlorides, bromides, iodides	NaCl	Those containing Ag^+, Pb^{2+}, or Hg_2^{2+}	AgCl
Sulfates	K_2SO_4	Those containing Pb^{2+}, Sr^{2+}, Ba^{2+}, or Hg_2^{2+}	$PbSO_4$
Water insoluble			
Hydroxides	$Mg(OH)_2$	Those containing alkali metal cations, Ca^{2+}, Sr^{2+}, or Ba^{2+}	NaOH
Phosphates	$FePO_4$	Those containing NH_4^+ or alkali metal cations	Li_3PO_4
Carbonates	$PbCO_3$	Those containing NH_4^+ or alkali metal cations	K_2CO_3

SAMPLE PROBLEM 7.2

Predicting the water solubility of ionic compounds

Indicate whether each ionic compound is soluble or insoluble in water.

a. KNO_3 **b.** $BaSO_4$ **c.** LiI

Strategy

To answer part a, note that Table 7.1 shows that all nitrate-containing ionic compounds are soluble in water. Parts b and c can also be answered by referring to this table.

Solution

a. soluble **b.** insoluble **c.** soluble

PRACTICE PROBLEM 7.2

Write the formula of each ionic compound and indicate whether it is soluble or insoluble in water.

a. barium sulfate **b.** lithium nitrate **c.** sodium carbonate

Precipitation Reactions

The information in Table 7.1 can be used to decide whether or not a chemical reaction will take place when a given pair of ionic compounds is mixed in water. Consider the balanced equation for the combination of cobalt(II) chloride with calcium hydroxide:

$$CoCl_2 + Ca(OH)_2 \longrightarrow Co(OH)_2 + CaCl_2$$

In this double replacement reaction the cations and anions switch places, with Co^{2+} being paired with OH^- and Ca^{2+} with Cl^- in the products. From Table 7.1, of the four ionic compounds represented in this equation, all are water soluble except for $Co(OH)_2$. Using this knowledge, the reaction equation can be rewritten, this time showing physical states.

$$CoCl_2(aq) + Ca(OH)_2(aq) \longrightarrow Co(OH)_2(s) + CaCl_2(aq)$$

In this reaction, the reactants are homogeneous mixtures and one of the products, $Co(OH)_2$, is a solid (a precipitate).

Knowing the physical state of reactants and products allows us to write an **ionic equation**, one in which the formulas of electrolytes are *written as individual ions.* If, for example, NaCl(*aq*) is a reactant, this electrolyte (dissolves in water to form ions) is written Na^+ (*aq*) and Cl^- (*aq*). For the reaction above between $CoCl_2$ and $Ca(OH)_2$, the ionic equation is

$$Co^{2+}(aq) + 2Cl^-(aq) + Ca^{2+}(aq) + 2OH^-(aq) \longrightarrow Co(OH)_2(s) + Ca^{2+}(aq) + 2Cl^-(aq)$$

Ionic equation

Note that this ionic equation is balanced in terms of atoms and with respect to charges (the net charge is zero on both the reactant and the product side).

If you look at the reactants and products in the ionic reaction equation above, you see that the ions Cl^- and Ca^{2+} remain unchanged. Ions such as these, which *appear identically on both sides of an ionic equation*, are called **spectator ions.** When the spectator ions are removed from an ionic equation, a **net ionic equation** results. The net ionic equation represents the actual change that takes place during a reaction. For the reaction between $CoCl_2$ and $Ca(OH)_2$, the net ionic equation is

- Spectator ions appear identically on both sides of an ionic reaction equation. Removing the spectator ions gives the net ionic equation.
- In an ionic equation, electrolytes are written as individual ions.

$$Co^{2+}(aq) + 2OH^-(aq) \longrightarrow Co(OH)_2(s)$$

Net ionic equation

A precipitation reaction (one in which a precipitate forms) takes place between sodium carbonate (Na_2CO_3) and magnesium nitrate ($Mg(NO_3)_2$). The balanced equation for this reaction is

$$Na_2CO_3 + Mg(NO_3)_2 \longrightarrow 2NaNO_3 + MgCO_3$$

To write the corresponding ionic equation, we must know which reactants and products are soluble in water. From Table 7.1, we learn that all nitrates are water soluble, as are carbonates that contain NH_4^+ or alkali metal cations, such as Na^+. All other carbonates are water insoluble. This information allows us to rewrite the reaction equation, this time specifying which reactants are aqueous and which are not.

$$Na_2CO_3(aq) + Mg(NO_3)_2(aq) \longrightarrow 2NaNO_3(aq) + MgCO_3(s)$$

From this, the ionic equation and net ionic equations can be written.

$$2Na^+(aq) + CO_3^{2-}(aq) + Mg^{2+}(aq) + 2NO_3^-(aq) \longrightarrow 2Na^+(aq) + 2NO_3^-(aq) + MgCO_3(s)$$

Ionic equation

$$CO_3^{2-}(aq) + Mg^{2+}(aq) \longrightarrow MgCO_3(s)$$

Net ionic equation

Precipitation reactions play many important roles in living things. Antibodies produced by your immune system, for example, combine with unwanted molecules to form large water-insoluble precipitates that are easily captured and destroyed.

SAMPLE PROBLEM 7.3

Precipitation reactions

In water, iron(III) bromide reacts with potassium phosphate to form potassium bromide and iron(III) phosphate, one of which is a solid. Write a balanced equation, an ionic equation, and a net ionic equation for this reaction.

Strategy

To solve this problem, you must first determine the formula of each reactant and product. Next, refer to Table 7.1 to decide which product is water insoluble.

Solution

From Table 7.1 we can determine that $FeBr_3$, K_3PO_4, and KBr are water soluble, while $FePO_4$ is not.

$$FeBr_3(aq) + K_3PO_4(aq) \longrightarrow FePO_4(s) + 3KBr(aq)$$

Balanced equation

$$Fe^{3+}(aq) + 3Br^-(aq) + 3K^+(aq) + PO_4^{3-}(aq) \longrightarrow FePO_4(s) + 3K^+(aq) + 3Br^-(aq)$$

Ionic equation

$$Fe^{3+}(aq) + PO_4^{3-}(aq) \longrightarrow FePO_4(s)$$

Net ionic equation

PRACTICE PROBLEM 7.3

Write the balanced equation, ionic equation, and net ionic equation for the precipitation reaction that takes place when an aqueous solution of lead(II) nitrate is mixed with one of ammonium chloride. Identify the precipitate.

Gas Producing Reactions

In an aqueous solution, ions will sometimes react to form gases. For example, hydrogen chloride (HCl) combines with sodium sulfide (Na_2S) to produce hydrogen sulfide (H_2S), the gas that gives rotten eggs their awful odor.

$$2HCl(aq) + Na_2S(aq) \longrightarrow H_2S(g) + 2NaCl(aq)$$

In this equation, the compounds on the reactant side (HCl and Na_2S) exist as dissolved, aqueous ions, as does NaCl on the product side. Although HCl is a molecule and not an ionic compound, most of the HCl in an aqueous solution is found in an ionic form as $H^+(aq)$ and $Cl^-(aq)$ because HCl is a strong acid (the chemistry of acids is introduced in Chapter 9).

As with precipitation reactions, we can write an ionic and a net ionic equation for the reaction between HCl and Na_2S. While Table 7.1 does not provide information about the water solubility of molecules (HCl and H_2S) or of ionic sulfides (Na_2S), the information that we need for writing these two equations is provided in the paragraph above.

$$2H^+(aq) + 2Cl^-(aq) + 2Na^+(aq) + S^{2-}(aq) \longrightarrow H_2S(g) + 2Na^+(aq) + 2Cl^-(aq)$$

Ionic equation

$$2H^+(aq) + S^{2-}(aq) \longrightarrow H_2S(g)$$

Net ionic equation

The reaction that takes place between hydrogen chloride and sodium bicarbonate ($NaHCO_3$) produces carbonic acid (H_2CO_3), which rapidly falls apart to give carbon dioxide gas and water (Figure 7.5). This reaction is important in baking, because sodium bicarbonate (baking soda) reacts with acids present in dough to produce the CO_2 gas that causes the dough to rise.

FIGURE 7.5

$$HCl(aq) + NaHCO_3(aq) \rightleftharpoons H_2CO_3(aq) + NaCl(aq)$$
$$H_2CO_3(aq) \rightleftharpoons CO_2(g) + H_2O(l)$$

A gas-forming reaction
The reaction that takes place when aqueous HCl (left) is mixed with aqueous $NaHCO_3$ (right) results in the formation of CO_2 gas (bubbles).

HARD WATER

The water supplied to many cities comes into underground contact with minerals such as limestone ($CaCO_3$), magnesite ($MgCO_3$), siderite ($FeCO_3$), and dolomite ($CaCO_3$ and $MgCO_3$). While these minerals are insoluble in water, when exposed to the carbon dioxide typically dissolved in water, each reacts to produce aqueous ions.

$$CaCO_3(s) + CO_2(aq) + H_2O(l) \longrightarrow Ca^{2+}(aq) + 2HCO_3^-(aq)$$
$$MgCO_3(s) + CO_2(aq) + H_2O(l) \longrightarrow Mg^{2+}(aq) + 2HCO_3^-(aq)$$
$$FeCO_3(s) + CO_2(aq) + H_2O(l) \longrightarrow Fe^{2+}(aq) + 2HCO_3^-(aq)$$

The resulting hard water (water containing Ca^{2+}, Mg^{2+}, and Fe^{2+}) is unsuitable for many home and industrial uses. One problem with hard water is that the dissolved metal ions react with soaps to form water-insoluble precipitates (soap scum). This reduces the cleansing ability of soaps (Figure 7.6).

Another problem with hard water is that heating it causes the reactions shown above to run in reverse, resulting in the formation of solids. These precipitates, called scale, form on the inside of water pipes, water heaters, and tea kettles.

$$Ca^{2+}(aq) + 2HCO_3^-(aq) \longrightarrow CaCO_3(s) + CO_2(aq) + H_2O(l)$$
$$Mg^{2+}(aq) + 2HCO_3^-(aq) \longrightarrow MgCO_3(s) + CO_2(aq) + H_2O(l)$$
$$Fe^{2+}(aq) + 2HCO_3^-(aq) \longrightarrow FeCO_3(s) + CO_2(aq) + H_2O(l)$$

Besides restricting the flow of water through pipes (Figure 7.7), scale makes heat transfer less efficient. The failure of heating elements in water heaters is often caused by the buildup of scale.

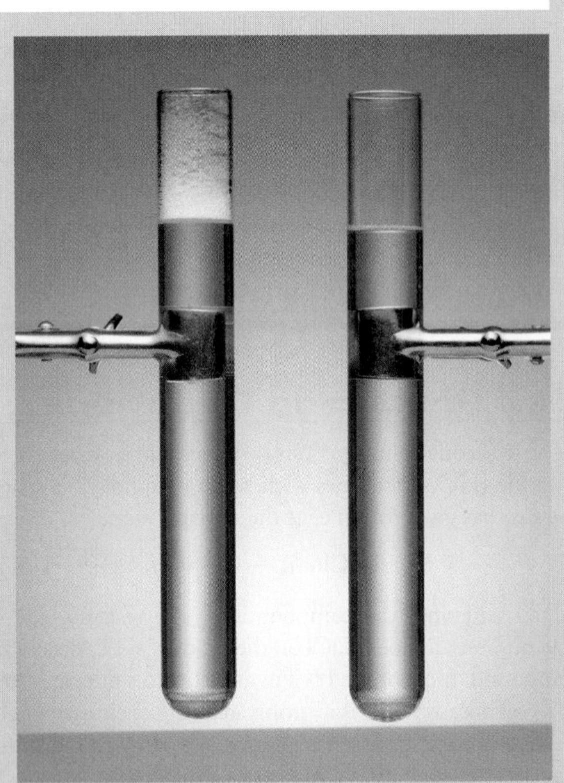

FIGURE 7.6

Hard water
(*Left*) The sudsing ability of soap in soft water. (***Right***) Soap in hard water.

Source: Andy Washnik.

FIGURE 7.7

Scale
When hard water is heated, scale forms on the inside of water pipes.

Source: Tony Freeman/PhotoEdit.

7.3 SOLUBILITY OF GASES IN WATER

In Section 7.1 we saw that gases can be solutes in aqueous solutions and that the higher the temperature, the lower the solubility of a gas. The English chemist William Henry (1775–1836) studied the effect of pressure on the solubility of gases. His observations are summarized by **Henry's law**, which states that *the solubility of a gas in a liquid is proportional to the pressure of the gas over the liquid.* This law predicts that increasing the pressure of a gas increases its solubility by the same proportion—doubling the pressure doubles the solubility, tripling the pressure triples the solubility, and so on. Since carbon dioxide (CO_2) has a solubility of 0.169 g/100 g H_2O at 1 atm and 20°C, the solubility will increase to 0.338 g/100 g at 2 atm and 20°C.

■ The higher the pressure of a gas above a liquid, the greater its solubility.

Carbonated beverages provide an example of how Henry's law relates to everyday life. The gas inside an unopened bottle of carbonated water is at a pressure slightly above atmospheric pressure. When the bottle is opened you hear the hiss of escaping pressurized gas, and bubbles of CO_2 immediately form in the liquid. This happens because the pressure of CO_2 over the liquid has been reduced, making CO_2 less soluble. Any CO_2 in excess of the solubility limit leaves the solution as bubbles (Figure 7.8).

FIGURE 7.8

Gas solubility
When a bottle of carbonated water is opened, the pressure of CO_2 over the liquid is reduced to that of the surrounding atmospheric pressure. CO_2 has a lower solubility at this reduced pressure and it rapidly leaves the solution as bubbles.

Source: Charles D. Winters/ Photo Researchers, Inc.

Some gases have a very high solubility in water because they react with this solvent and are converted into a water-soluble product. For example, the following reactions contribute to the fact that water holds over 25 times more CO_2 and over 80,000 times more ammonia (NH_3) than nitrogen gas (N_2) at 25°C and 1 atm.

$$CO_2(g) + H_2O(l) \rightleftharpoons H_2CO_3(aq) \rightleftharpoons H^+(aq) + HCO_3^-(aq)$$

$$NH_3(g) + H_2O(l) \rightleftharpoons NH_4^+(aq) + OH^-(aq)$$

The ability of a liquid to absorb a gas can also be enhanced by interactions between the gas and other substances present in the liquid. Blood, for example, takes up about 66 times more O_2 than does water, due to the interaction of O_2 with hemoglobin, an oxygen-transporting compound that is present in red blood cells (Figure 7.9).

The solubility of gases in blood is a factor in how quickly gaseous anesthetic drugs take effect. The lower the solubility of a gaseous drug in blood, the faster it moves out of the blood and into the appropriate tissue. Of the two commonly used gaseous anesthetics nitrous oxide (N_2O) and halothane ($C_2HBrClF_3$), nitrous oxide is about five times less soluble in blood and has a faster onset of action.

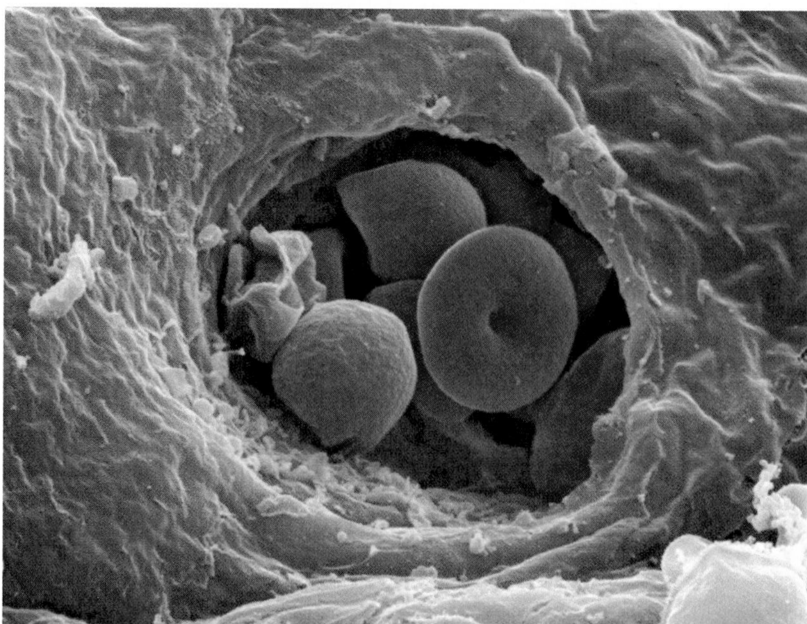

FIGURE 7.9

Red blood cells moving through a capillary
Each red blood cell contains about 300 million hemoglobin molecules.

Source: Science Photo Library/ Photo Researchers, Inc.

HEALTH Link — *Hyperbaric Medicine*

Breathing $O_2(g)$ under hyperbaric (*high pressure*) conditions causes higher than normal levels of this gas to dissolve in the blood. As shown in Section 7.3, the amount of gas that dissolves in a liquid is proportional to the pressure of the gas over the liquid. Hyperbaric O_2 treatment is an effective way to treat carbon monoxide (CO) poisoning because elevated O_2 concentrations help to displace CO from hemoglobin, the protein that binds and carries O_2. Hyperbaric O_2 may also be used to treat gas gangrene and tetanus, because the bacteria that cause these conditions cannot survive in an oxygen-rich environment. This treatment is also used to speed the healing of some skin grafts and wounds. Slow decompression from hyperbaric conditions treats the decompression sickness sometimes experienced by divers.

Modern hyperbaric chambers can hold 16 or more patients and have many of the comforts of home, including televisions, radios, and CD and DVD players (Figure 7.10). Pet-sized chambers are even available for veterinary use.

While hyperbaric oxygen treatment has many benefits, it is also potentially harmful. A large amount of O_2 dissolved in the blood can result in the formation of harmful reaction products that damage important biological compounds. This may cause nervous system damage and other severe health problems.

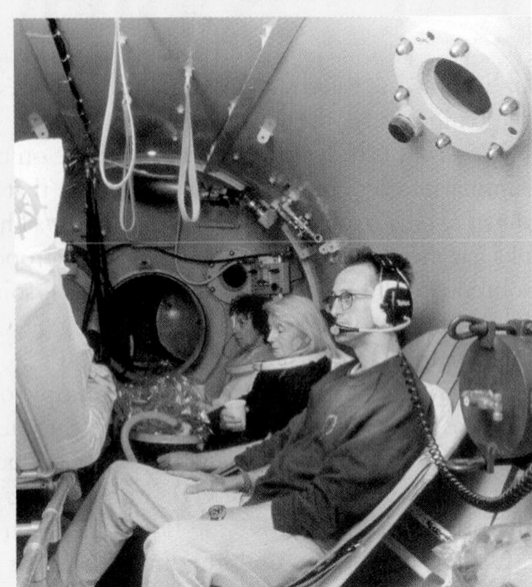

FIGURE 7.10

A hyperbaric (high pressure) chamber
High pressure moves more O_2 into the blood.

Source: James King-Holms/Science Photo Library/Photo Researchers, Inc.

7.4 ORGANIC COMPOUNDS

Alkane molecules readily mix with one another, so a solution can be prepared by adding pentane (solute) to hexane (solvent). As pure substances (Figure 7.11*a*) these molecules are held to their neighbors by London forces. When pentane is mixed with hexane, the molecules intermingle and new London force interactions take place between pentane and hexane molecules (Figure 7.11*b*). Pentane is soluble in hexane because the solute and solvent molecules are about the same size and interact with one another. As explained in Section 7.1, *like dissolves like* and in this case the "like" molecules are hydrocarbons that interact through London forces.

Hydrocarbons do not dissolve in water. These nonpolar molecules do not interact effectively with the polar water molecules, so when the two are mixed, hydrocarbon molecules will stay associated with one another (London forces) and water molecules will do the same (hydrogen bonding).

Although hydrocarbons are not water soluble, many organic compounds are. Ethanol (CH_3CH_2OH), for example, mixes with water in all proportions. In pure ethanol and pure water (Figure 7.12*a*), the molecules are held to one another primarily by hydrogen bonds. When ethanol is mixed with water, some of the hydrogen bonds in the pure substances are replaced by hydrogen bonds between alcohol and water molecules (Figure 7.12*b*).

The more carbon atoms an alcohol molecule has, the less water soluble it becomes. While ethanol is soluble in water in all amounts, only very small amounts of 1-octanol will dissolve in this solvent.

$$CH_3CH_2CH_2CH_2CH_2CH_2CH_2CH_2OH$$

1-Octanol

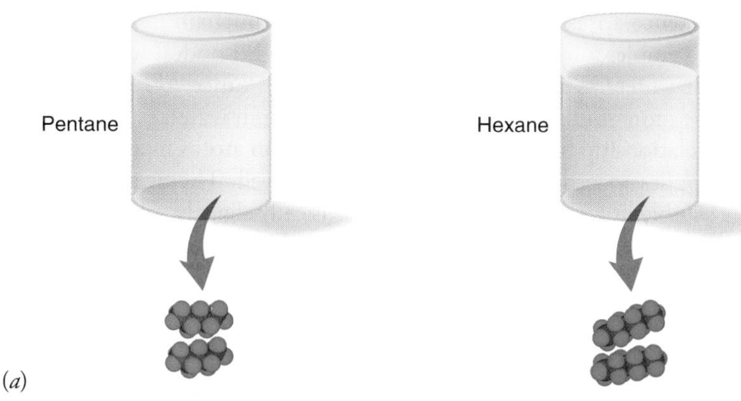

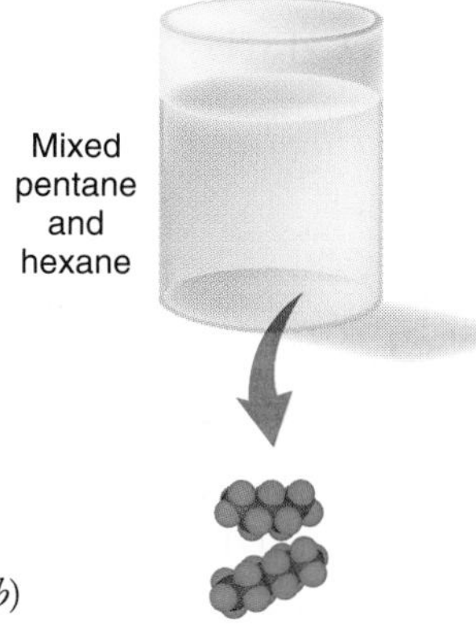

FIGURE 7.11

Solubility and London forces
(*a*) In pure pentane ($CH_3CH_2CH_2CH_2CH_3$) and pure hexane ($CH_3CH_2CH_2CH_2CH_2CH_3$), the molecules are held to one another by London forces. (*b*) When the two hydrocarbons are mixed, London forces are still at work.

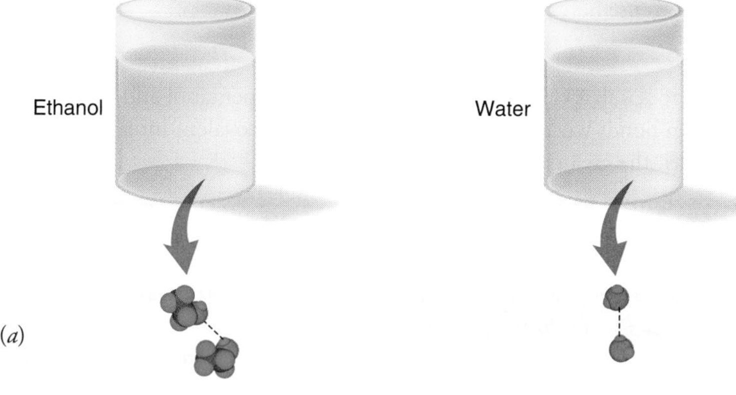

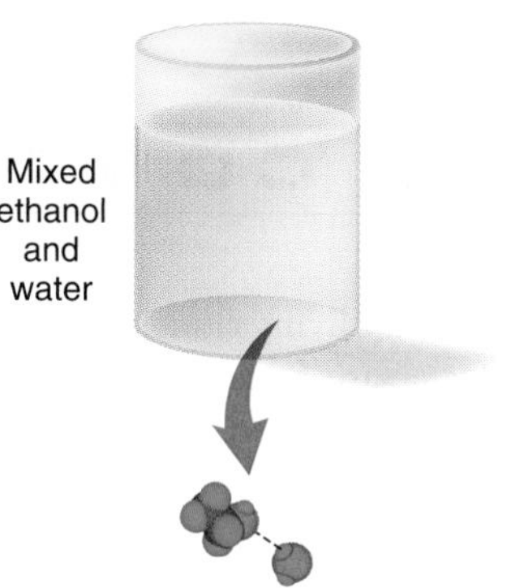

FIGURE 7.12

Solubility and hydrogen bonding
(*a*) In pure ethanol (CH_3CH_2OH) and pure water, the molecules are held to one another primarily by hydrogen bonds. (*b*) When the two compounds are mixed, hydrogen bonds form between alcohol and water molecules.

This difference arises because the longer the carbon chain in the molecule, the more the alcohol resembles a hydrocarbon (interacts through London forces) and the less it resembles water (interacts through hydrogen bonding). Since *like dissolves like,* large alcohols tend to be more soluble in nonpolar solvents than in water.

This relationship between the number of carbon atoms in a molecule and its water solubility applies to other organic compounds as well. The acetic acid in vinegar is completely soluble in water, while octanoic acid is only slightly soluble. Small esters, like ethyl methanoate, are slightly soluble in water, while larger esters are insoluble.

$CH_3C(=O)-OH$
Acetic acid

$CH_3CH_2CH_2CH_2CH_2CH_2CH_2C(=O)-OH$
Octanoic acid

$H-C(=O)-OCH_2CH_3$
Ethyl methanoate

$CH_3CH_2CH_2CH_2CH_2CH_2C(=O)-OCH_2CH_2CH_2CH_3$
Butyl heptanoate

SAMPLE PROBLEM 7.4

The solubility of organic compounds

Pentanoic acid and 1-pentanol have the same number of carbon atoms, but pentanoic acid has a greater solubility in water. Explain.

$CH_3CH_2CH_2CH_2C(=O)-OH$
Pentanoic acid

$CH_3CH_2CH_2CH_2CH_2OH$
1-Pentanol

Strategy

The rule to use here is *like dissolves like.*

Solution

Pentanoic acid has one more oxygen atom than 1-pentanol and can form more hydrogen bonds with water. The stronger the noncovalent interactions between a solute and water, the greater the solubility of the solute.

PRACTICE PROBLEM 7.4

Which molecule is more soluble in water, $CH_3CH_2CH_2OH$ or $CH_3CH_2CH_2SH$?

7.5 BIOCHEMICAL COMPOUNDS

The compounds found in living things can be placed into one of three solubility classes. Those *soluble in water* are **hydrophilic** (water loving), those *insoluble in water* are **hydrophobic** (water fearing), and those that fall somewhere in between are **amphipathic** (have both hydrophilic and hydrophobic parts). A few compounds from each of these groups will be introduced in the paragraphs that follow, with more detailed discussions saved for later chapters.

Hydrophilic Compounds

Since *like dissolves like*, to be hydrophilic a molecule must resemble water. For the large molecules typically found in living things, this means that the water solubility of polar groups and groups able to form hydrogen bonds must be more significant than the water insolubility due to carbon chains, which interact through London forces. Glucose, fructose, and other simple sugars (Figure 7.13) are hydrophilic because they contain a large number of —OH groups that are able to interact with water.

Amino acids, the building blocks of proteins, are also hydrophilic (Figure 7.14), due to the presence of nitrogen, oxygen, and hydrogen atoms able to form hydrogen bonds with water. The water solubility of amino acids varies somewhat from one to the next, depending on the other atoms and groups that are present in the molecule. Phenylalanine and valine, having more carbon atoms than glycine and alanine, are slightly less soluble in water.

■ Hydrophilic compounds are water soluble.

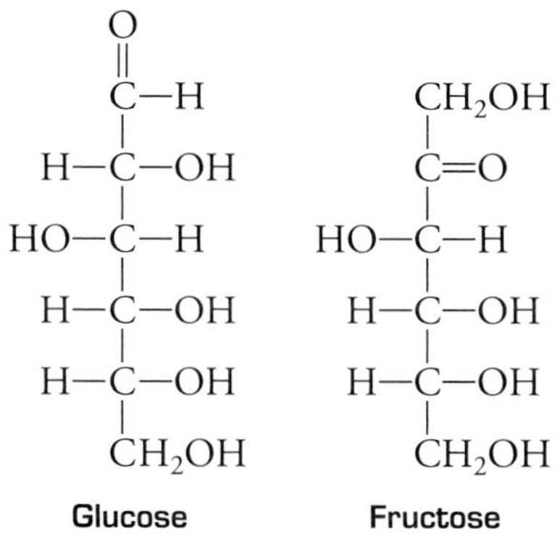

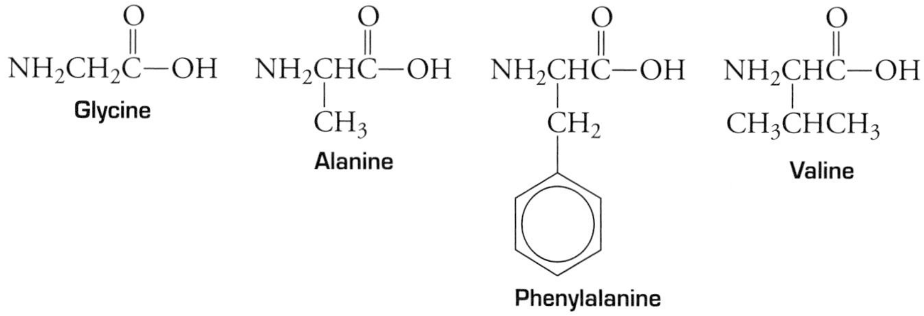

Glucose: O=C—H, H—C—OH, HO—C—H, H—C—OH, H—C—OH, CH_2OH

Fructose: CH_2OH, C=O, HO—C—H, H—C—OH, H—C—OH, CH_2OH

Glucose **Fructose**

FIGURE 7.13

Simple sugars
Glucose, fructose, and other simple sugars are hydrophilic (soluble in water).

$NH_2CH_2C(=O)—OH$ **Glycine**

$NH_2CHC(=O)—OH$ with CH_3 **Alanine**

$NH_2CHC(=O)—OH$ with CH_2—phenyl ring **Phenylalanine**

$NH_2CHC(=O)—OH$ with CH_3CHCH_3 **Valine**

FIGURE 7.14

Amino acids
All amino acids are hydrophilic, but some (glycine and alanine) are more soluble in water than others (phenylalanine and valine).

Hydrophobic Compounds

Stearic acid, oleic acid, and linolenic acid are fatty acids—carboxylic acids having between 12 and 20 carbon atoms (Figure 7.15). A quick inspection of the structure of stearic acid shows why this and other fatty acids are hydrophobic (water insoluble). While one end of the molecule has the polar carboxyl group, the molecule consists mostly of a chain of carbon atoms. This makes a fatty acid very similar to a water-insoluble hydrocarbon molecule.

■ Hydrophobic compounds are water insoluble.

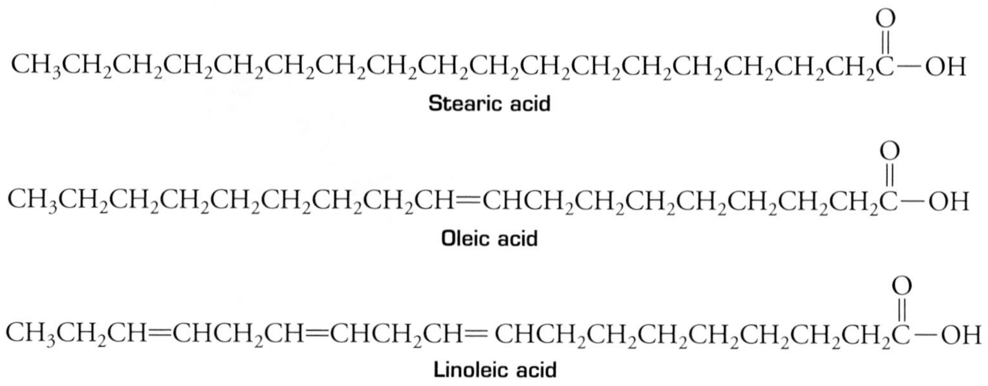

$CH_3CH_2CH_2CH_2CH_2CH_2CH_2CH_2CH_2CH_2CH_2CH_2CH_2CH_2CH_2CH_2CH_2C(=O)—OH$

Stearic acid

$CH_3CH_2CH_2CH_2CH_2CH_2CH_2CH_2CH{=}CHCH_2CH_2CH_2CH_2CH_2CH_2CH_2C(=O)—OH$

Oleic acid

$CH_3CH_2CH{=}CHCH_2CH{=}CHCH_2CH{=}CHCH_2CH_2CH_2CH_2CH_2CH_2CH_2C(=O)—OH$

Linoleic acid

FIGURE 7.15

Fatty acids
Stearic acid, oleic acid, and linolenic acid are typical fatty acids.

Amphipathic Compounds

In the presence of aqueous sodium hydroxide (NaOH), fatty acids lose H^+ to form carboxylate anions, which combine with Na^+ from NaOH to form the ionic compounds that we call soap. Through this reaction, oleic acid can be converted into the soap named sodium oleate (Figure 7.16). The negative charge makes the polar end of the oleate anion very hydrophilic, enough so that the soap is amphipathic—it has both a hydrophilic and a hydrophobic nature.

- Amphipathic compounds have a hydrophobic part and a hydrophilic part.

FIGURE | 7.16

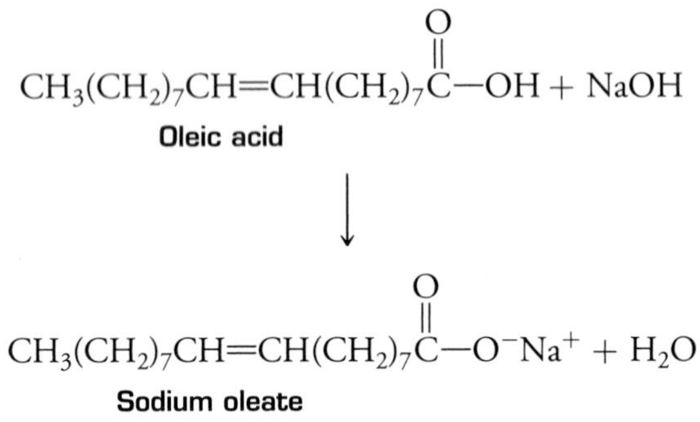

Soaps
Reacting oleic acid with NaOH produces sodium oleate, a typical soap. In these condensed structures, $(CH)_7$ indicates seven repeats of CH_2.

$$CH_3(CH_2)_7CH{=}CH(CH_2)_7\overset{\overset{\displaystyle O}{\|}}{C}{-}OH + NaOH$$

Oleic acid

$$\downarrow$$

$$CH_3(CH_2)_7CH{=}CH(CH_2)_7\overset{\overset{\displaystyle O}{\|}}{C}{-}O^-Na^+ + H_2O$$

Sodium oleate

When a drop of soap is carefully spread across the surface of water a monolayer (single layer) forms, with the hydrophilic end of each soap anion pointing down into the water and the hydrophobic end pointing up into the air (Figure 7.17*a*). Soaps also form micelles, tiny spheres in which hydrophilic ends make up the outer surface and hydrophobic ends cluster in the center (Figure 7.17*b*). The interaction of the nonpolar micelle interior with grease and other water-insoluble substances is what gives soaps their cleansing abilities. Interaction with the Ca^{2+}, Mg^{2+}, and Fe^{2+} ions present in hard water converts the amphipathic soap anion into a water-insoluble ionic compound (see Hard Water).

FIGURE | 7.17

Soaps in water
When added to water, soaps can form (*a*) a monolayer or (*b*) a micelle.

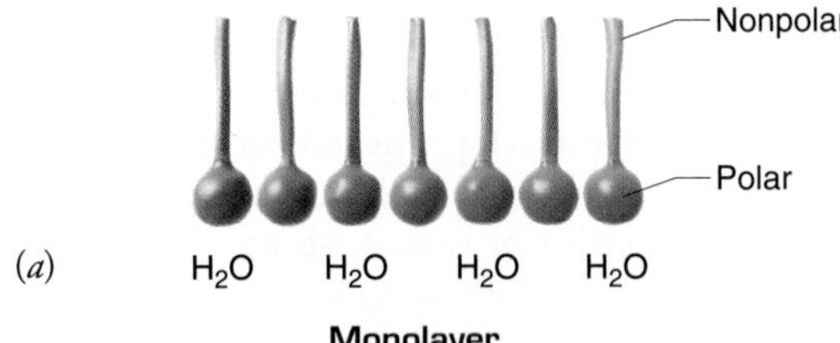

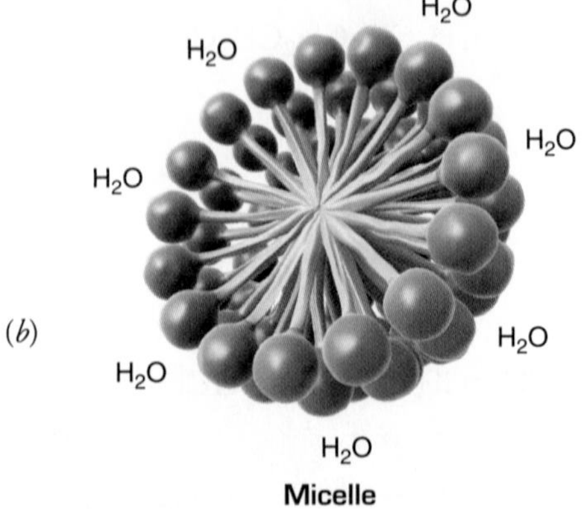

SAMPLE PROBLEM 7.5

Identifying hydrophilic, hydrophobic, and amphipathic compounds

Based on its structure, do you expect vitamin C to be hydrophobic or hydrophilic? Explain.

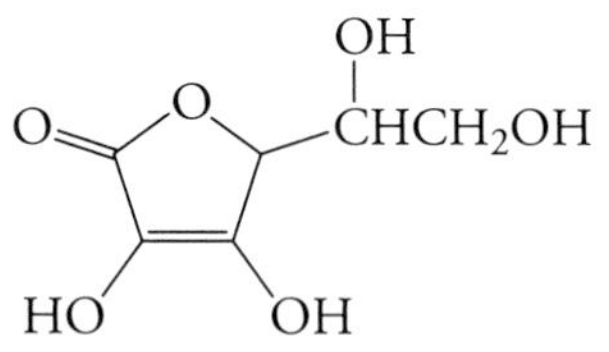

OH
|
$CHCH_2OH$

O=, O, HO, OH

Vitamin C

Strategy

In a hydrophobic compound the structure is predominantly nonpolar. In a hydrophilic compound, groups capable of interacting with water molecules predominate.

Solution

Hydrophilic. Vitamin C has a a high proportion of polar groups that are able to form hydrogen bonds with water.

PRACTICE PROBLEM 7.5

Sodium lauryl sulfate is a detergent often present in shampoos. Identify the hydrophilic and hydrophobic parts of this amphipathic compound.

$$CH_3CH_2CH_2CH_2CH_2CH_2CH_2CH_2CH_2CH_2CH_2CH_2O{-}\overset{\overset{\displaystyle O}{\|}}{\underset{\underset{\displaystyle O}{\|}}{S}}{-}O^-Na^+$$

Sodium lauryl sulfate

HEALTH Link: *Prodrugs*

Some drugs are administered as prodrugs, inactive compounds that, in the body, are converted to fully active drugs. In many cases, prodrugs are used to overcome problems with solubility. For example, adrenaline is effective at treating glaucoma, but its use is limited because it is not hydrophobic enough to pass into the eye when administered as eyedrops. A prodrug form of adrenaline, called dipivefrin, is more hydrophobic and readily dissolves in eye tissue. Once in the eye, enzymes catalyze the hydrolysis (Section 6.4) of the two ester groups in dipivefrin, releasing adrenaline (Figure 7.18*a*).

Another example involves the antibiotic chloramphenicol. This drug has an extremely unpleasant taste when administered orally. Chloramphenicol palmitate, a prodrug ester made by combining chloramphenicol with a fatty acid (palmitic acid), has a much lower water solubility than chloramphenicol. Because it does not dissolve in saliva, this prodrug has little taste. Once in the body, it is hydrolyzed to yield chloramphenicol and palmitate ion (Figure 7.18*b*).

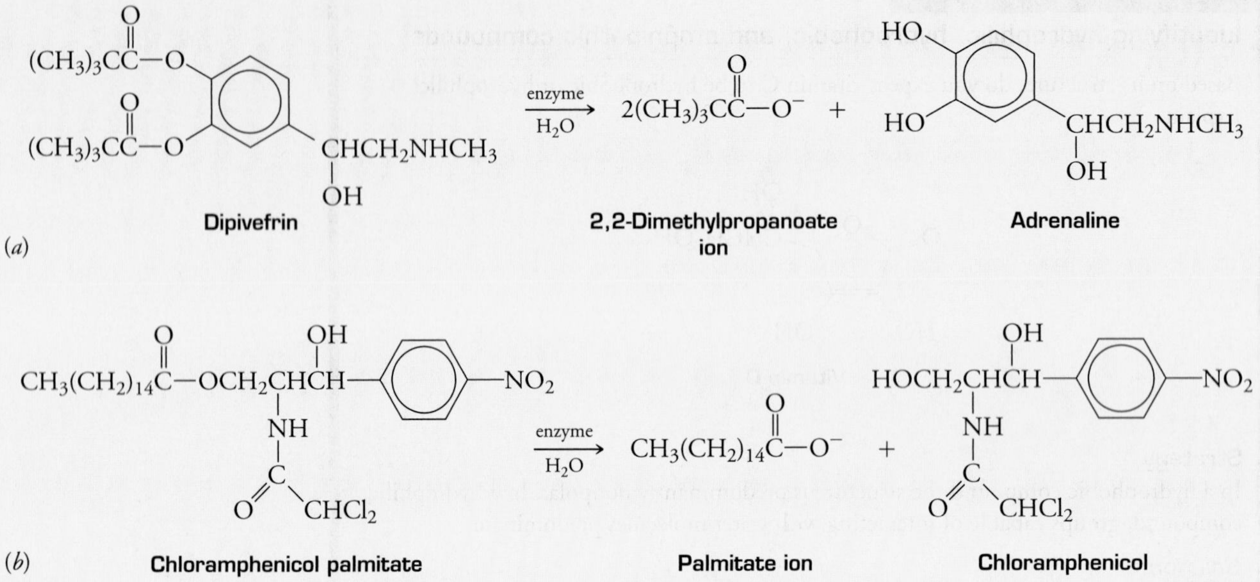

$$\text{Dipivefrin} \xrightarrow[H_2O]{\text{enzyme}} 2(CH_3)_3C\overset{O}{\overset{\|}{C}}{-}O^- + \text{Adrenaline}$$

(*a*) Dipivefrin — 2,2-Dimethylpropanoate ion — Adrenaline

$$\text{Chloramphenicol palmitate} \xrightarrow[H_2O]{\text{enzyme}} CH_3(CH_2)_{14}\overset{O}{\overset{\|}{C}}{-}O^- + \text{Chloramphenicol}$$

(*b*) Chloramphenicol palmitate — Palmitate ion — Chloramphenicol

FIGURE 7.18 **Prodrugs**
In the body, the prodrugs (*a*) dipivefrin and (*b*) chloramphenicol palmitate are hydrolyzed to produce active drugs.

7.6 CONCENTRATION

When dealing with solutions, it is often important to known how much solute is present. Blood tests are run, for example, to check the levels of various blood serum solutes. These solutes include glucose (normally 80–100 milligrams per deciliter of serum), lactate (normally 0.6–1.8 millimoles per liter), and testosterone (in men, normally 300–1000 nanograms per milliliter).

The term **concentration** is used to refer to the *amount of solute that is dissolved in a solvent.* If we are not concerned with exact amounts of solute that are present, the terms **saturated** (*holds the maximum amount of solute that can be dissolved at a particular temperature*) and **unsaturated** (*holds less than a saturating amount of solute*) may be used to indicate concentration. In the case of sodium chloride (NaCl), which has a solubility of 35.9 g/100 g H_2O at 20°C, an aqueous solution that contains this amount of NaCl is saturated, while one that holds any amount less than 35.9 g/100 g H_2O is unsaturated (Figure 7.4).

■ Concentration is the amount of solute dissolved in a solvent.

The terms dilute and concentrated are used to indicate relative concentration—a dilute solution contains less solute than a more concentrated one. If a tea bag is dipped into a cup of hot water two or three times, a dilute tea solution has been prepared. If the tea bag is put back in the cup and allowed to stand for a few minutes, the solution will be more concentrated.

To be specific about the amount of solute present in a solvent, concentration units must be used. Depending on the application, any number of different units are available. In the paragraphs that follow we will look at units that you are most likely to encounter.

Percent

Percent means the same thing as "parts per hundred," so when percent is used as a concentration unit, the number of parts of solute present in every 100 parts of solution is being specified. The three commonly used percent measurements are **weight/volume percent**, **volume/volume percent**, and **weight/weight percent**.

Weight/volume percent or % (w/v) indicates the mass of solute (in grams) per volume of solution (in milliliters).

$$\text{Weight/volume percent} = \frac{\text{g of solute}}{\text{mL of solution}} \times 100$$

Volume/volume percent or % (v/v) specifies the volume of solute (in milliliters) per volume of solution (in milliliters).

$$\text{Volume/volume percent} = \frac{\text{mL of solute}}{\text{mL of solution}} \times 100$$

Weight/weight percent or % (w/w) is the mass of solute (in grams) per mass of solution (in grams).

$$\text{Weight/weight percent} = \frac{\text{g of solute}}{\text{g of solution}} \times 100$$

Let us see an example of how % (w/v) can be calculated. Potassium bitartrate, an ionic compound present in wine sediment, can be used as a laxative. What is the % (w/v) of 80.0 mL of solution that contains 0.250 g of potassium bitartrate?

To solve this problem we use the weight/volume percent equation above, insert the values for grams of solute and milliliters of solution, and solve.

$$\begin{aligned}\text{Weight/volume percent} &= \frac{\text{g of solute}}{\text{mL of solution}} \times 100 \\ &= \frac{0.250\text{ g potassium bitartrate}}{80.0\text{ mL solution}} \times 100 \\ &= 0.313\%\text{ (w/v)}\end{aligned}$$

SAMPLE PROBLEM 7.6

Calculating percent concentration

Potassium iodide (KI) is used to treat iodine deficiencies. What is the % (w/v) of a 75 mL solution containing 2.0 g of KI?

Strategy

Weight/volume percent is calculated by dividing grams of solute by milliliters of solution and multiplying by 100.

Solution

$$\begin{aligned}\text{Weight/volume percent} &= \frac{\text{g of solute}}{\text{mL of solution}} \times 100 \\ &= \frac{2.0\text{ g KI}}{75\text{ mL of solution}} \times 100 \\ &= 2.7\%\text{ (w/v)}\end{aligned}$$

PRACTICE PROBLEM 7.6

If 25 mL of ethanol are combined with enough water to give 150 mL of solution, what is the % (v/v)?

Parts per Thousand, Parts per Million, and Parts per Billion

For very dilute solutions, the concentration units **parts per thousand** (ppt), **parts per million** (ppm), and **parts per billion** (ppb) are sometimes used (Figure 7.19). As is the case for percent (parts per hundred), each of these concentration units can be defined in terms of weight or volume of solute per given weight or volume of solution. We will use definitions analogous to weight/volume percent. Also, it is worth noting that ppt can be used refer to parts per trillion. In this text, ppt stands only for parts per thousand.

$$\text{Parts per hundred or \% (w/v)} = \frac{\text{g of solute}}{\text{mL of solution}} \times 10^2$$

$$\text{Parts per thousand} = \frac{\text{g of solute}}{\text{mL of solution}} \times 10^3$$

$$\text{Parts per million} = \frac{\text{g of solute}}{\text{mL of solution}} \times 10^6$$

$$\text{Parts per billion} = \frac{\text{g of solute}}{\text{mL of solution}} \times 10^9$$

Many cities add sodium fluoride to their drinking water to help reduce dental cavities. If 25 L of city water contains 0.018 g of sodium fluoride, what is the concentration in parts per thousand? This problem is solved by using the parts per thousand equation shown above. Before doing so, however, the volume of solution must be converted from liters into the milliliters specified in the equation.

The volume of the solution is 25 L, which is equivalent to 2.5×10^4 mL.

$$25\ \cancel{\text{L}} \times \frac{1\ \text{mL}}{1 \times 10^{-3}\ \cancel{\text{L}}} = 2.5 \times 10^4\ \text{mL}$$

Inserting grams of solute and milliliters of solution into the parts per thousand equation gives an answer of 0.00072 ppt.

$$\begin{aligned}\text{Parts per thousand} &= \frac{\text{g of solute}}{\text{mL of solution}} \times 10^3 \\ &= \frac{0.018\ \text{g NaF}}{2.5 \times 10^4\ \text{mL of solution}} \times 10^3 \\ &= 0.00072\ \text{ppt}\end{aligned}$$

(*a*)

(*b*)

FIGURE 7.19 **Parts per thousand, parts per million, and parts per billion** A 30 mg quantity of sucrose dissolved in a swimming pool full of water is equivalent to 0.000001 ppt, 0.001 ppm, and 1 ppb.

Source: (*a*) Andy Washnik; (*b*) Photodisc Red/Getty Images.

SAMPLE PROBLEM 7.7

Calculating parts per thousand and parts per billion

For the NaF solution just described, what is the concentration in parts per billion?

Strategy

To calculate this concentration, grams of solute is divided by milliliters of solution and then multiplied by 10^9.

Solution

As shown previously, the volume of the solution is 2.5×10^4 mL. The concentration of the solution in parts per billion is

$$\begin{aligned}\text{Parts per billion} &= \frac{\text{g of solute}}{\text{mL of solution}} \times 10^9 \\ &= \frac{0.018 \text{ g NaF}}{2.5 \times 10^4 \text{ mL of solution}} \times 10^9 \\ &= 720 \text{ ppb } (7.2 \times 10^2 \text{ ppb})\end{aligned}$$

PRACTICE PROBLEM 7.7

What is the concentration of the solution mentioned directly above, in (a) ppm and (b) % (w/v)?

Molarity

Another widely used concentration unit is **molarity** (M), which is defined as *the number of moles of solute present in each liter of solution*, or

$$\text{Molarity} = \frac{\text{moles of solute}}{\text{liters of solution}}$$

To calculate the molarity of a solution, the number of moles of solute and the volume of the solution must be known (Figure 7.20). The concentration of a solution that contains 0.12 mol of glucose in 0.25 L is

$$\text{Molarity} = \frac{\text{moles of solute}}{\text{liters of solution}} = \frac{0.12 \text{ mol}}{0.25 \text{ L}} = 0.48 \text{ M}$$

FIGURE 7.20

Molarity

Molarity is calculated by dividing the moles of solute by the volume of solution (in liters). Each of the following recipes gives a 1 M solution: (*a*) 0.1 mol of $CuSO_4$ is dissolved in enough water to give 0.1 L of solution, (*b*) 0.5 mol of $CuSO_4$ is dissolved in enough water to give 0.5 L of solution, and (*c*) 1 mol of $CuSO_4$ is dissolved in enough water to give 1 L of solution.

Source: Andy Washnik.

SAMPLE PROBLEM 7.8

Calculating molarity

A solution is prepared by dissolving 0.010 mol of the amino acid alanine in enough water to give a final volume of 75 mL. What is the molarity of the solution?

Strategy

To calculate molarity, you must know the number of moles of solute and the volume of the solution (in L). Here you must convert from milliliters into liters, using a conversion factor based on the equality 1 mL = 1×10^{-3} L.

Solution

The volume of the solution is 0.075 L.

$$75 \text{ \sout{mL}} \times \frac{1 \times 10^{-3} \text{ L}}{1 \text{ \sout{mL}}} = 0.075 \text{ L}$$

Molarity is calculated by dividing moles of solute by liters of solution.

$$\text{Molarity} = \frac{\text{moles of solute}}{\text{liters of solution}} = \frac{0.010 \text{ mol}}{0.075 \text{ L}} = 0.13 \text{ M}$$

PRACTICE PROBLEM 7.8

Veterinarians can use sodium iodide (NaI) to treat animals that have ringworm. What is the molarity of 95.0 mL of an aqueous solution that contains 40.0 g of NaI?

A variety of concentration units are used in medicine. Table 7.2 lists the concentration of six different blood serum solutes, each reported using its typical concentration unit. Most of these reported concentrations are given in terms of mass of solute in a particular volume of serum (mg/dL, ng/L, ng/mL, and g/dL).

TABLE 7.2 CONCENTRATION RANGES FOR SOME BLOOD SERUM SOLUTES

Solute	Normal Range
Glucose (fasting)	80–100 mg/dL
Vitamin B_{12}	160–170 ng/L
Testosterone	300–1000 ng/mL (men)
	30–70 ng/mL (women)
Total protein	6.5–8.3 g/dL
Lactate	0.6–1.8 mmol/L
Na^+	136–146 mEq/L

Let us take a look at the mg/dL unit. Suppose that analysis of a patient's blood shows that 1.00 mL of serum contains 0.00220 g of fibrinogen (a protein involved in blood coagulation). Is this concentration within the normal range of 105–350 mg/dL?

To solve this problem the mass of fibrinogen must be converted to milligrams and the volume of serum converted to deciliters (dL) (100 mL = 1 dL).

$$0.00220 \text{ g} \times \frac{1 \text{ mg}}{1 \times 10^{-3} \text{ g}} = 2.20 \text{ mg}$$

$$1.00 \text{ mL} \times \frac{1 \text{ dL}}{100 \text{ mL}} = 0.0100 \text{ dL}$$

The concentration of fibrinogen, calculated by dividing mg of solute by dL of solution, is within the normal range.

$$\frac{2.20 \text{ mg}}{0.0100 \text{ dL}} = 220 \text{ mg/dL}$$

SAMPLE PROBLEM 7.9

Calculating micrograms per liter

Higher than normal serum levels of troponin I (a protein found in heart tissue) can be used to diagnose a myocardial infarction (heart attack). In a myocardial infarction, heart cells are damaged or destroyed and their contents leak into blood serum. If 5.0×10^{1} μL of a patient's serum contains 4.1×10^{-11} g of troponin I, is this level higher than the normal range of 0–0.4 μg/L?

Strategy

To solve this problem, the mass of solute must first be converted into micrograms and the volume of solution into liters.

Solution

$$4.1 \times 10^{-11} \text{ g} \times \frac{1 \text{ μg}}{1 \times 10^{-6} \text{ g}} = 4.1 \times 10^{-5} \text{ μg}$$

$$5.0 \times 10^{1} \text{ μL} \times \frac{1 \times 10^{-6} \text{ L}}{1 \text{ μL}} = 5.0 \times 10^{-5} \text{ L}$$

$$\frac{4.1 \times 10^{-5} \text{ μg}}{5.0 \times 10^{-5} \text{ L}} = 0.82 \text{ μg/L}$$

A troponin I concentration of 0.82 μg/L is higher than normal.

PRACTICE PROBLEM 7.9

According to Table 7.2, a total protein serum concentration of 6.9 g/dL is within the normal range. Express this concentration as % (w/v).

Two of the units in Table 7.2 pertain to the number of moles of solute in a volume of solution. The first of these, mmol/L, is very similar to the molarity (mol/L) unit described above. The second, mEq/L, uses milliequivalents (mEq) to indicate the amount of solute. An **equivalent (Eq)** is the *number of moles of charges that one mole of a solute contributes to a solution.* For example, 1 mol of Na^+ is equal to 1 Eq of Na^+, because 1 mol of Na^+ (6.02×10^{23} Na^+ ions) contributes 1 mol of positive charges.

One mole of Ca^{2+} is equal to 2 Eq of Ca^{2+}, because 1 mol of Ca^{2+} contributes 2 mol of positive charges (each Ca^{2+} carries a 2+ charge). Similarly, 1 mol of Cl^- equals 1 Eq Cl^- and 1 mol PO_4^{3-} equals 3 Eq PO_4^{3-}.

$$1 \text{ mol } Na^+ = 1 \text{ Eq } Na^+$$

$$1 \text{ mol } Ca^{2+} = 2 \text{ Eq } Ca^{2+}$$

$$1 \text{ mol } Cl^- = 1 \text{ Eq } Cl^-$$

$$1 \text{ mol } PO_4^{3-} = 3 \text{ Eq } PO_4^{3-}$$

If blood serum has a phosphate ion (PO_4^{3-}) concentration of 6.6×10^{-4} M, is this within the normal range of 1.8–2.6 mEq/L? Solving this problem involves converting from moles to equivalents and then from equivalents to milliequivalents. The conversion factor to use for converting moles to equivalents is based on the equality 1 mol PO_4^{3-} = 3 Eq PO_4^{3-}.

$$\frac{6.6 \times 10^{-4} \cancel{\text{mol } PO_4^{3-}}}{\text{L}} \times \frac{3 \text{ Eq } PO_4^{3-}}{1 \cancel{\text{mol } PO_4^{3-}}} = \frac{2.0 \times 10^{-3} \text{ Eq } PO_4^{3-}}{\text{L}}$$

$$\frac{2.0 \times 10^{-3} \cancel{\text{Eq}} \ PO_4^{3-}}{\text{L}} \times \frac{1 \text{ mEq}}{1 \times 10^{-3} \cancel{\text{Eq}}} = \frac{2.0 \text{ mEq } PO_4^{3-}}{\text{L}}$$

A phosphate concentration of 2.0 mEq/L is within the normal range.

SAMPLE PROBLEM 7.10

Calculating milliequivalents per liter

Sometimes, different concentration units are used to describe levels of the same blood solute. The serum concentration of Ca^{2+}, for example, is often reported in either mmol/L or in mEq/L. The normal range of Ca^{2+} in blood serum is 1.15–1.30 mmol/L. Convert this range into mEq/L.

Strategy

Solving this problem requires you to use a conversion factor that relates millimoles of Ca^{2+} to milliequivalents of Ca^{2+}.

Solution

Above, we saw that 1 mol Ca^{2+} = 2 Eq Ca^{2+}. It is also true that 1 mmol Ca^{2+} = 2 mEq Ca^{2+}.

$$\frac{1.15 \cancel{\text{mmol } Ca^{2+}}}{\text{L}} \times \frac{2 \text{ mEq } Ca^{2+}}{1 \cancel{\text{mmol } Ca^{2+}}} = \frac{2.30 \text{ mEq } Ca^{2+}}{\text{L}}$$

$$\frac{1.30 \cancel{\text{mmol } Ca^{2+}}}{\text{L}} \times \frac{2 \text{ mEq } Ca^{2+}}{1 \cancel{\text{mmol } Ca^{2+}}} = \frac{2.60 \text{ mEq } Ca^{2+}}{\text{L}}$$

In serum, Ca^{2+} concentrations normally range from 2.30 to 2.60 mEq/L.

PRACTICE PROBLEM 7.10

Convert the normal serum concentration range of Ca^{2+} (1.15–1.30 mmol/L) into mg/dL.

One important use of concentration is in letting you calculate the amount of solute present in a particular amount of solution. To carry out such calculations, concentration units can be used as conversion factors. Suppose, for example, that you are asked to calculate the number of moles of glucose present in 4.0 L of 0.17 M glucose solution. To convert from liters of solution to moles of glucose, we use one of two conversion factors that relate moles and liters.

The glucose concentration is 0.17 M

$$\text{conversion factors: } \frac{0.17 \text{ mol glucose}}{1 \text{ L of solution}} \quad \text{and} \quad \frac{1 \text{ L of solution}}{0.17 \text{ mol glucose}}$$

The number of moles of glucose in 4.0 L of solution is calculated as follows:

$$4.0 \cancel{\text{L}} \text{ of solution} \times \frac{0.17 \text{ mol glucose}}{1 \cancel{\text{L}} \text{ of solution}} = 0.68 \text{ mol glucose}$$

Calculations involving other concentration units work the same way. For example, suppose that blood serum has a fibrinogen concentration of 109 mg/dL. How many milligrams of fibrinogen are present in a 0.0750 L sample? The conversion factor to use is based on the concentration.

The fibrinogen concentration is 109 mg/dL

$$\text{conversion factors: } \frac{109 \text{ mg fibrinogen}}{1 \text{ dL of solution}} \quad \text{and} \quad \frac{1 \text{ dL of solution}}{109 \text{ mg fibrinogen}}$$

Before converting 0.0750 L of solution into milligrams of fibrinogen, the initial volume (liters) must be switched into the volume unit used in the conversion factor (deciliters).

$$0.0750 \cancel{\text{L}} \text{ of solution} \times \frac{1 \text{ dL}}{1 \times 10^{-1} \cancel{\text{L}} \text{ of solution}} = 0.750 \text{ dL}$$

$$0.750 \cancel{\text{dL}} \times \frac{109 \text{ mg fibrinogen}}{\cancel{\text{dL}}} = 81.8 \text{ mg fibrinogen}$$

SAMPLE PROBLEM 7.11

Using concentration as a conversion factor

A sample of blood serum is tested and found to contain testosterone at a concentration of 625 ng/mL. How many grams of testosterone are present in 50.0 mL of this serum?

Strategy

The concentration of 625 ng/mL is used as a conversion factor. Check to be sure that your answer is reported in grams.

Solution

Multiplying the volume (50.0 mL) by the concentration gives an answer with units of nanograms.

$$50.0 \cancel{\text{mL}} \times \frac{625 \text{ ng testosterone}}{\cancel{\text{mL}}} = 3.13 \times 10^4 \text{ ng testosterone}$$

Multiplying this value by a conversion factor that relates grams and nanograms gives the final answer.

$$3.13 \times 10^4 \cancel{\text{ng}} \text{ testosterone} \times \frac{1 \times 10^{-9} \text{ g}}{\cancel{\text{ng}}} = 3.13 \times 10^{-5} \text{ g testosterone}$$

PRACTICE PROBLEM 7.11

A sample of blood serum is tested and found to contain a total uric acid concentration of 4.2 mg/dL. How many grams of uric acid are present in 0.025 L of this serum?

7.7 DILUTION

When *more solvent is added* to a solution, it has been diluted. **Dilution** is an easy way to reduce the concentration of a solution and is often a much easier way to prepare a solution of known concentration than starting from scratch by measuring out solute. Dilution is important in health care because some drugs must be diluted to the proper concentration before being administered (Figure 7.21).

A useful equation to use when doing dilutions is

$$V_{\text{original}} \times C_{\text{original}} = V_{\text{final}} \times C_{\text{final}}$$

where V is volume and C is concentration. It does not make any difference which volume and concentration units are used, as long as the same units are used for the original solution (the one being diluted) and the final solution (the new one being made).

For example, if you have 0.50 L of a 0.24 M aqueous glucose solution and add enough water to give a final volume of 1.0 L, the new diluted solution has a concentration of 0.12 M.

$$V_{\text{original}} \times C_{\text{original}} = V_{\text{final}} \times C_{\text{final}}$$

$$C_{\text{final}} = \frac{V_{\text{original}} \times C_{\text{original}}}{V_{\text{final}}}$$

$$C_{\text{final}} = \frac{0.50\ \cancel{\text{L}} \times 0.24\ \text{M}}{1.0\ \cancel{\text{L}}} = 0.12\ \text{M}$$

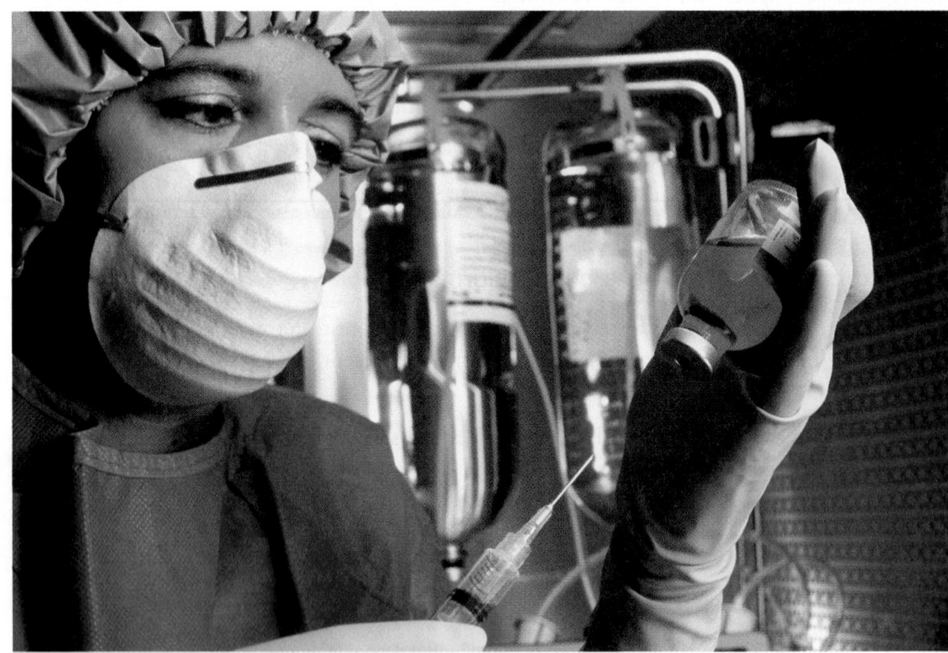

FIGURE | 7.21

Dilution
Before being used, many injectable drugs must be diluted.

Source: Stone/Getty Images.

SAMPLE PROBLEM 7.12

Dilution calculations involving molarity

You begin with 25 mL of a 1.8 M aqueous LiCl solution and add enough water to give a final volume of 35 mL. What is the new concentration?

Strategy

This problem gives you values for three of the four variables in the equation below. You must solve for the fourth.

$$V_{\text{original}} \times C_{\text{original}} = V_{\text{final}} \times C_{\text{final}}$$

Solution

The unknown quantity in this problem is the new concentration, C_{final}. Rearranging the equation above to solve for C_{final} gives

$$C_{final} = \frac{V_{original} \times C_{original}}{V_{final}}$$

Substituting the given values,

$$C_{final} = \frac{25\ \cancel{mL} \times 1.8\ M}{35\ \cancel{mL}} = 1.3\ M$$

Practice Problem 7.12

Another 25 mL of a 1.8 M aqueous LiCl solution is diluted to a final volume of 350 mL. What is the new concentration?

Sample Problem 7.13

Dilution calculations involving % (w/v)

A drug is sold at a concentration of 5.0% (w/v). Before use, 0.50 mL of the drug must be diluted to give a solution with a concentration of 0.83% (w/v). What is the final volume of the diluted solution?

Strategy

Use the same strategy here as in Sample Problem 7.12. All that is different is the concentration unit.

Solution

The unknown quantity in this problem is the final volume, V_{final}.

$$V_{final} = \frac{V_{original} \times C_{original}}{C_{final}}$$

Substituting and solving give

$$V_{final} = \frac{0.50\ mL \times 5.0\ \cancel{\%\ (w/v)}}{0.83\ \cancel{\%\ (w/v)}} = 3.0\ mL$$

Practice Problem 7.13

When 0.50 mL of a water sample was diluted to 1.0 L, the lead (Pb) content in this new solution was determined to be 7.5 ppb. What was the lead concentration in the original water sample?

7.8 COLLOIDS AND SUSPENSIONS

We have seen that a solution consists of a solute dissolved in a solvent and that in liquid solutions the similar-sized solute particles and solvent molecules interact through noncovalent interactions. Because of their size and their interactions with solvent, solute particles do not settle out of solution, and because you cannot visibly distinguish solutes from solvents, solutions are usually clear.

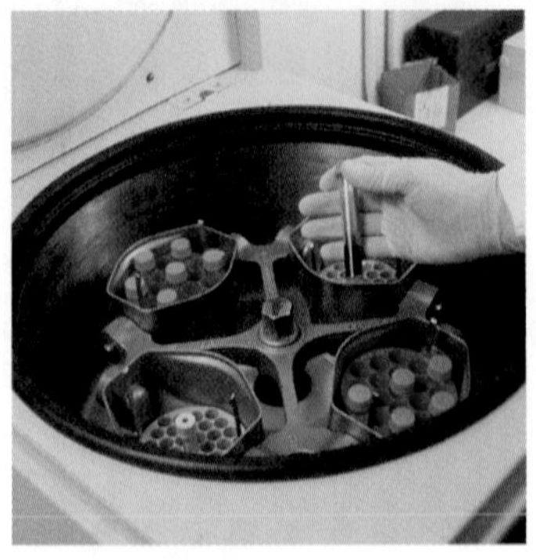

FIGURE | 7.22

Suspensions
Blood cells and other suspended particles in blood are forced to the bottom of a test tube during centrifugation.

Source: Chris Priest/Photo Researchers.

Colloids and suspensions are mixtures that resemble solutions, but they have some important differences. A **suspension** contains *large particles suspended in a liquid.* Usually these particles are large enough to see with the naked eye or with a simple light microscope. The suspended particles are too large to dissolve in the liquid and, unless the suspension is stirred or mixed, they will settle due to gravity. Their vastly different sizes make it possible to use a filter or centrifuge to separate the suspended particles from the molecules that make up the liquid (Figure 7.22). Solutes and solvents that make up a solution cannot be separated by either of these means. Unlike solutions, suspensions have a cloudy appearance. Milk of magnesia, Pepto Bismol, and some other medications are suspensions. The directions for these say to shake the contents before using, because the active ingredients settle over time.

In the mixture called a **colloid**, the particles are larger than the solutes in a solution but smaller than the particles that make up a suspension. Typically, colloid particles cannot be distinguished by the naked eye or by a light microscope. Because of their intermediate size, colloid particles cannot form a solution, but they also will not separate out upon standing. Colloid particles can, however, usually be separated by special filtration or centrifugation techniques. Soapy water, fog, shaving cream, mayonnaise, and milk are examples of colloids. A few of the properties of solutions, suspensions, and colloids are summarized in Table 7.3.

- The solute particles in a solution are approximately the same size as solvent molecules. The particles in a colloid are larger and those in a suspension are larger yet.

SAMPLE PROBLEM 7.14

Solution, colloid, or suspension?

If you wade into a clear lake and the silt on the bottom gets stirred up, what has been produced, a suspension, a colloid, or a solution?

Strategy
One thing that distinguishes a suspension from a colloid or a solution is that the particles in a suspension will settle.

Solution
A suspension. The silt particles will eventually settle back to the bottom of the lake.

PRACTICE PROBLEM 7.14

Identify each of the following as being a solution, a colloid, or a suspension.

a. salt water **b.** cheese **c.** orange juice **d.** air

TABLE | 7.3 SOLUTIONS, COLLOIDS, AND SUSPENSIONS

Mixture	Particle Size	Appearance	Particle Settling?	Separation of Particles
Solution	Small (<1 nm)	Clear	No	Particles cannot be separated out by filtration or centrifugation
Colloid	Intermediate (1 nm–1 μm)	Usually cloudy	No	Particles can usually be separated out by special filtration or centrifugation techniques
Suspension	Large (>1 μm)	Cloudy	Yes	Particles can be separated out by filtration or centrifugation

HEALTH Link

Saliva

In 1982, police collected evidence from a murder scene in Seattle, Washington. Because DNA technology was still fairly new at the time, forensic experts were unable to analyze the DNA sample. Ten years later, with improved techniques, they were able to carry out DNA fingerprinting on the evidence (Figure 7.23). However, because the DNA did not match any of that in DNA databases, the police were unable to identify a suspect.

The Seattle police decided to try to obtain a DNA sample from the prime suspect in the murder case. They sent him a letter that appeared to be from a law firm, inviting him to join a class action lawsuit. He signed the paperwork and mailed it back. When investigators tested the glue on the envelope that he had licked, they found DNA matching that from the crime scene. The suspect was arrested, convicted, and is now serving a ten to twenty year sentence for second degree manslaughter.

From this story, it is clear that saliva contains DNA. But what else does it contain? Saliva is about 98% water, with the other 2% a mix of a variety of substances, including simple cations and anions, enzymes and other proteins, and nucleic acids (DNA and RNA). The ions present in saliva in the greatest amounts are Na^+, K^+, Mg^{2+}, Ca^{2+}, Cl^-, HCO_3^-, and PO_4^{3-}. The calcium and phosphate ions are believed to play an important role in remineralization, the maintenance of tooth enamel (Chapter 3). By acting as a buffer (Section 9.10) to prevent the mouth from becoming too acidic, bicarbonate ion helps to prevent tooth decay.

The primary enzymes found in saliva are amylase, lingual lipase, and lysozyme. The first two of these begin the digestion process. In the mouth, amylase takes some of the starch present in food and breaks it down into smaller molecules. This action of amylase stops once the food has been swallowed, because the enzyme is deactivated by the acidic conditions in the stomach. The digestion of starch and other carbohydrates continues in the small intestine (Section 15.3). Lingual lipase, on the other hand, is not activated until it reaches the stomach. When a food/saliva mix is swallowed, the lingual lipase present in saliva, along with gastric lipase released in the stomach, begins to break down some fats into smaller molecules. Most of the digestion of fats, however, takes place in the small intestine. Lysozyme, an enzyme that breaks down bacteria, helps to prevent infection in the mouth. Saliva also contains other antibiotic (bacteria fighting) compounds, including immunoglobulin A (an antibody), defensins (a type of protein), hydrogen peroxide (H_2O_2), and thiocyanate ion (SCN^-).

Mucins are another important component of saliva. When mixed with water, these glycoproteins (Section 12.4) lubricate the mouth and teeth, and form a thick mucus that helps bind food into a slippery mass that is easily swallowed.

Some of the other substances present in saliva can be used for medical or forensic testing. As mentioned above, DNA in saliva can be used for identification purposes. Recent studies have shown that the presence of certain RNAs (Section 14.8) can indicate the onset of oral cancer. Additionally, as an alternative to testing urine or blood, saliva can be used to detect drug use. The recent scientific literature reports the ability to use saliva to identify the presence of heroin, THC, methamphetamine, cocaine, PCP, and many other drugs.

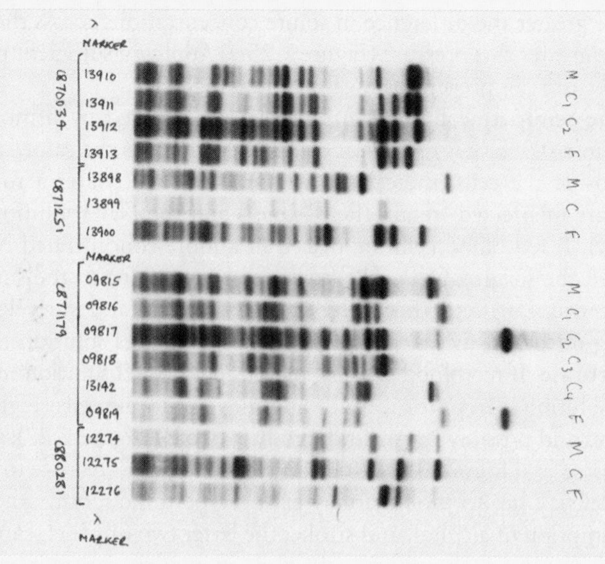

FIGURE 7.23

DNA fingerprinting

DNA fingerprints resemble barcodes, with each line representing a particular fragment of DNA. Because DNA varies slightly from one individual to the next, each person has his or her own particular DNA fingerprint (see Chapter 14, Biochemistry Link, DNA Fingerprinting).

Source: David Parker/Photo Researchers, Inc.

7.9 DIFFUSION AND OSMOSIS

Substances *move from areas of higher concentration to those of lower concentration* in a process called **diffusion** (Figure 7.24). For example, when you chop onions some of the organic molecules that give onions their odor are released into the air, forming a solution (molecules from the onion are solutes and air is the solvent). Initially the solute molecules are located near the onion, but within a few minutes you can smell them throughout the kitchen. The high concentration of solutes near the onion has been redistributed, or diluted, into the rest of the air in the room.

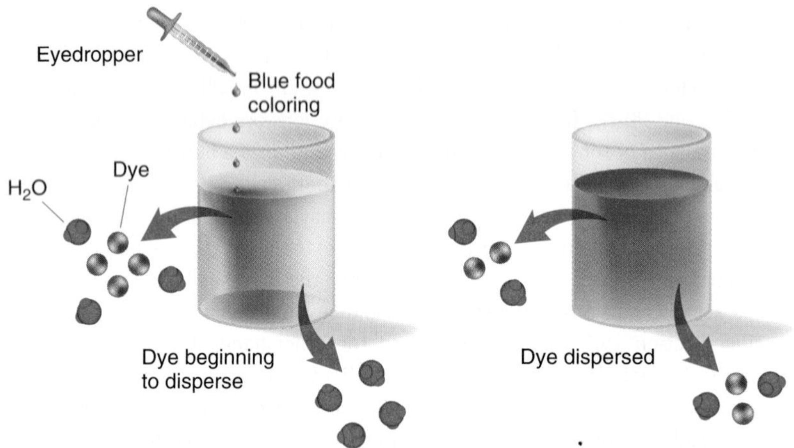

FIGURE | 7.24

Diffusion
As a drop of food coloring dissolves in water, the dye molecules diffuse (move to areas of lower concentration) until they are evenly distributed throughout the solution.

Certain membranes, called semipermeable membranes, are barriers to diffusion because they allow solvents but not all solutes to pass through. If a semipermeable membrane separates solutions with different concentrations, the solvent molecules move from the side with the lower solute concentration to the one with the higher concentration (Figure 7.25*a*,*b*). The effect is the same as what takes place during diffusion—as solvent molecules move in this direction, the solution with higher concentration is diluted. This *net movement of water across a membrane from the solution of lower concentration to one of higher concentration* is called **osmosis**. Seawater has a higher concentration of solutes (Na^+, Cl^-, and others) than freshwater. If these two solutions are separated by a semipermeable membrane, osmosis will move water from the freshwater side to the seawater side.

- In osmosis, solvent moves across a semipermeable membrane from the solution with lower concentration into the solution with higher concentration.

The pressure exerted by water as it flows through a semipermeable membrane is called **osmotic pressure**. The greater the difference in solute concentrations across the semipermeable membrane, the greater this pressure (Figure 7.25*c*). Applying sufficient pressure to counteract osmotic pressure can prevent osmosis from taking place.

Cell membranes are semipermeable and are, therefore, affected by osmosis. If red blood cells are placed in a 0.9% (w/v) solution of NaCl, which has the same solute concentration as found inside the cells, there is no net movement of water across the cell membrane. The cells are unaffected (retain their "tone") and the NaCl solution is called **isotonic** (Figure 7.26*a*). If red blood cells are placed in a more concentrated NaCl solution, water moves from the inside of the cells (lower concentration) to the surrounding solution (higher concentration) and the cells become wrinkled and shriveled (Figure 7.26*b*). This wrinkling of the cell is called **crenation** and the NaCl solution that causes this shrinkage is **hypertonic**. If red blood cells are placed in an NaCl solution more dilute than 0.9% (w/v), the solution is **hypotonic** and water moves into the cells (Figure 7.27*c*), causing them to expand and possibly burst (undergo **hemolysis**)(Figure 7.27).

If blood serum is tested and found to be hypertonic or hypotonic (relative to the solute concentration inside cells), a health problem may exist. Hypertonic serum can be caused by dehydration, consumption of alcohol, and stroke, the latter two of which can affect the release of ADH (antidiuretic hormone) used by the body to regulate the tonicity of serum. Hypotonic blood serum can be caused by excess intake of fluids.

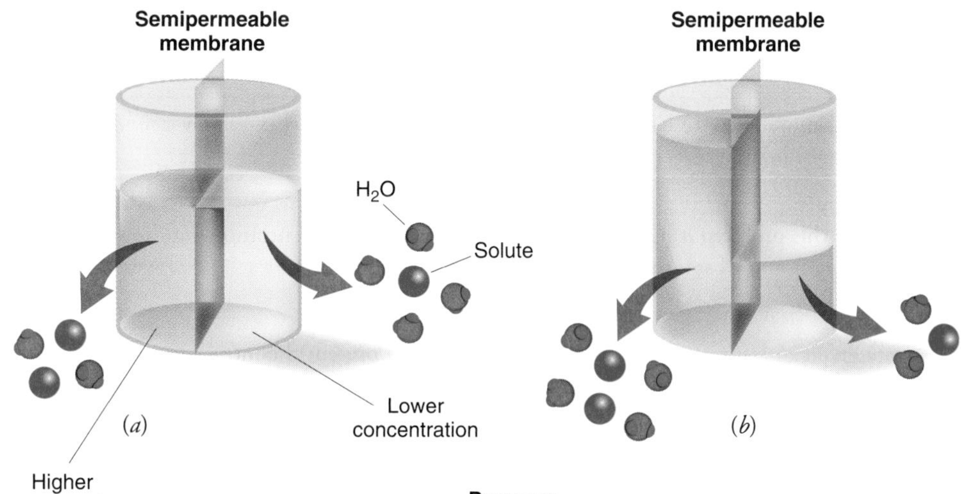

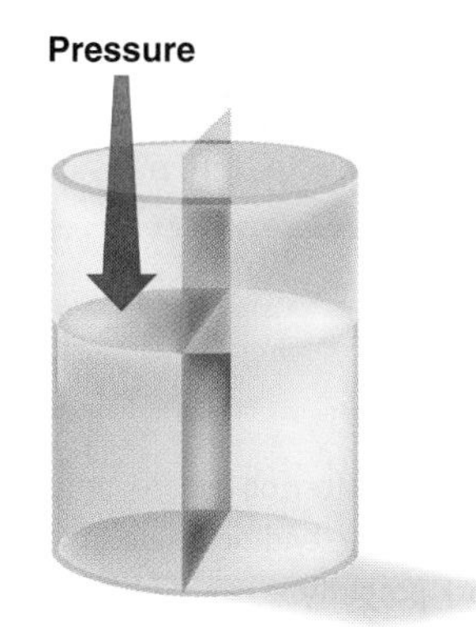

FIGURE | 7.25

Osmosis
When a semipermeable membrane separates two solutions, solvent particles (H_2O in this case) move from the solution with lower solute concentration to the one with higher concentration. (*a*) Before osmosis begins. (*b*) After osmosis has taken place. (*c*) Applying pressure to one side can reverse the effects of osmosis.

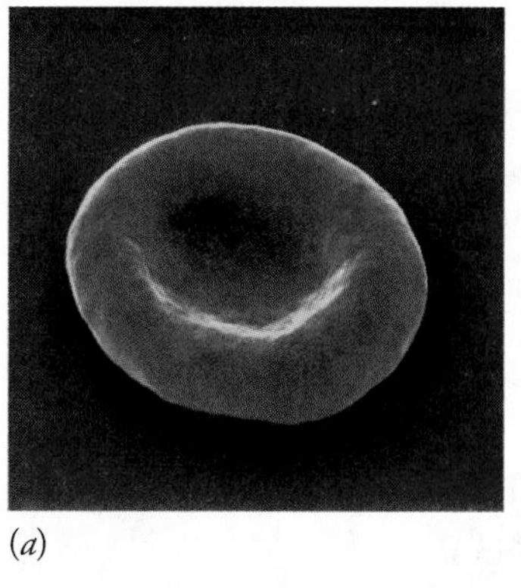
(*a*)

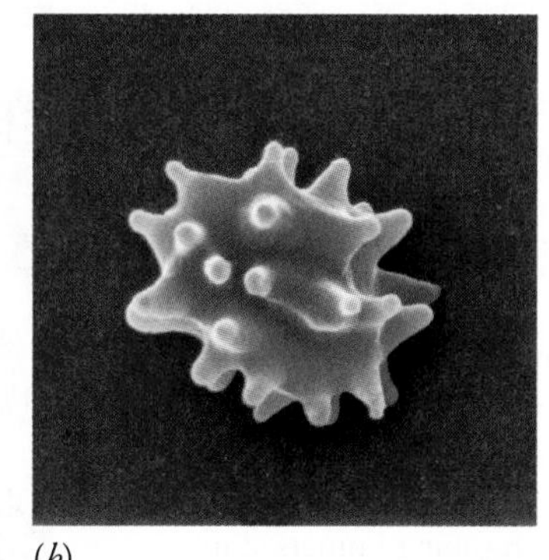
(*b*)

FIGURE | 7.26

The effect of osmosis on red blood cells
(*a*) In an isotonic solution cells retain their normal shape and size. (*b*) In a hypertonic solution osmosis pulls water from the cells, making them shrink. (*c*) In a hypotonic solution osmosis causes water to move into the cells, making them swell or burst.

Source: (*a*) Dr. Stanley Flegler/Visuals Unlimited; (*b*) Dr. David M. Phillips/Visuals Unlimited; (*c*) Dr. Stanley Flegler/Visuals Unlimited.

(*c*)

(*a*) (*b*)

FIGURE | 7.27

The effect of osmosis
(*a*) When placed in salt water, a carrot shrinks as a net movement of water from the carrot cells to the solution takes place. (*b*) When placed in pure water, a carrot swells, as a net movement of water into the cells takes place.

Source: Charles D. Winters.

SAMPLE PROBLEM 7.15

Osmosis

Which is more likely to happen to red blood cells if a person is extremely dehydrated—crenation or hemolysis?

Strategy

To solve this problem it might help to think in terms of concentration and how it indicates the amount of solute in a particular volume of solution.

Solution

For a given amount of solute, the less solvent (water) present the more concentrated a solution (blood serum). The higher the concentration of blood solutes, the more hypertonic the blood serum. Cells will undergo crenation.

PRACTICE PROBLEM 7.15

Explain why consuming alcohol (ethanol) can make blood serum hypertonic.

HEALTH Link

Diffusion and the Kidneys

Your kidneys serve the important function of filtering waste products, unwanted ions, and excess water from your blood, sending them off to be released in urine. If your kidneys fail to work properly it can be life threatening, because toxic solutes will build up in the blood. One treatment option for those with kidney failure is hemodialysis, in which blood is passed through an artificial kidney that removes excess water, ions, and waste (Figure 7.28).

In hemodialysis, the patient's blood is pumped into selectively permeable tubing that is immersed in dialyzing solution, an isotonic aqueous solution of electrolytes and glucose. Because the dialyzing solution is isotonic with blood, no net movement of desired solutes takes place. Waste products and unwanted ions are at a higher concentration in the blood than in the dialyzing solution, however, so they move out of the blood and into the dialyzing solution. No blood cells or proteins are lost because they are too large to move across the membrane.

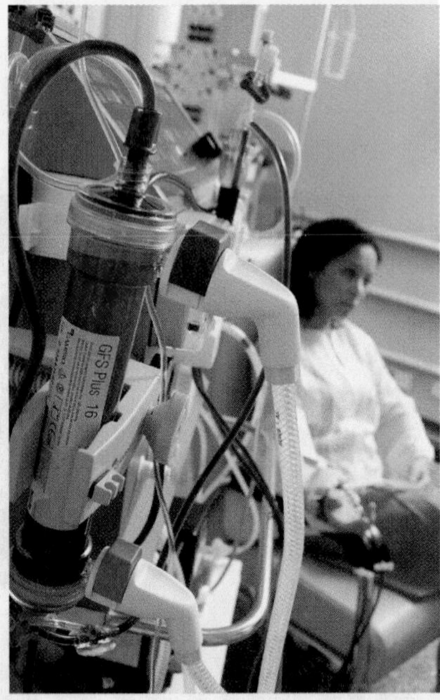

FIGURE 7.28

Kidney dialysis
In kidney dialysis, blood is pumped through tubing immersed in dialyzing solution. Unwanted blood solutes are removed by diffusion.

Source: AJPhoto/Photo Researchers, Inc.

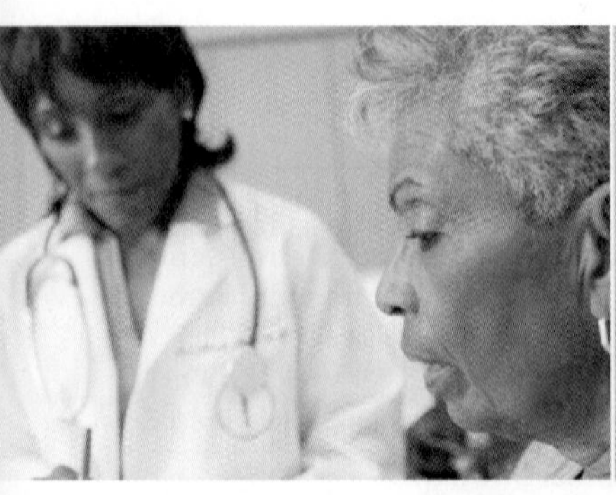

The link between gout and some forms of cancer chemotherapy is that certain drugs can cause the rapid death of large numbers of cancer cells. This leads to a sharp increase in the blood serum concentration of molecules called purines, which are present in both cancer cells and normal cells. Ordinarily, the purines formed by the breakdown of cells or by digestion of the food that you eat are converted into uric acid, which is pulled from the blood by the kidneys and released in urine.

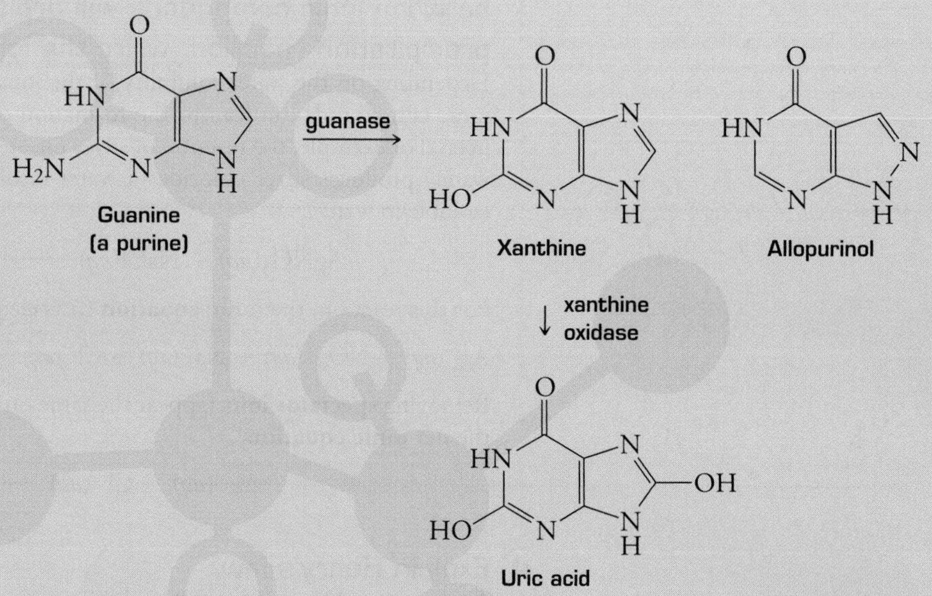

The rapid release of purines by chemotherapy can lead to the overpro-duction of uric acid. At high concentrations, this blood solute reaches its solubility limit and begins to crystallize. This crystal formation, the source of gout, typically happens in the toes, joints, and kidneys, and is the source of severe pain.

Gout can be treated using allopurinol, a drug that slows the production of uric acid from purines.

summary of objectives

1 Define the term pure substance and describe what is meant by the terms homogeneous mixture and heterogeneous mixture. For a homogeneous mixture, explain what differentiates solutes from a solvent.

Pure substances consist of just one element or compound. Mixtures of pure substances can be homogeneous (uniformly distributed) or heterogeneous (not uniformly distributed). In a homogeneous mixture (or **solution**), the **solvent** is present in the greatest amount and **solutes** in lesser amounts.

2 Describe the effect that temperature has on the solubility of gases, liquids, and solids in water.

As temperature is increased, the **solubility** (amount of solute that will dissolve in a solvent at a given temperature) of gases in water decreases. For most liquids and solids, a rise in temperature leads to an increased solubility in water.

3 Use solubility rules to predict whether or not a reaction between ionic compounds will produce a precipitate. For a precipitation reaction, write the ionic and net ionic equation.

Depending on the water solubility of the ions involved, some reactions of aqueous ionic compounds can lead to the formation of **precipitates** (water-insoluble products). For example, the reaction of silver nitrate and sodium chloride (both soluble in water) produces silver chloride (a water-insoluble precipitate) and sodium nitrate (soluble in water).

$$AgNO_3(aq) + NaCl(aq) \longrightarrow AgCl(s) + NaNO_3(aq)$$

For this reaction, the **ionic equation** (lists electrolytes as individual ions) is

$$Ag^+(aq) + NO_3^-(aq) + Na^+(aq) + Cl^-(aq) \longrightarrow AgCl(s) + Na^+(aq) + NO_3^-(aq)$$

Removing **spectator ions** (appear the same on both sides of the reaction arrow) gives the **net ionic equation**.

$$Ag^+(aq) + Cl^-(aq) \longrightarrow AgCl(s)$$

4 Explain Henry's law.

According to **Henry's law**, the solubility of a gas in a liquid is proportional to the pressure of the gas over the liquid. This means that the higher the pressure, the greater the solubility of the gas in a liquid.

5 Explain the terms hydrophilic, hydrophobic, and amphipathic and give examples of compounds that belong to each category.

Water-soluble substances are classed as **hydrophilic** (water loving), water-insoluble substances as **hydrophobic** (water fearing), and substances that contain both water-soluble and water-insoluble parts as **amphipathic** (has both). Amino acids and simple sugars are hydrophilic, fatty acids are hydrophobic, and soaps (sodium salts of carboxylic acids) are amphipathic.

6 List some of the commonly encountered concentration units and perform calculations involving concentration.

Concentration units include **weight/volume percent**, **volume/volume percent**, **weight/weight percent**, **parts per thousand**, **parts per million**, **parts per billion**, and **molarity**. Additional units commonly used in medical applications are milligrams per deciliter, nanograms per liter, millimoles per liter, and milliequivalents per liter. Given a known amount of solute and solution, concentration can be calculated. For example, a solution prepared by dissolving 0.012 g of NaF in water to give a final volume of 250.0 mL has a concentration of 48 ppm.

$$\begin{aligned}\text{Parts per million} &= \frac{\text{g of solute}}{\text{mL of solution}} \times 10^6 \\ &= \frac{0.012 \text{ g NaF}}{250.0 \text{ mL of solution}} \times 10^6 \\ &= 48 \text{ ppm}\end{aligned}$$

Given a concentration and a volume of solution, the amount of solute present can be calculated. For example, a 75.0 mL (0.0750 L) blood serum sample with a vitamin B_{12} concentration of 132 ng/L contains 9.90 ng of vitamin B_{12}.

$$\frac{132 \text{ ng vitamin } B_{12}}{\cancel{L}} \times 0.0750\ \cancel{L} = 9.90 \text{ ng vitamin } B_{12}$$

7 Calculate the concentration of a solution after it has been diluted.

The equation $V_{original} \times C_{original} = V_{final} \times C_{final}$ may be used in calculations related to dilution. If you have 0.25 mL of 0.15 M aqueous glucose and add enough water to give a final volume of 1.00 mL, the new diluted solution has a concentration of 0.038 M.

$$V_{original} \times C_{original} = V_{final} \times C_{final}$$

$$C_{final} = \frac{V_{original} \times C_{original}}{V_{final}}$$

$$C_{final} = \frac{0.25\ \cancel{mL} \times 0.15\ M}{1.00\ \cancel{mL}} = 0.038\ M$$

8 Distinguish solutions, suspensions, and colloids.

In a liquid solution, the solute particles and solvent molecules are about the same size and solute particles do not settle out. In a **suspension** particles are too large to dissolve and will settle due to gravity. In a **colloid** the particles are intermediate in size between those in a solution and those in a suspension. Colloid particles will not settle.

9 Describe the principles of diffusion and osmosis.

The movement of substances from areas of higher concentration to areas of lower concentration is called **diffusion**. Semipermeable membranes can be barriers to diffusion and, often, only solvent molecules can pass through them. The movement of water across semipermeable membranes from areas of lower concentration to areas of higher concentration is called **osmosis**. When cells are placed in a **hypotonic** solution, osmosis causes a net movement of water from outside to inside and the cells swell. In a **hypertonic** solution cells shrink, while in an **isotonic** solution no change in cell size takes place.

summary of reactions

Section 7.2

Precipitation (formation of a solid)

$$CoCl_2(aq) + Ca(OH)_2(aq) \longrightarrow Ca(OH)_2(s) + CaCl_2(aq)$$

Balanced equation

$$Co^{2+}(aq) + 2Cl^-(aq) + Ca^{2+}(aq) + 2OH^-(aq) \longrightarrow Co(OH)_2(s) + Ca^{2+}(aq) + 2Cl^-(aq)$$

Ionic equation

$$Co^{2+}(aq) + 2OH^-(aq) \longrightarrow Co(OH)_2(s)$$

Net ionic equation

Gas formation

$$2HCl(aq) + Na_2S(aq) \longrightarrow H_2S(g) + 2NaCl(aq)$$

Balanced equation

$$2H^+(aq) + 2Cl^-(aq) + 2Na^+(aq) + S^{2-}(aq) \longrightarrow H_2S(g) + 2Na^+(aq) + 2Cl^-(aq)$$

Ionic equation

$$2H^+(aq) + S^{2-}(aq) \longrightarrow H_2S(g)$$

Net ionic equation

END OF CHAPTER PROBLEMS

Answers to problems whose numbers are printed in color are given in Appendix D. More challenging questions are marked with an asterisk. Problems within colored rules are paired. **ILW** = Interactive Learning Ware solution is available at *www.wiley.com/college/raymond.*

7.1 Solutions

7.1 A bottle of 190 proof alcohol is 95% ethanol and 5% water. What is the solvent in this solution?

7.2 You add sugar to a pan of cold water and obtain a saturated solution. When you heat the pan on the stove, all of the sugar dissolves. Explain why this happens.

7.3 A pan full of hot salt water, NaCl(*aq*), is cooled and NaCl(*s*) precipitates. Explain why this happens.

7.4 Which are solutions?
a. sand
b. vinegar
c. oxygen (O_2)
d. mayonnaise

7.5 Give an example of a solution in which
a. the solute is a solid and the solvent is a liquid
b. the solute is a gas and the solvent is a gas

7.6 Give an example of a solution in which
a. the solute is a gas and the solvent is a liquid
b. the solute is a liquid and the solvent is a liquid

7.2 Reactions of Ions in Aqueous Solutions

7.7 When aqueous copper(II) sulfate is mixed with aqueous sodium sulfide, a precipitate forms. Write a balanced equation for this reaction.

7.8 When aqueous barium chloride is mixed with aqueous sodium sulfate, a precipitate forms. Write a balanced equation for this reaction.

7.9 Predict whether each ionic compound is soluble or insoluble in water.
a. $(NH_4)_2SO_4$
b. K_2SO_4
c. $CaCO_3$
d. $NaNO_3$

7.10 Predict whether each ionic compound is soluble or insoluble in water.
a. $(NH_4)_3PO_4$
b. $Ba(NO_3)_2$
c. AgCl
d. K_2CO_3

7.11 Complete and balance each precipitation reaction.
a. $CaCl_2(aq) + Li_2CO_3(aq) \longrightarrow$
b. $Pb(NO_3)_2(aq) + NaCl(aq) \longrightarrow$
c. $CaBr_2(aq) + K_3PO_4(aq) \longrightarrow$

7.12 **a.** Complete and balance the precipitation reaction.

$$Na_2CO_3 + Pb(NO_3)_2 \longrightarrow$$

b. Using Table 7.1, come up with your own precipitation reaction.

7.13 **a.** Complete and balance the precipitation reaction.

$$CuCl_2 + AgNO_3 \longrightarrow$$

b. Using Table 7.1, come up with your own precipitation reaction.

7.14 An aqueous solution of nickel(II) sulfide reacts with aqueous hydrogen chloride to form aqueous nickel(II) chloride and hydrogen sulfide gas. Write a balanced equation for this reaction.

***7.15** In water, ammonium carbonate reacts with hydrogen chloride to produce ammonium chloride and carbonic acid, which falls apart to form water and carbon dioxide gas. Write balanced equations for these reactions.

7.16 Write the ionic equation and the net ionic equation for the reaction in Problem 7.7.

7.17 Write the ionic equation and the net ionic equation for the reaction in Problem 7.8.

7.18 Write the ionic equation and the net ionic equation for each reaction in Problem 7.11.

7.19 Write the ionic equation and the net ionic equation for the reaction in Problem 7.12a.

7.20 Write the ionic equation and the net ionic equation for the reaction in Problem 7.13a.

7.21 Write the ionic equation and the net ionic equation for the reaction in Problem 7.14.

7.3 Solubility of Gases in Water

7.22 What happens to the solubility of carbon dioxide gas (CO_2) in water in each situation? (Answer as increase, decrease, or no change.)
a. The pressure of CO_2 over the solution is increased.
b. The temperature is increased.

7.23 What happens to the solubility of oxygen gas (O_2) in water in each situation? (Answer as increase, decrease, or no change.)
- **a.** The pressure of O_2 over the solution is decreased.
- **b.** The temperature is decreased.

*__7.24__ Two unopened bottles of carbonated water are at the same temperature. If one is opened at the top of a mountain and the other at sea level, which will produce more bubbles? Explain.

7.25 Which are correct statements?
- **a.** The higher the pressure of a gas over a liquid, the more soluble the gas.
- **b.** At a given pressure, all gases are equally soluble in a liquid.
- **c.** The lower the temperature, the greater the solubility of a gas in a liquid.

7.26 Explain why an opened bottle of carbonated water keeps its fizz longer if kept in a refrigerator than if kept at room temperature.

7.4 Organic Compounds

7.27 When your body metabolizes amino acids, one of the final end products is urea, a water-soluble compound that is removed from the body in urine. Why is urea soluble in water, when hexanamide, a related compound, is not?

$NH_2-C(=O)-NH_2$ — Urea

$CH_3CH_2CH_2CH_2CH_2C(=O)-NH_2$ — Hexanamide

7.28 Which of the compounds is more soluble in water?

$CH_3CH_2CH_2CH_2CH_2C(=O)-OH$ — Hexanoic acid

$H-C(=O)-OH$ — Formic acid

7.29 Explain why $CH_3CH_2OCH_2CH_3$ is less soluble in water than its constitutional isomer $CH_3CH_2CH_2CH_2OH$.

7.30 Explain why propylene glycol is more soluble in water than 2-propanol.

$CH_3CH(OH)CH_2OH$ — Propylene glycol

$CH_3CH(OH)CH_3$ — 2-Propanol

7.31 Most general anesthetics have low water solubility, which is sometimes a factor in how long-acting the drugs are. Why do water-insoluble drugs tend to stay in the body for a longer time than water-soluble drugs?

7.32 **a.** Skunk spray contains the two molecules below. Based on the structures, explain why skunk spray is hard to rinse off with water.

$CH_3CH(CH_3)CH_2CH_2SH$ $\quad$ $CH_3CH{=}CHCH_2SH$

b. Washing with tomato juice is one of the ways to remove skunk odor. For this to be the case, what type of molecules must be present in tomato juice?

7.5 Biochemical Compounds

7.33 Vitamin D is produced in the skin upon exposure to sunlight. Based on its structure, is vitamin D hydrophilic, hydrophobic, or amphipathic?

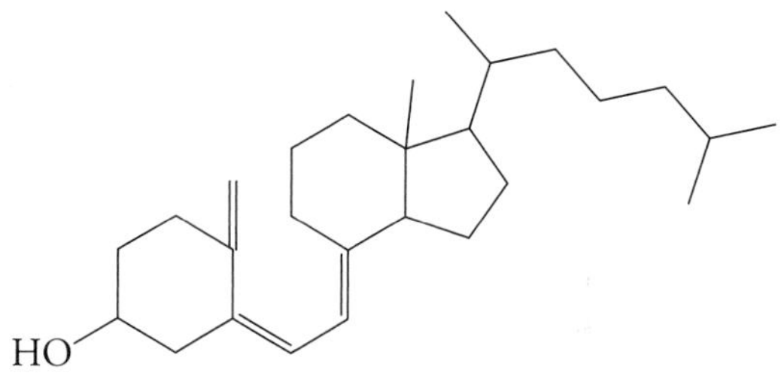

HO

Vitamin D

7.34 Vitamin A is not water soluble. Why?

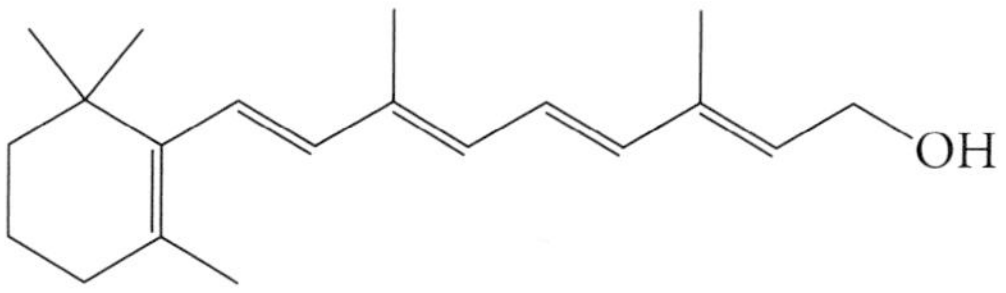

OH

Vitamin A

7.35 Some swimmers, football players, sprinters, and other athletes make illegal use of anabolic steroids to increase muscle strength and endurance. One of these drugs, nandrolone decanoate, is slowly released from its site of injection and is metabolized. Six to twelve months after its use, nandrolone decanoate can still be detected in urine. Why does this anabolic steroid remain in the fatty tissues of the body for such a long period of time?

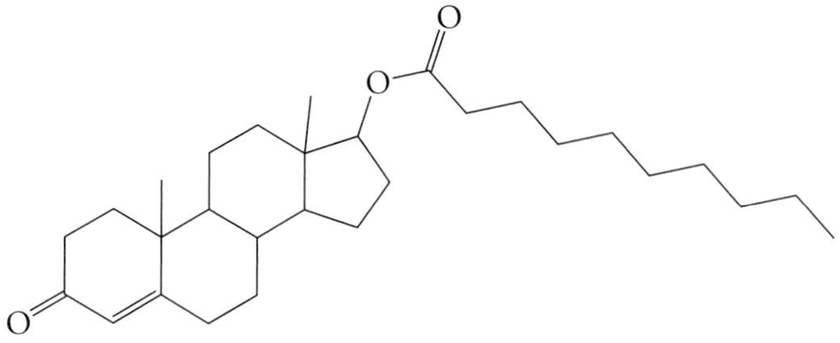

O O O

Nandrolone decanoate

7.6 Concentration

7.36 Explain how you would prepare a saturated aqueous solution of baking soda ($NaHCO_3$).

7.37 Explain how you would prepare a saturated aqueous solution of table sugar.

7.38 A 0.50 L bottle of wine contains 60 mL of ethanol. What is the % (v/v) of ethanol in this aqueous solution?

*7.39 You carefully measure 1.0 tsp of milk into some coffee and end up with exactly 1.0 cup of liquid. What is the % (v/v) of milk in this aqueous solution? (See Tables 1.1 and 1.2 for conversion factors.)

7.40 If 100 mL of blood serum contains 5.0 mg of thyroxine, a hormone released by the thyroid gland, thyroxine levels are within the normal range for an adult. Express this concentration of thyroxine in parts per million and parts per billion.

7.41 The normal concentration range of vitamin B_{12} in blood serum is 160–170 ng/L. Express this concentration range in parts per million and parts per billion.

***7.42** If 0.30 mg of KCl is present in 1.0 L of aqueous solution, what is the concentration in terms of the following?

a. molarity
b. weight/volume percent
c. parts per thousand
d. parts per million
e. parts per billion

*7.43 If 15.0 g of $CaCl_2$ are present in 250 mL of aqueous solution, what is the concentration in terms of the following?

a. molarity
b. weight/volume percent
c. parts per thousand
d. parts per million
e. parts per billion

7.44 Calculate the molarity of each.

a. 0.33 mol of NaCl in 2.0 L of solution
b. 55.0 g of NaCl in 125 mL of solution

7.45 Calculate the molarity of each.

a. 1.75 mol of NaCl in 15.2 L of solution
b. 270 mg of NaCl in 1.00 mL of solution

***7.46** For women, normal levels of uric acid in blood serum range from 26 to 60 ppm. If a female patient has 1.2 mg of uric acid in 10.0 mL of blood serum, is she within the normal range?

7.47 Lead is a toxic metal that can delay mental development in babies. The Environmental Protection Agency's action level (that level where remedial action must be taken to clean the water) is 15 ppb. Does a 50 mL sample of water that contains 0.35 μg of lead fall above or below this action level?

***7.48** Albumin, a protein, is present in normal blood serum at concentrations of 3.5–5.5 g/dL. What is the % (w/v) of albumin in serum that contains 4.0 g of albumin per deciliter?

7.49 Thyroxine, a thyroid hormone, is present in normal blood serum at 58–167 nmol/L. What is the molar concentration of thyroxine in serum that contains 150 nmol of thyroxine per liter?

7.50 The normal concentration range of lactate in blood serum is 0.6–1.8 mmol/L. What is the molar concentration of lactate in serum that contains 1.2 mmol of lactate per liter?

*7.51 The serum concentration of cortisol, a hormone, is expected to be in the range 8–20 mg/dL. What is the part per million concentration of cortisol in serum that contains 10 mg of cortisol per deciliter?

***7.52** The serum concentration of glucose is expected to be in the range 80–100 mg/dL. What is the part per million concentration of glucose in serum that contains 95 mg of glucose per deciliter?

7.53 The normal serum concentration of potassium ion (K^+) is 3.5–4.9 mEq/L. Convert this concentration range into mmol/L.

7.54 The normal serum concentration of phosphate ion (PO_4^{3-}) is 2.8–4.5 mEq/L. Convert this concentration range into mmol/L.

7.55 The normal serum concentration of chloride ion (Cl^-) is 95–107 mmol/L. Convert this concentration range into mEq/L.

7.56 The normal serum concentration of magnesium ion (Mg^{2+}) is 1.2–2.5 mmol/L. Convert this concentration range into mEq/L.

7.57 How many milliequivalents of bicarbonate (HCO_3^-) are present in a 75.0 mL blood serum sample with a concentration of 25 mEq/L HCO_3^-?

7.58 How many equivalents of calcium ion (Ca^{2+}) are present in a 25.0 mL blood serum sample with a Ca^{2+} concentration of 5.2 mEq/L?

7.59 How many moles of sodium ion (Na^+) are present in a 10.0 mL blood serum sample with a Na^+ concentration of 132 mEq/L?

7.60 How many grams of vitamin B_{12} are present in a 100.0 mL blood serum sample with a vitamin B_{12} concentration of 168 ng/L?

7.61 **a.** How many milliliters of 25 mEq/L HCO_3^- are required to obtain 95 mEq of HCO_3^-?

b. How many milliliters of 25 mEq/L HCO_3^- are required to obtain 0.055 mol of HCO_3^-?

c. How many liters of 25 mEq/L HCO_3^- are required to obtain 1.3 g of HCO_3^-?

7.62 **a.** How many liters of 72.8 nM thyroxine are required to obtain 126 nmol of thyroxine?

b. How many milliliters of 72.8 nM thyroxine are required to obtain 9.22×10^{-8} mol of thyroxine?

c. How many microliters of 72.8 nM thyroxine are required to obtain 31.9 μmol of thyroxine?

7.7 Dilution

7.63 If 15.0 mL of 3.0 M HCl are diluted to a final volume of 100.0 mL, what is the new concentration?

7.64 If 3.0 mL of 0.15 M HCl are diluted to a final volume of 250.0 mL, what is the new concentration?

7.65 A 10.0% (w/v) solution of ethanol is diluted from 50.0 mL to 200.0 mL. What is the new weight/volume percent?

7.66 A 5.0% (w/v) solution of ethanol is diluted from 50.0 mL to 75.0 mL. What is the new weight/volume percent?

7.67 How many milliliters of 2.00 M NaOH are needed to prepare 300.0 mL of 1.50 M NaOH?

7.68 How many milliliters of 1.50 M NaOH are needed to prepare 225.0 mL of 0.150 M NaOH?

7.69 Calculate the final volume required to prepare each solution.

a. Starting with 100.0 mL of 1.00 M KBr, prepare 0.500 M KBr.

b. Starting with 50.0 mL of 0.250 M alanine (an amino acid), prepare 0.110 M alanine.

7.70 Calculate the final volume required to prepare each solution.

a. Starting with 10 mL of 2.0% (w/v) KI, prepare 0.020% (w/v) KI.

b. Starting with 1.0 mL of 60 ppm Cu^{2+}, prepare 18 ppm Cu^{2+}.

***7.71** When 1.0 mL of a water sample was diluted to 5.0 L, the new solution was found to have a lead concentration of 5.3 ppb. What was the concentration of lead (in ppm) in the original sample?

7.8 Colloids and Suspensions

7.72 How is a colloid different from a suspension?

7.73 How is a colloid different from a solution?

7.74 What do you end up with if you pour dirt into water and stir—a solution, a colloid, or a suspension?

7.75 Give an example of each.

a. a solution
b. a suspension
c. a colloid

7.9 Diffusion and Osmosis

7.76 To make pickles, you soak cucumbers in a concentrated salt solution called brine. Describe how this process is related to osmosis.

7.77 A process called active transport moves certain ions and compounds across cell membranes from areas of lower concentration to areas of higher concentration. Does active transport involve diffusion? Explain.

7.78 Taking a sitz bath (sitting in a bath of dissolved Epsom salts) is a home remedy that can help to reduce the swelling and discomfort of hemorrhoids. Why is this hypertonic solution of $MgSO_4$ able to reduce the swelling of hemorrhoidal tissue, while pure water is not?

Hard Water

7.79 Scale can be removed from a tea kettle by adding some vinegar (a source of H^+) and heating. Write a balanced equation for the reaction of scale (calcium carbonate) and H^+ to produce aqueous calcium ion, carbon dioxide gas, and water.

HEALTH Link | *Hyperbaric Medicine*

7.80 Dry air is approximately 78.0% N_2, 21.1% O_2, and 0.9% other gases. What is the partial pressure of O_2 in a hyperbaric chamber filled with dry air at a pressure of 1900 torr? (See Section 5.4).

7.81 Why is breathing 100% O_2 in a pressurized hyperbaric chamber more effective at treating gangrene than breathing 100% O_2 at atmospheric pressure?

HEALTH Link | *Prodrugs*

7.82 Why is the prodrug chloramphenicol palmitate (Figure 7.18) less water soluble than chloramphenicol?

***7.83** Phenacetin is a prodrug that, in the liver, is converted into acetaminophen, a commonly used pain and fever reducer. While available in many parts of the world, phenacetin is not sold in the U. S. due to concerns about its toxicity. What is the theoretical yield of acetaminophen (in grams) upon reaction of the 325 mg of phenacetin contained in one tablet?

CH_3CH_2O–C_6H_4–NH–C(=O)CH_3 ⟶ HO–C_6H_4–NH–C(=O)CH_3

Phenacetin **Acetaminophen**

HEALTH *Link* | *Saliva*

7.84 Saliva has some characteristics of a solution. Explain.

7.85 Saliva has some characteristics of a colloid. Explain.

7.86 Salivary amylase is not present (or is present only in very small amounts) in the saliva of carnivores. Why do you suppose this is the case?

***7.87** Draw the line-bond structure of H_2O_2 and SCN^-, showing all nonbonding electrons and formal charges.

HEALTH *Link* | *Diffusion and the Kidneys*

7.88 Why is it important that dialyzing solution be isotonic with blood?

7.89 Kidney disease can lead to nephrotic syndrome, which is characterized, in part, by a low blood serum concentration of albumin (a protein). Explain why low serum albumin levels result in edema, swelling caused by the movement of fluid from the blood into tissue.

7.90 "Kidney Bones" describes how patients with kidney disease often experience bone decay, and attributes this bone loss to reduced levels or the absence of the protein BMP7. After reading the Health Link "Diffusion and the Kidneys," do you think it is likely that kidney dialysis plays a role in lowering BMP7 concentrations in the body?

THINKING IT THROUGH

***7.91** When given to a patient, the drug allopurinol is converted to oxypurinol. Oxypurinol blocks the action of xanthine oxidase, an enzyme that catalyzes one of the steps in the production of uric acid from purines.

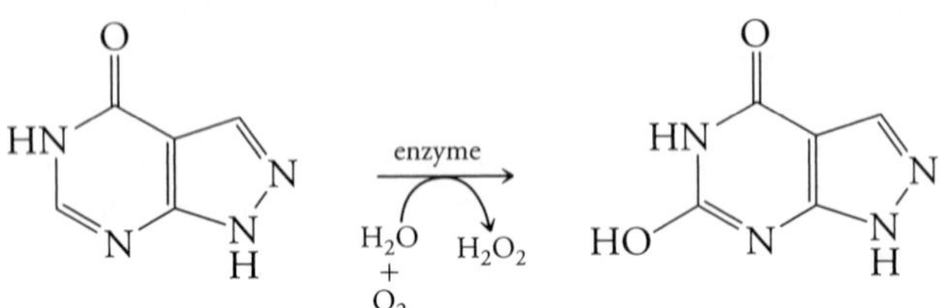

a. Is allopurinol oxidized or is it reduced when it is converted into oxypurinol?
b. Might allopurinol be considered a prodrug? (Refer to Health Link: Prodrugs.)
c. Suggest an explanation for the fact that oxypurinol is better at blocking xanthine oxidase than is allopurinol. (Hint: Look at the structures in the reactions presented just before the chapter summary.)

INTERACTIVE LEARNING PROBLEMS

ILW **7.92** How many grams of solute are needed to make 250 ml of 0.100 M K_2SO_4?

SOLUTIONS TO PRACTICE PROBLEMS

7.1 The lower the temperature of water, the greater the solubility of a gas in it. The fish may be gathering where more O_2 is available to them.

7.2 **a.** $BaSO_4$ (insoluble); **b.** $LiNO_3$ (soluble); **c.** Na_2CO_3 (soluble)

7.3 $Pb(NO_3)_2(aq) + 2NH_4Cl(aq) \longrightarrow PbCl_2(s) + 2NH_4NO_3(aq)$

Balanced equation

$$Pb^{2+}(aq) + 2NO_3^-(aq) + 2NH_4^+(aq) + 2Cl^-(aq) \longrightarrow PbCl_2(s) + 2NH_4^+(aq) + 2NO_3^+(aq)$$

Ionic equation

$$Pb^{2+}(aq) + 2Cl^-(aq) \longrightarrow PbCl_2(s)$$

Net Ionic equation

The product $PbCl_2$ is the precipitate.

7.4 $CH_3CH_2CH_2OH$

7.5 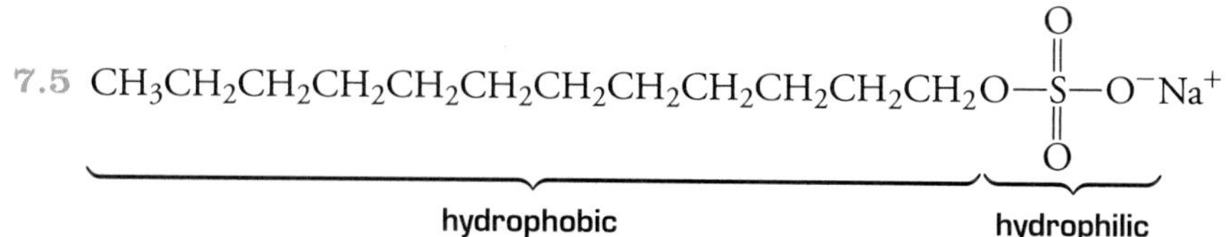

7.6 17% (v/v)

7.7 **a.** 0.72 ppm; **b.** 7.2×10^{-5}% (w/v)

7.8 2.81 M

7.9 6.9 g/dL = 6.9% (w/v)

7.10 4.61–5.21 mg/dL

7.11 0.0011 g

7.12 0.13 M

7.13 15,000 ppb (1.5×10^4 ppb)

7.14 **a.** solution (clear, does not settle); **b.** colloid (cloudy, does not settle); **c.** colloid (cloudy, does not settle) or a suspension (if pulp settles to the bottom or floats to the top); **d.** solution (clear—hopefully!)

7.15 Alcohol affects the release of antidiuretic hormone. Additionally, adding alcohol to blood serum will cause the total solute concentration to rise, making the serum hypertonic.

In the spring of 2003 a track coach told the U.S. Anti-Doping Agency, an organization involved with drug testing and drug education related to sports, that some world-class athletes were using a new performance-enhancing synthetic steroid. To support his statement, he provided a used syringe that had contained the drug. After several months of work, scientists identified the substance, a compound called tetrahydrogestrinone (THG), and had developed a test to detect this previously unknown steroid. Since then, a number of well-known athletes have tested positive for THG. Sports organizations have reported that they will retest athletes' urine samples saved from earlier sporting events. Those who competed after taking THG could have medals taken away, could have their names removed from the record books, and could be banned from future competitions.

Photo source: David Madison/The Image Bank/Getty Images.

8 LIPIDS AND MEMBRANES

In Chapter 4 we saw that molecules can be placed into families based on the functional groups that they possess—alkenes have a C═C group, alkynes a C≡C group, alcohols an —OH group, and so on. In this chapter we will consider the family of biochemical compounds called **lipids**. Compounds are not classified as lipids based on functional groups, but because they are *largely water insoluble*. The principles of solubility described in Chapter 7 will be applied here, for the larger molecules found in living things. Biologically, lipids are put to a wide range of uses, including as cell membrane components, as energy storage molecules, for insulation, and as hormones. As we will see, THG has a chemical structure similar to certain members of the class of lipids called steroid hormones.

objectives

After completing this chapter, you should be able to:

1. Explain what makes a compound a lipid.
2. Describe the structure of fatty acids and explain how saturated, monounsaturated, and polyunsaturated fatty acids differ from one another.
3. Identify the primary biological function of waxes and describe the esters that predominate in them.
4. Describe the makeup of triglycerides and list their biological functions.
5. Describe the structure of phospholipids and glycolipids.
6. Identify the basic steroid structure and list important members of this class of lipids.
7. Name the three types of eicosanoids and describe their biological function.
8. Describe the makeup of a cell membrane and explain how various compounds cross the membrane.

8.1 FATTY ACIDS

The first class of lipids that we will consider is the **fatty acids**, *carboxylic acids that typically contain between 12 and 20 carbon atoms.* When describing the structure of many lipids, the term *head* is used to represent a polar group and *tail* is used for a nonpolar hydrocarbon chain. In fatty acids, the head is the carboxyl group and the tail is the long chain of carbon atoms. Fatty acids typically have an even number of carbon atoms.

$$\underbrace{CH_3CH_2CH_2CH_2CH_2CH_2CH_2CH_2CH_2CH_2CH_2CH_2CH_2CH_2CH_2}_{\text{tail}}\underbrace{\overset{\overset{\displaystyle O}{\|}}{C}-OH}_{\text{head}}$$

A fatty acid

■ Unsaturated fatty acids contain carbon–carbon double bonds. Saturated fatty acids do not.

Fatty acids differ from one another in the number of carbon atoms that they contain and in their number of carbon–carbon double bonds (Table 8.1). **Saturated** fatty acids, like stearic acid, have only *single bonds* joining the carbon atoms in their hydrocarbon tails. **Monounsaturated** fatty acids, such as oleic acid, have just *one carbon–carbon double bond*, and **polyunsaturated** fatty acids (linoleic acid, linolenic acid, and others) have *two or more double bonds* (Figure 8.1).

■ TABLE 8.1 COMMON FATTY ACIDS

Number of Carbon Atoms	Number of Double Bonds	Name	Formula	Melting Point (°C)	Source
12	0	Lauric acid	$CH_3(CH_2)_{10}CO_2H$	43	Coconut
14	0	Myristic acid	$CH_3(CH_2)_{12}CO_2H$	54	Nutmeg
16	0	Palmitic acid	$CH_3(CH_2)_{14}CO_2H$	62	Palm
16	1	Palmitoleic acid	$CH_3(CH_2)_5CH{=}CH(CH_2)_7CO_2H$	−0.5	Macadamia nuts
18	0	Stearic acid	$CH_3(CH_2)_{16}CO_2H$	69	Lard
18	1	Oleic acid	$CH_3(CH_2)_7CH{=}CH(CH_2)_7CO_2H$	13	Olives
18	2	Linoleic acid	$CH_3(CH_2)_4(CH{=}CHCH_2)_2(CH_2)_6CO_2H$	−9	Safflower
18	3	Linolenic acid	$CH_3CH_2(CH{=}CHCH_2)_3(CH_2)_6CO_2H$	−17	Flax

The hydrocarbon tails of saturated fatty acids are able to pack closely together with one another, interacting through London forces. Because London forces depend on the surface area of the molecules involved, the longer the hydrocarbon tail in a fatty acid, the stronger the interaction between molecules (Figure 8.2*a*). This is why the melting points and boiling points of saturated fatty acids increase with increasing length (Table 8.1). Most saturated fatty acids are solids at room temperature (about 25°C). (Table 8.1 lists only melting points for fatty acids because the boiling points are high enough that they are not typically of concern.)

In unsaturated fatty acids, the carbon–carbon double bonds are usually *cis*, which means that there is a kink in the hydrocarbon tail. The *cis* double bonds hold the hydrocarbon tails of unsaturated fatty acids farther apart from one another than is the case for saturated fatty acids. This reduces the contact between the molecules and weakens the (*a*) $CH_3CH_2CH_2CH_2CH_2CH_2CH_2CH_2CH_2CH_2CH_2CH_2CH_2CH_2CH_2CH_2CH_2\overset{O}{\overset{\|}{C}}{-}OH$

Stearic acid

(*b*) $CH_3CH_2CH_2CH_2CH_2CH_2CH_2CH_2{-}\underset{H}{C}{=}\underset{H}{C}{-}CH_2CH_2CH_2CH_2CH_2CH_2CH_2\overset{O}{\overset{\|}{C}}{-}OH$

Oleic acid

(*c*) $CH_3CH_2CH_2CH_2CH_2{-}\underset{H}{C}{=}\underset{H}{C}{-}CH_2{-}\underset{H}{C}{=}\underset{H}{C}{-}CH_2CH_2CH_2CH_2CH_2CH_2CH_2\overset{O}{\overset{\|}{C}}{-}OH$

Linoleic acid

$CH_3CH_2{-}\underset{H}{C}{=}\underset{H}{C}{-}CH_2{-}\underset{H}{C}{=}\underset{H}{C}{-}CH_2{-}\underset{H}{C}{=}\underset{H}{C}{-}CH_2CH_2CH_2CH_2CH_2CH_2CH_2\overset{O}{\overset{\|}{C}}{-}OH$

Linolenic acid

FIGURE 8.1 **Saturated and unsaturated fatty acids**
(*a*) Stearic acid, a saturated fatty acid, has no carbon–carbon double bonds in its hydrocarbon tail. (*b*) The hydrocarbon tail of oleic acid, a monounsaturated fatty acid, has one C=C. (*c*) Linoleic acid and linolenic acid, polyunsaturated fatty acids, have more than one C=C.

(*a*)

Stearic acid

(*b*)

Linolenic acid

FIGURE 8.2 **Interactions between fatty acids**
(*a*) The hydrocarbon tails of saturated fatty acids interact through London forces. These fatty acids have relatively high melting and boiling points. (*b*) The *cis* double bonds in unsaturated fatty acids do not allow their hydrocarbon tails to interact with one another as effectively as is the case for saturated fatty acids. As a result, unsaturated fatty acids melt and boil at lower temperatures. In this figure, fatty acids are shown using space-filling drawings. A space-filling drawing shows atoms with their approximate relative sizes.

London forces between them (Figure 8.2*b*). For fatty acids with the same number of carbon atoms, the more unsaturated they are, the lower their melting points and boiling points. Unsaturated fatty acids are liquids at room temperature.

Fatty acids are water insoluble (hydrophobic) because their carboxyl head groups are not polar enough to counteract the effect of their long, nonpolar hydrocarbon tails. The solubility properties of fatty acids change when they react with sodium hydroxide (NaOH) to form ionic compounds consisting of Na^+ and fatty acid anions (carboxylate ions), in which the singly bonded oxygen atom has a 1− charge (Figure 8.3). Because the charge on the head group increases their hydrophilic nature, fatty acid anions are amphipathic (have a hydrophilic end and a hydrophobic end). Under the conditions most often found in the body, fatty acids exist as anions.

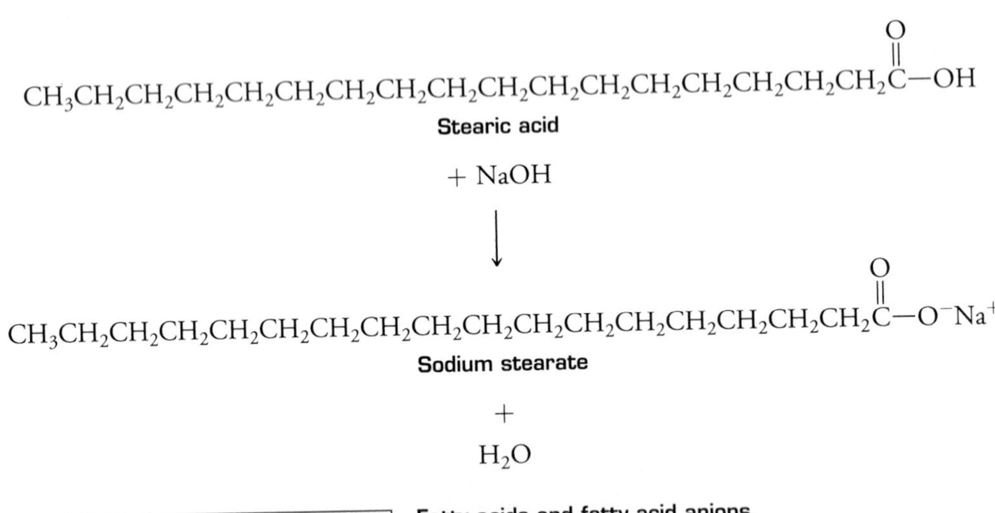

FIGURE 8.3

Fatty acids and fatty acid anions
Like all fatty acids, stearic acid is hydrophobic. When treated with NaOH, fatty acids are converted into salts (ionic compounds) containing Na^+ and amphipathic fatty acid anions. Here, stearic acid is converted into sodium stearate.

As we will see while progressing through this chapter, fatty acids are a structural component of many other lipids. In Chapter 15 we will also see that fatty acids are a significant source of energy. Polyunsaturated fatty acids play key roles in controlling some of what goes on within cells, for example, in regulating the formation of enzymes involved in fatty acid synthesis, breakdown, and transport. A great deal of interest has been focused on the impact that a particular group of polyunsaturated fatty acids known as omega-3s have on health (see Health Link: Omega-3 fatty acids).

SAMPLE PROBLEM 8.1

The relationship between fatty acid structure and melting point

Explain why stearic acid has a higher melting point than lauric acid.

Strategy

To account for the difference, consider the effect that increasing the length of a hydrocarbon tail has on the London force interactions between fatty acids.

Solution

Stearic acid has a longer hydrocarbon tail than lauric acid, so stearic acid molecules interact more strongly with one another through London forces than do lauric acid molecules. The stronger the London forces, the higher the temperature required for melting.

PRACTICE PROBLEM 8.1

Linolenic acid has a lower melting point than linoleic acid. Explain why.

HEALTH Link: Omega-3 fatty acids

In recent years, scientific studies have indicated that there are health benefits to be gained from eating foods that are high in omega-3 fatty acids. This class of polyunsaturated fatty acids gets its name from the last letter of the Greek alphabet—omega (ω). In one naming system, the carbon atoms in fatty acids are numbered beginning with the omega carbon atom—the last one in the hydrocarbon tail (Figure 8.4). For omega-3 fatty acids, the first double bond begins three carbons from the ω end. Examples include linolenic acid, eicosapentaenoic acid, and docosahexaenoic acid.

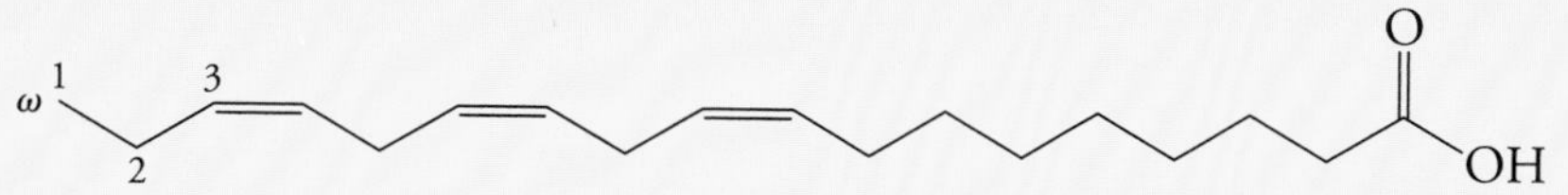

ω 1 2 3 O OH

Linolenic acid

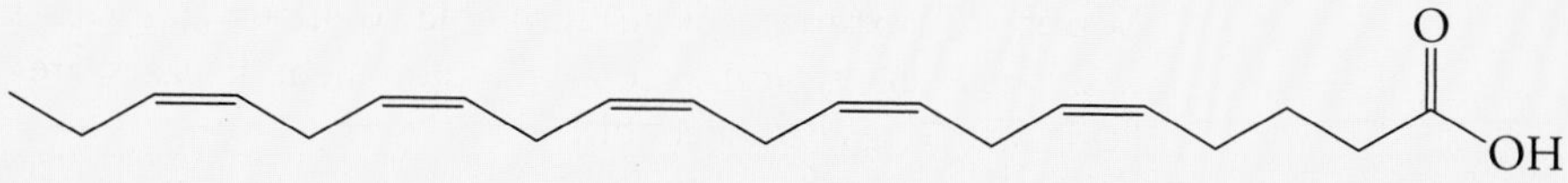

O OH

Eicosapentaenoic acid

O OH

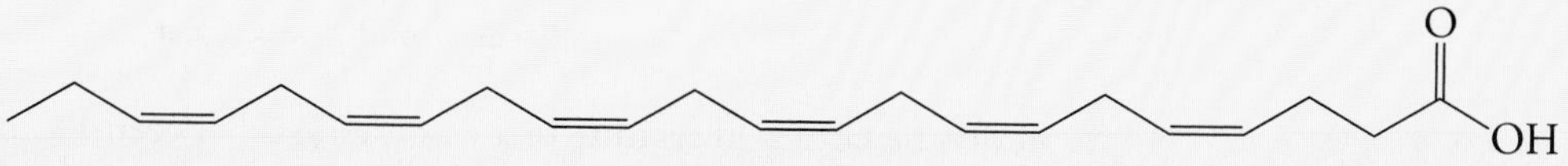

Docosahexaenoic acid

FIGURE 8.4

Omega-3 fatty acids are important in the diet

Linolenic acid, eicosapentaenoic acid, and docosahexaenoic acid are three omega-3 fatty acids present in fish. The last carbon atom in the hydrocarbon tail is the omega (ω) carbon. Note that the linolenic acid molecule in Figure 8.1 is drawn using a condensed formula, while here it is represented by a skeletal structure.

The health advantages attributed to a diet that contains omega-3 fatty acids include increased HDL and lowered LDL and triglyceride levels (refer to Section 8.5 for details on HDLs and LDLs), decreased blood pressure, a lowered risk of heart disease, and a reduction of inflammation caused by rheumatoid arthritis.

Omega-3 fatty acids are essential nutrients. This means that they are not produced in the body and must be obtained in (are essential to) the diet. Some plant-derived oils, including flax and canola, are good sources of these fatty acids, as are salmon, herring, sardines, anchovies, and other cold water fish. To meet dietary guidelines related to omega-3 fatty acids, the U.S. Food and Drug Administration (FDA) recommends that the average adult eat two six-ounce servings of fish per week.

One concern about eating a diet high in seafood is that pollutants work their way into the food chain and can become concentrated in fish. These pollutants include mercury compounds, dioxins, and PCBs, all of which are toxic. Mercury, for example, is harmful to fetuses and young children. As part of their recommendation, the FDA advises that women who are nursing or pregnant or who may become pregnant, as well as young children, should not eat Shark, Swordfish, King Mackerel, or Tilefish, all of which contain relatively high levels of mercury (Figure 8.5). Albacore Tuna is also high in mercury. If tuna is eaten, then only six total ounces of fish should be eaten per week.

FIGURE 8.5

Mackerels
Although they contain omega-3 fatty acids, Mackerels also accumulate relatively high levels of the toxin mercury.

Digital Archive Japan/Punchstock.

8.2 WAXES

Waxes are mixtures of water insoluble compounds, including esters, alcohols, and alkanes. Wax esters, typically the major component of waxes, are produced by combining fatty acids with long chain alcohols.

- A wax ester is the combination of a fatty acid and a long-chain alcohol.

Fatty acid — Alcohol

Structure of a wax ester

Typically the fatty acid **residue** in a wax ester has between 14 and 36 carbon atoms and the alcohol residue has between 16 and 30 carbon atoms (Figure 8.6). A residue is *that part of a reactant molecule that remains* when it has been incorporated into a product. For the wax ester shown in Figure 8.6, the fatty acid residue is a fatty acid molecule, minus the —OH on the polar head group. The alcohol residue is an alcohol molecule, minus the H atom of the —OH group.

$CH_3(CH_2)_{14}C(=O)—OH$

A fatty acid

$HOCH_2(CH_2)_{28}CH_3$

A long chain alcohol

$CH_3(CH_2)_{14}C(=O)—OCH_2(CH_2)_{28}CH_3$

A wax ester

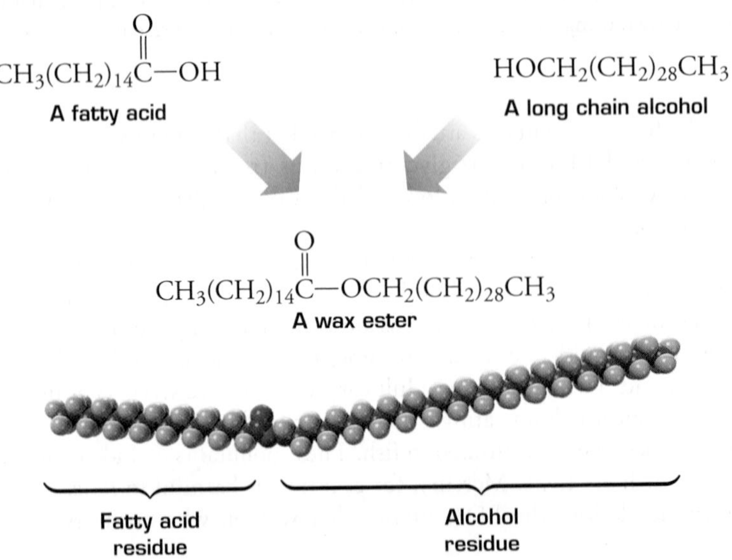

FIGURE 8.6

Waxes
Waxe esters consist of a fatty acid residue attached to a long chain alcohol residue by an ester bond. The ester shown here, formed by combining palmitic acid with an alcohol containing 30 carbon atoms, is the major constituent of beeswax.

Waxes primarily serve a protective function, in many cases to keep water either in or out of an organism. For example, the layer of wax on the leaves of most plants prevents the evaporation of water, while the wax present in feathers acts as a water repellant. Without the protective wax layer, ducks and other waterfowl would be unable to float (Figure 8.7). Other biological uses of waxes include protecting skin and hair by keeping it soft and waterproof and, in some microorganisms, serving as energy storage molecules.

FIGURE 8.7

Solubility of waxes
Being hydrophobic, waxes are not soluble in water. They are, however, soluble in nonpolar solvents, including the hydrocarbons that make up petroleum. When waterfowl are caught in an oil spill, the protective layer of wax dissolves from their feathers and the birds are unable to keep dry and buoyant.

Source: Ben Osborne/Stone/Getty Images.

Table 8.2 identifies principal esters in a few common waxes. Note that some of the fatty acid residues found in wax esters have more than the 12–20 carbon atoms typical for most fatty acids.

TABLE 8.2 KEY ESTERS FOUND IN SOME WAXES

Name	Formula	Source	Use
Beeswax	$CH_3(CH_2)_{14}CO_2CH_2(CH_2)_{28}CH_3$	Honeycomb	Candles
Carnauba wax	$CH_3(CH_2)_{24}CO_2CH_2(CH_2)_{28}CH_3$	Palm	Furniture wax
Insect wax	$CH_3(CH_2)_{24}CO_2CH_2(CH_2)_{48}CH_3$	Insects	Polish

8.3 TRIGLYCERIDES

Animal fats and vegetable oils are **triglycerides** or triacylglycerides, in which *three fatty acid residues are joined to a glycerol residue by ester bonds.*

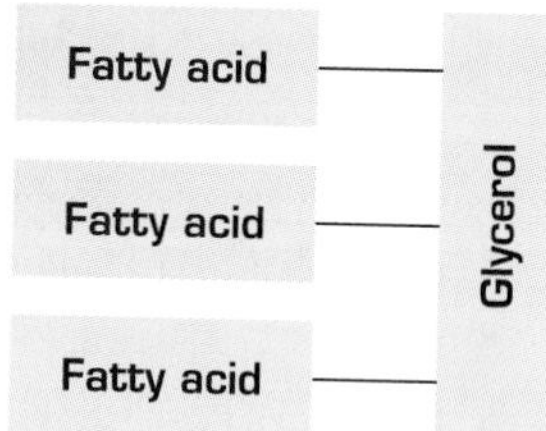

Structure of a triglyceride

■ Saturated triglycerides contain more saturated than unsaturated fatty acid residues. For unsaturated triglycerides the reverse is true.

Triglycerides usually contain two or three different fatty acid residues (Figure 8.8). Triglycerides like lard and beef fat are saturated because they contain more saturated than unsaturated fatty acid residues, while vegetable oils are unsaturated because they contain mostly monounsaturated and polyunsaturated ones. The main fatty acid residues in beef fat are palmitic acid (saturated) and oleic acid (monounsaturated). Olive oil contains mostly oleic acid residues, while safflower oil has mostly polyunsaturated linoleic acid residues. Fish oils contain relatively high amounts of omega-3 and other polyunsaturated fatty acid residues (see Health Link: Omega-3 fatty acids).

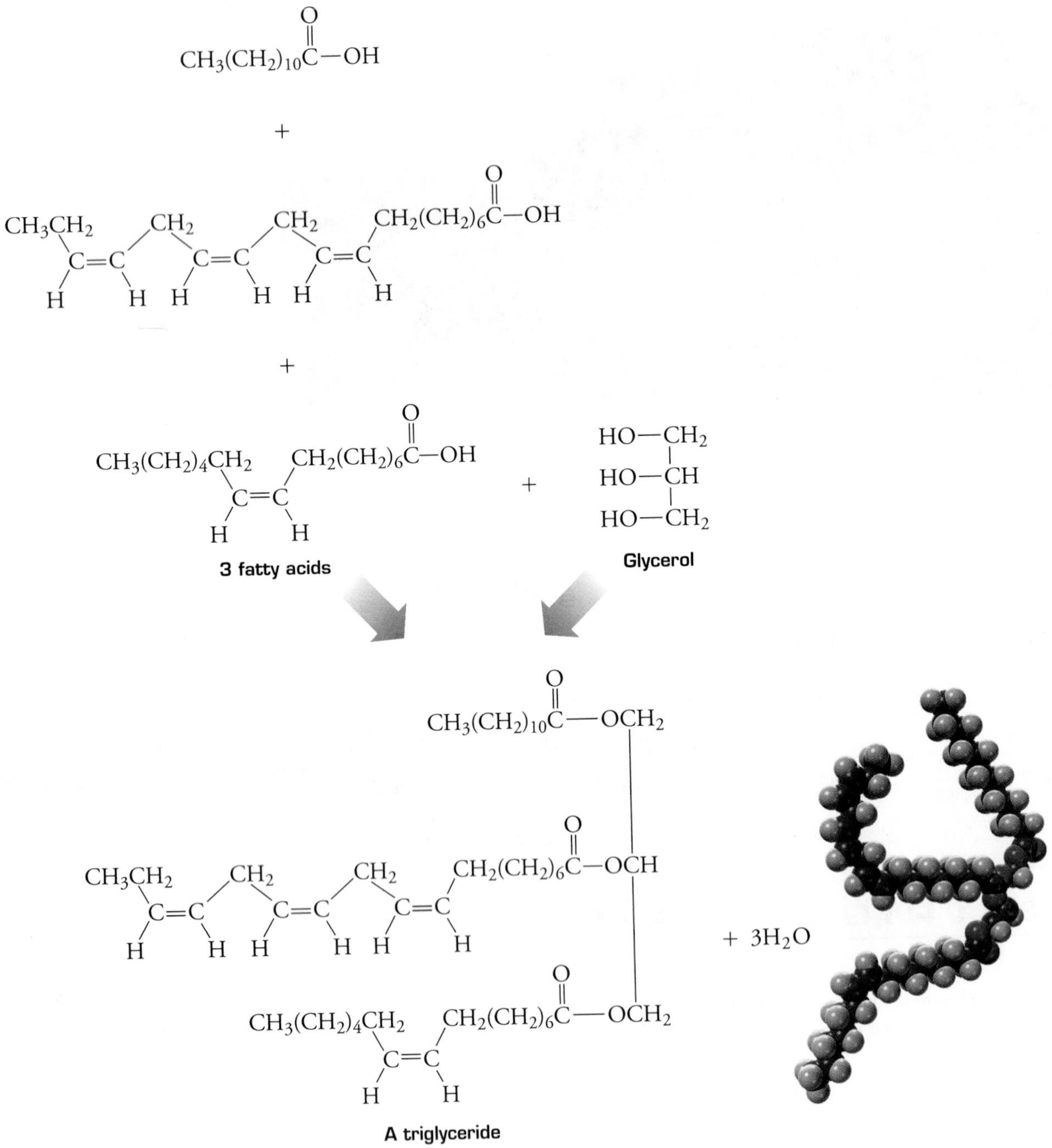

$CH_3(CH_2)_{10}C(=O)-OH$

+

$CH_3CH_2CH=CHCH_2CH=CHCH_2CH=CHCH_2(CH_2)_6C(=O)-OH$

+

$CH_3(CH_2)_4CH_2CH=CHCH_2(CH_2)_6C(=O)-OH$ + $HO-CH_2-CH(OH)-CH_2-OH$

3 fatty acids

Glycerol

$CH_3(CH_2)_{10}C(=O)-OCH_2$

$CH_3CH_2CH=CHCH_2CH=CHCH_2CH=CHCH_2(CH_2)_6C(=O)-OCH$

$CH_3(CH_2)_4CH_2CH=CHCH_2(CH_2)_6C(=O)-OCH_2$

+ $3H_2O$

A triglyceride

FIGURE 8.8

Triglycerides
Triglycerides are formed by joining three fatty acids to a glycerol molecule by ester bonds. The triglyceride pictured here is unsaturated because it contains more unsaturated fatty acid residues than saturated ones.

SAMPLE PROBLEM 8.2

The structure of triglycerides

While grocery shopping, you walk past the meat section and look at the various cuts of beef that are displayed. Draw a possible structure of a triglyceride that is present in the beef fat that you see.

Strategy

A triglyceride is constructed from three fatty acids and a glycerol molecule. Beef fat contains more saturated than unsaturated fatty acid residues. Palmitic and oleic acid residues predominate.

Solution

Several answers are possible. One structure is

$$\begin{array}{r} CH_3(CH_2)_{14}\overset{O}{\overset{\|}{C}}{-}OCH_2 \\ CH_3(CH_2)_7CH{=}CH(CH_2)_7\overset{O}{\overset{\|}{C}}{-}OCH \\ CH_3(CH_2)_{14}\overset{O}{\overset{\|}{C}}{-}OCH_2 \end{array}$$

PRACTICE PROBLEM 8.2

Draw two other possible structures for beef fat.

Section 8.1 described how the *cis* double bonds in unsaturated fatty acids cause them to have lower melting points and boiling points than saturated ones. The same holds true for unsaturated and saturated triglycerides. The *cis* configuration in the hydrocarbon tails does not allow the triglyceride molecules to pack together very well, so the more unsaturated fatty acid residues a triglyceride has, the lower its melting point. Fats are solids because they contain more saturated fatty acid residues and have melting points above room temperature. Vegetable oils, on the other hand, are liquids because they contain a high percentage of unsaturated fatty acid residues and have melting points below room temperature. Coconut oil is an exception—although it contains mostly saturated fatty acid residues, this oil has a melting point characteristic of unsaturated triglycerides. The lower melting point is the result of coconut oil containing significant amounts of lauric acid, a relatively short saturated fatty acid. As is the case for fatty acids, the shorter the hydrocarbon tails the weaker the London forces between molecules and the lower the melting point.

One of the primary biological roles of triglycerides is to provide energy. On a gram-per-gram basis, triglycerides provide more than twice as many calories (energy) as do carbohydrates and proteins. In animals these molecules are stored in adipose tissue (fat) for future use (Figure 8.9). The layers of fat in seals, penguins, and other aquatic animals provide thermal insulation and bouyancy. Triglycerides also serve a protective function as a form of padding—the layers of fat surrounding our internal organs act as a cushion to protect against injury.

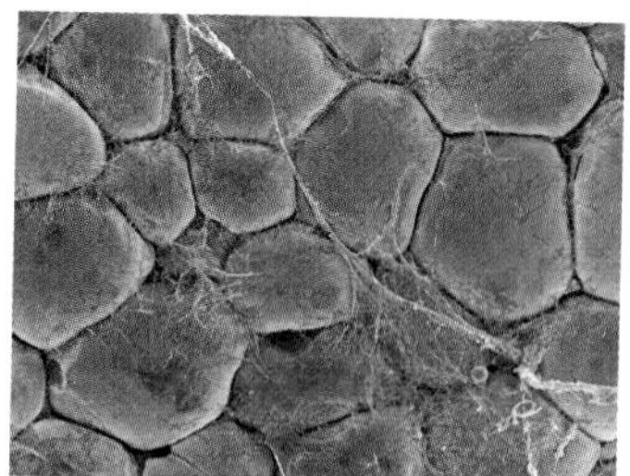

FIGURE 8.9

Adipocytes
Fat is stored in adipocytes (fat cells) for future use.

Source: SPL/Photo Researchers, Inc.

Important Reactions of Triglycerides

Section 6.3 showed that reacting alkenes with H_2 and platinum (Pt), a catalyst, converts them into alkanes. The carbon–carbon double bonds in unsaturated triglycerides can undergo the same catalytic reduction reaction (Figure 8.10*a*). In a commercial application called partial hydrogenation, vegetable oil is treated with H_2 and Pt, but the reaction is halted before all of the double bonds have been removed (Figure 8.10*b*). This converts liquid vegetable oil into a semisolid product called partially hydrogenated vegetable oil, which has a consistency that is good for cooking but does not contain the cholesterol that is present in solid triglycerides (fats), such as butter and lard, that come from animal sources. Partially hydrogenated vegetable oil is used as shortening, in solid and liquid margarine, and in a great number of other foods (Figure 8.11). Having fewer carbon–carbon double bonds than vegetable oil, partially hydrogenated vegetable oils do not spoil (oxidize) as rapidly (see below).

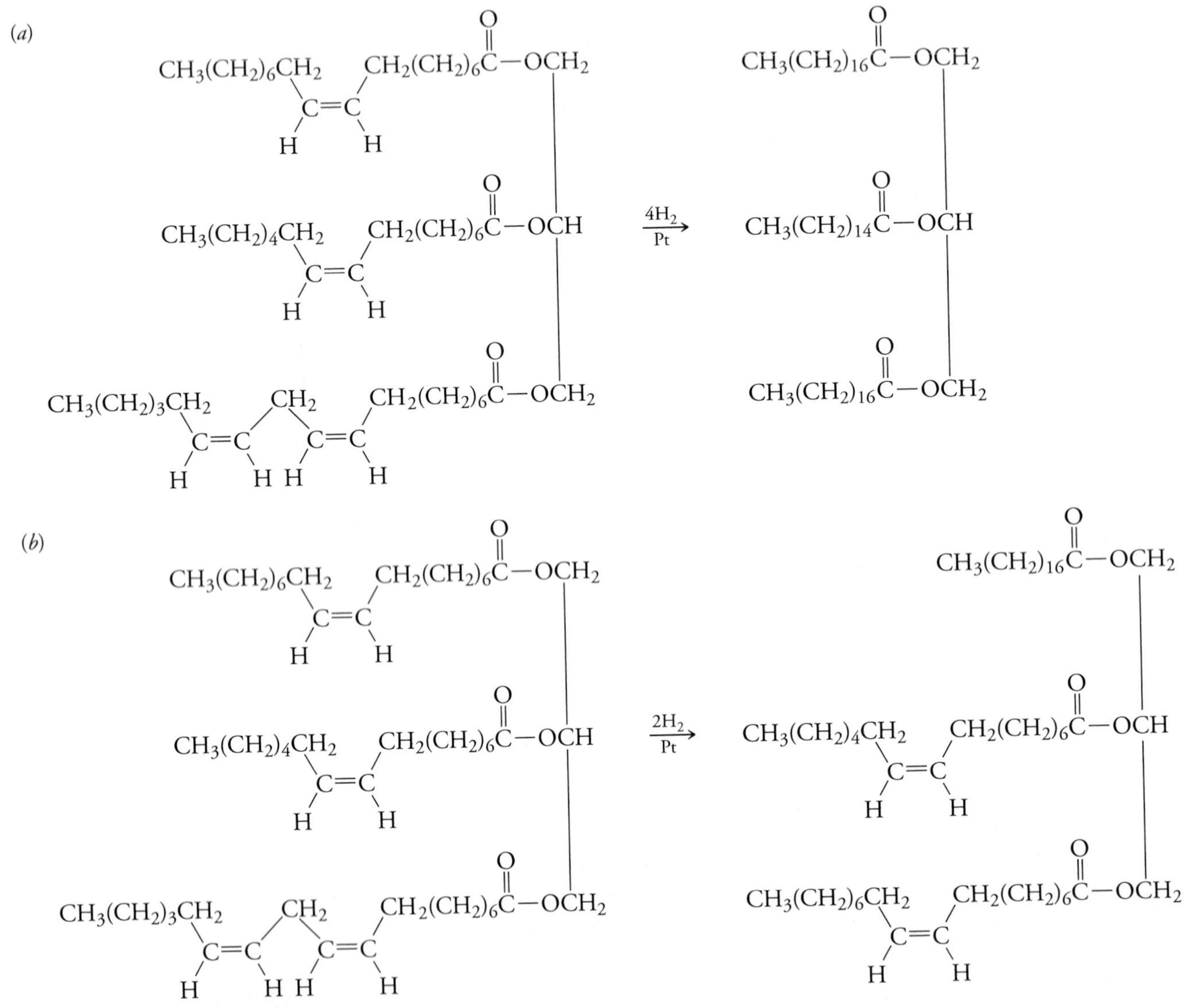

(*a*)

$CH_3(CH_2)_6CH_2$ $CH_2(CH_2)_6\overset{O}{\overset{\|}{C}}-OCH_2$

C=C

H H

$CH_3(CH_2)_4CH_2$ $CH_2(CH_2)_6\overset{O}{\overset{\|}{C}}-OCH$

C=C

H H

$CH_3(CH_2)_3CH_2$ CH_2 $CH_2(CH_2)_6\overset{O}{\overset{\|}{C}}-OCH_2$

C=C C=C

H H H H

$\xrightarrow[Pt]{4H_2}$

$CH_3(CH_2)_{16}\overset{O}{\overset{\|}{C}}-OCH_2$

$CH_3(CH_2)_{14}\overset{O}{\overset{\|}{C}}-OCH$

$CH_3(CH_2)_{16}\overset{O}{\overset{\|}{C}}-OCH_2$

(*b*)

$CH_3(CH_2)_6CH_2$ $CH_2(CH_2)_6\overset{O}{\overset{\|}{C}}-OCH_2$

C=C

H H

$CH_3(CH_2)_4CH_2$ $CH_2(CH_2)_6\overset{O}{\overset{\|}{C}}-OCH$

C=C

H H

$CH_3(CH_2)_3CH_2$ CH_2 $CH_2(CH_2)_6\overset{O}{\overset{\|}{C}}-OCH_2$

C=C C=C

H H H H

$\xrightarrow[Pt]{2H_2}$

$CH_3(CH_2)_{16}\overset{O}{\overset{\|}{C}}-OCH_2$

$CH_3(CH_2)_4CH_2$ $CH_2(CH_2)_6\overset{O}{\overset{\|}{C}}-OCH$

C=C

H H

$CH_3(CH_2)_6CH_2$ $CH_2(CH_2)_6\overset{O}{\overset{\|}{C}}-OCH_2$

C=C

H H

FIGURE | 8.10

Catalytic hydrogenation
(*a*) Reaction with H_2 and Pt converts an unsaturated triglyceride into a saturated one.
(*b*) In partial hydrogenation, only some of the carbon–carbon bonds are removed.

■ Triglycerides can be reduced, oxidized, and saponified.

The carbon–carbon double bonds in unsaturated triglycerides can be oxidized. When exposed to the oxygen (O_2) in air, for example, the double bonds can be broken to produce small organic molecules that have unpleasant odors (Figure 8.12). When this happens, the triglyceride has spoiled or has gone rancid. Having a greater degree of unsaturation, vegetable oils spoil more quickly than do fats. To make oils last longer you can keep them refrigerated (the lower the temperature, the slower the oxidation reaction) and in tightly capped bottles (less O_2 exposure). Removing some of the double bonds by partial hydrogenation also slows spoilage. Sometimes antioxidants, molecules that slow oxidation, are added to foods.

FIGURE 8.11

Partially hydrogenated vegetable oils
Many foods contain partially hydrogenated vegetable oil.

Source: Justine Sullivan/Stone/Getty Images.

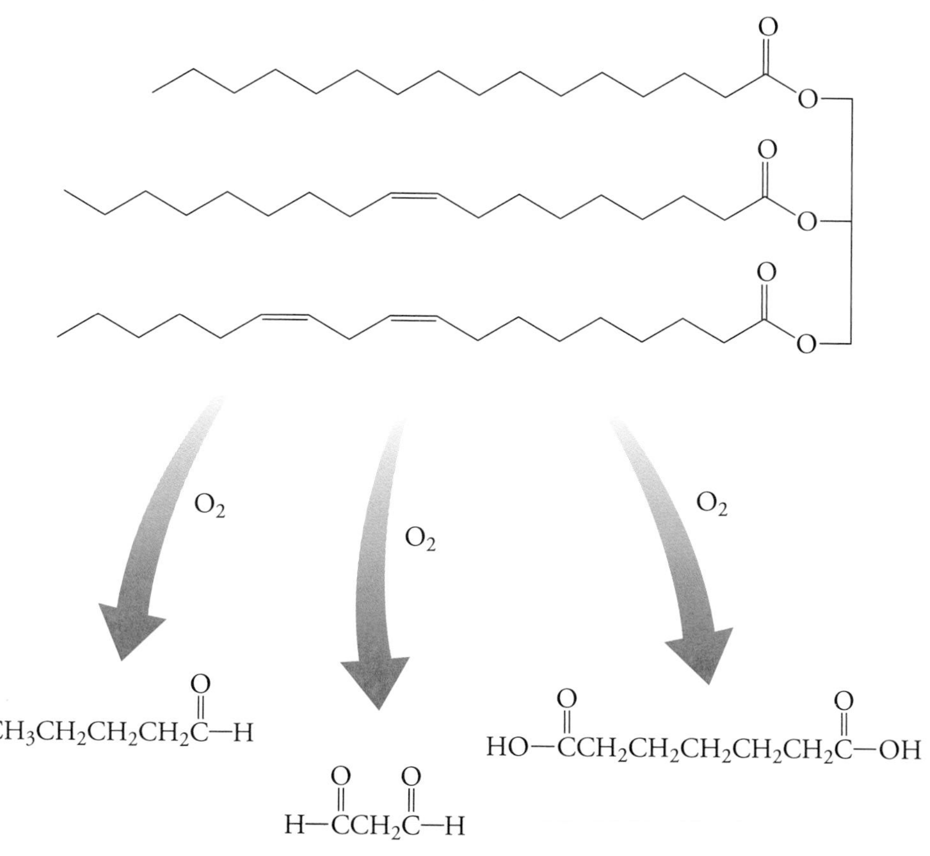

FIGURE 8.12

Oxidation of triglycerides
Oxygen (O_2) can oxidize the hydrocarbon tails of unsaturated triglycerides, producing small organic molecules that have unpleasant odors.

A third important reaction of triglycerides is hydrolysis of their ester groups in the presence of OH^-. This hydrolysis reaction, first introduced in Section 6.4, is also known as **saponification** (soap making). When triglycerides are saponified, glycerol and fatty acid salts (soap) are produced, (Figure 8.13). As we saw in Section 7.5, soaps are amphipathic compounds that form monolayers and micelles (Figure 7.17).

The particular source of OH^- used in saponification can affect the properties of the soap that is produced. Sodium hydroxide (NaOH) gives a solid soap, while potassium hydroxide (KOH) results in a liquid soap. The degree of unsaturation in the triglyceride is also important; greater amounts of unsaturation result in softer or liquid soaps. Glycerol, the other product of saponification, is used in lotions and creams.

Hydrolysis plays a role in the digestion of fats and oils, a process that takes place in the large intestine. Digestive enzymes catalyze the hydrolysis of triglycerides into their component parts, which move through the walls of the intestine and are absorbed. We will discuss this process in more detail in Chapter 15.

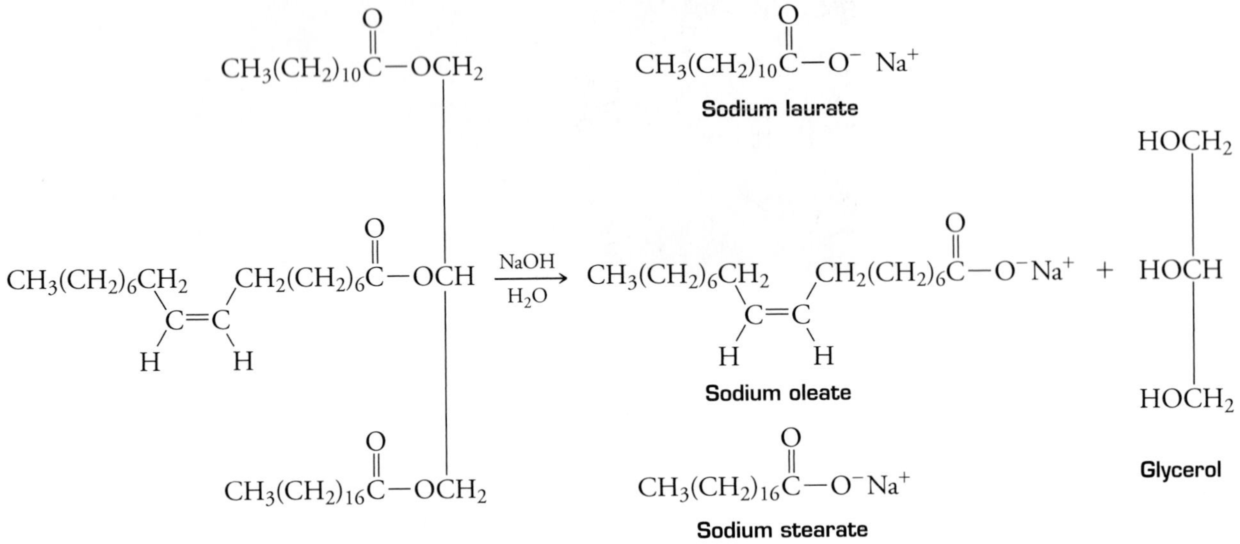

$$CH_3(CH_2)_{10}C(=O){-}OCH_2$$
$$CH_3(CH_2)_6CH{=}CHCH_2(CH_2)_6C(=O){-}OCH \quad (cis,\ H\ and\ H\ on\ same\ side)$$
$$CH_3(CH_2)_{16}C(=O){-}OCH_2$$

$$\xrightarrow[H_2O]{NaOH}$$

$$CH_3(CH_2)_{10}C(=O){-}O^- Na^+$$

Sodium laurate

$$CH_3(CH_2)_6CH{=}CHCH_2(CH_2)_6C(=O){-}O^- Na^+$$

Sodium oleate

$$CH_3(CH_2)_{16}C(=O){-}O^- Na^+$$

Sodium stearate

$$+ \quad HOCH_2{-}HOCH{-}HOCH_2$$

Glycerol

FIGURE 8.13

Saponification
Saponification (hydrolysis of fats in the presence of OH^-) converts triglycerides into fatty acid salts (soaps) and glycerol. The ionic compound NaOH is the source of the OH^- used in the reaction shown here.

SAMPLE PROBLEM 8.3

Properties of triglycerides

a. Draw a triglyceride that would probably exist as a liquid at room temperature.
b. Draw another that would probably be a solid at room temperature.

Strategy
Consider the effect of unsaturation on the London force interactions that can take place between triglycerides.

Solution
Triglycerides that contain mostly unsaturated fatty acids are expected to be liquids and those that contain mostly saturated fatty acids should be solids. Possible answers include the following triglycerides:

a.

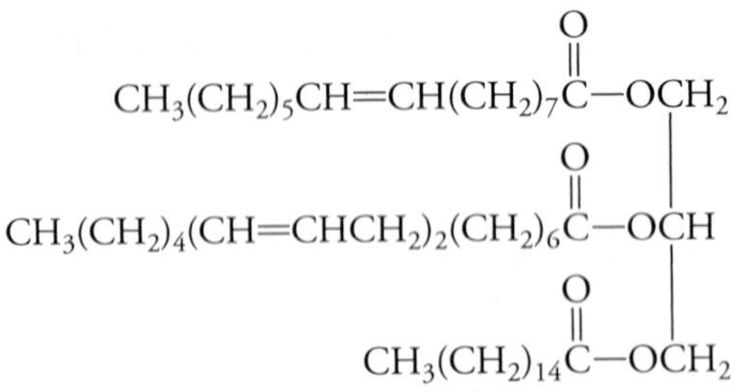

$$CH_3(CH_2)_5CH{=}CH(CH_2)_7C(=O){-}OCH_2$$
$$CH_3(CH_2)_4(CH{=}CHCH_2)_2(CH_2)_6C(=O){-}OCH$$
$$CH_3(CH_2)_{14}C(=O){-}OCH_2$$

b.

$$CH_3(CH_2)_{14}C(=O){-}OCH_2$$
$$CH_3(CH_2)_{14}C(=O){-}OCH$$
$$CH_3(CH_2)_{16}C(=O){-}OCH_2$$

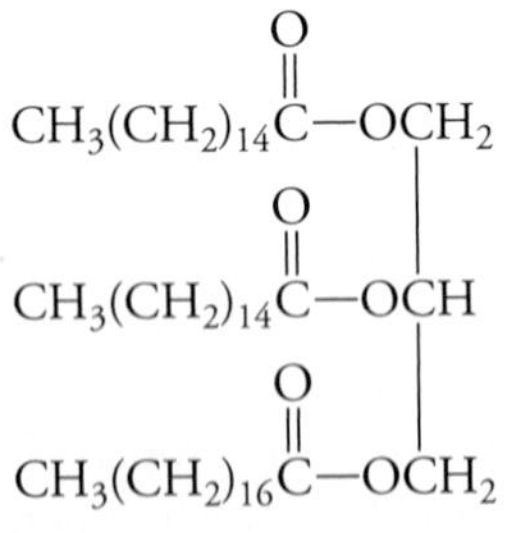

PRACTICE PROBLEM 8.3

Draw and name the products obtained when the triglyceride in Figure 8.13 is saponified using KOH.

HEALTH Link — *Trans* Fats

When unsaturated vegetable oils are partially hydrogenated (Section 8.3), more chemistry takes place than just the conversion of carbon–carbon double bonds into single bonds. As unsaturated vegetable oils, which contain *cis*-double bonds to begin with, come into contact with the catalyst, some of the double bonds are converted to the *trans* stereoisomer (Figure 8.14). This means that any food containing partially hydrogenated vegetable oils also contains some *trans* fats. Diets high in *trans* fats have been linked to lowered HDL levels (Section 8.5), an increased risk of heart disease, and changes in membrane structure (Section 8.7).

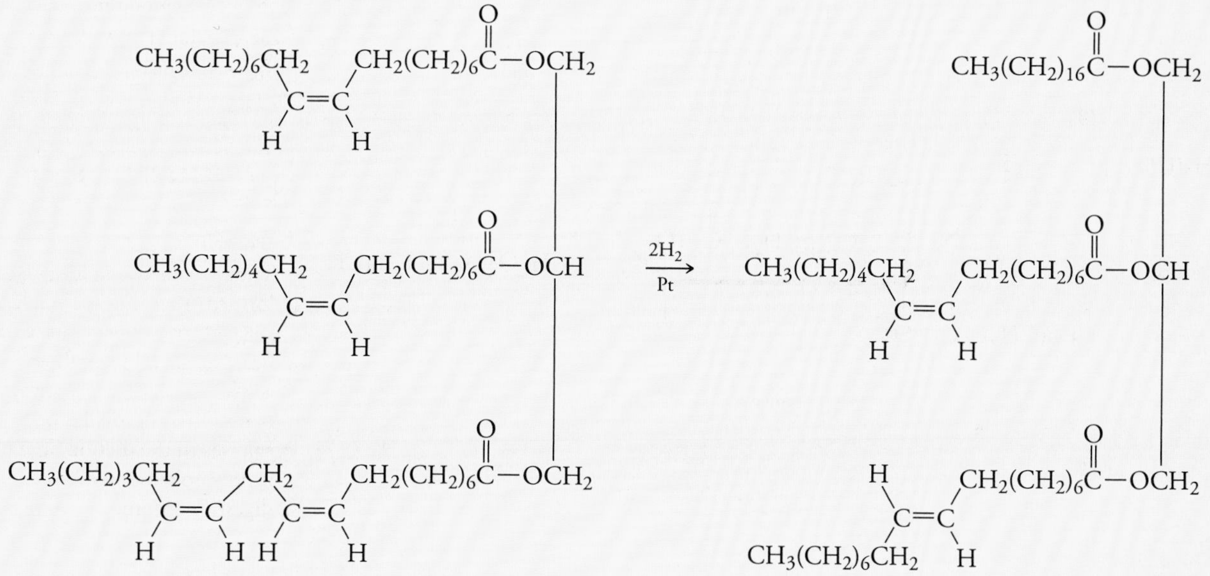

$CH_3(CH_2)_6CH_2$ and $CH_2(CH_2)_6C(=O)-OCH_2$ on C=C, with H and H (cis)

$CH_3(CH_2)_4CH_2$ and $CH_2(CH_2)_6C(=O)-OCH$ on C=C, with H and H (cis)

$CH_3(CH_2)_3CH_2$, CH_2, $CH_2(CH_2)_6C(=O)-OCH_2$ on C=C–C=C, with H, H H, H

$\xrightarrow[Pt]{2H_2}$

$CH_3(CH_2)_{16}C(=O)-OCH_2$

$CH_3(CH_2)_4CH_2$ and $CH_2(CH_2)_6C(=O)-OCH$ on C=C, with H and H

H and $CH_2(CH_2)_6C(=O)-OCH_2$ on C=C, with $CH_3(CH_2)_6CH_2$ and H

FIGURE 8.14

***Trans* fats**
Partial hydrogenation of unsaturated vegetable oil converts some of the carbon–carbon double bonds from *cis* into *trans* isomers.

Recognizing the health risks posed by *trans* fats in the diet, the U.S. Food and Drug Administration now requires that the Nutrition Facts label on packaged foods include information on the amount of *trans* fats present. In late 2006, the New York City Board of Health imposed a ban on restaurant use of *trans* fats. By July 2008, food served by restaurants within the city will not be allowed to contain any artificial *trans* fats. Other cities are considering similar bans, and some fast food restaurants have stopped using *trans* fats.

Some foods contain natural *trans* fats, which explains New York City's use of the word "artificial" in their ban. *Trans* fatty acids are naturally present in beef and in dairy products, including milk, butter, and cheese. These *trans* isomers are produced during the biohydrogenation that takes place in the digestive system of cattle. In this process, bacteria convert unsaturated fatty acids into saturated ones. As is the case for *trans* fats that arise during incomplete catalytic hydrogenation of vegetable oils (Section 8.3), incomplete biohydrogenation gives rise to natural *trans* fatty acids.

HEALTH Link — *Olestra*

Because triglycerides (fats and oils), on a gram-per-gram basis, contain more than twice as many calories as do carbohydrates and proteins, there has been interest in developing a low-calorie fat substitute. In 1996, the U.S. Food and Drug Administration (FDA) approved the fat substitute olestra for use in foods (Figure 8.15). Although olestra, like triglycerides, is a fatty acid ester, digestive enzymes are unable to catalyze its breakdown. As a result, this "fake fat" passes through the digestive system untouched and provides no calories to the diet. Some health professionals are concerned that use of this fat substitute may reduce the body's ability to absorb vitamin A, vitamin E, and other fat-soluble (water-insoluble) vitamins. Some who eat foods containing olestra experience what is euphemistically described as lower gastrointestinal distress. Until 2003, the FDA required products containing olestra to carry a label that warned of these potential vitamin- and abdomen-related side effects.

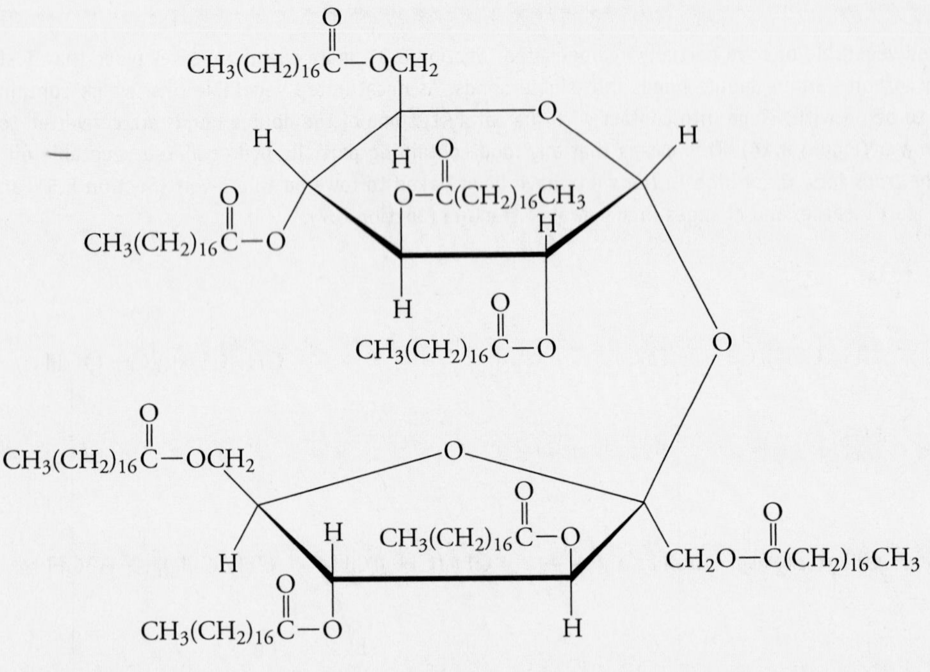

Olestra

FIGURE 8.15

Olestra
The fat substitute olestra is formed when sucrose (table sugar) forms ester bonds with between six and eight fatty acids. The octaester is shown here. Olestra provides no calories to the diet because it is not broken down by digestive enzymes.

8.4 PHOSPHOLIPIDS AND GLYCOLIPIDS

Cell **membranes** are *a double layer of amphipathic lipids* that block the movement of polar molecules and ions. Figure 8.16 shows the basic layout of a membrane, with the

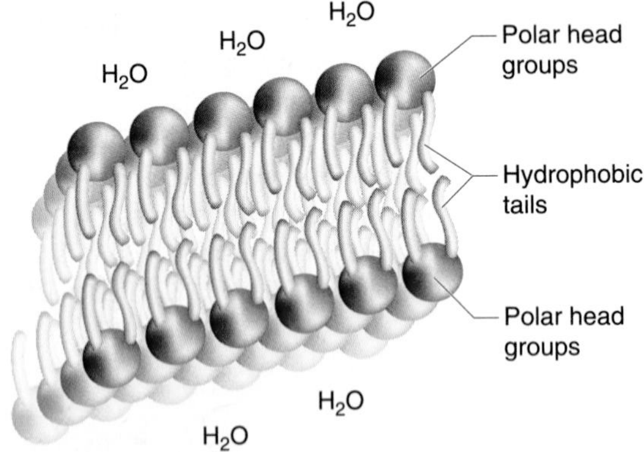

FIGURE 8.16

Membranes
Membranes are double layers of amphipathic lipids.

hydrophilic head group of each lipid on the membrane surface and the hydrophobic tail groups pointing inward. We will read more about membranes in Section 8.7, but for now let us focus our attention on the structure of phospholipids and glycolipids, the amphipathic lipids commonly found in membranes.

Phospholipids

Phospholipids get their name from the fact that *phosphate ion* (PO_4^{3-}) is one of the components used in their formation. The two classes of phospholipids, called glycerophospholipids and sphingolipids, will be considered separately.

Glycerophospholipids are made by combining *glycerol, two fatty acids, one phosphate group, and one alcohol-containing compound.*

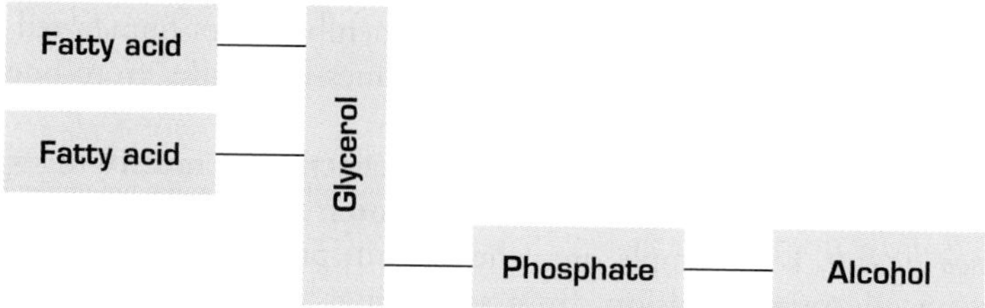

Structure of a glycerophospholipid

Fatty acid residues are attached to the glycerol backbone by ester bonds. Phosphoester bonds, which closely resemble carboxylic ester bonds, connect the phosphate group to both glycerol and the alcohol residue (Figure 8.17). As shown in this figure, different alcohol components can be incorporated into glycerophospholipids.

$$CH_3(CH_2)_7CH{=}CH(CH_2)_7\overset{\overset{\displaystyle O}{\|}}{C}{-}OCH_2$$

$$CH_3(CH_2)_{14}\overset{\overset{\displaystyle O}{\|}}{C}{-}OCH$$

$$CH_2O{-}\overset{\overset{\displaystyle O}{\|}}{\underset{\underset{\displaystyle O^-}{|}}{P}}{-}OX$$

Glycerophospholipid	X
Phosphatidylcholine	$-CH_2CH_2\overset{+}{N}(CH_3)_3$
Phosphatidylethanolamine	$-CH_2CH_2NH_3^+$
Phosphatidylserine	$-CH_2\underset{\underset{\displaystyle {}^+NH_3}{|}}{C}H\overset{\overset{\displaystyle O}{\|}}{C}{-}O^-$

FIGURE 8.17

Common glycerophospholipids Phosphatidylcholines, phosphatidylethanolamines, and phosphatidylserines are three families of glycerophospholipids. They are considered families of compounds because fatty acid residues other than the ones shown here can be present in the glycerophospholipids.

Phosphatidylcholine (lecithin) is the phospholipid usually found in the highest percentages in plants and animals. It is the major phospholipid in cell membranes and is also the main lipid component of the lipoproteins that are responsible for transporting lipids through the blood (Section 8.5).

Phosphatidylethanolamine (cephalin) is the next most common phospholipid in plants and animals, and is also a major contributor to membrane structure. In *E. coli* bacteria this phospholipid plays an important role in active transport (Section 8.7) of the sugar lactose—if phosphatidylethanolamine is not present, the enzyme lactose permease is unable to move lactose across the membrane and into the cell.

Phosphatidylserine is less common than the two phospholipids just mentioned, but it is also important to the structure of membranes. This lipid is found at its highest levels in myelin (the sheath around nerve cells) of brain tissue.

- Phospholipids are formed by combining glycerol, two fatty acids, phosphate, and an alcohol. In some phospholipids, sphingosine replaces glycerol and one fatty acid.

Each of the phospholipids just discussed is considered a family of glycerophospholipids because their fatty acid components can vary. The membranes of human red blood cells, for example, contain 21 different phosphatidylcholines that differ from one another in the particular fatty acid residues that they contain.

Because they are amphipathic, phospholipids are effective as **emulsifying agents**, compounds that make or stabilize emulsions. An **emulsion** is a *colloid* (Section 7.8) *formed by combining two liquids*. For example, it is the lecithin present in egg whites that keeps mayonnaise, an oil–water emulsion, from separating. Lecithin is also widely used in chocolate, baked goods, and cosmetics.

Sphingolipids are the other class of phospholipids. All sphingolipids contain the alcohol sphingosine, and sphingolipids that belong to the phospholipid family contain phosphate attached to both sphingosine and an alcohol residue.

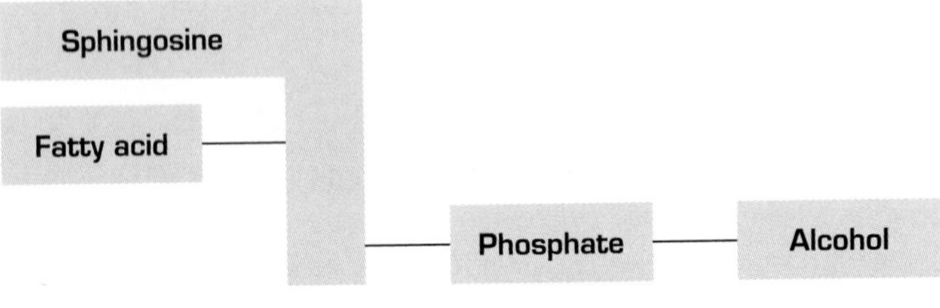

Structure of a sphingolipid

As implied by their name, the sphingomyelin family of sphingolipids (Figure 8.18) are prevalent in the myelin sheath that surrounds nerve cells.

Glycolipids

Glycolipids are lipids that contain a sugar residue. In many cases this residue is attached to a sphingosine backbone.

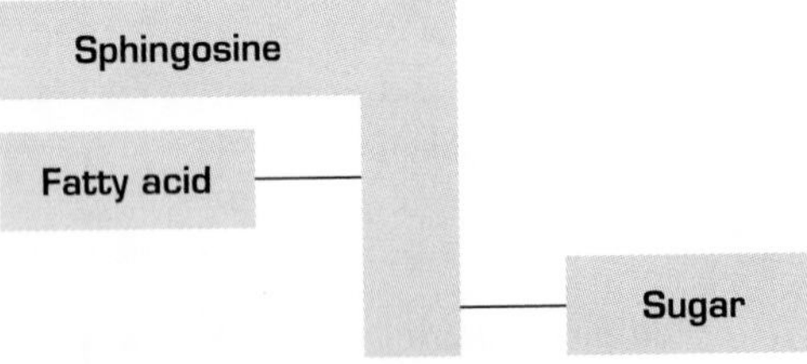

Structure of a glycolipid

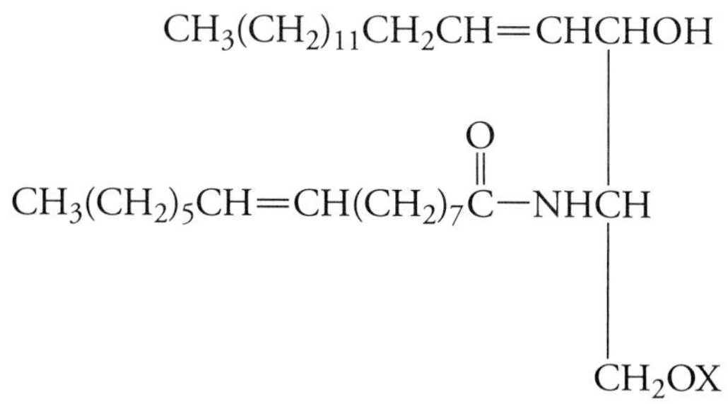

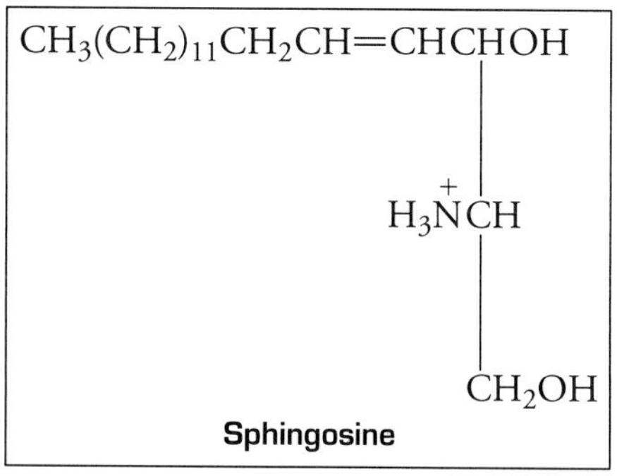

$$\begin{array}{l} CH_3(CH_2)_{11}CH_2CH{=}CHCHOH \\ CH_3(CH_2)_5CH{=}CH(CH_2)_7\overset{O}{\overset{\|}{C}}{-}NHCH \\ CH_2OX \end{array}$$

$$\begin{array}{l} CH_3(CH_2)_{11}CH_2CH{=}CHCHOH \\ H_3\overset{+}{N}CH \\ CH_2OH \end{array}$$

Sphingosine

Sphingolipid

X

Sphingomyelin $-\overset{O}{\overset{\|}{P}}(O^-){-}OCH_2CH_2\overset{+}{N}(CH_3)_3$ or $-\overset{O}{\overset{\|}{P}}(O^-){-}OCH_2CH_2\overset{+}{N}H_3$

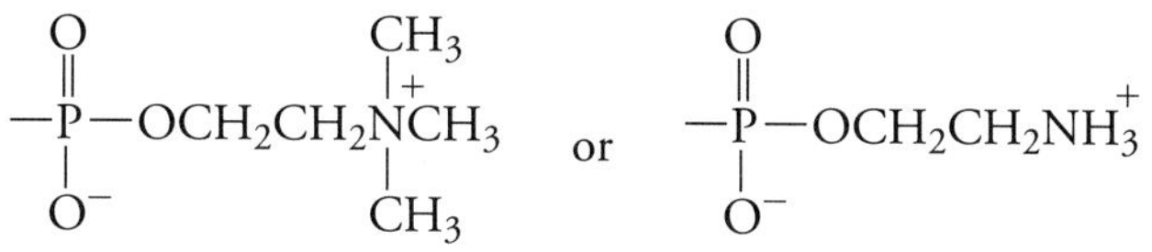

Glycolipid

Cerebroside

Ganglioside

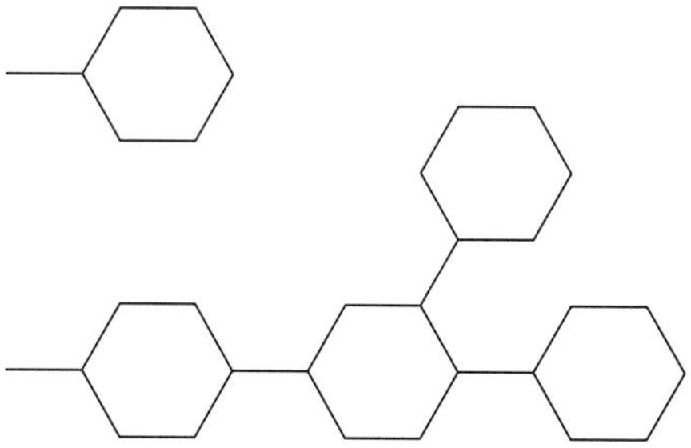

FIGURE 8.18

Common sphingolipids and glycolipids
Sphingolipids and many glycolipids contain a sphingosine backbone. These families of compounds can contain other fatty acid residues than the one shown. Hexagons are used to represent simple sugars. We will study these compounds in Chapter 12.

Simple sugars (consisting of only one ring), such as glucose, are used to produce the glycolipids called **cerebrosides**. **Gangliosides**, another class of glycolipids, are made using a chain of simple sugars (Figure 8.18). Cerebrosides are found at nerve synapses and in the brain, while gangliosides are important in nerve membranes and act as cell surface receptors for hormones and drugs. The sugar molecules in some gangliosides help determine blood type (Chapter 12, Health Link: Blood Type).

- Glycolipids are formed by combining sphingosine, a fatty acid, and a sugar.

SAMPLE PROBLEM 8.4

The structure of phospholipids

How many products are obtained when the phosphatidylethanolamine below is saponified? Hint: Each of the ester bonds in the molecule is hydrolyzed.

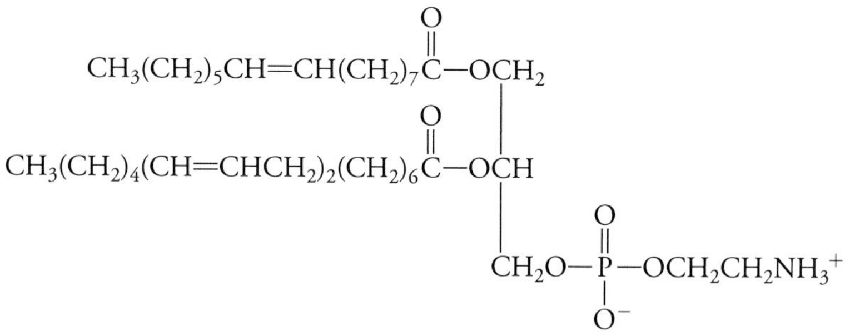

$$\begin{array}{l} CH_3(CH_2)_5CH{=}CH(CH_2)_7\overset{O}{\overset{\|}{C}}{-}OCH_2 \\ CH_3(CH_2)_4(CH{=}CHCH_2)_2(CH_2)_6\overset{O}{\overset{\|}{C}}{-}OCH \\ CH_2O{-}\overset{O}{\overset{\|}{P}}(O^-){-}OCH_2CH_2NH_3^+ \end{array}$$

Strategy

Phospholipids contain two different types of ester bonds, those formed by combining a carboxylic acid with an alcohol and those formed by combining phosphate with an alcohol. When hydrolyzed, the ester groups are split to give the components from which they are formed.

Solution

Five. When the glycerophospholipid is broken apart into its components, two fatty acid anions, glycerol, phosphate, and ethanolamine ($HOCH_2CH_2NH_3^+$), are obtained.

PRACTICE PROBLEM 8.4

Draw three different sphingomyelin molecules that might be found in the membranes of nerve cells.

8.5 STEROIDS

Steroids are a class of lipids that share the same basic fused ring structure—three 6-carbon atom rings and one 5-carbon atom ring.

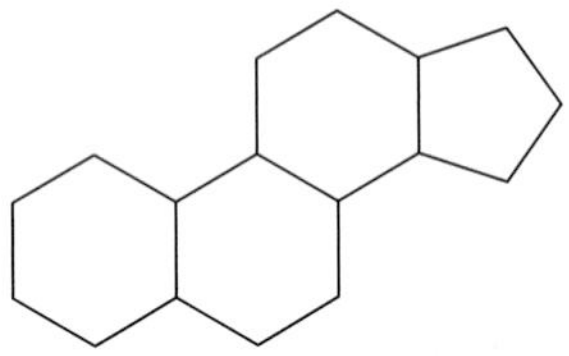

Ring structure of a steroid

In this section we will consider cholesterol, steroid hormones, and bile salts, three of the important types of steroids.

- Cholesterol, steroid hormones, and bile salts all contain the characteristic steroid ring structure.

Cholesterol is the steroid found most often in humans and other animals. It is not present in plants, so your choice of diet can affect the amount of cholesterol that you take in. Regardless of what you eat, your body will contain some cholesterol because it is manufactured in the liver.

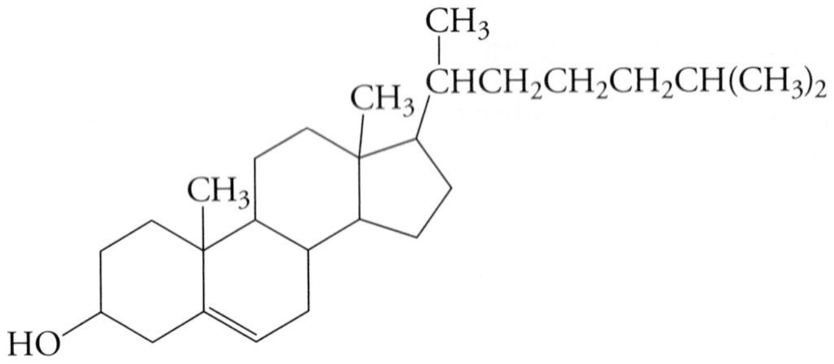

CH_3, $CHCH_2CH_2CH_2CH(CH_3)_2$, CH_3, CH_3, HO

Cholesterol

In cholesterol, the nonpolar rings and hydrocarbon chain are hydrophobic and the —OH group, which makes up a much smaller part of the molecule, is hydrophilic. Overall, this makes the molecule somewhat amphipathic and, like phospholipids and glycolipids, cholesterol can be a component of cell membranes. The primary biological use of cholesterol, however, is as a starting material for the synthesis of other steroids.

Cholesterol is not freely soluble in water, so it and other lipids are transported through the blood as suspensions by lipid–protein complexes called lipoproteins (Figure 8.19). Lipoproteins called chylomicrons mainly carry triglycerides, while very low density lipoproteins (VLDLs) carry triglycerides, phospholipids, and cholesterol. The major

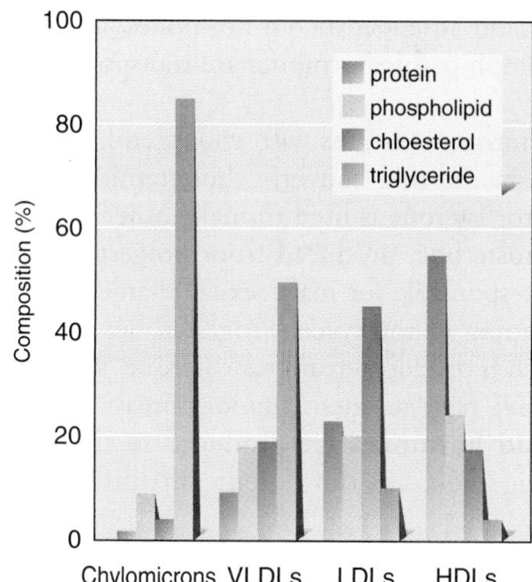

FIGURE 8.19

Lipoproteins

Chylomicrons, VLDLs (very low density lipoproteins), LDLs (low density lipoproteins), and HDLs (high density lipoproteins) are responsible for transporting various lipids through the blood.

function of low density lipoproteins (LDLs) is to transport cholesterol and phospholipids from the liver to the cells, where they are incorporated into membranes or, in the case of cholesterol, transformed into other steroids. High density lipoproteins (HDLs) transport cholesterol and phospholipids from the cells back to the liver.

Low HDL and high LDL levels in the blood are warning signs of atherosclerosis, the buildup of cholesterol-containing deposits in arteries (Figure 8.20). This deposit reduces the diameter of the arteries that supply the heart and organs with blood, making the arteries more easily blocked. When reduced blood flow to the heart or brain deprives cells of O_2 and vital nutrients, heart attacks or strokes can occur.

If you have a cholesterol screening test done, the report will typically list your HDL, LDL, and total cholesterol levels, as well as your cholesterol ratio. To be in the normal

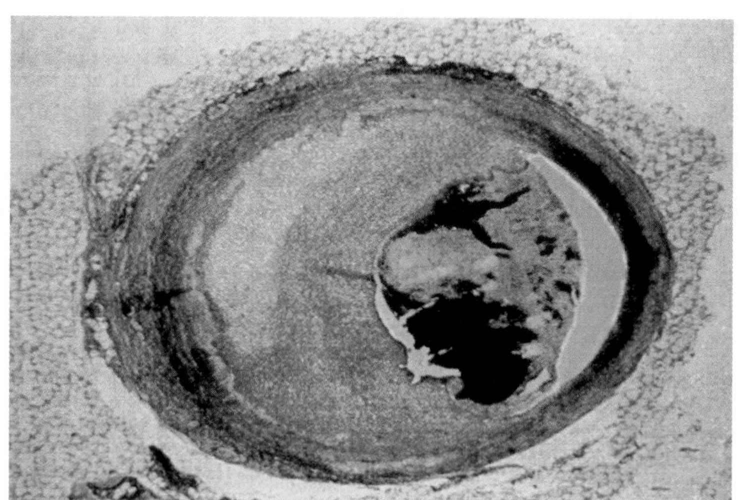

FIGURE 8.20

Atherosclerosis

Atherosclerosis involves the buildup of cholesterol-containing deposits in arteries. In this cross-section of a human coronary artery, the only unblocked portion is shown in blue at the right.

Source: GJLP/Photo Researchers, Inc.

range, HDL serum concentrations should be greater than 40 mg/dL, LDL concentrations less than 100 mg/dL, and total cholesterol (cholesterol present in HDL, LDL, and VLDL) less than 200 mg/dL. A healthy person's cholesterol ratio (LDL/HDL) should be less than 2.5.

Steroid Hormones

Hormones, *molecules that regulate the function of organs and tissues*, come in a variety of forms. Some, such as sex hormones and adrenocorticoid hormones, are steroids (Figure 8.21). After being produced, these amphipathic hormones are transported to their target tissues by protein carriers.

The manufacture of the steroid hormones begins with cholesterol. Shortening of the hydrocarbon chain and alterations on the ring converts cholesterol into progesterone, one of the female **sex hormones**. Progesterone is used to make other sex hormones and the adrenocorticoid hormones. Testosterone, produced from progesterone, is the most important male sex hormone. It is responsible for male sexual characteristics, including the growth of facial hair and deepening of the voice. Progesterone, which has already been mentioned, and estradiol, which is produced from testosterone, are two of the more important female sex hormones. They regulate menstruation, breast development, and other female traits. **Adrenocorticoid hormones** are produced in the adrenal glands, starting from progesterone. Among these are the potent anti-inflammatory agents cortisol and cortisone.

Bile Salts

Bile salts, produced from cholesterol, are amphipathic. Glycocholate, taurocholate (Figure 8.21), and the other bile salts are released from the gallbladder into the small intestine, where they aid digestion by forming emulsions with dietary lipids.

8.6 EICOSANOIDS

The lipids called **eicosanoids** are hormones that are derived from arachidonic acid (Figure 8.22) and other essential twenty-carbon fatty acids (the prefix *eicos* means 20). The term "essential" is used to refer to nutrients that are essential in the diet because they are not produced in the body. When hydrolyzed from a certain phospholipid by hormone action, arachidonic acid undergoes reactions that transform it into the various eicosanoids—prostaglandins, thromboxanes, and leukotrienes. Prostaglandins have a wide range of biological effects, including causing pain, inflammation, and fever; affecting blood pressure; and, in the case of the prostaglandin PGE_2, inducing labor. Thromboxanes, such as thromboxane A_2, are involved in blood clotting, while the leukotrienes, including leukotriene A_4, induce muscle contractions in the lungs and are linked to asthma attacks. Some anti-asthma drugs block the production of leukotrienes (Figure 8.23).

Nonsteroidal anti-inflammatory drugs (NSAIDs) such as aspirin and ibuprofen reduce pain, fever, and inflammation by blocking the action of an enzyme involved in the conversion of arachidonic acid into prostaglandins and thromboxanes. There are two forms of this enzyme, COX-1 and COX-2 (COX stands for cyclooxygenase). COX-2, which is activated when you are injured or ill, catalyzes the production of eicosanoids that cause pain, fever, and inflammation. By blocking the action of COX-2, NSAIDs alleviate these symptoms. NSAIDs also inhibit the COX-1 enzyme, which is responsible for the production of the prostaglandins and thromboxanes that control blood pressure and blood clotting. Because NSAIDs block COX-1, use of these drugs can result in ulcers and kidney damage, side effects not seen with COX-2 specific

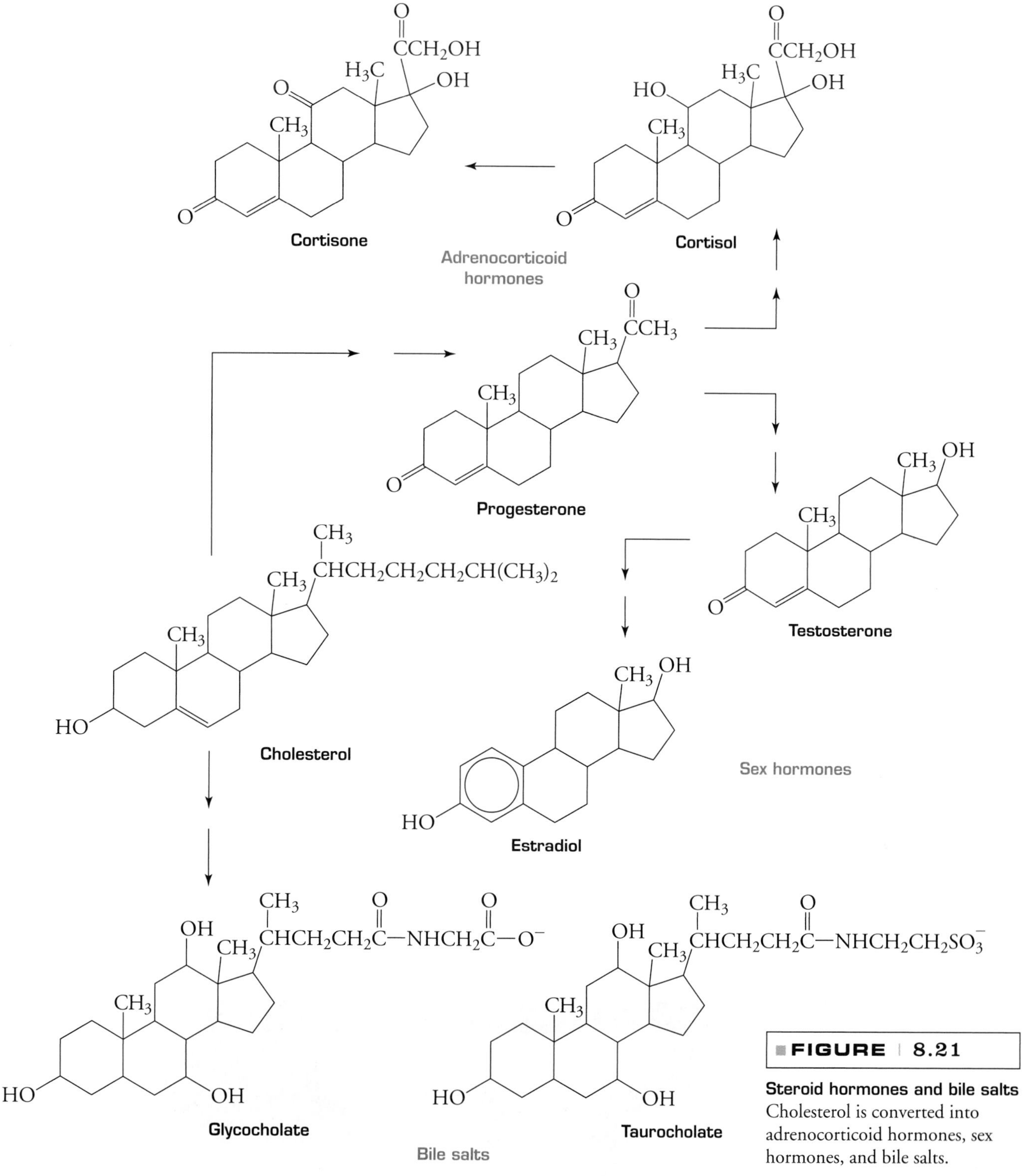

Cortisone

Cortisol

Adrenocorticoid hormones

Progesterone

Testosterone

Cholesterol

Estradiol

Sex hormones

Glycocholate

Taurocholate

Bile salts

FIGURE 8.21

Steroid hormones and bile salts Cholesterol is converted into adrenocorticoid hormones, sex hormones, and bile salts.

drugs, Vioxx, Bextra, and Celebrex. Use of these COX-2 inhibitors is not risk-free, however. Studies have shown that users of these drugs may have an increased risk of heart attack. As of early 2005, Vioxx and Bextra had been pulled off the market. Celebrex comes with a warning that its use may lead to an increased risk of heart problems and stroke.

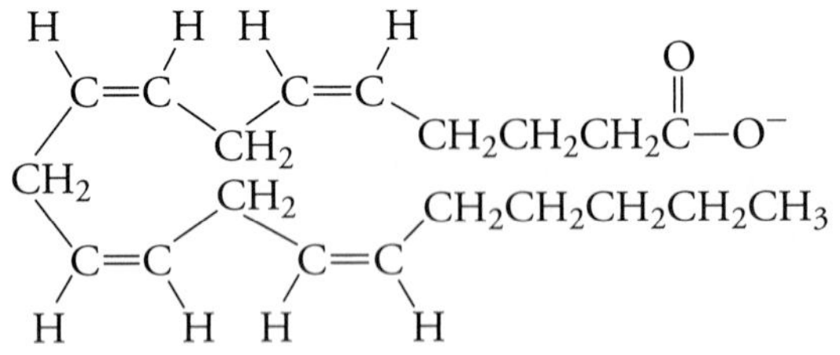

Arachidonic acid

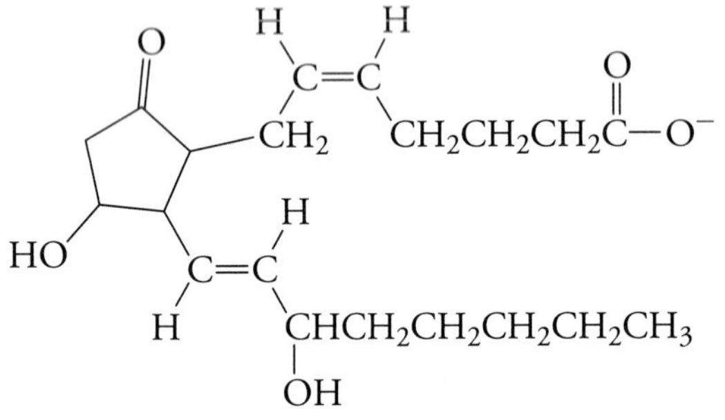

PGE_2
(a prostaglandin)

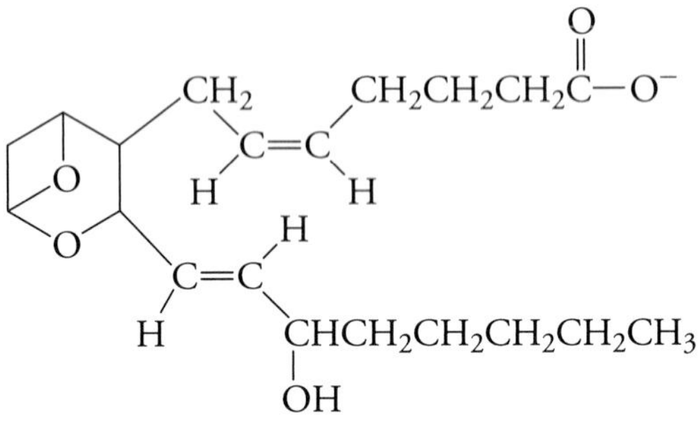

Thromboxane A_2

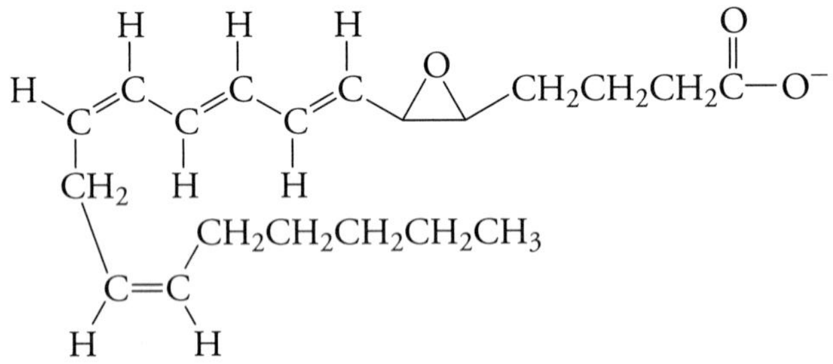

Leukotriene A_4

FIGURE 8.22

Eicosanoids
Arachidonic acid, a polyunsaturated fatty acid, is converted into prostaglandins, thromboxanes, and leukotrienes.

FIGURE 8.23

Anti-asthma drugs
Some of the drugs used to treat asthma block the formation of leukotrienes.

Source: Paul Windsor/Taxi/Getty.

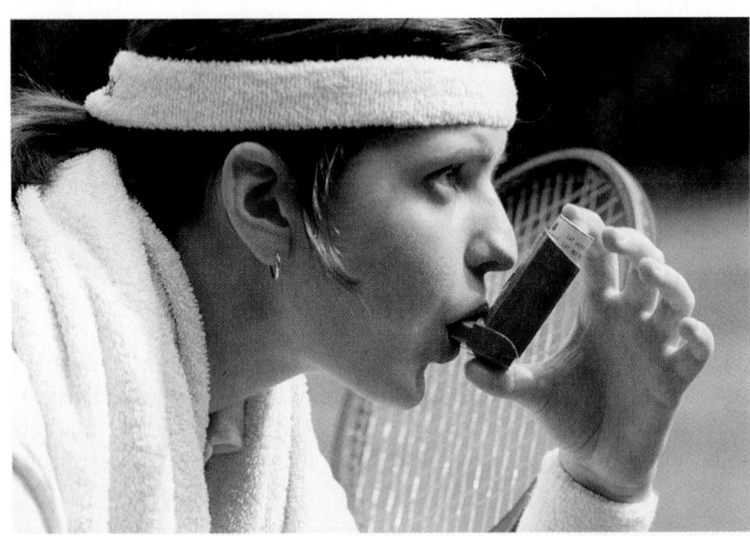

HEALTH Link: Anabolic Steroids

The term anabolic refers to synthesis or growth, and an anabolic hormone is one that stimulates this process. The male sex hormone testosterone (Figure 8.21) is anabolic, and muscle building is one of its anabolic effects. Some athletes get a jump on the competition by taking performance-enhancing anabolic hormones to increase their muscle mass (Figure 8.24).

When considering the use of these drugs, athletes might want to ask themselves whether competing with what many consider to be an unfair advantage is worth the potential harm that they may do to themselves. There are many health risks associated with the use of anabolic steroids, including acne, high blood pressure, increased aggressiveness, testicular atrophy, fluid retention, sleep disturbances, and irreversible liver damage. Women who use anabolic steroids risk the growth of facial hair, baldness, deepening of the voice, and menstrual irregularities.

Besides the health risks, athletes who use these restricted drugs may lose medals or be banned from competition. Not all anabolic steroids are banned by all sports associations or committees, however. Professional baseball, for example, only recently prohibited the use of androstenedione, a naturally occurring steroid that the body converts into testosterone. In 1998, Mark McGwire hit his record-breaking 70 home runs while taking this supplement.

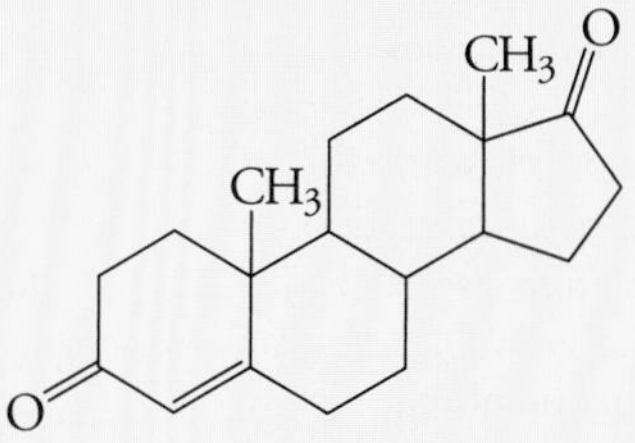

CH_3 O CH_3 O

Androstenedione

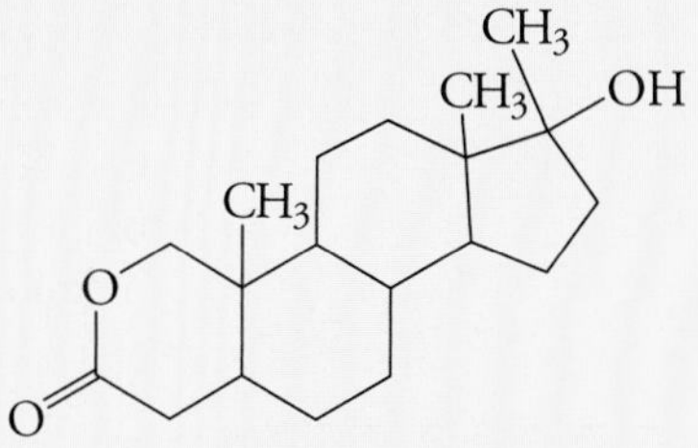

CH_3 CH_3 OH CH_3 O O

Oxandrolone

O CH_3 $O-\overset{\|}{C}(CH_2)_{10}CH_3$ O

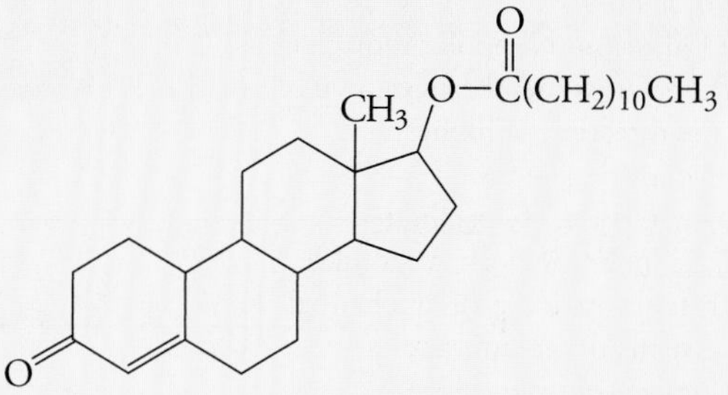

Nandrolone laurate

FIGURE 8.24

Anabolic steroid hormones
Muscle building is one of the effects of testosterone and related steroid hormones, such as androstenedione, oxandrolone, and nandrolone laurate.

8.7 MEMBRANES

Membranes, barriers that surround cells or that separate one part of a cell from another, are a bilayer of amphipathic lipids—usually phospholipids, glycolipids, and cholesterol (Figure 8.25). The lipids are arranged so that their hydrophilic heads interact with one another and with water at the surface of the membrane, and so that their hydrophobic tails interact with one another at the center of the membrane. Many of the fatty acid residues present in membrane lipids are unsaturated, so a significant number of the hydrophobic tails have the *cis* double bonds that do not allow tight packing of the membrane structure. Because the loosely packed membrane lipids have room to move around, the membranes are fluid. The fused ring portion of cholesterol is fairly inflexible and the presence of this lipid is thought to add rigidity to membrane structure.

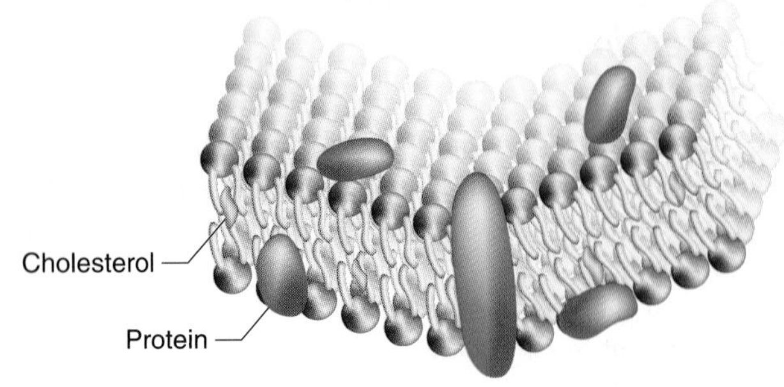

FIGURE | 8.25

The fluid mosaic model of membrane structure
Membranes are made up of glycerophospholipids, sphingolipids, glycolipids, cholesterol, and proteins. The unsaturated hydrocarbon tails on the interior of the membrane are in motion (fluid) because the kinks caused by unsaturation do not allow them to pack tightly together. Cholesterol gives some rigidity to the membrane structure. Proteins are embedded in the membrane, like tiles in a mosaic.

Proteins are another important membrane component. We will discuss these molecules in some detail in Chapter 13, but for now let us just say that proteins are large biomolecules that can have hydrophilic and hydrophobic parts. Protein molecules are inlayed into the membrane, either on the membrane surface, in the interior, or stretching from one side to the other. This view of membrane structure, which talks in terms of mobile phospholipids and embedded proteins, is called the **fluid mosaic model** of membrane structure (Figure 8.25).

Membranes are selectively permeable. Water, small nonpolar molecules (including O_2, N_2, and CO_2), and amphipathic steroids are able to diffuse directly through membranes (Figure 8.26*a*). As discussed in Chapter 7, the movement of water through cell membranes gives rise to osmosis.

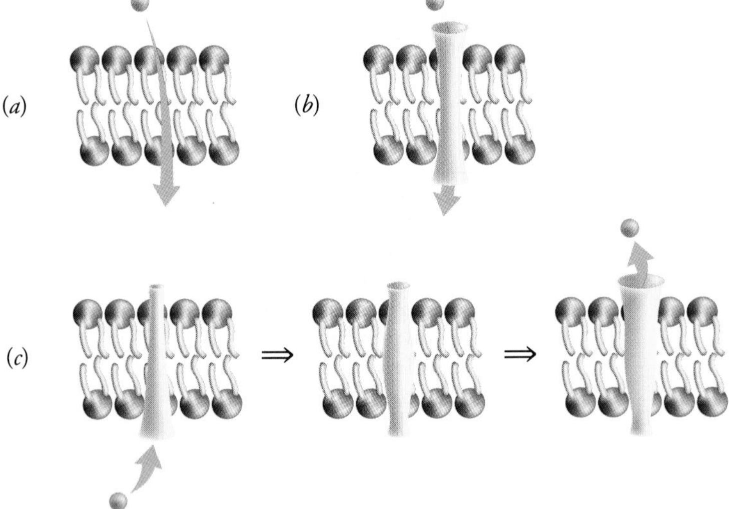

FIGURE | 8.26

Movement across cell membranes
(*a*) Diffusion. Nonpolar molecules diffuse across membranes moving in the direction of higher concentration to lower concentration. (*b*) Facilitated diffusion. Some polar molecules diffuse through a protein channel that spans the membrane. (*c*) Active transport. Some molecules and ions are transported across membranes in the direction of lower to higher concentration in an energy-requiring process.

In most cases, membranes do not allow ions, small polar molecules (water is an exception), and large biochemical molecules (including DNA and RNA) to diffuse through. Those that do diffuse through cell membranes, including glucose and small ions (K^+, Na^+, Cl^-, and others), are assisted by proteins in a process called **facilitated diffusion** (Figure 8.26*b*) that, like diffusion, moves solutes from higher concentration to lower concentration. In some cases, the proteins involved in facilitated diffusion act as highly selective membrane-spanning tubes with a hydrophilic interior. The particular molecules or ions that pass through these tubes never come into contact with hydrophobic membrane interiors. Some of the movement of water across cell membranes involves facilitated diffusion with the help of channel-forming proteins called aquaporins. In another type of facilitated diffusion, membrane proteins bind ions and undergo a conformational (shape) change that transports the ions through the membrane. Ionophores (Chapter 3, Biochemistry Link: Ionophores and Biological Ion Transport) function in this way.

Movement against diffusion, from lower to higher concentration, is called **active transport** (Figure 8.26*c*). In active transport, a protein moves compounds (amino acids and others) and ions (K^+, Na^+, etc.) across membranes in a process that requires the input of energy. Diffusion and facilitated diffusion, in contrast, do not require an energy input.

SAMPLE PROBLEM 8.5

Membrane structure

Plant membranes contain a higher proportion of unsaturated fatty acids than do animal membranes. Which type of membrane would you expect to be more fluid, plant or animal?

Strategy

To solve this problem, consider the effect of *cis* double bonds on membrane fluidity.

Solution

Plant. The presence of unsaturated fatty acids gives membranes their fluidity, so plant membranes are more fluid than animal membranes. Another factor that contributes to the greater rigidity of animal membranes is that they contain cholesterol, while plant membranes do not.

PRACTICE PROBLEM 8.5

Why are triglycerides not found in cell membranes?

Like all steroids, the basic ring structure of THG (tetrahydrogestrinone) consists of three fused 6-carbon atom rings and one 5-carbon atom ring. THG is similar to trenbolone, another anabolic steroid that is used by ranchers to promote the growth of cattle and by some weight lifters and bodybuilders. THG, trenbolone, and other anabolic steroids are related to the male sex hormone testosterone which, among its many effects, can increase muscle mass.

It is believed that THG is synthesized by reducing the carbon–carbon triple bond of gestrinone, a synthetic steroid used for the treatment of endometriosis.

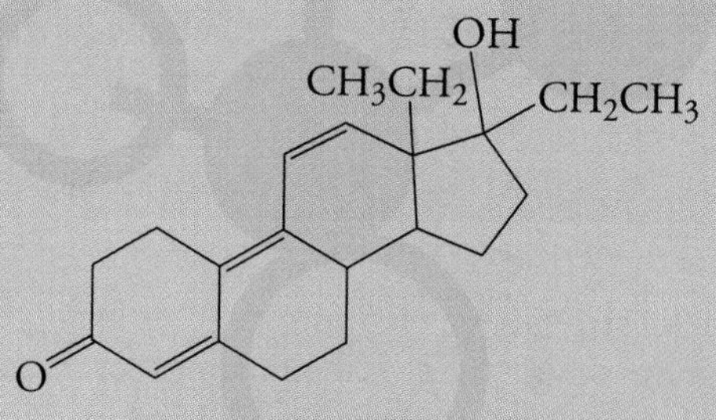

OH, CH_3CH_2, CH_2CH_3, O

Tetrahydrogestrinone (THG)

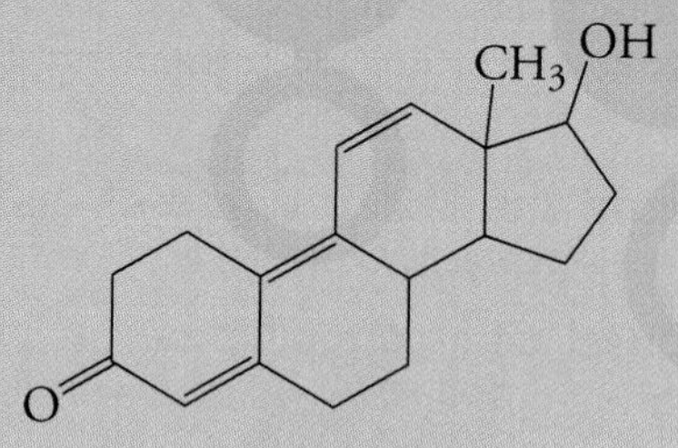

CH_3, OH, O

Trenbolone

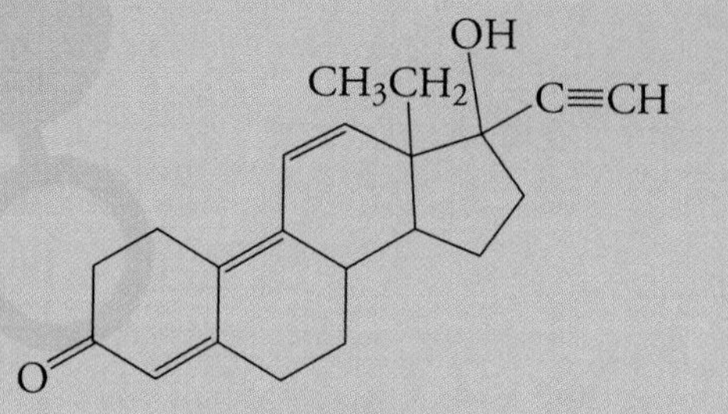

OH, CH_3CH_2, $C{\equiv}CH$, O

Gestrinone

summary of objectives

1 Explain what makes a compound a lipid.

A **lipid** is a water-insoluble biochemical compound.

2 Describe the structure of fatty acids and explain how saturated, monounsaturated, and polyunsaturated fatty acids differ from one another.

The class of lipids called **fatty acids** consists of long chain carboxylic acids having between 12 and 20 carbon atoms. Those containing only single carbon–carbon bonds in their hydrocarbon tail are **saturated**. **Monounsaturated** fatty acids have one carbon–carbon double bond and **polyunsaturated** ones have two or more. The double bonds in naturally occurring fatty acids are usually *cis*.

3 Identify the primary biological function of waxes and describe the esters that predominate in them.

The lipids called **waxes** contain esters formed by combining long chain fatty acids with long chain alcohols. Waxes primarily serve a protective function by keeping water either in or out of an organism.

4 Describe the makeup of triglycerides and list their biological functions.

A **triglyceride** is a triester formed by combining one glycerol molecule with three fatty acids. Triglycerides are saturated if they contain more saturated than unsaturated fatty acid residues, and unsaturated if unsaturated fatty acid residues predominate. Fats (saturated triglycerides) are solids at room temperature, while vegetable oils (unsaturated triglycerides) are liquids. In living things, triglycerides serve as energy storage molecules and provide thermal insulation, padding, and bouyancy.

5 Describe the structure of phospholipids and glycolipids.

Phospholipids and **glycolipids** are amphipathic compounds important to membrane structure. The two classes of phospholipids are the **glycerophospholipids**, consisting of residues of glycerol, two fatty acids, phosphate, and an alcohol, and **sphingolipids**, consisting of residues of **sphingosine**, one fatty acid, phosphate, and an alcohol. Many glycolipids are constructed of sphingosine, one fatty acid, and either a simple sugar or a chain of simple sugars.

6 Identify the basic steroid structure and list important members of this class of lipids.

All **steroids** have the same basic fused ring structure—three six-membered rings and one five-membered ring. The steroid **cholesterol** is used to make other steroids, including **sex hormones** (progesterone, testosterone, estradiol, etc.), **adrenocorticoid hormones** (cortisol and cortisone), and **bile salts** (taurocholate and glycocholate).

7 Name the three types of eicosanoids and describe their biological function.

Prostaglandins, **thromboxanes**, and **leukotrienes** are **eicosanoids** (they are derived from 20-carbon fatty acids). Prostaglandins cause pain, inflammation, fever, and a wide range of other effects. Thromboxanes are involved in blood clotting and leukotrienes affect muscle contractions.

8 Describe the makeup of a cell membrane and explain how various compounds cross the membrane.

Membranes are bilayers containing phospholipids, glycolipids, cholesterol, and proteins. In the **fluid mosaic model**, proteins are viewed as being inlayed into the membrane like a mosaic, and membrane flexibility or fluidity is provided by the hydrocarbon tails of unsaturated fatty acid residues. Some substances can move across cell membranes by diffusion. Others can diffuse across membranes with the assistance of specific proteins in a process called **facilitated diffusion. Active transport** is an energy-requiring process in which molecules and ions are moved across a membrane in the opposite direction of diffusion.

END OF CHAPTER PROBLEMS

Answers to problems whose numbers are printed in color are given in Appendix D. More challenging questions are marked with an asterisk. Problems within colored rules are paired. **ILW** = Interactive Learning Ware solution is available at *www.wiley.com/college/raymond.*

8.1 Fatty Acids

8.1 Why is the melting point of lauric acid (Table 8.1) lower than that of myristic acid?

8.2 Why is the melting point of palmitoleic acid (Table 8.1) lower than that of oleic acid?

8.3 Draw line-bond structures of the following (see Table 8.1).
a. lauric acid **b.** linolenic acid

8.4 Draw line-bond structures of the following (see Table 8.1).
a. palmitic acid **b.** palmitoleic acid

***8.5** Sodium palmitate, $CH_3(CH_2)_{14}CO_2$ has a higher melting point than palmitic acid, $CH_3(CH_2)_{14}CO_2H$. Why?

8.6 Define the terms.
a. hydrophobic **c.** amphipathic
b. hydrophilic

8.7 In terms of structure, what distinguishes fatty acids from the carboxylic acids that were discussed in Chapter 4?

8.8 Fatty acids are carboxylic acids that typically contain 12-20 carbon atoms.
a. Some biological molecules contain carboxylic acid residues longer than 20 carbon atoms. Would these fatty acids with more than 20 carbon atoms in length be considered lipids? Explain.
b. Some biological molecules contain carboxylic acid residues shorter than 12 carbon atoms. Would these fatty acids with less than 12 carbon atoms in length be considered lipids? Explain.

8.9 Fatty acid synthesis in cells (Section 15.9) involves the repeated addition of an acyl group to a growing fatty acid chain. How does this help to explain one of the structural differences between some of the fatty acids listed in Table 8.1?

$$CH_3\overset{\overset{\displaystyle O}{\|}}{C}-$$

Acyl group

8.2 Waxes

8.10 Draw skeletal structures for the products formed when the beeswax ester (Table 8.2) is saponified (hydrolyzed).

8.11 Draw skeletal structures for the products formed when the carnauba wax ester (Table 8.2) is saponified (hydrolyzed).

8.12 List the biological functions of waxes.

8.13 The fragrance of spermaceti, a wax produced by whales, once made it important to the perfume industry. One of the main constituents of spermaceti is cetyl palmitate, which is formed from palmitic acid and cetyl alcohol, $CH_3(CH_2)_{14}CH_2OH$. Draw the condensed structure of cetyl palmitate.

8.14 One of the esters present in oil of jojoba, a liquid wax obtained from the seeds of a particular desert shrub, is formed from eicosanoic acid [$CH_3(CH_2)_{18}CO_2H$] and docosanol [$CH_3(CH_2)_{20}CH_2OH$]. Draw the condensed structure of this ester.

8.15 Draw the fatty acid and alcohol from which the following kingfisher green wax ester is made.

$$CH_3CH_2CH_2CH(CH_3)CH_2CH(CH_3)CH_2CH(CH_3)CH_2CH(CH_3)C(=O){-}OCH_2CH(CH_3)CH_2CH(CH_3)CH_2CH(CH_3)CH_2CH(CH_3)CH_2CH_2CH_3$$

8.16 Draw the fatty acid and alcohol from which the ester in insect wax (Table 8.2) is made.

8.17 Some of the ester molecules present in beeswax have an —OH group in the ω-3 position of the alcohol residue. Draw a possible structure for one of these esters. Refer to Health Link: Omega-3 Fatty Acids for the meaning of ω-3.

8.18 The virulence (disease causing ability) of the bacterium that causes tuberculosis is partly due to the presence of a particular class of wax ester in its outer cell membrane. This ester contains two fatty acid residues and one alcohol residue. Draw the structure of one of these esters, formed from the carboxylic acid and alcohol molecules shown.

$$CH_3(CH_2)_{18}(CH_2CH(CH_3))_4C(=O){-}OH$$

$$CH_3(CH_2)_{20}CH(OH)CH_2CH(OH)(CH_2)_4CH(CH_3)CH(OCH_3)CH_3$$

8.3 Triglycerides

8.19 List four biological functions of triglycerides.

8.20 True or false?

a. Saturated fats contain only saturated fatty acids.

b. Unsaturated fats contain only unsaturated fatty acids.

c. Triglycerides are carboxylic acids.

8.21 Write a balanced reaction equation for the complete hydrogenation of palmitoleic acid (Table 8.1).

8.22 Write a balanced reaction equation for the complete hydrogenation of linolenic acid (Table 8.1).

8.23 Draw a triglyceride made from glycerol, myristic acid, palmitic acid, and oleic acid. Would you expect this triglyceride to be a liquid or a solid at room temperature? Explain.

8.24 Draw a triglyceride made from glycerol, linoleic acid, and two linolenic acid molecules. Would you expect this triglyceride to be a liquid or a solid at room temperature? Explain.

8.25 Draw the products formed when the triglyceride is saponified.

$$CH_3(CH_2)_{14}C(=O){-}OCH_2$$
$$CH_3CH_2(CH{=}CHCH_2)_3(CH_2)_6C(=O){-}OCH$$
$$CH_3(CH_2)_7CH{=}CH(CH_2)_7C(=O){-}OCH_2$$

8.26 Draw the products formed when the triglyceride is saponified.

$$CH_3(CH_2)_5CH{=}CH(CH_2)_7C(=O){-}OCH_2$$
$$CH_3(CH_2)_{12}C(=O){-}OCH$$
$$CH_3(CH_2)_7CH{=}CH(CH_2)_7C(=O){-}OCH_2$$

8.27 Vegetable oils tend to become rancid more rapidly than do animal fats. Why?

8.28 **a.** What does it mean to partially hydrogenate vegetable oil?

b. Why is this done?

8.29 From a molecular standpoint, explain why vegetable oil and vinegar, a 5% (w/v) aqueous solution of acetic acid (ethanoic acid, Table 4.7), do not mix.

8.4 Phospholipids and Glycolipids

8.30 Which two phospholipids are most prevalent in plants and animals?

8.31 True or false?

a. All phospholipids contain a phosphate residue.
b. All phospholipids contain two fatty acid residues.
c. All phospholipids contain a glycerol residue.

8.32 a. To which class of phospholipids does the compound belong?

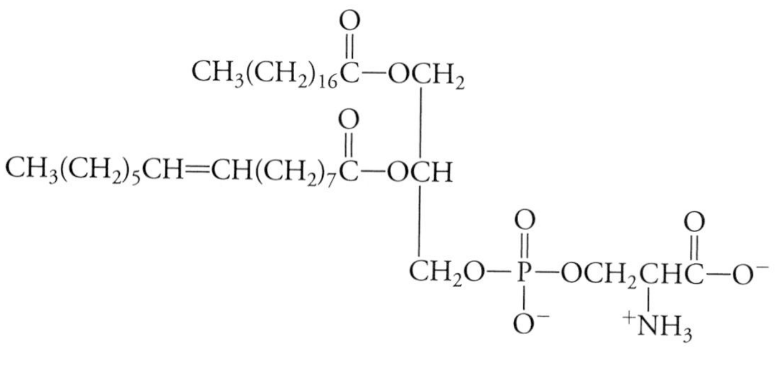

$$CH_3(CH_2)_{16}\overset{O}{\overset{\|}{C}}-OCH_2$$
$$CH_3(CH_2)_5CH{=}CH(CH_2)_7\overset{O}{\overset{\|}{C}}-OCH$$
$$CH_2O-\underset{O^-}{\underset{|}{\overset{O}{\overset{\|}{P}}}}-OCH_2\underset{^+NH_3}{\underset{|}{CH}}\overset{O}{\overset{\|}{C}}-O^-$$

b. Draw the products obtained when the compound is saponified into its component parts.

8.33 a. To which class of phospholipids does the compound belong?

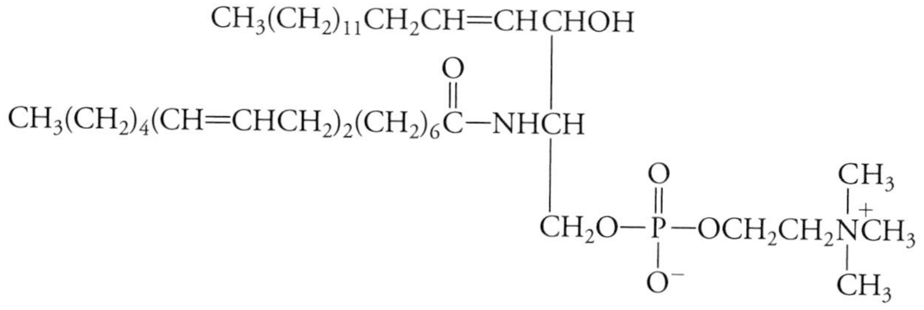

$$CH_3(CH_2)_{11}CH_2CH{=}CHCHOH$$
$$CH_3(CH_2)_4(CH{=}CHCH_2)_2(CH_2)_6\overset{O}{\overset{\|}{C}}-NHCH$$
$$CH_2O-\underset{O^-}{\underset{|}{\overset{O}{\overset{\|}{P}}}}-OCH_2CH_2\underset{CH_3}{\underset{|}{\overset{CH_3}{\overset{|\,+}{N}}}}CH_3$$

b. Draw the products obtained when the molecule is saponified into its component parts.

8.34 How do phosphatidylethanolamines differ from phosphatidylcholines?

8.35 Lecithin (phosphatidylcholine in Figure 8.17) is an emulsifying agent, but triglycerides are not. Account for the difference.

8.36 Draw a sphingomyelin that contains oleic acid.

8.37 Draw a sphingomyelin that contains palmitoleic acid.

8.38 In certain membranes sphingomyelin serves as a replacement for phosphatidylcholine. What structural similarities between these classes of phospholipids makes this a good substitution?

8.39 Which is better at forming hydrogen bonds, phosphatidylethanolamine or phosphatidylcholine? Explain.

8.40 The sphingomyelins present in cow brain include residues of saturated fatty acids having between 16 and 32 carbon atoms and/or residues of unsaturated fatty acids with 18 or 22 carbon atoms. Draw one of these sphingomyelins.

8.41 The membrane of an animal cell nucleus contains phosphatidylcholine in which a significant number of the fatty acid residues are saturated. After reading Section 8.7, would you expect this to increase or decrease the fluidity of the membrane structure?

8.5 Steroids

8.42 How is cholesterol used in the human body?

8.43 Label the hydrophobic and hydrophilic parts of taurocholate (Figure 8.21).

8.44 Label the hydrophobic and hydrophilic parts of glycocholate (Figure 8.21).

8.45 Which molecule is the starting point for the synthesis of sex hormones and bile salts?

8.46 Bile salts aid in the digestion of triglycerides by acting as emulsifiers. Explain.

8.47 Name one male sex hormone and one female sex hormone.

8.48 In the body, the steroid hormone cortisone is prepared from cortisol (Figure 8.21). In this reaction is cortisol oxidized or is it reduced?

8.49 **a.** What are the biological functions of HDL and LDL?

b. In the news, you often hear HDL referred to as "good cholesterol" and LDL as "bad cholesterol." In terms of the structure of HDL and LDL, is use of the term "cholesterol" totally correct? Explain.

c. What makes HDL "good" and LDL "bad"?

***8.50** A cholesterol screening finds that person's blood serum has 35 mg/dL HDL and 150 mg/dL LDL.

a. Are these values within the normal range?

b. How many grams of HDL are present in 1.0 mL of the serum?

c. How may micrograms of LDL are present in 2.5×10^{-4} L of the serum?

d. What is the LDL/HDL ratio? Is this value considered to be healthy?

e. Can total cholesterol be calculated from the given HDL and LDL concentrations? Explain.

8.6 EICOSANOIDS

8.51 **a.** How do nonsteroidal anti-inflammatory drugs (NSAIDs) act to reduce pain, fever, and swelling?

b. How is the action of Celebrex different from that of NSAIDs?

***8.52** Cortisol, cortisone, and other anti-inflammatory steroids block the action of an enzyme that catalyzes the hydrolysis of unsaturated fatty acids, including arachidonic acid, from membrane lipids. How does this result in reduced inflammation?

8.53 Low doses of aspirin can reduce the risk of heart attack and stroke, but can also increase the risk of gastrointestinal bleeding. Aspirin's effect on the synthesis of which type of eicosanoid leads to the greater risk of bleeding?

8.54 Dexamethasone, an alternative to COX-2 inhibitors, blocks the production of COX-2.

a. What are the benefits of taking a drug that prevents formation of COX-2, but has no effect on COX-1?

b. To which class of lipid does dexamethasone belong?

O C CH_2OH CH_3 OH HO CH_3 CH_3 F O

Dexamethasone

8.7 MEMBRANES

8.55 How are facilitated diffusion and active transport different?

8.56 How are facilitated diffusion and active transport similar?

8.57 How are facilitated diffusion and diffusion different?

8.58 How are facilitated diffusion and diffusion similar?

***8.59** To function properly, membranes must be flexible or fluid. In light of this fact, propose an explanation of why the cell membranes in the feet and legs of a reindeer contain a higher percentage of unsaturated fatty acids than do the cell membranes in the interior of its body.

HEALTH*Link* | *Omega-3 Fatty Acids*

8.60 Arachidonic acid (Figure 8.22) is not an omega-3 fatty acid. Which type of omega fatty acid is it?

8.61 **a.** In the name "eicosapentaenoic acid," the prefix *eicos* indicates that there are twenty carbon atoms in the molecule. To what does the *pentaen* part of the name refer?

b. In the name "docosahexaenoic acid", what does the prefix *docos* indicate? To what does the *hexaen* part of the name refer?

8.62 What are the health benefits of consuming omega-3 fatty acids?

8.63 Why is it recommended that some of the fish that are a good source of omega-3 fatty acids not be eaten?

HEALTH*Link* | Trans *Fats*

8.64 When unsaturated fatty acids were discussed in Section 8.1, it was said that their carbon–carbon double bonds are usually *cis*. Name one food source that is a natural source of unsaturated fatty acids with *trans* double bonds.

8.65 Draw a *trans* fat that might form if the following triglyceride is subjected to partial hydrogenation.

```
                          O
                          ||
          CH3(CH2)12C—OCH2
                                   |
                          O        |
                          ||       |
CH3(CH2)4CH2    CH2(CH2)6C—OCH
         \      /                  |
          C=C                      |
         /    \                    |
        H      H                   |
                          O        |
                          ||       |
CH3(CH2)4CH2    CH2(CH2)6C—OCH2
         \      /
          C=C
         /    \
        H      H
```

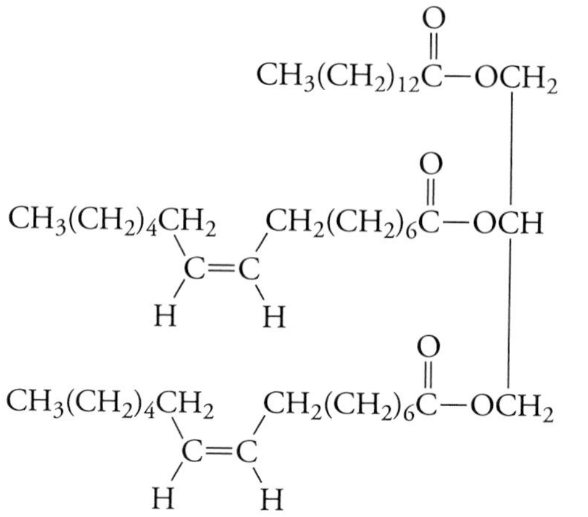

8.66 Draw a *trans* fat that might form if the following triglyceride is subjected to partial hydrogenation.

```
                                        O
                                        ||
          CH3(CH2)6CH2    CH2(CH2)6C—OCH2
                   \      /                  |
                    C=C                      |
                   /    \                    |
                  H      H                   |
                                   O         |
                                   ||        |
                      CH3(CH2)16C—OCH
                                             |
                                   O         |
                                   ||        |
CH3(CH2)3CH2     CH2       CH2(CH2)6C—OCH2
         \      /    \      /
          C=C          C=C
         /    \      /    \
        H      H    H      H
```

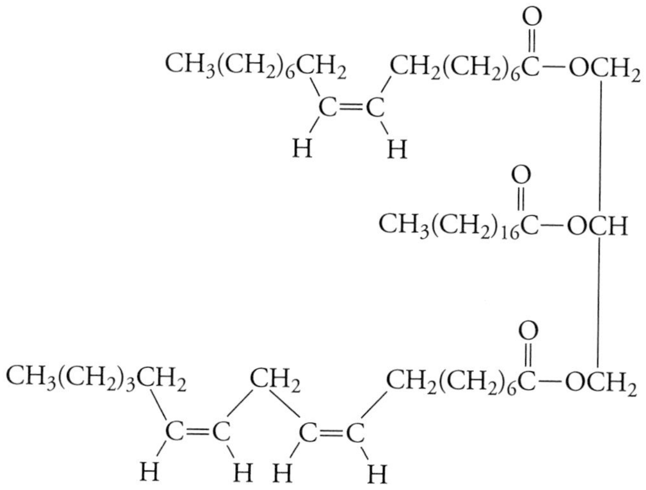

8.67 Which is likely to have the most *trans* fats, vegetable oil, margarine, or butter? Explain.

HEALTH*Link* | *Olestra*

*__8.68__ Draw the products formed if olestra (Figure 8.15) is saponified.

8.69 Olestra passes through the digestive system untouched, taking some dietary vitamin A, vitamin E, and other fat-soluble (water-insoluble) vitamins with it. Which noncovalent interaction allows these vitamins to be associated with olestra?

HEALTH*Link* | *Anabolic Steroids*

8.70 Draw the hydrolysis products obtained when nandrolone laurate (Figure 8.29) is saponified.

8.71 In the body, androstenedione is converted into testosterone. Is androstenedione oxidized or is it reduced in this process?

8.72 **a.** "Teen Steroids" describes a study that involves giving steroids to hamsters. What side effect was observed and what is its duration?

b. Based on this study, what are the implications for teens who take steroids?

THINKING IT THROUGH

8.73 What is the chemical difference between vegetable oil and motor oil?

8.74 Suppose that you are an athlete and that you have just been told that a new performance-enhancing synthetic steroid has been developed. If this steroid would never be detectable in drug tests would you take it? Why or why not?

INTERACTIVE LEARNING PROBLEMS

ILW

8.75 What fatty acid is predominantly involved in the biosynthesis of eicosanoids, compounds often involved in inflammatory processes?

SOLUTIONS TO PRACTICE PROBLEMS

8.1 A linolenic acid molecule has three carbon–carbon double bonds, compared to linoleic acid's two carbon–carbon double bonds. The more *cis* double bonds a fatty acid has, the farther apart the hydrocarbon tails of different fatty acids are held from one another. This results in weaker London force interactions and a lower melting point.

8.2

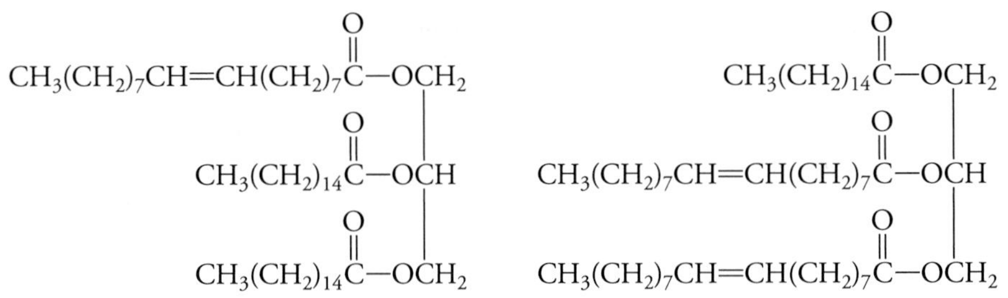

$$\begin{array}{l} CH_3(CH_2)_7CH{=}CH(CH_2)_7\overset{O}{\overset{\|}{C}}{-}OCH_2 \\ \qquad\qquad CH_3(CH_2)_{14}\overset{O}{\overset{\|}{C}}{-}OCH \\ \qquad\qquad CH_3(CH_2)_{14}\overset{O}{\overset{\|}{C}}{-}OCH_2 \end{array}$$

$$\begin{array}{l} \qquad\qquad CH_3(CH_2)_{14}\overset{O}{\overset{\|}{C}}{-}OCH_2 \\ CH_3(CH_2)_7CH{=}CH(CH_2)_7\overset{O}{\overset{\|}{C}}{-}OCH \\ CH_3(CH_2)_7CH{=}CH(CH_2)_7\overset{O}{\overset{\|}{C}}{-}OCH_2 \end{array}$$

8.3

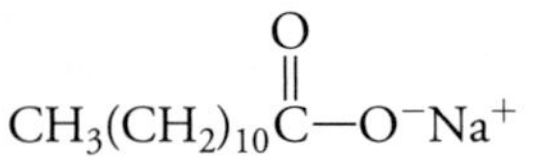

$$CH_3(CH_2)_{10}\overset{O}{\overset{\|}{C}}{-}O^-Na^+$$

Potassium laurate

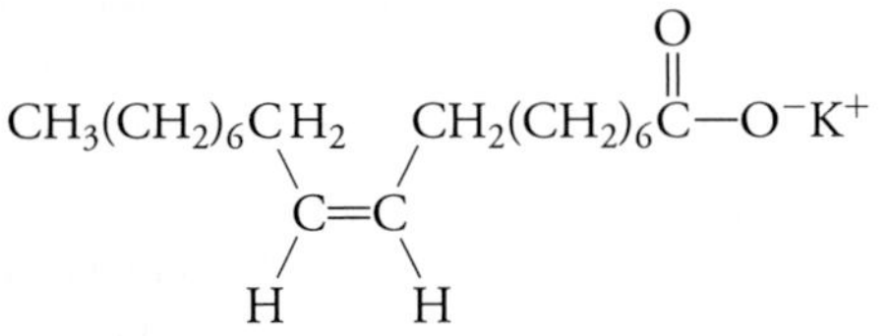

$$\begin{array}{c} CH_3(CH_2)_6CH_2 \qquad CH_2(CH_2)_6\overset{O}{\overset{\|}{C}}{-}O^-K^+ \\ C{=}C \\ H \qquad H \end{array}$$

Potassium oleate

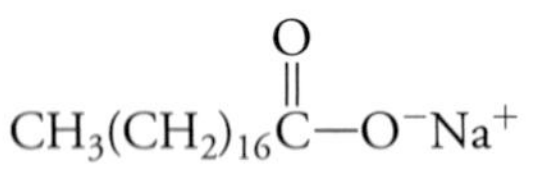

$$CH_3(CH_2)_{16}\overset{O}{\overset{\|}{C}}{-}O^-Na^+$$

Potassium stearate

$$\begin{array}{c} HOCH_2 \\ | \\ HOCH \\ | \\ HOCH_2 \end{array}$$

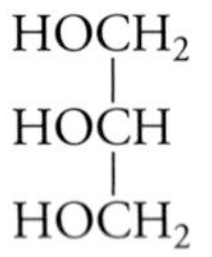

Glycerol

8.4

$$
\begin{array}{l}
\qquad\qquad CH_3(CH_2)_{11}CH_2CH{=}CHCHOH \\
CH_3(CH_2)_7CH{=}CH(CH_2)_7\overset{O}{\overset{\|}{C}}{-}NHCH \\
\qquad\qquad\qquad\qquad\quad CH_2O{-}\underset{O^-}{\underset{|}{\overset{O}{\overset{\|}{P}}}}{-}OCH_2CH_2NH_3^+
\end{array}
$$

$$
\begin{array}{l}
\qquad\qquad CH_3(CH_2)_{11}CH_2CH{=}CHCHOH \\
CH_3(CH_2)_7CH{=}CH(CH_2)_7\overset{O}{\overset{\|}{C}}{-}NHCH \\
\qquad\qquad\qquad\qquad\quad CH_2O{-}\underset{O^-}{\underset{|}{\overset{O}{\overset{\|}{P}}}}{-}OCH_2CH_2\underset{CH_3}{\underset{|}{\overset{CH_3}{\overset{|+}{N}}}}CH_3
\end{array}
$$

$$
\begin{array}{l}
\qquad\qquad\qquad CH_3(CH_2)_{11}CH_2CH{=}CHCHOH \\
CH_3(CH_2)_4(CH{=}CHCH_2)_2(CH_2)_6\overset{O}{\overset{\|}{C}}{-}NHCH \\
\qquad\qquad\qquad\qquad\qquad\quad CH_2O{-}\underset{O^-}{\underset{|}{\overset{O}{\overset{\|}{P}}}}{-}OCH_2CH_2\underset{CH_3}{\underset{|}{\overset{CH_3}{\overset{|+}{N}}}}CH_3
\end{array}
$$

8.5 The lipids present in membranes are amphipathic. Triglycerides are not.

At the library you run into a classmate who has put off studying for the upcoming chemistry exam. Between the stress of trying to learn the material at the last minute and drinking way too much coffee, she complains of an upset stomach. She reaches into her book bag, pulls out a roll of antacids, and takes a few. Seeing this reminds you of an advertisement that talked about how antacids consume excess stomach acid.

Photo source: PNC/Brand X Pictures/Jupiter Images.

9 ACIDS, BASES, AND EQUILIBRIUM

Acids and bases are related classes of compounds that are important to much of the chemistry that goes on around us everyday. Vinegar is an acid, as is the citric acid found in oranges, lemons, grapefruit, and other citrus fruits. Bases commonly found around the home include baking soda, baking powder, ammonia, antacids, and drain cleaners. Your body also produces and uses acids and bases, and maintaining a proper balance of them is essential for maintaining good health.

objectives

By the time you finish this chapter, you should be able to:

1. List the common characteristics of acids and bases.
2. Describe Brønsted–Lowry acids and bases and explain how they differ from their conjugates. Relate acid strength to conjugate base strength.
3. Write the equilibrium constant for a reversible reaction and use Le Châtelier's principle to explain how an equilibrium responds to being disturbed.
4. Use H_3O^+ concentration and pH to identify a solution as being acidic, basic, or neutral.
5. Describe the processes of neutralization and titration.
6. Explain how the pH of a solution can affect the relative concentrations of an acid and its conjugate base and describe buffers.
7. Describe the role of buffers, respiration, and the kidneys in maintaining a stable blood serum pH.

9.1 ACIDS AND BASES

Some compounds are acids and others are bases. Determining whether a particular compound belongs in one category or the other can sometimes be as simple as making a direct observation. **Acids** have a *sour taste* (such as that from the citric acid in citrus fruits), will *dissolve some metals*, and will turn the plant pigment called litmus *to a pink color* (Figure 9.1). **Bases**, on the other hand, have a *bitter taste* (such as that of caffeine or antihistamines), *feel slippery or soapy*, and cause litmus *to turn blue*.

In an earlier era, tasting was a routine way of identifying chemicals. This technique is no longer practiced because the potentially harmful side effects of doing so are well understood. This is especially true of acids and bases—many will cause extensive tissue damage and can be fatal, even in small doses.

Table 9.1 lists a number of common acids and bases and their uses. Several of the acids in this table are named differently than might be expected, based on the rules for naming binary compounds that were presented in Section 3.6. For example, Table 9.1 identifies HCl as hydrochloric acid, not hydrogen chloride. Although both are correct official names, when acidity is being emphasized, the acid name is used instead of the binary compound name.

TABLE 9.1 SOME COMMON ACIDS AND BASES

Name	Formula	Uses
Acids		
Hydroiodic acid	HI	Disinfectant
Hydrobromic acid	HBr	Veterinary sedative
Hydrochloric acid (muriatic acid)	HCl	Household cleaning products, swimming pool maintenance, and metal cleaning
Sulfuric acid	H_2SO_4	Fertilizers, explosives, dyes, and glues
Nitric acid	HNO_3	Fertilizers, explosives, and dyes
Bases		
Sodium hydroxide	NaOH	Drain cleaners, soap manufacture
Potassium hydroxide	KOH	Paint and varnish removers
Ammonia	NH_3	Fertilizers, cleaning

(*a*)

(*b*)

FIGURE 9.1

Some properties of acids and bases

(*a*) Some metals react with acids. Here, an iron nail reacts with sulfuric acid and forms hydrogen (H_2) gas. (*b*) In the presence of an acid (lemon juice) litmus, a pH indicator, is pink. In the presence of a base (baking soda), it is blue.

Source: (*a*) OPC, Inc.; (*b*) © Leonard Lessin/Peter Arnold, Inc.

9.2 BRØNSTED–LOWRY ACIDS AND BASES

One of the early challenges that chemists faced was developing a theory that would connect the structure of a compound to its acid or base characteristics. In the 1880s the Swedish chemist Svante Arrhenius identified acids as compounds that produce H^+ in water and bases as compounds that produce OH^-. By this definition HCl is an Arrhenius acid and NaOH is an Arrhenius base.

$$HCl \xrightarrow{H_2O} H^+ + Cl^-$$

$$NaOH \xrightarrow{H_2O} Na^+ + OH^-$$

In the 1920s, Johannes Brønsted, a Danish chemist, and Thomas Lowry, an English chemist, proposed a more general definition, one which described acids and bases in terms of the transfer of H^+. In the **Brønsted–Lowry** definition, *acids release* H^+ and bases accept H^+. Consider the equation for the reaction that takes place between HCN and water:

■ A Brønsted–Lowry acid releases H^+ and a Brønsted–Lowry base accepts H^+.

$$\underset{\text{Acid}}{HCN} + \underset{\text{Base}}{H_2O} \rightleftharpoons \underset{\text{Base}}{CN^-} + \underset{\text{Acid}}{H_3O^+}$$

The double arrows indicate that the reaction is reversible—goes from left to right and from right to left. In the forward direction of this reaction HCN is the acid (it releases H^+ to become CN^-) and H_2O is the base (it accepts H^+ to become H_3O^+). An acid and a base can also be identified for the reverse of this reaction: H_3O^+ is the acid (it releases H^+ to become H_2O) and CN^- is the base (it accepts H^+ to become HCN).

■ When released in water, H^+ rapidly associates with H_2O to form a hydronium ion (H_3O^+). H_3O^+ and H^+ are often used interchangeably.

The household ammonia used for cleaning purposes is an aqueous solution of NH_3. In this solution NH_3 acts as a base (it accepts H^+ to become NH_4^+) and H_2O serves as the acid (it releases H^+ to become OH^-). For the reverse reaction, NH_4^+ is the acid and OH^- is the base.

$$\underset{\text{Base}}{NH_3} + \underset{\text{Acid}}{H_2O} \rightleftharpoons \underset{\text{Acid}}{NH_4^+} + \underset{\text{Base}}{OH^-}$$

SAMPLE PROBLEM 9.1

Identifying acids and bases

For the forward and reverse directions of the reaction, identify the Brønsted–Lowry acids and bases.

$$HCN + CO_3^{2-} \rightleftharpoons HCO_3^- + CN^-$$

Strategy

To solve this problem you must look across the reaction arrow to see how a particular reactant changes. CO_3^{2-}, for example, gains H^+ to become HCO_3^-.

Solution

$$\underset{\text{Acid}}{HCN} + \underset{\text{Base}}{CO_3^{2-}} \rightleftharpoons \underset{\text{Acid}}{HCO_3^-} + \underset{\text{Base}}{CN^-}$$

In the forward reaction, H^+ moves from HCN to CO_3^{2-}. In the reverse reaction, H^+ moves from HCO_3^- to CN^-.

PRACTICE PROBLEM 9.1

For the forward and reverse directions of each reaction, identify the Brønsted–Lowry acids and bases.

a. $HSO_4^- + HPO_4^{2-} \rightleftharpoons H_2SO_4 + PO_4^{3-}$

b. $HPO_4^{2-} + OH^- \rightleftharpoons H_2O + PO_4^{3-}$

■ An acid and its conjugate base differ by the presence of H^+.

Compounds, such as HCN and CN^- or NH_3 and NH_4^+, which *differ only in the presence or absence of H^+*, are called **conjugates.** For the reaction of acetic acid (CH_3CO_2H) in water, CH_3CO_2H and $CH_3CO_2^-$ are conjugates, as are H_3O^+ and H_2O.

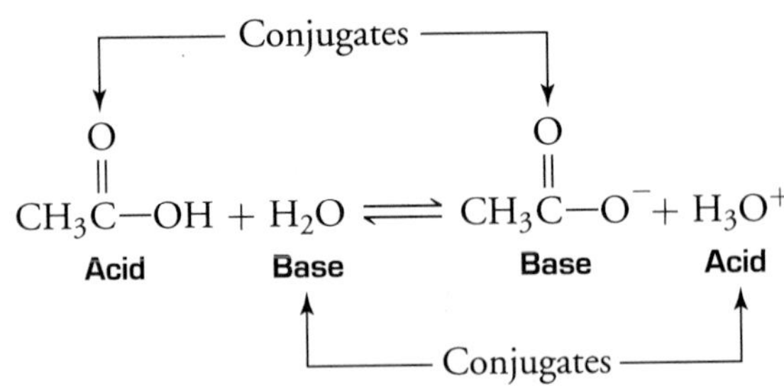

Note that an acid (CH_3CO_2H, for example) and its conjugate base ($CH_3CO_2^-$) are always on opposite sides of the equilibrium reaction arrows.

Water plays two different roles in the reactions described above. With HCN and CH_3CO_2H, water acts as a base, while in the reaction with NH_3, it is an acid. Compounds that *can act as acids or as bases* are called **amphoteric.** Other amphoteric compounds include hydrogen carbonate (HCO_3^-), dihydrogen phosphate ($H_2PO_4^-$), and the amino acids used to make proteins.

9.3 EQUILIBRIUM

The acid–base reactions introduced above are reversible, which means that reactants can be converted into products and products can be converted back into reactants. Let us use the decomposition of dinitrogen tetroxide (N_2O_4) to form nitrogen dioxide (NO_2), a reversible reaction that does not involve acids and bases, to introduce the concept of equilibrium.

$$N_2O_4(g) \rightleftharpoons 2NO_2(g)$$

Use of the terms *reactant* and *product* when describing this or any other reversible reaction can be rather vague, since N_2O_4 is the reactant of the forward reaction and the product of the reverse reaction, while NO_2 is the product of the forward reaction and the reactant of the reverse reaction. When dealing with reversible reactions, we will define reactants as the compounds that appear to the left of the reaction equation arrows and products as the compounds that appear to the right.

If N_2O_4 is placed in a sealed tube and warmed, a decomposition reaction begins to take place. Early in the reaction, the concentration of N_2O_4 is at its highest and the concentration of NO_2 is at its lowest (Figure 9.2). Since, in this case, reactant concentration

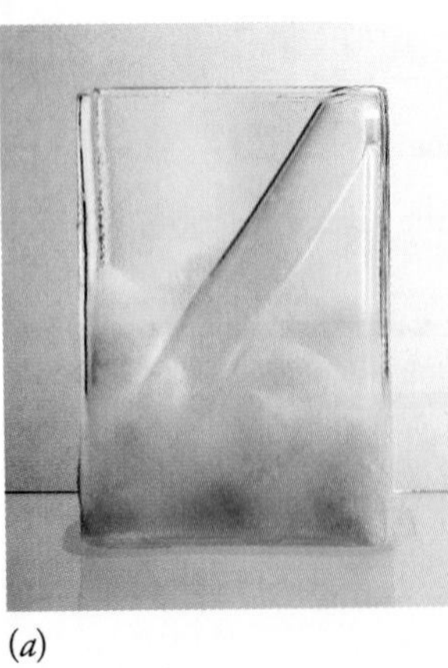

(*a*)

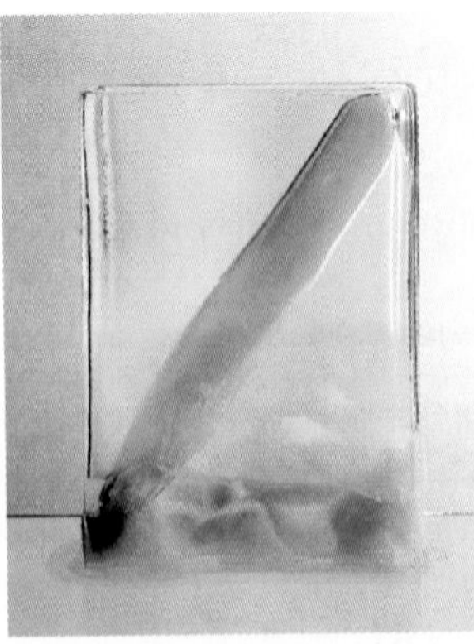

(*b*)

(*c*)

FIGURE 9.2

Reaching equilibrium
(*a*) N_2O_4 is a colorless gas. (*b*) As N_2O_4 is warmed it is converted into brown NO_2 gas, as represented by the equation $N_2O_4(g) \rightleftharpoons 2NO_2(g)$. (*c*) At equilibrium, the color in the tube stops changing because the forward and reverse reactions occur at the same rate and the concentrations of $N_2O_4(g)$ and $NO_2(g)$ no longer vary.

Source: © 1990 Richard Megna/ Fundamental Photographs, NYC.

has a direct effect on reaction rate (Section 6.6), when the reaction is started the rate of the forward reaction is at its greatest and the rate of the reverse reaction is at its lowest. As more and more N_2O_4 is converted into NO_2, the rate of the forward reaction slows and the rate of the reverse reaction increases. At the point where *the rate of the forward reaction and the rate of the reverse reaction are equal*, **equilibrium** has been reached. At equilibrium reactants and products are produced as rapidly as they are consumed and there is no change in their concentrations.

■ At equilibrium the rate of the forward and reverse reactions are the same and concentrations of reactants and products do not change.

Equilibrium can be reached from any set of starting conditions. For the reaction described above, this includes: only N_2O_4 being present at the beginning of the reaction, only NO_2 being present, or any combination of N_2O_4 and NO_2 being present. Being at equilibrium does not mean that the concentrations of reactants and products are equal—each reversible reaction has its own characteristic set of equilibrium concentrations of reactants and products.

Equilibrium Constants

If N_2O_4 and NO_2 are allowed to reach equilibrium and the equilibrium concentrations of each are measured, then the following will always be true:

$$K_{eq} = \frac{[NO_2]^2}{[N_2O_4]} = 4.6 \times 10^{-3}$$

where the square brackets stand for concentration in units of molarity (moles per liter, Section 7.6). Regardless of the starting concentration of N_2O_4 and NO_2, the equilibrium NO_2 concentration squared, divided by the equilibrium concentration of N_2O_4, will always equal 4.6×10^{-3}. This value is the **equilibrium constant** (K_{eq}) for the reaction.

■ Equilibrium constant values vary with temperature. Examples used in this chapter will assume a temperature of 25°C.

While we will not be doing equilibrium calculations, it will be useful to see how these expressions are derived. For the generalized reversible reaction

$$aA + bB \rightleftharpoons cC + dD$$

where A and B are reactants, C and D are products, and *a*, *b*, *c*, and *d* are coefficients (Section 6.1), the equilibrium constant expression takes the form:

$$K_{eq} = \frac{[C]^c[D]^d}{[A]^a[B]^b}$$

In this equation, the numerator is obtained by multiplying the concentrations of products together, with each raised to a power equal to its coefficient (if the coefficient is 2, the concentration is squared; if it is 3, the concentration is cubed). The denominator is obtained in the same way, using reactant concentrations.

Consider, for example, the reaction that takes place between nitrogen gas (N_2) and hydrogen gas (H_2) to produce the ammonia (NH_3) used in fertilizer.

$$N_2(g) + 3H_2(g) \rightleftharpoons 2NH_3(g)$$

The equilibrium constant for this reversible reaction has a value of 6.9×10^5 and the equilibrium constant expression is written:

$$K_{eq} = \frac{[NH_3]^2}{[N_2][H_2]^3} = 6.9 \times 10^5$$

SAMPLE PROBLEM 9.2

Writing equilibrium constant expressions

Balance the reaction equation and then write the corresponding equilibrium constant expression.

$$CO(g) + O_2(g) \rightleftharpoons CO_2(g)$$

■ Some of the reactions introduced in previous chapters were reversible. To simplify those earlier discussions, reversibility was largely ignored.

Strategy

Reaction equations are balanced by changing the coefficients on reactants and/or products. Using the balanced reaction equation, the equilibrium constant expression is written with product concentrations in the numerator and reactant concentrations in the denominator, each raised to a power equal to its coefficient.

Solution

$$2CO(g) + O_2(g) \rightleftharpoons 2CO_2(g)$$

$$K_{eq} = \frac{[CO_2]^2}{[CO]^2[O_2]}$$

PRACTICE PROBLEM 9.2

Write the equilibrium constant expression for each reaction.

a. $H_2(g) + CO_2(g) \rightleftharpoons CO(g) + H_2O(g)$

b. $2N_2O_5(g) \rightleftharpoons 4NO_2(g) + O_2(g)$

FIGURE | 9.3

An equilibrium involving a solid Solid PbI_2, which has a yellow color, is in equilibrium with colorless $Pb^{2+}(aq)$ and $I^-(aq)$. The equilibrium constant expression for this reaction does not include the concentration of PbI_2.

Source: Andy Washnik/Wiley Archive.

- Solvents and solids are not included in equilibrium constant equations.

In some reversible reactions, reactants and products can be present in different phases. For example, when lead(II) iodide, a solid, is dissolved in water, aqueous lead(II) ion and aqueous iodide ion are produced (Figure 9.3).

$$PbI_2(s) \rightleftharpoons Pb^{2+}(aq) + 2I^-(aq) \qquad K_{eq} = [Pb^{2+}][I^-]^2 = 7.1 \times 10^{-9}$$

It may seem that there is an error in the equilibrium constant expression just written—the concentration of PbI_2 does not appear in the denominator. There is no mistake, however, because when an equilibrium constant expression for a reaction is written, *only concentrations that can change are included.* The concentration of a solid is the number of moles present in a given volume, a value that never changes at a given temperature, so the concentration of solid reactants and products is omitted. Solvents are also left out of equilibrium constant expressions because their concentration is very high and does not change significantly during a reaction. In the reaction of hydrogen cyanide (HCN) and water to produce cyanide ion (CN^-) and hydronium ion (H_3O^+), for example, water is the solvent and does not appear in K_{eq}.

$$HCN(aq) + H_2O(l) \rightleftharpoons CN^-(aq) + H_3O^+(aq)$$

$$K_{eq} = \frac{[CN^-][H_3O^+]}{[HCN]} = 4.9 \times 10^{-10}$$

The size of equilibrium constants can vary greatly from one reaction to the next. For the reaction just described, K_{eq} has a value of 4.9×10^{-10}. (Because this is an acid–base reaction, the equilibrium constant, K_{eq}, is also known as the **acidity constant,** K_a.) In terms of the math involved, the only way that K_{eq} can have a value as small as 4.9×10^{-10} is if the denominator of the expression is much larger than the numerator. In this particular case it means that the concentration of HCN is much higher than the concentration of the two products. It is always true that *when an equilibrium constant has a value of less than 1, the denominator* (related to reactant concentrations) of the equilibrium constant expression *is greater than the numerator* (related to product concentrations) (Table 9.2).

For another acid–base reaction, that of hydrochloric acid (HCl) and water, the equilibrium constant has a value of 1.0×10^7.

$$HCl(aq) + H_2O(l) \rightleftharpoons Cl^-(aq) + H_3O^+(aq) \qquad K_{eq} = \frac{[Cl^-][H_3O^+]}{[HCl]} = 1.0 \times 10^7$$

TABLE 9.2 INTERPRETING EQUILIBRIUM CONSTANTS (K_{eq})

Relative Concentrations at Equilibrium	Value of K_{eq}
[Reactants] < [products]	$K_{eq} > 1$
[Reactants] > [products]	$K_{eq} < 1$
[Reactants] ≈ [products]	$K_{eq} \approx 1$

When an equilibrium constant has a value greater than 1, the denominator of the equilibrium constant expression *is smaller than the numerator*, which means that at equilibrium there are more products than reactants. In this example, the product concentrations (Cl^- and H_3O^+) far outweigh the reactant concentration (HCl).

SAMPLE PROBLEM 9.3

Using K_{eq} to predict relative concentrations

In which reaction below are equilibrium product concentrations greater than reactant concentrations?

$$HI + H_2O \rightleftharpoons I^- + H_3O^+ \qquad K_{eq} = \frac{[I^-][H_3O^+]}{[HI]} = 2.5 \times 10^{10}$$

$$HF + H_2O \rightleftharpoons F^- + H_3O^+ \qquad K_{eq} = \frac{[F^-][H_3O^+]}{[HF]} = 6.6 \times 10^{-4}$$

Strategy

K_{eq} is a direct indication of the relative amounts of reactants and products. If, at equilibrium, the concentration of products is higher than that of reactants, then the value of K_{eq} will be greater than 1. If the reverse is true, K_{eq} will be less than 1.

Solution

$$HI + H_2O \rightleftharpoons I^- + H_3O^+$$

In this reaction, the numerator of the K_{eq} expression is 2.5×10^{10} (25 billion) times larger than the denominator, so the equilibrium concentration of products is greater than the concentration of reactant. For the other reaction, the numerator is 6.6×10^{-4} (about 1/2000) times as large as the denominator.

PRACTICE PROBLEM 9.3

For the reaction of N_2O_4 (described earlier in this section), K_{eq} has a value of 4.6×10^{-3}. Which is larger at equilibrium, reactant concentration or product concentration?

$$N_2O_4(g) \rightleftharpoons 2NO_2(g)$$

9.4 LE CHÂTELIER'S PRINCIPLE

As we saw in the previous section, when a reversible reaction is at equilibrium, the rate of the forward reaction is equal to the rate of the reverse reaction and the concentrations of reactants and products do not change. It turns out that equilibrium conditions are actually quite precarious and can be easily disturbed by changes in conditions. The response to a loss of equilibrium is predicted by **Le Châtelier's principle**, which states that *when a reversible reaction is pushed out of equilibrium, the reaction responds to reestablish equilibrium.*

■ When an equilibrium is disturbed, the reaction responds to reestablish equilibrium.

Varying the concentration of a reactant or a product is one way to upset an equilibrium. A good example of this involves carbon dioxide, a product of metabolism. In red blood cells the enzyme carbonic anhydrase catalyzes the reversible reaction between H_2O and CO_2 to form H_2CO_3. (Chapter 6, Health Link: Carbonic Anhydrase).

$$H_2O(l) + CO_2(g) \overset{\text{carbonic anhydrase}}{\rightleftharpoons} H_2CO_3(aq)$$

At equilibrium, the concentration of reactants and products remains constant and the rates of the forward and reverse reactions are equal. When CO_2 is added to an equilibrium mixture (for example, when CO_2 moves from the cells into blood), the equilibrium is lost and the forward reaction will respond to reestablish equilibrium. The response involves an increase in the rate of the forward reaction, $H_2O(l) + CO_2(g) \longrightarrow H_2CO_3(aq)$, which reduces $[CO_2]$ and increases $[H_2CO_3]$. The rate of the forward reaction increases because the reactant concentration rises when CO_2 is added. As H_2CO_3 concentrations rise, the rate of the reverse reaction also increases until, eventually, the rates of the forward and reverse reaction match and equilibrium is reestablished. While equilibrium concentrations of reactants and products in the reestablished equilibrium will be different from that in the initial equilibrium state, the value of K_{eq} remains constant.

$$H_2O(l) + CO_2(g) \rightleftharpoons H_2CO_3(aq)$$

$$\longrightarrow$$

Increasing $[CO_2]$ upsets the equilibrium and a net forward reaction takes place.

If the concentration of CO_2 is decreased (for example, when CO_2 moves from blood into the lungs), the forward reaction slows, and the net effect is that the reverse reaction produces H_2O and CO_2 until equilibrium is reestablished.

$$H_2O(l) + CO_2(g) \rightleftharpoons H_2CO_3(aq)$$

$$\longleftarrow$$

Decreasing $[CO_2]$ upsets the equilibrium and a net reverse reaction takes place.

If the concentration of H_2CO_3 is increased, the rate of the reverse reaction ($H_2O + CO_2 \leftarrow H_2CO_3$) will increase, which lowers the concentration of H_2CO_3 and raises the concentration of CO_2.

$$H_2O(l) + CO_2(g) \rightleftharpoons H_2CO_3(aq)$$

$$\longleftarrow$$

Increasing $[H_2CO_3]$ upsets the equilibrium and a net reverse reaction takes place.

SAMPLE PROBLEM 9.4

Applying Le Châtelier's principle

For the equilibrium $2SO_2(g) + O_2(g) \rightleftharpoons 2SO_3(g)$, predict which reaction (forward or reverse) will be the faster one until equilibrium is reestablished when

a. the concentration of SO_2 is increased
b. the concentration of SO_2 is decreased
c. the concentration of SO_3 is decreased
d. the concentration of O_2 is increased

Strategy

Changing the concentration of SO_2 or O_2 will influence the rate of the forward reaction. Changing the concentration of SO_3 will influence the rate of the reverse reaction.

Solution

a. forward
b. reverse
c. forward
d. forward

PRACTICE PROBLEM 9.4

One step in the metabolism of glucose involves aldolase, an enzyme that catalyzes the cleavage of the molecule fructose 1,6-bisphosphate into dihydroxyacetone phosphate and glyceraldehyde 3-phosphate.

$$\text{Fructose 1,6-bisphosphate} \xrightleftharpoons{\text{aldolase}} \text{Dihydroxyacetone phosphate} + \text{Glyceraldehyde 3-phosphate}$$

For an equilibrium mixture, predict which direction (forward or reverse) will be the faster one until equilibrium is reestablished when

a. the concentration of dihydroxyacetone phosphate is increased
b. the concentration of glyceraldehyde 3-phosphate is decreased
c. the concentration of fructose 1,6-bisphosphate is decreased

Catalysts

Catalysts increase reaction rates by lowering the activation energy (Section 6.7). When a catalyst is used in a reversible reaction the lowered activation energy speeds up the forward and reverse reactions to the same extent. The overall result is that a catalyst has no effect on an equilibrium or on the value of K_{eq}.

BIOCHEMISTRY  Link

Diving Mammals, Oxygen, and Myoglobin

Muscle tissues contain the O_2 storage protein called myoglobin, which is found in two forms: deoxymyoglobin (Mb), which does not carry O_2, and oxymyoglobin (MbO_2), which does. In muscle tissue, these two forms of myoglobin exist in equilibrium with one another.

$$Mb + O_2 \rightleftharpoons MbO_2$$

As predicted by Le Châtelier's principle (Section 9.4), an increase in the concentration of O_2 in muscle tissue will upset an existing equilibrium and cause the forward reaction ($Mb + O_2 \longrightarrow MbO_2$) to speed up until equilibrium is reestablished. This results in a greater amount of O_2 being stored in muscle as MbO_2. When muscles are active and the body is unable to keep up with the O_2 demands of muscle tissue, the O_2 concentration begins to drop. In response to the loss of equilibrium a net reverse reaction takes place ($Mb + O_2 \longleftarrow MbO_2$), which releases O_2 to tissues.

Diving mammals, such as whales, dolphins, and seals, are able to stay underwater for much longer periods of time than humans. Some seals, for example, can remain submerged for over an hour (Figure 9.4). One of the factors responsible for this ability is that the muscles of many diving mammals contain three to ten times more myoglobin than found in human muscle tissue. When under water, MbO_2 in muscles is a major source of O_2 for diving mammals, while humans must depend mostly on the oxygen present in the lungs and blood.

FIGURE 9.4 **A harbor seal**
Muscle tissue of harbor seals has higher myoglobin concentrations than that of humans and, as a result, their tissues can bind greater amounts of O_2.

Source: PhotoDisc Blue/Getty Images.

9.5 IONIZATION OF WATER

Water is amphoteric because it can act as either an acid or base, depending on what it reacts with (Section 9.2). For example, in the presence of HCl water is a base and in the presence of NH_3 it is an acid.

$$\underset{\text{Acid}}{HCl} + \underset{\text{Base}}{H_2O} \rightleftharpoons \underset{\text{Base}}{Cl^-} + \underset{\text{Acid}}{H_3O^+}$$

$$\underset{\text{Base}}{NH_3} + \underset{\text{Acid}}{H_2O} \rightleftharpoons \underset{\text{Acid}}{NH_4^+} + \underset{\text{Base}}{OH^-}$$

This amphoteric nature of water even extends to a reaction that involves only water molecules as reactants (Figure 9.5). For the forward direction of this reaction, one H_2O molecule releases H^+ (is the acid) and the other accepts H^+ (is the base), while in the reverse reaction, H_3O^+ is the acid and OH^- is the base. In pure water this **ionization** (*formation of ions*) is spontaneous, but only involves a small fraction (less than 2 ppb) of the molecules at any one time.

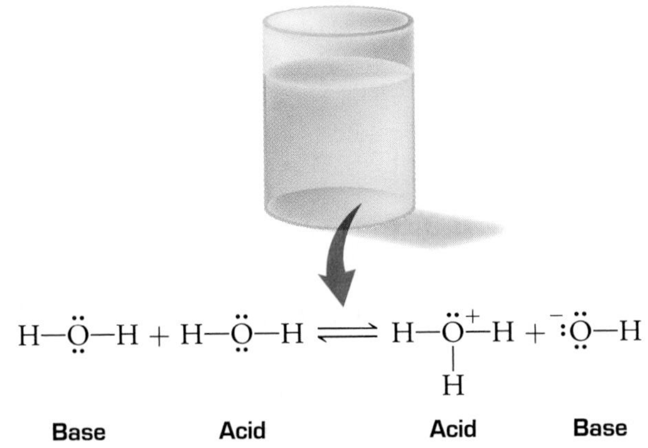

FIGURE 9.5

The self-ionization of water
One H_2O molecule (the acid) releases H^+ and another H_2O molecule (the base) accepts H^+. In the reverse reaction, H_3O^+ is the acid and OH^- is the base.

The equilibrium constant expression for the ionization of water is written

$$2H_2O(l) \rightleftharpoons H_3O^+(aq) + OH^-(aq) \qquad K_w = [H_3O^+][OH^-] = 1.0 \times 10^{-14}$$

■ $K_w = [H_3O^+][OH^-] = 1.0 \times 10^{-14}$

$\boldsymbol{K_w}$, the equilibrium constant for this reaction, has a value of 1.0×10^{-14}, so the hydronium ion (H_3O^+) concentration multiplied by the hydroxide ion (OH^-) concentration always equals 1.0×10^{-14}. This means that, in an aqueous solution, if the concentration of either one of these ions is known, the concentration of the other can be calculated. For example, if $[OH^-] = 1.0 \times 10^{-9}$ M, then $[H_3O^+] = 1.0 \times 10^{-5}$ M.

$$K_w = [H_3O^+][OH^-]$$

$$[H_3O^+] = \frac{K_w}{[OH^-]} = \frac{1.0 \times 10^{-14}}{1.0 \times 10^{-9}} = 1.0 \times 10^{-5}\ \text{M}$$

SAMPLE PROBLEM 9.5

Using K_w to solve for $[H_3O^+]$ or $[OH^-]$

What is the concentration of OH^- in an aqueous solution when

a. $[H_3O^+] = 1.0 \times 10^{-3}$ M
b. $[H_3O^+] = 1.0 \times 10^{-7}$ M
c. $[H_3O^+] = 5.0 \times 10^{-11}$ M

Strategy

Solving this type of problem involves solving for the unknown $[OH^-]$ in the equation $[H_3O^+][OH^-] = 1.0 \times 10^{-14}$.

Solution

$$K_w = [H_3O^+][OH^-]$$

$$[OH^-] = \frac{K_w}{[H_3O^+]}$$

a. $[OH^-] = \dfrac{1.0 \times 10^{-14}}{1.0 \times 10^{-3}} = 1.0 \times 10^{-11}\ M$

b. $[OH^-] = \dfrac{1.0 \times 10^{-14}}{1.0 \times 10^{-7}} = 1.0 \times 10^{-7}\ M$

c. $[OH^-] = \dfrac{1.0 \times 10^{-14}}{5.0 \times 10^{-11}} = 2.0 \times 10^{-4}\ M$

PRACTICE PROBLEM 9.5

What is the concentration of H_3O^+ in an aqueous solution when

a. $[OH^-] = 1.0 \times 10^{-5}$ M

b. $[OH^-] = 8.5 \times 10^{-11}$ M

c. $[OH^-] = 3.8 \times 10^{-1}$ M

d. $[OH^-] = 4.4 \times 10^{-9}$ M

9.6 THE pH SCALE

The concentration of H_3O^+ determines whether or not a solution is acidic, basic, or neutral. An aqueous solution is **acidic** when *the concentration of H_3O^+ is greater than 1×10^{-7} M*, and the higher the concentration the more acidic it is. A solution is **basic** when *the concentration of H_3O^+ is less than 1×10^{-7} M*, and the lower the concentration, the more basic it is. A solution in which *the concentration of H_3O^+ equals 1×10^{-7} M* is **neutral**.

As an alternative to using $[H_3O^+]$, the term **pH** is used, where

$$pH = -\log[H_3O^+]$$

A solution is acidic when $pH < 7$, basic when $pH > 7$, and neutral when $pH = 7$ (Figure 9.6). Table 9.3 lists the pHs of some common solutions.

To convert from $[H_3O^+]$ to pH, you take the logarithm of the H_3O^+ concentration (Math Support—Logs and Antilogs) and change the sign. For example, a solution containing 8×10^{-4} M H_3O^+ has a pH of 3.1.

$$pH = -\log[H_3O^+] = -\log(8 \times 10^{-4}) = -(-3.1) = 3.1$$

SAMPLE PROBLEM 9.6

Calculating pH

What is the pH of the following aqueous solutions?

a. $[H_3O^+] = 6.3 \times 10^{-1}$ M

b. $[OH^-] = 6.3 \times 10^{-1}$ M

FIGURE | 9.6

$[H_3O^+]$, $[OH^-]$, and pH
In aqueous solutions, $[H_3O^+][OH^-] = 1.0 \times 10^{-14}$, so if the concentration of one is large, the concentration of the other is small. In a neutral solution pH = 7, in an acidic solution pH < 7, and in a basic solution pH > 7.

	pH	$[H_3O^+]$	$[OH^-]$
	14	1×10^{-14}	1×10^{0}
	13	1×10^{-13}	1×10^{-1}
	12	1×10^{-12}	1×10^{-2}
Basic	11	1×10^{-11}	1×10^{-3}
	10	1×10^{-10}	1×10^{-4}
	9	1×10^{-9}	1×10^{-5}
	8	1×10^{-8}	1×10^{-6}
Neutral	7	1×10^{-7}	1×10^{-7}
	6	1×10^{-6}	1×10^{-8}
	5	1×10^{-5}	1×10^{-9}
	4	1×10^{-4}	1×10^{-10}
Acidic	3	1×10^{-3}	1×10^{-11}
	2	1×10^{-2}	1×10^{-12}
	1	1×10^{-1}	1×10^{-13}
	0	1×10^{0}	1×10^{-14}

TABLE | 9.3 pH VALUES OF SOME COMMON SOLUTIONS

Solution	pH
Battery acid	0.5
Gastric fluid in the stomach	1.5
Soft drinks	2.0–4.0
Vinegar	2.4–3.4
Grapefruit juice	3.2
Urine	4.8–7.5
Rainwater (unpolluted)	6.2
Milk	6.3–6.6
Pure water	7.0
Blood	7.35–7.45
Soap	8.0–10.0
Antacids	10.5
Household ammonia	11.5
Lye (a 1 M NaOH solution)	14.0

Strategy

The equation pH $= -\log[H_3O^+]$ is used to calculate pH. To solve part b, you must first convert from $[OH^-]$ into $[H_3O^+]$, using the equation $[H_3O^+][OH^-] = 1.0 \times 10^{-14}$.

Solution

a. $pH = -\log[H_3O^+] = -\log(6.3 \times 10^{-1}) = -(-0.20) = 0.20$
b. $[H_3O^+][OH^-] = 1.0 \times 10^{-14}$

$$[H_3O^+] = \frac{K_w}{[OH^-]} = \frac{1.0 \times 10^{-14}}{6.3 \times 10^{-1}} = 1.6 \times 10^{-14}\text{ M}$$

Once $[H_3O^+]$ is known, pH can be calculated.

$$pH = -\log[H_3O^+] = -\log(1.6 \times 10^{-14}) = -(-13.80) = 13.80$$

PRACTICE PROBLEM 9.6

Calculate the pH of a solution in which

a. $[H_3O^+] = 5.0 \times 10^{-8}$ M
b. $[H_3O^+] = 1 \times 10^{-13}$ M
c. $[OH^-] = 1 \times 10^{-13}$ M
d. $[OH^-] = 4.7 \times 10^{-3}$ M

Sometimes it is necessary to reverse this calculation, converting pH into $[H_3O^+]$. The equation to use for this is $[H_3O^+] = 10^{-pH}$ (Math Support—Logs and Antilogs). For example, if pH = 2.0, $[H_3O^+]$ is calculated as follows:

$$[H_3O^+] = 10^{-pH} = 10^{-2.0} = 0.01$$

The pH of a solution can be measured by using a pH meter, a device that consists of a pair of electrodes that can detect H_3O^+ (Figure 9.7), or by using a pH indicator, a compound whose appearance changes with variations in pH. Most pH indicators change from one color to another, as shown in Figure 9.1*b* for litmus and in Figure 9.8 for purple cabbage.

FIGURE 9.7

A pH meter
Some pH meters, built to be rugged, can be used to test pH out in the field.

Source: © Greg Smith/Corbis.

MATH SUPPORT—LOGS AND ANTILOGS

On your scientific calculator you will find a button labeled "log." If you try a few calculations, you will see that

$$\log 100 = 2$$
$$\log 1000 = 3$$
$$\log 0.0001 = -4$$

Converting these three numbers (100, 1000, and 0.0001) into scientific notation (Section 1.4) should help explain what the log or logarithm of a number is.

$$\log 100 = \log 10^2 = 2$$
$$\log 1000 = \log 10^3 = 3$$
$$\log 0.0001 = \log 10^{-4} = -4$$

The log or logarithm of a number *is the power to which ten must be raised to equal the number* ($\log 10^n = n$). The log of 100 is 2, because 10^2 equals 100 and the log of 0.0001 is -4 because $10^{-4} = 0.0001$.

When a number has the value 1×10^n, where n is an integer (-2, 5, etc.), its log can be determined without using a calculator. In these cases, the log is equal to the value of n.

$$\log 1 \times 10^{-2} = -2$$
$$\log 1 \times 10^{5} = 5$$

In all other cases (7.9×10^2, 2.2×10^{-5}, etc.), a calculator will be required. To calculate the log of a number, enter the number and press the "log" button.

$$\log 7.9 \times 10^2 = 2.90$$
$$\log 2.2 \times 10^{-5} = -4.66$$

Reversing this process gives the antilog or antilogarithm of a number ($\text{antilog } n = 10^n$).

$$\text{log of } 10^2 = 2, \text{ so antilog } 2 = 10^2$$
$$\text{log of } 10^{-8} = -8, \text{ so antilog } -8 = 10^{-8}$$

When n is an integer (2, -8, etc.), its antilog can be determined without using a calculator, because the antilog is equal to 10^n (see the two examples directly above). At other times (3.5, -1.3, etc.) a calculator must be used. Enter the number and press the "inv" (inverse) or "2nd" (second function) button, followed by "log."

$$\text{antilog } 3.5 = 10^{3.5} = 3.16 \times 10^3$$
$$\text{antilog } -1.3 = 10^{-1.3} = 5.01 \times 10^{-2}$$

Significant Figures

In the examples just given the numbers were exact, which means that they have an unlimited number of significant figures. The answers to the calculations performed using a calculator were arbitrarily reported with two digits to the right of the decimal point.

When taking the logarithm of a measured value, it is important to report the correct number of significant figures. In a log value, the digits to the left of the decimal place are not significant—they only indicate the power to which 10 is raised in the original number. Only the digits to the right of the decimal point in a logarithm are significant (shown in bold type).

$\mathbf{1} \times 10^2$	$\log(1 \times 10^2) = 2.\mathbf{0}$
$\mathbf{1} \times 10^{-13}$	$\log(1 \times 10^{-13}) = -13.\mathbf{0}$
$\mathbf{2.1} \times 10^{-8}$	$\log(2.1 \times 10^{-8}) = -7.\mathbf{68}$

The same approach is used in reverse when taking antilogs—the number of significant figures must be maintained.

$-6.\mathbf{8}$	$10^{-6.8} = \mathbf{2} \times 10^{-7}$
$-3.\mathbf{21}$	$10^{-3.21} = \mathbf{6.2} \times 10^{-4}$

9.7 ACID AND BASE STRENGTH

Acids

The stronger an acid, the more H_3O^+ it produces and the lower the pH of the solution that it forms. A strong acid such as HCl ionizes almost completely in water,

$$HCl(aq) + H_2O(l) \rightleftharpoons Cl^-(aq) + H_3O^+(aq)$$

so if a 0.1 M aqueous solution of HCl is prepared, it contains 0.1 M H_3O^+, 0.1 M Cl^-, and almost no HCl. The pH of this solution will be 1.0:

$$pH = -\log[H_3O^+] = -\log(0.1) = 1.0$$

Weak acids ionize much less. In HF, for example, only about 8% of the molecules are ionized at any given time.

$$HF(aq) + H_2O(l) \rightleftharpoons F^-(aq) + H_3O^+(aq)$$

A 0.1 M solution of HF contains about 0.008 M H_3O^+, 0.008 M F^-, and 0.09 M HF and has a pH of approximately 2.0.

When dealing with acids, the equilibrium constant (K_{eq}) is also known as the acidity constant (K_a). For the two reactions just mentioned, the acidity constant expressions are written

$$HCl(aq) + H_2O(l) \rightleftharpoons Cl^-(aq) + H_3O^+(aq) \qquad K_a = \frac{[Cl^-][H_3O^+]}{[HCl]} = 1.0 \times 10^7$$

$$HF(aq) + H_2O(l) \rightleftharpoons F^-(aq) + H_3O^+(aq) \qquad K_a = \frac{[F^-][H_3O^+]}{[HF]} = 6.6 \times 10^{-4}$$

As indicated by their respective K_a values, the equilibrium positions for these two acids are very different. For HCl, with a K_a of 1.0×10^7, products predominate at equilibrium, while for HF, with a K_a of 6.6×10^{-4}, reactants predominate. Another way of looking at this is to say that HCl is a stronger acid (is better at donating H^+) than is HF. *The larger the* K_a *for an acid, the stronger an acid it is.*

BIOCHEMISTRY Link

Plants as pH Indicators

Purple cabbage and many other plants contain natural pH indicators. If you add a few milliliters of purple cabbage indicator (prepared by boiling a few leaves of purple cabbage in water) to test tubes containing solutions of various pHs, a broad range of colors will be produced (Figure 9.8). With a variation in H_3O^+ concentrations, slight changes take place in the structure of the pigments that are present in purple cabbage leaves, changes that affect how they absorb light and the color that they appear. Other plant parts that contain pH-sensitive pigments include grape skins and eggplant peels.

FIGURE 9.8

Purple cabbage indicator
Purple cabbage juice contains a pH indicator. The indicator is red in acidic solutions, purple in neutral solutions, and blue, violet, green, and yellow in increasingly basic solutions.

Source: © Richard Megna/Fundamental Photographs.

TABLE 9.4 K_a AND pK_a VALUES FOR SELECTED ACIDS

Name	Formula	K_a	pK_a
Hydrochloric acid	HCl	1.0×10^{7}	−7.00
Phosphoric acid	H_3PO_4	7.5×10^{-3}	2.12
Hydrofluoric acid	HF	6.6×10^{-4}	3.18
Lactic acid	$CH_3CH(OH)CO_2H$	1.4×10^{-4}	3.85
Acetic acid	CH_3CO_2H	1.8×10^{-5}	4.74
Carbonic acid	H_2CO_3	4.4×10^{-7}	6.36
Dihydrogenphosphate ion	$H_2PO_4^-$	6.2×10^{-8}	7.21
Ammonium ion	NH_4^+	5.6×10^{-10}	9.25
Hydrocyanic acid	HCN	4.9×10^{-10}	9.31
Hydrogencarbonate ion	HCO_3^-	5.6×10^{-11}	10.25
Methylammonium ion	$CH_3NH_3^+$	2.4×10^{-11}	10.62
Hydrogenphosphate ion	HPO_4^{2-}	4.2×10^{-13}	12.38

Instead of using K_a we can also indicate the strength of an acid using **pK_a ($pK_a = -\log K_a$)**—*the lower the pK_a, the stronger the acid.* For HCl and HF, the pK_a values are calculated as follows.

■ Strong acids have large K_a and small pK_a values.

$$\text{HCl} \qquad pK_a = -\log(1.0 \times 10^{7}) = -7.00$$

$$\text{HF} \qquad pK_a = -\log(6.6 \times 10^{-4}) = 3.18$$

Table 9.4 lists K_as, and pK_as for a number of common acids.

Bases

The stronger a base, the more OH^- it produces and the higher the pH of the solution. A strong base like NaOH **dissociates** *(falls apart)* completely in water,

$$NaOH(s) \xrightarrow{H_2O} Na^+(aq) + OH^-(aq)$$

so if a 0.1 M NaOH solution is prepared, it will contain 0.1 M OH^- and 0.1 M Na^+, and the pH of the solution will be 13.0.

$$[H_3O^+] = \frac{K_w}{[OH^-]} = \frac{1.0 \times 10^{-14}}{0.1} = 1 \times 10^{-13}\ \text{M}$$

$$pH = -\log[H_3O^+] = -\log 1 \times 10^{-13} = 13.0$$

Some weak bases, like $Mg(OH)_2$, have an extremely low solubility in water (Table 7.1). Because they do not dissolve well, the OH^- concentrations remain low and the solution does not become very basic. For other weak bases, such as ammonia (NH_3), it is not a matter of solubility, but just that they are poor H^+ acceptors. When ammonia reacts with water, the concentration of products (NH_4^+ and OH^-) is very low and the solution is not strongly basic.

$$NH_3(aq) + H_2O(l) \rightleftharpoons NH_4^+(aq) + OH^-(aq)$$

To predict the strength of a base, it helps to know something about the acidity of its conjugate acid (Section 9.2), because *the stronger an acid, the weaker its conjugate base.* HCl, for example, is a strong acid, so Cl^- (its conjugate) is a weak base. H_2O is a weak acid, so OH^- (its conjugate) is a strong base. Table 9.5 shows the relationship between acid and conjugate base strength.

■ Strong acids have weak conjugate bases.

9.8 NEUTRALIZING ACIDS AND BASES

Neutralization takes place when *an acid and a base react to form water and a salt* (an ionic compound). For example, HCl and NaOH react with one another according to the equation

$$\underset{\text{Acid}}{HCl(aq)} + \underset{\text{Base}}{NaOH(aq)} \longrightarrow \underset{\text{Salt}}{NaCl(aq)} + \underset{\text{Water}}{H_2O(l)}$$

The term "neutralization" comes from the fact that if proper amounts of an acid and a base are combined, they will cancel one another to produce a neutral solution. To see how this comes about, let us take a closer look at the reaction between HCl and NaOH. HCl is a strong acid that completely ionizes in water to form H_3O^+ and Cl^-, NaOH is a strong base that completely dissociates in water to form Na^+ and OH^-, and NaCl is a water-soluble ionic compound. The ionic equation for this double replacement reaction is

$$H_3O^+(aq) + Cl^-(aq) + Na^+(aq) + OH^-(aq) \longrightarrow Na^+(aq) + Cl^-(aq) + 2H_2O(l)$$

Ionic equation

TABLE 9.5 Relative Strengths of Some Acids and Their Conjugate Bases

Acid	Name	Base	Name
Strong acids		**Extremely weak bases**	
HI	Hydroiodic acid	I^-	Iodide ion
HBr	Hydrobromic acid	Br^-	Bromide ion
HCl	Hydrochloric acid	Cl^-	Chloride ion
H_2SO_4	Sulfuric acid	HSO_4^-	Hydrogensulfate ion
HNO_3	Nitric acid	NO_3^-	Nitrate ion
Weak acids		**Very weak bases**	
H_3O^+	Hydronium ion	H_2O	Water
HSO_4^-	Hydrogensulfate ion	SO_4^{2-}	Sulfate ion
H_3PO_4	Phosphoric acid	$H_2PO_4^-$	Dihydrogenphosphate ion
HF	Hydrofluoric acid	F^-	Fluoride ion
CH_3CO_2H	Acetic acid	$CH_3CO_2^-$	Acetate ion
Very weak acids		**Weak bases**	
$H_2PO_4^-$	Dihydrogenphosphate ion	HPO_4^{2-}	Hydrogenphosphate ion
H_2CO_3	Carbonic acid	HCO_3^-	Hydrogencarbonate ion
NH_4^+	Ammonium ion	NH_3	Ammonia
$CH_3NH_3^+$	Methylammonium ion	CH_3NH_2	Methylamine
HCN	Hydrocyanic acid	CN^-	Cyanide ion
HCO_3^-	Hydrogencarbonate ion	CO_3^{2-}	Carbonate ion
HPO_4^{2-}	Hydrogenphosphate ion	PO_4^{3-}	Phosphate ion
Extremely weak acid		**Strong base**	
H_2O	Water	OH^-	Hydroxide

Removing the spectator ions gives the net ionic equation.

$$H_3O^+(aq) + OH^-(aq) \longrightarrow 2H_2O(l)$$

Net ionic equation

This shows that the neutralization of HCl by NaOH involves a reaction between H_3O^+ and OH^- to form two water molecules.

- When an acid is neutralized by a base, the solution has a pH of 7.0.

In a neutralization reaction the concentrations of H_3O^+ and OH^- never reach zero, because water self-ionizes to form these two ions (Section 9.5). If an acid and base completely react with one another, the dissociation of water leaves final H_3O^+ and OH^- concentrations of 1×10^{-7} M and a pH of 7.0.

Titration

- Titration is used to determine the concentration of an unknown acid or base solution.

A technique called **titration** can be used to *determine the concentration of an acid or base solution.* In the example that follows, we will consider the titration of an acid of unknown concentration. A buret (Figure 9.9) is used to add just enough base of a known concentration (a standard base solution) to totally consume any acid that is present. The volume of base needed to consume all of the acid is called the end point of the titration. End points can be determined using pH indicators or a pH meter.

Suppose that it takes 0.025 L of 0.20 M NaOH to neutralize 0.10 L of an HCl solution of unknown concentration. Calculating the HCl concentration (Figure 9.10) begins with determining how many moles of NaOH solution were added. To do this, we use concentration as a conversion factor.

$$0.025\ \cancel{L} \times \frac{0.20 \text{ mol NaOH}}{\cancel{L}} = 5.0 \times 10^{-3} \text{ mol NaOH}$$

Next, the number of moles of HCl initially present is calculated, making use of the fact that in the balanced equation for the reaction of HCl with NaOH, one mole of NaOH is required to completely react with one mole of HCl.

$$5.0 \times 10^{-3}\ \cancel{\text{mol NaOH}} \times \frac{1 \text{ mol HCl}}{1\ \cancel{\text{mol NaOH}}} = 5.0 \times 10^{-3} \text{ mol HCl}$$

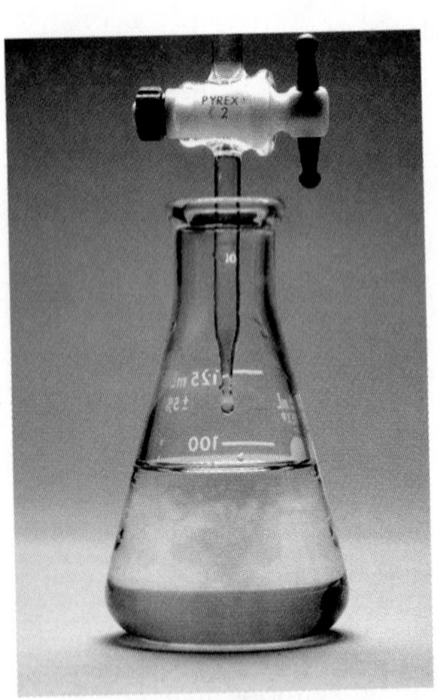

FIGURE 9.9

Titration
A buret is used to measure the amount of standard solution used during a titration and a pH indicator can be used to detect the end point. An HCl solution containing a few drops of phenolphthalein is colorless. When enough standard NaOH has been added to neutralize the HCl, phenolphthalein turns pink.

Source: Ken Karp/Wiley Archive.

FIGURE 9.10

Titration calculations
Calculations involving titrations must be done in terms of the number of moles of acid and base. An approach to use for determining the concentration of an unknown acid is shown here.

Source: BROWN, THEODORE E; LEMAY, H. EUGENE; BURSTEN, BRUCE E, CHEMISTRY: THE CENTRAL SCIENCE, 10th edition, © 2006, p. 741. Adapted permission of Pearson Education, Inc., Upper Saddle River, NJ.

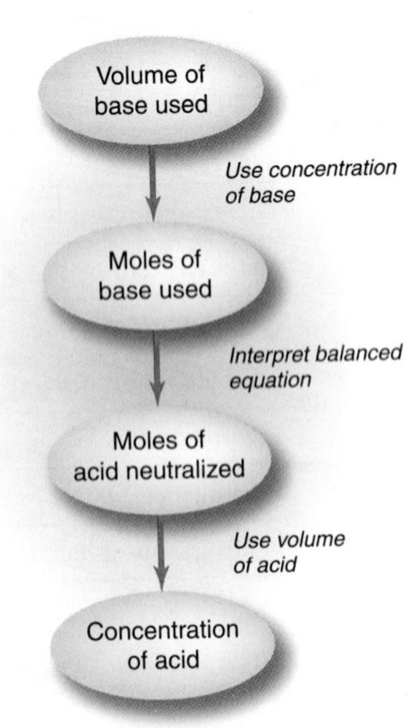

To obtain the concentration of HCl, moles of HCl are divided by the initial volume of solution.

$$\frac{5.0 \times 10^{-3} \text{ mol HCl}}{0.10 \text{ L}} = 5.0 \times 10^{-2} \text{ M HCl}$$

SAMPLE PROBLEM 9.7

Titration calculations

It requires 15 mL (0.015 L) of 0.25 M NaOH to reach the end point in a titration of 75 mL (0.075 L) of an HCl solution of unknown concentration. What is the initial concentration of HCl?

Strategy

Calculating the concentration of the HCl solution involves determining the number of moles of HCl present in the initial 75 mL of solution. Moles of HCl can be calculated from the number of moles of NaOH used in the titration (1 mol of NaOH and neutralizes 1 mol of HCl). You should begin by calculating the number of moles of NaOH in 15 mL of 0.25 M NaOH.

Solution

The first step in solving this problem involves calculating the number of moles of NaOH that were added. Recall that molarity (M) is moles/liter. The volume of NaOH added must, therefore, be expressed in L.

$$0.015 \cancel{\text{L}} \times \frac{0.25 \text{ mol NaOH}}{1 \cancel{\text{L}}} = 3.8 \times 10^{-3} \text{ mol NaOH}$$

The balanced equation for the reaction ($HCl + NaOH \longrightarrow NaCl + H_2O$) tells us that one mole of HCl reacts with one mole of NaOH, so

$$3.8 \times 10^{-3} \cancel{\text{mol NaOH}} \times \frac{1 \text{ mol HCl}}{1 \cancel{\text{mol NaOH}}} = 3.8 \times 10^{-3} \text{ mol HCl}$$

The initial HCl concentration is obtained by dividing the number of moles of HCl by the volume (in liters) of the initial solution.

$$\frac{3.8 \times 10^{-3} \text{ mol HCl}}{0.075 \text{ L}} = 5.1 \times 10^{-2} \text{ M HCl}$$

PRACTICE PROBLEM 9.7

It requires 26 mL of 0.30 M NaOH to titrate 15 mL of an HCl solution of unknown concentration. What is the HCl concentration?

9.9 EFFECT OF pH ON ACID AND CONJUGATE BASE CONCENTRATIONS

We have seen that K_a and pK_a provide information about the relative concentrations of products and reactants in acid–base reactions. For example, the K_a for HF is 6.6×10^{-4} (pK_a = 3.18), so at equilibrium, the concentration of HF is greater than the product of the concentrations of F^- and H_3O^+.

$$HF + H_2O \rightleftharpoons F^- + H_3O^+ \qquad K_a = \frac{[F^-][H_3O^+]}{[HF]} = 6.6 \times 10^{-4}$$

We have also seen that Le Châtelier's principle allows us to predict how changing the concentration of a reactant or product will be followed by a change in the rate of the forward or reverse reaction as equilibrium is reestablished. For the reaction above, increasing the concentration of H_3O^+ (lowering the pH) will cause the reverse reaction to speed up, which consumes F^- and produces HF. Reducing the concentration of H_3O^+ (raising the pH) will cause the forward reaction to speed up, which consumes HF and produces F^-. Thus, *the relative amounts of HF and F^- present in a solution depend on the pH.*

While the math will not be shown here, an interesting relationship exists between pH and the relative concentrations of an acid and its conjugate base:

- When the pH is adjusted to have the same value as the pK_a ($pH = pK_a$), the concentration of acid equals the concentration of its conjugate base. For the acid HF ($pK_a = 3.18$), this means that *[HF]* = *[F^-]* when the pH is 3.18 (Table 9.6).
- When the pH is lower than the pK_a ($pH < pK_a$), the concentration of the acid is higher than that of the conjugate base. At a pH of 1.0, which is lower than the pK_a of 3.18 for HF, *[HF]* > *[F^-]*.
- When the pH is higher than the pK_a ($pH > pK_a$), the concentration of the conjugate base is higher than that of the acid. At pH 9.0, which is higher than the pK_a of 3.18 for HF, *[HF]* < *[F^-]*.

TABLE 9.6 ACID (HA) AND CONJUGATE BASE (A^-) CONCENTRATIONS, AS A FUNCTION OF pH

If	$pH = pK_a$ of HA	then	$[HA] = [A^-]$
If	$pH < pK_a$ of HA	then	$[HA] > [A^-]$
If	$pH > pK_a$ of HA	then	$[HA] < [A^-]$

This explains some of the chemistry related to carboxylic and fatty acids discussed in earlier chapters. Depending on the pH of a solution, a carboxylic acid or fatty acid will be found as either the acid or as the conjugate base.

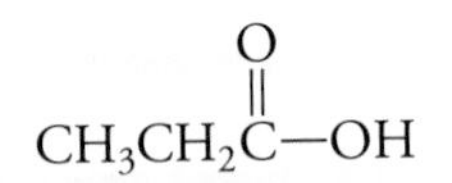

$CH_3CH_2C(=O)\text{—}OH$

Carboxylic acid

When pH < pK_a, this form predominates

$CH_3CH_2C(=O)\text{—}O^-$

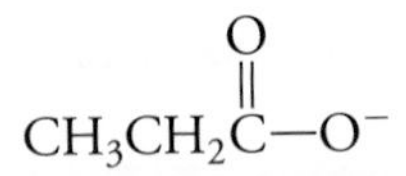

Conjugate base

When pH > pK_a, this form predominates

The typical carboxylic acid has a pK_a of about 5, so under the basic conditions of ester hydrolysis (Section 6.4), the product carboxylic acid exists as the conjugate base. The typical fatty acid also has a pK_a of about 5, so at pH 7, conditions commonly found in the body, fatty acids exist as amphipathic fatty acid anions (Sections 7.5 and 8.1).

9.10 BUFFERS

■ Buffers resist changes in pH.

A **buffer** is a solution that *resists changes in pH* when small amounts of acid or base are added. Here only buffers prepared from weak acids and their conjugate bases will be considered, one example being a mixture of acetic acid (CH_3CO_2H) and acetate ion ($CH_3CO_2^-$) in water.

$$\underset{\text{Weak acid}}{CH_3CO_2H} + H_2O \rightleftharpoons \underset{\text{Conjugate base}}{CH_3CO_2^-} + H_3O^+$$

If a small amount of H_3O^+ is added to this buffer, Le Châtelier's principle predicts that the rate of the reverse reaction will increase in order to reestablish equilibrium. In the process, excess H_3O^+ will be consumed by $CH_3CO_2^-$ and the pH will remain mostly unchanged.

$$CH_3CO_2H + H_2O \leftarrow CH_3CO_2^- + H_3O^+$$

If a small amount of OH^- is added, this base will react with CH_3CO_2H to produce its conjugate base, $CH_3CO_2^-$. This process consumes OH^-, so the pH remains relatively stable.

$$CH_3CO_2H + OH^- \longrightarrow CH_3CO_2^- + H_2O$$

You might wonder what this reaction has to do with the equilibrium reaction shown at the beginning of this section ($CH_3CO_2H + H_2O \rightleftharpoons CH_3CO_2^- + H_3O^+$). To explain, let us use a different approach to describe what happens when hydroxide ion is added to this buffer: The reaction of OH^- with H_3O^+ reduces the concentration of each.

$$H_3O^+ + OH^- \longrightarrow 2H_2O$$

This causes a slowing of the reverse reaction between $CH_3CO_2^-$ and H_3O^+, which allows the forward reaction to predominate until equilibrium is reestablished. As a result, lost H_3O^+ will be mostly replaced and the pH will remain fairly constant.

$$CH_3CO_2H + H_2O \longrightarrow CH_3CO_2^- + H_3O^+$$

Interestingly, when the two reactions above are added together, and when reactants and products appearing unchanged on each side of the reaction arrow are removed, the result is a reaction equation identical to the one at the beginning of this paragraph.

$$\begin{array}{rcl} \cancel{H_3O^+} + OH^- & \longrightarrow & \cancel{2}\,H_2O \\ CH_3CO_2H + \cancel{H_2O} & \longrightarrow & CH_3CO_2^- + \cancel{H_3O^+} \\ \hline CH_3CO_2H + OH^- & \longrightarrow & CH_3CO_2^- + H_2O \end{array}$$

Regardless of which approach is used to explain the effect of adding OH^- to this buffer, the result is the same. There is a net conversion of CH_3CO_2H into $CH_3CO_2^-$ and the pH remains constant.

Buffers are most resistant to pH changes when the pH equals the pK_a of the weak acid and are effective when the pH is within one unit of the pK_a ($pH = pK_a \pm 1$). The pK_a of acetic acid is 4.74, so an acetic acid and acetate ion buffer functions well over the pH range 3.74 to 5.74 and is most effective at a pH of 4.74.

Sample Problem 9.8

Understanding buffers

A buffer can be prepared using $H_2PO_4^-$ and its conjugate base.

a. Write the acid–base reaction equation for $H_2PO_4^-$ and H_2O reacting to form the conjugate base of $H_2PO_4^-$ and H_3O^+.

b. Write a reaction equation that shows what takes place when H_3O^+ is added to this buffer.

Strategy

Part a gives the formula of the two reactants and one of the two products. The conjugate base of $H_2PO_4^-$ has one less H^+ than $H_2PO_4^-$. In part b, think about what happens to the rate of the forward or reverse reaction if the concentration of a reactant or product is changed.

Solution

a. $H_2PO_4^- + H_2O \rightleftharpoons HPO_4^{2-} + H_3O^+$

b. An increase in the concentration of H_3O^+ causes the rate of the reverse reaction to increase.

$$H_2PO_4^- + H_2O \leftarrow HPO_4^{2-} + H_3O^+$$

PRACTICE PROBLEM 9.8

a. For the buffer described in Sample Problem 9.8, write a reaction equation that shows what takes place when OH^- is added to this buffer.

b. Over what pH range is this buffer effective? (See Table 9.4.)

9.11 MAINTAINING THE pH OF BLOOD SERUM

The pH of blood serum normally falls between 7.35 and 7.45. It is important that the pH not move out of this range, because of the effect that pH variations have on the ability of enzymes to function properly. Enzymes, which catalyze chemical reactions that take place in the body, are proteins consisting of chains of amino acid residues, some of which contain acidic or basic groups. Altering the pH can change the structure of these groups (Figure 9.11) in a way that affects the noncovalent interactions (Figure 4.8) that are so important for maintaining enzyme shape and function. The adverse health effects associated with blood serum pHs that fall outside of the 7.35–7.45 range largely result from changes to enzyme structure.

Maintaining the appropriate blood serum pH involves dealing with acids that are produced by metabolism (the reactions that take place within living things). These acids include fatty acids, lactic acid, phosphoric acid, and certain members of a group of compounds called ketone bodies. Some acids are also present in the foods that we eat.

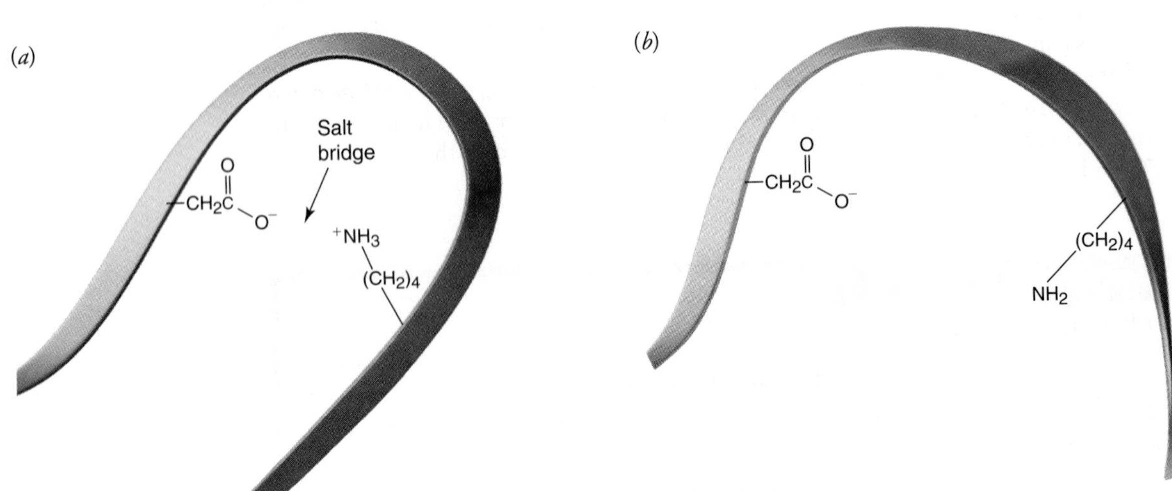

FIGURE 9.11 **The effect of changing pH on enzyme structure**
In enzymes and other proteins, a change in pH can switch groups between their acidic and conjugate base forms. This can disrupt noncovalent interactions, including salt bridges (ionic bonds), that are essential for maintaining function. (a) Near pH 7 a salt bridge holds a portion of the protein chain in the shape required for the enzyme to function. (b) Above pH 9, $—NH_3^+$ (acid) is converted into $—NH_2$ (conjugate base). The salt bridge no longer exists and the shape and catalytic ability of the enzyme change.

Another significant source of acid is the CO_2 produced by cells. In moving from the cells where it is produced to the lungs where it is exhaled (Figure 5.27), most of the CO_2 is carried in the blood as H_2CO_3 or HCO_3^-. H_2CO_3 is produced when CO_2 reacts with water, and HCO_3^- is formed in the acid-base reaction that takes place between H_2CO_3 and water.

$$H_2O + CO_2 \rightleftharpoons H_2CO_3$$

$$H_2CO_3 + H_2O \rightleftharpoons HCO_3^- + H_3O^+$$

To deal with these acids and to maintain a constant serum pH, the body relies on two general approaches. One involves using buffers to minimize pH changes. The other is to control the H_3O^+ concentration in the blood through respiration and by action of the kidneys.

Let us begin by considering the buffers present in serum. The most important of these consists of H_2CO_3 and its conjugate base HCO_3^-.

$$H_2CO_3 + H_2O \rightleftharpoons HCO_3^- + H_3O^+$$

As shown by the equations below, small amounts of H_3O^+ or OH^- can be consumed by this buffer:

$$\text{Addition of } H_3O^+\text{:} \quad H_2CO_3 + H_2O \longleftarrow HCO_3^- + H_3O^+$$

$$\text{Addition of } OH^-\text{:} \quad H_2CO_3 + OH^- \longrightarrow HCO_3^- + H_2O$$

For this buffer system to work, both H_2CO_3 and HCO_3^- must be available. As we saw above, H_2CO_3 is produced when CO_2 formed in cells reacts with H_2O. Some HCO_3^- is formed by loss of H^+ from H_2CO_3. The kidneys, by releasing or taking up HCO_3^-, also control blood concentrations of this buffer component.

Dihydrogenphosphate ion ($H_2PO_4^-$) and hydrogenphosphate ion (HPO_4^{2-}) are another buffer system that is present in blood (see Sample Problem 9.8). Proteins, which are weak acids, provide yet another buffer system that helps to maintain a relatively constant blood pH.

The respiratory system also plays an important role in maintaining the pH of blood serum. If, in spite of the action of buffers, the pH of the blood becomes too acidic, the respiratory center in the brain signals for faster and deeper breathing. This leads to more CO_2 being exhaled and a drop in its partial pressure (P_{CO_2}) (Figure 5.27). In terms of Le Châtelier's principle, loss of CO_2 leads to a net reduction in the concentration of H_2CO_3, which leads to a drop in the concentration of H_3O^+ (the pH becomes less acidic).

$$H_2O + CO_2 \longleftarrow H_2CO_3$$

$$H_2CO_3 + H_2O \longleftarrow HCO_3^- + H_3O^+$$

A good example of this is what happens when you exercise. Exercise increases metabolism, which leads to an increased cellular production of CO_2. Subsequent to this, the concentration of H_2CO_3 rises and the blood buffer systems are unable to deal with the excess H_3O^+ that forms. In response, you start to breathe faster. As described above, this helps to move the serum pH back into the normal range.

The respiratory system can also deal with serum that is too basic. In this case the respiratory center in the brain signals for slower and shallower breathing, which causes P_{CO_2} in the blood to rise. This leads to the production of more H_2CO_3 and, subsequently, a drop in the serum pH.

The kidneys also play a role in maintaining an appropriate pH. If serum is too acidic, the kidneys release HCO_3^- into the blood. This component of the H_2CO_3/HCO_3^- buffer system helps to reduce the H_3O^+ concentration. Additionally, the kidneys remove H_3O^+ from the body by releasing it into urine. If the pH of blood serum becomes too high, the kidneys do the opposite: HCO_3^- is removed from the blood and H_3O^+ is added. The effects of the H_2CO_3/HCO_3^- buffer system, respiration, and the kidneys on blood pH are summarized in Figure 9.12.

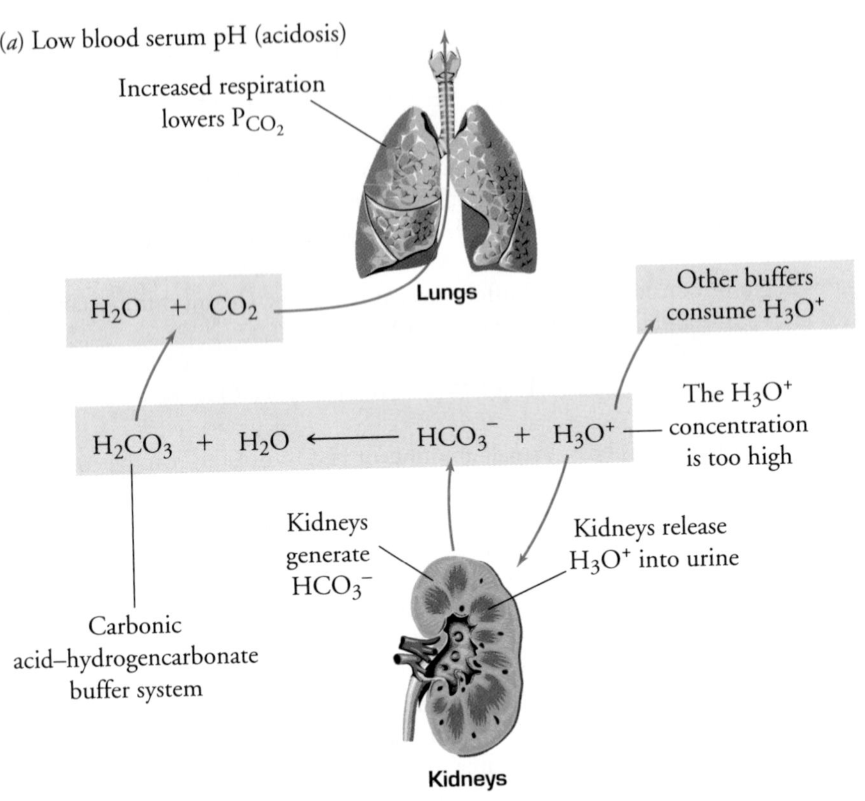
(a) Low blood serum pH (acidosis)
Increased respiration lowers P_{CO_2}
Lungs
$H_2O + CO_2$
Other buffers consume H_3O^+
$H_2CO_3 + H_2O \longleftarrow HCO_3^- + H_3O^+$
The H_3O^+ concentration is too high
Kidneys generate HCO_3^-
Kidneys release H_3O^+ into urine
Carbonic acid–hydrogencarbonate buffer system
Kidneys

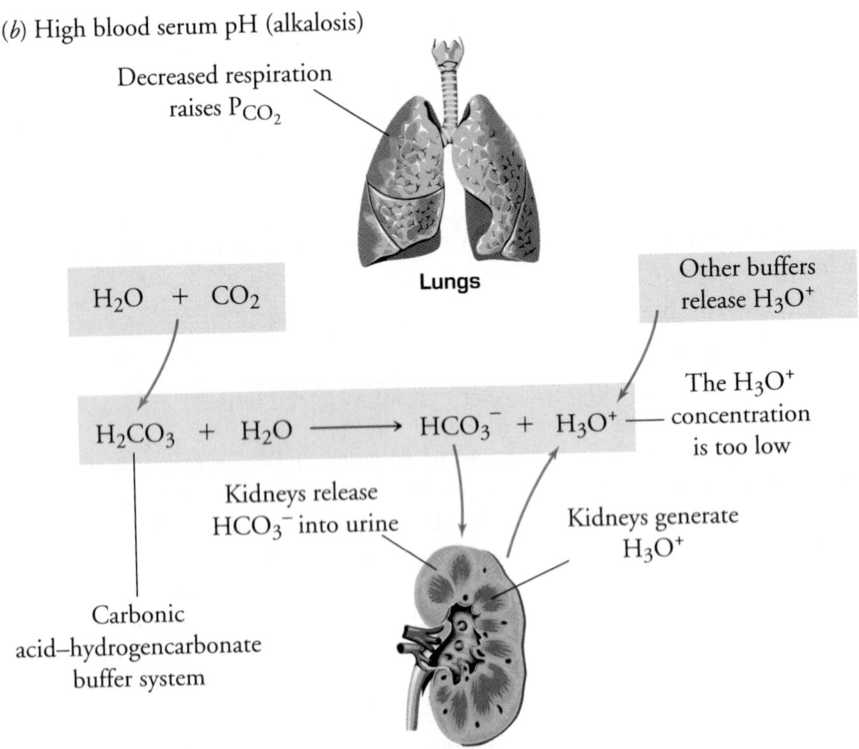
(b) High blood serum pH (alkalosis)
Decreased respiration raises P_{CO_2}
Lungs
$H_2O + CO_2$
Other buffers release H_3O^+
$H_2CO_3 + H_2O \longrightarrow HCO_3^- + H_3O^+$
The H_3O^+ concentration is too low
Kidneys release HCO_3^- into urine
Kidneys generate H_3O^+
Carbonic acid–hydrogencarbonate buffer system
Kidneys

FIGURE | 9.12

Controlling plasma pH
The H_2CO_3/HCO_3^- buffer system, respiration, the kidneys, and other buffer systems all play a role in controlling the pH of blood plasma. (a) The response to a drop in pH. (b) The response to a rise in pH.

Source: Figure 27.11, p. 1034 from FUNDAMENTALS OF ANATOMY AND PHYSIOLOGY, 6th ed. by Frederic H. Martini. © 2004 by Frederic H. Martini, Inc. Reprinted by permission of Pearson Education, Inc.

■ **TABLE | 9.7 ACID/BASE DISORDERS**

Condition	pH	P_{CO_2}	HCO_3^- concentration
Normal range in blood plasma	7.35–7.45	35–45 mm Hg	22–26 mEq/L
Respiratory acidosis	lower	higher	higher, if compensating
Metabolic acidosis	lower	lower, if compensating	lower
Respiratory alkalosis	higher	lower	lower, if compensating
Metabolic alkalosis	higher	higher, if compensating	higher

Source: Page 1055, HUMAN ANATOMY AND PHYSIOLOGY, 6th ed. By Elaine N. Marieb. © 2004 by Pearson, Education, Inc. Reprinted by permission.

Even with all of these controls, the pH of blood can still sometimes fall outside of the normal range. These acid/base disorders are classified as being either respiratory or metabolic in nature (Table 9.7). Respiratory disorders occur when exhaling does not remove CO_2 from the blood at the same rate that it is produced in cells. Metabolic disorders arise from an inability to deal with acids produced by metabolism, problems with controlling the blood concentration of HCO_3^-, ingestion of compounds that are acids or bases, or ingestion of compounds that can be converted into acids or bases.

Mild cases of **acidosis**, a *low blood serum pH*, can result in light-headedness. Severe cases produce a depression of the central nervous system that can lead to coma and death. Respiratory acidosis, characterized by a low pH, an elevated blood P_{CO_2}, and normal or higher than normal HCO_3^- concentrations, is caused by any condition that interferes with the ability of the lungs to exchange gases—specifically, to remove CO_2 from the blood. Pneumonia, emphysema, cystic fibrosis, shallow breathing, or holding your breath can cause this form of acidosis. As described above, a rise in P_{CO_2} produces higher concentrations of H_2CO_3, which can upset the H_2CO_3/HCO_3^- buffer system enough to lower the pH of blood. The serum concentration of HCO_3^- is sometimes higher than normal in respiratory acidosis because the kidneys are trying to compensate for the inability of the lungs to control pH. Releasing more HCO_3^- into the blood provides a buffer component to help remove unwanted H_3O^+.

Metabolic acidosis, distinguished by low pH, low HCO_3^- concentrations, and a P_{CO_2} that is normal, or lower if the lungs are compensating, can be caused by ketone bodies (produced from starvation or poorly controlled diabetes), lactic acid (produced by strenuous exercise), excessive alcohol consumption (alcohol is metabolized into acetic acid), diarrhea (excessive loss of HCO_3^-), and certain kidney problems.

The symptoms of **alkalosis**, a *high blood serum pH*, include headaches, nervousness, cramps, and, in severe cases, convulsions and death. Respiratory alkalosis (high pH, low P_{CO_2}, and HCO_3^- concentrations that are normal, or lower if the kidneys are compensating) occurs when CO_2 is exhaled from the body more quickly than it is produced by cells. A net loss of CO_2 causes a reduction in the concentration of H_2CO_3 present in the blood, and too large of a decrease leads to a rise in pH. Hyperventilation, which can be brought on by anxiety, central nervous system damage, aspirin poisoning, fever, and other factors, are the usual causes.

Metabolic alkalosis (high pH, high HCO_3^-, and a P_{CO_2} that is normal, or higher if the lungs are compensating) can result from excessive use of antacids (weak bases) and from constipation (greater than normal amounts for HCO_3^- end up in the blood).

To help with the digestion of food and to activate digestive enzymes, your stomach produces gastric juice, a pH 1.5 solution that contains HCl. Overeating or stress can cause too much gastric juice to be released, which can lead to stomach upset. For temporary relief, many people take antacids.

The active ingredients in antacids are weak bases, most commonly $CaCO_3$, $MgCO_3$, $NaHCO_3$, $Mg(OH)_2$, and $Al(OH)_3$. Some examples include

Antacid	Weak Base
Alka Seltzer	$NaHCO_3$
Di-Gel, Mylanta, Maalox	$Al(OH)_3$ and $Mg(OH)_2$
Milk of magnesia	$Mg(OH)_2$
Tums, Alka-2	$CaCO_3$
Gaviscon	$Al(OH)_3$ and $MgCO_3$
Rolaids	$AlNa(OH)_2CO_3$
Riopan	$AlMg(OH)_5$

The carbonate, hydrogencarbonate, or hydroxide ions provided by these bases react with and neutralize some of the H_3O^+ that is present in gastric juice.

$$CO_3^{2-}(aq) + H_3O^+(aq) \longrightarrow HCO_3^-(aq) + H_2O(l)$$

$$HCO_3^-(aq) + H_3O^+(aq) \longrightarrow H_3CO_3(aq) + H_2O(l)$$

$$OH^-(aq) + H_3O^+(aq) \longrightarrow 2H_2O(l)$$

Excessive use of antacids can lead to alkalosis, a higher than normal blood pH (Section 9.11). Besides alkalosis, some of the bases used as antacids can have other side effects—Al^{3+} and Ca^{2+} in large doses cause constipation, while Mg^{2+} has a laxative effect.

summary of objectives

1 List the common characteristics of acids and bases.

Acids dissolve some metals, they turn litmus pink, and those that are safe to ingest have a sour taste. **Bases** feel slippery or soapy, turn litmus blue, and have a bitter taste.

2 Describe Brønsted–Lowry acids and bases and explain how they differ from their conjugates. Relate acid strength to conjugate base strength.

According to the **Brønsted–Lowry** definition, acids release H^+ and bases accept H^+. In the equation for a Brønsted–Lowry acid–base reaction, an acid and a base appear on each side of the reaction arrow. Each acid has a corresponding **conjugate** base on the other side of the reaction equation. The stronger an acid (the greater the value of K_a or the smaller the value of pK_a), the more it undergoes ionization to produce H_3O^+. Also, the stronger an acid, the weaker is its conjugate base.

3 Write the equilibrium constant for a reversible reaction and use Le Châtelier's principle to explain how an equilibrium responds to being disturbed.

When the rates of the forward and reverse reactions are the same, a reversible reaction has reached **equilibrium**. At equilibrium, there is no net change in reactant or product concentrations. The equilibrium position of a reversible reaction is described by the **equilibrium constant (K_{eq})**. For the generalized reaction

$$aA + bB \rightleftharpoons cC + dD$$

the equilibrium constant expression takes the form

$$K_{eq} = \frac{[C]^c[D]^d}{[A]^a[B]^b}$$

Because they do not change, concentration terms for solids and solvents are not included in the expression.

Le Châtelier's principle states that when a reversible reaction is pushed out of equilibrium, the reaction responds to restore equilibrium. Increasing the concentration of reactant(s), for example, will increase the rate of the forward reaction, producing more product. As product concentration climbs, the reverse reaction speeds up, until the rates of the forward and reverse reactions are identical and equilibrium has been reestablished. Catalysts have no effect on an equilibrium.

4 Use H_3O^+ concentration and pH to identify a solution as being acidic, basic, or neutral.

In a **neutral** solution, $[H_3O^+] = 1.0 \times 10^{-7}$, in an **acidic** solution, $[H_3O^+] > 1.0 \times 10^{-7}$, and in a **basic** solution, $[H_3O^+] < 1.0 \times 10^{-7}$. Alternatively, in a neutral solution, pH = 7 (**$\mathbf{pH = -\log[H_3O^+]}$**), in an acidic solution, pH < 7, and in a basic solution, pH > 7.

5 Describe the processes of neutralization and titration.

Neutralization involves reacting an acid with a base to produce water, a salt, and a neutral solution. The concentration of unknown acidic solution can be determined through **titration**, in which just enough of a basic solution of known concentration is added to neutralize the acid.

6 Explain how the pH of a solution can affect the relative concentrations of an acid and its conjugate base and describe buffers.

In a solution containing an acid (HA) and its conjugate base (A^-), pH has an effect on the relative concentrations of each. When $pH = pK_a$, $[HA] = [A^-]$; when $pH < pK_a$, $[HA] > [A^-]$; and when $pH > pK_a$, $[HA] < [A^-]$. **Buffers**, solutions that resist changes in pH when small amounts of acid or base are added, can be made by preparing solutions containing a weak acid and its conjugate base. Buffers are most effective when the pH of the solution is within one unit of the pK_a of the weak acid ($pH = pK_a \pm 1$).

7 Describe the role of buffers, respiration, and the kidneys in maintaining a stable blood serum pH.

Carbonic acid (H_2CO_3) and hydrogencarbonate (HCO_3^-) make up the most important of the blood serum buffers.

$$H_2CO_3 + H_2O \rightleftharpoons HCO_3^- + H_3O^+$$

Following Le Châtelier's principle, a rise in the concentration of H_3O^+ produces a net reverse reaction (reducing H_3O^+ levels) and a rise in the concentration of OH^- causes a net forward reaction (increasing H_3O^+ levels).

$$\text{Addition of } H_3O^+\text{:} \quad H_2CO_3 + H_2O \longleftarrow HCO_3^- + H_3O^+$$

$$\text{Addition of } OH^-\text{:} \quad CH_3CO_2H + OH^- \longrightarrow CH_3CO_2^- + H_2O$$

The hydrogen carbonate present in blood serum is formed when CO_2 molecules produced by metabolism react with water.

$$H_2O + CO_2 \rightleftharpoons H_2CO_3$$

Respiration helps to determine serum concentrations CO_2 and, indirectly, the pH of blood. Slow, shallow breathing allows the partial pressure of CO_2 (P_{CO_2}) in serum to increase, which leads to higher H_2CO_3 concentrations and a lower pH. Rapid, deep breathing lowers P_{CO_2} and the concentration of H_2CO_3, and raises the pH. The kidneys help control the pH of serum by regulating the amount of HCO_3^- present and by either adding H_3O^+ to serum or moving it into urine.

END OF CHAPTER PROBLEMS

Answers to problems whose numbers are printed in color are given in Appendix D. More challenging questions are marked with an asterisk. Problems within colored rules are paired. **ILW** = Interactive Learning Ware solution is available at *www.wiley.com/college/raymond.*

9.1 ACIDS AND BASES

9.1 Based on taste, is it possible to tell whether a food contains acids or bases? Explain.

9.2 Name each of the following as an acid and as a binary compound.
a. HCl **b.** HBr

9.3 Name each of the following as an acid and as a binary compound.
a. HF **b.** HI

9.2 BRØNSTED–LOWRY ACIDS AND BASES

9.4 Give the formula for the conjugate base of each acid.
a. H_2CO_3 **c.** H_2SO_4
b. HCO_3^- **d.** $H_2PO_4^-$

9.5 Give the formula for the conjugate base of each acid.
a. HSO_4^- **c.** HPO_4^{2-}
b. H_3PO_4 **d.** HNO_3

9.6 Define the term *amphoteric.*

***9.7** HCO_3^- is amphoteric.
a. Write the equation for the reaction that takes place between HCO_3^- and the acid HCl.
b. Write the equation for the reaction that takes place between HCO_3^- and the base OH^-.

***9.8** HSO_4^- is amphoteric.
a. Write the equation for the reaction that takes place between HSO_4^- and the acid HCl.
b. Write the equation for the reaction that takes place between HSO_4^- and the base OH^-.

9.9 Identify the Brønsted–Lowry acids and bases for the forward and reverse reactions of each.
a. $F^-(aq) + HCl(aq) \rightleftharpoons HF(aq) + Cl^-(aq)$
b. $CH_3CO_2H(aq) + NO_3^-(aq) \rightleftharpoons CH_3CO_2^-(aq) + HNO_3(aq)$

9.10 Identify the Brønsted–Lowry acids and bases for the forward and reverse reactions of each.
a. $PO_4^{3-}(aq) + NH_4^+(aq) \rightleftharpoons NH_3(aq) + HPO_4^{2-}(aq)$

b. $HCN(aq) + H_2PO_4^-(aq) \rightleftharpoons CN^-(aq) + H_3PO_4(aq)$

9.11 Complete each acid-base reaction. For the forward and reverse reactions, identify each acid and its conjugate base.

a. $NH_4^+ + SO_4^{2-} \rightleftharpoons$

b. $CN^- + HI \rightleftharpoons$

9.12 Complete each acid-base reaction. For the forward and reverse reactions, identify each acid and its conjugate base.

a. $OH^- + HNO_3 \rightleftharpoons$

b. $HPO_4^{2-} + CO_3^{2-} \rightleftharpoons$

9.3 Equilibrium

*__9.13__ Which of the following statements are correct at equilibrium?

a. The concentration of reactants is always equal to the concentration of products.

b. No reactants are converted into products.

c. The rate of the forward reaction is equal to the rate of the reverse reaction.

*__9.14__ Which of the following statements are correct at equilibrium?

a. The forward and reverse reactions stop.

b. The concentration of reactants and products does not change.

c. The rate of the forward reaction is greater than the rate of the reverse reaction.

9.15 Write the equilibrium constant expression for each reaction. In part b, H_2O is a solvent.

a. $C(s) + CO_2(g) \rightleftharpoons 2CO(g)$

b. $NH_3(aq) + H_2O(l) \rightleftharpoons NH_4^+(aq) + OH^-(aq)$

9.16 Write the equilibrium constant expression for each reaction. In part b, H_2O is a solvent.

a. $PCl_5(g) \rightleftharpoons PCl_3(g) + Cl_2(g)$

b. $H_2SO_4(aq) + 2H_2O(l) \rightleftharpoons 2H_3O^+(aq) + SO_4^{2-}(aq)$

9.17 Explain why the concentrations of solids do not appear in equilibrium constant expressions.

9.18 Explain why the concentrations of solvents do not appear in equilibrium constant expressions.

9.19 Write the reaction equation from which each equilibrium constant expression is derived (assume that no solids or solvents are present).

a. $K_{eq} = \dfrac{[NOCl]^2}{[NO]^2[Cl_2]}$ **b.** $K_{eq} = \dfrac{[HBr]^2}{[H_2][Br_2]}$

9.20 Write the reaction equation from which each equilibrium constant expression is derived (assume that no solids or solvents are present).

a. $K_{eq} = \dfrac{[CH_4][H_2O]}{[CO][H_2]^3}$ **b.** $K_{eq} = \dfrac{[SO_3]^2}{[O_2][SO_2]^2}$

9.21 K_{eq} for the reaction below has a value of 4.2×10^{-4}.

$CH_3NH_2(aq) + H_2O(l) \rightleftharpoons CH_3NH_3^+(aq) + OH^-(aq)$

Which are there more of at equilibrium, products or reactants?

9.22 When steam is passed over coal and allowed to come to equilibrium at 800 °C, the equilibrium constant for the reaction has a value of 8.1×10^{-2}.

$$C(s) + H_2O(g) \rightleftharpoons CO(g) + H_2(g)$$

a. Explain why the concentration of water appears in the equilibrium constant expression for this reaction, when it is usually omitted from such equations.

b. Write the equilibrium constant expression for the reaction.

c. Which are there more of at equilibrium, reactants or products?

9.23 For each reaction, which is there more of at equilibrium, reactants or products?

a. $HPO_4^{2-} + CN^- \rightleftharpoons PO_4^{3-} + HCN$ $K_{eq} = 8.6 \times 10^{-4}$

b. $H_3PO_4 + CN^- \rightleftharpoons H_2PO_4^- + HCN$ $K_{eq} = 1.5 \times 10^7$

9.24 For each reaction, which is there more of at equilibrium, reactants or products?

a. $HF + HCO_3^- \rightleftharpoons F^- + H_2CO_3$ $K_{eq} = 1.5 \times 10^3$

b. $NH_4^+ + HCO_3^- \rightleftharpoons NH_3 + H_2CO_3$ $K_{eq} = 1.5 \times 10^{-3}$

9.4 Le Châtelier's Principle

9.25 The enzyme carbonic anhydrase catalyzes the rapid conversion of CO_2 and H_2O into H_2CO_3.

$$CO_2(g) + H_2O(l) \rightleftharpoons H_2CO_3(aq)$$

a. Write the equilibrium constant expression for this reaction.

b. What effect, if any, does doubling the amount of carbonic anhydrase have on an equilibrium mixture of H_2CO_3, CO_2, and H_2O? Explain.

9.26 Carbon disulfide (CS_2), used in the manufacture of some synthetic fabrics, is prepared by heating sulfur (S_2) and charcoal (C).

$$S_2(g) + C(s) \rightleftharpoons CS_2(g)$$

a. For the equilibrium above, what is the effect of increasing $[CS_2]$? Of decreasing $[CS_2]$? Of decreasing $[S_2]$?
b. What would be the effect of continually removing CS_2 from the reaction?

9.27 When carbon monoxide reacts with hydrogen gas, methanol (CH_3OH) is formed.

$$CO(g) + 2H_2(g) \rightleftharpoons CH_3OH(g)$$

a. For the equilibrium above, what is the effect of increasing [CO]? Of decreasing $[H_2]$? Of increasing $[CH_3OH]$?
b. What would be the effect of continually removing CH_3OH from the reaction?

9.28 One step in glycolysis (Section 15.4) involves the reversible conversion of glucose 6-phosphate into fructose 6-phosphate.

$$\text{glucose 6-phosphate} \rightleftharpoons \text{fructose 6-phosphate}$$

a. For the equilibrium above, what is the effect of increasing the concentration of glucose 6-phosphate?
b. What is the effect of increasing the concentration of fructose 6-phosphate?
c. What is the effect of decreasing the concentration of glucose 6-phosphate?

9.29 Starvation, severe diets, and poorly controlled diabetes can result in the production of excess ketone bodies. The reversible reaction shown here involves the reduction of the ketone body acetoacetate by NADH to form the ketone body 3-hydroxybutyrate (Figure 15.19).

$$\text{Acetoacetate} + \text{NADH} + H^+ \rightleftharpoons \text{3-hydroxybutyrate} + \text{NAD}^+$$

a. For the equilibrium above, what is the effect of decreasing the concentration of acetoacetate?
b. What is the effect of increasing the concentration of 3-hydroxybutyrate?
c. What is the effect of increasing the concentration of NADH?

9.5 Ionization of Water

9.30 Calculate the H_3O^+ concentration present in water when
a. $[OH^-] = 8.4 \times 10^{-3}$ M
b. $[OH^-] = 2.9 \times 10^{-9}$ M
c. $[OH^-] = 5.8 \times 10^{-1}$ M

9.31 Calculate the H_3O^+ concentration present in water when
a. $[OH^-] = 4.8 \times 10^{-8}$ M
b. $[OH^-] = 6.6 \times 10^{-2}$ M
c. $[OH^-] = 1.5 \times 10^{-12}$ M

9.32 In Problem 9.30, indicate whether each solution is acidic, basic, or neutral.

9.33 In Problem 9.31, indicate whether each solution is acidic, basic, or neutral.

9.34 Calculate the OH^- concentration present in water when
a. $[H_3O^+] = 9.1 \times 10^{-7}$ M
b. $[H_3O^+] = 1.3 \times 10^{-3}$ M
c. $[H_3O^+] = 8.8 \times 10^{-2}$ M

9.35 Calculate the OH^- concentration present in water when
a. $[H_3O^+] = 6.2 \times 10^{-4}$ M
b. $[H_3O^+] = 8.5 \times 10^{-8}$ M
c. $[H_3O^+] = 1.9 \times 10^{-11}$ M

9.36 In Problem 9.34, indicate whether each solution is acidic, basic, or neutral.

9.37 In Problem 9.35, indicate whether each solution is acidic, basic, or neutral.

9.6 The pH Scale

9.38 Calculate the pH of a solution in which
a. $[H_3O^+] = 1 \times 10^{-5}$ M
b. $[H_3O^+] = 3.9 \times 10^{-2}$ M
c. $[OH^-] = 6.8 \times 10^{-7}$ M
d. $[OH^-] = 1 \times 10^{-7}$ M

9.39 Calculate the pH of a solution in which
a. $[H_3O^+] = 1 \times 10^{-7}$ M
b. $[H_3O^+] = 7.0 \times 10^{-5}$ M
c. $[OH^-] = 7.0 \times 10^{-5}$ M
d. $[OH^-] = 1 \times 10^{-1}$ M

9.40 In Problem 9.38, indicate whether each solution is acidic, basic, or neutral.

9.41 In Problem 9.39, indicate whether each solution is acidic, basic, or neutral.

9.42 What is the concentration of H_3O^+ in a solution if the pH is
a. 7.00
b. 1.74
c. 9.37

9.43 What is the concentration of H_3O^+ in a solution if the pH is
a. 5.54
b. 13.8
c. 2.94

*__9.44__ The alkali metals and alkaline earth metals are so named because they produce alkaline (basic) solutions. When Na, an alkali metal, is added to water, NaOH and $H_2(g)$ are rapidly formed. Write a balanced equation for this reaction.

9.45 A 1.00 mL sample of blood serum has a pH of 7.35.
a. What is the concentration of H_3O^+?
b. What is the concentration of OH^-?
c. How many moles of H_3O^+ are present?
d. How many moles of OH^- are present?
e. How many H_3O^+ ions are present?
f. How many OH^- ions are present?

9.46 A 5.00 mL sample of blood serum has a pH of 7.45.
a. What is the concentration of H_3O^+?
b. What is the concentration of OH^-?
c. How many moles of H_3O^+ are present?
d. How many moles of OH^- are present?
e. How many H_3O^+ ions are present?
f. How many OH^- ions are present?

9.7 Acid and Base Strength

9.47 Write the chemical equation for the reaction of each weak acid with water. Write the corresponding acidity constant expression.
a. NH_4^+ **b.** HPO_4^{2-}

9.48 Write the chemical equation for the reaction of each weak acid with water. Write the corresponding acidity constant expression.
a. $CH_3NH_3^+$ **b.** HCO_3^-

9.49 Calculate the pK_a of each acid and indicate which is the stronger acid.
a. HClO, $K_a = 3.0 \times 10^{-8}$
b. $C_2O_4H^-$, $K_a = 6.4 \times 10^{-5}$

9.50 Calculate the pK_a of each acid and indicate which is the stronger acid.
a. $C_2O_4H_2$, $K_a = 5.9 \times 10^{-2}$
b. $C_6O_2H_8$, $K_a = 1.7 \times 10^{-5}$

***9.51** 0.10 M solutions of each of the following acids are prepared: acetic acid ($K_a = 1.8 \times 10^{-5}$) and hydrofluoric acid ($K_a = 6.6 \times 10^{-4}$). Which acid solution will have the lowest pH? Explain.

9.52 Write the formula of the conjugate base of each acid in Problem 9.49. Which of the two is the stronger base?

9.53 Write the formula of the conjugate base of each acid in Problem 9.50. Which of the two is the stronger base?

9.8 Neutralizing Acids and Bases

9.54 a. How many moles of NaOH are present in 31.7 mL of 0.155 M NaOH?
b. How many moles of HCl are present in a 15.0 mL sample that is neutralized by the 31.7 mL of 0.155 M NaOH?
c. What is the molar concentration of the HCl solution described in part b of this question?

9.55 a. How many moles of NaOH are present in 71.3 mL of 0.551 M NaOH?
b. How many moles of HCl are present in a 25.0 mL sample that is neutralized by the 71.3 mL of 0.551 M NaOH?
c. What is the molar concentration of the HCl solution described in part b of this question?

9.56 It requires 17 mL of 0.100 M KOH to titrate 75 mL of an HCl solution of unknown concentration. Calculate the initial HCl concentration.

9.57 It requires 35.0 mL of 0.250 M KOH to titrate 50.0 mL of an HCl solution of unknown concentration. Calculate the initial HCl concentration.

***9.58** It requires 45 mL of 0.20 M HCl to titrate 25 mL of a $Ca(OH)_2$ solution of unknown concentration. What is the $Ca(OH)_2$ concentration?

9.9 Effect of pH on Acid and Conjugate Base Concentrations

9.59 HF is added to water and the solution is allowed to come to equilibrium.

$$HF + H_2O \rightleftharpoons F^- + H_3O^+$$

a. Based on this reaction equation, is HF an acid or is it a base?
b. What happens to the pH of the solution if NaF (a water-soluble ionic compound) is added to the solution?

9.60 An equilibrium mixture of H_2CO_3 and HCO_3^- has a pH of 6.5. What happens to the H_2CO_3 and HCO_3^- concentrations when the pH of the solution is adjusted to 8.5?

9.61 An equilibrium mixture of CH_3CO_2H and $CH_3CO_2^-$ has a pH of 4.8. What happens to the CH_3CO_2H and $CH_3CO_2^-$ concentrations when the pH of the solution is adjusted to 2.8?

9.62 Explain how the relative values of pH and pK_a influence the relative concentrations of an acid and its conjugate base.

9.63 a. At pH 7.0 does lactic acid appear predominantly in its acid form ($CH_3CH(OH)CO_2H$) or in its conjugate base form ($CH_3CH(OH)CO_2^-$)? See Table 9.4.
b. At pH 1.0 does lactic acid appear predominantly in its acid form or in its conjugate base form?

9.64 a. At pH 7.0 does methylammonium ion appear predominantly in its acid form ($CH_3NH_3^+$) or in its conjugate base form (CH_3NH_2)? See Table 9.4.
b. At pH 1.0 does methylammonium ion appear predominantly in its acid form or in its conjugate base form?

9.10 Buffers

9.65 Which weak acid or weak acids from Table 9.4 would be the best choice if you wished to prepare buffers with the following pH values?
a. 7.0 **b.** 10.0 **c.** 4.0

9.66 A buffer can be prepared using NH_4^+ and NH_3.
a. Write an equation for the reaction that takes place when H_3O^+ is added to this buffer.
b. Write an equation for the reaction that takes place when OH^- is added to this buffer.

9.67 A buffer can be prepared using lactic acid $[CH_3CH(OH)CO_2H]$ and its conjugate base, lactate $[CH_3CH(OH)CO_2^-]$.
a. Write an equation for the reaction that takes place when H_3O^+ is added to this buffer.
b. Write an equation for the reaction that takes place when OH^- is added to this buffer.

9.68 If you wish to maintain a pH of 6.0, which is the better buffer—H_2CO_3 and HCO_3^- or $H_2PO_4^-$ and HPO_4^{2-}?

9.11 Maintaining the pH of Blood Serum

9.69 Explain how the H_2CO_3/HCO_3^- buffer system helps maintain blood serum at a constant pH.

9.70 At which pH is the H_2CO_3/HCO_3^- buffer system most effective?

9.71 Explain how respiration helps maintain blood serum at a constant pH.

9.72 Explain how the kidneys help maintain blood serum at a constant pH.

9.73 In respiratory alkalosis, the serum concentration of HCO_3^- can be lower than normal. Explain this form of compensation by the kidneys.

9.74 In metabolic acidosis, the serum P_{CO_2} levels can be lower than normal. Explain this form of compensation by the respiratory system.

9.75 One treatment for respiratory alkalosis is to breathe into a paper bag. Explain how this helps restore the pH of the blood to its normal value.

9.76 **a.** According to the Health Link "Respiratory and Metabolic Imbalances," how does cystic fibrosis upset the blood buffer system to cause acidosis?
b. "Cystic Fibrosis Mucus" states that mucus is not what clogs the lungs of cystic fibrosis patients. What, instead, does clog the lungs?

BIOCHEMISTRY*Link* | *Diving Mammals, Oxygen, and Myoglobin*

9.77 In terms of Le Châtelier's principle, describe how the equilibrium $Mb + O_2 \rightleftharpoons MbO_2$ responds to increases in the concentration of
a. Mb **b.** O_2 **c.** MbO_2

9.78 Studies have shown that the K_{eq} for the reaction in the previous problem is identical for humans and diving animals. Why is the MbO_2 in diving animals able to provide more O_2 than the MbO_2 in humans?

Math Support—Logs and Antilogs

9.79 Calculate the logarithm of each number. Assume that each is a measured quantity.
a. 1×10^{-9} **c.** 3.4×10^{-2}
b. 1×10^{12} **d.** 9.7×10^4

9.80 Calculate the logarithm of each number. Assume that each is a measured quantity.
a. 1×10^3 **c.** 2.2×10^{-13}
b. 1×10^{-6} **d.** 8.7×10^9

9.81 Calculate the antilogarithm of each number.
a. 8 **c.** 7.9
b. −3 **d.** 15.3

9.82 Calculate the antilogarithm of each number.
a. −1 **c.** 3.5
b. 12 **d.** 10.9

BIOCHEMISTRY*Link* | *Plants as pH Indicators*

9.83 When you pick a particular flower and immerse it in a pH 7 solution, there is no change in the color. If you place the same flower in a pH 9 solution, the color changes. Why does this happen?

THINKING IT THROUGH

9.84 Suggest an explanation for the fact that the antacids Di-Gel, Mylanta, and Maalox contain both $Al(OH)_3$ and $Mg(OH)_2$.

9.85 The amino acid called glycine contains a basic amino group and an acidic carboxyl group. At pH 7, the amino group exists in its conjugate acid form and the carboxyl group exists in its conjugate base form.

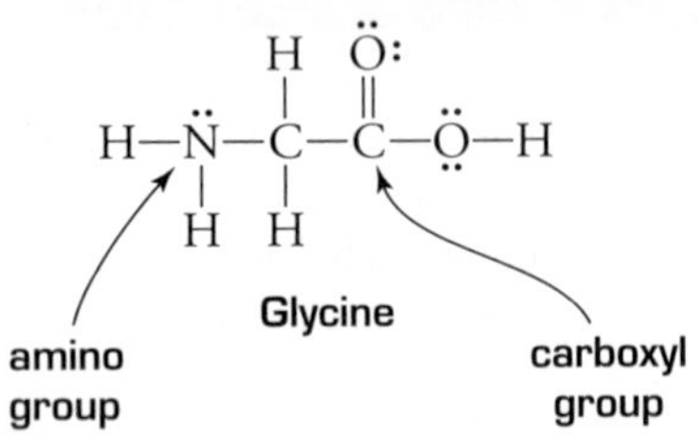

Draw glycine as it exists at pH 7.

INTERACTIVE LEARNING PROBLEMS

9.86 Calculate the pH of a solution in which:

a. $[H_3O^+] = 1 \times 10^{-3}$ M
b. $[H_3O^+] = 9.5 \times 10^{-6}$ M
c. $[OH^-] = 3.7 \times 10^{-4}$ M
d. $[OH^-] = 1 \times 10^{-12}$ M

9.87 Ammonia gas and oxygen gas react as described by the equation

$$4NH_3(g) + 5O_2(g) \rightleftharpoons 4NO(g) + 6H_2O(g)$$

a. For the equilibrium above, what is the effect of increasing $[NH_3]$?
b. What is the effect of decreasing $[O_2]$?
c. What is the effect of continually removing NO from the reaction?

SOLUTIONS TO PRACTICE PROBLEMS

9.1 **a.** $HSO_4^- + HPO_4^{2-} \rightleftharpoons H_2SO_4 + PO_4^{3-}$
Base Acid Acid Base

b. $HPO_4^{2-} + OH^- \rightleftharpoons H_2O + PO_4^{3-}$
Acid Base Acid Base

9.2 **a.** $K_{eq} = \dfrac{[CO][H_2O]}{[H_2][CO_2]}$; **b.** $K_{eq} = \dfrac{[NO_2]^4[O_2]}{[N_2O_5]^2}$

9.3 Reactant concentration.

9.4 **a.** reverse; **b.** forward; **c.** reverse

9.5 **a.** 1.0×10^{-9}; **b.** 1.2×10^{-4}; **c.** 2.6×10^{-14}; **d.** 2.3×10^{-6}

9.6 **a.** 7.30; **b.** 13.0; **c.** 1.0; **d.** 11.67

9.7 0.52 M

9.8 **a.** $H_2PO_4^- + OH^- \longrightarrow HPO_4^{2-} + H_2O$

b. 6.21–8.21

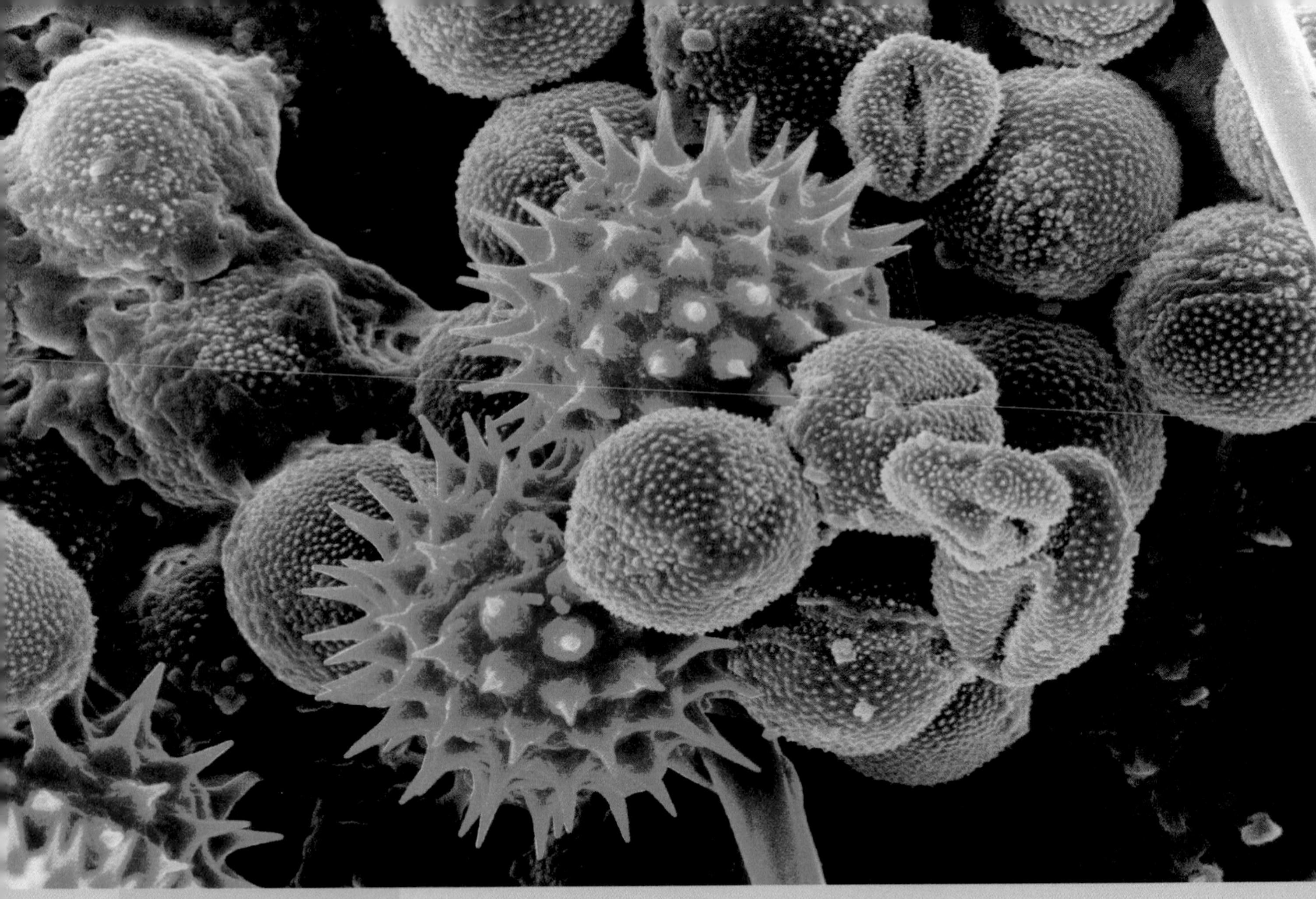

It is early spring and the air is filled with pollen. You are acutely aware of this because your eyes are watery and itchy, your nose is running, and you feel just awful. On a trip to the drug store, you stop and read the labels on antihistamines, hoping to find one that will alleviate your allergy symptoms. Some antihistamines claim to be non-drowsy, others warn you to not operate heavy machinery while taking them, and all contain drugs with very long names. You suspect that antihistamines work against histamine, but do not know what histamine is.

Photo source: Derek Berwin/The Image Bank/Getty Images.

10 CARBOXYLIC ACIDS, PHENOLS, AND AMINES

Having introduced the chemistry of acids and bases in Chapter 9, we will now turn our attention to carboxylic acids and phenols (families of organic acids) and amines (a family of organic bases).

objectives

After completing this chapter, you should be able to:

1. List the intermolecular forces that attract carboxylic acids, phenols, or amine molecules to one another and describe the effect that these forces have on boiling point.
2. Describe what takes place when carboxylic acids and phenols react with water or with strong bases. Explain how the conjugate bases of carboxylic acids and phenols are named.
3. Describe what takes place when amines react with water or with strong acids. Explain how the conjugate acids of amines are named.
4. Explain how carboxylic acids can be converted into esters and amides. Describe how esters and amides are named.
5. Explain the difference in the products obtained when an ester is hydrolyzed under acidic conditions and under basic conditions. Describe the products formed when an amide is hydrolyzed under acidic conditions.
6. Explain the terms chiral molecule and chiral carbon atom, distinguish between enantiomers and diastereomers, and define the terms dextrorotatory and levorotatory.

Carboxylic acids contain a carboxyl group ($—CO_2H$), which is the combination of a hydroxyl (—OH) group and a carbonyl (C=O) group. **Phenols** are compounds in which *a hydroxyl group is attached to an aromatic ring.* **Amines** have *a nitrogen atom directly attached to one or more carbon atoms* (excluding C=O carbon atoms).

Members of these three families of organic compounds are significant in biochemistry. As we have seen in earlier chapters, carboxylic acids are important to the chemistry of fatty acids, triglycerides, and phospholipids. Some phenols are associated with colors, flavors, and fragrances, including malvidin, which gives blueberries their color, and eugenol, which gives cloves their characteristic odor and flavor (Figure 10.1). Amines are found throughout nature and most have at least some physiological effect, such as the caffeine present in coffee and tea and the nicotine present in tobacco. Diphenhydramine and other antihistamines are also amines. As part of our discussion of amines later in this chapter, we will also take a brief look at a related family of compounds called **amides**, in which *a nitrogen atom is attached to the carbon atom of a carbonyl.*

$$CH_3\overset{O}{\overset{\|}{C}}—OH$$

Acetic acid

(*a*) $$CH_3CH_2CH_2CH_2CH_2CH_2CH_2CH_2CH_2CH_2CH_2CH_2CH_2CH_2CH_2CH_2CH_2\overset{O}{\overset{\|}{C}}—OH$$

Stearic acid

(*b*)

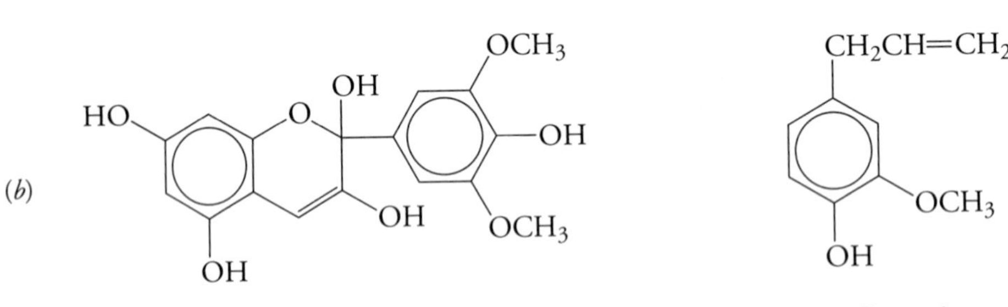

Malvidin **Eugenol**

(*c*)

Caffeine **Nicotine**

$$C_6H_5—\underset{\underset{C_6H_5}{|}}{CH}OCH_2CH_2\underset{\underset{CH_3}{|}}{N}CH_3$$

Diphenhydramine

FIGURE 10.1

Carboxylic acids, phenols, and amines
(*a*) Carboxylic acids, including acetic acid (in vinegar) and stearic acid (a fatty acid), contain an —OH that is attached to a carbonyl (C=O) group. (*b*) In malvidin, eugenol, and other phenols, an —OH is attached to an aromatic ring. (*c*) Amines, such as caffeine, nicotine, and diphenhydramine, have a nitrogen atom attached to one or more carbon atoms (excluding C=O carbon atoms).

10.1 CARBOXYLIC ACIDS

We will begin our look at carboxylic acids by discussing the naming and properties of the members of this family of organic compounds. When naming a carboxylic acid according to the IUPAC rules, the parent is the longest continuous carbon chain that includes the carboxyl group. Numbering begins at the carboxyl group, and alkyl groups are identified by name, position, and number of appearances. IUPAC names for carboxylic acid parent chains are formed by dropping the final "e" on the name of the corresponding hydrocarbon and adding "oic acid." For example, the alkane that contains just one carbon atom (CH_4) is called *methane* and the carboxylic acid with just one carbon atom (HCO_2H) is called *methanoic acid* (Figure 10.2). Similarly, the parent chains of propanoic acid, 3-methylbutanoic acid, and 3-ethyl-4-methylhexanoic acid have, respectively, 3, 4, and 6 carbon atoms. Although we will not consider specific examples, if a carboxylic acid parent chain contains a carbon–carbon double bond it is an *enoic acid* and if it contains a carbon–carbon triple bond it is an *ynoic acid.* Note, as shown in Figure 10.2, that the position of the carboxyl carbon atom (always 1) is not specified in the name.

Some carboxylic acids have well-known common names. The two common names that you are most likely to encounter are those of methanoic acid (known also as formic acid) and ethanoic acid (known also as acetic acid). Formic comes from *formica,* Latin for *ant* (the bite of a red ant contains formic acid, which is used to immobilize prey), and acetic comes from *acetum,* Latin for *vinegar* [a 5% (w/v) aqueous solution of acetic acid].

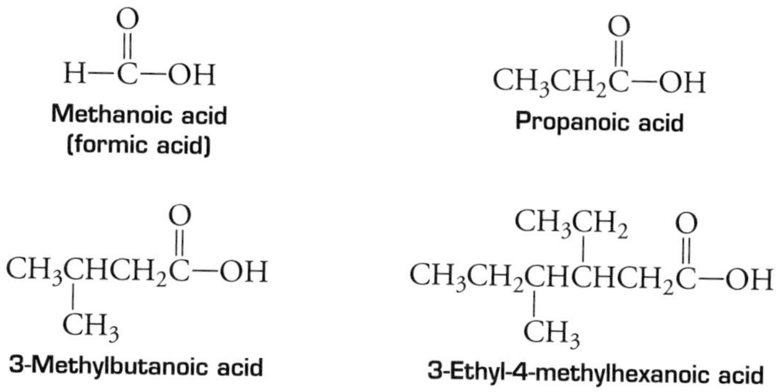

$H{-}C({=}O){-}OH$
Methanoic acid (formic acid)

$CH_3CH_2C({=}O){-}OH$
Propanoic acid

$CH_3CH(CH_3)CH_2C({=}O){-}OH$
3-Methylbutanoic acid

$CH_3CH_2CH(CH_3)CH(CH_2CH_3)CH_2C({=}O){-}OH$
3-Ethyl-4-methylhexanoic acid

FIGURE 10.2

Carboxylic acids
In the IUPAC name of a carboxylic acid, the parent chain (the longest carbon chain that contains the carboxyl group) is numbered starting at the carboxyl group. The common name of methanoic acid is given in parentheses.

SAMPLE PROBLEM 10.1

Naming carboxylic acids

Give the IUPAC name of each carboxylic acid.

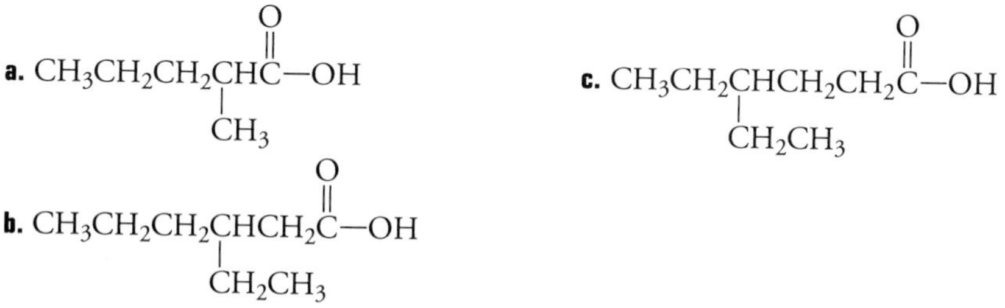

a. $CH_3CH_2CH_2CH(CH_3)C({=}O){-}OH$

b. $CH_3CH_2CH_2CH(CH_2CH_3)CH_2C({=}O){-}OH$

c. $CH_3CH_2CH(CH_2CH_3)CH_2CH_2C({=}O){-}OH$

Strategy
Following the IUPAC rules, the parent chain is numbered beginning at the carboxyl group of each molecule.

Solution

a. 2-methylpentanoic acid
b. 3-ethylhexanoic acid
c. 4-ethylhexanoic acid

PRACTICE PROBLEM 10.1

Draw each molecule. Hint: When named as a substituent, —OH is hydroxy, —Cl is chloro, and —Br is bromo.

a. 4-hydroxybutanoic acid
b. 2-bromopropanoic acid
c. 4-chlorohexanoic acid

■ The relatively high boiling points of carboxylic acids are due to hydrogen bonding.

Compared to other organic compounds with a similar molecular weight, carboxylic acids have relatively high boiling points (Table 10.1) due to their ability to form hydrogen bonds with one another. As is the case for all organic compounds, the more carbon atoms that a carboxylic acid contains, the higher its boiling point. This is because greater surface area leads to increased London forces. The ability to form hydrogen bonds, in addition to the presence of polar C=O, C—O, and O—H bonds, gives small carboxylic acids a significant water solubility. As we have seen before, an increasing number of carbon atoms leads to a reduction in water solubility.

Carboxylic acids that contain from 3 to 10 carbon atoms have particularly unpleasant odors. Butanoic acid is responsible for the odor of stale perspiration, rank tennis shoes, and rancid butter, while hexanoic, heptanoic, and octanoic acids are secreted from the skin of goats and give them their distinctive odor. Carboxylic acids with 11 or more carbon atoms have low vapor pressures, so they have little odor.

■ **TABLE 10.1 PHYSICAL PROPERTIES OF SOME SMALL CARBOXYLIC ACIDS**

Formula	IUPAC Name	Common Name[a]	Boiling Point (°C)	Water Solubility (g/100 mL)
HCO_2H	Methanoic acid	Formic acid	100	Very large
CH_3CO_2H	Ethanoic acid	Acetic acid	118	Very large
$CH_3CH_2CO_2H$	Propanoic acid		141	Very large
$CH_3(CH_2)_2CO_2H$	Butanoic acid		164	Very large
$CH_3(CH_2)_3CO_2H$	Pentanoic acid		187	3.7
$CH_3(CH_2)_4CO_2H$	Hexanoic acid		205	1.0

[a] Only selected common names are listed.

10.2 PHENOLS

In the simplest phenol, named *phenol*, the —OH group is attached to a benzene ring. When substituents are added to this molecule, the carbon atom carrying the —OH is designated carbon 1. Examples of phenol naming include 2-chlorophenol and 4-ethylphenol (Figure 10.3*a*). As described in Section 4.8, the position of substituents in disubstituted benzenes can also be indicated using *ortho* (*o*), *meta* (*m*), and *para* (*p*), so 2-chlorophenol is also known as *o*-chlorophenol and 4-ethylphenol as *p*-ethylphenol.

Catechol, resorcinol, and hydroquinone are phenols that are common constituents of biomolecules (Table 10.2). Urushiol, a catechol-containing phenol present in poison ivy and poison oak, contributes to the itching and burning that occurs when someone comes into contact with these plants (Figure 10.3*b*). Catechin, a catechol- and resorcinol-containing phenol used for dyeing and tanning, is present in mahogany. Gentisic acid is a hydroquinone-containing anti-rheumatism drug.

(a) OH — Phenol; OH, Cl — 2-Chlorophenol (o-chlorophenol); OH, CH_2CH_3 — 4-Ethylphenol (p-ethylphenol)

(b) OH, OH, $CH_2(CH_2)_{13}CH_3$ — Urushiol; HO, O, OH, OH, OH, OH — Catechin; OH, O, C—OH, OH — Gentisic acid

FIGURE 10.3

Phenol names

(*a*) The smallest phenol is named phenol. The ring position of substituents may be indicated by number or, in the case of disubstituted benzene rings, by using *ortho* (*o*), *meta* (*m*), and *para* (*p*). (*b*) Urushiol, catechin, and gentisic acid are examples of more structurally complex phenols.

Source: Wally Eberhart/Botanica/Getty Images.

The —OH group of phenols allows these molecules to form hydrogen bonds with one another and, as a result, phenols tend to have relatively high boiling points and many are liquids at room temperature (Table 10.2). The polar covalent C—O and O—H bonds and the ability to form hydrogen bonds give phenols with low molecular weight some degree of water solubility.

HEALTH Link: A Chili Pepper Painkiller

The fiery taste of chili peppers is due to the phenol called capsaicin, which is also present in cayenne and paprika (Figure 10.4). This phenol is an irritant that affects the cardiovascular and respiratory systems. When applied to the skin, capsaicin is an effective painkiller because it desensitizes the nerves, reducing the perception of pain.

CH_3O, HO, $CH_2NH—\overset{O}{\overset{\|}{C}}CH_2CH_2CH_2CH_2$, H, C=C, H, $CH(CH_3)_2$

Capsaicin

FIGURE 10.4

Capsaicin

This phenol, which is responsible for the hot taste of chili peppers, can be used as a painkiller.

Source: PhotoDisc, Inc./Getty Images.

TABLE 10.2 PHYSICAL PROPERTIES OF SELECTED PHENOLS

Formula	Common Name	Boiling Point (°C)	Water Solubility (g/100 mL)
C_6H_5OH	Phenol	182	9.3
1,2-$C_6H_4(OH)_2$	Catechol	246	45
1,3-$C_6H_4(OH)_2$	Resorcinol	281	123
1,4-$C_6H_4(OH)_2$	Hydroquinone	286	8

10.3 CARBOXYLIC ACIDS AND PHENOLS AS WEAK ORGANIC ACIDS

As implied by the name, members of the carboxylic acid family are acids (they can donate H^+). For example, acetic acid (CH_3CO_2H) reacts with water to form acetate ion ($CH_3CO_2^-$) and H_3O^+.

$$\underset{\text{Acetic acid}}{CH_3\overset{O}{\overset{\|}{C}}{-}OH} + H_2O \rightleftharpoons \underset{\text{Acetate ion}}{CH_3\overset{O}{\overset{\|}{C}}{-}O^-} + H_3O^+ \quad K_a = \frac{[CH_3CO_2^-][H_3O^+]}{[CH_3CO_2H]} = 1.8 \times 10^{-5}$$

Having a K_a of 1.8×10^{-5} ($pK_a = 4.74$), acetic acid is a weak acid.

In Section 9.9 we saw that for an acid (HA) and its conjugate base (A^-), *there is more A^- when the pH is greater than the pK_a*. Since most carboxylic acids have a pK_a with a value near 5, this means that at physiological pH (approximately pH 7), carboxylate ions (*conjugate bases of carboxylic acids*) predominate.

A biochemical significance of this chemistry relates to fatty acids. In their acidic form, fatty acids are hydrophobic. Under physiological pHs, however, they exist as amphipathic ions that can mix with water, to some extent. For example, water-insoluble palmitic acid is found as the amphipathic palmitate ion at pH 7.

$$CH_3CH_2CH_2CH_2CH_2CH_2CH_2CH_2CH_2CH_2CH_2CH_2CH_2CH_2CH_2\overset{\overset{\Large O}{||}}{C}{-}OH$$

Palmitic acid (hydrophobic; predominates at pH < 5)

$$CH_3CH_2CH_3CH_2CH_2CH_2CH_2CH_2CH_2CH_2CH_2CH_2CH_2CH_2CH_2\overset{\overset{\Large O}{||}}{C}{-}O^-$$

Palmitate ion (amphipathic; predominates at pH > 5)

In general, carboxylate ions with 12 or more carbon atoms, like palmitate ion, are amphipathic compounds, while those with fewer than 12 carbon atoms are water soluble.

At a number of points in this and previous chapters, carboxylate ion names have been given without a description of how the naming rules work. The reason for delaying this explanation was that we had not yet seen how to name carboxylic acids, the starting point for naming carboxylate ions. To name a carboxylate ion, the ending on the name (IUPAC or common) of the related carboxylic acid is changed from "ic acid" to "ate ion." Palmitic acid becomes palmitate ion, propanoic acid becomes propanoate ion, and 2-methylbutanoic acid becomes 2-methylbutanoate ion.

$$CH_3CH_2\overset{\overset{\Large O}{||}}{C}{-}OH \qquad\qquad CH_3CH_2\overset{\overset{\Large O}{||}}{C}{-}O^-$$

Propanoic acid **Propanoate ion**

$$CH_3CH_2\underset{\underset{\Large CH_3}{|}}{C}H\overset{\overset{\Large O}{||}}{C}{-}OH \qquad\qquad CH_3CH_2\underset{\underset{\Large CH_3}{|}}{C}H\overset{\overset{\Large O}{||}}{C}{-}O^-$$

2-Methylbutanoic acid **2-Methylbutanoate ion**

Although phenols are also considered organic acids, they are considerably weaker acids than carboxylic acids. Phenol, for example, has a K_a value of 1.1×10^{-10} ($pK_a = 9.96$).

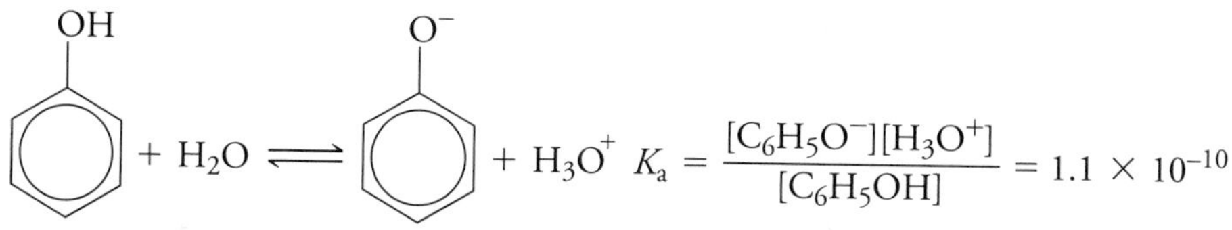

$$C_6H_5OH + H_2O \rightleftharpoons C_6H_5O^- + H_3O^+ \quad K_a = \frac{[C_6H_5O^-][H_3O^+]}{[C_6H_5OH]} = 1.1 \times 10^{-10}$$

Phenol
C_6H_5OH
(predominates when pH < pK_a)

Phenoxide ion
$C_6H_5O^-$
(predominates when pH > pK_a)

Phenols are less effective at releasing H_3O^+ than are carboxylic acids and, having pK_a values near 10, they exist mainly in their acidic form at physiological pH. The conjugate base of a phenol is named by dropping the "ol" ending of the phenol name and adding "oxide ion."

Like carboxylic acids and phenols, alcohols (CH_3CH_2OH, for example) contain an —OH group. Alcohols are not acids, however. This difference in the ability of —OH groups to donate H^+ is directly linked to the structure of the molecule involved. When attached to a carbonyl (C=O), as in carboxylic acids, the —OH group is weakly acidic. When attached to an aromatic ring, as in phenols, the —OH group is very weakly acidic and, when attached to an alkane-type carbon atom, as in alcohols, the —OH group is not at all acidic.

FIGURE | 10.5

Carboxylic acids and phenols as acids

(*a*) In the presence of a strong base, a carboxylic acid reacts to form a carboxylate salt.
(*b*) Phenols react with a strong base to form a phenoxide salt.

(*a*) $C_6H_5-\overset{O}{\overset{\|}{C}}-OH + NaOH \longrightarrow C_6H_5-\overset{O}{\overset{\|}{C}}-O^-Na^+ + H_2O$

Benzoic acid → **Sodium benzoate**

(*b*) $HO-C_6H_4-CH_3 + NaOH \longrightarrow Na^+\,{}^-O-C_6H_4-CH_3 + H_2O$

4-Methylphenol → **Sodium 4-methylphenoxide**

- A strong base converts a carboxylic acid or a phenol into its conjugate base.

When it comes to removing H^+ from carboxylic acids, strong bases are more effective than water. Sodium hydroxide, for example, can completely convert benzoic acid into sodium benzoate, a salt (Figure 10.5*a*). The salts of some carboxylic acids are commonly used as food additives—sodium benzoate and sodium propanoate as preservatives and potassium sorbate as a mold inhibitor (Figure 10.6). As described above, converting a carboxylic acid into a carboxylate salt can alter solubility characteristics—benzoic acid is insoluble in water, while sodium benzoate is soluble.

As with carboxylic acids, reacting a phenol with a strong base, such as sodium hydroxide, converts all of the reactant into a phenoxide salt (Figure 10.5*b*). Formation of a phenoxide salt can dramatically alter water solubility: 2,4-dinitrophenol (used in the manufacture of dyes) has a very low solubility in water, while its sodium salt, sodium 2,4-dinitrophenoxide, is completely water soluble.

- Sodium salts of carboxylic acids are more soluble in water than carboxylic acids. Sodium salts of phenols are more soluble in water than phenols.

$HO-C_6H_3(NO_2)_2$ **2,4-Dinitrophenol**

$Na^+\,{}^-O-C_6H_3(NO_2)_2$ **Sodium 2,4-Dinitrophenoxide**

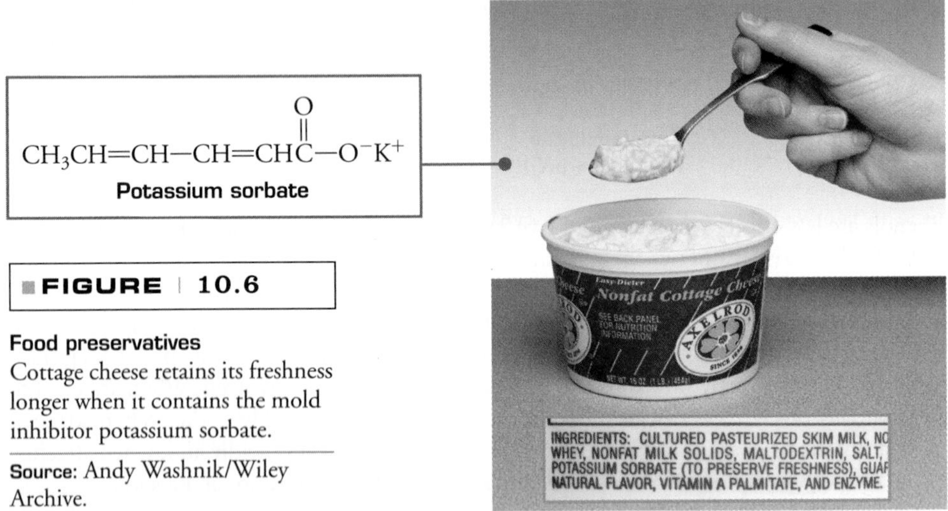

FIGURE | 10.6

Food preservatives

Cottage cheese retains its freshness longer when it contains the mold inhibitor potassium sorbate.

Source: Andy Washnik/Wiley Archive.

SAMPLE PROBLEM 10.2

Reacting carboxylic acids and phenols with a strong base

Draw the product of the reaction between KOH and

a. butanoic acid **b.** *p*-bromophenol

Strategy

Reacting a carboxylic acid or a phenol with a strong base converts each into a salt that contains its conjugate base.

Solution

a. $CH_3CH_2CH_2\overset{\overset{\large O}{\|}}{C}{-}O^-K^+$

b. p-$BrC_6H_4{-}O^-K^+$ (benzene ring with O^-K^+ at top and Br at bottom)

PRACTICE PROBLEM 10.2

Name the product of the reaction between NaOH and

a. 2,3-dimethylpentanoic acid **b.** 2,4-dichlorophenol

10.4 OTHER REACTIONS OF CARBOXYLIC ACIDS

In Section 4.9 we saw that esters contain a C—O—C linkage in which one of the carbon atoms belongs to a carbonyl (C=O) group. Every ester molecule contains a carboxylic acid residue and an alcohol residue. This is a reflection of how esters are formed — by the reaction of a carboxylic acid with an alcohol under acidic conditions. In addition to an ester, this double replacement reaction produces a water molecule (Figure 10.7a). In living things esters are not prepared as shown in Figure 10.7*a*.

$$\underbrace{CH_3CH_2\overset{\overset{\large O}{\|}}{C}}_{\text{Carboxylic acid residue}}{-}\underbrace{OCH_2CH_3}_{\text{Alcohol residue}}$$

$$(a)\quad CH_3CH_2\overset{\overset{\large O}{\|}}{C}{-}OH + HOCH_3 \overset{H^+}{\rightleftharpoons} CH_3CH_2\overset{\overset{\large O}{\|}}{C}{-}OCH_3 + H_2O$$

$$CH_3CH_2\overset{\overset{\large O}{\|}}{C}{-}OH + HOCH_2CH_3 \overset{H^+}{\rightleftharpoons} CH_3CH_2\overset{\overset{\large O}{\|}}{C}{-}OCH_2CH_3 + H_2O$$

$$C_6H_5{-}\overset{\overset{\large O}{\|}}{C}{-}OH + HO\underset{\underset{\large CH_3}{|}}{C}HCH_3 \overset{H^+}{\rightleftharpoons} C_6H_5{-}\overset{\overset{\large O}{\|}}{C}{-}O\underset{\underset{\large CH_3}{|}}{C}HCH_3 + H_2O$$

$$(b)\quad CH_3CH_2\overset{\overset{\large O}{\|}}{C}{-}OCH_3 + H_2O \overset{H^+}{\rightleftharpoons} CH_3CH_2\overset{\overset{\large O}{\|}}{C}{-}OH + HOCH_3$$

$$(c)\quad CH_3CH_2\overset{\overset{\large O}{\|}}{C}{-}OCH_3 + H_2O \xrightarrow{OH^-} CH_3CH_2\overset{\overset{\large O}{\|}}{C}{-}O^- + HOCH_3$$

FIGURE 10.7

Esters

(*a*) In the presence of H^+, a carboxylic acid reacts with an alcohol to form an ester plus water. (*b*) In the presence of water and H^+, an ester is hydrolyzed to produce a carboxylic acid and an alcohol. (*c*) In the presence of water and OH^-, a carboxylate ion and an alcohol are formed.

Instead, the—OH group on the carboxylic acid is first replaced by functional groups that alcohol molecules have an easier time replacing.

To name an ester, you first list the name of the alkyl group contained in the alcohol residue. The second part of the name is formed by changing the ending on the name of the carboxylic acid that provides the carboxylic acid residue from "ic acid" to "ate" (Figure 10.8). In Figure 10.8, the name ethyl acetate is a combination of *ethyl* (from the alcohol residue) and *acetate* (from the acetic acid residue). Methyl benzoate is a combination of *methyl* (from the alcohol residue) and *benzoate* (from the benzoic acid residue).

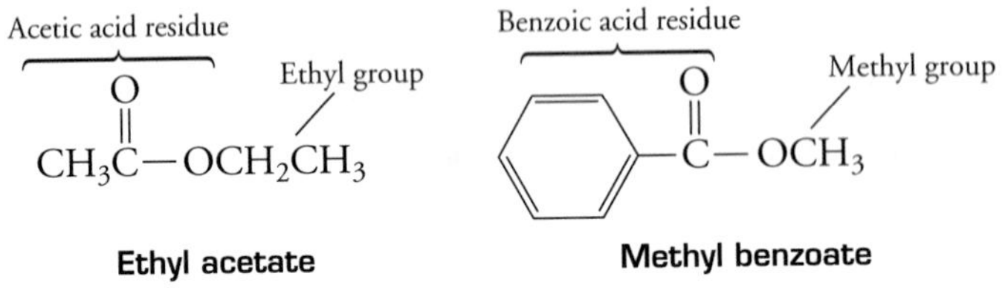

FIGURE 10.8

Naming esters
Esters are named by placing the name of the alkyl group in the alcohol residue before the name of the carboxylic acid (change the "ic acid" ending to "ate") in the carboxylic acid residue.

Acetic acid residue — Ethyl group

$CH_3C(=O)—OCH_2CH_3$

Ethyl acetate

Benzoic acid residue — Methyl group

$C_6H_5—C(=O)—OCH_3$

Methyl benzoate

SAMPLE PROBLEM 10.3

Ester formation

Draw the products of the reaction. Name the ester that forms.

$$CH_3CH_2CH_2C(=O)—OH + HOCH_2CH_2CH_3 \overset{H^+}{\rightleftharpoons}$$

Strategy

Ester formation is a double replacement reaction, and the product ester contains a carboxylic acid residue and an alcohol residue. Each of these residues is considered when assigning a name to the ester.

Solution

$$CH_3CH_2CH_2C(=O)—OH + HOCH_2CH_2CH_3 \overset{H^+}{\rightleftharpoons} CH_3CH_2CH_2C(=O)—OCH_2CH_2CH_3 + H_2O$$

Butanoic acid residue — Propyl group

Propyl butanoate

PRACTICE PROBLEM 10.3

Draw and name the organic product of the reaction.

$$CH_3CH(CH_3)CH_2C(=O)—OH + HOCH(CH_3)CH_2CH_3 \overset{H^+}{\rightleftharpoons}$$

As shown in Figure 10.7*a*, ester formation is a reversible process. This means that in the presence of H^+, a carboxylic acid plus an alcohol will react to yield an ester plus water, and an ester plus water will react to give a carboxylic acid plus an alcohol. To push the reaction one direction (making esters) or another (hydrolyzing esters), Le Châtelier's principle (Section 9.4) may be applied. To drive the reactions in Figure 10.7*a* forward, an excess of carboxylic acid and/or alcohol is used. To favor the reverse reactions, high concentrations of ester and/or water are used.

In hydrolysis reactions, esters are broken apart by water to form a carboxylic acid plus an alcohol. The reaction in Figure 10.7b shows an example of ester hydrolysis under acidic conditions. Note that this is the reverse of one of the reactions that appears in Figure 10.7*a*.

Earlier we saw that esters can be hydrolyzed under basic conditions (Section 6.4 and 8.3). This reaction initially produces a carboxylic acid and an alcohol, but the basic conditions immediately convert the product carboxylic acid into its conjugate base (Figure 10.7*c*).

Decarboxylation is a decomposition reaction that results in the *loss of carbon dioxide* (CO_2) from a carboxylic acid. In biochemistry, decarboxylation is seen in carboxylic acids that contain a keto (C=O) functional group at alpha (α) or beta (β) carbon atoms. (The term "keto" comes from "ketone," a family of organic compounds that will be introduced in Chapter 11.) An **α carbon atom** *is directly attached to the carbon atom of a carboxyl group* and a **β carbon atom** is *two atoms removed from the carbon atom of a carboxyl group.*

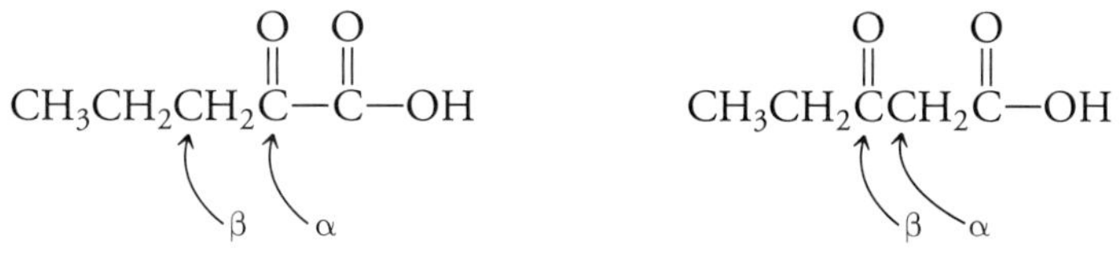

$$CH_3CH_2CH_2\overset{\beta}{\overset{O}{\overset{\|}{C}}}-\overset{\alpha}{\overset{O}{\overset{\|}{C}}}-OH$$

An α-keto acid

$$CH_3CH_2\overset{\beta}{\overset{O}{\overset{\|}{C}}}\overset{\alpha}{C}H_2\overset{O}{\overset{\|}{C}}-OH$$

A β-keto acid

In a laboratory setting, heating β-keto acids is often all that is required for decarboxylation to occur. When acetoacetic acid (a β-keto acid) is heated, CO_2 is released (Figure 10.9*a*). A biochemical example of β-keto acid decarboxylation follows from metabolic changes that take place as a result of starvation or poorly controlled diabetes. These conditions lead to the overproduction of molecules called ketone bodies, one of which is the acetoacetic acid reactant in Figure 10.9*a*. When this β-keto acid undergoes enzyme-catalyzed decarboxylation, the sweet-smelling acetone product can be detected on the breath.

■ Decarboxylation involves the loss of CO_2 from a carboxylic acid.

α-Keto acids can also be decarboxylated. Removal of CO_2 from an α-keto acid requires the addition of specific reactants (not discussed in this text) or the catalytic action of an enzyme. Brewer's yeast contains pyruvate decarboxylase, an enzyme that catalyzes the decarboxylation of pyruvic acid, an α-keto acid, transforming it into the molecule named acetaldehyde (Figure 10.9*b*). This particular reaction is one step in alcoholic fermentation, a process that ultimately converts pyruvic acid into ethanol (CH_3CH_2OH).

$$(a)\quad \underset{\textbf{Acetoacetic acid}}{CH_3\overset{O}{\overset{\|}{C}}CH_2\overset{O}{\overset{\|}{C}}-OH} \xrightarrow{\text{heat}} \underset{\textbf{Acetone}}{CH_3\overset{O}{\overset{\|}{C}}CH_3} + CO_2$$

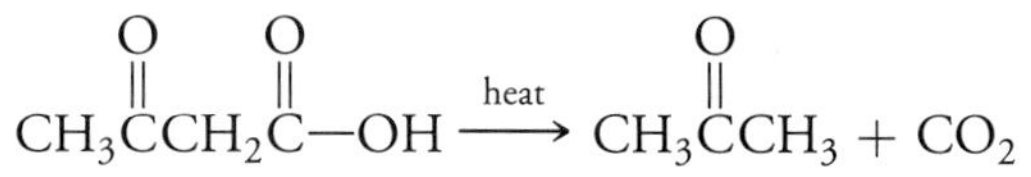

$$(b)\quad \underset{\textbf{Pyruvic acid}}{CH_3\overset{O}{\overset{\|}{C}}-\overset{O}{\overset{\|}{C}}-OH} \xrightarrow{\text{pyruvate decarboxylase}} \underset{\textbf{Acetaldehyde}}{CH_3\overset{O}{\overset{\|}{C}}-H} + CO_2$$

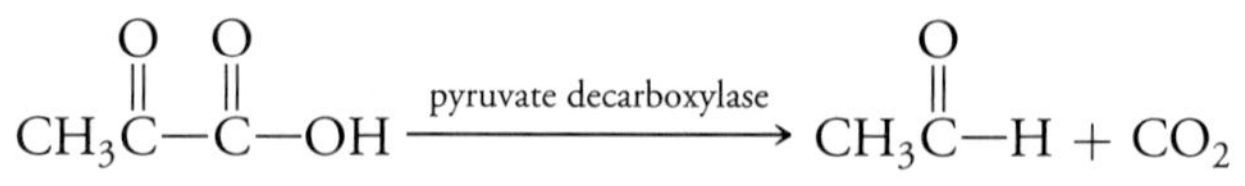

FIGURE 10.9

Decarboxylation

(*a*) β-Keto acids lose CO_2 when heated. (*b*) α-Keto acids lose CO_2 when in the presence of specific enzymes.

SAMPLE PROBLEM 10.4

Decarboxylation reactions

Draw the products of each decarboxylation reaction.

a. 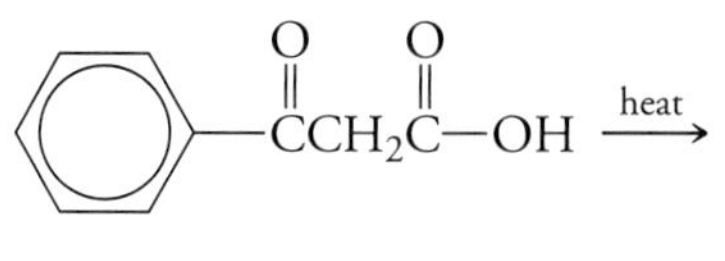 $C_6H_5-\overset{O}{\overset{\|}{C}}CH_2\overset{O}{\overset{\|}{C}}-OH \xrightarrow{\text{heat}}$

b. 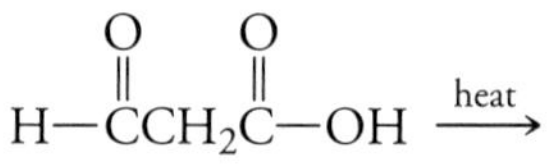 $H-\overset{O}{\overset{\|}{C}}CH_2\overset{O}{\overset{\|}{C}}-OH \xrightarrow{\text{heat}}$

Strategy

In a decarboxylation reaction, CO_2 is released from a carboxylic acid reactant.

Solution

a. 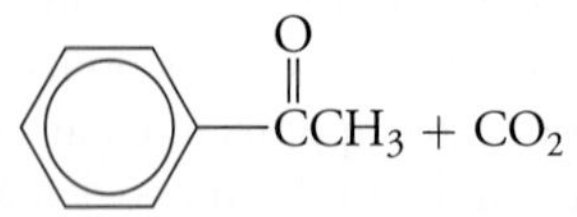 $C_6H_5-\overset{O}{\overset{\|}{C}}CH_3 + CO_2$ **b.** $H-\overset{O}{\overset{\|}{C}}CH_3 + CO_2$

PRACTICE PROBLEM 10.4

Draw the missing product of each reaction.

a. $CH_3CH_2\overset{O}{\overset{\|}{C}}CH_2\overset{O}{\overset{\|}{C}}-OCH_3 \xrightarrow[H_2O]{H^+} \qquad + CH_3OH$

b. Product of a $\xrightarrow{\text{heat}} \qquad + CO_2$

10.5 OXIDATION OF PHENOLS

Section 6.3 showed us that an organic molecule has been oxidized if carbon atoms gain oxygen and/or lose hydrogen. The oxidizing agent potassium dichromate ($K_2Cr_2O_7$) can be used to oxidize a number of different organic compounds, including phenols. Hydroquinone, for example, is oxidized to benzoquinone (Figure 10.10*a*). In this reaction, removal of hydrogen atoms from hydroquinone is evidence that oxidation has taken place.

FIGURE 10.10

Phenol oxidation

The oxidizing agent $K_2Cr_2O_7$ converts (*a*) hydroquinone into benzoquinone and (*b*) other hydroquinone-containing compounds into quinone-containing ones.

(*a*) Hydroquinone + $K_2Cr_2O_7$ ⟶ Benzoquinone

(*b*) + $K_2Cr_2O_7$ ⟶ Dichlone

(*c*) + $K_2Cr_2O_7$ ⟶ 1-Aminoanthraquinone

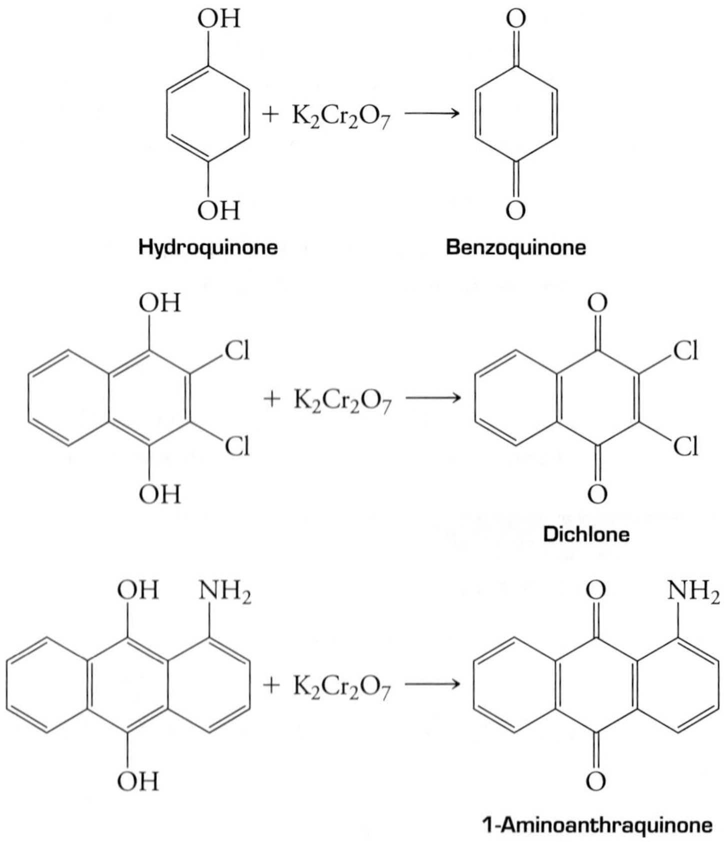

Other quinones (molecules that contain the benzoquinone structure) can be prepared in the same way. Dichlone, a fungicide and herbicide, and 1-aminoanthraquinone (amino is $-NH_2$), used in the manufacture of dyes and pharmaceuticals, can be synthesized from the appropriate hydroquinones (Figure 10.10*b*). Phenols that possess just one —OH group are also oxidized to quinones by $K_2Cr_2O_7$.

Quinones are widely distributed throughout nature. One example, vitamin K_1, plays an important role in blood clotting and in calcium metabolism. Some quinones are highly colored. Mordant red, which can be obtained from the roots of the madder plant, was used as a dye in ancient Egypt, Persia, and India. Lawsone, an orange-colored quinone, is found in henna, a plant whose leaves provide a dye that can be used to give hair an auburn color.

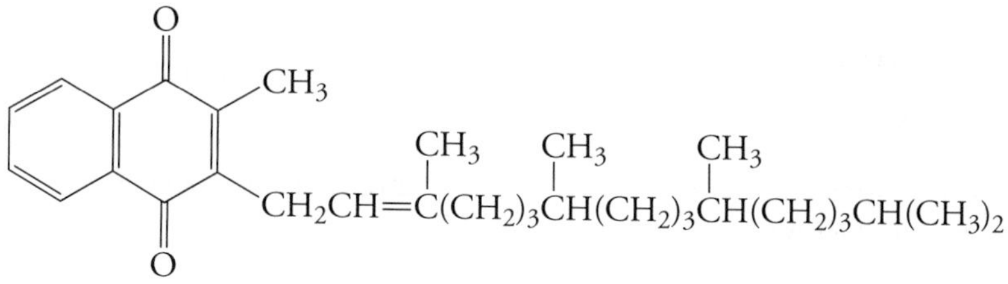

Vitamin K_1

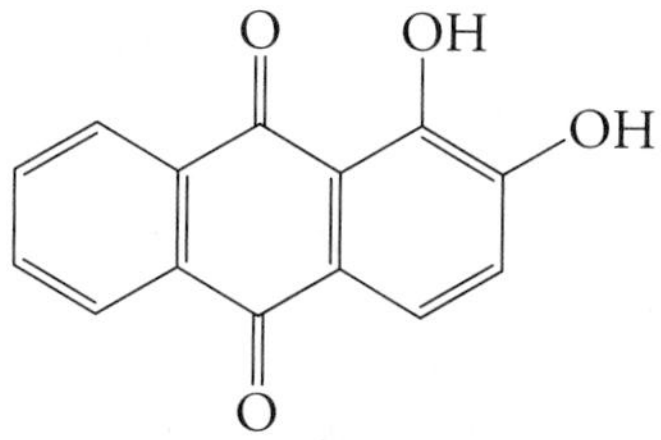

Mordant red

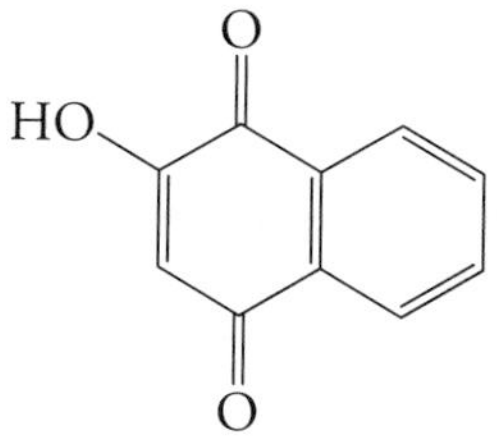

Lawsone

10.6 AMINES

Amines are classified based on the number of carbon atoms directly attached to the nitrogen atom. If *only one carbon atom is attached to the amine nitrogen atom*, the amine is **primary (1°)**. A **secondary (2°)** amine *has two attached carbon atoms* and a **tertiary (3°)** amine *has three attached carbon atoms*. **Quaternary (4°) ammonium ions** are a class of amine-related compounds that are organic derivatives of ammonium ion (NH_4^+). In a 4° ammonium ion, *four carbon atoms are directly attached to the nitrogen atom and the nitrogen carries a 1+* charge (Figure 10.11).

To name a 1°, 2°, or 3° amine using the IUPAC rules, the parent, the longest chain of carbon atoms attached to the amine nitrogen atom, is numbered from the end nearer the point of attachment of the nitrogen. The parent chains of amines are named by dropping "e" from the name of the corresponding hydrocarbon and adding "amine." For example, the amine $CH_3CH_2NH_2$ is named ethanamine (Figure 10.11). If an amine parent chain contains a carbon–carbon double bond it is an *enamine* and if it contains a carbon–carbon triple bond it is an *ynamine*. If an amine is 2° or 3°, the carbon-containing groups

FIGURE | 10.11

Classifying amines
The number of carbon atoms directly attached to the nitrogen atom of an amine determines its classification as 1°, 2°, or 3°. In 4° ammonium ions, four carbon atoms are attached to the nitrogen atom. Here, IUPAC names and common names (in parentheses) are given.

Primary (1°) amine	Secondary (2°) amine
$CH_3CH_2NH_2$ Ethanamine (ethylamine)	$CH_3CH_2CH_2NHCH_3$ *N*-Methyl-1-propanamine (methylpropylamine)
Tertiary (3°) amine	**Quaternary (4°) ammonium ion**
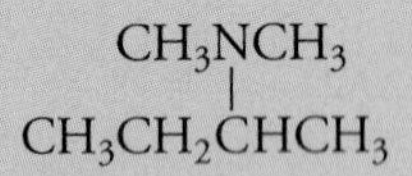 CH_3NCH_3 (N bonded to C-2 of $CH_3CH_2CHCH_3$) *N,N*-Dimethyl-2-butanamine (*sec*-butyldimethylamine)	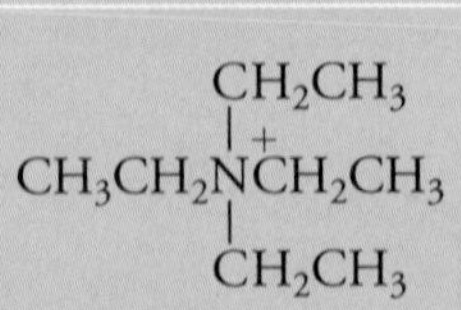 $CH_3CH_2\overset{+}{N}CH_2CH_3$ with CH_2CH_3 above and below N Tetraethylammonium ion

attached to the nitrogen atom that are not part of the parent chain are substituents and *N* is used to indicate their location (*N*-methyl, *N,N*-dimethyl, etc.). As shown in the name *N,N*-dimethyl-2-butanamine, when a parent chain contains more than two carbon atoms, the position of the nitrogen atom must be specified.

Simple amines, those with a relatively small number of carbon atoms, are often identified by common names formed by placing "amine" after the names of the groups attached to the nitrogen. Examples include methylamine (CH_3NH_2), dipropylamine ($CH_3CH_2CH_2NHCH_2CH_2CH_3$), and triethylamine [$N(CH_2CH_3)_3$]. To name a 4° ammonium ion, the names of the four groups attached to the nitrogen are combined with "ammonium ion," as for tetraethylammonium ion in Figure 10.11.

SAMPLE PROBLEM 10.5

Naming amines

Match each IUPAC and common name to the correct structural formula:

a. $CH_3CH_2CH_2NHCH_2CH_3$

b. $CH_3CH_2NCH_2CH_3$ with $CH_2CH_2CH_3$ on N

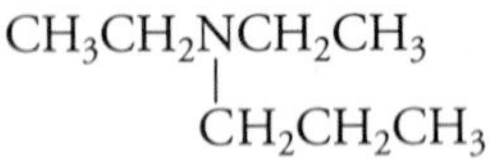

c. $H_2NCH_2CH_2CH_2CH_3$

1-butanamine	butylamine
N,N-diethyl-1-propanamine	diethylpropylamine
ethylpropylamine	*N*-ethyl-1-propanamine

Strategy

Begin by distinguishing the IUPAC names from the common names. While both end in "amine," common names only include the names of the alkyl groups attached to the nitrogen atom (methylamine, dipropylamine, etc.).

Solution

a. *N*-ethyl-1-propanamine and ethylpropylamine
b. *N,N*-diethyl-1-propanamine and diethylpropylamine
c. 1-butanamine and butylamine

PRACTICE PROBLEM 10.5

Give an IUPAC and common name for each molecule.

a. $CH_3CHNHCH_2CH_3$ with CH_3 on the second carbon

b. $CH_3CH_2CHCH_3$ with NH_2 on the third carbon

c. $CH_3CH_2CHCH_3$ with CH_3NH on the third carbon

In one particular class of amines called heterocyclic amines, one or more nitrogen atoms are part of a ring. Pyridine, pyrimidine, and purine are three heterocyclic amines important to biochemistry (Figure 10.12*a*). Thymine, one of the amines present in DNA, is a pyrimidine because it contains an arrangement of atoms similar to pyrimidine (Figure 10.12*b*). Adenine, another amine found in DNA, is a purine. Nicotine, shown in Figure 10.1, is a pyridine.

Hydrogen bonding, or the lack of it, plays a role in determining many of the properties of a particular amine. Although the polar covalent C—N bonds and pyramidal shape about the nitrogen atom make all amines polar, *only 1° and 2° amines form hydrogen bonds with like amines.* Tertiary amines do not, since they do not have a hydrogen atom directly attached to the nitrogen atom.

1° and 2° amines form hydrogen bonds with like amines. 3° amines do not.

1°	2°	3°
$CH_3—\ddot{N}—H$ (with H on N)	$CH_3—\ddot{N}—CH_3$ (with H on N)	$CH_3—\ddot{N}—CH_3$ (with CH_3 on N)
$CH_3—\ddot{N}—H$ (with H on N)	$CH_3—\ddot{N}—CH_3$ (with H on N)	$CH_3—\ddot{N}—CH_3$ (with CH_3 on N)

hydrogen bonds

Because the hydrogen bonds that attract one 1° or 2° amine to another are stronger than the dipole–dipole interactions that attract one 3° amine to another, 1° and 2° amines tend to have higher boiling points than do 3° amines with similar molecular weights (Table 10.3). For example, trimethylamine has a low enough boiling point that it is a gas at room temperature, while one of its constitutional isomers, propylamine, is a liquid at this temperature. The cations and anions that make up 4° ammonium salts are held to one another by strong ionic bonds and, for this reason, 4° ammonium salts have very high melting and boiling points.

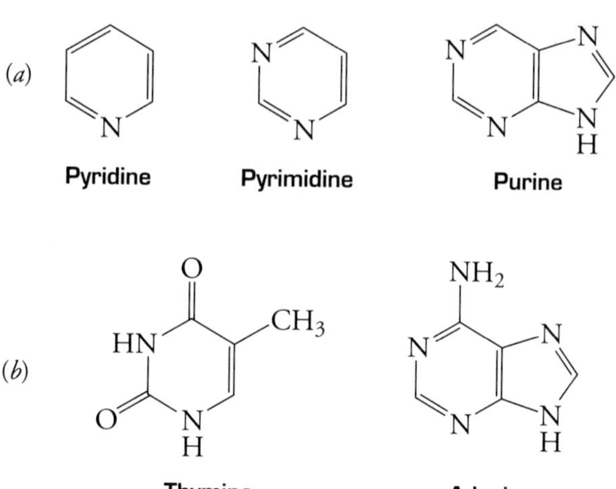

FIGURE 10.12

Heterocyclic amines
(*a*) Pyridine, pyrimidine, and purine rings are common in biochemical molecules.
(*b*) Thymine is a pyrimidine and adenine is a purine.

TABLE 10.3 PHYSICAL PROPERTIES OF SELECTED AMINES

Formula	IUPAC Name	Common Name	Boiling Point (°C)	Water Solubility (g/100 mL)
CH_3NH_2	Methanamine	Methylamine	−7.5	Very large
$CH_3CH_2NH_2$	Ethanamine	Ethylamine	17	Very large
$CH_3CH_2CH_2NH_2$	1-Propanamine	Propylamine	49	Very large
$CH_3CH_2NHCH_3$	*N*-Methylethanamine	Ethylmethylamine	36–37	Very large
$N(CH_3)_3$	*N,N*-Dimethylmethanamine	Trimethylamine	3	91
$N(CH_2CH_3)_3$	*N,N*-Diethylethanamine	Triethylamine	89	14

Amines can form hydrogen bonds with water molecules and are polar, so small amines are highly soluble in water. The more carbon atoms in an amine, the lower its solubility in water.

An additional property of many amines is their strong, unpleasant odor. Small amines have an odor reminiscent of ammonia and some have a fishy odor. The stench of some larger amines makes them quite unpleasant to work with, including putrescine ($NH_2CH_2CH_2CH_2CH_2NH_2$) and cadaverine ($NH_2CH_2CH_2CH_2CH_2CH_2NH_2$), compounds produced during the decomposition of tissue.

10.7 AMINES AS WEAK ORGANIC BASES

Amines are weak bases (H^+ acceptors), a characteristic that distinguishes them from the other classes of organic compounds that we will consider. When methylamine (CH_3NH_2) is added to water, H^+ is transferred from a water molecule to the amine (Figure 10.13*a*). The products of this acid–base reaction are hydroxide ion (makes the solution basic) and methylammonium ion, the conjugate acid of the amine. Because amines are weak bases, in the reversible reaction with water, reactant concentrations are higher than product concentrations and the solution does not get especially basic. The conjugate acids of many amines have pK_a values ranging between 9 and 11, so at pHs of less than 9 amines are found as their conjugate acids, while at pHs above 11 they exist as amines.

$CH_3NH_3^+$	CH_3NH_2
Methylammonium ion [predominates when pH < pK_a]	**Methylamine [predominates when pH > pK_a]**

Amine conjugate acids are named in the same way as quaternary ammonium ions. For example, $CH_3NH_3^+$ is methylammonium ion and $(CH_3)_3NH^+$ is trimethylammonium ion.

While primary, secondary, and tertiary amines are bases, quaternary ammonium ions are not, because they have no nonbonding electrons with which to form bonds to H^+. In other words, the nitrogen of the 4° ammonium ion is already directly bonded to four different carbon atoms and has no space available to accept an H^+.

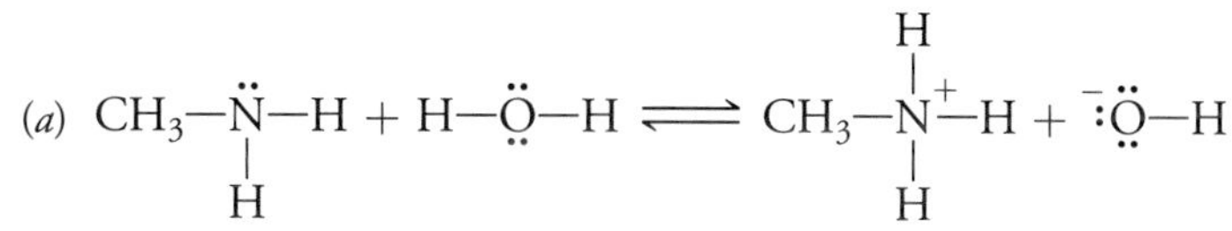

$$(a)\ CH_3{-}\ddot{N}H_2 + H{-}\ddot{O}{-}H \rightleftharpoons CH_3{-}\overset{+}{N}H_3 + {}^{-}{:}\ddot{O}{-}H$$

Methylamine — Methylammonium ion

$$(b)\ CH_3CH_2NH_2 + HCl \longrightarrow CH_3CH_2\overset{+}{N}H_3\ Cl^-$$

Ethylamine — Ethylammonium chloride

FIGURE 10.13

Amines as bases
(*a*) Methylamine reacts with water to form methylammonium ion and OH^-. (*b*) Ethylamine reacts with hydrochloric acid to form the salt ethylammonium chloride.

Just as amines can accept H^+ from water, they can also accept H^+ from acids. When ethylamine reacts with HCl, the acid–base reaction produces the salt ethylammonium chloride (Figure 10.13*b*). Salts formed from amines are important to the drug industry because they have a greater water solubility than do the amines themselves, making amine salt drugs easy to administer. The decongestant ephedrine, for example, is sold as ephedrine hydrochloride or ephedrine sulfate.

■ A strong acid converts an amine into an ammonium salt.

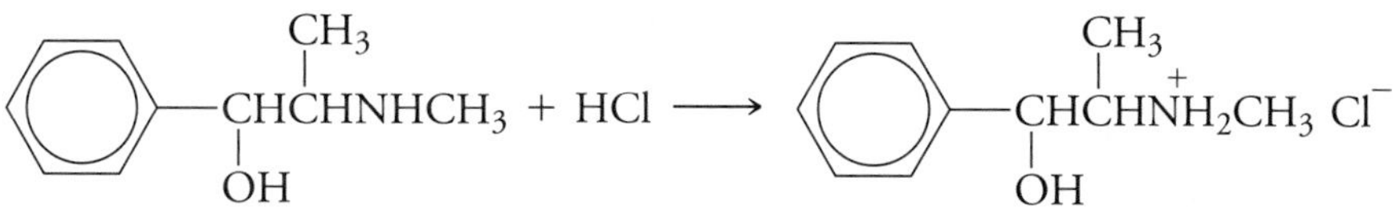

$$C_6H_5{-}CH(OH)CH(CH_3)NHCH_3 + HCl \longrightarrow C_6H_5{-}CH(OH)CH(CH_3)\overset{+}{N}H_2CH_3\ Cl^-$$

Ephedrine — Ephedrine hydrochloride

$$C_6H_5{-}CH(OH)CH(CH_3)NHCH_3 + H_2SO_4 \longrightarrow \left(C_6H_5{-}CH(OH)CH(CH_3)\overset{+}{N}H_2CH_3\right)_2 SO_4^{2-}$$

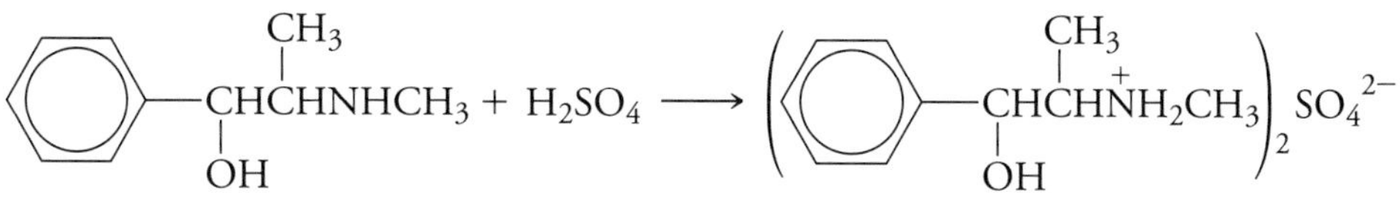

Ephedrine sulfate

The salts of amine-containing drugs, such as ephedrine, are usually named by combining the common name of the drug with the name of the anion. When HCl is used to form the salt, *hydrochloride* is used instead of *chloride*.

Although 4° ammonium ions are not basic, they can exist as salts. Muscarine, a highly poisonous 4° ammonium ion found in the mushroom *Amanita muscaria*, can be isolated as an iodide salt.

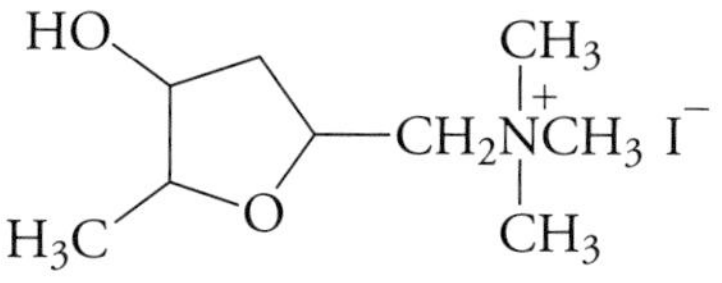

(Structure: tetrahydrofuran ring bearing HO, H_3C, and $-CH_2\overset{+}{N}(CH_3)_3\ I^-$)

Muscarine iodide

10.8 AMIDES

When a carboxylic acid reacts with ammonia or with an amine, the product of the acid–base reaction is a salt composed of a carboxylate ion and the conjugate acid of ammonia or an amine. If this salt is heated, an amide is produced (Figure 10.14*a*).

■ Amides can be formed by heating carboxylate–ammonium ion salts.

Amides, all of which have a nitrogen atom directly attached to a carbonyl carbon, consist of a carboxylic acid residue attached to an ammonia or amine residue. In the chapters that follow, we will see the important roles that amides play in the chemistry of living things.

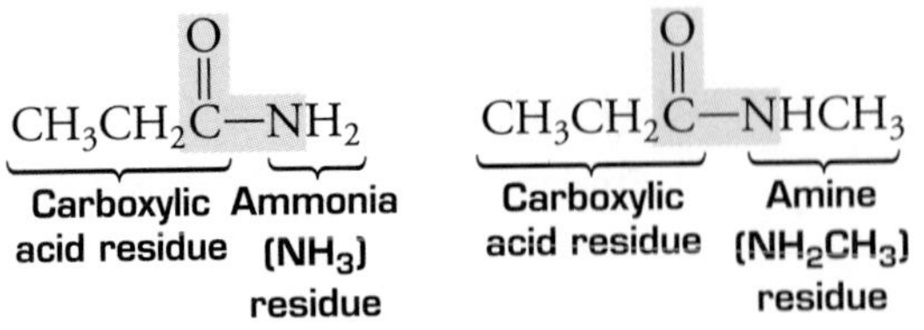

$$\underbrace{CH_3CH_2\overset{O}{\overset{\|}{C}}}_{\text{Carboxylic acid residue}}\!-\!\underbrace{NH_2}_{\text{Ammonia }(NH_3)\text{ residue}} \qquad \underbrace{CH_3CH_2\overset{O}{\overset{\|}{C}}}_{\text{Carboxylic acid residue}}\!-\!\underbrace{NHCH_3}_{\text{Amine }(NH_2CH_3)\text{ residue}}$$

There are ways of producing amides other than the reaction just shown, most of which involve replacing the —OH group on the carboxylic acid with a functional group that ammonia or amine molecules have an easier time displacing. In the case of the amide bonds formed during protein synthesis in cells, the —OH is replaced by a large molecule called tRNA (Chapter 14).

Amides are named in much the same way as esters. In the simplest amides, where only an $—NH_2$ is attached to the carbonyl carbon atom, names are obtained by changing the name of the carboxylic acid in the carboxylic acid residue. The "oic acid" ending on IUPAC names or the "ic acid" ending on common names is changed to "amide" (Figure 10.15*a*). When an amine residue is present, any substituents attached to the nitrogen atom must be identified and their position specified using *N* (Figure 10.15*b*).

Like esters, amides can be hydrolyzed in a double replacement reaction. When heated in the presence of water and acid, amides react to form carboxylic acids and the conjugate acids of ammonia or amines (Figure 10.14*b*).

$$(a)\quad CH_3CH_2\overset{O}{\overset{\|}{C}}-OH + NH_3 \longrightarrow CH_3CH_2\overset{O}{\overset{\|}{C}}-O^-NH_4^+ \xrightarrow{\text{heat}} CH_3CH_2\overset{O}{\overset{\|}{C}}-NH_2 + H_2O$$

$$CH_3CH_2\overset{O}{\overset{\|}{C}}-OH + H_2NCH_3 \longrightarrow CH_3\overset{O}{\overset{\|}{C}}-O^-H_3\overset{+}{N}CH_3 \xrightarrow{\text{heat}} CH_3CH_2\overset{O}{\overset{\|}{C}}-NHCH_3 + H_2O$$

$$(b)\quad CH_3\overset{O}{\overset{\|}{C}}-NH_2 + H_2O \xrightarrow[\text{heat}]{H^+} CH_3\overset{O}{\overset{\|}{C}}-OH + NH_4^+$$

$$CH_3\overset{O}{\overset{\|}{C}}-NHCH_2CH_3 + H_2O \xrightarrow[\text{heat}]{H^+} CH_3\overset{O}{\overset{\|}{C}}-OH + H_3\overset{+}{N}CH_2CH_3$$

FIGURE | 10.14

Amides
(*a*) Reaction of a carboxylic acid (an acid) with ammonia or an amine (a base) produces a salt. Heating converts the salt into an amide. (*b*) When heated in the presence of water and H^+, an amide is hydrolyzed to a carboxylic acid and the conjugate acid of ammonia or an amine.

(a)

Pentanoic acid residue

$$CH_3CH_2CH_2CH_2\overset{\overset{\displaystyle O}{\|}}{C}-NH_2$$

Pentanamide

Acetic acid residue

$$CH_3\overset{\overset{\displaystyle O}{\|}}{C}-NH_2$$

Acetamide

(b)

3-methylbutanoic acid residue; Ethyl group

$$CH_3\underset{\underset{\displaystyle CH_3}{|}}{C}HCH_2\overset{\overset{\displaystyle O}{\|}}{C}-NHCH_2CH_3$$

***N*-ethyl-3-methylbutanamide**

Formic acid residue; Methyl groups

$$H\overset{\overset{\displaystyle O}{\|}}{C}-\underset{\underset{\displaystyle CH_3}{|}}{N}CH_3$$

***N,N*-dimethylformamide**

FIGURE 10.15

Naming amides (*a*) For amides that contain an ammonia residue, names are assigned by changing the ending of the name of the carboxylic acid in the carboxylic acid residue from "oic acid" (IUPAC names) or "ic acid" (common names) to "amide". (*b*) When an amine residue is present, the amide name also identifies the substituents attached to the nitrogen atom, using *N* to indicate their position. Note that, unlike ester names, the names of amides contain no spaces.

HEALTH Link: *Biofilms*

Have you ever had a head of lettuce sitting in the refrigerator and found, overnight, that it had become slimy and started to spoil? If so, you have observed the results of a type of bacterial action known as quorum sensing.

Bacteria are single-celled organisms that, until recent years, were thought to lead a solitary existence. Now many bacteria are known to release molecules whose function is to say "Here I am!" Other bacteria have receptors that detect these signaling molecules, called autoinducers. When a population of bacteria reaches a high enough level, the concentration of autoinducers reaches the point where bacterial metabolism changes from producing compounds that benefit an individual bacterium to producing those that help bacteria to act as a group. This switch from individual to group activity is called quorum sensing. In human terms, a quorum is the minimum attendance required at a meeting for business to take place. For bacteria, a quorum is a population of sufficient size to make group activity successful.

Biofilms are one of the common outcomes of quorum sensing. Bacteria begin to manufacture a slime that sticks large numbers of them together. These biofilms keep nutrients close at hand and have channels that allow for the uptake of water and the removal of waste products. For bacteria in some biofilms, genes that provide resistance to antibiotics or that produce toxins are switched on.

Biofilms are widespread. They have been shown to play a role in the spoilage of food, including vegetables, meat, and fish. They also impact health. Biofilms form on contact lenses, inside catheters, on surgical sutures, on toothbrushes, and in the mouth (Figure 10.16). Bacteria in biofilms are one factor in the onset of gum disease and dental cavities. The bad breath associated with poor oral hygiene comes from sulfur compounds produced in biofilms.

Quorum-sensing autoinducers come in a number of forms. Many bacteria release variations of what are called acyl homoserine lactones (AHLs). AHLs consist of two parts: a homoserine residue and an acyl group (Figure 10.17). Homoserine is an amino acid that contains a carboxylic acid, an

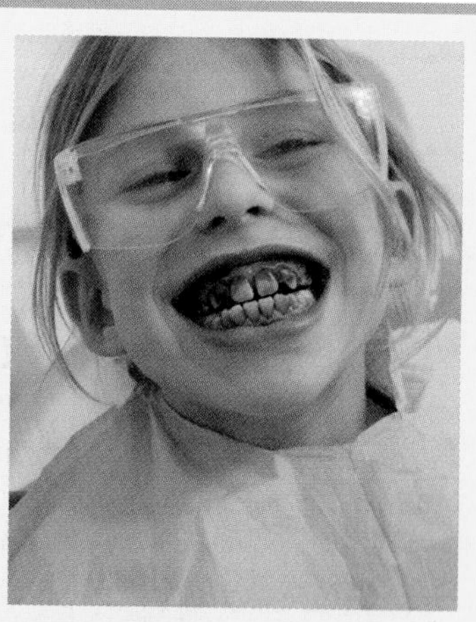

FIGURE 10.16

Biofilm
The purple color on the teeth is due to a stain used to show dental plaque, a biofilm of oral bacteria.

Source: Michael Donne/Photo Researchers, Inc.

alcohol, and an amine group. In AHLs the carboxylic acid and alcohol group have reacted to form a lactone (a cyclic ester). The nitrogen atom of the amino group is joined with an acyl group (a carboxylic acid residue) to form an amide. The acyl group has an even number of carbon atoms and, depending on the bacterial species, between four and eighteen carbon atoms.

FIGURE 10.17

An acyl homoserine lactone
Acyl homoserine lactones (AHLs) are one class of molecule that triggers quorum sensing in bacteria. (*a*) Three families of organic compounds are present in the amino acid homoserine: carboxylic acid, alcohol, and amine. (*b*) The carboxylic acid and alcohol group in homoserine react to form a cyclic ester (a lactone). (*c*) AHLs consist of a homoserine lactone residue coupled to an acyl group (a carboxylic acid residue) by an amide bond.

(*a*)

NH_2—CH—C(=O)—OH, with CH_2—CH_2—OH on CH

Amine

Carboxylic acid

Alcohol

Homoserine

(*b*)

NH_2

Lactone

Homoserine lactone

(*c*)

Acyl group

$CH_3CH_2CH_2C(=O)$—NH

An acyl homoserine lactone

BIOCHEMISTRY Link

A Cure for Fleas

Anyone who owns a cat or dog has, at some point, probably had to deal with fleas (Figure 10.18). Although fleas live only for about one month, they reproduce very efficiently—a single female flea can lay up to 2,000 eggs in its lifetime. It doesn't take long for a pet, a sofa, or an entire house to be infested with these tiny pests.

The Ciba-Geigy company markets an amide-containing drug called Lufenuron that, when administered on a monthly basis, helps to control flea problems. This drug is hydrophobic (it dissolves in a pet's fatty tissues) and is slowly released into the bloodstream. When fleas bite a lufenuron-treated pet, they ingest some of this drug. Adult fleas are not affected but, because the drug interferes with the proper formation of insect exoskeletons, eggs laid by fleas that have been exposed to lufenuron fail to develop.

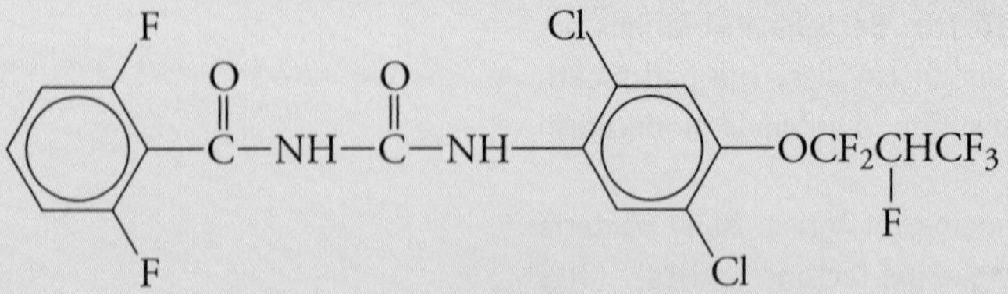

Lufenuron

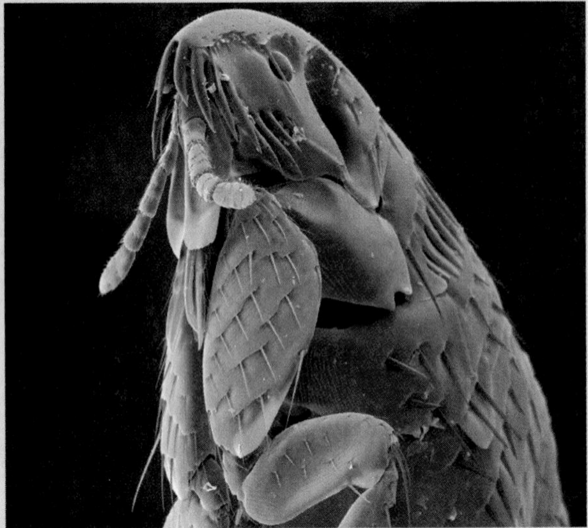

FIGURE 10.18

Ctenocephalides felis
The cat flea.

Source: Eye of Science/Photo Researchers, Inc.

10.9 STEREOISOMERS

In our earlier discussion of cycloalkanes and alkenes (Chapter 4), we saw that limited rotation of the carbon–carbon single bonds in cycloalkanes and the lack of rotation about carbon–carbon double bonds in alkenes can produce stereoisomers, molecules that

- have the same molecular formula
- have the same atomic connections
- have a different three-dimensional shape
- are interchanged only by breaking bonds

Cis and *trans* stereoisomers that arise from hindered bond rotation are called geometric isomers.

Members of another class of stereoisomers, called **enantiomers**, are *nonsuperimposable mirror image forms of a molecule.* The amine molecule shown in Figure 10.19 exhibits this form of stereoisomerism, since only bond breaking will convert one 2-butanamine molecule into its mirror image.

■ Enantiomers are nonsuperimposable mirror image molecules.

The term **chiral** (from *cheir*, Greek for *hand*) is used to describe objects that *cannot be superimposed on their mirror image.* Your hands are chiral, because your left hand is not superimposable on your right hand, its mirror image. Figure 10.20 shows examples of objects that are chiral and objects that are not.

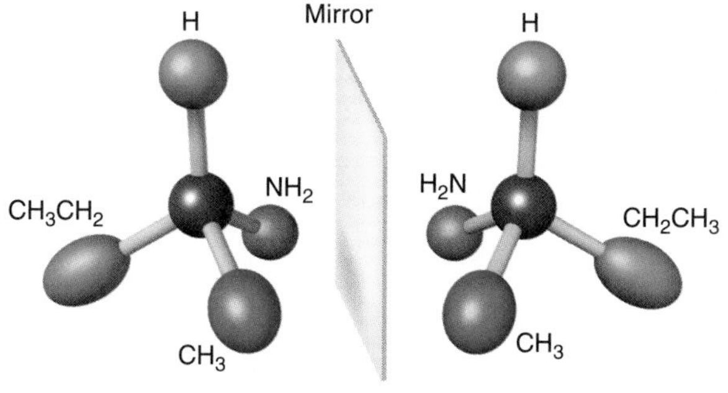

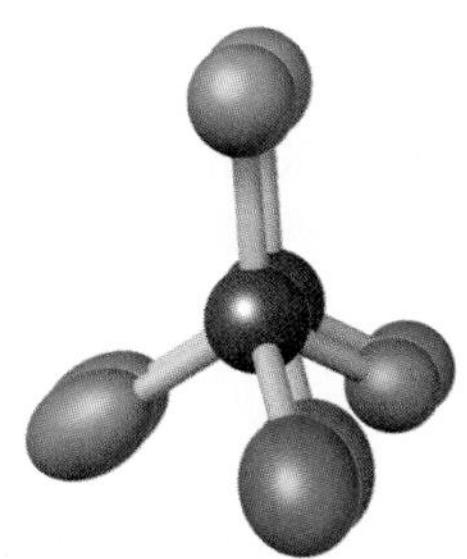

FIGURE 10.19

Enantiomers
This amine (2-butanamine) and its mirror image are stereoisomers known as enantiomers, because they cannot be superimposed.

(*a*)

(*b*)

Chiral objects
(*a*) Each battery is chiral because it is different from its mirror image. (*b*) Each ball is not chiral because it is identical to its mirror image.

Source: Andy Washnik/Wiley Archive.

A molecule is usually chiral if it contains one or more **chiral carbon atoms**, *carbons with four different attached atoms or groups of atoms.* In Figure 10.21 the chiral carbon atom of the molecule is marked with an asterisk. The four atoms or groups attached to this chiral atom are $-H$, $-CH_3$, $-OH$, and $-CO_2H$. When mirror images of this molecule are drawn in three dimensions, they are not superimposable and are, therefore, enantiomers.

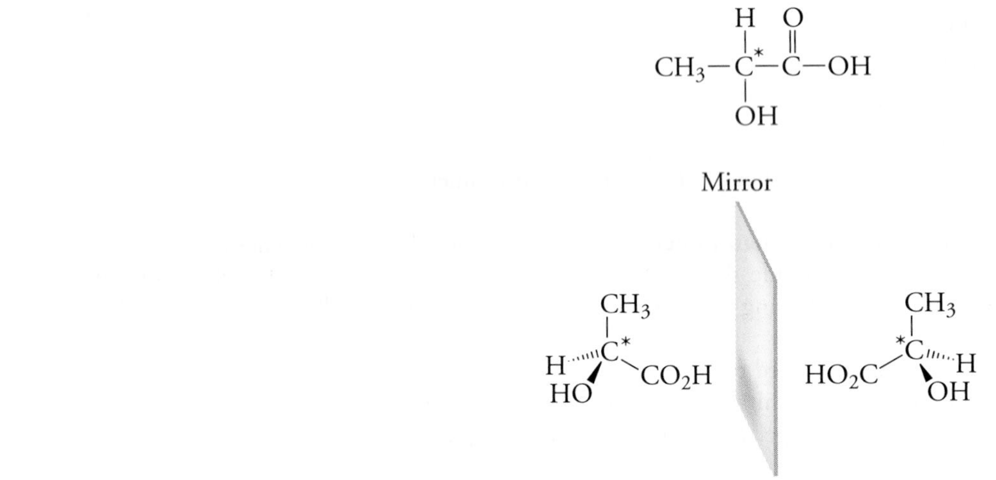

$$CH_3-\overset{H}{\underset{OH}{\overset{|}{\underset{|}{C^*}}}}-\overset{O}{\overset{\|}{C}}-OH$$

Mirror

H····C*(CH₃)(HO)—CO₂H | HO₂C—*C(CH₃)····H(OH)

FIGURE 10.21

Chiral carbon atoms
A chiral carbon atom (*) carries four different atoms or groups of atoms, and molecules that contain a chiral carbon atom are almost always chiral. The chiral carbon atom in this molecule, 2-hydroxypropanoic acid (lactic acid), carries $-H$, $-CH_3$, $-OH$, and $-CO_2H$.

SAMPLE PROBLEM 10.6

Identifying chiral carbon atoms

Use an asterisk to label the chiral carbon atom in each molecule.

a. 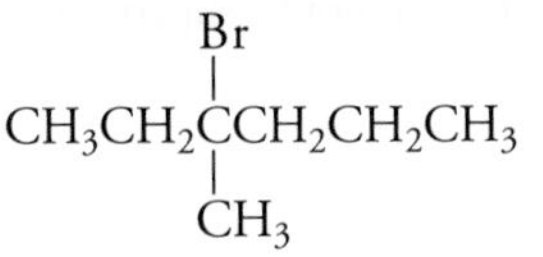 $CH_3CH_2\overset{Br}{\overset{|}{\underset{CH_3}{\underset{|}{C}}}}CH_2CH_2CH_3$

b. 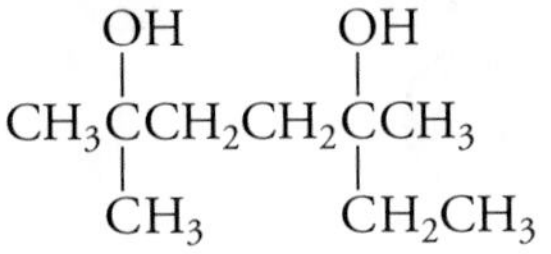 $CH_3\overset{OH}{\overset{|}{\underset{CH_3}{\underset{|}{C}}}}CH_2CH_2\overset{OH}{\overset{|}{\underset{CH_2CH_3}{\underset{|}{C}}}}CH_3$

Strategy
Remember that a chiral carbon atom is attached to four different groups and that $-CH_3$, $-CH_2CH_3$, and $-CH_2CH_2CH_3$ are different groups.

Solution

a. 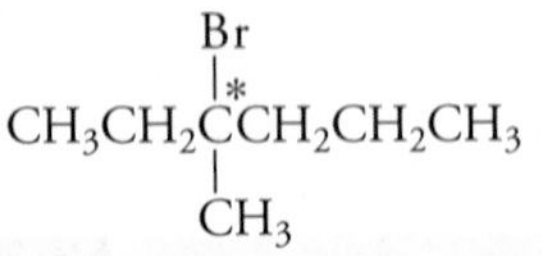 $CH_3CH_2\overset{Br}{\overset{|*}{\underset{CH_3}{\underset{|}{C}}}}CH_2CH_2CH_3$

b. 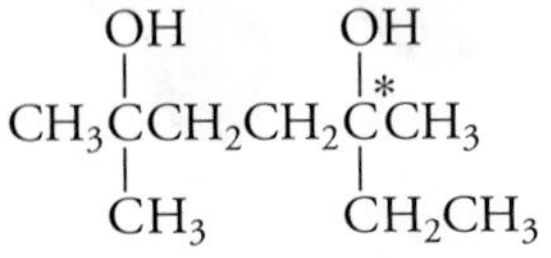 $CH_3\overset{OH}{\overset{|}{\underset{CH_3}{\underset{|}{C}}}}CH_2CH_2\overset{OH}{\overset{|*}{\underset{CH_2CH_3}{\underset{|}{C}}}}CH_3$

PRACTICE PROBLEM 10.6

Label all chiral carbon atoms in the molecule.

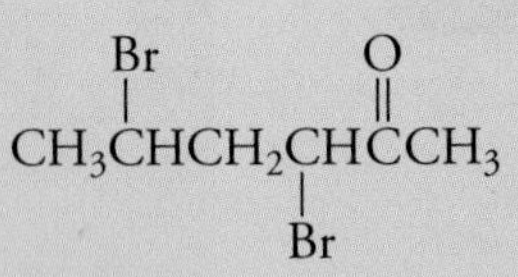

$$CH_3\overset{Br}{\overset{|}{C}}HCH_2\underset{Br}{\underset{|}{C}}H\overset{O}{\overset{\|}{C}}CH_3$$

Plane-Polarized Light

Pairs of enantiomers are identical in almost every way. Their boiling points, melting points, solubilities, and other physical properties match, which makes it difficult to tell an enantiomer from its mirror image. One property that can be used to distinguish enantiomers is the difference in how they interact with **plane-polarized light**, light that has passed through a filter that blocks all light waves except those vibrating in *one particular orientation* (Figure 10.22).

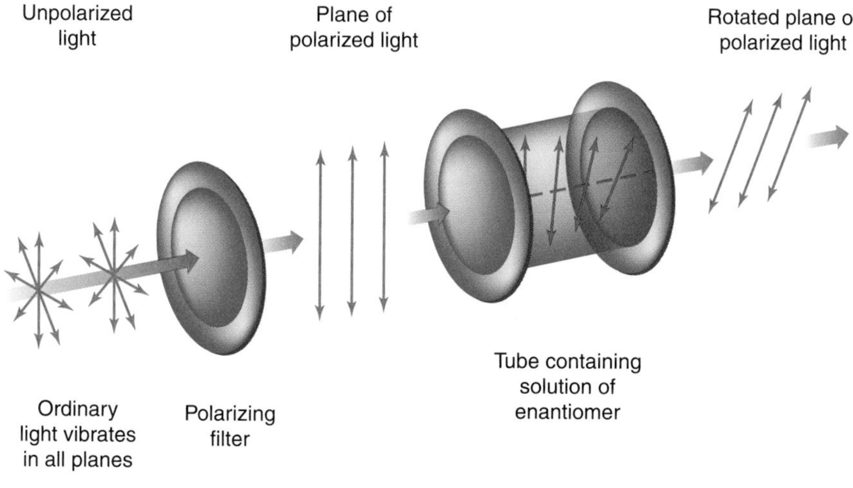

FIGURE 10.22

Plane-polarized light
When light is sent into a polarizing filter, only light that vibrates in one orientation is allowed to pass through. One enantiomer of a molecule rotates plane-polarized light in a clockwise direction (viewed as light travels toward the observer), while the other enantiomer rotates it in a counterclockwise direction.

Source: From Introductory Chemistry for Today (with InfoTrac), 4th edition by Seager/Slabaugh, © 2000. Reprinted with permission of Brooks/Cole, a division of Thomson Learning: www.thomsonrights.com. Fax 800 730-2215.

When plane-polarized light is passed through a solution of one enantiomer, the light is rotated in a clockwise direction. The other enantiomer rotates the plane-polarized light to the same extent, but in a counterclockwise direction. The enantiomer that *rotates the light in a clockwise direction* is called the **dextrorotatory** or (+) isomer, and the enantiomer that *rotates the light counterclockwise* is the **levorotatory** or (−) isomer. These terms come from the Latin words *dexter* (right) and *laevus* (left).

As we will see in the chapters that follow, living things typically produce either the (+) or the (−) enantiomer of a given chiral molecule, but not both. When chiral molecules are produced in laboratories, however, *a 50:50 mixture of enantiomers* (a **racemic mixture**) often results. In a racemic mixture the rotation of plane-polarized light due to one enantiomer cancels that of the other enantiomer and no net rotation of light occurs.

In addition to their different abilities at rotating plane-polarized light, enantiomers affect biological systems differently. In some cases, enantiomers have different odors or flavors. The (−) enantiomer of carvone, for example, gives caraway its odor and flavor, while the (+) enantiomer is responsible for the odor and flavor of spearmint (Figure 10.23*a*). The (−) enantiomer of the amino acid asparagine tastes sweet, while the (+) enantiomer tastes bitter.

FIGURE | 10.23

Stereoisomers and biochemistry
(*a*) The enantiomers of carvone and asparagine have very different flavors. (*b*) One stereoisomer of a drug may have a very different biological effect than its enantiomer. The (+) enantiomers pictured here are either more effective drugs or are safer to use than their (−) counterparts.

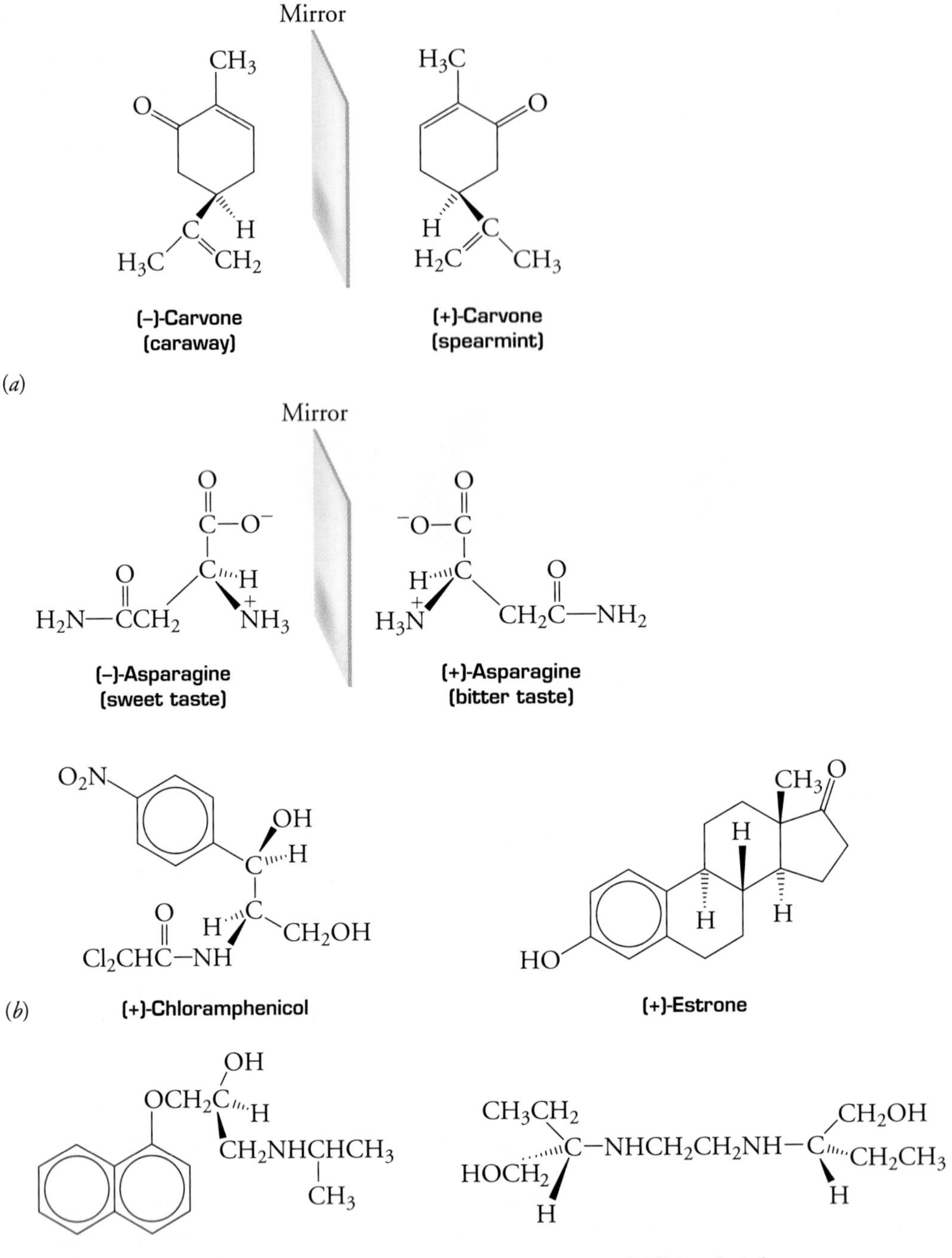

(*a*)

Mirror

(−)-Carvone (caraway)

(+)-Carvone (spearmint)

Mirror

(−)-Asparagine (sweet taste)

(+)-Asparagine (bitter taste)

(*b*)

(+)-Chloramphenicol

(+)-Estrone

(+)-Propranolol

(+)-Ethambutol

Drug effectiveness can also be a function of the particular enantiomer that is present. The (+) enantiomer of chloramphenicol is used as an antibiotic, while the (−) enantiomer has no antibiotic properties whatsoever (Figure 10.23*b*). Similarly, (+)-estrone has hormonal activity, but (−)-estrone does not. When it comes to treating irregular heartbeat (+)-propranolol is 100 times more effective than the (−) enantiomer and, while (−)-propranolol is effective as a contraceptive, the (+) enantiomer is not. (+)-Ethambutol can be used to treat tuberculosis, but not the (−) enantiomer—it causes blindness. The varied biological responses to enantiomers are important in the biochemistry of carbohydrates (Chapter 12) and amino acids, peptides, and proteins (Chapter 13).

Molecules with More than One Chiral Carbon Atom

As the number of chiral carbon atoms in a molecule increases, so does the number of stereoisomers that can exist. A general formula for predicting the maximum number of stereoisomers possible for a molecule is

$$\text{Number of stereoisomers} = 2^n$$

where n is the number of chiral carbon atoms.

2-Bromo-3-chlorobutane (Figure 10.24) has two chiral carbons and can exist as four different stereoisomers ($2^2 = 4$). As can be seen in this figure, the four stereoiosmers consist of two pairs of enantiomers, molecules A and B and molecules C and D. Being enantiomers, A and B are nonsuperimposable mirror images that have identical properties. C and D are also a set of nonsuperimposable mirror images with identical properties, but not necessarily the same properties as A and B. Molecules A and B, for example, rotate plane polarized light to a different extent than do molecules C and D.

Stereoisomers of a molecule that are not enantiomers are called **diastereomers**. In Figure 10.24, molecule A is an enantiomer of B and is a diastereomer of molecules C and D. The distinction between enantiomers and diasteromers is important in biochemistry, as we will see in Chapter 12, when we consider the chemistry of sugars and other carbohydrates.

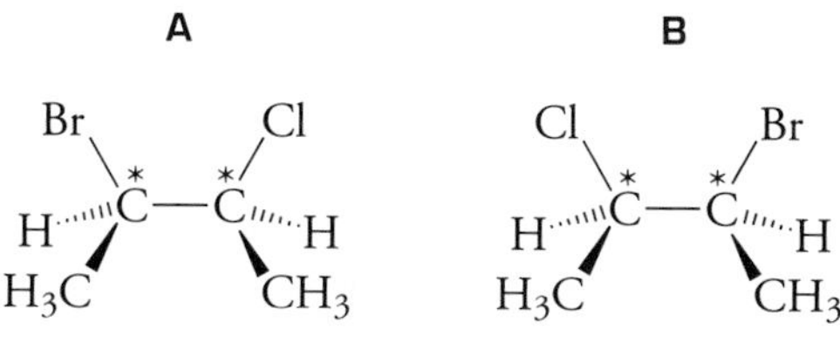

A pair of enantiomers

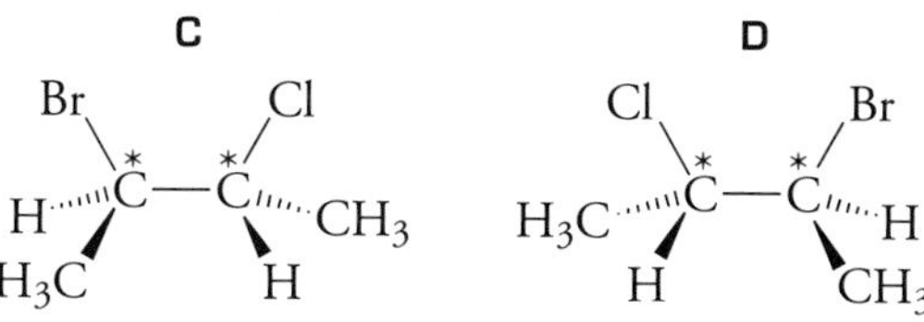

A pair of enantiomers

FIGURE 10.24

Enantiomers and diastereomers
The four stereoisomers of 2-bromo-3-chlorobutane can be grouped as two pairs of enantiomers. The diastereomers of this molecule are those stereoisomers that are not mirror images—molecule D is a diastereomer of molecules A and B.

SAMPLE PROBLEM 10.7

Counting stereoisomers

a. How many chiral carbon atoms does 3-bromo-4-methylhexane have?
b. How many stereoisomers are possible for this molecule?

Strategy
To identify the chiral carbon atoms in a molecule, you should look for carbon atoms that are attached to four different groups. Remember that $—CH_3$ and $—CH_2CH_3$ are different groups.

Solution

a. Two chiral carbon atoms. Carbon 3 is attached to $-Br$, $-H$, $-CH_2CH_3$, and $-CH(CH_3)CH_2CH_3$. Carbon 4 is attached to $-CH_3$, $-H$, $-CH_2CH_3$, and $-CH(Br)CH_2CH_3$.

$$CH_3CH_2\overset{*}{C}H(Br)\overset{*}{C}H(CH_3)CH_2CH_3$$

b. 4 $(2^2 = 4)$

PRACTICE PROBLEM 10.7

a. Draw the four stereoisomers of 3-bromo-4-methylhexane (Sample Problem 10.7).
b. Label pairs of enantiomers and identify diastereomers.

HEALTH Link — Adrenaline and Related Compounds

If you are startled, frightened, or angry, your adrenal gland releases the hormone epinephrine, also known as adrenaline (Figure 10.25). Although this hormone causes your pulse rate and blood pressure to climb, one of its major functions is to produce a rapid jump in the blood concentration of glucose, the main fuel supply for biochemical reactions. One of the immediate effects of epinephrine production is a burst of energy that comes from increased levels of blood glucose.

The drug called amphetamine is structurally similar to epinephrine. Like epinephrine, it increases pulse rate, blood pressure, and acts as a stimulant. Amphetamine has also been used as a weight reduction drug and to treat depression. A related compound, methamphetamine, has similar but more potent effects. Amphetamine and methamphetamine are relatively easy to synthesize, and they have been widely produced and sold as illegal street drugs.

Amphetamine and methamphetamine each have one chiral carbon atom (Section 10.9), and each exists as a pair of enantiomers. The (+) enantiomer of amphetamine, known as Dexedrine, is more biologically active than is the (−) enantiomer. Similarly, the (+) enantiomer of methamphetamine, called methedrine, is more active than the (−) isomer.

Phenylpropanolamine, a stimulant that is used in over-the-counter (nonprescription) appetite suppressants, and ephedrine, a nasal decongestant, are two other structural relatives of epinephrine.

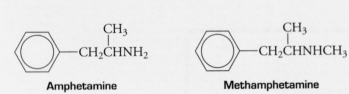

$(HO)_2C_6H_3-CH(OH)CH_2NHCH_3$

Epinephrine

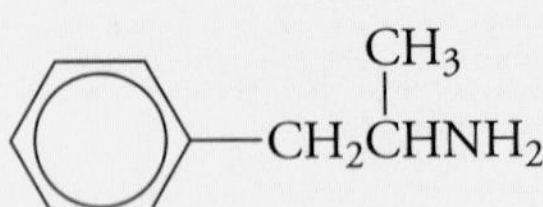

$C_6H_5-CH_2CH(CH_3)NH_2$

Amphetamine

$C_6H_5-CH_2CH(CH_3)NHCH_3$

Methamphetamine

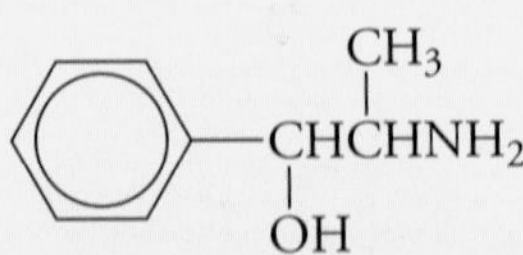

$C_6H_5-CH(OH)CH(CH_3)NH_2$

Phenylpropanolamine

$C_6H_5-CH(OH)CH(CH_3)NHCH_3$

Ephedrine

FIGURE 10.25

Epinephrine and some related compounds

Allergies result from an extreme sensitivity to dust, pollen, foods, or other allergens. If you are exposed to an allergen, your immune system responds by releasing histamine, prostaglandins and leukotrienes (Section 8.6), and other compounds. The effects of histamine include headaches, itching, watery eyes, a runny nose, and bronchoconstriction (narrowing of the airways to the lungs). These responses to histamine begin when histamine binds to particular cell receptors.

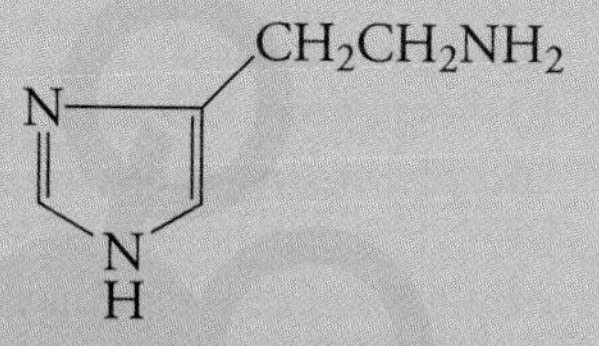

Histamine

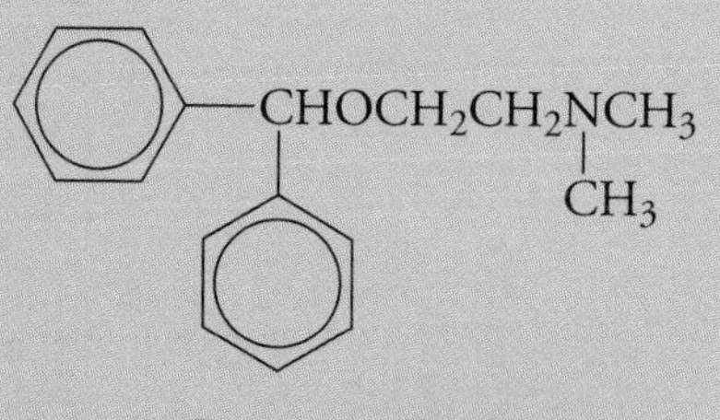

Diphenhydramine

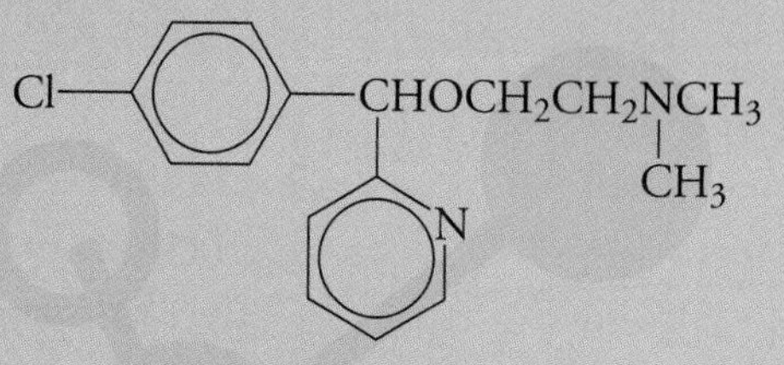

Chlorpheniramine

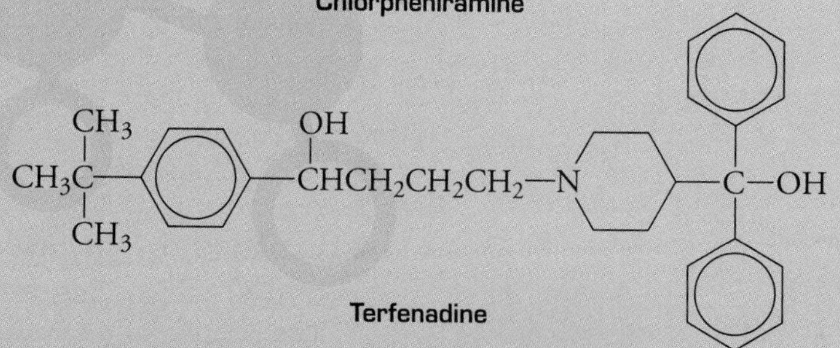

Terfenadine

Diphenhydramine (Benadryl) and chlorpheniramine (Chlor-Trimeton) are two of the many antihistamines currently available. These and other antihistamines act by attaching to histamine cell receptors, which prevents histamine from doing so. The drowsiness that accompanies the use of some antihistamines is the result of these drugs also blocking a particular histamine receptor found in the brain. Antihistamines that do not cause drowsiness, including terfenadine (Seldane), do not interact with this particular type of receptor.

summary of objectives

1 List the intermolecular forces that attract carboxylic acids, phenols, or amine molecules to one another and describe the effect that these forces have on boiling point.

Because they are able to form hydrogen bonds with one another, carboxylic acids have higher boiling points than do many other organic compounds of similar molecular weight. Increased size leads to an increase in the boiling point (stronger London forces). Like carboxylic acids, **phenols** can form hydrogen bonds with one another. This results in relatively high boiling points. All amines are polar due to the presence of polar covalent bonds, but only **1°** and **2°** amines can form hydrogen bonds with like amines. As a result, 1° and 2° amines tend to have higher boiling points than **3°** amines of similar molecular weight. **Quaternary (4°)** ammonium ions are ionic and have high melting and boiling points.

2 Describe what takes place when carboxylic acids and phenols react with water or with strong bases. Explain how the conjugate bases of carboxylic acids and phenols are named.

In the presence of water, carboxylic acids and phenols react to form **carboxylate ions** and **phenoxide ions**, respectively. Being weak acids, the concentration of conjugate base tends to be low at equilibrium. Reacting carboxylic acids and phenols with strong bases converts them entirely into their conjugate base form. The sodium salts (ionic compounds) formed from the conjugate bases of carboxylic acids and phenols are more soluble in water than are their acid counterparts. The conjugate base of a carboxylic acid is named by changing the ending on the name of the carboxylic acid from "ic acid" to "ate ion." The conjugate base of a phenol is named by changing the ending on the name of the phenol from "ol" to "oxide ion."

3 Describe what takes place when amines react with water or with strong acids. Explain how the conjugate acids of amines are named.

Amines are weak bases that, in the presence of water, react to form low concentrations of conjugate acid and OH^-. In the presence of strong acids, amines are converted entirely into their salt form. The salts formed from the conjugate acids of amines are more soluble in water than are their base counterparts. The conjugate acid of an amine is named by changing the ending on the name of the amine from "ine" to "onium ion."

4 Explain how carboxylic acids can be converted into esters and amides. Describe how esters and amides are named.

The reaction of a carboxylic acid and an alcohol, under acidic conditions, produces an ester. The reaction of a carboxylic acid with ammonia or an amine produces a carboxylate ion–ammonium ion salt. An **amide** is formed when this salt is heated. An ester is named by giving the name of the alkyl group contained in the alcohol residue, followed by the name of the carboxylic acid present in the carboxylic acid residue, with the ending on its name changed from "ic acid" to "ate." An amide that contains an ammonia residue is named by changing the ending of the name of the carboxylic acid in the carboxylic acid residue from "oic acid" (IUPAC names) or "ic acid" (common names) to "amide." The name is derived in the same way if an amine residue is present in an amide, but the name begins with the name of each substituent attached to the nitrogen, with *N* used to indicate its position.

5 Explain the difference in the products obtained when an ester is hydrolyzed under acidic conditions and under basic conditions. Describe the products formed when an amide is hydrolyzed under acidic conditions.

When hydrolyzed under acidic conditions, an ester yields a carboxylic acid plus an alcohol, while under basic conditions, a carboxylate ion plus an alcohol are produced. Under acidic conditions amides are hydrolyzed to produce carboxylic acids and conjugate acids of ammonia or amines.

6 Explain the terms chiral molecule and chiral carbon atom, distinguish between enantiomers and diastereomers, and define the terms dextrorotatory and levorotatory.

Chiral molecules are not superimposable on their mirror image. Most chiral molecules contain one or more **chiral carbon atoms**, carbon atoms attached to four different atoms or groups of atoms. Stereoisomers of a molecule that are nonsuperimposable mirror images are called **enantiomers** and those that are not mirror images are called **diastereomers**. A molecule with *n* chiral carbon atoms can have up to 2^n stereoisomers. **Plane-polarized light** can be used to distinguish enantiomers—one enantiomer rotates the light clockwise (**dextrorotatory** or +) and the other rotates it counterclockwise (**levorotatory** or −).

summary of reactions

Section 10.3

Carboxylic acids as weak acids

$$CH_3\overset{O}{\overset{\|}{C}}{-}OH + H_2O \rightleftharpoons CH_3\overset{O}{\overset{\|}{C}}{-}O^- + H_3O^+$$

$$CH_3\overset{O}{\overset{\|}{C}}{-}OH + NaOH \longrightarrow CH_3\overset{O}{\overset{\|}{C}}{-}O^-Na^+ + H_2O$$

Phenols as weak acids

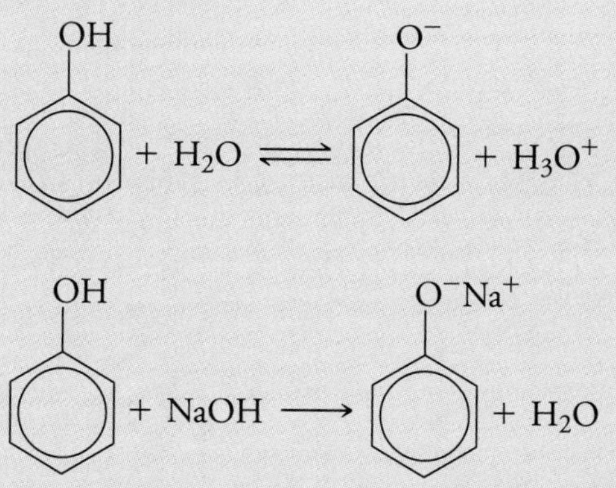

$$C_6H_5OH + H_2O \rightleftharpoons C_6H_5O^- + H_3O^+$$

$$C_6H_5OH + NaOH \longrightarrow C_6H_5O^-Na^+ + H_2O$$

Section 10.4

Ester formation

$$CH_3\overset{O}{\overset{\|}{C}}{-}OH + HOCH_2CH_3 \overset{H^+}{\rightleftharpoons} CH_3\overset{O}{\overset{\|}{C}}{-}OCH_2CH_3 + H_2O$$

Ester hydrolysis

$$CH_3\overset{O}{\overset{\|}{C}}{-}OCH_2CH_3 + H_2O \overset{H^+}{\rightleftharpoons} CH_3\overset{O}{\overset{\|}{C}}{-}OH + HOCH_2CH_3$$

$$CH_3\overset{O}{\overset{\|}{C}}{-}OCH_2CH_3 + H_2O \xrightarrow{OH^-} CH_3\overset{O}{\overset{\|}{C}}{-}O^- + HOCH_2CH_3$$

Decarboxylation

$$CH_3\overset{O}{\overset{\|}{C}}CH_2\overset{O}{\overset{\|}{C}}{-}OH \xrightarrow{\text{heat}} CH_3\overset{O}{\overset{\|}{C}}CH_3 + CO_2$$

$$CH_3CH_2\overset{O}{\overset{\|}{C}}{-}\overset{O}{\overset{\|}{C}}{-}OH \xrightarrow{\text{enzyme}} CH_3CH_2\overset{O}{\overset{\|}{C}}{-}H + CO_2$$

Section 10.5

Phenol oxidation

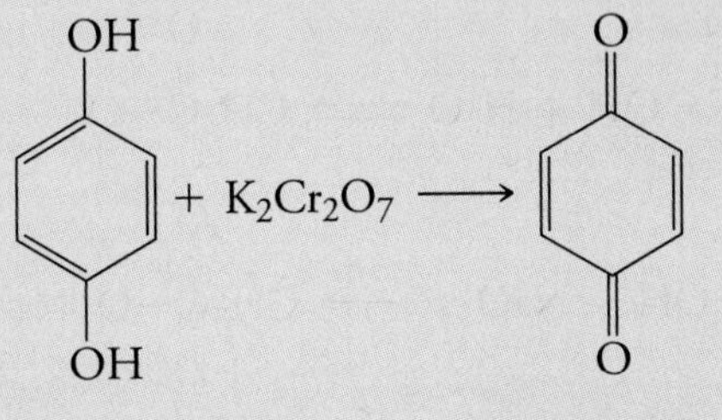

$$\text{HO}{-}C_6H_4{-}\text{OH (hydroquinone)} + K_2Cr_2O_7 \longrightarrow \text{O}{=}C_6H_4{=}\text{O (}p\text{-benzoquinone)}$$

Section 10.7

Amines as weak bases

$$CH_3NH_2 + H_2O \rightleftharpoons CH_3NH_3^+ + OH^-$$

$$CH_3NH_2 + HCl \longrightarrow CH_3NH_3^+Cl^-$$

Section 10.8

Amides from carboxylic acids

$$CH_3\overset{O}{\overset{\|}{C}}{-}OH + NH_3 \longrightarrow CH_3\overset{O}{\overset{\|}{C}}{-}O^-NH_4^+ \xrightarrow{\text{heat}} CH_3\overset{O}{\overset{\|}{C}}{-}NH_2 + H_2O$$

Amide hydrolysis

$$CH_3\overset{O}{\overset{\|}{C}}{-}NH_2 + H_2O \xrightarrow[\text{heat}]{H^+} CH_3\overset{O}{\overset{\|}{C}}{-}OH + NH_4^+$$

END OF CHAPTER PROBLEMS

Answers to problems whose numbers are printed in color are given in Appendix D. More challenging questions are marked with an asterisk. Problems within colored rules are paired. **ILW** = Interactive Learning Ware solution is available at *www.wiley.com/college/raymond*.

10.1 Carboxylic Acids

10.1 Match each structure to the correct IUPAC name: 3-methylbutanoic acid, 2-methylpentanoic acid, 2-methylbutanoic acid.

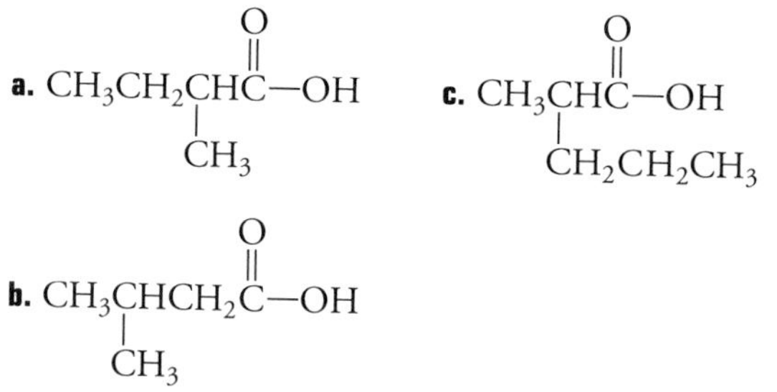

a. $CH_3CH_2CH(CH_3)C(=O)—OH$

b. $CH_3CH(CH_3)CH_2C(=O)—OH$

c. $CH_3CH(CH_2CH_2CH_3)C(=O)—OH$

10.2 Match each structure to the correct IUPAC name: *o*-bromobenzoic acid, *m*-bromobenzoic acid, *p*-bromobenzoic acid.

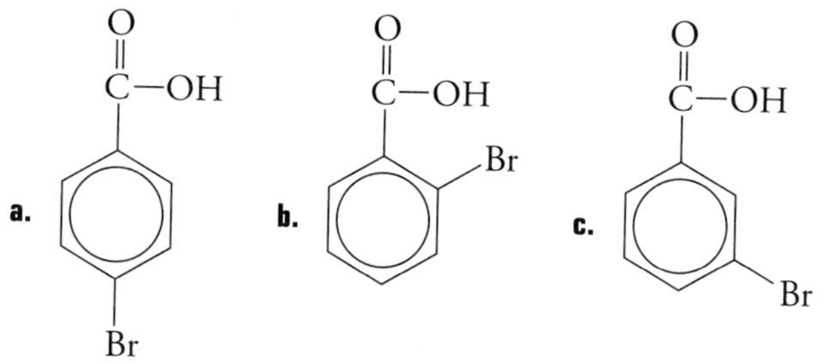

a. benzene ring with C(=O)—OH and Br para

b. benzene ring with C(=O)—OH and Br ortho

c. benzene ring with C(=O)—OH and Br meta

10.3 Draw each carboxylic acid.
a. octanoic acid
b. 3,3-dimethylheptanoic acid
c. 3-isopropylhexanoic acid
d. 2-bromopentanoic acid

10.4 Draw each carboxylic acid.
a. 4-methylheptanoic acid
b. 2-propylhexanoic acid
c. 3-chlorobutanoic acid

10.5 Acetic acid (CH_3CO_2H) and propyl alcohol ($CH_3CH_2CH_2OH$) have the same molecular weight, but their boiling points differ by about 20°C (acetic acid, 118°C; propyl alcohol, 97.4°C). Account for this difference.

10.6 Butanoic acid is more soluble in water than hexanoic acid. Account for this difference.

10.2 Phenols

10.7 Draw each phenol.
a. 2,4-dimethylphenol
b. 3,5-dimethylphenol
c. 4-propylphenol
d. *m*-propylphenol

10.8 Draw each phenol.
a. *p*-*t*-butylphenol
b. 2,6-dichlorophenol
c. 4-isopropylphenol
d. *o*-isopropylphenol

10.9 If you come into contact with urushiol (Figure 10.3*b*), the phenol partially responsible for the itching and burning caused by poison ivy, you cannot wash it off with water. Explain.

10.10 Erbstatin has shown antitumor activity in combating skin, breast, and esophageal cancer.

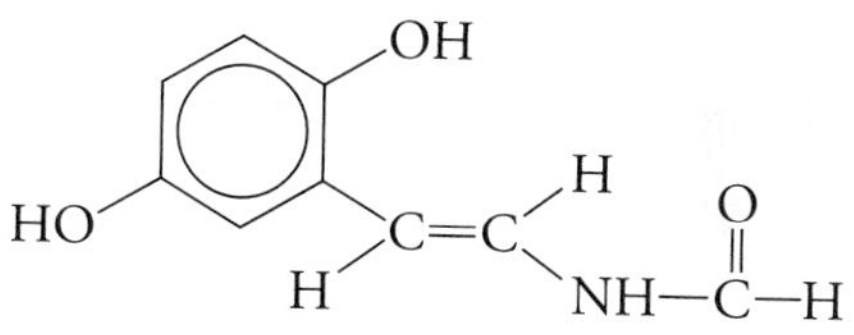

HO—(benzene ring, OH)—CH=CH—NH—C(=O)—H

Erbstatin

a. Is erbstatin a catechol-, a resorcinol-, or a hydroquinone-containing phenol?
b. Is the nitrogen atom present in this compound part of an amine functional group?
c. Which stereoisomer (*cis* or *trans*) is the carbon–carbon double bond?

10.11 Hexylresorcinol (4-hexyl-3-hydroxyphenol) is an antiseptic used in some mouthwashes and throat lozenges. Draw this compound.

10.12 Is malvidin (Figure 10.1b) a catechol-, a resorcinol-, or a hydroquinone-containing compound?

10.13 Is mordant red (see the last structural formulas in Section 10.5) a catechol-, a resorcinol-, or a hydroquinone-containing compound?

10.3 Carboxylic Acids and Phenols as Weak Organic Acids

10.14 Draw octanoic acid and its conjugate base.

10.15 Draw *m*-chlorobenzoic acid and its conjugate base.

10.16 Draw the conjugate base of each carboxylic acid.
a. propanoic acid
b. hexanoic acid
c. 3-chloropentanoic acid

10.17 Draw the conjugate base of each carboxylic acid.
a. 3-methylhexanoic acid
b. formic acid
c. 3,5-dibromobenzoic acid

10.18 Name each of the conjugate bases in your answer to Problem 10.16.

10.19 Name each of the conjugate bases in your answer to Problem 10.17.

10.20 Draw and name the conjugate base of each carboxylic acid reactant shown in Figure 10.9.

10.21 Draw the organic product of each reaction.

a. 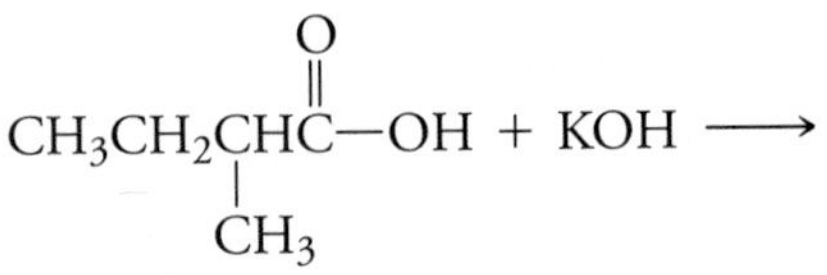 $CH_3CH_2CH(CH_3)C(=O)-OH + KOH \longrightarrow$

b. 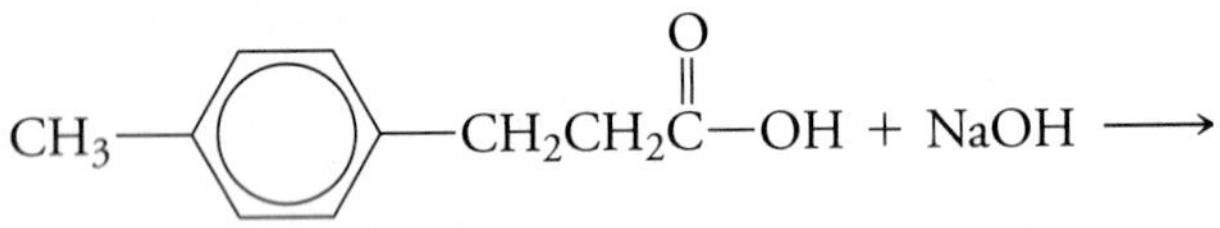 $CH_3-C_6H_4-CH_2CH_2C(=O)-OH + NaOH \longrightarrow$

10.22 Draw the organic product of each reaction.

a. 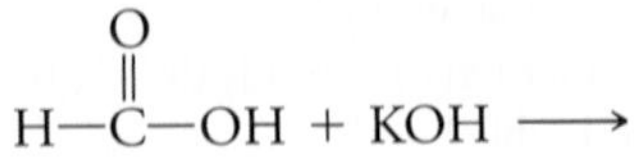 $H-C(=O)-OH + KOH \longrightarrow$

b. 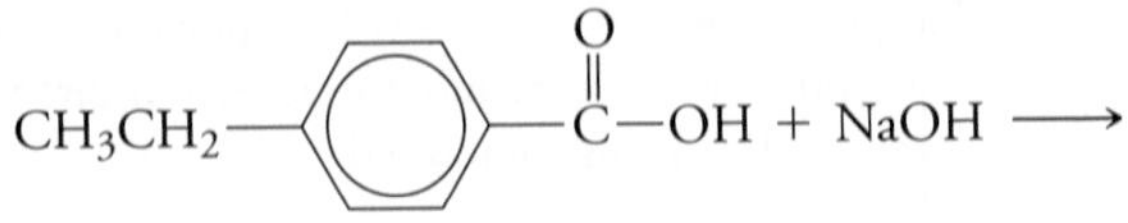 $CH_3CH_2-C_6H_4-C(=O)-OH + NaOH \longrightarrow$

10.23 Dinoseb, an herbicide and insecticide, is sold as a water-soluble ammonium salt. Draw this ammonium salt, which is produced by reacting Dinoseb with ammonia.

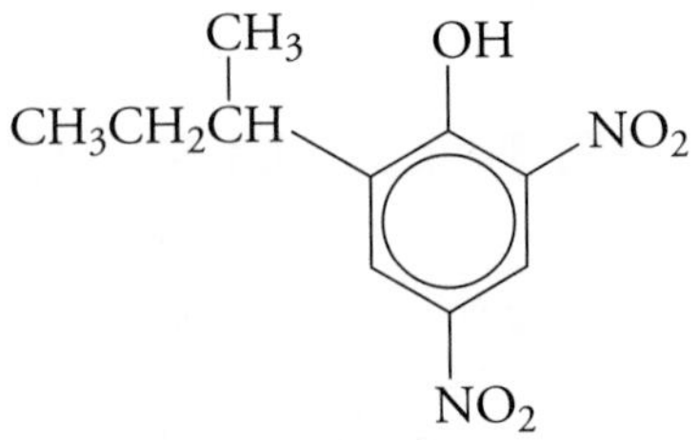

$CH_3CH_2CH(CH_3)$, OH, NO_2, NO_2

Dinoseb

*__10.24__ In 1875 sodium salicylate, the sodium salt of salicylic acid, was introduced as an analgesic (pain killer).

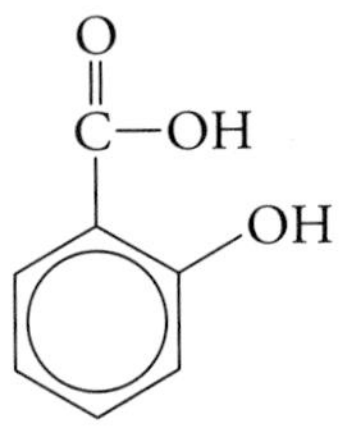

C(=O)—OH, OH

Salicyclic acid

a. Which functional group of salicylic acid is the most acidic, the phenol group or the carboxyl group?
b. Draw the ionic compound sodium salicylate. (Hint: Only one H^+ is removed from salicylic acid.)

*__10.25__ Gentisic acid (Figure 10.3*b*) can be used as an anti-rheumatism drug.
a. Which functional group of gentisic acid is the most acidic, one of the phenol groups or the carboxyl group?
b. Draw the ionic compound potassium gentisate. (Hint: Only one H^+ is removed from gentisic acid.)

10.26 *p*-Ethylphenoxide ion is the conjugate base of *p*-ethylphenol. Draw both compounds.

10.27 2,4-Dichlorophenoxide ion is the conjugate base of 2,4-dichlorophenol. Draw both compounds.

10.28 Draw the salt that forms when eugenol (Figure 10.1*b*) reacts with NaOH.

10.29 **a.** Draw 4-chloropentanoic acid.
b. Draw and name the conjugate base of 4-chloropentanoic acid.
c. Which predominates in water at pH 7, 4-chloropentanoic acid or its conjugate base?

10.30 a. Draw 2,4-dichlorophenol.

b. Draw and name the conjugate base of 2,4-dichlorophenol.

c. Which predominates in water at pH 7, 2,4-dichlorophenol or its conjugate base?

10.4 Other Reactions of Carboxylic Acids

10.31 Draw the organic product of each reaction.

a. 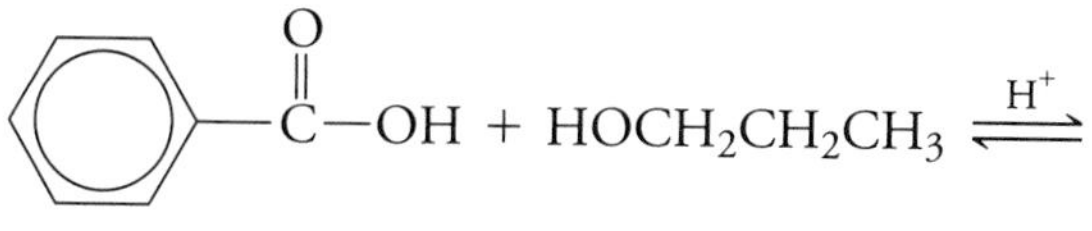 $C_6H_5-\overset{O}{\overset{\|}{C}}-OH + HOCH_2CH_2CH_3 \xrightleftharpoons{H^+}$

b. 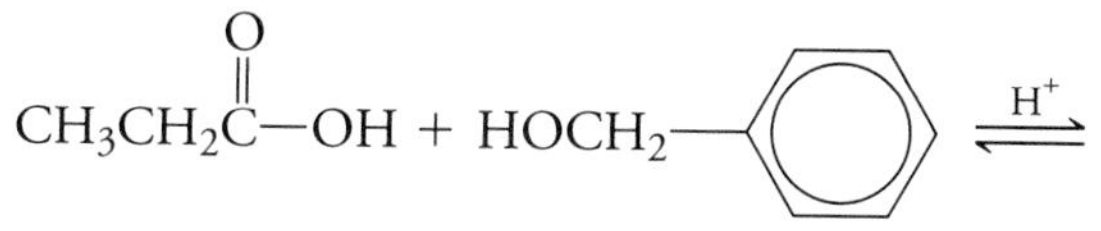 $CH_3CH_2\overset{O}{\overset{\|}{C}}-OH + HOCH_2-C_6H_5 \xrightleftharpoons{H^+}$

10.32 Draw the organic product of each reaction.

a. $CH_3\underset{CH_3}{\underset{|}{C}}HCH_2\overset{O}{\overset{\|}{C}}-OH + HOCH_3 \xrightleftharpoons{H^+}$

b. 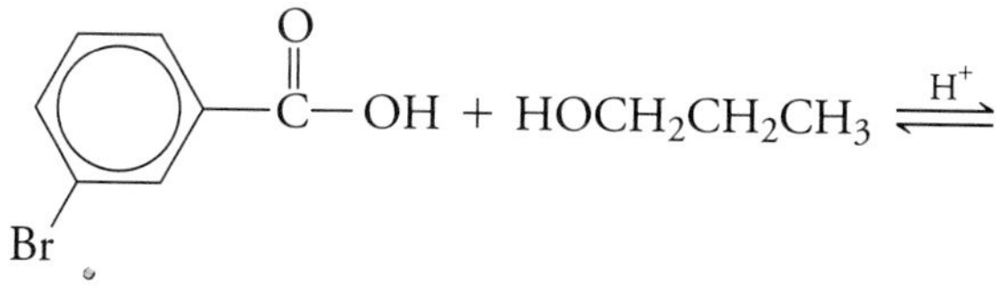 (3-bromophenyl)$-\overset{O}{\overset{\|}{C}}-OH + HOCH_2CH_2CH_3 \xrightleftharpoons{H^+}$ (Br on ring)

10.33 Name the organic products in Problem 10.31. Hint: $-CH_2C_6H_5$ is a benzyl group.

10.34 Name the organic products in Problem 10.32.

10.35 Draw and name the ester formed when *p*-ethylbenzoic acid reacts with isopropyl alcohol, $CH_3CH(OH)CH_3$, in the presence of H^+.

10.36 Draw and name the ester formed when benzoic acid reacts with pentyl alcohol, $HOCH_2CH_2CH_2CH_2CH_3$, in the presence of H^+.

10.37 Draw the products obtained from each reaction.

a. $CH_3\underset{CH_3}{\underset{|}{C}}HCH_2CH_2\overset{O}{\overset{\|}{C}}-OCH_3 + H_2O \xrightleftharpoons{H^+}$

b. 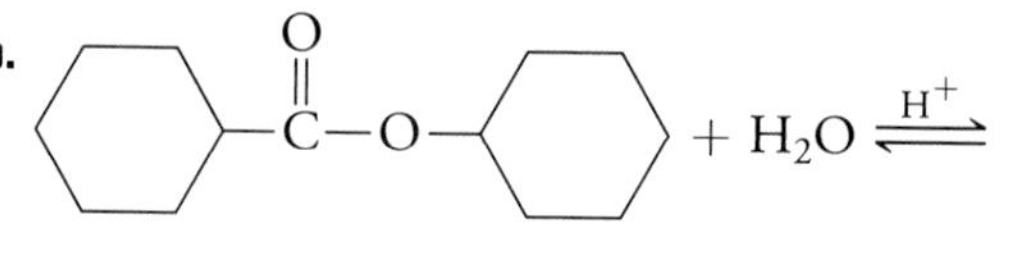 $C_6H_{11}-\overset{O}{\overset{\|}{C}}-O-C_6H_{11} + H_2O \xrightleftharpoons{H^+}$

10.38 Draw the products obtained from each reaction.

a. $CH_3\underset{CH_3}{\underset{|}{C}}HCH_2CH_2\overset{O}{\overset{\|}{C}}-OCH_3 + H_2O \xrightarrow{OH^-}$

b. 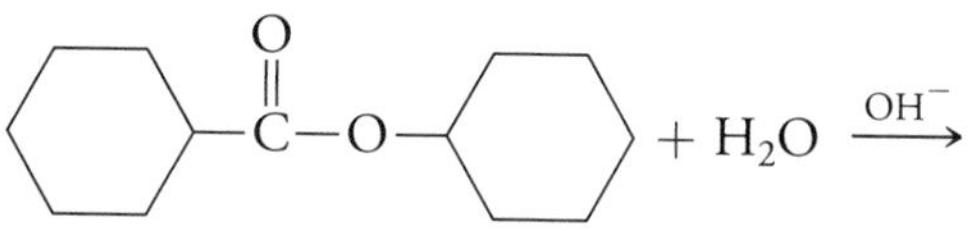 $C_6H_{11}-\overset{O}{\overset{\|}{C}}-O-C_6H_{11} + H_2O \xrightarrow{OH^-}$

10.39 Draw the products obtained when

a. cyclopentyl acetate reacts with H_2O in the presence of H^+.

b. cyclopentyl acetate reacts with H_2O in the presence of OH^-.

10.40 Draw the products obtained when

a. methyl 3,3-dimethyl hexanoate reacts with H_2O in the presence of H^+.

b. methyl 3,3-dimethyl hexanoate reacts with H_2O in the presence of OH^-.

10.41 Draw the organic product formed when each β-keto acid is decarboxylated.

a. 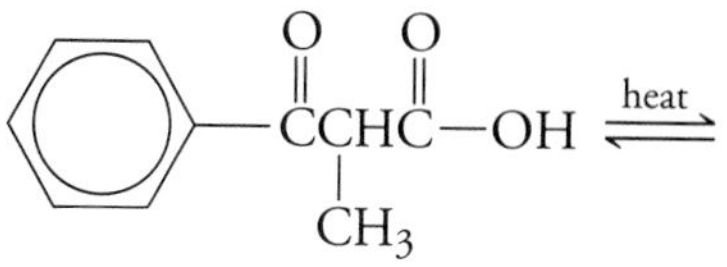 $C_6H_5-\overset{O}{\overset{\|}{C}}\underset{CH_3}{\underset{|}{C}}H\overset{O}{\overset{\|}{C}}-OH \xrightleftharpoons{heat}$

b. $CH_3\underset{CH_3}{\underset{|}{C}}H\overset{O}{\overset{\|}{C}}CH_2\overset{O}{\overset{\|}{C}}-OH \xrightleftharpoons{heat}$

10.42 Draw the organic product formed when each β-keto acid is decarboxylated.

a. $H-\overset{O}{\overset{\|}{C}}CH_2\overset{O}{\overset{\|}{C}}-OH \xrightarrow{heat}$

b. $CH_3CH_2CH_2\overset{O}{\overset{\|}{C}}CH_2\overset{O}{\overset{\|}{C}}-OH \xrightarrow{heat}$

10.43 Draw the organic product formed when the α-keto acid undergoes an enzyme-catalyzed decarboxylation.

$H_3C-C_6H_4-CH_2\overset{O}{\overset{\|}{C}}-\overset{O}{\overset{\|}{C}}-OH$

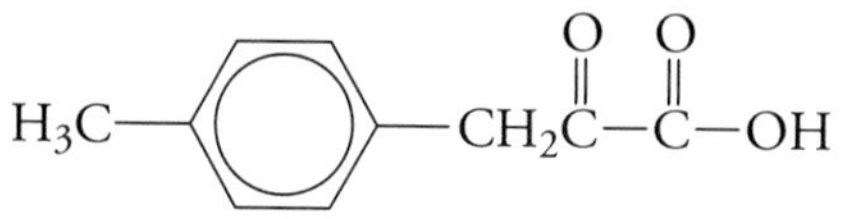

10.44 Draw the organic product formed when the α-keto acid undergoes an enzyme-catalyzed decarboxylation.

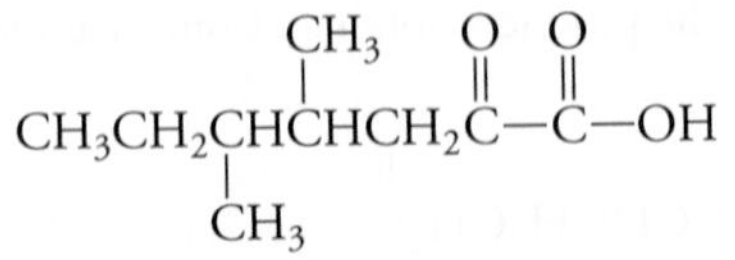
CH3 O O
CH3CH2CHCHCH2C—C—OH
CH3

10.5 Oxidation of Phenols

10.45 Geranylhydroquinone is an experimental radioprotective drug, one that protects against the harmful effects of radiation. Draw the organic compound formed when geranylhydroquinone is treated with $K_2Cr_2O_7$.

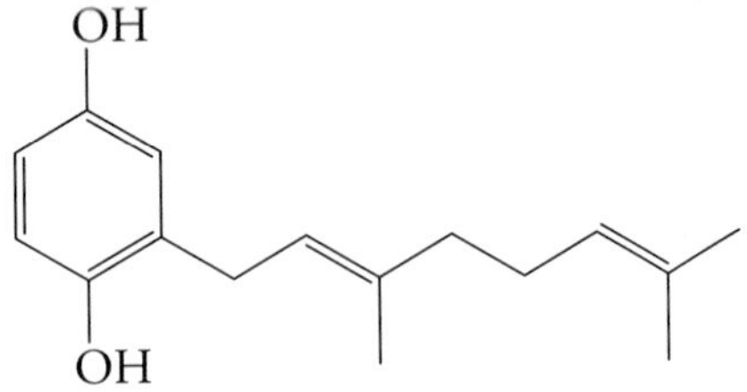
OH
OH

Geranylhydroquinone

10.46 Anthraquinone is used in the manufacture of some dyes and is also used by farmers to make seeds distasteful to birds. Draw the hydroquinone-containing compound from which anthraquinone can be produced by oxidation.

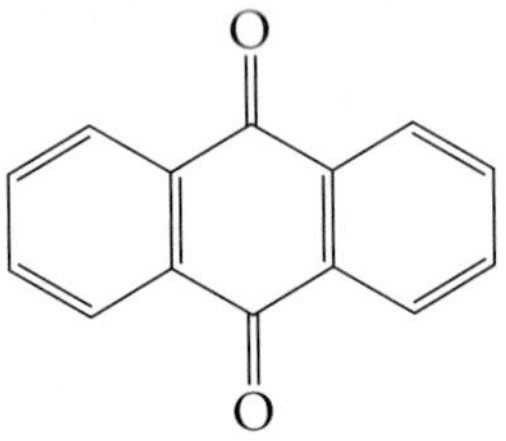
O
O

Anthraquinone

10.6 Amines

10.47 Label each amine nitrogen atom in nicotine as being 1°, 2°, or 3° (Figure 10.1*c*).

10.48 Label each nitrogen atom in purine as being 1°, 2°, or 3° (Figure 10.12*a*).

10.49 Give the IUPAC name of each amine.

a. CH_3NH_2

b.
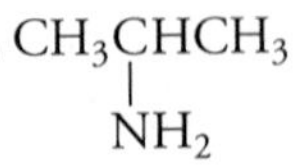
CH3CHCH3
NH2

c.
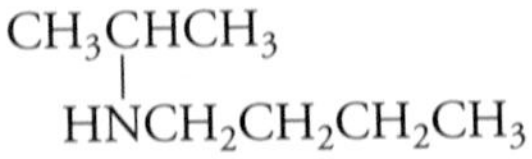
CH3CHCH3
HNCH2CH2CH2CH3

d.
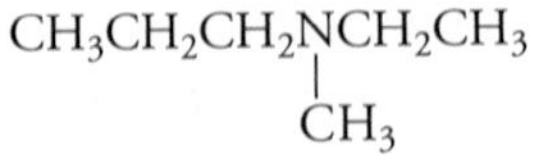
CH3CH2CH2NCH2CH3
CH3

10.50 Give the IUPAC name of each amine.

a. CH_3NHCH_3

b.
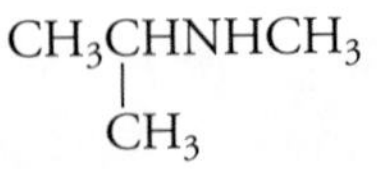
CH3CHNHCH3
CH3

c.
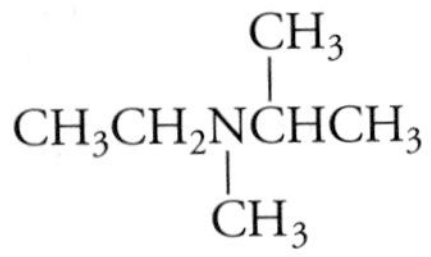
CH3
CH3CH2NCHCH3
CH3

d.
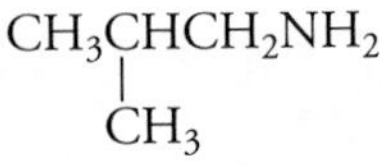
CH3CHCH2NH2
CH3

10.51 Give the common name of each amine in the Problem 10.49.

10.52 Give the common name of each amine in the Problem 10.50.

10.53 Identify each compound as a pyridine, a pyrimidine, or a purine (Figure 10.12*a*).

a.
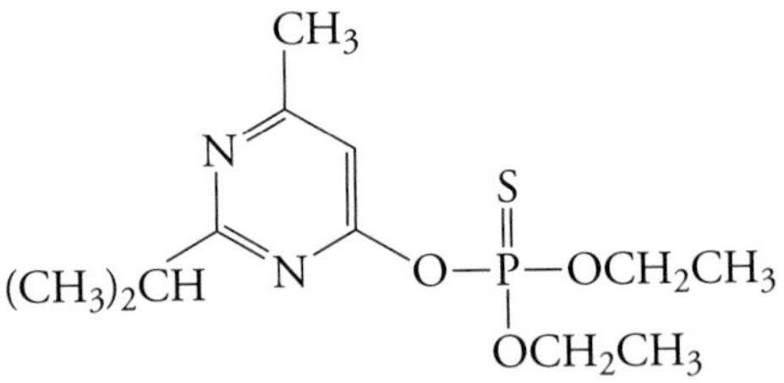
CH3
N
N
(CH3)2CH
S
O—P—OCH2CH3
OCH2CH3

Diazinon (an insecticide)

b.
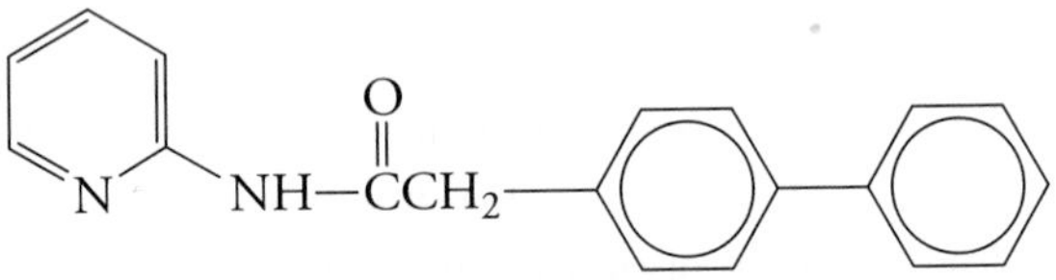
N
O
NH—CCH2

Difenpiramide (an anti-inflammatory drug)

c.
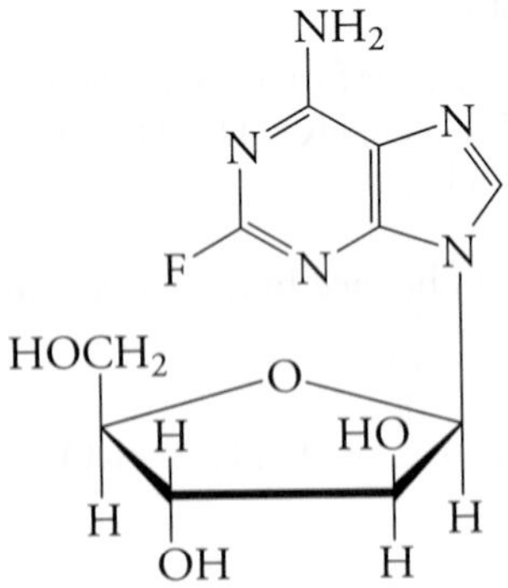
NH2
N
N
N
N
F
HOCH2
O
H
HO
H
H
OH
H

Fludarabine (an anti-cancer drug)

10.54 Account for the fact that ethylamine is more water soluble than hexylamine.

10.55 Account for the fact that propylamine is more water soluble than trimethylamine.

10.56 Draw each compound.
- **a.** 1-pentanamine
- **b.** *N*-isopropyl-1-pentanamine
- **c.** *N*-ethyl-*N*-methyl-2-hexanamine
- **d.** dimethyldipropylammonium ion

10.57 Draw each compound.
- **a.** 2-hexanamine
- **b.** *N*-methyl-3-pentanamine
- **c.** *N*,*N*-dimethyl-1-butanamine
- **d.** tetramethylammonium ion

10.7 Amines As Weak Organic Bases

10.58 Draw and name the conjugate acid of
- **a.** ammonia
- **b.** propylamine
- **c.** methylethylamine
- **d.** triethylamine

10.59 Draw and name the conjugate acid of
- **a.** butylamine
- **b.** dipropylamine
- **c.** trimethylamine
- **d.** *t*-butylamine

10.60 Anabasine hydrochloride, the salt of anabasine (an amine present in tobacco), is sold as an insecticide. Draw this salt. (Hint: The nitrogen atom in the pyridine ring is not basic.)

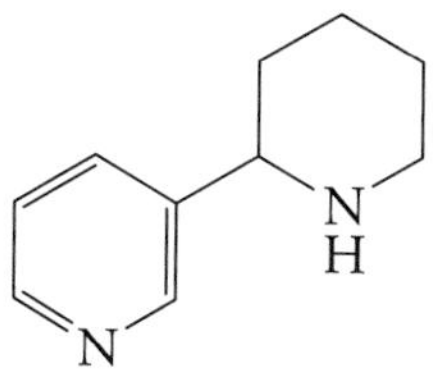

N
H
N

Anabasine

10.61 In the early 1980s a street drug sold as synthetic heroin appeared in southern California. This drug contained an impurity called MPTP, which caused irreversible symptoms of Parkinson's disease (including immobility, slurred speech, and tremors). The MPTP existed as a hydrochloride salt. Draw this salt.

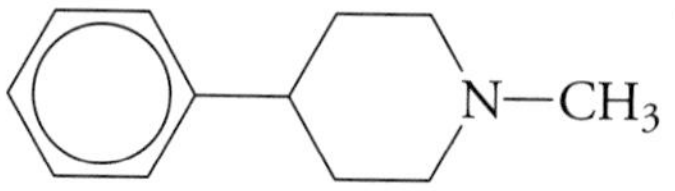

N—CH_3

MPTP

10.8 Amides

10.62 Caffeine (Figure 10.1*c*) has four nitrogen atoms, two of which are amines, and the other two of which are amides. Label the amide and amine nitrogen atoms in this compound.

10.63 Difenpiramide (Problem 10.53) has two nitrogen atoms (one amine and one amide). Label the amide and amine nitrogen atoms in this compound.

10.64 Draw the amide that will be produced by each reaction.

a. 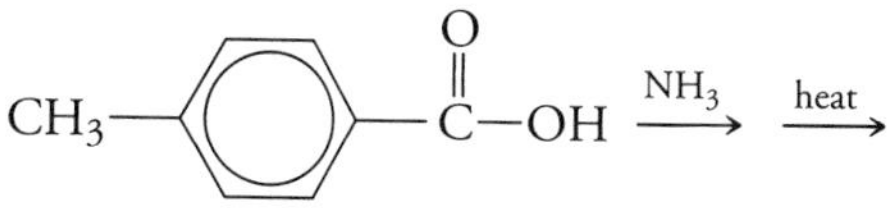 $CH_3-C_6H_4-\overset{O}{\overset{\|}{C}}-OH \xrightarrow{NH_3} \xrightarrow{heat}$

b. $CH_3CH_2CH_2CH_2\overset{O}{\overset{\|}{C}}-OH \xrightarrow{NH_2CH_3} \xrightarrow{heat}$

10.65 Draw the amide that will be produced by each reaction.

a. $CH_3CH_2CH_2\overset{O}{\overset{\|}{C}}-OH \xrightarrow{NH_2CH_3} \xrightarrow{heat}$

b. $CH_3CH_2CH_2\overset{O}{\overset{\|}{C}}-OH \xrightarrow{CH_3NHCH_3} \xrightarrow{heat}$

10.66 Name the amide product of each reaction in Problem 10.64.

10.67 Name the amide product of each reaction in Problem 10.65.

10.68 Draw each amide.
- **a.** hexanamide
- **b.** *N*-propylacetamide
- **c.** *N*-butyl-*N*-methylbenzamide

10.69 Draw each amide.
- **a.** *N*-isopropylacetamide
- **b.** *N*,*N*-dimethyl-3,4-dimethylhexanamide
- **c.** *N*-ethyl-*N*-methylbutyramide

10.70 Draw the products formed when each amide in Problem 10.68 is reacted with H_2O in the presence of H^+.

10.71 Draw the products formed when each amide in Problem 10.69 is reacted with H_2O in the presence of H^+.

10.72 Draw the products obtained when erbstatin (Problem 10.10) is hydrolyzed under acidic conditions.

10.73 The boiling point of propanamide is 213°C and that of methyl acetate is 57.5°C. Account for this difference in boiling points for these two molecules with very similar molecular weights.

$CH_3CH_2C(=O)-NH_2$ — **Propanamide**

$CH_3C(=O)-OCH_3$ — **Methyl acetate**

10.74 The boiling point of propanamide is 213°C and that of pentanamide is 232°C. Account for this difference in boiling points.

$CH_3CH_2C(=O)-NH_2$ — **Propanamide**

$CH_3CH_2CH_2CH_2C(=O)-NH_2$ — **Pentanamide**

10.9 Stereoisomers

10.75 Label the chiral carbon atom in each molecule.

a. $CH_3CH_2CH(Br)CH_2CH_2CH_3$

b. $C_6H_5-CH_2CH(NH_2)C(=O)-OH$

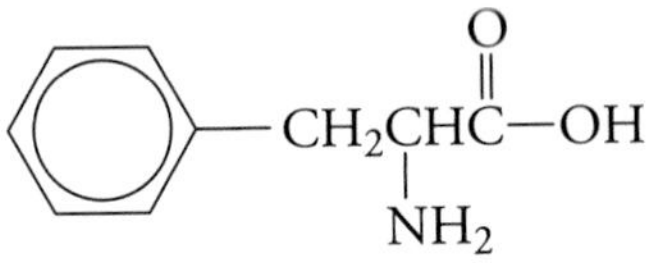

10.76 Label the chiral carbon atom in each molecule.

a. $CH_3CH(CH_3)CH(CH_3)CH_2Br$

b. 3,5-dimethylphenyl (CH_3, H_3C on ring) $-C(OH)(CH_3)CH_2CH_3$

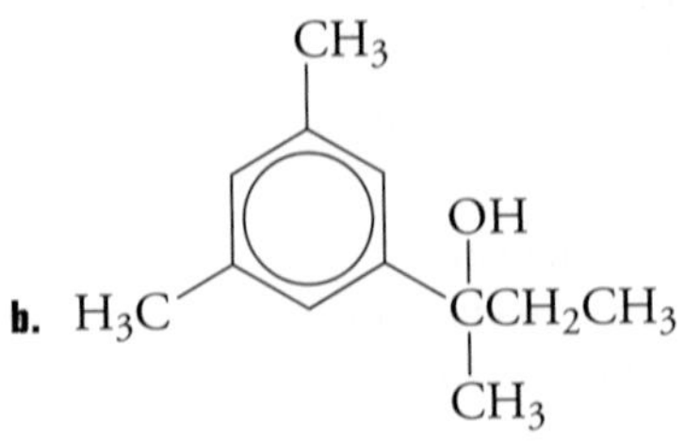

10.77 Draw both enantiomers of each molecule in Problem 10.75, using wedge and dashed line notation to show the three-dimensional shape about each chiral carbon atom.

10.78 Draw both enantiomers of each molecule in Problem 10.76, using wedge and dashed line notation to show the three-dimensional shape about each chiral carbon atom.

***10.79** **a.** Draw the enantiomer of each molecule in Figure 10.23*b*.
b. Draw one diastereomer of (+)-ethambutol (Figure 10.23*b*).

10.80 (+)-Propoxyphene is an analgesic (a painkiller).

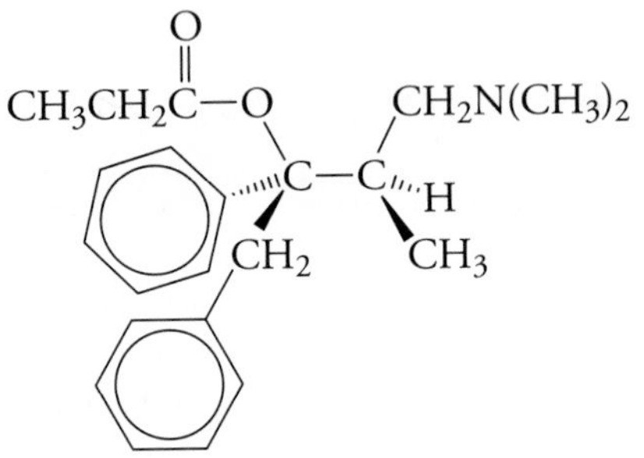

$CH_3CH_2C(=O)-O$, $CH_2N(CH_3)_2$, C—C, H, CH_2, CH_3

(+)-Propoxyphene

a. Label the chiral carbon atom(s) in this molecule, using an asterisk.
b. How many other stereoisomers of propoxyphene exist? Explain.
c. The enantiomer of (+)-propoxyphene is an antitussive (a cough suppressant). Draw this enantiomer.

10.81 Draw two diastereomers of the molecule shown in the previous problem.

10.82 When chiral molecules are synthesized in the laboratory, equal amounts of each enantiomer often form. A solution that contains equal concentrations of a pair of enantiomers does not rotate plane-polarized light. Explain why.

HEALTH*Link* | A Chili Pepper Painkiller

10.83 Olvanil, a compound structurally related to capsaicin (Figure 10.4), has been studied as a potential analgesic. Locate the phenol, amide, and alkene functional groups in the molecule.

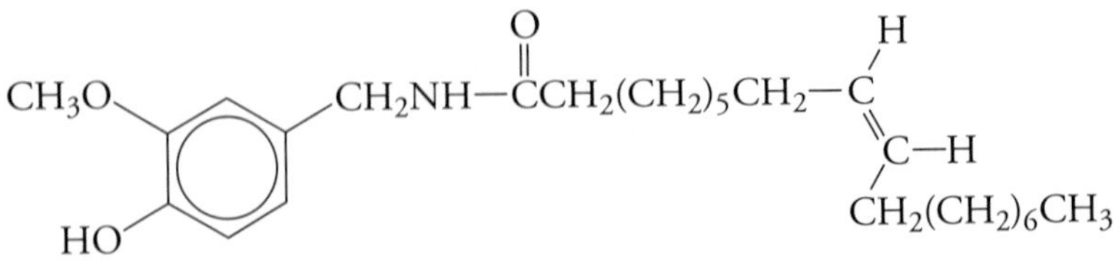

CH_3O, HO (on ring) $-CH_2NH-C(=O)CH_2(CH_2)_5CH_2-C(H)=C(H)-CH_2(CH_2)_6CH_3$

Olvanil

HEALTH*Link* | *Biofilms*

10.84 **a.** What is the function of quorum sensing?
b. What role do autoinducers play in quorum sensing?
c. How does biofilm formation benefit bacteria?

10.85 Homoserine (Figure 10.17) forms a lactone, but serine, a related amino acid, does not. Explain why.

$$NH_2-\underset{\underset{\underset{OH}{|}}{\underset{CH_2}{|}}}{CH}-\overset{\overset{O}{\|}}{C}-OH$$

Serine

BIOCHEMISTRY*Link* | *A Cure for Fleas*

10.86 Draw the products from the hydrolysis of Lufenuron under acidic conditions.

*__10.87__ Precor, used in flea collars and animal sprays, is a hormone that prevents flea pupae (the stage of life between larval and adult forms) from developing.

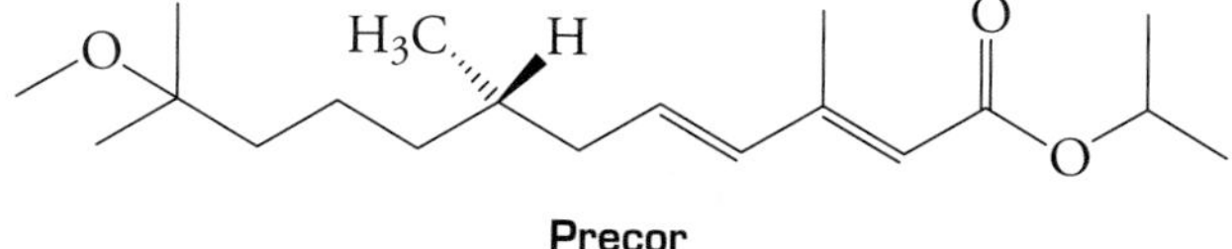

Precor

a. Label the chiral carbon atom with an asterisk.
b. Precor has low water solubility. Explain why.
c. Draw the products obtained when Precor is hydrolyzed under basic conditions.
d. Draw the products obtained when Precor is hydrolyzed under acidic conditions.

HEALTH*Link* | *Adrenaline and Related Compounds*

10.88 **a.** How many different stereoisomers are possible for phenylpropanolamine (Figure 10.25)?
b. Should all of the stereoisomers of phenylpropanolamine be expected to have the same physiological activity? Explain.

10.89 **a.** According to the Health Link "Adrenaline and Related Compounds," what are some of the biological effects of methamphetamine?
b. What additional effects are described in "Meth and the Brain," and how do these effects pose challenges for treating meth addiction?

THINKING IT THROUGH

*__10.90__ *N*-Methylhistamine is one of the products formed during the metabolism of histamine (see the chapter summary). This compound is formed by replacing one of the hydrogen atoms on histamine's primary amine with a methyl group. Draw *N*-methylhistamine.

INTERACTIVE LEARNING PROBLEMS

ILW **10.91** What is the IUPAC name for the following compound?

$$(CH_3CH_2)_3CCO_2H$$

ILW **10.92** Name this compound:

$$\begin{array}{l} CH_3CH_2CHCH_3 \\ \qquad\quad\;| \\ CH_3CH_2NCHCH_3 \\ \qquad\qquad| \\ \qquad\quad\; CH \\ \qquad\qquad| \\ \qquad\quad\; CH_3 \end{array}$$

ILW **10.93** Triethylamine is insoluble in water, but soluble in 1 M hydrochloric acid. What is the name of the product of triethylamine and HCl? Why is this product soluble in water? Amines are known for their bad smells. When HCl is added to triethylamine, the odor disappears. Why?

ILW **10.94** There are some undesirable side reactions to the frequently prescribed drug "Albuterol," a drug which is often used to relieve certain types of breathing difficulty. When this happens, "Xopenex" (levalbuterol HCl) is sometimes substituted. The patient insert claims that levalbuterol is a more purified form of the drug. Based on the prefix "lev," what do you think is the difference between "Albuterol" and "Xopenex?"

SOLUTIONS TO PRACTICE PROBLEMS

10.1 **a.** $HOCH_2CH_2CH_2\overset{\overset{O}{\|}}{C}{-}OH$ **b.** $CH_3\underset{\underset{Br}{|}}{C}H\overset{\overset{O}{\|}}{C}{-}OH$ **c.** $CH_3CH_2\underset{\underset{Cl}{|}}{C}HCH_2CH_2\overset{\overset{O}{\|}}{C}{-}OH$

10.2 **a.** sodium 2,3-dimethylpentanoate; **b.** sodium 2,4-dichlorophenoxide

10.3

$$CH_3\underset{\underset{CH_3}{|}}{C}HCH_2\overset{\overset{O}{\|}}{C}{-}O\underset{\underset{CH_3}{|}}{C}HCH_2CH_3$$

***sec*-butyl 3-methylbutanoate**

10.4 **a.** $CH_3CH_2\overset{\overset{O}{\|}}{C}CH_2\overset{\overset{O}{\|}}{C}{-}OH$ **b.** $CH_3CH_2\overset{\overset{O}{\|}}{C}CH_3$

10.5 **a.** IUPAC name *N*-ethyl-2-propanamine, common name ethylisopropylamine;

b. IUPAC name 2-butanamine, common name *sec*-butylamine; **c.** IUPAC name *N*-methyl-2-butanamine, common name, *sec*-butylmethylamine

10.6 $CH_3\underset{*}{\overset{\overset{Br}{|}}{C}}HCH_2\underset{\underset{Br}{|}}{\overset{*}{C}}H\overset{\overset{O}{\|}}{C}CH_3$

Mirror

10.7 a.

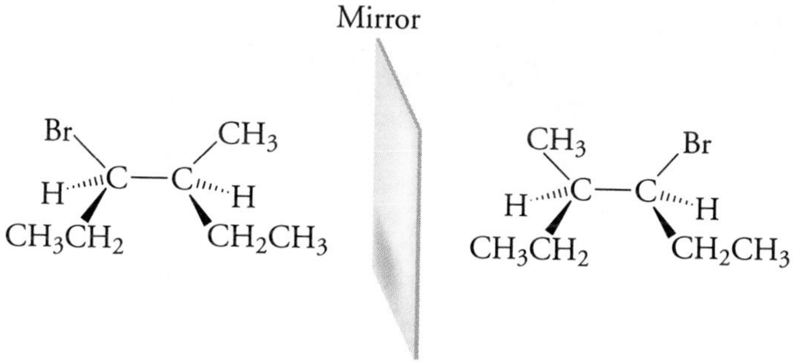

Br, CH_3, H, H, CH_3CH_2, CH_2CH_3

A

CH_3, Br, H, H, CH_3CH_2, CH_2CH_3

B

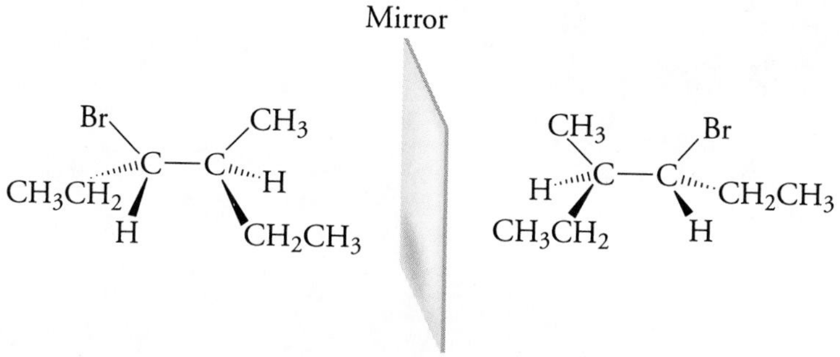

Mirror

Br, CH_3, CH_3CH_2, H, H, CH_2CH_3

C

CH_3, Br, H, CH_2CH_3, CH_3CH_2, H

D

b. enantiomers: A and B, C and D;

diastereomers: A is a diastereomer of C and D,
B is a diastereomer of C and D,
C is a diastereomer of A and B,
D is a diastereomer of A and B

While on vacation last summer, a student visited a friend whose family has several horses. A veterinarian had come to take a look at one of the horses that had developed a limp. Seeing the veterinarian put on a pair of gloves and apply a pasty material to the horse's sore leg, the student asked her what the paste did. The veterinarian told him that the paste contained DMSO, an organic compound that acts as an anti-inflammatory agent.

Photo source: George Skene/Orlando Sentinel/NewsCom.

11 ALCOHOLS, ETHERS, ALDEHYDES, AND KETONES

In this chapter we will expand on our earlier discussions of the chemistry of alcohols and will be introduced to a few new families of oxygen-containing compounds: ethers, aldehydes, and ketones. As part of our discussion, we will include a look at some related sulfur-containing compounds, including the DMSO mentioned on the facing page.

objectives

After completing this chapter, you should be able to:

1. Describe the structure of molecules that belong to the alcohol, ether, thiol, sulfide, disulfide, aldehyde, and ketone families, describe how they are named, and identify the intermolecular forces that attract similar molecules from this list to one another.
2. Distinguish 1°, 2°, and 3° alcohols.
3. Describe the nucleophilic substitution reactions that can be used to prepare alcohols, ethers, thiols, and sulfides.
4. Predict the major product of the addition reaction between an alkene and H_2O/H^+ and the major product for the elimination reaction between an alcohol and H^+/heat.
5. Describe the oxidation reactions of alcohols, thiols, sulfides, and aldehydes.
6. Explain what happens when an aldehyde or ketone is reacted with H_2 and Pt and when one of these compounds is reacted with one or two alcohol molecules, in the presence of H^+.

In alcohols an —OH group is attached to an alkane-type carbon atom, and in **ethers**, *an oxygen atom is attached to two alkane-type or aromatic carbon atoms (C—O—C)*. The three sulfur-containing families that we will consider in this chapter are **thiols** (*contain an —SH group attached to an alkane-type or aromatic carbon atom*), **sulfides** (*contain a C—S—C linkage—a sulfur atom attached to two alkane-type or aromatic carbon atoms*), and **disulfides** (*contain a C—S—S—C linkage—each sulfur atom is attached to one other sulfur atom and one alkane-type or aromatic carbon atom*). These sulfur compounds have been included because oxygen and sulfur belong to the same group in the periodic table and have some properties in common.

Alcohols, ethers, thiols, sulfides, and disulfides appear widely in nature and often occur in compounds that contain more than one functional group. The steroid cholesterol contains the alcohol functional group, as do menthol, used in flavorings, and citronellol, a component of rose oil (Figure 11.1*a*). Kadsurenone and nuciferine are ethers

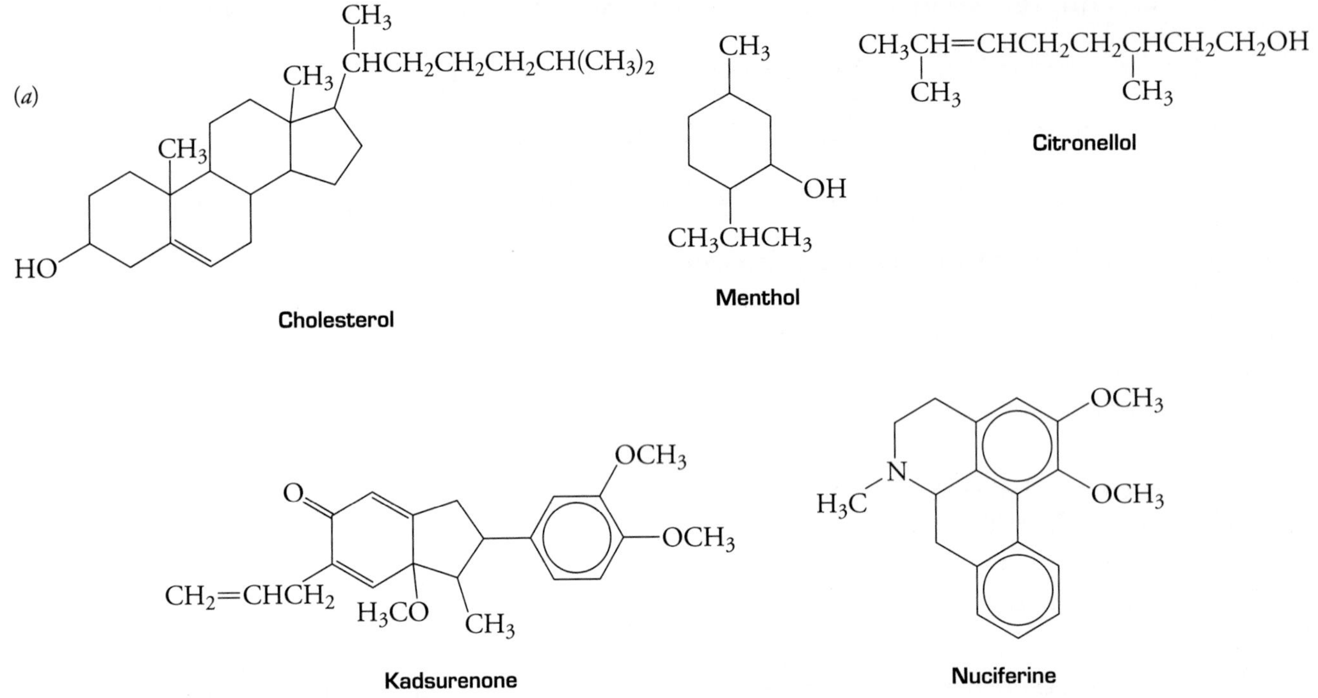

(*a*)

$CH_3CH{=}CHCH_2CH_2CH(CH_3)CH_2CH_2OH$ (with CH_3 on C-2)

Cholesterol

Menthol

Citronellol

Kadsurenone

Nuciferine

(*b*)

$CH_3CH(CH_3)CH_2CH_2SH$

3-Methyl-1-butanethiol

$CH_3CH{=}CHCH_2SH$

2-buten-1-thiol

$\overset{+}{N}H_3CH(CH_2SH)C(=O)O^-$

Cysteine

$\overset{+}{N}H_3CH(CH_2CH_2SCH_3)C(=O)O^-$

Methionine

$CH_2{=}CHCH_2SSCH_2CH{=}CH_2$

Diallyl disulfide

■ From this point on in the text, biochemical compounds will often be shown as they exist at pH 7. Carboxyl groups will appear in their basic form ($—CO_2^-$) and amino groups in their acidic form ($—NH_3^+$).

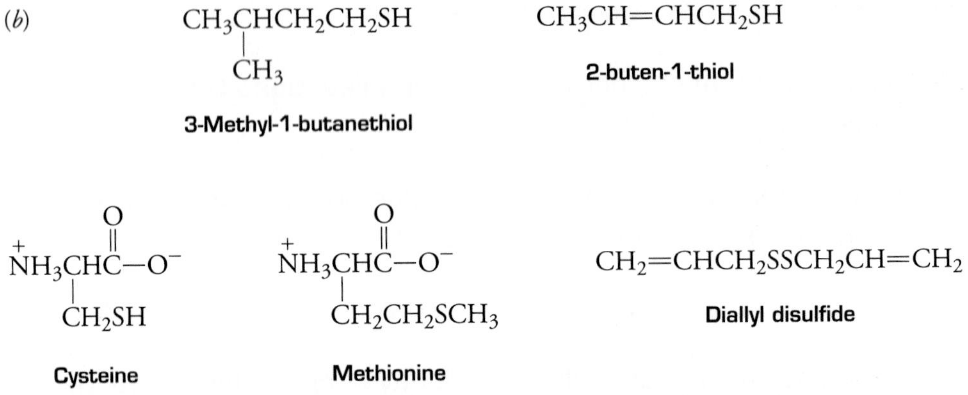

FIGURE 11.1 **Alcohols, ethers, and related sulfur compounds**
(*a*) Alcohols contain an —OH group and ethers contain a C—O—C linkage. (*b*) Thiols contain an —SH group, sulfides a C—S—C linkage, and disulfides a C—S—S—C linkage.

found in herbs used in traditional Chinese medicine. Two of the compounds responsible for the strong odor of skunk spray, 3-methyl-1-butanethiol and 2-buten-1-thiol, are thiols, as is the amino acid cysteine, a building block for protein synthesis (Figure 11.1*b*). Methionine, another amino acid, is among the many sulfides found in nature. Diallyl disulfide, one of the compounds that give garlic its characteristic odor, is a disulfide.

Ketones and aldehydes contain a carbonyl (C=O) group. In **ketones** the carbonyl carbon atom is attached to *two other carbon atoms* and in **aldehydes** it is attached to *one carbon atom and one hydrogen atom* or to *two hydrogen atoms*. Aldehydes and ketones are of interest to chemists because they are part of the structure of many key biomolecules. Some steroid hormones, including testosterone and progesterone, contain the ketone functional groups (Figure 11.2). Other aldehydes and ketones are associated with the pleasant odors and flavors of specific plants, including vanillin, from vanilla beans; cinnamaldehyde, from cinnamon; and jasmone, from jasmine. Simple sugars, such as glucose and fructose, are either aldehydes or ketones (Chapter 12).

Testosterone

Progesterone

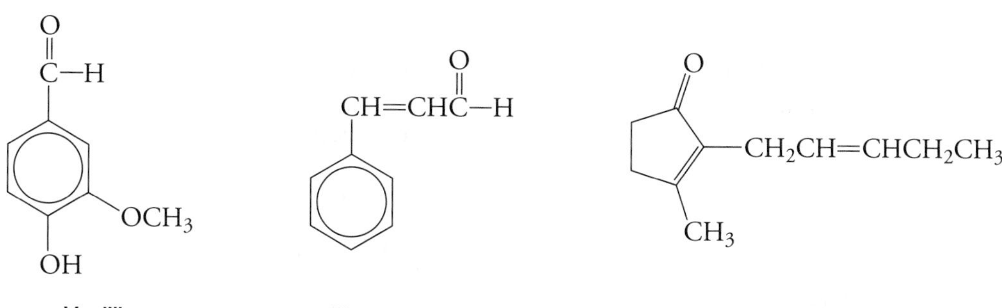

FIGURE 11.2 **Examples of ketones and aldehydes found in nature**
In ketones the C=O carbon atom is attached to two other carbon atoms. In aldehydes the C=O carbon atom is attached to one carbon atom and one hydrogen atom or to two hydrogen atoms.

11.1 ALCOHOLS, ETHERS, AND RELATED COMPOUNDS

When the IUPAC rules are used to name an alcohol, the parent (the longest carbon chain carrying the —OH group) is numbered from the end nearer the —OH and named by dropping the "e" ending on the name of the corresponding hydrocarbon and adding "ol." When a parent chain contains more than two carbon atoms, the position of the —OH group must be specified. Any alkyl groups attached to the parent chain are identified by name, position, and number of appearances (Figure 11.3*a*). Smaller alcohols are often known by their common names, which are obtained by combining the name of the alkyl group attached to the —OH with the word "alcohol." The common name of ethanol is ethyl alcohol (grain alcohol) and that of 2-propanol is isopropyl alcohol (known also as rubbing alcohol).

The IUPAC rules for naming thiols closely follow those for naming alcohols. The parent chain contains the —SH (thiol) group, is numbered from the end nearer this group, and is named by adding "thiol" to the name of the corresponding hydrocarbon (Figure 11.3*a*). The common names of thiols consist of the name of the alkyl group that is present plus "mercaptan," a term that comes from the fact that thiols readily react with mercury (mercaptan for *mer*cury *capt*urer).

In this text, only the common names of ethers, sulfides, and disulfides will be described. The common names of these compounds are formed by placing the name of the organic family (ether, sulfide, or disulfide) after the names of the attached groups (Figure 11.3*b*).

- The carbon atom carrying the —OH group is attached to just one other carbon atom in a 1° alcohol, to two other carbon atoms in a 2° alcohol, and to three other carbon atoms in a 3° alcohol.

Alcohols are categorized according to the nature of the carbon atom that is bonded directly to the —OH group. The hydroxy-carrying carbon atom is attached to only one other carbon atom in a **primary (1°)** alcohol, to two other carbon atoms in a **secondary (2°)** alcohol, and to three other carbon atoms in a **tertiary (3°)** alcohol. In Figure 11.3*a* ethanol and 1-propanol are primary alcohols, 2-propanol is a secondary alcohol, and 2-methyl-2-propanol is a tertiary alcohol. Note that this definition of 1°, 2°, and 3° alcohols differs from that used for amines (Section 10.6).

Compared to hydrocarbons with a similar molecular weight, alcohols have relatively high boiling points (Table 11.1). Methanol (CH_3OH), for example, has a boiling point that is more than 150°C higher than that of ethane, although their molecular weights differ by only 2.0 amu. This large difference in boiling points is due to the ability of alcohol molecules to form hydrogen bonds with one another (Figure 11.4).

FIGURE 11.3

Naming alcohols, thiols, ethers, sulfides, and disulfides
Examples of (*a*) IUPAC and common names (in parentheses) of alcohols and thiols and (*b*) common names of ethers, sulfides, and disulfides.

(*a*)

CH_3CH_2OH
Ethanol
(ethyl alcohol)

$CH_3CH_2CH_2OH$
1-Propanol
(propyl alcohol)

$CH_3CH(OH)CH_3$
2-Propanol
(isopropyl alcohol)

$CH_3C(OH)(CH_3)CH_3$
2-Methyl-2-propanol
(*t*-butyl alcohol)

CH_3SH
Methanethiol
(methyl mercaptan)

$CH_3CH(SH)CH_2CH_3$
2-Butanethiol
(*s*-butyl mercaptan)

(*b*)

$CH_3CH_2OCH_2CH_3$
(Diethyl ether)

$CH_3CH_2OCH_3$
(Ethyl methyl ether)

CH_3SCH_3
(Dimethyl sulfide)

$CH_3CH_2CH_2SSCH_2CH_2CH_3$
(Dipropyl disulfide)

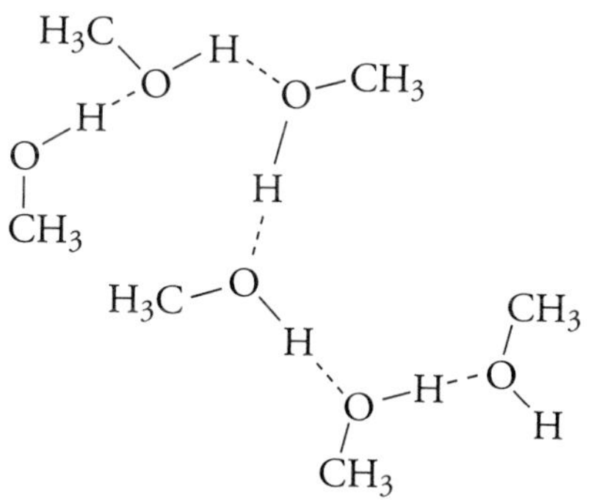

FIGURE 11.4

Hydrogen bonding
The dashed lines represent hydrogen bonds between alcohol molecules. Their ability to form hydrogen bonds contributes to the relatively high boiling points of alcohols and to the ability of small alcohols to dissolve in water.

TABLE 11.1 Physical Properties of Selected Alcohols, Ethers, Thiols, Sulfides, and Hydrocarbons

Formula	IUPAC Name	Common Name	Boiling Point (°C)	Water Solubility (g/100 mL)
Alcohols				
CH_3OH	Methanol	Methyl alcohol	65.0	Miscible[a]
CH_3CH_2OH	Ethanol	Ethyl alcohol	78.5	Miscible
$CH_3CH_2CH_2OH$	1-Propanol	Propyl alcohol	97.4	Miscible
$CH_3CH_2CH_2CH_2OH$	1-Butanol	Butyl alcohol	117.3	8.0
$CH_3CH_2CH_2CH_2CH_2OH$	1-Pentanol	Pentyl alcohol	138	2.2
Ethers				
CH_3OCH_3		Dimethyl ether	−24.9	Slight
$CH_3CH_2OCH_2CH_3$		Diethyl ether	34.6	Very slight
$CH_3CH_2CH_2OCH_2CH_2CH_3$		Dipropyl ether	90.5	~0
Thiols				
CH_3SH	Methanethiol	Methyl mercaptan	6	2.3
CH_3CH_2SH	Ethanethiol	Ethyl mercaptan	35	0.7
Sulfides				
CH_3SCH_3		Dimethyl sulfide	38	~0
$CH_3CH_2SCH_2CH_3$		Diethyl sulfide	90–92	~0
$CH_3CH_2CH_2SCH_2CH_2CH_3$		Dipropyl sulfide	142–143	~0
Alkanes				
CH_3CH_3	Ethane		−89	~0
$CH_3CH_2CH_3$	Propane		−42	~0
$CH_3CH_2CH_2CH_3$	Butane		0	~0
$CH_3CH_2CH_2CH_2CH_3$	Pentane		36	~0

[a] Can be mixed in all proportions.

As we have seen for other families of organic compounds, the longer the carbon chain of an alcohol, the higher its boiling point, due to increased London force interactions between the hydrocarbon parts of the molecules (Table 11.1).

The boiling points of ethers, thiols, sulfides, and disulfides are lower than those of alcohols with similar molecular weights, because none of these compounds is able to form hydrogen bonds to like molecules. Ether molecules are slightly polar as a consequence of the C—O—C linkage, but the dipole–dipole attractions that occur between ether molecules are not strong enough to raise boiling points much above those of similarly sized hydrocarbons. The relatively weak London forces that hold thiol, sulfide, and disulfide molecules to one another are responsible for their lower boiling points.

SAMPLE PROBLEM 11.1

Intermolecular forces

Which of the following molecules can form hydrogen bonds with other molecules of the same type?

a. H_2O

b. $CH_3\overset{\overset{\large O}{\|}}{C}CH_3$

c. CH_3CH_2OH

d. $CH_3\overset{\overset{\large O}{\|}}{C}—OH$

Strategy

The key to solving this problem is to review the definition of hydrogen bonding. As we saw in Section 4.3, a hydrogen bond is the interaction of a nitrogen, oxygen, or fluorine atom with a hydrogen atom that is covalently bonded to a different nitrogen, oxygen, or fluorine atom.

Solution

a, c, and d. Although the molecule in part b can form hydrogen bonds when different molecules provide the hydrogen atom (a hydrogen atom from H_2O, for example), two of these ketone molecules cannot hydrogen bond with one another. Neither has a hydrogen atom covalently attached to N, O, or F.

PRACTICE PROBLEM 11.1

Which molecule has the higher boiling point?

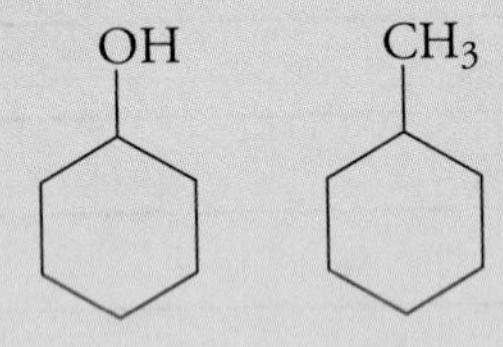

OH CH₃ (cyclohexanol) CH_3 (methylcyclohexane)

The *like dissolves like* rule (Section 7.1) can be used to roughly predict the water solubility of a molecule. Small alcohol molecules are significantly soluble in water because, like water, they are polar and are able to form hydrogen bonds. Larger alcohol molecules have a lower solubility in water because the nonpolar hydrocarbon portion makes up an increasingly greater share of the molecular structure. Ethers are less polar than alcohols and form fewer hydrogen bonds with water molecules, so ethers are less water soluble. Thiols, sulfides, and disulfides, all of which are nonpolar, have very low solubility in water. Many thiols, sulfides, and disulfides have a strong, unpleasant odor, as is true of the two thiols present in skunk spray (Figure 11.1). The low water solubility of these compounds is a factor to consider if you are sprayed by a skunk—the smell does not wash off easily.

■ Hydrogen bonding makes alcohols more soluble in water than are ethers, thiols, sulfides, and disulfides.

11.2 PREPARATION

In this section we will take a look at a few of the reactions that chemists use to prepare alcohols, ethers, and related sulfur compounds. When living things manufacture these compounds, they tend to use enzyme-catalyzed versions of these same reactions.

Alcohols can be prepared using a **nucleophilic substitution reaction**, in which *an electron-rich atom or group of atoms*, called a **nucleophile**, replaces a **leaving group**, *an easily replaced atom or group of atoms* that is held to a carbon atom by a relatively weak covalent bond. Chlorine, bromine, and iodine are common leaving groups used in organic chemistry.

One example of a nucleophilic substitution reaction used to prepare an alcohol is that of OH^- with methyl chloride (CH_3Cl). The electrons in the new covalent bond between carbon and oxygen are provided by OH^-, the nucleophile. The electrons in what was the carbon–chlorine bond remain with the leaving group, Cl, giving it a negative charge (Figure 11.5*a*). Whenever OH^- is used as the nucleophile in a substitution reaction, an alcohol is produced and the structure of the product alcohol depends on the **alkyl halide** reactant. (In an alkyl halide *a halogen atom is attached to an alkane-type carbon atom.*) The key to understanding nucleophilic substitution reactions is recognizing that the leaving group (Cl, Br, or I) is always replaced by the nucleophile.

(*a*)

$$:\ddot{\underset{..}{O}}H^- + CH_3—\ddot{\underset{..}{Cl}}: \longrightarrow CH_3—\ddot{\underset{..}{O}}H + :\ddot{\underset{..}{Cl}}:^-$$

Methyl chloride → Methyl alcohol

$$:\ddot{\underset{..}{O}}H^- + CH_3—\ddot{\underset{..}{Br}}: \longrightarrow CH_3—\ddot{\underset{..}{O}}H + :\ddot{\underset{..}{Br}}:^-$$

Methyl bromide → Methyl alcohol

$$:\ddot{\underset{..}{O}}H^- + CH_3CH_2CH_2—\ddot{\underset{..}{I}}: \longrightarrow CH_3CH_2CH_2—\ddot{\underset{..}{O}}H + :\ddot{\underset{..}{I}}:^-$$

Propyl iodide → Propyl alcohol

(*b*)

$$:\ddot{\underset{..}{O}}CH_3^- + CH_3—\ddot{\underset{..}{Br}}: \longrightarrow CH_3—\ddot{\underset{..}{O}}CH_3 + :\ddot{\underset{..}{Br}}:^-$$

Methyl bromide → Dimethyl ether

$$:\ddot{\underset{..}{S}}H^- + CH_3CH_2—\ddot{\underset{..}{Cl}}: \longrightarrow CH_3CH_2—\ddot{\underset{..}{S}}H + :\ddot{\underset{..}{Cl}}:^-$$

Ethyl chloride → Ethyl mercaptan

$$:\ddot{\underset{..}{S}}CH_2CH_3^- + CH_3CH_2—\ddot{\underset{..}{I}}: \longrightarrow CH_3CH_2—\ddot{\underset{..}{S}}CH_2CH_3 + :\ddot{\underset{..}{I}}:^-$$

Ethyl iodide → Diethyl sulfide

FIGURE 11.5

Nucleophilic substitution
In a nucleophilic substitution reaction, a nucleophile (blue) replaces a leaving group (red). This reaction may be used to prepare (*a*) alcohols and (*b*) ethers, thiols, and sulfides.

By changing the nucleophile, ethers, thiols, and sulfides can be prepared. In place of OH^-, SH^- is used to form thiols, a nucleophile with a $C—O^-$ linkage is used to form ethers, and one with a $C—S^-$ linkage to form sulfides (Figure 11.5*b*).

Nucleophilic reactions occur in biochemical systems too, but the leaving groups are larger and more complex than the halogen atoms frequently used in organic reactions. For example, in the biosynthesis of epinephrine (adrenaline) from norepinephrine, the nitrogen atom of norepinephrine (the nucleophile) forms a bond to the $—CH_3$ group of *S*-adenosylmethionine, replacing the large leaving group in the process (Figure 11.6).

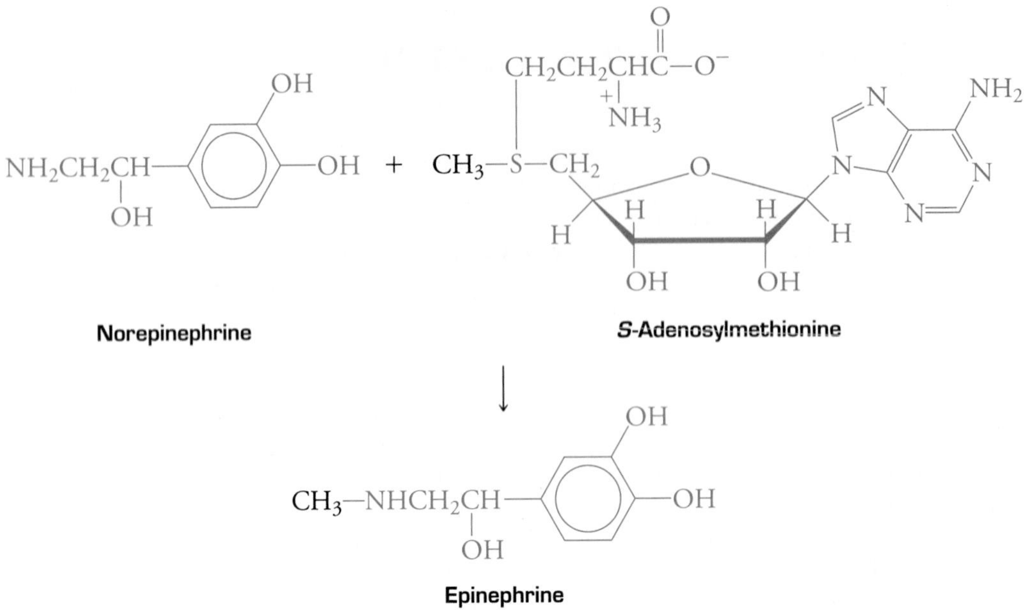

Norepinephrine + S-Adenosylmethionine → Epinephrine

FIGURE | 11.6

A biochemical nucleophilic substitution reaction
Adrenaline is produced in the body when the nucleophilic nitrogen atom of norepinephrine forms a covalent bond with the CH_3 of *S*-adenosylmethionine and displaces the leaving group.

Sample Problem 11.2

Predicting the product of a substitution reaction

Draw the organic product of each nucleophilic substitution reaction.

a. $^{-}OH + CH_3CH_2Br \longrightarrow$

b. $^{-}SH + CH_3CH(Br)CH_3 \longrightarrow$

c. $^{-}OCH_3 + CH_3CH_2CH_2CH_2Br \longrightarrow$

Strategy

To arrive at the product of each reaction you must decide which reactant is the nucleophile and which reactant carries the leaving group. In a nucleophilic substitution reaction, a nucleophile replaces a leaving group.

Solution

a. CH_3CH_2OH

b. $CH_3CH(SH)CH_3$

c. $CH_3CH_2CH_2CH_2OCH_3$

Practice Problem 11.2

What is the organic product of each reaction in Sample Problem 11.2, if iodine replaces bromine in each reactant?

Section 6.4 showed us that alcohols can be prepared from alkenes when they are reacted with water in the presence of H^+, a catalyst. From asymmetric alkenes, those with *different groups on each side of the carbon–carbon double bond,* two different addition products are possible. For example, the addition of H_2O to propene produces the constitutional isomers 1-propanol and 2-propanol (Figure 11.7*a*). This addition reaction does not give a 50:50 mixture of the two products. According to **Markovnikov's rule,** the major product (the product obtained in the greater amount) is the one in which *a hydrogen atom has been added to the double-bonded carbon atom that carries more hydrogen atoms.* This rule, developed in 1869, is named after Victor Markovnikov, the Russian chemist involved in some of the early investigations of alkene addition reactions. Figure 11.7*b* shows two additional applications of Markovnikov's rule.

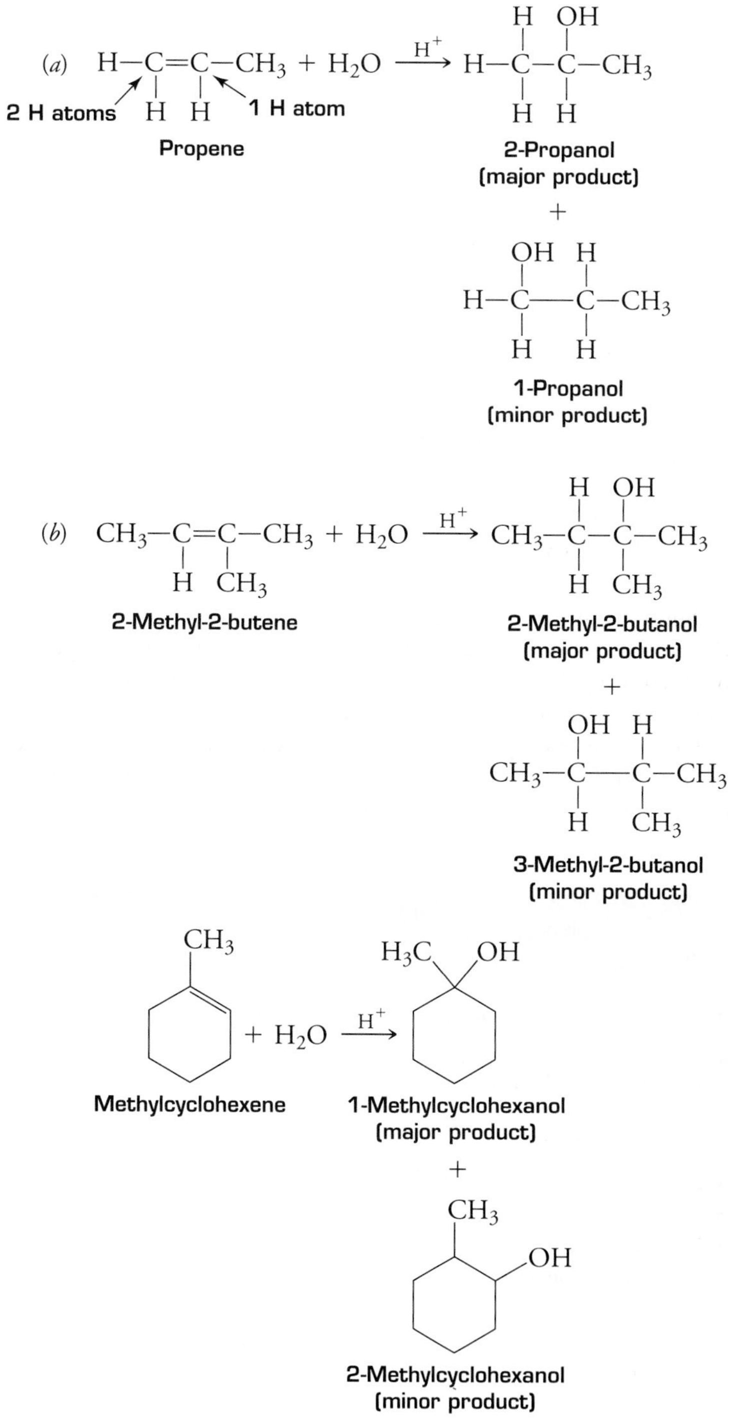

(*a*) H—C=C—CH$_3$ + H$_2$O $\xrightarrow{H^+}$ H—C(H)(H)—C(OH)(H)—CH$_3$

2 H atoms ↗ H H ↖ 1 H atom

Propene

2-Propanol
(major product)

+

H—C(OH)(H)—C(H)(H)—CH$_3$

1-Propanol
(minor product)

(*b*) CH$_3$—C(H)=C(CH$_3$)—CH$_3$ + H$_2$O $\xrightarrow{H^+}$ CH$_3$—C(H)(H)—C(OH)(CH$_3$)—CH$_3$

2-Methyl-2-butene

2-Methyl-2-butanol
(major product)

+

CH$_3$—C(OH)(H)—C(H)(CH$_3$)—CH$_3$

3-Methyl-2-butanol
(minor product)

CH$_3$ / H$_3$C, OH / + H$_2$O $\xrightarrow{H^+}$

Methylcyclohexene

1-Methylcyclohexanol
(major product)

+

CH$_3$ / OH

2-Methylcyclohexanol
(minor product)

FIGURE 11.7

Alkene hydration and Markovnikov's rule

When adding H_2O to an alkene, the major product is that formed when the hydrogen atom adds to the double-bonded carbon atom that carries more hydrogen atoms. As written, the reaction equations are not balanced.

11.3 REACTIONS

Oxidation

In Section 10.5 we saw that potassium dichromate ($K_2Cr_2O_7$) oxidizes hydroquinones into quinones. In a similar fashion, this oxidizing agent can be used to oxidize certain alcohols to produce ketones or carboxylic acids (Figure 11.8). During the oxidation of an alcohol, a carbonyl group (C=O) is formed when one hydrogen atom is removed from the —OH group and another is removed from the carbon atom that carries the —OH.

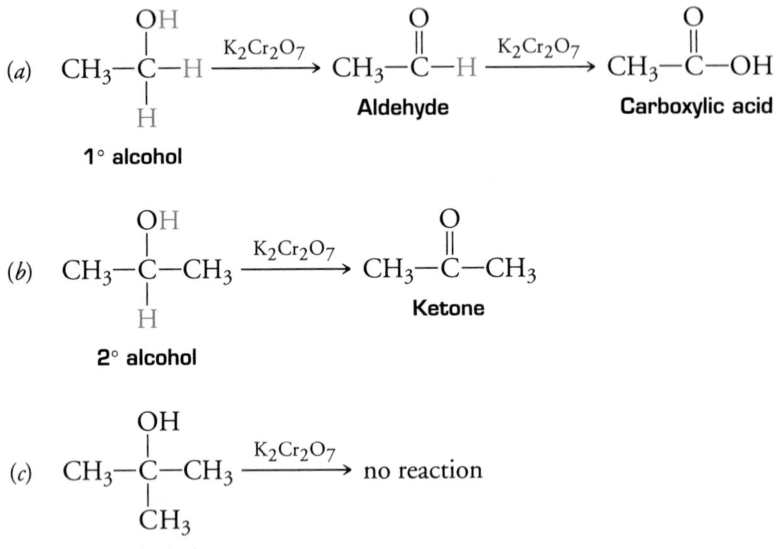

FIGURE 11.8

Alcohol oxidation
Two hydrogen atoms are lost from an alcohol molecule when it is oxidized by $K_2Cr_2O_7$: one from the —OH group and one from the carbon atom that carries the —OH. (*a*) Primary alcohols are oxidized to aldehydes, which are then oxidized to carboxylic acids. (*b*) Secondary alcohols are oxidized to ketones. (*c*) Tertiary alcohols are not oxidized.

(*a*) $CH_3-CH(OH)-H \xrightarrow{K_2Cr_2O_7} CH_3-C(=O)-H \xrightarrow{K_2Cr_2O_7} CH_3-C(=O)-OH$

1° alcohol → **Aldehyde** → **Carboxylic acid**

(*b*) $CH_3-CH(OH)-CH_3 \xrightarrow{K_2Cr_2O_7} CH_3-C(=O)-CH_3$

2° alcohol → **Ketone**

(*c*) $CH_3-C(OH)(CH_3)-CH_3 \xrightarrow{K_2Cr_2O_7}$ no reaction

3° alcohol

Since one of the hydrogen atoms removed during alcohol oxidation must come from the carbon that carries the —OH, there are limitations to the types of alcohols that can be oxidized using this reaction. A primary alcohol has two of the necessary C—H bonds and can be oxidized to an aldehyde, which is immediately oxidized to a carboxylic acid by $K_2Cr_2O_7$. Secondary alcohols also carry the required C—H bond and are oxidized to ketones. Tertiary alcohols have no hydrogen atom attached to the carbon atom holding the —OH group, so they cannot be oxidized by $K_2Cr_2O_7$.

SAMPLE PROBLEM 11.3

Oxidation of alcohols

Draw the product expected from each reaction.

a. $CH_3CH_2CH_2OH \xrightarrow{K_2Cr_2O_7}$

b. 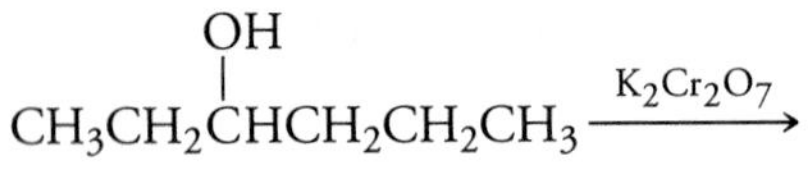 $CH_3CH_2CH(OH)CH_2CH_2CH_3 \xrightarrow{K_2Cr_2O_7}$

Strategy

The place to begin is determining which type of alcohol (1°, 2°, or 3°) appears in each reaction. Primary alcohols are oxidized to aldehydes, which are immediately oxidized to carboxylic acids. Secondary alcohols are oxidized to ketones and tertiary alcohols are not oxidized.

Solution

a. $CH_3CH_2C(=O){-}OH$

b. $CH_3CH_2C(=O)CH_2CH_2CH_3$

PRACTICE PROBLEM 11.3

In Chapter 15 we will study the citric acid cycle, a series of reactions involved in making compounds that can be used in a separate process to manufacture an energy-rich compound called ATP. A reaction early in the citric acid cycle involves the oxidation of an alcohol. Of the two reactants shown below (each is a reactant somewhere in the cycle), which has an alcohol group that can be oxidized?

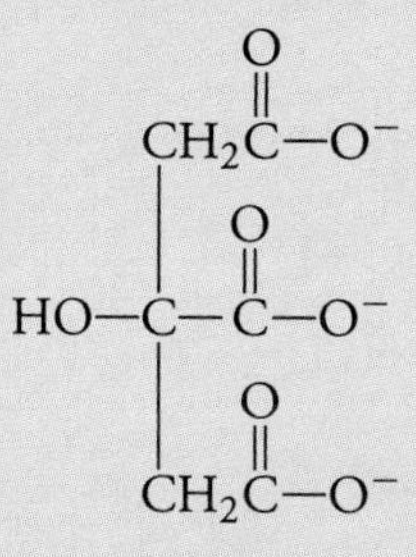

Citrate: $CH_2C(=O){-}O^-$ / $HO{-}C{-}C(=O){-}O^-$ / $CH_2C(=O){-}O^-$

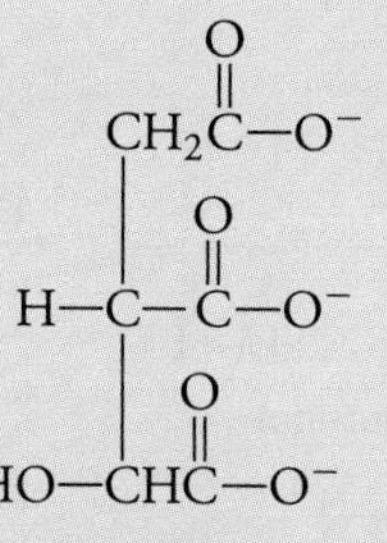

Citrate

Isocitrate: $CH_2C(=O){-}O^-$ / $H{-}C{-}C(=O){-}O^-$ / $HO{-}CHC(=O){-}O^-$

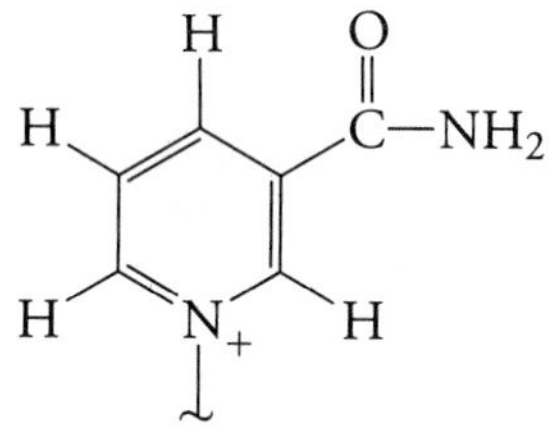

Isocitrate

Nicotinamide adenine dinucleotide (NAD^+), an oxidizing agent used by living things, works in conjunction with certain enzymes that catalyze the oxidation of alcohols. NAD^+ assists in the oxidation of an alcohol molecule by accepting one of its hydrogen atoms, becoming NADH in the process (Figure 11.9). The second hydrogen atom released from the alcohol becomes H^+. For example, the first step in the metabolism of the ethanol present in beer, wine, and other alcoholic beverages takes place in the liver and is catalyzed by an NAD^+-requiring enzyme (Health Link: Aldehyde Dehydrogenase).

The oxidation of thiols produces a different type of product than obtained from the oxidation of alcohols. On treatment with the oxidizing agent I_2, two thiol molecules combine to form a disulfide. The loss of a hydrogen atom by each thiol is evidence that oxidation has taken place. (As is often the case in organic chemistry, this and many of the other reaction equations shown in this chapter are not balanced.)

$$CH_3CH_2SH + HSCH_2CH_2 \xrightarrow{I_2} CH_3CH_2SSCH_2CH_3$$

Ethanethiol **Diethyl disulfide**

When —SH groups in a protein are oxidized to form disulfides, the resulting bond is called a disulfide bridge. Disulfide bridges play an important role in holding many proteins in the three-dimensional shape required for biological activity.

Sulfides can be oxidized to form **sulfoxides**, *molecules containing charged sulfur and oxygen atoms joined by a single bond* ($S^+{-}O^-$). Dimethyl sulfoxide (DMSO), known as a "universal solvent" because so many different compounds dissolve in it, can be formed by reacting dimethyl sulfide with the oxidizing agent hydrogen peroxide (H_2O_2).

$$CH_3SCH_3 \xrightarrow{H_2O_2} CH_3\overset{O^-}{\overset{|+}{S}}CH_3$$

Dimethyl sulfide **Dimethyl sulfoxide (DMSO)**

NAD⁺ structure: pyridine ring with H atoms, $C(=O){-}NH_2$ substituent, N^+

NAD⁺

NADH structure: ring with H, H on one carbon, $C(=O){-}NH_2$ substituent, N

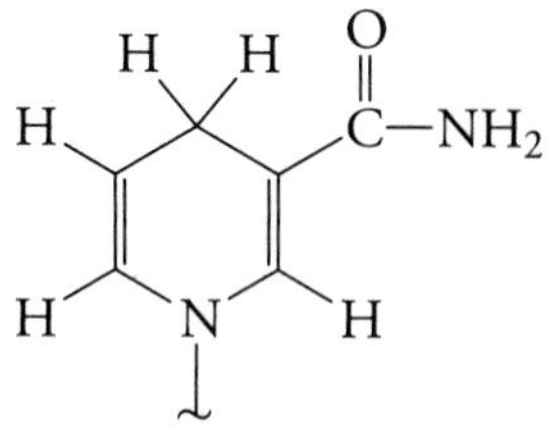

NADH

FIGURE 11.9

NAD^+ and NADH
Nicotinamide adenine dinucleotide (NAD^+)—a partial structure is shown—is a common biochemical oxidizing agent. In the process of oxidizing a compound, NAD^+ is converted to NADH. The structure of NADH differs from that of NAD^+ only in the pyridine ring. For complete structures of these two compounds, see Figure 15.2.

Unlike alcohols, thiols, and sulfides, members of the ether family are quite unreactive. Apart from their being highly flammable, it is quite difficult to involve ethers in chemical reactions. This makes them good solvents for carrying out chemical reactions, since they will not take part in the reactions. Although ethers are generally unreactive, many oxidize slowly in air to form explosive peroxides, compounds that contain two oxygen atoms joined by a single covalent bond (O—O). Diethyl ether ($CH_3CH_2OCH_2CH_3$), for example, is oxidized to form diethyl ether hydroperoxide.

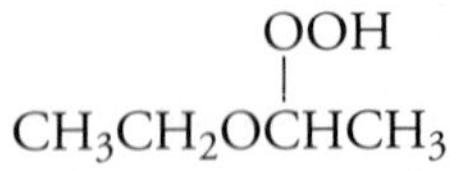

$$CH_3CH_2OCH(OOH)CH_3$$

Diethyl ether hydroperoxide

FIGURE | 11.10

Dehydration of alcohols
When alcohols are dehydrated in the presence of H^+ and heat, the major product is that formed by removal of —OH from one carbon atom and removal of —H from the neighboring C atom that carries fewer H atoms. As is common for organic reaction equations, only the organic products are shown.

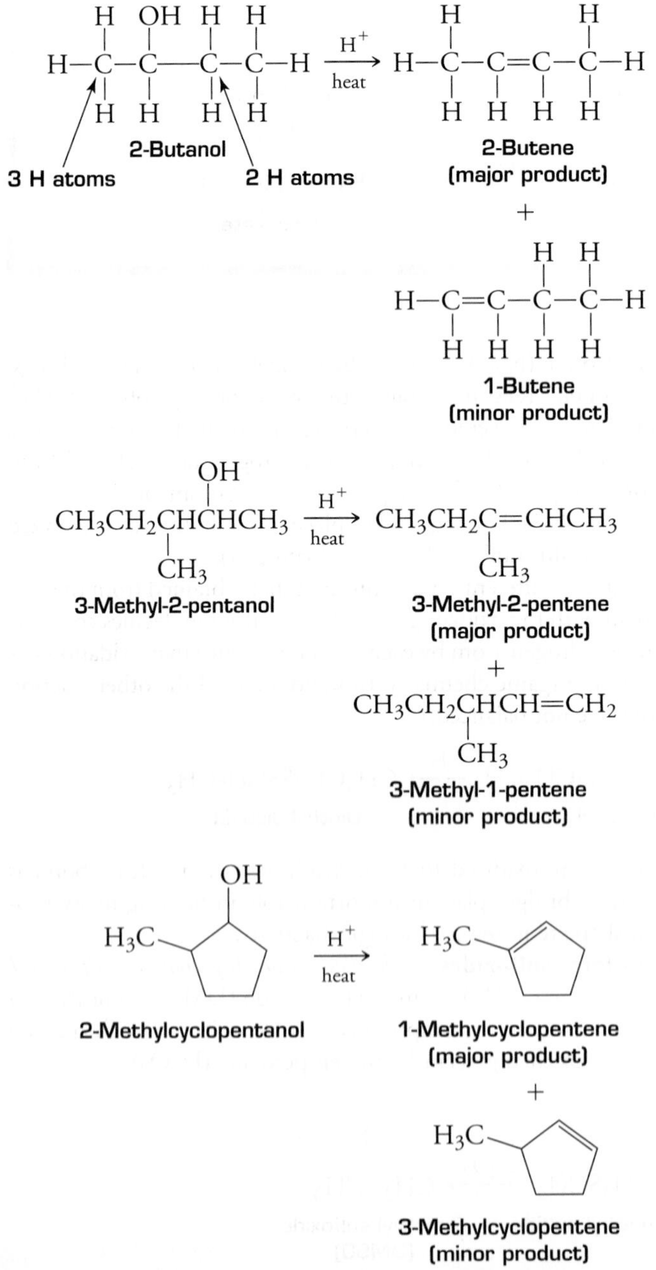

$$CH_3CH(OH)CH_2CH_3 \xrightarrow[\text{heat}]{H^+} CH_3CH{=}CHCH_3 + CH_2{=}CHCH_2CH_3$$

2-Butanol (3 H atoms; 2 H atoms) → 2-Butene (major product) + 1-Butene (minor product)

$$CH_3CH_2CH(CH_3)CH(OH)CH_3 \xrightarrow[\text{heat}]{H^+} CH_3CH_2C(CH_3){=}CHCH_3 + CH_3CH_2CH(CH_3)CH{=}CH_2$$

3-Methyl-2-pentanol → 3-Methyl-2-pentene (major product) + 3-Methyl-1-pentene (minor product)

2-Methylcyclopentanol $\xrightarrow[\text{heat}]{H^+}$ 1-Methylcyclopentene (major product) + 3-Methylcyclopentene (minor product)

Dehydration of Alcohols

In the presence of an H^+ catalyst and heat, alcohol molecules undergo dehydration (Section 6.4), in which loss of —OH and —H from neighboring carbon atoms yields an alkene. For alcohols, such as 2-butanol, where more than one alkene can be formed, the major product is the one produced by *removal of H from the neighboring carbon atom that carries fewer H atoms.* For example, when 2-butanol undergoes dehydration 2-butene is the major product (Figure 11.10).

11.4 ALDEHYDES AND KETONES

When naming aldehydes and ketones according to the IUPAC rules, the carbonyl group (C=O) must be part of the parent chain, which is numbered from the end nearer this group. Since the carbonyl carbon atom of an aldehyde is always in position number 1, its position is not specified in the name. For ketones, however, the position of the carbonyl carbon is given, unless the molecule is small enough that there is no question as to carbonyl placement. Parent chains are named by dropping the final "e" from the name of the corresponding hydrocarbon and adding "al" for aldehydes or "one" for ketones (Figure 11.11).

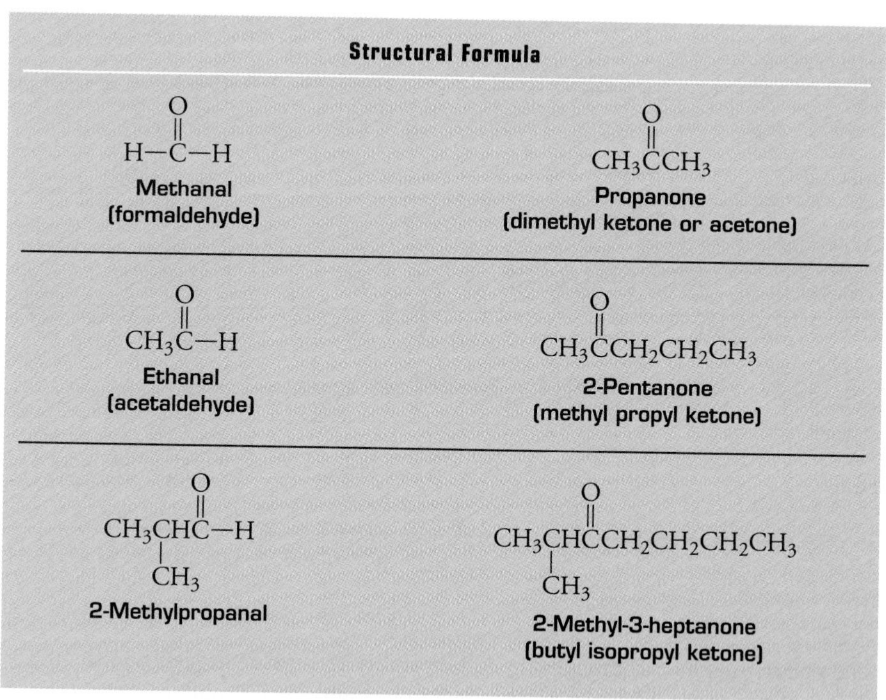

FIGURE 11.11

Naming aldehydes and ketones IUPAC names and common names (in parentheses) are given for selected aldehydes and ketones.

The common names of ketones are formed by placing "ketone" after the names of the alkyl groups attached to the carbonyl carbon atom. For aldehydes and for some ketones, other common names are assigned. Three important ones to know are formaldehyde (methanal), acetaldehyde (ethanal), and acetone (propanone) (Figure 11.11).

Aldehydes and ketones have much lower boiling points than alcohols with a similar molecular weight. Ethanol, for example, has a boiling point of 78.5°C, while ethanal has a boiling point of 20°C (Table 11.2). The difference in boiling points is due to differences in how the molecules are attracted to one another. While alcohol molecules can form hydrogen bonds with a neighbor, aldehydes and ketones are unable to do so because hydrogen bonding requires that a hydrogen atom be covalently bonded to a nitrogen, oxygen, or fluorine atom. Aldehdyes and ketones interact with like molecules through London forces, as well as through dipole–dipole forces—the carbonyl group is slightly polar. These interactions are weaker, however, than the intermolecular forces that attract alcohol molecules to one another, so aldehydes and ketones have lower boiling points than do alcohols of similar size.

TABLE 11.2 Physical Properties of Selected Aldehydes and Ketones

Formula	IUPAC Name	Common Name	Boiling Point (°C)	Water Solubility (g/100 mL)
Aldehydes				
$H-C(=O)-H$	Methanal	Formaldehyde	−21	Miscible[a]
$CH_3C(=O)-H$	Ethanal	Acetaldehyde	20	Miscible
$CH_3CH_2C(=O)-H$	Propanal		49	16
$CH_3CH_2CH_2C(=O)-H$	Butanal		76	~0
Ketones				
$CH_3C(=O)CH_3$	Propanone	Acetone	56	Miscible
$CH_3C(=O)CH_2CH_3$	Butanone	Methyl ethyl ketone	80	26
$CH_3C(=O)CH_2CH_2CH_3$	2-Pentanone	Methyl propyl ketone	102	6
$CH_3C(=O)CH_2CH_2CH_2CH_3$	2-Hexanone	Methyl butyl ketone	150	2

[a] Can be mixed in all proportions.

The polarity of the carbonyl group and its ability to form hydrogen bonds with water molecules allows small aldehydes and ketones to be highly water soluble (*like dissolves like*). The larger an aldehyde or ketone, the less water soluble it is, because the water-insoluble hydrocarbon portion of the molecule contributes a greater share to the overall structure.

11.5 OXIDATION OF ALDEHYDES

As we saw in Section 11.3, $K_2Cr_2O_7$ will oxidize secondary alcohols into ketones and primary alcohols into aldehydes, which are further oxidized to carboxylic acids. Carboxylic acids can also be formed directly from aldehydes using $K_2Cr_2O_7$. The gain of an oxygen atom by the aldehyde is indication that oxidation has taken place (Figure 11.12*a*).

■ Benedict's reagent oxidizes aldehydes but not alcohols.

Special reagents have been developed for oxidizing aldehydes without affecting alcohols. One of these, called **Benedict's reagent**, contains Cu^{2+} as the oxidizing agent. An aldehyde will be oxidized to a carboxylic acid by Benedict's reagent, but alcohols remain unoxidized

(Figure 11.12*b*). As Benedict's reagent oxidizes an aldehyde, its initial blue color (due to Cu^{2+}) changes as Cu^{2+} is reduced to Cu^{+}. The final color of the solution depends on the amount of aldehyde initially present (Figure 11.13). Benedict's reagent has been used to monitor for diabetes by testing urine for the presence of glucose, an aldehyde-containing sugar. We will discuss other ways to measure glucose levels in Chapter 12.

(*a*)

$$CH_3\overset{O}{\overset{\|}{C}}-H \xrightarrow{K_2Cr_2O_7} CH_3\overset{O}{\overset{\|}{C}}-OH$$

$$HOCH_2-C_6H_4-\overset{O}{\overset{\|}{C}}-H \xrightarrow{K_2Cr_2O_7} HO-\overset{O}{\overset{\|}{C}}-C_6H_4-\overset{O}{\overset{\|}{C}}-OH$$

$$CH_3\overset{OH}{\overset{|}{C}}HCH_2\overset{O}{\overset{\|}{C}}-H \xrightarrow{K_2Cr_2O_7} CH_3\overset{O}{\overset{\|}{C}}CH_2\overset{O}{\overset{\|}{C}}-OH$$

(*b*)

$$CH_3\overset{O}{\overset{\|}{C}}-H \xrightarrow{Cu^{2+}} CH_3\overset{O}{\overset{\|}{C}}-OH$$

$$HOCH_2-C_6H_4-\overset{O}{\overset{\|}{C}}-H \xrightarrow{Cu^{2+}} HOCH_2-C_6H_4-\overset{O}{\overset{\|}{C}}-OH$$

$$CH_3\overset{OH}{\overset{|}{C}}HCH_2\overset{O}{\overset{\|}{C}}-H \xrightarrow{Cu^{2+}} CH_3\overset{OH}{\overset{|}{C}}HCH_2\overset{O}{\overset{\|}{C}}-OH$$

FIGURE 11.12

Aldehyde oxidation

(*a*) Potassium dichromate ($K_2Cr_2O_7$) oxidizes aldehydes and 1° alcohols into carboxylic acids and 2° alcohols into ketones. (*b*) Benedict's reagent (contains Cu^{2+}) oxidizes aldehydes into carboxylic acids, but does not oxidize 1° or 2° alcohols.

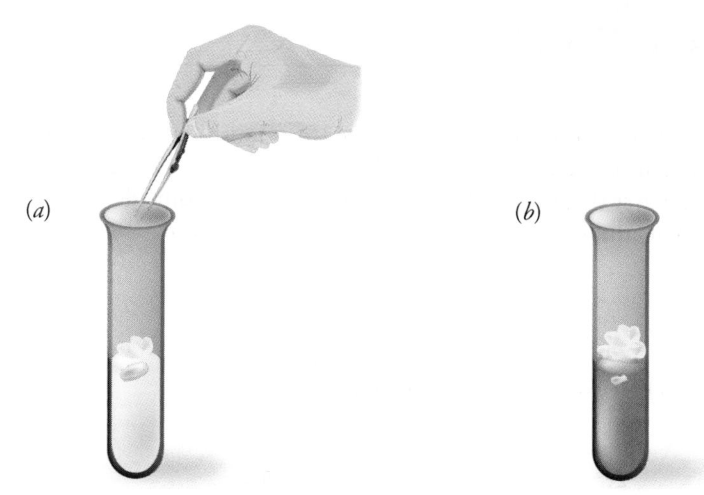

(*c*)

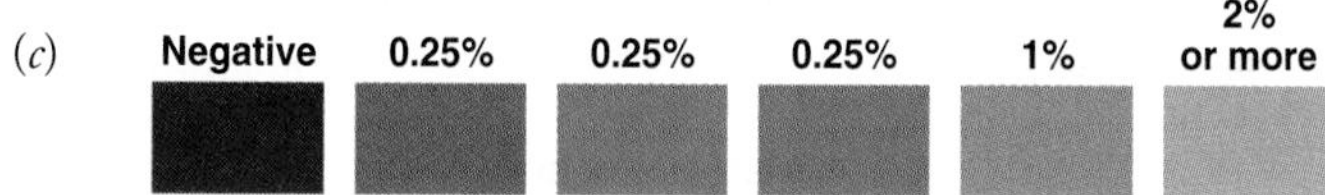

FIGURE 11.13

Benedict's reagent

(*a*) When Benedict's reagent is added to a test tube containing a glucose solution, Cu^{2+} gives the initial solution a blue color. (*b*) Oxidation of glucose's aldehyde group causes a color change to take place after a few minutes. (*c*) The final color of the solution depends on the initial % (w/v) concentration of glucose.

HEALTH Link

Aldehyde Dehydrogenase

When someone drinks beer, wine, or any other alcoholic beverage, the ethanol that it contains ends up in their liver, where it is metabolized. The first step in the metabolism of this alcohol is its oxidation to acetaldehyde, a reaction that is catalyzed by the enzyme alcohol dehydrogenase. Acetaldehyde is subsequently oxidized to acetic acid, with the help of a different enzyme called aldehyde dehydrogenase (Figure 11.14).

Acetaldehyde is toxic, so aldehyde dehydrogenase plays the important role of catalyzing the removal of any acetaldehyde that appears in the body as a result of ethanol oxidation. If, for some reason, the acetaldehyde remains in the body for long, its effects range from an intense flushing of the face and neck to nausea and vomiting.

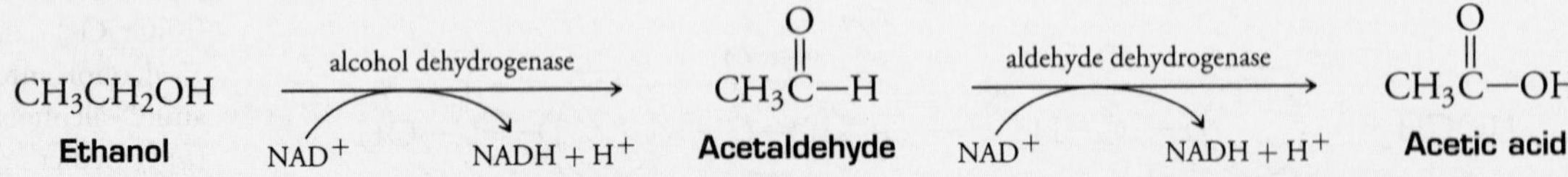

FIGURE | 11.14

Metabolizing ethanol
The first two steps in the metabolism of ethanol within the body are oxidation reactions. Alcohol dehydrogenase catalyzes the oxidation of ethanol to acetaldehyde and then aldehyde dehydrogenase catalyzes the oxidation of acetaldehyde to acetic acid. Each of these reactions takes place in the presence of NAD^+, a biochemical oxidizing agent.

One of the drugs that have been used to treat chronic alcoholism interferes with the action of aldehyde dehydrogenase. If a person drinks ethanol after taking this drug (disulfiram), the ethanol is oxidized to acetaldehyde in a normal fashion, but the oxidation of acetaldehyde is considerably slowed. This causes acetaldehyde concentrations to climb, and the person soon experiences nausea and vomiting, which, in some cases, can be a deterrent to further drinking.

$$CH_3CH_2N(CH_2CH_3)-\overset{S}{\overset{\|}{C}}-SS-\overset{S}{\overset{\|}{C}}-N(CH_2CH_3)CH_2CH_3$$

Disulfiram

11.6 REDUCTION OF ALDEHYDES AND KETONES

Catalytic hydrogenation, introduced previously in Sections 6.3 and 8.3, involves the reaction of the carbon–carbon double bond of an alkene with H_2, in the presence of a Pt catalyst (Figure 11.15*a*). Because a hydrogen atom adds to each of the double-bonded carbon atoms, hydrogenation is a reduction reaction.

FIGURE | 11.15

Catalytic hydrogenation
(*a*) Catalytic hydrogenation of an alkene produces an alkane.
(*b*) Catalytic hydrogenation of an aldehyde or a ketone produces an alcohol.

(*a*) $$\underset{\text{Ethylene}}{H_2C=CH_2} + H_2 \xrightarrow{Pt} \underset{\text{Ethane}}{H_3C-CH_3}$$

(*b*) $$\underset{\text{Formaldehyde}}{H-\overset{O}{\overset{\|}{C}}-H} + H_2 \xrightarrow{Pt} \underset{\text{Methyl alcohol}}{H-C(OH)H-H}$$

$$\underset{\text{Acetone}}{CH_3-\overset{O}{\overset{\|}{C}}-CH_3} + H_2 \xrightarrow{Pt} \underset{\text{Isopropyl alcohol}}{CH_3-CH(OH)-CH_3}$$

In the same way, catalytic hydrogenation can reduce the carbon–oxygen double bond of an aldehyde or ketone. In the presence of H_2 and Pt, the carbon–oxygen double bond of a carbonyl is converted to a single bond (Figure 11.15*b*) forming an —OH group in the process. When formaldehyde is reduced using H_2 and Pt, methanol is the product. Reduction converts all other aldehydes into 1° alcohols and all ketones into 2° alcohols.

SAMPLE PROBLEM 11.4

Predicting reduction products

Draw the alcohol product expected from each reduction reaction.

a. $CH_3CH_2CH_2\overset{\overset{\displaystyle O}{\|}}{C}{-}H + H_2 \xrightarrow{Pt}$

b. $CH_3CH_2\overset{\overset{\displaystyle O}{\|}}{C}CH_2CH_3 + H_2 \xrightarrow{Pt}$

c. 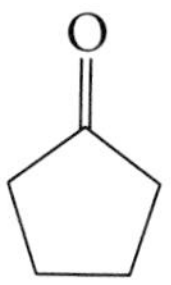 cyclopentanone (C=O on a five-membered ring) $+ H_2 \xrightarrow{Pt}$

Strategy

As with the reduction of a carbon–carbon double bond by H_2 and Pt, in the reduction of the carbon–oxygen double bond of an aldehyde or ketone a hydrogen atom adds to each double-bonded atom. The product of each reaction will be an alcohol.

Solution

a. $CH_3CH_2CH_2\overset{\overset{\displaystyle OH}{|}}{C}H_2$

b. $CH_3CH_2\overset{\overset{\displaystyle OH}{|}}{C}HCH_2CH_3$

c. 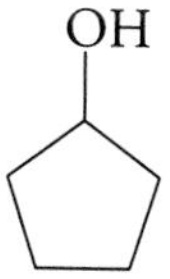 cyclopentanol (OH on a five-membered ring)

PRACTICE PROBLEM 11.4

Provide the missing reactant for each reaction.

a. $\quad + H_2 \xrightarrow{Pt} CH_3\overset{\overset{\displaystyle OH}{|}}{C}HCH_2CH_2CH_3$

b. $\quad + H_2 \xrightarrow{Pt} CH_3CH_2CH_2CH_2CH_2OH$

c. $\quad + H_2 \xrightarrow{Pt}$ cyclohexane ring carbon bearing H and OH, with CH_2CH_3 on the ring

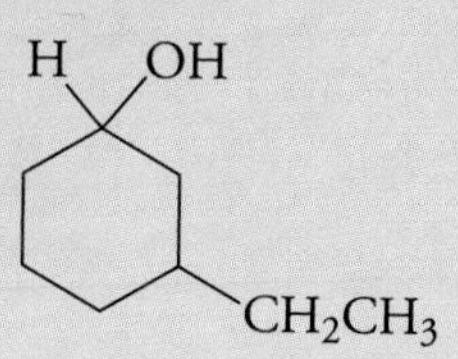

In the presence of the biochemical reducing agent NADH, particular enzymes in your body catalyze the reduction of aldehydes and ketones into alcohols. Hydroxysteroid dehydrogenases (HSDHs), for example, are a class of enzymes that catalyze reactions

involved in the biochemical synthesis of steroids. There are a number of different HSDH enzymes, each specific for a different region of a steroid carbon skeleton. Among these is 3α-HSDH, which reduces a ketone at position 3 of a steroid ring system. This enzyme catalyzes one of the reactions in the multistep biosynthesis of bile acids from cholesterol (Figure 11.16).

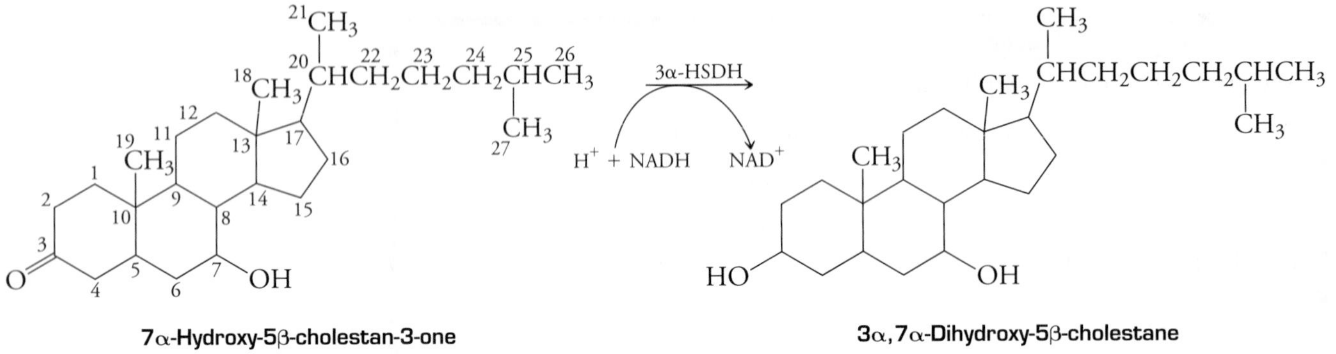

7α-Hydroxy-5β-cholestan-3-one $\xrightarrow[H^+ + NADH \;\to\; NAD^+]{3\alpha\text{-HSDH}}$ 3α,7α-Dihydroxy-5β-cholestane

FIGURE 11.16

Hydroxysteroid dehydrogenases
Hydroxysteroid dehydrogenases (HSDHs) catalyze the reduction of ketone groups in steroid carbon skeletons. 3α-HSDH acts on steroids that have a carbonyl group at ring position 3. The reaction shown is one step in the biosynthesis of bile acids from cholesterol.

HEALTH Link — *Protective Enzymes*

Oxidation and reduction reactions are a normal part of metabolic activity. As a by-product of these reactions, O_2 is sometimes reduced to superoxide ion (O_2^-) or hydrogen peroxide (H_2O_2), two extremely toxic compounds that can damage proteins, nucleic acids, and other vitally important biomolecules. As a protective measure, the body manufactures superoxide dismutase, an enzyme that catalyzes the conversion of two superoxide ions into a hydrogen peroxide molecule and an oxygen molecule. Catalase, another enzyme, catalyzes the breakdown of two hydrogen peroxide molecules into two water molecules and an oxygen molecule.

$$2O_2^- + 2H^+ \xrightarrow{\text{superoxide dismutase}} H_2O_2 + O_2$$

$$2H_2O_2 \xrightarrow{\text{catalase}} 2H_2O + O_2$$

Some white blood cells use the high toxicity of superoxide and hydrogen peroxide as a way to protect the body from bacteria and other foreign substances. In a process called phagocytosis (Greek: *phago* "to eat" + *cyto* "cell"), white blood cells engulf foreign material and destroy it by releasing O_2^- and H_2O_2 (Figure 11.17).

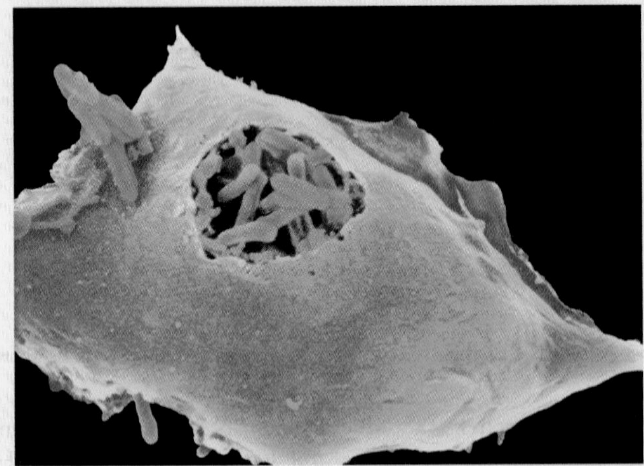

FIGURE 11.17

Phagocytosis
A white blood cell engulfs and digests bacteria by releasing superoxide ion and hydrogen peroxide.

Source: SPL/Photo Researchers, Inc.

11.7 REACTIONS OF ALCOHOLS WITH ALDEHYDES AND KETONES

Hemiacetals

In the presence of an H^+ catalyst, alcohols can undergo an addition reaction with the carbonyl group of an aldehyde or ketone. As with the reduction reaction described in Section 11.6, this reaction involves adding an atom or group of atoms to each of the double-bonded atoms of the C=O group. In the reaction of an alcohol with an aldehyde or ketone, the alcohol's —OH hydrogen atom attaches to the carbonyl oxygen atom and the remainder of the alcohol attaches to the carbonyl carbon atom (Figure 11.18*a*). The product that forms when an aldehyde or ketone is reacted with one alcohol molecule is called a **hemiacetal**. (Hemiacetals formed from ketones are also known as hemiketals.) A hemiacetal consists of *a carbon atom that is attached to both —OH and —OC.* Although the —OH group is found in alcohols and the C—O—C linkage is present in ethers, the distinct chemical behavior of hemiacetals results in their not being considered members of either family.

■ A hemiacetal carbon atom is attached to —OH and —OC.

To produce a hemiacetal, an alcohol must react with an aldehyde or a ketone. Molecules that contain both an alcohol and an aldehyde or a ketone functional group can form cyclic hemiacetals when the —OH reacts with a C=O in the same molecule. The cyclic hemiacetal shown in Figure 11.18*b* is formed in the same way as any other hemiacetal—the carbonyl oxygen atom gains a hydrogen atom and the carbonyl carbon atom gains an —OC group. As we will see in Chapter 12, the formation of cyclic hemiacetals is a very important part of the chemistry of sugars.

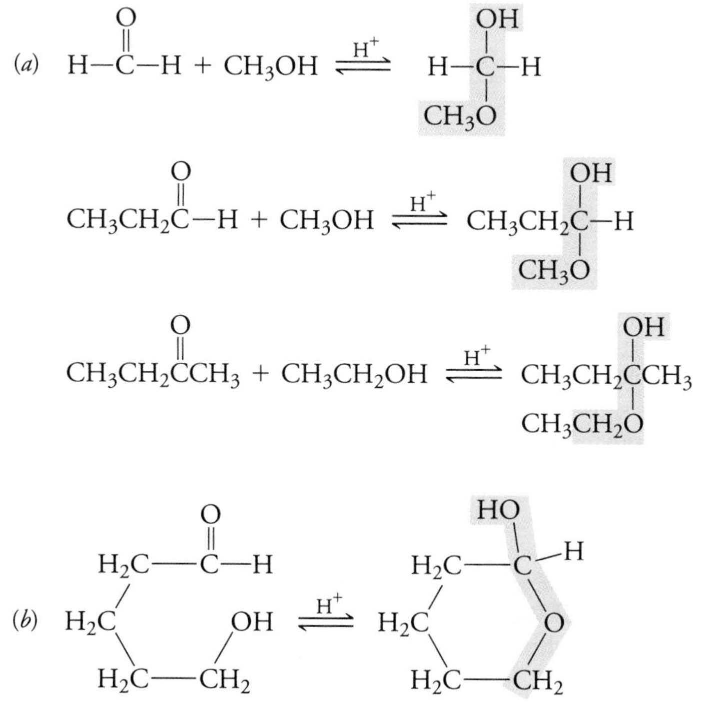

(*a*)
$$H{-}\overset{\overset{\displaystyle O}{\|}}{C}{-}H + CH_3OH \xrightleftharpoons{H^+} H{-}\underset{\underset{\displaystyle CH_3O}{|}}{\overset{\overset{\displaystyle OH}{|}}{C}}{-}H$$

$$CH_3CH_2\overset{\overset{\displaystyle O}{\|}}{C}{-}H + CH_3OH \xrightleftharpoons{H^+} CH_3CH_2\underset{\underset{\displaystyle CH_3O}{|}}{\overset{\overset{\displaystyle OH}{|}}{C}}{-}H$$

$$CH_3CH_2\overset{\overset{\displaystyle O}{\|}}{C}CH_3 + CH_3CH_2OH \xrightleftharpoons{H^+} CH_3CH_2\underset{\underset{\displaystyle CH_3CH_2O}{|}}{\overset{\overset{\displaystyle OH}{|}}{C}}CH_3$$

(*b*) H_2C—$C(=O)$—H, H_2C, OH, H_2C—CH_2 (open chain) $\xrightleftharpoons{H^+}$ H_2C—C(HO)(H), H_2C, O, H_2C—CH_2 (ring)

FIGURE 11.18

Hemiacetals
(*a*) The addition of an alcohol molecule to an aldehyde or ketone molecule produces a hemiacetal, in which —OH and —OC are attached to a single carbon atom. (*b*) When an alcohol group is present in the same molecule as an aldehyde or a ketone group, a cyclic hemiacetal can form.

Acetals

When two alcohol molecules react with an aldehyde or ketone in the presence of H^+, an **acetal** forms (Figure 11.19*a*). An acetal consists of *a carbon atom that is attached to two —OC groups.* (Acetals formed from ketones are also known as ketals.) Even though they contain C—O—C linkages, acetals are not considered to be ethers. The chemistry of these two groups is entirely different. Varying the alcohol used allows formation of a wide range of acetals from a given aldehyde or ketone. As shown in Figure 11.19*b*,

■ An acetal carbon atom is attached to two —OC groups.

FIGURE 11.19

Acetals
Acetals, in which two OC groups are attached to a single carbon atom, can be produced (*a*) by the reaction of an aldehyde or a ketone with two alcohol molecules or (*b*) by the reaction of a hemiacetal with one alcohol molecule.

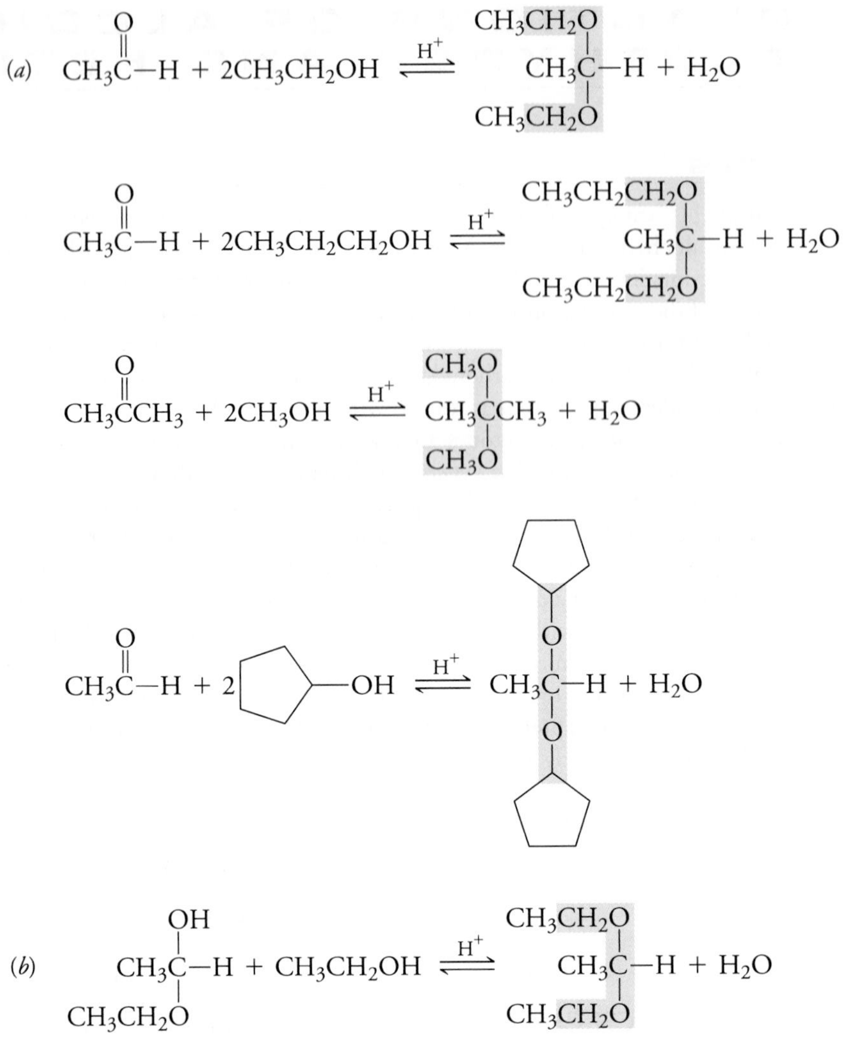

(*a*)

$$CH_3\overset{O}{\overset{\|}{C}}-H + 2CH_3CH_2OH \underset{}{\overset{H^+}{\rightleftharpoons}} CH_3\underset{OCH_2CH_3}{\overset{OCH_2CH_3}{C}}-H + H_2O$$

$$CH_3\overset{O}{\overset{\|}{C}}-H + 2CH_3CH_2CH_2OH \overset{H^+}{\rightleftharpoons} CH_3\underset{OCH_2CH_2CH_3}{\overset{OCH_2CH_2CH_3}{C}}-H + H_2O$$

$$CH_3\overset{O}{\overset{\|}{C}}CH_3 + 2CH_3OH \overset{H^+}{\rightleftharpoons} CH_3\underset{OCH_3}{\overset{OCH_3}{C}}CH_3 + H_2O$$

$$CH_3\overset{O}{\overset{\|}{C}}-H + 2\,(\text{cyclopentyl})-OH \overset{H^+}{\rightleftharpoons} CH_3\underset{O-\text{cyclopentyl}}{\overset{O-\text{cyclopentyl}}{C}}-H + H_2O$$

(*b*)

$$CH_3\underset{OCH_2CH_3}{\overset{OH}{C}}-H + CH_3CH_2OH \overset{H^+}{\rightleftharpoons} CH_3\underset{OCH_2CH_3}{\overset{OCH_2CH_3}{C}}-H + H_2O$$

acetals also can be produced directly from hemiacetals. In Chapter 12 we will see that acetals play a key role in the structure of starch, glycogen, cellulose, and many other carbohydrates.

As shown in the reaction equations in this section, hemiacetal and acetal formation is reversible. Adding water will increase the rate of the reverse reaction and convert acetals into alcohols and aldehydes or ketones.

SAMPLE PROBLEM 11.5

Forming hemiacetals and acetals

Draw the organic product of each reaction.

a. $CH_3CH_2\overset{O}{\overset{\|}{C}}-H + CH_3CH_2CH_2OH \overset{H^+}{\rightleftharpoons}$

b. 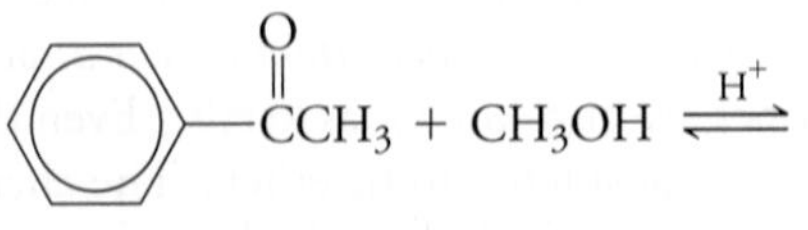 $C_6H_5-\overset{O}{\overset{\|}{C}}CH_3 + CH_3OH \overset{H^+}{\rightleftharpoons}$

c.  $H{-}\overset{\overset{\displaystyle O}{\|}}{C}{-}H + 2CH_3CH_2CH_2OH \xrightleftharpoons{H^+}$

d. 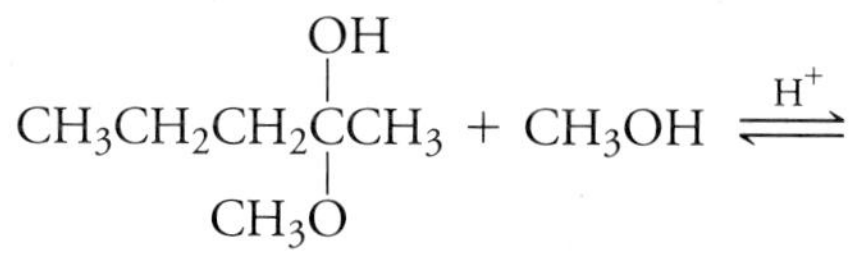 $CH_3CH_2CH_2\overset{\overset{\displaystyle OH}{|}}{\underset{\underset{\displaystyle CH_3O}{|}}{C}}CH_3 + CH_3OH \xrightleftharpoons{H^+}$

Strategy

The product formed depends on the reactant used and the number of alcohol molecules added. An aldehyde or a ketone molecule plus one alcohol molecule gives a hemiacetal. Adding two alcohol molecules gives an acetal. A hemiacetal plus one alcohol molecule produces an acetal.

Solution

a. 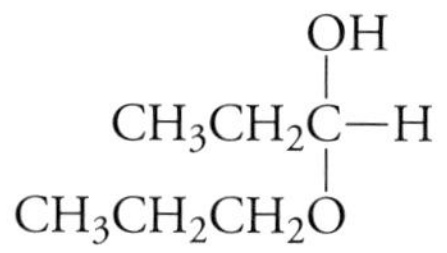 $CH_3CH_2\overset{\overset{\displaystyle OH}{|}}{\underset{\underset{\displaystyle CH_3CH_2CH_2O}{|}}{C}}{-}H$

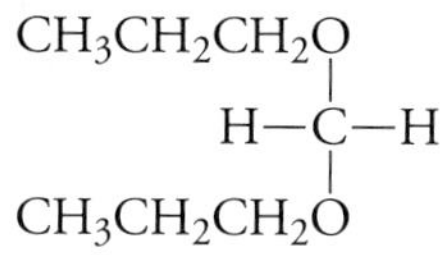

b. 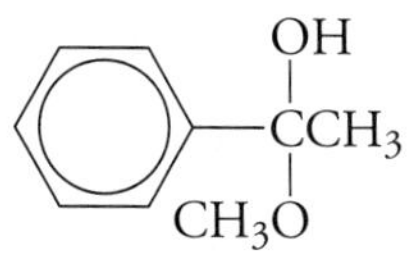 $C_6H_5{-}\overset{\overset{\displaystyle OH}{|}}{\underset{\underset{\displaystyle CH_3O}{|}}{C}}CH_3$

c. $H{-}\overset{\overset{\displaystyle CH_3CH_2CH_2O}{|}}{\underset{\underset{\displaystyle CH_3CH_2CH_2O}{|}}{C}}{-}H$

d. 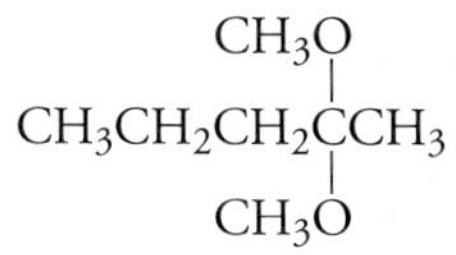 $CH_3CH_2CH_2\overset{\overset{\displaystyle CH_3O}{|}}{\underset{\underset{\displaystyle CH_3O}{|}}{C}}CH_3$

PRACTICE PROBLEM 11.5

Draw the missing reactant for each reaction.

a. 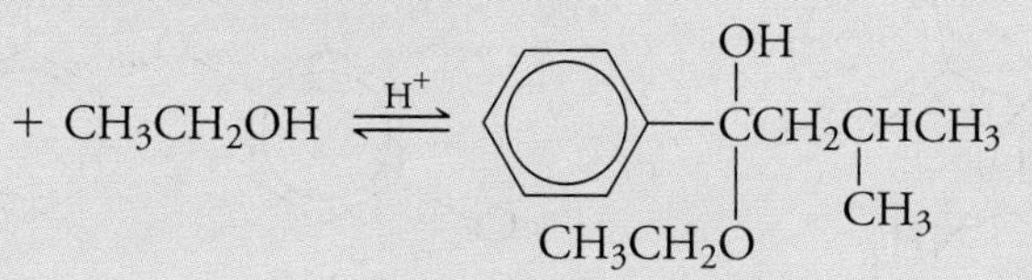 $+ CH_3CH_2OH \xrightleftharpoons{H^+} C_6H_5{-}\overset{\overset{\displaystyle OH}{|}}{\underset{\underset{\displaystyle CH_3CH_2O}{|}}{C}}CH_2\underset{\underset{\displaystyle CH_3}{|}}{C}HCH_3$

b. $CH_3CH_2CH_2\overset{\overset{\displaystyle O}{\|}}{C}CH_3 + \quad \xrightleftharpoons{H^+} CH_3CH_2CH_2\overset{\overset{\displaystyle OH}{|}}{\underset{\underset{\displaystyle CH_3CH_2CH_2O}{|}}{C}}CH_3$

c. $+ CH_3OH \xrightleftharpoons{H^+}$ 1-methoxy-3-methylcyclopentan-1-ol (ring carbon bearing CH_3O and OH; CH_3 on ring)

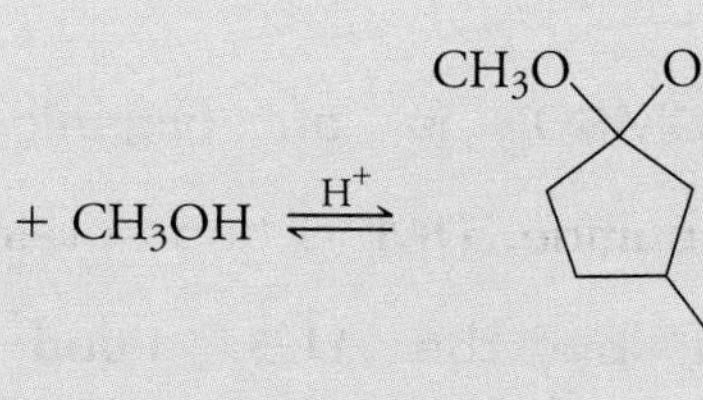

d. 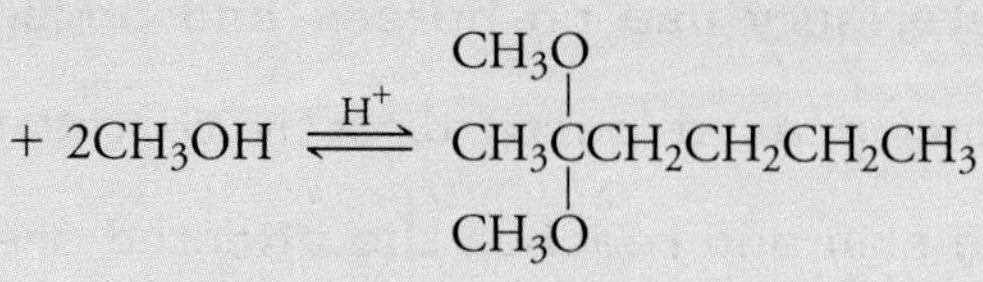 $+ 2CH_3OH \xrightleftharpoons{H^+} CH_3\overset{\overset{\displaystyle CH_3O}{|}}{\underset{\underset{\displaystyle CH_3O}{|}}{C}}CH_2CH_2CH_2CH_3$

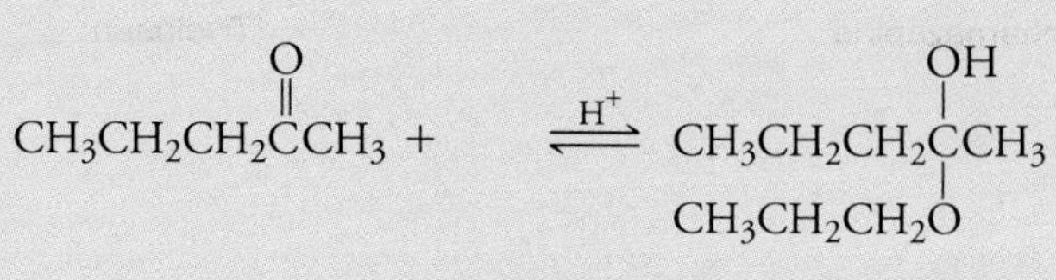

HEALTH Link: Drugs in the Environment

Each day, millions of gallons of liquid and solid human waste arrive at sewage treatment plants. After removal of solids, the liquid waste is cleaned to remove disease-causing organisms such as bacteria and viruses, and is then released into surface or ground water. The sanitized solids end up as biosolids (sludge), about half of which are used as fertilizer for gardens and farmland.

Research has found that not all traces of human activity are removed from these treated liquids and solids and that, through them, pharmaceuticals and other chemicals are introduced into the environment. When a drug is flushed down the toilet, either directly as a means of quick disposal or indirectly in human waste, some of that drug is likely to end up being present in the water or biosolids that leave a sewage treatment plant.

Among the pharmaceuticals that have been detected in rivers, lakes, and biosolids are the painkiller hydrocodone, the anti-inflammatory ketoprofen, the antiepileptic carbamazepine (Figure 11.20), and the antidepressant fluoxetine (Prozac; see Problem 11.77). Acetaminophen (a pain reliever; see Problem 11.76), the disinfectant triclosan (used in antibacterial soaps), the steroid hormone estrone (used in birth control pills; see problem 11.78), and caffeine and nicotine (Figure 10.1) are among the other chemicals that have been detected.

Scientists question whether these compounds introduced into the environment via sewage treatment plants have any adverse effects. At this time, the concentration of these drugs and other chemicals is below that expected to be harmful to organisms in the environment.

FIGURE 11.20

Compounds present in water and biosolids released from sewage treatment plants Hydrocodone (painkiller), ketoprofen (anti-inflammatory), carbamazepine (antiepileptic), and triclosan (disinfectant) are among the compounds that make it through sewage treatment and into the environment.

Hydrocodone

Ketoprofen

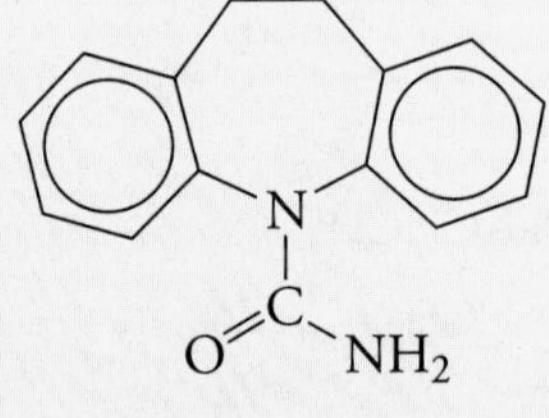

Carbamazepine

Triclosan

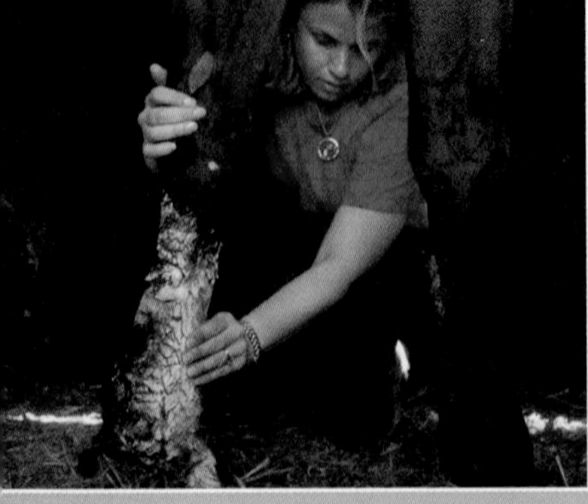

Dimethyl sulfoxide (DMSO) is an organic compound produced by oxidizing dimethyl sulfide. DMSO has anti-inflammatory properties and is approved by the U.S. Food and Drug Administration for veterinary use on horses and dogs. On horses, for example, it is applied as a paste directly to the legs—DMSO rapidly penetrates the skin and reaches the affected area.

Some thought has been given to using DMSO as a solvent to dissolve therapeutic drugs and to transport them directly through the skin of humans. Studies using laboratory animals, however, have shown that DMSO can damage vision, so its use on humans is not approved.

One unpleasant side effect of using DMSO is that once this odorless compound is absorbed by the body of an animal (humans included) it is readily reduced to dimethyl sulfide, which has a very strong garlic-like odor. To overcome this drawback, veterinarians have found that DMSO's oxidation product, methylsulfonylmethane (MSM), is a good replacement for DMSO. MSM has the same anti-inflammatory effect as DMSO, but in the body it is not reduced to dimethyl sulfide.

$$\underset{\text{Dimethyl sulfide}}{CH_3SCH_3} \longrightarrow \underset{\text{DMSO}}{CH_3\overset{\overset{O^-}{|+}}{S}CH_3} \longrightarrow \underset{\text{MSM}}{CH_3\overset{\overset{O^-}{|2+}}{\underset{\underset{O^-}{|}}{S}}CH_3}$$

summary of objectives

1 Describe the structure of molecules that belong to the alcohol, ether, thiol, sulfide, disulfide, aldehyde, and ketone families, describe how they are named, and identify the intermolecular forces that attract similar molecules from this list to one another.

Alcohols have the —OH functional group, **ethers** the C—O—C functional group, **thiols** the —SH group, **sulfides** the C—S—C group, and **disulfides** the C—S—S—C group. When assigning IUPAC names to alcohols and thiols, the parent chain contains the functional group and is numbered from the end nearer that group. IUPAC names of alcohols end in "ol" and those of thiols end in "thiol." Common names of alcohols, ethers, sulfides, and disulfides are put together by giving the names of the group or groups attached to O or S, followed by the name of the family (ethyl methyl sulfide, for example). For purposes of assigning common names, thiols are called mercaptans (ethyl mercaptan, etc.). Aldehydes and ketones contain a **carbonyl** (C=O) group. In **aldehydes** the carbonyl carbon atom is attached to two hydrogen atoms or one hydrogen atom and one carbon atom, and in **ketones** the carbonyl carbon atom is attached to two other carbon atoms. When assigning IUPAC names to aldehydes and ketones, the parent chain contains the functional group and is numbered from the end nearer that group. IUPAC names of aldehydes end in "al" and those of ketones end in "one." Common names of ketones can be formed by giving the names of the groups attached to the C=O, followed by "ketone." Alcohol molecules are held to one another primarily by hydrogen bonds. London forces also contribute, with the extent being determined by the size of the molecule. Aldehydes, ketones, and ethers are attracted to one other by dipole–dipole and London forces, while thiols, sulfides, and disulfides interact with like molecules only through London forces.

2 Distinguish 1°, 2°, and 3° alcohols.

In **1°** alcohols the carbon atom carrying the —OH is attached to just one other carbon atom. In **2°** and **3°** alcohols, this carbon atom is attached, respectively, to two and three other carbon atoms.

3 Describe the nucleophilic substitution reactions that can be used to prepare alcohols, ethers, thiols, and sulfides.

In a **nucleophilic substitution reaction**, an electron-rich **nucleophile** displaces a **leaving group** from a carbon atom. When certain alkyl halides (RCl, RBr, or RI) are reacted (a) with OH^- an alcohol is produced, (b) with RO^- an ether is produced, (c) with SH^- a thiol is produced, and (d) with RS^- a sulfide is produced.

4 Predict the major product of the addition reaction between an alkene and H_2O/H^+ and the major product for the elimination reaction between an alcohol and H^+/heat.

Reacting an alkene with H_2O in the presence of H^+ produces an alcohol. From an asymmetric alkene, the major product (as predicted by **Markovnikov's rule**) is the alcohol formed by adding a hydrogen atom to the double-bonded carbon atom that carries more hydrogen atoms. In the presence of H^+ and heat, alcohols are dehydrated to form alkenes. When two different alkenes can be produced from one alcohol, the major product is the one formed by removing a hydrogen atom from the carbon atom adjacent to the C—OH that carries the fewer hydrogen atoms.

5 Describe the oxidation reactions of alcohols, thiols, sulfides, and aldehydes.

When reacted with $K_2Cr_2O_7$, 1° alcohols are converted into aldehydes, which are then oxidized to carboxylic acids, and 2° alcohols are converted into ketones. Tertiary (3°) alcohols are not oxidized by this reagent. In the oxidation of thiols by I_2, two thiol molecules react to form a disulfide. The oxidation of a sulfide by H_2O_2 produces a sulfoxide. When reacted with $K_2Cr_2O_7$ or **Benedict's reagent** (Cu^{2+}), aldehydes are oxidized to carboxylic acids. Alcohols are not oxidized by Benedict's reagent.

6 Explain what happens when an aldehyde or ketone is reacted with H_2 and Pt and when one of these compounds is reacted with one or two alcohol molecules, in the presence of H^+.

When treated with H_2 and Pt, an aldehyde or ketone is reduced to an alcohol. In the presence of H^+, an aldehyde or ketone reacts with one alcohol molecule to form a **hemiacetal** and with two alcohol molecules to form an **acetal**. Hemiacetal and acetal formation is reversible.

summary of reactions

Section 11.2

Nucleophilic substitution

$$^{-}OH + CH_3Br \longrightarrow CH_3OH + Br^-$$

$$^{-}OCH_3 + CH_3Br \longrightarrow CH_3OCH_3 + Br^-$$

$$^{-}SH + CH_3Br \longrightarrow CH_3SH + Br^-$$

$$^{-}SCH_3 + CH_3Br \longrightarrow CH_3SCH_3 + Br^-$$

Hydration of alkenes

$$CH_2{=}CHCH_3 + H_2O \xrightarrow{H^+} CH_3\overset{\overset{\Large OH}{|}}{C}HCH_3 + \overset{\overset{\Large OH}{|}}{C}H_2CH_2CH_3$$

Major product — Minor product

Section 11.3

Oxidation of alcohols

$$CH_3CH_2OH \xrightarrow{K_2Cr_2O_7} CH_3\overset{\overset{O}{\|}}{C}{-}OH$$

$$CH_3\overset{\overset{OH}{|}}{C}HCH_3 \xrightarrow{K_2Cr_2O_7} CH_3\overset{\overset{O}{\|}}{C}CH_3$$

Oxidation of thiols

$$CH_3SH + HSCH_3 \xrightarrow{I_2} CH_3SSCH_3$$

Oxidation of sulfides

$$CH_3SCH_3 \xrightarrow{H_2O_2} CH_3\overset{\overset{O^-}{|+}}{S}CH_3$$

Dehydration of alcohols

$$CH_3CH_2\overset{\overset{OH}{|}}{C}HCH_3 \xrightarrow[heat]{H^+} CH_3CH{=}CHCH_3 + CH_3CH_2CH{=}CH_2$$

Major product — Minor product

Section 11.5

Oxidation of aldehydes

$$CH_3\overset{\overset{O}{\|}}{C}{-}H \xrightarrow{K_2Cr_2O_7} CH_3\overset{\overset{O}{\|}}{C}{-}OH$$

$$CH_3\overset{\overset{O}{\|}}{C}{-}H \xrightarrow{Cu^{2+}} CH_3\overset{\overset{O}{\|}}{C}{-}OH$$

Section 11.6

Reduction of aldehydes and ketones

$$CH_3\overset{\overset{O}{\|}}{C}{-}H + H_2 \xrightarrow{Pt} CH_3CH_2OH$$

Section 11.7

Hemiacetal formation

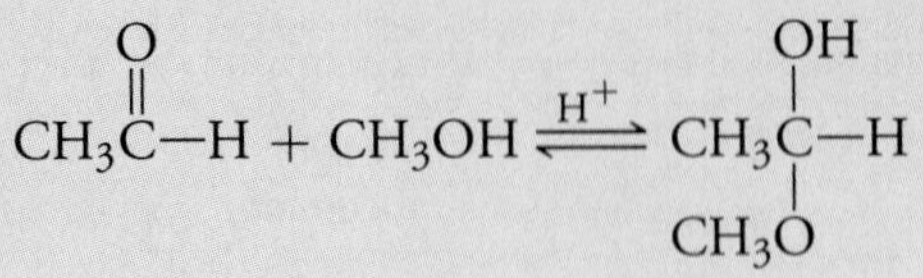

$$\mathrm{CH_3\overset{\overset{\displaystyle O}{\|}}{C}{-}H + CH_3OH \overset{H^+}{\rightleftharpoons} CH_3\underset{\underset{\displaystyle CH_3O}{|}}{\overset{\overset{\displaystyle OH}{|}}{C}}{-}H}$$

Acetal formation

$$\mathrm{CH_3\overset{\overset{\displaystyle O}{\|}}{C}{-}H + 2CH_3OH \overset{H^+}{\rightleftharpoons} CH_3\underset{\underset{\displaystyle CH_3O}{|}}{\overset{\overset{\displaystyle CH_3O}{|}}{C}}{-}H}$$

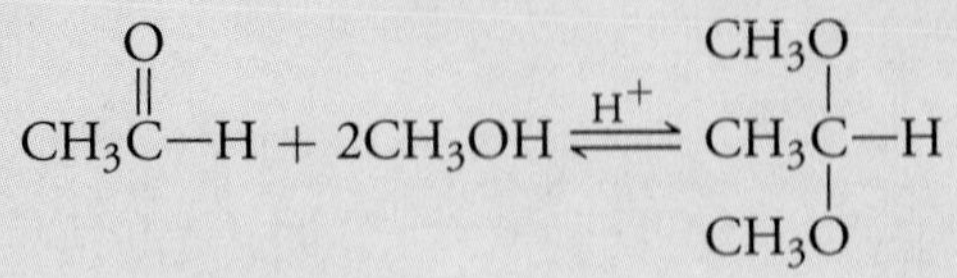

END OF CHAPTER PROBLEMS

Answers to problems whose numbers are printed in color are given in Appendix D. More challenging questions are marked with an asterisk. Problems within colored rules are paired. **ILW** = Interactive Learning Ware solution is available at *www.wiley.com/college/raymond.*

11.1 Alcohols, Ethers, and Related Compounds

11.1 Describe the structure of each type of molecule.
a. alcohol
b. ether
c. thiol
d. sulfide
e. disulfide

11.2 List the intermolecular forces that can hold each molecule to an identical one.
a. alcohol
b. ether
c. thiol
d. sulfide
e. disulfide

11.3 Identify each alcohol as 1°, 2°, or 3°.

a. 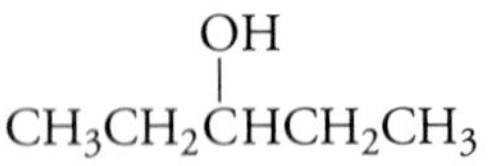 $\mathrm{CH_3CH_2\overset{\overset{\displaystyle OH}{|}}{C}HCH_2CH_3}$

b. $\mathrm{CH_3\underset{\underset{\displaystyle CH_2CH_3}{|}}{\overset{\overset{\displaystyle OH}{|}}{C}}CH_2CH_3}$

c. $\mathrm{HOCH_2\overset{\overset{\displaystyle CH_3}{|}}{C}HCH_2CH_3}$

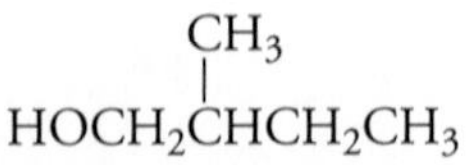

11.4 Identify each alcohol as 1°, 2°, or 3°.

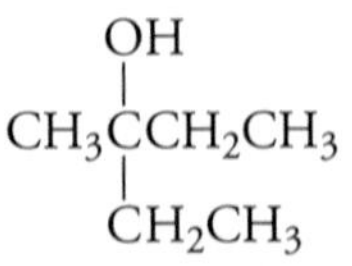

a. 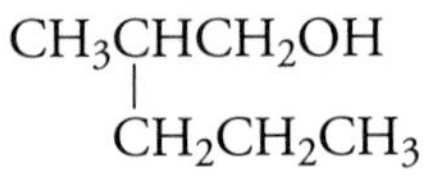 $\mathrm{CH_3\underset{\underset{\displaystyle CH_2CH_2CH_3}{|}}{C}HCH_2OH}$

b. $\mathrm{CH_3CH_2\overset{\overset{\displaystyle CH_3CHCH_3}{|}}{C}HOH}$

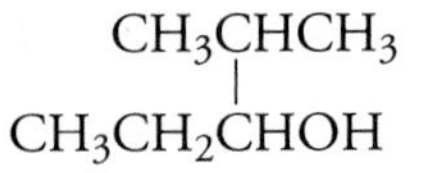

c. (cyclopentane ring bearing H_3C and OH on the same carbon)

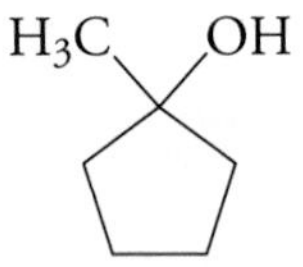

11.5 Give the IUPAC name of each alcohol in Problem 11.3.

11.6 Give the IUPAC name of each alcohol in Problem 11.4.

11.7 Draw each alcohol molecule.
a. 2-hexanol
b. 3-methyl-1-pentanol
c. 4-isopropylcyclohexanol

11.8 Draw each alcohol molecule.
a. 2,3-dimethyl-3-hexanol
b. 4,4-dimethyl-2-hexanol
c. *cis*-3-methylcyclohexanol

11.9 **a.** In some arid parts of the United States, a thin layer of 1-octadecanol (octadec is the IUPAC prefix for 18) is floated on the top of reservoirs to slow water evaporation. Write the condensed molecular formula of this alcohol.

b. Why is this alcohol insoluble in water?

11.10 The molecule $CH_3CH_2CH_2SH$ is partly responsible for the strong odor that is associated with onions. Give its IUPAC and common name.

11.11 Give the IUPAC and common name of the molecule.

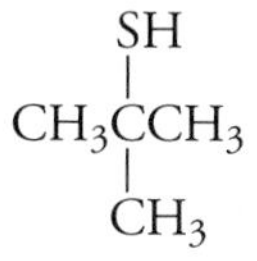

$$CH_3\underset{\displaystyle CH_3}{\overset{\displaystyle SH}{C}}CH_3$$

11.12 Give the common name of each ether.

a. 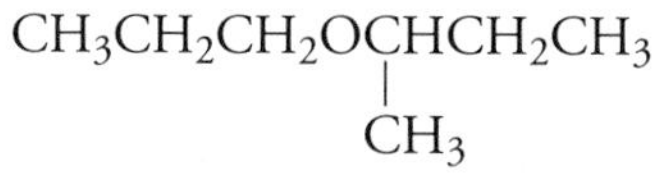 $CH_3CH_2CH_2O\underset{\displaystyle CH_3}{C}HCH_2CH_3$

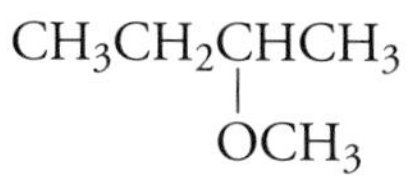

b. $CH_3CH_2\underset{\displaystyle OCH_3}{C}HCH_3$

c. cyclohexyl—OCH_3

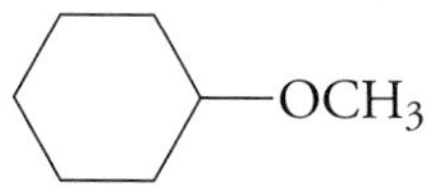

11.13 Give the common name of each ether.

a. 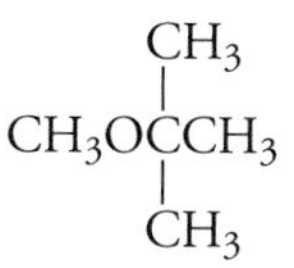 $CH_3O\underset{\displaystyle CH_3}{\overset{\displaystyle CH_3}{C}}CH_3$

b. 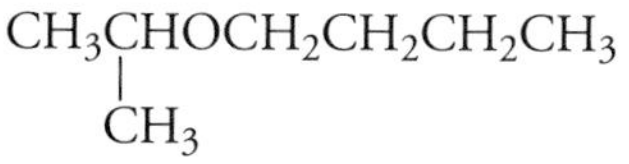 $CH_3\underset{\displaystyle CH_3}{C}HOCH_2CH_2CH_2CH_3$

c. $CH_3CH_2CH_2CH_2CH_2OCH_2CH_2CH_2CH_2CH_3$

*__11.14__ Which of the following is a liquid at STP: CH_3CH_2OH or CH_3OCH_3?

11.15 Account for the fact that dipropyl ether has a higher boiling point than diethyl ether.

11.16 Explain why $CH_3CH_2OCH_2CH_3$ is less soluble in water than its constitutional isomer $CH_3CH_2CH_2CH_2OH$.

11.17 Account for the fact that dipropyl ether has a lower solubility in water than diethyl ether.

11.18 Ethanol (molecular weight = 46.0 amu) has a boiling point of 75.5°C and propane (molecular weight = 44.0 amu) has a boiling point of −42°C. Account for the difference in boiling point.

11.19 Ethanol (molecular weight = 46.0 amu) is miscible in water, while propane (molecular weight = 44.0 amu) is insoluble. Account for the difference in water solubility.

11.20 1-Propanol has a boiling point of 97.4°C, and 1-butanol has a boiling point of 117.3°C. Account for the difference in boiling point.

11.21 1-Propanol is miscible in water and 1-butanol has a water solubility of 8.0 g/100 mL. Account for this difference in water solubility.

11.22 Methanethiol (molecular weight = 48.0 amu) has a boiling point of 6°C and ethanol (molecular weight = 46.0 amu) has a boiling point of 78.5°C. Account for the difference in boiling point.

11.23 Methanethiol (molecular weight = 48.0 amu) has a water solubility of 2.3 g/100 mL and ethanol (molecular weight = 46.0 amu) is miscible in water. Account for the difference in water solubility.

11.24 Isoimpinellin, a naturally occurring ether, is found at very low levels in celery. Tests have shown that at high concentrations this compound can be a carcinogen (cancer-causing agent). This is not cause to avoid eating celery, however, because many foods naturally contain toxic substances at very low levels.

a. How many ether groups does isoimpinellin contain?

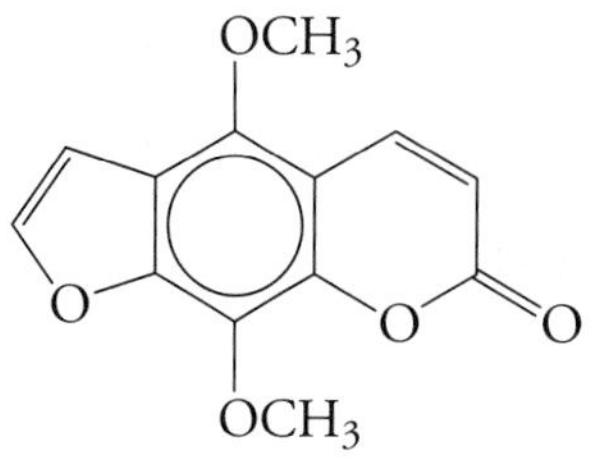

OCH_3 O O O OCH_3

Isoimpinellin

* **b.** What other functional groups are present in this molecule?

11.25 Thiotic acid is a growth factor for many bacteria. Which functional groups does this molecule contain?

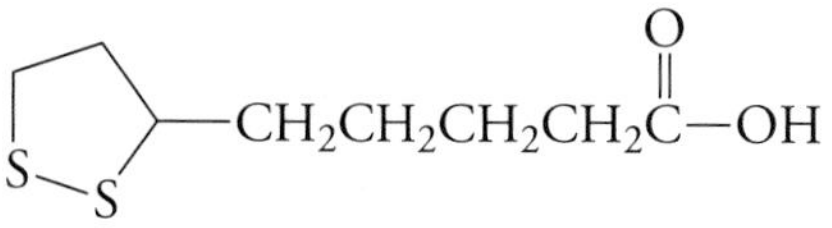

S S —$CH_2CH_2CH_2CH_2\overset{\displaystyle O}{\overset{\|}{C}}—OH$

Thiotic acid

11.2 Preparation

11.26 Draw the organic product of each nucleophilic substitution reaction.

a. $^{-}OH + CH_3CH(CH_3)CH_2CH_2I \longrightarrow$

b. $^{-}OCH_2CH_2CH_3 + CH_3Br \longrightarrow$

c. $^{-}SCH_3 + CH_3CH_2CH(Cl)CH_3 \longrightarrow$

d. $^{-}SH + CH_3CH_2CH_2CH_2Br \longrightarrow$

11.27 Draw the organic product of each nucleophilic substitution reaction.

a. $^{-}OH + CH_3CH(Br)CH_3 \longrightarrow$

b. $^{-}OCH_3 + CH_3CH_2CH_2CH_2CH_2Br \longrightarrow$

c. $^{-}SH + CH_3Cl \longrightarrow$

d. $^{-}SCH_2CH_2CH_3 + CH_3CH_2CH_2I \longrightarrow$

11.28 Name each of the products in Problem 11.26.

11.29 Name each of the products in Problem 11.27.

11.30 Draw the missing alkyl bromide (RBr) for each reaction.

a. $^{-}OCH_3 + \qquad \longrightarrow CH_3CH_2OCH_3$

b. $^{-}SH + \qquad \longrightarrow CH_3CH_2CH_2SH$

c. $^{-}SCH_2CH_3 + \qquad \longrightarrow CH_3CH_2SCH_2CH_3$

11.31 Draw the missing alkyl bromide (RBr) for each reaction.

a. $^{-}OCH_2CH_3 + \qquad \longrightarrow CH_3CH_2OCH_2CH_3$

b. $^{-}SCH_2CH_2CH_3 + \qquad \longrightarrow CH_3SCH_2CH_2CH_3$

c. $^{-}SH + \qquad \longrightarrow CH_3CH_2CH_2CH(SH)CH_3$

11.32 Write a chemical equation that shows how each compound can be produced from an alkyl bromide (RBr).

a. $CH_3CH_2CH_2OH$

b. $CH_3OCH_2CH_2CH_3$

11.33 Write a chemical equation that shows how each compound can be produced from an alkyl bromide (RBr).

a. $CH_3CH_2SCH_2CH_2CH_2CH_3$

b. $CH_3CH(CH_3)CH_2CH_2SH$

11.34 Draw the major product of each reaction.

a. $CH_3CH_2C(CH_3){=}CH_2 + H_2O \xrightarrow{H^+}$

b. 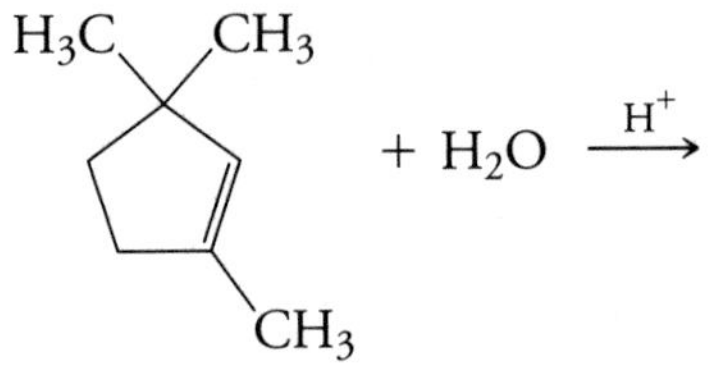 3,3-dimethyl-1-methylcyclopentene (ring with H_3C and CH_3 on one carbon, CH_3 on the double-bond carbon) $+ H_2O \xrightarrow{H^+}$

c. $CH_2{=}C(CH_3)CH_2CH_2CH_3 + H_2O \xrightarrow{H^+}$

11.35 Draw the major product of each reaction.

a. $CH_3CH{=}C(CH_2CH_3)CH_2CH_3 + H_2O \xrightarrow{H^+}$

b. 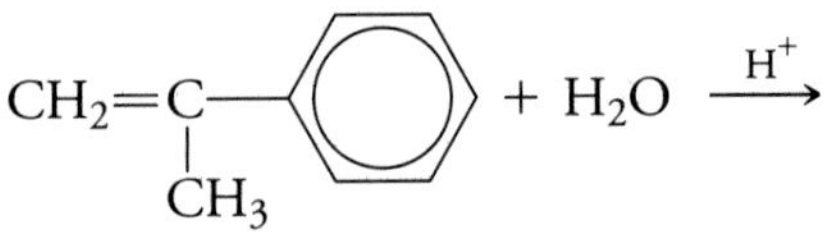 $CH_2{=}C(CH_3){-}C_6H_5 + H_2O \xrightarrow{H^+}$

c. cyclohexene $+ H_2O \xrightarrow{H^+}$

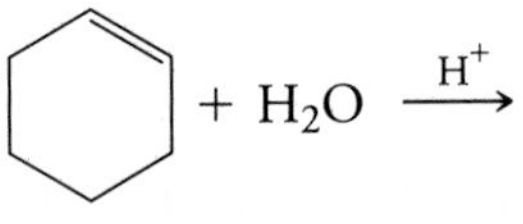

11.36 The molecule CH_3Cl is named methyl chloride. Name each of the following molecules:

a. CH_3CH_2Br

b. CH_3CHFCH_3

c. $CH_3CH_2CH_2CH_2I$

d. $CH_3CH(CH_3)CH_2Cl$

11.37 Methyl chloride has the formula CH_3Cl. Draw each of the following molecules:

a. propyl chloride

b. *s*-butyl bromide

c. isopropyl fluoride

d. *t*-butyl iodide

11.3 Reactions

11.38 Draw the organic product (if any) expected from each reaction.

a. 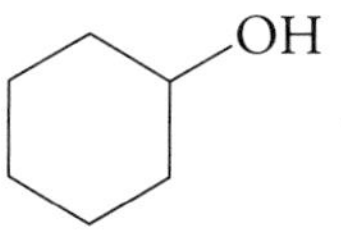 (cyclohexanol) + $K_2Cr_2O_7$ ⟶

b. $HOCH_2CH_2CH_2CH_2CH_3 + K_2Cr_2O_7$ ⟶

c. 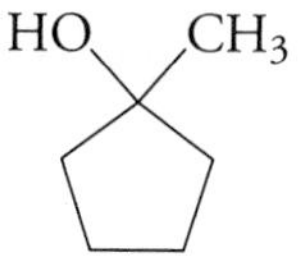 (1-methylcyclopentanol) + $K_2Cr_2O_7$ ⟶

d. $2CH_3SH + I_2$ ⟶

e. $2CH_3CH(SH)CH_3 + I_2$ ⟶

11.39 Draw the organic product (if any) expected from each reaction.

a. $CH_3CH_2CH_2CH_2OH + K_2Cr_2O_7$ ⟶

b. $CH_3CH_2CH(OH)CH_3 + K_2Cr_2O_7$ ⟶

c. 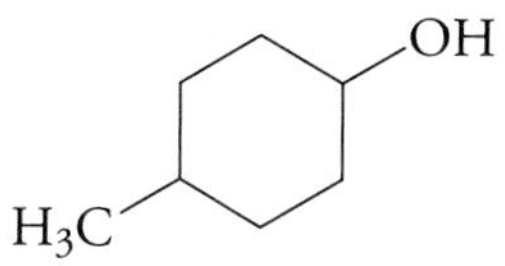 (4-methylcyclohexanol) + $K_2Cr_2O_7$ ⟶

d. $2CH_3CH_2CH_2SH + I_2$ ⟶

e. 2 (cyclopentanethiol) + I_2 ⟶

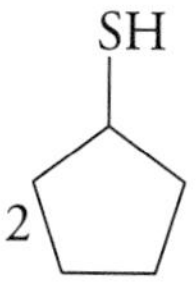

11.40 a. Lactate, which builds up in muscle cells during exercise, is sent to the liver where the enzyme lactate dehydrogenase catalyzes the oxidation of this 2° alcohol to pyruvate. Draw pyruvate.

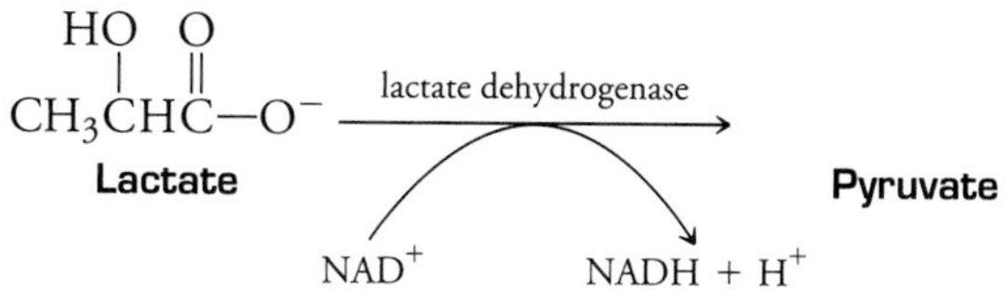

$$CH_3CH(OH)C(=O)-O^- \xrightarrow{\text{lactate dehydrogenase}} \text{Pyruvate}$$

Lactate ⟶ Pyruvate (NAD^+ ⟶ $NADH + H^+$)

b. Lactate and pyruvate are the conjugate bases of lactic acid and pyruvic acid, respectively. Draw these carboxylic acids.

***11.41** Draw each molecule named below and draw the major organic product expected when each is reacted with H^+ and heat.

a. 2,3-dimethyl-2-butanol

b. 2,3-dimethyl-3-hexanol

c. 2-methylcyclopentanol

***11.42** Draw each molecule named below and draw the major organic product expected when each is reacted with H^+ and heat.

a. 2,3-dimethyl-3-pentanol

b. 2-hexanol

c. 1-methylcyclopentanol

11.43 Draw the major product of each reaction.

a. $CH_3CH(CH_3)CH(CH_3)CH(OH)CH_3 \xrightarrow[\text{heat}]{H^+}$

(CH₃ above the second carbon; CH₃ and OH below the third and fourth carbons)

b. $CH_3C(CH_3)(OH)CH_2CH_3 \xrightarrow[\text{heat}]{H^+}$

11.44 Draw the major product of each reaction.

a. $CH_3CH_2CH(OH)CH_2CH_3 \xrightarrow[\text{heat}]{H^+}$

b. 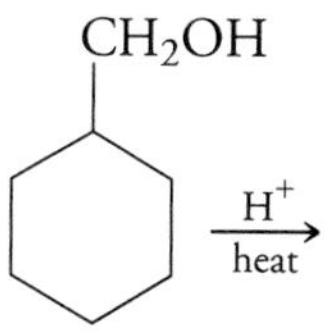 (cyclohexyl)$CH_2OH \xrightarrow[\text{heat}]{H^+}$

11.45 Describe the difference in the products obtained when 1-butanol and 1-butanethiol are oxidized.

11.46 Describe the difference in the products obtained when diethyl sulfide and diethyl ether are oxidized.

11.4 Aldehydes and Ketones

11.47 a. Prednisone is often used to reduce inflammation. Which functional groups does this molecule contain?

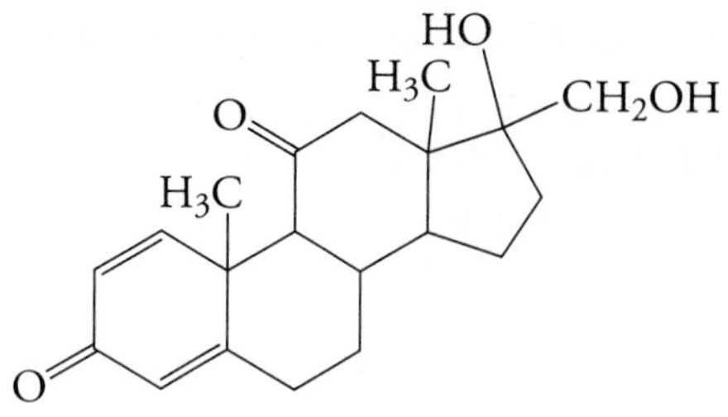

HO H_3C CH_2OH O H_3C O

Prednisone

b. To which class of lipids does prednisone belong?

11.48 Name each of the following aldehydes and ketones.

a. $CH_3CH(CH_3)CH(CH_3)\overset{O}{\overset{\|}{C}}{-}H$

b. $CH_3CH_2\overset{O}{\overset{\|}{C}}CH(CH_2CH_3)_2$

c. $CH_3C(CH_2CH_3)_2\overset{O}{\overset{\|}{C}}{-}H$

d. $CH_3\overset{O}{\overset{\|}{C}}C(CH_3)_3$

11.49 Name each of the following aldehydes and ketones.

a. $CH_3C(CH_3)_2CH_2\overset{O}{\overset{\|}{C}}{-}H$

b. $CH_3\overset{O}{\overset{\|}{C}}CH_2CH(CH_3)_2$

c. $CH_3CH(CH_3)\overset{O}{\overset{\|}{C}}CH(CH_3)_2$

d. $CH_3CH_2CH(CH_2CH_2CH_3)\overset{O}{\overset{\|}{C}}{-}H$

11.50 Draw each molecule.
a. pentanal
b. 3-bromohexanal
c. dipropyl ketone
d. 2,5-dibromocyclohexanone

11.51 Draw each molecule.
a. octanal
b. 2,5-dimethylcyclopentanone
c. 2-octanone
d. 2-isopropylpentanal

11.52 2-Hexanone has a boiling point of 150°C and 2-pentanone has a boiling point of 102°C.
a. Draw each compound.
b. Account for the difference in boiling point.

11.53 Acetone has a boiling point of 56°C and isopropyl alcohol has a boiling point of 82°C.
a. Draw each molecule.
b. Account for the difference in boiling point.

11.54 Propanone (molecular weight = 58.0 amu) has a boiling point of 56°C and 1-propanol (molecular weight = 60.0 amu) has a boiling pont of 97.4°C. Account for the difference in boiling point.

11.55 Propanone (molecular weight = 58.0 amu) has a boiling point of 56°C and butane (molecular weight = 58.0 amu) has a boiling point of 0°C. Account for the difference in boiling point.

11.56 Formaldehyde is infinitely soluble in water, while butanal is only slightly soluble.
a. Draw each molecule.
b. Account for the difference in water solubility.

11.5 Oxidation of Aldehydes

11.57 Draw each molecule named below and draw the product (if any) obtained when each is reacted with $K_2Cr_2O_7$. When named as a substituent, —Cl is chloro and —OH is hydroxy.
a. propanal
b. butanone
c. 1-butanol
d. 3-chloro-3-methylpentanal
e. 3-hydroxybutanal

11.58 Draw each molecule named below and draw the product (if any) obtained when each is reacted with Benedict's reagent. When named as a substituent, —Cl is chloro and —OH is hydroxy.
a. propanal
b. butanone
c. 1-butanol
d. 3-chloro-3-methylpentanal
e. 3-hydroxybutanal

11.59 Draw the product (if any) of each reaction.

a. $CH_3CH_2\underset{CH_3}{\underset{|}{C}}HCH_2\overset{O}{\overset{\|}{C}}{-}H \xrightarrow{Cu^{2+}}$

b. $HOCH_2CH_2CH_2\overset{CH_3}{\overset{|}{\underset{CH_3}{\underset{|}{C}}}}CH_3 \xrightarrow{Cu^{2+}}$

c. $CH_3\overset{OH}{\overset{|}{C}}H\underset{CH_3}{\underset{|}{C}}H\overset{O}{\overset{\|}{C}}{-}H \xrightarrow{Cu^{2+}}$

*__11.60__ Virgin olive oil contains vanillic acid, 3,4-dihydroxycinnamic acid, and syringic acid, compounds that can be prepared by oxidizing the appropriate aldehydes. Complete each reaction equation by drawing the structures of the reactants. To simplify this question, ignore the possible oxidation of phenol groups by $K_2Cr_2O_7$.

a. $\xrightarrow{K_2Cr_2O_7}$ HO–(benzene ring, CH_3O ortho to HO)–C(=O)–OH

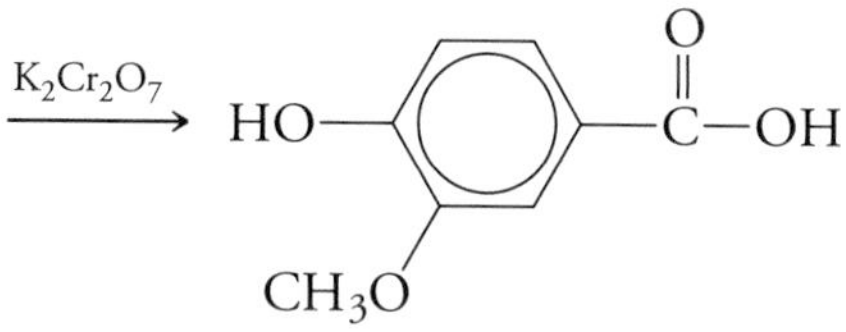

Vanillic acid

b. $\xrightarrow{K_2Cr_2O_7}$ (benzene ring bearing CH=CH–C(=O)–OH, with OH and OH at positions 3 and 4)

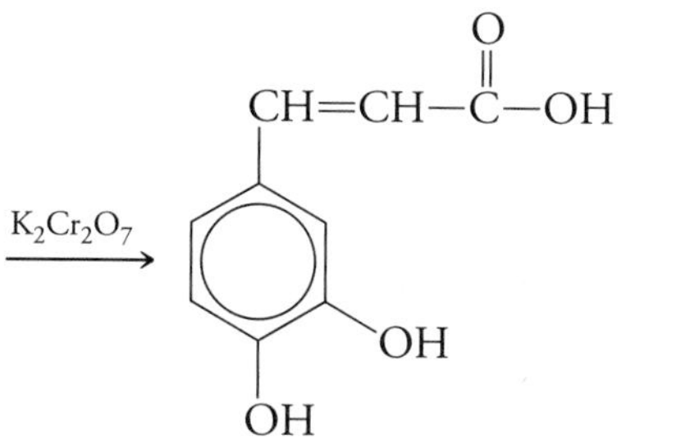

3,4-Dihydroxycinnamic acid

c. $\xrightarrow{K_2Cr_2O_7}$ HO–(benzene ring with two CH_3O groups flanking HO)–C(=O)–OH

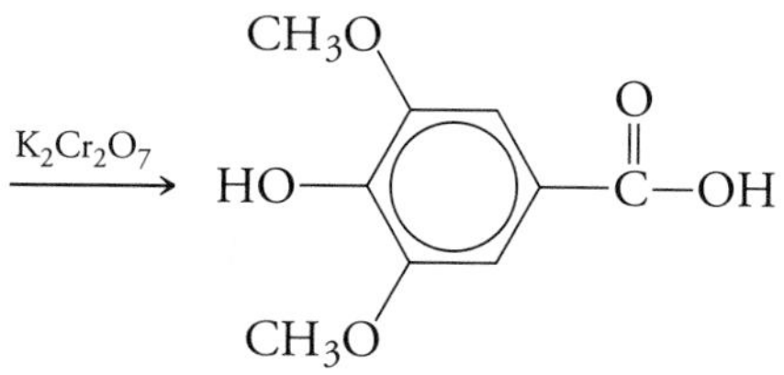

Syringic acid

11.6 Reduction of Aldehydes and Ketones

11.61 Draw the product of each reaction.

a. $CH_3CH_2C(=O)CH(CH_2CH_3)CH_3 + H_2 \xrightarrow{Pt}$

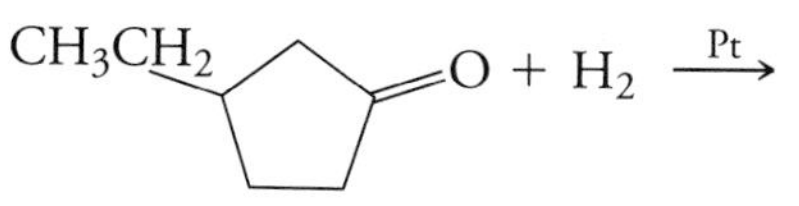

b. 3-ethylcyclopentanone (CH_3CH_2–cyclopentane=O) $+ H_2 \xrightarrow{Pt}$

c. $CH_3C(CH_3)_2CH_2C(=O)–H + H_2 \xrightarrow{Pt}$

11.62 Draw the product of each reaction.

a. $CH_3C(=O)CH_2CH(CH_3)CH_3 + H_2 \xrightarrow{Pt}$

b. $CH_3CH_2CH_2CH_2C(=O)–H + H_2 \xrightarrow{Pt}$

c. 2,5-dimethylcyclopentanone (H_3C–cyclopentanone–CH_3) $+ H_2 \xrightarrow{Pt}$

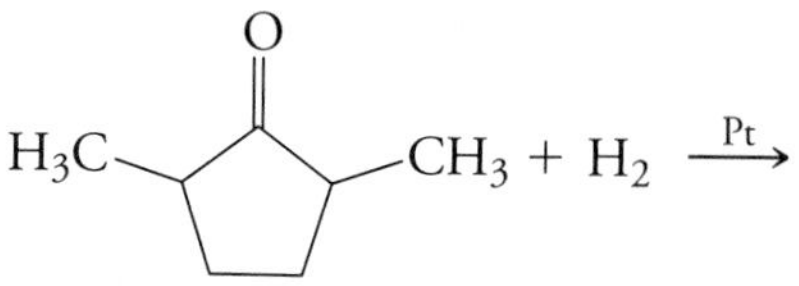

*__11.63__ The first step in the biochemical breakdown of the steroid progesterone is a reaction involving 20α-hydroxysteroid dehydrogenase (20α-HSDH), an enzyme which catalyzes the reduction of a ketone group at position 20 of steroids.

a. Draw the product that forms when progesterone is reduced by the action of 20α-HSDH and NADH.

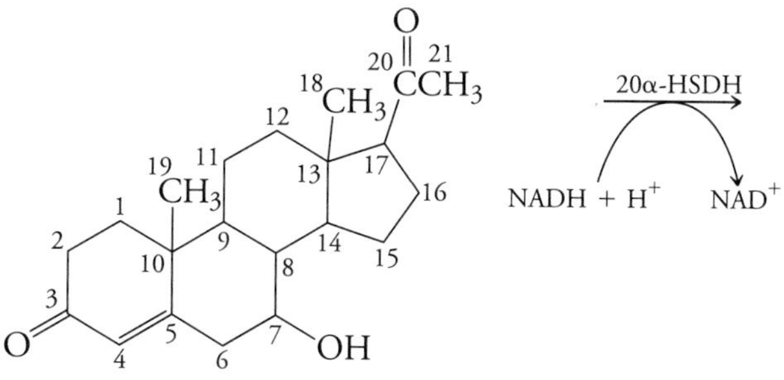

(Progesterone structure: steroid ring system numbered 1–21, with 3-keto (O at C3), C4=C5 double bond, 18-CH_3, 19-CH_3, C20 ketone $C(=O)CH_3$ (21); OH shown near C7/C8) $\xrightarrow[\text{NADH + H}^+ \;\rightarrow\; \text{NAD}^+]{20\alpha\text{-HSDH}}$

Progesterone

b. Draw the product that forms when 1 mol progesterone is reacted with 3 mol of H_2 in the presence of Pt.

11.7 Reactions of Alcohols with Aldehydes and Ketones

11.64 Draw the product of each reaction.

a. $CH_3CH_2CH_2C(=O)–H + CH_3CH(OH)CH_3 \overset{H^+}{\rightleftharpoons}$

b. cyclohexanone $+ 2CH_3CH_2CH_2CH_2OH \overset{H^+}{\rightleftharpoons}$

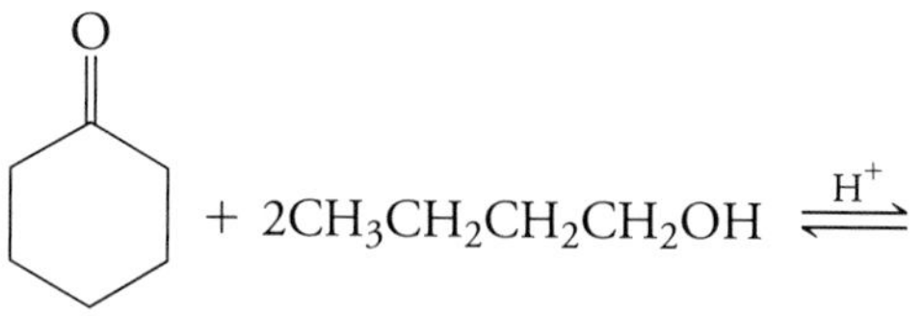

c. $CH_3CH_2CH(CH_3)C(=O)CH_2CH_2CH_3$ + cyclopentyl–OH $\xrightleftharpoons{H^+}$

11.65 Draw the product of each reaction.

a. $CH_3CH_2C(=O)CH_2CH_3 + CH_3OH \xrightleftharpoons{H^+}$

b. C_6H_5–C(=O)–H + $2CH_3CH_2OH \xrightleftharpoons{H^+}$

c. $CH_3C(=O)CH_2CH_3$ + C_6H_5–$CH_2OH \xrightleftharpoons{H^+}$

11.66 Draw the hemiacetal that forms when muscone (a deer sex attractant) is reacted with CH_3CH_2OH and H^+. Muscone is the "musk" scent used in some perfumes.

Muscone + $CH_3CH_2OH \xrightleftharpoons{H^+}$

Muscone

11.67 Draw the missing reactant for each reaction.

a. + $CH_3CH(OH)CH_3 \xrightleftharpoons{H^+} CH_3CH_2C(OH)(OCH(CH_3)CH_3)CH_2CH_3$

b. + $2CH_3CH(OH)CH_3 \xrightleftharpoons{H^+}$ cyclopentane with $CH_3CH(CH_3)O$ and $OCH(CH_3)CH_3$ on the same carbon

c. + $2CH_3CH(OH)CH_3 \xrightleftharpoons{H^+} CH_3C(H)(OCH(CH_3)CH_3)(OCH(CH_3)CH_3)$

11.68 Draw the missing reactant for each reaction.

a. + $CH_3CH_2CH_2OH \xrightleftharpoons{H^+} CH_3C(OH)(OCH_2CH_2CH_3)CH_3$

b. $CH_3CH_2C(=O)$–H + $\xrightleftharpoons{H^+} CH_3CH_2C(OH)(OCH_2CH_2CH_3)$–H

c. + $2CH_3OH \xrightleftharpoons{H^+}$ cyclopentane with CH_3O and OCH_3 on the same carbon

*11.69 a. How are cyclic hemiacetals similar to lactones (Figure 10.17)?
b. How are cyclic hemiacetals different from lactones?

11.70 What product is formed in the presence of H^+ when an alcohol molecule is reacted with each of the following?
a. an aldehyde
b. a carboxylic acid

HEALTH*Link* | Aldehyde Dehydrogenase

11.71 The toxicity of methanol is mainly due to the aldehyde produced when it is oxidized in the liver. Draw methanol, then draw and name the aldehyde formed on its oxidation.

11.72 Describe the physiological action of disulfiram.

HEALTH*Link* | Protective Enzymes

11.73 Is O_2 reduced or is it oxidized when it is converted into O_2^- or H_2O_2?

*11.74 In the breakdown of superoxide, which product results from the reduction of O_2^- and which product results from the oxidation of O_2^-? Explain.

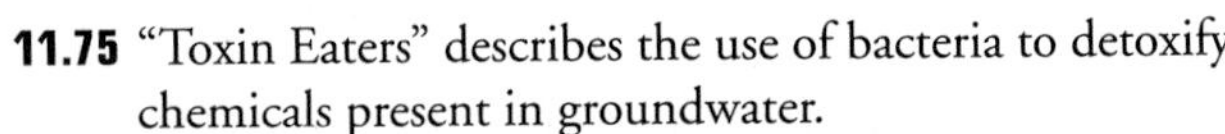

HEALTH*Link* | Drugs in the Environment

11.75 "Toxin Eaters" describes the use of bacteria to detoxify chemicals present in groundwater.
a. The chemical mentioned in this video, vinyl chloride, is also known as chloroethene. Draw vinyl chloride.

b. From where did scientists obtain bacteria that can break down vinyl chloride?
c. How could this technique for removing vinyl chloride from groundwater be modified to allow removal of pharmaceuticals and other compounds?

11.76 Acetaminophen, which has been detected in groundwater, is an amide that can be formed by reacting acetic acid with 4-hydroxyaniline. Draw acetaminophen.

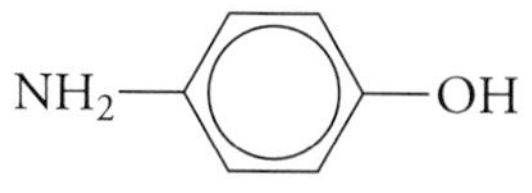

4-hydroxyaniline

11.77 The antidepressant fluoxetine, which has been detected in biosolids, is typically sold as the hydrochloride salt. Draw this salt.

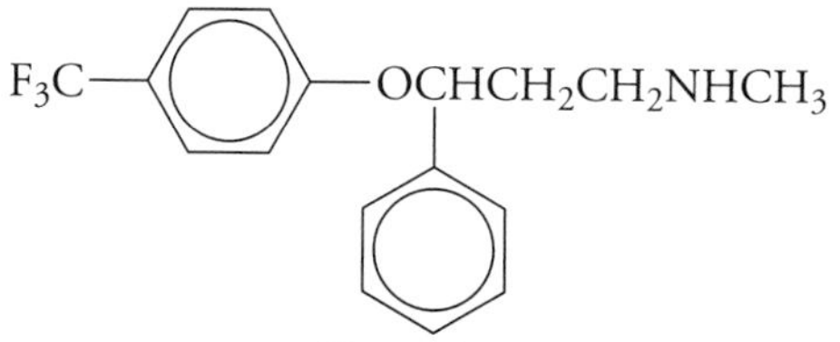

Fluoxetine

11.78 The steroid hormone estrone has been detected in river, stream, and well water. Draw the product formed when the ketone group in estrone is reduced by H_2 and Pt.

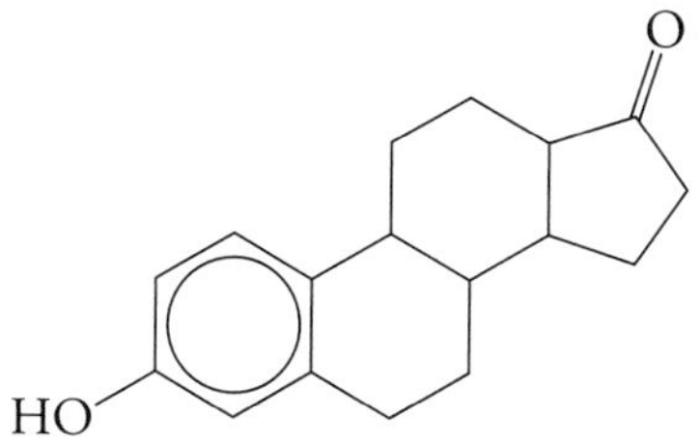

Estrone

11.79 The antidepressant venlafaxine has been detected in ground water at concentrations of 50 ng/L.
a. At this concentration, how many grams of this drug are present in 3.0 L of water?
b. Convert 50 μg/L into parts per million.
c. Convert 50 μg/L into parts per billion.

11.80 Paraxanthine, a product of caffeine metabolism in humans, has been detected in ground water at concentrations of 150 μg/L.
a. At this concentration, how many grams of paraxanthine are present in 25.0 mL of water?
b. Convert 150 μg/L into parts per million.
c. Convert 150 μg/L into parts per billion.

THINKING IT THROUGH

***11.81** After reviewing the definition of the terms hemiacetal and acetal, identify any hemiacetal or acetal carbons in lactose (also known as milk sugar).

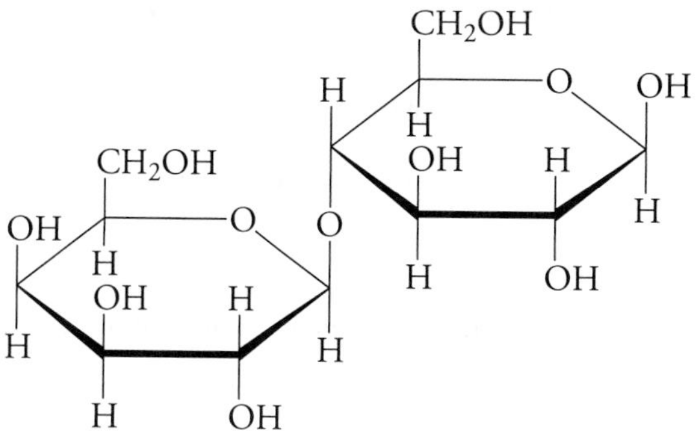

Lactose

***11.82 a.** How many chiral carbon atoms does lactose (Problem 11.61) contain?
b. Lactose is one of how many stereoisomers?

INTERACTIVE LEARNING PROBLEMS

ILW **11.83** Name this compound:

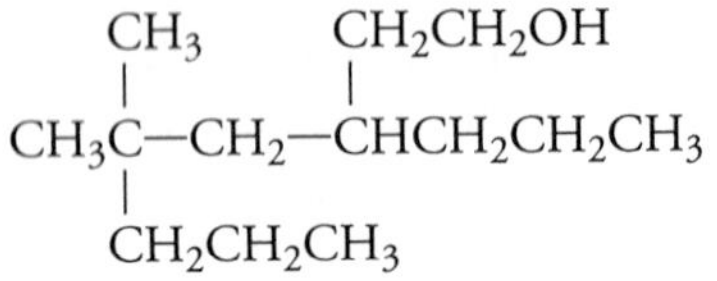

$$\begin{array}{c} \quad CH_3 \qquad\quad CH_2CH_2OH \\ CH_3\overset{|}{\underset{|}{C}}-CH_2-\overset{|}{C}HCH_2CH_2CH_3 \\ CH_2CH_2CH_3 \end{array}$$

ILW **11.84** Give the IUPAC name for the following compound:

$$(CH_3)_3C-\overset{\overset{\displaystyle O}{\|}}{C}-CH_2C(CH_3)_3$$

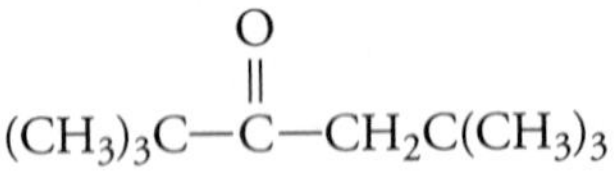

ILW **11.85** Name the products of the hydrolysis of the following compound:

(cyclopentane ring bonded to OCH_2CH_2 / OCH_2CH_2 — a cyclic ketal)

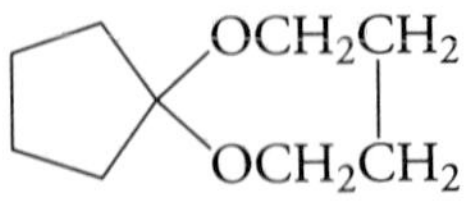

SOLUTIONS TO PRACTICE PROBLEMS

11.1 The alcohol molecule.

11.2 **a.** CH_3CH_2OH **b.** 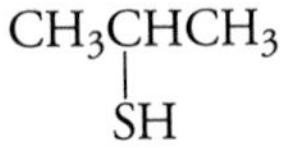 $CH_3\underset{\underset{\displaystyle SH}{|}}{C}HCH_3$ **c.** $CH_3CH_2CH_2CH_2OCH_3$

11.3 Citrate contains a 3° alcohol and cannot be oxidized. Isocitrate contains a 2° alcohol and can.

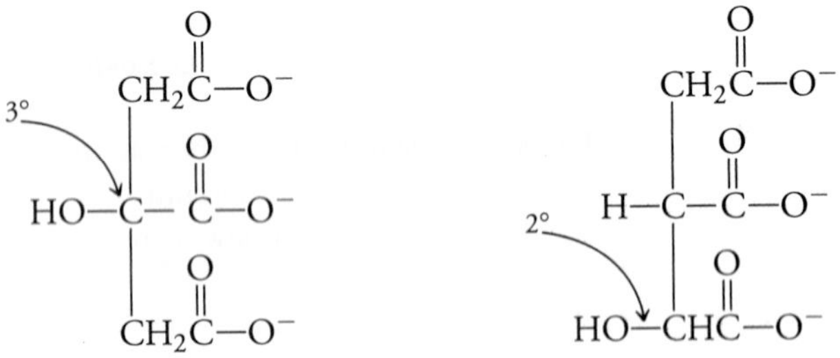

Citrate: $CH_2C(=O)-O^-$ / $HO-C-C(=O)-O^-$ (3°) / $CH_2C(=O)-O^-$

Citrate

Isocitrate: $CH_2C(=O)-O^-$ / $H-C-C(=O)-O^-$ / $HO-CHC(=O)-O^-$ (2°)

Isocitrate

11.4 **a.**  $CH_3\overset{\overset{\displaystyle O}{\|}}{C}CH_2CH_2CH_3$

b. $CH_3CH_2CH_2CH_2\overset{\overset{\displaystyle O}{\|}}{C}-H$

c. (cyclohexanone ring with CH_3 substituent at position 3)

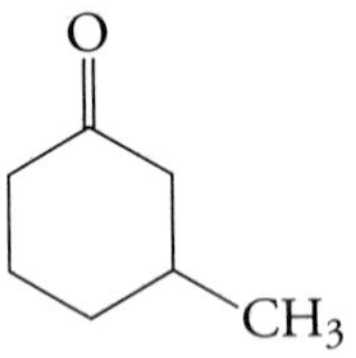

11.5 **a.** 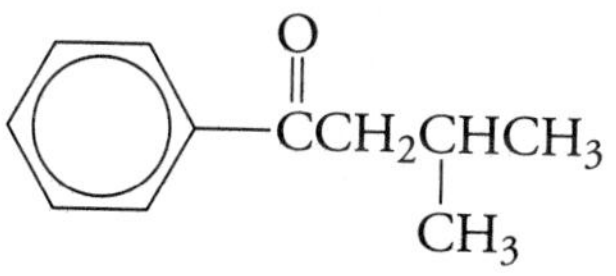 $C_6H_5-\overset{O}{\overset{\|}{C}}CH_2\underset{CH_3}{\underset{|}{C}}HCH_3$

b. $CH_3CH_2CH_2OH$

c. 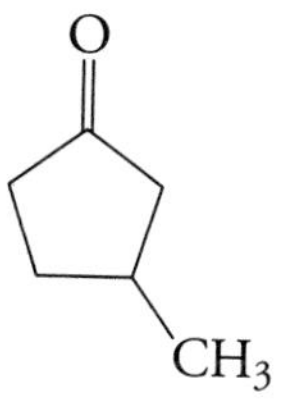 3-methylcyclopentanone (cyclopentane ring with C=O, CH_3 substituent)

d. $CH_3\overset{O}{\overset{\|}{C}}CH_2CH_2CH_2CH_3$

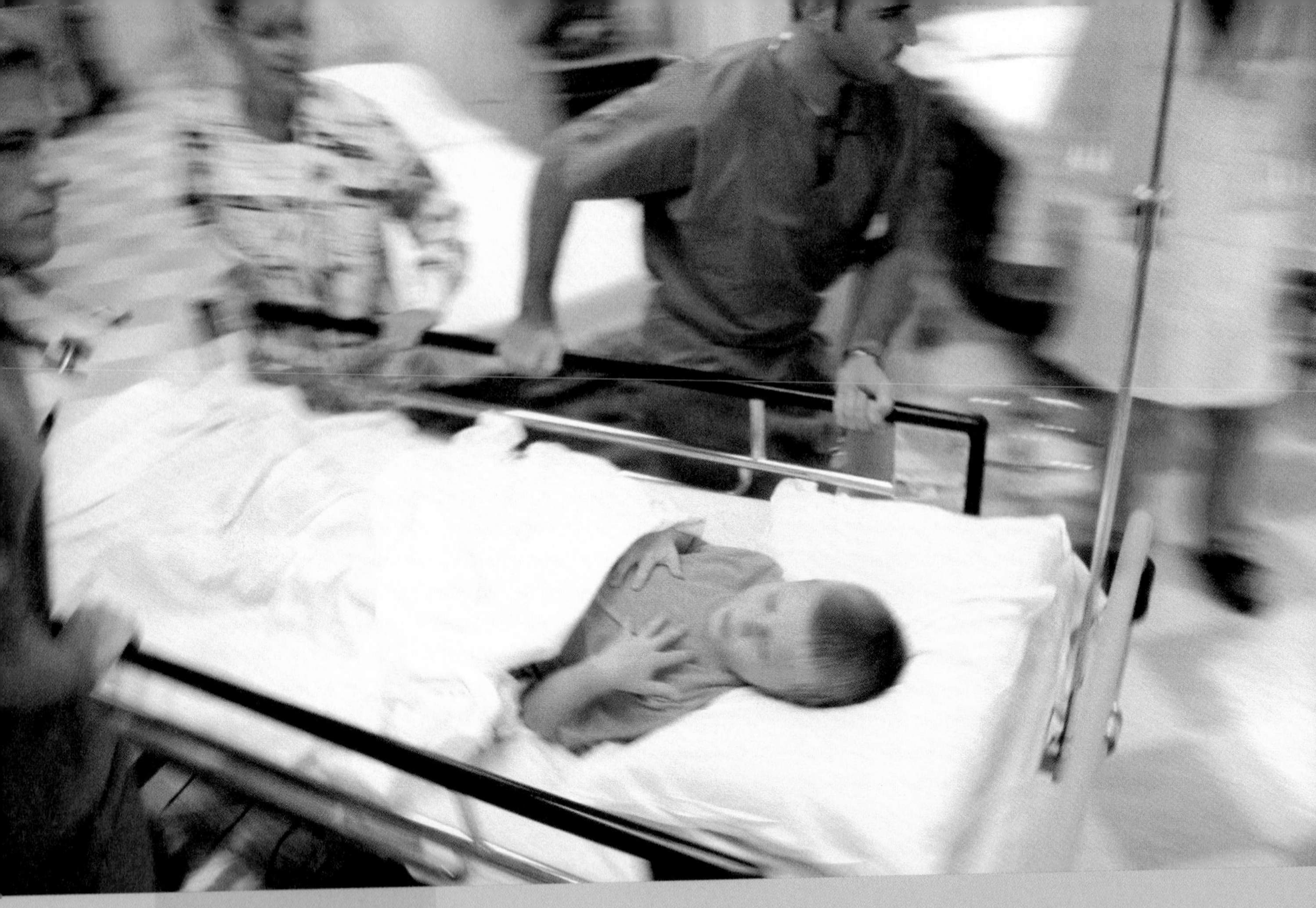

While walking across campus a student meets a classmate, a premed student currently doing volunteer work in a hospital emergency room. He tells of a case that happened the previous day, when a child was brought in who had eaten some leaves from a foxglove plant. After initially complaining of a headache, the child soon began vomiting and displaying an irregular heartbeat. Quick action by the emergency room staff prevented the onset of convulsions.

Photo source: Arthur Tilley/Taxi/Getty Images.

12

CARBOHYDRATES

Early studies on the sugar named glucose showed that it has the molecular formula $C_6H_{12}O_6$. Rearranging this formula into $C_6(H_2O)_6$ made it appear that glucose is a hydrate of carbon—six carbon atoms attached to six water molecules—so the name *carbohydrate* was adopted for glucose and related compounds. Although we now know that carbohydrates are not hydrates of carbon, the name has been retained.

objectives

After completing this chapter, you should be able to:

1. Describe the difference between mono-, oligo-, and polysaccharides and explain the classification system used to categorize monosaccharides.
2. Identify the four common types of monosaccharide derivatives.
3. Explain what happens when a monosaccharide is reacted with H_2 and Pt (a catalyst) or with Benedict's reagent.
4. Define the term anomer and describe the difference between pyranoses and furanoses and the difference between α and β anomers. Explain what is meant by the term mutarotation.
5. Identify four important disaccharides and describe how the monosaccharide residues in them are joined to one another.
6. Distinguish homopolysaccharides from heteropolysaccharides and give examples of each.

On the earth, more than half of the carbon atoms tied up in organic compounds are found in carbohydrates. Plants produce most of the carbohydrates, doing so through photosynthesis, in which energy from the sun is converted into chemical energy that is, in turn, used to produce carbohydrates from atmospheric CO_2 (Figure 12.1).

Carbohydrates play many important biological roles. Some, such as starch and glycogen, are used to store energy. Others, including glucose and fructose, can be oxidized to provide the energy needed for many biochemical reactions to take place. Carbohydrates are also found in cell membranes, in the lubricant that surrounds joints, and in the nucleic acids RNA and DNA. In this chapter we will take a look at the structure, properties, and biochemical uses of some important carbohydrates.

$$6CO_2 + 6H_2O \longrightarrow C_6H_{12}O_6 + 6O_2$$

FIGURE 12.1

Carbohydrate factories
Foxgloves and all other plants indirectly use the energy of sunlight to make carbohydrates from CO_2.

Source: © Corbis Digital Stock.

12.1 MONOSACCHARIDES

■ Oligosaccharides contain 2–10 monosaccharide residues. Polysaccharides contain more than 10.

Carbohydrate molecules can be placed into one of three categories: monosaccharides, oligosaccharides, and polysaccharides. **Monosaccharides** are used as *building blocks to produce* **oligosaccharides**, *which contain from 2 to 10 monosaccharide residues*, and **polysaccharides**, *which contain more than 10 monosaccharide residues*. As we saw in Section 8.2, a residue is that part of a reactant molecule that remains when it has been incorporated into a product. The term saccharide comes from the Greek word *sakcharon* ("sugar").

From an organic chemistry perspective, monosaccharides are *polyhydroxy aldehydes or ketones containing three or more carbon atoms*. In simpler terms, this means that each monosaccharide molecule is either an aldehyde or a ketone and that each of the other carbon atoms usually carries a hydroxyl (—OH) group (Figure 12.2).

FIGURE 12.2

Monosaccharides
Monosaccharides are polyhydroxy aldehydes (aldoses) or ketones (ketoses). An aldopentose is an aldose that contains five carbon atoms. A ketotetrose is a ketose that contains four carbon atoms.

```
    H  H  H  H  O                 H  H  O  H
    |  |  |  |  ||                |  |  ||  |
H—C—C—C—C—C—H            H—C—C—C—C—H
    |  |  |  |                    |  |     |
   OH OH OH OH                   OH OH    OH
```

An aldopentose | A ketotetrose

Monosaccharides can be classified based on functional group, on number of carbon atoms, or both (Table 12.1). Those that *contain an aldehyde group* are **aldoses** and *those with a ketone group* are **ketoses**. **Trioses** have *three carbon atoms*, **tetroses** have *four carbon atoms*, **pentoses** have *five*, and so on. An **aldohexose** is an aldehyde sugar with six carbon atoms. The "ose" ending indicates that the molecule being named is a carbohydrate.

As a family, monosaccharides have a sweet taste (Health Link: Relative Sweetness), have high melting points, and are hydrophilic (water soluble). The latter two properties are primarily due to the ability of monosaccharides to form hydrogen bonds.

TABLE 12.1 MONOSACCHARIDES

Type	Functional Group	Number of Carbon Atoms
Aldose	Aldehyde	
Ketose	Ketone	
Triose		3
Tetrose		4
Pentose		5
Hexose		6
Heptose		7
Octose		8
Nonose		9
Aldotriose	Aldehyde	3
Aldopentose	Aldehyde	5
Ketoheptose	Ketone	7

SAMPLE PROBLEM 12.1

Classifying monosaccharides

Classify each monosaccharide in terms of functional group and number of carbon atoms.

a. 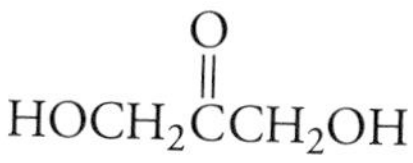 $HOCH_2\overset{O}{\overset{\|}{C}}CH_2OH$

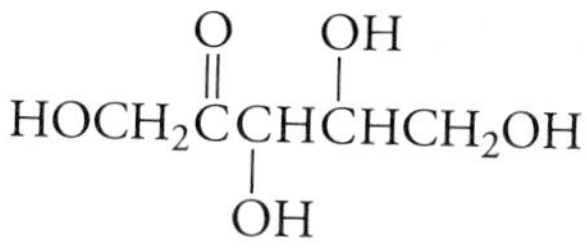

b. 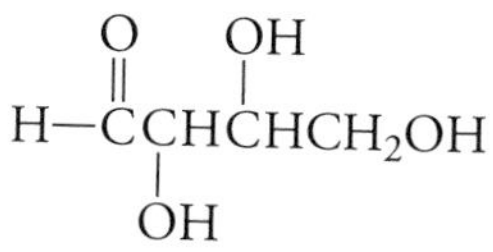 $H{-}\overset{O}{\overset{\|}{C}}\underset{OH}{\underset{|}{C}}H\overset{OH}{\overset{|}{C}}HCH_2OH$

c. $HOCH_2\overset{O}{\overset{\|}{C}}\underset{OH}{\underset{|}{C}}H\overset{OH}{\overset{|}{C}}HCH_2OH$

d. 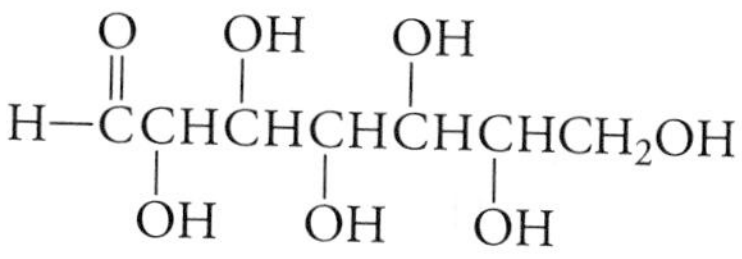 $H{-}\overset{O}{\overset{\|}{C}}\underset{OH}{\underset{|}{C}}H\overset{OH}{\overset{|}{C}}H\underset{OH}{\underset{|}{C}}H\overset{OH}{\overset{|}{C}}H\underset{OH}{\underset{|}{C}}HCH_2OH$

Strategy

To classify each, you must determine whether it is an aldehyde or a ketone and you must count the number of carbon atoms.

Solution

a. ketotriose
b. aldotetrose
c. ketopentose
d. aldoheptose

PRACTICE PROBLEM 12.1

Draw an example of each of the following.

a. an aldotriose
b. an aldopentose
c. a ketohexose
d. a ketotetrose

Stereoisomers

One important characteristic of carbohydrates is that they contain chiral carbon atoms (carbon atoms with four different attached atoms or groups of atoms) and have stereoisomers. The aldotriose called glyceraldehyde (Figure 12.3) contains one chiral carbon atom and, as described in Section 10.9, we can calculate that it is one of two stereoisomers with this formula ($2^1 = 2$). The three-dimensional representations in this figure show that the stereoiosmers are nonsuperimposable mirror images, or enantiomers. In Figure 12.3*b*, the three-dimensional structure of each glyceraldehyde enantiomer is given using a **Fischer projection**. In Fischer projections, each chiral carbon atom (not shown) sits at the intersection of a vertical and a horizontal line. *The horizontal lines represent bonds pointing toward the viewer and the vertical lines are for bonds pointing away from the viewer.* For monosaccharides, Fischer projections are drawn with the carbon atoms running vertically and the aldehyde or ketone group at or near the top. Drawing them in this particular orientation makes it easy to classify monosaccharides using D and L notation. In **D sugars**, the —OH attached to the chiral carbon atom farthest from the C=O *points to the right*, and in **L sugars** this —OH *points to the left.* Monosaccharide enantiomers have the same name, except for the D or L designation, as seen for D-glyceraldehyde and L-glyceraldehyde in Figure 12.3. In nature, most monosaccharides are D sugars, so D-glyceraldehyde is commonly found, while L-glyceraldehyde is not.

■ When drawn in a Fischer projection, the —OH attached to the chiral carbon atom farthest from the C=O points to the right in a D-monosaccharide and to the left in an L-monosaccharide.

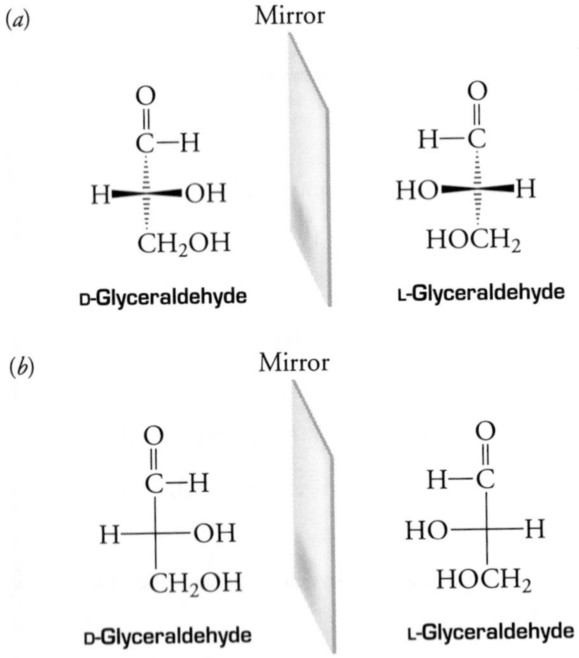

FIGURE 12.3

Glyceraldehyde enantiomers Glyceraldehyde, an aldotriose, has one chiral carbon atom and exists in two stereoisomeric forms. The two enantiomers are represented (*a*) in wedge and dashed line notation and (*b*) as Fischer projections.

(*a*) Mirror

O=C—H, H—C—OH, CH_2OH — **D-Glyceraldehyde**

O=C—H, HO—C—H, $HOCH_2$ — **L-Glyceraldehyde**

(*b*) Mirror

O=C—H, H—C—OH, CH_2OH — **D-Glyceraldehyde**

O=C—H, HO—C—H, $HOCH_2$ — **L-Glyceraldehyde**

As the number of chiral carbon atoms in a monosaccharide increases, so do the number of possible stereoisomers. Glucose (Figure 12.4) has four chiral carbon atoms and is one of sixteen different aldohexose stereoisomers ($2^4 = 16$). In Figure 12.4, which shows Fischer projections for three aldohexose stereoisomers, the chiral carbon atoms are numbered 2, 3, 4, and 5. (Monosaccharides are numbered beginning at the end of the molecule nearer the carbon–oxygen double bond.) In D-glucose three of the —OH groups attached to chiral carbon atoms point to the right and one —OH group points to the left. The D designation is arrived at by looking at the orientation of the chiral carbon atom farthest from the aldehyde group (carbon atom 5 in this case).

D-Glucose and L-glucose are enantiomers, so they have the same common name (glucose). D-Galactose is a diastereomer (non-mirror image stereoisomer) of the other two.

D-Glucose	L-Glucose	D-Galactose
$_1$C(=O)–H	C(=O)–H	C(=O)–H
H–$_2$–OH	HO–H	H–OH
HO–$_3$–H	H–OH	HO–H
H–$_4$–OH	HO–H	HO–H
H–$_5$–OH	HO–H	H–OH
$_6$$CH_2OH$	CH_2OH	CH_2OH

FIGURE 12.4

Aldohexose stereoisomers Aldohexoses have four chiral carbon atoms and exist as sixteen ($2^4 = 16$) stereoisomers. D-Glucose and L-glucose are enantiomers and D-galactose is a diastereomer of the other two.

SAMPLE PROBLEM 12.2

Drawing monosaccharide stereoisomers

a. How many 2-ketopentose stereoisomers are possible? (The "2-" specifies that the C=O is at carbon atom 2.)

b. Draw a Fischer projection for each of the D-2-ketopentoses.

Strategy

For part a, you might begin by drawing a 2-ketopentose and counting the number of chiral carbon atoms. This will allow you to calculate the number of stereoisomers.

Solution

a. Four. A 2-ketopentose has two chiral carbon atoms and a total of four ($2^2 = 4$) stereoisomers are possible.

$$HOCH_2C(=O)C^*H(OH)C^*H(OH)CH_2OH$$

b. Two of the stereoisomers are D and two are L. Only the D sugars are shown.

CH_2OH	CH_2OH
C=O	C=O
H–OH	HO–H
H–OH	H–OH
CH_2OH	CH_2OH

PRACTICE PROBLEM 12.2

Figure 12.4 shows two of the eight possible D-aldohexoses. Draw the other six.

Important Monosaccharides

While monosaccharides of various sizes appear in nature, pentoses and hexoses are the most abundant. D-Ribose and D-2-deoxyribose are aldopentoses that are often incorporated into larger biomolecules, including the biochemical reducing agent NADPH

(Section 6.3), the biochemical oxidizing agent NAD^+ (Section 11.3), and the nucleic acids RNA and DNA (Chapter 14).

O
‖
C—H
H—C—OH
H—C—OH
H—C—OH
CH_2OH

D-Ribose

O
‖
C—H
H—C—H
H—C—OH
H—C—OH
CH_2OH

D-2-Deoxyribose

The structural difference between these two monosaccharides is that 2-deoxyribose lacks an —OH group on carbon atom 2, hence the "2-deoxy" part of the name. This is just one example of the variations in monosaccharide structure that are observed in nature.

D-Glucose (Figure 12.4), also known as dextrose or blood sugar, is one of the most important monosaccharides in human biochemistry. It is released when the polysaccharides starch and glycogen are broken down and is a key reactant in the series of reactions that produce molecules used to drive otherwise nonspontaneous reactions. One form of energy storage in the body involves combining glucose molecules to produce glycogen (Section 12.5).

D-Galactose (Figure 12.4) is combined with glucose to produce lactose (Section 12.4), a disaccharide (one type of oligosaccharide) that gives milk its sweetness. When lactose is digested, the galactose released is transformed into a glucose derivative used for energy production or storage. Galactose and glucose are diastereomers, and one of the key steps in the conversion of galactose into a glucose derivative involves flipping the orientation of the chiral carbon atom at position 4.

When infants are fed a milk diet, they depend on the ability to convert galactose into a glucose derivative to supply some of their energy needs. Galactosemias, a group of genetic disorders related to the inability to metabolize galactose, can pose great health risks in newborns. The buildup of galactose and related compounds can lead to liver problems, mental retardation, and damage to the central nervous system. Galactosemias cannot be cured, so treatment involves maintaining a galactose-free diet.

D-Fructose, or fruit sugar, is the ketose found most often in nature. This monosaccharide is present in fruit, as the name *fruit sugar* implies, and makes up about 40% of honey. As is the case for glucose, fructose plays a key role in energy production.

CH_2OH
C=O
HO—C—H
H—C—OH
H—C—OH
CH_2OH

D-Fructose

Monosaccharide Derivatives

■ Deoxy, amino, alcohol, and carboxylic acid sugars are common monosaccharide derivatives.

A number of monosaccharide derivatives appear in nature, and many are incorporated into oligo- and polysaccharides. The four major classes of monosaccharide derivatives that will be considered here are deoxy sugars, amino sugars, alcohol sugars, and carboxylic acid sugars.

In **deoxy sugars** *a hydrogen atom replaces one or more of the —OH groups* in a monosaccharide. D-2-Deoxyribose is an example that we have already seen. L-Fucose (6-deoxy-L-galactose) and L-rhamnose (6-deoxy-L-mannose) are two others that are found in the oligo- and polysaccharides of plants, animals, and bacteria.

L-Fucose **L-Rhamnose**

Above, it was stated that D sugars are the stereoisomers that predominate in nature. L-Fucose and L-rhamnose, two of the naturally occurring L sugars, are produced from D-monosaccharides.

In **amino sugars** *an —OH group of a monosaccharide has been replaced by an amino (—NH_2) group*. In some amino sugars, the amino group has been converted to an amide. D-Glucosamine and *N*-acetyl-D-glucosamine are two amino sugars that will be important in the discussion of polysaccharides that follows.

D-Glucosamine ***N*-Acetyl-D-glucosamine**

In **alcohol sugars** *the carbonyl group of a monosaccharide has been reduced to an alcohol group*. Sorbitol, derived from glucose (Figure 12.5), is used as a sweetener

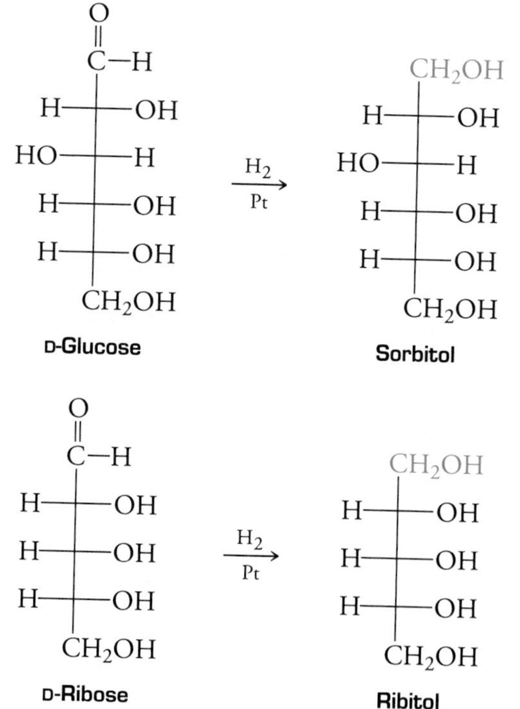

FIGURE 12.5

Alcohol sugars
When D-glucose and D-ribose are treated with H_2 and the catalyst Pt, they are converted into the corresponding alcohol sugars.

(Health Link: Relative Sweetness). Ribitol, from ribose, is present in FAD, a biochemical oxidizing agent that will be discussed in Chapter 15.

In **carboxylic acid sugars** *an aldehyde or alcohol group of a monosaccharide has been oxidized to form a carboxyl group.* D-Gluconic acid and D-glucuronic acid, both produced from glucose, are carboxylic acid sugars present in many oligosaccharides and polysaccharides.

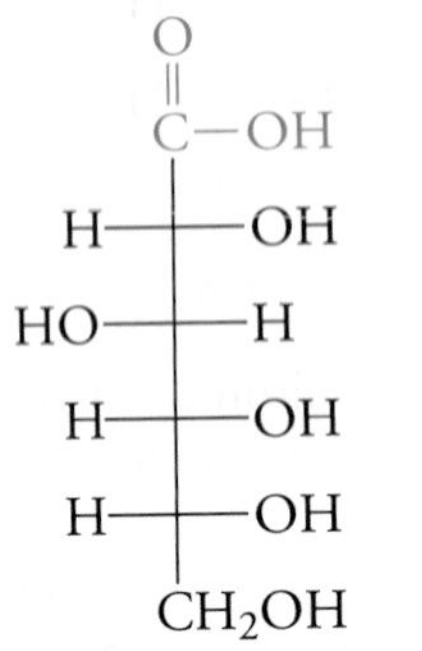

```
      O
      ‖
      C—OH
      |
  H———+———OH
      |
 HO———+———H
      |
  H———+———OH
      |
  H———+———OH
      |
     CH2OH
```

D-Gluconic acid

```
      O
      ‖
      C—H
      |
  H———+———OH
      |
 HO———+———H
      |
  H———+———OH
      |
  H———+———OH
      |
      C—OH
      ‖
      O
```

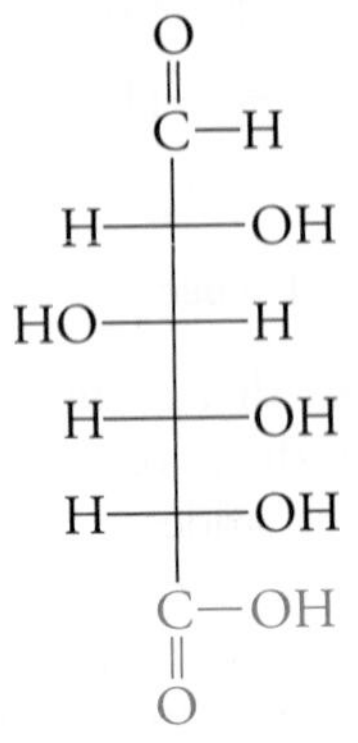

D-Glucuronic acid

SAMPLE PROBLEM 12.3

Drawing monosaccharide derivatives

In D-galactosamine, the —OH at carbon atom 2 of D-galactose is replaced by an $—NH_2$ group. Draw this amino sugar.

Strategy

Refer to Figure 12.4 for the structure of galactose and then modify its structure at carbon atom 2.

Solution

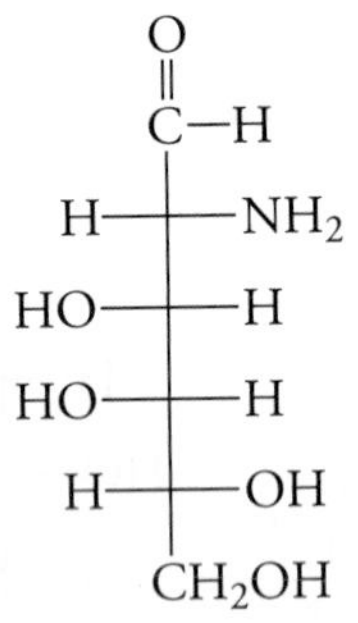

```
      O
      ‖
      C—H
      |
  H———+———NH2
      |
 HO———+———H
      |
 HO———+———H
      |
  H———+———OH
      |
     CH2OH
```

PRACTICE PROBLEM 12.3

D-Quinovose is another name for 6-deoxy-D-glucose. Draw this deoxy sugar.

12.2 REACTIONS OF MONOSACCHARIDES

In this section we will consider a few of the important reactions of monosaccharides. Monosaccharides contain alcohol and aldehyde or ketone groups, so the reactions discussed will be ones that were presented in Chapter 11.

Reduction

When the carbon–oxygen double bond of an aldehyde or ketone is treated with H_2 and the catalyst platinum (Pt) or encounters the appropriate enzyme, it is reduced to an alcohol (Section 11.6). Alcohol sugars formed by reduction of aldoses and ketoses (Figure 12.5) appear widely in nature and are also produced by the food industry for use as sweeteners.

Sorbitol and the other alcohol sugars used to sweeten some sugarless gum cannot be utilized by bacteria in the mouth to produce the acidic waste products that lead to dental cavities and the formation of plaque.

■ Aldehyde and ketone groups of monosaccharides can be reduced. Aldehyde and alcohol groups of monosaccharides can be oxidized.

Oxidation

As described above, carboxylic acid sugars are produced by oxidizing either the aldehyde group or an alcohol group of monosaccharide molecules. In nature, these oxidations are catalyzed by enzymes.

Benedict's reagent (Section 11.5) oxidizes aldehydes, but not alcohols. *Sugars that give a positive Benedict's test* (react with Benedict's reagent) are called **reducing sugars** because, in the process of being oxidized, they reduce the Cu^{2+} present in the reagent. Interestingly enough, fructose and many other ketoses are also reducing sugars, even though ketones are not oxidized by Benedict's reagent. What happens to these ketoses is that, in the presence of the basic Benedict's test solution, they rearrange to become aldoses. D-Fructose, for example, is converted into its constitutional isomers D-glucose and D-mannose, each of which is a reducing sugar (Figure 12.6).

D-Fructose		D-Glucose		D-Mannose
CH_2OH		O=C—H		O=C—H
C=O		H—C—OH		HO—C—H
HO—C—H	Benedict's solution ⇌	HO—C—H	+	HO—C—H
H—C—OH		H—C—OH		H—C—OH
H—C—OH		H—C—OH		H—C—OH
CH_2OH		CH_2OH		CH_2OH

FIGURE 12.6

D-Fructose is a reducing sugar Although it does not contain an aldehyde group, D-fructose gives a positive Benedict's test (is a reducing sugar). Under the basic conditions of the Benedict's test, D-fructose rearranges to become D-glucose or D-mannose, each of which is an aldehyde.

Its ability to detect the presence of aldoses has made Benedict's reagent useful for monitoring diabetes, through the testing of urine for the presence of glucose (Figure 11.13). Benedict's testing of urine can give false-positive readings, however. This test is not specific for glucose because any molecule that contains an aldehyde group (or contains a ketone group that can be converted into an aldehyde under the test conditions) can give a positive result. A newer glucose-detection system uses an enzyme-based method to overcome this limitation. Test strips are coated with glucose oxidase, an enzyme that catalyzes the oxidation of the aldehyde group in glucose, but no other compounds. When glucose is oxidized by this enzyme, an indicator produces a color change. The degree of color change depends on the amount of glucose that is present (Figure 12.7).

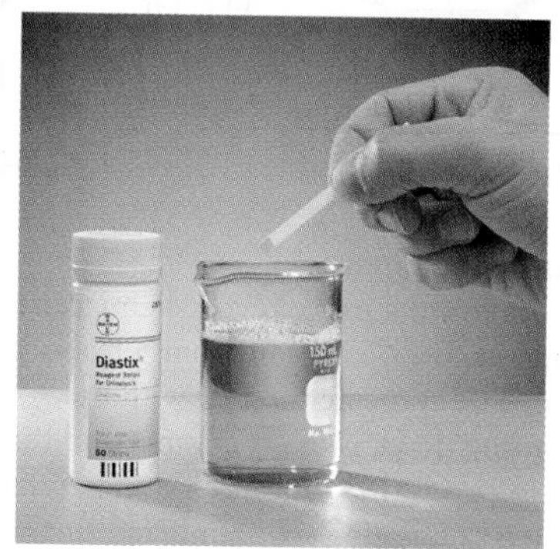

(*a*)

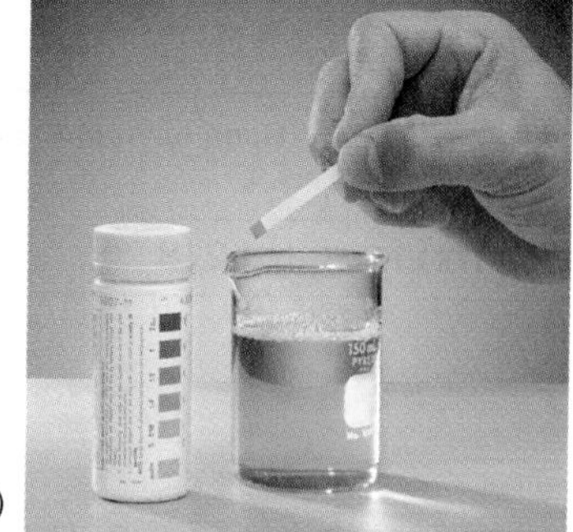
(*b*)

FIGURE 12.7

Detecting glucose
(*a*) Test strips are coated with glucose oxidase and an indicator. (*b*) The strip changes color if dipped in a solution that contains glucose. The final color depends on the initial glucose concentration.

Source: Andy Washnik/Wiley Archive.

Hemiacetal Formation

Hemiacetal formation (Section 11.7) is very important to the structure of monosaccharides. In Section 12.3 we will take a detailed look at this reaction between the alcohol and aldehyde or ketone groups present in monosaccharides.

12.3 MONOSACCHARIDES IN THEIR CYCLIC FORM

When the carbonyl group of an aldehyde or ketone reacts with the hydroxyl group of an alcohol, a hemiacetal forms. A hemiacetal carbon atom is one that is attached to —OH and —OC. Aldehydes or ketones that also contain an alcohol group (aldoses and ketoses fall into this category) can form cyclic hemiacetals (Figure 11.18b). D-Glucose, for example, has one aldehyde and five alcohol groups, which opens up the possibility for the existence of a number of different cyclic hemiacetals. The one most often seen is that formed between the C=O and the —OH on carbon atom 5, the chiral carbon atom farthest from the C=O. This cyclic hemiacetal has a six-membered ring containing five carbon atoms (carbons 1–5 of glucose) and one oxygen atom (from the —OH on carbon 5).

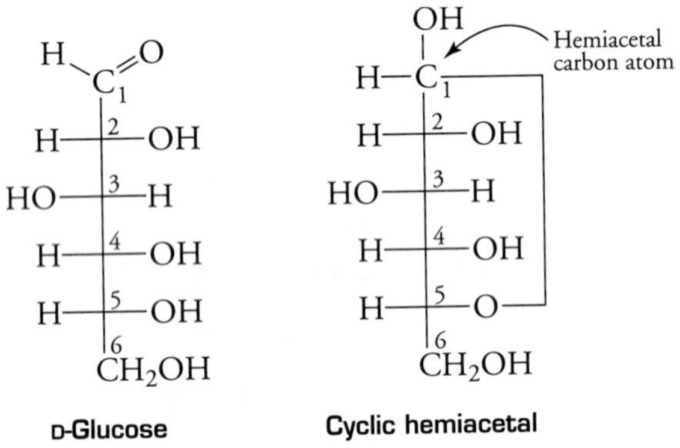

Hemiacetal carbon atom

D-Glucose

Cyclic hemiacetal

Typically, the six-membered ring forms of monosaccharides are not drawn as shown above, but instead as in Figure 12.8, with the oxygen atom of the ring at the back right and the hemiacetal carbon on the right side.

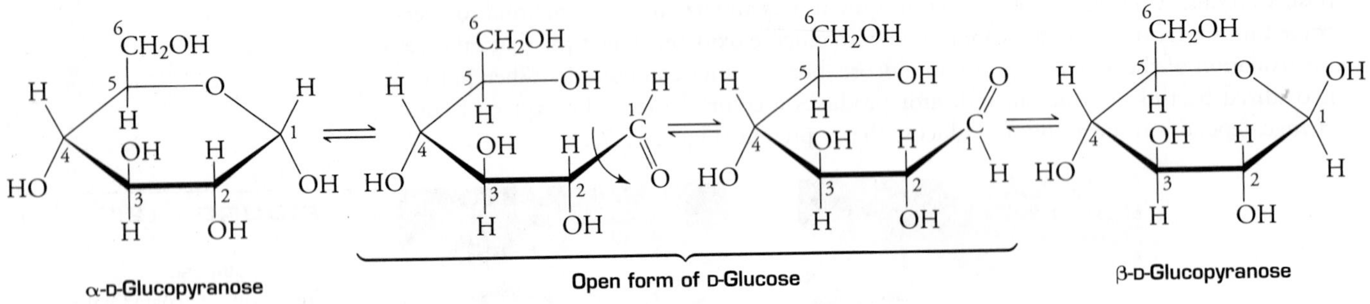

α-D-Glucopyranose

Open form of D-Glucose

β-D-Glucopyranose

FIGURE 12.8 D-Glucose anomers
Glucose forms a pair of cyclic hemiacetals when the —OH group attached to carbon atom 5 reacts with the C=O group. The cyclic hemiacetal with the —OH on carbon atom 1 pointing down is the α anomer and that with the —OH pointing up is the β anomer. Monosaccharides that form anomers containing six-membered rings are called pyranoses.

Figure 12.8 shows D-glucose reacting to form two different cyclic hemiacetals. The C=O group points down in the noncyclic (open) form of the D-glucose drawn to the left of center, so when the ring forms the newly created —OH of the hemiacetal carbon atom also points down. Rotation about the single bond between carbons 1 and 2 converts the open form at the left of center to that at the right of center. When this new conformation reacts to form a hemiacetal, the hemiacetal —OH points up.

The open form of D-glucose has four chiral carbon atoms, while the two cyclic forms have five chiral carbon atoms each because the carbon atom of the newly formed hemiacetal is chiral. *The two cyclic forms of glucose* in Figure 12.8 are called **anomers**. The one with the hemiacetal —OH *pointing down* is called the **alpha (α) anomer** and that with the —OH *pointing up* is the **beta (β) anomer**.

■ When drawn in the orientation shown in Figures 12.8 and 12.9, the hemiacetal —OH points down for an α anomer and up for a β anomer.

The complete name for a monosaccharide anomer must specify the anomer (α or β), the monosaccharide, and the ring size. Five- and six-membered monosaccharide rings predominate and the common names of two cyclic ethers, furan and pyran, are used to indicate ring size. A **furanose** has a *five-membered ring* and a **pyranose**, a *six-membered ring*. The two glucose anomers in Figure 12.8 are pyranoses and their full names are α-D-glucopyranose and β-D-glucopyranose.

■ Organic chemists and biochemists use α and β differently, so the α- and β-keto acids described in Section 10.4 are completely unrelated to the α and β anomers described here.

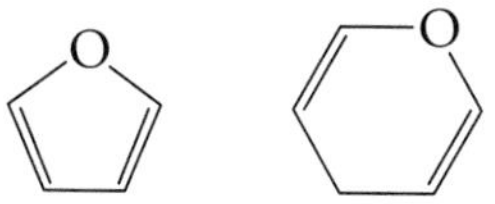

Furan   Pyran

An idea to keep in mind when it comes to drawing pyranose anomers of D-glucose and related monosaccharides is that when the six-membered ring anomers are drawn as in Figure 12.8 (oxygen atom to the back right, hemiacetal carbon atom to the right), the —CH_2OH group on carbon atom 5 always points up for a D sugar and down for an L sugar.

As can be seen in Figure 12.4, D-galactose is a diastereomer of glucose that differs only in the orientation at carbon atom 4. D-Galactose forms α and β pyranose isomers that are identical to those of glucose except that the —OH on carbon atom 4 points up instead of down.

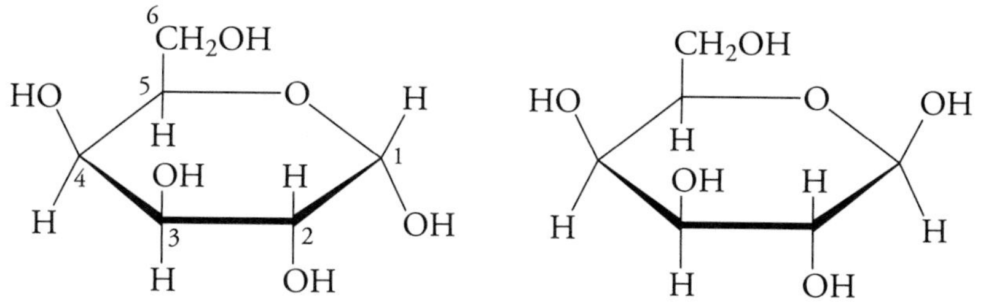

α-D-Galactopyranose   β-D-Galactopyranose

The amino sugar D-glucosamine differs from D-glucose only at carbon atom 2, by having an —NH_2 group instead of an —OH group. This difference is maintained in the pyranose anomers of D-glucosamine, in which —NH_2 points down on carbon atom 2.

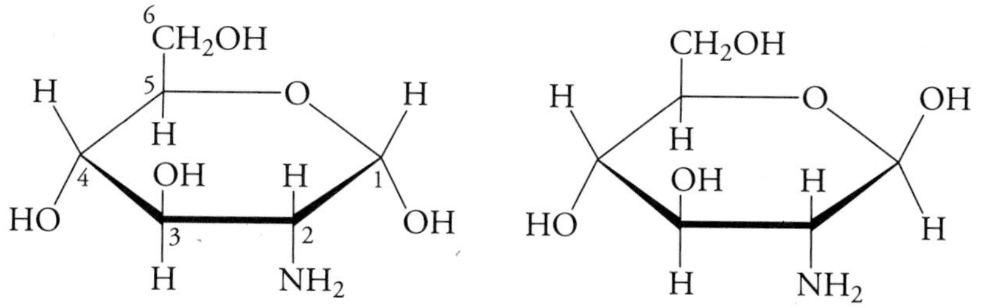

α-D-Glucosaminopyranose   β-D-Glucosaminopyranose

D-Ribose, an aldopentose, forms five-membered cyclic hemiacetals. As is the case for the three aldohexoses just discussed, the cyclic hemiacetal is formed when the —OH group

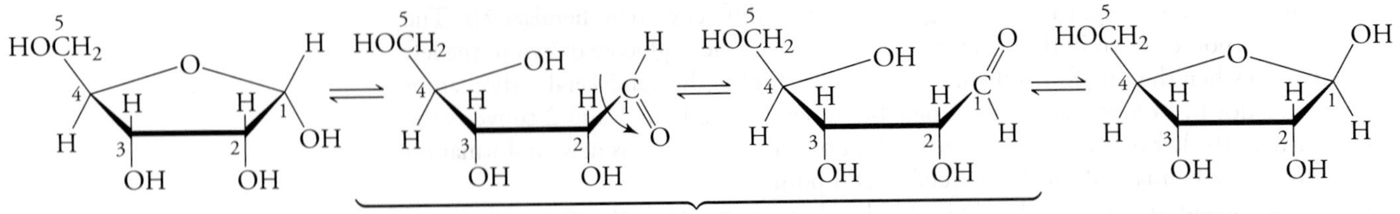

α-D-Ribofuranose

Open form of D-Ribose

β-D-Ribofuranose

FIGURE 12.9 D-Ribose anomers
In its cyclic hemiacetal forms, D-ribose is a furanose—it contains a five-membered ring.

attached to the chiral carbon atom farthest from the C=O reacts with the aldehyde group. By convention, the five-membered furanose ring is drawn with the ring oxygen atom at the rear and the hemiacetal carbon atom at the right (Figure 12.9). When a furanose is drawn in this form, the $-CH_2OH$ on the left side will always point up for a D sugar and down for an L sugar.

D-2-Deoxyribose is identical to D-ribose, except that —H replaces —OH at carbon atom 2. D-2-deoxyribose is found in α and β furanose forms.

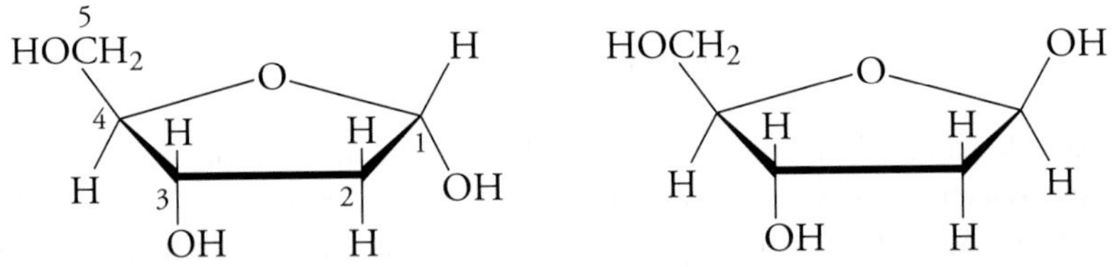

α-D-2-Deoxyribofuranose

β-D-2-Deoxyribofuranose

D-Fructose, the only ketose that we will consider in this section, appears as α and β furanose anomers (Figure 12.10). Unlike the aldoses just considered, the hemiacetal carbon atom of fructose (position 2) holds $-CH_2OH$ and —OH rather than —H and —OH. The difference arises because in the open form of fructose the C=O is on carbon atom 2. When the ring forms, carbon atom 1 is not involved and it remains as $-CH_2OH$.

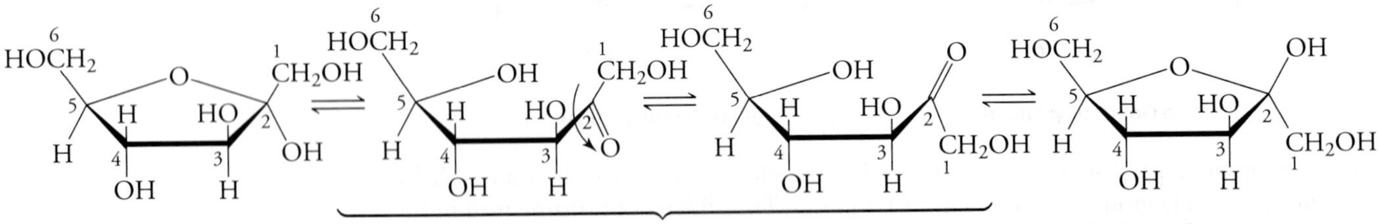

α-D-Fructofuranose

Open form of D-Fructose

β-D-Fructofuranose

FIGURE 12.10 D-Fructose anomers
D-Fructose is a furanose in its cyclic hemiacetal forms.

Mutarotation

In solution, the cyclic hemiacetal groups of α and β anomers undergo continuous change. For example, if α-D-glucopyranose (Figure 12.8) is dissolved in water, all three forms (α, β, and open) will appear. The same would be true beginning with pure β-D-glucopyranose or the open form of glucose. The *process of converting back and forth from an α anomer to the open form to the β anomer* is called **mutarotation**. The term mutarotation is based on the observation that, for glucose, converting from the α to the β anomer produces a change in the rotation of plane-polarized light (Latin: *mutare*, "to change").

Although, in solution, monosaccharides spend most of their time in a cyclic form, there is still enough open form present for aldehyde and ketone reactions to take place. All of the cyclic monosaccharides mentioned above, for example, are reducing sugars: they can be oxidized by Benedict's reagent.

■ Mutarotation is the interconversion of α and β anomers.

SAMPLE PROBLEM 12.4

Drawing α and β anomers

Draw and name the α and β furanose anomers of D-xylose.

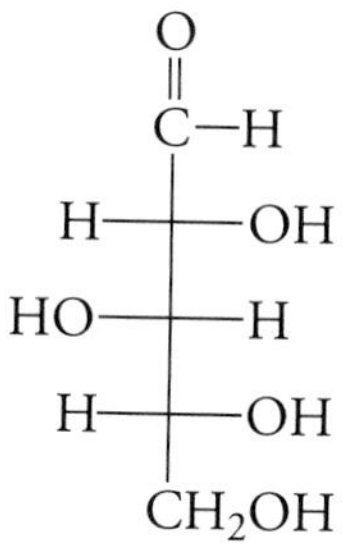

D-Xylose

Strategy

One approach to use is to determine the difference in the structures of the open forms of D-xylose and D-ribose. The same difference will be maintained in the anomers. Figure 12.9 shows the two furanose anomers of ribose.

Solution

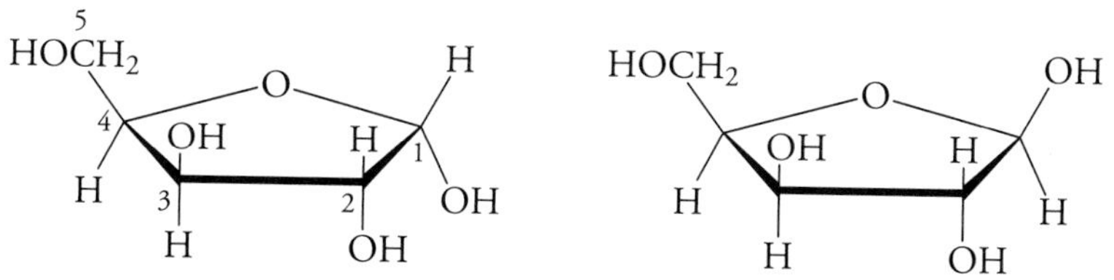

α-D-Xylofuranose

β-D-Xylofuranose

D-Xylose differs from D-ribose only at carbon atom 3, so the furanose forms of these two monosaccharides also differ only at this position.

PRACTICE PROBLEM 12.4

Draw and name the α and β pyranose anomers of D-abqueose, a dideoxy aldohexose found in the membranes of some bacteria.

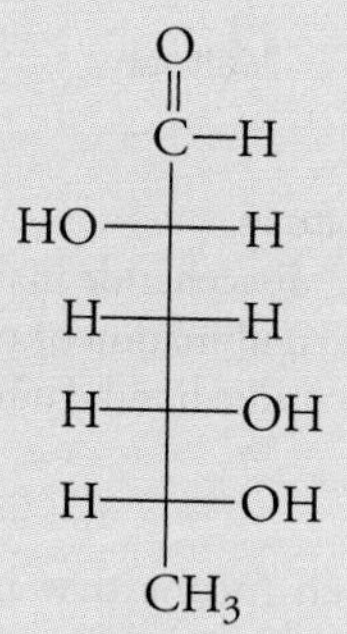

D-Abqueose

12.4 OLIGOSACCHARIDES

■ Glycosidic bonds connect monosaccharide residues to one another.

Figure 11.19*b* showed that one way to form an acetal (contains a carbon atom that is attached to two —OC groups) is to react a hemiacetal with another molecule that contains an —OH group. When the hemiacetal in question is the α or β anomer of a monosaccharide, biochemists call the resulting *acetal* a **glycoside**, and *the bond that connects the acetal carbon to the newly added —OC group* is a **glycosidic bond**. Oligosaccharides are formed when 2 to 10 monosaccharide residues are joined to one another by glycosidic bonds.

Disaccharides, which *contain two monosaccharide residues*, are the oligosaccharides found most widely in nature. Maltose, cellobiose, lactose, and sucrose are four important disaccharides. Maltose is the combination of two α-D-glucopyranoses. As pictured in Figure 12.11, the glucose molecule on the left provides the hemiacetal carbon atom (carbon atom 1) used to make the glycosidic bond and the glucose molecule on the right provides the —OH group (that on carbon atom 4). The bond that joins the two α-D-glucopyranoses in maltose is called an α-(1 → 4) glycosidic bond, because it connects carbon atom 1 (α orientation) of one glucose residue to carbon atom 4 of the other glucose residue. In this notation, the arrow points from the number of the carbon atom that provides the hemiacetal group to the number of the atom that provides the alcohol group.

Maltose is a product of the digestion of starch and glycogen. In the intestines, the hydrolysis of maltose—the reverse of the reaction shown in Figure 12.11—is catalyzed by the enzyme maltase to produce glucose molecules, which pass through the intestinal lining and into the bloodstream.

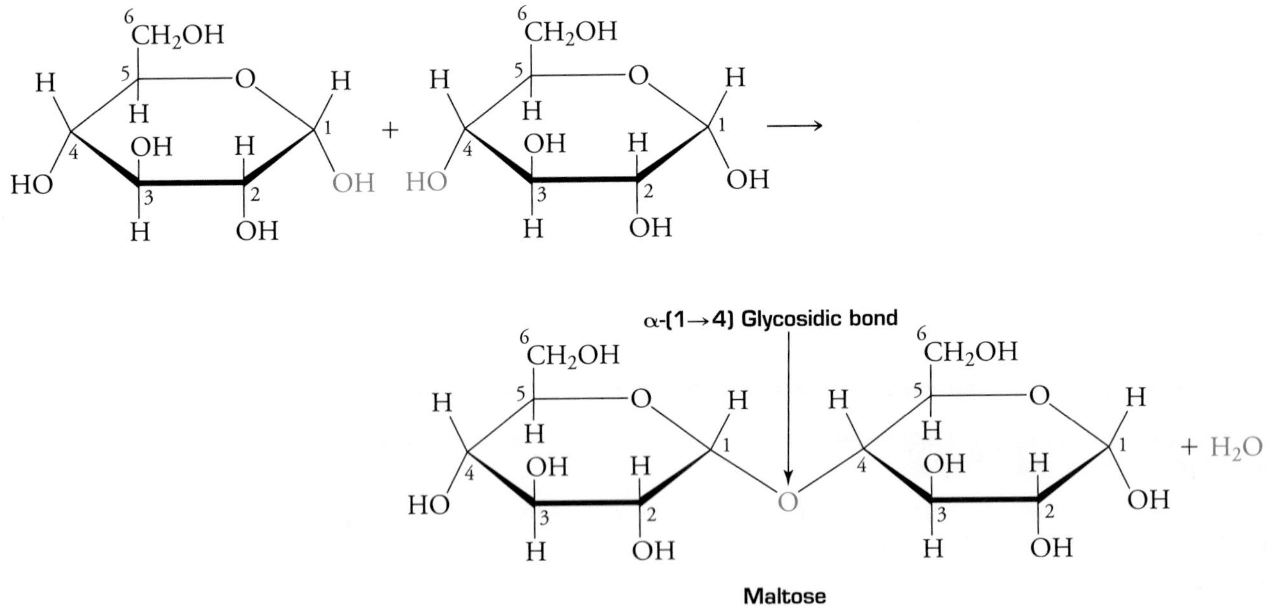

α-(1→4) Glycosidic bond

$+ H_2O$

Maltose

■ FIGURE | 12.11

Maltose
The disaccharide maltose is an acetal (glycoside) formed by the combination of two D-glucose molecules. An α-(1 → 4) glycosidic bond connects the two glucose residues.

Maltose is a reducing sugar, which means that in one of its solution forms it has an aldehyde group that is oxidized by Cu^{2+} (Benedict's reagent). As typically drawn, it is the glucose residue on the right side of maltose that is oxidized, because this anomer can undergo mutarotation (Figure 12.12).

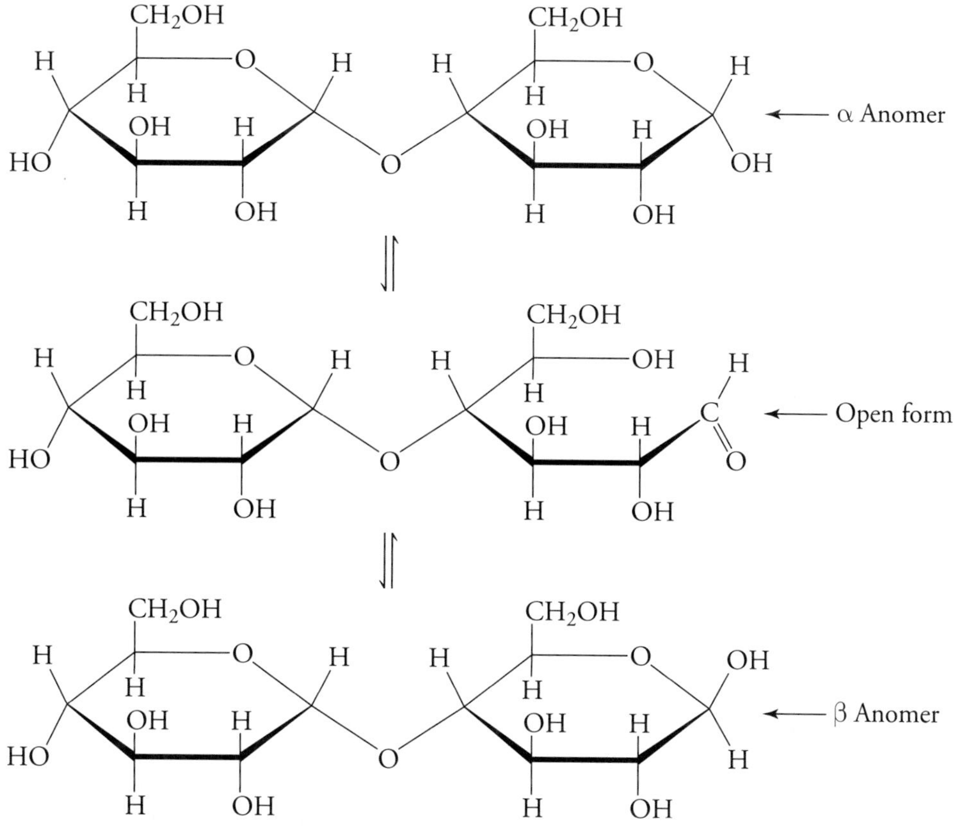

FIGURE 12.12

Maltose is a reducing sugar In maltose, the glucose residue with the hemiacetal carbon atom (the residue drawn on the right-hand side) can undergo mutarotation. When in the open form, it reacts as an aldehyde and can be oxidized by Benedict's reagent. This makes maltose a reducing sugar.

Cellobiose contains two glucose residues and differs from maltose only in that its glycosidic bond is β-(1 → 4). The oxygen atom that connects the two residues points up from the glucose on the left (β orientation) and down from the glucose on the right (carbon atom 4).

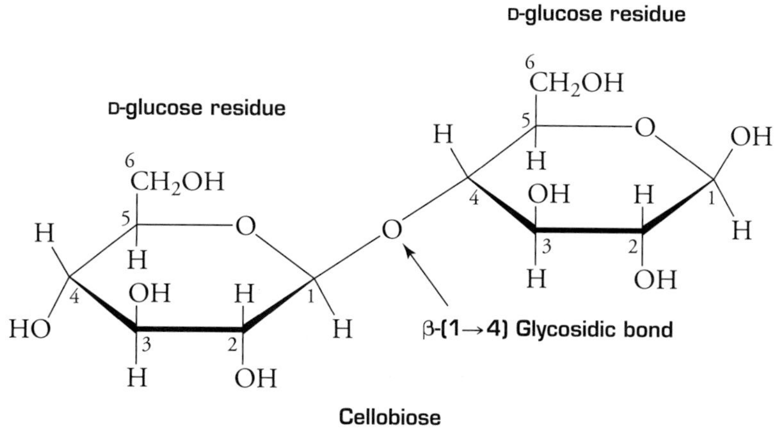

Cellobiose

Cellobiose is formed during the breakdown of cellulose, and when hydrolyzed cellobiose yields two glucoses. While cellobiose gives the same hydrolysis products as maltose, humans are unable to use cellobiose as a fuel source because we do not produce the enzyme necessary to catalyze this particular hydrolysis reaction. Like maltose, cellobiose is a reducing sugar.

Lactose, which makes up about 5% (w/v) of cow's milk and about 7% (w/v) of human milk, is an important fuel source for infants (bovine, human, or otherwise). It consists of a β-galactopyranose residue joined to a glucopyranose residue by a β-(1 → 4) glycosidic bond. The only difference between lactose and cellobiose is that in lactose the —OH on carbon atom 4 of the monosaccharide residue on the left (galactose) points up and in cellobiose the —OH on this same carbon points down (glucose).

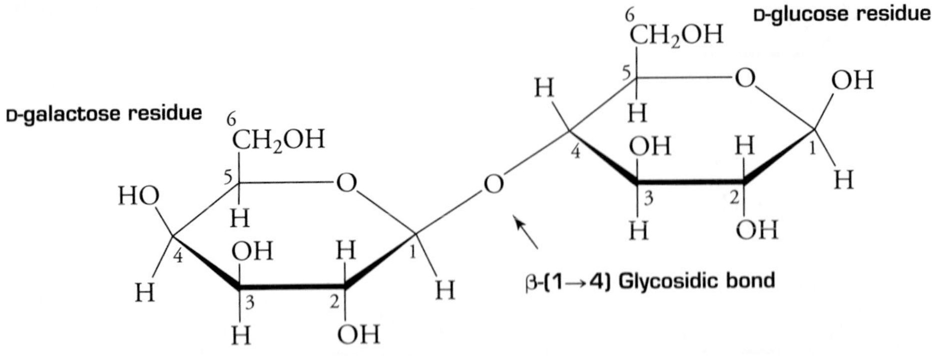

D-galactose residue
D-glucose residue
β-(1→4) Glycosidic bond

Lactose

When hydrolyzed in the intestines with the assistance of the enzyme β-galactosidase (known also as lactase), lactose yields one galactose and one glucose molecule. These monosaccharides pass through the intestinal lining and move into the bloodstream. Several health problems are associated with lactose and its digestion. Lactose intolerance, the inability to hydrolyze lactose, is a common ailment (Health Link: Indigestible Oligosaccharides) and the presence of galactose from the breakdown of lactose poses a problem for those individuals with galactosemia (Section 12.1). Like maltose and cellobiose, lactose is a reducing sugar.

Sucrose, known also as table sugar, consists of an α-glucopyranose residue joined to a β-fructofuranose by an α,β-(1 ↔ 2) glycosidic bond. A double-headed arrow is used here because each monosaccharide supplies a hemiacetal group to the glycosidic bond. Unlike the disaccharides mentioned above, sucrose is not a reducing sugar—it contains no hemiacetal group and is unable to mutarotate. When sucrose forms, the hemiacetal group of fructose combines with the hemiacetal —OH of glucose to produce the glycosidic (acetal) bond.

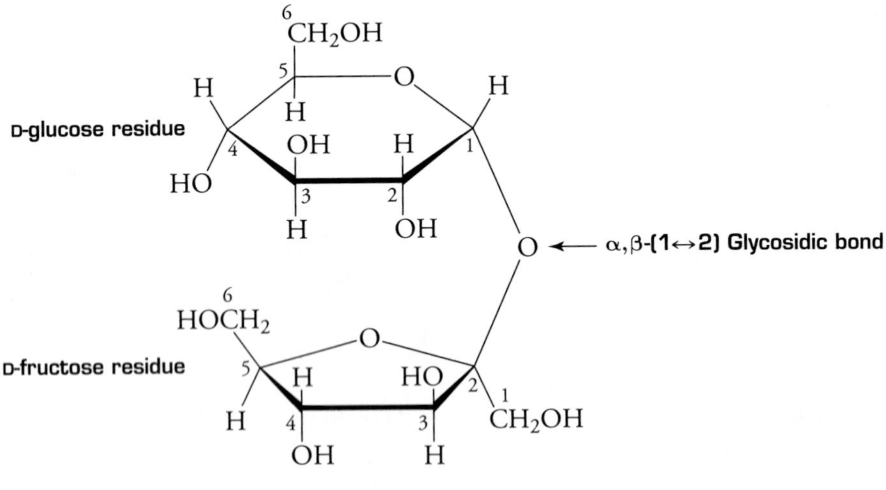

D-glucose residue
α,β-(1↔2) Glycosidic bond
D-fructose residue

Sucrose

Sucrose is an energy and carbon storage molecule for many plants, including the sugar beets and sugar cane that are the source of most table sugar available in stores. When hydrolyzed, sucrose yields one fructose and one glucose molecule, each of which passes into the bloodstream.

While disaccharides are the most commonly found oligosaccharides, those with 3 or more monosaccharide units are also known. Peas and beans, for example, contain the oligosaccharides shown in Figure 12.13—raffinose (a trisaccharide), stachyose (a tetrasaccharide), and verbascose (a pentasaccharide). Many people have problems digesting these oligosaccharides (Health Link: Indigestible Oligosaccharides).

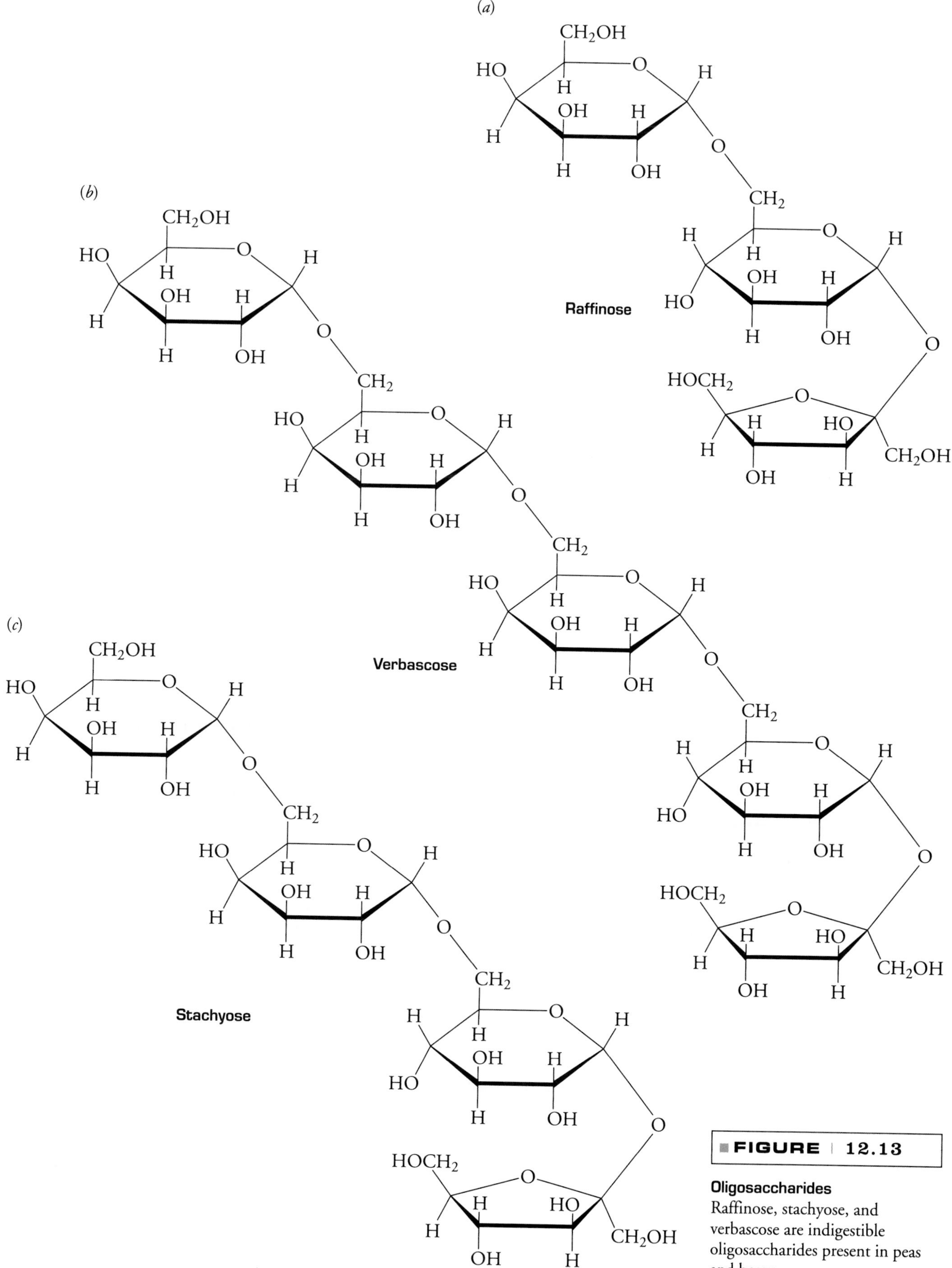

FIGURE 12.13

Oligosaccharides
Raffinose, stachyose, and verbascose are indigestible oligosaccharides present in peas and beans.

Glycolipids are sugar-containing lipids present in nerve cell membranes and that serve as identifying markers on cell surfaces. In Section 8.4 glycolipids were described as containing a sugar residue. More carefully defined, a glycolipid consists of a mono-, oligo-, or polysaccharide attached to an alcohol group of a lipid by a glycosidic bond.

In cerebrosides, glycolipids present in nerve and brain membranes, the attached sugar residue is usually glucose or galactose (Figure 12.14). The sugar part of membrane glycolipids called gangliosides is usually an oligosaccharide. Many gangliosides serve as markers for organ and tissue specificity. Blood type is determined by the oligosaccharide portion of glycolipids or glycoproteins attached to the surface of red blood cells (Health Link: Blood Type).

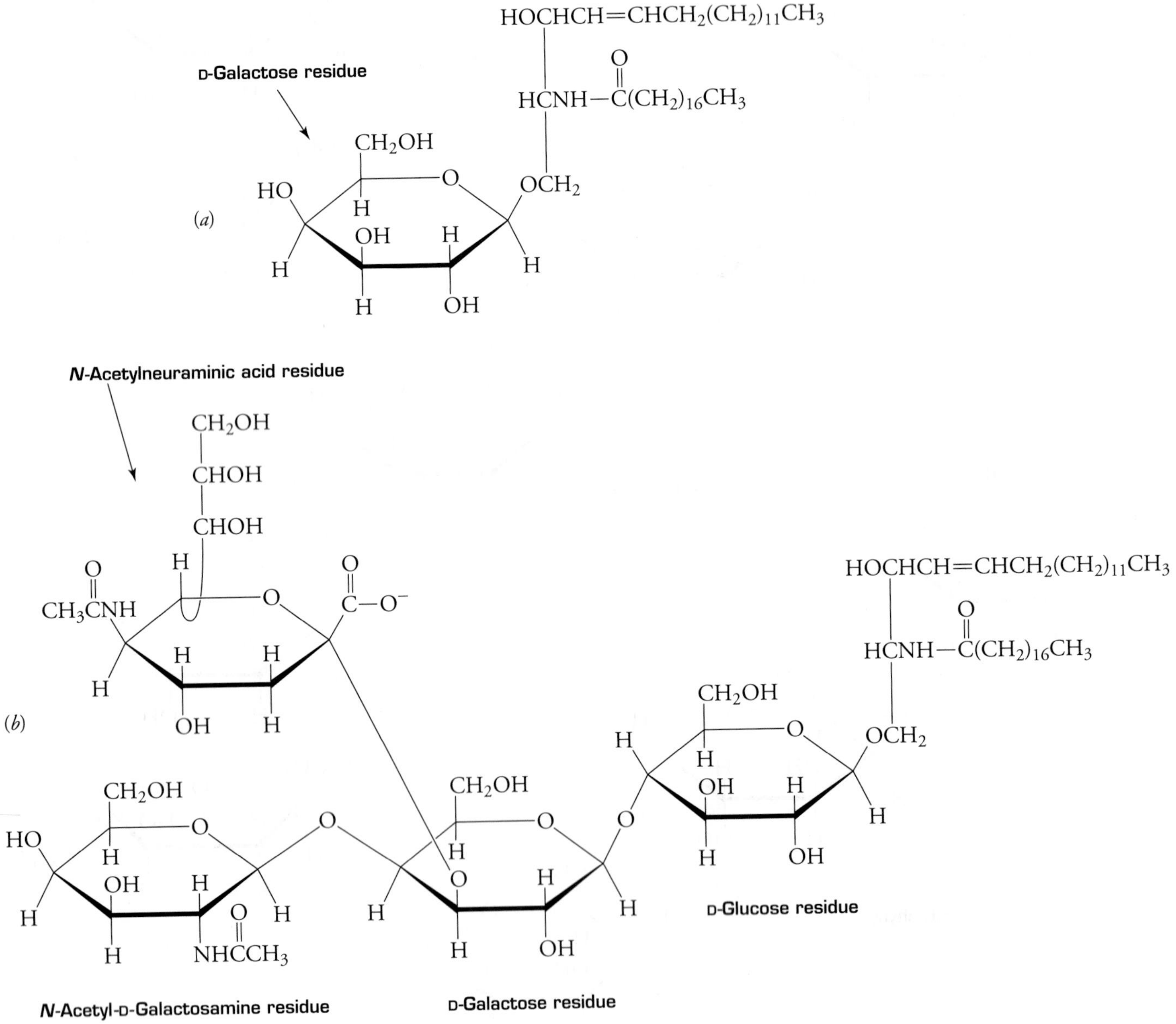

(*a*)

$HOCHCH{=}CHCH_2(CH_2)_{11}CH_3$

$HCNH{-}C(=O)(CH_2)_{16}CH_3$

OCH_2

D-Galactose residue

(*b*)

N-Acetylneuraminic acid residue

$HOCHCH{=}CHCH_2(CH_2)_{11}CH_3$

$HCNH{-}C(=O)(CH_2)_{16}CH_3$

OCH_2

D-Glucose residue

N-Acetyl-D-Galactosamine residue

D-Galactose residue

FIGURE 12.14

Glycolipids
(*a*) Cerebrosides often consist of a fatty acid residue and a glucose or galactose residue attached to a sphingosine backbone. (*b*) The carbohydrate portion of gangliosides is typically an oligosaccharide. Ganglioside G_{M2} is shown here. Glycolipids can contain other fatty acid residues than the stearic acid residues pictured here.

HEALTH Link

Relative Sweetness

Many carbohydrates present in our diet are there because of their sweetness. Relative sweetness is determined by assigning sucrose (table sugar) a sweetness of 100 and then comparing other sweeteners to this standard. A sugar with a relative sweetness of 200, for example, would be twice as sweet as an equal amount of sucrose. Table 12.2 gives some typical relative sweetness values.

TABLE 12.2 RELATIVE SWEETNESS

Sweetener	Relative Sweetness	Details
Natural sweeteners		
Fructose	170	Fruit sugar, a ketohexose, one component of sucrose
Invert sugar	120	Hydrolyzed sucrose
Sucrose	100	Table sugar, a disaccharide containing one glucose and one fructose residue
Xylitol	100	An alcohol sugar
Glucose	75	Corn syrup, an aldohexose, one component of sucrose and lactose
Sorbitol	55	An alcohol sugar
Maltose	32	A disaccharide containing two glucose residues
Galactose	30	An aldohexose, one component of lactose
Lactose	15	A disaccharide containing one glucose and one galactose residue
Artificial sweeteners		
Sodium cyclamate	3,000	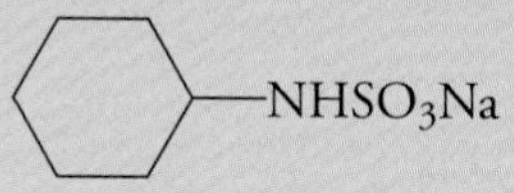 (cyclohexyl)—$NHSO_3Na$
Acesulfame-K	20,000	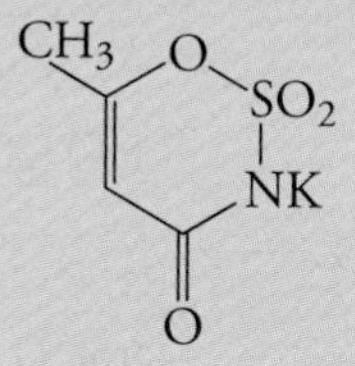 CH_3, O, SO_2, NK, O (ring structure)
Aspartame	18,000	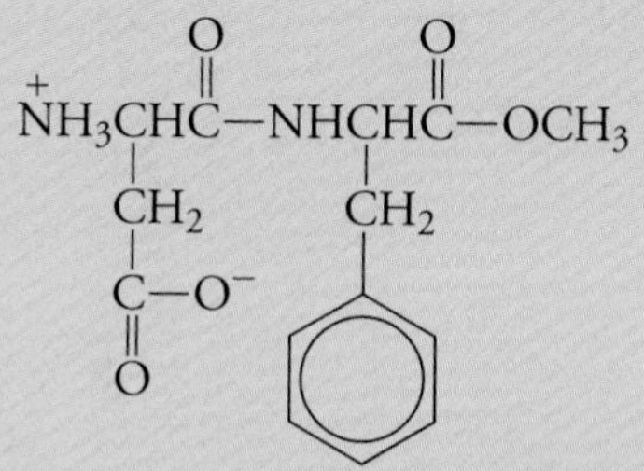 $\overset{+}{N}H_3CH(CH_2COO^-)C(=O)$—$NHCH(CH_2C_6H_5)C(=O)$—$OCH_3$
Saccharin	30,000	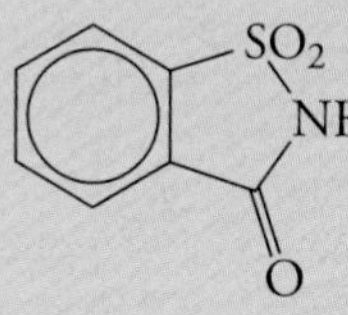 (benzo ring) SO_2, NH, C=O

Continues on next page

■ TABLE | 12.2 CONTINUED

Sweetener	Relative Sweetness	Details
Artificial sweeteners		
Sucralose	60,000	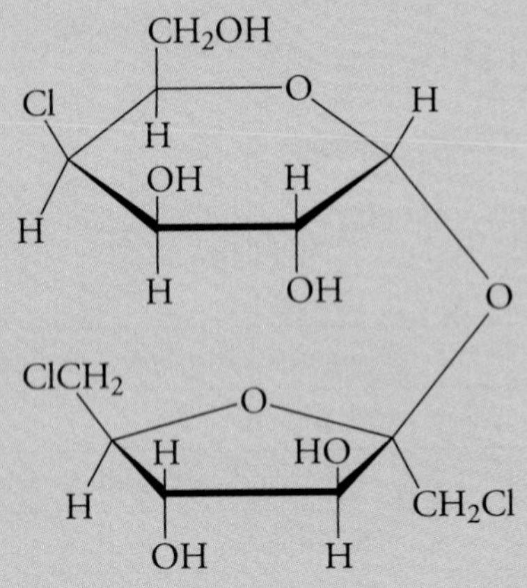

The list of ingredients on candy bars, cookies, and other sweet treats typically includes invert sugar and/or high fructose corn syrup. Invert sugar is produced by hydrolyzing sucrose, a disaccharide, into its component monosaccharides, glucose and fructose. This sweetener gets its name from the fact that sucrose rotates plane-polarized light in a clockwise (+) direction, but the product mixture of glucose and fructose rotates plane-polarized light in a counterclockwise (−) direction; thus the rotation is inverted. Invert sugar is used because it is a cost-effective way of increasing sweetness. Together, the products of sucrose hydrolysis (glucose and fructose) are sweeter than sucrose.

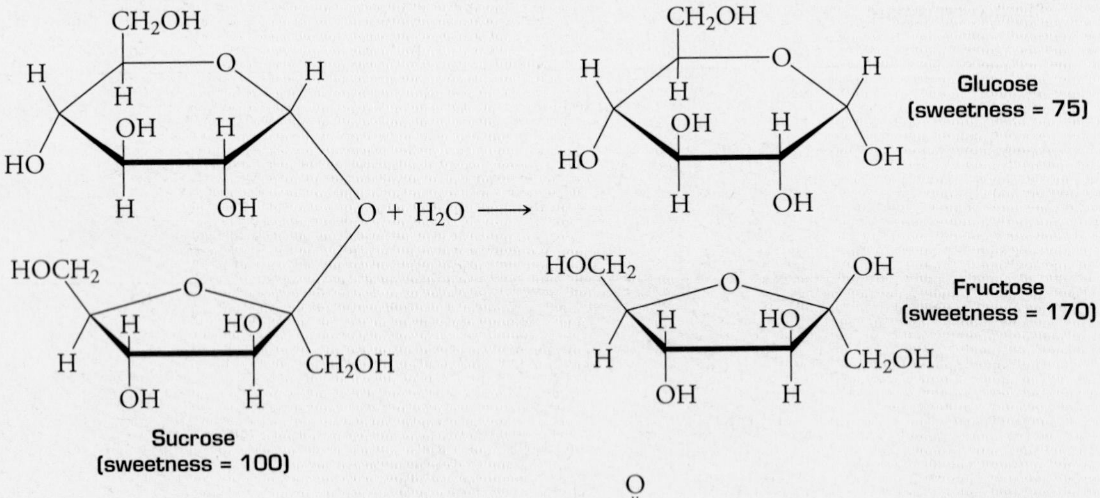

$$\text{Sucrose} + H_2O \longrightarrow \text{Glucose} + \text{Fructose}$$

Sucrose (sweetness = 100)

Glucose (sweetness = 75)

Fructose (sweetness = 170)

Use of high fructose corn syrup is another cost-effective way of increasing sweetness. Corn syrup consists of glucose, which has a relative sweetness of 75. When corn syrup is treated with the enzyme glucose isomerase, some of the D-glucose is converted into its constitutional isomer D-fructose. The resulting fructose–glucose mixture is sweeter than the original corn syrup.

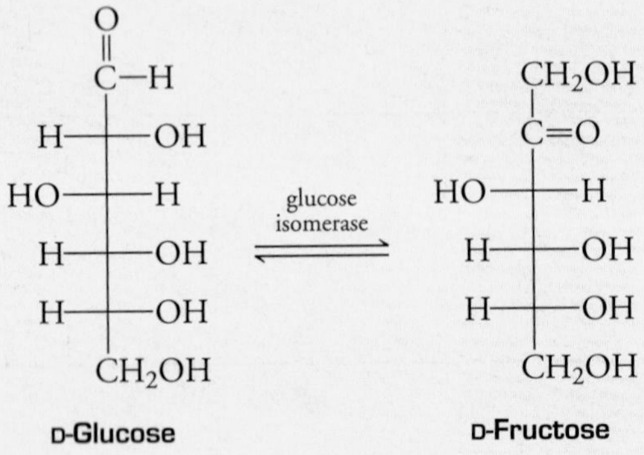

$$\text{D-Glucose} \xrightleftharpoons{\text{glucose isomerase}} \text{D-Fructose}$$

D-Glucose

D-Fructose

Sorbitol (Figure 12.5) is often used as the sweetener in foods prepared for diabetics. This alcohol sugar is not absorbed well by the intestines and, therefore, has minimal effects on blood levels of glucose and insulin.

Artificial sweeteners, noncarbohydrate molecules that provide a great deal of sweetness with few or no calories, are also available. Among those in use today are saccharin (in Sweet' N Low), aspartame (NutraSweet), acesulfame-K (Sunette), and sucralose (Splenda). To enhance their water solubility, some are sold as ionic compounds, including cyclamate as the sodium (Na) salt and acesulfame-K as the potassium (K) salt.

These artificial sweeteners do not give an exact "sugar experience," because they either have a taste that is slightly off or have an aftertaste. Although all currently in use have been approved, there have been concerns about the long-term health consequences of using some of these products. The U.S. Food and Drug Administration banned the use of sodium cyclamate in 1969, after studies showed that large amounts of this artificial sweetner cause cancer in some animals. Beginning in 1977, saccharin-containing products sold in the United States were required to carry a label warning that saccharin is a cancer-suspect agent. In late 2000 the U.S. Congress repealed this requirement after saccharin was removed from the list of suspected carcinogens.

HEALTH Link

Indigestible Oligosaccharides

The digestion of oligosaccharides takes place in the intestines, where enzymes catalyze the hydrolysis of glycosidic bonds to produce monosaccharides. Lactose intolerance is caused by a deficiency in β-galactosidase, the enzyme that catalyzes the hydrolysis of the β-(1 $\rightarrow$ 4) glycosidic bond in this disaccharide. Infants normally produce this enzyme, but by the time people reach the age of 5 a significant number have either stopped producing it or make it only at reduced levels. Worldwide, more than 70% of adult humans are deficient in β-galactosidase, but these numbers vary widely between population groups. For example, lactose intolerance is common in those with African, Asian, Hispanic, and Native American backgrounds and is less common for those with northern European ancestry.

When a lactose-intolerant person consumes dairy products, the lactose is metabolized by intestinal bacteria instead of by β-galactosidase. This results in the production of gases such as CO_2, H_2, and CH_4, which cause bloating. Small carboxylic acids that also form cause diarrhea by drawing water into the intestines through osmosis.

Lactose intolerance cannot be cured, so one way that lactose-intolerant people can avoid its symptoms is to avoid dairy products altogether. Other options are to buy lactose-free milk, yogurt, and cheese, or to use tablets, such as Lactaid, that contain a bacterial β-galactosidase which catalyzes the breakdown of lactose.

The same type of digestion problem is associated with some of the oligosaccharides present in beans, peas, and other vegetables. Raffinose, stachyose, and verbascose (Figure 12.13) all contain galactopyranose residues involved in α-(1 $\rightarrow$ 6) glycosidic bonds. Humans do not produce the α-galactosidase enzyme necessary to hydrolyze these bonds, so when foods containing these carbohydrates are eaten, intestinal bacteria break them down and produce gases. Beano and similar products contain an enzyme that catalyzes the hydrolysis of the α-(1 $\rightarrow$ 6) glycosidic bonds (Figure 12.15).

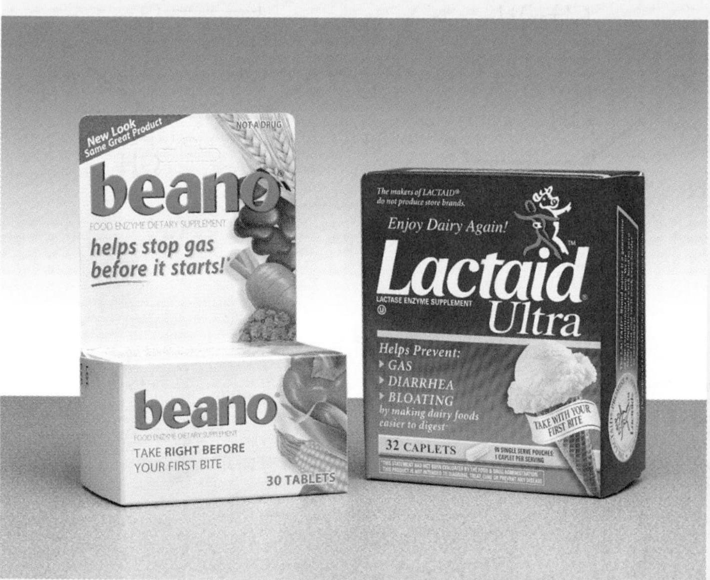

FIGURE 12.15

Digestive aids

Lactaid contains an enzyme that catalyzes the breakdown of lactose present in dairy products. Beano contains an enzyme that catalyzes the hydrolysis of indigestible oligosaccharides found in peas, beans, and other vegetables.

Source: Andy Washnik/Wiley Archive.

12.5 POLYSACCHARIDES

- Homopolysaccharides contain one type of monosaccharide residue. Heteropolysaccharides contain more than one.

Carbohydrates containing more than 10 monosaccharide residues are called polysaccharides. In living things, polysaccharides help provide structure and also act as energy storage molecules. **Homopolysaccharides** *are composed of just one type of monosaccharide* and **heteropolysaccharides** *are built from more than one type*. Typically, a heteropolysaccharide molecule will be constructed from very few different monosaccharides.

Homopolysaccharides

The homopolysaccharide cellulose is of major importance to the structure of plants (wood is about 50% cellulose) because it provides support to stems and stalks and provides a tough and water-insoluble protective barrier. Cellulose consists of a long chain of β-glucopyranose residues joined by β-(1 → 4) glycosidic bonds. The strength of cellulose comes from the way in which cellulose molecules form hydrogen bonds with neighboring ones, producing durable sheets of polysaccharide (Figure 12.16).

Animals do not produce the enzyme required to hydrolyze cellulose into cellobiose and individual glucose molecules, so cellulose is not typically an animal food source. The ability of horses, cows, and other ruminants to survive on grass, hay, and other cellulose-containing plants is due to the presence of symbiotic bacteria in their digestive tract that catalyze the breakdown of cellulose.

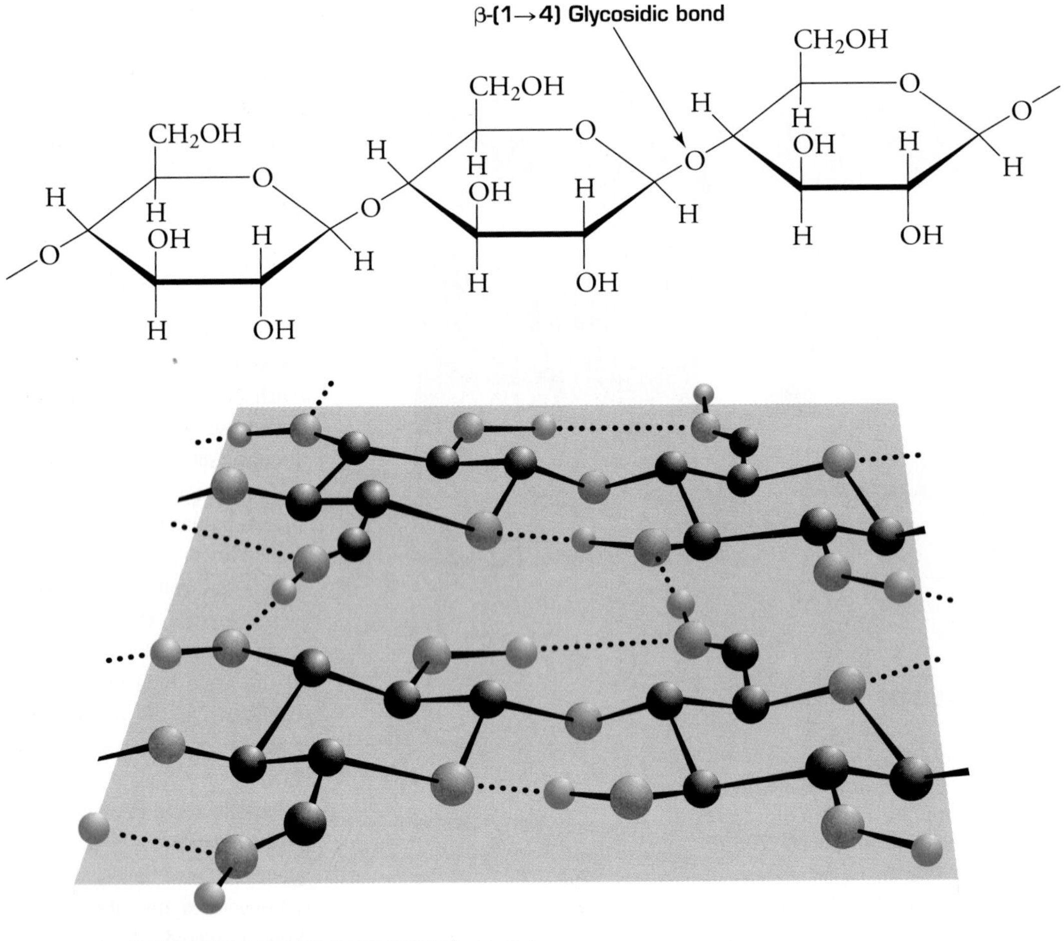

β-(1→4) Glycosidic bond

CH_2OH O H H OH H H OH O H CH_2OH O H H OH H H OH O H CH_2OH O H H OH H H OH O H

FIGURE 12.16 **Cellulose**
In cellulose, glucose residues are joined by β-(1 → 4) glycosidic bonds. Hydrogen bonds that form between neighboring cellulose strands give cellulose its strength.

Starch is produced by plants as a way to store energy. Some plants, such as potatoes, store large amounts of this polysaccharide. Starch consists of two different homopolysaccharides, amylose and amylopectin, each of which is composed entirely of glucose building blocks. Amylose, a stereoisomer of cellulose, is a chain of α-glucopyranose residues joined by α-(1 → 4) glycosidic bonds (Figure 12.17*a*). The α arrangement of atoms gives amylose a very different three-dimensional structure than cellulose. Instead of forming hydrogen-bonded sheets of polysaccharide molecules as in cellulose, amylose molecules coil up into flexible helical shapes.

(*a*)

α-(1→4) Glycosidic bond

(*b*)

α-(1→6) Glycosidic bond

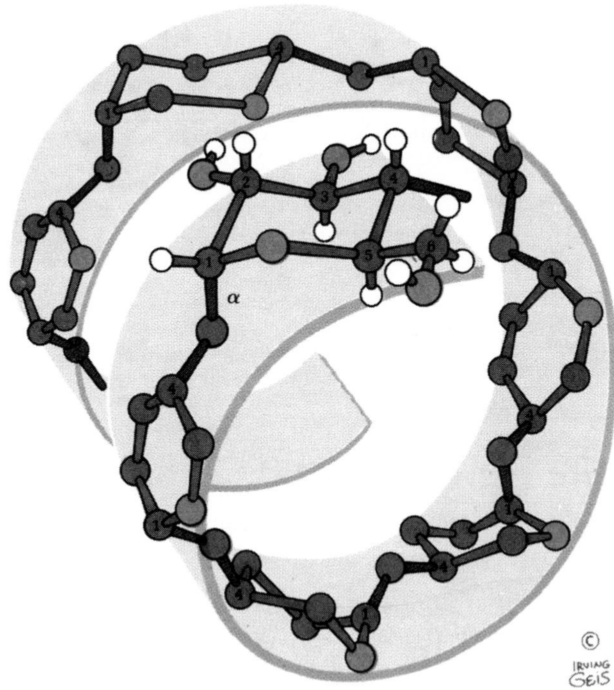

FIGURE | 12.17

Starch
Starch consists of two different homopolysaccharides:
(*a*) amylose, in which glucose residues are connected by α-(1 → 4) glycosidic bonds, and (*b*) amylopectin, in which glucose residues are joined by α-(1 → 4) and α-(1 → 6) bonds. Amylose molecules coil into flexible helical shapes.

Source: Illustration, Irving Geis.

Amylopectin, the other component of starch, consists of α-glucopyranose residues joined by α-(1 → 4) glycosidic bonds, with an α-(1 → 6) glycosidic bond at about every 25 or 30 glucose units. The α-(1 → 6) linkages produce branching in the chain (Figure 12.17*b*).

In animals and plants, the hydrolysis of α-(1 → 4) bonds in amylose and amylopectin is catalyzed by amylases and yields glucose and the disaccharide maltose which is, in turn, hydrolyzed with the assistance of maltase. In amylopectin, hydrolysis of the α-(1 → 6) glycosidic bonds is catalyzed by debranching enzymes.

Glycogen is to animals what starch is to plants—an energy storage molecule. Like amylopectin, glycogen consists of a chain of α-(1 → 4) linked glucose residues, with occasional α-(1 → 6) glycosidic bonds (Figure 12.18). In glycogen, however, these branch points occur every 8 to 12 glucose residues, rather than every 25 to 30 as in amylopectin. A variety of enzymes are responsible for catalyzing the hydrolysis of this homopolysaccharide.

Chitin makes up the exoskeleton of crustaceans and insects, and is present in the cell walls of some algae and fungi. This homopolysaccharide consists of *N*-acetyl-D-glucosamine residues joined to one another by β-(1 → 4) glycosidic bonds. The rigidity of this polysaccharide comes from the way in which separate strands are held to one another by hydrogen bonds, much as is the case for cellulose.

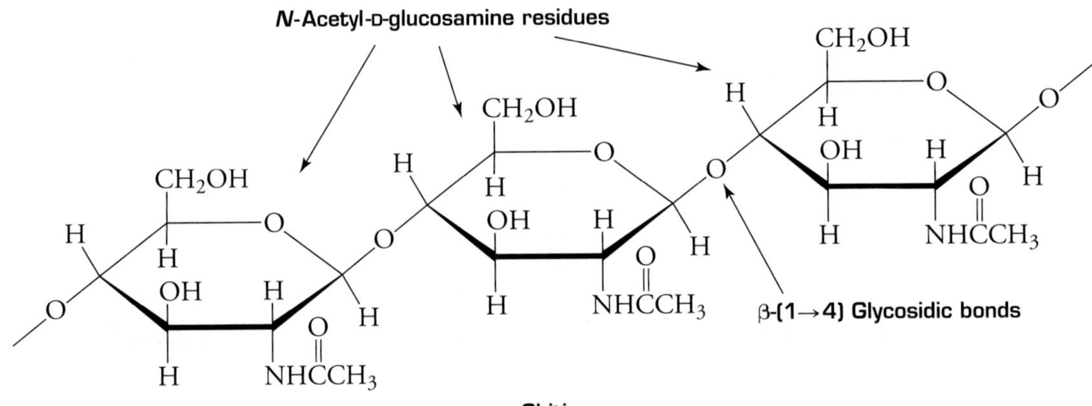

Chitin

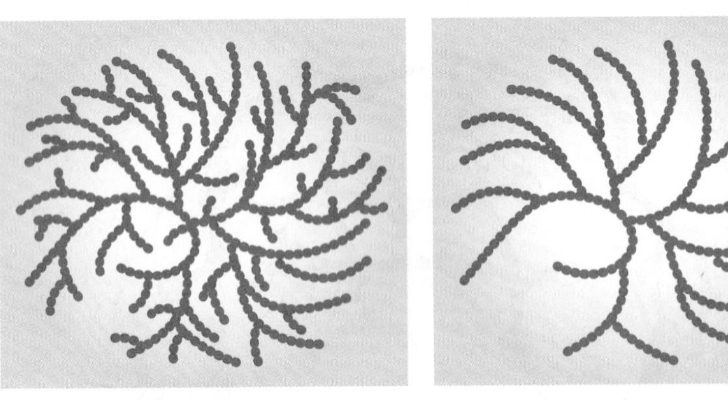

FIGURE 12.18 **Glycogen** Glycogen and amylopectin (Figure 12.17*b*) have similar structures except that glycogen is more highly branched.

Heteropolysaccharides

Hyaluronic acid, found in the lubricating fluid that surrounds joints and in the vitreous humor (clear gel) present inside the eye, is made up of alternating residues of *N*-acetyl-D-glucosamine and D-glucuronate (the conjugate base of D-glucuronic acid, a carboxylic acid sugar) connected to one another, respectively, by β-(1 $\rightarrow$ 4) and β-(1 $\rightarrow$ 3) bonds.

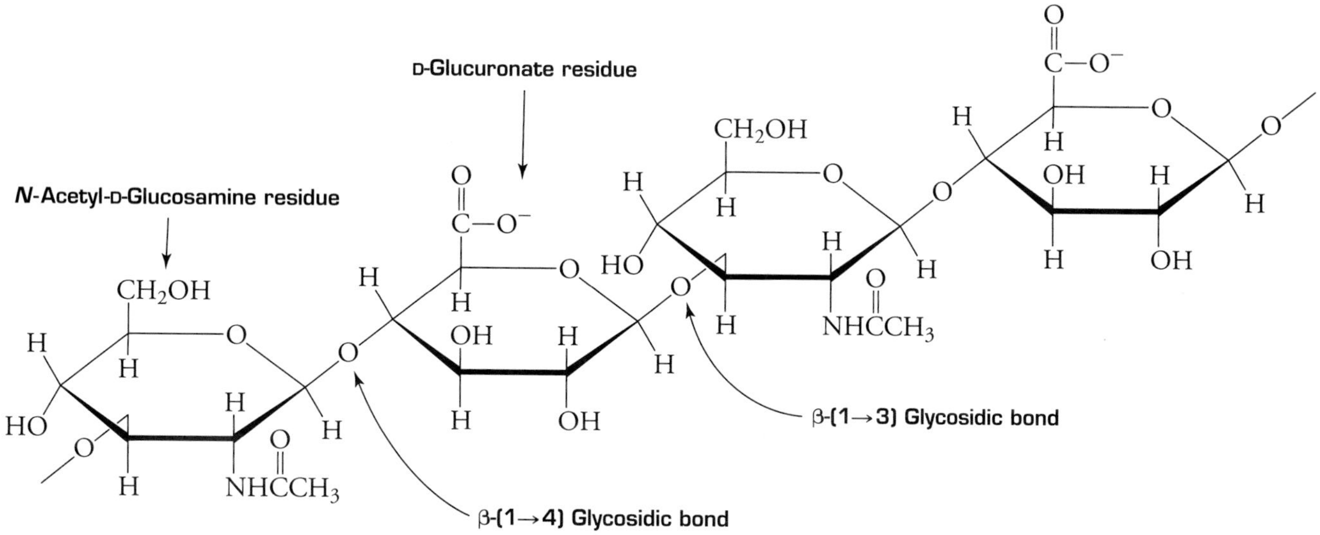

Hyaluronic acid

Chondroitin 4-sulfate consists of *N*-acetyl-D-galactosamine-4-sulfate residues alternating with D-glucuronate residues, attached, respectively, by β-(1 $\rightarrow$ 4) and β-(1 $\rightarrow$ 3) glycosidic bonds. This heteropolysaccharide is present in connective tissue, and can be purchased in over-the-counter (nonprescription) forms to assist in the repair of damaged cartilage.

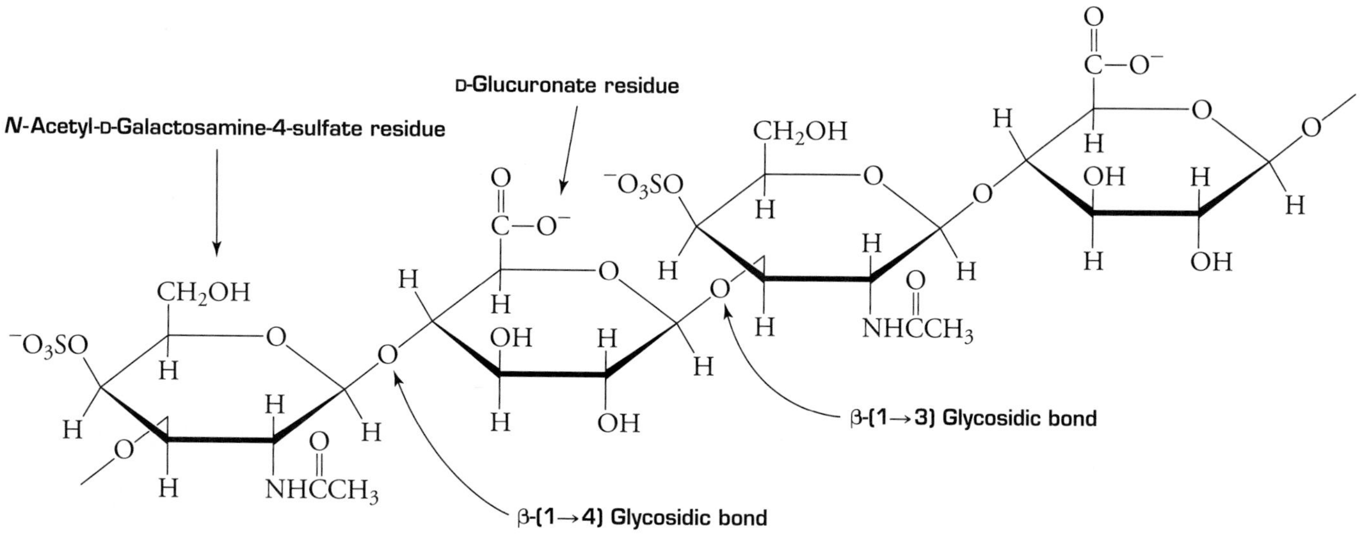

Chondroitin 4-sulfate

HEALTH Link — Blood Type

In 2003, a 17-year-old girl who underwent a heart and lung transplant at the Duke University Hospital died after being given organs from a donor whose blood type was not a proper match. When organ transplants or blood transfusions are done it is vitally important that blood type be known, because certain blood types are incompatible. This incompatibility produces an immune response that causes red blood cells to clump together, which can lead to organ damage and death.

Blood type is determined by the oligosaccharides present in glycolipids and glycoproteins (carbohydrate–protein combinations) attached to the surface of red blood cells. There are over 30 different types of red blood cell antigens (substances that cause an immune response), but the so-called ABO antigens are the source of most of the problems that take place with blood transfusions and transplants. What these three glycosphingolipids have in common is that all consist of an oligosaccharide chain attached to sphingosine (Figure 12.19). The difference between them is that type A and B group antigens have an additional monosaccharide unit attached to the galactose residue near the end of the hexasaccharide chain—galactose in the case of blood type B and *N*-acetylgalactosamine for blood type A.

N-Acetyl-D-Galactosamine residue

D-Galactose residue

N-Acetyl-D-Glucosamine residue

— remainder of glycosphingolipid

L-Fucose residue

FIGURE 12.19 **ABO blood group antigens**
A portion of the A blood group antigen is shown. For type B blood, the *N*-acetyl-D-galactosamine residue at the upper left is replaced by D-galactose. In type O blood, the monosaccharide at the upper left is missing.

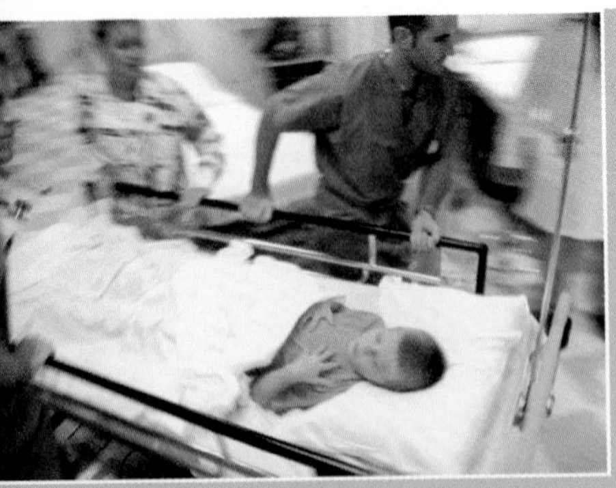

Foxglove is the common name for plants that belong to the genus (grouping) called *Digitalis*. The digitalis heart medicines contain digitoxin or digoxin, glycosides that are obtained from foxglove. Each of these glycosides (alcohol-containing steroids attached to a trisaccharide) has a direct effect on the heart muscle of animals and will produce a slower, stronger, and steadier heartbeat. Digitalis medicines are used to treat congestive heart failure and arrhythmias (irregular heartbeat).

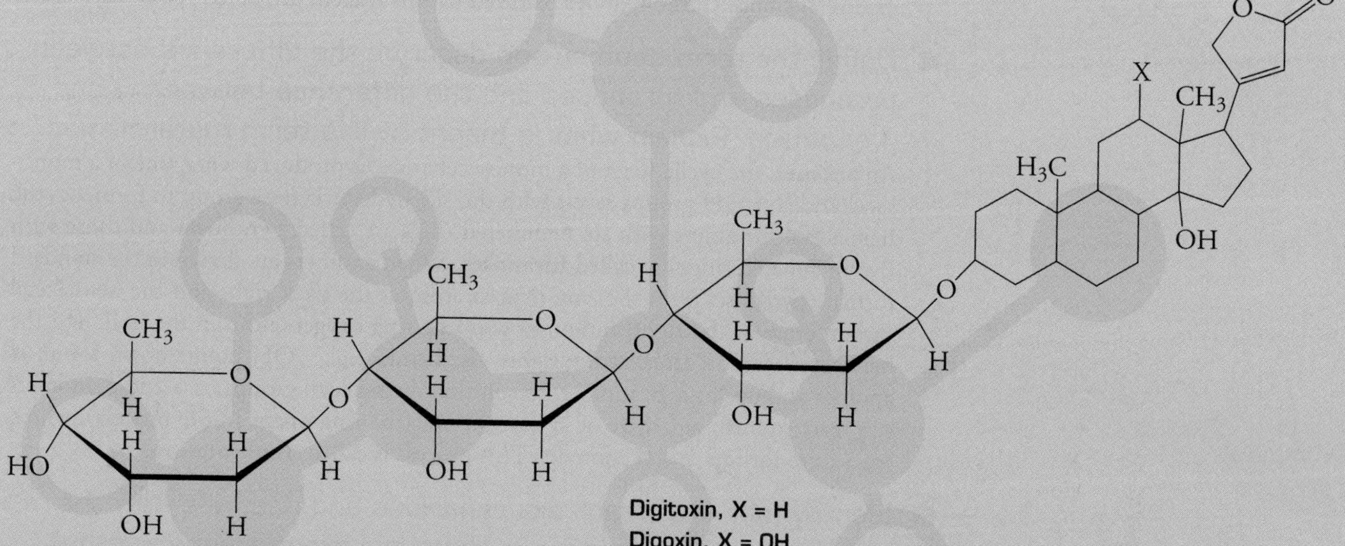

Digitoxin, X = H
Digoxin, X = OH

Digitoxin and digoxin are also quite toxic (centuries ago the Romans used foxglove as both a heart medication and a rat poison). The symptoms of poisoning include an upset stomach, mental confusion, convulsions, an increased heart rate, and heart failure. Treatment for foxglove poisoning includes gastric lavage (stomach pumping) and the administration of digitoxin-specific antibodies.

summary of objectives

1 Describe the difference between mono-, oligo-, and polysaccharides and explain the classification system used to categorize monosaccharides.

Monosaccharides are polyhydroxy aldehydes and ketones that are classified by functional group (**aldose**, **ketose**), by number of carbon atoms (**triose**, **tetrose**, etc.), or both (**aldopentose**, **ketohexose**, etc.). Monosaccharides are building blocks of **oligosaccharides** (2–10 monosaccharide residues) and **polysaccharides** (more than 10 residues).

2 Identify the four common types of monosaccharide derivatives.

The commonly observed monosaccharide derivatives are the **deoxy sugars** (—H replaces an —OH), **amino sugars** ($-NH_2$ replaces an —OH), **alcohol sugars** (C=O has been reduced to an alcohol group), and **carboxylic acid sugars** (an aldehyde group or an —OH group has been oxidized to a carboxylic acid).

3 **Explain what happens when a monosaccharide is reacted with H_2 and Pt (a catalyst) or with Benedict's reagent.**
Reacting a monosaccharide with H_2 in the presence of Pt converts the aldose or ketose into an alcohol sugar. Certain enzymes also catalyze this reduction. Benedict's reagent, which contains Cu^{2+}, oxidizes aldehydes to carboxylic acids, but will not oxidize alcohols. Carbohydrates oxidized by this reagent are called **reducing sugars**.

4 **Define the term anomer and describe the difference between pyranoses and furanoses and the difference between α and β anomers. Explain what is meant by the term mutarotation.**
An **anomer**, the cyclic form of a monosaccharide, is produced when one of a monosaccharide's —OH groups reacts with the aldehyde or ketone group to form a cyclic hemiacetal. Anomers with six-membered rings are called **pyranoses** and those with five-membered rings are called **furanoses**. When anomers are drawn in the standard format (pyranoses with the ring oxygen atom at the back right and the hemiacetal carbon atom at the right; furanoses with the ring oxygen atom in the back and the hemiacetal carbon atom at the right), the hemiacetal —OH points down for an ***α*** anomer and up for a ***β*** anomer. In solution, hemiacetal groups of α and β anomers can react to form aldehyde or ketone groups (the ring opens in the process), which react to re-form α and β anomers. This process is called **mutarotation**.

5 **Identify four important disaccharides and describe how the monosaccharide residues in them are joined to one another.**
When the hemiacetal group of one monosaccharide reacts with an alcohol or with the hemiacetal —OH of another monosaccharide, a **glycosidic** (acetal) bond forms. In an α-(1 → 4) glycosidic bond, carbon atom 1 of the α anomer of one monosaccharide provides the hemiacetal group for the bond and carbon atom 4 of the other provides the —OH group. Important disaccharides include maltose [two α-(1 → 4) linked glucose residues], cellobiose [two β-(1 → 4) linked glucose residues], lactose [a galactose residue linked β-(1 → 4) to a glucose residue], and sucrose [an α-glucose residue linked α,β-(1 ↔ 2) to a β-fructose residue].

6 **Distinguish homopolysaccharides from heteropolysaccharides and give examples of each.**
Homopolysaccharides consist of one type of monosaccharide residue, while **heteropolysaccharides** consist of more than one type. Important homopolysaccharides include cellulose [β-(1 → 4) linked glucose residues], chitin [β-(1 → 4) linked *N*-acetylglucosamine residues], amylose [α-(1 → 4) linked glucose residues], and amylopectin and glycogen [α-(1 → 4) and α-(1 → 6) linked glucose residues]. The heteropolysaccharides hyaluronic acid, chondroitin sulfate, and heparin consist of glucose and galactose derivatives linked by various glycosidic bonds.

END OF CHAPTER PROBLEMS

Answers to problems whose numbers are printed in color are given in Appendix D. More challenging questions are marked with an asterisk. Problems within colored rules are paired. **ILW** = Interactive Learning Ware solution is available at *www.wiley.com/college/raymond.*

12.1 MONOSACCHARIDES

12.1 How do oligosaccharides differ from polysaccharides?

12.2 How are oligosaccharides similar to polysaccharides?

12.3 Draw an example of each type of monosaccharide.
a. an aldoheptose **b.** a ketononose

12.4 Draw an example of each type of monosaccharide.
a. an aldononose **b.** a ketooctose

12.5 **a.** How many chiral carbon atoms does D-lyxose contain?

O
||
C—H
HO—|—H
HO—|—H
H—|—OH
CH_2OH

D-Lyxose

b. How many total stereoisomers are possible for this aldopentose?
c. Draw and name the enantiomer of D-lyxose.
d. D-Ribose is a diastereomer of D-lyxose. Draw two additional D-diastereomers of this monosaccharide.

12.6 **a.** How many chiral carbon atoms does D-glucose contain?

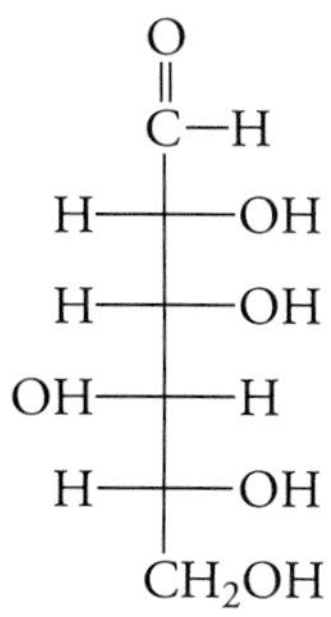

D-Gulose

b. How many total stereoisomers are possible for this aldohexose?
c. Draw and name the enantiomer of D-glucose.
d. Draw two D-diastereomers of this monosaccharide.

12.7 Draw each of the following glucose derivatives.
a. D-2-deoxyglucose
b. D-glucosamine
c. D-glucuronic acid
d. sorbitol

12.8 For the molecules in Problem 12.7,
a. to which class of monosaccharide derivative does each belong?
b. how many chiral carbon atoms does each have?
c. how many stereoisomers are possible for each?

12.9 Why, do you suppose, does sorbitol (Figure 12.5) not have a D or an L designation?

12.10 Draw and name the two products obtained when *N*-acetyl-D-glucosamine is hydrolyzed (reacted with H_2O in the presence of H^+).

12.11 When D-gluconic acid is produced from D-glucose, which functional group is oxidized?

12.12 When D-glucuronic acid is produced from D-glucose, which functional group is oxidized?

12.13 **a.** Draw all possible aldotetroses.
b. Which are D and which are L sugars?

12.14 **a.** Draw all possible 2-ketopentoses.
b. Which are D and which are L sugars?

12.15 Draw each galactose derivative (see Figure 12.4).
a. D-2-deoxygalactose
b. galactitol (the alcohol sugar derived from galactose)
c. D-galacturonic acid (carbon atom 6 of galactose is oxidized to a carboxylic acid)

12.16 Draw each monosaccharide derivative of D-lyxose (see Problem 12.5).
a. D-2-deoxylyxose
b. lyxitol (the alcohol sugar derived from lyxose)
c. D-lyxonic acid (carbon atom 1 of lyxose is oxidized to a carboxylic acid)

12.2 Reactions of Monosaccharides

12.17 Draw the product obtained when
a. D-mannose is reduced (see Figure 12.6)
b. carbon atom 1 of D-mannose is oxidized to a carboxylic acid
c. carbon atom 6 of D-mannose is oxidized to a carboxylic acid

12.18 D-Talose is an aldohexose. Draw the product obtained when

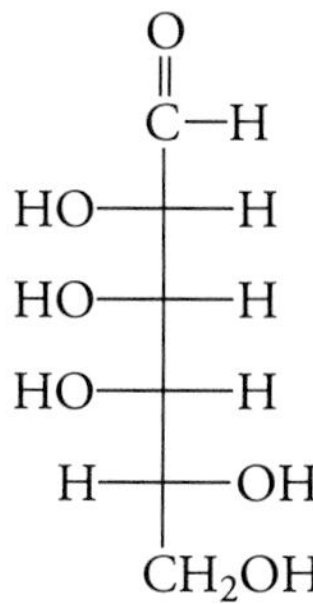

D-Talose

a. D-talose is reduced
b. carbon atom 1 of D-talose is oxidized to a carboxylic acid
c. carbon atom 6 of D-talose is oxidized to a carboxylic acid

12.19 Define the term reducing sugar.

12.20 D-Fructose does not contain an aldehyde group, yet it is a reducing sugar. Explain why.

12.21 Draw arabinitol, the alcohol sugar formed when D-arabinose is reacted with H_2 and Pt.

O
||
C—H
HO—H
H—OH
H—OH
CH_2OH

D-Arabinose

12.22 Draw D-araburonic acid, the alcohol sugar formed when carbon atom 5 of D-arabinose (see the previous question) is oxidized to a carboxylic acid.

***12.23** **a.** Draw the D-aldohexose that gives the alcohol sugar below, when treated with H_2 and Pt.

CH_2OH
H—OH
HO—H
HO—H
H—OH
CH_2OH

b. Is the aldohexose a reducing sugar?
c. Is the aldohexose a deoxy sugar?
d. Is the aldohexose an amino sugar?
e. Draw the L-aldohexose that also gives the alcohol sugar above, when treated with H_2 and Pt.

***12.24** **a.** Draw the D-2-ketopentose that gives the alcohol sugar below, when treated with H_2 and Pt.

CH_2OH
H—OH
H—OH
H—OH
CH_2OH

b. Is the ketopentose a reducing sugar?
c. Is the ketopentose a deoxy sugar?
d. Is the ketopentose an amino sugar?
e. Draw the L-2-ketopentose that also gives the alcohol sugar above, when treated with H_2 and Pt.

12.3 Monosaccharides in Their Cyclic Form

12.25 **a.** What is an anomer?
b. How does an α anomer differ from a β anomer?
c. How does a pyranose differ from a furanose?

12.26 Draw each molecule.
a. β-D-glucopyranose
b. α-D-ribofuranose
c. β-D-galactopyranose
d. α-D-arabinopyranose (see Problem 12.21)

12.27 Draw each molecule.
a. α-D-glucopyranose
b. β-D-ribofuranose
c. α-D-galactopyranose
d. β-D-arabinopyranose (see Problem 12.21)

12.28 **a.** Each molecule in Problem 12.26 has how many chiral carbon atoms and is one of how many total stereoisomers?
b. Each molecule in Problem 12.27 has how many chiral carbon atoms and is one of how many total stereoisomers?

12.29 D-Lyxose (see Problem 12.5) is a diastereomer of D-ribose. Draw α-D-lyxofuranose.

12.30 D-Gulose (see Problem 12.6) is a diastereomer of D-glucose. Draw β-D-glucopyranose.

12.31 Draw β-D-mannopyranose (see Figure 12.6).

12.32 Draw α-D-allopyranose.

O
||
C—H
H—OH
H—OH
H—OH
H—OH
CH_2OH

D-Allose

12.33 **a.** Draw the α pyranose anomer of D-2-deoxyglucose.
b. Draw the β pyranose anomer of D-glucuronic acid

12.34 a. Draw the α pyranose anomer of *N*-Acetyl-D-glucosamine.

b. Draw the β pyranose anomer of D-galactosamine (identical to D-glucosamine except that the positions of —H and —OH on carbon atom #4 are switched).

12.35 Define the term mutarotation.

12.36 Is the following statement true or false? When β-D-glucopyranose is dissolved in water, α-D-glucopyranose and the open form of glucose soon appear.

12.4 Oligosaccharides

12.37 Cellobiose is a reducing sugar. Write a reaction equation that shows why.

12.38 Lactose is a reducing sugar. Write a reaction equation that shows why.

12.39 a. Gentiobiose consists of two D-glucopyranose residues joined by a β-(1 → 6) glycosidic bond. Draw this disaccharide.

b. Is gentiobiose a reducing sugar? Explain.

12.40 a. *N*-Acetyllactosamine, found in various glycoproteins, consists of a β-D-galactopyranose residue joined to an *N*-acetyl-D-glucosamine residue by a β-(1 → 4) glycosidic bond. Draw this disaccharide.

b. Is *N*-acetyllactosamine a reducing sugar? Explain.

***12.41** Vanillin β-D-glucoside gives vanilla extract its flavor. Draw the two products obtained when this acetal is hydrolyzed.

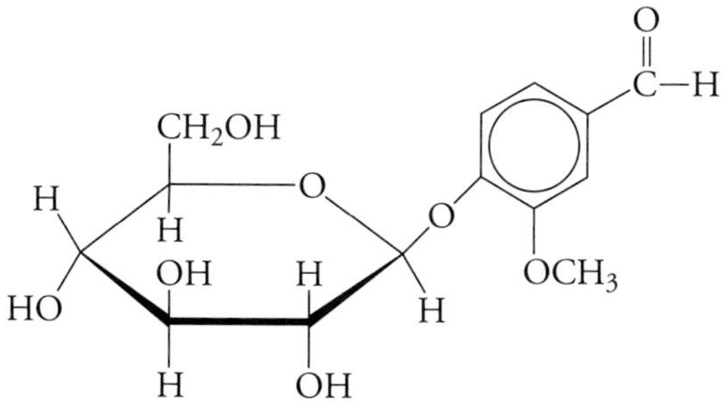

Vanillin β-D-Glucoside

***12.42** Daunorubicin is a glycoside prodrug that shows promise for the treatment of certain forms of cancer.

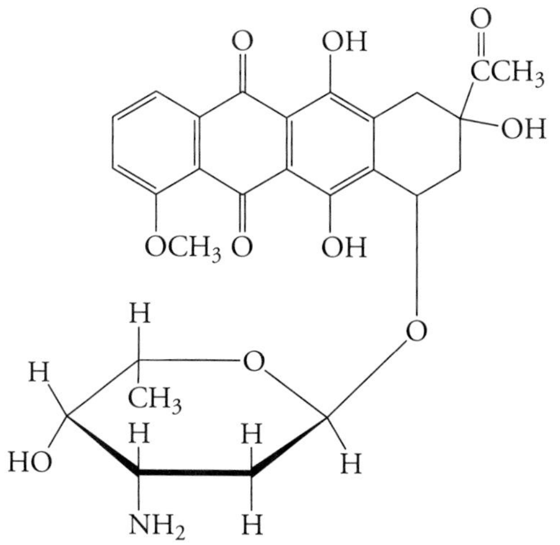

Daunorubicin

a. Draw the two products obtained when this acetal is hydrolyzed.

b. Identify the functional groups present in daunorubicin.

12.43 Draw a disaccharide consisting of two D-glucose residues that is not a reducing sugar.

12.44 Draw a disaccharide consisting of two D-galactose residues that is a reducing sugar.

12.45 a. Is trehalose, a disaccharide found in a wide range of living things, a reducing sugar?

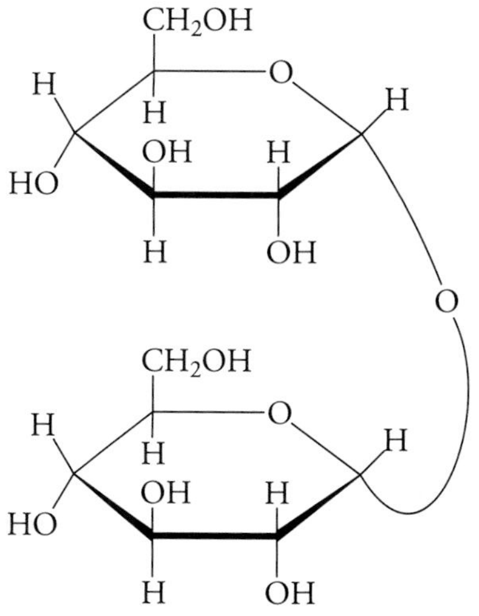

Trehalose

b. Which describes the glycosidic bond in this disaccharide: α,α-(1 ↔ 1), α-(1 → 2), α-(1 → 3), α-(1 → 4), α-(1 → 5), or α-(1 → 6)?

12.46 **a.** Is raffinose (Figure 12.13) a reducing sugar?
b. Identify the type of glycosidic bonds [α-(1 $\rightarrow$ 4), β-(1 $\rightarrow$ 4), etc.] in this trisaccharide.

12.47 **a.** Draw the product obtained when lactose is reacted with Benedict's reagent.
b. Draw the product obtained when cellobiose is reacted with Benedict's reagent.

12.48 True or false?
a. Lactose and cellobiose are stereoisomers.
b. Lactose and cellobiose are enantiomers.
c. Lactose and cellobiose are diastereomers.

***12.49** Human milk contains lactose at a concentration of 7% (w/v).
a. How many grams of lactose are present in 15 mL of human milk?
b. What is the molar concentration of lactose in human milk?
c. What is the parts per million concentration of lactose in human milk?

***12.50** Cow's milk contains lactose at a concentration of 5% (w/v).
a. How many grams of lactose are present in 150 μL of cow's milk?
b. What is the molar concentration of lactose in cow's milk?
c. What is the parts per million concentration of lactose in cow's milk?

12.51 **a.** What three products are obtained when the cerebroside in Figure 12.14a is hydrolyzed?
b. What products are obtained when the ganglioside in Figure 12.14b is hydrolyzed?

***12.52** Saliva contains glycoproteins called mucins (Chapter 7 Health Link, Saliva). Draw the following oligosaccharides that have been detected in mucins present in human saliva.
a. A β-D-galactopyranose residue joined to an *N*-acetyl-D-galactosamine residue (Figure 12.14) by a β-(1 $\rightarrow$ 3) glycosidic bond.
b. An *N*-acetylneuramic acid residue (Figure 12.14) joined α-(2 $\rightarrow$ 3) to a D-galactopyranose residue, which is connected β-(1 $\rightarrow$ 3) to an *N*-acetyl-D-galactosamine residue.

12.5 Polysaccharides

12.53 **a.** How is the structure of cellulose similar to that of amylose?
b. How is the structure of cellulose different from that of amylose?

12.54 Why can humans use starch as a food source, but not cellulose?

12.55 How is the structure of glycogen similar to that of amylopectin?

12.56 How is the structure of glycogen different from that of amylopectin?

12.57 A particular lichen produces the polysaccharide below.

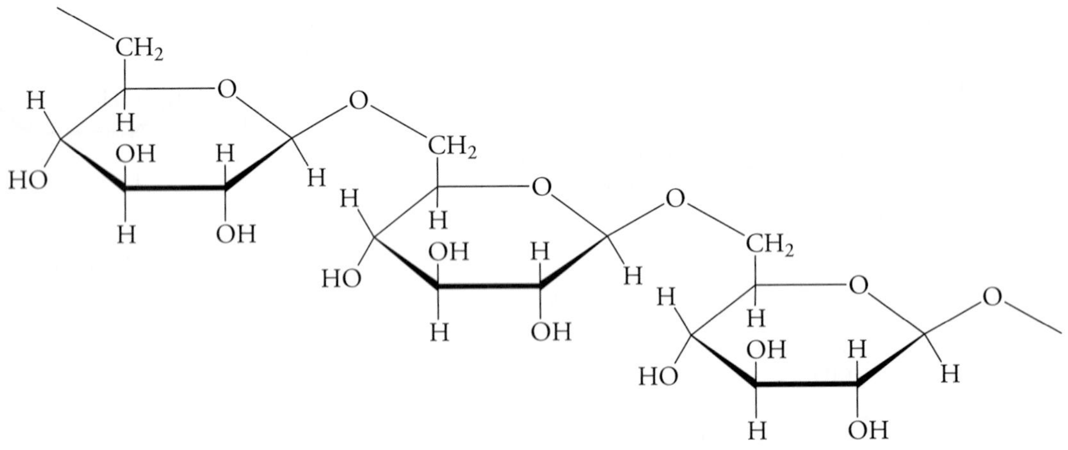

CH_2 O H H OH H HO H OH H O CH_2 O H H OH H HO H OH H O CH_2 O H H OH H HO H OH H O

a. Is the carbohydrate a homopolysaccharide or a heteropolysaccharide?
b. Name the glycosidic bond present, as α-(1 $\rightarrow$ 4) β-(1 $\rightarrow$ 2), etc.

12.58 Suppose that a newly discovered heteropolysaccharide consists of alternating α-D-glucopyranose and β-D-galactopyranose residues, with glucose joined to galactose by an α-(1 $\rightarrow$ 4) glycosidic bond and the galactose joined to glucose by a β-(1 $\rightarrow$ 6) glycosidic bond. Draw a segment of this polysaccharide.

12.59 Chondroitin 6-sulfate, identical to chondroitin 4-sulfate (Section 12.5) with the exception of where the sulfate group is attached, is another of the heteropolysaccharides present in connective tissue. Draw chondroitin 6-sulfate.

HEALTH*Link* | *Relative Sweetness*

12.60 Sucralose (Table 12.2) is a derivative of which naturally occurring oligosaccharide?

12.61 Olestra (Figure 8.13) is a derivative of which naturally occurring oligosaccharide?

12.62 What is invert sugar and why is it used in foods?

12.63 How is high fructose corn syrup made?

12.64 In addition to being found in certain plants, the sweetener xylitol (Table 12.2) can be made by reducing D-xylose (Sample Problem 12.4). Draw xylitol.

12.65 What does the "K" in acesulfame-K (Table 12.2) stand for? Why is acesulfame-K sold in this form?

12.66 As described in "Sweet Spot", scientists have discovered the gene for the sweet receptor — a molecule in taste cells of the tongue that tells the brain that something is sweet. Scientists plan to use what they discover about this gene and the receptor that it codes for to develop compounds that could replace artificial sweeteners currently in use. How would the action of these compounds differ from those of artificial sweeteners such as aspartame and saccharin?

HEALTH*Link* | *Indigestible Oligosaccharides*

12.67 What is lactose intolerance and what are some of the options for dealing with it?

12.68 When the enzyme lactase is used to hydrolyze lactose present in milk, lactose-free milk is obtained. Explain why lactose-free milk is sweeter than regular milk (see Table 12.2).

HEALTH*Link* | *Blood Type*

12.69 The blood group antigens described in Figure 12.19 also appear in another form, in which the D-galactose drawn in the center is attached to the *N*-acetyl-D-glucosamine on the right by a β-(1 $\rightarrow$ 3) glycosidic bond. Draw the oligosaccharide portion of this alternate structure.

12.70 To which monosaccharide derivative group (Section 12.1) does L-fucose (Figure 12.19) belong?

THINKING IT THROUGH

12.71 Suppose that a new artificial sweetener is approved for use by the Food and Drug Administration, but some public interest groups question its safety (this happened when aspartame was put on the market). Would you use the sweetener?

12.72 **a.** Which functional groups are present in digitoxin? (See the chapter conclusion.)

b. To what lipid family does the complex ring drawn to the right of the structure belong?

c. Identify the type of glycosidic bond that joins each of the monosaccharide residues in the molecule.

d. Draw the open form of the monosaccharide used to produce the trisaccharide portion of digitoxin.

INTERACTIVE LEARNING PROBLEMS

12.73 ILW Which of the following is not a disaccharide and does not contain a glycosidic linkage? What are the constituents of each of the disaccharides?

a. Maltose **c.** Lactose

b. Sucrose **d.** Galactose

SOLUTIONS TO PRACTICE PROBLEMS

12.1 **a.** 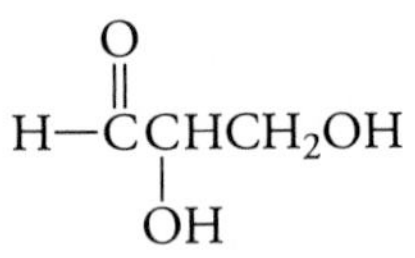 $H{-}\overset{\overset{O}{\|}}{C}\underset{\underset{OH}{|}}{C}HCH_2OH$

b. $HOCH_2\underset{\underset{OH}{|}}{C}H\overset{\overset{OH}{|}}{C}H\underset{\underset{OH}{|}}{C}H\overset{\overset{O}{\|}}{C}{-}H$

c. $HOCH_2\underset{\underset{OH}{|}}{C}H\overset{\overset{OH}{|}}{C}H\underset{\underset{OH}{|}}{C}H\overset{\overset{O}{\|}}{C}CH_2OH$

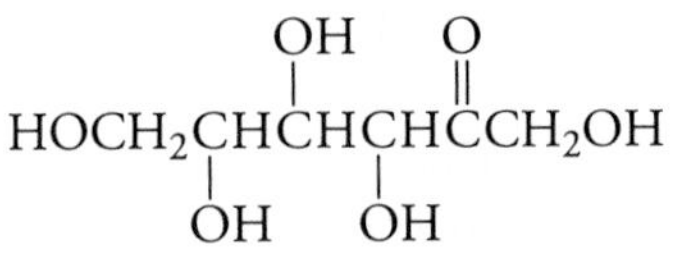

d. 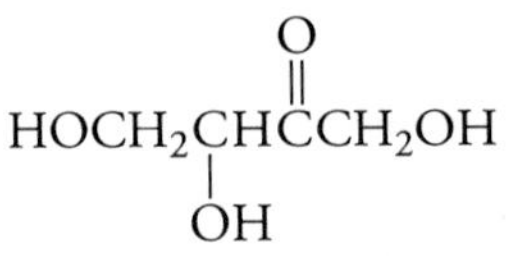 $HOCH_2\underset{\underset{OH}{|}}{C}H\overset{\overset{O}{\|}}{C}CH_2OH$

12.2

$$\begin{array}{ccc} & \overset{\overset{O}{\|}}{C}{-}H & \\ H & \text{—}\!\!+\!\!\text{—} & OH \\ H & \text{—}\!\!+\!\!\text{—} & OH \\ H & \text{—}\!\!+\!\!\text{—} & OH \\ H & \text{—}\!\!+\!\!\text{—} & OH \\ & CH_2OH & \end{array} \qquad \begin{array}{ccc} & \overset{\overset{O}{\|}}{C}{-}H & \\ HO & \text{—}\!\!+\!\!\text{—} & H \\ H & \text{—}\!\!+\!\!\text{—} & OH \\ H & \text{—}\!\!+\!\!\text{—} & OH \\ H & \text{—}\!\!+\!\!\text{—} & OH \\ & CH_2OH & \end{array} \qquad \begin{array}{ccc} & \overset{\overset{O}{\|}}{C}{-}H & \\ H & \text{—}\!\!+\!\!\text{—} & OH \\ H & \text{—}\!\!+\!\!\text{—} & OH \\ OH & \text{—}\!\!+\!\!\text{—} & H \\ H & \text{—}\!\!+\!\!\text{—} & OH \\ & CH_2OH & \end{array}$$

$$\begin{array}{ccc} & \overset{\overset{O}{\|}}{C}{-}H & \\ HO & \text{—}\!\!+\!\!\text{—} & H \\ HO & \text{—}\!\!+\!\!\text{—} & H \\ H & \text{—}\!\!+\!\!\text{—} & OH \\ H & \text{—}\!\!+\!\!\text{—} & OH \\ & CH_2OH & \end{array} \qquad \begin{array}{ccc} & \overset{\overset{O}{\|}}{C}{-}H & \\ HO & \text{—}\!\!+\!\!\text{—} & H \\ H & \text{—}\!\!+\!\!\text{—} & OH \\ HO & \text{—}\!\!+\!\!\text{—} & H \\ H & \text{—}\!\!+\!\!\text{—} & OH \\ & CH_2OH & \end{array} \qquad \begin{array}{ccc} & \overset{\overset{O}{\|}}{C}{-}H & \\ HO & \text{—}\!\!+\!\!\text{—} & H \\ HO & \text{—}\!\!+\!\!\text{—} & H \\ HO & \text{—}\!\!+\!\!\text{—} & H \\ H & \text{—}\!\!+\!\!\text{—} & OH \\ & CH_2OH & \end{array}$$

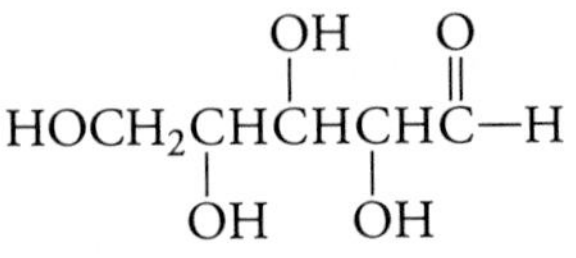

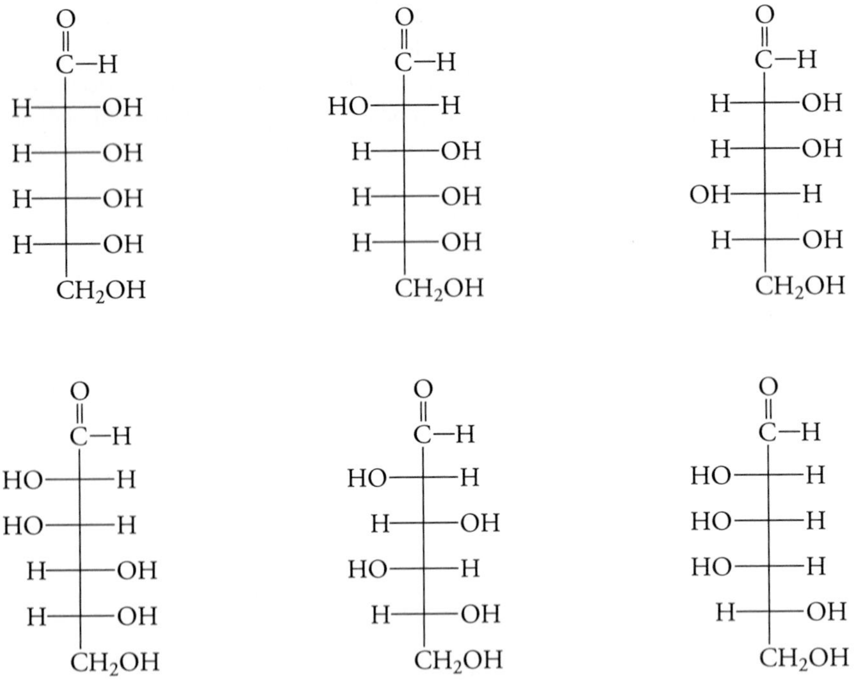

12.3

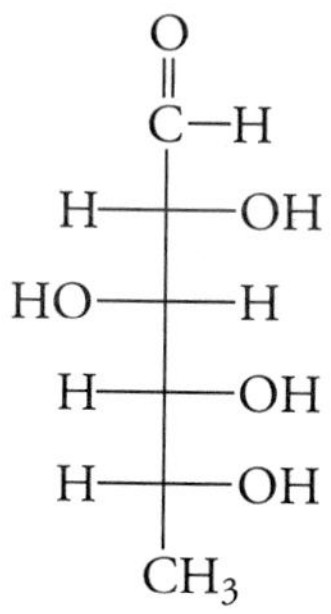

$$\begin{array}{c} \mathrm{O} \\ \| \\ \mathrm{C-H} \\ \mathrm{H-C-OH} \\ \mathrm{HO-C-H} \\ \mathrm{H-C-OH} \\ \mathrm{H-C-OH} \\ \mathrm{CH_3} \end{array}$$

12.4

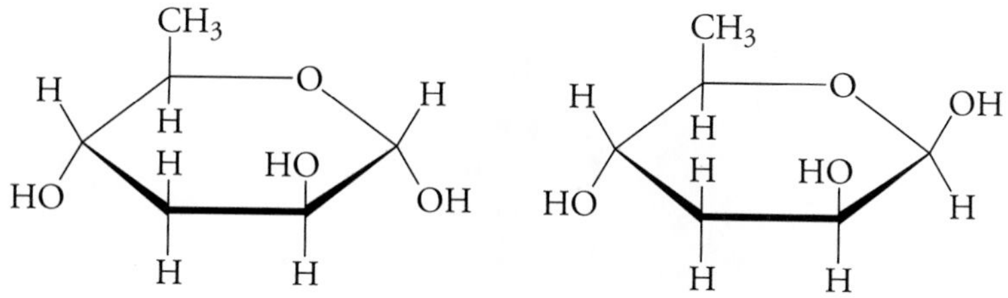

α-D-Abqueopyranose

β-D-Abqueopyranose

You have arranged to meet a friend from class at a local coffee shop to study for the upcoming chemistry exam. After buying a cup of coffee and a peanut butter cookie, you go and sit at a table to reserve space for you and your study partner. After her arrival you offer to share the cookie, but she declines because of a peanut allergy. She goes on to explain that even though her allergy is relatively minor, exposure to certain proteins found in peanuts causes her to break out in a rash and makes her lips and mouth tingle.

13 PEPTIDES, PROTEINS, AND ENZYMES

Peptides and proteins play more diverse roles in living things than any other class of compounds. They can act as catalysts, hormones, and transport molecules and are a key part of the structure of skin, tendons, cartilage, and bone. Still other proteins are involved in muscle action, the visual process, the immune response, and the operation of the nervous system. We begin this chapter with a look at amino acids, the building blocks from which peptides and proteins are constructed.

objectives

After completing this chapter, you should be able to:

1. Describe the structure of amino acids and the system used to classify amino acids.
2. Distinguish between oligopeptides, polypeptides, and proteins and describe the bond that joins amino acid residues in these compounds.
3. Define primary, secondary, tertiary, and quaternary structure. Name the covalent and noncovalent forces responsible for each level of structure.
4. Explain what is meant by the term denaturation, and list some of the ways to denature a protein.
5. Distinguish between absolute specificity, relative specificity, and stereospecificity.
6. Identify the steps required for a typical Michaelis–Menten enzyme to convert a substrate into a product. Explain how competitive and noncompetitive inhibitors affect these enzymes, and distinguish between reversible and irreversible inhibitors.
7. Describe allosteric enzymes and the role that effectors play in the ability of these enzymes to catalyze reactions.
8. Explain what is meant by feedback inhibition and covalent modification.

13.1 AMINO ACIDS

From the perspective of organic chemistry, amino acids are molecules that contain at least one carboxyl ($—CO_2H$) group and one amino ($—NH_2$) group. The 20 amino acids used to make peptides and proteins (Table 13.1) are **α-amino acids**, named for the fact that *the amino group is attached to the carbon atom α to the carboxyl group.* Only differences in the side chain or R group, the atom or group of atoms also attached to the α carbon atom, distinguish one amino acid from another.

```
        H  :Ö:
        |   ||
  H—N̈—C— C—ÖH
     |   |
     H   R   ← α carbon atom
         ↑
     side chain
```

An α-amino acid

TABLE | 13.1 α-AMINO ACIDS PRESENT IN PROTEINS

Common Name	Formula	Common Name	Formula
Nonpolar			
Glycine (Gly)	$H—N^+H_2—CH_2—C(=O)—O^-$ (H H O / H—N⁺—C—C—O⁻ / H H)	Methionine (Met)	H H O / H—N⁺—C—C—O⁻ / H CH₂ / CH₂ / S / CH₃
Alanine (Ala)	H H O / H—N⁺—C—C—O⁻ / H CH₃	Proline (Pro)	H H O / H—N⁺——C—C—O⁻ / CH₂ CH₂ / CH₂
Valine (Val)	H H O / H—N⁺—C—C—O⁻ / H CHCH₃ / CH₃	Phenylalanine (Phe)	H H O / H—N⁺—C—C—O⁻ / H CH₂ / [benzene ring]
Leucine (Leu)	H H O / H—N⁺—C—C—O⁻ / H CH₂ / CHCH₃ / CH₃	Tryptophan (Trp)	H H O / H—N⁺—C—C—O⁻ / H CH₂ / [indole ring] NH
Isoleucine (Ile)	H H O / H—N⁺—C—C—O⁻ / H CHCH₃ / CH₂ / CH₃		

Common Name	Formula
Polar-acidic	
Aspartic acid (Asp)	H H O $H-\overset{+}{N}-C-C-O^-$ H CH_2 $C-O^-$ O
Glutamic acid (Glu)	H H O $H-\overset{+}{N}-C-C-O^-$ H CH_2 CH_2 $C-O^-$ O
Polar-basic	
Lysine (Lys)	H H O $H-\overset{+}{N}-C-C-O^-$ H CH_2 CH_2 CH_2 CH_2 $H-\overset{+}{N}-H$ H
Arginine (Arg)	H H O $H-\overset{+}{N}-C-C-O^-$ H CH_2 CH_2 CH_2 NH $H-\overset{+}{N}=C$ H NH_2
Histidine (His)	H H O $H-\overset{+}{N}-C-C-O^-$ H CH_2 NH $H\overset{+}{N}$=

Common Name	Formula
Polar-neutral	
Serine (Ser)	H H O $H-\overset{+}{N}-C-C-O^-$ H CH_2OH
Threonine (Thr)	H H O $H-\overset{+}{N}-C-C-O^-$ H $CHCH_3$ OH
Cysteine (Cys)	H H O $H-\overset{+}{N}-C-C-O^-$ H CH_2SH
Tyrosine (Tyr)	H H O $H-\overset{+}{N}-C-C-O^-$ H CH_2 OH
Asparagine (Asn)	H H O $H-\overset{+}{N}-C-C-O^-$ H CH_2 $C-NH_2$ O
Glutamine (Gln)	H H O $H-\overset{+}{N}-C-C-O^-$ H CH_2 CH_2 $C-NH_2$ O

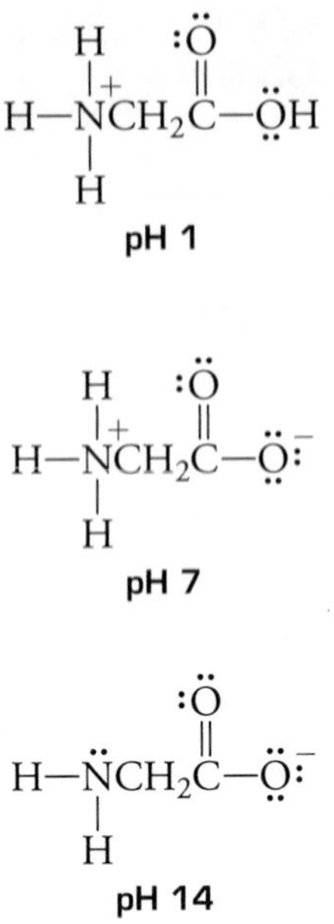

$$H_3N^+CH_2C(=O)OH$$

pH 1

$$H_3N^+CH_2C(=O)O^-$$

pH 7

$$H_2NCH_2C(=O)O^-$$

pH 14

FIGURE 13.1

The effect of pH on glycine
The structure of the carboxyl and amino groups in an amino acid vary as a function of pH. Glycine has a net charge of 1+ at pH 1, a net charge of 0 at pH 7, and a net charge of 1− at pH 14.

- An amino acid has no net charge at its isoelectric point.

For the amino acid drawn above Table 13.1, the effect of pH on structure has been ignored. Because amino acids contain carboxyl and amino groups, their structure is as dependent on pH as the structure of carboxylic acids and amines. The pK_a values for the carboxyl groups in amino acids range between 1.8 and 4.3, so at pH 7 the carboxylic acid portion of an amino acid is found in its conjugate base form ($—CO_2^-$). The pK_a values for most of the amine conjugate acids present in amino acids fall between 9.1 and 12.5, so at pH 7 the amine portion of an amino acid will usually be found as the conjugate acid ($—NH_3^+$). For the amino acid glycine, this means that at pH 7 the molecule carries both a 1+ and a 1− charge (Figure 13.1). This form of an amino acid, which *has one positive and one negative charge*, is called the **zwitterion** (German: *zwei*, "two").

As pH changes, so does the net charge on an amino acid. At a pH of 1, a value less than the pK_a of both $—CO_2H$ and $—NH_3^+$, glycine has a net 1+ charge and at a pH of 14, a value greater than the pK_a of each of these functional groups, glycine has a net 1− charge (Figure 13.1). In living things pH does not vary over the range 1 to 14, but even over much narrower pH ranges the net charge on amino acids can vary.

Slight differences in pK_a values make each amino acid a bit different from the others, so at a given pH the net charge will vary from amino acid to amino acid. The pH at which an amino acid has *a net charge of zero* is called the **isoelectric point** (**pI**). As we will see later in this chapter, the dependence of structure on pH is important in the biochemical function of peptides and proteins.

SAMPLE PROBLEM 13.1

The effect of pH on amino acid structure

Draw the amino acid valine as it would appear at pH 1, 7, and 14 (see Table 13.1).

Strategy

As explained in Section 9.9, when the pH equals the pK_a for an acid, the concentration of acid and its conjugate base are equal. When the pH is less than the pK_a the acid predominates and when the pH is greater than pK_a the conjugate base predominates.

Solution

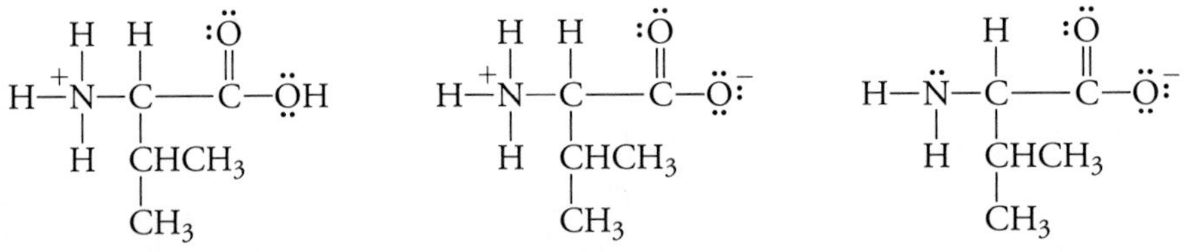

$$H_3N^+—CH(CH(CH_3)_2)—C(=O)—OH$$

pH 1

$$H_3N^+—CH(CH(CH_3)_2)—C(=O)—O^-$$

pH 7

$$H_2N—CH(CH(CH_3)_2)—C(=O)—O^-$$

pH 14

PRACTICE PROBLEM 13.1

Draw the amino acid aspartic acid as it would appear at pH 1, 7, and 14.

Classifying Amino Acids

All amino acids are hydrophilic (Section 7.5) due to the presence of polar covalent bonds, to the presence of N, O, and H atoms that are capable of forming hydrogen bonds with water, and to the fact that they can carry charges ($—CO_2^-$ and/or $—NH_3^+$). The water solubility and the chemistry of amino acids vary to some extent, depending on the makeup of the side chain, so it is useful to use side chains to classify amino acids (Table 13.1).

The side chain of a **nonpolar** amino acid is usually *an alkyl group, an aromatic ring, or a nonpolar collection of atoms.* The nonpolar amino acids include alanine, which carries a methyl group; glycine, whose side chain consists only of a hydrogen atom; and proline, whose side chain forms a ring containing the α-amino group.

■ Amino acids are nonpolar, polar-acidic, polar-basic, or polar-neutral, depending on the nature of their side chain.

The side chain of a **polar-acidic** amino acid contains *a carboxyl group.* At pH 7, the carboxyl group is found in its conjugate base form ($—CO_2^-$), which means that the side chains on the two polar-acidic amino acids, aspartic acid and glutamic acid, carry a negative charge at this pH.

The **polar-basic** side chain contains *an amino group*, as is the case for lysine and its $—CH_2CH_2CH_2CH_2NH_2$ group, *or another basic nitrogen-containing group.* At pH 7 the amines exist in their conjugate acid form, so polar-basic amino acids carry a positive charge at this pH.

■ Unlike the other polar-basic amino acids, histidine's side chain carries a positive charge only when the pH is less than 6.

The side chain of a **polar-neutral** amino acid is usually *an alcohol, a phenol, or an amide.* None of these functional groups is acidic or basic enough to carry a charge at pH 7. Examples include serine, tyrosine, and asparagine.

Stereoisomers

With the exception of glycine, all of the α-amino acids in Table 13.1 are chiral (the α carbon atom in each is attached to four different groups). This means that each amino acid is nonsuperimposable on its mirror image, as shown in Figure 13.2 for alanine. The D and L designations of the two enantiomers are based on the relative position of the amino group and the hydrogen atom attached to the α carbon. When amino acids are drawn in Fischer projections with the carboxyl group pointing up and the side chain pointing down, a **D-amino acid** has the amino group on the *right side* and an **L-amino acid** has it on the *left side.*

■ With the exception of glycine, amino acids are D or L.

In living things only L-amino acids are used to produce proteins. In some cases, an organism will produce an enzyme that catalyzes the conversion of an L-amino acid into a D-amino acid, once a protein has been assembled. Examples of proteins that contain some D-amino acid residues include those used by bacteria to manufacture cell walls and a number of bacteria-produced antibiotics.

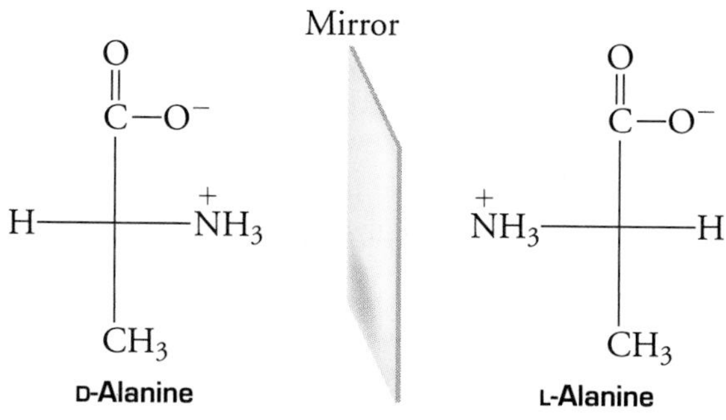

FIGURE 13.2

Fischer projections
Alanine contains one chiral carbon atom and can exist as a pair of enantiomers, D-alanine and L-alanine. In nature, L-amino acids predominate.

13.2 THE PEPTIDE BOND

In Chapter 12 we saw that oligo- and polysaccharides are chains of monosaccharide residues attached to one another by glycosidic bonds. In a similar fashion, peptides and proteins are chains of amino acid residues connected to one another by peptide bonds. **Peptide bond** is the name that biochemists give the amide bond (Section 10.8) that joins one amino acid residue to another.

■ Peptides and proteins consist of amino acid residues joined by peptide (amide) bonds.

The structure in Figure 13.3*a*, which consists of *two amino acid residues*, is called a **dipeptide**. Similar to the system used for carbohydrates, a combination of *three* amino acid units is a **tripeptide**, a combination of *four* is a **tetrapeptide**, and so on. **Oligopeptides** contain from *2 to 10 amino acid residues* and **polypeptides**, *more than 10.* Oligo- and polypeptides are referred to, collectively, as **peptides**. At some point, not

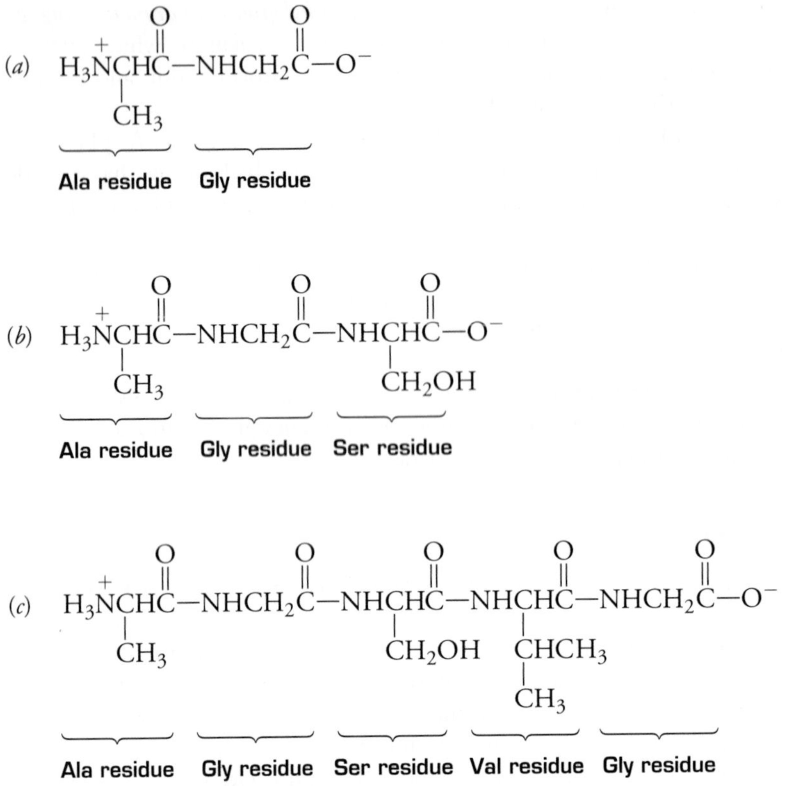

FIGURE | 13.3

Oligopeptides
Peptide (amide) bonds connect one amino acid residue to another. The amino acid sequence of a peptide is drawn or listed from the N-terminus to the C-terminus. (*a*) The dipeptide Ala-Gly. (*b*) The tripeptide Ala-Gly-Ser. (*c*) The pentapeptide Ala-Gly-Ser-Val-Gly.

(*a*) $$H_3\overset{+}{N}\underset{\large CH_3}{CH}\overset{O}{\overset{\|}{C}}{-}NHCH_2\overset{O}{\overset{\|}{C}}{-}O^-$$

Ala residue Gly residue

(*b*) $$H_3\overset{+}{N}\underset{\large CH_3}{CH}\overset{O}{\overset{\|}{C}}{-}NHCH_2\overset{O}{\overset{\|}{C}}{-}NH\underset{\large CH_2OH}{CH}\overset{O}{\overset{\|}{C}}{-}O^-$$

Ala residue Gly residue Ser residue

(*c*) $$H_3\overset{+}{N}\underset{\large CH_3}{CH}\overset{O}{\overset{\|}{C}}{-}NHCH_2\overset{O}{\overset{\|}{C}}{-}NH\underset{\large CH_2OH}{CH}\overset{O}{\overset{\|}{C}}{-}NH\underset{\large CHCH_3(CH_3)}{CH}\overset{O}{\overset{\|}{C}}{-}NHCH_2\overset{O}{\overset{\|}{C}}{-}O^-$$

Ala residue Gly residue Ser residue Val residue Gly residue

generally agreed upon, polypeptides contain enough amino acid units to be called **proteins**. For our purposes, a protein will be defined as a polypeptide chain having *more than 50 amino acid residues.*

The peptide bond for the dipeptide shown in Figure 13.3*a* resulted from the alanine on the left donating the C=O of its carboxyl group and the glycine on the right donating an NH from its amino group. As we will see in Chapter 14, the process of joining amino acids to one another is a complex process in living things.

Because each amino acid has both a carboxyl and an α-amino group, each amino acid can take part in two peptide bonds. If, for example, the glycine in this dipeptide contributes its $—CO_2H$ to a peptide bond with the $—NH_2$ from a serine, a tripeptide is formed (Figure 13.3*b*). The addition of successive amino acids leads to the formation of a longer oligopeptide or a polypeptide (Figure 13.3*c*).

It is customary to draw peptides and proteins with the **N-terminus** (*the end of the peptide chain with the unreacted amino group*) on the left and the **C-terminus** (*the end of the peptide chain with the free carboxyl group*) on the right. For the tripeptide in Figure 13.3*b*, alanine is at the N-terminus and serine is at the C-terminus.

Peptides and proteins are named by listing the amino acid residues, in order, from the N- to the C-terminus. The dipeptide in Figure 13.3 is Ala-Gly and the tripeptide is Ala-Gly-Ser. Names of larger oligopeptides, polypeptides, and proteins arrived at using this method can be quite lengthy, so these molecules are sometimes known by common names (Figure 13.4).

For the amino acid residues in a peptide or protein, hydrolysis of the peptide bond restores their amino and carboxyl groups to their original state (Figure 13.5). This hydrolysis, which is identical to amide hydrolysis described in Section 10.8, is a key reaction in the digestion of proteins.

Biochemically active peptides and proteins come in all sizes. Thyrotropin-releasing factor (3 amino acid residues) controls the release of another hormone, thyrotropin; somatostatin (14 amino acid residues) inhibits the release of growth hormone, while

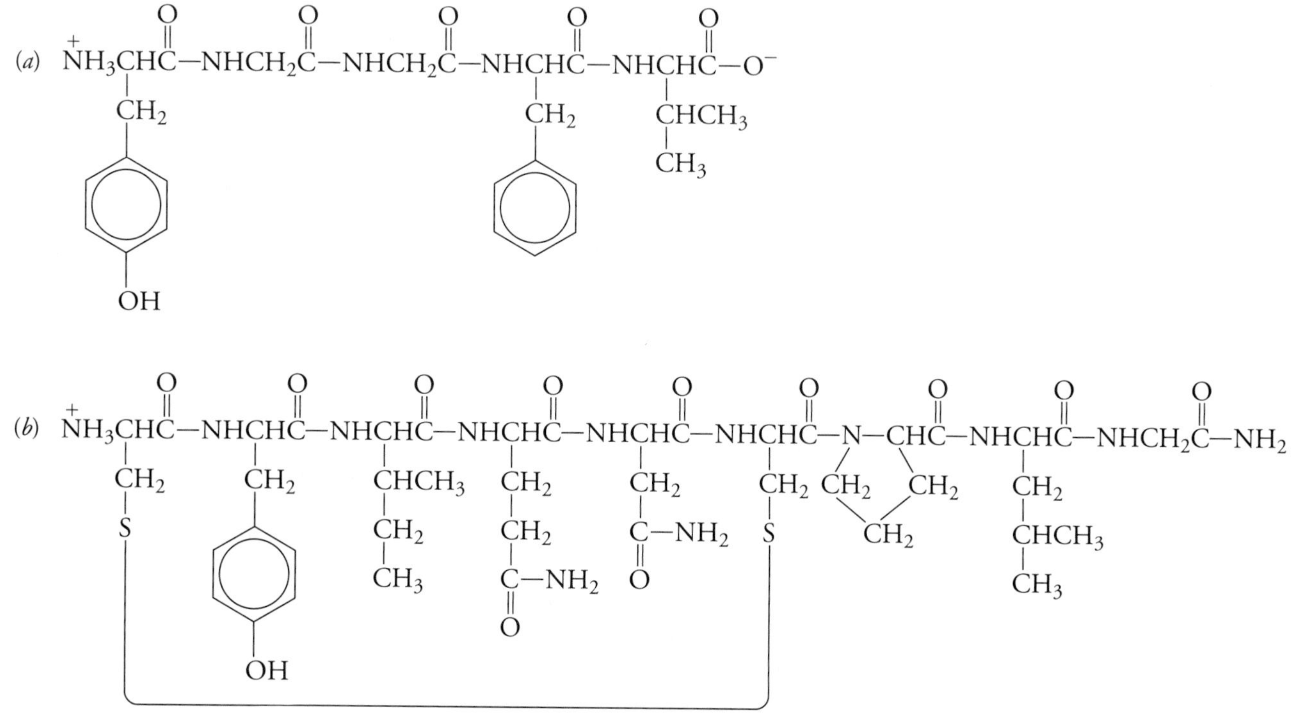

(a) $\overset{+}{N}H_3CH(CH_2C_6H_4OH)C(=O)-NHCH_2C(=O)-NHCH_2C(=O)-NHCH(CH_2C_6H_5)C(=O)-NHCH(CH_2CH(CH_3)CH_3)C(=O)-O^-$

(b) $\overset{+}{N}H_3CH(CH_2S-)C(=O)-NHCH(CH_2C_6H_4OH)C(=O)-NHCH(CHCH_3CH_2CH_3)C(=O)-NHCH(CH_2CH_2C(=O)NH_2)C(=O)-NHCH(CH_2C(=O)NH_2)C(=O)-NHCH(CH_2S-)C(=O)-N(CH_2CH_2CH_2)CHC(=O)-NHCH(CH_2CHCH_3CH_3)C(=O)-NHCH_2C(=O)-NH_2$ (the two S atoms are joined by a disulfide bond)

FIGURE 13.4

Larger oligopeptides are known by common names
(*a*) Leucine-enkephalin, a pentapeptide, is a naturally occurring pain reliever found in the brain. (*b*) Oxytocin, a nonapeptide, is a hormone that induces labor during pregnancy.

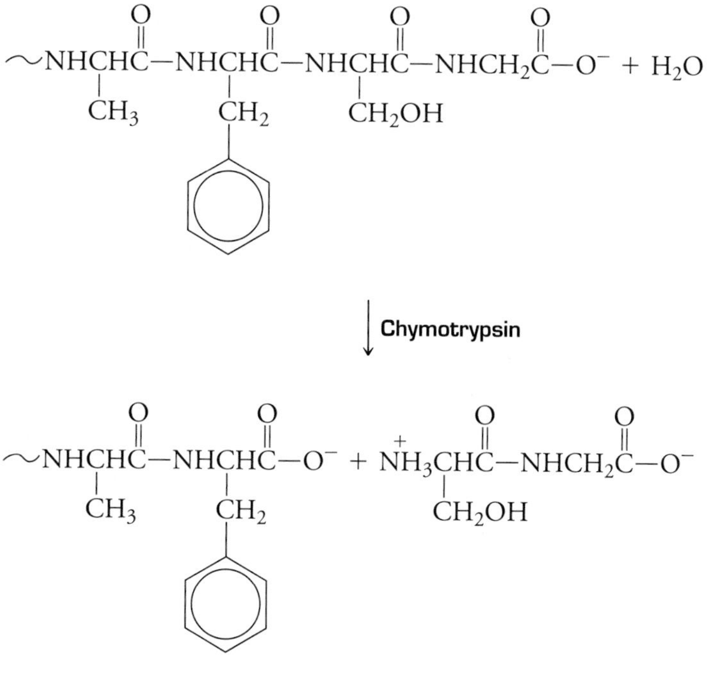

$$\sim NHCH(CH_3)C(=O)-NHCH(CH_2C_6H_5)C(=O)-NHCH(CH_2OH)C(=O)-NHCH_2C(=O)-O^- + H_2O$$

$$\downarrow \text{Chymotrypsin}$$

$$\sim NHCH(CH_3)C(=O)-NHCH(CH_2C_6H_5)C(=O)-O^- + \overset{+}{N}H_3CH(CH_2OH)C(=O)-NHCH_2C(=O)-O^-$$

FIGURE 13.5

Peptide bond hydrolysis
Hydrolysis of a peptide bond restores the carboxyl and amino groups on the amino acid residues held together by the bond. Chymotrypsin is a digestive enzyme that catalyzes the hydrolysis of peptide bonds on the C-terminal side of amino acid residues that contain aromatic groups. The partial structure of a peptide is shown here.

Source: MORAN, LAURENCE A.; SCRIMGEOUR, K. GRAY; HORTON, H. ROBERT; OCHS, RAYMOND S.; RAWM, J. DAVID, BIOCHEMISTRY, 2nd edition, © 1994, p. 5.15. Adapted by permission of Pearson Education, Inc., Upper Saddle River, NJ.

growth hormone-releasing factor (44 residues) stimulates its release; chymotrypsinogen (245 residues), produced in the pancreas, is a biologically inactive protein that is split into smaller pieces to become the digestive enzyme chymotrypsin; and titin (approximately 27,000 residues) is responsible for muscle's ability to spring back into shape after being stretched.

13.3 PEPTIDES, PROTEINS, AND pH

Peptide and protein structure is just as sensitive to pH is as amino acid structure, and the same principles apply when it comes to predicting whether carboxyl and amino groups are found in their acidic or basic forms. In acidic solutions, amino and carboxyl groups appear in their acidic forms, $-NH_3^+$ and $-CO_2H$, respectively. In neutral solutions, amino groups remain in their acidic form, but carboxyl groups are found in their basic form ($-CO_2^-$), and in basic solutions each appears in its basic form, $-NH_2$ and $-CO_2^-$. Peptide bonds are not acidic or basic, so their structure does not vary with changes in pH.

The tripeptide Lys-Ser-Asp contains one polar-basic amino acid residue and one polar-acidic amino acid residue so, in all, four functional groups may change structure with variations in pH (Figure 13.6*a*). At low pH values, the tripeptide has a net charge of 2+, at pH 7 a net charge of 0, and at high pH values a net charge of 2−. The —OH group in serine is not acidic, so the structure of its side chain does not vary with pH.

Making changes in the tripeptide can alter the net charge at a particular pH. Lys-Lys-Ala (Figure 13.6*b*), for example, has a net charge of 3+ at pH 1, a net charge of 2+ at pH 7, and a net charge of 1− at pH 14. Net charge affects the water solubility of a peptide or a protein. When peptides carry either a net positive or a net negative charge, the like-charged molecules tend to repel one another and stay dissolved in water. When the net charge is zero, these repulsive forces are reduced and the molecules are more likely to aggregate and precipitate out of solution. Lys-Ser-Asp is the least soluble in water near pH 7, because at this pH its net charge is zero. For Lys-Lys-Ala, the lowest water solubility comes at a basic pH.

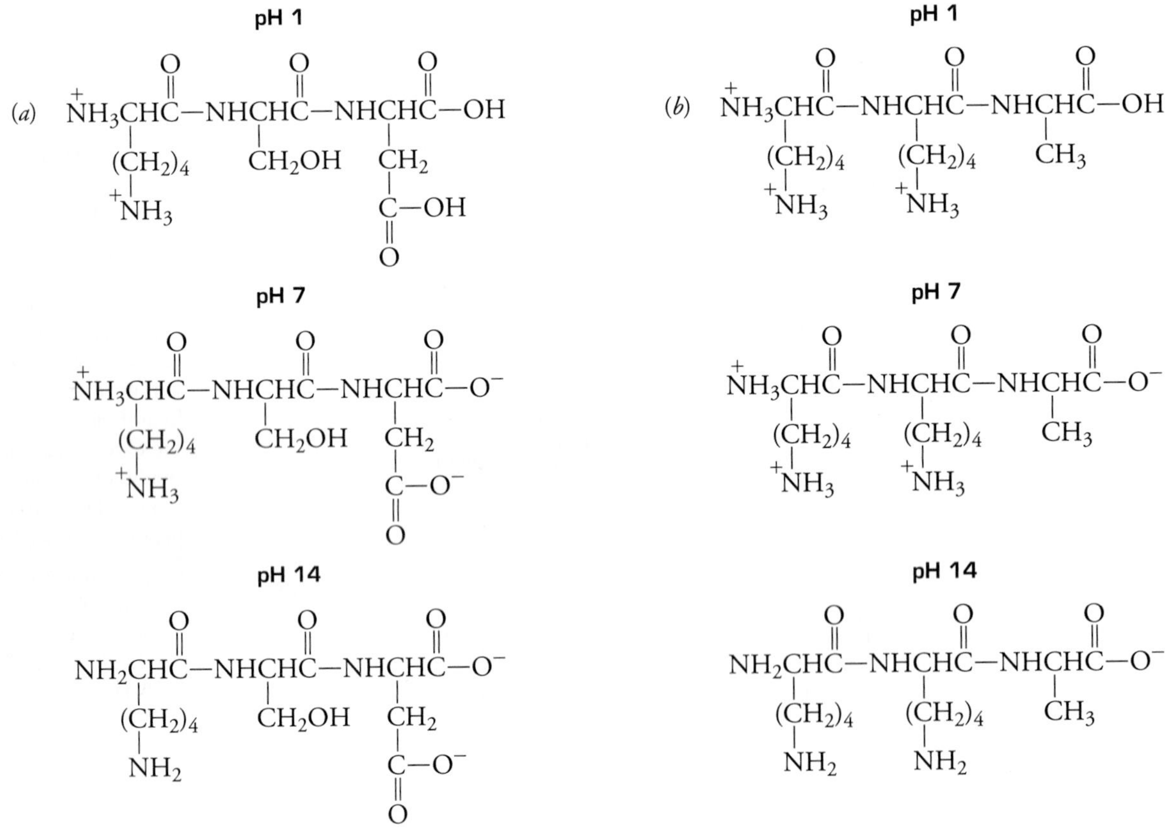

(*a*)

pH 1

$$\overset{+}{N}H_3CH(C{=}O)-NHCH(C{=}O)-NHCH(C{=}O)-OH$$

Side chains: $(CH_2)_4\overset{+}{N}H_3$; CH_2OH; $CH_2C(=O)-OH$

pH 7

$$\overset{+}{N}H_3CH(C{=}O)-NHCH(C{=}O)-NHCH(C{=}O)-O^-$$

Side chains: $(CH_2)_4\overset{+}{N}H_3$; CH_2OH; $CH_2C(=O)-O^-$

pH 14

$$NH_2CH(C{=}O)-NHCH(C{=}O)-NHCH(C{=}O)-O^-$$

Side chains: $(CH_2)_4NH_2$; CH_2OH; $CH_2C(=O)-O^-$

(*b*)

pH 1

$$\overset{+}{N}H_3CH(C{=}O)-NHCH(C{=}O)-NHCH(C{=}O)-OH$$

Side chains: $(CH_2)_4\overset{+}{N}H_3$; $(CH_2)_4\overset{+}{N}H_3$; CH_3

pH 7

$$\overset{+}{N}H_3CH(C{=}O)-NHCH(C{=}O)-NHCH(C{=}O)-O^-$$

Side chains: $(CH_2)_4\overset{+}{N}H_3$; $(CH_2)_4\overset{+}{N}H_3$; CH_3

pH 14

$$NH_2CH(C{=}O)-NHCH(C{=}O)-NHCH(C{=}O)-O^-$$

Side chains: $(CH_2)_4NH_2$; $(CH_2)_4NH_2$; CH_3

FIGURE 13.6

The effect of pH on peptide structure

In addition to the N-terminal amino group and the C-terminal carboxyl group, side chains containing amino and carboxyl groups are affected by changes in pH. (*a*) The tripeptide Lys-Ser-Asp has a net charge of 2+ at pH 1, a net charge of 0 at pH 7, and a net charge of 2− at pH 14. (*b*) The tripeptide Lys-Lys-Ala has a net charge of 3+ at pH 1, a net charge of 2+ at pH 7, and a net charge of 1− at pH 14.

SAMPLE PROBLEM 13.2

The effect of pH on peptide structure

The tripeptide glutathione, γ-Glu-Cys-Gly, stimulates tissue growth. Draw this tripeptide as it appears at pH 7. (A γ-glutamic acid residue results when glutamic acid forms a peptide bond using the carboxyl group of its side chain, rather than the carboxyl group attached to its α carbon atom.)

Strategy

Begin by drawing the tripeptide without worrying about charges. The structure of the amino group at the N-terminus and that of the carboxyl group at the C-terminus will vary with pH, as will the side chains of any polar-acidic or polar-basic amino acids. Table 13.1 will help you classify Glu, Cys, and Gly.

Solution

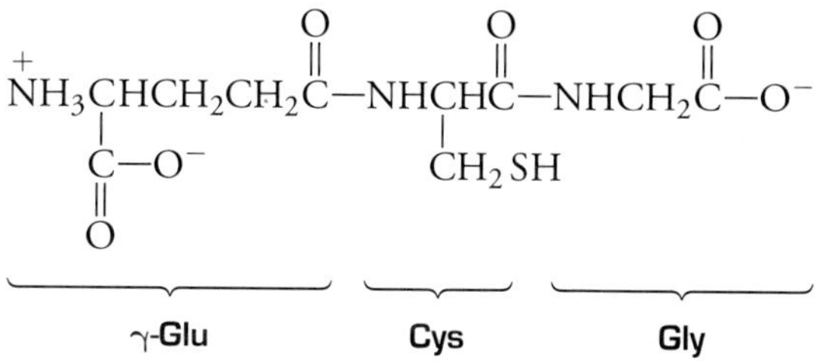

$$\overset{+}{N}H_3\underset{\underset{\underset{O}{\|}}{C-O^-}}{\underset{|}{C}}HCH_2CH_2\overset{O}{\overset{\|}{C}}-NH\underset{CH_2SH}{\underset{|}{C}}H\overset{O}{\overset{\|}{C}}-NHCH_2\overset{O}{\overset{\|}{C}}-O^-$$

γ-Glu | Cys | Gly

PRACTICE PROBLEM 13.2

Draw the tripeptide above as it would appear at pH 1.

13.4 PROTEIN STRUCTURE

In general terms, a protein can be classified as being either fibrous or globular. **Fibrous proteins** *exist as long fibers or strings.* These proteins, including collagen (in skin) and keratin (in hair), are usually tough and water insoluble. **Globular proteins** are *spherical in shape,* highly folded, and tend to be water soluble.

The structure of proteins is understood in terms of four levels of organization: primary, secondary, tertiary, and quaternary structure.

■ The sequence of amino acid residues in a peptide or protein is its primary structure.

Primary Structure

The *order of amino acid residues* in a peptide or protein is referred to as its **primary structure**. The name Ala-Gly-Val indicates primary structure and, reading from the N-terminus, this tripeptide consists of an alanine residue attached to a glycine residue, which is attached to a valine residue. Five other tripeptides, each with a different primary structure, can be constructed from one each of these three amino acids (Figure 13.7).

The primary structure of peptides and proteins is analogous to the arrangement of letters in a word. By rearranging the sequence of letters in a word, you can entirely change its meaning—“edit” becomes “diet,” “tide,” or “tied.” Similarly, rearranging the amino acid residues in a peptide or protein changes its function. One example that illustrates this point is the artificial sweetener aspartame (Table 12.2), whose primary structure is Asp-Phe-OMe, where the OMe indicates that the carboxyl group of the phenylalanine residue exists as the methyl ester. Aspartame is 180 times sweeter than sucrose, but its constitutional isomer with a different primary structure, Phe-Asp-OMe, is not sweet at all.

Ala–Gly–Val
Ala–Val–Gly
Gly–Ala–Val
Gly–Val–Ala
Val–Ala–Gly
Val–Gly–Ala

FIGURE 13.7

Primary structure
The primary structure of an oligo- or polypeptide is the order in which the amino acid residues appear. Six tripeptides, each with a different primary structure, contain one residue each of the amino acids alanine, glycine, and valine.

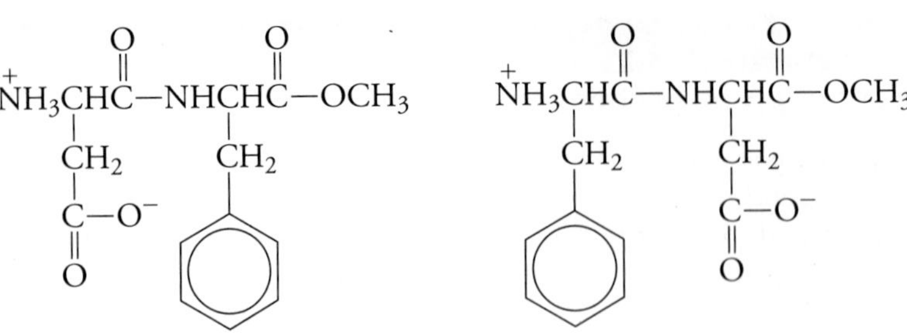

Asp-Phe-OMe **Phe-Asp-OMe**

■ The α-helix and the β-sheet are two types of secondary structure.

To stretch the analogy that relates letters in a word to amino acid residues in a peptide or protein, consider the huge number of words there are in an English language dictionary, all of which are formed by combining the 26 letters of the alphabet in different ways. With 20 basic building blocks to work with (the 20 common amino acids), the number of primary structures possible is enormous. Insulin, a protein involved in glucose metabolism, contains 51 amino acid residues. It is 1 of 20^{51} (1 of about 2.3×10^{66}) proteins of this length that can be constructed from the 20 amino acid building blocks.

Secondary Structure

The properties of proteins depend not only on their sequence of amino acid residues, but also on how they are folded, twisted, and bent. The two types of **secondary structure** commonly observed are *the α-helix and the β-sheet.* Secondary structure results from hydrogen bonding that takes place between amide N—H and C=O groups along the **polypeptide backbone.** The backbone is the *series of alternating α carbon atoms and peptide (amide) groups.* Amino acid side chains are attached to the backbone.

Peptide backbone

$\overset{+}{N}H_3CHC(=O)-NHCHC(=O)-NHCHC(=O)-NHCHC(=O)-NHCHC(=O)-O^-$

Side chains: CH_2OH, CH_3, $CH_2C(=O)O^-$, CH_2SH, $(CH_2)_4\overset{+}{N}H_3$

Side chains

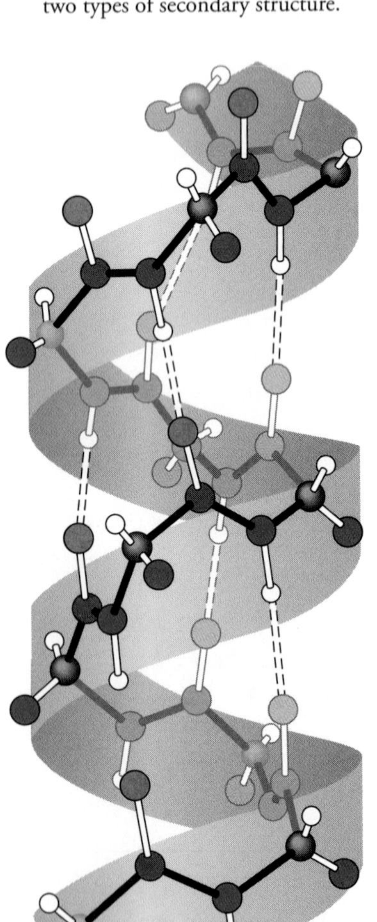

■ **FIGURE 13.8**

The α-helix
In the form of secondary structure called the α-helix, the polypeptide coils into a helical shape that is held in place by hydrogen bonds between amide N—H and C=O groups from different parts of the polypeptide backbone.

Source: Illustration, Irving Geis. Rights owned by Howard Hughes Medical Institute. Not to be reproduced without permission.

When a part of a polypeptide folds into an **α-helix** it assumes a shape that looks much like *a coiled spring.* The helical shape is maintained through hydrogen bonding between the N—H group of one amino acid residue and the C=O group of an amino acid four residues farther along the chain (Figure 13.8). Although a single hydrogen bond is fairly weak, the combination of the hydrogen bonds between the various amino acid residues in an α-helix stabilize the structure. Because of the spring-like nature of α-helices, proteins with a high α-helix content are flexible and stretchy. Keratins, the proteins that make up wool, are a good example.

The term **β-sheet** or β-pleated sheet refers to a sheet-like arrangement that *forms when different segments of a polypeptide chain align side by side* (Figure 13.9). In **parallel sheets** all of the interacting *strands run in the same N-terminal to C-terminal direction,* while in **antiparallel sheets** adjacent *segments run in opposite directions,* one N-terminal to C-terminal and the other C-terminal to N-terminal. In all β-sheets, the polypeptide segments are held together by hydrogen bonds between the N—H group of one amino acid residue and the C=O group of another. As with the α-helix, many hydrogen bonds cooperate to stabilize the β-sheet. Silk and other proteins with a high β-sheet content are strong but not stretchy.

■ The overall three-dimensional shape of a protein is its tertiary structure.

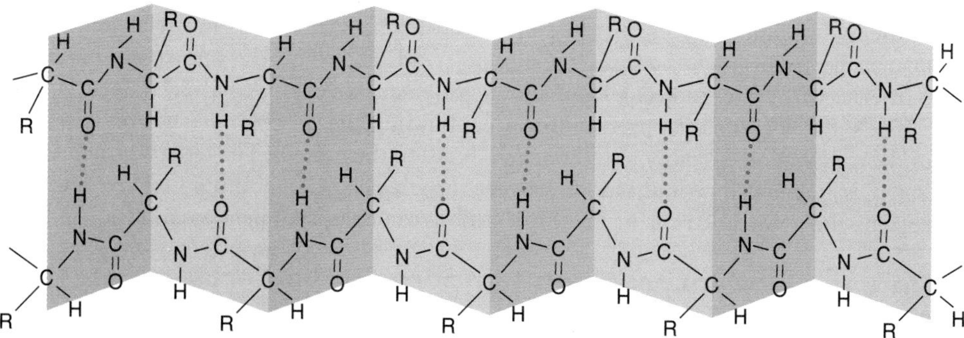

FIGURE 13.9

The β-sheet
In the form of secondary peptide structure called the β-sheet, portions of a polypeptide chain line up side by side. As with the α-helix, this form of secondary structure is held together by hydrogen bonds between amide N—H and C=O groups along the backbone. In antiparallel β-sheets (shown here) the strands run in opposite N-terminal to C-terminal directions. In parallel β-sheets, adjacent strands all run in the same direction. The term β-pleated sheet is sometimes used to describe a β-sheet because of the pleated appearance of its structure.

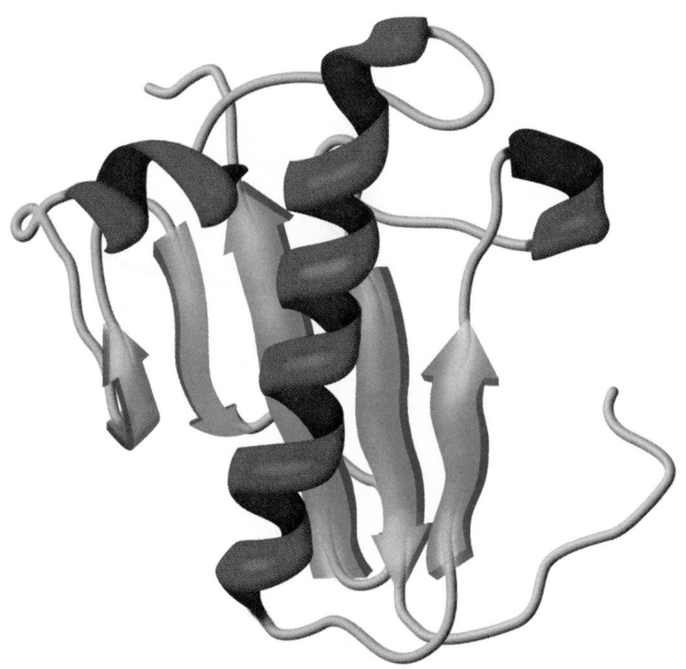

FIGURE 13.10

Tertiary structure
The overall three-dimensional shape of a protein is referred to as its tertiary structure. Here the tertiary structure of a protein is shown, with the blue arrows representing β-sheets, red coils representing α-helices, and the yellow rope representing other twists and turns of the polypeptide chain.

Source: Lima, C.D., Klein, M.G., Hendrickson, W.A.: Structure-based analysis of catalysis and substrate definition in the HIT protein family. *Science* 278, pp. 286 (1997). H.M. Berman, J. Westbrook, Z., Feng, G., Gilliland, T.N. Bhat, H. Weissig, I.N. Shindyalov, P.E. Bourne. The Protein Data Bank. *Nucleic Acids Research*, 28, pp. 235–242 (2000), *http://www.pdb.org.*

Tertiary Structure

Tertiary structure refers to *the overall three-dimensional shape* of a protein, including the folding of α-helices or β-sheets with respect to one another (Figure 13.10). Of the many folding patterns (conformations) possible for a protein, there is usually only one that leads to a **native** (*biologically active*) molecule. The sequence of amino acids (primary structure) ultimately determines which folding pattern is selected, so both secondary structure and tertiary structure depend on primary structure.

In an aqueous environment, the native form of a globular protein typically has its nonpolar amino acid side chains folded into the hydrophobic interior and its polar side chains on the hydrophilic surface. This folding rule, known as "nonpolar in, polar out," is the most stable arrangement because it allows polar side chains on the surface of the protein to interact with water molecules and allows nonpolar side chains to avoid water.

The surface of a globular protein usually has **binding sites**, *indentations or clefts* important for the ability of other compounds to attach (bind) to the protein. The tertiary structure of a protein is maintained by interactions between the side chains of the amino acid residues that make up the polypeptide chain. These interactions can be noncovalent or covalent.

Noncovalent interactions important to tertiary structure

Hydrogen bonds, the hydrophobic effect, and salt bridges contribute to stabilizing the tertiary structure of proteins (Figure 13.11). Of these, hydrogen bonds and the hydrophobic effect make the most important contributions.

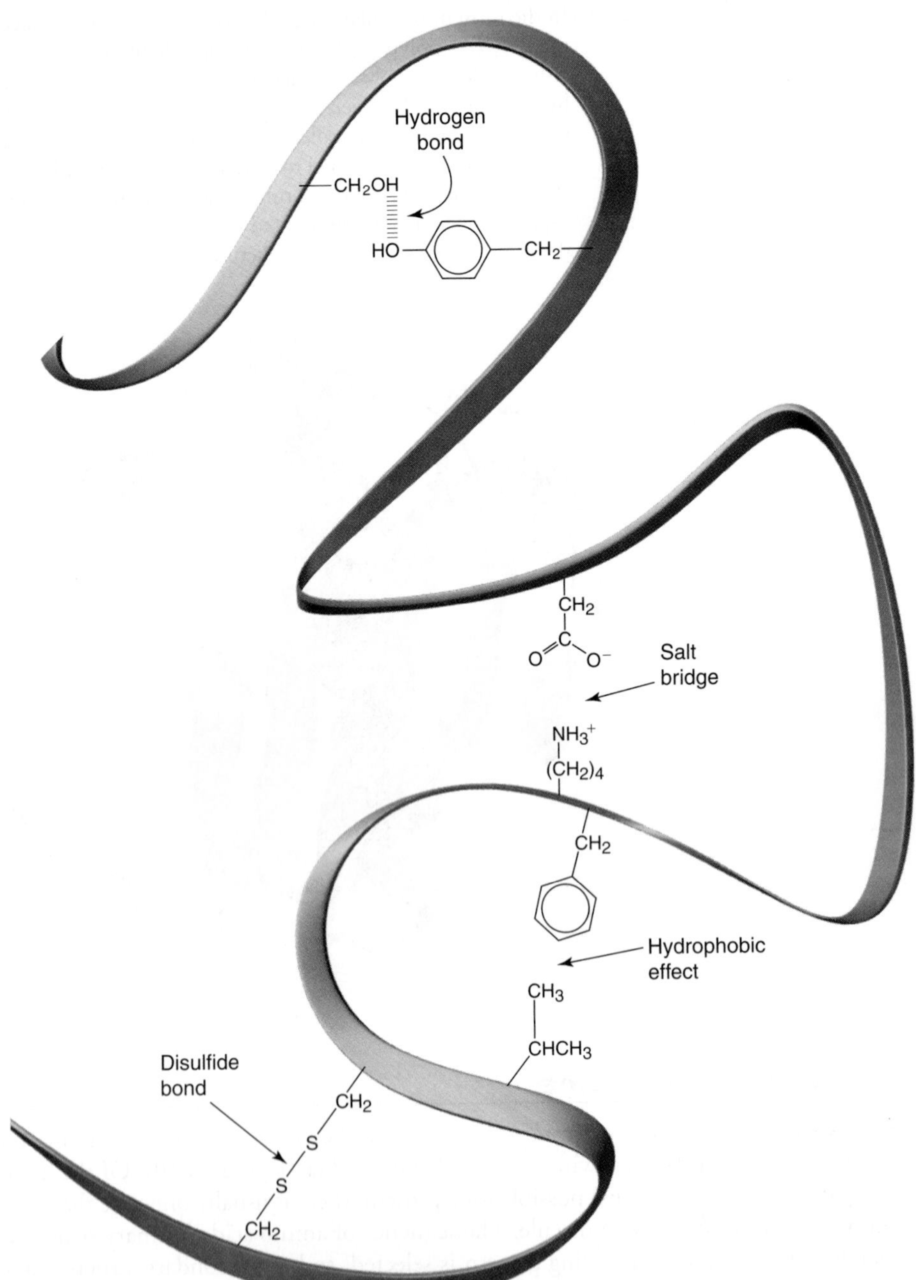

FIGURE 13.11

Forces that maintain tertiary structure
The interactions between side chains that are most important to maintaining the overall three-dimensional shape of a polypeptide are hydrogen bonds, salt bridges, the hydrophobic effect, and disulfide bonds.

For hydrogen bonds to form between side chains, one amino acid residue must carry either an N—H or O—H and the other a nitrogen or an oxygen atom. With the large number of polar, polar-basic, and polar-acidic amino acid residues typically present in a protein, there are many opportunities for hydrogen bonding to take place between side chains.

As described in Section 7.5, hydrophobic compounds are "water fearing" and do not mix with water. The term **hydrophobic effect** is used to describe the way that *nonpolar (hydrophobic) compounds are drawn to one another and avoid water*, when placed in an aqueous solution. This arrangement is favorable because it allows the nonpolar compounds to interact through London forces and does not disrupt hydrogen bonding between water molecules. When applied to tertiary structure, the hydrophobic effect is displayed by the way that nonpolar amino acid side chains become folded into the interior of the protein, avoiding exposure to water.

A salt bridge (ionic bond) is formed when a positively charged side chain is attracted to a negatively charged side chain. Amino acid residues that carry charged side chains belong to the polar-basic and polar-acidic classifications and, being polar, the ionic interactions between their side chains usually take place on the protein surface.

Covalent interactions important to tertiary structure

When molecules containing thiol (—SH) groups are oxidized, they combine to form disulfides (Section 11.3). When the $—CH_2SH$ side chains of two cysteine residues undergo the same reaction, the resulting disulfide bond is called a disulfide bridge (Figure 13.11).

Disulfide bonds are easily reduced to re-form thiol groups. Since the cytoplasm of a cell tends to contain reducing agents (NADH, NADPH, and others), disulfide bonds are less common in intracellular (*within a cell*) proteins than in extracellular (*outside of a cell*) ones. Because they are covalent, disulfide bonds are relatively strong and often play a key role in helping extracellular proteins maintain their native structure. For example, the structure of hair depends on disulfide bonds that form between strands of protein. Insulin is another protein whose tertiary structure depends on the formation of disulfide bridges.

For some proteins, tertiary structure also includes **prosthetic groups** (*nonpeptide components* tightly bound to the polypeptide chain), such as the heme group (Figure 13.12) and the metal ions carried by some enzymes (Section 13.6). Proteins that *require a prosthetic group* for biological activity are called **complex**, while those that *function without a prosthetic group* are **simple**.

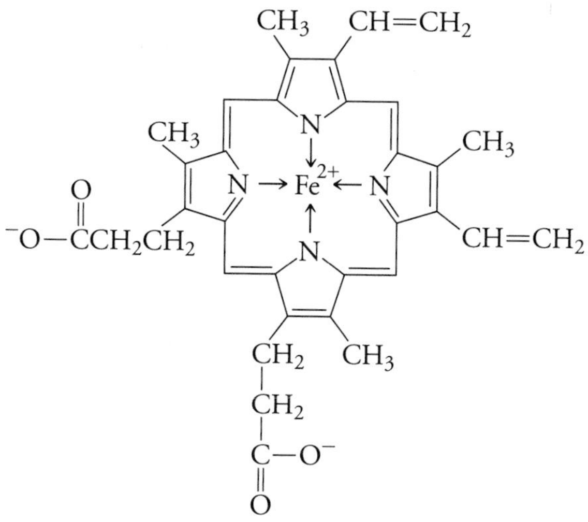

CH_3 $CH{=}CH_2$ CH_3 CH_3 $N \rightarrow Fe^{2+} \leftarrow N$ $^{-}O—\overset{O}{\overset{\|}{C}}CH_2CH_2$ $CH{=}CH_2$ CH_2 CH_3 CH_2 $C—O^{-}$ O

Heme

FIGURE 13.12

A prosthetic group
Tertiary structure sometimes includes nonpeptide components called prosthetic groups. The heme group (part of hemoglobin, the O_2-carrying protein present in red blood cells) is a prosthetic group.

Quaternary Structure

A large number of native proteins are a combination of more than one polypeptide chain. Quaternary structure refers to the arrangement of these chains in a functioning protein (Figure 13.13). The number of polypeptide subunits varies widely from protein to protein, as does the number of different polypeptide chains. Hemoglobin (Biochemistry Link: Hemoglobin, a Globular Protein) is a tetramer (four subunits) containing two copies of one polypeptide and two copies of another.

- Quaternary structure is the arrangement of multiple polypeptide chains in a functioning protein.

Quaternary structure is largely maintained through noncovalent attractions such as hydrogen bonding, the hydrophobic effect, and salt bridges, and by disulfide bonds. Quaternary structure is important because cooperation between subunits is often a factor in the regulation of protein activity. The binding of a compound to one subunit can alter the tertiary structure of another subunit, greatly affecting the ability of other subunits to bind compounds. The discussions of hemoglobin and allosteric enzymes that follow illustrate this point.

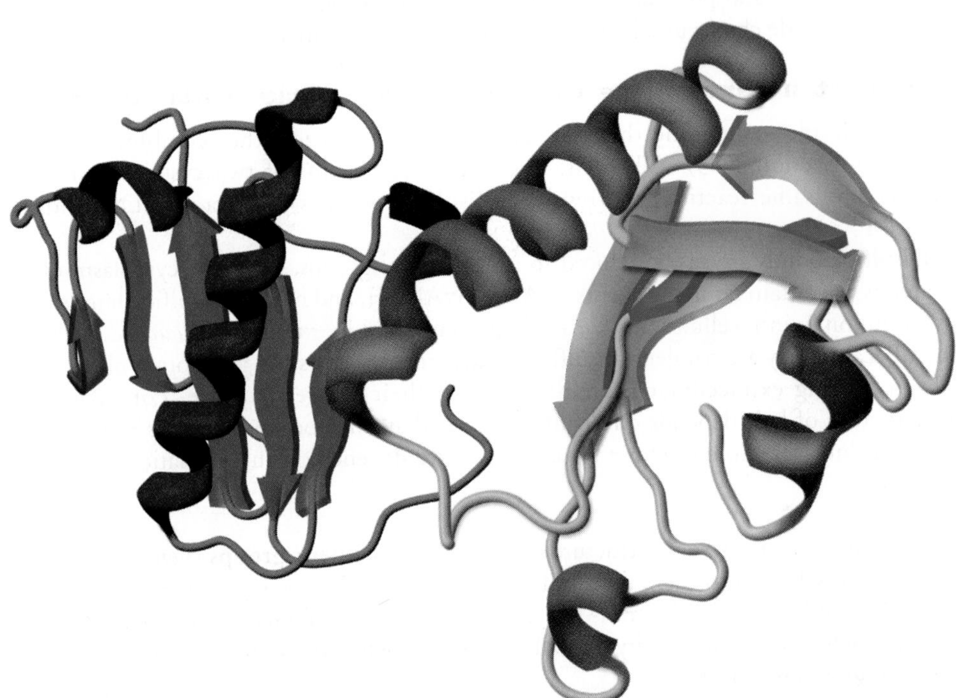

FIGURE | 13.13

Quaternary structure
Some proteins require more than one polypeptide chain to be biologically active. Quaternary structure refers to the arrangement of these chains. Here a protein dimer (two subunits) is pictured.

Source: Lima, C.D., Klein, M.G., Hendrickson, W.A.: Structure-based analysis of catalysis and substrate definition in the HIT protein family. *Science* 278, p. 286 (1997). H.M. Berman, J. Westbrook, Z., Feng, G., Gilliland, T.N. Bhat, H. Weissig, I.N. Shindyalov, P.E. Bourne. The Protein Data Bank. *Nucleic Acids Research*, 28, pp. 235–242 (2000), *http://www.pdb.org*.

SAMPLE PROBLEM 13.3

In which of the following levels of protein structure can hydrogen bonding play a role?

a. 1° structure **b.** 2° structure **c.** 3° structure **d.** 4° structure

Strategy
To solve this problem, you will want to think about how each level of protein structure is defined.

Solution
b, c, and d. Secondary structure arises from hydrogen bonding along a polypeptide chain. Tertiary protein structure (overall three-dimensional shape) and quaternary structure (association of separate polypeptide chains) is due, in part, to hydrogen bonding.

PRACTICE PROBLEM 13.3

In which level(s) of protein structure do disulfide bonds play a role?

BIOCHEMISTRY Link

Collagen and α-Keratin, Two Fibrous Proteins

Collagen, which represents almost one-third of the total protein in the human body, is the major structural protein found in tendon, cartilage, bone, blood vessels, and teeth. Because of this biological role, collagen needs to be strong, water insoluble, and relatively resistant to chemical change.

The basic building blocks for all collagen molecules are proteins that contain about 1000 amino acid residues, the majority of which are glycine, alanine, glutamic acid, arginine, proline, and hydroxyproline. Hydroxyproline (Hyp), which is not one of the 20 common amino acids, is produced by the enzyme-catalyzed addition of a hydroxyl group to a proline residue, once a protein chain has been formed.

$$\sim N-CH\overset{O}{\overset{\|}{C}}\sim \ (ring: CH_2, CH_2, CH_2) \xrightarrow{\text{enzyme}} \sim N-CH\overset{O}{\overset{\|}{C}}\sim \ (ring: CH_2, CH_2, CH-OH)$$

Proline residue → **Hydroxyproline residue**

In collagen, every third amino acid residue is usually a glycine, and the sequence Gly-Pro-Hyp reoccurs many times. This primary structure causes collagen proteins to twist into helices, three of which wrap together to form a coil known as a triple helix (Figure 13.14). Parallel triple helices often pack together to form collagen fibers. Initially, hydrophobic interactions and hydrogen bonds are the major forces that hold the proteins to one another, but as collagen ages, covalent cross-links form between neighboring protein chains. As a result of their structure, collagen fibers are very strong—some fibers 1 mm in diameter can hold over 20 pounds. The presence of proline, alanine, and other nonpolar amino acid residues on the surface of collagen helps account for its insolubility in water (the "nonpolar in, polar out" rule applies only to globular proteins in an aqueous environment).

Abnormal collagens cause a number of different medical problems. Scurvy, which results from a vitamin C deficiency, produces skin lesions, fragile blood vessels, and bleeding gums. These symptoms have been traced to hydroxyproline-deficient collagen, which is relatively weak and easily pulled apart. The skeletal deformities caused by another disease, homocystinuria, are a consequence of incomplete cross-linking of collagen. The replacement of glycine residues with serine residues leads to Ehlers-Danlos syndrome, a disorder characterized by loose joints.

The proteins in hair, fingernails, wool, claws, quills, horns, hooves, and feathers are similar to collagen in their overall construction. The basic building blocks for these α-keratins are protofibrils, which consist of four polypeptide α helices coiled around one another. Protofibrils combine to form microfibrils that are held together, in part, by disulfide bonds (Figure 13.15). Microfibrils combine to make macrofibrils, which are packed into even larger bundles.

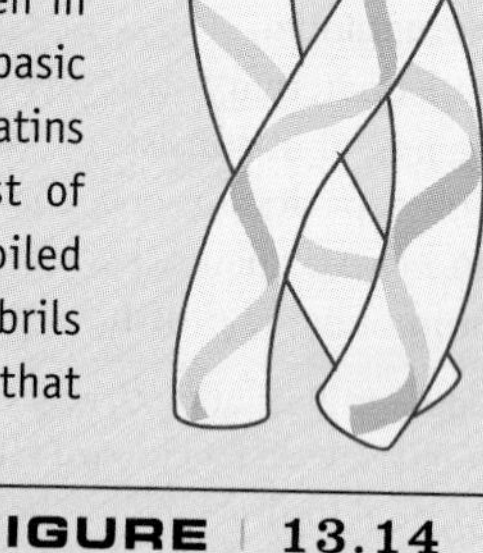

FIGURE 13.14

Collagen
Collagen, an important structural protein, exists as a triple helix.

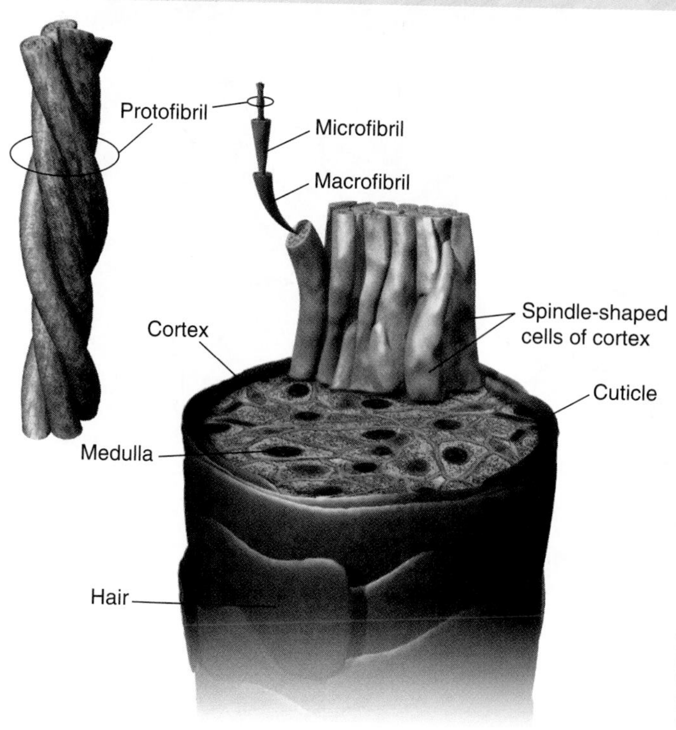

FIGURE 13.15

Keratin
Keratin, the protein found in hair, claws, and fingernails, consists of bundles of protofibrils, which are formed by wrapping four polypeptide chains together.

Source: MORAN, LAURENCE A.; SCRIMGEOUR, K. GRAY; HORTON, H. ROBERT; OCHS, RAYMOND S.; RAWN, J. DAVID, BIOCHEMISTRY, 2nd Edition, © 1994, p. 515. Reprinted by permission of Pearson Education, Inc., Upper Saddle River, NJ.

BIOCHEMISTRY Link

Hemoglobin, a Globular Protein

Most of the O_2 carried from the lungs to cells is transported by hemoglobin, a transport protein found only in red blood cells. The CO_2 produced by cells enters the bloodstream, where some of it binds to hemoglobin for transport back to the lungs.

A single human hemoglobin molecule is a tetramer consisting of four separate polypeptide chains—two identical α chains and two identical β chains (Figure 13.16). Each α chain contains 141 amino acid residues that coil into seven α-helical regions, and each β chain contains 146 amino acid residues that coil to form eight α-helical regions.

Each α and β chain carries a heme prosthetic group (Figure 13.12) held in place by side chains of polar amino acids that interact with the heme's two charged carboxylate groups and with Fe^{2+}. The surface of each α and β chain is covered with hydrophobic side chains and the quaternary structure of hemoglobin is formed when the hydrophobic regions in the α and β chains pack together in order to escape from surrounding water molecules. Although hydrophobic forces are primarily responsible for the packing of the four subunits, hydrogen bonds and salt bridges account for about one-third of the noncovalent bonds that maintain hemoglobin's quaternary structure.

Hemoglobin has four O_2 binding sites because each heme–polypeptide subunit can bind one O_2 to its Fe^{2+}. The binding of an O_2 to any one subunit in nonoxygenated hemoglobin (deoxyhemoglobin) leads to changes in the shape of the other three subunits which, in the process, increases their ability to bind O_2. The binding of a second O_2 increases the O_2 affinity of the remaining nonoxygenated subunits in a similar fashion and the binding of a third O_2 has a comparable effect. After the first O_2 has bound to hemoglobin, each of the next three O_2 molecules binds more and more readily.

In molecules with multiple binding sites, binding is said to be cooperative if binding at one site impacts the binding at other sites. Cooperative binding is positive if the binding of one ligand (an atom, ion, or molecule that binds to a protein) enhances the binding of subsequent ligands, and is negative when the binding of one ligand makes it more difficult for additional ligands to bind. The positive cooperative binding of O_2 by hemoglobin accounts for its tendency to carry either four O_2 molecules or none.

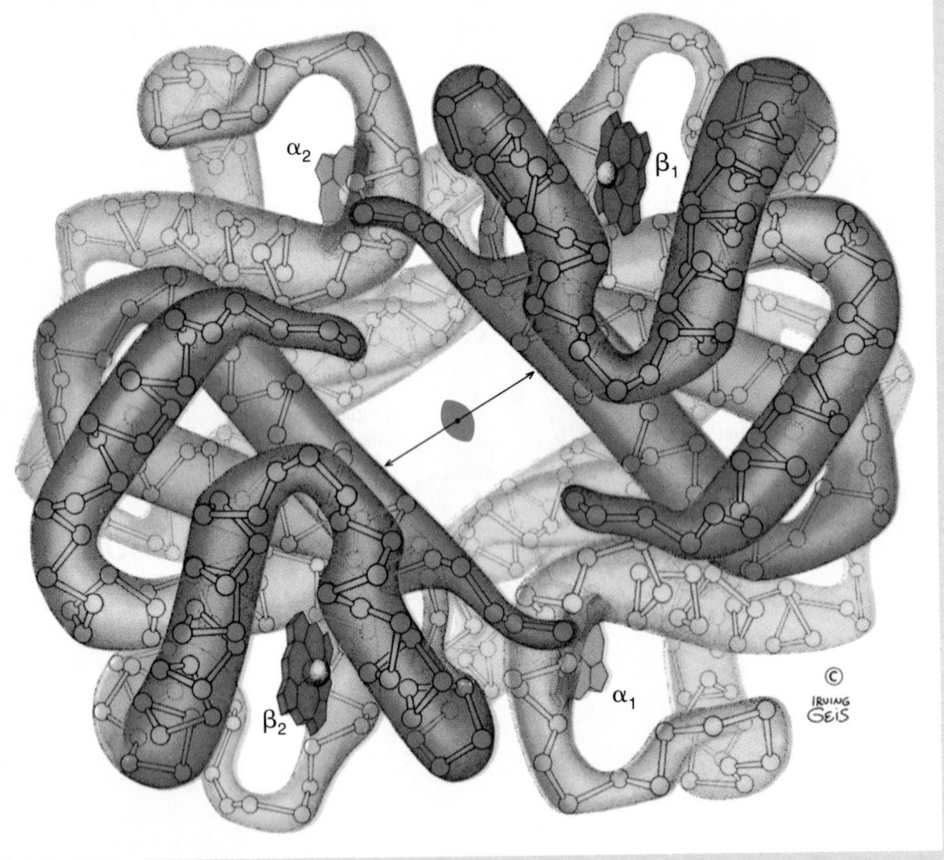

FIGURE 13.16

Hemoglobin
The quaternary structure of hemoglobin consists of four polypeptides—two α chains and two β chains. Each chain holds a heme prosthetic group.

13.5 DENATURATION

As we saw above, the native conformation (shape) of a protein is the biologically active one. **Denaturation** is any *change in protein conformation* caused by disruption of the noncovalent forces and disulfide bonds responsible for maintaining 2°, 3°, and 4° structure (Figure 13.17). A loss of biological activity normally accompanies denaturation, and this process is reversible only if the changes that take place are minor.

■ When the three-dimensional shape of an active protein is changed, the protein has been denatured.

Denaturation can be caused by a variety of factors, including changes in temperature or pH, agitation, and the use of detergents or soaps. An increase in temperature is associated with an increase in kinetic energy, and this increased motion can be enough to disrupt hydrogen bonds and other noncovalent interactions. Varying pH affects protein shape, in part, because the charges on amino acid side chains involved in salt bridges may disappear. Detergents and soaps, which are amphipathic, interfere with hydrophobic interactions.

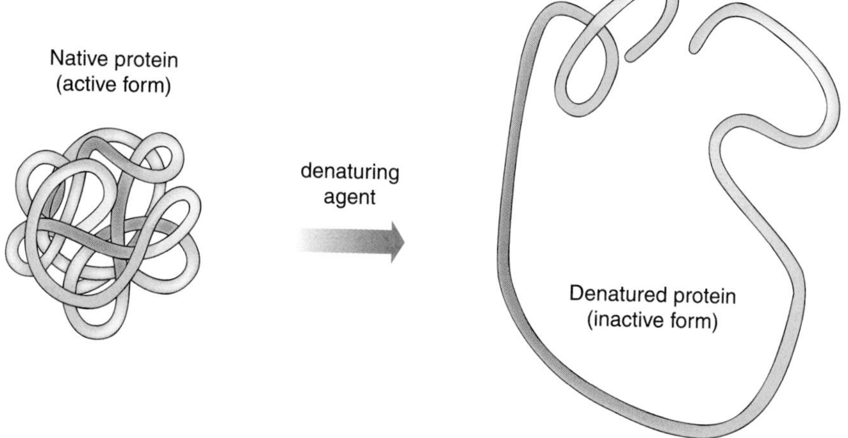

■ FIGURE 13.17 **Denaturation**

Heat, changes in pH, agitation, and denaturing agents such as soaps and detergents can disrupt the shape of a protein.

Source: From *Biochemistry* 4th edition by Campbell/Farrell. © 2003. Reprinted with permission of Brooks/Cole, a division of Thomson Learning: *www.thomsonrights.com*. Fax 800 730-2215.

13.6 ENZYMES

Thousands of chemical reactions are taking place within your body at any given time. Most of these would not occur at a biologically significant rate without the help of biological catalysts called enzymes. While most enzymes are globular proteins, a few newly discovered ones are RNA molecules (Section 14.8). We will focus our attention on the protein catalysts.

A number of different systems have been developed for naming enzymes. One commonly used system identifies enzymes by recommended names, which usually specify the reaction catalyzed or the **substrate** (*reactant*) that is modified during the reaction. For example, aldehyde dehydrogenase (most recommended enzyme names end in "ase") catalyzes the dehydrogenation (oxidation) of aldehydes, and maltase catalyzes the hydrolysis of maltose (Figure 13.18). Some recommended names, such as chymotrypsin, which catalyzes the hydrolysis of polypeptides (Figure 13.5), give no clue as to the nature of the reaction catalyzed by the enzyme.

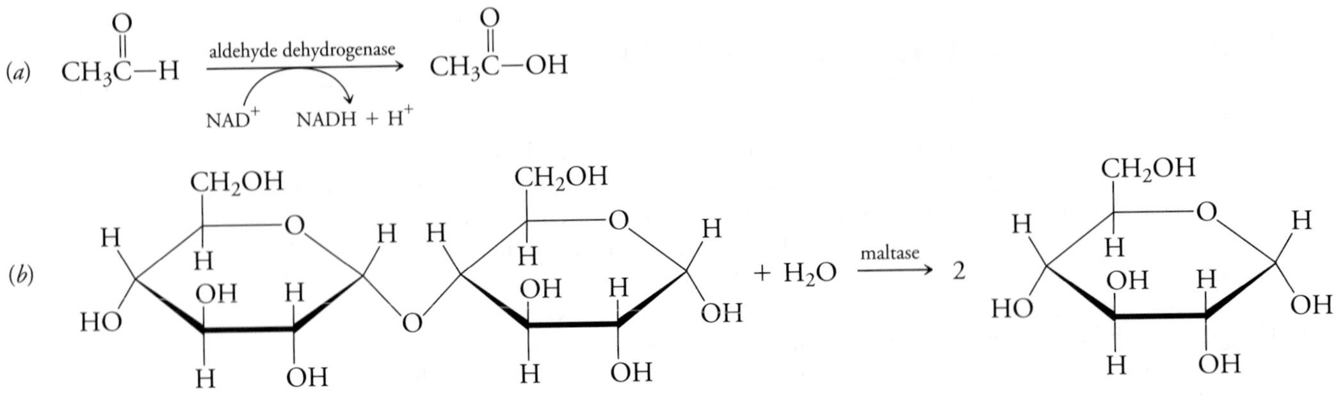

(a) $CH_3\overset{O}{\overset{\|}{C}}-H \xrightarrow[NAD^+ \quad NADH + H^+]{\text{aldehyde dehydrogenase}} CH_3\overset{O}{\overset{\|}{C}}-OH$

(b) Maltose + H_2O $\xrightarrow{\text{maltase}}$ 2 Glucose

FIGURE | 13.18

Recommended enzyme names
Generally, recommended names indicate the reaction catalyzed or the substrate recognized by the enzyme. (*a*) Aldehyde dehydrogenase oxidizes (removes hydrogen atoms from) aldehydes. (*b*) Maltase catalyzes the hydrolysis of the disaccharide maltose.

Specificity

■ Enzymes can display stereospecificity and absolute or relative specificity.

Depending on the enzyme, the degree of specificity (*selectivity*) shown for substrates can vary widely. Some enzymes display **absolute specificity**, which means that *the enzyme accepts only one specific substrate.* Urease, which catalyzes the hydrolysis of urea to produce CO_2 and NH_3, shows absolute specificity because it catalyzes the hydrolysis of urea but not any other molecule (Figure 13.19). Most enzymes are much less specific and will react with *a range of substrates having the same functional groups or similar structures.* Alcohol dehydrogenase, which catalyzes the oxidation of a wide range of alcohols, is an enzyme that displays this **relative specificity**.

FIGURE | 13.19

Enzyme specificity
(*a*) Urease shows absolute specificity because it acts as a catalyst for only one specific substrate, urea. (*b*) Alcohol dehydrogenase displays relative specificity because it catalyzes the oxidation of a wide range of alcohols (here R represents a carbon atom-containing group). (*c*) Stereospecific enzymes react with or produce a particular stereoisomer. Stearoyl-CoA 9-desaturase converts a saturated fatty acid derivative into a 9-*cis* unsaturated one. (Some of the reactants for this reaction are not shown.)

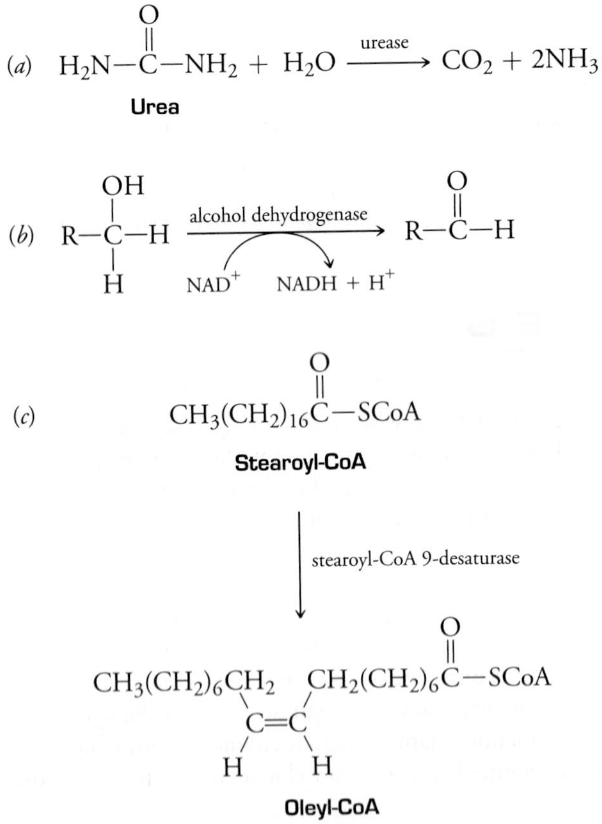

(a) $H_2N-\overset{O}{\overset{\|}{C}}-NH_2 + H_2O \xrightarrow{\text{urease}} CO_2 + 2NH_3$

Urea

(b) $R-\overset{OH}{\underset{H}{C}}-H \xrightarrow[NAD^+ \quad NADH + H^+]{\text{alcohol dehydrogenase}} R-\overset{O}{\overset{\|}{C}}-H$

(c) $CH_3(CH_2)_{16}\overset{O}{\overset{\|}{C}}-SCoA$

Stearoyl-CoA

↓ stearoyl-CoA 9-desaturase

$CH_3(CH_2)_6CH_2(H)C{=}C(H)CH_2(CH_2)_6\overset{O}{\overset{\|}{C}}-SCoA$ (cis)

Oleyl-CoA

Many enzymes show **stereospecificity**, because they only *react with or produce a particular stereoisomer.* Stearoyl-CoA 9-desaturase converts a derivative of stearic acid (a saturated fatty acid) into a derivative of oleic acid (an unsaturated fatty acid), producing only the *cis* stereoisomer in the process. Other enzymes are stereospecific in that they selectively catalyze the reaction or formation of only L or only D isomers of amino acids and carbohydrates.

Catalysis

Section 6.7 discussed the factors that can influence the rate of a chemical reaction: temperature, concentration, and catalysts. Raising the temperature increases the kinetic energy of reactants, which leads to more frequent collisions with a greater average energy. This, in turn, means that more reactants will have sufficient energy to overcome the activation barrier of a reaction, so its rate will increase. Increasing the concentration of reactants can also lead to an increased reaction rate, because the more reactants that are packed into a given volume the greater the chances of reactants colliding properly and moving over the activation barrier. Catalysts increase reaction rates by lowering the activation energy.

The activation energy for a reaction has no connection to whether the reaction is spontaneous ($\Delta G < 0$) or nonspontaneous ($\Delta G > 0$). This means that an uncatalyzed reaction that is nonspontaneous is still nonspontaneous in the presence of a catalyst. As we will see in Chapter 15, many of the reactions necessary for life are nonspontaneous. One approach that living things use to get these reactions to take place is to change them by having high energy molecules (they have high chemical potential energy) supply the needed energy.

For an enzyme, catalysis is the direct result of interactions that take place between it and the substrate. Substrates attach to specific binding sites called **active sites.**

■ Substrates (reactants) attach to active sites (crevices) on an enzyme's surface.

The first step in enzyme catalysis involves the formation of an enzyme–substrate complex (ES in Figure 13.20) due to binding of the substrate to an enzyme's active site. The amino acid side chains present in the active site interact with the substrate and hold it in place. A reaction takes place faster in the presence of the appropriate enzyme than in its absence because the substrate is held in the proper orientation and near other atoms that are involved in the bond making or bond breaking process that goes on during the reaction. Once the product has formed, it leaves the enzyme (in Figure 13.20, ES becomes E + P), which is restored to its original form, ready for another cycle of catalysis.

The extent of catalysis varies, depending on the enzyme and on the reaction being catalyzed. The peptide bond hydrolysis catalyzed by papain, an enzyme used in meat tenderizers, runs only 10 times faster than the uncatalyzed reaction. Acetylcholinesterase, which assists in the hydrolysis of an ester involved in the transmission of nerve impulses, gives a rate enhancement of about 1000, while catalase, involved in the decomposition of H_2O_2, speeds the reaction by a factor of 10 million.

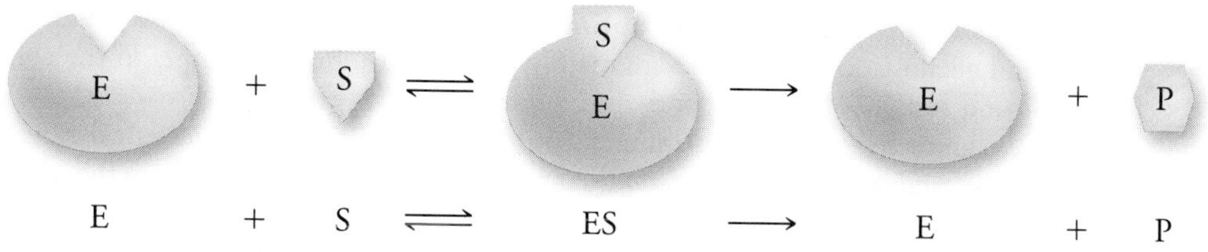

FIGURE 13.20

An enzyme reaction

For an enzyme-catalyzed reaction to take place, an enzyme (E) must bind substrate (S) and form an enzyme–substrate complex (ES). Once substrate has been converted to product (P), the product leaves, restoring the catalyst (E) to its original form. K_M, the Michaelis constant, describes the first step in this process ($E + S \rightleftharpoons ES$) and V_{max}, the maximum velocity, describes the second step ($ES \longrightarrow E + P$).

Cofactors and Coenzymes

For some enzymes, the polypeptide chain is all that is required for catalytic action. Others require the presence of **cofactors**, which are either *inorganic ions*, such as Fe^{2+}, Fe^{3+}, Cu^{2+}, Zn^{2+}, Na^{+}, and K^{+}, *or organic compounds* called **coenzymes** (Table 13.2). Most coenzymes are derived from water-soluble vitamins.

TABLE 13.2 SELECTED ENZYME COFACTORS

Ions	An enzyme that requires the ion	Dietary source
Zn^{2+}	Alcohol dehydrogenase	Trace element[a]
Fe^{2+}	Peroxidase	Trace element
Ni^{2+}	Urease	Trace element
Coenzymes[b]	**Reaction type that uses coenzyme**	**Dietary source**
NADH	Oxidation and reduction	Niacin
$FADH_2$	Oxidation and reduction	Riboflavin (B_2)
CoA	Group transfer	Pantothenic acid

[a] See Figure 2.4.
[b] The structure of these coenzymes can be seen in Figure 15.2.

In about one-third of known enzymes the metal ions are tightly bound to the enzyme, which means that these cofactors are prosthetic groups (discussed earlier in this chapter as part of tertiary protein structure). For other enzymes the cofactors are not considered prosthetic groups because they are loosely held or come and go during the course of the reaction.

The Effect of pH and Temperature

Enzymes are globular proteins, and their ability to act as catalysts depends on maintaining their native (biologically active) conformation. For this reason pH and temperature have an effect on enzyme activity.

Protein conformation is sensitive to pH changes because pH influences the charges on the amino acid side chains that are involved in maintaining tertiary and quaternary protein structure. Most human enzymes are efficient catalysts at pH values near 7 (often called physiological pH because most human body fluids have a pH near this value), but they are denatured and inactive at extremes in pH (Figure 13.21*a*). The pH at which an enzyme *is most active* is called its **pH optimum**. The potentially lethal consequences of acidosis and alkalosis (Section 9.11) are mainly due to pH-induced changes in the conformation and activity of enzymes and other proteins.

■ An enzyme operates best at its optimum pH and temperature.

Some enzymes have a pH optimum far outside of the usual physiological range. Pepsin, which catalyzes the hydrolysis of dietary proteins in the stomach where the pH is normally close to 1.5, exhibits its highest activity at that pH.

A rise in temperature leads to an increase in the rate of enzyme-catalyzed reactions—at least until a temperature is reached where the protein begins to denature and the enzyme ceases to operate as a catalyst (Figure 13.21*b*). Deaths due to high fevers are typically a result of the denaturation of enzymes and other proteins.

Although most human enzymes have a **temperature optimum** (*the temperature at which an enzyme is most active*) near 37°C and are inactivated well before they reach 90°C, the enzymes in some bacteria found living in Yellowstone hot springs function quite effectively at temperatures close to the boiling point of water (100°C) (Figure 13.22). One of these temperature-resistant enzymes plays a key role in the polymerase chain reaction

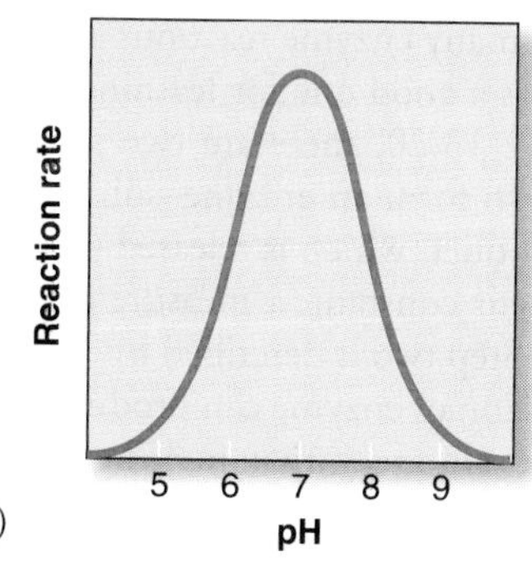

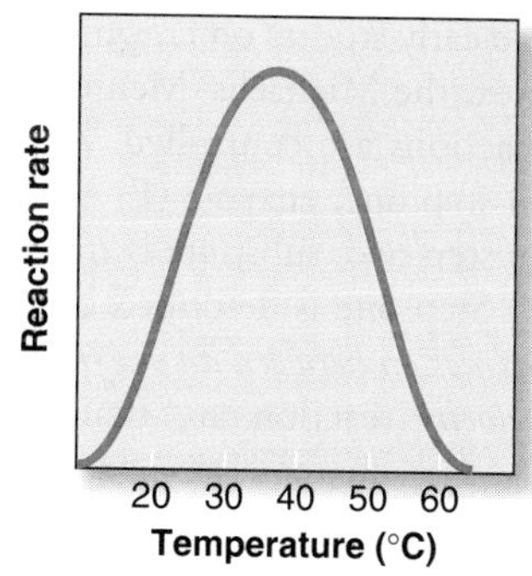

FIGURE 13.21

Effect of pH and temperature on enzymes
(*a*) Each enzyme has its own particular pH optimum, the pH at which it has maximum catalytic activity. Changing the pH can alter the charges on amino acid side chains, which leads to varying degrees of denaturation. (*b*) Each enzyme has its own particular temperature optimum. Increasing the temperature leads to an increased reaction rate, but too high a temperature can result in denaturation and loss of catalytic ability.

(PCR), a process that is used to increase the amount of DNA in a sample (Chapter 14, Biochemistry Link: DNA Fingerprinting).

FIGURE 13.22

High-temperature enzymes
Yellowstone hot springs contain thermophilic (heat loving) bacteria whose enzymes remain active at temperatures close to the boiling point of water.

Source: © Corbis Digital Stock.

13.7 CONTROL OF ENZYME-CATALYZED REACTIONS

The rate of a chemical reaction, whether enzyme catalyzed or not, is a measure of how quickly reactants are consumed or products are formed. As we saw in Section 6.7, raising the concentration of substrate is one way to increase the rate of a reaction. This also works for enzyme-catalyzed reactions, but only up to a point. Once the substrate concentration reaches levels high enough that an enzyme is working at top speed, adding more substrate will not increase the rate any further.

Many enzymes fall into one of two well-studied groups with respect to varying substrate concentrations: Michaelis–Menten enzymes and allosteric enzymes.

Michaelis–Menten Enzymes

The generalized enzyme reaction shown in Figure 13.20 is for what is often known as a **Michaelis–Menten enzyme**, named after two of the scientists who carried out some of the early studies on enzyme reactions. While many enzyme reactions are much more complex, the Michaelis–Menten enzyme system is a good one for learning how some enzyme reactions are controlled. As shown in Figure 13.20, there are two steps to the process. In step one, enzyme (E) binds substrate (S) to form an enzyme–substrate complex (ES). In step two, substrate is transformed into product, which is released (ES $\rightarrow$ E + P).

Step one is described by $\boldsymbol{K_M}$, the Michaelis constant, a measure of *the strength of the attraction between an enzyme and a substrate.* Step two is described by $\boldsymbol{V_{max}}$, the *maximum velocity* (reaction rate) that a given concentration of enzyme can produce. K_M and V_{max} are used to characterize enzymes in much the same way that melting points and boiling points are used to characterize simpler organic compounds. Each enzyme has a unique K_M and V_{max} under a particular set of reaction conditions (pH, temperature, etc.).

Consider, for example, a particular alcohol dehydrogenase found in human liver cells. This enzyme, which displays relative specificity (Figure 13.19), catalyzes the oxidation of a wide range of primary and secondary alcohols. At pH 10 and a temperature of 25°C, the K_M values for 1-octanol ($CH_3CH_2CH_2CH_2CH_2CH_2CH_2CH_2OH$), ethanol ($CH_3CH_2OH$), and methanol ($CH_3OH$) vary 2000-fold. The enzyme binds 1-octanol 200 times more effectively than ethanol, and it binds ethanol 10 times more effectively than methanol. V_{max} values for these substrates vary by a factor of 20. Once substrate is attached to the enzyme, ethanol and 1-octanol are converted to product and released 20 times faster than methanol.

To better understand K_M and V_{max}, let us use an example. Suppose that you are given a box of oranges and are asked to peel all of them. You reach in the box with your left hand and pick up an orange. With your right hand, you peel it. Then you set the peeled orange down and reach into the box with your left hand to start the process over again.

How does this relate to K_M and V_{max}? As shown in Figure 13.23*a*, if you imagine yourself as being the enzyme and the orange as being the substrate, then picking up the orange is related to K_M, formation of the enzyme–substrate complex. Peeling the orange and putting it down corresponds to V_{max}, the enzyme–substrate complex being converted to enzyme plus product.

Inhibition

■ Some enzyme inhibitors are reversible. Others are irreversible.

Any *compound that reduces an enzyme's ability to act as a catalyst* is called an **inhibitor** and the effect of an inhibitor can be irreversible or reversible. **Irreversible inhibitors** react with enzymes, whether Michaelis–Menten enzymes or not, to *produce a protein that is not a catalyst.* The ability of aspirin to reduce pain, fever, and inflammation is due to its irreversible inhibition of COX enzymes responsible for the production of prostaglandins (Section 8.6). The transfer of an acetyl group from aspirin to a serine side chain at the COX active site renders the enzyme useless (Figure 13.24).

Reversible inhibitors bind to an enzyme and interfere with its catalytic ability, but *the inhibitory effect is not permanent* because the inhibitors are loosely bound to the enzyme and can dissociate, restoring the enzyme to its original state. For Michaelis–Menten enzymes, competitive and noncompetitive inhibition are two important types of reversible inhibition.

A **competitive inhibitor** *competes with a substrate at the active site of an enzyme.* These inhibitors usually have a structure similar to that of the substrate and are mistaken for a substrate by an enzyme. Some drugs act as competitive inhibitors, including allopurinol, which inhibits an enzyme involved in the production of uric acid, and the antibiotic sulfanilamide, which inhibits bacterial production of folate, a growth factor (Figure 13.25).

As we have seen, catalysis by a Michaelis–Menten enzyme can be divided into two steps. The first step, described by K_M, involves the binding of substrate to the enzyme. The second step, described by V_{max}, is the subsequent conversion of substrate into product, followed by its release (Figure 13.23*a*). If either of these steps is affected by an inhibitor, the rate of an enzyme-catalyzed reaction will slow.

To explain how a competitive inhibitor affects the values of K_M and V_{max}, let us return to the example used above. You are given another box of oranges, but this time it also contains orange plastic balls that look and feel very much like real oranges. If you pick up

FIGURE 13.23

Competitive and noncompetitive inhibition
(*a*) In the absence of an inhibitor, enzyme binds substrate to form an enzyme–substrate complex ($E + S \rightleftharpoons ES$). After product (P) forms, it leaves the enzyme ($ES \rightarrow E + P$). (*b*) *Competitive inhibition.* E binds competitive inhibitor (I) at the active site to form an enzyme–inhibitor complex ($E + I \rightleftharpoons EI$), which cannot react to form product. The presence of I affects the binding of S to E, and K_M is affected. If ES forms, it reacts the same in the presence or the absence of I, so V_{max} is unaffected.
(*c*) *Noncompetitive inhibition.* E binds noncompetitive inhibitor (I) at a different binding site than substrate. I has no effect on the binding of S (K_M is not affected) but when I is attached to E, the reaction cannot proceed (V_{max} is affected).

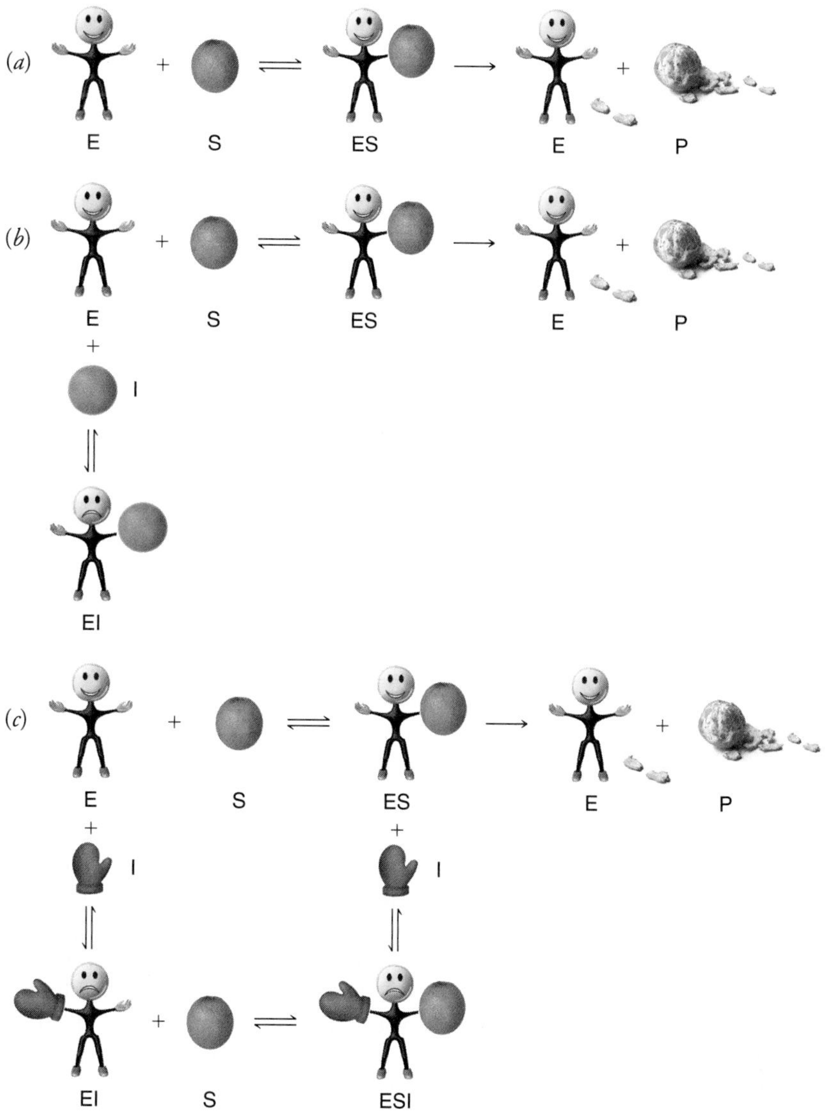

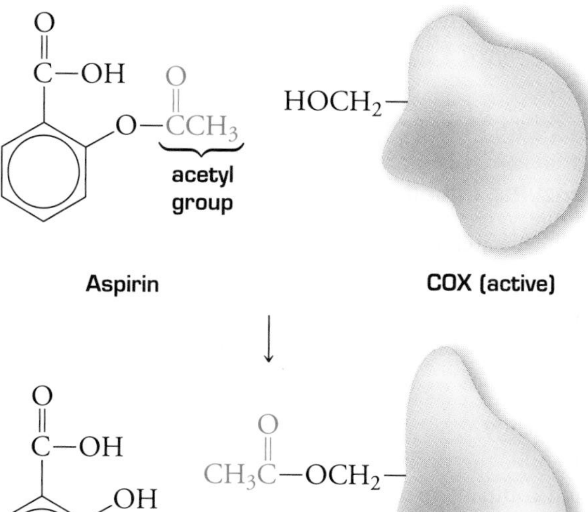

FIGURE 13.24

Irreversible inhibition
An irreversible enzyme inhibitor permanently deactivates an enzyme. Transfer of an acetyl group from aspirin to a COX enzyme, for example, causes irreversible inhibition. Since COX enzymes catalyze the production of prostaglandins responsible for pain, fever, and inflammation, their irreversible inhibition by aspirin blocks these symptoms.

FIGURE | 13.25

Competitive inhibitors
(*a*) Allopurinol is a competitive inhibitor of an enzyme involved in the formation of uric acid. Allopurinol resembles the substrate xanthine (*b*) The antibiotic sulfanilamide is a competitive inhibitor of an enzyme that catalyzes the formation of a growth factor in some harmful bacteria. Sulfanilamide resembles the substrate *p*-aminobenzoate ion.

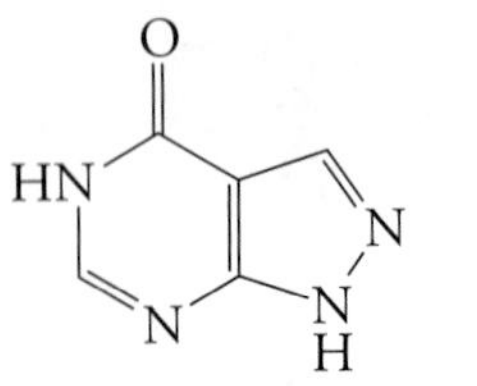

Allopurinol

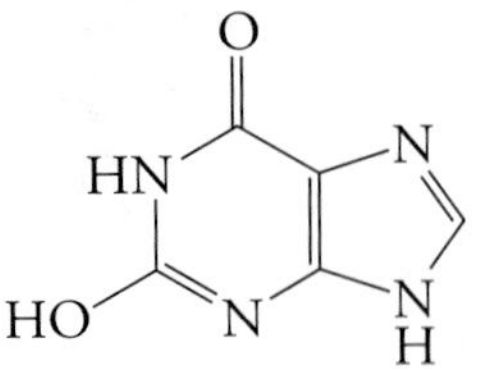

Xanthine

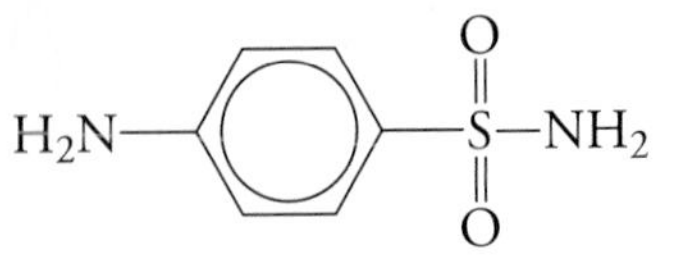

Sulfanilamide

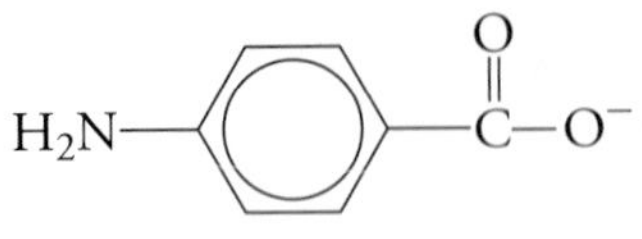

***p*-Aminobenzoate ion**

a ball instead of an orange, you must put the ball back in the box and try again. This slows you down (Figure 13.23*b*). In terms of enzyme kinetics, the presence of a competitive inhibitor (the orange balls) interferes with the formation of the enzyme–substrate complex (your ability to pick up an orange with your left hand), so K_M is changed. The competitive inhibitor does not interfere with conversion of the enzyme–substrate complex into enzyme plus product (your ability to peel the orange with your right hand, once you are holding an orange in your left hand), so V_{max} is unaffected. To summarize, *a competitive inhibitor affects* K_M, *but not* V_{max}.

In Michaelis–Menten enzymes controlled by **noncompetitive inhibition**, *inhibitor and substrate attach to different sites on the enzyme.* The separate binding sites are a reflection of the fact that noncompetitive inhibitors usually have a very different structure than substrates. Some heavy metal ions, including Pb^{2+} and Hg^{2+}, are noncompetitive inhibitors. The toxicity of these common environmental contaminants is primarily due to their reversible interactions with the thiol groups (—SH) present in many enzymes.

Once again, we will return to orange peeling to explain the effects of a noncompetitive inhibitor. Suppose that you are given a box containing only oranges. This time, at random intervals, a person standing next to you puts a mitten onto your right hand. This slows you down because you need your right hand to do the peeling (Figure 13.23*c*). In terms of enzyme kinetics, in the presence of a noncompetitive inhibitor (the mitten on your right hand), an enzyme's ability to bind substrate (your ability to pick up an orange with your left hand) is unaffected, so K_M is unchanged. The noncompetitive inhibitor slows down the conversion of the enzyme–substrate complex into enzyme plus product (your ability to peel the orange with your right hand), so V_{max} is changed. To summarize, *a noncompetitive inhibitor affects* V_{max}, *but not* K_M.

SAMPLE PROBLEM 13.4

Enzyme inhibitors

Which statements accurately describe a competitive inhibitor?

a. It denatures the enzyme.
b. It changes V_{max} for the enzyme.
c. It catalyzes the hydrolysis of the normal substrate for the inhibited enzyme.
d. It changes K_M for the enzyme.

Strategy

To solve this problem it would be useful to reread the paragraphs above on denaturation and competitive inhibition.

Solution

d. Competitive inhibitors of an enzyme compete with the substrate at the active site. K_M changes and V_{max} does not. Competitive inhibitors do not denature an enzyme.

PRACTICE PROBLEM 13.4

Which statements accurately describe a noncompetitive inhibitor?

a. It denatures the enzyme.
b. It changes V_{max} for the enzyme.
c. It catalyzes the hydrolysis of the normal substrate for the inhibited enzyme.
d. It changes K_M for the enzyme.

Allosteric Enzymes

■ Allosteric enzymes display cooperativity and are regulated by positive and negative effectors.

Allosteric enzymes, the second category of enzymes that we will consider, play a key role in controlling the thousands of reactions that maintain life. Typically, allosteric enzymes consist of more than one polypeptide chain. While many Michaelis–Menten enzymes also have quaternary structure, what distinguishes allosteric enzymes is their display of **cooperativity**, in which *the binding of substrate to the active site on one subunit affects the binding of substrate at other subunits.*

Allosteric enzymes *are regulated by ions or molecules* called **allosteric effectors.** Effectors, which bind at sites different from substrate binding sites, usually have very different structures than the substrates for the enzymes that they regulate. The attachment of an allosteric effector to a binding site alters the three-dimensional shape of enzyme subunits and influences the ability of the enzyme to bind substrates and catalyze their conversion to products.

A **positive effector** *enhances substrate binding and increases the speed of the enzyme-catalyzed reaction*, while a **negative effector** *reduces binding and slows the reaction.* Allosteric control is an important form of self-regulation for an enzyme because the effectors act as "on" and "off" switches.

The terms V_{max} and K_M were developed to describe the behavior of Michaelis–Menten enzymes and are not used when dealing with allosteric enzymes. K_M, for example, is not useful for describing allosteric enzymes because the interaction of enzyme with substrate varies depending on how many other substrate molecules or which effectors are already attached to the enzyme. Similarly, application of the terms "competitive inhibition" and "noncompetitive inhibition" do not translate well to allosteric enzymes.

Some allosteric enzymes are involved in **feedback inhibition**. In this form of inhibition the product of a metabolic pathway (a series of reactions) *prevents its own overproduction by inhibiting one or more of the enzymes responsible for its synthesis.* This product usually functions as a negative effector and it targets an allosteric enzyme placed early in the metabolic pathway. In the process, the buildup of intermediates and final products is slowed (Figure 13.26).

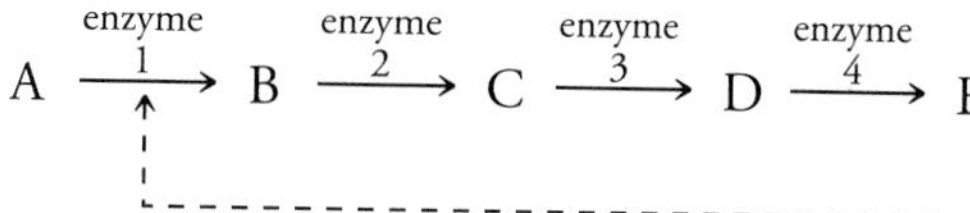

■ **FIGURE 13.26**

Feedback inhibition

In the metabolic pathway (series of reactions) shown here, compound A is converted into compound E. The pathway is under feedback control, because product E is a negative effector of enzyme 1. When the concentration of E is low, enzyme 1 functions normally and the metabolic pathway produces E. As concentrations of E rise, enzyme 1 is inhibited and production of E drops.

Covalent Modification

The activity of many enzymes is controlled by **covalent modification**, *the enzyme-catalyzed making or breaking of covalent bonds within the enzyme.* Phosphorylation and dephosphorylation, common forms of covalent modification, involve the *formation of phosphate esters (phosphorylation) or their hydrolysis (dephosphorylation).* Phosphorylation is catalyzed by a class of enzymes called kinases and involves the transfer of a phosphate group from ATP (Chapter 14) to the —OH group on the side chain of a serine, threonine, or tyrosine residue in a protein. Dephosphorylation is catalyzed by phosphatase enzymes. Depending on the enzyme in question, one form (phosphorylated or dephosphorylated) will be a better catalyst than the other.

Glycogen phosphorylase, a dimeric enzyme involved in the breakdown of glycogen, is controlled by phosphorylation/dephosphorylation. Glycogen phosphorylase *b* is the less active form of this enzyme (Figure 13.27). When there is a metabolic demand for the breakdown of glycogen, a kinase produces the highly active form of glycogen phosphorylase (the *a* form) by phosphorylating the *b* form. When the breakdown of glycogen needs to be slowed, a phosphatase converts the *a* form into the less active *b* form. In addition to being controlled by this covalent modification, glycogen phosphorylase *a* and *b* are allosteric enzymes, each of which has different positive and negative effectors.

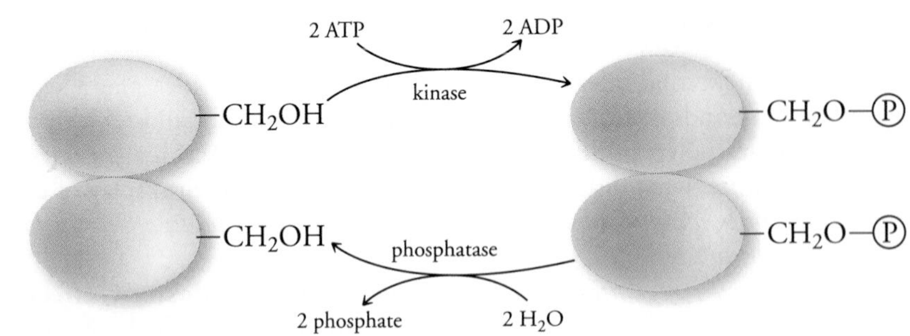

FIGURE 13.27

Phosphorylation and dephosphorylation
Some enzymes are controlled by this form of covalent modification, which involves the addition of a phosphate group (phosphorylation) to the side chain of an amino acid residue or its removal (dephosphorylation). Addition of two phosphate residues (Ⓟ) switches glycogen phosphorylase on by converting the less active form (glycogen phosphorylase *b*) into the more active form (glycogen phosphorylase *a*). Removal of the phosphate residues by hydrolysis reverses the process, switching the enzyme off.

- Zymogens are inactive enzyme precursors.

Another type of covalent modification used to control enzyme action is the formation of **zymogens**, *inactive enzyme precursors*. Trypsin and chymotrypsin, digestive enzymes that catalyze the hydrolysis of peptide bonds in proteins, are examples of enzymes that begin as zymogens. If produced as active enzymes, they would damage the pancreatic cells that produce them. Instead, each is initially produced as an inactive precursor (trypsinogen and chymotrypsinogen, respectively), neither of which is activated (converted to an active protein) until it has entered the small intestine. The premature activation of trypsinogen and chymotrypsinogen within the pancreas leads to pancreatitis, a potentially fatal disease.

The activation of zymogens usually involves the hydrolysis of peptide bonds, followed by changes in conformation that either create or expose an active site (Figure 13.28). The polypeptide chain of trypsinogen, which has 246 amino acid residues, is cut by the intestinal

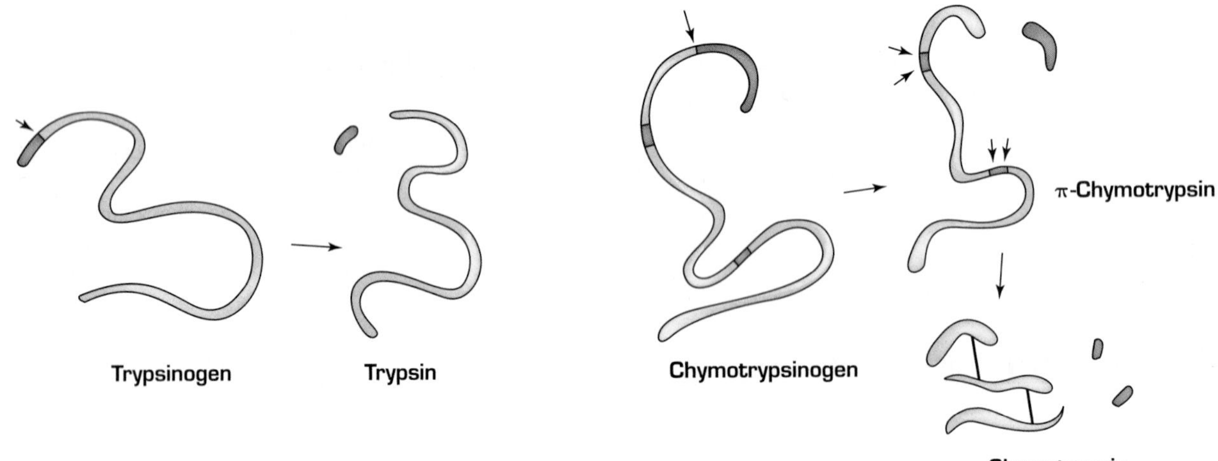

FIGURE 13.28

Zymogens
The digestive enzymes trypsin and chymotrypsin are initially formed as zymogens (inactive enzyme precursors). (*a*) Removal of a hexapeptide converts trypsinogen (inactive) into trypsin (active). (*b*) Removal of a polypeptide fragment (15 amino acid residues) converts chymotrypsinogen (inactive) into π-chymotrypsin (active), which then cuts itself into smaller polypeptide fragments to become chymotrypsin (active).

enzyme enteropeptidase between residues 6 and 7 (counting from the N-terminus), releasing a hexapeptide. The remaining polypeptide chain takes a conformation that is active trypsin. Once formed, trypsin can catalyze the activation of trypsinogen and chymotrypsinogen molecules.

The polypeptide chain of chymotrypsinogen contains 245 amino acid residues. Trypsin hydrolyzes the peptide bond in chymotrypsinogen between residues 15 and 16 to produce the enzyme π-chymotrypsin. π-Chymotrypsin then hydrolyzes several of its own peptide bonds, removing two dipeptides, consisting of residues 14 and 15 and residues 147 and 148. The remaining three polypeptide chains, held to one another by disulfide bridges, are the active form of chymotrypsin.

HEALTH Link — Proteins in Medicine

Many diseases, including those of the heart, liver, and kidney, can be diagnosed by the presence of higher than normal concentrations of particular proteins in blood serum. One example is the detection of a myocardial infarction (heart attack), injury to heart muscle that is usually caused by a restricted flow of blood to the heart. When heart cells are damaged, they dump their contents into the blood, which leads to elevated serum levels of heart cell proteins. These include creatine kinase and troponin.

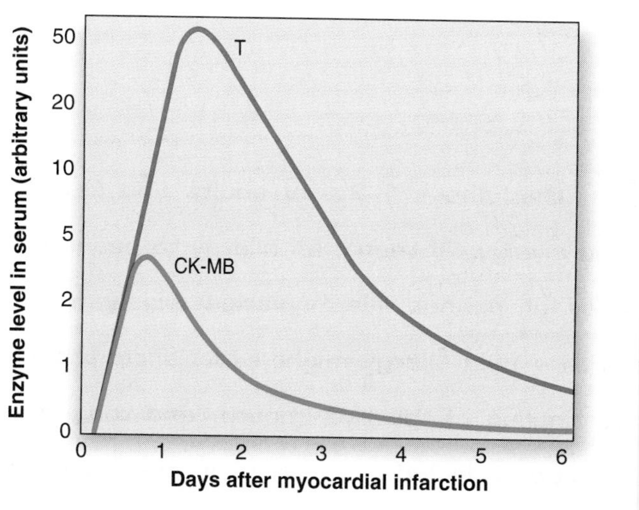

Creatine kinase catalyzes the reversible formation of creatine and ATP from phosphocreatine and ADP. This enzyme is especially important in muscle, brain, heart, and other cells that consume a great deal of energy. By being able to catalyze the rapid production of ATP, creatine kinase can prevent rapid decreases in the concentration of this compound that is so vitally important to many of the reactions that take place within cells (Section 15.2). Creatine kinase has quaternary structure, and two polypeptide chains, each with a different primary structure, are the building blocks used to form the active enzyme. One of these polypeptides is designated M (for muscle) and the other is called B (for brain). In skeletal muscle, the predominant form of creatine kinase is CK-MM (creatine kinase produced from two M polypeptide chains). In brain cells, most creatine kinase is of the CK-BB variety (produced from two B polypeptide chains). In heart cells, CK-MB predominates (produced from one M and one B polypeptide chain).

Enzyme level in serum (arbitrary units)
50
20
10
5
2
1
0
T
CK-MB
0 1 2 3 4 5 6
Days after myocardial infarction

FIGURE 13.29

Detecting heart attacks
Following a heart attack (myocardial infarction) the concentration of heart cell proteins found in blood serum increases, followed by a return to normal levels. Creatine kinase (CK-MB) and heart-specific troponin I (T) levels rise and fall over different time periods.

When someone has a heart attack, his/her serum level of CK-MB begins to rise within a few hours, reaching a maximum within about a day (Figure 13.29). Elevated levels of CK-MB do not result only from myocardial infarction, however. For example, because some CK-MB is present in skeletal muscle cells, this enzyme's appearance in blood serum can be the result of muscle injury. Due to this uncertainty in interpreting CK-MB results, additional factors, such as medical history and the results of electrocardiograms (ECGs), are taken into account when diagnosing a heart attack.

The presence of heart-specific troponin I in blood serum is a much better indicator than CK-MB that a myocardial infarction has occurred. Troponin, a protein that binds to muscle fibers and helps to control muscle contraction, consists of three different polypeptide chains. One of these chains, called troponin I, has different primary structures in heart muscle and in skeletal muscle. The presence of the heart-specific troponin I in blood serum is a clear indication that damage to the heart has taken place.

The therapeutic uses of enzymes vary widely and diseases treated using enzymes include cancer, blood clotting disorders, genetic defects, inflammation, digestive problems, and kidney failure. The treatment of some forms of leukemia with asparaginase provides a specific example.

Asparaginase catalyzes hydrolysis of the side chain of the amino acid asparagine (Asn), producing aspartic acid (Asp) and ammonia (NH_3).

$$\underset{\textbf{Asn}}{H_3\overset{+}{N}CH(CH_2C(=O)NH_2)C(=O)O^-} \xrightarrow{\text{asparaginase}} \underset{\textbf{Asp}}{H_3\overset{+}{N}CH(CH_2C(=O)O^-)C(=O)O^-}$$

Normal cells can synthesize all of the asparagine that they require to manufacture proteins. Some leukemia cells, however, cannot synthesize enough asparagine to meet their needs, so their survival depends on the asparagine that is normally supplied in blood.

In one leukemia treatment, asparaginase is administered intravenously in conjunction with one or more chemotherapy drugs. The asparaginase destroys the asparagine in the blood, which deprives the cancer cells of this essential nutrient. Since the injected asparaginase cannot pass through cell membranes, normal cells do not lose their self-made supply of asparagine. Patients undergoing this particular treatment must be closely monitored because the asparaginase, which is obtained from bacteria (*Escherichia coli*), can sometimes cause allergic reactions.

It is estimated that about 1–2% of adults and 5–6% of children have at least one food allergy. Of these, an allergy to peanuts is the most common. The severity of peanut allergy varies widely from person to person, and those with a mild allergy might experience itchiness, a rash, stomach cramps, or tingling of the lips, mouth, and tongue after eating peanuts. Severe reactions include swelling of the face, difficulty in breathing, anaphylactic shock, and death. Some children and adults are so allergic to peanuts that just touching someone who has eaten peanuts can be enough to cause a reaction.

Three proteins (Ara h1, Ara h2, and Ara h3) present in peanuts are the cause of this allergy. These three proteins belong to a class of storage proteins and one of them, Ara h2, is a trypsin inhibitor. By inhibiting trypsin, an enzyme that catalyzes the hydrolysis of peptide bonds, Ara h2 prevents stored proteins from being broken down.

Allergies are an unusually strong response of the immune system to compounds (in this case, Ara h1, h2, and h3) that it identifies as outsiders. While the biochemistry of this response is quite complex, it results in the release of large amounts of histamine, prostaglandins, and other inflammation-producing molecules.

summary of objectives

1 Describe the structure of amino acids and the system used to classify amino acids.

Amino acids contain an amino group and a carboxyl group. In **α-amino acids**, the amino group is attached to the carbon atom α to the carboxyl group. Because both amines and carboxylic acids have acidic and basic forms, amino acid structure is pH dependent. The nature of the side chain (the atom or group of atoms also attached to the α carbon) is used to classify amino acids as one of the following: **nonpolar**, **polar-acidic**, **polar-basic**, or **polar-neutral**.

2 Distinguish between oligopeptides, polypeptides, and proteins and describe the bond that joins amino acid residues in these compounds.

Amino acids are the building blocks used to produce **oligopeptides** (2–10 amino acid residues) and **polypeptides** (more than 10 residues). A **protein** is a polypeptide with more than 50 amino acid residues. In an oligo- or polypeptide the amino acid residues are joined to one another by **peptide** (amide) **bonds**.

3 Define primary, secondary, tertiary, and quaternary structure. Name the covalent and noncovalent forces responsible for each level of structure.

Primary structure refers to the sequence of amino acid residues in a peptide or protein and **secondary structure** to common structural features (**α-helix** and **β-sheet**). **Tertiary structure** refers to the overall three-dimensional shape of a protein, including secondary structural features (**globular proteins** are spherical and water soluble, as opposed to **fibrous proteins**, which are long fibers that are tough and water insoluble). **Quaternary structure** is the combination of more than one polypeptide chain that may be required to produce an active protein. Tertiary structure can also include **prosthetic groups**—nonpeptide components attached to the polypeptide. Secondary structure is held together by hydrogen bonds between atoms in the polypeptide **backbone**, tertiary structure by interactions between side chains (hydrogen bonding, the hydrophobic effect, salt bridges, and disulfide bridges), and quaternary structure by the same forces responsible for tertiary structure.

4 Explain what is meant by the term denaturation, and list some of the ways to denature a protein.

Denaturation is any disruption of the noncovalent forces and disulfide bonds responsible for maintaining the native (biologically active) form of a protein. Changes in temperature or pH, agitation, or the addition of soaps or detergents are among the ways that a protein can be denatured.

5 Distinguish between absolute specificity, relative specificity, and stereospecificity.

Enzymes can display **absolute specificity** (limited to one particular substrate) or **relative specificity** (can use substrates containing the same functional group or similar structures). The term **stereospecificity** applies to enzymes that react with or produce a particular stereoisomer.

6 Identify the steps required for a typical Michaelis–Menten enzyme to convert a substrate into a product. Explain how

competitive and noncompetitive inhibitors affect these enzymes, and distinguish between reversible and irreversible inhibitors.

For catalysis by a **Michaelis–Menten** enzyme to occur, substrate (S) must bind to enzyme (E) and produce an enzyme–substrate complex (ES). Once the reaction has taken place, the product (P) dissociates and the enzyme is restored to its original form. Overall, these steps are summarized as $E + S \leftrightharpoons ES \rightarrow E + P$. $\boldsymbol{K_M}$, which is related to the first step in catalysis ($E + S \leftrightharpoons ES$), is a measure of how well the enzyme and substrate interact, and $\boldsymbol{V_{max}}$, related to the second step ($ES \rightarrow E + P$), is the maximum velocity (reaction rate) that a given concentration of enzyme can produce. **Competitive inhibitors** of Michaelis–Menten enzymes resemble the substrate, bind to the same site on the enzyme surface, change the value of K_M, and have no effect on V_{max}. **Noncompetitive inhibitors** do not resemble the substrate, bind to a different site, change V_{max}, and leave K_M unchanged. Unlike the effect of **irreversible inhibitors** (enzyme function is permanently destroyed), the action of competitive and noncompetitive inhibitors can be reversed.

7 Describe allosteric enzymes and the role that effectors play in the ability of these enzymes to catalyze reactions.

Allosteric enzymes usually consist of more than one polypeptide chain, and the binding of substrate to the active site on one subunit affects the binding of substrate at other subunits. Allosteric enzymes are controlled by **positive effectors** (increase reaction velocity) and **negative effectors** (decrease reaction velocity).

8 Explain what is meant by feedback inhibition and covalent modification.

These terms relate to the control of enzyme reactions. In **feedback inhibition** the product of a metabolic pathway prevents its own overproduction by inactivating one or more of the enzymes responsible for its synthesis. In **covalent modification** an enzyme is switched "on" or "off" by the making or breaking of covalent bonds to or within the enzyme. Phosphorylation/dephosphorylation and **zymogens** are examples of covalent modification.

END OF CHAPTER PROBLEMS

Answers to problems whose numbers are printed in color are given in Appendix D. More challenging questions are marked with an asterisk. Problems within colored rules are paired. **ILW** = Interactive Learning Ware solution is available at *www.wiley.com/college/raymond.*

13.1 Amino Acids

13.1 Draw methionine as it would appear at each of the following pHs.
a. pH 1
b. pH 7
c. pH 14

13.2 Draw proline as it would appear at each of the following pHs.
a. pH 1
b. pH 7
c. pH 14

13.3 **a.** What is the net charge on arginine at pH 1?
b. What is the net charge on arginine at pH 14?

13.4 **a.** What is the net charge on valine at pH 1?
b. What is net charge on valine at pH 14?

13.5 Using Fischer projections, draw each amino acid as it would appear at pH 7.
a. L-isoleucine
b. D-aspartic acid
c. L-tyrosine
d. D-phenylalanine

13.6 Using Fischer projections, draw each amino acid as it would appear at pH 7.

a. D-methionine
b. L-leucine
c. L-lysine
d. D-asparagine

13.7 Using a Fischer projection, draw each amino acid from Problem 13.5 as it appears at pH 1.

13.8 Using a Fischer projection, draw each amino acid from Problem 13.6 as it appears at pH 1.

13.9 Two of the amino acids in Table 13.1 have two chiral carbon atoms. Which ones?

13.10 One of the amino acids in Table 13.1 has no chiral carbon atoms. Which one?

*__13.11__ Monosodium glutamate (MSG), used to enhance the flavor of certain foods, carries a net charge of zero. It can be formed by reacting glutamic acid as it appears at low pH (below) with sufficient NaOH. Draw MSG by showing the molecule below as it would appear at pH 7, and then attaching Na^+ to the side chain.

$$H_3\overset{+}{N}CH(CH_2CH_2COOH)C(=O)OH$$

13.12 One metabolic fate of glutamic acid is its conversion into γ-aminobutyric acid. Draw γ-aminobutyric acid. [Hint: A gamma (γ) carbon atom is attached to a beta (β) carbon atom.]

*__13.13__ Other amino acids than the twenty common ones shown in Table 13.1 appear in proteins. One of these, selenocysteine (Sec), is identical to cysteine except that the sulfur atom in Cys is replaced by a selenium atom.

a. The side chain in Sec is as acidic as the side chain in Asp. Draw Sec as it would appear at pH 7.
b. To which classification of amino acids (nonpolar, polar-acidic, polar-basic, polar-neutral) does Sec belong?

*__13.14__ Pyrrolysine is a naturally occurring amino acid used by some bacteria.

$$\overset{+}{N}H_3CH((CH_2)_4NH-C(=O)-C_4H_6N(CH_3))C(=O)-O^-$$

Pyrrolysine

a. Hydrolysis of the amide bond in pyrrolysine yields which amino acid?
b. Assuming that the side chain in pyrrolysine is as basic as the side chain in arginine, draw pyrrolysine as it would appear at pH 7. Recall that amide nitrogen atoms do not act as bases.
c. To which classification of amino acids (nonpolar, polar-acidic, polar-basic, polar-neutral) does pyrrolysine belong?

13.2 The Peptide Bond

13.15 Which specific class of bonds holds one amino acid residue to the next in the primary structure of a protein? Are these bonds covalent or noncovalent?

13.16 How does an amino acid residue in a protein differ from an amino acid?

13.17 **a.** Is Ala-Phe-Thr-Ser an oligopeptide or a polypeptide?
b. How many peptide bonds does the molecule contain?
c. Which is the N-terminal amino acid?

13.18 Circle the backbone for the pentapeptide shown in Figure 13.4*a*.

13.19 Circle the backbone for the nonapeptide shown in Figure 13.4*b*.

13.20 Name the amino acid residues in the oligopeptide shown in Figure 13.4*a*.

13.21 Name the amino acid residues in the oligopeptide shown in Figure 13.4*b*.

13.3 Peptides, Proteins, and pH

*__13.22__ Draw Gly-Phe-Lys-Lys as it would appear at

a. pH 1
b. pH 7
c. pH 14

*__13.23__ Draw Asp-Ser-Lys-Val as it would appear at

a. pH 1
b. pH 7
c. pH 14

13.24 What is the net charge on Asp-Lys at each pH?

a. pH 1
b. pH 7
c. pH 14

13.25 What is the net charge on Phe-Asp at each pH?

a. pH 1
b. pH 7
c. pH 14

13.4 Protein Structure

13.26 List the primary structure of all possible tripeptides containing one residue each of aspartic acid, phenylalanine, and valine.

13.27 How many tetrapeptides can be produced that contain one residue each of serine, methionine, arginine, and tyrosine?

13.28 The rule "nonpolar in, polar out" assumes that a protein exists in an aqueous environment. What is the folding rule for a protein in a nonpolar environment? Explain.

13.29 In an aqueous environment will the following peptide fragment more likely be buried inside a globular protein or located on its surface? Explain.

```
       O          O           O         O
       ‖          ‖           ‖         ‖
~NHCHC—NHCHC—NHCHC—NHCHC~
     |          |           |          |
CH3CHCH3   CH2        CHCH3   CH2
                |           |          |
            (C6H5)      CH2      CH2
                            |          |
                           CH3       S
                                       |
                                      CH3
```

13.30 In an aqueous environment will the following peptide fragment more likely be buried inside a globular protein or located on its surface? Explain.

```
       O          O           O          O
       ‖          ‖           ‖          ‖
~NHCHC—NHCHC—NHCHC—NHCHC~
     |          |           |           |
 CH2OH    CH2   CH3CHOH     CH2
                |                       |
               C—O⁻                   C—NH2
               ‖                       ‖
               O                       O
```

13.31 List the chemical bonds or forces that are primarily responsible for maintaining

a. the primary structure of a protein

b. the secondary structure of a protein

c. the tertiary structure of a protein

d. the quaternary structure of a protein

13.32 Describe the two types of secondary structure.

13.33 Which amino acids have side chains that can participate in salt bridge formation (ionic bonds) at pH 7?

13.34 Which nonpolar amino acid has a side chain that can participate in hydrogen bonding?

13.35 Explain why many proteins have no quaternary structure.

13.36 What distinguishes a simple protein from a complex protein?

13.37 Why do extracellular proteins tend to contain more disulfide bonds than do intracellular ones?

13.38 Draw structures that show

a. hydrogen bonding between His and Glu side chains

b. hydrogen bonding between Thr and Ser side chains

13.39 Draw structures that show

a. a salt bridge between Asp and Arg side chains

b. hydrophobic interactions between Trp and Val side chains

*__13.40__ Which is more likely to lead to a change in the biological activity of a protein, the replacement of a leucine residue with a valine residue or the replacement of the same leucine residue with a lysine? Explain.

13.5 Denaturation

13.41 What term is used to describe an enzyme whose tertiary structure has been unfolded?

*__13.42__ Denaturing a native protein tends to decrease its water solubility. Why?

13.43 Which is/are true for a denatured globular protein?

a. is biologically inactive

b. contains no peptide bonds

c. has an abnormal primary structure

13.44 Which types of bonds or interactions are disrupted during the denaturation of a protein?

13.45 List some of the ways to denature a protein.

13.6 Enzymes

13.46 Is it possible for an enzyme to show both absolute specificity and relative specificity?

13.47 Is it possible for an enzyme to show both relative specificity and stereospecificity?

*__13.48__ **a.** Draw an energy diagram for the exergonic ($\Delta G < 0$) reaction catalyzed by urease (Figure 13.19*a*). Label reactants, products, activation energy, and ΔG.

b. How will the energy diagram change if the reaction is run in the absence of enzyme?

13.49 The enzyme trypsin, found in the small intestine, operates best at pH 8. Draw a graph for trypsin, similar to that found in Figure 13.21*a*.

13.50 Why does a change in pH usually produce a change in the ability of an enzyme to act as a catalyst?

13.51 A thermophilic bacterium found in a hot spring at Yellowstone Park produces an enzyme with a temperature optimum of 98°C. For this enzyme, draw a graph similar to that found in Figure 13.21*b*. Assume that the enzyme denatures at 100°C.

*__13.52__ The coenzyme NADH is a cofactor but is not a prosthetic group. Explain.

13.7 Control of Enzyme-Catalyzed Reactions

13.53 A change in pH is likely to produce a change in the K_M and V_{max} of an enzyme. Explain why.

13.54 An inhibitor is added to a Michaelis–Menten enzyme. How could you distinguish between the inhibitor being reversible or irreversible?

13.55 An inhibitor is added to a Michaelis–Menten enzyme. How could you distinguish between the inhibitor being competitive or noncompetitive?

13.56 Explain why most competitive inhibitors are structurally similar to a substrate for the enzyme they inhibit, while most noncompetitive inhibitors are not.

13.57 Some bacterial proteins contain D-amino acid residues, which are produced from L-amino acids once the polypeptide chain has been produced. One such case involves the conversion of L-alanine into D-alanine by the enzyme alanine racemase.

a. Using Fischer projections, write a reaction equation for this transformation.

b. To what does the term "racemase" apply? (Hint: See Section 10.9.)

c. Scientists are investigating the possible use of L-fluoroalanine as an antibiotic. By inhibiting alanine racemase, this compound would slow bacterial growth. Is L-fluoroalanine more likely to be a competitive inhibitor or a noncompetitive inhibitor of alanine racemase? Explain.

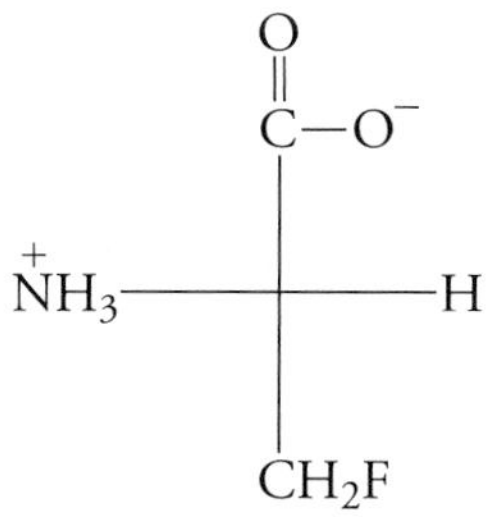

O
||
C—O^-
$\overset{+}{N}H_3$ —+— H
CH_2F

L-fluoroalanine

d. To which classification of amino acids (nonpolar, polar-acidic, polar-basic, polar-neutral) does alanine belong?

e. To which classification of amino acids (nonpolar, polar-acidic, polar-basic, polar-neutral) does fluoroalanine belong?

*****13.58** **a.** To which reaction type (synthesis, decomposition, single replacement, double replacement) does the reaction in Figure 13.24 belong?

b. Due to its action on COX enzymes, aspirin is considered to be a "suicide inhibitor." Explain.

c. If transfer of an acetyl group from aspirin to the side chain of a serine residue in a COX enzyme caused a change in the tertiary structure of the enzyme, which interaction was most likely disrupted: a hydrogen bond, a salt bridge, or a disulfide bridge?

*****13.59** By inhibiting COX enzymes, low doses of aspirin can reduce the risk of heart attack and stroke. However, if aspirin and acetaminophen are taken at the same time, some of the beneficial effects of aspirin are lost. After reviewing Figure 13.23, explain why acetaminophen (a competitive inhibitor of COX enzymes) has this effect.

*****13.60** Celecoxib is an inhibitor of COX-2 (Section 8.6). Given that arachidonic acid (Figure 8.22) is a substrate for COX-2, is celecoxib more likely to be a competitive inhibitor or a noncompetitive inhibitor?

CH_3 CF_3 N N NH_2SO_2

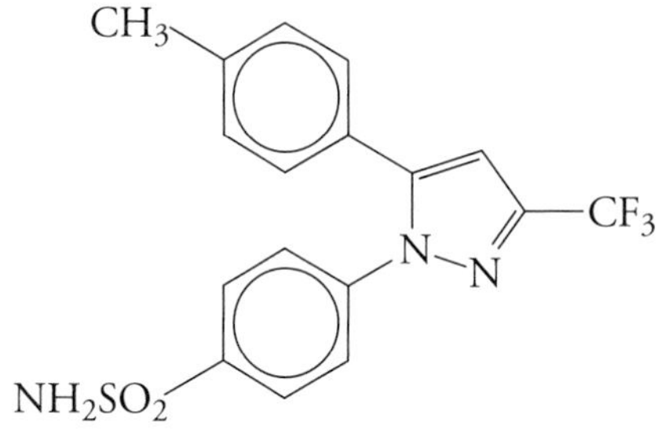

Celecoxib

13.61 Methanol (CH_3OH) is poisonous because, once it is ingested, the enzyme alcohol dehydrogenase catalyzes its oxidation into formaldehyde (CH_2O). This aldehyde damages tissues and often causes blindness. Ethanol (CH_3CH_2OH) given intravenously is a common treatment for methanol poisoning. Alcohol dehydrogenase becomes so "busy" oxidizing ethanol that methanol gets metabolized in a way that does not produce formaldehyde.

a. Which of these two alcohols binds more effectively to alcohol dehydrogenase? (See Section 13.7.)

b. Which of these two alcohols, once bound to the enzyme, is converted to product more quickly?

c. In terms of its ability to act as a substrate for alcohol dehydrogenase, would 1-octanol be as effective as ethanol when it comes to treating methanol poisoning?

d. Besides its ability to act as a substrate, what additional information might you want about 1-octanol before using it to treat methanol poisoning?

13.62 Describe the effect of positive and negative effectors on allosteric enzymes.

13.63 Explain how addition or removal of a phosphate group can be used to control the activity of an enzyme.

13.64 Why are some enzymes produced initially as zymogens?

*****13.65** A person is found to produce trypsinogen molecules that have an abnormal primary structure. The trypsin produced from this zymogen is perfectly normal, both structurally and functionally. How is this possible?

13.66 What is feedback inhibition?

BIOCHEMISTRY*Link* | *Collagen and α-Keratin, Two Fibrous Proteins*

13.67 List one unusual amino acid residue found in collagen. Is this residue present in a newly assembled collagen polypeptide? Explain.

***13.68** Give a molecular explanation for the fact that steak from an old steer tends to be tougher than steak from a young one.

BIOCHEMISTRY*Link* | *Hemoglobin, a Globular Protein*

13.69 Describe the quaternary structure of hemoglobin.

13.70 The binding of oxygen to hemoglobin is cooperative. Explain this term.

HEALTH*Link* | *Proteins in Medicine*

13.71 True or false? If liver cells are the only cells in the human body that produce enzyme *x*, an increase in serum levels of enzyme *x* following a viral infection probably indicates that the infection has resulted in some liver damage. Explain.

13.72 Explain how asparaginase treatment kills certain leukemia cells, while having no impact on normal cells.

***13.73** Would orally administered asparaginase be as effective as intravenously injected asparaginase in reducing serum asparagine levels? Explain.

13.74 a. According to "Young Hearts", what does the adult form of the protein troponin I regulate?

b. How does adult troponin I respond to a heart attack?

c. In mice, how does adult troponin I that has a portion of the fetal troponin I molecule incorporated into its primary structure respond to a heart attack?

THINKING IT THROUGH

13.75 As we will see in the next chapter, some of the information carried by DNA specifies the primary structure of proteins. Can a change in the structure of DNA (a change in the information) result in the altered tertiary structure of a protein that it carries the code for?

13.76 Most drugs that act as reversible enzyme inhibitors are competitive inhibitors. If you were trying to discover a new drug to inhibit a particular enzyme, why would it be easier to find a competitive inhibitor than a noncompetitive one?

INTERACTIVE LEARNING PROBLEMS

ILW **13.77** What is the name of this pentapeptide? (Use the three-letter abbreviations)

$H_2N-CH(CH_2CH_2C(=O)NH_2)-C(=O)-NH-CH(CH_2C(=O)OH)-C(=O)-NH-CH_2-C(=O)-N$ (pyrrolidine ring) $-C(=O)-NH-CH(CH_2$-indole (HN)$)-C(=O)-OH$

ILW **13.78** The structure for serine is the following. The isoelectric point for serine is 5.7.

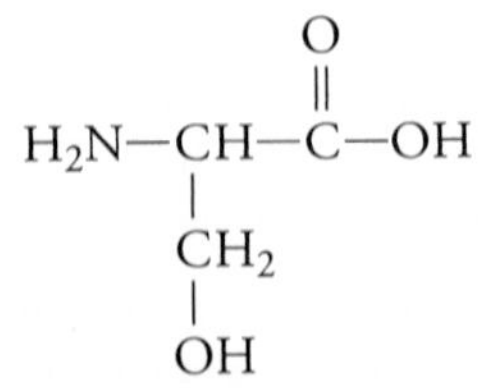

Show the structure of serine at pH 5.7.

ILW **13.79** Which of the following amino acids would you expect to find in a hydrophilic region of a peptide?

a. valine
b. isoleucine
c. glycine
c. alanine
d. aspartic acid

ILW **13.80** An enzyme is a ________ (protein/nucleic acid/polysaccharide) which ________ (raises/lowers) activation energy needed for some biochemical reaction and thereby acts as a ________ .

SOLUTIONS TO PRACTICE PROBLEMS

13.1

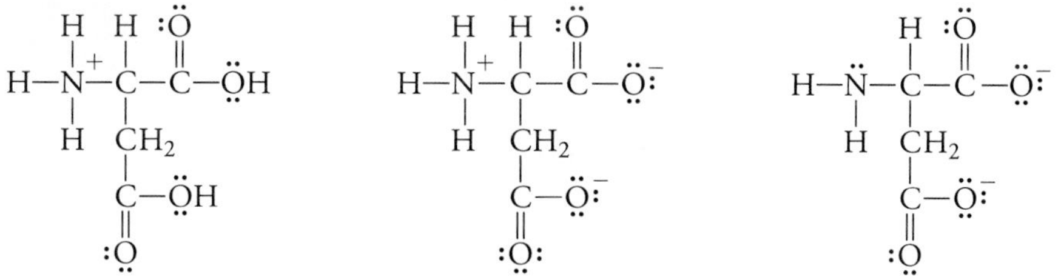

```
   H  H :O:            H  H :O:              H :O:
   |+ |  ||            |+ |  ||              |  ||
H—N—C—C—OH        H—N—C—C—O:⁻         H—N—C—C—O:⁻
   |  |                |  |                |  |
   H  CH2              H  CH2              H  CH2
      |                   |                   |
      C—OH                C—O:⁻               C—O:⁻
      ||                  ||                  ||
     :O:                 :O:                 :O:
```

pH 1	pH 7	pH 14

13.2

```
 +                    O          O           O
                      ||         ||          ||
NH3CHCH2CH2C—NHCHC—NHCH2C—OH
     |                   |
     C—OH             CH2SH
     ||
     O
```

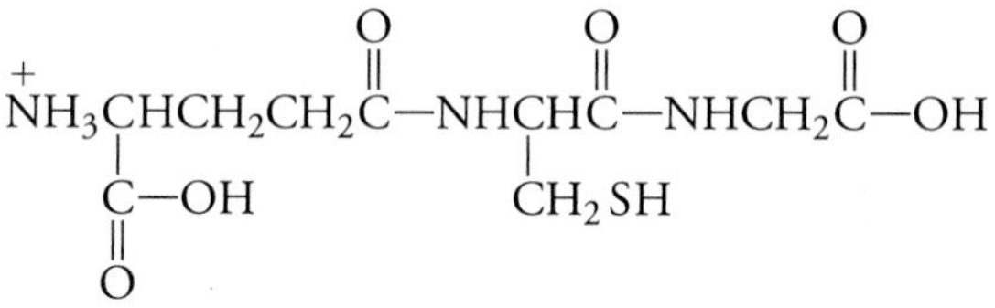

pH 7

13.3 Tertiary and quaternary structure.

13.4 b

While watching the TV news one night you see a story on genetically modified (GM) foods that are currently on the market. According to the report, the first GM food to be sold was Flavr-Savr tomatoes. These tomatoes, available for only a few years during the 1990s, were modified in a way that caused them to ripen slowly. Unlike normal tomatoes, which must be picked while still green if they are to reach stores unspoiled, Flavr-Savrs can be left on the vine until they are ripe and will still reach the market in good shape.

14 NUCLEIC ACIDS

The hereditary information passed from one generation to the next is carried in genes, stretches of deoxyribonucleic acid (DNA) from which chromosomes are made. In a process that involves ribonucleic acid (RNA), the information present in each gene is used to produce a given peptide. By specifying which enzymes and other proteins are made, DNA determines the identity of each living thing. In this chapter we will consider the structure of nucleic acids, the processes by which genetic information is transmitted, and the techniques that scientists have developed for modifying the DNA of various organisms, including those sold as genetically modified foods.

objectives

After completing this chapter, you should be able to:

1. Describe the makeup of nucleosides, nucleotides, oligonucleotides, and polynucleotides.
2. Describe the primary structure of DNA and RNA and the secondary and tertiary structure of DNA.
3. Explain how replication takes place and describe the roles of DNA polymerase in this process.
4. Explain how transcription takes place and describe the role of RNA polymerase in this process.
5. Name the three types of RNA required for the synthesis of proteins and identify the role of each in translation.
6. Describe one way that *E. coli* control the expression of genes.
7. List the steps involved in producing recombinant DNA.
8. Explain the term mutation.

14.1 NUCLEIC ACID BUILDING BLOCKS

■ Nucleotides contain a monosaccharide, phosphate, and a base.

Nucleic acids consist of chains of nucleotide residues. Each **nucleotide** is put together from three building blocks: phosphoric acid, a monosaccharide, and an organic base.

Phosphoric Acid

Phosphoric acid contains three acidic hydrogen atoms so, depending on the pH of a solution, it can be found in one of four different forms—as phosphoric acid (H_3PO_4), dihydrogenphosphate ion ($H_2PO_4^-$), hydrogenphosphate ion (HPO_4^{2-}), or phosphate ion (PO_4^{3-}). At physiological pH (near pH 7), dihydrogenphosphate ($H_2PO_4^-$) and hydrogenphosphate (HPO_4^{2-}) predominate.

Present at low pH | Present at physiological pH | Present at high pH

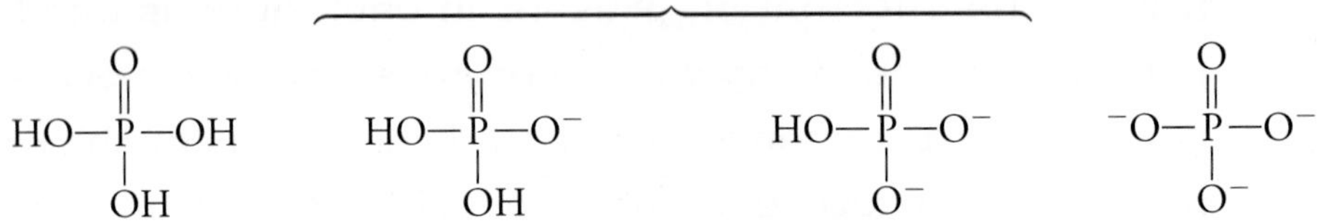

$$\underset{\text{Phosphoric acid}}{HO{-}\overset{\overset{\displaystyle O}{\|}}{\underset{\underset{\displaystyle OH}{|}}{P}}{-}OH} \qquad \underset{\text{Dihydrogenphosphate ion}}{HO{-}\overset{\overset{\displaystyle O}{\|}}{\underset{\underset{\displaystyle OH}{|}}{P}}{-}O^-} \qquad \underset{\text{Hydrogenphosphate ion}}{HO{-}\overset{\overset{\displaystyle O}{\|}}{\underset{\underset{\displaystyle O^-}{|}}{P}}{-}O^-} \qquad \underset{\text{Phosphate ion}}{^-O{-}\overset{\overset{\displaystyle O}{\|}}{\underset{\underset{\displaystyle O^-}{|}}{P}}{-}O^-}$$

The chemistry of phosphoric acid important to the structure of nucleotides is its ability to form phosphate esters, which closely resemble esters made from carboxylic acids (Section 10.4). When phosphoric acid in one of its four forms (we will use phosphate ion in the discussions that follow) reacts with one alcohol molecule, a phosphate monoester forms (Figure 14.1). Reaction with two alcohol molecules produces a phosphate diester. Like phosphoric acid, the mono- and diesters can exist in acid or base forms and carry negative charges at physiological pH.

You may have noticed in the structures shown here that the phosphorus atoms have five covalent bonds. While this does violate the octet rule (Section 3.3), it turns out that this rule does not always apply when atoms of elements beyond the second period (Li, Be, B, C, N, O, F, and Ne) are considered.

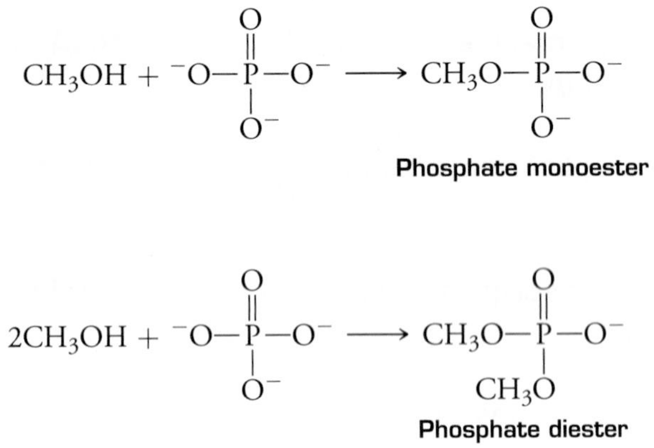

$$CH_3OH + {^-O}{-}\overset{\overset{\displaystyle O}{\|}}{\underset{\underset{\displaystyle O^-}{|}}{P}}{-}O^- \longrightarrow \underset{\text{Phosphate monoester}}{CH_3O{-}\overset{\overset{\displaystyle O}{\|}}{\underset{\underset{\displaystyle O^-}{|}}{P}}{-}O^-}$$

$$2CH_3OH + {^-O}{-}\overset{\overset{\displaystyle O}{\|}}{\underset{\underset{\displaystyle O^-}{|}}{P}}{-}O^- \longrightarrow \underset{\text{Phosphate diester}}{CH_3O{-}\overset{\overset{\displaystyle O}{\|}}{\underset{\underset{\displaystyle CH_3O}{|}}{P}}{-}O^-}$$

FIGURE 14.1 **Phosphate esters**
When phosphoric acid or any of its basic forms (PO_4^{3-} is shown here) reacts with one alcohol molecule, a phosphate monoester forms. Reaction with two alcohol molecules produces a phosphate diester. (The reaction equations are not balanced.)

Monosaccharides

D-Ribose and D-2-deoxyribose are the two monosaccharides incorporated into nucleotide structures. In nucleotides, each appears in its β-furanose form (Section 12.3).

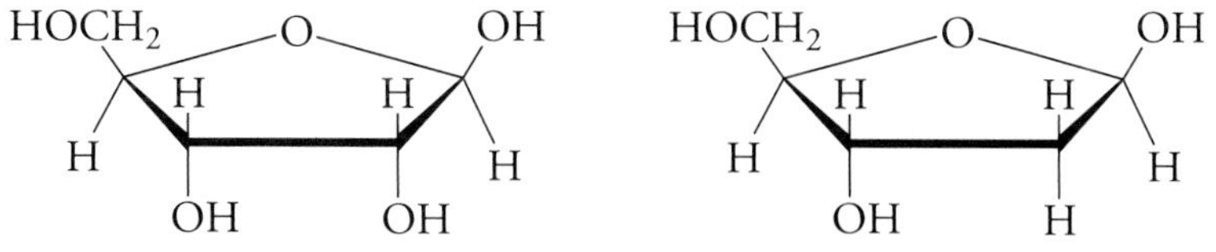

β-D-Ribofuranose **β-D-2-Deoxyribofuranose**

The two reactions of monosaccharides that are most important to nucleotide structure are the formation of phosphate esters and *N*-glycosides. As just described, a phosphate monoester forms when phosphate ion reacts with an alcohol. In nucleotides, the alcohol groups are provided by ribose or 2-deoxyribose (Figure 14.2*a*).

Glycosides form when the hemiacetal group of an α or β monosaccharide anomer reacts with a hydroxyl group from a different molecule (Section 12.4), as is the case when β-2-deoxyribofuranose reacts with methanol (Figure 14.2*b*). A related compound, an ***N*-glycoside**, is formed when this same reaction takes place with an amine instead of an alcohol (Figure 14.2*c*).

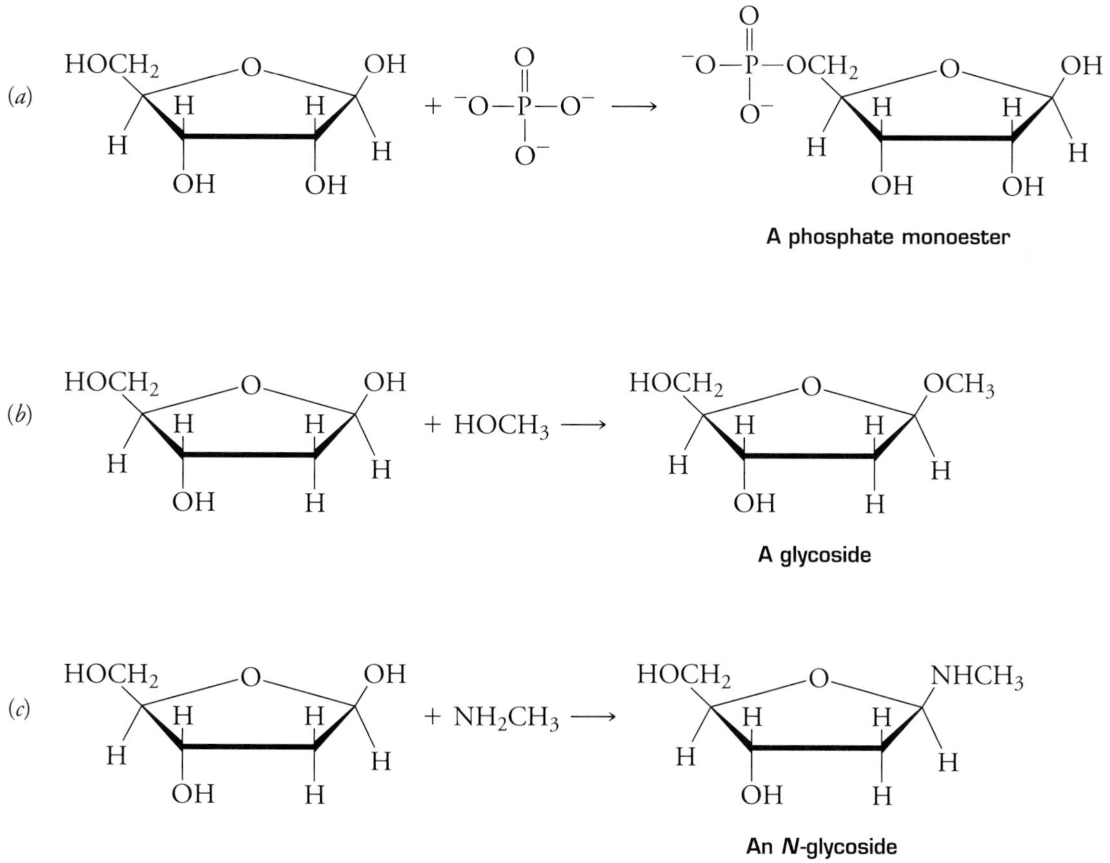

(*a*) + $^-O-P(O)(O^-)-O^-$ ⟶ **A phosphate monoester**

(*b*) + $HOCH_3$ ⟶ **A glycoside**

(*c*) + NH_2CH_3 ⟶ **An *N*-glycoside**

FIGURE 14.2 **Monosaccharide derivatives**
(*a*) The —OH groups of ribose and 2-deoxyribose can react with phosphate to form phosphate monoesters. (*b*) When an alcohol reacts at the hemiacetal carbon atom of a monosaccharide, a glycoside forms. (*c*) When an amine reacts at the hemiacetal carbon atom, an *N*-glycoside forms.

Organic Bases

There are two types of organic bases (amines) that are incorporated into nucleotides: purines and pyrimidines (Figure 14.3). The **major bases** found in nucleotides are the two purine bases, adenine (A) and guanine (G) and the three pyrimidine bases, cytosine (C), thymine (T), and uracil (U).

Besides the five major bases listed above, more than 50 other bases are known (Figure 14.4). These **minor bases** *are formed by the enzyme-catalyzed modification of the major bases* after they have been incorporated into nucleic acid strands.

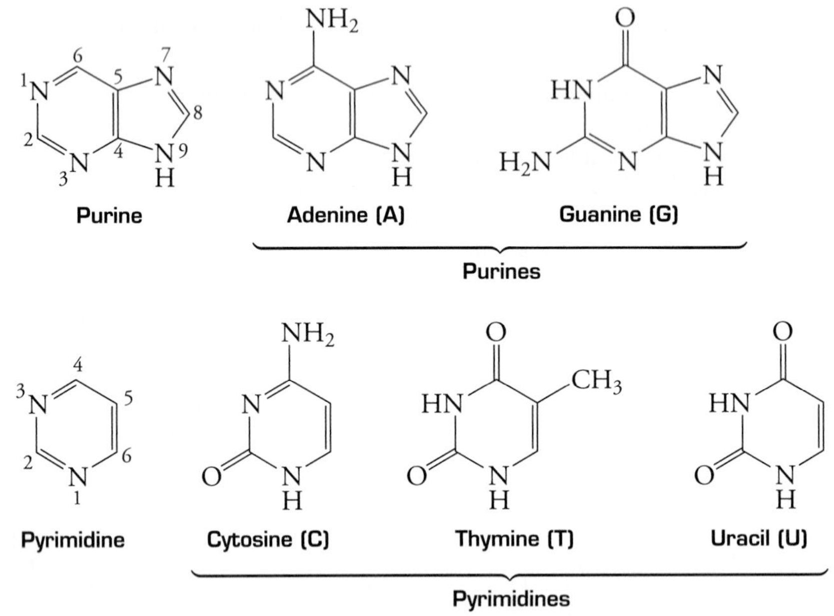

Purine — Adenine (A) — Guanine (G)

Purines

Pyrimidine — Cytosine (C) — Thymine (T) — Uracil (U)

Pyrimidines

FIGURE 14.3

Purines and pyrimidines
Nucleic acids contain the purine bases adenine (A) and guanine (G) and the pyrimidine bases cytosine (C), thymine (T), and uracil (U). T is present in DNA, U in RNA, and A, C, and G in both.

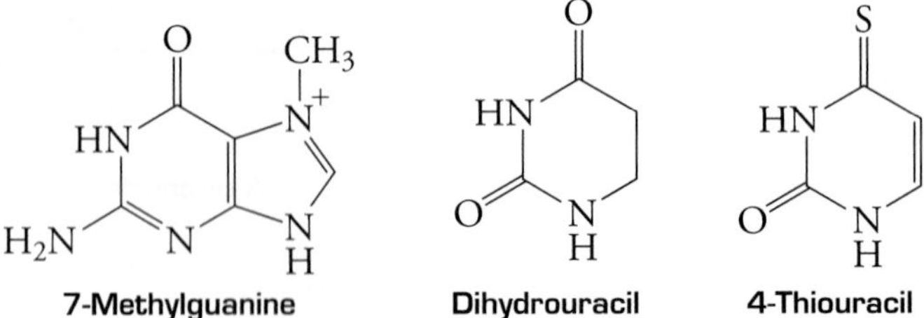

FIGURE 14.4

Minor bases
Some nucleic acids contain minor bases, including the three shown here. Minor bases are formed by modification of bases in existing nucleic acid strands.

Nucleosides

When *ribose or 2-deoxyribose is combined with a purine or pyrimidine base*, a **nucleoside** is formed. Nucleosides containing *ribose* are called **ribonucleosides**, and those containing *2-deoxyribose* are **deoxyribonucleosides**. The monosaccharide and base residues are connected by a β *N*-glycosidic bond involving carbon 1′ of the monosaccharide and one of the nitrogen atoms in the base (nitrogen atom 1 of pyrimidines and nitrogen atom 9 of purines) (Figure 14.5). The enzymes responsible for making nucleosides produce only certain combinations of sugar and base: ribose is paired with A, G, C, or U and 2-deoxyribose is paired with A, G, C, or T.

Ribonucleosides | **Deoxyribonucleosides**

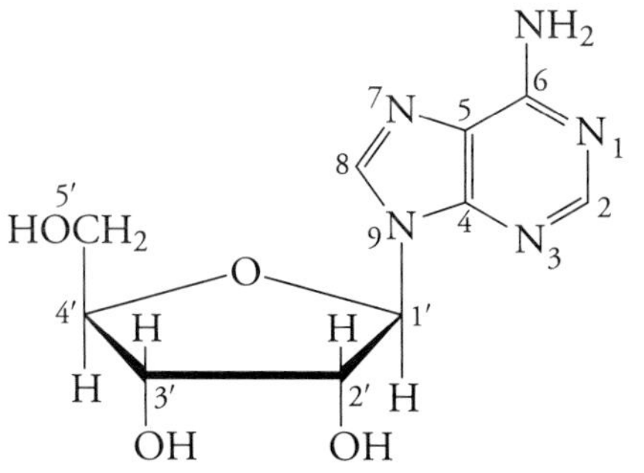

Adenosine

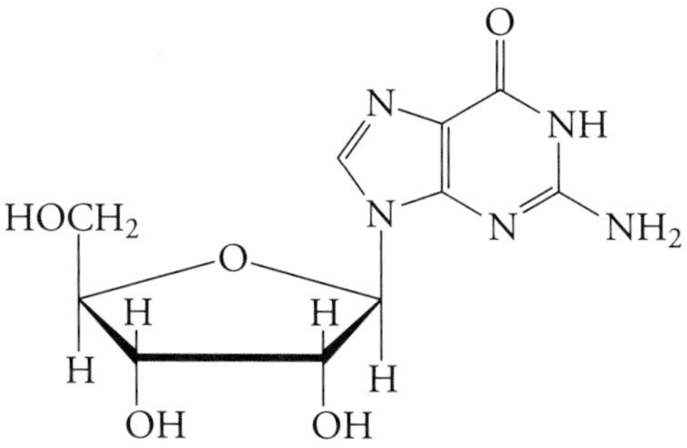

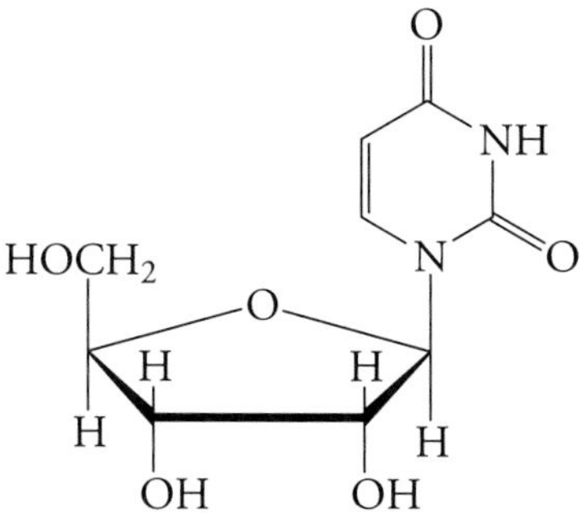

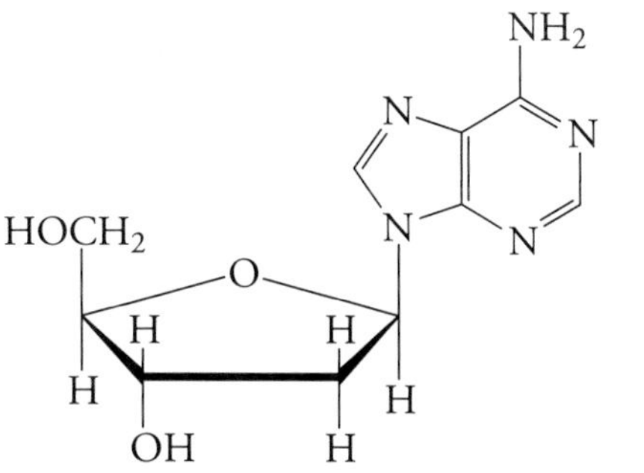

Deoxyadenosine

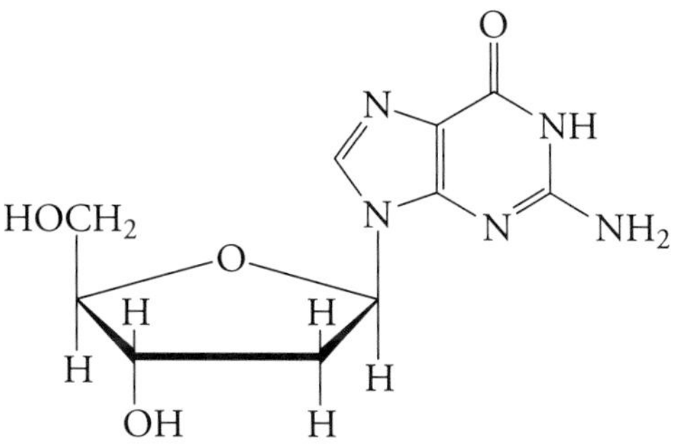

Guanosine

Deoxyguanosine

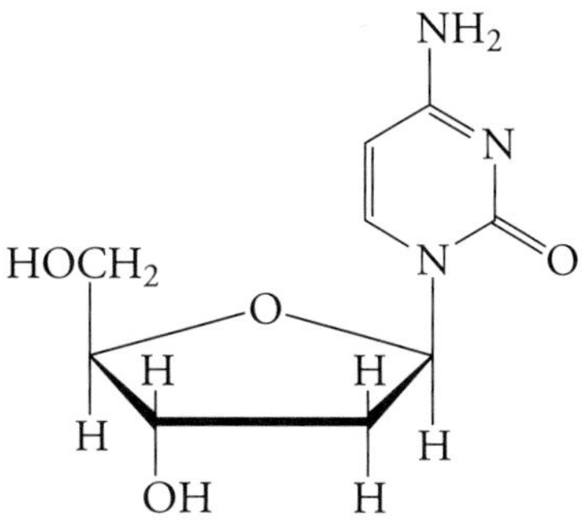

Cytidine

Deoxycytidine

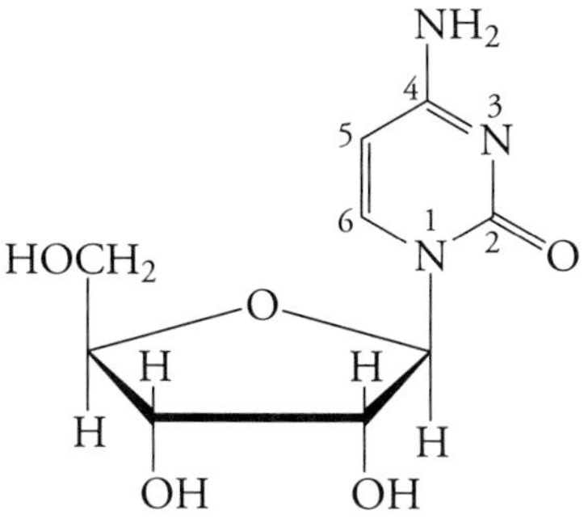

Uridine

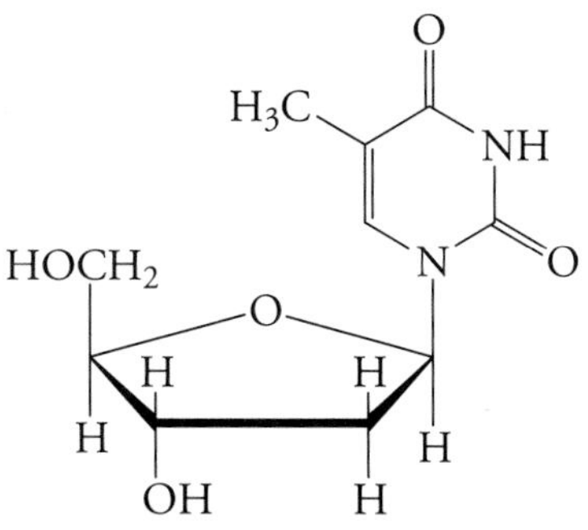

Deoxythymidine

FIGURE 14.5

Nucleosides

A nucleoside is an *N*-glycoside formed by combining a monosaccharide with a purine or a pyrimidine base. Atom numbers in the sugar residue are "primed" (as in the 1′ on the sugar residue in adenosine) to distinguish them from atom numbers in the bases. Ribonucleosides contain β-ribofuranose and either A, G, C, or U, while deoxyribonucleosides contain β-2-deoxyribofuranose and either A, G, C, or T.

Nucleotides

When phosphate ion reacts with one of the —OH groups on the sugar residue of a nucleoside to form a phosphate monoester, a nucleotide is produced (Figure 14.6). In a nucleotide, the sugar and phosphate residues are joined by a phosphoester bond. Ribonucleosides react to form **ribonucleotides** and deoxyribonucleosides react to form **deoxyribonucleotides**. The 5′ nucleotides, including cytidine 5′-monophosphate (CMP) and deoxyadenosine 5′-monophosphate (dAMP), are most common in nature, although ribonucleosides can also react to form 2′ and 3′ nucleotides and deoxyribonucleosides to form 3′ nucleotides.

FIGURE 14.6

Nucleotides
A nucleotide is a nucleoside monophosphate. Nucleoside 5′-monophosphates predominate in nature.

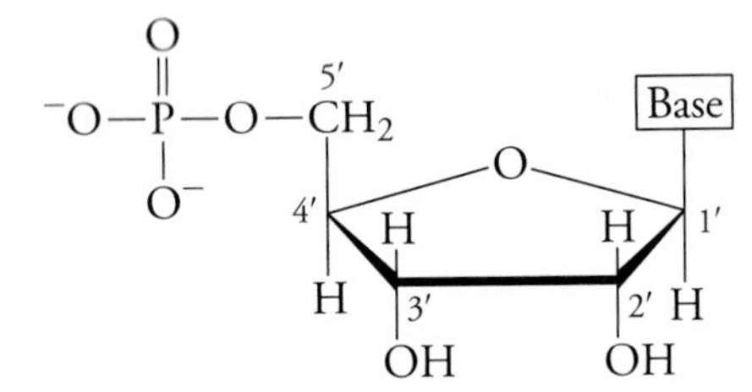

General structure of a ribonucleotide

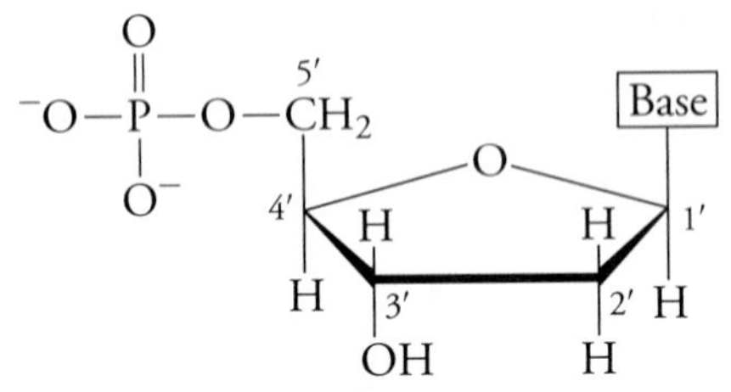

Name	Base
Adenosine 5′-monophosphate (AMP)	Adenine
Guanosine 5′-monophosphate (GMP)	Guanine
Cytidine 5′-monophosphate (CMP)	Cytosine
Uridine 5′-monophosphate (UMP)	Uracil

General structure of a deoxyribonucleotide

Name	Base
Deoxyadenosine 5′-monophosphate (dAMP)	Adenine
Deoxyguanosine 5′-monophosphate (dGMP)	Guanine
Deoxycytidine 5′-monophosphate (dCMP)	Cytosine
Deoxythymidine 5′-monophosphate (dTMP)	Thymine

SAMPLE PROBLEM 14.1

Drawing nucleotides

Draw adenosine 3′-monophosphate. Label the phosphoester bond and *N*-glycosidic bond.

Strategy

The name *adenosine 3′-monophosphate* contains clues about the nucleotide—which monosaccharide residue is present, which base residue is attached at the hemiacetal carbon, and the location of the phosphate monoester.

Solution

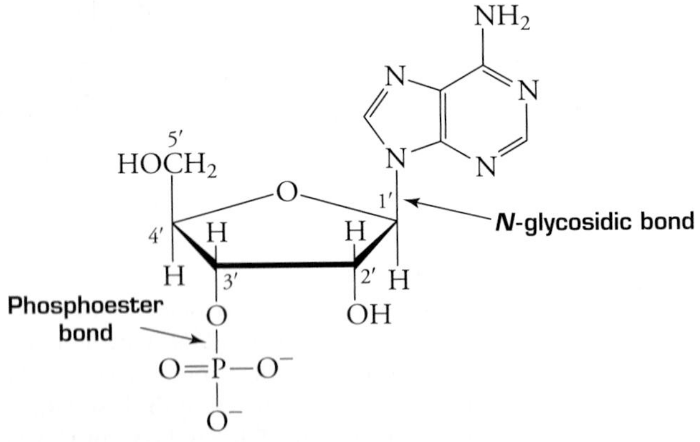

Phosphoester bond

N-glycosidic bond

PRACTICE PROBLEM 14.1

Why is deoxyuridine 5′-monophosphate not listed in Figure 14.6?

14.2 NUCLEOSIDE DI- AND TRIPHOSPHATES, CYCLIC NUCLEOTIDES

As we saw directly above, a nucleotide (nucleoside monophosphate) consists of a nucleoside residue and a phosphate residue joined by a phosphoester bond. Attaching a *second phosphate* to the first generates a **nucleoside diphosphate** and addition of a *third phosphate* forms a **nucleoside triphosphate** (Figure 14.7*a*). The connections between the phosphate groups are called phosphoanhydride bonds.

As we will see in Chapter 15, nucleoside triphosphates, such as adenosine triphosphate (ATP), are the immediate source of energy for most of the energy-requiring (nonspontaneous) processes that take place in living things. Nucleoside triphosphates are also used as substrates in the enzyme-catalyzed synthesis of polynucleotides (Section 14.3) and cyclic nucleotides.

The ability of phosphate ion to form a diester allows one phosphate residue to attach at two different positions on the same nucleoside. In **cyclic nucleotides**, such as adenosine 3′,5′-cyclic monophosphate (cyclic AMP or cAMP), *one phosphate residue has two phosphoester connections to the same monosaccharide residue* (Figure 14.7*b*). Cyclic nucleotides help regulate a large number of biochemical processes by controlling the activities of specific proteins to which they bind. In most cases, they function as effectors for allosteric enzymes (Section 13.7). cAMP, cGMP, and other cyclic nucleotides are produced from nucleoside triphosphates by the action of enzymes called cyclases.

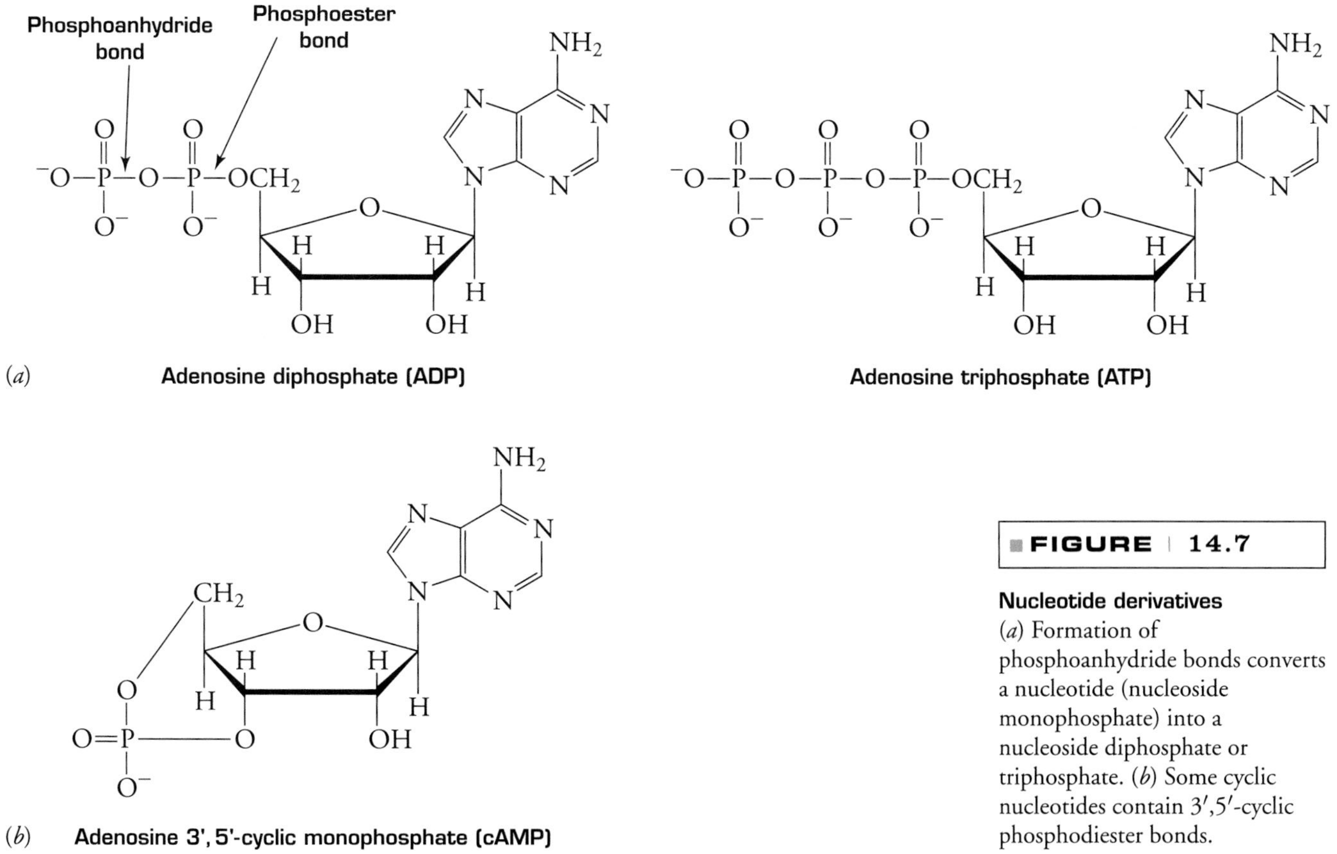

Phosphoanhydride bond

Phosphoester bond

(*a*) Adenosine diphosphate (ADP)

Adenosine triphosphate (ATP)

(*b*) Adenosine 3′,5′-cyclic monophosphate (cAMP)

FIGURE 14.7

Nucleotide derivatives
(*a*) Formation of phosphoanhydride bonds converts a nucleotide (nucleoside monophosphate) into a nucleoside diphosphate or triphosphate. (*b*) Some cyclic nucleotides contain 3′,5′-cyclic phosphodiester bonds.

14.3 POLYNUCLEOTIDES

Nucleotides can be connected to one another to form **oligonucleotides** (*2 to 10 nucleotide residues*) and **polynucleotides** (*more than 10 nucleotide residues*). In these molecules, ribonucleotide (in RNA) or deoxyribonucleotide (in DNA) residues are joined to one another by 3′,5′-phosphodiester bonds, in which a phosphate residue is attached at the 3′ position of one nucleotide residue and the 5′ position of another (Figure 14.8*a*). The sequence of nucleotide residues in an oligo- or polynucleotide is listed from the **5′-terminus** to the **3′-terminus** (the —OH groups on either end are not attached to nucleotide residues). Typically, the 5′ end is a phosphate monoester and the 3′ end a free —OH group.

(*a*)

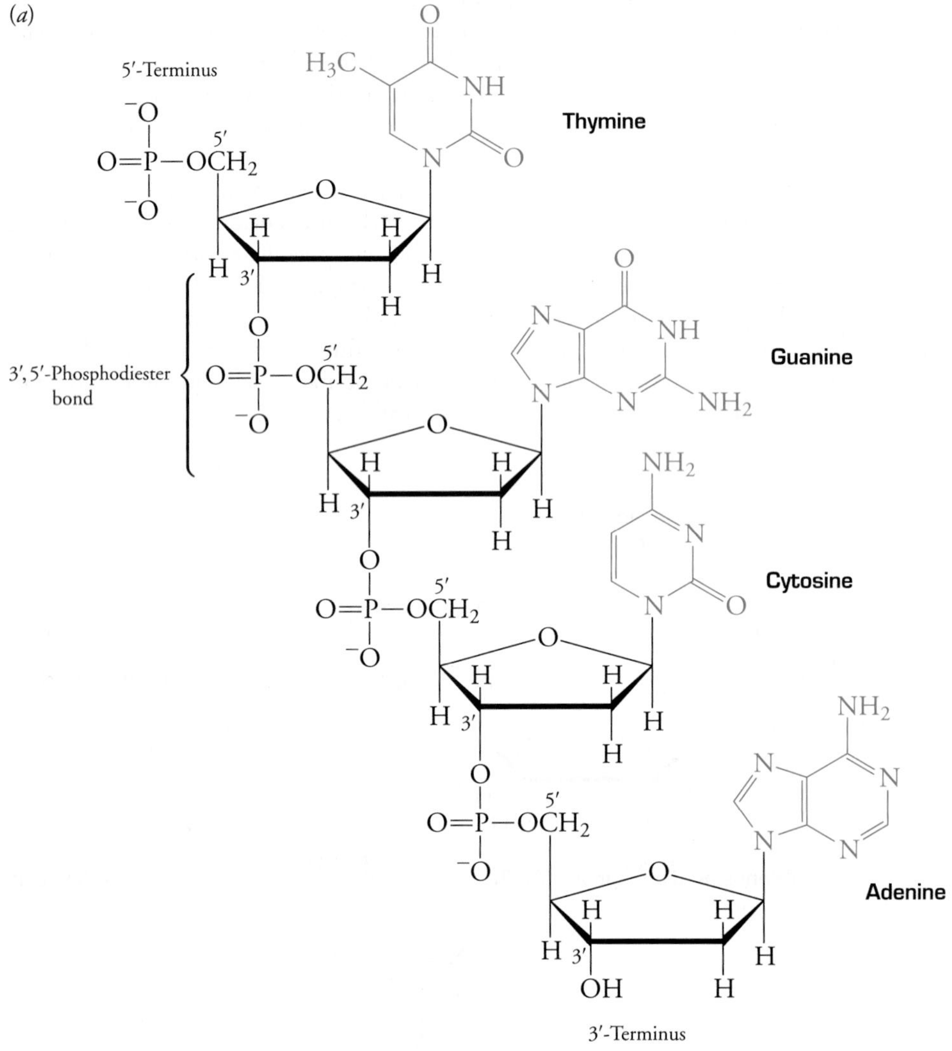

5′-Terminus

Thymine

3′,5′-Phosphodiester bond

Guanine

Cytosine

Adenine

3′-Terminus

FIGURE 14.8

Polynucleotides
(*a*) In deoxyribonucleic acid (DNA), deoxyribonucleotide residues are joined by 3′,5′-phosphodiester bonds. (*b*) The same bonds connect the ribonucleotide residues in ribonucleic acid (RNA). Both DNA and RNA consist of a sugar–phosphate backbone (in black) to which purines and pyrimidines are attached. The nucleotide residue at each end of a polynucleotide (the 3′ end and the 5′ end) has no phosphodiester bond.

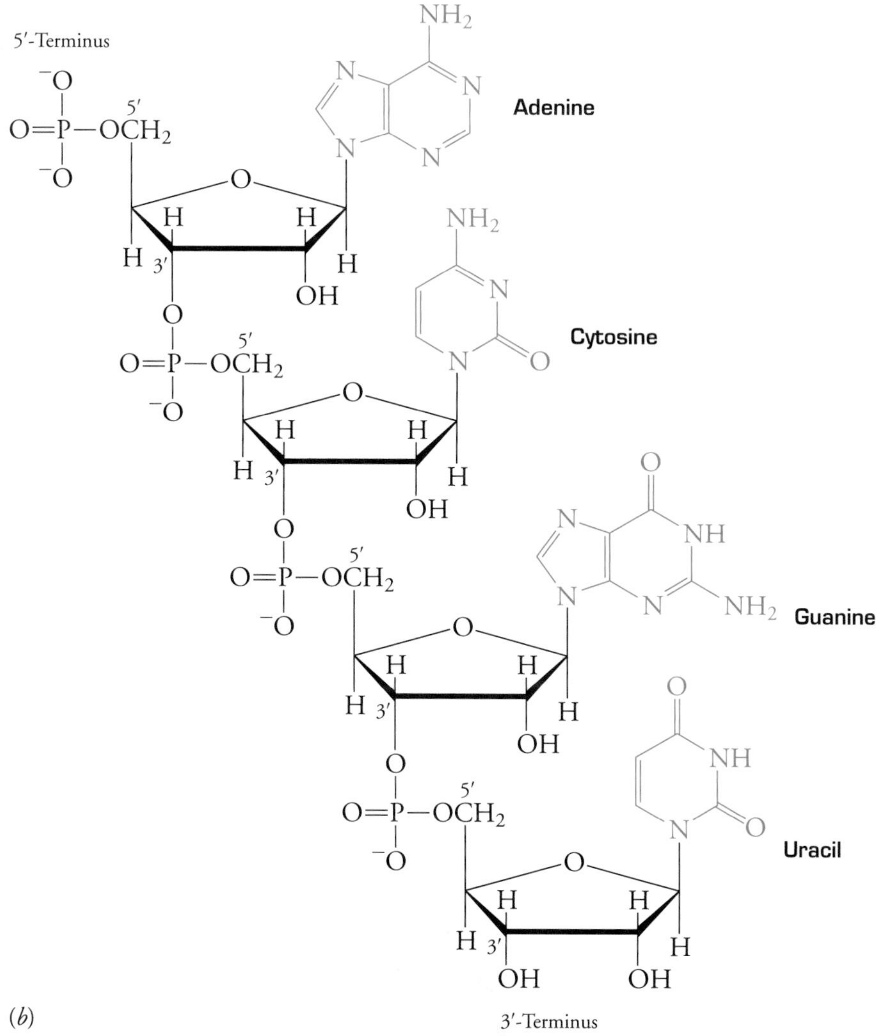

(*b*)

FIGURE | 14.8

Continued.

Figure 14.8*a* shows a short segment of **deoxyribonucleic acid (DNA)**. The portion of the structure drawn in black represents the **sugar–phosphate backbone**, which consists of *repeating units of phosphate attached to 2-deoxyribose attached to phosphate*, and so on. At C-1′ of each 2-deoxyribose residue, one of four bases common to deoxyribonucleotides is attached—A, G, C, or T. The charge on each phosphate residue varies with pH (Section 14.1), and at physiological pH the sugar–phosphate backbone of a polynucleotide carries negative charges.

■ DNA consists of a sugar–phosphate backbone to which A, G, C, and T are attached.

Figure 14.8*b* shows a strand of **ribonucleic acid (RNA)**. RNA differs from DNA in the sugar residue (ribose instead of 2-deoxyribose) and the combination of bases (A, G, C, and U instead of A, G, C, and T).

14.4 DNA STRUCTURE

Every cell in a particular living thing contains the exact same DNA, and in plant and animal cells most of the DNA is found in the cell nucleus. Human DNA, which contains 3 billion paired deoxyribonucleotide residues (see Secondary Structure, below), carries an estimated 25,000 **genes**, *stretches of DNA that carry the codes for protein production.* Even this many genes accounts for only about 5% of the total DNA. The remaining "noncoding" DNA serves functions not yet understood in any detail.

■ Genes code for the production of proteins.

Primary Structure

As is the case for proteins, the nucleic acids can be identified in terms of different levels of structure. The primary structure of a nucleic acid refers to the *sequence of its nucleotide residues*. The primary structure of the DNA segment shown in Figure 14.8*a*, reading from the 5′ to the 3′ end, is dTGCA and that of the RNA segment in Figure 14.8*b* is ACGU. In this abbreviated naming system, only the order of the base residues attached to the sugar–phosphate backbone is specified. The "d" in dTGCA indicates that the polynucleotide is DNA (contains 2-deoxyribose). Alterations in primary structure can change the way in which a nucleic acid functions.

Secondary Structure

When applied to proteins, secondary structure refers to the α-helix and the β-sheet. In DNA, secondary structure pertains to the *helix formed by the interaction of two DNA strands* (Figure 14.9). In the most commonly found form of DNA, two single strands lie side by side in an antiparallel arrangement, with *one running 5′ to 3′ and the other running 3′ to 5′*. The two DNA strands are held to one another by **base pairing**, *hydrogen bonding between the bases attached to the sugar–phosphate backbone*. This base pairing is **complementary**, which means that *A forms hydrogen bonds with T and G forms hydrogen bonds with C* (Figure 14.10). The two hydrogen-bonded strands are complementary along their entire length—each A in one strand is across from a T in the other and each G is across from a C.

■ Double-stranded DNA is held together by hydrogen bonding between complementary bases.

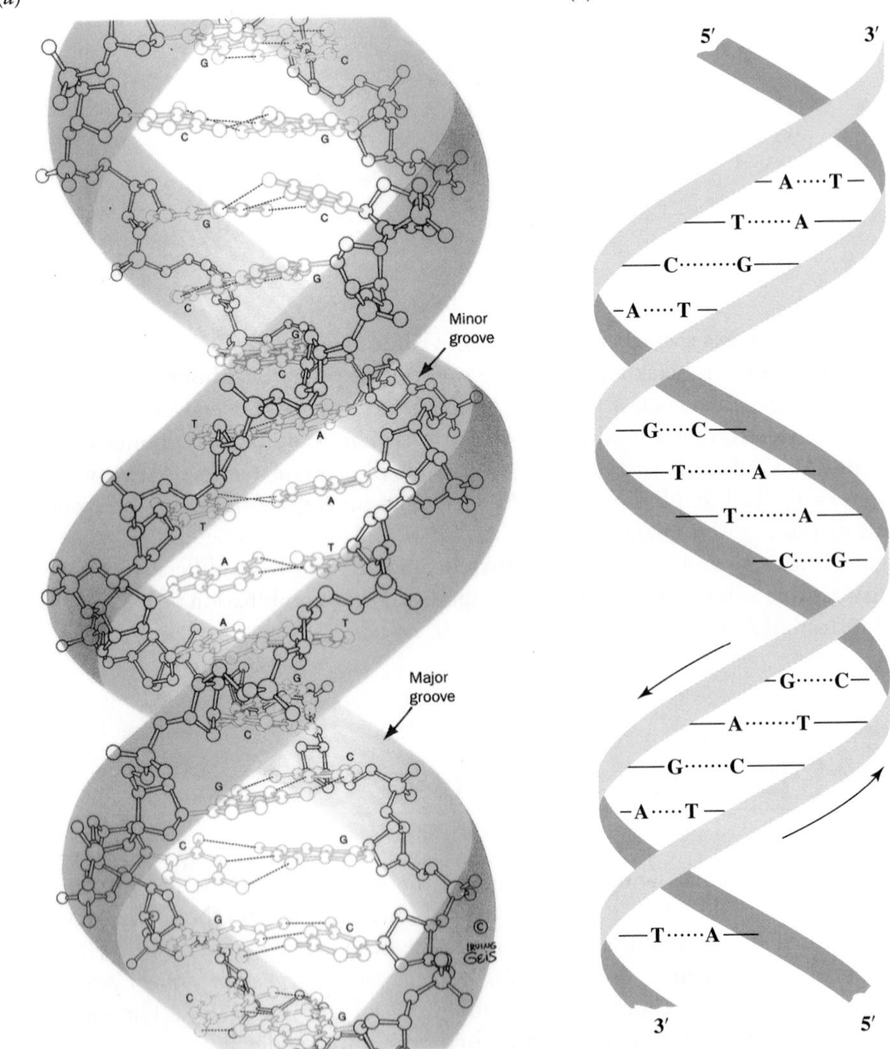

■ **FIGURE 14.9**

Double helix

DNA exists as a double strand that is twisted into a helical shape. The double strand is held together by base pairing, hydrogen bonds that form between A and T or G and C residues.

Source: (*a*) Illustration, Irving Geis. Rights owned by Howard Hughes Medical Institute. Not to be reproduced without permission. (*b*) D.S. Page, Principles of Biological Chemistry, Second Edition.

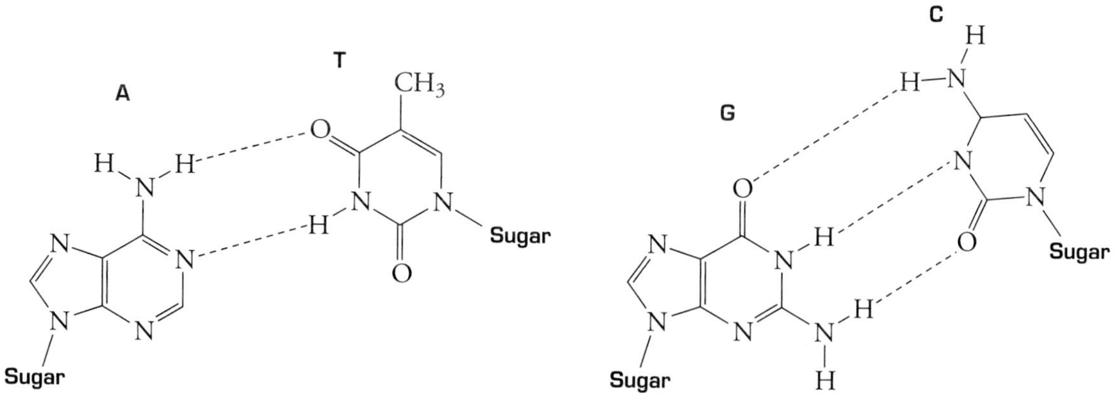

FIGURE 14.10

Complementary bases
The structures of the bases present in DNA (A, G, C, and T) are such that specific hydrogen bonding interactions take place between them. Adenine forms two hydrogen bonds with thymine and guanine forms three hydrogen bonds with cytosine.

The double strand of DNA twists into a helix that resembles a spiral staircase. The sugar–phosphate backbones with their negatively charged phosphate groups are on the outside and the bases are on the inside, where base pairing occurs. Hydrophobic interactions between neighboring bases on the same DNA strand help to stabilize this secondary structure.

Tertiary Structure

For nucleic acids, tertiary structure refers to *overall three-dimensional shape*, including the contribution of secondary structure. In DNA, tertiary structure arises from **supercoiling**, which involves double *helices being twisted into tighter, more compact shapes*. A relatively simple example of supercoiling in DNA is that which takes place with plasmids, a particular form of circular DNA found in bacteria. Enzymes called topoisomerases catalyze the conversion of the plasmids into a more compact, supercoiled state (Figure 14.11).

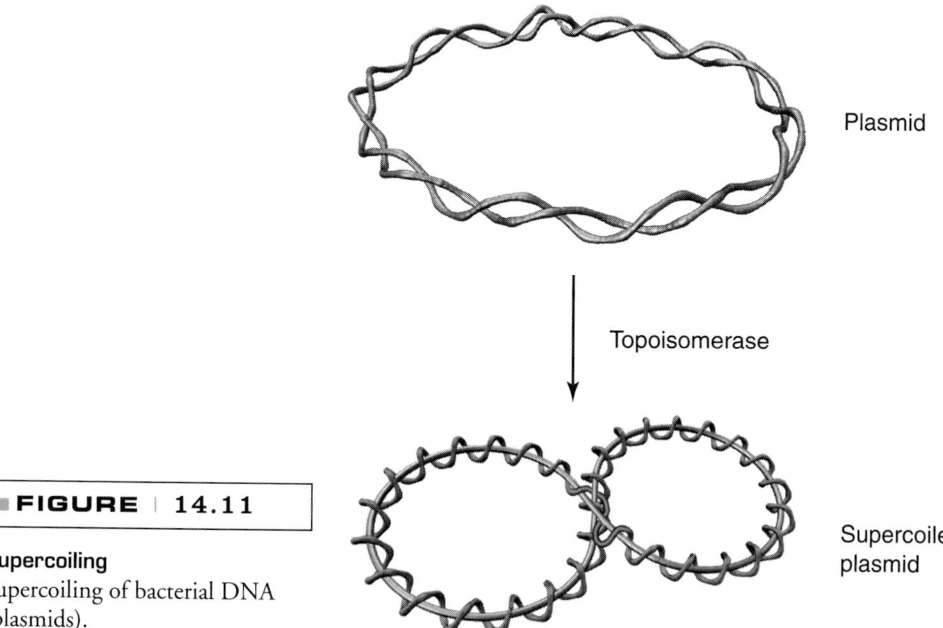

FIGURE 14.11

Supercoiling
Supercoiling of bacterial DNA (plasmids).

The tertiary structure of the DNA found in the nucleus of plants and animals is more complex, due to the involvement of proteins called histones (Figure 14.12) that interact with the phosphate groups of the DNA backbone. The DNA double helix is wrapped around a group of histones to form a nucleosome, and the strands of DNA that connect the nucleosomes twist further to produce a coiled structure called chromatin. Prior to cell division, each chromatin molecule coils and folds to become a chromosome. Human cells normally contain 23 pairs of chromosomes with one member of each pair contributed by each parent.

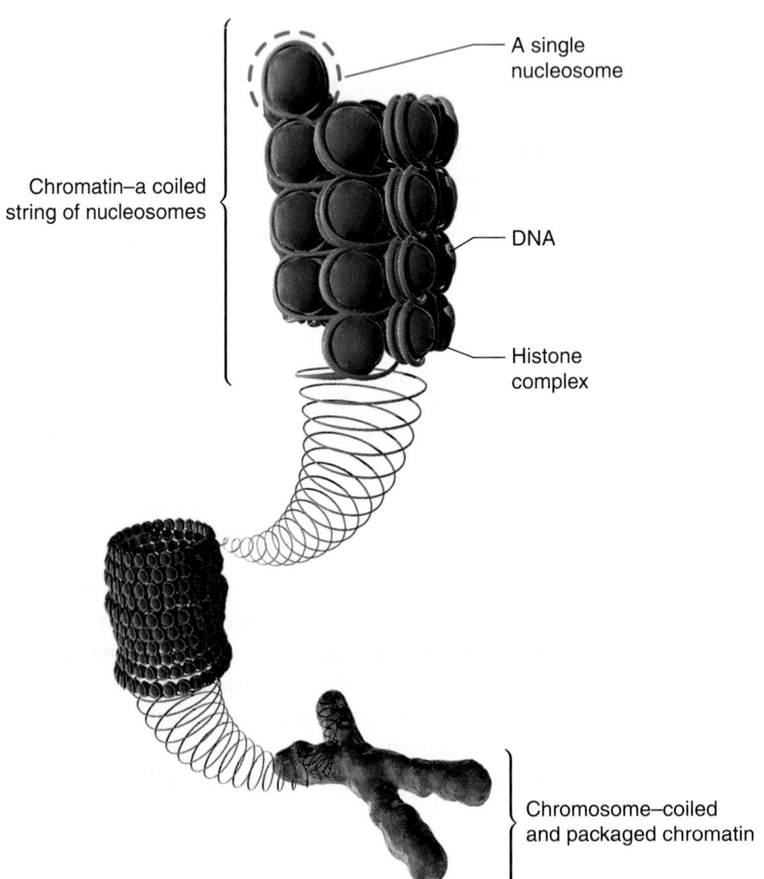

FIGURE 14.12

Larger structures involving DNA

In plants and animals, the DNA double helix wraps around a complex of histone proteins to form a nucleosome. Nucleosomes coil to produce chromatin which, prior to cell division, coils and folds to become a chromosome.

Source: From Biochemistry: A Foundation, 1st edition by Ritter. © 1996. Reprinted with permission of Brooks/Cole, a division of Thomson Learning: *www.thomsonrights.com*. Fax 800 730-2215.

14.5 DENATURATION

As we saw in Section 14.4, the double helix of DNA is held together by hydrogen bonding between complementary bases. When these interactions are disrupted, the DNA strand has been denatured.

Nucleic acids can be denatured by the same conditions that denature proteins (Section 13.5). Depending on the amount of heat added, for example, a double helix may unwind or even separate entirely, forming two single strands of DNA. Upon cooling, DNA can be "renatured" (Figure 14.13). Denaturation by heat has applications in experimental work involving DNA (Biochemistry Link: DNA Fingerprinting).

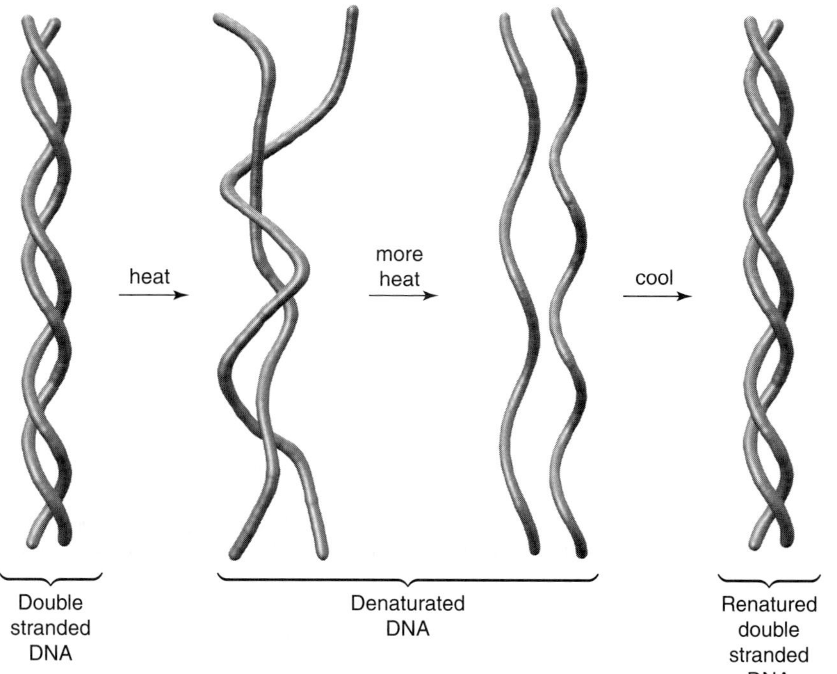

FIGURE 14.13

Denaturing DNA
Heating denatures DNA by disrupting the hydrogen bonds that hold the two strands to one another. With sufficient heat, the two strands separate entirely. Cooling can reverse the process, renaturing the DNA.

14.6 NUCLEIC ACIDS AND INFORMATION FLOW

DNA is often called the blueprint for life because it contains all of the information necessary for making the proteins required by living things. This information is carried in the primary structure of DNA.

To pass the information stored in DNA to a new generation of cells, **DNA replication** (Figure 14.14) must take place. This process involves *using existing DNA as a template for making new DNA molecules* before a cell divides (Section 14.7). For daughter cells to have the same characteristics as the parent cell, it is important that replication be as error free as possible.

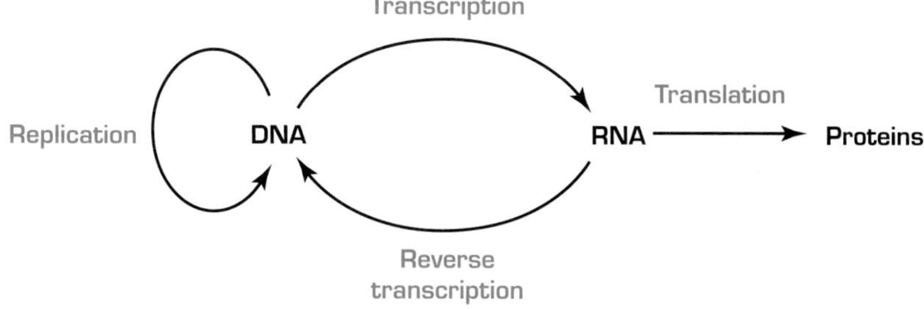

FIGURE 14.14

Information flow
DNA carries the information needed to make proteins. DNA replication (copying) allows the genetic information to be passed to a new generation of cells. Decoding DNA to produce proteins first involves transcription (transcribing the information in DNA by making RNA molecules) and translation (translating the information in RNA to make protein). Some viruses rely on reverse transcription to use the information present in RNA to make new DNA.

Processing the information in DNA to make proteins begins with *mapping portions of the primary structure of DNA onto various RNA molecules.* This procedure, which uses DNA as a template, is called **transcription** (Section 14.8). Some viruses, such as HIV (the virus that causes AIDS), are capable of **reverse transcription**, in which *viral RNA is read to produce new DNA that carries information for the manufacture of viral proteins and new virus particles* (Health Link: Viruses).

Once RNA has been produced from DNA, protein synthesis can begin. In a process called **translation** (Section 14.9), *the primary structure of a particular form of RNA is read to generate the correct primary structure of the desired protein.*

14.7 DNA REPLICATION

When DNA is replicated, each strand of the double helix serves as a template for the manufacture of a new strand of DNA. *In each of the daughter DNA strands, one strand from the parent DNA is present.* This is called **semiconservative replication.**

DNA replication begins when particular enzymes alter the secondary and tertiary structure of DNA by separating two strands and forming a bubble or **origin**, *where single DNA strands are exposed* (Figure 14.15). Only one origin is formed when a bacterial DNA molecule is replicated. During replication in plants and animals, many origins are opened along a double-stranded DNA helix at one time.

FIGURE | 14.15

DNA replication

(*a*) At an origin, DNA polymerases attach to the single strands of DNA and catalyze the formation of new DNA that is complementary to the original. As the origin expands, the replication forks move and more and more of the original DNA is replicated. (*b*) DNA polymerase catalyzes the attachment of a deoxyribonucleotide residue to the 3′ end of a growing DNA strand.

Source: From Essentials of General, Organic, and Biochemistry 1st edition by Armold/Amend/Mundy/Thomson Learning Inc. 2001. Reprinted with permission of Brooks/Cole, a division of Thomson Learning: *www.thomsonrights.com.* Fax 800 730-2215.

(*a*)

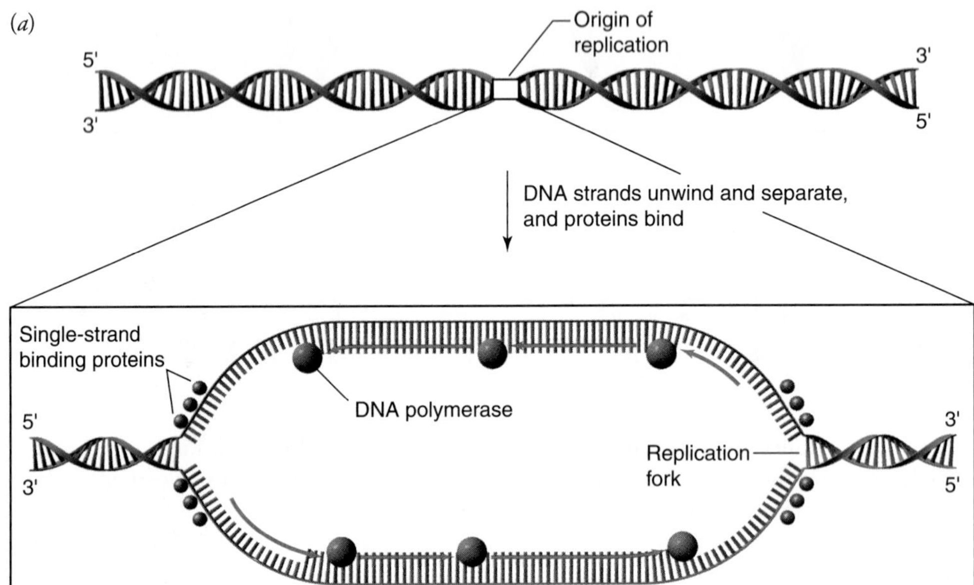

(*b*)

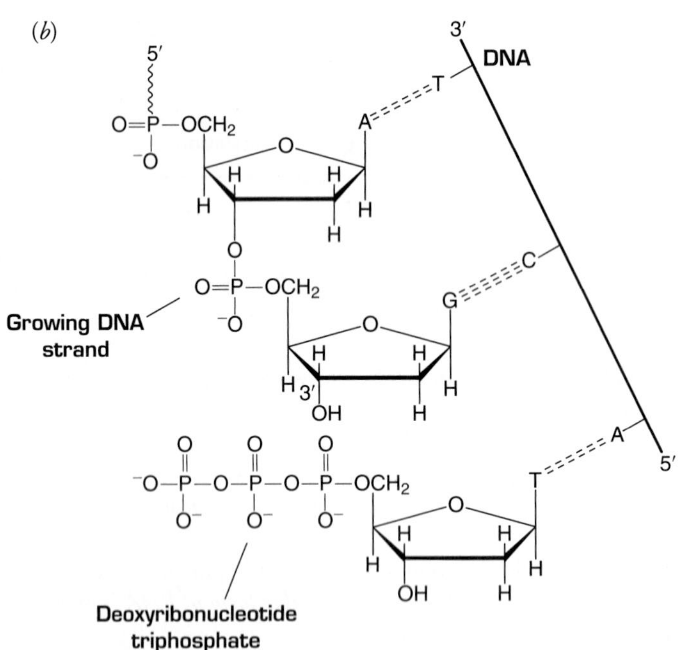

When an origin is formed, proteins called single-strand binding proteins (SSBs) attach to the newly exposed single strands of DNA. The function of these SSBs is to protect the DNA strands from endonucleases, protective enzymes that destroy foreign DNA. Any single-stranded DNA, regardless of its source, is a target for these enzymes. Each end of the origin has a **replication fork** *where replication takes place.*

The *production of new DNA* is carried out by **DNA polymerases.** These enzymes attach to each of the existing single strands of DNA at the replication fork and, using dATP, dGTP, dCTP, and dTTP (Section 14.2) as substrates, match each base in the parent strand with the proper complementary base. When the correct match has been made, a new deoxyribonucleotide residue is added to the growing strand and the polymerase moves along to the next base in the parent strand. DNA polymerases move along an existing DNA strand in a $3'$ to $5'$ direction and synthesize new DNA in a $5'$ to $3'$ direction.

■ DNA polymerase catalyzes the addition of deoxyribonucleotide residues to a growing DNA strand.

As all of this is going on, the double helix continues to be unwound at the replication forks. With the exposure of new single strands of DNA, additional SSBs and DNA polymerases attach and, eventually, the entire DNA is replicated. Because more than one DNA polymerase is attached to DNA at any given time during replication, there are some breaks in the newly formed DNA. Enzymes called ligases seal up these breaks by forming the necessary phosphoester bonds.

Proofreading and Repair

When adding new deoxyribonucleotides to a growing DNA strand, DNA polymerase inserts the wrong residue slightly less than once every 10,000 times. The enzyme proofreads its work to see if the correct deoxyribonucleotide residue has been added; if a mistake has been made, DNA polymerase clips the residue off and tries again.

If proofreading does not catch an error, then other DNA repair enzymes are likely to catch the problem. After proofreading and repair, the error rate during replication falls to less than one in 1 billion bases.

SAMPLE PROBLEM 14.2

Proofreading and repair

After proofreading and repair, how often does an error occur in the replication of human DNA (3 billion base pairs)?

Strategy

To solve this problem you must know the error rate. Remember that both strands of DNA are copied.

Solution

After proofreading and repair, the error rate of replication is less than 1 in a billion, so there are less than 6 errors.

$$3 \text{ billion } \cancel{\text{base pairs}} \times \frac{2 \cancel{\text{ bases}}}{\cancel{\text{base pair}}} \times \frac{1 \text{ error}}{1 \text{ billion } \cancel{\text{bases}}} = 6 \text{ errors}$$

PRACTICE PROBLEM 14.2

Without proofreading and repair, how many errors would occur during replication of human DNA?

14.8 RNA

RNA differs from DNA in that it contains a different monosaccharide residue (ribose instead of 2-deoxyribose); it contains the bases A, G, C, and U instead of A, G, C, and T; and it usually exists as a *single strand* instead of a double strand. The first step in using the information stored in DNA to produce proteins is transcription—using DNA as a template to make RNA.

Transcription

■ RNA polymerase catalyzes the addition of nucleotide residues to a growing RNA strand.

Transcription begins when **RNA polymerase** attaches to a strand of DNA at a sequence of bases called a **promoter site**, disrupting the double helix so that the sequence of deoxyribonucleotides in this strand can be used as a template for making RNA. *The DNA strand that RNA polymerase reads* is called the **template strand** (Figure 14.16).

FIGURE 14.16

Transcription
(*a*) At a promoter site on DNA, RNA polymerase binds and catalyzes the formation of an RNA strand that is complementary to the template DNA strand. (*b*) RNA polymerase catalyzes the attachment of a ribonucleotide residue to the 3′ end of a growing RNA strand.

(*a*)

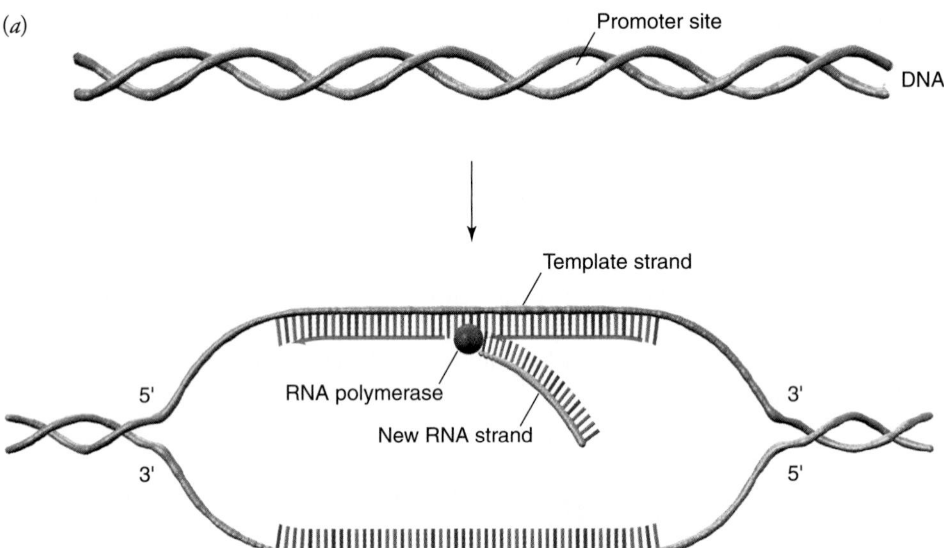

(*b*)

3′ DNA 5′

Growing RNA strand

Ribonucleotide triphosphate

RNA polymerase uses ATP, GTP, CTP, and UTP as substrates and catalyzes the addition of ribonucleotides to the growing RNA strand, with the base in each incoming ribonucleotide being complementary to the corresponding base in the template DNA strand (G is complementary to C and A is complementary to T and U—note the similarity in the structures of T and U in Figure 14.3). As RNA polymerase moves down the template strand 3′ to 5′ and makes RNA 5′ to 3′, the DNA double helix re-forms after it. When RNA polymerase reaches the **termination sequence**, *a series of bases that signals the end of the stretch of DNA that it is to copy*, the enzyme detaches and releases the RNA.

In plants and animals transcription takes place in the nucleus. After being produced, RNA is moved from the nucleus into the cytoplasm or the endoplasmic reticulum where it is used in the synthesis of proteins (Section 14.9).

Unlike DNA replication, in which all of the DNA is copied, only genes (the portions of DNA that code for the specific RNA needed to make a particular protein) and those portions of DNA that make additional RNA used in protein synthesis or for other purposes are copied during transcription.

Types of RNA

Three different types of RNA are required for the manufacture of proteins: transfer RNA (tRNA), messenger RNA (mRNA), and ribosomal RNA (rRNA). **Transfer RNAs** (Figure 14.17), the smallest of the three types (73–93 nucleotide residues), carry the correct amino acid to the site of protein synthesis. **Messenger RNAs** (variable size, depending on

■ Three types of RNA (tRNA, mRNA, and rRNA) are involved in protein synthesis.

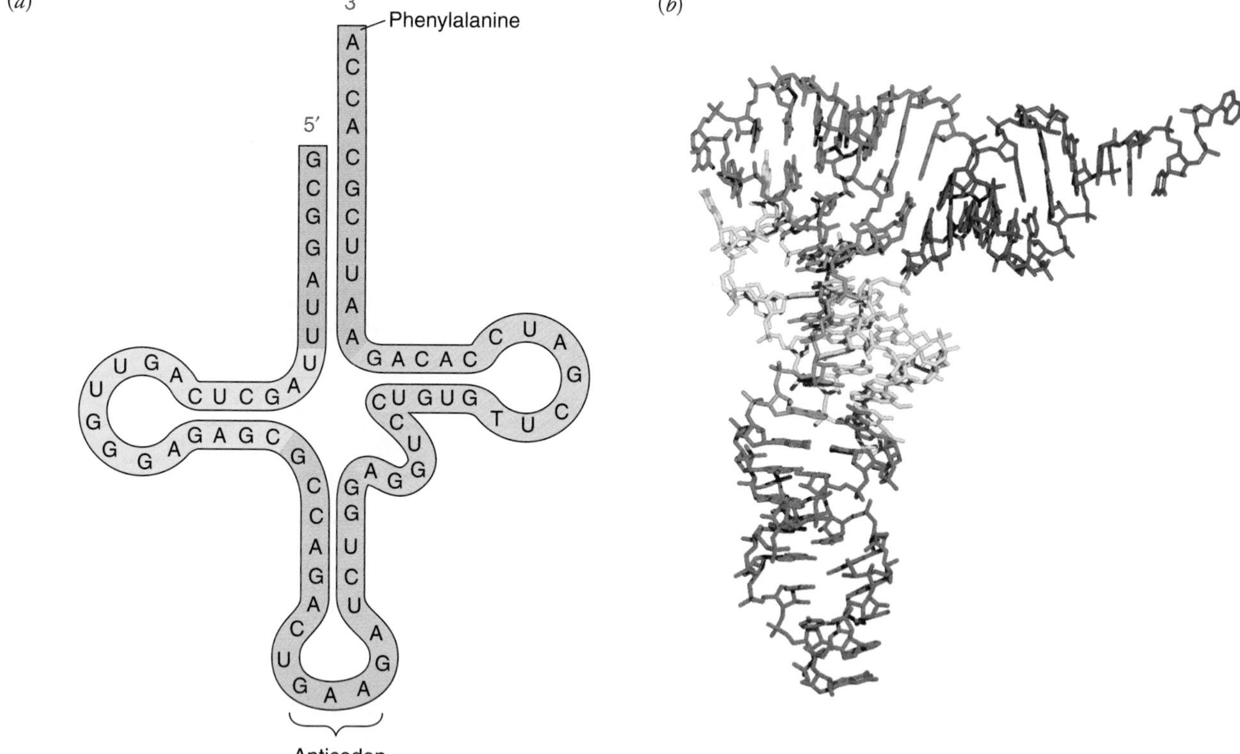

FIGURE 14.17 **Yeast tRNAPhe**

(*a*) Primary and secondary structure. tRNAs are held in a cloverleaf shape by hydrogen bonds between complementary bases. Although not shown here, a number of the base residues in every tRNA molecule have been converted into minor bases. Each tRNA holds an amino acid residue at its 3′ end. The yeast tRNA shown here is specific for phenylalanine. An anticodon, across the cloverleaf from where the amino acid residue is held, ensures that the correct amino acid is transferred to the site of protein synthesis. (*b*) Tertiary structure. The secondary structure of tRNA folds to form tertiary structure. The color coding is the same in parts a and b.

the protein to be manufactured) carry the information that specifies which protein should be made. This message is carried as a sequence of nucleotides that is complementary to the template strand of DNA. **Ribosomal RNAs** are relatively long RNA strands (hundreds or thousands of nucleotide residues) that *combine with proteins to form* **ribosomes**, *the multisubunit complexes in which protein synthesis takes place.*

In all organisms, with the exception of bacteria, *RNA undergoes modification following transcription* (**post-transcriptional modification**). Transfer RNA is shortened and some of the major bases (A, G, C, and U) are converted into minor bases (Section 14.1). The modifications to mRNA may involve altering the 3′ and 5′ ends and removing sections of mRNA, both from the middle and the ends, whose code is not used in protein production. Ribosomal RNA is clipped into smaller pieces that become active rRNA.

The Genetic Code

The information that mRNA carries regarding the primary structure of a protein to be produced is read three bases at a time. Each *series of three bases*, or **codon**, specifies a particular amino acid in a peptide chain. There are 64 ways to combine bases in this manner (Table 14.1) and only 20 amino acids used to make proteins, so in the **genetic code**, known also as the **triplet code**, many amino acids have two or more codons. For example, the codons AGU and AGC specify serine, while GGU, GGC, GGA, and GGG all code for glycine. The genetic code is universal because, with few exceptions, all living things use the same set of codons to specify the same amino acids.

One codon, AUG, has two different meanings. It codes for the amino acid methionine and also, if preceded by a specific series of bases, indicates the place to start making protein. When a protein is being synthesized, the first amino acid residue in the chain is always *N*-formylmethionine (fMet), a modified form of methionine.

TABLE 14.1 Codons in the 5′ to 3′ Sequence of mRNA

	Position 2 (Middle)				
Position 1 (5′ End)	**U**	**C**	**A**	**G**	**Position 3 (3′ End)**
U	Phe	Ser	Tyr	Cys	U
	Phe	Ser	Tyr	Cys	C
	Leu	Ser	(Stop)	(Stop)	A
	Leu	Ser	(Stop)	Trp	G
C	Leu	Pro	His	Arg	U
	Leu	Pro	His	Arg	C
	Leu	Pro	Gln	Arg	A
	Leu	Pro	Gln	Arg	G
A	Ile	Thr	Asn	Ser	U
	Ile	Thr	Asn	Ser	C
	Ile	Thr	Lys	Arg	A
	Met (Start)	Thr	Lys	Arg	G
G	Val	Ala	Asp	Gly	U
	Val	Ala	Asp	Gly	C
	Val	Ala	Glu	Gly	A
	Val	Ala	Glu	Gly	G

N-Formylmethionine is not the first residue in all native proteins, however, because it is usually removed before the protein is completed. Three codons, UAA, UAG, and UGA, signal the place to stop making protein.

Each tRNA carries an **anticodon**, *a series of three bases that is complementary to a codon.* For example, one of the codons (on mRNA) for serine is AGC (listed 5′ to 3′). The tRNA that carries serine to this codon has the anticodon UCG (listed 3′ to 5′). The base pairing between the codon and anticodon ensures that the proper amino acid is added to a growing protein chain at the correct time. Amino acids are matched with their respective tRNAs by synthetase enzymes.

SAMPLE PROBLEM 14.3

a. Refer to Table 14.1 to determine which amino acid is specified by the codon CGA (listed 5′ to 3′).
b. Which tRNA anticodon (listed 3′ to 5′) is complementary to this codon?

Strategy

Complementary bases form base pairs with one another (C to G, etc.)

Solution

a. Arginine
b. GCU. (Remember that in RNA A base pairs with U.)

PRACTICE PROBLEM 14.3

Which other codons specify arginine? List the corresponding anticodons.

14.9 TRANSLATION

Protein synthesis, or translation, takes place in three steps: **initiation**, *where a ribosome, mRNA, and tRNA come together to form a complex*; **elongation**, in which *amino acids are joined to the growing polypeptide chain*; and **termination**, *when the protein has been synthesized and the ribosome–mRNA–tRNA complex dissociates.*

- mRNA, rRNA, and tRNA must all come together for translation to take place.

Initiation

The first step in protein synthesis takes place when an mRNA and a tRNA combine with ribosome subunits. This happens near the 5′ end of the mRNA at the start codon (AUG). The tRNA that associates with the start codon has the anticodon UAC and carries the amino acid fMet (Figure 14.18). The fMet will be the N-terminal amino acid residue in the protein being manufactured.

Elongation

Once the initiation complex has formed, a second tRNA–amino acid binds beside the first. The particular tRNA is determined by the codon at this site on the mRNA. The ribosome catalyzes a reaction between the two amino acids, forming a new peptide bond and breaking the bond between the first amino acid and its tRNA.

The first tRNA dissociates from the ribosome, and the *ribosome shifts (5′ to 3′)* three nucleotides down the mRNA to the next codon, in a process called translocation. A third tRNA–amino acid attaches at the free binding site and the process takes place all over again, with formation of a peptide bond, breaking of a tRNA–amino acid bond, and translocation. As the ribosome moves down the mRNA the polypeptide grows in length.

FIGURE | 14.18

Translation
Initiation. Ribosome subunits combine with mRNA and a tRNA carrying fMet. *Elongation.* Another tRNA–amino acid, whose anticodon is complementary to the next codon on mRNA, attaches and an enzyme in the ribosome transfers fMet to the amino acid on the second tRNA, forming a peptide bond. The ribosome undergoes translocation (shifts one codon down the mRNA) and the first tRNA is released. *Termination.* Elongation repeats until a stop codon is reached, at which time the newly synthesized protein is released and the mRNA–ribosome–tRNA complex dissociates.

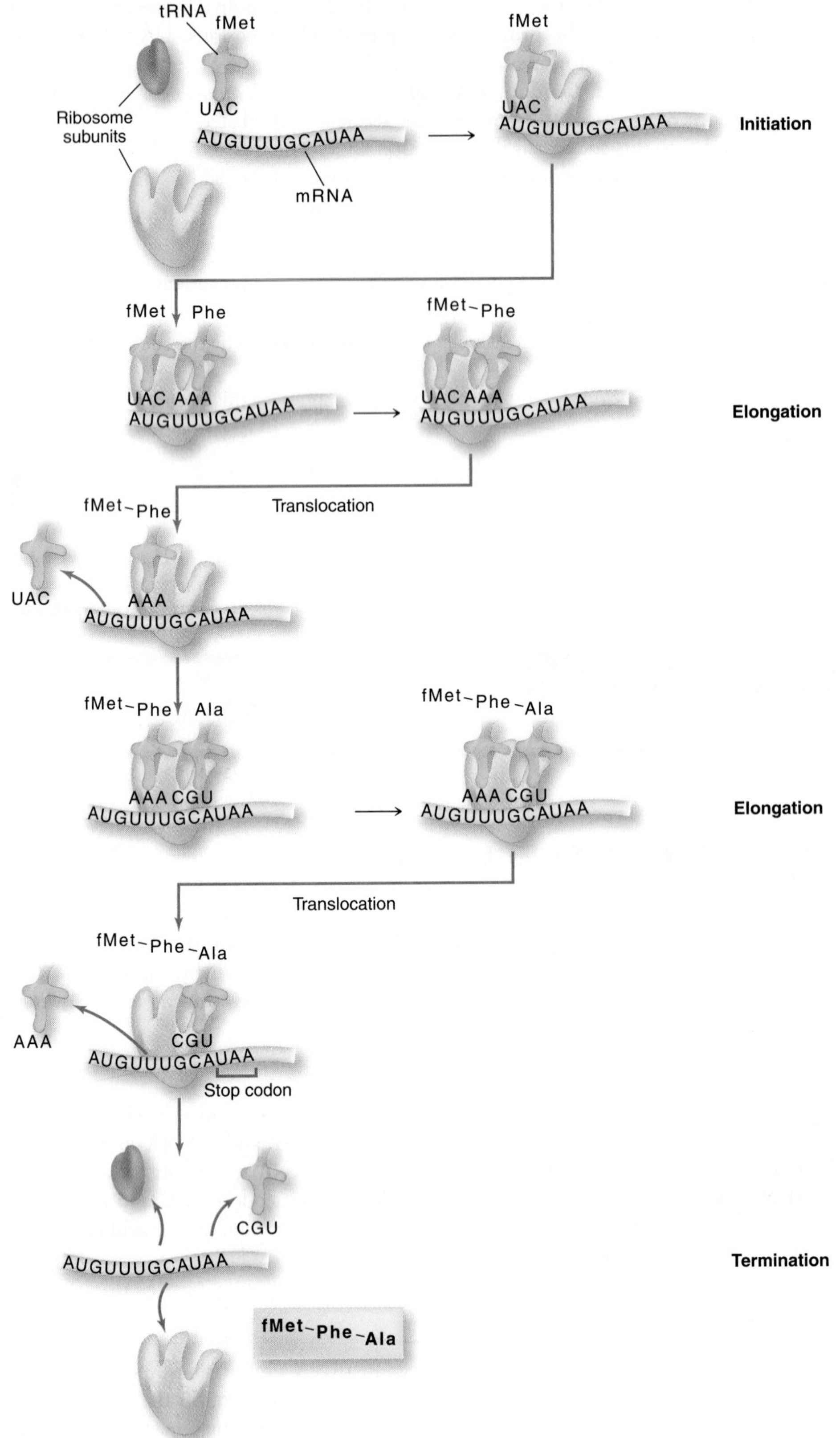

Termination

When the ribosome hits a stop codon, protein synthesis is halted. An enzyme catalyzes the removal of the tRNA from the last amino acid residue in the polypeptide which, along with the mRNA and tRNA, dissociates from the ribosome, which then separates.

Post-Translational Modification of Proteins

After a protein dissociates from the ribosome it is *processed before becoming active.* Among the **post-translational modifications** that a protein undergoes are removal of fMet at the N-terminus, trimming and cutting of the polypeptide chain, folding of the protein into its proper three-dimensional shape, the addition of prosthetic groups (heme, metal ions, etc.), formation of disulfide bridges, modification of amino acid residues (converting L into D amino acids, proline into hydroxyproline, and so on), and association with other polypeptide strands (quaternary structure).

HEALTH Link — Viruses

Viruses consist of a protein coat surrounding DNA or RNA and, sometimes, enzymes or other proteins (Figure 14.19). When a virus infects a cell, it does so by attaching to the cell membrane and inserting its nucleic acid. The virus reproduces by taking over the host cell's machinery used to make nucleic acids and proteins. In the case of DNA viruses, the genetic information that they provide (DNA) is in the form that the host cell can directly use. Retroviruses (RNA-containing viruses) provide an enzyme called reverse transcriptase, which catalyzes the reactions necessary to make a DNA copy of the viral RNA. Viral DNA is replicated and then transcribed to make viral RNA, which is then translated into viral proteins. All of the newly made viral components assemble into new viruses, the cell bursts, and the infection spreads.

Colds, smallpox, herpes, and pneumonia are caused by DNA viruses, while influenza, mumps, polio, and AIDS (acquired immune deficiency syndrome) are caused by RNA viruses. One difficulty that arises in treating retroviruses is that reverse transcriptase is not effective at proofreading and repair (Section 14.7), so the viruses rapidly mutate (Section 14.11) into new forms.

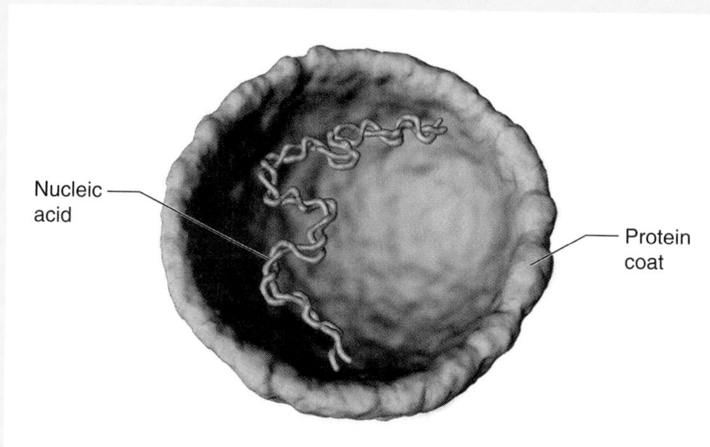

FIGURE 14.19 **Viruses**
Viruses consist of DNA or RNA surrounded by a protein coat.

14.10 CONTROL OF GENE EXPRESSION

The DNA of each living thing contains thousands of genes. These genes are not continually **expressed** (*read to make proteins*), because the production of unneeded proteins would be an inefficient use of resources. Organisms use a variety of different techniques to manage gene expression, including switching transcription on and off, controlling the rate of post-transcriptional modification of RNA, and controlling the rate of post-translational modification of proteins. In this section we will take a look at how *Escherichia coli* bacteria control the transcription of one set of genes.

■ Control of gene expression prevents the manufacture of unwanted proteins.

β-Galactosidase is an enzyme that catalyzes the hydrolysis of the disaccharide lactose to form glucose and galactose. Only very small amounts of this enzyme are produced in *E. coli* when no lactose is present. When lactose is introduced and glucose (an alternative energy source) is absent, the manufacture of β-galactosidase is switched on.

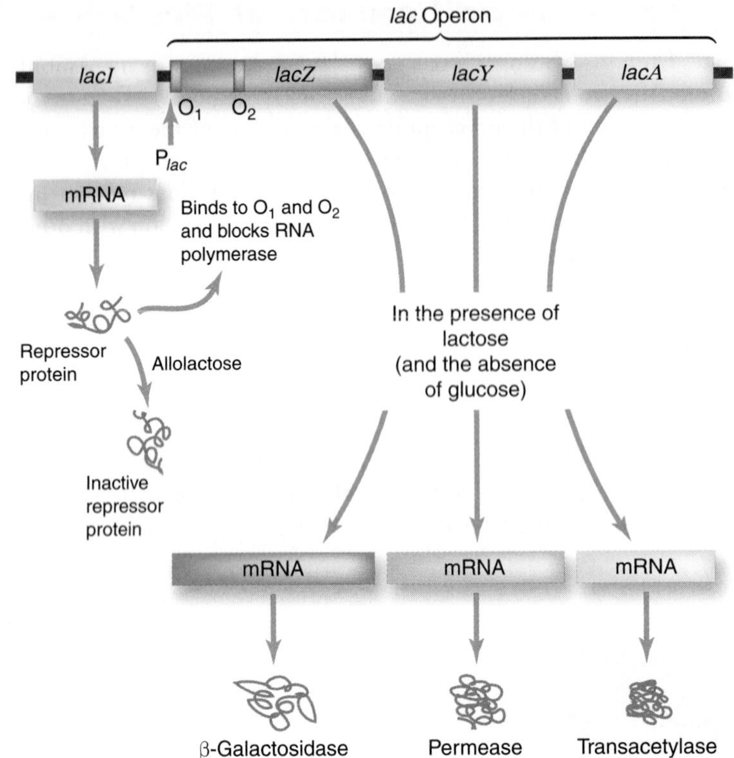

FIGURE | 14.20

Control of the *lac* operon
The *lac* operon contains three genes all under control of one promoter site (P_{lac}). The genes, *lacZ*, *lacY*, and *lacA*, code for the production of enzymes involved in lactose metabolism. The *lacI* regulator gene produces a repressor protein that binds at operator sites (O_1 and O_2) to prevent RNA polymerase from transcribing the three *lac* genes. When lactose is added, one of its metabolites, allolactose, interferes with the repressor protein's ability to bind to the operator sites. This allows RNA polymerase to transcribe the *lac* genes, resulting in the production of enzymes able to catalyze the metabolism of lactose.

Sources: HORTON, H. ROBERT; MORAN, LAURENCE A.; OCHS, RAYMOND S.; RAWN, DAVID J.; SCRIMGEOUR, K. GRAY, PRINCIPLES OF BIOCHEMISTRY, 3rd edition, © 2002, p. 683. Adapted by permission of Pearson Education.

The gene that codes for β-galactosidase is part of the *lac* operon (Figure 14.20). An **operon** is *a group of genes under the control of one promoter site*, and this particular operon is named after the fact that it codes for enzymes involved in the breakdown of *lac*tose. The *lac* operon consists of three genes, *lacZ*, which codes for β-galactosidase, *lacY*, which codes for lactose permease (transports lactose into the bacterial cell), and *lacA*, which codes for transacetylase (function unknown). Transcription of these three genes, which is initiated at one promoter site, called P_{lac}, is under the control of a repressor protein, a tetrameric allosteric protein coded for by the *lacI* regulator gene. The repressor protein binds to operator sites O_1 and O_2, control sites on the DNA near or at the promoter site. With repressor protein attached to the operator sites, transcription of the *lac* operon is blocked because RNA polymerase is unable to move down the DNA template strand.

If lactose is added, β-galactosidase (small amounts are always present in the cell) catalyzes the conversion of some of it into allolactose, in which the β-(1 → 4) glycosidic bond of lactose has been rearranged into a β-(1 → 6) glycosidic bond. Allolactose binds to the repressor protein and changes its shape in such a way that it can no longer attach to the operator sites. This allows RNA polymerase to bind to the promoter site and move down the DNA, transcribing the *lacZ*, *lacY*, and *lacA* genes.

This is not quite the end of the story, though. In addition to the requirement that lactose be present for transcription to begin, glucose must also be absent. For reasons that we will not go into here, the absence of glucose stimulates the production of cAMP (Section 14.2) which attaches to a protein called CAP (catabolite activator protein). The cAMP–CAP complex combines with the promoter site (P_{lac}), making it easier for RNA polymerase to attach there. So, when glucose is absent, RNA polymerase binds effectively to P_{lac} and transcription of the *lac* operon takes place. If glucose is present, the cAMP–CAP complex does not form and the binding of RNA polymerase to P_{lac} is less effective, and transcription of the *lac* operon slows.

HEALTH Link — RNA Interference

In this chapter we have seen how the information carried in genes is transcribed into RNA, which is then translated into proteins. The extent to which a particular gene is expressed can be controlled at a number of different stages, including transcription (Section 14.10), post-transcriptional modification of RNA (Section 14.8), and post-translational modification of proteins (Section 14.9). Another way that gene expression is controlled is through a process called RNA interference (RNAi). In RNAi the mRNA transcribed from a particular gene is destroyed, which prevents the gene product (protein) from being produced.

RNA interference begins with particular RNA strands folding back on themselves to produce regions of double-stranded structure (G pairs with C and A pairs with U). The double-stranded RNA is cut by a nuclease (hydrolyzes RNA) into small 21–23 base pair fragments (Figure 14.21). In these short sections of double-stranded RNA, called small interfering RNA, one strand (the guide strand) is complementary to a portion of the mRNA targeted for destruction.

The small interfering RNAs are separated into single strands, and the guide strand is combined with a number of proteins to form RISC (RNA-Induced Silencing Complex). Through base pairing, the guide RNA strand attracts the appropriate mRNA to RISC, where the mRNA is broken up. As this process repeats, more and more mRNA is destroyed and its translation into protein stops.

There is a great deal of interest in applying RNA interference to the treatment of disease. The plan is to silence the genes associated with particular diseases, by synthesizing small interfering RNA that contains a primary structure complementary to part of a given mRNA. Clinical trials of RNAi drugs to be used for the treatment of age-related macular degeneration and various viral infections are planned.

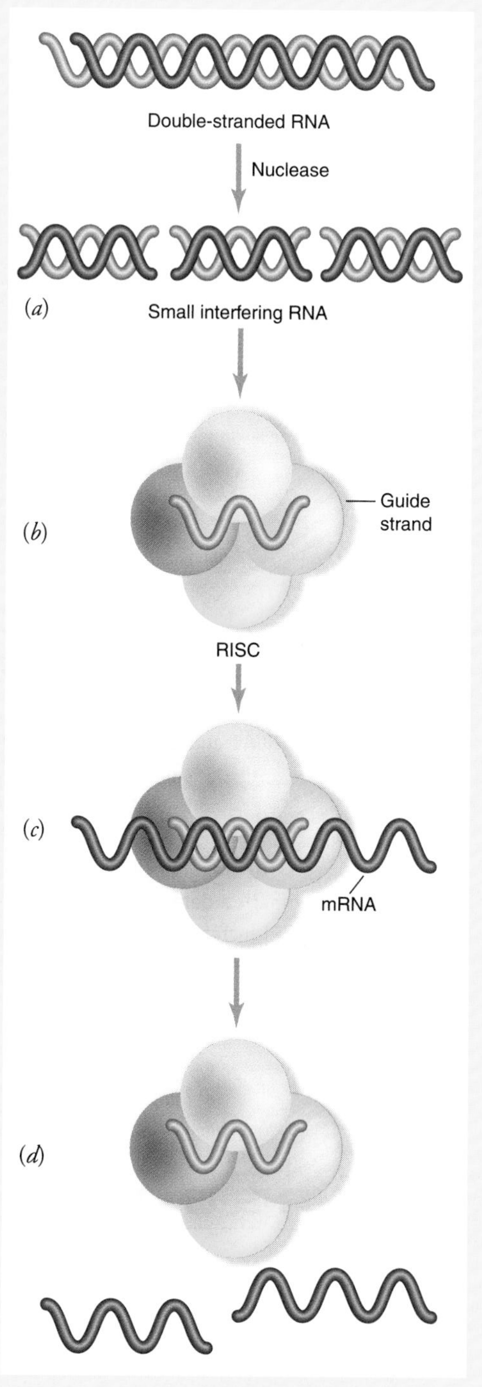

FIGURE 14.21

RNA interference (RNAi)
(*a*) A double-stranded RNA molecule is cut into 21–23 base pair fragments called small interfering RNA. (*b*) One strand of small interfering RNA, the guide strand, is complementary to a portion of the mRNA targeted for destruction. This guide strand is incorporated into a protein complex called RISC (RNA-Induced Silencing Complex). (*c*) mRNA forms base pairs with the guide strand. (*d*) RISC splits the mRNA, which prevents translation of the mRNA's coded information.

14.11 MUTATION

Any *permanent change in the primary structure* of (sequence of nucleotide residues in) DNA is called a **mutation**. Mutations might involve the switching of one base pair for another or the addition or deletion of base pairs. Errors in replication and exposure to

■ Mutations are changes in the primary structure of DNA.

mutagens (mutation-causing agents, including x rays, UV radiation, nuclear radiation, and chemicals) are the common causes of mutations. Many of the changes that take place in DNA are repaired. During replication, for example, DNA polymerase proofreads the growing DNA strand and corrects any mistakes that have been made. In addition, DNA repair enzymes continually inspect existing DNA and repair damage that is found.

As we saw earlier in this chapter, the primary structure of DNA is transcribed to produce RNA, whose message is translated to make proteins. In humans, genes occupy only 5% of the total DNA structure, so mutations that take place in the other 95% of DNA's structure have little or no effect on gene expression. Even if a mutation takes place in a gene, there is no guarantee that an individual will be harmed, because some changes will have no effect on the protein specified by a gene. This is because several different codons (Table 14.1) code for the same amino acid. For example, GGU, GGC, GGA, and GGG all specify glycine.

While some mutations affect only one individual, those mutations that take place in egg or sperm cells can be inherited. The list of **genetic diseases** (*diseases caused by inherited mutations*) is long—over 4,000 have been identified in humans. Susceptibility to colon cancer and breast cancer (Health Link: Breast Cancer Genes) is linked to mutations in DNA repair enzymes, cystic fibrosis to deletion of a phenylalanine residue in a membrane protein, and sickle cell anemia to replacement of a glutamic acid residue with a valine residue in the β chain of hemoglobin.

HEALTH Link

Breast Cancer Genes

BRCA1 and BRCA2 are genes that code for proteins indirectly involved in a form of DNA repair called recombinational repair. Defects in either of these genes increase the risk of getting breast cancer (BRCA stands for BReast CAncer) and other forms of cancer. While a person, either female or male, who inherits a mutated form of BRCA1 or BRCA2 is predisposed to breast cancer, the disease does not always appear.

In the general population, 13% of women will have breast cancer at some point in their life. For those who inherit a defective copy of BRCA1 or BRCA2, the percentage increases to between 36% and 85%, a three- to sevenfold increase. A similar increase occurs for ovarian cancer, which occurs in 1.7% of all women and in between 16% and 60% of those with a mutated form of either of these genes. An increased risk of colon and prostate cancer may also be associated with defects in BRCA1 or BRCA2.

14.12 RECOMBINANT DNA

Gaining an understanding of the processes involved in DNA replication has given scientists the ability to create **recombinant DNA**, which *contains DNA from two or more sources*. Combining DNA from different sources is not a new idea. For thousands of years people have been crossbreeding plants, a technique that involves exchanging DNA between plants. The modern techniques described here give quicker results and a more predictable outcome.

The first steps in the process of making recombinant DNA are to identify a gene of interest and cut it from the DNA strand in such a way that it can be combined with DNA from another source. This cutting can be done using restriction enzymes (endonucleases), bacterial protective enzymes that degrade DNA that is not recognized as belonging to the cell. The particular restriction enzymes used to cut DNA create sticky ends—strands of complementary single-stranded DNA (Figure 14.22).

Through base pairing, the sticky ends allow DNA fragments from different sources to attach to one another. For example, the gene that codes for human growth hormone

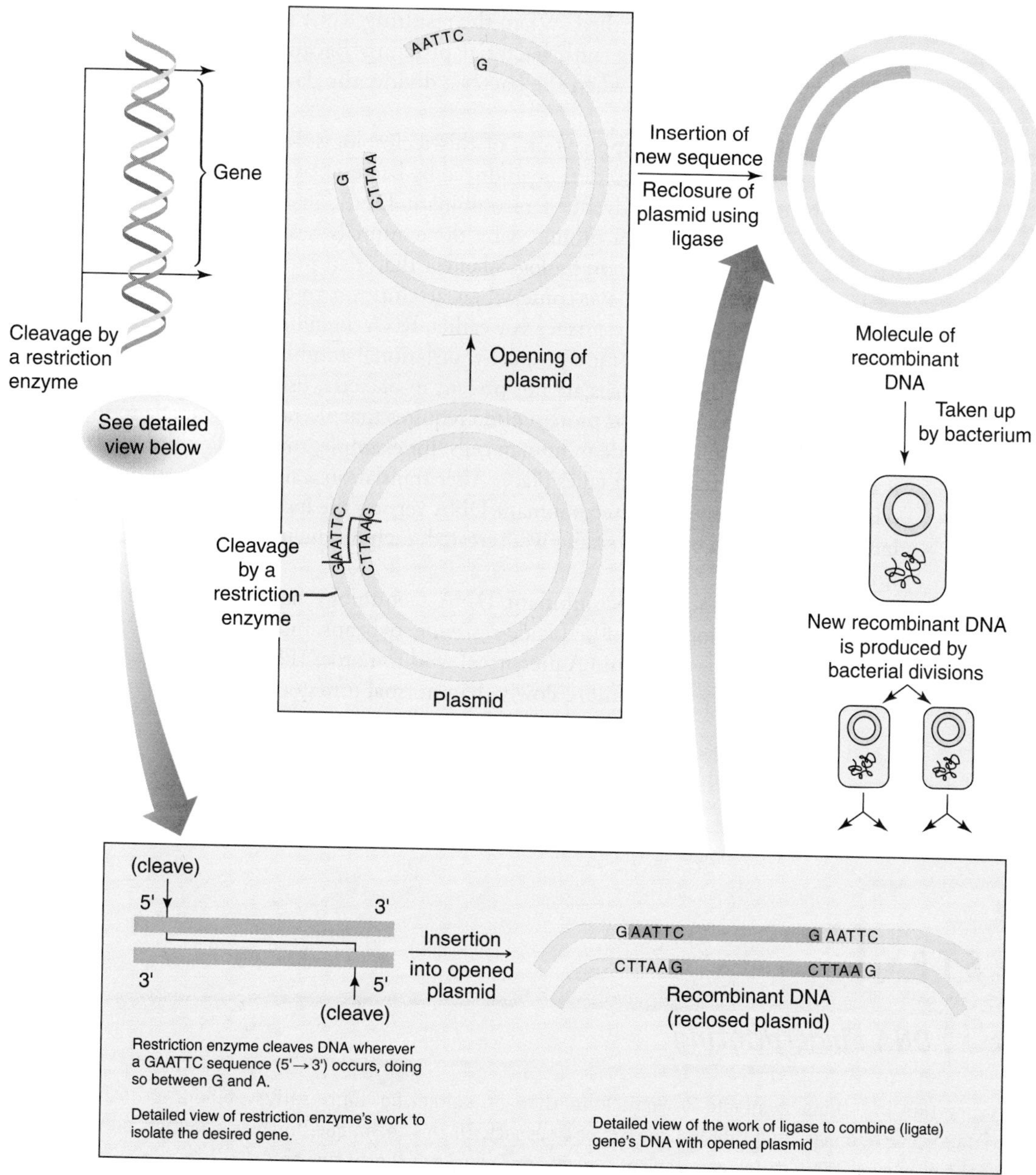

FIGURE 14.22 **Recombinant DNA**
The first step in creating recombinant DNA is to identify and determine the primary structure of the gene of interest. This allows a suitable restriction enzyme to be selected to cleave the DNA at either end of the gene. A bacterial plasmid (circular DNA) is cut by the same restriction enzyme, the DNA fragments from both sources are mixed, and their sticky ends left by the restriction enzyme form base pairs. Ligases connect the sugar–phosphate backbones, forming recombined (recombinant) DNA. The new plasmid is reinserted into bacteria, which reproduce to produce new copies of the DNA, whose gene products include those from the foreign gene.

can be snipped from a strand of human DNA and combined with a bacterial plasmid (small circular DNA) that has been treated with the same restriction enzyme. When the combined sections of DNA attach to one another and are treated with ligase (Section 14.7) to connect the sugar–phosphate backbones, recombinant DNA is produced. If the plasmid containing the combined bacteria–human DNA is reintroduced into bacteria, the DNA is transcribed. When the resulting RNA is translated, human growth hormone is produced along with bacterial proteins. Because the human gene is now part of the bacterial DNA, when the bacteria divide, the daughter cells also contain the human gene.

There have been many applications of this technology. The human growth hormone now used to treat growth disorders is produced by bacteria, as is the human insulin used to treat diabetes. Before the advent of recombinant DNA, insulin used by diabetics came from cattle and pigs. Insulin produced by these animals has a slightly different primary structure and a slightly different biological effect than that produced by humans.

There are challenges to overcome when attempting to manufacture proteins using recombinant DNA techniques. As we saw earlier, RNA often undergoes post-transcriptional modification and proteins undergo post-translational modifications. Since the enzymes required to make these changes are not present in bacteria, using recombinant DNA techniques to produce a particular protein often requires that alterations be made to the gene of interest. When insulin is made in human cells, for example, the middle third of the mRNA is removed before translation takes place. After translation a significant portion of the protein is trimmed away. In the recombinant DNA version the insulin gene was split—two different plasmids and bacterial strains were created, each producing a different fragment of the finished protein.

Many applications of recombinant DNA technology have appeared in agriculture, including the development of herbicide-resistant soybeans, insect-resistant cotton, rice that produces β-carotene (a vitamin A precursor), and tomatoes (Flavr-Savrs) that have a longer shelf life because they ripen more slowly than normal tomatoes.

BIOCHEMISTRY Link: DNA Fingerprinting

One of the modern crime-fighting tools available to police is that of collecting hair, saliva, blood, or other DNA-containing samples at a crime scene and matching the DNA to that of a suspect. In this technique, forensic scientists do not study the entire 3 billion base pairs of human DNA but, instead, look at a number of relatively short fragments.

As we saw in Section 14.4, genes make up about 5% of human DNA and the remaining 95% is noncoding. Much of the noncoding DNA consists of repeating series of base pairs—more than 30% in humans. The latest form of DNA fingerprinting analyzes short tandem repeats (STRs), relatively small stretches of DNA that contain short repeating sequences of bases. In the 13 STRs selected by the FBI for use in DNA fingerprinting, all contain four-base repeats and are less than 500 bases in length. The STR called TH01, for example, has between 7 and 11 repeats of the sequence dAATG (Figure 14.23).

dAATGAATGAATGAATGAATGAATGAATG
dAATGAATGAATGAATGAATGAATGAATGAATG
dAATGAATGAATGAATGAATGAATGAATGAATGAATG
dAATGAATGAATGAATGAATGAATGAATGAATGAATGAATG
dAATGAATGAATGAATGAATGAATGAATGAATGAATGAATGAATG

FIGURE 14.23

Short tandem repeats
Short tandem repeats (STRs) are regions of noncoding DNA that contain a variable number of repeats of a short series of bases. TH01, one of the STRs used in DNA fingerprinting, has between 7 and 11 repeats of the series dAATG.

Carrying out a DNA analysis requires a certain minimum amount of DNA. If a sample has too little DNA, additional DNA copies can be made using the polymerase chain reaction (PCR). The entire DNA is not copied in this process; instead only those parts needed for further study are duplicated. The first step in PCR involves denaturing the DNA by heating it to 98°C. At this temperature the two strands of DNA separate, which allows the addition of primers, short segments of single-stranded DNA that are complementary to the DNA at the 3′ end of the region to be copied (amplified). Lowering the temperature to 60°C allows the primers to form base pairs (hybridize) with the single-stranded DNA. At this point, DNA polymerase and deoxyribonucleoside triphosphates (dATP, dGTP, dCTP, and dTTP) are added, the temperature is raised, and a complementary strand of DNA is "grown" along each of the existing strands (Figure 14.24). After this single PCR cycle, the amount of DNA of interest has been doubled. By repeating this process over and over, in an afternoon it is possible to produce millions of copies.

We have previously seen that increasing the temperature will lead to an increase in reaction rate. Enzyme-catalyzed reactions are an exception to this, because high temperatures usually denature enzymes. Ordinarily, the temperatures used for PCR would denature DNA polymerase, but researchers have solved this problem by using a thermostable DNA polymerase isolated from bacteria that live in hot springs, where the temperatures are near 100°C.

Once a DNA sample of sufficient size has been obtained, restriction enzymes are used to cut the DNA strands into small segments. Then, using techniques not described here, the number of repeats in each of the STRs is determined.

As shown in Table 14.2, the probability of two individuals chosen at random having the same STR types has been determined. For example, the probability of two African-Americans having the same TH01 type is about 1 in 10. The more STRs analyzed, the lower the odds that two people chosen at random will match. The probability of finding identical STR matches for Caucasians in each of the first four STRs in Table 14.2 is 1 in 60,000 (13 × 15 × 28 × 11). The probability of matching all 13 of the STRs is 1 in 400 trillion.

The fact that STRs are obtained from noncoding regions means that DNA fingerprinting does not pose the same sort of confidentiality problems that analyzing genes might cause. Those doing DNA fingerprinting learn nothing about genetic diseases, predisposition to cancer (Health Link: Breast Cancer Genes), or other genetic information that individuals might wish to keep to themselves.

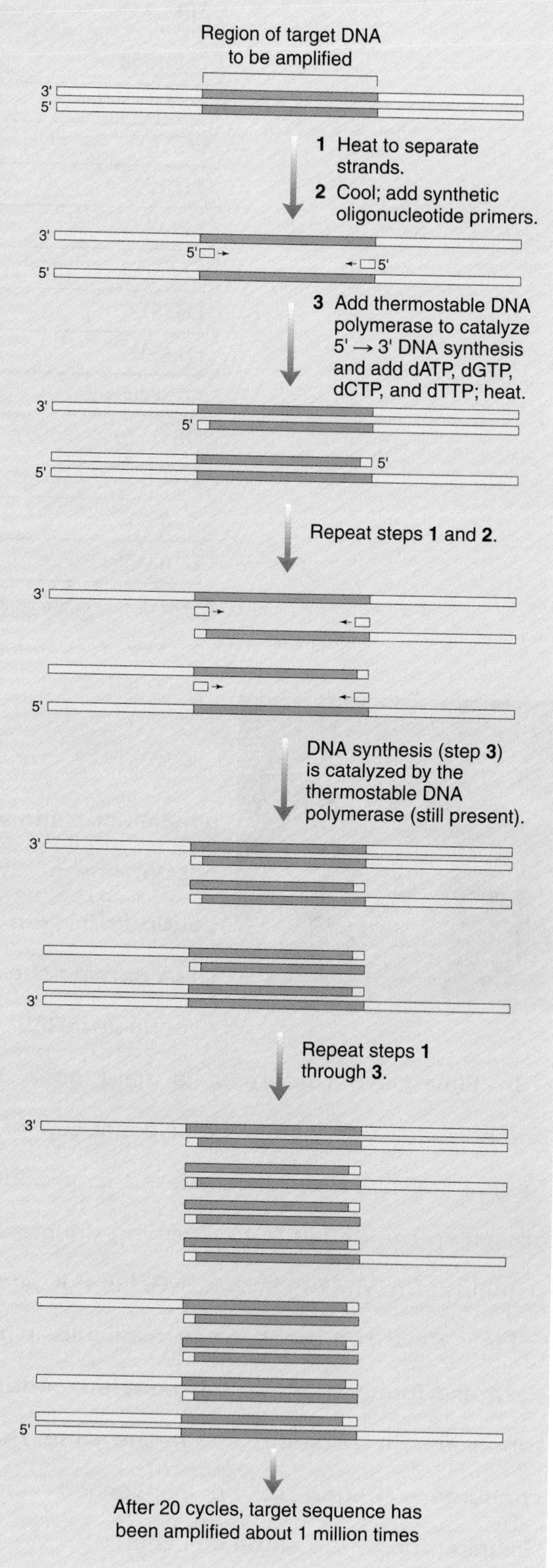
Region of target DNA to be amplified
3'
5'
1 Heat to separate strands.
2 Cool; add synthetic oligonucleotide primers.
3 Add thermostable DNA polymerase to catalyze 5' → 3' DNA synthesis and add dATP, dGTP, dCTP, and dTTP; heat.
Repeat steps 1 and 2.
DNA synthesis (step 3) is catalyzed by the thermostable DNA polymerase (still present).
Repeat steps 1 through 3.
After 20 cycles, target sequence has been amplified about 1 million times

FIGURE 14.24

Polymerase chain reaction

The polymerase chain reaction (PCR), used to increase (amplify) a DNA sample, has three steps: (1) the DNA is heat denatured; (2) primers, which tell DNA polymerase where to begin DNA synthesis, are added and the DNA is cooled to allow primers to associate with the DNA strands; (3) DNA polymerase and its substrates are added and new DNA is produced. Steps 1–3 are repeated. After 10 cycles, the original DNA will have been amplified 1000-fold and, at 20 cycles, about 1 million-fold.

TABLE 14.2 SHORT TANDEM REPEATS (STRs) AND THE PROBABILITY OF THEIR OCCURRENCE

STR	African-American	U.S. Caucasian
D3S1358	1 in 10[a]	1 in 13
vWA	1 in 17	1 in 15
FGA	1 in 29	1 in 28
TH01	1 in 10	1 in 11
TPOX	1 in 12	1 in 5
CSF1PO	1 in 14	1 in 8
D5S818	1 in 10	1 in 7
D13S317	1 in 8	1 in 14
D7S820	1 in 12	1 in 16
D8S1179	1 in 13	1 in 15
D21S11	1 in 30	1 in 22
D18S51	1 in 36	1 in 33
D16S539	1 in 15	1 in 10

[a] The chances of two randomly selected African-Americans having the same D3S1358 STR type is 1 in 10.

During ripening, tomatoes produce an enzyme called polygalacturonase (PGU). This enzyme catalyzes the hydrolysis of pectin, one of the polysaccharides present in tomato cell walls. As pectin is broken down, a ripening tomato begins to soften. Tomato DNA carries the PGU gene that, when switched on and transcribed, produces mRNA that carries the codons required to produce PGU.

In Flavr-Savr tomatoes, a new gene has been inserted into the tomato DNA. When transcribed, this added gene produces an "antisense" mRNA that is complementary to the "sense" mRNA that is translated to produce PGU. Base pairing causes the antisense RNA to attach to the sense RNA, forming double-stranded RNA. Because double-stranded RNA is not translated to make protein, PGU is not produced and ripening is slowed.

Flavr-Savr tomatoes were available in the United States only between 1995 and 1997. Farmers found them to be more susceptible to disease and to give lower yields than normal tomatoes. In addition, about the time that these tomatoes appeared on the market many consumers began to be concerned about the safety of genetically modified foods, or "Frankenfoods" as some call them.

summary of objectives

1 Describe the makeup of nucleosides, nucleotides, oligonucleotides, and polynucleotides.

Nucleotides are a combination of ribose (in RNA) or 2-deoxyribose (in DNA); phosphate; and one of the bases **adenine** (A), **guanine** (G), **cytosine** (C), **thymine** (T—present in DNA only), or **uracil** (U—present in RNA only). If the phosphate group is absent, the sugar–base combination is called a **nucleoside**. If a phosphate group is added to the phosphate of a nucleotide, a **nucleoside diphosphate** is formed. Addition of a third phosphate produces a **nucleoside triphosphate**. Nucleoside triphosphates are energy-rich compounds that are used to synthesize oligo- and polynucleotides and to produce **cyclic nucleotides**, such as cAMP, that are important in the regulation of enzyme activity. **Oligo-** and **polynucleotides** are held together by 3′,5′-phosphodiester bonds linking one nucleotide residue to another and are described as having a **sugar–phosphate backbone** to which bases are attached.

2 Describe the primary structure of DNA and RNA and the secondary and tertiary structure of DNA.

The sequence of nucleotide residues in a DNA or RNA strand is its primary structure. Two strands of DNA combine as a result of **base pairing**—hydrogen bonding between the complementary bases A and G or C and T—to form antiparallel **double-stranded DNA**. This double strand twists into a helix (secondary structure) which then **supercoils** (tertiary structure). In plants and animals, DNA wraps around histones to form **nucleosomes** and **chromatin**.

3 Explain how replication takes place and describe the roles of DNA polymerase in this process.

DNA **replication** (copying DNA) is **semiconservative**—each double-stranded daughter DNA contains one DNA strand from the parent DNA. Replication begins when double-stranded DNA is separated to form an **origin** (a bubble). **DNA polymerase** attaches to the exposed single strands of DNA and, using dATP, dGTP, dCTP, and dTTP as substrates, moves along the strand matching each base in the parent strand with its complementary base in the growing DNA strands. DNA polymerase **proofreads** the result and corrects any errors that may have taken place.

4 Explain how transcription takes place and describe the role of RNA polymerase in this process.

Transcription (using DNA as a template to make RNA) begins at **promoter sites**. Using ATP, CTP, GTP, and UTP as substrates, **RNA polymerase** produces RNA by matching incoming bases with their **complementary bases** on DNA (C and G, A and U or T). When a **termination sequence** is reached, RNA polymerase and the product RNA detach from DNA. In all organisms but bacteria, RNA is modified following transcription.

5 Name the three types of RNA required for the synthesis of proteins and identify the role of each in translation.

The three types of RNA are **transfer RNA** (tRNA), which transports amino acids; **messenger RNA** (mRNA), which carries information for the primary structure of the protein to be made; and **ribosomal RNA** (rRNA), which combines with proteins to make ribosomes. **Translation** (reading the RNA to make proteins) begins when a tRNA and the ribosome subunits combine with an mRNA at the start **codon** (AUG). Then, one by one, tRNAs bring in the appropriate amino acid, which is added to the growing protein chain, followed by a shift of the ribosome down the mRNA to the next codon. The anticodon carried by each tRNA is complementary to a codon on mRNA, ensuring that the proper amino acid is brought to the site of protein synthesis. When the ribosome reaches a stop codon the protein is released. Many proteins are modified after transcription takes place.

6 Describe one way that *E. coli* control the expression of genes.

It is necessary for living things to have some control over gene expression (transcribing the DNA in a gene and translating the resulting mRNA to make protein). In *E. coli,* genes important to lactose metabolism are part of the same **operon** (under control of one promoter site). In the absence of lactose a repressor protein binds to DNA and prevents RNA polymerase from making the RNA needed to produce lactose metabolizing enzymes. In the presence of lactose, the repressor protein is altered in a way that prevents it from attaching to DNA, so RNA polymerase can make the necessary RNA and the enzymes to metabolize lactose are produced. Glucose also plays a role in the process. If glucose is present, RNA polymerase does not bind effectively to the promoter site and the enzymes needed to utilize lactose are not produced—glucose is metabolized instead. In the absence of glucose, RNA polymerase binds to the promoter site and the genes are expressed.

7 List the steps involved in producing recombinant DNA.

Recombinant DNA contains DNA from two or more sources. Making recombinant DNA involves using restriction enzymes to cut the DNA, combining the DNA fragments with other DNA cut in a similar manner, using ligases to link the sugar–phosphate backbones of the DNA fragments into one chain, and introducing the recombinant DNA into a host cell.

8 Explain the term mutation.

A **mutation**, any permanent change to the primary structure of DNA, can be caused by mutagens. Not all mutations are harmful to an individual. Mutations can be inherited if they take place in egg or sperm cells.

END OF CHAPTER PROBLEMS

Answers to problems whose numbers are printed in color are given in Appendix D. More challenging questions are marked with an asterisk. Problems within colored rules are paired. **ILW** = Interactive Learning Ware solution is available at *www.wiley.com/college/raymond.*

14.1 Nucleic Acid Building Blocks

14.1 Which three components go into making a nucleotide?

14.2 **a.** Which monosaccharide is used to make DNA?
b. Draw this monosaccharide in its β-furanose form.

14.3 **a.** Which monosaccharide is used to make RNA?
b. Draw this monosaccharide in its β-furanose form.

14.4 Name the four bases that are present in DNA.

14.5 Name the four bases that are present in RNA.

14.6 Which of the bases present in nucleotides are purines?

14.7 Which of the bases present in nucleotides are pyrimidines?

14.8 The structure of phosphoric acid is pH dependent.
a. Draw the four forms in which phosphate appears.
b. Which form(s) appear at physiological pH?

***14.9** Figure 14.1 shows the phosphorus atom in a phosphate ion as having five covalent bonds. Draw a phosphate ion in which the phosphorus atom has an octet of valence electrons. Specify formal charges.

14.10 At physiological pH, nucleotides are negatively charged. Explain.

14.11 Draw the phosphate monoester in Figure 14.1 as it would appear at low pH.

14.12 Draw the phosphate diester in Figure 14.1 as it would appear at low pH.

14.13 How is a phosphate monoester similar to an ester made from a carboxylic acid?

14.14 **a.** Draw β-D-ribofuranose.
b. Draw α-D-ribofuranose.

14.15 **a.** Draw β-D-2-deoxyribofuranose.
b. Draw α-D-2-deoxyribofuranose.

14.16 **a.** How many different phosphate monoesters can β-D-ribofuranose form?
b. When reacted with NH_2CH_3, how many different *N*-glycosides can β-D-ribofuranose form?

14.17 **a.** How many different phosphate monoesters can β-D-2-deoxyribofuranose form?
b. When reacted with NH_2CH_3, how many different *N*-glycosides can β-D-2-deoxyribofuranose form?

14.18 **a.** What structural feature makes a base found in nucleic acids a purine?
b. Of the bases adenine, thymine, guanine, cytosine, and uracil, which are purines?
c. What structural feature makes a base found in nucleic acids a pyrimidine?
d. Of the bases adenine, thymine, guanine, cytosine, and uracil, which are pyrimidines?

14.19 Draw the complete structure of guanosine 5′-monophosphate (see Figure 14.6).

14.20 Draw the complete structure of thymidine 5′-monophosphate (see Figure 14.6).

14.2 Nucleoside Di- and Triphosphates, Cyclic Nucleotides

14.21 How many phosphoester bonds and how many phosphoanhydride bonds are present in the following?
a. a nucleotide
b. a nucleoside diphosphate

14.22 How many phosphoester bonds and how many phosphoanhydride bonds are present in the following?
a. a nucleoside triphosphate
b. a cyclic nucleotide

14.23 **a.** Draw 3′-dATP. **b.** Draw 3′-ATP.

14.24 **a.** Draw 5′-dGTP. **b.** Draw 5′-GTP.

14.25 **a.** Name the types of bonds broken when ATP is completely hydrolyzed.
b. What products are obtained when ATP is completely hydrolyzed?
c. Name the types of bonds broken when cAMP is completely hydrolyzed.
d. What products are obtained when cAMP is completely hydrolyzed?

14.26 **d.** Name the types of bonds broken when the molecule in Problem 14.23a is completely hydrolyzed.
b. What products are obtained when the molecule in Problem 14.23a is completely hydrolyzed?
c. Name the types of bonds broken when the molecule in Problem 14.24a is completely hydrolyzed.
d. What products are obtained when the molecule in Problem 14.24a is completely hydrolyzed?

14.3 Polynucleotides

14.27 In the sugar–phosphate backbone of DNA
a. which sugar residue is present?
b. which type of bonds connect the sugar and phosphate residues?
c. where are bases attached?

14.28 In the sugar–phosphate backbone of RNA
a. which sugar residue is present?
b. which type of bonds connect the sugar and phosphate residues?
c. where are bases attached?

14.29 In terms of DNA structure, what do the terms 3′-terminus and 5′-terminus mean?

14.30 In terms of RNA structure, what do the terms 3′-terminus and 5′-terminus mean?

14.31 How does a single strand of DNA differ from a single strand of RNA?

14.4 DNA Structure

14.32 The term *primary structure* refers to the sequence of what in DNA?

14.33 Describe the secondary and tertiary structure of DNA.

14.34 **a.** What does the term *base pairing* mean?
b. In DNA, which bases are complementary to one another?

14.35 What force holds double-stranded DNA together?

14.36 **a.** Draw the complete structure of the DNA dinucleotide dGC.
b. Label the 3′ and 5′ ends of the molecule.

14.37 **a.** Draw the complete structure of the DNA trinucleotide dTAT.
b. Label the 3′ and 5′ ends of the molecule.

14.38 Double-stranded DNA is antiparallel. Explain this term.

14.39 **a.** What are histones?
b. What are nucleosomes?
c. What is chromatin?

14.40 **a.** Salt bridges play a role in the interaction between DNA and histones. Amino acid side chains in histones must carry what net charge?
b. Are the histone amino acid residues polar-acidic or are they polar-basic?

*__14.41__ Compare the primary, secondary, and tertiary structure of proteins to that of DNA.

14.5 DENATURATION

14.42 When DNA is denatured, which of the following is disrupted?
a. primary structure
b. secondary structure
c. tertiary structure

14.43 What does it mean to renature DNA?

14.6 NUCLEIC ACIDS AND INFORMATION FLOW

14.44 What molecule is made during the following processes?
a. DNA replication
b. transcription
c. translation
d. reverse transcription

14.45 The primary structure of what molecule is read during the following processes?
a. DNA replication
b. transcription
c. translation
d. reverse transcription

14.7 DNA REPLICATION

14.46 Explain the term *semiconservative replication.*

14.47 **a.** What is an origin?
b. During DNA replication in plants and animals does just one origin form?

14.48 Which nucleoside triphosphates are substrates for DNA polymerase?

*__14.49__ **a.** In which direction along an existing DNA strand does DNA polymerase move?
b. In which direction does DNA polymerase synthesize a new DNA strand?
c. If an origin opens and a DNA polymerase attaches to each of the exposed single strands of DNA, do the polymerases move in the same direction or in opposite directions as they make new DNA?

14.50 In addition to adding nucleotides to a growing DNA strand, what is another important function of DNA polymerases?

14.8 RNA

14.51 Name the three types of RNA and describe their function.

14.52 Which nucleoside triphosphates are substrates for RNA polymerase?

*__14.53__ Enzymes called aminoacyl tRNA synthetases catalyze the combination of amino acids with the proper tRNAs. The first step in this process is the reaction of an amino acid with ATP to form an aminoacyl-AMP. An aminoacyl-AMP consists of an amino acid residue attached through its carboxyl group to the 5′ phosphate group of AMP (Figure 14.6) by an anhydride bond. Draw the aminoacyl-AMP that involves the amino acid alanine.

*__14.54__ The second of the reactions catalyzed by aminoacyl tRNA synthetases (Problem 14.53) is reaction of an aminoacyl-AMP with a tRNA molecule to form an ester. The ester consists of a carboxylic acid residue (supplied by the amino acid) and an alcohol residue (supplied by an —OH group attached to the ribose on the 3′-terminus of the tRNA).
a. Class 1 aminoacyl tRNA synthetases attach the amino acid to the 2′ —OH group of the terminal nucleotide residue. Draw the ester formed when alanine is attached to the 2′ —OH group at the 3′-terminus of the tRNA below.

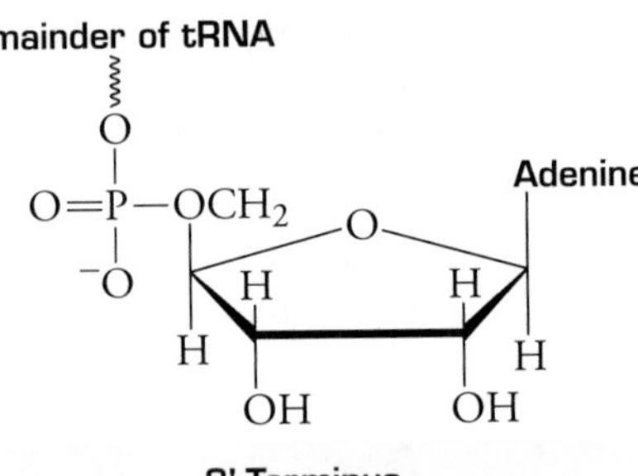

b. Class 2 aminoacyl tRNA synthetases attach the amino acid to the 3′ —OH group of the terminal nucleotide residue. Draw the ester formed when serine is attached to the 3′ —OH group at the 3′-terminus of the tRNA above.

*__14.55__ The sequence dGGCAT appears in a template strand of DNA.
a. Which base is at the 3′-terminus of this primary structure?
b. What is the sequence in the DNA strand that is complementary to this sequence? (Label the 3′ and 5′ ends.)
c. What is the sequence in the RNA strand that is synthesized from dGGCAT? (Label the 3′ and 5′ ends.)

*14.56 The sequence dATCCCC appears in a template strand of DNA.

a. Which base is at the 3′-terminus of this primary structure?

b. What is the sequence in the DNA strand that is complementary to this sequence? (Label the 3′ and 5′ ends.)

c. What is the sequence in the RNA strand that is synthesized from dATCCCC? (Label the 3′ and 5′ ends.)

14.57 a. What are codons?

b. Which type of RNA has codons?

14.58 a. What are anticodons?

b. Which type of RNA has anticodons?

14.59 Which amino acid is specified by each codon (listed 5′ to 3′)?

a. CCU
b. AGU
c. GUU

14.60 Which amino acid is specified by each codon (listed 5′ to 3′)?

a. UUU
b. CCC
c. AAA
d. GGG

14.61 Which codon(s) specify each amino acid?

a. Phe
b. Lys
c. Asp

14.62 Which codon(s) specify each amino acid?

a. Asn
b. Tyr
c. Gly

14.63 What types of post-transcriptional modifications do RNAs undergo?

14.64 Which anticodons are complementary to the codons in Problem 14.59? (Label the 3′ and 5′ ends.)

14.65 Which anticodons are complementary to the codons in Problem 14.60? (Label the 3′ and 5′ ends.)

14.9 Translation

14.66 During translation, fMet is always the N-terminal amino acid residue in the peptide being formed. Explain how proteins can exist that do not have fMet as their N-terminal amino acid residue.

14.67 Describe the role of mRNA, tRNA, and rRNA in protein synthesis.

*14.68 A tripeptide has the primary structure Ser-Lys-Asp.

a. Assuming that no post-translational modification took place, what is a sequence of bases in the mRNA that would code for this tripeptide? (Ignore start and stop codons and label the 3′ and 5′ ends.)

b. Assuming that no post-transcriptional modification took place, what is the sequence of bases in the template DNA strand used to make the mRNA? (Label the 3′ and 5′ ends.)

*14.69 a. For the tripeptide in Problem 14.68, again assuming that no post-translational modification took place, what is a different sequence of bases in the mRNA that would code for this tripeptide? (Ignore start and stop codons and label the 3′ and 5′ ends.)

b. Assuming that no post-transcriptional modification took place, what is the sequence of bases in the template DNA strand used to make the mRNA? (Label the 3′ and 5′ ends.)

14.70 How are rRNAs and zymogens similar?

14.71 Describe each step in translation:

a. initiation
b. elongation
c. termination

14.10 Control of Gene Expression

14.72 What is an operon?

14.73 a. What are the gene products of the *lac* operon?

b. How does the repressor protein prevent the *lac* operon from being expressed?

c. How does the presence of lactose act as an “on” switch for the *lac* operon?

14.74 Explain how allolactose is able to influence transcription of the *lacZ*, *lacY*, and *lacA* genes.

14.75 a. Draw lactose and allolactose.

b. What products are obtained when allolactose is hydrolyzed?

14.11 Mutation

14.76 Define the term *mutation*.

14.77 a. Are all mutations harmful to an individual? Explain.

b. Are all mutations passed on to offspring? Explain.

14.12 Recombinant DNA

14.78 Describe the general procedure used to make recombinant DNA and include the role of restriction enzymes, plasmids, and ligases.

HEALTH*Link* | *RNA Interference*

14.79 Describe the role of each in RNA interference.

a. double-stranded RNA
b. nuclease
c. small interfering RNA
d. guide strand
e. RISC

14.80 Although each involves double-stranded RNA, RNA interference is different than the use of antisense mRNA (see the discussion of Flavr-Savr tomatoes directly before the summary of chapter objectives). Explain.

14.81 Using RNA interference to treat disease requires synthesizing the appropriate small interfering RNA.
- **a.** True or false? Knowing the primary structure of the mRNA targeted for destruction would allow the appropriate small interfering RNA to be produced.
- **b.** True or false? Knowing the primary structure of the gene that codes for the mRNA targeted for destruction would allow the appropriate small interfering RNA to be produced.
- **c.** True or false? Knowing the primary structure of the native protein coded for by the mRNA targeted for destruction would allow the appropriate small interfering RNA to be produced.

HEALTH *Link* | *Viruses*

14.82 Why do retroviruses have a high mutation rate?

14.83 What is reverse transcriptase? Why do some viruses produce this enzyme?

HEALTH *Link* | *Breast Cancer Genes*

14.84 Why do mutations in the BRCA1 or BRCA2 genes result in increased cancer risks?

14.85 What is the normal function of the BRCA1 and BRCA2 genes?

14.86
- **a.** According to “Cancer Screening”, what percentage of breast cancer is caused by gene mutations?
- **b.** If genetic screening shows that a woman has a BRAC mutation, what lifestyle choices may delay the onset of cancer?
- **c.** If she has a gene mutation, what surgical options can greatly reduce a woman's chances of getting avarian and breast cancer?

BIOCHEMISTRY *Link* | *DNA Fingerprinting*

14.87 What are short tandem repeats and how are they used in DNA fingerprinting?

14.88 If a person's DNA is “fingerprinted” and the STR information is saved in a database, should he/she be concerned that information regarding genetic diseases that he/she might carry will be released to insurance companies? Explain.

14.89 Beginning with one double strand of DNA, how many double-stranded DNAs will be present after 15 cycles of PCR?

14.90 Which primer would be appropriate if you wanted to initiate copying of the single strand of DNA that has the primary structure dCCTAGGCGAATCCG? Recall that the primary structure of nucleotides is listed 5′ to 3′.
- **a.** dCCTAG
- **b.** dGGATC
- **c.** dCGGAT
- **d.** dGGCGC
- **e.** dGCCTA

THINKING IT THROUGH

14.91 While the Flavr-Savr tomato is no longer on the market, many other genetically modified foods are.
- **a.** Do you think that genetically modified foods should be labeled as such?
- **b.** If you knew that a food had been genetically modified and was approved by the FDA, would you eat it? Why or why not?

INTERACTIVE LEARNING PROBLEMS

ILW

14.92 Messenger RNA, mRNA, is often described, when examined under an electron microscope, as a fuzzy Christmas tree or a string of pearls. Why?

ILW

14.93 What is the name of the molecule shown? This is one of four nucleotides that are precursors for ________ (RNA/DNA). These strands ________ (are/are not) normally found as dual complementary strands.

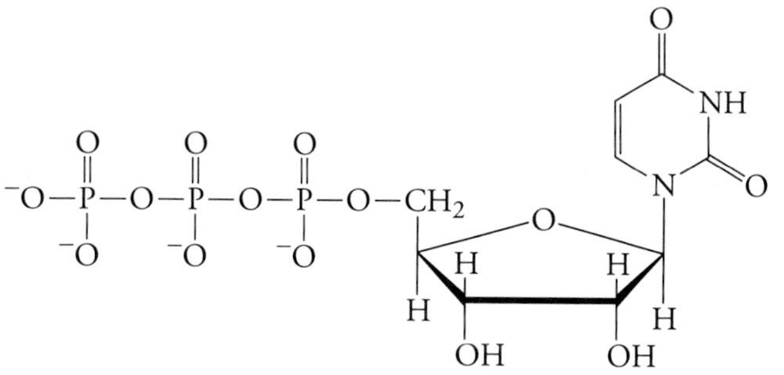

SOLUTIONS TO PRACTICE PROBLEMS

14.1 Uracil can be combined with ribose, but not with 2-deoxyribose.

14.2 600,000.

14.3 codons (listed 5′ to 3′): CGU, CGC, CGG, AGA, AGG; corresponding anticodons (listed 5′ to 3′): GCA, GCG, GCC, UCU, UCC.

At a recent family gathering, word got around that you are taking a chemistry class. One relative, an avid fisherman, said that he had a chemistry-related question for you. His favorite fishing lake had been recently closed by the Fish and Wildlife Department so that unwanted fish could be removed and replaced by trout. This is done by using a chemical called rotenone to kill all of the fish in the lake, after which the lake is restocked with desired fish species. Your relative has heard that rotenone suffocates fish and wonders if it affects people in the same way.

Photo source: © Corbis Digital Stock.

15 METABOLISM

In earlier chapters we looked at the structure and properties of lipids, carbohydrates, proteins, and nucleic acids, the four major classes of compounds important in biochemistry. Now we will turn our attention to the processes by which these molecules are either broken down or synthesized by living things. Learning about these processes can help us understand the health consequences of being exposed to certain types of compounds. In fish, for example, rotenone blocks an essential process called the electron transport chain.

objectives

After completing this chapter, you should be able to:

1. Define the terms metabolism, catabolism, and anabolism.
2. Describe three different types of metabolic pathways and explain what coupled reaction means.
3. Name the products formed during the digestion of polysaccharides, triglycerides, and proteins and state where the digestion of each takes place.
4. Identify the initial reactant and final products of glycolysis, describe how this pathway is controlled, and explain how gluconeogenesis differs from glycolysis. Describe how the manufacture and breakdown of glycogen are related to each of these pathways.
5. Give an overview of the citric acid cycle, explain how it is controlled, and describe how the products of this circular pathway are used by the electron transport chain and oxidative phosphorylation.
6. Describe the catabolism of triglycerides, the β oxidation spiral, and how the β oxidation spiral differs from fatty acid biosynthesis.
7. Explain the fate of the amino groups in amino acids.

15.1 METABOLIC PATHWAYS, ENERGY, AND COUPLED REACTIONS

■ Metabolic pathways can be linear, circular, or spiral.

To discuss how living things manufacture or break down carbohydrates, lipids, or members of any other biochemical class of compounds it is necessary to talk in terms of *groups of reactions* called **metabolic pathways.** The metabolic pathways discussed in this chapter will be **linear** (*a continuous series of reactions in which the product of one reaction is the reactant in the next*), **circular** (*a series of reactions where the final product is an initial reactant*), or **spiral** (*a series of repeated reactions is used to break down or build up a molecule*) (Figure 15.1).

(*a*) A ⟶ B ⟶ C ⟶ D ⟶ E

Linear

(*b*)

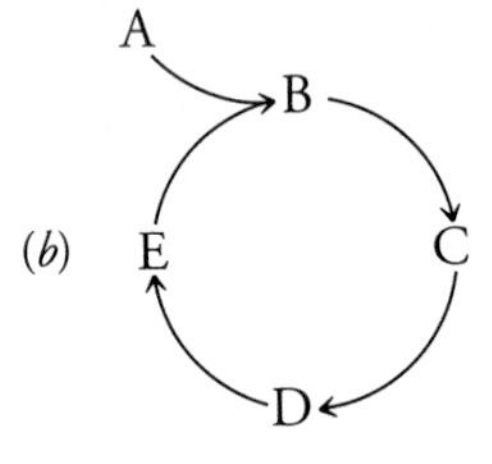

Circular

(*c*)

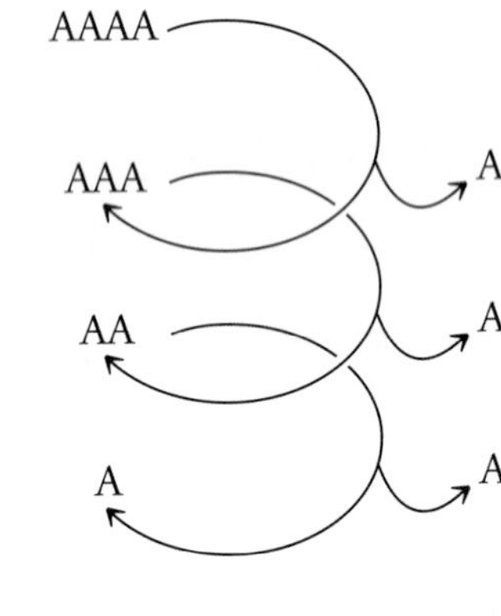

Spiral

FIGURE 15.1

Metabolic pathways
(*a*) Linear pathways are a continuous series of reactions in which the product of one reaction is the reactant in the next.
(*b*) In circular pathways the final product is one of the reactants used to begin the series again.
(*c*) In spiral pathways a series of repeated reactions is used to break down or build up a molecule.

As we saw in Section 6.7, free energy change (ΔG) can be used as a gauge of whether a reaction is spontaneous ($\Delta G < 0$) or nonspontaneous ($\Delta G > 0$). In the metabolic pathways introduced in this chapter we will see examples of how otherwise nonspontaneous reactions take place when coupled with an appropriate spontaneous reaction. In a **coupled reaction** *a spontaneous reaction provides the energy needed by a nonspontaneous one.*

The formation of adenosine 5′-triphosphate (ATP) from adenosine 5′-diphosphate (ADP) is an example of a coupled reaction (see Figure 15.2 for structures). This particular reaction, in which ADP is combined with phosphate ion (P_i), is nonspontaneous and will not take place without an input of energy.

$$\text{ADP} + \text{P}_i \longrightarrow \text{ATP} + \text{H}_2\text{O} \qquad \Delta G = +7.3 \text{ kcal}$$

In a number of different metabolic pathways, ATP is produced by coupling the reaction of ADP with P_i to a different reaction that releases enough energy to make the combined reaction spontaneous. In the first metabolic pathway that we will consider in depth, glycolysis (Section 15.4), one of the reactions used to generate ATP is the conversion of 1,3-bisphosphoglycerate into 3-phosphoglycerate. Individually, the reactions that take place are

$$\text{1,3-Bisphosphoglycerate} + \text{H}_2\text{O} \longrightarrow \text{3-phosphoglycerate} + \text{P}_i \qquad \Delta G = -11.8 \text{ kcal}$$
$$\text{ADP} + \text{P}_i \longrightarrow \text{ATP} + \text{H}_2\text{O} \qquad \Delta G = +7.3 \text{ kcal}$$

During glycolysis one enzyme catalyzes each of these reactions simultaneously, giving a coupled spontaneous reaction with a ΔG of -4.5 kcal/mol. The net reaction equation is obtained by adding the individual reaction equations and removing compounds that appear on both sides of the reaction arrow. The net free energy change is calculated by adding the ΔG values for the individual reactions that have been coupled.

$$\text{1,3-Bisphosphoglycerate} \cancel{+ \text{H}_2\text{O}} + \text{ADP} \cancel{+ \text{P}_i} \longrightarrow \text{3-phosphoglycerate} \cancel{+ \text{P}_i} + \text{ATP} \cancel{+ \text{H}_2\text{O}}$$
$$\Delta G = (-11.8 \text{ kcal}) + (+7.3 \text{ kcal})$$
$$\text{1,3-Bisphosphoglycerate} + \text{ADP} \longrightarrow \text{3-phosphoglycerate} + \text{ATP} \qquad \Delta G = -4.5 \text{ kcal}$$

The terms spontaneous and nonspontaneous can also be applied to an entire metabolic pathway. Making this determination is based on adding the ΔG values for all steps in the pathway. For the hypothetical pathway shown in Figure 15.1*a*, if ΔG values for the individual steps are -1 kcal/mol, $+5$ kcal/mol, -3 kcal/mol, and -6 kcal/mol, respectively, then the pathway is spontaneous, with $\Delta G = -5$ kcal/mol.

SAMPLE PROBLEM 15.1

Coupled reactions

One of the steps early in glycolysis involves the transformation of fructose 6-phosphate into fructose 1,6-bisphosphate in a reaction that is coupled to the hydrolysis of ATP.

$$\text{Fructose 6-phosphate} + P_i \longrightarrow \text{fructose 1,6-bisphosphate} + H_2O \qquad \Delta G = 3.9 \text{ kcal}$$

$$ATP + H_2O \longrightarrow ADP + P_i \qquad \Delta G = -7.3 \text{ kcal}$$

a. Is the conversion of fructose 6-phosphate into fructose 1,6-bisphosphate spontaneous or nonspontaneous in the absence of ATP?

b. Write the net coupled reaction equation and calculate ΔG for the overall reaction between fructose 6-phosphate and ATP.

c. Is the coupled reaction spontaneous or nonspontaneous?

Strategy

Recall that ΔG is negative for a spontaneous process and positive for a nonspontaneous one. You can obtain the net equation for the coupled reaction by placing all of the reactants to the left of a reaction arrow, all of the products to the right, and then canceling compounds that appear on both sides.

Solution

a. Nonspontaneous. ΔG has a positive value.

b. Fructose 6-phosphate + ATP $\longrightarrow$ fructose 1,6-bisphosphate + ADP
$\Delta G = -3.4$ kcal

c. Spontaneous. ΔG has a negative value.

PRACTICE PROBLEM 15.1

In a process called gluconeogenesis, fructose 1,6-bisphosphate is converted into fructose 6-phosphate. Which of the reaction equations below describe a spontaneous process? (Refer to Sample Problem 15.1.)

a. Fructose 1,6-bisphosphate + ADP $\longrightarrow$ fructose 6-phosphate + ATP

b. Fructose 1,6-bisphosphate + H_2O $\longrightarrow$ fructose 6-phosphate + P_i

15.2 OVERVIEW OF METABOLISM

In this chapter our discussion of **metabolism**, *the sum of all reactions that take place in a living thing*, will be divided into two parts, **catabolism** and **anabolism**. During catabolism compounds are *broken down into smaller ones in processes that, usually, release energy*. Anabolism involves the biosynthesis of *larger compounds from smaller ones in processes that, usually, require energy*.

- Nucleoside triphosphates (ATP, GTP, and UTP), coenzymes (NAD^+, $NADP^+$, and FAD), and acetyl-CoA are important to metabolism.

ATP (Figure 15.2) is a key player in metabolism. Energy released during catabolism is used to drive the formation of this compound. The energy obtained by hydrolyzing ATP

FIGURE | 15.2

Compounds important to metabolism

Energy released during catabolism is collected in ATP and, to some extent, GTP and UTP. This energy is released when ATP is hydrolyzed to ADP or AMP. (Note: *Adenosyl* refers to an adenosine residue.) The oxidized coenzymes NAD^+, $NADP^+$, and FAD are converted into their reduced counterparts during catabolism, and the reduced coenzymes NADH, NADPH, and $FADH_2$ play important roles in the generation of ATP or in anabolism. To activate many molecules for catabolism, a coenzyme A (CoA) residue is attached. Acetyl-CoA, one such activated molecule, is a product of the catabolism of carbohydrates, fatty acids, and some amino acids.

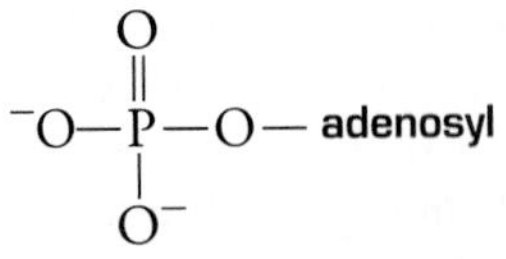

Adenosine 5'-monophosphate (AMP)

Adenosyl

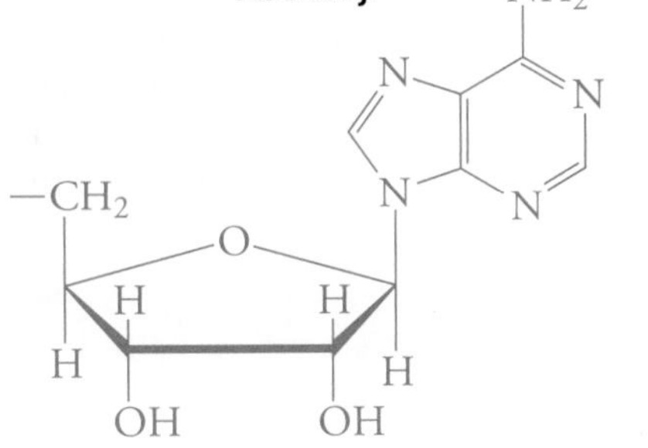

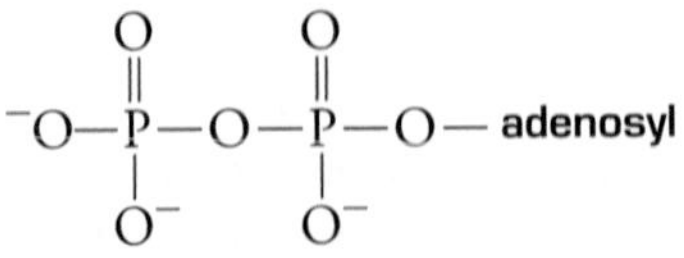

Adenosine 5'-diphosphate (ADP)

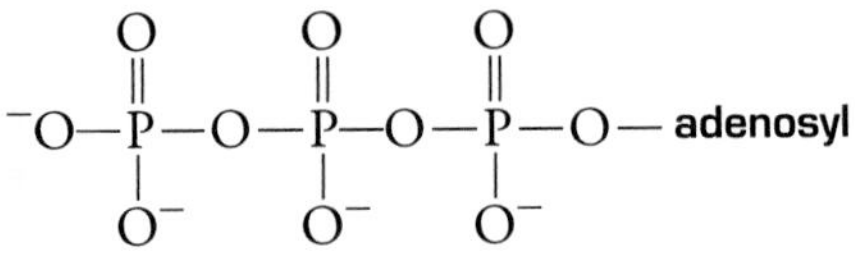

Adenosine 5'-triphosphate (ATP)

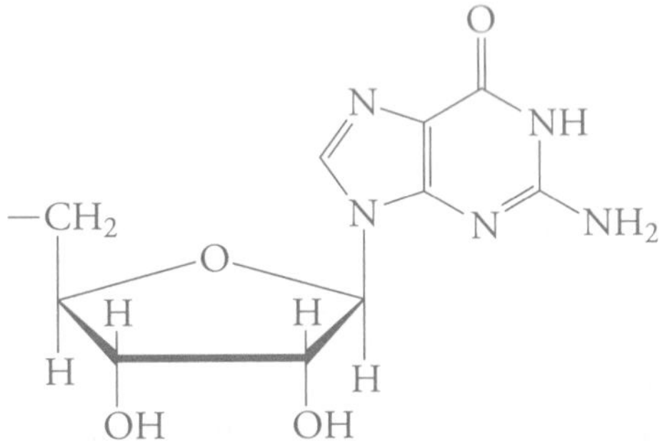

Guanosyl

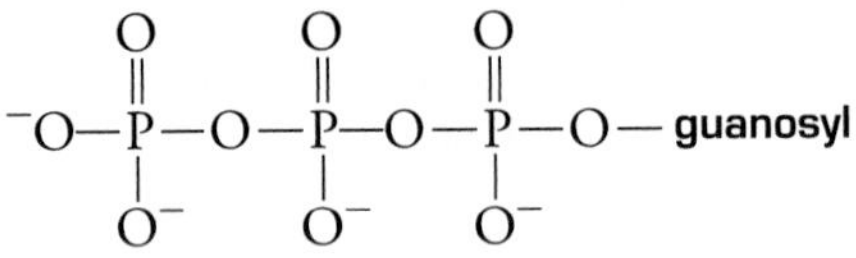

Guanosine 5'-triphosphate (GTP)

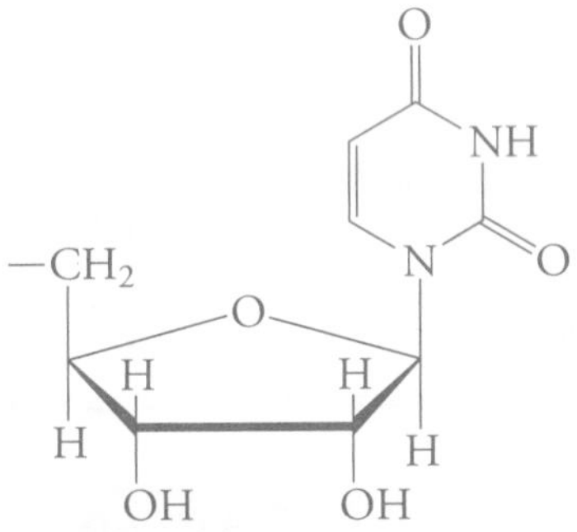

Uridinyl

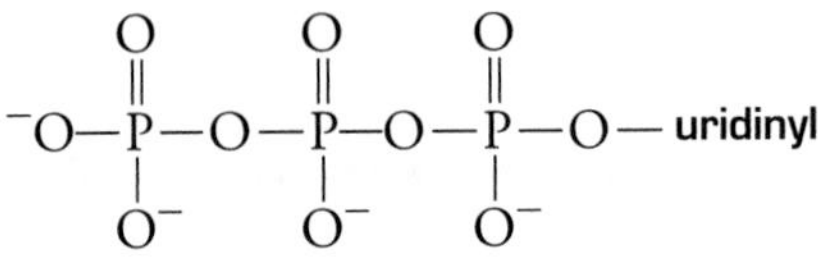

Uridine 5'-triphosphate (UTP)

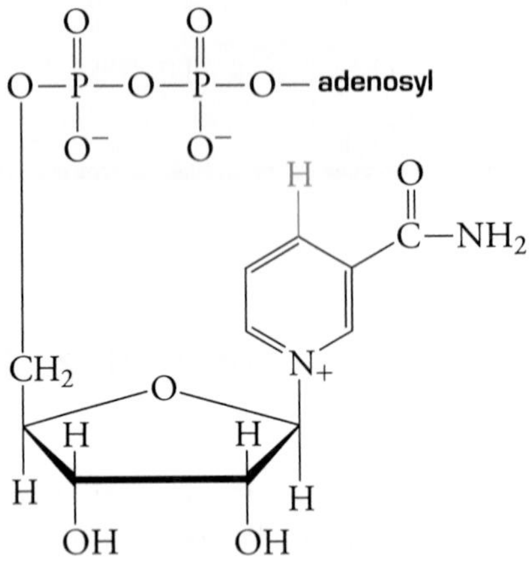

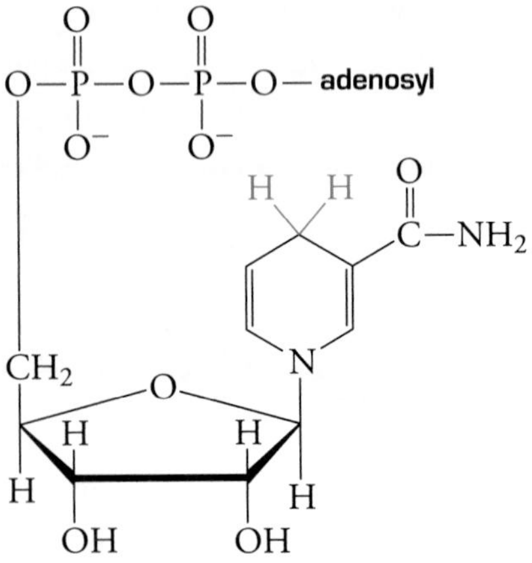

Nicotinamide adenine dinucleotide

NAD^+ (oxidized form)

NADH (reduced form)

FIGURE 15.2 Continued

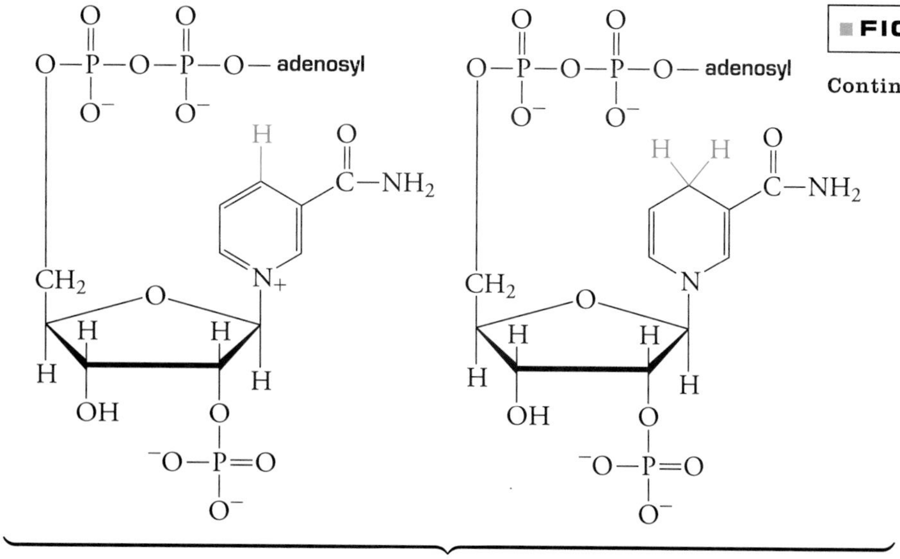

Nicotinamide adenine dinucleotide phosphate

NADP$^+$
(oxidized form)

NADPH
(reduced form)

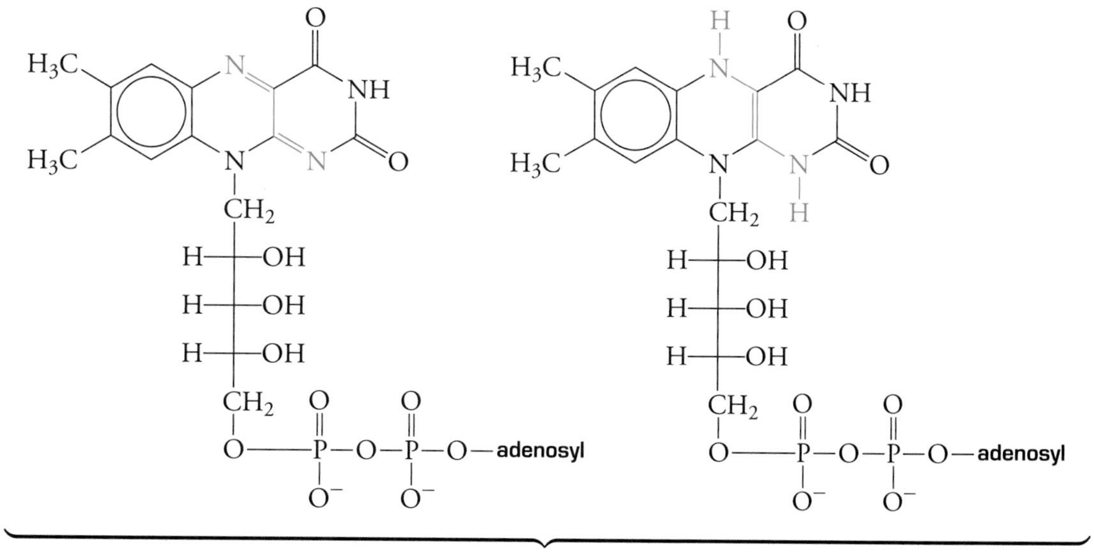

Flavin adenine dinucleotide

FAD
(oxidized form)

FADH$_2$
(reduced form)

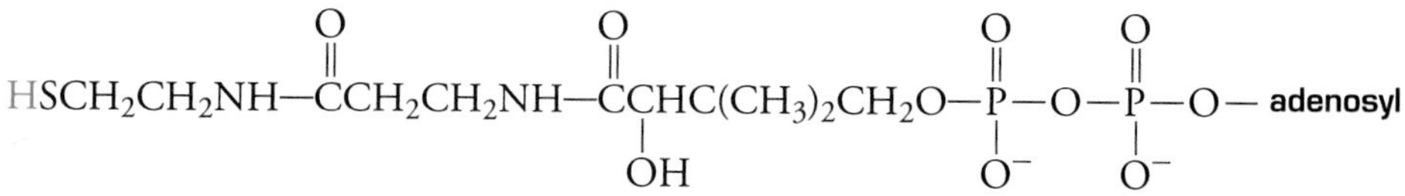

Coenzyme A (CoA)

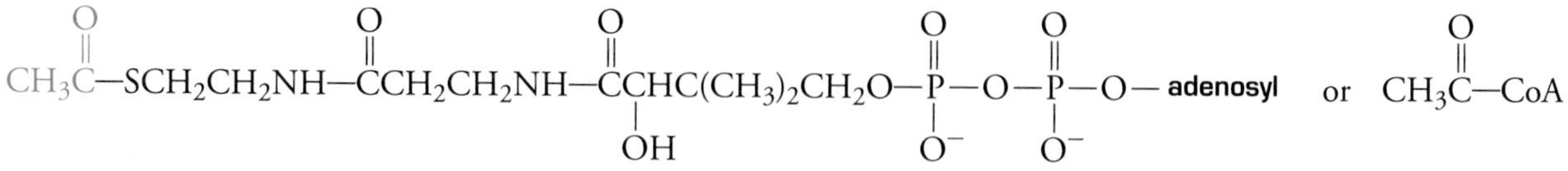

or $CH_3C(=O){-}CoA$

Acetyl-CoA

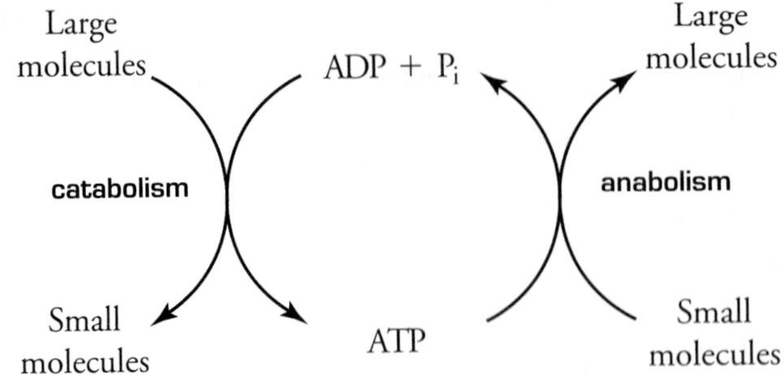

FIGURE 15.3

ATP and metabolism
ATP is produced using energy released during catabolism. Hydrolysis of ATP to produce ADP (or AMP, not shown) provides the energy used to drive anabolism.

can, in turn, be used for anabolism (Figure 15.3) or other energy-requiring processes, such as muscle contraction. The ATP hydrolysis reactions coupled to anabolism are

$$ATP + H_2O \longrightarrow ADP + P_i \qquad \Delta G = -7.3 \text{ kcal}$$
$$ATP + H_2O \longrightarrow AMP + PP_i \qquad \Delta G = -7.6 \text{ kcal}$$

where AMP is adenosine 5′-monophosphate (Figure 15.2) and PP_i is pyrophosphate.

$$^{-}O-\overset{\overset{\displaystyle O}{\|}}{\underset{\underset{\displaystyle ^{-}O}{|}}{P}}-O-\overset{\overset{\displaystyle O}{\|}}{\underset{\underset{\displaystyle O^{-}}{|}}{P}}-O^{-}$$

Pyrophosphate (PP_i)

Guanosine 5′-triphosphate (GTP) and uridine 5′-triphosphate (UTP) are two other nucleoside triphosphates involved in metabolism. The energy available from the hydrolysis of each is roughly equivalent to the energy supplied by ATP.

The coenzymes NAD^+, $NADP^+$, and FAD (Figure 15.2) are also important in metabolism. During oxidation reactions that take place in catabolism these oxidized coenzymes are converted to their reduced forms NADH, NADPH, and $FADH_2$, whose energy can be used to produce ATP or whose hydrogen atoms can be used during anabolism.

Coenzyme A (CoA) and acetyl-CoA are also central to metabolism. In some pathways, reactants are activated (their potential energy is raised) by the attachment of a coenzyme A residue. The energy released when this residue is hydrolyzed is used to drive otherwise nonspontaneous reactions. Acetyl-CoA is a product common to the breakdown of carbohydrates, fatty acids, and some amino acids and is the starting point for the citric acid cycle, a metabolic pathway that forms products involved in the generation of most of the ATP that originates from catabolism.

Catabolism

Catabolism begins with the digestion of food, during which triglycerides, carbohydrates, and proteins are split into their building blocks—fatty acids and glycerol, monosaccharides, and amino acids, respectively (Figure 15.4). Monosaccharides enter **glycolysis** (Section 15.4), which produces NADH and ATP during the *conversion of one hexose molecule into two pyruvates* (three carbon atoms each). Pyruvate can be converted into acetyl-CoA (two carbon atoms), which is the entry point for the **citric acid cycle** (Section 15.7), a series of reactions that *converts the carbon atoms in acetyl-CoA into CO_2*, generating GTP and reduced coenzymes in the process.

Fatty acids undergo fatty acid oxidation (Section 15.9), which produces NADH and $FADH_2$ as the *fatty acids are cut into two-carbon fragments to become acetyl-CoA.* Amino acids lose their amino groups (Section 15.10) and, depending on the amino acid, are converted into pyruvate, acetyl-CoA, intermediates in the citric acid cycle, or related compounds.

The NADH and $FADH_2$ produced in these processes can enter the **electron transport chain** (Section 15.8), where these *coenzymes are recycled to their oxidized forms* and the energy (in the form of electrons) released in the process is used during **oxidative phosphorylation** (Section 15.8) *to produce ATP.*

- Catabolism of food molecules results in the production of reduced coenzymes and ATP.

The catabolic pathways just mentioned do not all take place in the same part of a cell. Glycolysis occurs in the cytoplasm, while fatty acid oxidation, amino acid catabolism, the

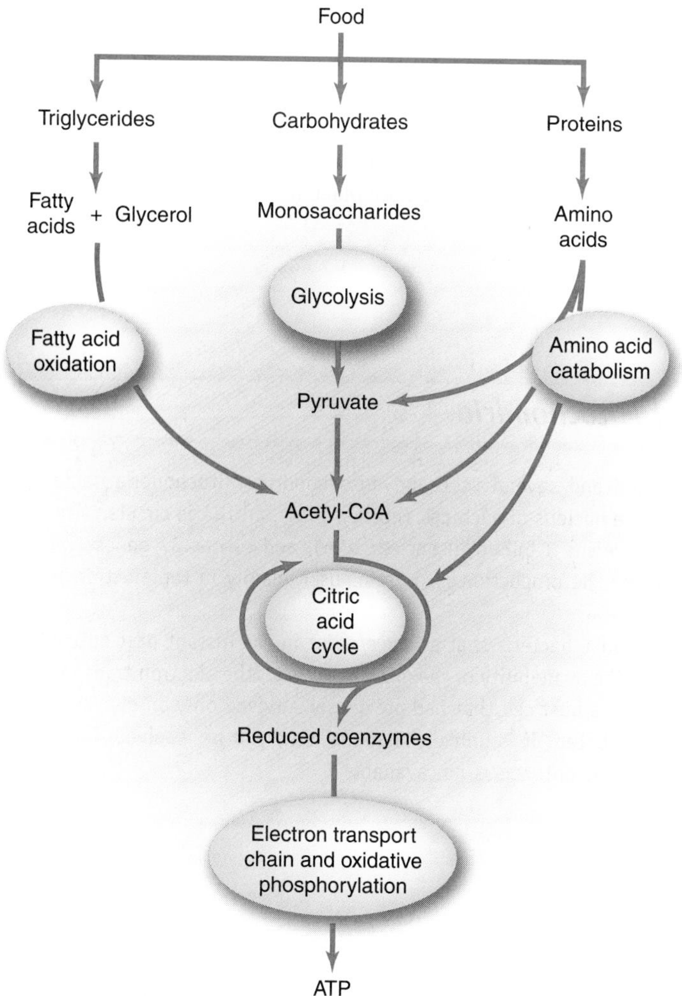

FIGURE 15.4

Catabolism
Fatty acids, glycerol, monosaccharides, and amino acids obtained from digestion enter various catabolic pathways, where reduced coenzymes and ATP are generated. Details on each pathway are presented later in this chapter.

citric acid cycle, the electron transport chain, and oxidative phosphorylation all take place in **mitochondria**, organelles found in plant and animal cells. These organelles, considered the powerhouse of cells because they *are where most of the ATP is produced*, have an outer membrane that is separated from a folded inner membrane by an intermembrane space. The outer membrane allows free passage of most ions and small molecules, but movement of substances across the inner membrane usually requires the action of transport proteins. The fluid inside a mitochondrion is called the matrix (Figure 15.5).

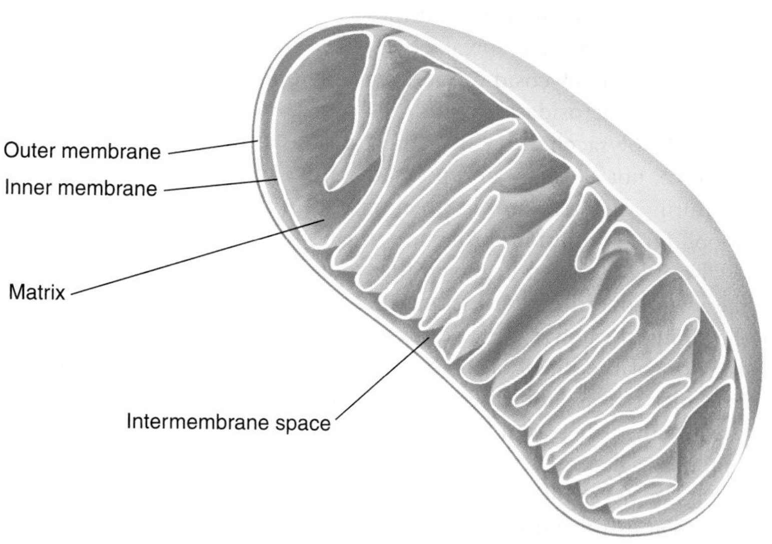

FIGURE 15.5

Mitochondrial structure
These organelles, found in animal and plant cells, contain the enzymes and other compounds required for fatty acid oxidation, amino acid catabolism, the citric acid cycle, the electron transport chain, and oxidative phosphorylation.

Source: AFTER D. VOET, J.G. VOET, AND C.W. PRATT, FUNDAMENTALS OF BIOCHEMISTRY: LIFE AT THE MOLECULAR LEVEL, SECOND EDITION ; COPYRIGHT 2006, DONALD VOET AND JUDITH VOET. REPRINTED BY PERMISSION OF JOHN WILEY AND SONS, INC.

Anabolism

During anabolism, small molecules such as pyruvate, acetyl-CoA, and intermediates in the citric acid cycle are used to make fatty acids, monosaccharides, and amino acids for incorporation into lipids, polysaccharides, and proteins. Anabolic pathways are not the exact reverse of the catabolic pathways shown in Figure 15.4—some make use of different enzymes and coenzymes, and some take place in different parts of a cell.

BIOCHEMISTRY Link

The Origin of Mitochondria

Every human cell contains between several hundred and several thousand mitochondria. Mitochondria contain their own DNA (mtDNA), which is different from the DNA found in the nucleus of each cell (nuclear DNA). mtDNA is circular, like bacterial plasmids (Section 14.12), contains about 17,000 base pairs (versus 3 billion in nuclear DNA), and carries 37 genes (versus approximately 25,000 in nuclear DNA). The genes in mtDNA code for the production of proteins used mainly in the electron transport chain and oxidative phosphorylation.

Scientists have proposed that mitochondria were once bacteria that at some point in the distant past entered into a symbiotic relationship with a different type of bacteria or cells. The availability of the citric acid cycle, the electron transport chain, and oxidative phosphorylation would have been an advantage to a host cell that had previously survived only on glycolysis.

This origin of mitochondria idea has been much debated. It is unlikely that the issue will be resolved, however, as the experimental evidence needed to support or to invalidate the hypothesis is not available.

15.3 DIGESTION

During **digestion** the *large molecules present in food are broken down into their respective building blocks*. Depending on the molecules involved, digestion takes place in the mouth, stomach, or small intestine (Figure 15.6).

The digestion of any starch present in food begins in the mouth, where salivary amylase catalyzes the hydrolysis of some of the α-(1→4) glycosidic bonds present in amylose and amylopectin. This enzyme does not act on any of the other oligo- or polysaccharides present in food. The action of salivary amylase stops soon after the food is swallowed, because the acidic conditions of the stomach rapidly denature the enzyme. After exiting the stomach and entering the small intestine, any remaining starch and some of the other oligo- and polysaccharides are split into monosaccharides. Starch is hydrolyzed to glucose by the action of amylase [breaks α-(1→4) glycosidic bonds] and debranching enzyme [breaks α-(1→6) glycosidic bonds in amylopectin] (Figure 15.7*a*). Other enzymes, including β-galactosidase (hydrolyzes lactose) and sucrase (hydrolyzes sucrose), split oligo- and polysaccharides into their monosaccharide building blocks. As noted in Chapter 12, not all oligo- and polysaccharides are digestible. The monosaccharides produced during digestion are absorbed by cells in the intestinal lining and are released into the blood for distribution throughout the body.

Triglyceride digestion begins in the stomach. Lingual lipase present in saliva (activated by the acidic conditions in the stomach, once saliva is swallowed) and gastric lipase secreted in the stomach catalyze the hydrolysis of triglycerides to yield diacylglycerides and fatty acids. Most of the digestion of triglycerides takes place in the small intestine. There bile acids, which act as emulsifiers (Section 8.5) to form lipid–water colloids, assist lipases secreted by the pancreas in breaking down diacylglycerides and the remaining triglycerides into monoacylglycerides, fatty acids, and glycerol (Figure 15.7*b*). These molecules cross into cells that line the small intestine and are eventually converted back into triglycerides, which are carried at first through lymph and then blood by lipoproteins (Section 8.5).

- Starch and triglycerides are primarily digested in the small intestine. Proteins are digested in the stomach and small intestine.

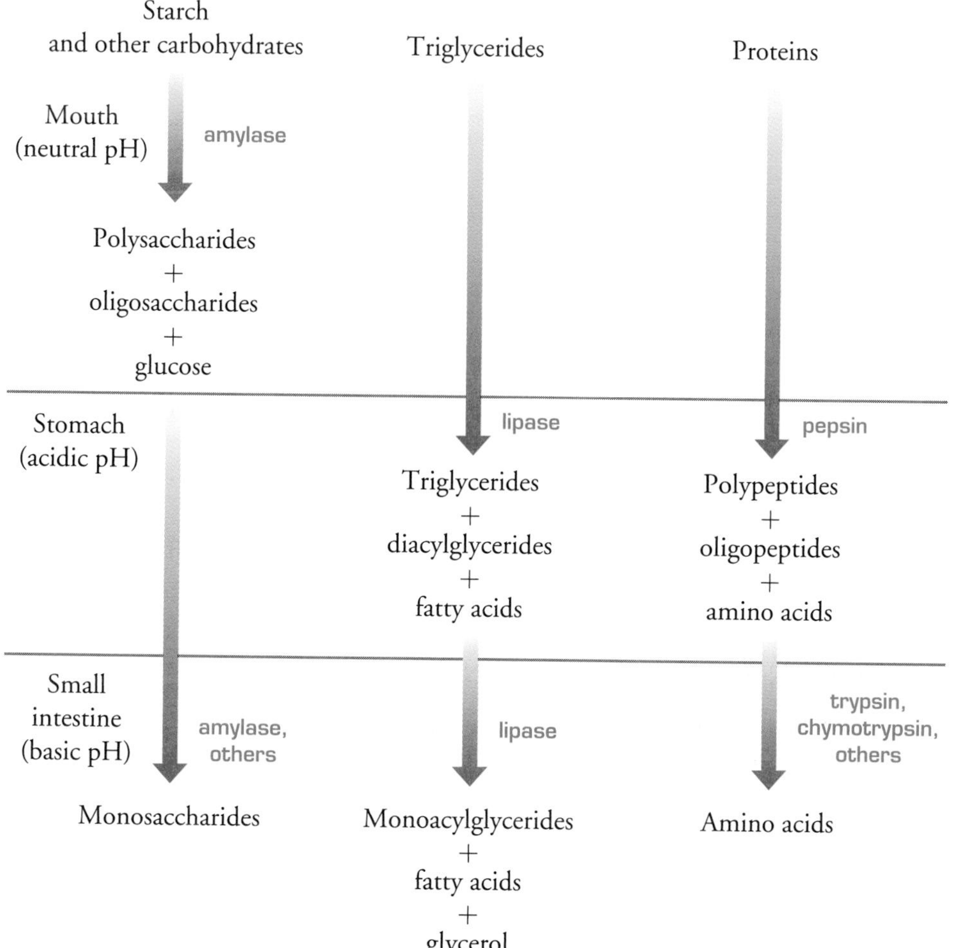

FIGURE 15.6

Digestion
The breakdown of starch to glucose takes place partly in the mouth and mostly in the small intestine. Some triglyceride digestion takes place in the stomach, but the majority occurs in the small intestine. Proteins are broken down to amino acids in the stomach and in the small intestine.

Protein digestion begins in the stomach, where pepsin catalyzes the hydrolysis of some of the peptide bonds that join amino acid residues (Figure 15.7*c*). When the mix of proteins, oligopeptides, and amino acids leaves the stomach and enters the small intestine, any unhydrolyzed peptide bonds are split by trypsin, chymotrypsin, and other protease enzymes. The resulting free amino acids are absorbed and move into blood.

FIGURE 15.7

Starch, triglyceride, and protein hydrolysis
(*a*) Hydrolysis of starch yields glucose molecules. The α-(1 → 4) glycosidic bonds in amylose and amylopectin are hydrolyzed by amylase and the α-(1 → 6) glycosidic bonds in amylopectin (not shown) are hydrolyzed by debranching enzyme. (*b*) Hydrolysis of triglycerides, catalyzed by lipases, yields fatty acids, glycerol, and monoacylglycerides (consisting of one fatty acid residue attached to glycerol). (*c*) Hydrolysis of proteins, catalyzed by proteases, gives amino acids.

Amylose

$\downarrow H_2O$

many Glucose

(*a*)

FIGURE 15.7

Continued

(b)

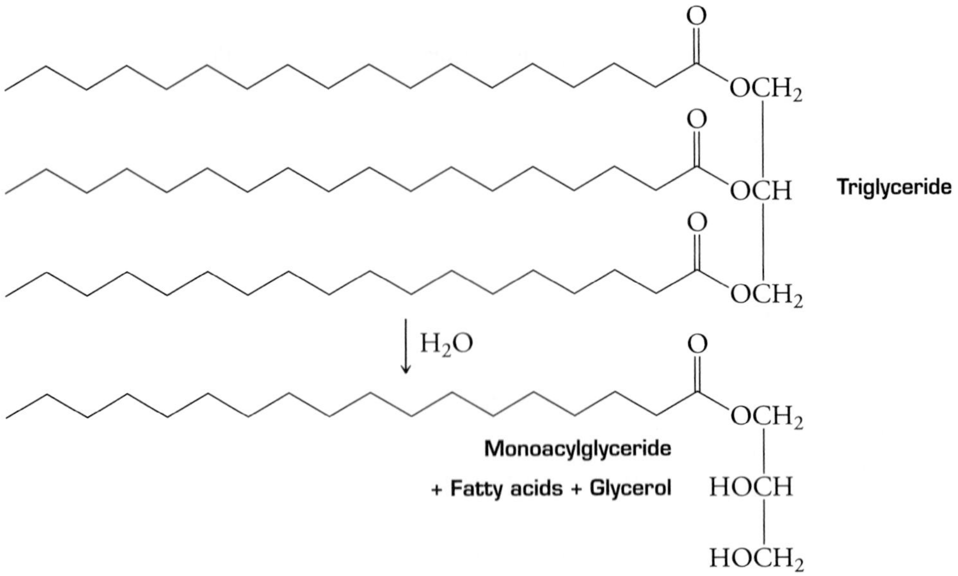

Triglyceride

$\downarrow H_2O$

Monoacylglyceride

+ Fatty acids + Glycerol

(c)

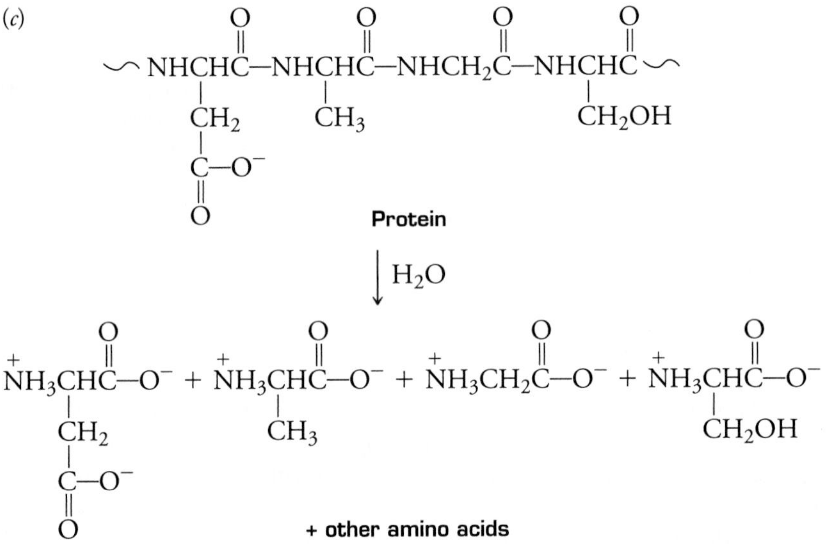

$$\sim NHCH(CH_2COO^-)C(=O)-NHCH(CH_3)C(=O)-NHCH_2C(=O)-NHCH(CH_2OH)C(=O)\sim$$

Protein

$\downarrow H_2O$

$$\overset{+}{N}H_3CH(CH_2COO^-)C(=O)O^- + \overset{+}{N}H_3CH(CH_3)C(=O)O^- + \overset{+}{N}H_3CH_2C(=O)O^- + \overset{+}{N}H_3CH(CH_2OH)C(=O)O^-$$

+ other amino acids

15.4 GLYCOLYSIS

In this section we focus on the catabolism of glucose via glycolysis (Figure 15.8). In this 10-step linear pathway, which takes place in the cytoplasm of a cell, one glucose molecule is converted into two pyruvate ions, with the accompanying production of 2 ATP and 2 NADH. Pyruvate is the conjugate base of pyruvic acid (recall that carboxylic acids exist in their conjugate base form at physiological pH).

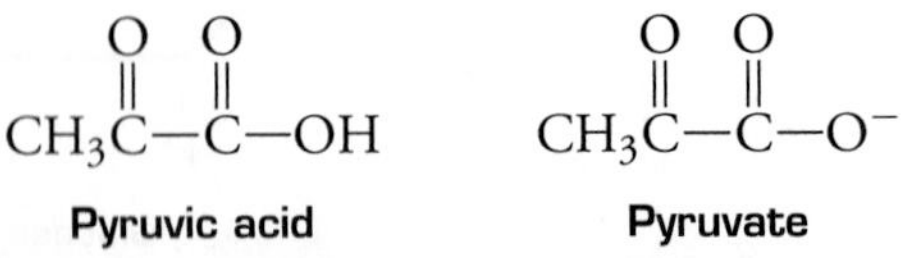

$$CH_3C(=O)-C(=O)-OH \qquad CH_3C(=O)-C(=O)-O^-$$

Pyruvic acid Pyruvate

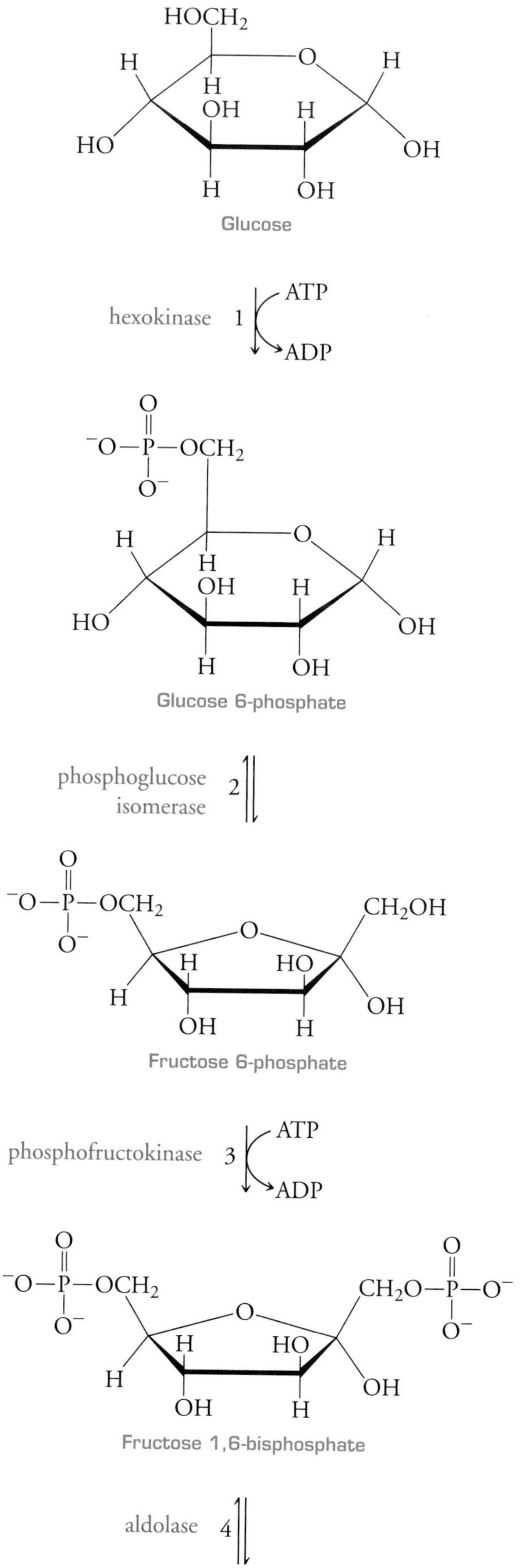

HOCH$_2$ H O H H OH H HO OH H OH

Glucose

hexokinase 1 ATP ADP

O $^-$O—P—OCH$_2$ O$^-$ H O H H OH H HO OH H OH

Glucose 6-phosphate

phosphoglucose isomerase 2

O $^-$O—P—OCH$_2$ O$^-$ O CH$_2$OH H HO H OH OH H

Fructose 6-phosphate

phosphofructokinase 3 ATP ADP

O $^-$O—P—OCH$_2$ O$^-$ O CH$_2$O—P—O$^-$ O O$^-$ H HO H OH OH H

Fructose 1,6-bisphosphate

aldolase 4

FIGURE 15.8

Glycolysis

Steps 1–5 of glycolysis convert one glucose molecule into two glyceraldehyde 3-phosphates at the expense of two ATP. Steps 6–10, which involve oxidation, rearrangement, and phosphate group transfer, produce 2 NADH and 4 ATP for each glucose that enters the pathway. Overall, the conversion of one glucose to two pyruvates produces a net 2 ATP and 2 NADH.

FIGURE 15.8

Continued

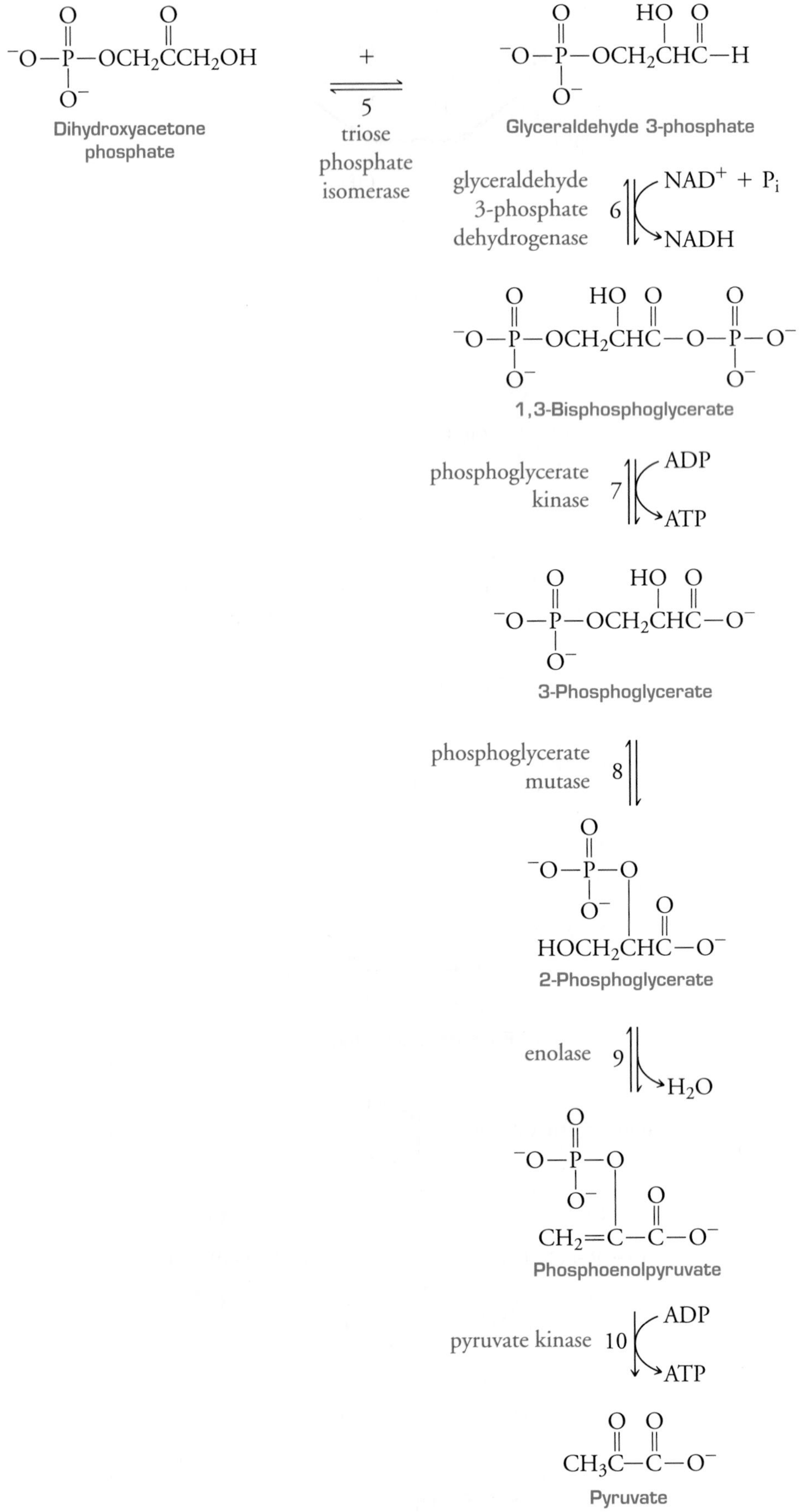

Dihydroxyacetone phosphate: $^{-}O{-}P(=O)(O^{-}){-}OCH_2C(=O)CH_2OH$

$\rightleftharpoons$ 5 triose phosphate isomerase

+ Glyceraldehyde 3-phosphate: $^{-}O{-}P(=O)(O^{-}){-}OCH_2CH(OH)C(=O){-}H$

6 glyceraldehyde 3-phosphate dehydrogenase ($NAD^+ + P_i$ → NADH)

1,3-Bisphosphoglycerate: $^{-}O{-}P(=O)(O^{-}){-}OCH_2CH(OH)C(=O){-}O{-}P(=O)(O^{-}){-}O^{-}$

7 phosphoglycerate kinase (ADP → ATP)

3-Phosphoglycerate: $^{-}O{-}P(=O)(O^{-}){-}OCH_2CH(OH)C(=O){-}O^{-}$

8 phosphoglycerate mutase

2-Phosphoglycerate: $HOCH_2CH(O{-}P(=O)(O^{-}){-}O^{-})C(=O){-}O^{-}$

9 enolase (→ H_2O)

Phosphoenolpyruvate: $CH_2{=}C(O{-}P(=O)(O^{-}){-}O^{-}){-}C(=O){-}O^{-}$

10 pyruvate kinase (ADP → ATP)

Pyruvate: $CH_3C(=O){-}C(=O){-}O^{-}$

The net reaction for glycolysis is

$$\text{Glucose} + 2\text{NAD}^+ + 2\text{ADP} + 2\text{P}_\text{i} \longrightarrow 2\text{pyruvate} + 2\text{NADH} + 2\text{ATP} + \text{energy}$$

In step 1, the hydrolysis of an ATP is coupled to the attachment of phosphate to glucose. After isomerization of glucose 6-phosphate to fructose 6-phosphate in step 2, ATP hydrolysis (step 3) is coupled to the addition of phosphate to fructose 6-phosphate.

In step 4, a six-carbon molecule (fructose 1,6-bisphosphate) is split into two three-carbon molecules (dihydroxyacetone phosphate and glyceraldehyde 3-phosphate), and in step 5, dihydroxyacetone phosphate is isomerized into glyceraldehyde 3-phosphate. Since one glucose molecule yields two glyceraldehyde 3-phosphates, steps 6–10 take place twice for each glucose molecule that enters glycolysis.

Steps 6–10 involve oxidation of glyceraldehyde 3-phosphate (makes NADH), removal of phosphate from 1,3-bisphosphoglycerate (makes ATP), rearrangement of 3-phosphoglycerate, dehydration of 2-phosphoglycerate, and transfer of phosphate from phosphoenolpyruvate (makes ATP). When one glucose molecule enters glycolysis, the early stages of the pathway consume 2 ATP and the later stages generate 4 ATP and 2 NADH (steps 6–10 occur twice). This gives a net gain of 2 ATP and 2 NADH from each glucose molecule.

Control of Glycolysis

The rate of glycolysis is controlled at steps 1, 3, and 10, with step 3—catalyzed by the enzyme phosphofructokinase—being the main control point. For this allosteric enzyme (Section 13.7), ATP and citrate are negative effectors (they switch the enzyme off) and ADP and AMP are positive effectors (they switch the enzyme on). When a cell is energy-rich, the concentrations of ATP and citrate ion, an intermediate in the citric acid cycle (Section 15.7), are high. Under these conditions, further production of ATP by glycolysis is inhibited. When a cell is energy-poor, the concentrations of ADP and AMP are high and phosphofructokinase is activated to allow glycolysis to proceed.

■ Glycolysis converts 1 glucose molecule into 2 pyruvates, 2 ATP, and 2 NADH.

Hexokinase, the allosteric enzyme that catalyzes step 1 of glycolysis, has glucose 6-phosphate as a negative effector. A rise in glucose 6-phosphate concentration, which can be caused by inhibition of step 3, by the manufacture of glucose (Section 15.5) or by the breakdown of glycogen (Section 15.6), causes step 1 of glycolysis to slow.

In step 10, the third control point of glycolysis, the allosteric enzyme pyruvate kinase has as negative effectors ATP, fructose 1,6-bisphosphate (the reactant in step 4), and alanine (an amino acid that can be produced when pyruvate concentrations are high; Section 15.10).

SAMPLE PROBLEM 15.2

Glycolysis

For glycolysis to occur an initial input of energy must take place. In which coupled reactions does this take place and which compound provides the energy used to drive these reactions?

Strategy

In a coupled reaction, a spontaneous reaction is used to drive a nonspontaneous one.

Solution

In steps 1 and 3. ATP is the immediate source of energy.

PRACTICE PROBLEM 15.2

Which step in glycolysis involves both oxidation and reduction? What is oxidized and what is reduced?

Pyruvate

Metabolism can take pyruvate in a number of different directions. In yeast, pyruvate undergoes **alcoholic fermentation** (Figure 15.9*a*). In this process *pyruvate is split into acetaldehyde plus CO_2, and acetaldehyde is reduced to ethanol.* These reactions serve to recycle NADH back into NAD^+, allowing glycolysis to continue. If NADH were not recycled, there would not be sufficient amounts of NAD^+ available for step 6 of glycolysis and the reactions of the pathway would come to a stop.

FIGURE 15.9

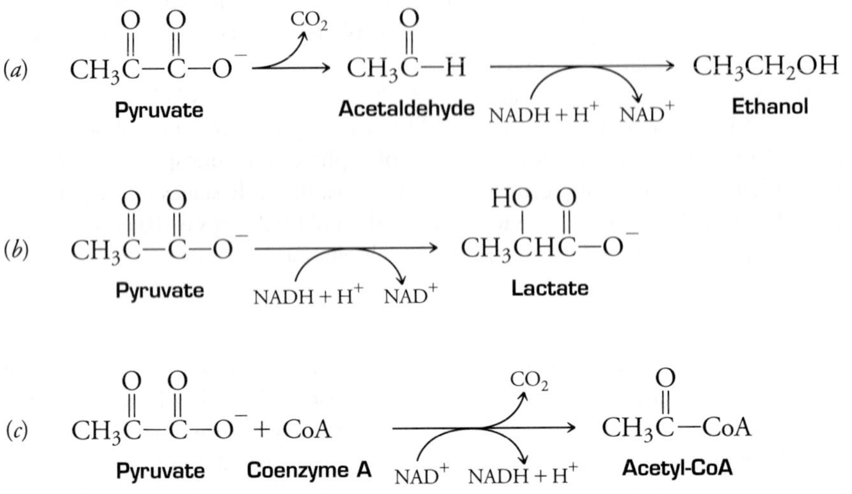

Reactions of pyruvate
(*a*) In yeast, pyruvate is decarboxylated to form acetaldehyde, which is reduced to ethanol. (*b*) Under anaerobic conditions pyruvate is reduced to lactate. (*c*) Under aerobic conditions, pyruvate moves to mitochondria where it is converted into acetyl-CoA.

(*a*) $CH_3C(=O)-C(=O)-O^- \xrightarrow{CO_2} CH_3C(=O)-H \xrightarrow[NADH + H^+ \;\to\; NAD^+]{} CH_3CH_2OH$

Pyruvate → Acetaldehyde → Ethanol

(*b*) $CH_3C(=O)-C(=O)-O^- \xrightarrow[NADH + H^+ \;\to\; NAD^+]{} CH_3CH(OH)C(=O)-O^-$

Pyruvate → Lactate

(*c*) $CH_3C(=O)-C(=O)-O^- + CoA \xrightarrow[NAD^+ \;\to\; NADH + H^+]{CO_2} CH_3C(=O)-CoA$

Pyruvate + Coenzyme A → Acetyl-CoA

■ Pyruvate can be converted into ethanol plus CO_2, lactate, or acetyl-CoA plus CO_2.

In humans, pyruvate is reduced to lactate when conditions are **anaerobic** (*O_2 deficient*), such as during sprinting or other vigorous exercise (Figure 15.9*b*). As with alcoholic fermentation, this reaction converts NADH into the NAD^+ needed for glycolysis to continue. Once produced, lactate is sent in blood to the liver, where it can be used to manufacture glucose (see gluconeogenesis, Section 15.5).

Under **aerobic conditions** (*O_2 is in sufficient supply*), pyruvate is converted into acetyl-CoA, the reactant for the first step in the citric acid cycle. Glycolysis takes place in a cell's cytoplasm, but the formation of acetyl-CoA and the citric acid cycle take place inside mitochondria. Before pyruvate can be converted into acetyl-CoA, it must move across the outer mitochondrial membrane by diffusion and across the inner mitochondrial membrane with the help of a transporter protein. Once inside the mitochondrial matrix, pyruvate encounters the pyruvate dehydrogenase complex (three different enzymes and five different coenzymes), which catalyzes the reactions required to convert pyruvate into acetyl-CoA (Figure 15.9*c*). During this reaction, NAD^+ is converted into NADH. Pyruvate that moves into the mitochondrion for conversion into acetyl-CoA cannot be used to recycle NADH in the cytoplasm, as is done under anaerobic conditions. However, NADH produced in the cytoplasm during glycolysis can be recycled to NAD^+ by the action of shuttles associated with the mitochondrial membrane (Section 15.8).

The pyruvate dehydrogenase complex is inhibited by ATP, NADH, and acetyl-CoA. When a cell is in an energy-rich state the concentrations of these compounds are high and the demand for them is reduced.

Energy and Glycolysis

In humans, when one mole of glucose enters glycolysis, under anaerobic conditions the final products of the pathway are two moles of lactate and two moles of ATP (NADH gets recycled into NAD^+ during the conversion of pyruvate into lactate). Because ΔG for this series of reactions has a negative value, the process is spontaneous.

$$\text{Glucose} + 2\text{ADP} + 2\text{P}_\text{i} \longrightarrow 2\text{lactate} + 2\text{ATP} \qquad \Delta G = -29.4 \text{ kcal}$$

The conversion of ADP into ATP has a ΔG of +7.3 kcal/mol, so the energetic cost of making two moles of ATP is +14.6 kcal/mol. After producing ATP, glycolysis is still spontaneous by 29.4 kcal/mol, so of the total free energy released by the conversion of one mole of glucose into two moles of lactate (29.4 + 14.6 = 44.0 kcal/mol) only 33.2% (14.6/44.0 × 100) is used to generate ATP. The remaining energy is released as heat.

Entry of Other Monosaccharides into Glycolysis

All of the common monosaccharides can enter glycolysis at some point, so glucose is not the only monosaccharide that can be catabolized by this pathway. Examples include the conversion of galactose to glucose 6-phosphate, mannose to fructose 6-phosphate, and fructose to fructose 6-phosphate or dihydroxyacetone phosphate. While the number of reactions required to convert a particular monosaccharide into an intermediate in glycolysis varies, the net gain of ATP and NADH from glycolysis is usually the same as for glucose.

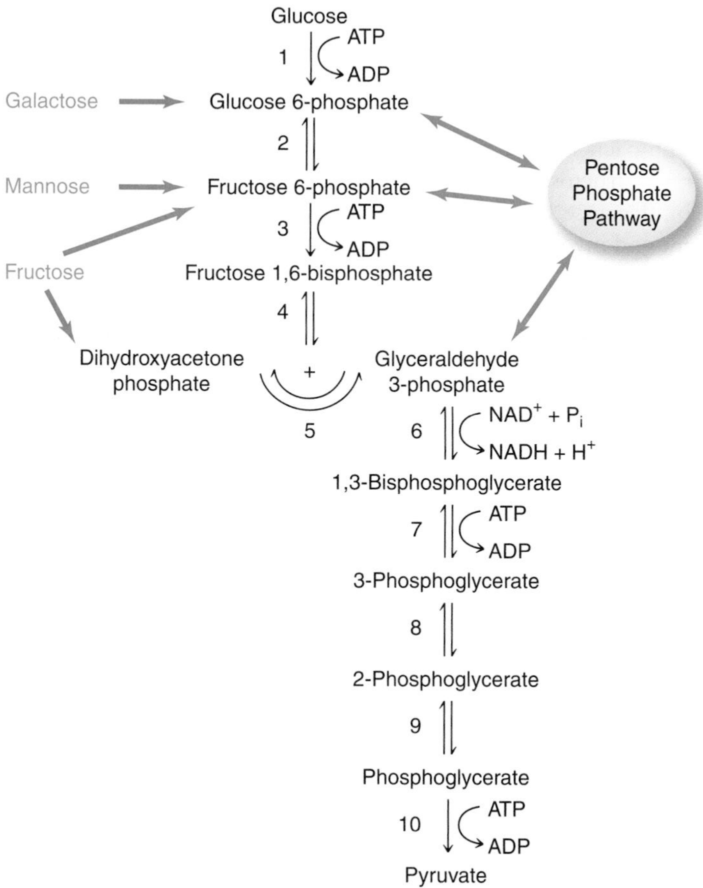

Some of the intermediates in glycolysis are also the starting point for the **pentose phosphate pathway**, a metabolic pathway that is *responsible for the production of ribose and 2-deoxyribose, other 3-, 4-, 5-, 6-, and 7-carbon sugar molecules, and NADPH,* a coenzyme required for the synthesis of fatty acids (Section 15.9). This pathway branches off from glycolysis at glucose 6-phosphate, fructose 6-phosphate, and glyceraldehyde 3-phosphate.

15.5 GLUCONEOGENESIS

Gluconeogenesis, the linear pathway involved in *making glucose from noncarbohydrate sources*, such as lactate, glycerol, and amino acids, takes place mostly in the liver. One important role of this process is the conversion of the lactate produced during anaerobic catabolism back into glucose, which either is transformed into glycogen (Section 15.6) or goes into the blood and is transported to other cells. In addition to recycling lactate, gluconeogenesis is a provider of glucose during fasting or in the early stages of starvation, when glucose and glycogen (a source of glucose) have been depleted. This supply of glucose is especially important to brain cells, which use only glucose to fuel metabolism, unlike other cells in the body which can also use lipids and proteins.

Gluconeogenesis is not simply the reverse of glycolysis. While these two pathways share many reactions and enzymes, three of the steps in glycolysis (steps 1, 3, and 10) are too spontaneous to be reversed. To overcome this free energy problem, in gluconeogenesis the reverse reactions of these three steps are catalyzed by different enzymes than the forward reactions, and follow different paths (Figure 15.10). In step 1 of glycolysis the hexokinase-catalyzed conversion of one glucose molecule to one glucose 6-phosphate is coupled with the consumption of one ATP.

$$\text{Glucose} + \text{ATP} \longrightarrow \text{glucose 6-phosphate} + \text{ADP} \qquad \Delta G = -4.0 \text{ kcal}$$

In gluconeogenesis the conversion of glucose 6-phosphate to glucose involves a hydrolysis reaction catalyzed by glucose 6-phosphatase. Removing ATP from the picture makes the reverse reaction spontaneous.

$$\text{Glucose 6-phosphate} \longrightarrow \text{glucose} + \text{P}_i \qquad \Delta G = -3.3 \text{ kcal}$$

The same approach is used at step 3 of glycolysis. The forward direction (fructose 6-phosphate $\longrightarrow$ fructose 1,6-bisphosphate) is coupled to the hydrolysis of ATP. The reverse direction (fructose 1,6-bisphosphate $\longrightarrow$ fructose 6-phosphate) is catalyzed by a different enzyme and involves only hydrolysis of a phosphate ester bond. ATP is not a participant in the reaction.

Reversing step 10 of glycolysis requires several steps. In the forward direction, the formation of pyruvate from phosphoenolpyruvate is coupled to the generation of ATP from ADP. To produce phosphoenolpyruvate from pyruvate, pyruvate is converted

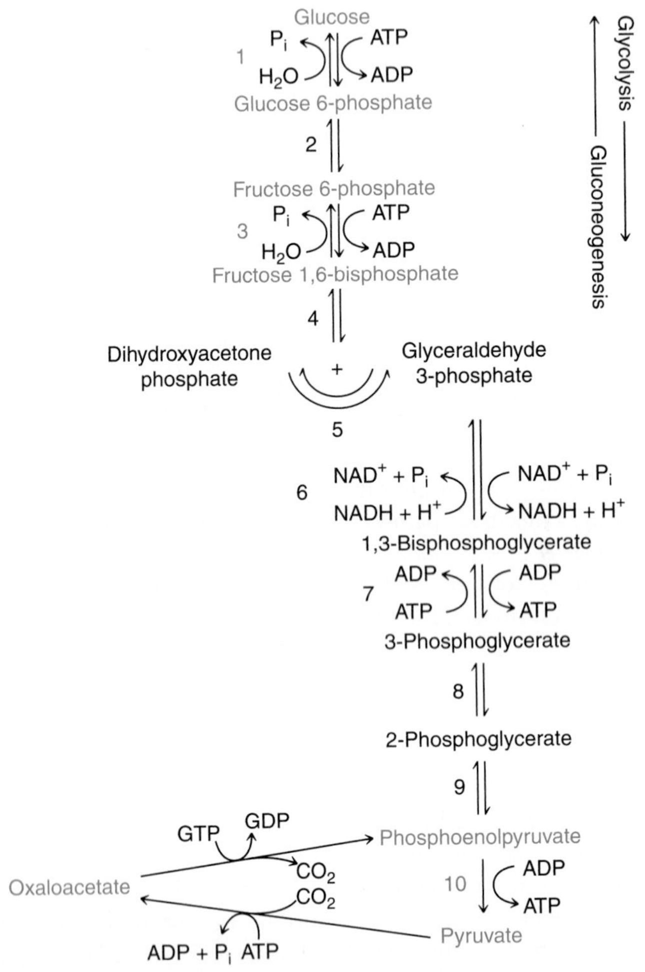

FIGURE 15.10

Gluconeogenesis
The formation of glucose from pyruvate is not the exact reverse of glycolysis. To reverse the changes that take place at steps 1, 3, and 10 of glycolysis, different enzymes are used and different reactions take place.

to oxaloacetate (an intermediate in the citric acid cycle) which is transformed into phosphoenolpyruvate.

As is the case for glycolysis, compounds can enter the gluconeogenesis pathway at a variety of steps. These include the conversion of lactate into pyruvate (the reverse of the reaction in Figure 15.9*b*), the conversion of glycerol into dihydroxyacetone phosphate (Section 15.9), and the conversion of amino acids into various compounds linked to gluconeogenesis (Section 15.10).

SAMPLE PROBLEM 15.3

Glucose metabolism

a. In the conversion of one glucose into two pyruvates during glycolysis, what is the net change in NADH and ATP (or ATP equivalents)?

b. In the conversion of two pyruvates into one glucose during gluconeogenesis, what is the net change in NADH and ATP or ATP equivalents (GTP is equivalent in energy to ATP)?

Strategy

Use Figure 15.10 as a map for these processes, and trace the change in NADH and ATP for each.

Solution

a. +2 NADH and +2 ATP

b. −2 NADH and −6 ATP (actually −4 ATP and −2 GTP)

PRACTICE PROBLEM 15.3

a. Is glycolysis catabolic or anabolic?

b. Is gluconeogenesis catabolic or anabolic?

15.6 GLYCOGEN METABOLISM

Glycogen, a highly branched homopolysaccharide consisting of α-(1→4) and α-(1→6) linked glucose residues (Section 12.5), is found mainly in liver and muscle cells. This carbohydrate is a glucose storage molecule that, when necessary, can be quickly broken down to release glucose.

The *synthesis of glycogen* (**glycogenesis**) begins with glucose 6-phosphate, a compound formed early in glycolysis or late in gluconeogenesis. To make glycogen, glucose 6-phosphate is isomerized into glucose 1-phosphate which is then reacted with uridine 5′-triphosphate (UTP) to form glucose-UDP, an activated molecule that can be used to add a glucose residue to the end of a growing glycogen chain. In the *breakdown of glycogen* (**glycogenolysis**), glucose residues are removed from the ends of the polymer, becoming glucose 1-phosphate, which is isomerized into glucose 6-phosphate (Figure 15.11).

■ Glycogen is synthesized from or is broken down into glucose 6-phosphate.

When the concentration of glucose is high, glycogenesis is triggered and excess glucose is stored as glycogen. When the concentration of glucose is low, glycogenolysis begins. In liver cells, the product glucose 6-phosphate is hydrolyzed to glucose, which enters the blood and is transported to cells throughout the body. In muscle cells the glucose 6-phosphate formed by glycogenolysis enters glycolysis.

The key enzymes involved in the making of glycogen (glycogen synthase) and the breakdown of glycogen (glycogen phosphorylase) are controlled by covalent modification (Section 13.7), and when one is switched on, the other is switched off. Which enzyme is active is indirectly determined by the pancreatic hormones insulin and

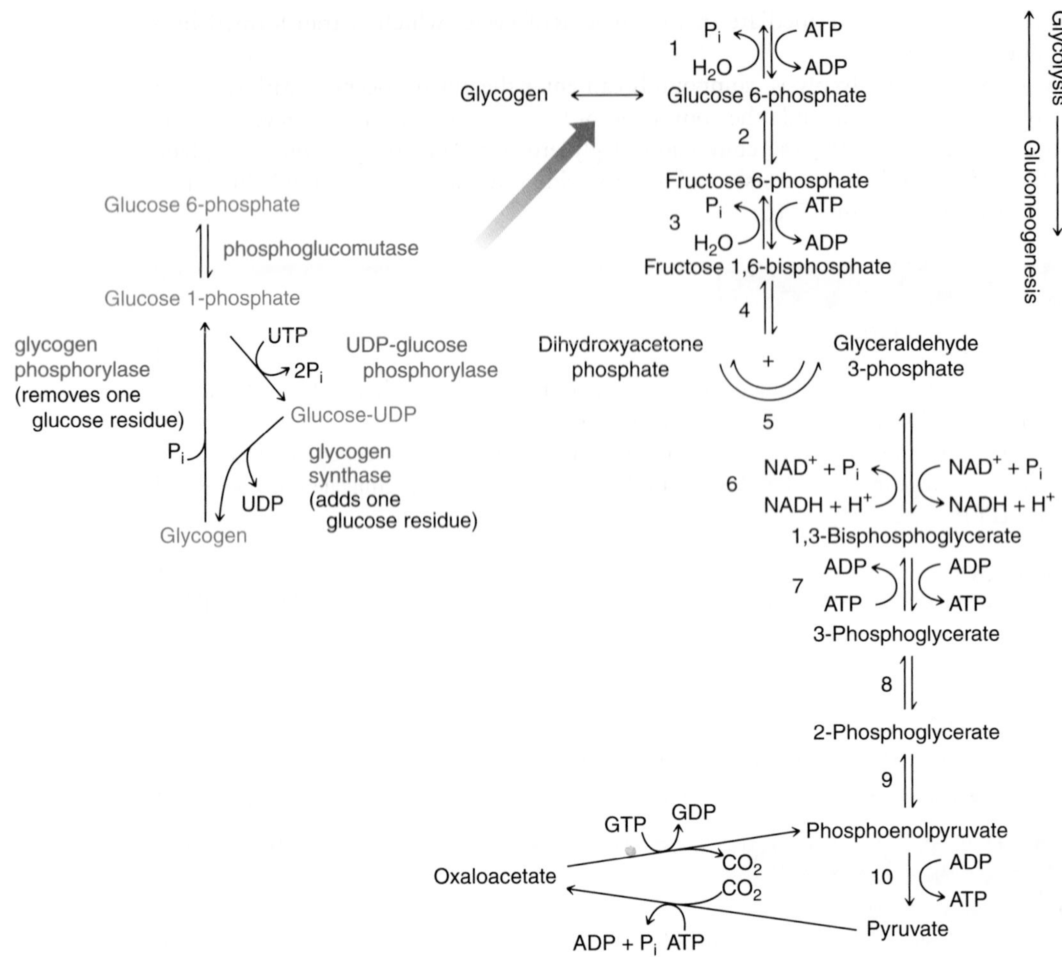

FIGURE 15.11 **Glycogen metabolism**
The synthesis of glycogen (glycogenesis) begins at glucose 6-phosphate and requires three steps. The removal of a glucose residue from glycogen (glycogenolysis) requires two steps and produces glucose 6-phosphate.

glucagon. When blood glucose levels rise, insulin is released to reduce the amount of glucose present. One way that insulin does so is by activating glycogen synthase (speeds up glycogen synthesis) and deactivating glycogen phosphorylase (slows down glycogen breakdown). Insulin also lowers blood glucose by causing an increase in the uptake of blood glucose by cells, an increase in the rate of glycolysis, and an increase in the synthesis of lipids and proteins.

Glucagon has the reverse effect. When blood glucose concentrations drop, glucagon is released to increase glucose levels. Glycogenolysis is switched on and glycogenesis is switched off by covalent modification of the appropriate enzymes, the movement of glucose from the blood into cells slows, and the breakdown of proteins and lipids is accelerated. The products of the latter two reactions provide amino acids and glycerol that can be used to synthesize glucose by gluconeogenesis.

15.7 CITRIC ACID CYCLE

During glycolysis some of the potential energy in a glucose molecule is extracted to form 2 pyruvates, 2 ATP, 2 NADH, and heat. The pyruvates can be catabolized further by converting them into acetyl-CoA (Figure 15.9*c*), which can enter the citric acid cycle. The citric acid cycle, which takes place in mitochondria, has eight steps and is circular because step 1 is the combination of acetyl-CoA with oxaloacetate, the product of the final reaction in the pathway (Figure 15.12). The products of one turn of the citric acid cycle are 2 CO_2 molecules, 3 NADH, 1 $FADH_2$, and 1 GTP. The net reaction is

$$\text{Acetyl-CoA} + 3\text{NAD}^+ + \text{FAD} + \text{GDP} + \text{P}_i \longrightarrow 2\text{CO}_2 + \text{CoA} + 3\text{NADH} + \text{FADH}_2 + \text{GTP}$$

$$\Delta G = -11 \text{ kcal}$$

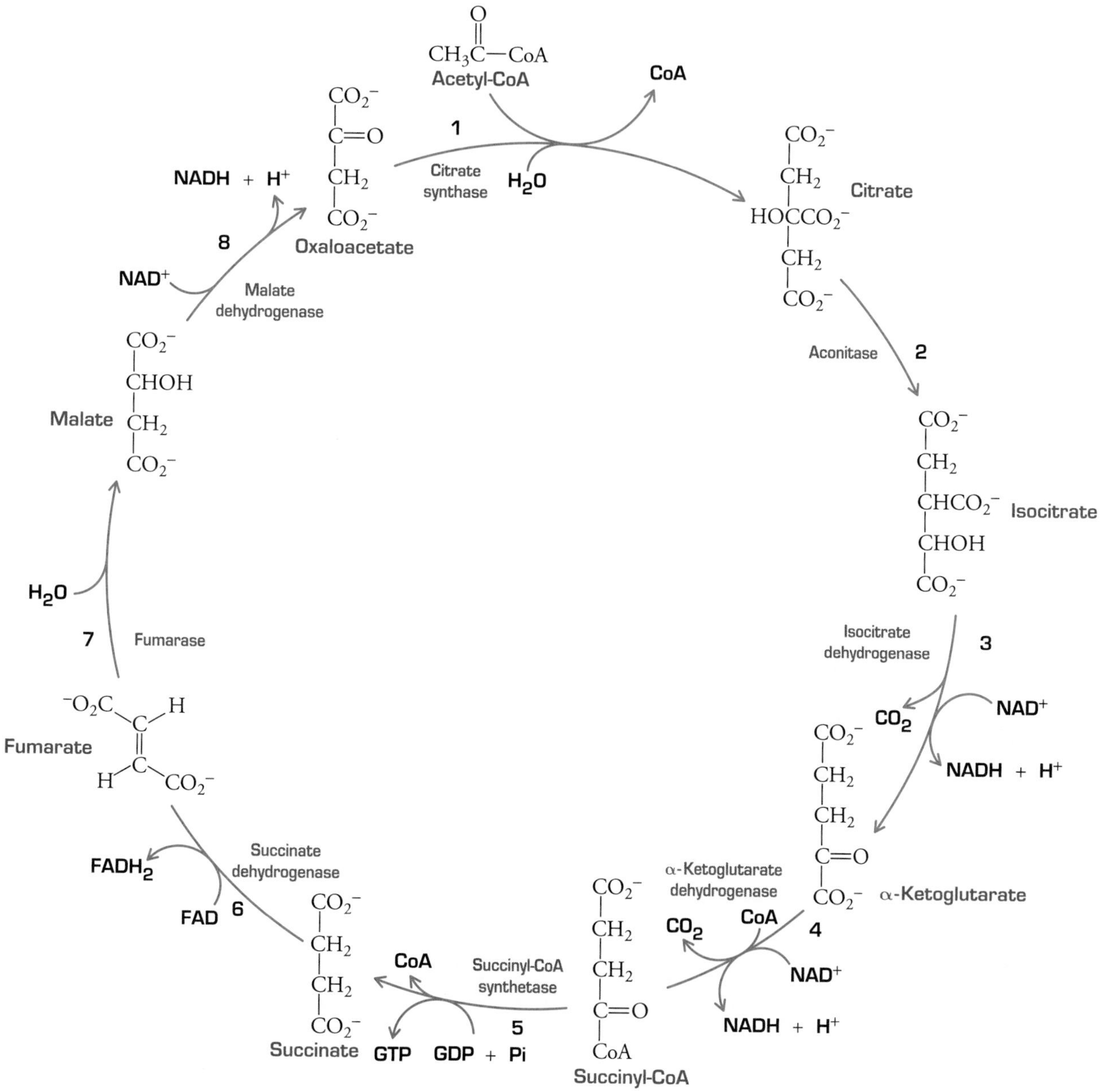

FIGURE 15.12 **The citric acid cycle**
In step 1, oxaloacetate (four carbon atoms) reacts with acetyl-CoA (two carbon atoms, ignoring those in CoA) to form citrate (six carbon atoms). In the remaining seven steps of the citric acid cycle, two carbon atoms are removed (as CO_2), one GTP is produced, and oxidation reactions account for the formation of 3 NADH and 1 $FADH_2$. The final product, oxaloacetate, combines with another acetyl-CoA to begin another cycle.

The citrate formed in step 1 rearranges in step 2 to become isocitrate. In step 3 isocitrate is oxidized to α-ketoglutarate, accompanied by the production of NADH and CO_2. NADH is also produced in step 4, where α-ketoglutarate reacts with CoA, is decarboxylated, and is converted to succinyl-CoA. In step 5 the removal of CoA from succinyl-CoA

- One turn of the citric acid cycle produces 2 CO_2, 3 NADH, 1 $FADH_2$, and 1 GTP.

is coupled to the production of GTP from GDP and P_i. The remaining steps of the cycle involve oxidation accompanied by the production of $FADH_2$ (step 6), hydration (step 7), and oxidation to produce oxaloacetate and NADH (step 8).

The reaction in step 5 of the citric acid cycle is the only one that directly generates ATP (GTP is equivalent in energy to ATP). The reduced coenzymes NADH and $FADH_2$ produced in the cycle can enter the electron transport chain (Section 15.8), where their energy is used to generate more ATP.

Control of the Citric Acid Cycle

Entry to the citric acid cycle is controlled by the conversion of pyruvate to acetyl-CoA. As previously noted, the allosteric enzyme complex that catalyzes this reaction has ATP, NADH, and acetyl-CoA as negative effectors. In the cycle itself, step 3—the conversion of isocitrate to α-ketoglutarate—is the main control point. Isocitrate dehydrogenase, the allosteric enzyme that catalyzes this reaction, has ATP and NADH as negative effectors and ADP and NAD^+ as positive effectors.

Sample Problem 15.4

Energy changes in the citric acid cycle

For the citric acid cycle, $\Delta G = -11$ kcal/mol.

a. Is the pathway spontaneous or is it nonspontaneous?

b. Must ΔG for each of the eight reactions in the cycle be negative for the overall ΔG to be negative?

Strategy

Think in terms of how the overall ΔG for a series of reactions is calculated.

Solution

a. Spontaneous.

b. No. The overall ΔG for the cycle is obtained by adding the ΔG values for the individual reactions. Some can have positive ΔG values and some have can have negative values.

Practice Problem 15.4

Not all of the energy released during the citric acid cycle is used to produce GTP and reduced coenzymes. What happens to the remainder?

15.8 ELECTRON TRANSPORT CHAIN AND OXIDATIVE PHOSPHORYLATION

In the previous section we saw that each turn of the citric acid cycle produces 2 CO_2, 3 NADH, 1 $FADH_2$, and 1 GTP. When the reduced coenzymes (NADH and $FADH_2$) encounter the electron transport chain, which is followed by oxidative phosphorylation, ATP is produced.

The electron transport chain is a group of proteins and other molecules embedded in the inner mitochondrial membrane (Figure 15.13). The first of these, complex I, accepts two electrons during the oxidation of NADH to NAD^+. $FADH_2$ donates two electrons at complex II, a different entry point into the electron transport chain. Pairs of electrons are passed from one molecule in the chain to the next, with energy being released at each step. The final acceptor of electrons is O_2, which undergoes the reaction

$$O_2 + 4H^+ + 4e^- \longrightarrow 2H_2O$$

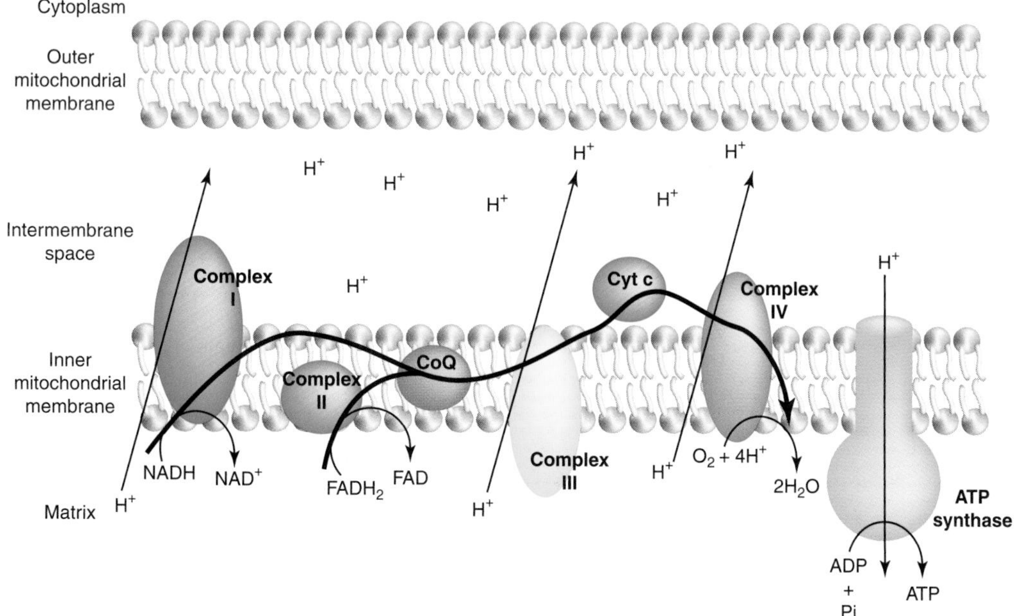

FIGURE 15.13 **Electron transport chain and oxidative phosphorylation** A series of proteins and other molecules make up the electron transport chain, which is involved in the removal of pairs of electrons from NADH and $FADH_2$. As the electron pairs are transferred down the chain, the energy released is used to move H^+ to the mitochondrial intermembrane space. As H^+ moves back through the inner mitochondrial membrane via an ATP synthase channel, the energy released is used to form ATP from ADP and P_i.

At several points in the electron transport chain the energy released during electron exchange is used for active transport (Section 7.9) to pump H^+ through the inner mitochondrial membrane into the intermembrane space. As a result, the concentration of H^+ becomes higher in the intermembrane space than in the mitochondrial matrix.

With a higher concentration of H^+ on one side of the inner mitochondrial membrane than the other, given the opportunity, H^+ will diffuse in the direction of higher concentration to lower concentration. The inner mitochondrial membrane does not usually allow H^+ to pass through, but the enzyme ATP synthase provides a channel for H^+ to do so. The energy released during this facilitated diffusion is coupled with the formation of ATP from ADP and P_i in a process called oxidative phosphorylation.

Because electrons from NADH enter the electron transport chain earlier than those from $FADH_2$, NADH provides more energy used for ATP production. On average, each NADH molecule is responsible for the formation of 2.5 ATPs and each $FADH_2$ for the formation of 1.5 ATPs. About 35% of the potential energy available in these reduced coenzymes is converted to ATP and the rest is released as heat.

- The electron transport chain and oxidative phosphorylation use the potential energy present in NADH and $FADH_2$ to make ATP.

The NADH and $FADH_2$ used by the electron transport chain are produced in mitochondria. The 2 NADH generated in the cytoplasm per glucose molecule during glycolysis can also provide electrons for this chain, but only indirectly because NADH cannot move across the inner mitochondrial membrane. Two different shuttle systems are used to move two electrons of NADH from the cytoplasm into mitochondria. One (in heart and liver cells) generates one NADH in the mitochondrial matrix for each NADH in the cytoplasm and the other (in other cells) generates one $FADH_2$ for each NADH (Figure 15.14).

If, in the last step of the electron transport chain, the reduction of O_2 to H_2O is incomplete, superoxide ion (O_2^-) or hydrogen peroxide (H_2O_2) can form. These toxic compounds are broken down by the protective enzymes superoxide dismutase and catalase (Chapter 11 Health Link: Protective Enzymes).

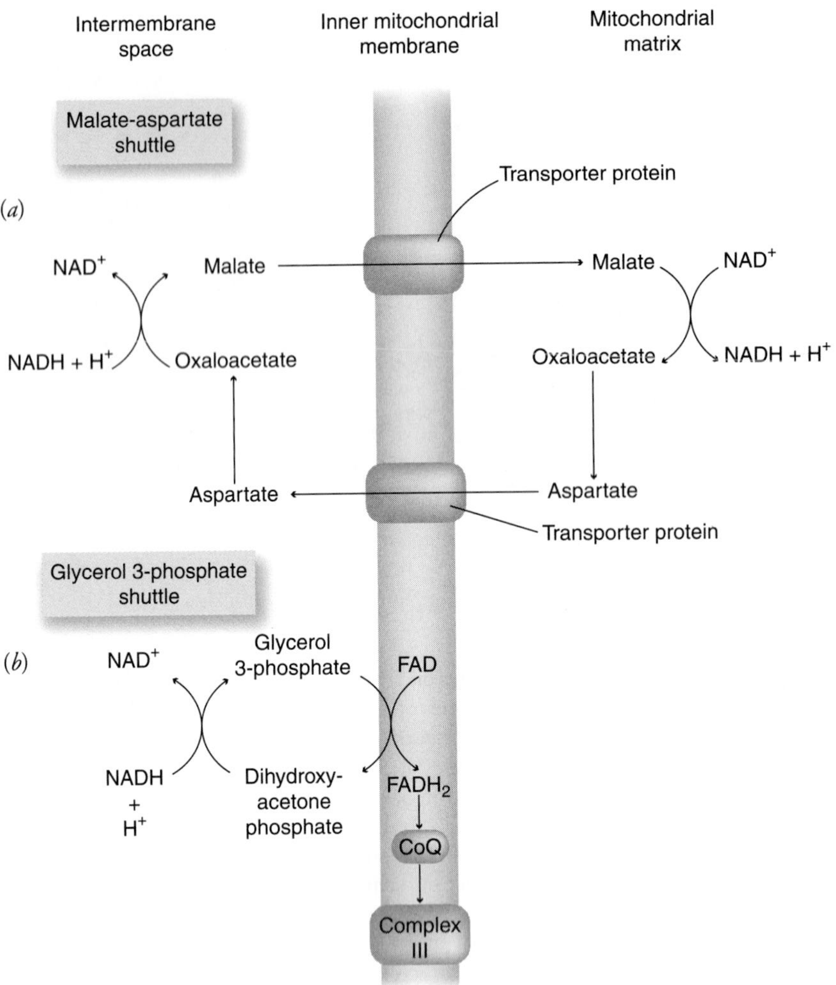

FIGURE | 15.14

NADH shuttles

NADH can move through the outer mitochondrial membrane into the intermembrane space but cannot cross the inner mitochondrial membrane. (*a*) The malate–aspartate shuttle (shown in a simplified form here) oxidizes NADH to NAD^+ in the intermembrane space and reduces a different NAD^+ to NADH in the mitochondrial matrix, in effect "moving" the NADH across the inner membrane. Note that two of the compounds involved in this shuttle are involved in other pathways discussed earlier—oxaloacetate in gluconeogenesis and both malate and oxaloacetate in the citric acid cycle. (*b*) The glycerol 3-phosphate shuttle oxidizes NADH to NAD^+ in the intermembrane space and reduces FAD to $FADH_2$ in the inner mitochondrial membrane. Electrons from $FADH_2$ generated in this way are transferred directly to coenzyme Q, a participant in the electron transport chain (Figure 15.13). The dihydroxyacetone phosphate involved in this shuttle is also an intermediate in glycolysis and gluconeogenesis. Glycerol 3-phosphate is an intermediate in glycerol metabolism (Figure 15.17*a*).

Summary of ATP Yield from Glucose

Glycolysis of one glucose molecule directly produces 2 ATPs and 2 NADH. For each 2 NADH generated in the cytoplasm, either 2 NADH or 2 $FADH_2$ are produced in mitochondria, depending on which system is used to shuttle the electrons from NADH across the inner mitochondrial membrane. The two pyruvates obtained from one glucose yield 2 NADH during the formation of acetyl-CoA. During the citric acid cycle the two acetyl-CoAs yield 6 NADH, 2 $FADH_2$, and 2 GTP (=2 ATP). After electron transport and oxidative phosphorylation, the net gain in ATP from one glucose molecule is 30–32 ATP (Figure 15.15).

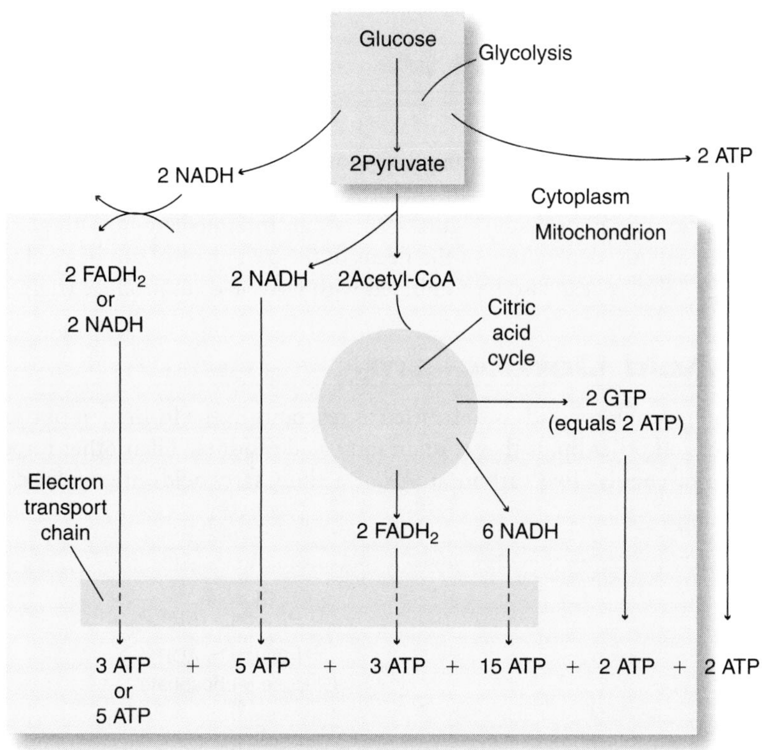

FIGURE 15.15

ATP generation
When one glucose molecule is catabolized by glycolysis, the citric acid cycle, and the electron transport chain, 30–32 ATPs are generated.

HEALTH Link — Brown Fat

Your body temperature is maintained by the heat given off during normal metabolism. Glycolysis, the citric acid cycle, the electron transport chain, and other spontaneous metabolic pathways give off heat in addition to generating reduced coenzymes and ATP. One of the ways that your body generates additional heat when you are cold is by shivering. The metabolism that results from the involuntary shuddering of skeletal muscle produces heat.

Newborn infants do not rely on shivering to generate heat to the same extent as adults. Instead, they use the oxidation of brown fat as the major pathway for rapid heat generation (Figure 15.16). Like all fat cells, brown fat cells contain triglycerides. What distinguishes brown fat from other fat is that it has a greater number of mitochondria, which is what gives brown fat cells their brown appearance. When an infant gets cold, hydrolysis of the triglycerides in brown fat is triggered and the resulting fatty acids are catabolized rapidly (due to the increased number of mitochondria) to form acetyl-CoA (Section 15.9). Acetyl-CoA enters the citric acid cycle, which is followed by the electron transport chain (Section 15.8).

Brown fat is especially good at generating heat because, in addition to the extra mitochondria, it contains a protein called thermogenin that serves as an uncoupler, a compound that makes the inner mitochondrial membrane porous to H^+. In the presence of thermogenin, H^+ can move directly through the inner membrane and into the mitochondrial matrix, so ATP synthase is not used to generate ATP and the energy available from the H^+ is released as heat.

Over the first year of life, shivering becomes more important for heat generation because the levels of brown fat decrease. Brown fat makes up only a small amount of the total fat in adults.

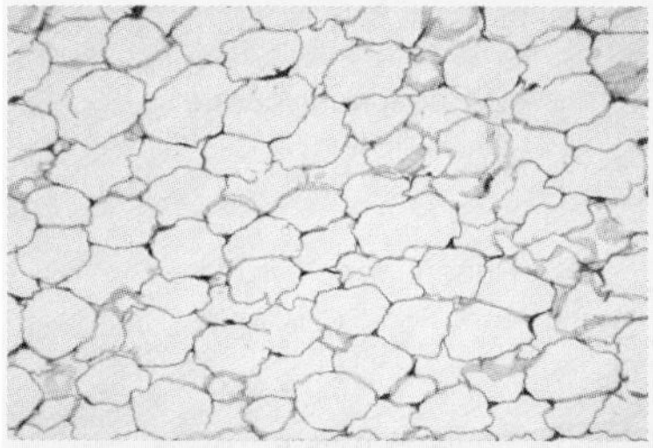

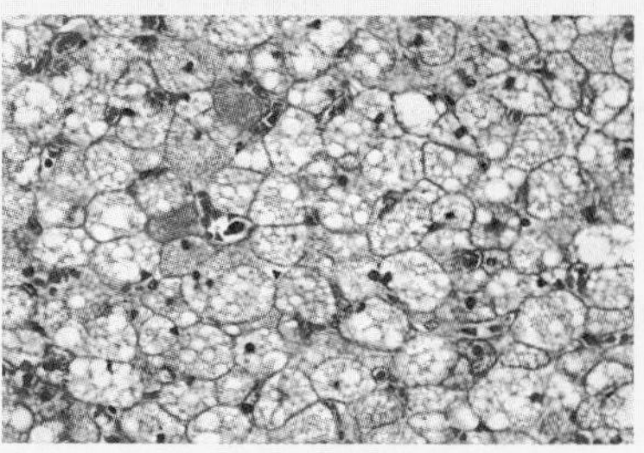

FIGURE 15.16

Brown fat
Having a higher concentration of mitochondria than white fat cells (top), brown fat cells (bottom) catabolize fatty acids very rapidly. The presence of thermogenin, a protein, uncouples the electron transport chain from oxidative phosphorylation. The energy of H^+ pumped into the inner mitochondrial membrane is not used to make ATP, but is released as heat.

Source: Gladden Willis, M. D./Visuals Unlimited.

15.9 LIPID METABOLISM

When triglycerides are hydrolyzed, the glycerol and fatty acids produced can be used in a number of ways. Glycerol can be converted into glycerol 3-phosphate, which is transformed into dihydroxyacetone phosphate (Figure 15.17*a*), an intermediate in both glycolysis and gluconeogenesis. Thus, depending on the needs of the cell, glycerol can be used to produce or store energy. Glycerol can also be combined with fatty acids to produce triglycerides.

Fatty Acid Catabolism

One common fate of fatty acids is to be used as reactants in the formation of triglycerides, sphingolipids, and other lipids that contain fatty acid residues. Their other important use is as a source of energy. Two carbon atoms at a time, fatty acids are oxidized to release

FIGURE | 15.17

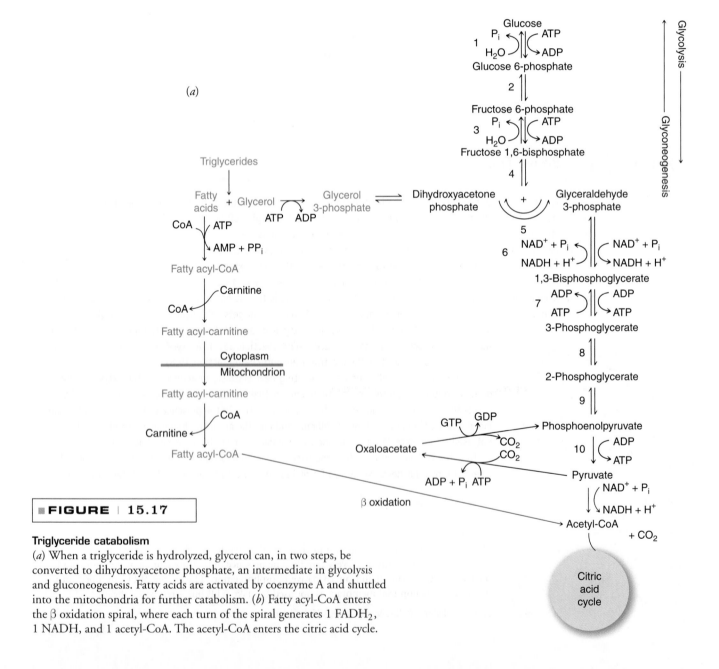

Triglyceride catabolism

(*a*) When a triglyceride is hydrolyzed, glycerol can, in two steps, be converted to dihydroxyacetone phosphate, an intermediate in glycolysis and gluconeogenesis. Fatty acids are activated by coenzyme A and shuttled into the mitochondria for further catabolism. (*b*) Fatty acyl-CoA enters the β oxidation spiral, where each turn of the spiral generates 1 $FADH_2$, 1 NADH, and 1 acetyl-CoA. The acetyl-CoA enters the citric acid cycle.

(*b*)

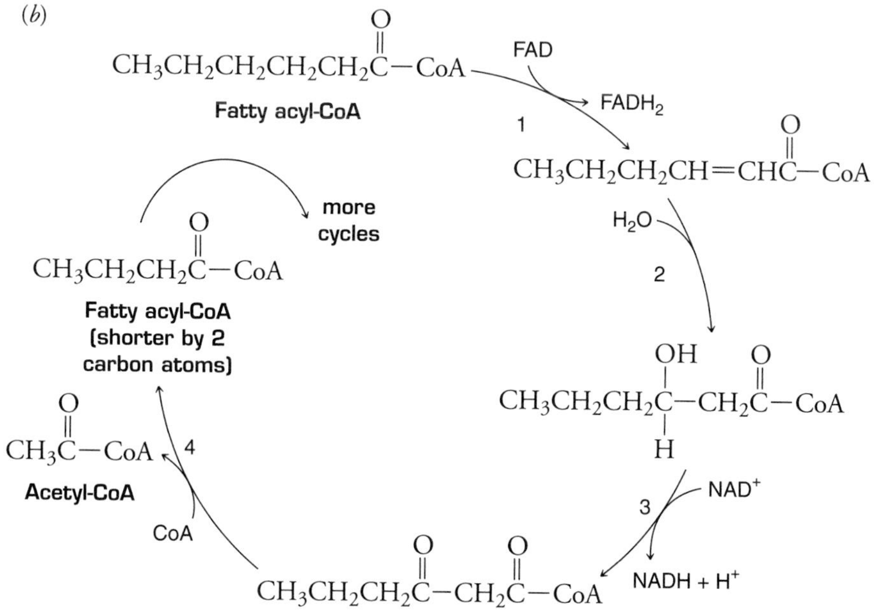

$CH_3CH_2CH_2CH_2CH_2C(=O){-}CoA$ Fatty acyl-CoA

1: FAD → $FADH_2$

$CH_3CH_2CH_2CH{=}CHC(=O){-}CoA$

2: H_2O

$CH_3CH_2CH_2CH(OH){-}CH_2C(=O){-}CoA$

3: NAD^+ → $NADH + H^+$

$CH_3CH_2CH_2C(=O){-}CH_2C(=O){-}CoA$

4: CoA

$CH_3C(=O){-}CoA$ Acetyl-CoA

$CH_3CH_2CH_2C(=O){-}CoA$ Fatty acyl-CoA (shorter by 2 carbon atoms)

more cycles

NADH, $FADH_2$, and acetyl-CoA. The acetyl-CoA can enter the citric acid cycle to produce GTP, NADH, and $FADH_2$, and the reduced coenzymes can be used by the electron transport chain to generate ATP.

Fatty acid catabolism begins in the cytoplasm, where the molecules are activated by the attachment of a coenzyme A (CoA) residue (Figure 15.17*a*). The energy required to drive this reaction is supplied by the hydrolysis of ATP to form AMP and pyrophosphate (PP_i). The newly formed pyrophosphate is immediately hydrolyzed to produce two phosphates. This reaction of pyrophosphate releases nearly the same amount of energy as does the hydrolysis of ATP and, for this reason, the activation of one fatty acid is considered to require 2 ATPs (actually, the energy equivalent of 2 ATPs).

■ Each turn of the β oxidation spiral produces 1 acetyl-CoA, 1 $FADH_2$, and 1 NADH.

$$\text{Fatty acid} + \text{CoA} + \text{ATP} \longrightarrow \text{Fatty acyl-CoA} + \text{AMP} + PP_i$$

$$PP_i + H_2O + \longrightarrow 2P_i$$

For fatty acid catabolism to continue, fatty acyl-CoA must be moved into mitochondria. This happens by replacing the CoA residue with the residue of a compound called carnitine and moving the fatty acyl-carnitine into the mitochondria via a transport protein. Once inside mitochondria, the fatty acyl-CoA is re-formed.

Fatty acid catabolism involves *a spiral metabolic pathway*, called the **β oxidation spiral**, where the same series of reactions is repeated on increasingly shorter reactants. This pathway has four steps (Figure 15.17*b*). In step 1 the activated fatty acid is oxidized by removal of hydrogen atoms at the α and β carbon atoms. At the same time, FAD is reduced to $FADH_2$. Step 2 involves hydration of the carbon–carbon double bond to place an alcohol group at the β carbon, which is then oxidized in step 3 to yield a C=O group (this step gives the β oxidation spiral its name). At the same time that the alcohol group is oxidized, NAD^+ is reduced to NADH. In the final step, acetyl-CoA is split from the fatty acid as another CoA is added, yielding an activated fatty acid that is two carbon atoms shorter than the original.

With each pass through the β oxidation spiral, one acetyl-CoA, one $FADH_2$, and one NADH are generated. As shown in Figure 15.18, a 12-carbon fatty acid can make five passes through the spiral, generating a total of 6 acetyl-CoA, 5 $FADH_2$, and 5 NADH. Each acetyl-CoA that enters the citric acid cycle gives 1 GTP, 3 NADH, and 1 $FADH_2$. After all of the reduced coenzymes are sent through the electron transport chain, the initial 12-carbon fatty acid yields a total of 78 ATPs. Triglycerides are an effective form of energy storage because each fatty acid has the potential of generating many ATPs.

FIGURE | 15.18

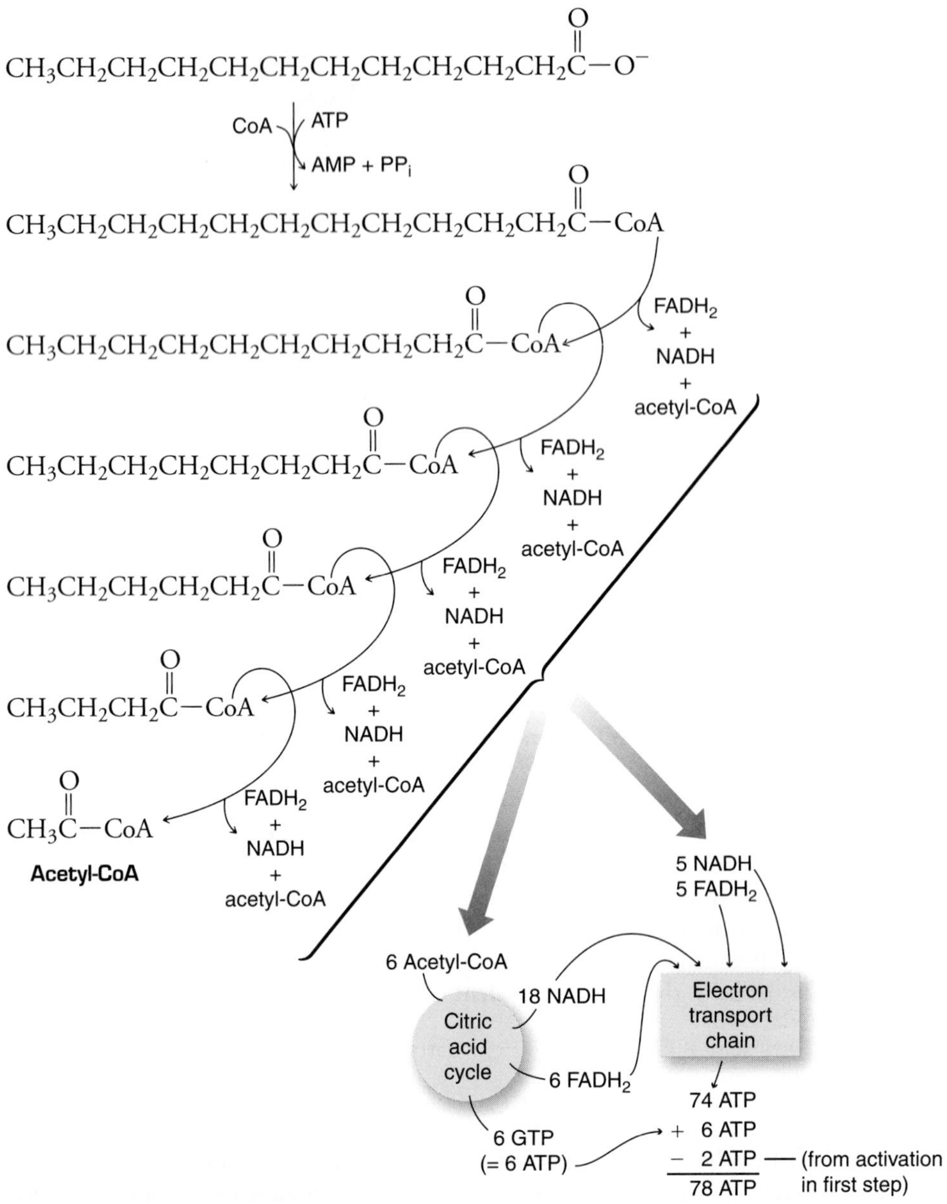

ATP generation from lauric acid
One lauric acid molecule can generate five passes through the β oxidation spiral, producing $FADH_2$, NADH, and acetyl-CoA on each pass. When GTP and reduced coenzymes generated from acetyl-CoA as it is sent through the citric acid cycle are taken into account, catabolism of lauric acid generates 78 ATPs.

$CH_3CH_2CH_2CH_2CH_2CH_2CH_2CH_2CH_2CH_2CH_2\overset{O}{\overset{\|}{C}}-O^-$

CoA, ATP → AMP + PP_i

$CH_3CH_2CH_2CH_2CH_2CH_2CH_2CH_2CH_2CH_2CH_2\overset{O}{\overset{\|}{C}}-CoA$

$CH_3CH_2CH_2CH_2CH_2CH_2CH_2CH_2CH_2\overset{O}{\overset{\|}{C}}-CoA$ + $FADH_2$ + NADH + acetyl-CoA

$CH_3CH_2CH_2CH_2CH_2CH_2CH_2\overset{O}{\overset{\|}{C}}-CoA$ + $FADH_2$ + NADH + acetyl-CoA

$CH_3CH_2CH_2CH_2CH_2\overset{O}{\overset{\|}{C}}-CoA$ + $FADH_2$ + NADH + acetyl-CoA

$CH_3CH_2CH_2\overset{O}{\overset{\|}{C}}-CoA$ + $FADH_2$ + NADH + acetyl-CoA

$CH_3\overset{O}{\overset{\|}{C}}-CoA$ + $FADH_2$ + NADH + acetyl-CoA

Acetyl-CoA

5 NADH, 5 $FADH_2$ → Electron transport chain

6 Acetyl-CoA → Citric acid cycle → 18 NADH, 6 $FADH_2$ → Electron transport chain

6 GTP (= 6 ATP)

74 ATP
+ 6 ATP
− 2 ATP — (from activation in first step)
78 ATP

SAMPLE PROBLEM 15.5

Fatty acid catabolism

a. Beginning with hexanoic acid, how many passes through the β oxidation spiral will take place?

b. From your answer to part a, what is the net change in NADH, $FADH_2$, acetyl-CoA, and ATP, when only activation and the passes through the β oxidation spiral are considered?

Strategy

Figure 15.18 tracks the catabolism of a fatty acid containing twelve carbon atoms. Using the parts of this figure that correspond to the catabolism of hexanoic acid (or its conjugate base), track the formation of NADH, $FADH_2$, acetyl-CoA, and ATP.

Solution

a. 2

b. +2 NADH, +2 $FADH_2$, +3 acetyl-CoA, −2 ATP (activation)

PRACTICE PROBLEM 15.5

a. From the answer to Sample Problem 15.5b, what will be the total yield of ATP once all of the acetyl-CoA has moved through the citric acid cycle and all of the reduced coenzymes produced (including those from the citric acid cycle) have been used for electron transport and oxidative phosphorylation?

b. How does the total yield of ATP obtainable from one hexanoic acid molecule compare with that from one glucose molecule?

Ketone Bodies

Up to this point in the chapter we have seen that acetyl-CoA can be produced from the pyruvate generated by glycolysis and by β oxidation of fatty acids. Acetyl-CoA in excess of what can be immediately used by the citric acid cycle is converted into the **ketone bodies** *acetoacetate, 3-hydroxybutyrate, and acetone* (Figure 15.19). Acetoacetate, which forms when two acetyl-CoA molecules are combined, undergoes spontaneous decarboxylation (Section 10.4) to form acetone or is enzymatically reduced to 3-hydroxybutyrate. Being water soluble, ketone bodies are easily transported from the liver cells, where they are typically produced, to other cells. Many cells can use acetoacetate and 3-hydroxybutyrate as reactants in metabolism.

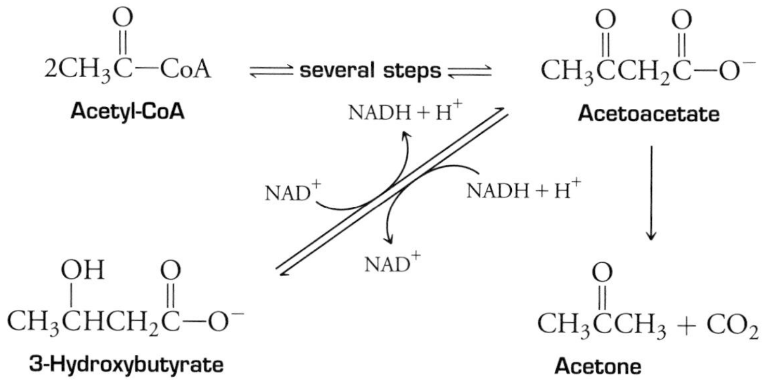

$2CH_3C(=O){-}CoA$ ⇌ several steps ⇌ $CH_3C(=O)CH_2C(=O){-}O^-$

Acetyl-CoA → Acetoacetate

Acetoacetate ⇌ 3-Hydroxybutyrate ($NADH + H^+$ → NAD^+; NAD^+ → $NADH + H^+$)

$CH_3CH(OH)CH_2C(=O){-}O^-$

3-Hydroxybutyrate

Acetoacetate → $CH_3C(=O)CH_3 + CO_2$

Acetone

FIGURE 15.19

Ketone bodies

Two acetyl-CoA molecules can combine to form the ketone body acetoacetate, which can be converted into the other two ketone bodies. Decarboxylation of acetoacetyl-CoA leads to acetone, and reduction produces 3-hydroxybutyrate.

Under starvation conditions or when glucose metabolism is impaired due to diabetes, glucose concentrations drop and the concentration of acetyl-CoA reaches high enough levels that ketone bodies begin to pose problems. Acetyl-CoA levels climb for several reasons. The first is that in the absence of glucose, cells must rely on other molecules for the production of ATP. As we just saw, the β oxidation of fatty acids is an excellent source of reduced coenzymes and acetyl-CoA. It might, at first glance, seem that higher concentrations of acetyl-CoA would allow the rate of the citric acid cycle to increase, generating even more reduced coenzymes and, ultimately, more ATP. This is not the case, however, due to another metabolic change associated with low glucose concentrations.

When glucose concentrations drop, gluconeogenesis is switched on to ensure that cells have a sufficient supply of this fuel. One of the steps early in this anabolic pathway is the conversion of oxaloacetate into phosphoenolpyruvate (Figure 15.10). Diverting oxaloacetate to gluconeogenesis means that there is less of this reactant to combine with acetyl-CoA to begin the citric acid cycle. If the citric acid cycle slows at the same time that fatty acid catabolism increases, the concentration of acetyl-CoA and ketone bodies climb. The *overproduction of ketone bodies* (**ketosis**) can be detected by the odor of acetone on the breath and the presence of ketone bodies in the urine and blood. High concentrations of ketone bodies in the blood can lead to ketoacidosis, a potentially fatal drop in blood pH caused by the presence of acetoacetate and 3-hydroxybutyrate.

Fatty Acid Anabolism

Fatty acid biosynthesis (**lipogenesis**) involves a spiral pathway that adds two carbon atoms at a time (Figure 15.20) to a growing fatty acid. Unlike fatty acid oxidation, which takes place in mitochondria, lipogenesis takes place in the cytoplasm. Other differences between fatty acid catabolism and biosynthesis are that the enzymes used in each pathway are different, as are the coenzymes. While the enzymes of β oxidation require NAD^+ and FAD, those of biosynthesis use NADPH (produced in the pentose phosphate pathway; Section 15.4). Finally, coenzyme A activates fatty acids for breakdown, while an acyl carrier protein (ACP) activates the growing fatty acid for addition of the next two-carbon fragment. In humans, palmitic acid is the largest fatty acid that can be produced by this process.

The formation of unsaturated fatty acids from saturated ones is catalyzed by desaturase enzymes, but human desaturase enzymes will not place a carbon–carbon double bond beyond carbon 10 in a fatty acid. This is why linoleic and linolenic acid (Table 8.1) cannot be produced in humans and are **essential fatty acids** (*they must be obtained in the diet*).

FIGURE 15.20

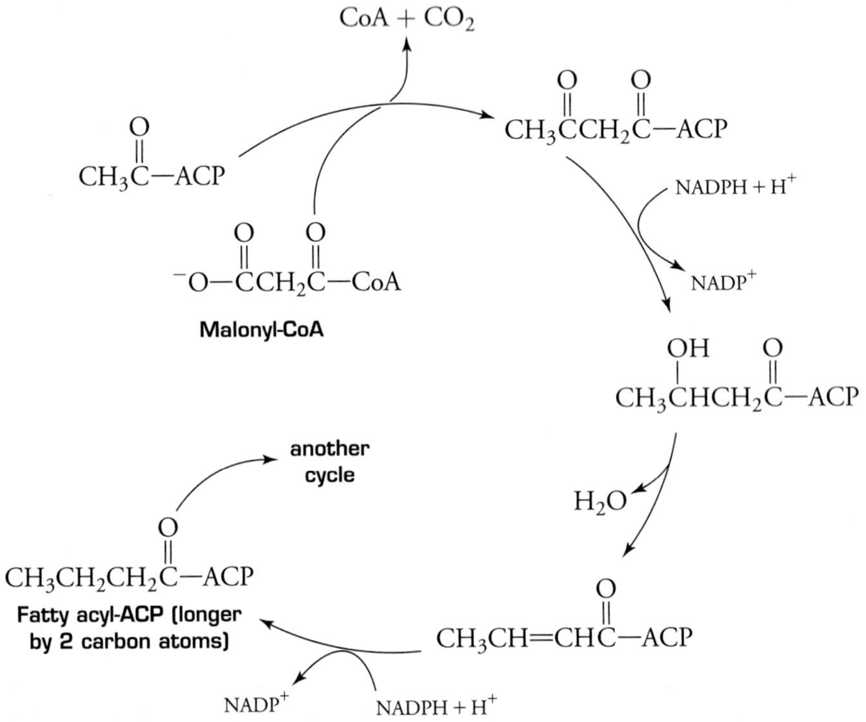

Fatty acid biosynthesis
Fatty acids are produced in a spiral anabolic pathway. Each turn of the spiral adds two carbon atoms to the growing fatty acid.

15.10 AMINO ACID METABOLISM

The digestion of dietary proteins in the stomach and small intestine releases amino acids for transport throughout the body. Some are used for the production of new peptides and proteins, but most are converted into compounds that take part in the citric acid cycle, gluconeogenesis, and other metabolic pathways.

■ Transamination and oxidative deamination are used to remove amino groups from amino acids.

Removal of the amino group is an important part of amino acid catabolism. The two reactions most often used to do this are transamination and oxidative deamination. In **transamination** reactions *the transfer of an amino group from an amino acid to an* α-keto acid is catalyzed by transaminase enzymes. Alanine, for example, reacts with α-ketoglutarate, a common amino group acceptor, to form pyruvate and glutamate—the basic form of glutamic acid—while aspartate reacts to form oxaloacetate, the product of the final step in the citric acid cycle (Figure 15.21*a*). In all, transamination is involved in catabolism of 11 of the 20 common amino acids.

(*a*)

$$\underset{\textbf{Alanine}}{CH_3CH(\overset{+}{N}H_3)\overset{O}{\overset{\|}{C}}{-}O^-} + \underset{\alpha\textbf{-Ketoglutarate}}{{}^-O{-}\overset{O}{\overset{\|}{C}}CH_2CH_2\overset{O}{\overset{\|}{C}}{-}\overset{O}{\overset{\|}{C}}{-}O^-} \xrightarrow{\text{transaminase}} \underset{\textbf{Pyruvate}}{CH_3\overset{O}{\overset{\|}{C}}{-}\overset{O}{\overset{\|}{C}}{-}O^-} + \underset{\textbf{Glutamate}}{{}^-O{-}\overset{O}{\overset{\|}{C}}CH_2CH_2CH(\overset{+}{N}H_3)\overset{O}{\overset{\|}{C}}{-}O^-}$$

$$\underset{\textbf{Aspartate}}{{}^-O{-}\overset{O}{\overset{\|}{C}}CH_2CH(\overset{+}{N}H_3)\overset{O}{\overset{\|}{C}}{-}O^-} + \underset{\alpha\textbf{-Ketoglutarate}}{{}^-O{-}\overset{O}{\overset{\|}{C}}CH_2CH_2\overset{O}{\overset{\|}{C}}{-}\overset{O}{\overset{\|}{C}}{-}O^-} \xrightarrow{\text{transaminase}} \underset{\textbf{Oxaloacetate}}{{}^-O{-}\overset{O}{\overset{\|}{C}}CH_2\overset{O}{\overset{\|}{C}}{-}\overset{O}{\overset{\|}{C}}{-}O^-} + \underset{\textbf{Glutamate}}{{}^-O{-}\overset{O}{\overset{\|}{C}}CH_2CH_2CH(\overset{+}{N}H_3)\overset{O}{\overset{\|}{C}}{-}O^-}$$

(*b*)

$$\underset{\textbf{Glutamate}}{{}^-O{-}\overset{O}{\overset{\|}{C}}CH_2CH_2CH(\overset{+}{N}H_3)\overset{O}{\overset{\|}{C}}{-}O^-} \xrightarrow[NADP^+ \;\; NADPH + H^+]{H_2O} \underset{\alpha\textbf{-Ketoglutarate}}{{}^-O{-}\overset{O}{\overset{\|}{C}}CH_2CH_2\overset{O}{\overset{\|}{C}}{-}\overset{O}{\overset{\|}{C}}{-}O^-} + NH_4^+$$

FIGURE 15.21

Amino acid catabolism
(*a*) In transamination reactions, the amino group of an amino acid is transferred to an α-keto acid, typically α-ketoglutarate. (*b*) During oxidative deamination, an amino acid loses its amino group to produce an α-keto acid plus NH_4^+.

In **oxidative deamination** *an amino group is replaced by a carbonyl (C=O) group.* Glutamate, for example, undergoes oxidative deamination to produce α-ketoglutarate and ammonium ion (NH_4^+). This reaction of glutamate is significant because it allows the amino groups previously transferred to α-ketoglutarate during transamination to be released (as ammonium ion, NH_4^+) for excretion (Figure 15.22). Due to its toxicity, NH_4^+ must be removed from the cell as rapidly as possible, and in mammals the NH_4^+ is converted into

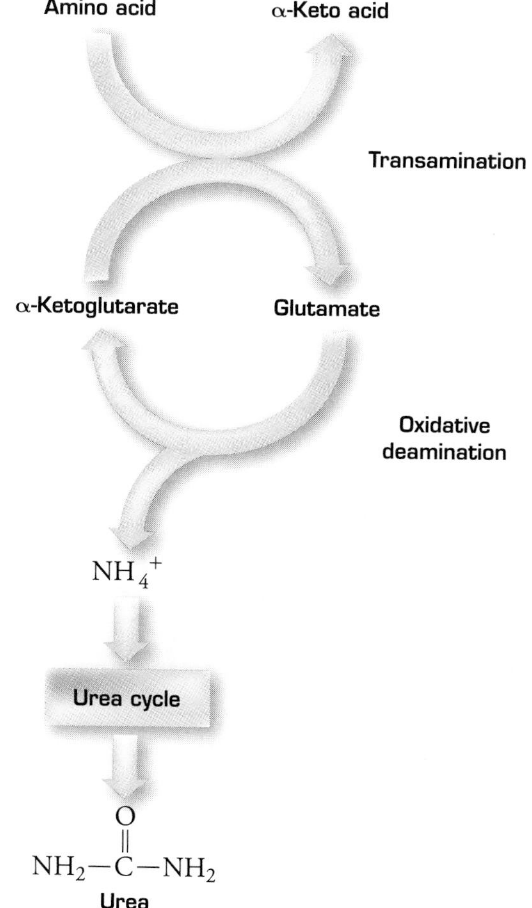

FIGURE 15.22

Excretion of nitrogen in mammals
Through transamination the amino group of an amino acid is transferred to α-ketoglutarate, which undergoes oxidative deamination to release NH_4^+. The urea cycle converts the NH_4^+ to urea, which is excreted in urine.

urea by the urea cycle. Birds and reptiles excrete nitrogen in a white paste that contains uric acid, while fish eliminate NH_4^+ directly into the water through their gills.

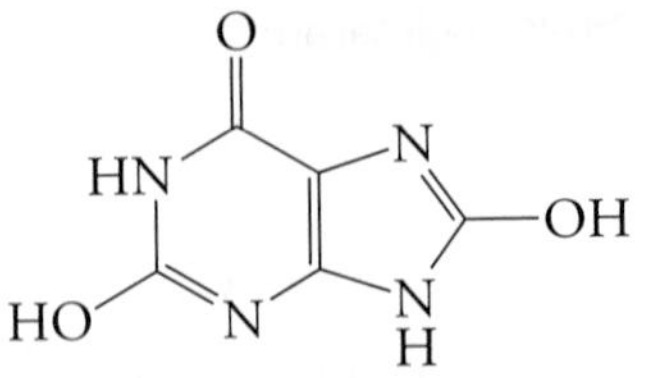

Uric acid

All of the amino acids can be converted into citric acid cycle intermediates or into compounds that can enter the cycle (Figure 15.23).

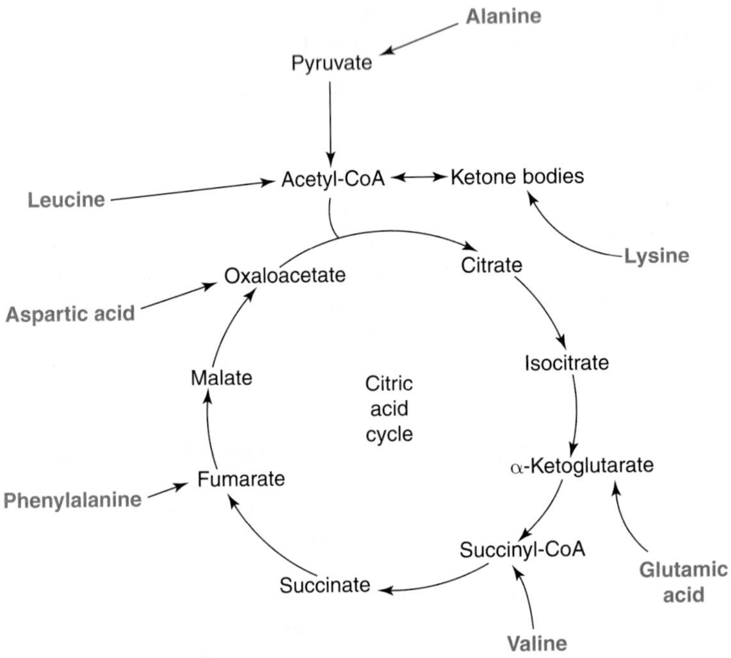

FIGURE | 15.23

Amino acid catabolism and the citric acid cycle
Amino acids can be converted into molecules important in ATP production, as illustrated here for 7 of the 20 amino acids present in peptides and proteins.

Amino Acid Anabolism

Of the 20 amino acids used to synthesize proteins, humans can make only half. The others, called **essential amino acids**, *must be obtained in the diet* (Table 15.1). In humans, synthesis of the nonessential amino acids (those that we can make) involves reversal of the steps that we looked at directly above in our discussion of amino acid catabolism.

TABLE | 15.1 AMINO ACIDS THAT ARE ESSENTIAL FOR HUMANS

Histidine	Phenylalanine
Isoleucine	Threonine
Leucine	Tryptophan
Lysine	Valine
Methionine	Arginine[a]

[a] Essential for children

Rotenone is a naturally occurring compound that can be obtained from the roots and stems of a number of tropical plants. When rotenone is dispersed in a lake, fish absorb it through their gills and die. Death does not occur by suffocation, but instead as the result of interference with the electron transport chain. Rotenone inhibits complex I (Figure 15.13), which prevents the transfer of electrons from NADH and affects the subsequent production of ATP during oxidative phosphorylation. The inability of NADH to be recycled back into its oxidized form shuts down the citric acid cycle, the β oxidation spiral, glycolysis, and other catabolic reactions that require NAD^+. Inhibition of complex I is fatal to fish.

At the levels used in lakes to kill fish, rotenone is not toxic to humans. Although it is an inhibitor of the human electron transport chain, enzymes in the liver detoxify rotenone before it can act as a poison.

Rotenone

summary of objectives

1 Define the terms metabolism, catabolism, and anabolism.
Metabolism, the sum of all reactions that take place in a living thing, can be divided into **catabolism** (reactions that break large molecules into small ones, usually releasing energy) and **anabolism** (reactions that biosynthesize large molecules from small ones, usually consuming energy).

2 Describe three different types of metabolic pathways and explain what coupled reaction means.
The series of reactions called **metabolic pathways** can be **linear**, **circular**, or **spiral**. Some reactions in metabolic pathways are **coupled**, where a spontaneous process supplies the energy required for a nonspontaneous one to take place.

3 Name the products formed during the digestion of polysaccharides, triglycerides, and proteins and state where the digestion of each takes place.

The digestion of carbohydrates converts oligo- and polysaccharides into monosaccharides. Starch digestion begins in the mouth and continues in the small intestine. All other carbohydrates are digested solely in the small intestine. Digestion of triglycerides, which produces glycerol, fatty acids, and monoacylglycerides, begins in the stomach, but most takes place in the small intestine. Digestion of proteins, which yields amino acids, takes place in the stomach and the small intestine.

4 Identify the initial reactant and final products of glycolysis, describe how this pathway is controlled, and explain how gluconeogenesis differs from glycolysis. Describe how the manufacture and breakdown of glycogen are related to each of these pathways.

Glycolysis, a 10-step linear pathway that takes place in the cytoplasm, converts one monosaccharide molecule into 2 pyruvates, producing 2 ATP and 2 NADH in the process. This pathway is controlled at steps 1, 3, and 10 by allosteric enzymes. Under **anaerobic** conditions pyruvate is converted to lactate or, in yeast, is converted to CO_2 and ethanol. Each of these reactions serves to recycle NADH back into NAD^+ so glycolysis can continue. Under **aerobic** conditions, pyruvate is converted into acetyl-CoA. **Gluconeogenesis**, a pathway in which glucose is biosynthesized from lactate and other noncarbohydrate sources, is the exact reverse of glycolysis, except at steps 1, 3, and 10 of glycolysis. Gluconeogenesis uses different enzymes and reactions to bypass these otherwise nonspontaneous steps. Glycogen is produced starting from and is broken down into glucose 6-phosphate, an intermediate in both glycolysis and gluconeogenesis.

5 Give an overview of the citric acid cycle, explain how it is controlled, and describe how the products of this circular pathway are used by the electron transport chain and oxidative phosphorylation.

The **citric acid cycle**, an eight-step circular pathway that takes place in mitochondria, begins with a reaction between oxaloacetate and acetyl-CoA and ends with the manufacture of oxaloacetate. The products of one pass through the citric acid cycle are 1 GTP, 3 NADH, 1 $FADH_2$, and 2 CO_2. The cycle is primarily controlled at step 3 by an allosteric enzyme. The **electron transport chain**, a series of proteins and other molecules embedded in the inner mitochondrial membrane, removes two electrons from NADH or $FADH_2$ and uses the energy released while transferring the electrons down the chain to pump H^+ into the intermembrane space. When the H^+ moves back into the mitochondrial matrix through channels provided by ATP synthase, ATP is manufactured from ADP and P_i in a process called **oxidative phosphorylation**. When 1 glucose molecule enters glycolysis, followed by the citric acid cycle, the electron transport chain, and oxidative phosphorylation, 30–32 ATP are manufactured.

6 Describe the catabolism of triglycerides, the β oxidation spiral, and how the β oxidation spiral differs from fatty acid biosynthesis.

Glycerol produced when triglycerides are hydrolyzed can be converted into dihydroxyacetone phosphate, an intermediate in glycolysis and gluconeogenesis. Fatty acids obtained from triglycerides can be oxidized in mitochondria by the β oxidation spiral, each cycle of which removes 2 carbon atoms from a fatty acid and produces 1 acetyl-CoA, 1 $FADH_2$, and 1 NADH. When glucose metabolism is impaired or starvation conditions exist, the resulting rise in acetyl-CoA concentrations leads to increased production of potentially harmful **ketone bodies**. Fatty acid biosynthesis involves a spiral pathway that takes place in the cytoplasm, uses NADPH, and uses different enzymes than the β oxidation spiral.

7 Explain the fate of the amino groups in amino acids.

Amino groups are removed from amino acids by **transamination**, in which an α-keto acid is the $-NH_2$ acceptor, or by **oxidative deamination**, in which NH_4^+ is released. Regardless of how the amino group is removed, the nitrogen is eventually excreted as urea (in mammals), uric acid (in birds and reptiles), or ammonium ion (in fish).

END OF CHAPTER PROBLEMS

Answers to problems whose numbers are printed in color are given in Appendix D. More challenging questions are marked with an asterisk. Problems within colored rules are paired. **ILW** = Interactive Learning Ware solution is available at *www.wiley.com/college/raymond*.

15.1 Metabolic Pathways, Energy, and Coupled Reactions

15.1 Define the term *metabolic pathway*.

15.2 After reviewing this chapter, name a metabolic pathway that
a. is linear
b. is circular
c. is a spiral

15.3 In the first step of glycolysis, the following two reactions are coupled

$$\text{Glucose} + P_i \longrightarrow \text{glucose 6-phosphate} + H_2O \qquad \Delta G = +3.3 \text{ kcal}$$

$$\text{ATP} + H_2O \longrightarrow \text{ADP} + P_i \qquad \Delta G = -7.3 \text{ kcal}$$

a. Is the reaction Glucose + $P_i \longrightarrow$ glucose 6-phosphate + H_2O spontaneous?
b. Write the net reaction equation and calculate ΔG for the coupled reaction.
c. Is the first step in glycolysis spontaneous?

***15.4** Suppose that an uncatalyzed reaction is spontaneous because ΔG has a value of -10 kcal/mol. An enzyme that catalyzes the reaction is identified. What effect will the enzyme have on the rate of the reaction?

***15.5** For the reaction described in Problem 15.4, what effect will the enzyme have on the value of ΔG for the reaction?

15.2 Overview of Metabolism

15.6 The abbreviation P_i is used to represent phosphate ion.
a. Draw phosphate ion.
b. Depending on the pH of a solution, phosphate ion can exist in three other forms. Draw and name them.

15.7 The abbreviation PP_i is used to represent pyrophosphate ion.
a. Draw pyrophosphate.
b. Draw pyrophosphate as it would appear at a very acidic pH.

15.8 With the exception of GTP and UTP, what structural feature do all of the molecules in Figure 15.2 have in common?

15.9 What provides the energy used to produce ATP from ADP and P_i, anabolism or catabolism?

15.10 What process does the energy provided by hydrolyzing ATP to ADP and P_i drive, anabolism or catabolism?

15.11 Define the term *reduction* and explain how it applies to the difference in the structures of NAD^+ and NADH.

15.12 Define the term *oxidation* and explain how it applies to the difference in the structures of FAD and $FADH_2$.

15.13 What type of bond is broken when acetyl-CoA is hydrolyzed?

15.14 Describe the role that acetyl-CoA has in catabolism.

15.15 How does anabolism differ from catabolism in terms of the relative size of the reactants that enter a particular pathway and the products that leave it?

15.16 Write a reaction equation for the hydrolysis of GTP to produce P_i.

15.17 Write a reaction equation for the hydrolysis of UTP to produce PP_i.

15.3 Digestion

15.18 What products are formed when each of the following undergoes digestion and is broken down to its building blocks?
a. a polysaccharide
b. a triglyceride
c. a protein

15.19 Where (in the mouth, the stomach, and/or the small intestine) does digestion of each take place?
a. polysaccharides
b. triglycerides
c. proteins

***15.20** Name the functional group hydrolyzed when each undergoes digestion.
a. a polysaccharide
b. a triglyceride
c. a protein

15.21 How do the structures of amylose and amylopectin, the two homopolysaccharides that make up starch, differ?

15.22 How are the structures of amylose and amylopectin, the two homopolysaccharides that make up starch, similar?

15.23 Only a small fraction of the starch in food is digested before it enters the small intestine. Why?

***15.24** Name three common disaccharides and name the monosaccharide(s) formed when they are hydrolyzed in the small intestine.

***15.25** What two types of glycosidic bonds must be hydrolyzed to break amylopectin down into individual glucose molecules? Does the same enzyme hydrolyze both types?

15.26 What is the role of bile acids in digestion?

15.27 To which class of lipids do bile acids belong?

15.28 Studies have shown that, in newborns, pancreatic lipases are poor catalysts. Lingual lipase, whose secretion is stimulated by feeding, is active, however. Explain how the digestion of triglycerides by newborns differs from that of adults.

15.4 Glycolysis

15.29 What is the role of glycolysis?

15.30 What is the net change in ATP and NADH from the passage of one glucose molecule through glycolysis?

15.31 a. Which step in glycolysis is the major control point?
b. What compounds act as positive effectors of the enzyme that catalyzes this reaction?
c. What compounds are negative effectors?

***15.32 a.** What is the relationship between glucose 6-phosphate and fructose 6-phosphate (step 1 of glycolysis)? Are they constitutional isomers, *cis/trans* isomers, stereoisomers, different conformations of the same compound, or identical compounds?
b. Are these two compounds reducing sugars? Explain.
c. Draw each as it appears in its open form.

15.33 a. Which class of reaction is involved in the conversion of 2-phosphoglycerate into phosphoenolpyruvate (step 9 of glycolysis): synthesis, decomposition, single replacement, or double replacement?
b. Draw each of these compounds as it would appear at low pH.

15.34 a. In humans, pyruvate is converted into what compound under anaerobic conditions?
b. What purpose does this reaction have?

15.35 a. What products are formed when pyruvate undergoes alcoholic fermentation?
b. What purpose does this reaction have?

15.36 What is the net change in ATP and NADH from the passage of one glucose molecule through glycolysis, followed by the conversion of pyruvates into lactates?

15.37 What is the net change in ATP and NADH from the passage of one glucose molecule through glycolysis, followed by the conversion of pyruvates into ethanol and CO_2 molecules?

15.38 Draw pyruvic acid, the acidic form of pyruvate. Which predominates at physiological pH, the acidic form or the conjugate base form of this acid?

15.39 Draw lactic acid, the acidic form of lactate. Which predominates at physiological pH, the acidic form or the conjugate base form of this acid?

15.40 **a.** Is the conversion of acetaldehyde to ethanol an oxidation or a reduction?
b. When this reaction takes place during alcoholic fermentation, what is the oxidizing agent?
c. What is the reducing agent?

15.41 **a.** Is the conversion of pyruvate to lactate an oxidation or a reduction?
b. When this reaction takes place in a cell, what is the oxidizing agent?
c. What is the reducing agent?

15.42 **a.** In what part of a cell does glycolysis take place?
b. In what part of a cell does the conversion of pyruvate into acetyl-CoA take place?

15.43 Is the conversion of pyruvate into acetyl-CoA an oxidation or a reduction?

15.44 Not all of the free energy (ΔG) released during glycolysis ends up in ATP. What happens to the remainder of the energy?

***15.45** Monosaccharides other than glucose are converted into compounds that are intermediates in glycolysis. Why is this more efficient than having a different catabolic pathway for the conversion of each monosaccharide into pyruvate?

15.46 **a.** Which steps in glycolysis involve a coupled reaction?
b. Are the reactions coupled to make the net reaction spontaneous or to make it nonspontaneous?
c. In the coupled reactions, is ΔG greater than zero or is it less than zero?

15.47 Glycolysis can be described as a process in which energy is invested "up front" in exchange for a greater return of energy later on. Which steps in glycolysis involve investment of energy?

15.5 Gluconeogenesis

15.48 **a.** Define the term *gluconeogenesis.*
b. What is the role of this pathway?

15.49 Which three steps in glycolysis cannot be directly reversed during gluconeogenesis?

15.50 Which step in gluconeogenesis involves a compound also used in the citric acid cycle?

15.6 Glycogen Metabolism

15.51 From which intermediate shared by glycolysis and gluconeogenesis is glycogen produced?

15.52 During glycogenolysis, the splitting of a glucose residue off of glycogen is spontaneous. During glycogenesis, the addition of a glucose residue to glycogen is also spontaneous. How is it possible for both processes to be spontaneous?

15.53 **a.** Describe the role that insulin plays in glycogenesis and glycogenolysis.
b. Describe the role that glucagon plays in glycogenesis and glycogenolysis.

***15.54** **a.** Describe the role that glycogen phosphorylase plays in glycogen metabolism.
b. Glycogen phosphorylase is controlled by covalent modification. Explain.
c. Glycogen phosphorylase is an allosteric enzyme controlled by positive and negative effectors. Explain.

15.7 Citric Acid Cycle

***15.55** The alcohol group in isocitrate can be oxidized, but that in citrate cannot. Explain.

15.56 In the first step of the citric acid cycle, citrate (6 carbon atoms) is formed. The final product of the cycle is oxaloacetate (four carbon atoms). Where do the missing two carbon atoms end up?

15.57 What is the net change in GTP, NADH, and $FADH_2$ from the passage of two acetyl-CoA through the citric acid cycle?

15.58 **a.** Which step is the control point of the citric acid cycle?
b. Name the enzyme that catalyzes the reaction that takes place at this step and name its positive and negative effectors.

15.8 Electron Transport Chain and Oxidative Phosphorylation

15.59 **a.** Which two compounds donate their electrons to the electron transport chain?
b. Which molecule is the final electron acceptor of the electron transport chain?

15.60 As electrons pass through the electron transport chain, H^+ is moved to the mitochondrial intermembrane space. Which term best describes this process: diffusion, facilitated diffusion, or active transport?

15.61 H^+ moves from the intermembrane space, through ATP synthase, and into the mitochondrial matrix. Which term best describes this process: facilitated diffusion or active transport?

15.62 Explain why an $FADH_2$ that enters the electron transport chain yields fewer ATP than an NADH entering the chain.

15.63 The electron transport chain and oxidative phosphorylation typically generate how many ATP from 1 NADH?

15.64 The electron transport chain and oxidative phosphorylation typically generate how many ATP from 1 $FADH_2$?

15.65 Account for the 30–32 ATPs generated when one glucose molecule is catabolized by glycolysis and the citric acid cycle.

15.9 Lipid Metabolism

15.66 Glycerol can be converted into a compound used in glycolysis and gluconeogenesis. Explain.

15.67 Describe how fatty acids are activated and then moved from the cytoplasm into the mitochondrion.

15.68 What happens to the potential energy of fatty acids when they become activated?

15.69 In one pass through the β oxidation spiral a fatty acyl-CoA is shortened by two carbon atoms. What other products are formed?

15.70 How does β oxidation get its name?

15.71 Calculate the net number of ATPs produced when one 14-carbon fatty acid is activated, enters the mitochondrion, and undergoes complete β oxidation to produce acetyl-CoA and reduced coenzymes.

15.72 Calculate the net number of ATPs produced when one 10-carbon fatty acid is activated, enters the mitochondrion, and undergoes complete β oxidation to produce acetyl-CoA and reduced coenzymes.

15.73 **a.** Name the three ketone bodies.
b. Are they all ketones?

15.74 Explain how impaired glucose metabolism can result in the production of ketone bodies.

15.75 Why is acidosis (low blood pH) associated with the overproduction of ketone bodies?

15.76 Where is the NADPH used in fatty acid biosynthesis produced?

***15.77** **a.** Draw linoleic and linolenic acid.
b. Which carbon–carbon double bonds in these molecules cannot be formed by human desaturase enzymes?

15.78 What makes a fatty acid "essential"?

15.79 Where in a cell does fatty acid biosynthesis take place?

15.80 Where in a cell does fatty acid oxidation take place?

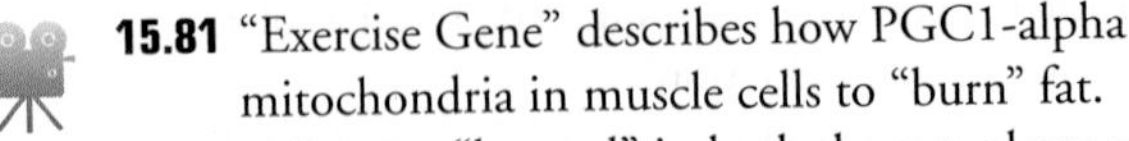

15.81 "Exercise Gene" describes how PGC1-alpha helps mitochondria in muscle cells to "burn" fat.
a. Fats are "burned" in both the cytoplasm and mitochondria of a cell. To what extent are fats broken down in the cytoplasm? What part of fat catabolism takes part in mitochondria?
b. Exercise leads PGC1-alpha to increase the production of mitochondria and of components of the electron transport chain. How does this lead to an improvement in how cells use fuel sources?

15.10 Amino Acid Metabolism

15.82 Which compound is used for nitrogen excretion in the following?
a. humans
b. birds
c. fish

15.83 What is the source of the NH_4^+ produced during amino acid catabolism?

15.84 During transamination, which molecule is a common amino group acceptor?

15.85 Draw the α-keto acid obtained when cysteine undergoes transamination.

15.86 Draw the α-keto acid obtained when phenylalanine undergoes transamination.

15.87 Oxidative deamination of glutamate produces which compound besides NH_4^+?

***15.88** When aspartate undergoes deamination, the reaction product can take part in the citric acid cycle to produce ATP or in gluconeogenesis to produce glucose. Explain.

BiochemistryLink | The Origin of Mitochondria

15.89 Why do some scientists believe that mitochondria were once free bacteria?

HealthLink | Brown Fat

15.90 How does the way that you generate some of your additional body heat when you are cold differ from the way that an infant does so?

15.91 Hibernating animals produce brown fat which helps them to survive during the winter months. Explain how brown fat makes better use of the animal's triglyceride supply during hibernation than other fat does.

THINKING IT THROUGH

15.92 Bakers use yeast because the CO_2 produced when it metabolizes sugar causes bread to rise. In yeast, what are the metabolic steps involved in the production of CO_2 from glucose?

15.93 Which functional groups are present in rotenone? (See the chapter conclusion.)

15.94 One effect of the "low carb" diet craze that swept the United States in 2004 was that a greater number of people had very bad breath, described as "sickeningly sweet." From the standpoint of metabolism, what molecule produced by a diet that is low in carbohydrates and high in fat and protein might cause bad breath?

INTERACTIVE LEARNING PROBLEMS

15.95 Cyanide is an acute poison in animals that targets the biochemical machinery involved in oxygen consumption. Which organelle would you expect cyanide to target? Would you expect ATP production to be interrupted?

15.96 During glucose breakdown, approximately what percentage of the available energy as ATP is sacrificed in the absence of oxygen?

15.97 During anaerobic metabolism, to recycle NADH back to NAD^+, which reaction is required?

SOLUTIONS TO PRACTICE PROBLEMS

15.1 **b** is spontaneous. Using the ΔG values from Sample Problem 15.1, the reaction described in part **a** has a positive value for ΔG and that in part **b** has a negative value.

a. Fructose 1,6-bisphosphate + ADP $\longrightarrow$ fructose 6-phosphate + ATP
$\Delta G = 3.4$ kcal/mol

b. Fructose 1,6-bisphosphate + H_2O $\longrightarrow$ fructose 6-phosphate + P_i
$\Delta G = -3.9$ kcal/mol

15.2 Step 6. Glyceraldehyde 3-phosphate is oxidized and NAD^+ is reduced.

15.3 **a.** catabolic; **b.** anabolic

15.4 It is released as heat.

15.5 **a.** 36 ATP; **b.** 36 ATP for hexanoic acid, 30 or 32 ATP for glucose

APPENDIX A

Scientific Calculators

This appendix describes the operation of a typical scientific calculator, the type of calculator that you may want to use to solve many of the math-related problems in this text. While many varieties are available, a scientific calculator should have, as a minimum, the following keys:

Key	Meaning
+	Addition
−	Subtraction
X	Multiplication
÷	Division
=	Equals
Exp	Exponential number
+/−	Change sign
Log	Logarithm
2nd	Second function

Not all calculators use the same labels for all keys. For example, EE is often used in place of Exp. Refer to the instructions that came with your calculator if you are unsure of the meaning of any keys.

On scientific calculators, it is typical for certain keys to be used for more than one mathematical operation. The second function key is used to make this switch. A common example is that of converting the log key into the 10^{x} key by first pressing the 2nd key.

One thing to consider is that the display area of a calculator can show up to 10 digits. This means that when carrying out calculations on measured quantities you should be sure to round the answer to the correct number of significant figures (see Section 1.5 and Chapter 9 Math Support). In the examples that follow, we will assume that all values are measured quantities.

Addition and Subtraction

To solve the equation

$74 + 5.7 =$

1. Enter 74	7 4
2. Press the plus key	+
3. Enter 5.7	5 . 7
4. Press the equals key	=

The display reads 79.7
which is rounded to 80.

To solve the equation

$0.12 - 0.0366 =$

1. Enter 0.12	0 . 1 2
2. Press the minus key	−
3. Enter 0.0366	0 . 0 3 6 6
4. Press the equals key	=

The display reads 0.0834
which is rounded to 0.08

To solve the equation

$6.2 + 3.1 - 1.23 =$

1. Enter 6.2	6 . 2
2. Press the plus key	+
3. Enter 3.1	3 . 1
4. Press the minus key	−
5. Enter 1.23	1 . 2 3
6. Press the equals key	=

The display reads 8.07
which is rounded to 8.1

Multiplication and Division

To solve the equation

$6.2 \times 3.1 =$

1. Enter 6.2	6 . 2
2. Press the multiply key	X
3. Enter 3.1	3 . 1
4. Press the equals key	=

The display reads 19.22
which is rounded to 19

To solve the equation

$14.01 \div 7 =$

1. Enter 14.01	1 4 . 0 1
2. Press the divide key	÷
3. Enter 7	7
4. Press the equals key	=

The display reads 2.001428571
which is rounded to 2

To solve the equation

$2.4 \times 3 \div 16.1 =$

1. Enter 2.4	2 . 4
2. Press the multiply key	X
3. Enter 3	3
4. Press the divide key	÷
5. Enter 16.1	1 6 . 1
6. Press the equals key	=

The display reads 0.447204969
which is rounded to 0.4

Scientific Notation

To solve the equation

$5.2 \times 10^3 \times 6.1 \times 10^1 =$

1. Enter 5.2	5 . 2
2. Press the exponential number key	Exp
3. Enter 3	3

The display reads 5.2 ^{03} (^{03}means $\times 10^3$)

4. Press the multiply key	X
5. Enter 6.1	6 . 1
6. Press the exponential number key	Exp
7. Enter 1	1

The display reads 6.01 ^{01}

8. Press the equals key	=

Depending on how your calculator is set, the display reads

317200 or 3.172 ^{05}

which is rounded to 3.2×10^5

To solve the equation

$1.022 \times 10^{-14} \div 8.49 \times 10^{12} =$

1. Enter 1.022	1 . 0 2 2
2. Press the exponential number key	Exp
3. Enter 14	1 4
4. Press the +/− key	+/−

The display reads 1.022^{-14}

1. Press the divide key	÷
2. Enter 8.49	8 . 4 9
3. Press the exponential number key	Exp
4. Enter 12	1 2

The display reads 8.49^{12}

9. Press the equals key	=

The display reads 1.20376914^{-27}
which is rounded to 1.20×10^{-27}

Logarithms and Antilogarithms

To solve the equation

$\log 56 =$

1. Enter 56	5 6
2. Press the logarithm key	Log

The display reads 1.748188027
which is rounded to 1.75

To solve the equation

$\log(7.22 \times 10^{-7}) =$

1. Enter 7.22	7 . 2 2
2. Press the exponential number key	Exp
3. Enter 7	7
4. Press the +/− key	+/−
5. Press the logarithm key	Log

The display reads −6.141462802
which is rounded to −6.141

To solve the equation

$\text{antilog}(-5.6) =$

1. Enter 5.6	5 . 6
2. Press the +/− key	+/−
3. Press the second function key	2nd
4. Press the logarithm key (typically, 2nd switches the log key to 10^x	10^x

The display reads
0.000002512
which is rounded to 3×10^{-6}

Important families of organic compounds

FAMILY	EXAMPLE	FUNCTIONAL GROUP	DESCRIPTION
Alkane	$CH_3CH_2CH_2CH_3$		Hydrocarbons in which carbon atoms are joined only by single bonds (Section 4.4).
Alkene	$CH_2{=}CHCH_2CH_3$	$-\overset{}{\underset{\mid}{C}}{=}\underset{\mid}{C}-$	Contain at least one carbon–carbon double bond (Section 4.8).
Alkyne	$HC{\equiv}CCH_2CH_3$	$-C{\equiv}C-$	Contain at least one carbon–carbon triple bond (Section 4.8).
Aromatic	$C_6H_4(CH_3)_2$ (ring with two CH_3 groups, meta)	benzene ring	Cyclic compounds in which rings have alternating double and single bonds (Section 4.8).
Alcohol	$CH_3CH_2CH_2CH_2OH$	$-\overset{\mid}{\underset{\mid}{C}}-OH$	Have an—OH group attached to an alkane-type carbon atom (a carbon atom that is singly bonded to carbon or hydrogen atoms) (Sections 4.9 and 11.1).
Carboxylic acid	$CH_3CH_2CH_2\overset{O}{\overset{\Vert}{C}}-OH$	$-\overset{O}{\overset{\Vert}{C}}-OH$	Contain a carboxyl group, which consists of an—OH group attached to a carbonyl (C=O group). The carbon atom of the carboxyl group is attached to a hydrogen atom, an alkane-type carbon atom, or an aromatic ring (Sections 4.9 and 10.1).

FAMILY	EXAMPLE	FUNCTIONAL GROUP	DESCRIPTION
Ester	$CH_3CH_2C(=O){-}OCH_2CH_3$	$-C(=O)-O-C$	Have a C—O—C linkage in which one of the carbon atoms belongs to a carboxylic acid residue and the other is part of an alcohol residue (Sections 4.9 and 10.4).
Phenol	OH, CH_3 on benzene ring (meta)	OH on benzene ring	Contain an—OH group attached to an aromatic ring (Section 10.2).
Amine	$CH_3CH_2NHCH_2CH_3$	$C-N\!<$	Contain a nitrogen atom attached to one, two, or three carbon atoms, excluding C=O carbon atoms (Section 10.6).
Amide	$CH_3CH_2C(=O){-}NHCH_2CH_3$	$-C(=O)-N\!<$	Have a nitrogen atom attached to a carbonyl group. The carbonyl belongs to a carboxylic acid residue and the nitrogen atom is part of an ammonia or amine residue (Section 10.8).
Thiol	$CH_3CH_2CH_2CH_2SH$	$C-SH$	Have an—SH group attached to an alkane-type or aromatic carbon atom (Section 11.1).
Ether	$CH_3CH_2OCH_2CH_3$	$C-O-C$	Contain a C—O—C linkage in which the carbon atoms are alkane-type or aromatic (Section 11.1).
Sulfide	$CH_3CH_2SCH_2CH_3$	$C-S-C$	Contain a C—S—C linkage in which the carbon atoms are alkane-type or aromatic (Section 11.1).
Disulfide	$CH_3CH_2SSCH_2CH_3$	$C-S-S-C$	Contain a C—S—S—C linkage in which the carbon atoms are alkane-type or aromatic (Section 11.1).
Alkyl halide	$CH_3CH_2CH_2CH_2Br$	$-\overset{\vert}{\underset{\vert}{C}}-X$ $X = F, Cl, Br, I$	Have a halogen atom attached to an alkane-type carbon atom (Section 11.2).

FAMILY	EXAMPLE	FUNCTIONAL GROUP	DESCRIPTION
Aldehyde	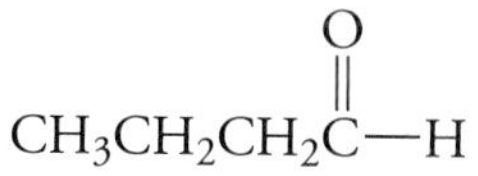 $CH_3CH_2CH_2C(=O){-}H$	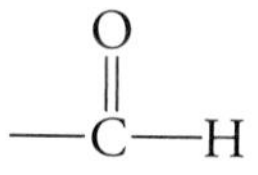 $-C(=O){-}H$	Contain a carbonyl group in which the carbonyl carbon is attached to one carbon atom and one hydrogen atom or to two hydrogen atoms (Section 11.4).
Ketone	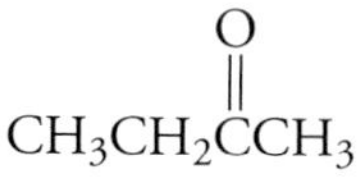 $CH_3CH_2C(=O)CH_3$	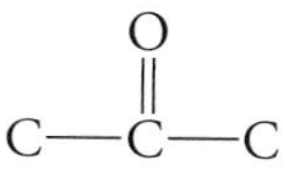 $C{-}C(=O){-}C$	Contain a carbonyl group in which the carbonyl carbon is attached to two other carbon atoms (Section 11.4).
Hemiacetal	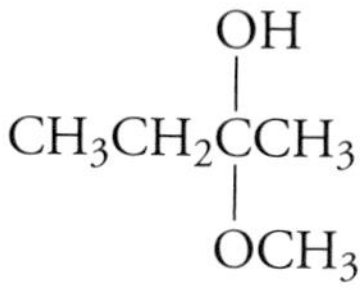 $CH_3CH_2C(OH)(OCH_3)CH_3$	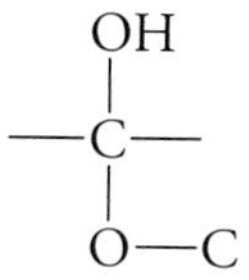 $-C(OH)(O{-}C)-$	Contain a hemiacetal carbon atom (one attached to both—OH and—OC) (Section 11.7). The cyclic hemiacetal formed by a monosaccharide is called an anomer (Section 12.3).
Acetal	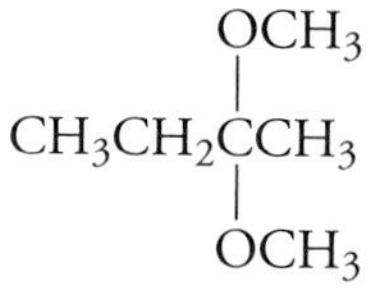 $CH_3CH_2C(OCH_3)_2CH_3$	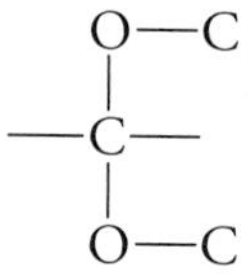 $-C(O{-}C)(O{-}C)-$	Contain an acetal carbon atom (one attached to two—OC—groups) (Section 11.7). The acetal formed when an anomer (see "hemiacetal") reacts with an–OH containing molecule is called a glycoside (Section 12.4).
Phosphate monoester	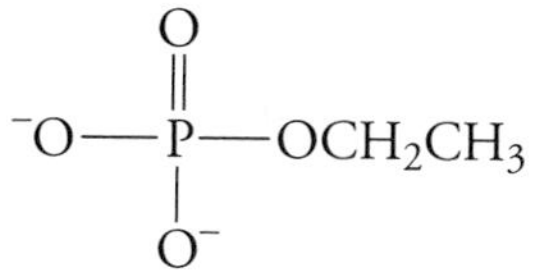 $^{-}O{-}P(=O)(O^{-}){-}OCH_2CH_3$	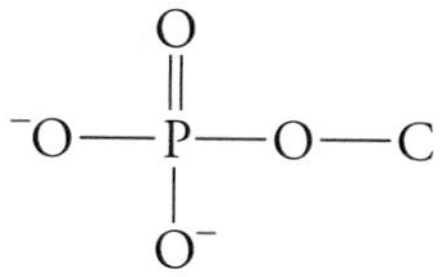 $^{-}O{-}P(=O)(O^{-}){-}O{-}C$	Have a P—O—C linkage in which the phosphorus atom belongs to a phosphate residue and the carbon atom is part of an alcohol residue (Section 14.1).
Phosphate diester	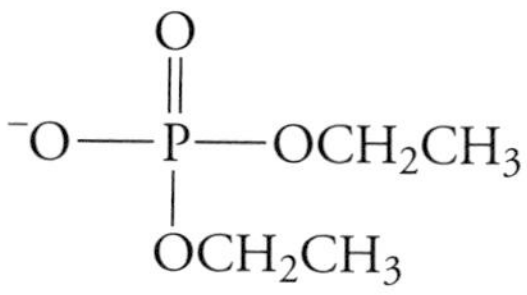 $^{-}O{-}P(=O)(OCH_2CH_3){-}OCH_2CH_3$	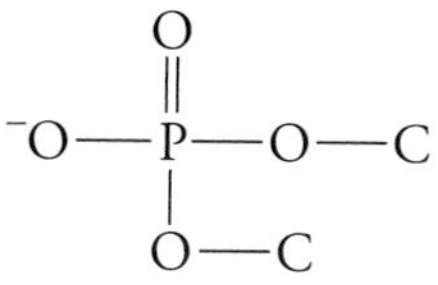 $^{-}O{-}P(=O)(O{-}C){-}O{-}C$	Have two P—O—C linkages in which the phosphorus atom belongs to a phosphate residue and the carbon atoms belong to alcohol residues (Section 14.1).
N-glycoside	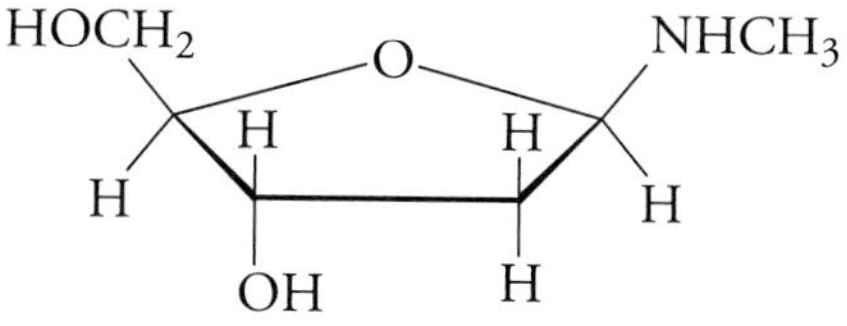 $HOCH_2$, O, $NHCH_3$, H, H, H, H, OH, H	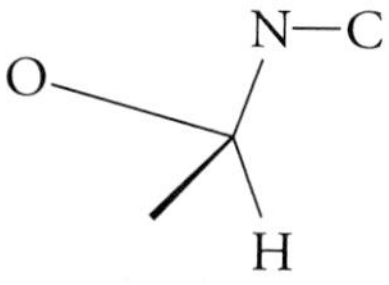 O, N—C, H	A glycoside (see "acetal") formed when an anomer reacts with an amine (Section 14.1).
Phosphate anhydride	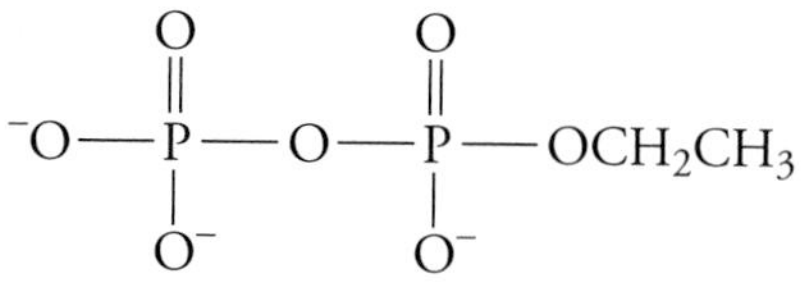 $^{-}O{-}P(=O)(O^{-}){-}O{-}P(=O)(O^{-}){-}OCH_2CH_3$	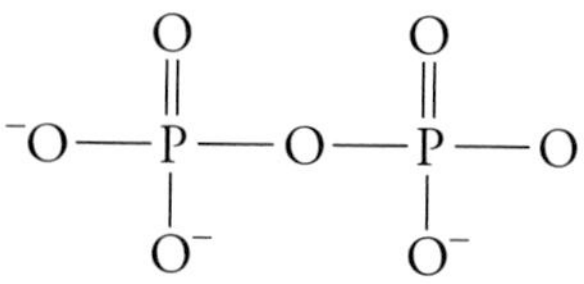 $^{-}O{-}P(=O)(O^{-}){-}O{-}P(=O)(O^{-}){-}O$	Have a P—O—P linkage in which the phosphorus atoms belong to phosphate residues (Section 14.2).

APPENDIX C

Naming ions, ionic compounds, binary molecules, and organic compounds

A. Naming ions

1. Monoatomic ions (Section 3.1)

a. Monoatomic cations (carry a positive charge) formed from representative elements (groups 1A–8A) are given the same name as the original element. Monoatomic anions (carry a negative charge) formed from representative elements are named by changing the ending of the element name to *ide* (Table A1).

TABLE A1. Names of monoatomic cations and anions formed from representative elements

Cation	Name	Anion	Name
Na^+	Sodium ion	F^-	Fluoride ion
Mg^{2+}	Magnesium ion	O^{2-}	Oxide ion
Al^{3+}	Aluminum ion	N^{3-}	Nitride ion

b. Because a given transition metal element (groups 1B–8B) can form cations of different charges, these cations are named by giving the name of the element, along with a Roman numeral indicating the charge. Alternatively, the suffix *ous* can be used for the cation with the smaller charge and the suffix *ic* used for the cation with the greater charge (Table A2).

TABLE A2. Names of monoatomic cations formed from transition elements

Cation	Name	Alternative name
Cu^+	Copper(I) ion	Cuprous ion
Cu^{2+}	Copper(II) ion	Cupric ion
Fe^{2+}	Iron(II) ion	Ferrous ion
Fe^{3+}	Iron(III) ion	Ferric ion
Sn^{2+}	Tin(II) ion	Stannous ion
Sn^{4+}	Tin(IV) ion	Stannic ion

2. Polyatomic ions (Section 3.1).

Table A3 lists the formulas and names of a number of important polyatomic ions (contain two or more atoms). Some polyatomic ion names indicate relative numbers of particular atoms. For example, carbonate ion and hydrogen carbonate ion differ by one hydrogen atom. When an element can form polyatomic ions by combining with oxygen in two different ways, the suffix *ite* is used for the ion with fewer oxygen atoms and the suffix *ate* is used for the ion with more oxygen atoms.

TABLE A3. Names of polyatomic ions

FORMULA	NAME	FORMULA	NAME
CO_3^{2-}	Carbonate ion	NO_2^-	Nitrite ion
HCO_3^-	Hydrogen carbonate	NO_3^-	Nitrate ion
PO_4^{3-}	Phosphate ion	SO_3^{2-}	Sulfite ion
HPO_4^{2-}	Hydrogen phosphate ion	SO_4^{2-}	Sulfate ion
$H_2PO_4^-$	Dihydrogen phosphate ion	OH^-	Hydroxide ion
H_3O^+	Hydronium ion	NH_4^+	Ammonium ion

B. Naming ionic compounds (Section 3.4).

Ionic compounds are named by specifying each ion present in the compound (drop "ion" from the name of each) and placing the cation first in the name. The number of times that a given ion appears in the formula of the ionic compound is not indicated—this can be determined from the charge on the individual ions (Table A4).

TABLE A4. Names of ionic compounds

IONIC COMPOUND	NAME	IONIC COMPOUND	NAME
NaCl	Sodium chloride	Na_2CO_3	Sodium carbonate
$CaSO_4$	Calcium sulfate	$Ca(OH)_2$	Calcium hydroxide
NH_4NO_3	Ammonium nitrate	$(NH_4)_2CO_3$	Ammonium carbonate
CuCl	Copper(I) chloride	$CuCl_2$	Copper(II) chloride
FeO	Iron(II) oxide	Fe_2O_3	Iron(III) oxide

C. Naming binary molecules (Section 3.6).

The names of binary molecules (contain atoms of just two different elements) indicate the atoms present and specify the number of times that an atom of a given element appears in the formula (Tables A5 and A6). The exception to this is that names should not begin with *mono*. For example, CO_2 is carbon dioxide, not monocarbon dioxide. When adding a prefix places two vowels together, the *a* or *o* ending on the prefix is dropped (CO is carbon monoxide, not carbon monooxide).

TABLE A5. Prefixes used for naming binary molecules

PREFIX	NUMBER OF ATOMS	PREFIX	NUMBER OF ATOMS
Mono	1	Hexa	6
Di	2	Hepta	7
Tri	3	Octa	8
Tetra	4	Nona	9
Penta	5	Deca	10

TABLE A6. Names of binary molecules

CATION	NAME	ANION	NAME
CO	Carbon monoxide	N_2O_3	Dinitrogen trioxide
CO_2	Carbon dioxide	PCl_3	Phosphorus trichloride
SO_2	Sulfur dioxide	PCl_5	Phosphorus pentachloride
SO_3	Sulfur trioxide	SF_6	Sulfur hexafluoride

D. Naming organic compounds.

The IUPAC rules are a widely used method of naming organic compounds. Using these rules involves identifying a parent chain (a particular chain of carbon atoms) and substituents (atoms or groups of atoms attached to the parent chain). When specifying the number of carbon atoms in a parent chain or substituent, numbering prefixes are used (Table A7).

TABLE A7. Prefixes used for IUPAC naming

Prefix	Number of Atoms	Prefix	Number of Atoms
Meth	1	Hex	6
Eth	2	Hept	7
Prop	3	Oct	8
But	4	Non	9
Pent	5	Dec	10

Formulas and names for the alkyl substituents (contain only carbon and hydrogen atoms) commonly found in organic compounds are given in Table A8.

The approach for using the IUPAC rules to name alkanes will be presented in some detail below. Modifications to the rules, needed to name members of other organic families, will be presented in the sections that follow.

TABLE A8. Formulas and names of alkyl groups

Fomula[a]	Name
$—CH_3$	Methyl
$—CH_2CH_3$	Ethyl
$—CH_2CH_2CH_3$	Propyl
$—CHCH_3$ with CH_3 on the CH	Isopropyl
$—CH_2CH_2CH_2CH_3$	Butyl
$—CH_2CHCH_3$ with CH_3 on the CH	Isobutyl
$—CHCH_2CH_3$ with CH_3 on the CH	*sec*-Butyl or *s*-butyl
$—C—CH_3$ with CH_3 above and CH_3 below the C	*tert*-Butyl or *t*-butyl

[a]The bond drawn to the left of each formula is the point of attachment to the parent chain.

1. Alkanes (Section 4.4).

The parent chain of an alkane (the longest continuous chain of carbon atoms in the molecule) is named by combining a numbering prefix (Table A7) with "ane," the last three letters in "alkane." Once the parent has been identified, alkyl substituents are identified and named (Table A8). Next, the carbon atoms in the parent chain are numbered, counting from the end nearer the first alkyl group. The IUPAC name is constructed by placing the names of the alkyl groups in alphabetical order and specifying the number of the parent carbon atom to which each is attached, followed by the name of the parent chain. The labels di, tri, tetra, etc., (Table A5) are added if two or more identical alkyl groups are present. For the examples in Table A9, note that commas separate numbers from numbers and that hyphens separate numbers from letters.

TABLE A9. Alkane IUPAC names

STRUCTURAL FORMULA	NAME
6 5 4 3 2 1 $CH_3CH_2CH_2CHCH_2CH_3$ | CH_3	3-Methylhexane
CH_2CH_3 | 1 2 3 4 5 6 7 $CH_3CHCHCH_2CH_2CH_2CH_3$ | CH_3	3-Ethyl-2-methylheptane
CH_3 CH_2CH_3 | | 10 9 8 7 6 5 4 3 2 1 $CH_3CH_2CH_2CH_2CCH_2CHCH_2CH_2CH_3$ | CH_3CCH_3 | CH_3	6-*t*-Butyl-4-ethyl-6-methyldecane
1 CH_3 CH_3 | | 6 5 4 3 2 $CH_2CH_2CHCHCH_2$ | | 7 CH_3 CH_3	3,4-Dimethylheptane
CH_3 | CH_3 CH_2 | | 1 2 3 4 5 6 7 8 $CH_3CH_2CCH_2CHCH_2CH_2CH_3$ | CH_3	5-Ethyl-3,3-dimethyloctane

2. Cycloalkanes (Section 4.7).

When naming cycloalkanes, the ring is usually the parent. The parent ring is named by combining "cyclo" with the appropriate numbering prefix (Table A7) and "ane." If the cycloalkane ring holds only one alkyl group, the point of attachment on the ring is designated carbon #1. This number is not included in the name. If more than one alkyl group is attached to the ring, it is numbered from the position and in the direction that gives the lowest numbers to substituents (Table A10).

3. Alkenes (Section 4.8).

When assigning IUPAC names to alkenes, the parent is the longest chain of carbon atoms that contains the carbon–carbon double bond. The parent is numbered from the end nearer the double bond and the position of the bond is specified using the lower number assigned to the double-bonded carbon atoms. The position of the double bond need not be included in the name of alkenes whose parent chain contains three or fewer carbon atoms. IUPAC names of alkene parent chains are formed by replacing the final "e" on the name of the corresponding alkane with "ene" (Table A11). Ethlyene and propylene are important alkene common names (names that are not assigned according to the IUPAC rules).

TABLE A10. Cycloalkane IUPAC names

STRUCTURAL FORMULA	NAME
(cyclopentane ring)	Cyclopentane
CH_3 on cyclopentane ring	Methylcyclopentane
H_3C, CH_3 on one carbon of cyclopentane ring	1,1-Dimethylcylcopentane
CH_2CH_3, CH_2CH_3, CH_3 on cyclohexane ring	2,4-Diethyl-1-methylcyclohexane

TABLE A11. Alkene names

STRUCTURAL FORMULA	NAME	COMMON NAME
$H-\underset{H}{C}=\underset{H}{C}-H$	Ethene	Ethylene
$H-\underset{H}{C}=\underset{H}{C}-CH_3$	Propene	Propylene
$CH_2=CHCH_2\underset{CH_3}{CH}CH_3$	4-Methyl-1-pentene	
$CH_3\overset{CH_3}{CH}CH_2CH=\underset{CH_2CH_3}{C}CH_2CH_2CH_2CH_3$	5-Ethyl-2-methyl-4-nonene	

4. Alkynes (Section 4.8).

When assigning IUPAC names to alkynes, the parent is the longest chain of carbon atoms that contains the carbon–carbon triple bond. The parent is numbered from the end nearer the triple bond and the position of the bond is specified using the lower number assigned to the triple-bonded carbon atoms. The position of the triple bond need not be included in the name of alkenes whose parent chain contains three or fewer carbon atoms. IUPAC names of alkyne parent chains are formed by replacing the final "e" on the name of the corresponding alkane with "yne" (Table A12). Acetylene is an important alkyne common name.

TABLE A12. Alkyne names

STRUCTURAL FORMULA	NAME	COMMON NAME
$H-C=C-H$ $\;\;\;\;\; \vert \;\;\;\; \vert$ $\;\;\;\;\; H \;\;\; H$	Ethyne	Acetylene
CH_3 $CH_3CHCHC\equiv C-H$ CH_3	3,4-Dimethyl-1-pentyne	
CH_3CHCH_3 $CH_3C\equiv CCH_2CHCH_2CH_2CH_3$	5-Isopropyl-2-octyne	
CH_3 $CH_3CHC=CCH_2CHCH_3$ CH_2CH_3	2,6-Dimethyl-3-octyne	

5. Aromatic hydrocarbons (Section 4.8).

When naming aromatic hydrocarbons, "benzene" is an accepted IUPAC name, and benzene-related compounds use benzene as the parent ring. When only one substituent is attached, its position is not specified in the name. In the presence of more than one substituent, the ring is numbered from the position and in the direction to give the lowest numbers to substituents. For common names of benzene-related compounds that have only two substituents attached to the ring, *ortho* or *o* refers to a 1,2 arrangement, *meta* or *m* to a 1,3 arrangement, and *para* or *p* to a 1,4 arrangement (Table A13). Toluene and xylene are important common names.

TABLE A13. Aromatic hydrocarbon names

STRUCTURAL FORMULA	NAME	COMMON NAME
	Benzene	
CH_3	Methylbenzene	Toluene
CH_3 CH_3	1,2-Dimethylbenzene	*o*-Xylene
CH_3 CH_3	1,4-Dimethylbenzene	*p*-Xylene
CH_2CH_3 CH_3 CH_3 $CH_2CH_2CH_2CH_3$	2-Butyl-5-ethyl-1,3-dimethylbenzene	

6. Carboxylic acids (Section 10.1) and carboxylate ions (Section 10.3).

When assigning IUPAC names to carboxylic acids, the parent is the longest chain of carbon atoms that contains the carboxyl (CO_2H) group. Numbering begins at the carboxyl group, but the position of the carboxyl group is not specified in the name. IUPAC names for carboxylic acid parent chains are formed by dropping the final "e" from the name of the corresponding alkane and adding "oic acid" (Table A14). Formic acid, acetic acid, and benzoic acid are important carboxylic acid common names.

Carboxylic acid names are the starting point for naming carboxylate ions (carboxylic acid conjugate bases). To name a carboxylate ion, the ending on the name (IUPAC or common) of the corresponding carboxylic acid is changed from "ic acid" to "ate ion" (Table A15).

TABLE A14. Carboxylic acid names

STRUCTURAL FORMULA	NAME	COMMON NAME
$H-C(=O)-OH$	Methanoic acid	Formic acid
$CH_3C(=O)-OH$	Ethanoic acid	Acetic acid
$CH_3CH(CH_3)CH_2CH_2CH_2C(=O)-OH$	5-Methylhexanoic acid	
$CH_3CH_2C(CH_2CH_3)_2CH_2CH_2CH_2C(=O)-OH$	5,5-Diethylheptanoic acid	
$C_6H_5-C(=O)-OH$		Benzoic acid[a]

[a]This text does not present the IUPAC naming rules for compounds in which a carboxyl group is attached to a ring.

TABLE A15. Carboxylate ion names

STRUCTURAL FORMULA	NAME	COMMON NAME
$H-C(=O)-O^-$	Methanoate ion	Formate ion
$CH_3C(=O)-O^-$	Ethanoate ion	Acetate ion
$CH_3CH(CH_3)CH_2CH_2CH_2C(=O)-O^-$	5-Methylhexanoate ion	
$CH_3CH_2C(CH_2CH_3)_2CH_2CH_2CH_2C(=O)-O^-$	5,5-Diethylheptanoate ion	
$C_6H_5-C(=O)-O^-$		Benzoate ion

7. Phenols (Section 10.2) and phenoxide ions (Section 10.3).

The simplest phenol (C_6H_5OH) is named phenol. When substituents are added to this molecule, the carbon atom carrying the —OH group is designated as carbon #1 and the ring is numbered in the direction that gives the lowest numbers to substituents. The position of the—OH group is not included in the name of a phenol. As for aromatic hydrocarbons, when only two groups (including the—OH) are attached to the benzene ring, *ortho*, *meta*, and *para* can be used (Table A16).

TABLE A16. Phenol names

STRUCTURAL FORMULA	NAME
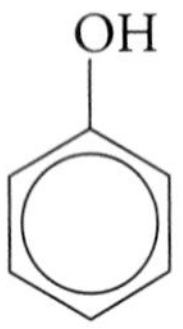 OH on benzene ring	Phenol
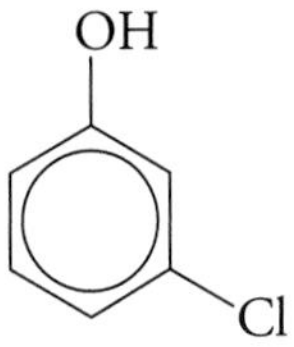 OH; Cl (meta) on benzene ring	3-Chlorophenol or *m*-chlorophenol
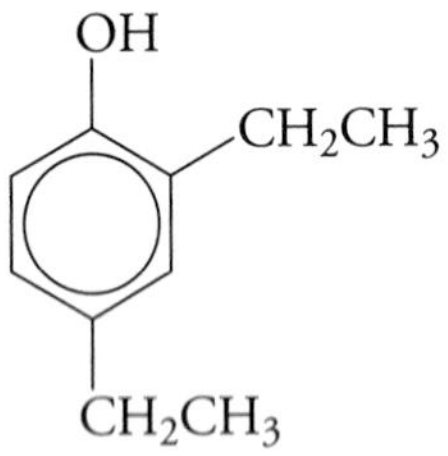 OH; CH_2CH_3; CH_2CH_3 on benzene ring	2,4-Diethylphenol

Phenol names are the starting point for naming phenoxide ions (phenol conjugate bases). To name a phenoxide ion, the ending on the name of the corresponding phenol is changed from "ol" to "oxide ion" (Table A17).

TABLE A17. Phenoxide ion names

STRUCTURAL FORMULA	NAME
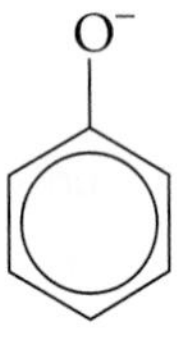 O^- on benzene ring	Phenoxide ion
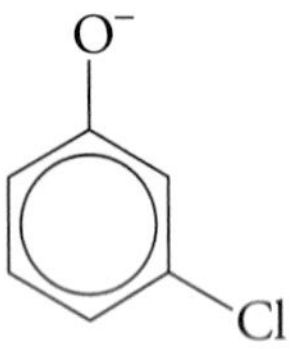 O^-; Cl (meta) on benzene ring	3-Chlorophenoxide ion or *m*-chlorophenoxide ion
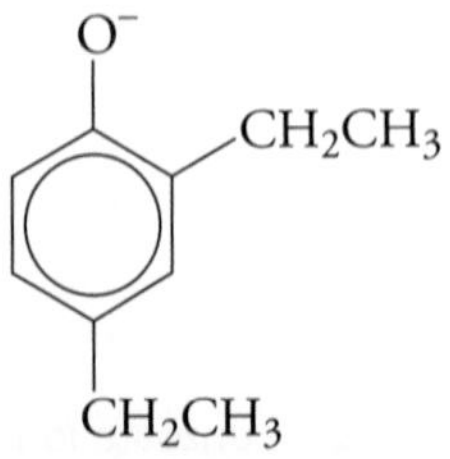 O^-; CH_2CH_3; CH_2CH_3 on benzene ring	2,4-Diethylphenoxide ion

8. Esters (Section 10.4).

Esters are named by specifying the name of the alkyl group contained in the alcohol residue, followed by the name of the carboxylic acid that supplies the carboxylic acid residue, with the ending on its name changed from "ic acid" to "ate" (Table A18).

TABLE A18. Ester names

STRUCTURAL FORMULA	NAME	COMMON NAME
$H-C(=O)-OCH_3$	Methyl methanoate	Methyl formate
$CH_3C(=O)-OCH_2CH_3$	Ethyl ethanoate	Ethyl acetate
$CH_3CH(CH_3)CH_2CH_2CH_2C(=O)-OCH(CH_3)CH_3$	Isopropyl 5-methylhexanoate	
$CH_3CH_2C(CH_2CH_3)_2CH_2CH_2CH_2C(=O)-OC(CH_3)_3$	*t*-Butyl 5,5-diethylheptanoate	
$C_6H_5-C(=O)-OCH_3$		Methyl benzoate

9. Amine (Section 10.6), quaternary ammonium ion (Section 10.6), and amine conjugate acid names (Section 10.7).

When naming a 1°, 2°, or 3° amine using the IUPAC rules, the parent is the longest chain of carbon atoms attached to the amine nitrogen atom. The parent is named by dropping "e" from the name of the corresponding hydrocarbon and adding "amine," and is numbered from the end of the parent nearer the point of attachment of the nitrogen atom. If the amine is 2° or 3°, the carbon-containing groups attached to the nitrogen atom, but not part of the parent, are substituents and their location is indicated using "*N*." Amine common names are formed by placing "amine" after the names of the groups attached to the nitrogen atom (Table A19). Quaternary (4°) ammonium ions and amine conjugate acids are named by giving the names of the groups attached to the nitrogen atom, followed by "ammonium ion" (Table A20).

TABLE A19. Amine names

Structural Formula	Name	Common Name
$CH_3CH_2NH_2$	Ethanamine	Ethylamine
$CH_3CH(CH_3)CH_2NH_2$	2-Methyl-1-propanamine	Isobutylamine
$CH_3CH_2CH(CH_3)NH_2$	2-Butanamine	*sec*-Butylamine
$CH_3CH_2CH_2NHCH_3$	*N*-Methyl-1-propanamine	Methylpropylamine
$CH_3CH_2CH_2N(CH_2CH_3)CH_2CH_3$	*N,N*-Diethyl-1-propanamine	Diethylpropylamine

TABLE A20. Quaternary ammonium ion and amine conjugate acid names

Structural Formula	Name
$CH_3CH_2\overset{+}{N}(CH_2CH_3)_2CH_2CH_3$	Tetraethylammonium ion
$CH_3CH_2CH_2\overset{+}{N}(CH_3)_2CH_3$	Trimethylpropylammonium ion
$CH_3CH_2\overset{+}{N}H_3$	Ethylammonium ion
$CH_3CH_2CH_2\overset{+}{N}H_2CH_3$	Methylpropylammonium ion
$CH_3CH_2CH_2\overset{+}{N}H(CH_2CH_3)CH_2CH_3$	Diethylpropylammonium ion

10. Amides (Section 10.8).

For amides that contain an ammonia (NH_3) residue, names are assigned by changing the ending on the name of the carboxylic acid that supplies the carboxylic acid residue from "oic acid" (IUPAC name) or "ic acid" (common name) to "amide." Amides that contain an amine residue are named in the same way, except that the name also indicates the substituents attached to the nitrogen atom, using *N* to indicate their position (Table A21).

TABLE A21. Amide names

Structural Formula	Name	Common Name
$H{-}\overset{\overset{O}{\parallel}}{C}{-}NH_2$	Methanamide	Formamide
$CH_3\overset{\overset{O}{\parallel}}{C}{-}NHCH_3$	*N*-Methylethanamide	*N*-Methylacetamide
$CH_3\underset{\underset{CH_3}{\mid}}{C}HCH_2CH_2CH_2\overset{\overset{O}{\parallel}}{C}{-}NH_2$	5-Methylhexanamide	
$CH_3CH_2\overset{\overset{CH_3}{\mid}}{C}HCH_2\overset{\overset{O}{\parallel}}{C}{-}\underset{\underset{CH_3}{\mid}}{N}CH_2CH_3$	*N*-Ethyl-*N*-methyl-3-methylpentanamide	
$C_6H_5{-}\overset{\overset{O}{\parallel}}{C}{-}NH_2$		Benzamide

11. Alcohols (Section 11.1).

When using the IUPAC rules to name alcohols, the parent (the longest carbon chain carrying the—OH group) is numbered from the end nearer the—OH. Alcohol names are derived by dropping the "e" ending on the name of the corresponding alkane and adding "ol." When a parent chain contains more than two carbon atoms, the position of the—OH group must be specified. Common names of alcohols are formed by combining the name of the alkyl group attached to the—OH with the word "alcohol" (Table A22).

12. Thiols (Section 11.1).

When using the IUPAC rules to name thiols, the parent (the longest carbon chain carrying the—SH group) is numbered from the end nearer the—SH. Thiol names are derived by adding "thiol" to the name of the corresponding alkane. When a parent chain contains more than two carbon atoms, the position of the—SH group must be specified. Common names of thiols are formed by combining the name of the group attached to the—SH with the word "mercaptan" (Table A23).

TABLE A22. Alcohol names

STRUCTURAL FORMULA	NAME	COMMON NAME
CH_3CH_2OH	Ethanol	Ethyl alcohol
OH | CH_3CHCH_3	2-Propanol	Isopropyl alcohol
$CH_3CH_2CHCH_2OH$ | CH_3	3-Methyl-1-butanol	
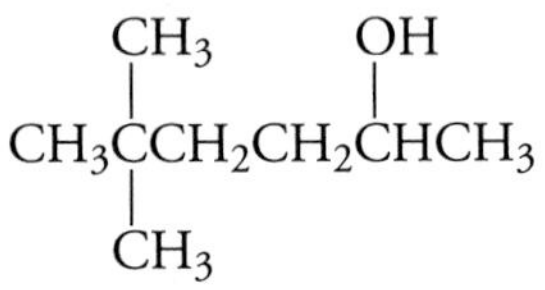 CH_3 OH | | $CH_3CCH_2CH_2CHCH_3$ | CH_3	5,5-Dimethyl-2-hexanol	
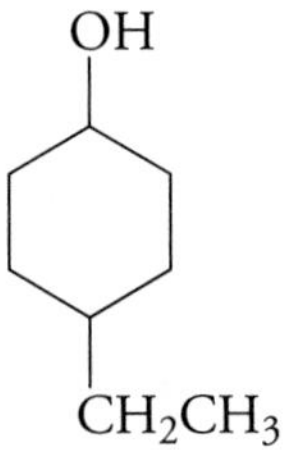 OH CH_2CH_3	3-Ethylcyclohexanol	

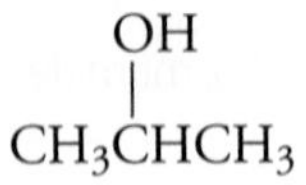

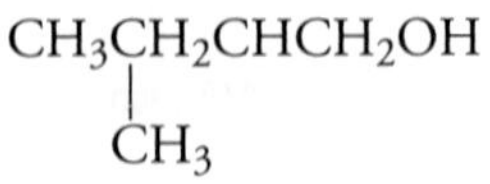

TABLE A23. Thiol names

STRUCTURAL FORMULA	NAME	COMMON NAME
CH_3CH_2SH	Ethanethiol	Ethyl mercaptan
SH | CH_3CHCH_3	2-Propanethiol	Isopropyl mercaptan
$CH_3CH_2CHCH_2OH$ | CH_3	3-Methyl-1-butanethiol	
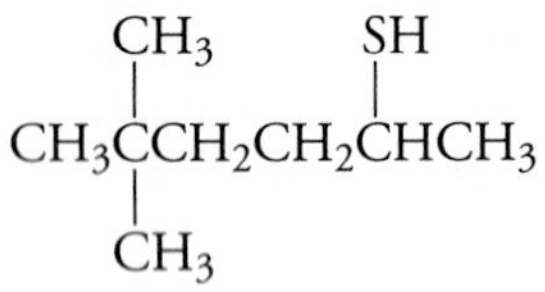 CH_3 SH | | $CH_3CCH_2CH_2CHCH_3$ | CH_3	5,5-Dimethyl-2-hexanethiol	
SH | $CH_3CHCHCH_2CHCH_2CH_2CH_3$ | | CH_3 CH_3CHCH_3	5-Isopropyl-2-methyl-3-octanethiol	

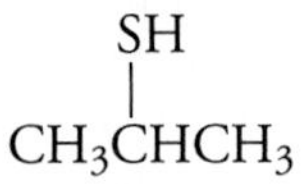

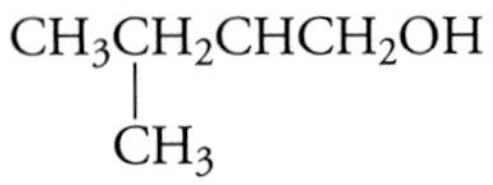

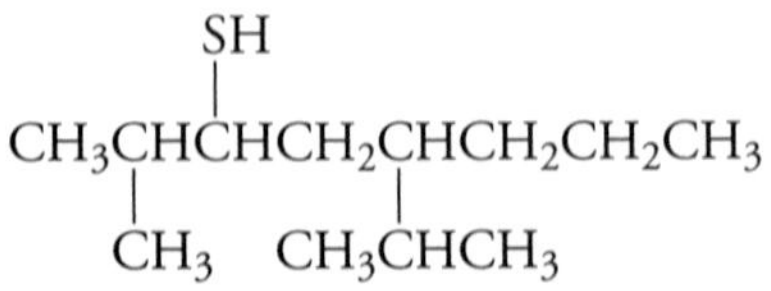

13. Ethers, sulfides, and disulfides (Section 11.1).

Common names of ethers are formed by placing "ether" after the names of the groups attached to the oxygen atom. Common names of sulfides are formed by placing "sulfide" after the names of the groups attached to the sulfur atom. Common names of disulfides are formed by placing "disulfide" after the names of the groups attached to the sulfur atoms (Table A24).

14. Aldehydes (Section 11.4).

TABLE A24. Ether, sulfide, and disulfide names

STRUCTURAL FORMULA	NAME
$CH_3CH_2OCH_2CH_3$	Diethyl ether
CH_3 | CH_3OCCH_3 | CH_3	Methyl *t*-butyl ether
$CH_3CH_2SCH_2CH_3$	Diethyl sulfide
CH_3 | $CH_3CHSCHCH_3$ | CH_3	Diisopropyl sulfide
$CH_3CH_2SSCH_2CH_3$	Diethyl disulfide
$CH_3CH_2CH_2SSCH_2CH_2CH_3$	Dipropyl disulfide

When assigning IUPAC names to aldehydes, the parent is the longest chain of carbon atoms that contains the carbonyl group. The parent is numbered from the end with the C=O and the position of this group (always carbon #1) is not specified in the name. Aldehyde names are derived by dropping the "e" ending on the name of the corresponding alkane and adding "al." Some aldehydes have common names that are based on the common name of the corresponding carboxylic acid: Compare formic acid (Table A14) and formaldehyde (Table A25).

15. Ketones (Section 11.4).

TABLE A25. Aldehyde names

STRUCTURAL FORMULA	NAME	COMMON NAME
O ‖ H—C—H	Methanal	Formaldehyde
O ‖ CH_3C—H	Ethanal	Acetaldehyde
O ‖ CH_3CHCH_2C—H | Br	3-Bromobutanal	
CH_3 O | ‖ $CH_3CHCHCH_2C$—H | CH_3	3,4-Dimethylpentanal	

When assigning IUPAC names to ketones, the parent is the longest chain of carbon atoms that contains the carbonyl group. The parent is numbered from the end nearer the C=O and for parents chains containing more than four carbon atoms, the position of this group must be specified in the name. Ketone names are derived by dropping the "e" ending on the name of the corresponding alkane and adding "one." Common names for ketones are obtained by placing "ketone" after the names of the groups attached to the carbonyl carbon atom (Table A26).

TABLE A26. Ketone names

Structural Formula	Name	Common Name
$CH_3C(=O)CH_3$	Propanone	Dimethyl ketone or Acetone
$CH_3C(=O)CH_2CH_3$	Butanone	Methyl ethyl ketone
$CH_3C(=O)CH(CH_3)CH_2CH_3$	3-Methyl-2-pentanone	Methyl *s*-butyl ketone
$CH_3CH_2CH(CH_2CH_2CH_3)C(=O)CH_3$	3-Ethyl-2-hexanone	

Answers to Odd-Numbered Problems

Chapter 1

1. A law. It describes what is observed but does not explain why it happens.
3. A hypothesis is a tentative explanation based on presently known facts while a theory is an experimentally tested explanation that is consistent with existing experimental evidence and accurately predicts the results of future experiments.
5. solid; liquid; gas.
7. **(a)** Kinetic energy increases as the gymnast accelerates across the mat. Some kinetic energy is converted to potential energy as the gymnast leaves the floor and starts upward into the flip. At the point where the upward motion stops and the downward motion begins, the potential energy is beginning to convert back to kinetic energy; **(b)** Potential energy increases as the gymnast leaves the floor until the maximum height is reached. At that point the potential energy decreases as the gymnast comes back down; **(c)** Work is done as the gymnast changes position.
9. **(a)** It is shiny, orange-brown in color, ductile, malleable, and conducts heat and electricity; **(b)** It can be stretched, pounded flat, and melted.
11. **(a)** Some of the potential (stored) energy in the body provides the heat used to melt the snow. **(b)** The body's energy reserves can be put to other uses, like keeping warm.
13. **(a)** 1 L; **(b)** 1°C; **(c)** 1 gal.
15. **(a)** 1 mg; **(b)** 1 gr; **(c)** 1 T; **(d)** 1 oz.
17. **(a)** milli or m; **(b)** mega or M; **(c)** centi or c.
19. **(a)** 1.5 km $= 1.5 \times 10^3$ m $=$ 1,500 m; **(b)** 5.67 mm $=$ 5.67×10^{-3} m $= 0.00567$ m; **(c)** 5.67 nm $= 5.67 \times 10^{-9}$ m $=$ 0.00000000567 m; **(d)** 0.3 cm $= 3 \times 10^{-3}$ m $=$ 0.003 m.
21. **(a)** 1 kcal; **(b)** 4.184 cal.
23. **(a)** 1×10^3 m; **(b)** 5×10^3 m; **(c)** 1×10^3 mm.
25. **(a)** 2×10^6mm; **(b)** 30 cL; **(c)** 4×10^{-6}kg.
27. **(a)** 8; **(b)** 5; **(c)** 3; **(d)** 2; **(e)** 2.
29. **(a)** 5.0×10^1; **(b)** 4.0×10^{-4}; **(c)** 47; **(d)** 7.9×10^2.
31. 35 mm.
33. **(a)** For a room 5.5 paces long and 5.0 paces wide, the area is 28 paces^2. **(b)** For a 31 inch stride (2.6 ft)) the conversion factor would be (2.6×2.6) ft^2/pace^2 and 28 paces^2 would be equivalent to 190 ft^2 (2 significant figures). **(c)** The room in part a is 172.0 inches (14.33 ft) long, 153.5 inches (12.79 ft) wide, and has an area of 183.3 ft^2. **(d)** The answers to parts b and c differ because the measuring techniques (counting paces versus using a yardstick) gave results with a different number of significant figures. The answer to part b is a better indicator of the area of the room.
35. **(a)** $\dfrac{12 \text{ eggs}}{1 \text{ dozen}}$ or $\dfrac{1 \text{ dozen}}{12 \text{ eggs}}$ **(b)** $\dfrac{1 \times 10^3}{1 \text{ km}}$ or $\dfrac{1 \text{ km}}{1 \times 10^3 \text{ m}}$ **(c)** $\dfrac{0.946 \text{ L}}{1 \text{ qt}}$ or $\dfrac{1 \text{ qt}}{0.946 \text{ L}}$
37. **(a)** 5.7 yd; **(b)** 442 in.
39. **(a)** 9.2×10^{-5} g; **(b)** 2.72×10^{-5} mg; **(c)** 3.3×10^5 mg; **(d)** 7.27×10^3 μg.
41. Your pound weight $\times \dfrac{1 \text{ kg}}{2.205 \text{ lb}}$
43. **(a)** 39 °C; **(b)** 77 °F; **(c)** 308 K; **(d)** 270 °F.
45. 2.2 Cal.
47. 25 g.
49. 1.7×10^2 mg.
51. 0.50 mL.
53. **(a)** 0.60 mL; **(b)** 13 gtt/min.
55. Insulin became available, the purity of insulin was improved, genetically engineered human insulin was put on the market, and oral drugs were developed.
59. 87 Ibs.
61. No. A normal body temperature of approximately 98.6 is based on the average of temperatures in normal healthy people. A difference of only 0.5 °F could easily be normal for you.
63. Tympanic temperature measurements can give false readings.

Chapter 2

1. It consists of protons (positive charge) and neutrons (no charge).
3. An atom is constructed from three subatomic particles: protons (1+ charge, 1 amu), neutrons (no charge, 1 amu), and electrons (1− charge, 1/2000 amu). Protons and neutrons make up the nucleus. Electrons are located outside of the nucleus.
5. **(a)** 6.07435×10^{23}amu; **(b)** 6.02214×10^{23} neutrons
7. **(a)** Li; **(b)** Br; **(c)** B; **(d)** Al; **(e)** F.
9. **(a)** beryllium; **(b)** neon; **(c)** magnesium; **(d)** phosphorus.
11. **(a)** 9 protons, 10 neutrons; **(b)** 11 protons, 12 neutrons; **(c)** 92 protons, 146 neutrons.
13. b.
15. $^{35}_{17}Cl$.
17. $^{148}_{65}Tb$; terbium−148.
19. **(a)** 12 protons, 12 neutrons, 12 electrons; **(b)** 25 protons, 30 neutrons, 25 electrons; **(c)** 30 protons, 34 neutrons, 30 electrons; **(d)** 34 protons, 40 neutrons, 34 electrons.
21.

helium	chlorine	iron
$^{3}_{2}He$	$^{34}_{17}Cl$	$^{56}_{26}Fe$
2	17	26
1	17	30
2	17	26
3	34	56

23. **(a)** Al; **(b)** Ca.
25. **(a)** I, Br, F; **(b)** O, F, Ne.
27. They are poor conductors of heat and electricity. As solids thay are non lusterous and brittle.
29. **(a)** nitrogen, N; phosphorus, P; astatine, As; antimony, Sb; bismuth, Bi. **(b)** N, nonmetal; P, nonmetal; As, semimetal; Sb, semimetal; Bi, metal.
31. 50.9415 amu (51 amu if rounded to 2 significant figures).
33. **(a)** 4.00 amu (rounded to two decimal places); **(b)** 4.00 g/mol; **(c)** 20.0 g He; **(d)** 0.400 g He; **(e)** 4.00 g He.
35. 1.20×10^{24} atoms.
37. **(a)** 32.1 amu (rounded to one decimal place); **(b)** 6.02×10^{23} atoms.
39. **(a)** 6.02×10^{23} atoms; **(b)** 7.34×10^{14} atoms; **(c)** 6.02×10^{23} atoms; **(d)** 2.26×10^{14} atoms.
41. **(a)** 0.00345 g; **(b)** 6.18×10^{-5}mol; **(c)** 3.72×10^{19} atoms.
43. **(a)** $^{187}_{80}Hg \longrightarrow {}^{183}_{78}Pt + {}^{4}_{2}\alpha$;
    (b) $^{226}_{88}Ra \longrightarrow {}^{222}_{86}Rn + {}^{4}_{2}\alpha$;
    (c) $^{238}_{92}U \longrightarrow {}^{234}_{90}Th + {}^{4}_{2}\alpha$.
45. **(a)** $^{14}_{8}O \longrightarrow {}^{14}_{7}N + {}^{0}_{1}\beta^{+}$;
    (b) $^{3}_{1}H \longrightarrow {}^{3}_{2}He + {}^{0}_{-1}\beta$;
    (c) $^{14}_{6}C \longrightarrow {}^{14}_{7}N + {}^{0}_{-1}\beta$.
47. **(a)** $^{35}_{16}S \longrightarrow {}^{31}_{14}Si + {}^{4}_{2}\alpha$;
    (b) $^{27}_{12}Mg \longrightarrow {}^{27}_{13}Al + {}^{0}_{-1}\beta$.
49. **(a)** $^{128}_{53}I + {}^{0}_{-1}e^{-} \longrightarrow {}^{128}_{52}Te + {}^{0}_{0}x$ ray
    (b) $^{128}_{53}I \longrightarrow {}^{128}_{52}Te + {}^{0}_{1}\beta^{+}$
    (c) No.
51. No. Since alpha particles travel only 4–5 cm in air and smoke detectors are usually on the ceiling, the risk of exposure to alpha radiation is small.
53. Alpha particles are relatively large and do not penetrate tissue very deeply. Radiation emissions must be able to pass through the body and reach a detector to be useful for diagnostic purposes.
55. **(a)** 0.0625 mg;
    (b) $^{197}_{80}Hg \longrightarrow {}^{196}_{80}Hg + {}^{1}_{0}n + {}^{0}_{0}\gamma$.
57. **(a)** $^{52}_{26}Fe \longrightarrow {}^{52}_{25}Mn + {}^{0}_{1}\beta^{+}$
    (b) When a positron and an electron collide, two gamma rays are produced
    (c) 2.1 hours
    (d) $^{52}_{25}Mn \longrightarrow {}^{52}_{24}Cr + {}^{0}_{1}\beta^{+}$
59. In fission, the nucleus of an atom splits to produce two smaller nuclei, neutrons, and energy. In a chain reaction, one or more of the products can initiate another cycle of the reaction. A critical mass is the minimum amount of radioactive material needed for a nuclear chain reaction to continue. Fissile material can undergo fission.
61. **(a)** Reprocessing can yield radioactive material suitable for use in nuclear weapons. If security at a reprocessing plant is lax, or if the spent fuel rods must be transported to a reprocessing site, it might be possible for some individual or group to steal enough material to make a nuclear weapon.
    (b) Reprocessing would reduce the amount of radioactive material stored at nuclear reactor sites.
63. $^{6}_{3}Li + {}^{1}_{0}n \longrightarrow {}^{4}_{2}He + {}^{3}_{1}H$
65. **(a)** AI (Adequate Intake) is used when RDA (Recommended Daily Allowance) values are not available.
    (b) RDA is a better measure, because it is defined as the daily intake of a nutrient that is sufficient to meet the needs of 97–98% of healthy people. AI, on the other hand, is the daily intake believed to be sufficient to meet the needs of people, but which percentage of people is unknown.
67. 330 mg (2 significant figures).
69. **(a)** Invasive refers to the fact that the imaging technique requires something (x rays in this case) to penetrate the body.
    (b) CT
    (c) This is a matter of personal choice. There are potential risks associated with exposure to x rays, so some people may not want a CT scan run. Many MRI devices are quite confining—some may not prefer that experience.
71. No. The patient contains no cobalt−60 and will not give off gamma radiation.

Chapter 3

1. **(a)** 19 protons, 18 electrons; **(b)** 12 protons, 10 electrons; **(c)** 15 protons, 18 electrons.
3. **(a)** 29 protons, 34 neutrons, 28 electrons; **(b)** 9 protons, 10 neutrons, 10 electrons; **(c)**17 protons, 20 neutrons, 18 electrons.
5. **(a)** fluoride ion; **(b)** oxide ion; **(c)** chloride ion; **(d)** bromide ion.
7. **(a)** carbonate ion; **(b)** nitrate ion; **(c)** sulfite ion; **(d)** acetate ion.

9. **(a)** HCO_3^-; **(b)** NO_2^-; **(c)** SO_4^{2-}.
11. H·
13. Yes. The emission spectrum is created when electrons move from excited state energy levels to more stable ones. A different energy separation between energy levels would produce different colors in an emission spectrum.
15. **(a)** n = 1 n = 2 n = 3 n = 4
    2 **3**
    (b) n = 1 n = 2 n = 3 n = 4
    2 **4**
    (c) n = 1 n = 2 n = 3 n = 4
    2 **8** **2**
17. **(a)** 1; **(b)** 4; **(c)** 3; **(d)** 8; **(d)** 5.
19. **(a)** 35 total electrons, 7 valence electrons held in level 4; **(b)** 36 total electrons, 8 valence electrons held in level 4; **(c)** 33 total electrons, 5 valence electrons held in level 4; **(d)** 53 total electrons, 7 valence electrons held in level 5.
21. **(a)** 118; **(b)** 8; **(c)** 7; **(d)** yes.
23. **(a)** potassium ion; **(b)** 1; **(c)** 1+.
25. **(a)** K·, K^+; **(b)** ·S̈e·, :S̈e:$^{2-}$; **(c)** Ca:, Ca^{2+}; **(d)** ·Ö·, :Ö:$^{2-}$.
27. **(a)** magnesium oxide; **(b)** sodium sulfate; **(c)** calcium fluoride; **(d)** iron (II) chloride.
29. **(a)** $CaHPO_4$; **(b)** $CuBr_2$; **(c)** $CuSO_4$; **(d)** $NaHSO_4$.
31. **(a)** Li_2SO_4; **(b)** $Ca(H_2PO_4)_2$; **(c)** $BaCO_3$.
33. **(a)** MgF_2; **(b)** KBr; **(c)** K_2S; **(d)** Al_2S_3.
35. **(a)** tin(II) chloride; **(b)** iron(II) oxide; **(c)** chromium(III) phosphate.
37. **(a)** 3; **(b)** 1; **(c)** 3.
39. **(a)**
```
    H     H
 H:C̤:Ö:C̤:H
    H     H
```
(b)
```
   H    H    H  Ö:
H:C̤  :  C̤  : C̤:C:H
   H  H:N̤:   H
        H
```
41.
```
    H  Ö: Ö:
    |  ‖  ‖
 H—C—C—C—Ö—H
    |
    H
```
43.
```
H:Ö: :Ö:H
H:C̤  :  C̤:H
   H     H
```
45. **(a)** nitrogen trichloride; **(b)** phosphorus trichloride; **(c)** phosphorus pentachloride.
47. phosphorus trihydride.
49. dinitrogen monoxide.
51. **(a)** H :F̤̈:
    (b) H:P̤̈:H
    H
    (c) H :B̤̈r:
53. **(a)** barium chloride; **(b)** oxygen dichloride; **(c)** carbon disulfide; **(d)** mercury(II) oxide; **(e)** dinitrogen trioxide; **(f)** copper(II) oxide.
55. **(a)** sulfur difluoride; **(b)** magnesium fluoride; **(c)** tin(II) chloride; **(d)** phosphorus pentafluoride; **(e)** nitrogen dioxide; **(f)** tin(II) oxide.
57. No. Binary molecules contain atoms of two different elements.
59. **(a)** 73.8 amu; **(b)** 9.82×10^{-3} g; **(c)** 0.996 mol.
61. **(a)** 278.1 amu; **(b)** 4.5×10^{18} Mg^{2+}; **(c)** 9.0×10^{18} I^-; **(d)** 9.7×10^{18} Mg^{2+}; **(e)** 1.9×10^{19} I^-.
63. 30 mmol.
65. **(a)** 154.0 amu; **(b)** 9.44×10^3 g; **(c)** 3.02×10^{-3} mol; **(d)** 2.15×10^{19} molecules.
67. 1.0×10^{21} molecules.
69. **(a)** 1.3×10^{-6}mol; **(b)** 7.8×10^{17} molecules.
71. When acted on by a particular enzyme in the presence of oxygen gas and ATP, the electrons in luciferin are pushed into an excited state. When the electrons return to ground state, light is emitted.
75. Since K^+ ions are larger than Na^+ ions, the cavity must be too large for the Na^+ ion.
77. A compound always has the same proportion of the same elements. In amalgam, the amounts of mercury, silver, tin, copper, and zinc vary depending on how it is prepared, so there is no one amalgam compound.

Chapter 4

1. **(a)**
```
             H
             |
           H—C—H
   H  H      |  H  H
   |  |      |  |  |
 H—C—C——C——C—C—H
   |  |   |   |  |
   H  H   H   H  H
```
(b)
```
   H  H
   |  |
 H—C—C—N̈—H
   |  |  |
   |  H  H
   |
 (benzene ring)
   H    C    H
    \ /  \\ /
     C     C
     ‖     |
     C     C
    / \\  / \
   H    C    H
        |
        H
```
3. **(a)** tetrahedral; **(b)** trigonal planar; **(c)** tetrahedral.
5. **(a)**
```
   H  H
   |  |
 H—C—C—H
   |  |
   H  H
```
(b)
```
 H—C=C—H
   |  |
   H  H
```
(c) H—C≡C—H
7. **(a)** :F̤̈—F̤̈: **(b)** :Ö=Ö: **(c)** :N≡N:
9. **(a)** each H atom (0); S atom (0); **(b)** C atom (0); double bonded O atom (0); each single bonded O atom (1−); **(c)** each Cl atom (0); N atom (0).
11. **(a)** $^-$:Ö—H **(b)**
```
     H
     |+
 H—N—H
     |
     H
```
(c) $^-$:C≡N:
13. **(a)**
```
        2+
 ⁻:Ö—S=Ö:
      |
     :Ö:⁻
```
(b)
```
        +
 ⁻:Ö—S—Ö:⁻
      |
     :Ö:⁻
```

15. **(a)** None. While H and S have different electronegativities, for our purposes the bonds are not considered to be polar covalent.

polar covalent bond

(b) 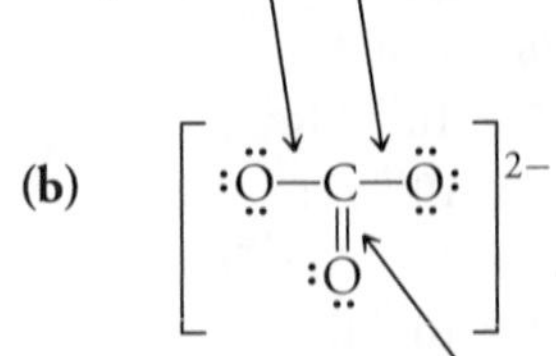

polar covalent bond

(c) None. While N and Cl have different electronegativities, for our purposes the bonds are not considered to be polar covalent.

17. **(a)** bent; **(b)** trigonal planar; **(c)** pyramidal
19. a and b.
21. False. Depending on the number of groups of electrons around an atom, a bent molecular shape can have a bond angle near 110° or near 120° (see Table 4.2).
23. No.
25. c and d.
27. b, c, and d.
29. b.
31. two $CH_3CH_2CH_2CH_2CH_3$ molecules.
33.
35. $CH_3CH_2CH_2CH_2CH_3$; Both molecules have the formula C_5H_{12}, but $CH_3CH_2CH_2CH_2CH_3$ is unbranched and has a greater surface area. This leads to stronger London force attractions between molecules.
37. decane; butane; propane.
39. **(a)**

(b)

41. **(a)** parent—pentane; name: 3-methylpentane; **(b)** parent—hexane; name: 3-methylhexane; **(c)** parent—heptane; name: 4-ethyl-4-methylheptane.
43. There are 9 constitutional isomers

heptane

2-methylhexane

3-methylhexane

2,3-dimethylpentane

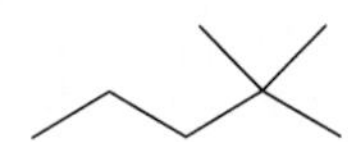

2,2-dimethylpentane

3,3-dimethylpentane

2,4-dimethylpentane

2,2,3-trimethylbutane

3-ethylpentane

45. constitutional isomers (a and c); identical (b).
47. b and c
49. **(a)** different conformations; **(b)** constitutional isomers
51. **(a)** In order to be constitutional isomers or conformations, molecules must have the same molecular formula; **(b)** Contitutional isomers have different atomic connections. Conformations have the same atomic connections, but different three-dimensional shapes that are interchanged by bond rotation.
53.

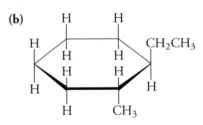

1-ethyl-2-methylcyclobutane

1-ethyl-3-methylcyclobutane

1-ethyl-1-methylcyclobutane

55. **(a)** methylcyclobutane; **(b)** 1,3-dimethylcyclohexane; **(c)** 1-ethyl-2-methylcyclopentane; **(d)** 1,1-dimethyl-2-propylcyclopropane.
57. b and c.
59. **(a)**

(b)

(c) 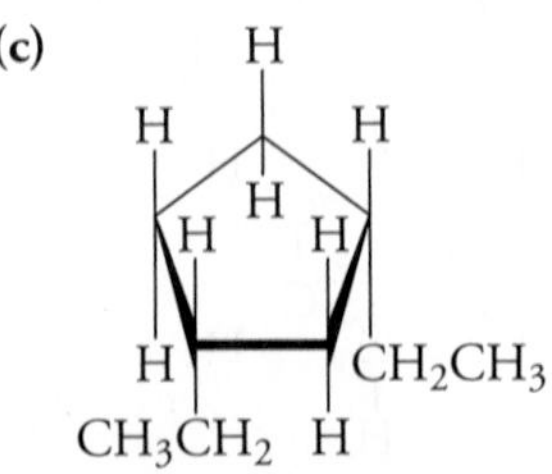

61. **(a)** *cis*-1,4-dipropylcyclohexane; **(b)** *trans*-1,3-diethylcyclopentane.

63. H, H, C=C, H, H, C, H, H

65. **(a)** 1-butene; **(b)** 2,4-dimethyl-2-pentene; **(c)** 4-methyl-2-pentyne.

67. **(a)** CH_3CHCH_3 / $CH_3CH_2CH_2CH_2CHCH{=}CH_2$

(b) $CH_3C{=}CCH_3$ / H_3C CH_3

(c) $CH_3CHCH_2CH_3$ / $CH_3CH_2CH_2CH_2CHC{\equiv}CCH_2CH_3$

69. Moving left to right across the molecule, the first double bond is *trans*, the second double bond is *cis*, and the third double bond is *cis*.

71. **(a)** 1-methyl-3-propylbenzene; **(b)** 1,2,4-trimethylbenzene; **(c)** 4-*t*-butyl-1,2-diethylbenzene.

73. **(a)** $CH_2CH_2CH_3$, $CH_2CH_2CH_3$

(b) CH_3CHCH_3, $CHCH_3$ / CH_3

(c) CH_3, CH_3, CH_3

75. *trans*.

77. **(a)** hydrogen bonding; **(b)** hydrogen bonding; **(c)** dipole-dipole.

79. **(a)** alcohol and alkene; **(b)** *trans*; **(c)** yes; **(d)** cyclohexane, cyclopentane and cyclopropane; **(e)** unsaturated.

81. Cattle feed should not contain parts from sheep, cattle, or other animals.

83. a, b, and c.

85. They are not volatile, so molecules do not reach the nose.

87. In the strictest interpretation of "organic", the sign would be interpreted as telling you that there are carbon containing compounds. The second meaning is that the foods were grown utilizing nutrients that came from decaying organic matter.

Chapter 5

1. **(a)** Heat of fusion is the energy required to melt a solid; **(b)** Heat of vaporization is the energy required to evaporate a liquid.

3. The energy goes into the ice and the energy is used in the melting process

5. **(a)** 1.6×10^3 cal; **(b)** 3.3×10^3 cal.

7. **(a)** 8.1×10^3 cal; **(b)** 3.5×10^3 cal.

9. 1.4×10^4 cal.

11. 67°C.

13. The process will not run by itself unless something keeps it going.

15. **(a)** spontaneous; **(b)** nonspontaneous; **(c)** spontaneous; ΔG is negative; **(d)** non spontaneous; ΔG is positive.

17. **(a)** 6.39×10^3cal; **(b)** 4.04×10^4cal

19. 233 g.

21. **(a)** iron; **(b)** mercury; **(c)** oxygen.

23. **(a)** 0.925 atm; **(b)** 703 torr.

25. **(a)** 0.45 atm; **(b)** 6.6 psi and 3.4×10^2 torr.

27. 765 torr.

29. **(a)** the mercury level is falling; **(b)** the air pressure is decreasing.

31. 430 kPa.

33. 0.46 L.

35. 1.65 L.

37. 835 torr.

39. 0.86 L.

41. 0.50 L.

43. 1.06 atm (greater than atmospheric pressure).

45. If the number of moles and the volume of a gas remain constant, then an increase in temperature results in an increase in pressure (PV = nRT).

47. **(a)** 3.21×10^3 atm; **(b)** 2.44×10^6 torr; **(c)** 4.72×10^4 psi.

49. 7.49 g.

51. **(a)** 0.0418 mol, **(b)** 0.84 g.

53. **(a)** 1.37×10^3torr; **(b)** 1.09×10^3torr; **(c)** 13.2 L; **(d)** 1.82×10^3torr; **(e)** 0.435 mol.

55. False. The same number of moles of each gas is present, but the molar mass of each is different.

57. **(a)** 0.86 atm N_2, 6.5×10^2 torr; 0.29 atm O_2, 2.2×10^2 torr; 0.29 atm He, 2.2×10^2 torr; **(b)** 1.44 atm; 1.09×10^3 torr.

59. 0.86 atm N_2, 6.5×10^2 torr; 0.29 atm O_2, 2.2×10^2 torr; 0.29 atm He, 2.2×10^2 torr; 0.58 atm CO_2, 4.4×10^2 torr; Total pressure =2.02 atm = 1.54×10^2 torr

61. Hot Coffee, Mississippi.

63. **(a)** 50. atm O_2, 22 atm He, 15 atm N_2; **(b)** 160 g O_2, 8.8 g He, 39 g N_2; **(c)** 2.9×10^{24} molecules O_2, 1.3×10^{24} atoms Ne, 8.4×10^{23} molecules N_2.

65. 1.06 g/mL.

67. 44.5 mL.

69. 12 mL.

71. 0.00143 g/ml.

73. 0.82.

75. 20 °C.

77. 120 °C.

79. **(a)** 300 cal; **(b)** 630 cal.

81. When the humidity is high, less water evaporates from the water soaked pads and less cooling takes place.

83. Ionic bonds are stronger than intermolecular forces.

85. Amorphous solids do not have the orderly arrangement of particles found in crystalline structures.

87. Molecular solids are held together by noncovalent interactions, which are weaker than the covalent bonds that hold covalent solids together.

89. A decrease in blood volume causes an increase in blood pressure.

91. Two hours might not be enough time to rehydrate and to recover from the effects of dehydration.

93. **(a)** 0.521 lbs/cup; **(b)** 2.9 lbs. No. Losing weight through dehydration can adversely affect endurance, strength, energy, and motivation. Extreme dehydration can lead to kidney failure, heart attack, and death.

95. Without sufficient hemoglobin, your cells will not be supplied with the O_2 that they require.

97. An increase in the number of red blood cells would allow for more O_2 transport to the cells.

Chapter 6

1. $2P + 3Cl_2 \longrightarrow 2PCl_3$

3. **(a)** the formation of a solid; **(b)** bubbling, indicating the formation of a gas.

5. **(a)** $2SO_2 + O_2 \longrightarrow 2SO_3$; **(b)** $2NO + O_2 \longrightarrow 2NO_2$.

7. **(a)** $2K + Cl_2 \longrightarrow 2KCl$; **(b)** $CH_4 + 2Cl_2 \longrightarrow CH_2Cl_2 + 2HCl$.

9. $CaC_2(s) + 2H_2O(l) \longrightarrow Ca(OH)_2(aq) + C_2H_2(g)$.

11. **(a)** double replacement; **(b)** decomposition.

13. **(a)** single replacement; **(b)** single replacement; **(c)** synthesis; **(d)** decomposition.

15. synthesis

17. synthesis

19. **(a)** oxidized; **(b)** reduced; **(c)** iron(II) ion; **(d)** magnesium atom.

21. **(a)** $4Al + 3O_2 \longrightarrow 2Al_2O_3$; **(b)** Al; **(c)** O_2; **(d)** O_2; **(e)** Al.

23. Problem 6.19, single replacement; Problem 6.21, synthesis,

25. $CH_3CH_2OH + 3O_2 \longrightarrow 2CO_2 + 3H_2O$.

27. **(a)** $2H_2 + O_2 \longrightarrow 2H_2O$; **(b)** CO_2 and H_2O; **(c)** Visit *http://www.wiley.com/college/raymond* to view this video.

29. oxidized.

31. $CH_3(CH_2)_{10}CH_2OH$.

33. **(a)** 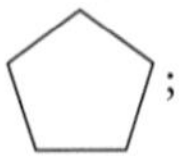 [cyclopentane structure]; **(b)** 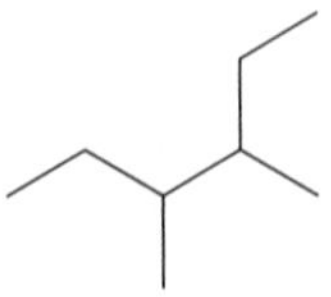 [skeletal structure]

35. **(a)** $H{-}\overset{O}{\overset{\|}{C}}{-}O^- + HOCH_2CH_3$

(b) $CH_3{-}\overset{O}{\overset{\|}{C}}{-}O^- + HOCH_2{-}C_6H_5$ (benzene ring)

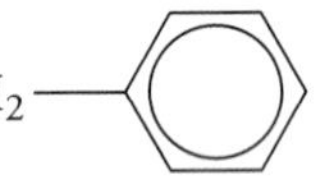

(c) $CH_3CH_2\overset{O}{\overset{\|}{C}}{-}O^- + HOCH_2CH_2CH_3$

37. **(a)** $CH_3CH_2CH_2\underset{}{\overset{OH}{\overset{|}{C}}}HCH_2CH_3$

(b) $CH_3CH_2CH_2\overset{OH}{\overset{|}{C}}HCH_2CH_3$

(c) 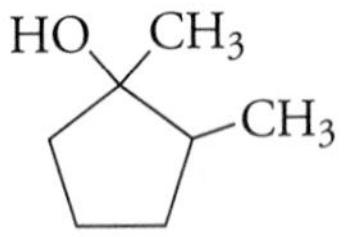 [cyclopentane ring bearing HO and CH_3 on one carbon, with CH_3]

39. **(a)** $CH_3\underset{CH_3}{\underset{|}{C}}{=}CH_2$

(b) 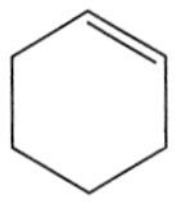 [cyclohexene structure]

(c) $CH_3CH_2CH{=}CHCH_3$

41. **(a)** 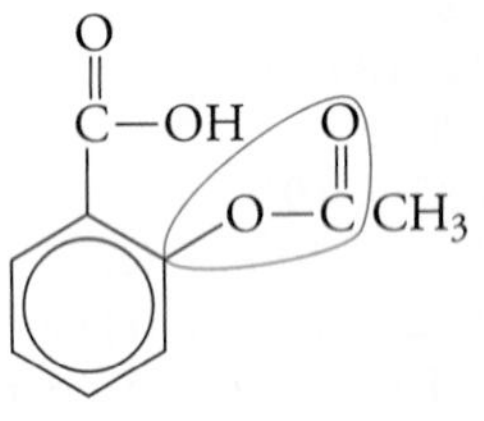 [benzene ring with $\overset{O}{\overset{\|}{C}}{-}OH$ and $O{-}\overset{O}{\overset{\|}{C}}CH_3$ (circled)]

(b) 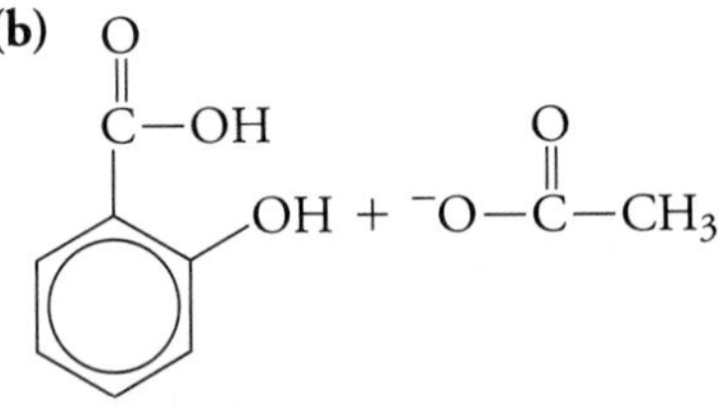 [benzene ring with $\overset{O}{\overset{\|}{C}}{-}OH$ and OH] $+ \ {}^-O{-}\overset{O}{\overset{\|}{C}}{-}CH_3$

The $-CO_2H$ and $-OH$ groups of the molecule on the left are actually found in their conjugate base forms (chapter 10).

43. **(a)** $CH_3\overset{OH}{\overset{|}{C}}HCH_3 = A$

(b) $CH_3CH{=}CH_2 = B$

(c) $CH_3\overset{OH}{\overset{|}{C}}HCH_3 = C$

45. **(a)** 6.30 mol; **(b)** 3.15 mol; **(c)** 6.30 mol.

47. **(a)** 55.7 mol; **(b)** 1.66 mol; **(c)** 1.12 mol; **(d)** 179 g.

49. **(a)** 281 g; **(b)** 135 g.

51. **(a)** $2Fe + O_2 \longrightarrow 2FeO$; **(b)** oxidized; **(c)** reduced.

53. **(a)** $P_4(s) + 5O_2(g) \longrightarrow P_4O_{10}(s)$; **(b)** 75.6 g.

55. **(a)** 113 g; **(b)** 44.2%

57. **(a)** oxidized.

(b) 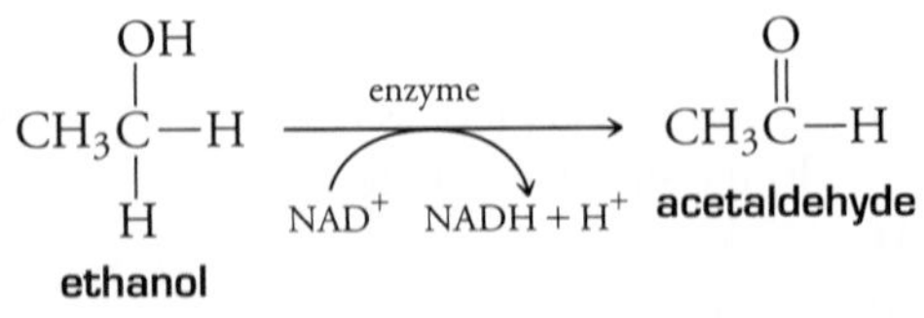 $CH_3\underset{H}{\underset{|}{\overset{OH}{\overset{|}{C}}}}{-}H$ (ethanol) $\xrightarrow[NAD^+ \ \ NADH + H^+]{enzyme}$ $CH_3\overset{O}{\overset{\|}{C}}{-}H$ (acetaldehyde)

(c) 9.55 g.

59. a.

61. **(a)** yes; **(b)** yes; **(c)** These are different reactions. The conversion of glucose 6-phosphate into glucose plus phosphate is spontaneous. By bringing ATP into play, the conversion of glucose into glucose 6-phosphate becomes spontaneous.

63. **(a)** decreases;
(b) decreases;
(c) increases.

65. $2C_{14}H_{10}O_4 + 29O_2 \longrightarrow 28CO_2 + 10H_2O$.

67. Approximately 1/3 of the CO_2 attaches to hemoglobin and the remaining part of the 95% is converted into carbonic acid in a hydration reaction catalyzed by the enzyme carbonic anhydrase.

Chapter 7

1. ethanol.
3. The amount of a substance that can dissolve in a solvent is temperature dependent. As the temperature is lowered the solubility of the salt decreases and the salt in excess settles out as a precipitate.
5. **(a)** Salt in water; **(b)** O_2 in the air
7. $CuSO_4(aq) + Na_2S(aq) \longrightarrow CuS(s) + Na_2SO_4(aq)$.
9. **(a)** soluble; **(b)** soluble;
(c) insoluble; **(d)** soluble.
11. **(a)** $CaCl_2(aq) + Li_2CO_3(aq) \longrightarrow CaCO_3(s) + 2\ LiCl(aq)$.
(b) $Pb(NO_3)_2(aq) + 2NaCl(aq) \longrightarrow PbCl_2(s) + 2NaNO_3(aq)$.
(c) $3CaBr_2(aq) + 2K_3PO_4(aq) \longrightarrow Ca_3(PO_4)_2(s) + 6KBr(aq)$.
13. **(a)** $CuCl_2(aq) + 2AgNO_3(aq) \longrightarrow Cu(NO_3)_2(aq) + 2AgCl(s)$
(b) $3Mg(NO_3)_2(aq) + 2Na_3PO_4(aq) \longrightarrow Mg_3(PO_4)_2(s) + 6NaNO_3(aq)$
15. $(NH_4)_2CO_3(aq) + 2HCl(aq) \longrightarrow H_2CO_3(aq) + 2NH_4Cl(aq)$
$H_2CO_3(aq) \longrightarrow H_2O(l) + CO_2(g)$
17. $BaCl_2(aq) + 2Na_2SO_4(aq) \longrightarrow BaSO_4(s) + 2NaCl(aq)$
balanced equation

$Ba^{2+}(aq) + 2Cl^-(aq) + 2Na^+(aq) + (SO_4)^{2-}(aq) \longrightarrow BaSO_4(s) + 2Na^+(aq) + 2Cl^-(aq)$
ionic equaton

$Ba^{2+}(aq) + SO_4^{2-}(aq) \longrightarrow BaSO_4(s)$
net ionic equation

19. **(a)** $2Na^+(aq) + CO_3^{2-}(aq) + Pb^{2+}(aq) + 2NO_3^-(aq) \longrightarrow PbCO_3(s) + 2Na^+(aq) + 2NO_3^-(aq)$
ionic equation

$Pb^{2+}(aq) + CO_3^{2-}(aq) \longrightarrow PbCO_3(s)$
net ionic equation

$3Ca^{2+}(aq) + 2PO_4^{3-}(aq) \longrightarrow Ca_3(PO_4)_2(s)$
net ionic equation

21. $Nis(aq) + 2HCl(aq) \longrightarrow NiCl_2(aq) + H_2S(g)$
balanced equation

$Ni^{2+}(aq) + S^{2-}(aq) + 2H^+(aq) + 2Cl^-(aq) \longrightarrow Ni^{2+}(aq) + 2Cl^-(aq) + H_2S(g)$
ionic equation

$S^{2-}(aq) + 2H^+(aq) \longrightarrow H_2S(g)$
net ionic equation

23. **(a)** decrease; **(b)** increase.
25. a, c.
27. Urea has more atoms capable of forming hydrogen bonds with water and does not contain the nonpolar chain of carbon atoms present in hexanamide.
29. $CH_3CH_2CH_2CH_2OH$ is able to form more hydrogen bonds with water.
31. Water insoluble drugs tend to stay dissolved in nonpolar tissues, while water-soluble ones are removed in urine.
33. hydrophobic.
35. The drug is hydrophobic and stays dissolved in the fatty (nonpolar) tissues of the body.

37. With stirring, add table sugar to water. Continue until no more will dissolve.
39. 2.1% (v/v).
41. 0.00016 – 0.00017 ppm. 0.16 – 0.17 ppb.
43. **(a)** 0.540 M; **(b)** 6.0% (w/v); **(c)** 60 ppt; **(d)** 6.0×10^4 ppm; **(e)** 6.0×10^7 ppb.
45. **(a)** 0.115 M; **(b)** 4.6 M.
47. 7 ppb lead; below the action level.
49. 1.5×10^{-7} M.
51. 100 ppm.
53. 3.5 – 4.9 mmol/L.
55. 95 – 107 mEq/L.
57. 1.9 mEq.
59. 1.32×10^{-3} mol.
61. **(a)** 3.8×10^3mL; **(b)** 2.2×10^3mL; **(c)** 0.85 L.
63. 0.45 M.
65. 2.50% (w/v).
67. 225 mL.
69. **(a)** 200 mL; **(b)** 114 mL.
71. 27 ppb.
73. The particles that make up a colloid are smaller than those present in a suspension. While suspensions settle, colloids do not.
75. **(a)** Salt in water; **(b)** Muddy water; **(c)** Soapy water.
77. No. In active transport solutes move in a direction opposite to diffusion.
79. $CaCO_3(s) + 2H^+(aq) \longrightarrow H_2O(l) + CO_2(g) + Ca^{2+}(aq)$.
81. The higher the pressure of a gas over a liquid, the its solubility in the liquid. In a hyperbaric chamber, more O_2 will be dissolved in the blood.
83. 0.274 g.
85. Saliva is cloudy and will not settle upon standing (see Table 7.3).
87. $\mathrm{H{-}\ddot{\underset{..}{O}}{-}\ddot{\underset{..}{O}}{-}H}$

    $\mathrm{:\ddot{\underset{..}{S}}{=}C{-}\ddot{\underset{..}{N}}:^-}$ or $\mathrm{^-:\ddot{\underset{..}{S}}{-}C{\equiv}N:}$
89. Osmosis causes the movement of water from blood serum (lower solute concentration) to tissues (higher solute concentration).
91. **(a)** Oxidized; **(b)** Yes; **(c)** The structure of oxypyrinol more closely resembles that of xanthine, so it is more likely to interact with xanthine oxidase.

Chapter 8

1. Lauric acid has a shorter hydrocarbon tail than myristic acid, so London force interactions between lauric acid molecules are weaker.
3. lauric acid

```
   H H H H H H H H H H H O
   | | | | | | | | | | | ||
 H—C—C—C—C—C—C—C—C—C—C—C—C—O—H
   | | | | | | | | | | |
   H H H H H H H H H H H
```

linolenic acid

```
   H H H H H       H H H H H H H H H H O
   | | | | |       | | | | | | | | | | ||
 H—C—C—C=C—C—C=C—C—C=C—C—C—C—C—C—C—C—C—O—H
   | |     | | | |     | | | | | | |
   H H     H H H H     H H H H H H H
```

5. The ionic bonds that hold the ions in sodium palmitate to one another are stronger than the noncovalent interactions that hold one palmitic acid molecule to another.
7. Fatty acids are carboxylic acids that typically have between 12 and 20 carbon atoms.
9. The fatty acids differ in length by an even number of carbon atoms.
11. 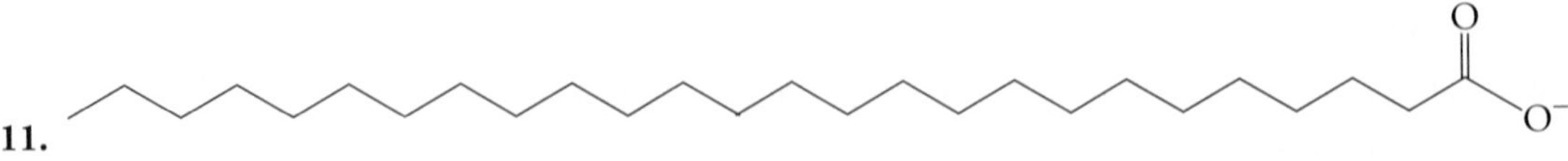

and

HO

13.  $CH_3(CH_2)_{14}\overset{O}{\overset{\|}{C}}-OCH_2(CH_2)_{14}CH_3$

15.  $CH_3CH_2CH_2\underset{CH_3}{CH}CH_2\underset{CH_3}{CH}CH_2\underset{CH_3}{CH}CH_2\underset{CH_3}{CH}\overset{O}{\overset{\|}{C}}-OH$ and $HOCH_2\underset{CH_3}{CH}CH_2\underset{CH_3}{CH}CH_2\underset{CH_3}{CH}CH_2\underset{CH_3}{CH}CH_2CH_2CH_3$

17. $CH_3(CH_2)_{14}CO_2CH_2(CH_2)_{26}CH(OH)CH_2CH_3$

19. energy source, thermal insulation, padding, buoyancy.

21. $CH_3(CH_2)_5CH{=}CH(CH_2)_7CO_2H \xrightarrow[Pt]{H_2} CH_3(CH_2)_5CH_2CH_2(CH_2)_7CO_2H$

23.

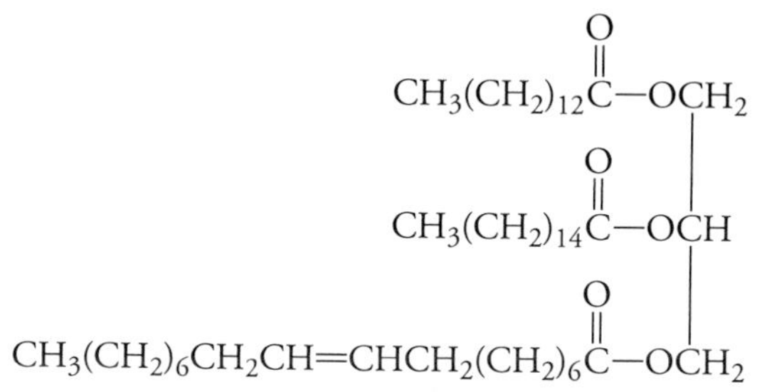

$CH_3(CH_2)_{12}\overset{O}{\overset{\|}{C}}-OCH_2$

$CH_3(CH_2)_{14}\overset{O}{\overset{\|}{C}}-OCH$

$CH_3(CH_2)_6CH_2CH{=}CHCH_2(CH_2)_6\overset{O}{\overset{\|}{C}}-OCH_2$

This triglyceride contains more satured than unsatured fatty acid residues, so it should be a solid at room temperature.

25. 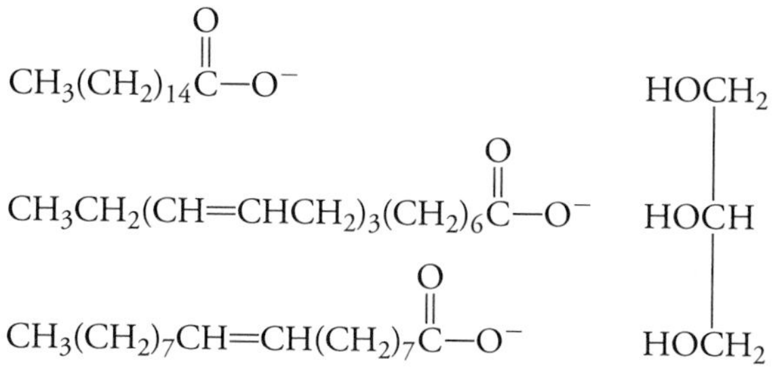 $CH_3(CH_2)_{14}\overset{O}{\overset{\|}{C}}-O^-$

$CH_3CH_2(CH{=}CHCH_2)_3(CH_2)_6\overset{O}{\overset{\|}{C}}-O^-$

$CH_3(CH_2)_7CH{=}CH(CH_2)_7\overset{O}{\overset{\|}{C}}-O^-$

$HOCH_2$

$HOCH$

$HOCH_2$

27. Vegetable oils contain more carbon-carbon double bonds than animal fats. It is the oxidation of these bonds that causes oils to become rancid.

29. The triglyceride molecules that make up vegetable oil are nonpolar, while the molecules present in vinegar (CH_3CO_2H and H_2O) are polar.

31. **(a)** true,
(b) false,
(c) false.

33. **(a)** sphingolipid;
(b) 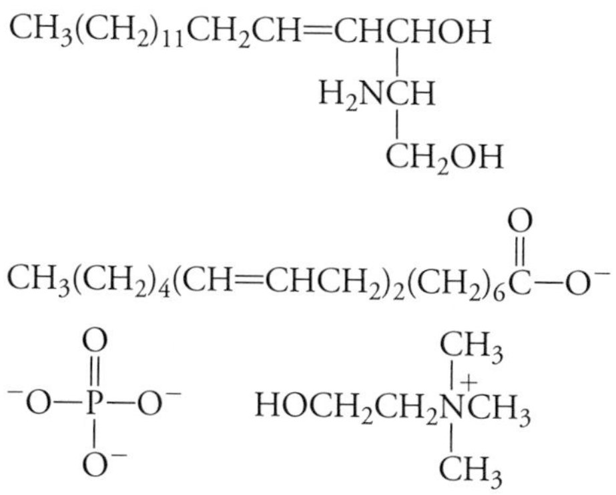 $CH_3(CH_2)_{11}CH_2CH{=}CHCHOH$

H_2NCH

CH_2OH

$CH_3(CH_2)_4(CH{=}CHCH_2)_2(CH_2)_6\overset{O}{\overset{\|}{C}}-O^-$

${}^-O-\overset{O}{\overset{\|}{P}}(-O^-)-O^-$

$HOCH_2CH_2\overset{+}{N}(CH_3)_3$

35. Lecithin is amphipathic and triglycerides are hydrophobic. Being amphipathic, lecithins can interact with both nonpolar compounds (oil) and polar ones (water) to keep an emulsion from separating.

37.

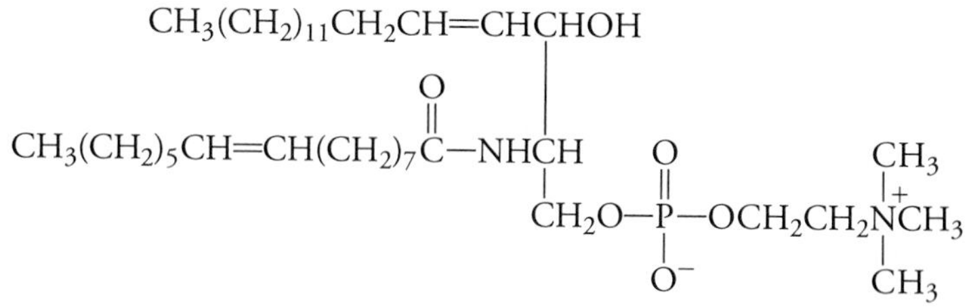

$CH_3(CH_2)_{11}CH_2CH{=}CHCHOH$

$CH_3(CH_2)_5CH{=}CH(CH_2)_7\overset{O}{\overset{\|}{C}}-NHCH$

$CH_2O-\overset{O}{\overset{\|}{P}}(O^-)-OCH_2CH_2\overset{+}{N}(CH_3)_3$

39. Phosphatidylethanolamine. Both compounds (Figure 8.17) contain O atom capable of forming hydrogen bonds. The—$CH_2CH_2NH_3^+$ part of phosphatidylethanolamine contains three N—H bonds able to form hydrogen bonds, while the—$CH_2CH_2N(CH_3)_3^+$ part of phosphatidylcholine does not.

41. decrease.

43. 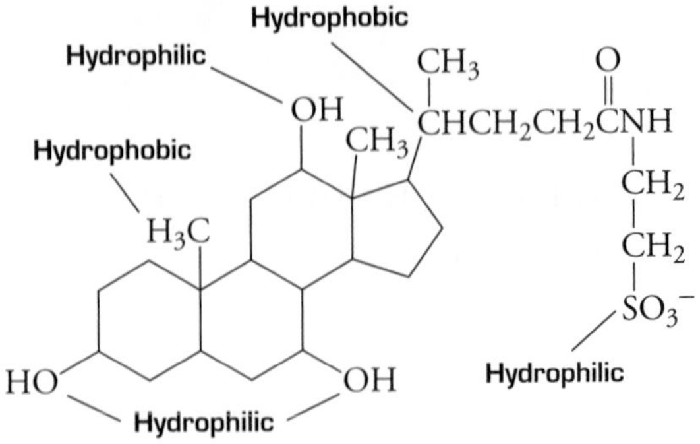 Structure (steroid with side chain CH_3–$CHCH_2CH_2C(=O)NH$–CH_2–CH_2–SO_3^-; ring substituents HO, H_3C, CH_3, OH, OH)
Labels: Hydrophobic; Hydrophilic; OH; Hydrophobic; H_3C; HO; Hydrophilic; OH; Hydrophilic

45. cholesterol.

47. male sex hormone-testosterone; female sex hormone-progesterone.

49. **(a)** HDL transports cholesterol and phospholipids from cells back to the liver. LDL transports cholesterol and phospholipids from the liver to cells. **(b)** Partially. Cholesterol is the major component of LDH. HDL contains more protein and phospholipids than cholesterol. **(c)** Having high LDH and low HDL levels is a warning sign for atherosclerosis and an increased risk of stroke and heart disease.

51. **(a)** They block the action of COX-1 and COX-2, enzymes involved in the conversion of arachidonic acid into prostaglandins and thromboxanes; **(b)** Celebrex only blocks the action of COX-2.

53. thromboxanes.

55. Facilitated diffusion moves substances from areas of higher concentration to areas of lower concentration, and does not require the input of energy. Active transport moves substances in the opposite direction and requires the input of energy.

57. In facilitated diffusion, diffusion across a membrane takes place with the assistance of proteins.

59. Reindeer are typically found in cold, snow covered regions. Since the melting point of fatty acids goes down as the degree of unsaturation goes up, having more unsaturated fatty acids in the feet and legs would reduce the likelihood that the fats would solidify. This would help keep the membranes more flexible.

61. **(a)** pentaen = 5 double bonds; **(b)** docos = 22, hexaen = 6 double bonds.

63. They may contain harmful levels of mercury.

65. 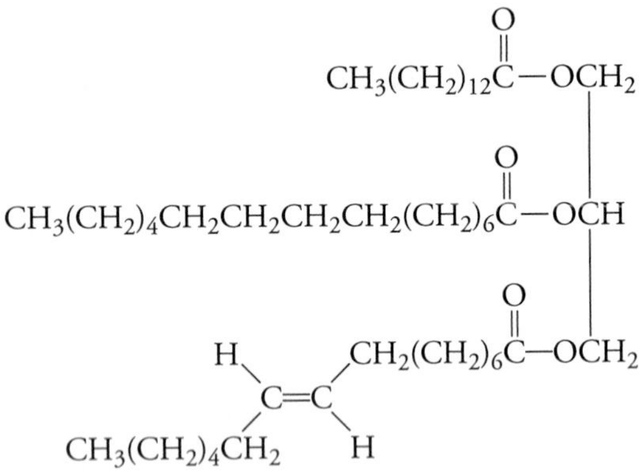
$$CH_3(CH_2)_{12}\overset{O}{\overset{\|}{C}}-OCH_2$$
$$CH_3(CH_2)_4CH_2CH_2CH_2CH_2(CH_2)_6\overset{O}{\overset{\|}{C}}-OCH$$
$$\underset{CH_3(CH_2)_4CH_2}{\overset{H}{\ }}\!\!\!>C{=}C<\!\!\!\underset{H}{\overset{CH_2(CH_2)_6\overset{O}{\overset{\|}{C}}-OCH_2}{\ }}$$

67. Margarine. It is produced from vegetable oil by partial hydrogenation, a process that forms *trans* fats.

69. London forces.

71. reduced.

73. Vegetable oil consists of triglycerides, while motor oil consists largely of hydrocarbon molecules.

Chapter 9

1. Yes. Acids have a sour taste and bases have a bitter taste.

3. **(a)** hydrofluoric acid and hydrogen fluoride; **(b)** hyrdroiodic acid and hydrogen iodide.

5. **(a)** SO_4^{2-}; **(b)** $H_2PO_4^-$; **(c)** PO_4^{3-}; **(d)** NO_3^-.

7. **(a)** $HCO_3^-(aq) + HCl(aq) \longrightarrow H_2CO_3(aq) + Cl^-(aq)$; **(b)** $HCO_3^-(aq) + OH^-(aq) \longrightarrow CO_3^{2-}(aq) + H_2O(l)$.

9. **(a)** $\underset{\text{base}}{F^-(aq)} + \underset{\text{acid}}{HCl(aq)} \rightleftharpoons \underset{\text{acid}}{HF(aq)} + \underset{\text{base}}{Cl^-(aq)}$
(b) $\underset{\text{acid}}{CH_3CO_2H(aq)} + \underset{\text{base}}{NO_3^-(aq)} \rightleftharpoons \underset{\text{base}}{CH_3CO_2^-(aq)} + \underset{\text{acid}}{HNO_3(aq)}$

11. **(a)** $NH_4^+ + SO_4^{2-} \rightleftharpoons NH_3 + HSO_4^-$
NH_4^+ (acid), NH_3(conjugate base); HSO_4^- (acid), SO_4^{2-} (conjugate base)
(b) $CN^- + HI \rightleftharpoons NCH + I^-$
HCN (acid), CN^- (conjugate base); HI (acid), I^- (conjugate base)

13. c.

15. **(a)** $K_{eq} = \dfrac{[CO]^2}{[CO_2]}$
(b) $K_{eq} = \dfrac{[NH_4^+][OH^-]}{[NH_3]}$

17. Only concentrations that change should be included in equilibrium constant expressions. The concentration of a solid is constant at a given temperature.

19. (a) $2NO + Cl_2 \longrightarrow 2NOCl$; (b) $H_2 + Br_2 \longrightarrow 2HBr$.

21. reactants.

23. (a) reactants; (b) products.

25. (a) $K_{eq} = \dfrac{[H_2CO_3]}{[CO_2]}$

(b) No effect. The catalyst increases the rate of the forward and reverse reactions to the same extent.

27. (a) Increasing [CO] increases the rate of the forward reaction. Decreasing $[H_2]$ slows the rate of the forward reaction. Increasing $[CH_3OH]$ increases the rate of the reverse reaction.
(b) The reaction would continually move to make CH_3OH and would never reach equilibrium.

29. (a) There is a net reverse reaction until equilibrium is reestablished; (b) There is a net reverse reaction until equilibrium is reestablished; (c) There is a net forward reaction until equilibrium is reestablished.

31. (a) 2.1×10^{-7} M; (b) 1.5×10^{-13} M; (c) 6.7×10^{-3} M.

33. (a) acidic; (b) basic; (c) acidic.

35. (a) 1.6×10^{-11} M; (b) 1.2×10^{-7} M; (c) 5.3×10^{-4} M.

37. (a) acidic; (b) basic; (c) basic.

39. (a) 7.0; (b) 4.15; (c) 9.85; (d) 13.0.

41. (a) neutral; (b) acidic; (c) basic; (d) basic.

43. (a) 2.9×10^{-6} M; (b) 2×10^{-4} M; (c) 1.1×10^{-3} M.

45. (a) 4.5×10^{-8} M; (b) 2.2×10^{-7} M; (c) 4.5×10^{-11} mol; (d) 2.2×10^{-10} mol; (e) 2.7×10^{13} H_3O^+; (f) 1.3×10^{14} OH^-

47. (a) $NH_4^+(aq) + H_2O(l) \rightleftharpoons H_3O^+(aq) + NH_3(aq)$;

$$K_a = \frac{[H_3O^+][NH_3]}{[NH_4^+]}$$

(b) $HPO_4^{2-}(aq) + H_2O(l) \rightleftharpoons H_3O^+(aq) + PO_4^{3-}(aq)$

$$K_a = \frac{[H_3O^+][PO_4^{3-}]}{[HPO_4^{2-}]}$$

49. (a) 7.52; (b) 4.19. $C_2O_4H^-$ is the stronger acid.

51. Hydrofluoric acid. Hydrofluoric acid is the stronger acid, so it is better at releasing H^+ (making an acidic solution) than acetic acid.

53. (a) $C_2O_4H^-$; (b) $C_6O_2H_7^-$; $C_6O_2H_8$ is the weaker acid, so $C_6O_2H_7^-$ is the stronger base.

55. (a) 4.91×10^{-3} mol NaOH; (b) 4.91×10^{-3} mol HCl; (c) 0.327 M HCl

57. 0.175 M.

59. (a) acid; (b) the pH increases.

61. The concentration of CH_3CO_2H will increase and the concentration of $CH_3CO_2^-$ will decrease.

63. (a) acid form; (b) conjugate base form

65. (a) H_2CO_3 or $H_2PO_4^-$; (b) NH_4^+, HCN, HCO_3^-, or $CH_3NH_3^+$; (c) HF, $CH_3CH(OH)CO_2H$, or CH_3CO_2H.

67. (a) $CH_3CH(OH)CO_2H + H_2O \longleftarrow CH_3CH(OH)CO_2^- + H_3O^+$
(b) $CH_3CH(OH)CO_2H + OH^- \longrightarrow CH_3CH(OH)CO_2^- + H_2O$

69. In response to an increase in $[H_3O^+]$ (a drop in pH) the equilibrium shifts to the left:
$H_2CO_3 + H_2O \longleftarrow HCO_3^- + H_3O^+$
In response to a decrease in $[H_3O^+]$ (a rise in pH) the equilibrium shifts to the right:
$H_2CO_3 + H_2O \longrightarrow HCO_3^- + H_3O^+$

71. In response to a drop in blood pH (a rise in $[H_3O^+]$) the rate of respiration increases. As more CO_2 is exhaled, $[H_2CO_3]$ drops, as does $[H_3O^+]$.
$H_2O + CO_2 \longleftarrow H_2CO_3$
$H_2CO_3 + H_2O \longleftarrow HCO_3^- + H_3O^+$
In response to a rise in blood pH (a drop in $[H_3O^+]$) the rate of respiration slows. As CO_2 accumulates, $[H_2CO_3]$ increases, as does $[H_3O^+]$.
$H_2O + CO_2 \longrightarrow H_2CO_3$
$H_2CO_3 + H_2O \longrightarrow HCO_3^- + H_3O^+$

73. Respiratory alkalosis occurs when CO_2 is exhaled more rapidly than it is produced by cells (too little CO_2 results in a drop in $[H_3O^+]$—see the answer to Problem 71.) The kidneys respond to a drop in $[H_3O^+]$ by taking up HCO_3^-. A drop in the concentration of HCO_3^- should lead to an increase in the concentration of H_3O^+.
$H_2CO_3 + H_2O \longrightarrow HCO_3^- + H_3O^+$

75. CO_2 exhaled into a paper bag is inhaled on the next breath. This leads to a rise in the concentrations of H_2CO_3 and H_3O^+.
$H_2O + CO_2 \longrightarrow H_2CO_3$
$H_2CO_3 + H_2O \longrightarrow HCO_3^- + H_3O^+$

77. (a) When [Mb] increases, the rate of the forward reaction increases and more MbO_2 is produced; (b) When $[O_2]$ increases, the rate of the forward reaction increases and more MbO_2 is produced; (c) When $[MbO_2]$ increases, the rate of the reverse reaction increases and more Mb and O_2 is produced.

79. (a) −9.0; (b) 12.0; (c) −1.47; (d) 4.99.

81. (a) 1×10^8; (b) 1×10^{-3}; (c) 7.9×10^7; (d) 2.0×10^{15}.

83. A variation in pH causes a change in the structure of the compound responsible for the color.

85.
```
    H  H  Ö:
    |  |  ||
H—N⁺—C—C—Ö:⁻
    |  |
    H  H
```

Chapter 10

1. (a) 2-methylbutanoic acid;
   (b) 3-methylbutanoic acid;
   (c) 2-methylpentanoic acid.

3. (a) 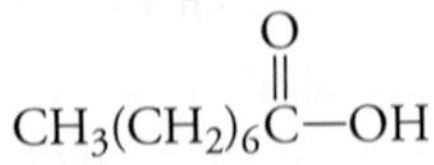 $CH_3(CH_2)_6C(=O)-OH$

   (b) 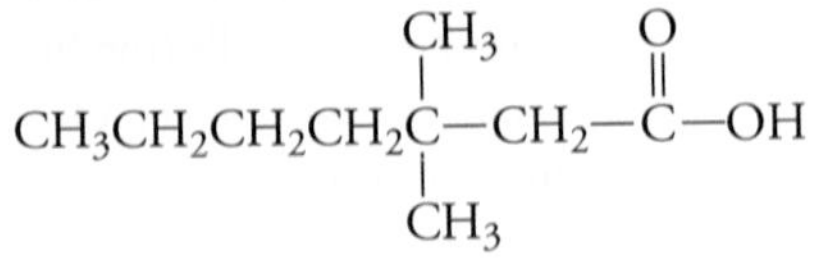 $CH_3CH_2CH_2CH_2C(CH_3)_2-CH_2-C(=O)-OH$

   (c) $CH_3CH_2CH_2CH(CH_3CHCH_3)CH_2-C(=O)-OH$

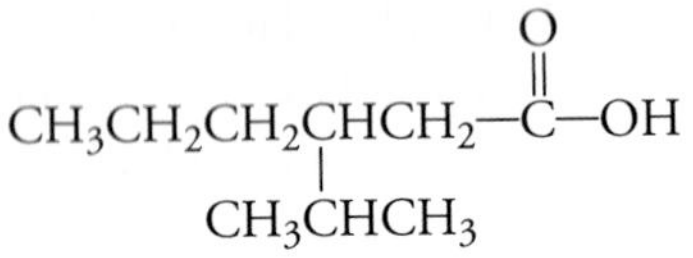

   (d) 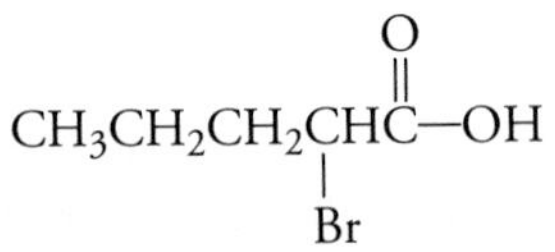 $CH_3CH_2CH_2CH(Br)C(=O)-OH$

5. Acetic acid molecules form more hydrogen bonds with one another than do propyl alcohol molecules.

7. (a) 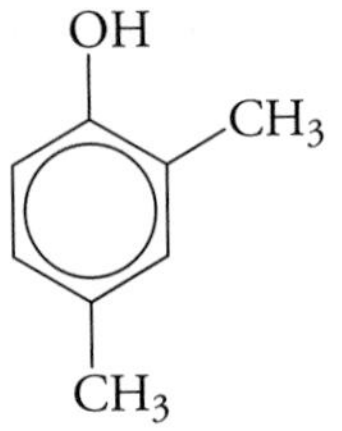 OH, CH_3, CH_3 (benzene ring structure)

   (b) 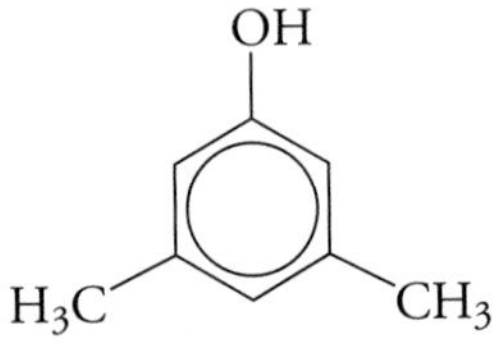 OH, H_3C, CH_3 (benzene ring structure)

   (c) 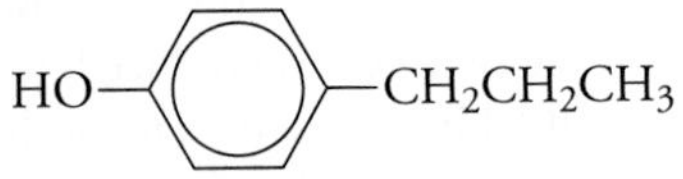 HO–(benzene ring)–$CH_2CH_2CH_3$

   (d) 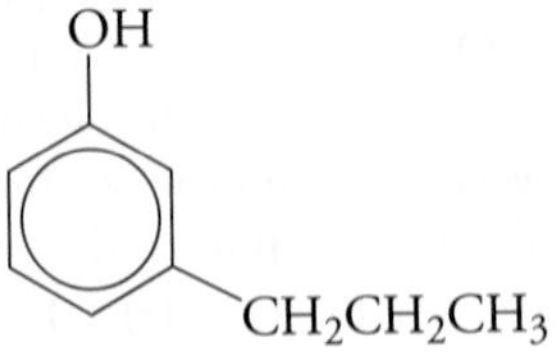 OH, $CH_2CH_2CH_3$ (benzene ring structure)

9. The urushiol molecule has a long hydrocarbon side chain that makes it hydrophobic (water-insoluble)

11. 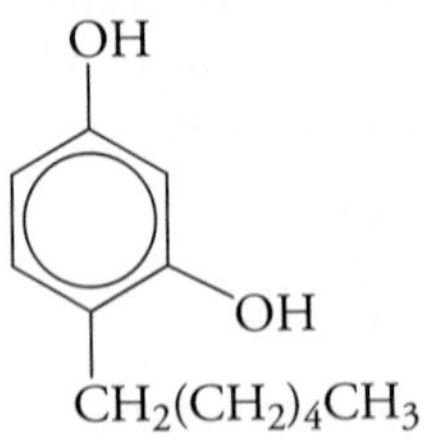 OH, OH, $CH_2(CH_2)_4CH_3$ (benzene ring structure)

13. catechol-containing.

15. *m*-chlorobenzoic acid: $C(=O)-OH$, Cl (benzene ring structure); conjugate base: $C(=O)-O^-$, Cl (benzene ring structure)

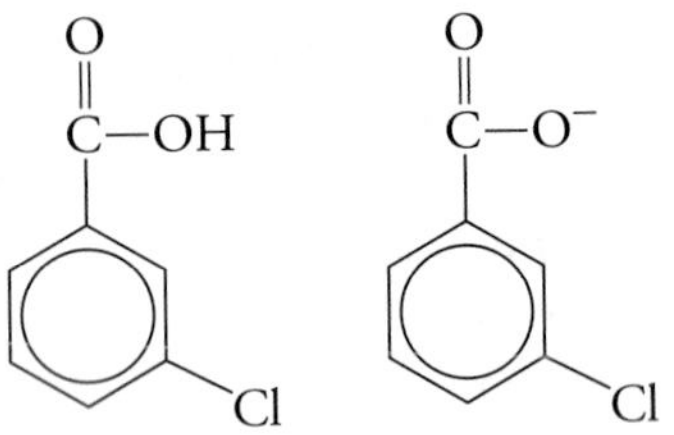

17. (a) 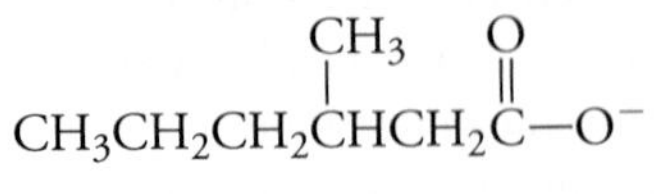 $CH_3CH_2CH_2CH(CH_3)CH_2C(=O)-O^-$

    (b) 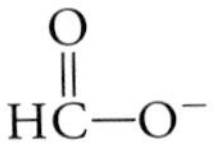 $HC(=O)-O^-$

    (c) 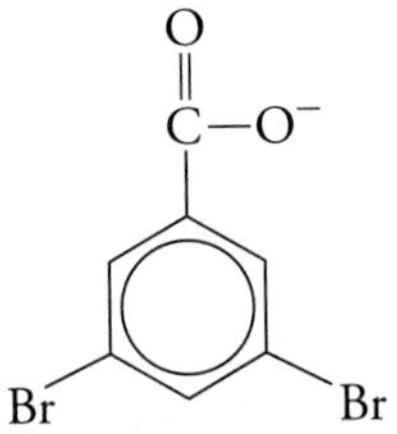 $C(=O)-O^-$, Br, Br (benzene ring structure)

19. (a) 3-methylhexanoate ion;
    (b) formate ion;
    (c) 3,5-dibromobenzoate ion.

21. (a) 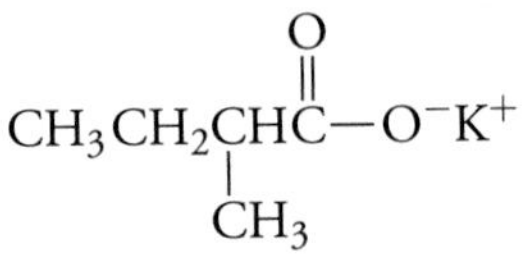 $CH_3CH_2CH(CH_3)C(=O)-O^-K^+$

    (b) 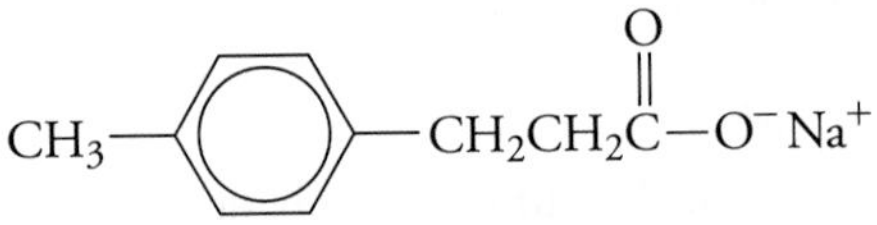 CH_3–(benzene ring)–$CH_2CH_2C(=O)-O^-Na^+$

23. 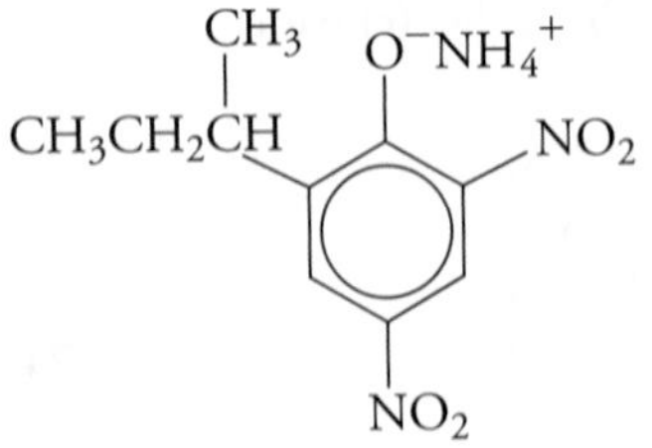 $CH_3CH_2CH(CH_3)$, $O^-NH_4^+$, NO_2, NO_2 (benzene ring structure)

25. (a) The carboxyl group;
    (b) 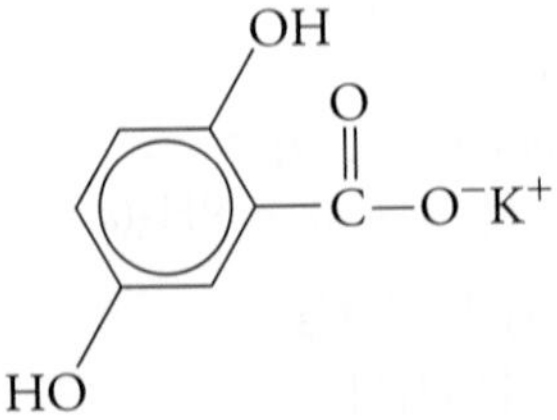 OH, $C(=O)-O^-K^+$, HO (benzene ring structure)

27. 2,4-dichlorophenol; 2,4-dichlorophenoxide ion

29. (a) $CH_3CH(Cl)CH_2CH_2C(=O)—OH$

(b) 4-chloropentanoate ion; $CH_3CH(Cl)CH_2CH_2C(=O)—O^-$

(c) conjugate base

31. (a) $C_6H_5—C(=O)—OCH_2CH_2CH_3$

(b) $CH_3CH_2C(=O)—OCH_2—C_6H_5$

33. (a) propyl benzoate; (b) benzyl propanoate

35. $CH_3CH_2—C_6H_4—C(=O)—OCH(CH_3)CH_3$

37. (a) $CH_3CH(CH_3)CH_2CH_2C(=O)—OH + HOCH_3$

(b) $C_6H_{11}—C(=O)—OH + HO—C_6H_{11}$

39. (a) $CH_3C(=O)—OH + HO—C_5H_9$

(b) $CH_3C(=O)—O^- + HO—C_5H_9$

41. (a) $C_6H_5—C(=O)CH_2CH_3$

(b) $CH_3CH(CH_3)C(=O)—CH_3$

43 $CH_3—C_6H_4—CH_2C(=O)—H$

45.

47. N 3°; CH_3; N 2°

49. (a) methanamine;
(b) 2-propanamine;
(c) *N*-isopropyl-1-butanamine;
(d) *N*-ethyl-*N*-methyl-1-propanamine

51. (a) methylamine; (b) isopropylamine; (c) butylisopropylamine; (d) ethylmethylpropylamine.

53. (a) pyrimidine; (b) pyridine; (c) purine.

55. Primary amines, like propylamine, have two N—H hydrogen atoms and one nitrogen atom available for forming hydrogen bonds with water. Tertiary amines, like trimethylamine, have only one nitrogen atom capable of forming hydrogen bonds.

57. (a) $CH_3CH_2CH_2CH_2CH(NH_2)CH_3$

(b) $CH_3CH_2CH(NHCH_3)CH_2CH_3$

(c) $CH_3CH_2CH_2CH_2N(CH_3)CH_3$

(d) $CH_3—N^+(CH_3)_2—CH_3$

59. (a) $CH_3CH_2CH_2CH_2NH_3^+$
butylammonium ion

(b) $CH_3CH_2CH_2\overset{+}{N}H_2CH_2CH_2CH_3$
dipropylammonium ion

(c) $CH_3\overset{+}{N}H(CH_3)CH_3$
trimethylammonium ion

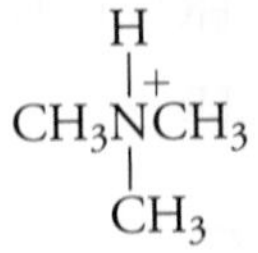

(d) $CH_3\overset{+}{C}(CH_3)(CH_3)NH_3$
t-butylammonium ion

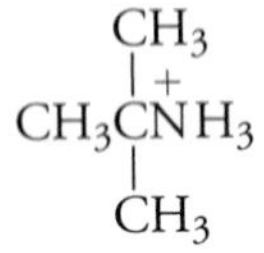

61. C_6H_5–(piperidine ring)–$\overset{+}{N}(CH_3)(H)$ Cl^-

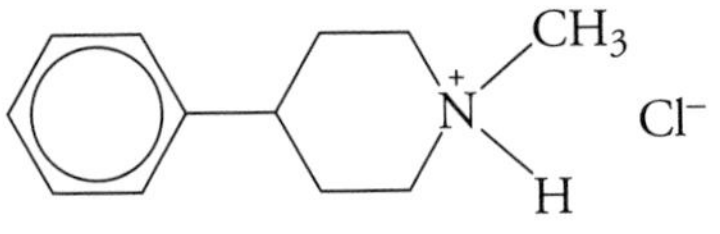

63. (pyridine ring N) amine –NH–$\overset{O}{\overset{\|}{C}}CH_2$–$C_6H_4$–$C_6H_5$
amide
amine

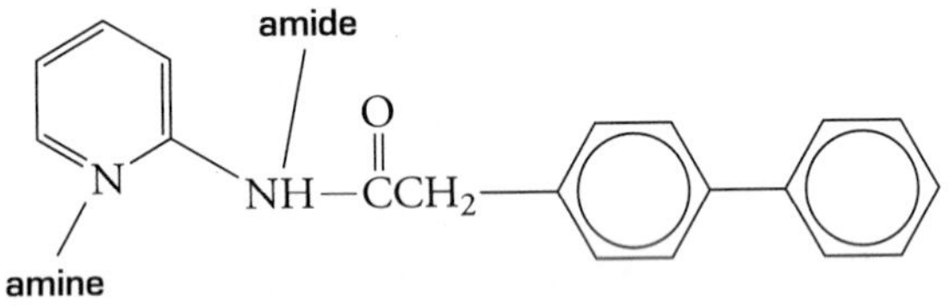

65. (a) $CH_3CH_2CH_2\overset{O}{\overset{\|}{C}}–NHCH_3$

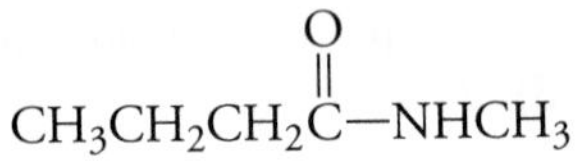

(b) $CH_3CH_2CH_2\overset{O}{\overset{\|}{C}}–N(CH_3)CH_3$

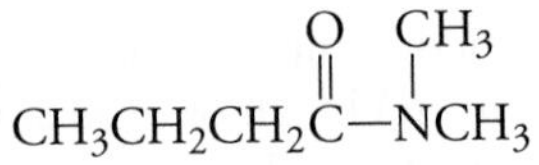

67. (a) *N*-methylbutanamide;
(b) *N,N*-dimethylbutanamide

69. (a) $CH_3\overset{O}{\overset{\|}{C}}–NHCH(CH_3)CH_3$

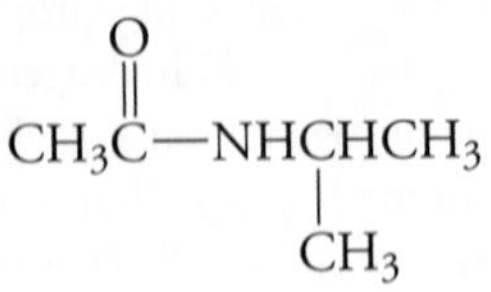

(b) $CH_3CH_2CH(CH_3)CH(CH_3)CH_2\overset{O}{\overset{\|}{C}}–N(CH_3)CH_3$

(c) $CH_3CH_2CH_2\overset{O}{\overset{\|}{C}}–N(CH_3)CH_2CH_3$

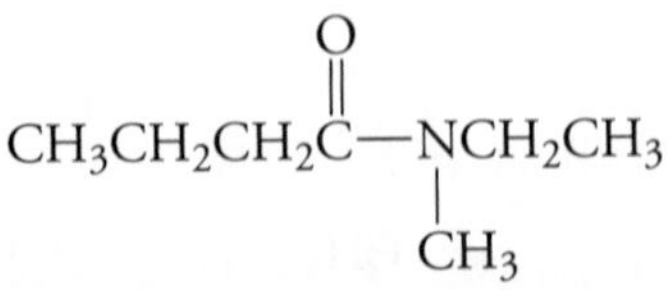

71. (a) $CH_3\overset{O}{\overset{\|}{C}}–OH + H_3\overset{+}{N}CH(CH_3)CH_3$

(b) $CH_3CH_2CH(CH_3)CH(CH_3)CH_2\overset{O}{\overset{\|}{C}}–OH + H_2\overset{+}{N}(CH_3)CH_3$

(c) $CH_3CH_2CH_2\overset{O}{\overset{\|}{C}}–OH + H_2\overset{+}{N}(CH_3)CH_2CH_3$

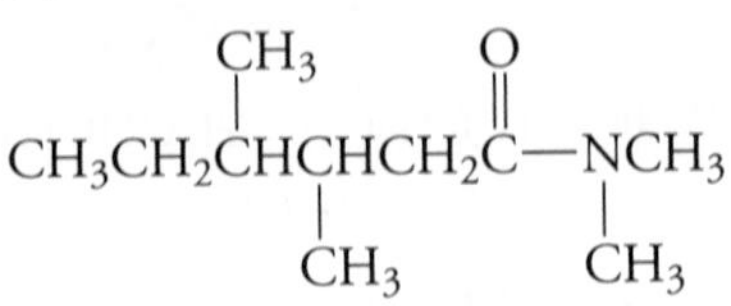

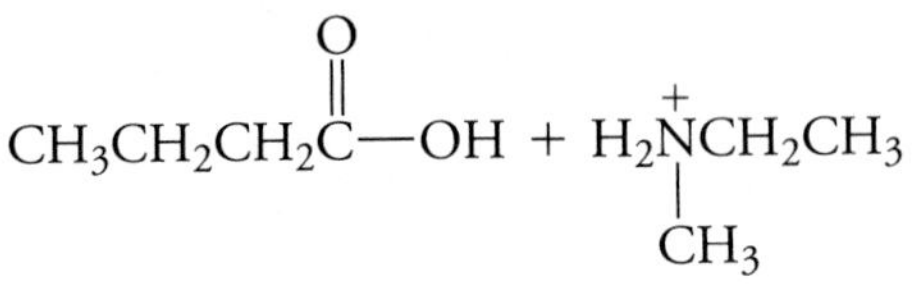

73. Propanamide interacts more effectively with water through hydrogen bonding than does methyl acetate.

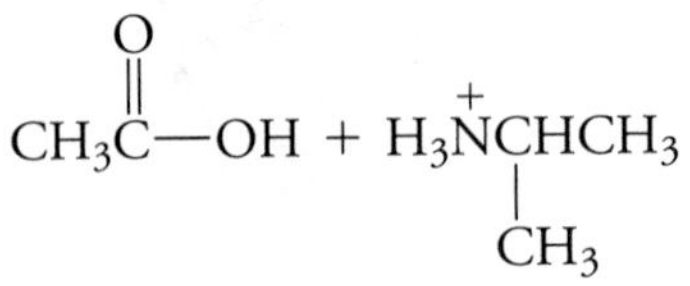

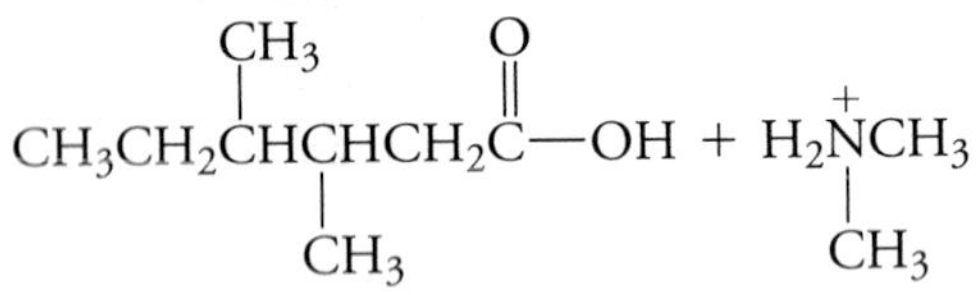

75. (a) $CH_3CH_2CH_2\overset{*}{C}H(Br)CH_2CH_3$

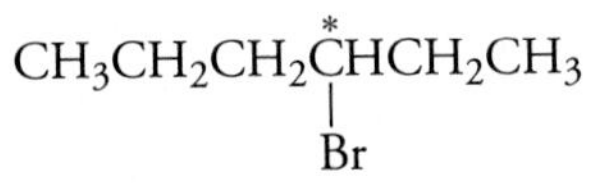

(b) C_6H_5–$CH_2\overset{*}{C}H(NH_2)\overset{O}{\overset{\|}{C}}–OH$

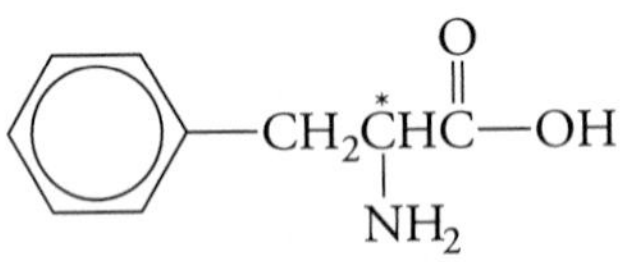

77. (a) CH_3CH_2, H, Br, $CH_3CH_2CH_2$ on C (H····C—Br) and mirror image CH_2CH_3, Br—C····H, $CH_2CH_2CH_3$

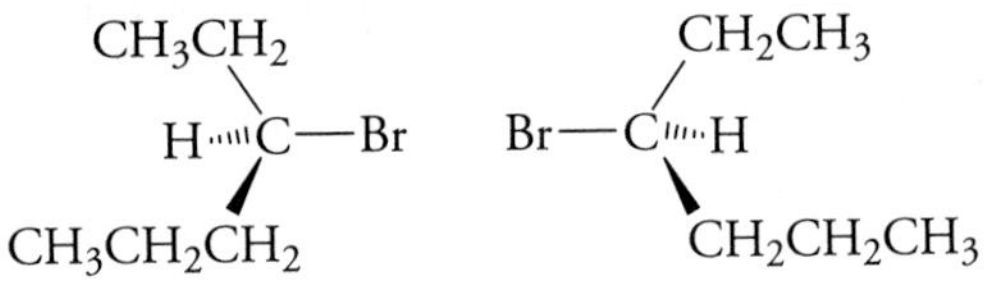

(b) $C_6H_5CH_2$, H····C, H_2N, CO_2H and mirror image $C_6H_5CH_2$, C····H, HO_2C, NH_2

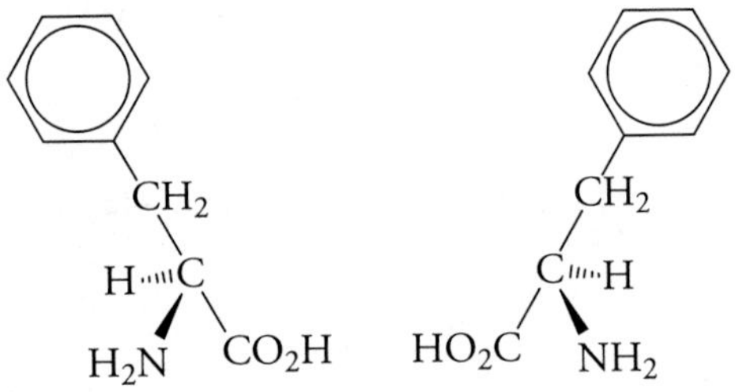

79. (a) HO, H····CCH_2–O–(naphthyl), $CH_3CH(CH_3)NHCH_2$

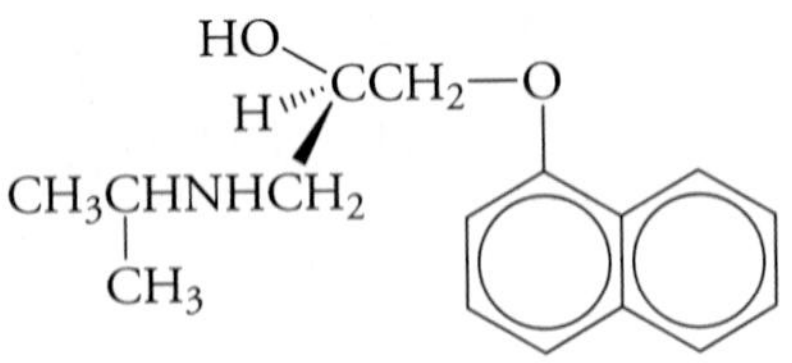

(–)-Propranolol

$HOCH_2$, CH_3CH_2····C(H)–$NHCH_2CH_2NH$–C(H)····CH_2OH, CH_2CH_3

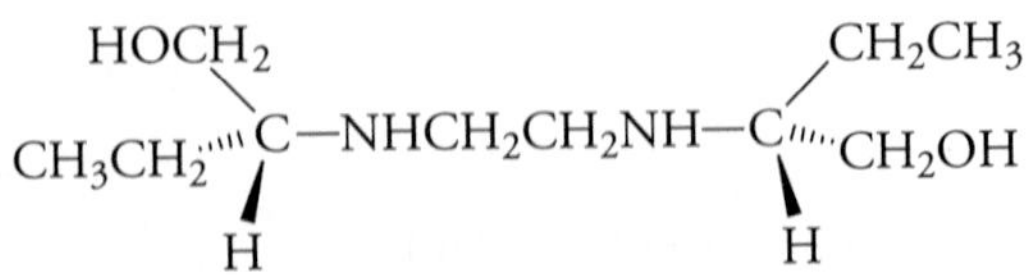

(–)-Ethambutol

(b) 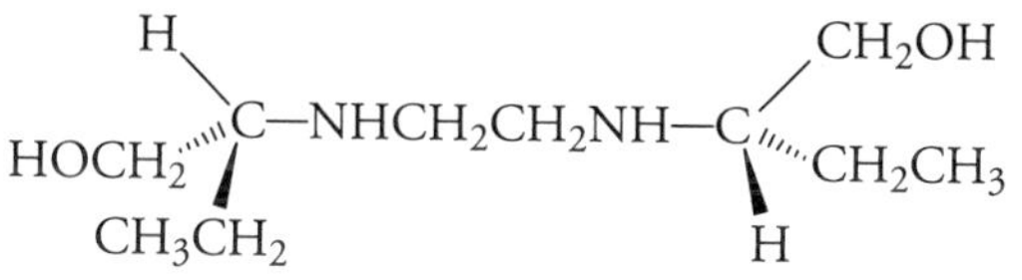 H / HOCH$_2$ C—NHCH$_2$CH$_2$NH—C CH$_2$OH / CH$_2$CH$_3$ / H; CH$_3$CH$_2$

or

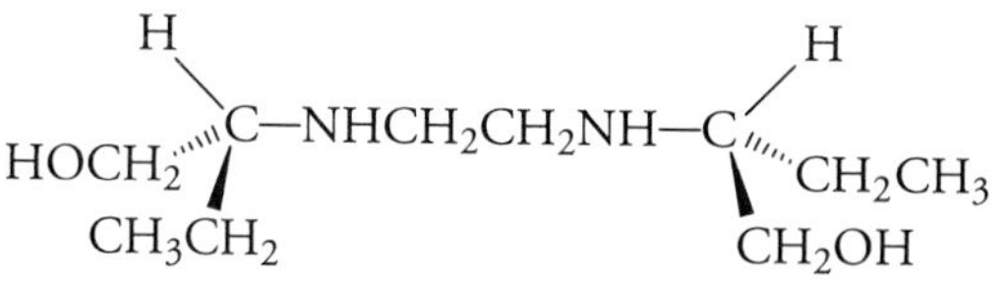

H / HOCH$_2$ C—NHCH$_2$CH$_2$NH—C H / CH$_2$CH$_3$ / CH$_2$OH; CH$_3$CH$_2$

81. 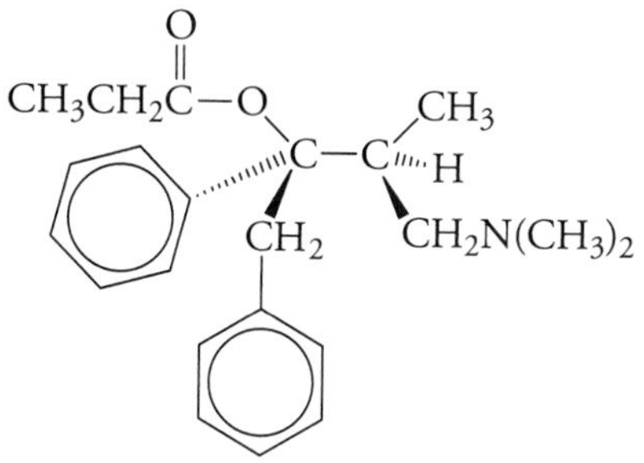 O ‖ CH$_3$CH$_2$C—O C—C CH$_3$ H CH$_2$ CH$_2$N(CH$_3$)$_2$

and

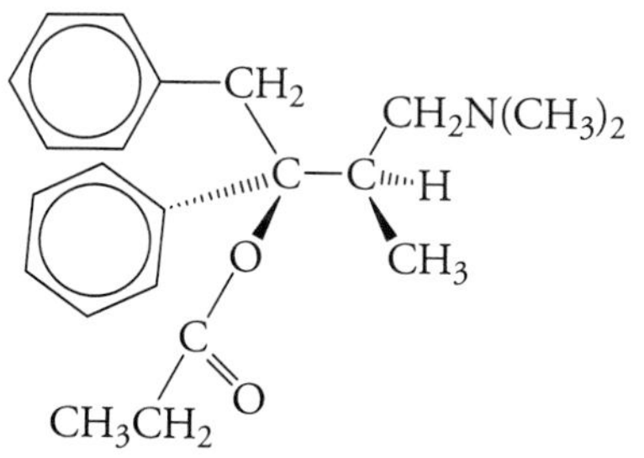

CH$_2$ CH$_2$N(CH$_3$)$_2$ C—C H O CH$_3$ C O CH$_3$CH$_2$

83. 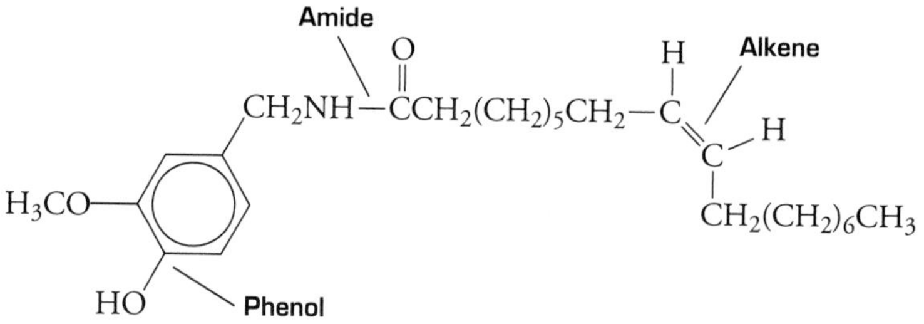 H$_3$CO— CH$_2$NH—CCH$_2$(CH$_2$)$_5$CH$_2$—C=C H H CH$_2$(CH$_2$)$_6$CH$_3$ HO

Amide

Alkene

Phenol

85. Because 4-membered rings are quite strained, serine would be unstable as a lactone.

87. (a) 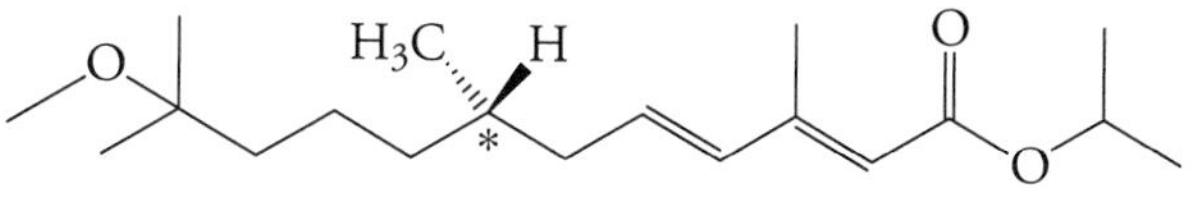 H$_3$C H O O O *

(b) The contribution of the polar ether and ester groups, which can form hydrogen bonds with water, is outweighed by the nonpolar nature of the rest of the molecule.

(c) 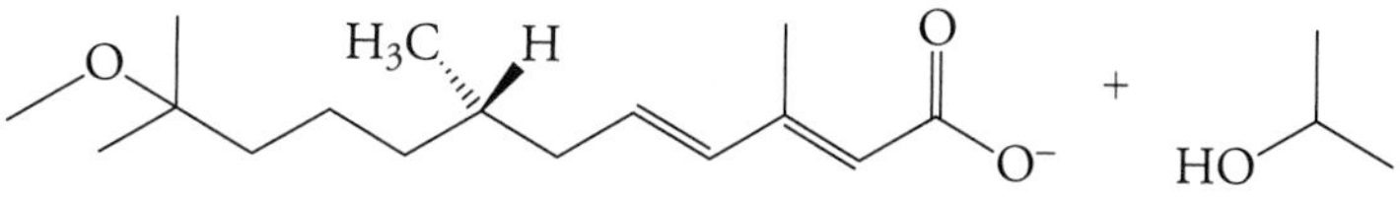 O H$_3$C H O O$^-$ + HO

(d) 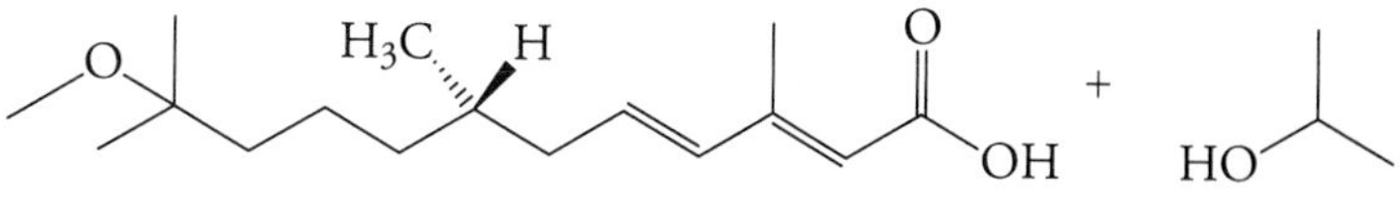 O H$_3$C H O OH + HO

Chapter 11

1. **(a)** Alcohols have an—OH group attached to an alkane-type carbon atom; **(b)** ethers have an oxygen atom attached to two alkane-type or aromatic carbon atoms; **(c)** thiols have an—SH group attached to an alkane-type or aromatic carbon atom; **(d)** sulfides have a sulfur atom attached to two alkane-type or aromatic carbon atoms; **(e)** disulfides contain a C—S—S—C linkages in which each sulfur atom is attached to an alkane-type or aromatic carbon atom.

3. **(a)** 2°; **(b)** 3°; **(c)** 1°.

5. **(a)** 3-pentanol; **(b)** 3-methyl-3-pentanol; **(c)** 2-methyl-1-butanol.

7. **(a)** CH$_3$CH$_2$CH$_2$CH$_2$CHCH$_3$ (OH on CH)

(b) CH$_3$CH$_2$CHCH$_2$CH$_2$OH (CH$_3$ on CH)

(c) OH

CH$_3$CHCH$_3$

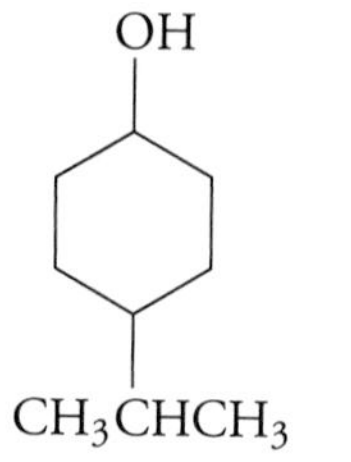

9. **(a)** CH$_3$(CH$_2$)$_{16}$CH$_2$OH

(b) The effects of the long nonpolar hydrocarbon chain outweigh those of the polar —OH group

11. IUPAC name: 2-methyl-2-propanethiol. Common name: *t*-butyl mercaptan.

13. **(a)** *t*-butyl methyl ether; **(b)** butyl isopropyl ether; **(c)** dipentyl ether.

15. Dipropyl ether has a longer hydrocarbon chain than diethyl ether and, therefore, stronger London force interactions.

17. Dipropyl ether has more carbon atoms (nonpolar, water insoluble) than diethyl ether.

19. Ethanol is polar and can form hydrogen bonds with water, while propane is nonpolar and cannot form hydrogen bonds.

21. Both can form hydrogen bonds with water molecules, but 1-butanol contains more carbon atoms and is more nonpolar.

23. Ethanol molecules can form hydrogen bonds with water molecules. Methanethiol cannot.

25. Disulfide and carboxylic acid.

27. **(a)** 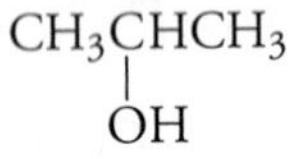 CH_3CHCH_3 (OH on the central C)

(b) $CH_3CH_2CH_2CH_2CH_2OCH_3$

(c) CH_3SH

(d) $CH_3CH_2CH_2SCH_2CH_2CH_3$

29. **(a)** isopropyl alcohol or 2-propanol; **(b)** methyl pentyl ether; **(c)** methyl mercaptan or methanethiol; **(d)** dipropyl sulfide.

31. **(a)** CH_3CH_2Br

(b) CH_3Br

(c) 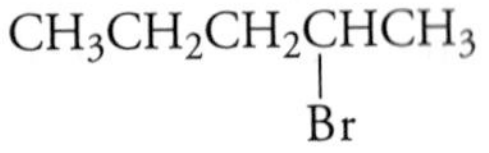 $CH_3CH_2CH_2CHCH_3$ (Br on the fourth C)

33. **(a)** $CH_3CH_2Br + {}^{-}SCH_2CH_2CH_2CH_3 \longrightarrow CH_3CH_2SCH_2CH_2CH_2CH_3 + Br^-$

(b) $CH_3CH(CH_3)CH_2CH_2Br + {}^{-}SH \longrightarrow CH_3CH(CH_3)CH_2CH_2SH + Br^-$

35. **(a)** 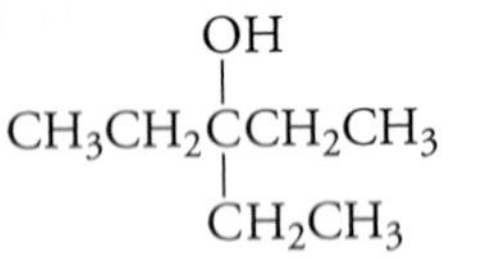 $CH_3CH_2CCH_2CH_3$ (OH and CH_2CH_3 on the central C)

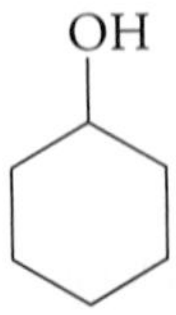

(b) CH_3C (OH and CH_3) bonded to a benzene ring

(c) OH on a cyclohexane ring

37. **(a)**

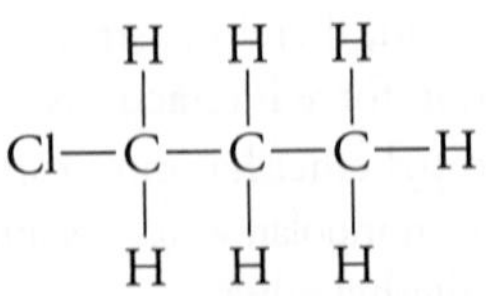

```
     H  H  H
     |  |  |
Cl—C—C—C—H
     |  |  |
     H  H  H
```

(b)

```
   H  H  H  H
   |  |  |  |
H—C—C—C—C—H
   |  |  |  |
   H  Br H  H
```

(c)

```
   H  H  H
   |  |  |
H—C—C—C—H
   |  |  |
   H  F  H
```

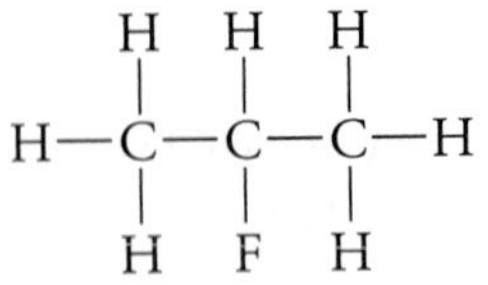

(d)

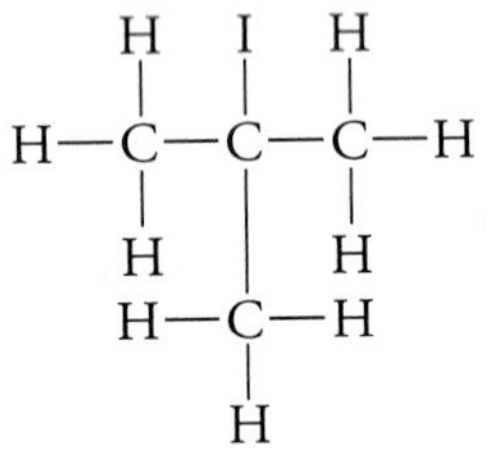

```
   H  I  H
   |  |  |
H—C—C—C—H
   |  |  |
   H  |  H
   H—C—H
      |
      H
```

39. **(a)** 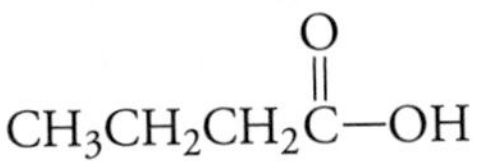 $CH_3CH_2CH_2\overset{O}{\overset{\|}{C}}—OH$

(b) $CH_3CH_2\overset{O}{\overset{\|}{C}}CH_3$

(c) cyclohexanone ring (C=O) with H_3C substituent

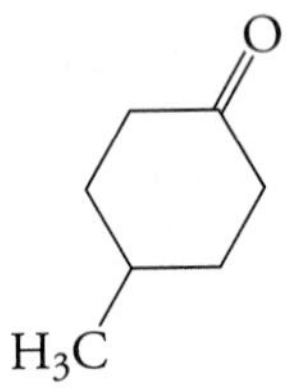

(d) $CH_3CH_2CH_2SSCH_2CH_2CH_3$

(e) 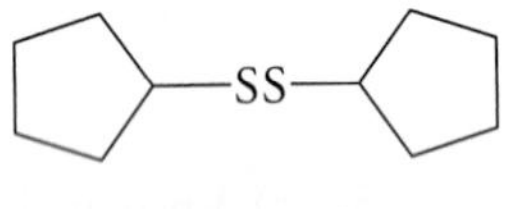 cyclopentyl—SS—cyclopentyl

41. **(a)** 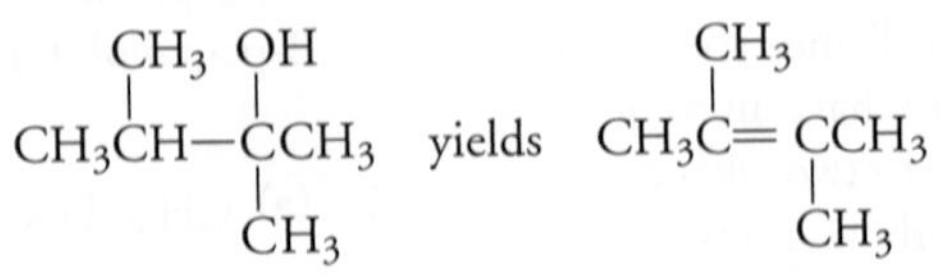 $CH_3\underset{}{C}H—CCH_3$ (CH_3 on the second C; OH and CH_3 on the third C) yields $CH_3C{=}CCH_3$ (CH_3 on each double-bonded C)

(b) $CH_3CH_2CH_2C—CHCH_3$ (OH and CH_3 on the fourth C; CH_3 on the fifth C) yields $CH_3CH_2CH_2C{=}CCH_3$ (CH_3 on each double-bonded C)

(c) cyclopentane with OH and adjacent CH_3 yields cyclopentene with CH_3 on the double-bonded C

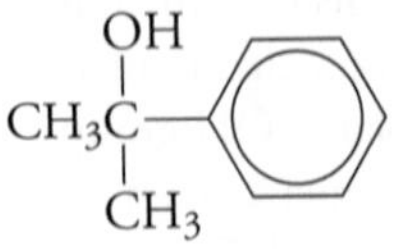

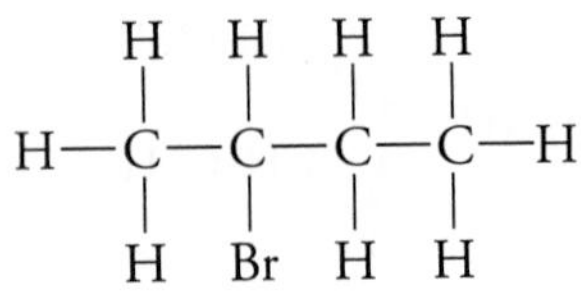

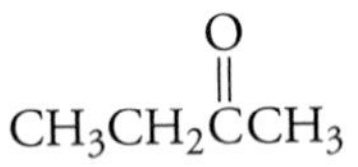

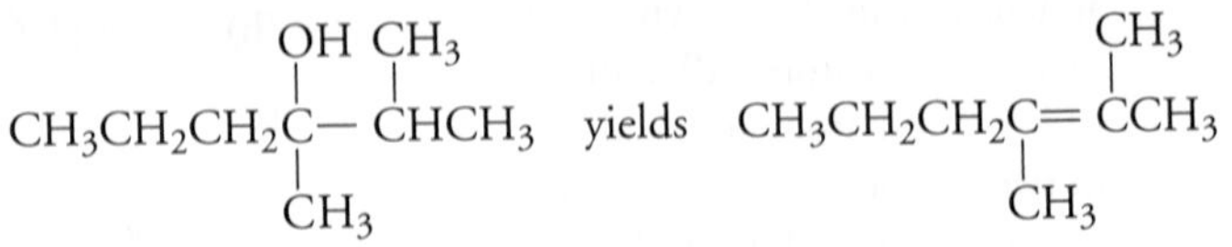

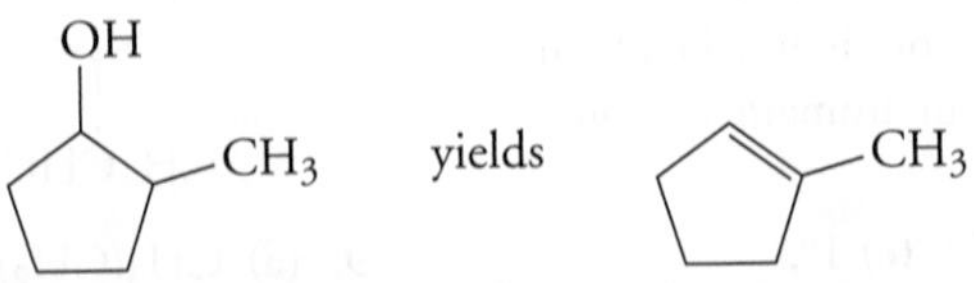

43. **(a)** 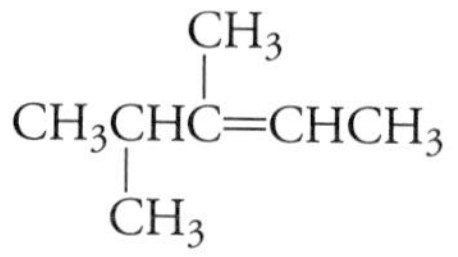 $CH_3CH(CH_3)C(CH_3){=}CHCH_3$

$$\begin{array}{c} CH_3 \\ | \\ CH_3CHC{=}CHCH_3 \\ | \\ CH_3 \end{array}$$

(b)

$$\begin{array}{c} CH_3 \\ | \\ CH_3C{=}CHCH_3 \end{array}$$

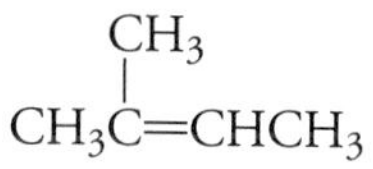

45. When 1-butanol is oxidized, a carboxylic acid is produced.
When 1-butanethiol is oxidized, a disulfide is produced.

47. **(a)** alcohol, alkene, and ketone;
(b) steroid.

49. **(a)** 3,3-dimethylbutanal;
(b) 4-methyl-2-pentanone;
(c) 2,4 -dimethyl-3-pentanone;
(d) 2-ethylpentanal.

51. **(a)** 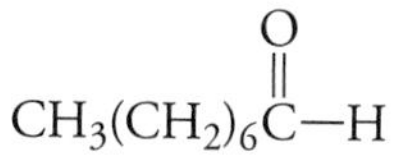 $CH_3(CH_2)_6C({=}O){-}H$

(b) 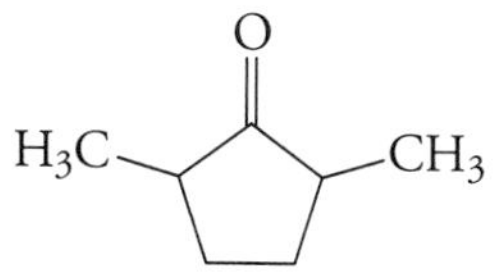 H_3C–cyclopentanone–CH_3 (2,5-dimethylcyclopentanone; O on ring carbon, H_3C and CH_3 on adjacent ring carbons)

(c) 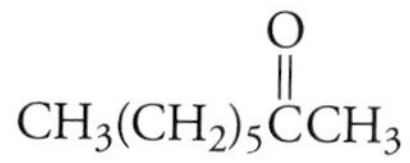 $CH_3(CH_2)_5C({=}O)CH_3$

(d) 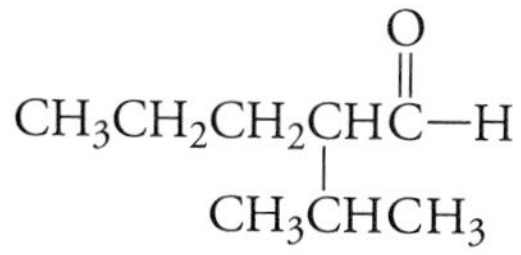 $CH_3CH_2CH_2CH(CH(CH_3)_2)C({=}O){-}H$

$$\begin{array}{l} \qquad\qquad\qquad\quad O \\ \qquad\qquad\qquad\quad \| \\ CH_3CH_2CH_2CHC{-}H \\ \qquad\qquad\quad | \\ \qquad\quad CH_3CHCH_3 \end{array}$$

53. **(a)** acetone isopropyl alcohol

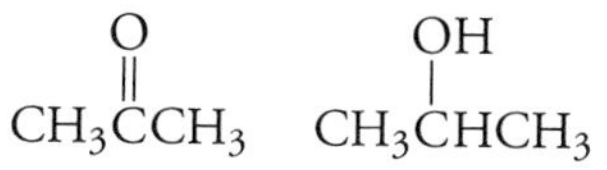

$CH_3C({=}O)CH_3$ $\quad CH_3CH(OH)CH_3$

(b) Alcohol molecules can interact through relatively strong hydrogen bonds, while ketone molecules interact through relatively weak dipole-dipole forces. The stronger the noncovalent interactions, the higher the boiling point.

55. Propanone molecules are primarily attracted to one another by dipole–dipole interactions. Butane molecules are attracted to one another by relatively weaker London forces.

57. **(a)** 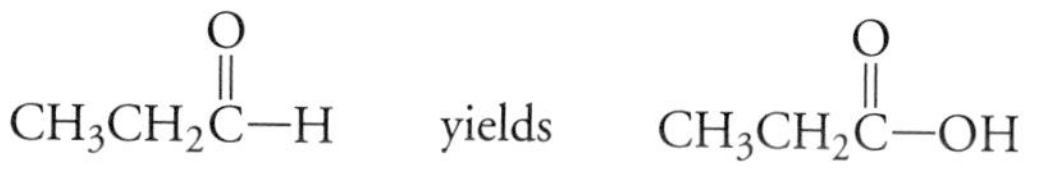 $CH_3CH_2C({=}O){-}H$ yields $CH_3CH_2C({=}O){-}OH$

(b) 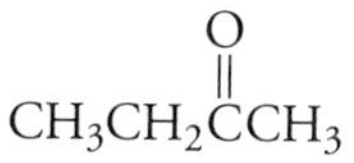 $CH_3CH_2C({=}O)CH_3$ no reaction

(c) 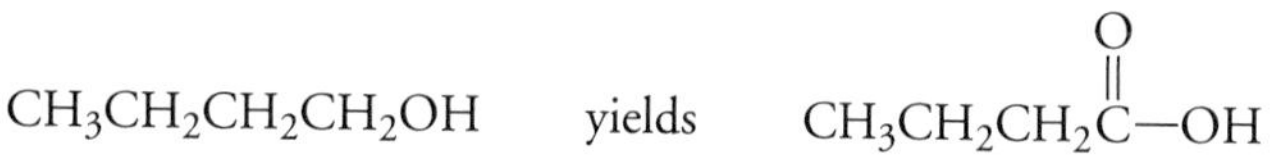 $CH_3CH_2CH_2CH_2OH$ yields $CH_3CH_2CH_2C({=}O){-}OH$

(d) 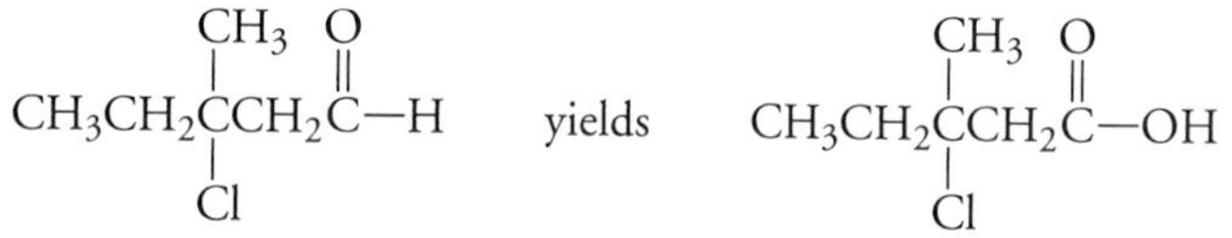 $CH_3CH_2C(CH_3)(Cl)CH_2C({=}O){-}H$ yields $CH_3CH_2C(CH_3)(Cl)CH_2C({=}O){-}OH$

(e)  $CH_3CH(OH)CH_2C({=}O){-}H$ yields $CH_3C({=}O)CH_2C({=}O){-}OH$

59. **(a)** 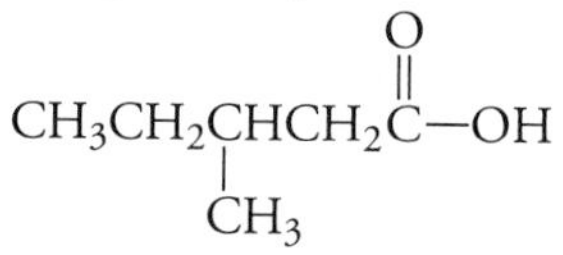 $CH_3CH_2CH(CH_3)CH_2C({=}O){-}OH$

(b) no reaction product

(c) 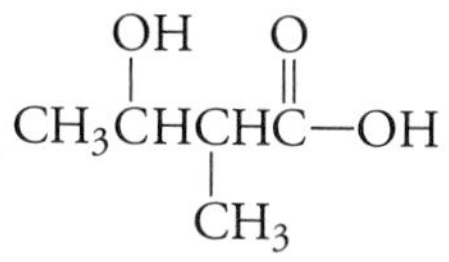 $CH_3C(OH)HCH(CH_3)C({=}O){-}OH$

61. **(a)** 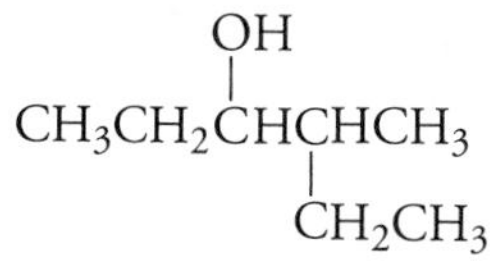 $CH_3CH_2CH(OH)CH(CH_2CH_3)CH_3$

(b) 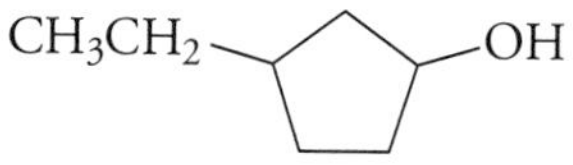 CH_3CH_2–cyclopentane–OH

(c) 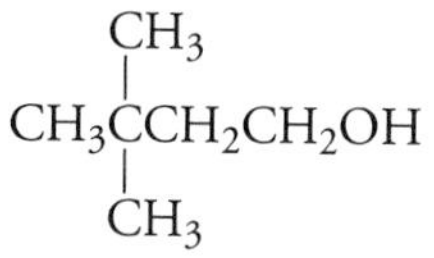 $CH_3C(CH_3)_2CH_2CH_2OH$

63. **(a)** 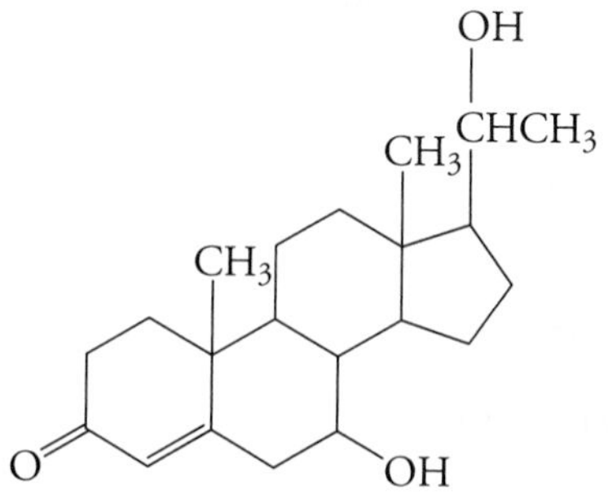 OH, CHCH$_3$, CH$_3$, CH$_3$, O, OH

(b) 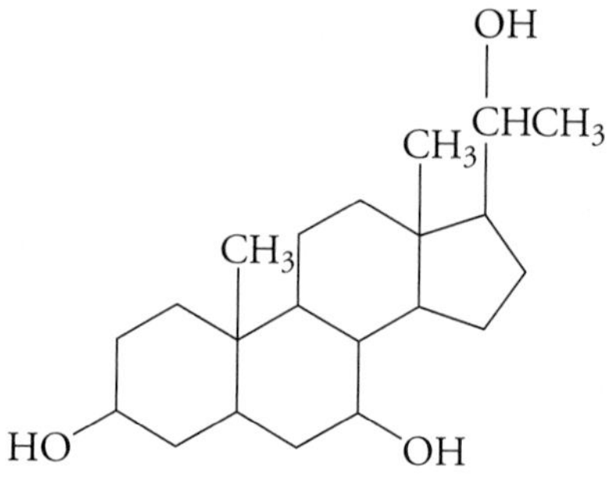 OH, CHCH$_3$, CH$_3$, CH$_3$, HO, OH

65. (a) 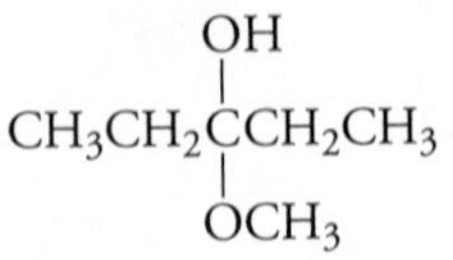

(b) 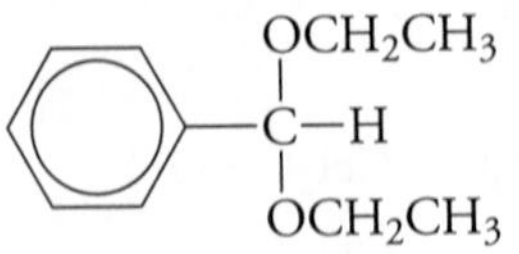

(c) 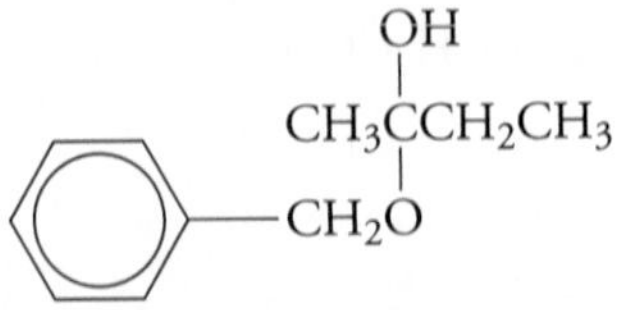

67. (a) 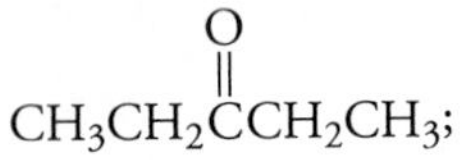;

(b) 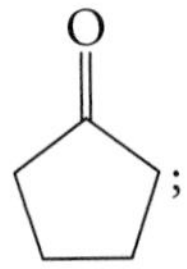;

(c) 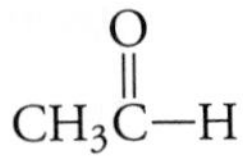

69. **(a)** Cyclic hemiacetals and lactones each have a ring that contains a C—O—C linkage; **(b)** In cyclic hemiacetals, one of the carbon atoms in the C—O—C linkage is also attached to an—OH group. In lactones, this carbon atom is double-bonded to an oxygen atom (C=O).

71. methanol methanal or formaldehyde

CH_3OH 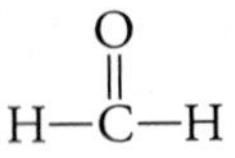

73. reduced.

77. 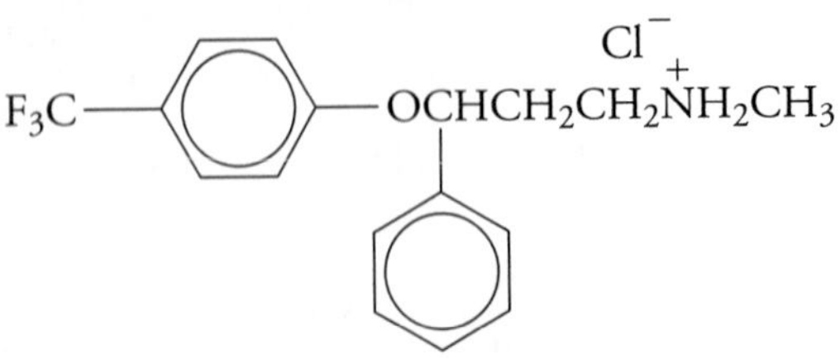

79. **(a)** 2×10^{-7}g; **(b)** 5×10^{-5}ppm; **(c)** 5×10^{-2}ppb

81. 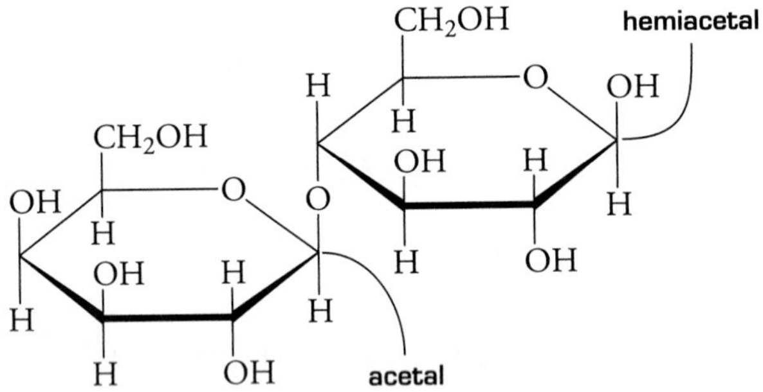

Chapter 12

1. Oligosaccharides contain from 2 to 10 monosaccharide residues while polysaccharides contain more than 10.

3. (a) 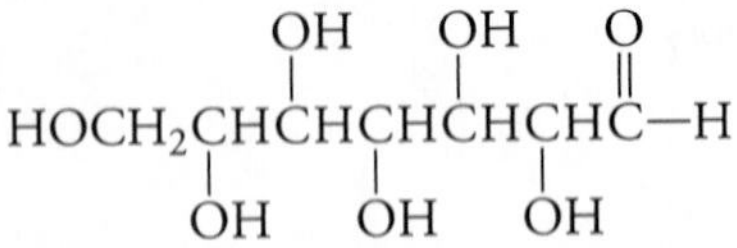

(b) 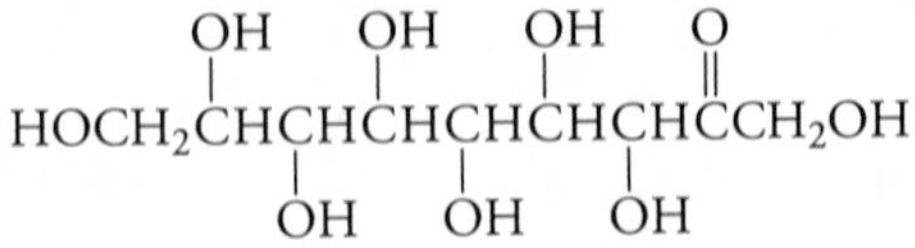

5. (a) 3; **(b)** 8; (c) 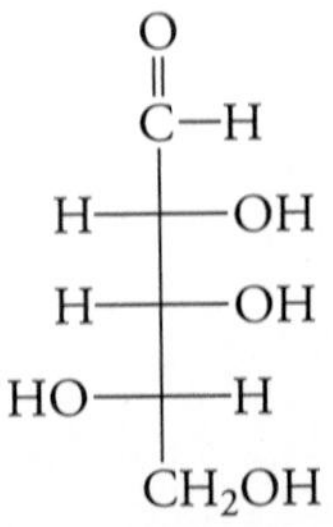

L-lyxose

(d) 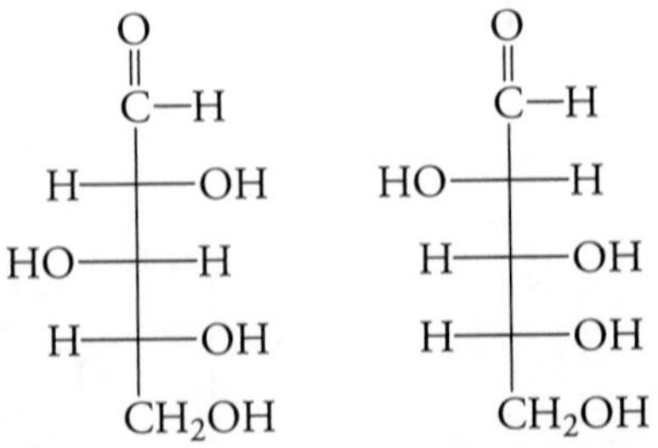

7. (a)

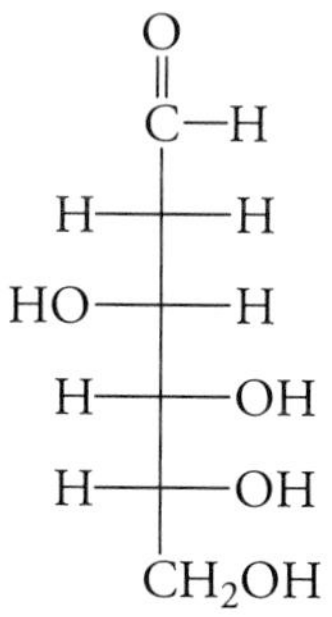

$$\begin{array}{rcl} & CHO & \\ H & - & H \\ HO & - & H \\ H & - & OH \\ H & - & OH \\ & CH_2OH & \end{array}$$

(b)

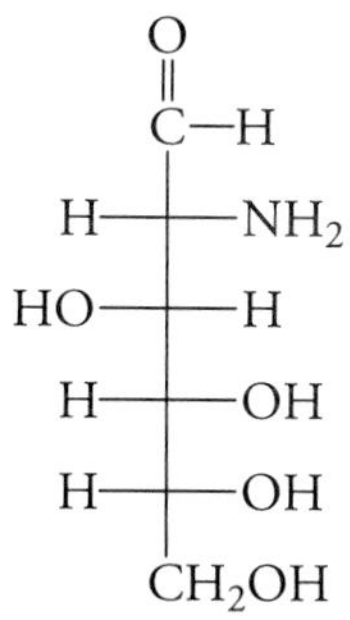

$$\begin{array}{rcl} & CHO & \\ H & - & NH_2 \\ HO & - & H \\ H & - & OH \\ H & - & OH \\ & CH_2OH & \end{array}$$

(c).

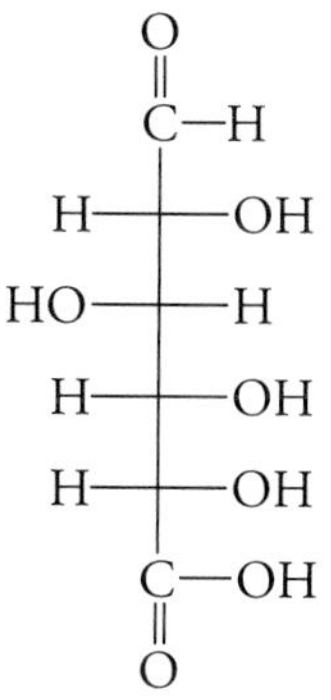

$$\begin{array}{rcl} & CHO & \\ H & - & OH \\ HO & - & H \\ H & - & OH \\ H & - & OH \\ & COOH & \end{array}$$

(d)

$$\begin{array}{rcl} & CH_2OH & \\ H & - & OH \\ HO & - & H \\ H & - & OH \\ H & - & OH \\ & CH_2OH & \end{array}$$

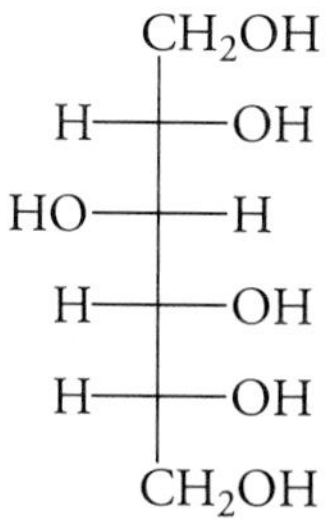

9. The D and L assignment is based on the orientation of the—OH group attached to the chiral carbon atom farthest away from the aldehyde or ketone group. Sorbitol has no aldehdye or ketone group.

11. aldehyde

13. (a)

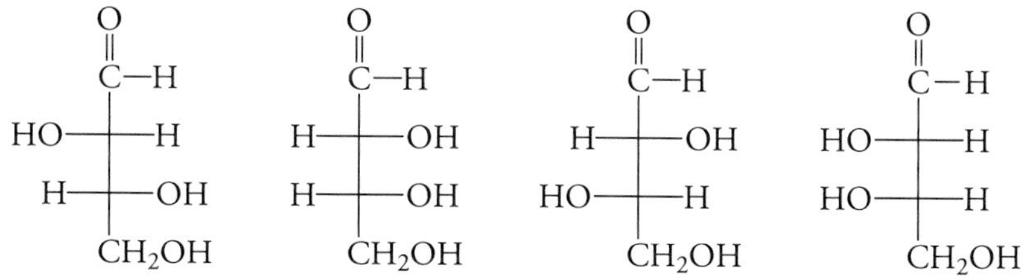

$$\begin{array}{rcl} & CHO & \\ HO & - & H \\ H & - & OH \\ & CH_2OH & \end{array} \quad \begin{array}{rcl} & CHO & \\ H & - & OH \\ H & - & OH \\ & CH_2OH & \end{array} \quad \begin{array}{rcl} & CHO & \\ H & - & OH \\ HO & - & H \\ & CH_2OH & \end{array} \quad \begin{array}{rcl} & CHO & \\ HO & - & H \\ HO & - & H \\ & CH_2OH & \end{array}$$

(b) **D** **D** **L** **L**

15. (a)

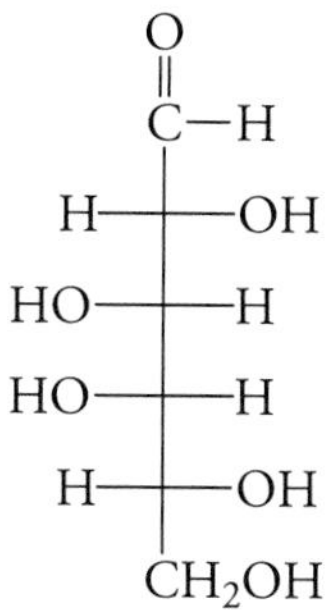

$$\begin{array}{rcl} & CHO & \\ H & - & OH \\ HO & - & H \\ HO & - & H \\ H & - & OH \\ & CH_2OH & \end{array}$$

(b)

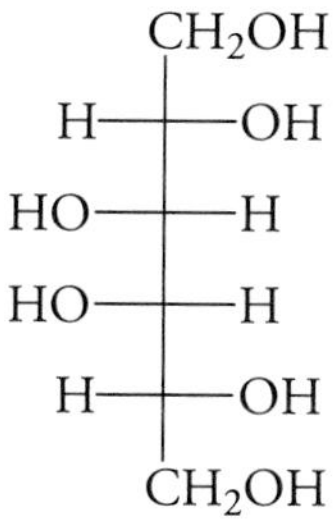

$$\begin{array}{rcl} & CH_2OH & \\ H & - & OH \\ HO & - & H \\ HO & - & H \\ H & - & OH \\ & CH_2OH & \end{array}$$

(c)

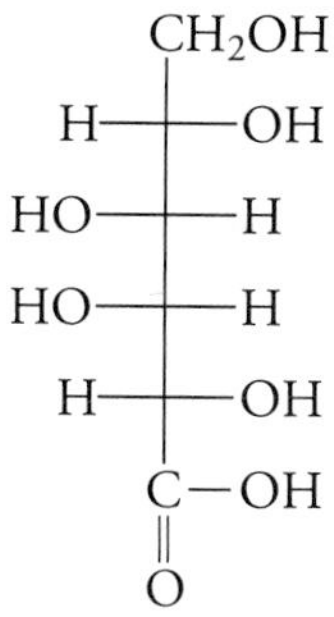

$$\begin{array}{rcl} & CH_2OH & \\ H & - & OH \\ HO & - & H \\ HO & - & H \\ H & - & OH \\ & COOH & \end{array}$$

17. (a)

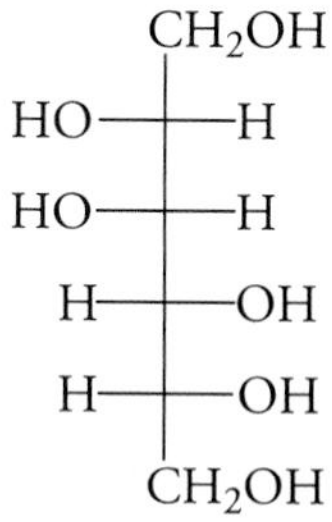

$$\begin{array}{rcl} & CH_2OH & \\ HO & - & H \\ HO & - & H \\ H & - & OH \\ H & - & OH \\ & CH_2OH & \end{array}$$

(b)

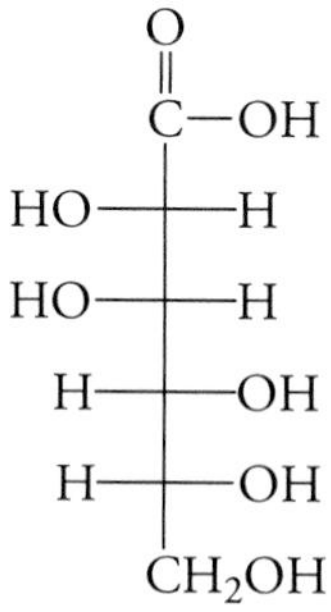

$$\begin{array}{rcl} & COOH & \\ HO & - & H \\ HO & - & H \\ H & - & OH \\ H & - & OH \\ & CH_2OH & \end{array}$$

(c)

$$\begin{array}{rcl} & CHO & \\ HO & - & H \\ HO & - & H \\ H & - & OH \\ H & - & OH \\ & COOH & \end{array}$$

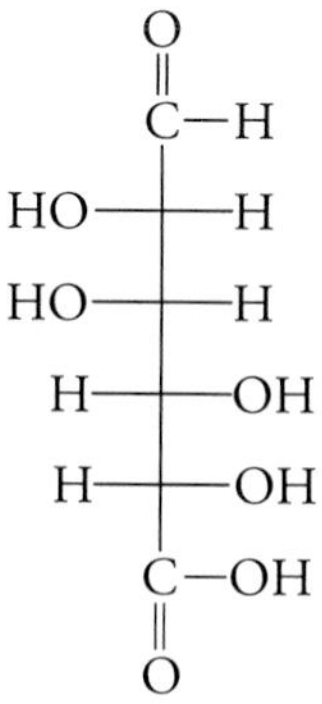

19. A reducing sugar is a carbohydrate that gives a positive Benedict's test (in the process of being oxidized, the sugar reduces Cu^{2+} ion present in the reagent).

21.

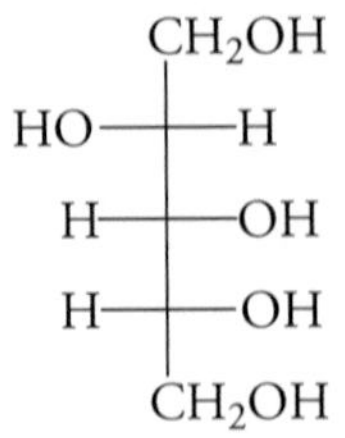

23. (a)

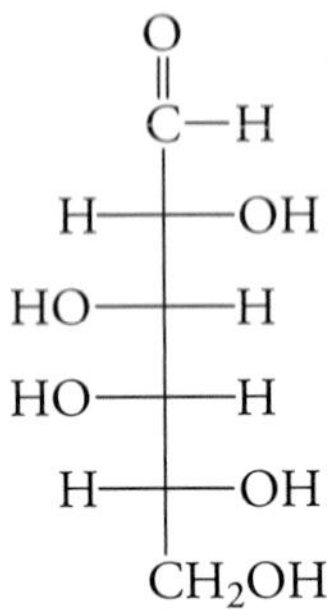

(b) yes; **(c)** no; **(d)** no.

(e)

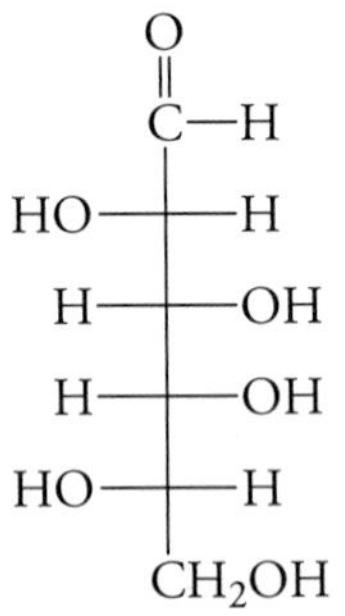

25. **(a)** An anomer is a cyclic hemiacetal form of a monosaccharide; **(b)** As pyranoses and furanoses are typically drawn (Figures 12.8 and 12.9), the hemiacetal—OH points down in an α anomer and up in a β anomer; **(c)** A pyranose contains a 6-membered cyclic hemiacetal ring and a furanose contains a 5-membered cyclic hemiacetal ring.

27. **(a)**

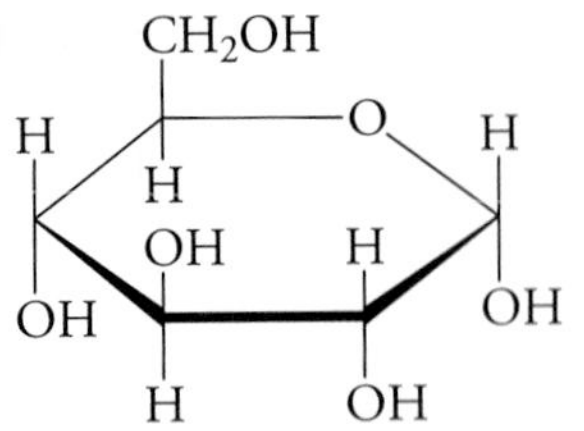

(b)

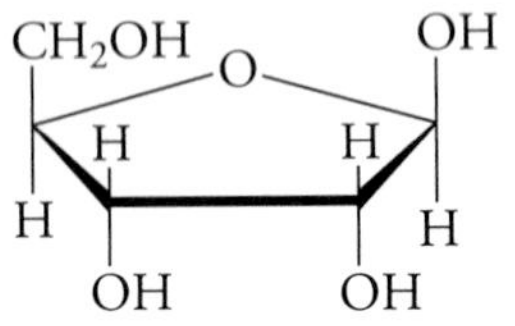

(c)

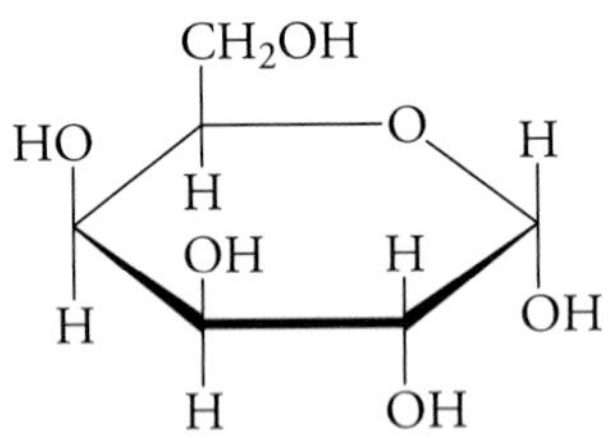

(d)

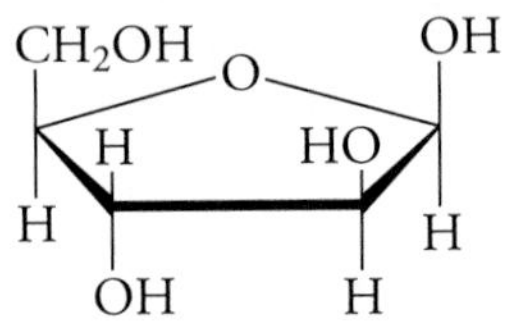

29.

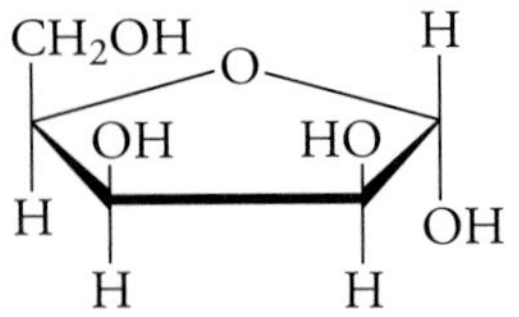

31.

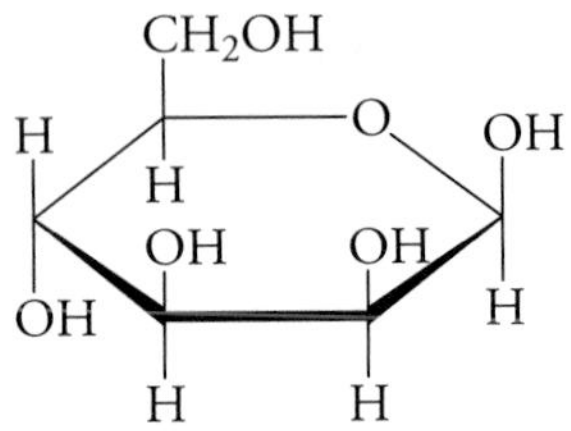

33.

(a)

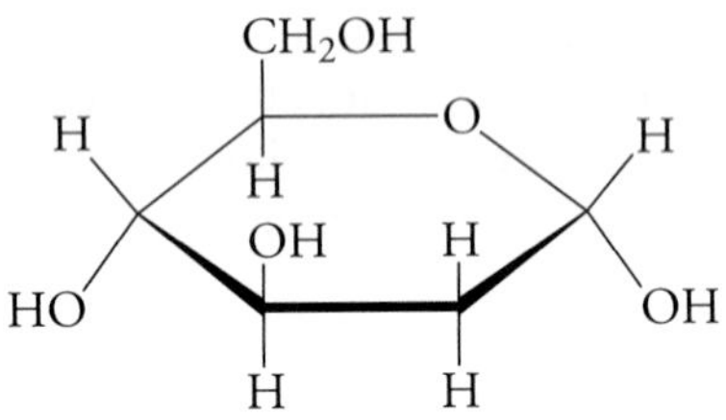

(b)

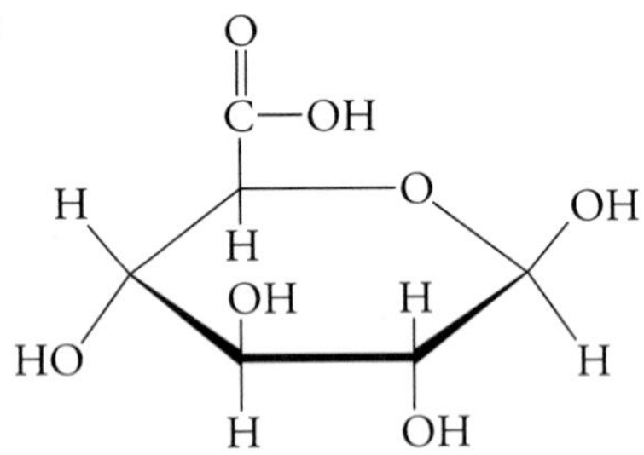

35. Mutarotation is the process of converting back and forth between an α anomer, the open form, and a β anomer.

37.

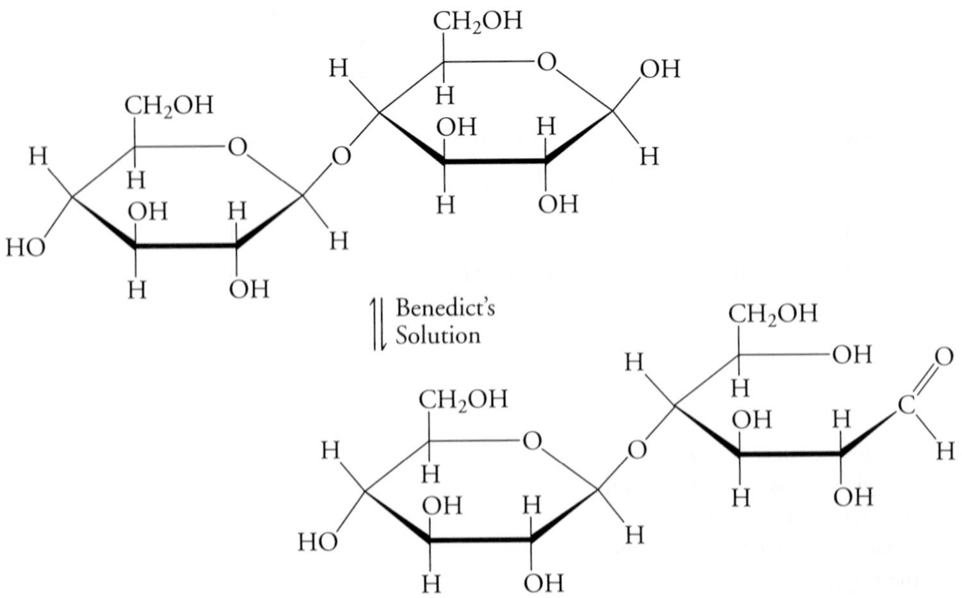

Benedict's Solution

39. **(a)**

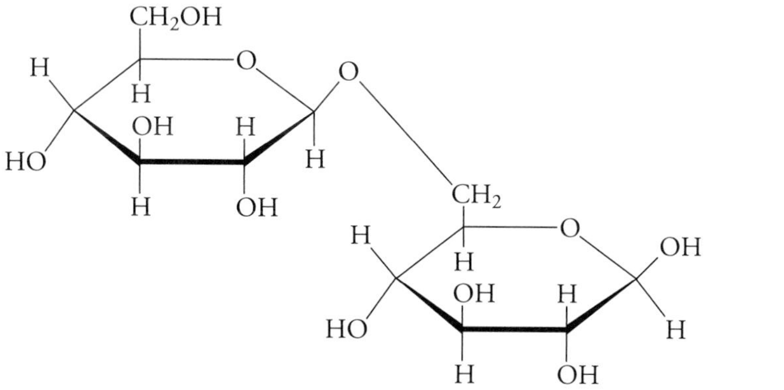

(b) Yes. When the β hemiacetal (at the right side of the molecule, as drawn) undergoes mutarotation, the resulting aldehyde group can be oxidized by Cu^{2+}

41.

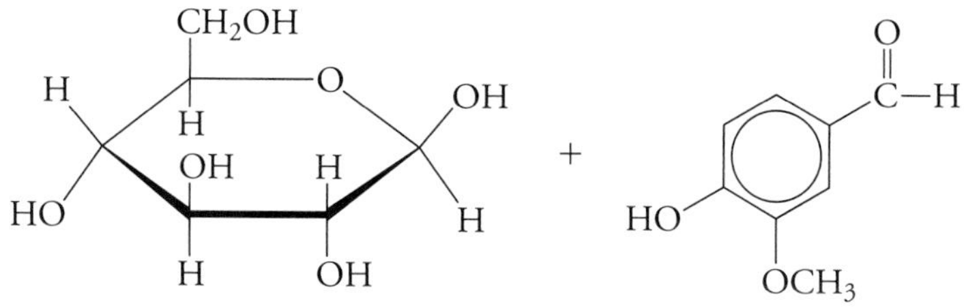

43.

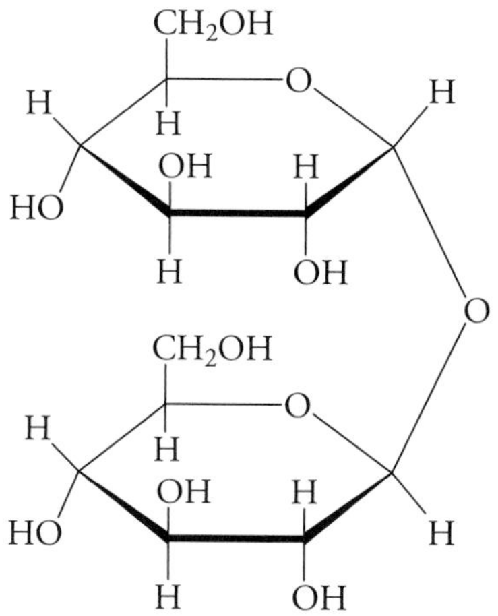

45. **(a)** no; **(b)** α, α–(1↔1)

47.

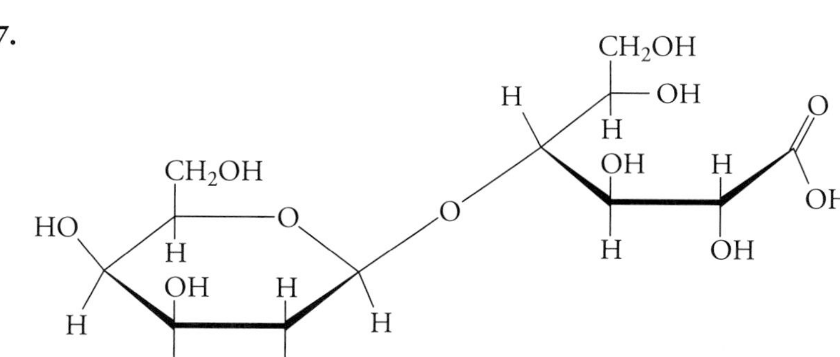

49. **(a)** 1 g; **(b)** 0.2 M; **(c)** 7×10^4ppm

51. **(a)** D-galactose, sphingosine, and stearic acid; **(b)** *N*-acetyl-D-galactosamine, D-galactose, *N*-acetylneuraminic acid, D-glucose, sphingosine, and stearic acid.

53. **(a)** Each is a homopolysaccharide consisting of glucose residues.
(b) Cellulose has β–(1→4) glycosidic bonds and amylose has α–(1→4) and α–(1→6) glycosidic bonds.

55. Glycogen and amylopectin each consist of glucose residues joined by α–(1→4) and α–(1→6) glycosidic bonds.

57. **(a)** homopolysaccharide; **(b)** β–(1→6)

59.

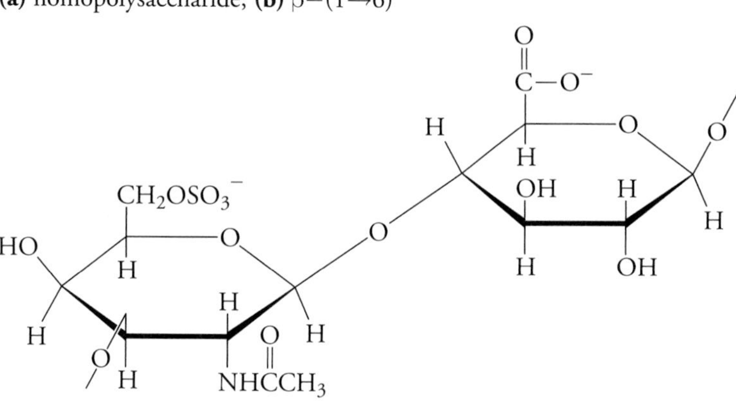

Chondroitin 6-sulfate

61. sucrose.

63. Corn syrup (glucose) is treated with glucose isomerase. This produces a fructose-glucose mixture known as high fructose corn syrup.

65. The "K" stands for potassium ion. In this form, Acesulfame has a greater solubility in water.

67. Lactose intolerance is a deficiency in β-galactosidase, the enzyme that catalyzes the hydrolysis of the β–(1→4) glycosidic bond in lactose. Persons with this disorder may deal with it by avoiding dairy products, by using dairy products from which lactose has been removed, or by taking tablets containing β-galactosidase.

69.

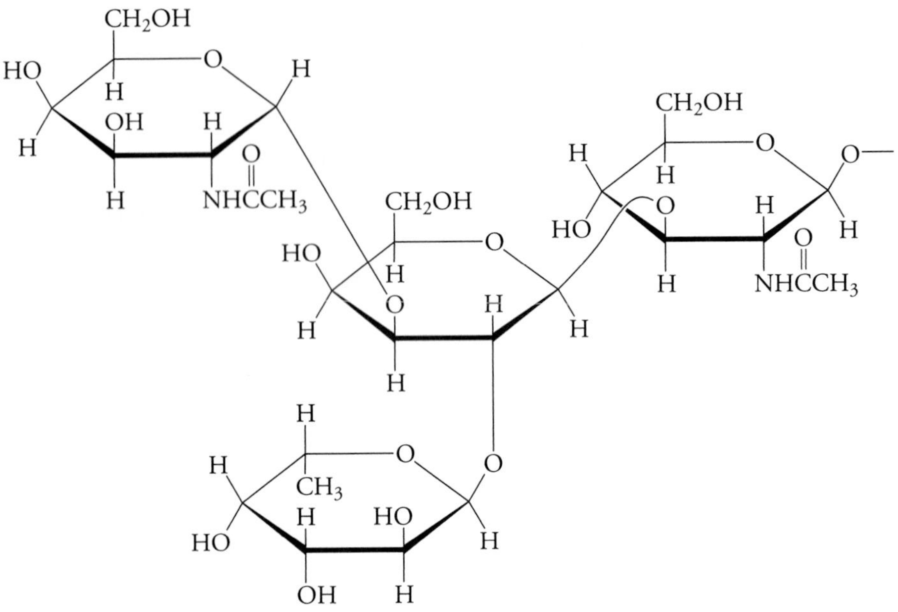

CH_2OH HO H O H OH H H O H $NHCCH_3$ CH_2OH HO H O H H H H H CH_3 O H HO HO O H OH H H CH_2OH H O O— H HO O H H H O H $NHCCH_3$

Chapter 13

1. **(a)** $H_3\overset{+}{N}CH\overset{O}{\overset{\|}{C}}{-}OH$ ($CH_2CH_2SCH_3$)

(b) $H_3\overset{+}{N}CH\overset{O}{\overset{\|}{C}}{-}O^-$ ($CH_2CH_2SCH_3$)

(c) $H_2NCH\overset{O}{\overset{\|}{C}}{-}O^-$ ($CH_2CH_2SCH_3$)

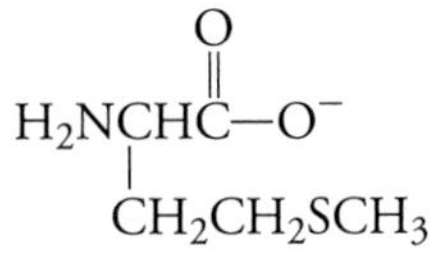

3. **(a)** 2+; **(b)** 1–.

5. **(a)** $\overset{O}{\overset{\|}{C}}{-}O^-$; ^+H_3N—C—H; $CHCH_3$; CH_2; CH_3

(b) $\overset{O}{\overset{\|}{C}}{-}O^-$; H—C—$NH_3^+$; CH_2; $\overset{O}{\overset{\|}{C}}{-}O^-$

(c) $\overset{O}{\overset{\|}{C}}{-}O^-$; ^+H_3N—C—H; CH_2; (benzene ring); OH

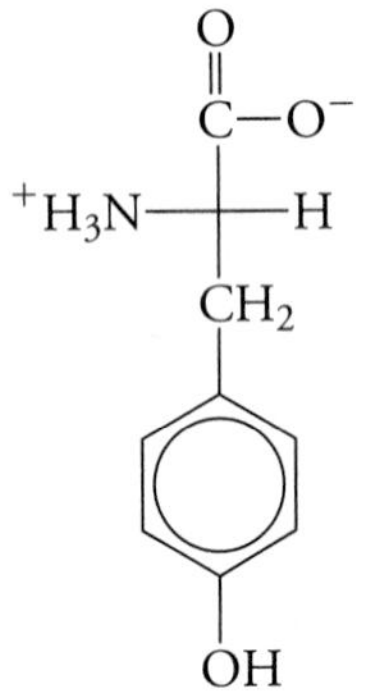

(d) $\overset{O}{\overset{\|}{C}}{-}O^-$; H—C—$NH_3^+$; CH_2; (benzene ring)

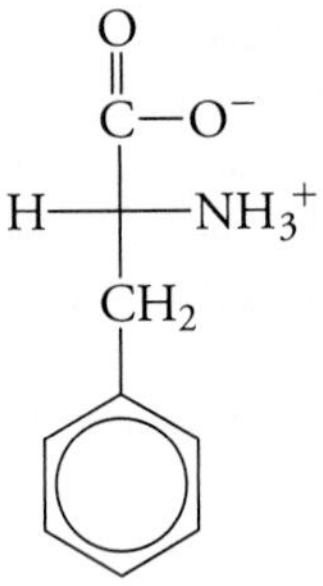

7. **(a)** $\overset{O}{\overset{\|}{C}}{-}OH$; ^+H_3N—C—H; $CHCH_3$; CH_2; CH_3

(b) $\overset{O}{\overset{\|}{C}}{-}OH$; H—C—$NH_3^+$; CH_2; $\overset{O}{\overset{\|}{C}}{-}OH$

(c) $\overset{O}{\overset{\|}{C}}{-}OH$; ^+H_3N—C—H; CH_2; (benzene ring); OH

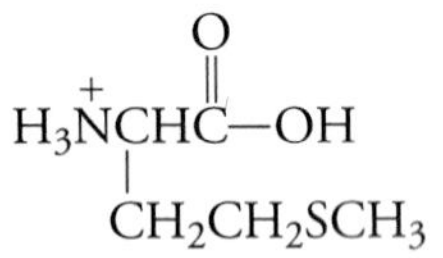

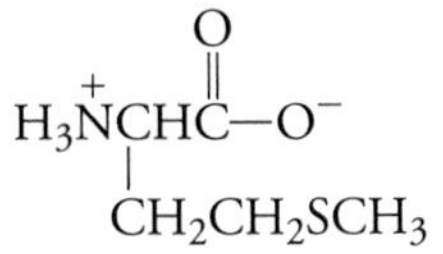

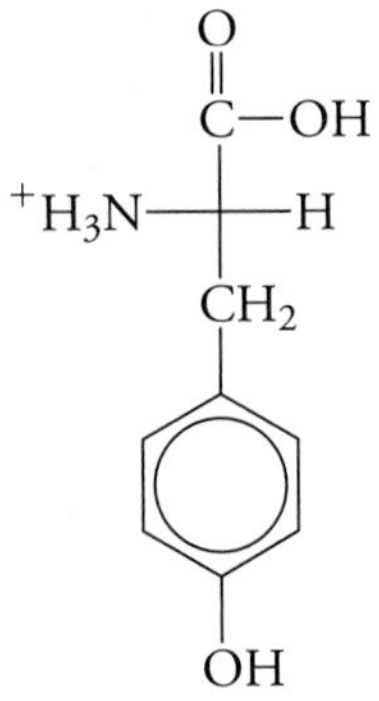

(d) $\overset{O}{\overset{\|}{C}}{-}OH$; H—C—$NH_3^+$; CH_2; (benzene ring)

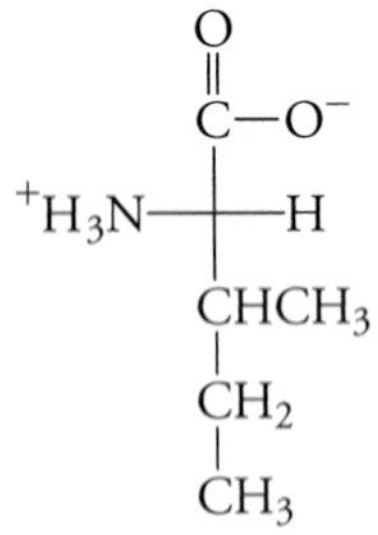

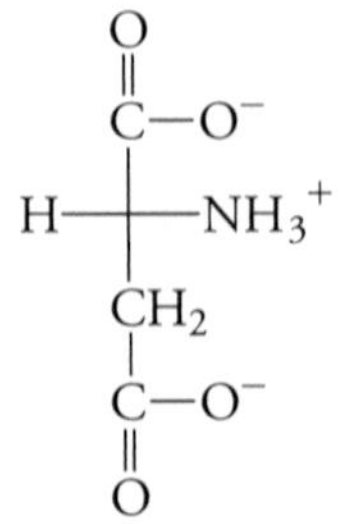

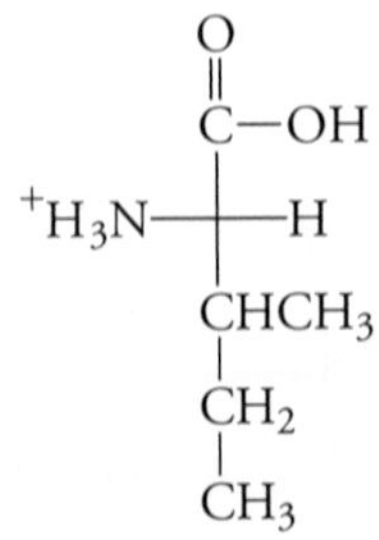

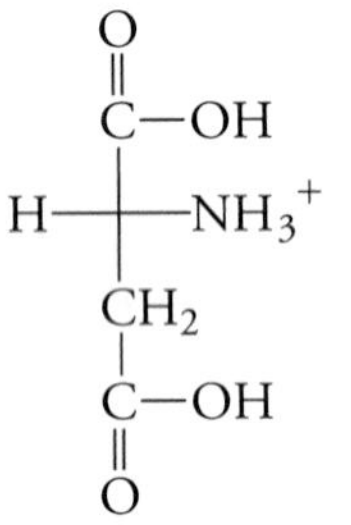

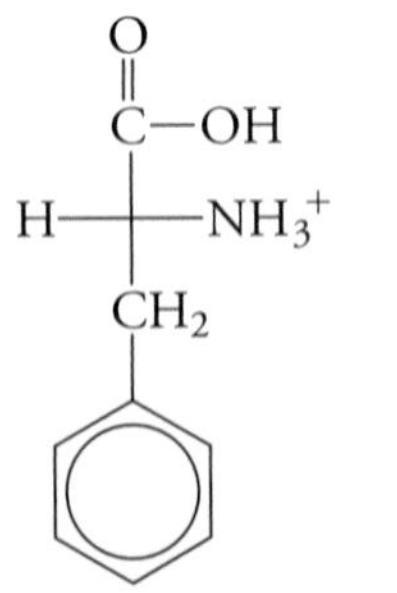

9. threonine and isoleucine.

11.

$$H_3\overset{+}{N}CH\underset{\displaystyle CH_2-CH_2-\underset{\displaystyle \|\atop O}{C}-O^-Na^+}{}\overset{O}{\overset{\|}{C}}-O^-$$

13.(a)

$$\overset{+}{N}H_3\underset{\displaystyle CH_2Se^-}{CH}\overset{O}{\overset{\|}{C}}-O^-$$

(b) polar acidic, just like Asp

15. peptide bonds, also known as amide bonds.

17. (a) oligopeptide; (b) three; (c) alanine.

19.

$$\overset{+}{N}H_3\underset{\displaystyle CH_2-S}{CH}\overset{O}{\overset{\|}{C}}-NH\underset{\displaystyle CH_2-C_6H_4-OH}{CH}\overset{O}{\overset{\|}{C}}-NH\underset{\displaystyle CHCH_3-CH_2-CH_3}{CH}\overset{O}{\overset{\|}{C}}-NH\underset{\displaystyle CH_2-CH_2-\overset{O}{\overset{\|}{C}}-NH_2}{CH}\overset{O}{\overset{\|}{C}}-NH\underset{\displaystyle CH_2-\overset{O}{\overset{\|}{C}}-NH_2}{CH}\overset{O}{\overset{\|}{C}}-NH\underset{\displaystyle CH_2-S}{CH}\overset{O}{\overset{\|}{C}}-N-\underset{(CH_2\ CH_2\ CH_2\ \text{ring})}{CH}\overset{O}{\overset{\|}{C}}-NH\underset{\displaystyle CH_2-CHCH_3-CH_3}{CH}\overset{O}{\overset{\|}{C}}-NHCH_2\overset{O}{\overset{\|}{C}}-NH_2$$

(the two S atoms are joined by a disulfide bond)

21. cysteine, tyrosine, isoleucine. glutamine, asparagine, cysteine, proline, leucine, glycine.

23. (a)

$$H_3\overset{+}{N}\underset{\displaystyle CH_2-\overset{O}{\overset{\|}{C}}-OH}{CH}\overset{O}{\overset{\|}{C}}-NH\underset{\displaystyle CH_2OH}{CH}\overset{O}{\overset{\|}{C}}-NH\underset{\displaystyle CH_2-CH_2-CH_2-CH_2-NH_3^+}{CH}\overset{O}{\overset{\|}{C}}-NH\underset{\displaystyle CHCH_3-CH_3}{CH}\overset{O}{\overset{\|}{C}}-OH$$

(b)

$$H_3\overset{+}{N}\underset{\displaystyle CH_2-\overset{O}{\overset{\|}{C}}-O^-}{CH}\overset{O}{\overset{\|}{C}}-NH\underset{\displaystyle CH_2OH}{CH}\overset{O}{\overset{\|}{C}}-NH\underset{\displaystyle CH_2-CH_2-CH_2-CH_2-NH_3^+}{CH}\overset{O}{\overset{\|}{C}}-NH\underset{\displaystyle CHCH_3-CH_3}{CH}\overset{O}{\overset{\|}{C}}-O^-$$

(c) $H_2NCHC(=O)—NHCHC(=O)—NHCHC(=O)—NHCHC(=O)—O^-$, with side chains (left to right): $CH_2—C(=O)—O^-$; CH_2OH; $CH_2CH_2CH_2CH_2NH_2$; $CHCH_3$ / CH_3

25. **(a)** 1+; **(b)** 1−; **(c)** 2−.
27. twenty-four.
29. Inside. This arrangement allows the nonpolar side chains to interact through London forces and does not disrupt hydrogen bonding between water molecules.
31. **(a)** peptide bonds; **(b)** hydrogen bonds; **(c)** hydrogen bonds, salt bridges, the hydrophobic effect, and disulfide bonds; **(d)** hydrogen bonds, salt bridges, the hydrophobic effect, disulfide bonds, and disulfide bonds.
33. aspartic acid, glutamic acid, lysine, arginine.
35. They consist of a single polypeptide chain.
37. Disulfide bonds of intracellular proteins tend to be reduced by NADH and other reducing agents present in the cytoplasm of the cell.
39. **(a)** $—CH_2C(=O)—O^-$ $\quad {}^{+}NH_2{=}C(H_2N)—NHCH_2CH_2CH_2—$

(b) $—CHCH_3$ / CH_3 $\qquad$ $—CH_2$–(indole ring, NH)

41. denatured.
43. a.
45. Change the temperature, change the pH, add detergent or soap, agitate the protein.
47. Yes.
49.

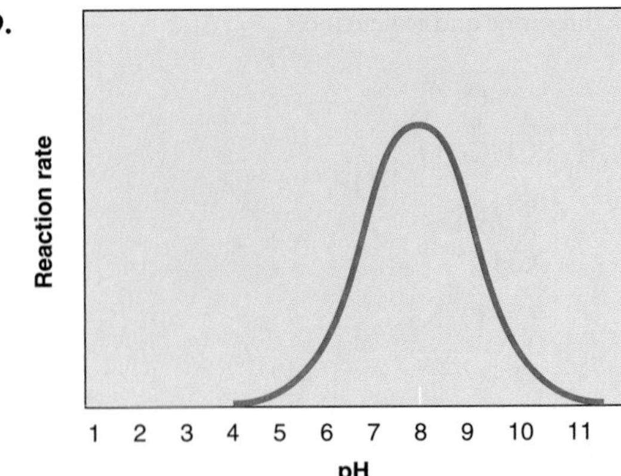

51.

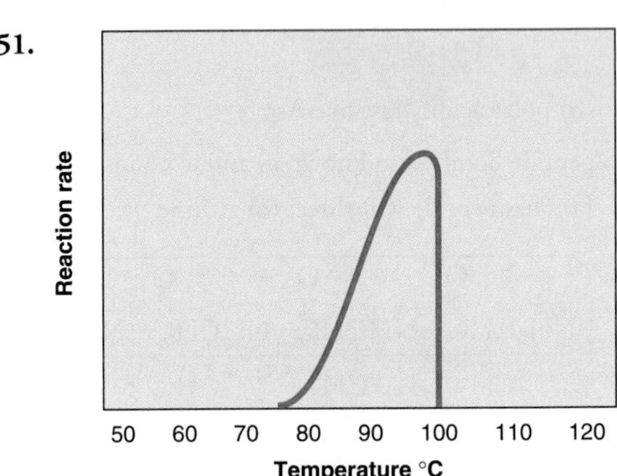

53. A change in PH will denature the enzyme. This will affect its ability to act as a catalyst and will change K_M and V_{max}.
55. A competitive inhibitor alters K_m but not V_{max}. A noncompetitive inhibitor alters V_{max} but not K_m.
57. **(a)** $\overset{+}{N}H_3—C(H)(CH_3)—C(=O)—O^- \longrightarrow H—C(\overset{+}{N}H_3)(CH_3)—C(=O)—O^-$

(b) A racemic mixture is a 50:50 mixture of enantiomers. As L-alanine is converted into its enantiomer D-alanine, the alanine mixture moves toward becoming racemic; **(c)** A competitive inhibitor because it resembles the substrate L-alanine; **(d)** nonpolar; **(e)** polar-neutral.
59. While acetaminophen is attached to the active site of a COX enzyme, aspirin is unable to irreversibly inhibit the enzyme. By the time that acetaminophen (a reversible competitive inhibitor) leaves the active site, the aspirin may no longer be present.
61. **(a)** Ethanol; **(b)** ethanol; **(c)** yes; **(d)** Is 1-octanol poisonous or is it converted into a poisonous compound?
63. Phosphorylation or dephosphorylation alters the structure of an enzyme in a way that activates or deactivates it.

65. The abnormal primary structure is in the part of the polypeptide chain removed during activation of trypsinogen.

67. Hydroxyproline (Hyp). No. Hyp is formed by the enzyme-catalyzed oxidation of a hydroxyl group of a proline residue.

69. A hemoglobin molecule is a tetramer consisting of four separate polypeptide chains (two identical α chains and two identical β chains). Each α chain consists of 141 amino-acid residues coiled into seven α-helical regions and each β chain consists of 146 amino acid residues coiled to form eight α-helical regions.

71. True. When cells die their contents are released into the bloodstream. If the enzyme x is detected in the bloodstream in elevated concentrations it probably indicates that liver cells have been damaged.

73. The asparaginase, like other proteins, would be broken down during digestion.

75. Yes. The tertiary structure of a protein depends on its primary structure.

Chapter 14

1. phosphoric acid, a monosaccharide, and an organic base.

3. **(a)** Ribose.
(b)

HOH$_2$C, O, OH, H, H, H, H, OH, OH

5. adenine, guanine, cytosine, and uracil.

7. cytosine, thymine, and uracil.

9.

$:\ddot{O}:^-$, $^-:\ddot{O}-P^+-\ddot{O}:^-$, $:\ddot{O}:^-$

11.

$CH_3O-P(=O)(OH)-OH$

13. A carboxylic acid ester contains a C—O—C linkage in which one of the carbon atoms belongs to a carbonyl group (C=O) and the other is an alkane-type or aromatic carbon atom. A phosphate monoester contains a P—O—C linkage in which the phosphorus atom is double-bonded to an oxgen atom (P=O).

15. **(a)**

HOCH$_2$, O, OH, H, H, H, H, OH, H

(b)

HOCH$_2$, O, H, H, H, H, OH, OH, H

17. **(a)** 3; **(b)** 1.

19.

O, $^-O-P-OCH_2$, O^-, O, H, H, H, H, OH, OH, N, N, NH, N, NH$_2$, O

21. **(a)** one phosphoester bond, no phosphoanhydride bonds.
(b) one phosphoester bond, one phosphoanhydride bond.

23.

HOCH$_2$, O, H, H, H, H, X, NH$_2$, N, N, N, N, O, $^-O-P=O$, O, $^-O-P=O$, O, $^-O-P=O$, O^-

(a) X = H
(b) X = OH

25. **(a)** phosphoanhydride, phosphoester, *N*-glycoside; **(b)** 3 phosphates, D-ribose, adenine; **(c)** phosphoester, *N*-glycoside; **(d)** 1 phosphate, D-ribose, adenine.

27. **(a)** 2-deoxyribose; **(b)** phosphoester bonds; **(c)** At C-1'.

29. At one end of a DNA strand (the 3′-terminus) the 3′ hydroxyl group has no attached nucleotide residue. At the other end (the 5′-terminus) the 5′ hydroxyl group has no attached nucleotide residue.

31. DNA contains 2-deoxyribose residues, while RNA contains ribose residues. The bases in DNA are A, G, C, and T, while those in RNA are A, G, C, and U.

33. The secondary structure of DNA consists of antiparallel double strands twisted into a helix. The tertiary structure of DNA involves supercoiling of the double helix.

35. hydrogen bonding between complementary bases.

37.

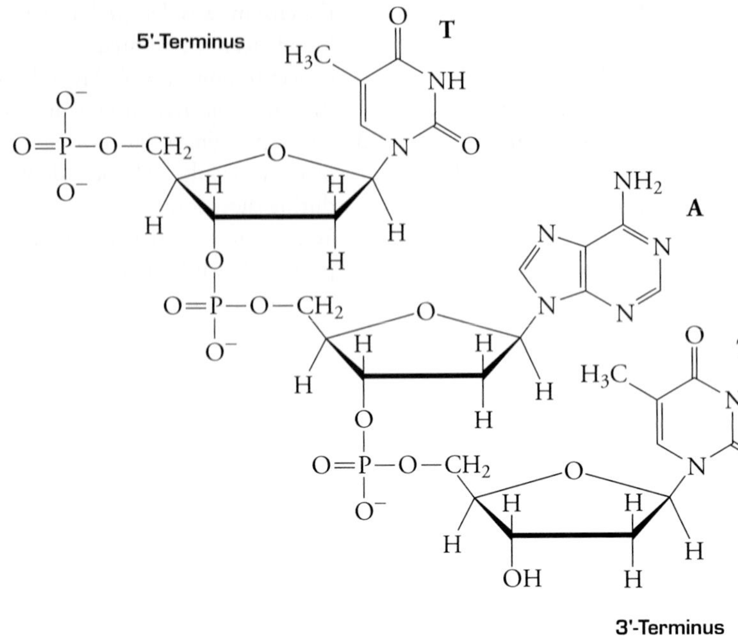

5′-Terminus

3′-Terminus

39. **(a)** Proteins that interact with the phosphate groups of the DNA backbone; **(b)** Groups of histones wrapped by DNA; **(c)** A coiled string of nucleosomes.

41. The primary structure of a protein is the sequence of amino acid residues. In DNA, primary structure is the sequence of nucleotide residues. Secondary structure is the arrangement of a protein into an α-helix or β-sheet, and the arrangement of DNA into a double helix. For both proteins and DNA, tertiary structure is the overall three-dimensional shape.

43. To renature DNA is to return denatured DNA to its original double stranded form.

45. **(a)** DNA; **(b)** DNA; **(c)** mRNA; **(d)** viral RNA.

47. **(a)** A site on DNA where the double strand has been pulled apart to expose single strands. DNA replication takes place at an origin; **(b)** No, during replication in plants and animals, many origins are opened at one time along a double-stranded DNA helix.

49. **(a)** 3′ to 5′; **(b)** 5′ to 3′; **(c)** Opposite directions.

51. Transfer RNAs carry the correct amino acid to the site of protein synthesis. Messenger RNAs carry the information that specifies which protein should be made. Ribosomal RNAs combine with proteins to form ribosomes, the multi-subunit complexes in which protein synthesis takes place.

53.

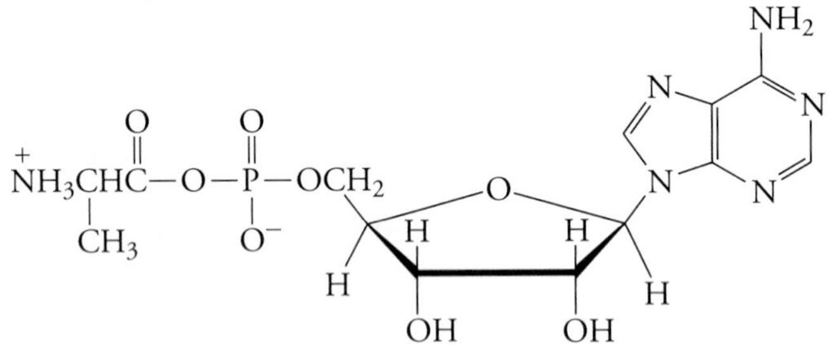

55. **(a)** Thymine; **(b)** dCCGTA (listed 3′ to 5′); **(c)** CCGUA (listed 3′ to 5′).

57. **(a)** A codon is a series of three bases that specifies a particular amino acid; **(b)** Messenger RNA

59. **(a)** Pro; **(b)** Ser; **(c)** Val.

61. **(a)** UUU and UUC; **(b)** AAA and AAG; **(c)** GAU and GAC.

63. Transfer RNA is shortened and some bases are converteed into minor bases. Messenger RNA may be trimmed at either end, the ends may be altered, and sections of the middle may be removed. Ribosomal RNA is cut into smaller pieces.

65. **(a)** (listed 3′ to 5′) AAA; **(b)** (listed 3′ to 5′) GGG; **(c)** (listed 3′ to 5′) UUU; **(d)** (listed 3′ to 5′)CCC.

67. rRNA combines with specific proteins to form a ribosome, the site of protein synthesis. mRNA carries the code that specifies the primary structure of the protein to be made. tRNA carries amino acid residues to where they are needed for addition to the growing protein.

69. **(a)** (listed 5′ to 3′) UCU-AAG-GAC; **(b)** (listed 3′ to 5′) dAGA-TTC-CTG.

71. **(a)** Ribosome subunits combine with an mRNA and a tRNA that carrys fMet; **(b)** A second tRNA attaches to the mRNA, a ribosome enzyme catalyzes the attachment of the amino acid carried by this tRNA to fMet, the ribosome shifts one codon down mRNA and the first tRNA leaves the ribosome, another tRNA attaches, its amino acid is added to the growing peptide chain, and the process repeats; **(c)** When a stop codon is reached, the newly synthesized peptide is released and the mRNA–ribosome–tRNA complex dissociates.

73. **(a)** β-galactosidase, permease, and transacetylase; **(b)** the repressor protein binds to operator sites O_1 and O_2, control sites on the DNA near or at the promoter site; **(c)** lactose is converted to allolactose by β-galactosidase. Allolactose binds to the repressor protein and changes its shape in such a

way that it can no longer attach to the operator sites. This allows RNA polymerase to bind to the promoter site, turning "on" transcription.

75. (a)

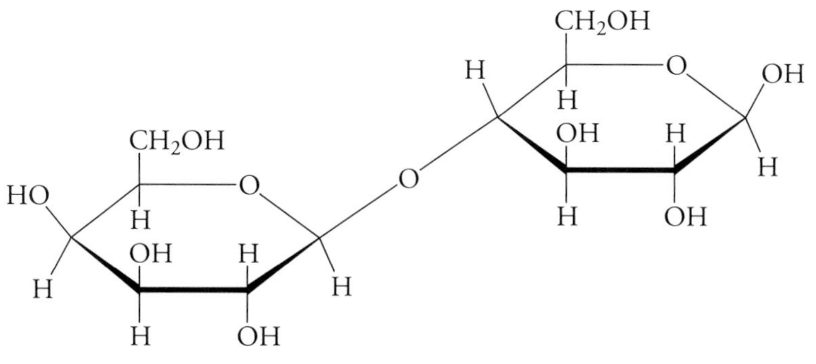

Lactose

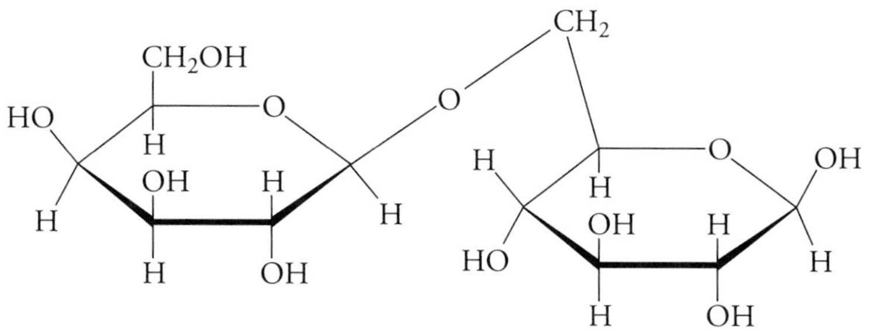

Allolactose

(b) one galactose and one glucose.

77. (a) No. Mutations that take place outside of genes will have little or no effect on gene expression. Because, in many cases more than one codon specifies the same amino acid residue, some mutations have no effect on the protein being coded for; (b) No. Only mutations that occur in the egg or sperm cells can be passed on to offspring.

79. (a) Double-stranded RNA is the source of small interfering RNA (siRNA); (b) A nuclease cuts double-stranded RNA into 21–23 base pair fragments; (c) siRNAs, the 21–23 base pair fragments created by the nuclease, are the source of guide strands; (d) A guide strand, one strand of each double-stranded siRNA, is complementary to the mRNA targeted for destruction; (e) RISC (RNA-induced silencing complex) binds a guide strand. When complementary mRNA binds to the guide strand, RISC cuts the mRNA, preventing it from undergoing translation and silencing the gene that produced the mRNA.

81. (a) True. One strand of small interfering RNA will be complementary to the mRNA; (b) False. mRNA undergoes post-transcriptional modifications, so the mRNA may not exactly reflect the primary structure of the gene that codes for it; (c) False. Proteins undergo post-translational modifications, so the native protein may not exactly reflect the primary structure of the mRNA from which it is produced.

83. Reverse transciptase is an enzyme that catalyzes the formation of DNA from viral RNA. Some viruses produce transcriptase because they contain RNA which, when injected into the host cell must be converted into DNA for viral proteins to be produced

85. The genes BRCA1 and BRCA2 code for proteins indirectly involved in a form of DNA repair called recombinational repair.

87. Short tandem repeats are relatively small stretches of DNA that contain short repeating sequences of bases. Restriction enzymes are used to cut the DNA into small segments and the number of base repeats of the STRs are determined.

89. 32, 768.

Chapter 15

1. A metabolic pathway is a group of biochemical reactions.

3. (a) No; (b) glucose + ATP $\longrightarrow$ glucose 6-phosphate + ADP $\Delta G = -4.0$ kcal (c) Yes

5. no effect.

7. (a) 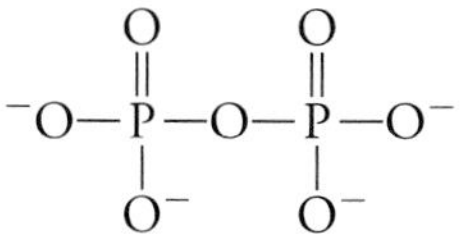 $^{-}O-\overset{\overset{O}{\|}}{\underset{\underset{O^-}{|}}{P}}-O-\overset{\overset{O}{\|}}{\underset{\underset{O^-}{|}}{P}}-O^{-}$

(b) 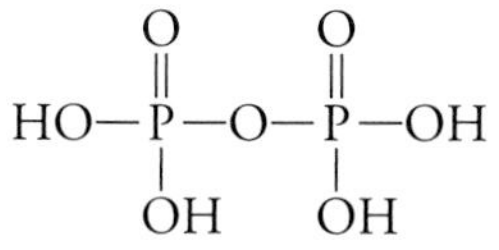 $HO-\overset{\overset{O}{\|}}{\underset{\underset{OH}{|}}{P}}-O-\overset{\overset{O}{\|}}{\underset{\underset{OH}{|}}{P}}-OH$

9. catabolism.

11. Reduction is the gain of electrons. In organic and biochemical molecules, the gain of hydrogen and/or the loss of oxygen is indication that reduction has taken place. NADH, the reduced form of the coenzyme has one more hydrogen atom than NAD^+, the oxidized form of the coenzyme.

13. an ester bond (specifically, a thioester bond).

15. During anabolism small molecules are used to make larger ones. In catabolism larger molecules are broken down into smaller ones.

17. UTP + $H_2O \longrightarrow$ UMP + PP_i

19. (a) mouth and small intestine; (b) stomach and small intestine; (c) stomach and small intestine.

21. Amylopectin contains $\alpha-(1 \rightarrow 6)$ glycosidic bonds. Amylose does not.

23. Salivary amylase, which initiates the digestion of carbohydrates in the mouth, is denatured when it reaches the stomach. Enzymes responsible for the majority of carbohydrate digestion are present in the small intestine.

25. α–(1 → 6) and α–(1 → 4) glycosidic bonds. No. α–(1 → 4) glycosidic bonds are hydrolyzed by amylase and α–(1 → 6) glycosidic bonds are hydrolyzed by a debranching enzyme.

27. steroid.

29. Glycolysis converts a glucose molecule into 2 pyruvates with accompanying production of 2 ATP and 2 NADH.

31. **(a)** Step 3; **(b)** ADP and AMP; **(c)** ATP and citrate.

33. **(a)** decomposition;

(b)

$$\mathrm{HOCH_2CH(O{-}P(=O)(OH){-}O^-)C(=O){-}OH}$$

2-Phosphoglycerate

$$\mathrm{CH_2{=}C(O{-}P(=O)(OH){-}O^-){-}C(=O){-}OH}$$

Phosphoenolpyruvate

35. **(a)** Ethanol and CO_2; **(b)** NADH is recycled back into NAD^+, which allows glycolysis to continue.

37. +2 ATP, 0 NADH.

39. $\mathrm{CH_3{-}CH(OH){-}C(=O){-}OH}$

The conjugate base form.

41. **(a)** reduction; **(b)** pyruvate; **(c)** NADH.

43. oxidation.

45. Each different pathway would require a different set of enzymes.

47. 1 and 3

49. 1, 3, and 10.

51. glucose 6-phosphate.

53. **(a)** Insulin activates glycogen synthetase (speeds us glycogenesis) and deactivates glycogen phosphorlase (slows down glycogenolysis); **(b)** Glucagon deactivates glycogen synthetase (slows down glycogenesis) and activates glycogen phosphorylase (speeds up glycogenolysis).

55. The alcohol group in isocitrate is secondary while, the alcohol group in citrate tertiary.

57. 2 GTP, 6 NADH, and 2 $FADH_2$

59. **(a)** NADH and $FADH_2$; **(b)** O_2

61. facilitated diffusion.

63. 2.5 ATP.

65. In glycolysis, one glucose molecule yields 2 pyruvates, 2 ATP and 2 NADH. Depending on the system used to shuttle the electrons from NADH into mitochondria, these 2 NADH become 2 NADH or 2 $FADH_2$. Converting the 2 pyruvates into 2 acetyl-CoAs produces 2 NADH. Two passes through the citric acid cycle starting with acetyl-CoA produces 2 GTP (equivalent to 2 ATP), 2 $FADH_2$, and 6 NADH. In the electron trasport chain, each $FADH_2$ generates 1.5 ATP and each NADH generates 2.5 ATP.

67. Fatty acids are activated by the attachment of a coenzyme A (CoA) residue. The CoA residue is replaced by a compound called carnitine and the fatty acyl-carnitine is moved into the mitochondria by active transport.

69. one acetyl-CoA, one $FADH_2$, and one NADH.

71. 92.

73. **(a)** acetoacetate, 3-hydroxybutyrate, and acetone; **(b)** no.

75. Two of the ketone bodies are carboxylic acids.

77. **(a)** $\mathrm{CH_3(CH_2)_4CH{=}CHCH_2CH{=}CH(CH_2)_7C(=O){-}OH}$

linoleic acid

$$\mathrm{CH_3CH_2CH{=}CHCH_2CH{=}CHCH_2CH{=}CH(CH_2)_7C(=O){-}OH}$$

linolenic acid

(b) The double bonds beyond carbon 10.

79. cytoplasm.

83. oxidative deamination.

85. $\mathrm{HSCH_2C(=O){-}C(=O){-}OH}$

87. α-ketogluterate.

89. Some scientists believe the mitochondria were once free bacteria because they have their own DNA that is different from that of the cell.

91. The presence of the uncoupler thermogenin allows energy released during catabolism to be diverted to the production of heat, rather than production of ATP.

93. ether, ketone, aromatic, and alkene.

APPENDIX E

Glossary

3′ terminus The end of an oligo- or polynucleotide whose 3′ carbon atom is not attached to another nucleotide residue.

5′ terminus The end of an oligo- or polynucleotide whose 5′ carbon atom is not attached to another nucleotide residue.

Absolute specificity Restriction of the activity of an enzyme to one specific substrate.

Accuracy An indication of how close a measured value is to the true value.

Acetal An organic compound in which a carbon atom is attached to two —OC groups.

Acid See Brønsted-Lowry acid.

Acidic solution A solution in which the concentration of H_3O^+ is greater than 1.0×10^{-7} M (pH < 7).

Acidity constant (K_a) The equilibrium constant for the reaction of an acid with water.

Acidosis A low blood serum pH.

Actinide elements Elements with atomic numbers 90–103.

Activation energy (E_{act}) The energy required to cross the energy barrier between reactants and products.

Active site An enzyme's substrate binding site.

Active transport An energy-requiring process in which molecules and ions are moved across a membrane in the direction opposite of diffusion.

Actual yield The amount of product obtained from a reaction.

Adrenocorticoid hormone A steroid hormone produced in the adrenal glands.

Aerobic conditions Conditions under which O_2 is available.

Alcohol An organic compound that contains a hydroxyl group attached to an alkane-type carbon atom.

Alcohol sugar A monosaccharide in which the aldehyde or ketone group has been reduced to an alcohol.

Alcoholic fermentation A metabolic process in which pyruvate ion is split into an acetaldehyde molecule and a CO_2 molecule, followed by the reduction of acetaldehyde into ethanol. During the reduction of acetaldehyde, NADH is oxidized to NAD^+.

Aldehyde An organic compound that contains a carbonyl group and in which the carbonyl carbon atom is attached to two hydrogen atoms or one hydrogen atom and one carbon atom.

Aldohexose A monosaccharide that contains an aldehyde group and six carbon atoms.

Aldose A monosaccharide that contains an aldehyde group.

Alkali metals Group 1A metals.

Alkaline earth metals Group 2A elements.

Alkane A hydrocarbon whose carbon atoms are joined only by single bonds.

Alkane-type carbon atom A carbon atom that is singly bonded to carbon or hydrogen atoms.

Alkene A hydrocarbon that contains at least one carbon–carbon double bond.

Alkyl group A substituent consisting only of carbon and hydrogen atoms and in which carbon atoms are joined to one another by single bonds.

Alkyl halide An organic compound in which a halogen atom is attached to an alkane-type carbon atom.

Alkyne A hydrocarbon that contains at least one carbon–carbon triple bond.

Allosteric effector An ion or molecule that regulates an allosteric enzyme. The attachment of an allosteric effector to a binding site alters the three-dimensional shape of enzyme subunits and influences the ability of the enzyme to bind substrates and to catalyze their conversion to products. Two general types of allosteric effectors exist: positive effectors and negative effectors.

Allosteric enzyme An enzyme that displays cooperativity between subunits and whose activity is controlled by positive and/or negative effectors.

Alpha (α) anomer One of two stereoisomers produced by formation of an anomer. When a pyranose is drawn with the ring oxygen atom at the back right and the hemiacetal carbon atom at the right or when a furanose is drawn with the ring oxygen atom in the back and the hemiacetal carbon atom at the right, the hemiacetal —OH points down in an alpha anomer.

Alpha particle ($^4_2\alpha$) A form of nuclear radiation that consists of two protons and two neutrons.

Alpha-amino acid An amino acid in which the amino ($—NH_2$) group is attached to an alpha carbon atom.

Alpha-helix A form of protein secondary structure in which a polypeptide chain is arranged into a helical (coiled) shape.

Alpha carbon atom A carbon atom that is directly attached to the carbon atom of a carboxyl group.

Amide An organic compound that contains a nitrogen atom that is directly attached to a carbonyl carbon atom.

Amine An organic compound that contains a nitrogen atom that is directly attached to one or more carbon atoms (excluding carbonyl carbon atoms).

Amino acid A compound that contains one or more carboxyl groups and one or more amino groups. Amino acids are the building blocks used to construct peptides and proteins.

Amino sugar A monosaccharide in which an amino ($—NH_2$) group has replaced one or more of the hydroxyl (—OH) groups. In some amino sugars, the amino group has been converted into an amide.

Amorphous solid A solid in which the particles are not arranged in an ordered fashion.

Amphipathic A term used to describe substances that have both hydrophilic and hydrophobic parts.

Amphoteric compound A compound that can act as an acid or a base.

Anabolism Metabolic reactions in which larger compounds are made from smaller ones. Anabolism usually requires energy.

Anaerobic conditions Conditions under which O_2 is not available.

Anion A negatively charged ion.

Anomer The cyclic form of a monosaccharide (or monosaccharide residue) created when an alcohol group reacts with the aldehyde or ketone group to form a hemiacetal.

Anticodon A series of three bases in tRNA that is complimentary to the a codon on mRNA. Base paring between the anticodon of a tRNA molecule and a codon on mRNA ensures that the proper amino acid residue will be added to a growing peptide chain.

Antiparallel When applied to protein secondary structure, antiparallel refers to interacting strands in a β-sheet that run in opposite N-terminal to C-terminal directions. When applied to DNA secondary structure, antiparallel refers to interacting strands in a double helix that run in opposite $3'$ to $5'$ directions.

Aromatic hydrocarbon A hydrocarbon that contains a ring with alternating double and single bonds.

Atom The building block of matter.

Atomic mass unit (amu) A unit used to describe the mass of atoms and subatomic particles. Protons and neutrons each have a mass of about 1 amu and electrons have a mass of about 1/2000 amu.

Atomic notation A shorthand technique used to represent the make-up of an atom. In atomic notation an atom's atomic number is shown as a subscript and its mass number as a superscript, both written to the left of the atomic symbol.

Atomic number The number of protons in an atom's nucleus.

Atomic orbital A three-dimensional region of space about an atom's nucleus, where there is a high probability of finding electrons.

Atomic symbol A one- or two-letter abbreviation of the name of an element.

Atomic weight The average mass of the atoms of an element, as it is found in nature.

Avogadro's law The law that states that at a given temperature and pressure, the volume occupied by a sample of a gas is directly proportional to the number of moles of gas that are present.

Avogadro's number The number of items in one mole (6.02×10^{23}).

Balanced equation A chemical equation in which the same number of atoms of each element appear on each side of the reaction arrow or a nuclear equation in which the sum of the mass numbers and the sum of the charges on atomic nuclei and subatomic particles is the same on each side of the arrow.

Barometer A device used to measure atmospheric pressure. A barometer consists of a tube (sealed at the top) placed in a mercury-filled dish. The greater the pressure of the atmosphere, the higher the mercury is pushed up into the tube.

Base See Brønsted-Lowry base.

Base pair The specific nucleic acid base combinations that arise through hydrogen bonding. In DNA, adenine pairs with thymine and guanine pairs with cytosine.

Basic solution A solution in which the concentration of H_3O^+ is less than 1.0×10^{-7} M (pH > 7).

Benedict's reagent An oxidizing agent that contains Cu^{2+}. Benedict's reagent will oxidize aldehydes into carboxylic acids, but will not oxidize alcohols.

Bent shape An arrangement of atoms in which (a) one atom can be imagined as being placed at the center of a tetrahedron and the two atoms to which it is attached placed at two of the corners of the same tetrahedron, or (b) one atom can be imagined as being placed at the center of an equilateral triangle and the two atoms to which it is attached placed at two of the corners of the same triangle.

Beta anomer One of two stereoisomers produced by formation of an anomer. When a pyranose is drawn with the ring oxygen atom at the back right and the hemiacetal carbon atom at the right or when a furanose is drawn with the ring oxygen atom in the back and the hemiacetal carbon atom at the right, the hemiacetal —OH points up in a beta anomer.

Beta oxidation spiral A spiral metabolic pathway in which fatty acids are broken down by the repeated removal of two carbon fragments in the form of acetyl-CoA. Each pass through the spiral results in the production of one acetyl-CoA, one $FADH_2$, and one NADH.

Beta particle ($_{-1}^{0}\beta$) A form of nuclear radiation that consists of an electron.

Beta-sheet A form of protein secondary structure in which a sheet-like arrangement forms between different segments of a polypeptide chain. β-sheets can be parallel or antiparallel. Also known as a β-pleated sheet.

Beta carbon atom A carbon atom that is two atoms removed from the carbon atom of a carboxyl group.

Bile salt An amphiphatic steroid released from the gallbladder into the small intestine, where it aids digestion by forming an emulsion with dietary lipids.

Binary ionic compound An ionic compound composed of just two different elements.

Binary molecule A molecule that contains atoms of only two different elements.

Binding site A crevice on the surface of a globular protein where molecules or ions attach.

Bohr's model of the atom An early model of the atom that viewed electrons as circling the nucleus in specific orbits.

Boiling point The temperature at which the vapor pressure of a liquid equals the atmospheric pressure.

Boyle's law The law that states that for a sample of a gas at a fixed temperature, the pressure is inversely proportional to the volume.

Brønsted-Lowry acid A compound that releases H^+.

Brønsted-Lowry base A compound that accepts H^+.

Buffer A solution that resists changes in pH when small amounts of acid or base are added.

Carbohydrate A term used to refer, collectively, to monosaccharides, oligosaccharides, and polysaccharides.

Carbonyl group A carbon atom and an oxygen atom joined by a double bond (C=O).

Carboxyl group A functional group that consists of a hydroxyl group attached to the carbon atom of a carbonyl group.

Carboxylate ion The conjugate base of a carboxylic acid.

Carboxylic acid An organic compound that contains a carboxyl group.

Carboxylic acid sugar A monosaccharide in which an aldehyde or alcohol group has been oxidized to a carboxylic acid.

Catabolism Metabolic reactions in which compounds are broken down into smaller ones. Catabolism usually releases energy.

Catalyst A substance that speeds up a reaction, without itself being altered.

Catalytic hydrogenation A reaction in which an unsaturated compound is reduced, in the presence of a catalyst, by H_2.

Cation A positively charged ion.

Cerebroside A glycolipid that contains a monosaccharide residue.

Chain reaction A reaction in which one or more products can initiate another cycle of the reaction.

Charles' law The law that states that for a sample of a gas at a fixed pressure, the volume is directly proportional to the temperature.

Chemical change A change in matter that involves changing its chemical composition.

Chemical composition The proportion of each type of atom in a substance.

Chemical equation An equation that represents the changes that take place during a chemical reaction. The reactants are written to the left of an arrow and products are written to the right.

Chemical properties The chemical changes that an element or a compound undergo.

Chemical reaction A chemical change in which the covalent or ionic bonds that hold elements or compounds together are broken and new bonds are formed.

Chemistry The study of matter and the changes it undergoes.

Chiral A term used to describe an object that cannot be superimposed on its mirror image. A molecule is usually chiral if it contains one or more chiral carbon atoms.

Chiral carbon atom A carbon atom that is attached to four different atoms or groups of atoms.

Circular metabolic pathway A metabolic pathway consisting of a series of reactions in which the final product is also an initial reactant.

Cis One of two possible geometric (*cis-trans*) isomers. For cycloalkanes, a *cis* geometric isomer has two substituents on the same face of the ring. For alkenes, a *cis* isomer has alkyl groups on the same side of a line connecting the two double-bonded carbon atoms.

Citric acid cycle A cyclic metabolic pathway that converts the carbon atoms in acetyl-CoA into CO_2, producing GTP and reduced coenzymes in the process.

Codon A series of three bases in mRNA that specifies the amino acid residue to be added to a growing peptide chain. A codon also indicates where translation should begin and end.

Coefficient A number that appears in front of the formula of a reactant or product in a chemical equation. The coefficient indicates how many of each reactant is used or product is formed during a reaction.

Coenzyme An organic cofactor.

Cofactor An inorganic ion or organic compound required by an enzyme for catalysis to take place.

Colloid A mixture that consists of particles smaller than found in a suspension, but larger than found in a solution. Upon standing, the particles that make up a colloid will not settle.

Combined gas law The law that relates the pressure (P), volume (V), and Kelvin temperature (T) of a gas.

$$\frac{P_1V_1}{T_1} = \frac{P_2V_2}{T_2}$$

Combustion An oxidation reaction that uses O_2 as an oxidizing agent.

Common name A molecule name that is not based on the IUPAC rules.

Common system A measurement system that uses the pound, the quart, and the foot to measure, respectively, mass, volume, and length.

Competitive inhibitor A reversible inhibitor that competes with a substrate at the active site of an enzyme, changes K_M, and has no effect on V_{max}.

Complex protein A protein that requires a prosthetic group for biological activity.

Complementary base The relationship, in a base pair, of one base to another. For example, cytosine is the complementary base of guanine.

Compound Matter constructed of two or more chemically combined elements.

Concentration The amount of solute dissolved in a solvent.

Condensed structural formula A structural formula that describes the attachment of atoms to one another without showing all of the bonds. A carbon atom attached to three hydrogen atoms can be written "CH_3," for example.

Conformations The various shapes that a molecule can assume through rotation of covalent bonds.

Conjugates An acid-base pair that differs only by the presence or absence of H^+.

Constitutional isomers Molecules that have the same molecular formula but different atomic connections.

Control rod A rod, made from a neutron-absorbing material, that is used to control the extent of fission in a nuclear reactor.

Conversion factor A ratio that relates two different units. A value expressed using one unit can be converted into another unit by multiplying by the appropriate conversion factor.

Coolant The substance in a nuclear reactor that removes heat generated by fission.

Cooperative binding Attachment of a ligand to a binding site on a protein that impacts binding at other sites.

Coordinate-covalent bond A noncovalent interaction between a metal cation and the nonbonding electrons of a nonmetal atom.

Counting unit A term, such as dozen or mole, that refers to a specific number of items.

Coupled reaction A reaction that is the combination of two others, one spontaneous and the other nonspontaneous. The spontaneous reaction provides the energy required for the nonspontaneous one to occur.

Covalent bond A bond in which two atoms share a pair of valence electrons.

Covalent modification A form of control over the activity of an enzyme, in which the making or breaking of covalent bonds within the enzyme activates or deactivates it.

Covalent solid A crystalline solid in which atoms are held to one another by an array of covalent bonds.

Crenation The wrinkling of a cell that takes place when it is placed in a hypertonic solution.

Critical mass The minimum amount of radioisotope needed to sustain a nuclear chain reaction.

Crystal lattice A three-dimensional array of alternating cations and anions.

Crystalline solid A solid in which the arrangement of particles is ordered.

C-terminus The end of a peptide chain that carries the unreacted carboxyl group.

Cyclic nucleotide A nucleotide in which the phosphate residue has two phosphoester connections to the one monosaccharide residue.

Cycloalkane An alkane that contains a ring of carbon atoms.

D amino acid An amino acid that belongs to one particular family of stereoisomers. When drawn in a Fischer projection, with the carboxyl group pointing up and the side chain pointing down, the alpha amino group points to the right in a D amino acid.

D monosaccharide A monosaccharide that belongs to one particular family of stereoisomers. When drawn as a Fischer projection, with carbon atoms running vertically and the aldehyde or ketone group at or near the top, the —OH attached to the chiral carbon atom farthest from the C=O points to the right in a D monosaccharide.

Dalton's atomic theory An early theory of atomic structure that viewed atoms as being indivisible. We now know that atoms are constructed of smaller subatomic particles (protons, neutrons, and electrons).

Dalton's law of partial pressure The law that states that the total pressure of a mixture of gases is the sum of the partial pressures of its components.

Decarboxylation A reaction in which CO_2 is removed from a molecule.

Decomposition reaction A reaction in which one compound is broken down to form elements or simpler compounds.

Dehydration A reaction in which an alkene is formed by removal of water from an alcohol.

Denaturation Any change in the three-dimensional shape of a protein or polynucleotide caused by disruption of the forces responsible for maintaining its shape.

Density The mass of substance divided by the volume of space that it occupies.

Deoxy sugar A monosaccharide in which a hydrogen atom has replaced one or more of the —OH groups.

Deoxyribonucleic acid (DNA) An oligo- or polynucleotide built from nucleotides that contain 2-deoxyribose and the bases adenine, guanine, cytosine, and thymine.

Deoxyribonucleoside A nucleoside that contains a 2-deoxyribose residue.

Deoxyribonucleotide A nucleotide that contains a 2-deoxyribose residue.

Deposition The conversion of a gas into a solid, bypassing the liquid phase.

Dextrorotatory (+) A chiral compound that rotates plane polarized light in a clockwise direction.

Diastereomers Stereoisomers that are not mirror images.

Diatomic molecule A molecule consisting of two atoms.

Diffusion The movement of substances from areas of higher concentration to those of lower concentration.

Digestion The process in which large molecules present in food are broken down into their building blocks.

Dilution A process in which the concentration of a solution is lowered by the addition of solvent.

Dipeptide An oligopeptide that consists of two amino acid residues.

Dipole–dipole force A noncovalent interaction that involves the attraction of neighboring polar groups for one another.

Disaccharide An oligosaccharide that consists of two monosaccharide residues.

Dissociation The process in which a compound separates to form ions when dissolved in a solvent.

Disulfide An organic compound that contains a C—S—S—C linkage.

DNA polymerase An enzyme that catalyzes the formation of DNA.

DNA replication See replication.

Double bond Two pairs of electrons shared by two atoms.

Double replacement reaction A reaction in which parts of two reactants trade places.

Eicosanoid A hormone derived from arachidonic acid or other essential fatty acids containing twenty carbon atoms (the prefix *eico* means 20). Prostaglandins, thromboxanes, and leukotrienes are examples of eicosanoids.

Electrolyte A compound that produces ions when dissolved in water.

Electromagnetic radiation Energy that travels as waves. The types of electromagnetic radiation include radiowaves, microwaves, infrared light, visible light, ultraviolet light, X rays, and gamma rays.

Electron A subatomic particle found outside of the nucleus of an atom. Electrons carry a 1− charge and have a mass of about 1/2000 amu.

Electron dot structure A representation of atomic structure in which dots are placed around an atomic symbol, with each dot representing a valence electron.

Electron transport chain A group of proteins and other molecules imbedded in the inner mitochondrial membrane, whose function is to pass electrons from NADH and $FADH_2$ to O_2, using the energy released to pump H^+ through the inner mitochondrial membrane into the intermembrane space.

Electronegativity An indication of the ability of an atom to attract bonding electrons.

Element A substance that contains only one type of atom.

Elongation A step in translation in which amino acid residues are joined to a growing peptide chain.

Emulsifying agent A compound that makes or stabilizes an emulsion.

Emulsion A colloid formed by the combination of two liquids.

Enantiomers Stereoisomers that are nonsuperimposable mirror images.

Endothermic process A process that absorbs heat.

Energy The ability to do work and to transfer heat.

Enthalpy change (ΔH) The heat released or absorbed during a process.

Entropy change (ΔS) The change in energy dispersal during a process.

Enzyme A biochemical catalyst (usually a protein).

Equilibrium A condition of a reversible reaction in which the rate of the forward reaction equals the rate of the reverse reaction. At equilibrium, there is no net change in reactant or product concentrations.

Equilibrium constant (K_{eq}) A constant which, for the generalized reaction $aA + bB \rightleftharpoons cC + dD$, takes the form

$$K_{eq} = \frac{[C]^c[D]^d}{[A]^a[B]^b}$$

Equivalent (Eq) The number of moles of charges that one mole of a solute contributes to a solution.

Essential amino acid An amino acid that cannot be produced in the body and must be obtained in the diet.

Essential fatty acid A fatty acid that cannot be produced in the body and must be obtained in the diet.

Ester An organic compound that contains a C—O—C linkage in which one of the carbon atoms belongs to a carbonyl group.

Ether An organic compound that contains a C—O—C linkage.

Exact numbers Numbers obtained by an exact count or by definition. Exact numbers have an unlimited number of significant figures.

Excited state Any arrangement of electrons about an atom's nucleus that is less stable than the ground state.

Exothermic process A process that releases heat.

Experiment An investigation designed to test a hypothesis.

Expression See gene expression.

Facilitated diffusion The aided movement of a solute through a membrane, from an area of higher concentration to one of lower concentration.

Factor-label method An approach for carrying out unit conversions that makes use of conversion factors.

Fatty acid A carboxylic acid typically having been twelve and twenty carbon atoms.

Feedback inhibition A form of control over the activity of an enzyme, in which the product of a metabolic pathway inhibits one or more of the enzymes responsible for the product's synthesis.

Fibrous protein A protein whose structure resembles long fibers or strings. Typically, fibrous proteins are tough and water insoluble.

Fischer projection An abbreviated method for showing a tetrahedral shape. Fischer projections are drawn as a large "+" sign with a carbon atom (not shown) sitting at the center. Horizontal lines represent bonds pointing toward the viewer and vertical lines represent bonds pointing away.

Fission A process in which an atom's nucleus splits to form two smaller nuclei, neutrons, and energy.

Fluid mosaic model A model of membrane structure in which the protein component of a membrane is viewed as being inlayed into the phospholipid, glycolipid, and cholesterol component.

Formal charge A count of the electron distribution around a covalently bonded atom. Formal charge is calculated by subtracting the number of valence electrons held by a neutral atom from the number of electrons that surround the atom in a molecule or ion, assuming that the electrons in covalent bonds are equally shared.

Formula weight The sum of the atomic weights of the elements in the formula of a compound.

Free energy change (ΔG) A combination of enthalpy change, entropy change, and temperature ($\Delta G = \Delta H - T\Delta S$) that can be used to indicate whether a process is spontaneous or nonspontaneous. By convention, ΔG for a spontaneous process has a negative value and ΔG for a nonspontaneous process has a positive value.

Fuel rods Rods in a nuclear reactor that contain the radioisotope that undergoes fission.

Functional group An atom, group of atoms, or bond that gives a molecule a particular set of chemical properties.

Furanose An anomer that contains a five-membered ring (one oxygen atom and four carbon atoms).

Fusion A nuclear process in which smaller nuclei combine to form a larger nucleus.

Gamma ray ($^{0}_{0}\gamma$) A form of nuclear radiation that consists of a particular type of electromagnetic radiation.

Ganglioside A glycolipid that contains an oligo- or polysaccharide residue.

Gas The form of matter that has a variable shape and volume.

Gas constant The constant (R) in the ideal gas law ($PV = nRT$). For an ideal gas, R has a value of 0.0821 L atm/mol K.

Gay-Lussac's law The law that states that for a sample of a gas with a fixed volume, the pressure is directly proportional to the temperature.

Gene A stretch of DNA that carrys a code for the production of a protein.

Gene expression The process of transcribing the DNA in a gene and translating the resulting mRNA to make a protein.

Genetic code The correlation between a codon and the corresponding amino acid residue.

Genetic disease A disease caused by an inherited mutation.

Geometric isomers Stereoisomers that result from restricted bond rotation. Cycloalkanes and alkenes are molecules that can exist as *cis* or *trans* geometric isomers.

Globular protein A protein whose structure is spherical and highly folded. Typically, a globular protein is water soluble.

Gluconeogenesis A metabolic pathway that produces glucose from non-carbohydrate sources, such as amino acids, glycerol, and lactate.

Glycerophospholipid A class of phospholipids whose members consist of a glycerol residue, two fatty acid residues, a phosphate residue, and an alcohol residue.

Glycogenesis The manufacture of glycogen.

Glycogenolysis The hydrolysis of glucose residues from glycogen.

Glycolipid A lipid that contains a carbohydrate residue.

Glycolysis A linear metabolic pathway that converts one glucose molecule into two pyruvate ions, producing 2 NADH and 2 ATP in the process.

Glycoside An acetal that is formed by reaction of an alcohol with the hemiacetal —OH of an anomer.

Glycosidic bond A name given to the acetal bond that connects the alcohol residue to the acetal carbon atom in a glycoside.

Ground state The most stable arrangement of electrons about an atom's nucleus.

Group Elements in the same vertical column of the periodic table.

Half life The time required for one-half of the atoms in a radioactive sample to decay (to undergo nuclear change by releasing nuclear radiation).

Halogens Group 7A elements.

Heat of fusion The energy required to melt a solid.

Heat of vaporization The energy required to vaporize a liquid.

Hemiacetal An organic compound in which a carbon atom is attached to —OC and —OH.

Hemolysis The bursting of a red blood cell that is caused by placing it in a hypotonic solution.

Henry's law A law pertaining to solubility, which states that the solubility of a gas in a liquid is proportional to the pressure of the gas over the liquid.

Heterogeneous mixture A mixture that is not evenly distributed.

Heteropolysaccharide A polysaccharide that contains more than one type of monosaccharide residue.

Homogeneous mixture A mixture that is uniformly distributed.

Homopolysaccharide A polysaccharide that contains just one type of monosaccharide residue.

Hormone A compound that regulates the function of organs and tissues.

Hydration A reaction in which an alcohol is formed by adding water to an alkene.

Hydrocarbon An organic compound that contains only carbon and hydrogen atoms.

Hydrogen bond A noncovalent interaction that involves the attraction between a nitrogen, oxygen, or fluorine atom and a hydrogen atom that is covalently bonded to a different nitrogen, oxygen, or fluorine atom.

Hydrolysis A reaction in which one reactant (water) is used to split another.

Hydrophilic A term used to describe substances that are soluble in water.

Hydrophobic A term used to describe substances that are insoluble in water.

Hydrophobic effect The attraction of nonpolar groups for one another when they are placed in an aqueous solution.

Hydroxyl group An —OH functional group.

Hypertonic Of two solutions on opposite sides of a membrane, the one with the higher solute concentration.

Hypothesis A tentative explanation of presently known facts.

Hypotonic Of two solutions on opposite sides of a membrane, the one with the lower solute concentration.

Ideal gas A gas that obeys the ideal gas law.

Ideal gas law The law ($PV = nRT$) that relates the pressure (P), volume (V), number of moles (n), and kelvin temperature of a gas. R is the gas constant.

Inert gases Group 8A elements.

Inhibitor Any substance that reduces an enzyme's ability to act as a catalyst.

Initiation The first step in translation, in which a ribosome, mRNA, and tRNA come together to form a complex.

Insoluble Unable to dissolve in (form a homogeneous mixture with) a given solute.

Ion A charged atom or group of atoms.

Ion–dipole interaction A noncovalent interaction that involves the attraction between ions and atoms that carry a partial charge.

Ionic bond The bond that holds anions and cations to one another. Ionic bonds result from the attraction of opposite charges.

Ionic compound A compound composed of cations and anions.

Ionic equation A reaction equation in which the formulas of electrolytes are written as individual ions.

Ionic solid A crystalline solid consisting of an array of oppositely charged ions held to one another by ionic bonds.

Ionization The formation of ions.

Irreversible inhibitor An enzyme inhibitor whose effect is permanent.

Isoelectric point The pH at which an amino acid has a net charge of zero.

Isoelectronic Describing ions and atoms that have identical arrangements of electrons.

Isotonic Describing either of two solutions on opposite sides of a membrane, when they have the same solute concentration.

Isotopes Atoms of a particular element that have a different number of neutrons.

IUPAC rules A set of rules created by the International Union of Pure and Applied Chemistry (IUPAC) that is used to name organic compounds.

Ketone An organic compound that contains a carbonyl group and in which the carbonyl carbon atom is attached to two other carbon atoms.

Ketone bodies Compounds (acetoacetate, 3-hydroxybutyrate, and acetone) produced from acetyl-CoA.

Ketose A monosaccharide that contains a ketone group.

Ketosis The overproduction of ketone bodies.

Kinetic energy Energy of motion.

K_{eq} See equilibrium constant.

K_M **(Michaelis constant)** A measure of the strength of the attraction between an enzyme and a substrate. K_M is related to the first step in the reaction catalyzed by a Michaelis-Menten enzyme: $E + S \rightleftharpoons ES$.

K_w The equilibrium constant for the ionization of water,

$$2H_2O(l) \rightleftharpoons H_3O^+(aq) + OH^-(aq)$$

which takes the form

$$K_w = [H_3O^+][OH^-] = 1.0 \times 10^{-14}$$

L amino acid An amino acid that belongs to one particular family of stereoisomers. When drawn in a Fischer projection with the carboxyl group pointing up and the side chain pointing down, the alpha amino group points to the left in an L amino acid.

L monosaccharide A monosaccharide that belongs to one particular family of stereoisomers. When drawn as a Fischer projection, with carbon atoms running vertically and the aldehyde or ketone group at or near the top, the —OH attached to the chiral carbon atom farthest from the C=O points to the left in an L monosaccharide.

Lanthanide elements Elements with the atomic numbers 58 through 71.

Law A statement that describes phenomena that are consistently and reproducibly observed.

Le Châtelier's principle A principle related to equilibrium, which states that when a reversible reaction is pushed out of equilibrium, the reaction responds to restore equilibrium.

Leaving group An atom or group of atoms that is displaced by a nucleophile during a nucleophilic substitution reaction.

Levorotatory (−) A chiral compound that rotates plane polarized light in a counterclockwise direction.

Ligand An atom, ion, or molecule that binds to a protein.

Like dissolves like A general solubility guideline based on the observation that one substance will usually dissolve in another if the particles from which they are made are of similar size and can interact through noncovalent forces.

Limiting reactant A reactant that determines the amount of product that can be formed in a reaction.

Linear shape An arrangement of atoms in which one atom can be imagined as being placed at the center of a line and the two atoms to which it is attached placed at the opposite ends of the line.

Linear metabolic pathway A metabolic pathway consisting of a continuous series of reactions in which the product of one reaction is the reactant in the next.

Line–bond method A method for drawing molecules, in which each pair of shared bonding electrons is represented by a line.

Lipid A water insoluble biochemical compound.

Lipogenesis A spiral metabolic pathway in which fatty acids are produced by the repeated addition of two carbon fragments.

Liquid The form of matter that has a variable shape and a fixed volume.

London force A noncovalent interaction resulting from temporary dipoles that arise from the continuous motion of electrons within an atom or molecule.

Major bases The five bases present in DNA and/or RNA: cytosine (C), thymine (T), uracil (U), adenine (A) and guanine (G).

Manometer A device used to measure gas pressure. In an open-end manometer, the pressure inside of a container is compared to atmospheric pressure.

Markovnikov's rule A rule that allows the major product of a reaction between an alkene and H_2O/H^+ to be predicted. This rule states that the major product is the alcohol formed by adding a hydrogen atom to the double bonded carbon atom that carries more hydrogen atoms and by adding —OH to the other double bonded carbon atom.

Mass number The total number of protons and neutrons in the nucleus of an atom.

Matter Anything that has mass and occupies space.

Membrane A bilayer containing phospholipids, glycolipids, cholesterol, and proteins.

Messenger RNA (mRNA) The form of RNA that holds the information (codons) transcribed from DNA regarding the primary structure of the peptide to be made.

Metabolic pathway A series of biochemical reactions.

Metabolism All of the reactions that take place in a living thing.

Metal An element that is a good conductor of electricity and heat and that, as a solid, is lustrous (it shines), malleable (it can be pounded without breaking), and ductile (it can be drawn into a wire).

Metallic solid A crystalline solid consisting of metal atoms. A metallic solid is considered to be an array of metal cations immersed in a cloud of electrons that spans the entire crystalline structure.

Metric system A measurement system that uses the gram, the liter, and the meter to measure, respectively, mass, volume, and length.

Michaelis-Menten enzyme An enzyme that catalyzes reactions following the scheme $E + S \rightleftharpoons ES \longrightarrow E + P$, where E is enzyme, S is substrate, ES is the enzyme–substrate complex, and P is product.

Minor bases Bases present in DNA and RNA other than adenine, guanine, cytosine, thymine, and uracil. Minor bases are formed by the enzyme-catalyzed modification of major bases after they have been incorporated into oligo- or polynucleotide strands.

Mitochondrion An organelle in plant and animal cells in which the citric acid cycle, electron transport (via the electron transport chain), oxidative phosphorylation, and fatty acid oxidation take place.

Mixture A combination of two or more pure substances.

Molarity (M) A concentration unit defined as follows:

$$\text{molarity} = \frac{\text{moles of solute}}{\text{liters of solution}}$$

Molar mass The mass of one mole of a substance. Molar mass (in grams per mole) is equivalent to the atomic weight, formula weight, or molecular weight of a substance (in amu).

Mole A counting unit that represents 6.02×10^{23} items.

Molecular formula A formula that lists the number of each type of atom present in a molecule.

Molecular solid A crystalline solid consisting of a ordered arrangement of molecules held to one another by noncovalent interactions.

Molecular weight The sum of the atomic weights of the elements in the formula of a molecule.

Molecule An uncharged group of atoms connected to one another by covalent bonds.

Monoatomic ion An ion formed from a single atom.

Monounsaturated fatty acid A fatty acid whose hydrocarbon tail has one carbon–carbon double bond.

Monsaccharide A polyhydroxy aldehyde or ketone containing three or more carbon atoms. Monosaccharides are the carbohydrate building blocks from which oligo- and polysaccharides are constructed.

Mutarotation Interconversion of the α and β anomers of a carbohydrate.

Mutation A permanent change in the primary structure of DNA.

Native protein The biologically active form of a protein.

Negative effector An allosteric effector that reduces substrate binding and slows the rate of an enzyme-catalyzed reaction.

Net ionic equation An ionic equation from which spectator ions have been removed.

Neutral solution A solution in which the concentration of H_3O^+ equals 1.0×10^{-7} M (pH = 7).

Neutralization The reaction of an acid with a base to produce water and a salt (ionic compound).

Neutron A subatomic particle found in the nucleus of an atom. Neutrons carry no charge and have a mass of about 1 amu.

***N*-glycoside** A glycoside formed by reaction of an amine with the hemiacetal group of an α or β monosaccharide anomer.

Nonbonding electrons Valence electrons not involved in covalent bonds.

Noncompetitive inhibitor A reversible inhibitor that attaches to an enzyme at a site other than the binding site, changes V_{max}, and has no effect on K_M.

Noncovalent interactions Interactions between different molecules or polyatomic ions or remote parts of the same that do not involve the sharing of valence electrons.

Nonmetal An element that is a poor conductor of electricity and heat and that, as a solid, is non-lustrous (not shiny) and brittle.

Nonpolar amino acid An amino acid whose side chain is an alkyl group, an aromatic ring, or another nonpolar collection of atoms.

Nonspontaneous process A process that will not run by itself unless something keeps it going.

***Normal* alkane** An alkane whose carbon atoms are joined in one continuous chain.

N-terminus The end of a peptide chain that carries the unreacted amino group

Nuclear change The change to atomic nuclei that takes place when nuclear radiation is released.

Nuclear equation An equation that represents the changes that take place during a nuclear reaction. The initial isotopes are written to the left of the arrow and products (final isotopes and radiation emitted) are written to the right.

Nuclear radiation High-energy particles (including alpha particles, beta particles, and positrons) and/or high-energy electromagnetic radiation (gamma rays) released from the nucleus of a radioactive isotope.

Nucleophile An electron rich atom or group of atoms that donates a pair of electrons to form a new covalent bond.

Nucleophilic substitution reaction A reaction in which a nucleophile replaces a leaving group.

Nucleoside A compound in which the β-furanose anomer of ribose (in RNA) or 2-deoxyribose (in DNA) is connected to one of the bases adenine (A), guanine (G), cytosine (C), thymine (T–present in DNA only), or uracil (U–present in RNA only) by an *N*-glycosidic bond.

Nucleotide A compound in which the β-furanose anomer of ribose (in RNA) or 2-deoxyribose (in DNA) is connected to a phosphate group by a phosphoester bond and to one of the bases adenine (A), guanine (G), cytosine (C), thymine (T–present in DNA only), or uracil (U–present in RNA only) by an *N*-glycoside bond.

Nucleotide diphosphate A nucleotide in which the phosphate residue is connected to another phosphate residue by a phosphoanhydride bond.

Nucleotide triphosphate A nucleotide in which the phosphate residue is connected to a second phosphate residue by a phosphoanhydride bond, and the second phosphate residue is connected to a third by another phosphoanhydride bond.

Nucleus The compact core of an atom where protons and neutrons are located.

Octet rule A rule stating that atoms gain, lose, or share valence electrons in order to have eight valence electrons.

Oligonucleotide A compound consisting of from two to ten nucleotide residues.

Oligopeptide A compound consisting of from two to ten amino acid residues.

Oligosaccharide A compound consisting of from two to ten monosaccharide residues.

Operon A group of genes under the control of one promoter site.

Organic compound A compound that contains carbon.

Origin A site on DNA where the two strands of DNA have been separated from one another, allowing replication to take place.

Osmosis The net movement of the solvent water across a membrane from the solution of lower concentration to the one of higher concentration.

Osmotic pressure The pressure due to water flowing through a membrane.

Oxidation The loss of electrons.

Oxidative deamination A reaction of amino acid metabolism in which an amino group is replaced by a carbonyl group.

Oxidative phosphorylation A metabolic process involving ATP synthase, in which the energy released when H^+ moves from the mitochondiral intermembrane space, through the inner mitochondrial membrane, is used to produce ATP from ADP and P_i.

Oxidizing agent A reactant that oxidizes (removes electrons from) another reactant.

Parallel When applied to protein secondary structure, parallel refers to interacting strands in a β-sheet that run in the same N-terminal to C-terminal direction.

Parent chain The continuous chain of carbon atoms used as the basis for assigning an IUPAC name to an organic molecule.

Partial pressure The pressure that each gas in a mixture exerts by itself.

Parts per billion (ppb) A concentration unit defined as follows:

$$\text{parts per billion} = \frac{\text{g of solute}}{\text{mL of solution}} \times 1{,}000{,}000{,}000$$

Parts per million (ppm) A concentration unit defined as follows:

$$\text{parts per million} = \frac{\text{g of solute}}{\text{mL of solution}} \times 1{,}000{,}000$$

Parts per thousand (ppt) A concentration unit defined as follows:

$$\text{parts per thousand} = \frac{\text{g of solute}}{\text{mL of solution}} \times 1{,}000$$

Pentose A monosaccharide that contains five carbon atoms.

Pentose phosphate pathway A metabolic pathway involved in the production of ribose, 2-deoxyribose, other 3-, 4-, 5-, 6-, and 7-carbon sugar molecules, and NADPH.

Peptide A general name given to oligo- and polypeptides.

Peptide bond Another name for the amide bond that connects amino acid residues in a peptide or a protein.

Percent yield

$$\text{percent yield} = \frac{\text{actual yield}}{\text{theoretical yield}} \times 100$$

Period Elements in the same horizontal row of the periodic table.

Periodic table of the elements A complete list of the elements arranged in order from smallest to largest atomic number, with elements having similar properties placed in the same group.

pH $\text{pH} = -\log[H_3O^+]$

pH optimum The pH at which an enzyme is most active.

Phenol An organic compound that contains a hydroxyl group which is attached to an aromatic ring.

Phospholipid A class of lipids whose members contain a phosphate residue.

Physical change A change in matter that does not involve varying its chemical composition.

Physical property A characteristic of matter that can be determined without changing its chemical composition.

pK_a $pK_a = -\log K_a$, where K_a is an acidity constant.

Plane polarized light Light that has been passed through a filter that blocks all light waves except those vibrating in one particular orientation (plane).

Polar molecule A molecule in which one side has a partial positive charge and the other has a partial negative charge.

Polar covalent bond A covalent bond in which bonding electrons are shared unequally by the two atoms involved.

Polar-acidic amino acid An amino acid whose side chain contains a carboxyl group.

Polar-basic amino acid An amino acid whose side chain contains an amino group or another basic nitrogen–containing group.

Polar-neutral amino acid An amino acid whose side chain is polar and neutral, typically an alcohol, a phenol, or an amide.

Polyatomic ion An ion consisting of two or more atoms.

Polycyclic aromatic hydrocarbon (PAH) An aromatic compound that contains benzene rings that are fused to one another (carbon–carbon bonds are shared by one or more rings).

Polynucleotide A compound consisting of more than ten nucleotide residues.

Polypeptide A compound consisting of more than ten amino acid residues.

Polypeptide backbone The series of alternating alpha carbon atoms and peptide groups present in a polypeptide.

Polysaccharide A compound consisting of more than ten monosaccharide residues.

Polyunsaturated fatty acid A fatty acid whose hydrocarbon tail has more than one carbon–carbon double bond.

Positive effector An allosteric effector that enhances substrate binding and increases the rate of an enzyme-catalyzed reaction.

Positron ($^{0}_{1}\beta^{+}$) A form of nuclear radiation that consists of a positively charged electron.

Post-transcriptional modification Changes made to RNA after its formation during transcription.

Post-translational modification Changes made to a peptide after its formation during translation.

Potential energy Stored energy.

Precipitate A solid formed during a chemical reaction.

Precision A measure of how close repeated measurements are to one another.

Pressure The force of the collisions that take place between the particles of a gas and an object.

Primary (1°) alcohol An alcohol in which the carbon atom carrying the —OH is attached to only one other carbon atom.

Primary (1°) amine An amine in which the nitrogen atom is attached to only one carbon atom.

Primary structure When applied to proteins, primary structure is the sequence of amino acid residues. When applied to oligo- or polynucleotides, primary structure is the sequence of deoxyribonucleotide (in DNA) or ribonucleotide (in RNA) residues.

Product A substance formed during a reaction.

Promoter site A site on DNA where transcription begins.

Prosthetic group A nonpeptide component that contributes to protein tertiary structure.

Protein A polypeptide containing more than fifty amino acid residues.

Proton A subatomic particle found in the nucleus of an atom. Protons carry a 1+ charge and have a mass of about 1 amu.

Pure substance A substance that consists of just one element or compound.

Pyramidal shape An arrangement of atoms in which one atom can be imagined as being placed at the center of a tetrahedron and the three atoms to which it is attached placed at three of the corners of the same tetrahedron.

Pyranose An anomer that contains a six-membered ring (one oxygen atom and five carbon atoms).

Quantum mechanics A theory that views electrons as residing in atomic orbitals.

Quaternary ammonium ion A positively charged polyatomic ion in which a nitrogen atom is attached to four carbon atoms.

Racemic mixture A 50:50 mixture of enantiomers.

Radioactive isotope (radioisotope) An atom that releases nuclear radiation.

Reactant A substance present at the start of a reaction.

Reaction energy diagram A diagram that shows energy on the *y*-axis and reaction progress (with reactants on the left and products on the right) on the *x*-axis.

Reaction rate A measure of the speed with which reaction products form.

Recombinant DNA A DNA molecule that contains DNA from two or more sources.

Reducing agent A reactant that reduces (adds electrons to) another reactant.

Reducing sugar A carbohydrate that reacts with (is oxidized by) Benedict's reagent.

Reduction The gain of electrons.

Relative specificity Restriction of the activity of an enzyme to a range of substrates with the same functional group or similar structure.

Replication A process that involves using existing DNA as a template for the production of new DNA molecules.

Replication fork The place, at either end of an origin, where replication takes place.

Representative elements Elements belonging to groups 1A–8A.

Reprocessing Chemical treatment of used nuclear fuel to separate out material capable of undergoing fission.

Residue That part of a reactant molecule that remains when it has been incorporated into a larger structure.

Reverse transcription A process in which the primary structure of RNA is used as a template for the production of DNA.

Reversible inhibitor An enzyme inhibitor whose effect is not permanent (is reversible) because the inhibitor is loosely bound to the enzyme and can dissociate, restoring the enzyme to its original state.

Ribonucleic acid (RNA) An oligo- or polynucleotide built from nucleotides that contain ribose and the bases adenine, guanine, cytosine, and uracil.

Ribonucleoside A nucleoside that contains a ribose residue.

Ribonucleotide A nucleotide that contains a ribose residue.

Ribosomal RNA (rRNA) The form of RNA that combines with proteins to form ribosomes.

Ribosome A multi-subunit complex composed of rRNA and proteins, where protein synthesis takes place.

RNA polymerase An enzyme that catalyzes the formation of RNA.

Salt bridge An ionic bond. This term is typically used to describe the ionic bonds that form between charged groups in protein molecules.

Saponification Hydrolysis of an ester in the presence of OH^-.

Saturated fatty acid A fatty acid whose hydrocarbon tail contains only single bonded carbon atoms.

Saturated hydrocarbon A hydrocarbon that contains only single bonded carbon atoms.

Saturated solution A solution that holds the most solute possible at a particular temperature.

Scientific method A process used to gather and interpret information about the world around us. Observation, experiment, and the development of laws, hypotheses, and theories are part of the scientific method.

Scientific notation Notation in which values are written as a number between 1 and 10 multiplied by a power of ten.

Secondary (2°) alcohol An alcohol in which the carbon atom carrying the —OH is attached to only two other carbon atoms.

Secondary (2°) amine An amine in which the nitrogen atom is attached to only two carbon atoms.

Secondary structure When applied to proteins, secondary structure is the folding of the chain that results from hydrogen bonding that takes place between amide N—H and amide C=O groups from different parts of the polypeptide chain. Two common forms of secondary structure are the α-helix and the β-sheet. When applied to polynucleotides, secondary structure is the helix formed by the interaction of two DNA strands.

Semiconservative replication DNA replication in which each of the two daughter DNA molecules (the new DNA) contains one strand of the original DNA.

Semimetal (metalloid) An element that has physical properties that are intermediate between those of metals and nonmetals.

Sex hormone A steroid hormone that controls the development of secondary sexual characteristics.

SI system A measurement system that uses the kilogram, the cubic meter, and the meter to measure, respectively, mass, volume, and length.

Significant figures Digits in a measurement that are reproducible when the measurement is repeated, plus the first doubtful digit.

Simple protein A protein that does not require a prosthetic group for biological activity.

Single bond One pair of electrons shared by two atoms.

Single replacement reaction A reaction in which one element switches places with a different element in a compound.

Skeletal structure An abbreviated structural formula that represents covalent bonds by lines, does not show carbon atoms, and shows hydrogen atoms only when they are attached to atoms other than carbon.

Solid The form of matter that has a fixed shape and volume.

Solubility A measure of the amount of solute that will dissolve in a solvent at a given temperature.

Soluble Able to dissolve in (form a homogeneous mixture with) a given solute.

Solute A solution component that is present in a lesser amount than the solvent.

Solution Another name for a homogeneous mixture.

Solvent The solution component that is present in the greatest amount.

Specific gravity The density of a substance divided by the density of water at the same temperature.

Specific heat The amount of heat required to raise the temperature of one gram of a substance by 1°C.

Spectator ion An ion that appears identically on both sides of an ionic equation.

Sphingolipid A class of lipids whose members contain a sphingosine residue.

Sphygmomanometer A device, consisting of an inflatable cuff attached to a manometer, that is used to measure blood pressure.

Spiral metabolic pathway A metabolic pathway consisting of a series of repeated reactions that are used to break down or produce a compound.

Spontaneous process A process that continues to occur once it has been started.

Standard temperature and pressure (STP) A temperature of 0°C and a pressure of 1 atm.

Stereoisomers Molecules that have the same molecular formula, the same atomic connections, different three-dimensional shapes, and are interchanged only by breaking bonds.

Stereospecificity Restriction of the activity of an enzyme to catalyzing the reaction of only one particular stereoisomer of a substrate or formation of only one stereoisomer of a product.

Steroid A class of lipids that share the same basic ring structure: three fused 6-carbon atom rings and one 5-carbon atom ring.

Structural formula A formula that shows which atoms are present in a compound and how they are connected to one another.

Subatomic particle A particle of matter that is smaller than an atom. Protons, neutrons, and electrons are the subatomic particles present in atoms.

Sublimation The conversion of a solid into a gas, bypassing the liquid phase.

Substituent An atom or group of atoms attached to the parent chain of an organic molecule.

Substrate The reactant in an enzyme-catalyzed reaction.

Sugar-phosphate backbone The portion of structure of oligo- and polynucleotides that consists of repeating units of phosphate attached to 2-deoxyribose (in DNA) or ribose (in RNA), attached to phosphate, and so on.

Sulfide An organic compound that contains a C—S—C linkage.

Sulfoxide An organic compound that contains a positively-charged sulfur atom joined to a negatively charged oxygen atom by a single bond (S^{+}—O^{-}).

Supercoiling The twisting or coiling of a helix into a tighter, more compact shape.

Suspension A mixture that consists of particles suspended in a liquid. Upon standing, the particles will settle.

Synthesis reaction A reaction in which two or more elements or compounds combine to form one more complex compound.

Temperature optimum The temperature at which an enzyme is most catalytically active.

Template strand The one strand of double-stranded DNA that RNA polymerase uses as a template to produce RNA.

Termination The last step in translation. Once a protein has been synthesized it leaves the ribosome, which dissociates.

Termination sequence A series of bases on DNA that indicates where transcription should end.

Tertiary (3°) alcohol An alcohol in which the carbon atom carrying the —OH is attached to three other carbon atoms.

Tertiary (3°) amine An amine in which the nitrogen atom is attached to three carbon atoms.

Tertiary structure Tertiary structure is the overall three-dimensional shape of a protein or a polynucleotide, including the contribution of secondary structure.

Tetrahedral shape An arrangement of atoms in which one atom can be imagined as being placed at the center of a tetrahedron and the four atoms to which it is attached placed at the four corners of the same tetrahedron.

Tetrapeptide An oligopeptide that contains four amino acid residues.

Tetrose A monosaccharide that consists of four carbon atoms.

Theoretical yield The theoretical maximum amount of product that can be obtained from a reaction.

Theory An experimentally tested explanation of an observed behavior. A theory is consistent with existing experimental evidence and can be used to make predictions.

Thiol An organic compound that contains an —SH group.

Titration A process used to determine the concentration of an acid or base solution.

Torr A unit of pressure.

$$760 \text{ torr} = 1 \text{ atm} = 14.7 \text{ psi.}$$

Trans One of two possible geometric (*cis-trans*) isomers. For cycloalkanes, a *trans* geometric isomer has two substituents on opposite faces of the ring. For alkenes, a *trans* isomer has alkyl groups on opposite sides of a line connecting the two double-bonded carbon atoms.

Transamination A reaction of amino acid metabolism in which an amino group from an amino acid is transferred to an α-keto acid, producing a new α-keto acid and a new amino acid.

Transcription A process that involves using part of the primary structure of DNA as a template for the production of RNA.

Transfer RNA (tRNA) The form of RNA that transports amino acid residues to the site of protein synthesis.

Transition elements (transition metals) Elements belonging to groups 1B–8B.

Translation A process in which the primary structure of mRNA is used as a template for the production of a protein.

Triglyceride A triester composed of three fatty acid residues and a glycerol residue.

Trigonal planar shape An arrangement of atoms in which one atom can be imagined as being placed at the center of an equilateral triangle and the three atoms to which it is attached placed at the three corners of the same triangle.

Triose A monosaccharide that consists of three carbon atoms.

Tripeptide An oligopeptide that consists of three amino acid residues.

Triple bond Three pairs of electrons shared by two atoms.

Triplet code See genetic code.

Unit A quantity used as a standard of measurement.

Unit Conversion Changing how a value is expressed by switching from one unit into another.

Unsaturated hydrocarbon A hydrocarbon that contains double or triple bonds.

Unsaturated solution A solution that holds less than the amount of solute required to produce a saturated solution.

Valence electron An electron held in an atom's valence shell.

Valence shell An atom's highest numbered energy level that holds electrons.

Vapor pressure The maximum pressure exerted by a gas formed by evaporation of a liquid.

Viscosity A liquid's resistance to flow.

$\boldsymbol{V_{max}}$ The maximum velocity (reaction rate) that a given concentration of enzyme can produce. V_{max} is related to the second step in the reaction catalyzed by a Michaelis-Menten enzyme: $ES \longrightarrow E + P$.

Volume/volume percent A concentration unit defined as follows:

$$\text{volume/volume percent} = \frac{\text{mL of solute}}{\text{mL of solution}} \times 100$$

Wax A lipid that is a mixture of water insoluble compounds. Wax esters, which typically contain 14–36 carbon fatty acid residues and 16–30 carbon alcohol residues, are a major constituent of waxes.

Weight/volume percent A concentration unit defined as follows:

$$\text{weight/volume percent} = \frac{\text{g of solute}}{\text{mL of solution}} \times 100$$

Weight/weight percent A concentration unit defined as follows:

$$\text{weight/weight percent} = \frac{\text{g of solute}}{\text{g of solution}} \times 100$$

Work A form of energy involved with making changes to matter.

Zwitterion The form of an amino acid that carries one positive and one negative charge.

Zymogen An inactive enzyme precursor.

INDEX

QUICK REFERENCE GUIDE

ACIDS AND BASES

Acid (HA) and conjugate base (A^-) concentrations, as a function of pH, Table 9.6
Interpreting equilibrium constants (K_{eq}), Table 9.2
K_a and pK_a values for selected acids, Table 9.4
Some common acids and bases, Table 9.1
pH values of some common solutions, Table 9.3
Relative strengths of some acids and their conjugate bases, Table 9.5

AMINO ACIDS, PROTEINS, AND ENZYMES

α-Amino acids present in proteins, Table 13.1
Amino acids that are essential for humans, Table 15.1
Enzyme cofactors, Table 13.2

ATOMS

Subatomic particles, Table 2.1
The ground state electron distribution for the first 20 elements, Table 3.4

BONDING

Bond types, Table 4.1

CARBOHYDRATES

Monosaccharides, Table 12.1
Relative sweetness, Table 12.2

ENERGY

Heat of fusion and heat of vaporization, Table 5.2
Specific heat, Table 5.1

GASES, LIQUIDS, AND SOLIDS

Density of common substances, Table 5.5
Solutions, colloids, and suspensions, Table 7.3
The solubility of ionic compounds in water, Table 7.1
The vapor pressure of water at various temperatures, Table 5.6

HEALTH

Blood pressure guidelines, Table 5.4
Concentration ranges for some blood serum solutes, Table 7.2
Dietary reference intakes (DRIs) for some essential elements, Table 2.3

IONS AND IONIC COMPOUNDS

Common polyatomic ions, Table 3.2
Some transition metal ions, Table 3.1
The uses of some ionic compounds, Table 3.5

LIPIDS

Common fatty acids, Table 8.1
Key esters found in some waxes, Table 8.2

MATH

Conversion factors and the factor label method, Section 1.6
Logs and antilogs, Chapter 9 Math Support
Measurements and significant figures, Section 1.5
Scientific notation and metric prefixes, Section 1.4
Significant figures, Table 1.5

NUCLEIC ACIDS

Codons in the 5′ to 3′ sequence of mRNA, Table 14.1
Short tandem repeats (STRs) and the probability of their occurrence, Table 14.2

ORGANIC COMPOUNDS

Common molecular shapes, Table 4.2
Formulas and names of alkyl groups, Table 4.6
Physical properties of selected alcohols, ethers, thiols, sulfides, and hydrocarbons, Table 11.1
Physical properties of selected aldehydes and ketones, Table 11.2
Physical properties of selected amines, Table 10.3
Physical properties of selected phenols, Table 10.2
Physical properties of some small carboxylic acids, Table 10.1
Structure, name, and properties of selected hydrocarbons, Table 4.4
The first ten numbering prefixes for IUPAC naming, Table 4.5

RADIOACTIVITY

Common forms of radioactivity, Table 2.4
Half lives and decay type for selected radioisotopes, Table 2.5
Some uses of radioisotopes in medicine, Table 2.6